Mammalian Evolutionary Morphology

Vertebrate Paleobiology and Paleoanthropology Series

Edited by

Eric Delson
Vertebrate Paleontology, American Museum of Natural History,
New York, NY 10024, USA
delson@amnh.org

Ross D. E. MacPhee
Vertebrate Zoology, American Museum of Natural History,
New York, NY 10024, USA
macphee@amnh.org

Focal topics for volumes in the series will include systematic paleontology of all vertebrates (from agnathans to humans), phylogeny reconstruction, functional morphology, paleolithic archaeology, taphonomy, geochronology, historical biogeography, and biostratigraphy. Other fields (e.g., paleoclimatology, paleoecology, ancient DNA, total organismal community structure) may be considered if the volume theme emphasizes paleobiology (or archaeology). Fields such as modeling of physical processes, genetic methodology, nonvertebrates or neontology are out of our scope.

Volumes in the series may either be monographic treatments (including unpublished but fully revised dissertations) or edited collections, especially those focusing on problem-oriented issues, with multidisciplinary coverage where possible.

Editorial Advisory Board

Published titles in this series are listed at the end of this volume

Mammalian Evolutionary Morphology

A Tribute to Frederick S. Szalay

Edited by

Eric J. Sargis
Yale University, Department of Anthropology
Peabody Museum of Natural History, Division of Vertebrate Zoology
New Haven, CT, USA

and

Marian Dagosto
Feinberg School of Medicine
Northwestern University,
Department of Cell and Molecular Biology,
Chicago, IL, USA

Springer

Library of Congress Control Number: 2008922957

ISBN 978-1-4020-6996-3 (HB)
ISBN 978-1-4020-6997-0 (e-book)

Published by Springer,
P.O. Box 17, 3300 AA Dordrecht, The Netherlands.

www.springer.com

Cover illustration:
Dryomomys szalayi, drawing by Doug M. Boyer.

Printed on acid-free paper

First published 2008, reprinted 2009

Justine A. Salton

12 March 1972–28 October 2005

This book is dedicated to the memory of Justine Salton, Fred Szalay's last student and a friend to many of those who contributed to this volume.

Figure 1. Justine A. Salton (left) with Frederick S. Szalay (right) in New York City after her dissertation defense on August 15, 2005.

Preface

Frederick S. Szalay is a commanding figure – one of those peerless inimitable people that leave a lasting impression however briefly they are encountered. Passionate and fearless, he approaches his work, as he does everything else in his life, with great gusto and verve and expects everyone around him to do the same. To have worked with him was alternately a terror and a blessing, but was in any case truly inspirational. Students and colleagues alike were apprehensive of his much renowned (but in reality rarely displayed and usually deserved) critiques, but therefore all the more appreciative of his generously given honest praise and unwavering confidence and support. His unbelievable breadth and depth of knowledge of all things mammalian and paleontological is due in no small part to his absurdly dense and complete library, compiled with the same ravenous collector's eye that he applies to souvenirs from foreign locales, abbreviation systems for tooth structures and joint surfaces, and dissectible road kill carcasses. Those brave readers prepared to work through the long philosophical preambles and the "very, very long sentences and creative grammar constructs" (in the words of one admirer) that distinguish Fred's insightful work from more mundane contributions are sure to learn something valuable from one of the most astute and creative practitioners of evolutionary morphology.

Equally at home with dental, cranial, or postcranial morphology, Fred made major contributions to the literature on mammalian evolutionary morphology, particularly in Primates, Archonta, and Metatheria, as will be detailed below. The esteem in which he is held by his colleagues can be partly measured by the number of taxa named after him in honor of his contributions to our knowledge of mammalian evolution. These include the primates *Jemezius szalayi* (Beard, 1987), *Szalatavus attricuspis* (Rosenberger et al., 1991), *Tatmanius szalayi* (Bown and Rose, 1991), *Dryomomys szalayi* (Bloch et al., 2007), and *Magnadapis fredi* (Godinot, this volume); the marsupials *Szalinia gracilis* (de Muizon and Cifelli, 2001), *Sinodelphys szalayi* (Luo et al., 2003), *Oklatheridium szalayi* (Davis et al., this volume), and *Fredszalaya hunteri* (Shockey and Anaya, this volume); and the multituberculate *Ectypodus szalayi* (Sloan, 1981).

Frederick Sigmund Szalay was born in Hungary on November 15, 1938. In many ways he was the product of the war-torn years of World War II where as a child he spent months forced to live in the cellars of Budapest while bombs were falling. Towards the end of the war this was followed by street combat between the German and Soviet forces, which he witnessed firsthand when he and other small rascals managed to sneak upstairs from the cellar. As a 6-year-old at the end of 1944, he helped his uncle and some friends coax an unexploded 500 lb bomb down the stairs from the third floor of the house where they lived. Having to scavenge for food with his beloved grandfather during the winter of 1945–1946 stands out as something he will never forget.

In addition to being a voracious reader of travel and natural history (and also a student of French and English), most of his high school years were spent playing a variety of sports (swimming, track and field, boxing, and rowing) and shooting photographs, with very little academic effort (but much mischief, and detailed planning with his friends on how to leave the Iron Curtain behind). Having the family background of a Jewish mother and a father from the titled nobility who was a feudal judge in pre-War Hungary nullified any chance of his attending university under the communists. Following the defeat of the 1956 uprising in Hungary, and after a previous attempt at escape which ended in capture, he and a good friend managed to reach Austria in late November 1956. He never finished his last year in Gymnasium (high school).

Oddly, Fred had no acquaintance with either vertebrate paleontology or evolutionary biology while attending college in the US. After reaching the United States in December 1956, he worked for nine months at odd jobs. Then a small Catholic college in Maryland offered him a refugee scholarship. Mt. St. Mary's College offered a straightforward premed curriculum with no opportunity for the study of geology. His consummate interests in mammalian natural history had to be satisfied with a biology major and chemistry minor, but his last two summers were spent in near bliss working at the Catskill Game Farm in New York State. The great variety of mammals that he worked with while living on the premises set the stage for his plans for a Mammalogy Ph.D. at the University of Massachusetts at Amherst, taking with him an NDEA (National Defense Education Act) Fellowship that he won following graduation from college (and naturalization in 1961). While at Amherst Fred took Albert Wood's year long course in Vertebrate Paleontology and a seminar with Lincoln Brower on Evolution and Ecology at Amherst College. Added to these pivotal experiences was the reading of George G. Simpson's *Meaning of Evolution* (followed by Simpson's other books on evolution and systematics), and as he often told his students, Simpson's writings were perhaps the most important reasons for shifting his interests from mammalogy to paleontology. A combination of these truly inspirational experiences at Amherst led to a quick trip to the American Museum of Natural History (AMNH) to interview with Malcolm McKenna, followed by a transfer to the Biology Department at Columbia University, and the AMNH.

In addition to McKenna's much valued tutelage (as Fred often stated), and the no less influential atmosphere of the Vertebrate Paleontology traditions at the AMNH, were the much treasured associations with fellow graduate students in Biology and Geology at Columbia; professors such as Bobb Schaeffer, Edwin Colbert, and John Imbrie; postdocs like Leigh Van Valen and Len Radinsky; and the hosts of perennial visitors that stream through the AMNH regularly. Yearly field work and field courses in geology rounded out the bases for his long continued dedication to understanding mammalian evolutionary history and macroevolutionary dynamics. After completing his doctorate in 1967, Fred stayed on as an NSF Postdoctoral fellow until taking a job in the Department of Anthropology at Hunter College, CUNY. He was a Research Associate in the Department of Vertebrate Paleontology of the AMNH until 1985, and on the Graduate Faculty of the City University of New York. He retired from Hunter College in 2003, and is now an Adjunct Professor in the Department of Biology, University of New Mexico. He is also Professor Emeritus in the Ecology and Evolutionary Biology Doctoral Program, City University of New York.

It is an honor and a privilege for all of us to have known him, to have learned from him, and to be able to contribute to this volume celebrating his career. This book acknowledges and celebrates the contributions of Dr. Frederick S. Szalay to the field of Mammalian Evolutionary Morphology. Fred Szalay has published about 200 articles, 6 monographs, and 6 books on this subject. His dissertation work was awarded the Newberry Prize in Vertebrate Zoology. He has received numerous grants from the National Science Foundation and the Wenner-Gren Foundation, and was awarded a Guggenheim Fellowship in 1980. Throughout his career, Fred has been a strong advocate for biologically and evolutionarily meaningful character analysis. In his view, this can be accomplished only through an integrated strategy of functional, adaptational, and historical analysis. Using this approach, he has made major contributions to the following areas of study:

Primate Evolutionary Morphology

1. ***Primate Origins***. Fred's dissertation work on the insectivore-primate transition set the groundwork for a career-long interest in the subject of primate origins both from a phylogenetic and ecological/adaptive perspective. His first monograph on the subject (Szalay, 1969, #10 in Szalay bibliography), the publication of his dissertation, concentrated on the dental evidence for the phylogenetic relationships of the still frustratingly difficult to interpret mixodectids and microsyopids. From this work he developed the hypothesis that the morphological changes in the dentition that distinguished the first primates (plesiadapiforms) from their predecessors was the result of a shift from a primarily insectivorous diet to a more herbivorous one (Szalay, 1968, #6). This work was followed by several papers that explored the dental, cranial, and postcranial evidence linking Plesiadapiformes to Euprimates and which developed a coherent explanation of the adaptive significance of primate synapomorphies. For example, Szalay et al. (1975, #52) used postcranial evidence to infer that plesiadapiforms were arboreal and closely related to euprimates. Although both of these proposals were initially challenged, they have subsequently been supported with evidence from new fossils (Bloch and Boyer, 2002; Bloch et al., 2007) and new phylogenetic analyses (Silcox, 2001; Bloch and Boyer, 2002; Bloch et al., 2007). Primate origins and the evolutionary morphology of plesiadapiforms are topics addressed in this volume by Silcox and Boyer and Bloch.

2. ***Phylogenetic relationships within Primates***. Fred Szalay also worked on the delineation of major taxa within Primates. Basicranial evidence was marshaled to understand the relationships within Strepsirhini (Szalay and Katz, 1973, #42) and to support the validity of Haplorhini (Szalay, 1975, #58). The latter paper, along with many to follow, argued that the fundamental division within Primates was Strepsirhini (adapids + lemuriforms) and Haplorhini (*Tarsius* + omomyids + anthropoids). These hypotheses, which are the best supported today, were defended by Fred against the rival hypotheses of "Plesitarsiiformes" (plesiadapiforms + tarsiiforms) and "Simiolemuriformes" (strepsirhines + anthropoids) favored by other paleontologists. Fred also addressed the origin and phylogenetic relationships of anthropoid primates (Szalay, 1975, #55; Rosenberger and Szalay, 1980, #75), a theme visited in this volume by Rosenberger et al. and Maier.

3. ***Major publications***. Fred has also described and named numerous Paleocene and Eocene primate taxa, a subject represented here by Godinot and Couette. Fred is the author or editor of several important books and monographs on the subject of primate evolution. These are:

1969: **Mixodectidae, Microsyopidae, and the insectivore-primate transition**. Bulletin of the American Museum of Natural History 140, 193–330.

1975: **Approaches to Primate Paleobiology**. Contributions to Primatology, Volume 5. Karger AG, Basel.

1975: **Phylogeny of the Primates: A Multi-disciplinary Approach**. Plenum, New York (Luckett, W. P. and F. S. Szalay, Eds.).

1976: **Systematics of the Omomyidae (Tarsiiformes, Primates): Taxonomy, Phylogeny, and Adaptations**. Bulletin of the American Museum of Natural History 156, 157–450.

1979: **Evolutionary History of the Primates**. Academic, New York (Szalay, F. S. and E. Delson, Eds.).

Szalay and Delson (1979, #72) is perhaps the most remarkable of all these volumes, as it was a huge undertaking that has never been replicated despite enormous interest in primates and a proliferation of primatologists since the late 1970s.

Mammalian Evolutionary Morphology

1. ***Archonta***. *The morphological evidence supporting the supraordinal grouping Archonta, and its adaptive significance*. Szalay (1977, #66) provided the first morphological support for McKenna's (1975) revised concept of Gregory's (1910) Archonta, a clade that includes Primates, Scandentia, Dermoptera, and Chiroptera. Szalay (1977, #66) used tarsal evidence to unite Primates, Scandentia, and Dermoptera, as well as previously cited similarities to include Chiroptera as well. Although Archonta (including Chiroptera) has not been subsequently supported, Euarchonta (excluding Chiroptera) has been strongly supported in molecular studies (e.g., Murphy et al., 2001). In other words, the grouping of Primates, Scandentia, and Dermoptera that he originally recognized based on tarsal evidence has now been supported in numerous other studies using different datasets. Szalay and Drawhorn (1980, #73) proposed that Archonta originated and diversified in an arboreal milieu, another hypothesis that has been supported in subsequent studies (e.g., Bloch and Boyer, 2002; Bloch et al., 2007). Szalay continued to work on this group throughout his career, including the publication of a monograph with S. G. Lucas in 1996 (#145).

2. ***Marsupialia***. After spending a sabbatical year in Australia in 1980, Szalay (1982, #80) proposed a completely novel hypothesis of marsupial relationships based on tarsal evidence. He hypothesized that the South American *Dromiciops* is more closely related to Australasian marsupials than to other South American marsupials. He formalized this by including *Dromiciops* with Australasian marsupials in Australidelphia, whereas other South American marsupials were placed in Ameridelphia. The classification of *Dromiciops* with Australasian taxa in Australidelphia was initially met with strong resistance and was highly criticized, but it has subsequently been supported in both morphological (e.g., Horovitz and Sanchez-Villagra, 2003) and molecular (e.g., Amrine-Madsen et al., 2003) analyses. Szalay's (1982, #80) novel hypothesis of marsupial relationships has major implications for the biogeographic history of this group. Szalay further developed his ideas on marsupial phylogeny, functional morphology, and biogeography in both a book (Szalay, 1994, #142) and a monograph (Szalay and Sargis, 2001, #198). This group is considered by Davis et al. and Kear et al. in this volume.

3. ***Other Mammals***. In 1990, Szalay, with co-editors M. J. Novacek and M. C. McKenna, organized an important conference on the subject of mammalian phylogeny and evolution, which resulted in the publication of two volumes (Szalay et al., 1993, #130–131). The themes of mammalian systematics and paleontology play into several contributions in this volume: e.g., Davis et al., Penkrot et al., Bergqvist, Shockey and Anaya, and O'Sullivan. Fred also published a monograph, with F. Schrenk in 1998 on "edentates" (#148). This study included an analysis of xenarthrans, a group discussed in this volume by Argot.

Theory and Practice of Phylogeny Reconstruction/Adaptive Scenarios

1. ***The integration of postcranial evidence into hypotheses of mammalian systematics***. Szalay's (1977, #66) phylogeny and classification of mammals were based completely on tarsal evidence, which was both novel and controversial at the time because such studies were typically based on teeth. Most were critical of this study, but George Gaylord Simpson (1978), probably the best known mammalian systematist in the history of the field, was supportive of Szalay's innovative analysis. In fact, Fred's analyses can fairly be seen as building on and refining the traditions of "total evidence" practiced by the best of the previous generation of mammalian paleontologists including Simpson, William K. Gregory, William D. Matthew, and Henry F. Osborne. The hegemony of dental evidence was based on the assumption that teeth reflected relationships better than the limb skeleton, which was thought to be more influenced by functional demands and thus more prone to parallelism. As anticipated by and demonstrated by Szalay, this assumption is faulty at best. Fred was able to use postcranial evidence to support controversial hypotheses on Primates (*sensu lato*; i.e., including plesiadapiforms), Euarchonta, and Marsupialia (specifically Australidelphia), as well as many other mammalian groups such as Glires, Xenarthra, and Mesozoic taxa. The majority of the contributions in this volume build on this aspect of Szalay's work, including those by Kear et al., Argot, Salton and Sargis, Penkrot et al., Bergqvist, Shockey and Anaya, O'Sullivan, Polly, Boyer and Bloch, Dagosto et al., Sargis et al., Harcourt-Smith et al., and Warshaw.

2. ***Phylogenetic and adaptational analysis***. In the 1970s the trend toward both numerical phenetic and cladistic methods of phylogenetic analysis was rapidly expanding. Fred Szalay was and is a vocal critic of the superficial character counting, distribution-based, algorithm driven solutions to phylogeny reconstruction advocated by some, particularly cladists. He advocates instead for the primacy of biologically informed

character analysis using functional, developmental, and adaptational criteria to both weight characters and test hypotheses of homology and polarity (Szalay, 1981, #76). Fred was also heavily influenced by the work of Walter Bock, and is a strong proponent of the logical inseparability of functional-adaptive and phylogenetic analysis; one is not primary to another, they are reciprocally illuminatory (if we might borrow that Hennigian phrase) (Szalay, 1981, #78; Szalay and Bock, 1991, #127; Szalay, 2000, #160). In fact, "The meeting of these two 'separate' disciplines is of course what is usually referred to as morphology" (Szalay, 1981, #78, p. 160). This point of view is represented in his concept of the "transformation series", a testable hypothesis of polarity based not on distribution, but on the fossil record and a functionally logical sequence of ancestor-descendant states. In Fred's view, the a-historical approaches, those that are phenetic, correlation based, and do not consider the phylogenetic history of the subject organism and its influence on the likely response to selection, are not adequate for analyzing adaptation. His 1981 (#78) paper outlined a historically informed approach for analyzing adaptations of fossil organisms. The influence of this point of view is clear in many of the contributions to this volume.

ERIC J. SARGIS
Department of Anthropology
Yale University
New Haven, CT 06520, USA
Eric.Sargis@yale.edu
Division of Vertebrate Zoology
Peabody Museum of Natural History

MARIAN DAGOSTO
Department of Cell and Molecular Biology
Feinberg School of Medicine
Northwestern University
Chicago, IL 60611, USA
m-dagosto@northwestern.edu

Bibliography of Frederick S. Szalay

1965

1. **First evidence of tooth replacement in the subclass Allotheria (Mammalia)**. American Museum Novitates 226, 1–12.

1966

2. **The tarsus of the Paleocene leptictid *Prodiacodon* (Insectivora, Mammalia)**. American Museum Novitates 2267, 1–13.
3. (Szalay, F. S. and S. J. Gould) **Asiatic Mesonychidae (Mammalia, Condylarthra)**. Bulletin of the American Museum of Natural History 132(2), 127–174.

1967

4. **The Affinities of *Apterodon* (Mammalia, Deltatheridia, Hyaenodontidae)**. American Museum Novitates 2293, 1–17.
5. (Van Valen, L., P. M. Butler, M. C. McKenna, F. S. Szalay, B. Patterson and A. S. Romer) ***Galeopithecus* Pallas, 1973 (Mammalia): proposed validation under the plenary powers Z.N. (SP 1792)**. Bulletin of Zoological Nomenclature 24(3), 190–191.

1968

6. **The beginnings of primates**. Evolution 22(1), 19–36.
7. **The Picrodontidae, a family of early primates**. American Museum Novitates 2329, 1–55.
8. (Mellett, J. S., and F. S. Szalay) ***Kennatherium shirensis* (Mammalia, Palaeoryctoidea) a new didymoconid from the Eocene of Asia**. American Museum Novitates 2342, 1–7.
9. **Origins of the Apatemyidae (Mammalia, Insectivora)**. American Museum Novitates 2352, 1–11.

1969

10. **Mixodectidae, Microsyopidae, and the insectivore-primate transition**. Bulletin of the American Museum of Natural History 140(4), 193–330.
11. **The Hapalodectinae and a phylogeny of the Mesonychidae (Mammalia, Condylarthra)**. American Museum Novitates 2361, 1–26.
12. **Uintasoricinae, a new subfamily of early Tertiary mammals (Primates)**. American Museum Novitates 2363, 1–36.
13. **Origin and evolution of function of the mesonychid feeding mechanism**. Evolution 23(4), 703–720.

1970

14. ***Amphipithecus* and the origin of catarrhine primates**. Nature 227(5256), 355–357.

1971

15. **On the cranium of the late Paleocene primate *Plesiadapis tricuspidens***. Nature 230(5292), 324–325.
16. **Artiodactyla**. *Encyclopedia of Science and Technology*. McGraw-Hill, New York, pp. 616–617.
17. **Astrapotheria**. *Encyclopedia of Science and Technology*. McGraw-Hill, New York, p. 649.
18. **Hyracoidea**. *Encyclopedia of Science and Technology*. McGraw-Hill, New York, p. 675–676.
19. **Mammalia**. *Encyclopedia of Science and Technology*. McGraw-Hill, New York, pp. 94–96.
20. **Monotremata**. *Encyclopedia of Science and Technology*. McGraw-Hill, New York, pp. 689–690.
21. **Perissodactyla**. *Encyclopedia of Science and Technology*. McGraw-Hill, New York, pp. 23–25.
22. **Proboscidea**. *Encyclopedia of Science and Technology*. McGraw-Hill, New York, pp. 731–732.
23. **Sirenia**. *Encyclopedia of Science and Technology*. McGraw-Hill, New York, pp. 396–397.
24. **Tooth (vertebrate)**. *Encyclopedia of Science and Technology*. McGraw-Hill, New York, pp. 699–701.
25. (Davis, D. D. and F. S. Szalay) **Allotheria**. *Encyclopedia of Science and Technology*. McGraw-Hill, New York, pp. 295–296.
26. (Davis, D. D. and F. S. Szalay) **Eutheria**. *Encyclopedia of Science and Technology*. McGraw-Hill, New York, p. 136.
27. (Davis, D. D. and F. S. Szalay) **Metatheria**. *Encyclopedia of Science and Technology*. McGraw-Hill, New York, p. 360.
28. (Davis, D. D. and F. S. Szalay) **Theria**. *Encyclopedia of Science and Technology*. McGraw-Hill, New York, p. 567.
29. (McKenna, M. C., J. S. Mellett and F. S. Szalay) **Affinities of the Cretaceous mammal *Deltatheridium***. Journal of Paleontology 145(3), 441–442.
30. (Szalay, F. S. and M. C. McKenna) **Beginnings of the Age of Mammals in Asia: the late Paleocene Gashato fauna, Mongolia**. Bulletin of the American Museum of Natural History 144(4), 269–318.

31. **The adapid primates *Agerina* and *Pronycticebus***. American Museum Novitates 2466, 1–19.
32. **Significance of the basicranium of early Tertiary primates for the phylogeny of the order**. (Abs.) American Journal of Physical Anthropology 35(2), 297.
33. **Biological level of organization of the Chesowanja robust australopithecine**. Nature 234, 229–230.
34. **Relationships of the alleged primate *Gesneropithex***. Journal of Mammalogy 52(4), 824–826.

1972

35. **Cranial morphology of the early Tertiary *Phenacolemur* and its bearing on primate phylogeny**. American Journal of Physical Anthropology 36(1), 56–76.
36. **Paleobiology of the earliest primates**. In: R. Tuttle (Ed.). The Functional and Evolutionary Biology of Primates. Aldine-Atherton, Chicago, pp.3–35.
37. ***Amphipithecus* revisited**. Nature 236(5343), 179.
38. **Review of** "Vertebrate Paleozoology" by E. Olson, American Journal of Physical Anthropology 36(3):449–450.
39. (Wilson, R. W. and F. S. Szalay) **New paromomyid primate from middle Paleocene beds, Kutz Canyon area, San Juan Basin, New Mexico**. American Museum Novitates 2499, 1–18.

1973

40. (Szalay, F. S. and A. Berzi) **Cranial anatomy of *Oreopithecus***. Science 180,183–185.
41. **New Paleocene primates and a diagnosis of the new suborder Paromomyiformes**. Folia Primatologica 19(2–3), 73–87.
42. (Szalay, F. S. and C. C. Katz) **Phylogeny of lemurs, galagos, and lorises**. Folia Primatologica 19(2–3), 88–103.

1974

43. (Szalay, F. S. and R. L. Decker) **Origins, evolution and function of the tarsus in late Cretaceous eutherians and Paleocene primates**. In: F.A. Jenkins, Jr. (Ed.). Primate Locomotion. Academic Press, New York, pp. 223–259.
44. (Decker, R. L. and F. S. Szalay) **Origins and function of the pes in the Eocene Adapidae (Lemuriformes, Primates)**. In: F.A. Jenkins, Jr. (Ed.). Primate Locomotion. Academic, New York, pp. 261–291.
45. **A review of some recent advances in paleoprimatology**. American Journal of Physical Anthropology Yearbook (1973), 17, 39–64.
46. **Review of** "Human evolution", Vol. II of Symposia of the Society for the Study of Human Biology (Day, M. H., Ed.), Taylor & Francis/Barnes and Noble Books, London/New York; Evolution 28(3), 507–508.
47. (Seligsohn, D. and F. S. Szalay) **Dental occlusion and the masticatory apparatus on *Lemur* and *Varecia*: their bearing on the systematics of living and fossil primates**. In: Martin, R. D., B.A. Doyle and A.C. Walker (Eds.). Prosimian Biology., Duckworth, London, pp. 543–561.
48. **Comments on** "New perspectives on ape and human evolution", article by A. Kortlandt. Current Anthropology 15(4), 438–439.
49. **New genera of European Eocene adapid primates**. Folia Primatologica 22(2–3),116–133.
50. **A new species and genus of early Eocene primate from North America**. Folia Primatologica 22(4), 243–250.

1975

51. (Szalay, F. S., Ed.) **Approaches to Primate Paleobiology**. Contributions to Primatology, Volume 5. Karger AG, Basel.
52. (Szalay, F. S., I. Tattersall and R. L. Decker) **Phylogenetic relationships of *Plesiadapis* – postcranial evidence**. In: F. S. Szalay (Ed.). Approaches to Primate Paleobiology. Contributions to Primatology, 5, 136–166.
53. **Early primates as a source for the taxon Dermoptera**. (Abs.); American Journal of Physical Anthropology 42(2), 332–333.
54. **Where to draw the nonprimate-primate taxonomic boundary**. Folia Primatologica 23, 158–163.
55. **Haplorhine relationships and the status of the Anthropoidea**. In: R. H. Tuttle (Ed.). Primate Functional Morphology and Evolution. Mouton Publishers, The Hague, The Netherlands, pp.3–22.
56. **Hunting-scavenging protohominids: a model for hominid origins**. Man 10, 420–429.
57. (Luckett, W. P. and F. S. Szalay, eds.) **Phylogeny of the Primates: a multi-disciplinary approach**. Plenum, New York.
58. **Phylogeny of primate higher taxa: the basicranial evidence**; In: W.P. Luckett and F.S. Szalay (Eds.). Phylogeny of the Primates: A Multidisciplinary Approach. Plenum, New York, pp. 357–404.
59. **Phylogeny, adaptations, and dispersal of the tarsiiform primates**; In: W.P. Luckett and F.S. Szalay (Eds.). Phylogeny of the Primates: A Multidisciplinary Approach. Plenum, New York, pp. 91–125.

1976

60. **Systematics of the Omomyidae (Tarsiiformes, Primates): taxonomy, phylogeny, and adaptations**. Bulletin American Museum Natural History 156, 157–450.
61. (Szalay, F. S. and J. A. Wilson) **Basicranial morphology of the early Tertiary tarsiiform *Rooneyia* from Texas**. Folia Primatologica 25, 88–293.
62. (Wilson, J. A. and F. S. Szalay) **New adapid primate of European affinities from Texas**. Folia Primatologica 25, 294–312.

1977

63. **Constructing primate phylogenies: a search for testable hypothesis with maximum empirical content**. Journal of Human Evolution 6(1), 3–18.
64. (Szalay, F. S. and D. Seligsohn) **Why did the strepsirrhine tooth comb evolve**? Folia Primatologica 27(1), 75–82.
65. **Ancestors, descendants, sister groups, and testing of phylogenetic hypotheses**. Systematic Zoology 26(1), 12–18.
66. **Phylogenetic relationships and a classification of the eutherian Mammalia**; In: M.K. Hecht, P.C. Goody, and B.M. Hecht (Eds.). Patterns of Vertebrate Evolution. Plenum, New York, pp. 315–374.
67. (Wilson, J.A. and F. S. Szalay) ***Mahgarita,* a new name for *Margarita* Wilson and Szalay, 1976**. Journal of Paleontology 51(3), 643.
68. (Seligsohn, D. and F. S. Szalay) **Relationship between natural selection and dental morphology: tooth function and diet in *Lepilemur* and *Hapalemur***. In: P. H. Butler, and K. A. Joysey (Eds.). Development, Function, and Evolution of Teeth. Academic, London. pp. 289–307.

1978

69. (Luckett, W. P. and F. S. Szalay) **Clades versus grades in primate phylogeny**. In: D. J. Chivers and K. A. Joysey (Eds.). Recent Advances in Primatology Volume 3: Evolution. Academic, London. pp. 227–237.

1979

70. **The problems of adaptation among Paleogene primates** (Abs.) VIIth Congress of the International Primatological Society, Bangalore, India.
71. (Dagosto, M. and F. S. Szalay) **The elbow joint of early primates** (Abs.) American Journal of Physical Anthropology 50(3), 431.
72. (Szalay, F. S. and E. Delson) **Evolutionary history of the Primates**. Academic, New York.

1980

73. (Szalay, F. S. and G. Drawhorn) **Evolution and diversification of the Archonta in an arboreal milieu**. In: W.P Luckett (Ed.). Comparative Biology and Evolutionary Relationships of Tree Shrews. Plenum, New York, pp. 133–169.
74. (Szalay, F. S. and M. Dagosto) **Locomotor adaptations as reflected on the humerus of Paleogene primates**. Folia Primatologica (34), 1–45.
75. (Rosenberger, A. L. and F. S. Szalay) **On the tarsiiform origins of the Anthropoidea**; In: R. L. Ciochon and A. B. Chiarelli (eds.) Evolutionary Biology of the New World Monkeys and Continental Drift. Plenum, New York, pp. 139–157.

1981

76. **Functional analysis and the practice of the phylogenetic method as reflected by some mammalian studies**. American Zoologist 21, 37–45.
77. **Review of** "Mesozoic Mammals" by J. A. Lillegraven, Z. Kielan-Jaworowska, and W. A. Clemens (Eds.). Journal of Mammalogy 62, 443–445.
78. **Phylogeny and the problem of adaptive significance: the case of the earliest primates**. Folia Primatologica 36, 157–182.

1982

79. **A critique of some recently proposed Paleogene primate taxa and suggested relationships**. Folia Primatologica 37, 152–182.
80. **A new appraisal of marsupial phylogeny and classification;** In: M. Archer (Ed.). Carnivorous Marsupials. Australian National University Press, Canberra, pp. 621–640.
81. (Flannery, T, and F. S. Szalay) ***Bohra paulae*, a new giant fossil tree kangaroo (Marsupialia: Macropodidae) from New South Wales, Australia**. Australian Mammalogy 5, 83–84.

1983

82. **Phylogenetic relationships of the marsupials**; Festschrift for R. Hoffstetter, Geobios, Memoire speciale 6, 177–190.
83. **An eco-ethological reassessment of the living mammals**. (A review of "The Mammalian Radiations" by John E. Eisenberg). Evolutionary Theory 6, 219–222.
84. **Review of** "Phylogenetic Patterns and Evolutionary Process" by N. Eldredge and J. Cracraft; American Journal of Physical Anthropology 61, 509–510.

1984

85. **Review of** "Mammalian Paleofaunas of the World" by D. E. Savage and D.E. Russell. American Journal of Physical Anthropology 64, 205–206.
86. **Arboreality: is it homologous in metatherian and eutherian mammals?** Evolutionary Biology 18, 215–258.

1985

87. **Rodent and lagomorph morphotype adaptations, origins, and relationships: some postcranial attributes analyzed**. In: W. P. Luckett and J.-L. Hartenberger (Eds.). Evolutionary relationships among Rodents. NATO ASI Series A: Life Sciences Volume. 92, Plenum, New York, pp. 83–132.
88. (Szalay, F. S. and J. Langdon) **Evolutionary morphology of the foot in *Oreopithecus***; (Abs.) American Journal of Physical Anthropology 66(2), 163.
89. (Delson, E. and F. S. Szalay) **Reconstruction of the 1958 cranium of *Oreopithecus***; (Abs.) American Journal of Physical Anthropology, 66(2), 163.
90. **Review of** "Vertebrate Zoogeography and Evolution in Australasia" by M. Archer and G. Clayton (Eds). American Scientist 73,173–174.

1986

91. (Szalay, F. S., C.-K. Li, and B.-Y. Wang) **Middle Paleocene omomyid primate from Anhui Province, China: *Decoredon anhuiensis* (Xu, 1976), new combination Szalay and Li, and the significance of *Petrolemur***. (Abs.) American Journal of Physical Anthropology 69(2), 269.
92. (Szalay, F. S., and C.-K. Li) **Middle Paleocene euprimate from southern China, and the distribution of primates in the Paleogene**. Journal of Human Evolution 15, 387–397.
93. (Szalay, F. S., and J. H. Langdon) **The foot of *Oreopithecus:* an evolutionary assessment**. Journal of Human Evolution 15, 585–621.
94. **Pedal grasping in primate evolution**. (Abs.) American Journal of Physical Anthropology 72(2), 260–261.

1987

95. **Review of** "Major Topics in Primate and Human Evolution" by B. Wood, L. Martin, and P. Andrews (Eds.). Quarterly Review of Biology 62, 353.
96. (Szalay, F. S., A. L. Rosenberger, and M. Dagosto) **Diagnosis and differentiation of the order Primates**. Yearbook of Physical Anthropology 30, 75–105.

1988

97. **A window into evolutionary primate systematics** (a review of "Comparative Primate Biology, Volume 1: Systematics, Evolution, and Anatomy", by D. R. Swindler and J. Erwin, Eds.), American Journal of Primatology, 14, 301–303.
98. (Szalay, F. S. and M. Dagosto) **Evolution of hallucial grasping in the Primates**. Journal of Human Evolution 17(1–2), 1–33.
99. **Anaptomorphinae**. In: I. Tattersall, E. Delson, and J. Van Couvering (Eds.). Encyclopedia of Human Evolution and Prehistory. Garland Publishing, New York/London, pp. 29–31.

100. **Apatemyidae**. In: I. Tattersall, E. Delson, and J. Van Couvering (Eds.). Encyclopedia of Human Evolution and Prehistory. Garland Publishing, New York/London, p. 39.
101. **Archonta**. In: I. Tattersall, E. Delson, and J. Van Couvering (Eds.). Encyclopedia of Human Evolution and Prehistory. Garland Publishing, New York/London, pp. 56–57.
102. **Carpolestidae**. In: I. Tattersall, E. Delson, and J. Van Couvering (Eds.). Encyclopedia of Human Evolution and Prehistory. Garland Publishing, New York/London, pp. 110–111.
103. ***Donrussellia***. In: I. Tattersall, E. Delson, and J. Van Couvering (Eds.). Encyclopedia of Human Evolution and Prehistory. Garland Publishing, New York/London, p. 166.
104. **Euprimates**. In: I. Tattersall, E. Delson, and J. Van Couvering (Eds.). Encyclopedia of Human Evolution and Prehistory. Garland Publishing, New York/London, pp. 184–185.
105. **Evolutionary Morphology**. In: I. Tattersall, E. Delson, and J. Van Couvering (Eds.). Encyclopedia of Human Evolution and Prehistory. Garland Publishing, New York/London, pp. 199–200.
106. **Evolutionary Systematics**. In: I. Tattersall, E. Delson, and J. Van Couvering (Eds.). Encyclopedia of Human Evolution and Prehistory. Garland Publishing, New York/London, pp. 200–201.
107. **Haplorhini**. In: I. Tattersall, E. Delson, and J. Van Couvering (Eds.). Encyclopedia of Human Evolution and Prehistory. Garland Publishing, New York/London, p. 242–243.
108. **Microchoerinae**. In: I. Tattersall, E. Delson, and J. Van Couvering (Eds.). Encyclopedia of Human Evolution and Prehistory. Garland Publishing, New York/London, pp. 339–340.
109. **Microsyopidae**. In: I. Tattersall, E. Delson, and J. Van Couvering (Eds.). Encyclopedia of Human Evolution and Prehistory. Garland Publishing, New York/London, p. 340.
110. **Omomyidae**. In: I. Tattersall, E. Delson, and J. Van Couvering (Eds.). Encyclopedia of Human Evolution and Prehistory. Garland Publishing, New York/London, pp. 396–397.
111. **Paromomyidae**. In: I. Tattersall, E. Delson, and J. Van Couvering (Eds.). Encyclopedia of Human Evolution and Prehistory. Garland Publishing, New York/London, pp. 443–444.
112. **Paromomyoidea**. In: I. Tattersall, E. Delson, and J. Van Couvering (Eds.). Encyclopedia of Human Evolution and Prehistory. Garland Publishing, New York/London, p. 444–445.
113. **Picrodontidae**. In: I. Tattersall, E. Delson, and J. Van Couvering (Eds.). Encyclopedia of Human Evolution and Prehistory. Garland Publishing, New York/London, p. 452.
114. **Plesiadapidae**. In: I. Tattersall, E. Delson, and J. Van Couvering (Eds.). Encyclopedia of Human Evolution and Prehistory. Garland Publishing, New York/London, pp. 464–466.
115. **Plesiadapiformes**. In: I. Tattersall, E. Delson, and J. Van Couvering (Eds.). Encyclopedia of Human Evolution and Prehistory. Garland Publishing, New York/London, pp. 466.
116. **Plesiadapoidea**. In: I. Tattersall, E. Delson, and J. Van Couvering (Eds.). Encyclopedia of Human Evolution and Prehistory. Garland Publishing, New York/London, p. 466–467.
117. **Saxonellidae**. In: I. Tattersall, E. Delson, and J. Van Couvering (Eds.). Encyclopedia of Human Evolution and Prehistory. Garland Publishing, New York/London, p. 504.
118. **Tarsiiformes**. In: I. Tattersall, E. Delson, and J. Van Couvering (Eds.). Encyclopedia of Human Evolution and Prehistory. Garland Publishing, New York/London, pp. 568–570.

1989

119. **Review of** "Vertebrates, Phylogeny, and Philosophy: A tribute to George Gaylord Simpson" (K. M. Flanagan and J. A. Lillegraven, Eds.). Journal of Human Evolution; 18(4), 415–416.
120. **Review of** "The Phylogenetic System. The systematization of organisms on the basis of their phylogenesis" by P. Ax; American Journal of Physical Anthropology 78(1), 124–126.
121. **On the evolutionary and interpretive frameworks of hominid lineage paleobiology**. In: Evolutionary Biology at the Crossroads (M. K. Hecht, Ed.), Queens College Press, Flushing, New York, pp. 145–148.

1990

122. (Szalay, F. S. and R. K. Costello) **Evolution of permanent estrus displays (PED) in the Hominidae**. (Abs.), American Journal of Physical Anthropology 81(2), 305.
123. **Evolution of the tarsal complex in Mesozoic mammals**. (Abs.), Journal of Vertebrate Paleontology 9(3), 45A.
124. (Szalay, F. S., and S. G. Lucas) **Postcranial skeleton of *Mixodectes* and a rediagnosis of the Primates**. (Abs.), Journal of Vertebrate Paleontology. 9(3), 45A.

1991

125. **Review of** "Functional morphology of the evolving hand and foot" by O. J. Lewis; Journal of Physical Anthropology 84(4), 493–495.
126. **Causal analysis in historical-narrative explanations**; (Abs.), American Journal of Physical Anthropology Supplement 12, 171.
127. (Szalay, F. S. and W. J. Bock) **Evolutionary theory and systematics: relationships between process and patterns**. Zeitschrift für zoologische Systematik und Evolutionsforschung 29, 1–39.
128. **The unresolved world between taxonomy and population biology: what is, and what is not, macroevolution?** (Lead review-essay of Genetics, Paleontology, and Macroevolution by Jeffrey Levinton) Journal of Human Evolution 20, 271–280.
129. (Szalay, F. S. and R. K Costello) **Evolution of permanent estrus displays in hominids**. Journal of Human Evolution 20, 439–464.

1993

130. (Szalay, F. S., M. J. Novacek, and M. C. McKenna, Eds.) **Mammal Phylogeny: Mesozoic differentiation, Monotremes, Therians, and Marsupials**. Springer, New York, 249 pp.
131. (Szalay, F. S., M. J. Novacek, and M. C. McKenna, Eds.) **Mammal Phylogeny: Placentals;** Springer, New York, 341 pp.
132. **Introduction**. In: F. S. Szalay, M. J. Novacek, and M. C. McKenna (Eds.). Mammal Phylogeny: Mesozoic Differentiation, Monotremes, Therians, and Marsupials. Springer, New York, pp. 1–3.
133. **Pedal evolution of mammals in the Mesozoic: tests for taxic relationships**. In: F. S. Szalay, M. J. Novacek, and M. C. McKenna (Eds.). Mammal Phylogeny: Mesozoic Differentiation, Monotremes, Therians, and Marsupials. Springer, New York, pp. 108–128.

134. **Metatherian taxon phylogeny: evidence and interpretation from the cranioskeletal system**. In: F. S. Szalay, M. J. Novacek, and M. C. McKenna (Eds.). Mammal Phylogeny: Mesozoic Differentiation, Monotremes, Therians, and Marsupials. Springer, New York, pp. 216–242.
135. **Introduction**. In: F. S. Szalay, M. J. Novacek, and M. C. McKenna (Eds.). Mammal Phylogeny: Placentals. Springer, New York, pp. 1–4.
136. **Species concepts: the tested, the untestable, and the redundant**. In: W. H. Kimbel, and L. B. Martin (Eds.). Species, Species Concepts, and Primate Evolution. Plenum, New York, pp. 21–41.
137. (Costello, R. K., C. Dickinson, A. L. Rosenberger, S. Boinski, and F. S. Szalay). **Squirrel Monkey (Genus *Saimiri*) taxonomy: a multidisciplinary study of the biology of species**. In: W. H. Kimbel, and L. B. Martin (Eds.). Species, Species Concepts, and Primate Evolution. Plenum, New York, pp. 177–210.
138. (Szalay, F. S. and S. G. Lucas) **Cranioskeletal morphology of archontans, and Diagnoses of Chiroptera, Volitantia, and Archonta**. In: R. D. E. MacPhee (Ed.). Primates and their Relatives in Phylogenetic Perspective. Plenum, New York, pp. 187–226.
139. (Trofimov, B. A., and F. S. Szalay) **New group of Asiatic marsupials (order Asiadelphia) from the Late Cretaceous of Mongolia**. Journal of Vertebrate Paleontology 13(Supplement 3), 60A.

1994

140. (Szalay, F. S., and F. Schrenk) **Middle Eocene *Eurotamandua* and the early differentiation of the Edentatata**. Journal of Vertebrate Paleontology 14 (Supplement 3), 48A.
141. (Trofimov, B. A., and Szalay, F. S.) **New Cretaceous marsupial from Mongolia and the early radiation of the Metatheria.** Proceedings of the National Academy of Sciences USA 91, 12569–12573.
142. **Evolutionary history of the marsupials and an analysis of osteological characters**. Cambridge University Press, New York, 481 pp.

1995

143. (Emry, R. J., S. G. Lucas, F. S. Szalay, and P. A. Tleuberdina) **A new herpetotheriine didelphid (Marsupialia) from the Oligocene of Central Asia**. Journal of Vertebrate Paleontology 15, 850–854.
144. **Review of** Evolution of the cercopithecoid Forelimb – Phylogenetic and Functional Implications from Morphometric Analyses, by R. L. Ciochon, 1994, University of California Publications, Geological Sciences, Volume 138, xxi + 251 pp., in Journal of Vertebrate Paleontology 15, 214–215.

1996

145. (Szalay, F. S., and S. G. Lucas) **The postcranial morphology of Paleocene *Chriacus* and *Mixodectes* and the phylogenetic relationships of archontan mammals**. Bulletin New Mexico Museum of Natural History 7, 1–47.
146. (Szalay, F. S., and B. A. Trofimov) **The Mongolian Late Cretaceous *Asiatherium*, and the early phylogeny and paleobiogeography of Metatheria**. Journal of Vertebrate Paleontology 16, 474–509.

1997

147. **The phylogenetic affinities of *Dromiciops* and the osteology of South American marsupials.** Noticiero de Biología (Organo oficial de sociedad de biología de Chile) 5(4), 48.

1998

148. (Szalay, F. S., and Schrenk, F.) **The middle Eocene *Eurotamandua* and a Darwinian phylogenetic analysis of the "edentates"**. Kaupia 7, 97–186.

1999

149. **Review of** *Classification of mammals above the species level*, by M. C. McKenna and S. K. Bell (with contributions from G.G. Simpson, R. H. Nichols, R. H. Tedford, K. F. Koopman, G. G. Musser, N. A. Neff, J. Shoshani, and D. M. McKenna), 1997, Columbia University Press, New York, 631 + xii pp., in Journal of Vertebrate Paleontology 19, 191–195.
150. **Paleontology and Macroevolution: On the theoretical conflict between an expanded synthesis and hierarchic punctuationism**. In: T. G. Bromage, and F. Schrenk (Eds.). African Biogeography, Climate Change, and Human Evolution. Oxford University Press, New York, pp. 35–56.
151. (Szalay, F. S., and Sargis, E. J.) **Paleocene marsupial postcranials from Itaboraí, Brazil: Model-based analysis of adaptations, phylogeny, & biogeography**. Journal of Vertebrate Paleontology 19(Supplement 3), 80A.

2000

152. **Review of** "Walker's Mammals of the World. Sixth Edition, Volumes I and II" (Nowak R. M.), 1999, Johns Hopkins University Press, Baltimore, MD. The Quarterly Review of Biology 75(1), 71–72.
153. **Marsupials**. In: R. Singer (Ed.). Encyclopedia of Paleontology. Fitzroy Dearborn Publishers pp. 703–714.
154. (Stafford, B. J., and F. S. Szalay). **Craniodental functional morphology and taxonomy of dermopterans**. Journal of Mammalogy 81(20), 360–385.
155. (Szalay, F. S., E. J. Sargis, and B. J. Stafford). **Small marsupial glider from the Paleocene of Itaboraí, Brazil**. Journal of Vertebrate Paleontology 20(Supplement 3), 73A.
156. (Schrenk, F., and F. S. Szalay). **Enigmatic new mammal (Dermoptera?) from the Messel middle Eocene, Germany**. Journal of Vertebrate Paleontology 20(Supplement 3), 68A.
157. (Warshaw, J., S. C. McFarlin, T. G. Bromage, and F. S. Szalay). **Some bone microstructure variables in extant therians, and their relationship to life history, locomotion, and phylogeny**. Journal of Vertebrate Paleontology 20(Supplement 3), 76A–77A.
158. **Review of** "The Atlas of European Mammals" (Mitchell-Jones, A. J. et al., Eds.) Academic, London. The Quarterly Review of Biology 75(4), 472.
159. (Goldman H. M, S. C. McFarlin, J. Warshaw, F. S. Szalay., and T.G. Bromage.). **Application of bone microstructural analysis to the comparative study of primate functional adaptation and life history**. American Journal of Physical Anthropology (Supplement 30), 160–161.
160. **Function and Adaptation in Paleontology and Phylogenetics: Why Do We Omit Darwin?** Palaeontologia Electronica 3(2), 25 pp., 372KB. *http://palaeo-electronica.org/2000_2/darwin/issue2_00.htm*

161. ***Altiatlasius***. In: E. Delson, I. Tattersall, J. A. Van Couvering and A. S Brooks (Eds.). Encyclopedia of Human Evolution and Prehistory (2nd Edition). Garland Publishing, New York, pp. 42–43.
162. **Anaptomorphinae**. In: E. Delson, I. Tattersall, J. A. Van Couvering and A. S Brooks (Eds.). Encyclopedia of Human Evolution and Prehistory (2nd Edition). Garland Publishing, New York, pp. 51–53.
163. **Apatemyidae**. In: E. Delson, I. Tattersall, J. A. Van Couvering and A. S Brooks (Eds.). Encyclopedia of Human Evolution and Prehistory (2nd Edition). Garland Publishing, New York, pp. 64–65.
164. **Archonta**. In: E. Delson, I. Tattersall, J. A. Van Couvering and A. S Brooks (Eds.). Encyclopedia of Human Evolution and Prehistory (2nd Edition). Garland Publishing, New York, p. 82.
165. **Carpolestidae**. In: E. Delson, I. Tattersall, J. A. Van Couvering and A. S Brooks (Eds.). Encyclopedia of Human Evolution and Prehistory (2nd Edition). Garland Publishing, New York, pp. 154–156.
166. ***Decoredon***. In: E. Delson, I. Tattersall, J. A. Van Couvering and A. S Brooks (Eds.). Encyclopedia of Human Evolution and Prehistory (2nd Edition). Garland Publishing, New York, p. 206.
167. **Dermoptera**. In: E. Delson, I. Tattersall, J. A. Van Couvering and A. S Brooks (Eds.). Encyclopedia of Human Evolution and Prehistory (2nd Edition). Garland Publishing, New York, pp. 207–208.
168. ***Donrussellia***. In: E. Delson, I. Tattersall, J. A. Van Couvering and A. S Brooks (Eds.). Encyclopedia of Human Evolution and Prehistory (2nd Edition). Garland Publishing, New York, p. 220.
169. **Ekgmowechashalinae**. In: E. Delson, I. Tattersall, J. A. Van Couvering and A. S Brooks (Eds.). Encyclopedia of Human Evolution and Prehistory (2nd Edition). Garland Publishing, New York, p. 233.
170. (Szalay, F. S. and J. A. Van Couvering). **Eocene**. In: E. Delson, I. Tattersall, J. A. Van Couvering and A. S Brooks (Eds.). Encyclopedia of Human Evolution and Prehistory (2nd Edition). Garland Publishing, New York, pp. 234–235.
171. **Eosimiidae**. In: E. Delson, I. Tattersall, J. A. Van Couvering and A. S Brooks (Eds.). Encyclopedia of Human Evolution and Prehistory (2nd Edition). Garland Publishing, New York, p. 235.
172. **Euprimates**. In: E. Delson, I. Tattersall, J. A. Van Couvering and A. S Brooks (Eds.). Encyclopedia of Human Evolution and Prehistory (2nd Edition). Garland Publishing, New York, p. 239.
173. **Evolutionary morphology**. In: E. Delson, I. Tattersall, J. A. Van Couvering and A. S Brooks (Eds.). Encyclopedia of Human Evolution and Prehistory (2nd Edition). Garland Publishing, New York, pp. 258–259.
174. **Evolutionary systematics**. In: E. Delson, I. Tattersall, J. A. Van Couvering and A. S Brooks (Eds.). Encyclopedia of Human Evolution and Prehistory (2nd Edition). Garland Publishing, New York, pp. 259–261.
175. **Flying primate hypothesis**. In: E. Delson, I. Tattersall, J. A. Van Couvering and A. S Brooks (Eds.). Encyclopedia of Human Evolution and Prehistory (2nd Edition). Garland Publishing, New York, pp. 271–272.
176. **Haplorhini**. In: E. Delson, I. Tattersall, J. A. Van Couvering and A. S Brooks (Eds.). Encyclopedia of Human Evolution and Prehistory (2nd Edition). Garland Publishing, New York, pp. 303–304.
177. ***Hoanghonius***. In: E. Delson, I. Tattersall, J. A. Van Couvering and A. S Brooks (Eds.). Encyclopedia of Human Evolution and Prehistory (2nd Edition). Garland Publishing, New York, p. 306.
178. ***Mahgarita***. In: E. Delson, I. Tattersall, J. A. Van Couvering and A. S Brooks (Eds.). Encyclopedia of Human Evolution and Prehistory (2nd Edition). Garland Publishing, New York, p. 403.
179. **Microchoerinae**. I In: E. Delson, I. Tattersall, J. A. Van Couvering and A. S Brooks (Eds.). Encyclopedia of Human Evolution and Prehistory (2nd Edition). Garland Publishing, New York, pp. 410–411.
180. **Microsyopidae**. In: E. Delson, I. Tattersall, J. A. Van Couvering and A. S Brooks (Eds.). Encyclopedia of Human Evolution and Prehistory (2nd Edition). Garland Publishing, New York, p. 411.
181. **Omomyidae**. In: E. Delson, I. Tattersall, J. A. Van Couvering and A. S Brooks (Eds.). Encyclopedia of Human Evolution and Prehistory (2nd Edition). Garland Publishing, New York, pp. 493–495.
182. **Omomyinae**. In: E. Delson, I. Tattersall, J. A. Van Couvering and A. S Brooks (Eds.). Encyclopedia of Human Evolution and Prehistory (2nd Edition). Garland Publishing, New York, pp. 495–497.
183. **Paleobiology**. In: E. Delson, I. Tattersall, J. A. Van Couvering and A. S Brooks (Eds.). Encyclopedia of Human Evolution and Prehistory (2nd Edition). Garland Publishing, New York, pp. 506–507.
184. **Paleocene**. In: E. Delson, I. Tattersall, J. A. Van Couvering and A. S Brooks (Eds.). Encyclopedia of Human Evolution and Prehistory (2nd Edition). Garland Publishing, New York, pp. 507–508.
185. **Paleogene**. In: E. Delson, I. Tattersall, J. A. Van Couvering and A. S Brooks (Eds.). Encyclopedia of Human Evolution and Prehistory (2nd Edition). Garland Publishing, New York, p. 512.
186. **Paromomyidae**. In: E. Delson, I. Tattersall, J. A. Van Couvering and A. S Brooks (Eds.). Encyclopedia of Human Evolution and Prehistory (2nd Edition). Garland Publishing, New York, pp. 548–550.
187. **Paromomyoidea**. In: E. Delson, I. Tattersall, J. A. Van Couvering and A. S Brooks (Eds.). Encyclopedia of Human Evolution and Prehistory (2nd Edition). Garland Publishing, New York, p. 550.
188. ***Petrolemur*** In: E. Delson, I. Tattersall, J. A. Van Couvering and A. S Brooks (Eds.). Encyclopedia of Human Evolution and Prehistory (2nd Edition). Garland Publishing, New York, p. 555.
189. **Picrodontidae**. In: E. Delson, I. Tattersall, J. A. Van Couvering and A. S Brooks (Eds.). Encyclopedia of Human Evolution and Prehistory (2nd Edition). Garland Publishing, New York, pp. 558–559.
190. **Plesiadapidae**. In: E. Delson, I. Tattersall, J. A. Van Couvering and A. S Brooks (Eds.). Encyclopedia of Human Evolution and Prehistory (2nd Edition). Garland Publishing, New York, pp. 570–572.

191. **Plesiadapiformes**. In: E. Delson, I. Tattersall, J. A. Van Couvering and A. S Brooks (Eds.). Encyclopedia of Human Evolution and Prehistory (2nd Edition). Garland Publishing, New York, pp. 572–573.
192. **Plesiadapoidea**. In: E. Delson, I. Tattersall, J. A. Van Couvering and A. S Brooks (Eds.). Encyclopedia of Human Evolution and Prehistory (2nd Edition). Garland Publishing, New York, pp. 573–574.
193. **Saxonellidae**. In: E. Delson, I. Tattersall, J. A. Van Couvering and A. S Brooks (Eds.). Encyclopedia of Human Evolution and Prehistory (2nd Edition). Garland Publishing, New York, p. 626.
194. ***Shoshonius***. In: E. Delson, I. Tattersall, J. A. Van Couvering and A. S Brooks (Eds.). Encyclopedia of Human Evolution and Prehistory (2nd Edition). Garland Publishing, New York, pp. 635–636.
195. **Tarsiiformes**. In: E. Delson, I. Tattersall, J. A. Van Couvering and A. S Brooks (Eds.). Encyclopedia of Human Evolution and Prehistory (2nd Edition). Garland Publishing, New York, pp. 691–693.
196. **Tarsioidea**. In: E. Delson, I. Tattersall, J. A. Van Couvering and A. S Brooks (Eds.). Encyclopedia of Human Evolution and Prehistory (2nd Edition). Garland Publishing, New York, pp. 693–694.
197. **Visual predation hypothesis**. In: E. Delson, I. Tattersall, J. A. Van Couvering and A. S Brooks (Eds.). Encyclopedia of Human Evolution and Prehistory (2nd Edition). Garland Publishing, New York, pp. 725–726.

2001

198. (Szalay, F. S., and E. J. Sargis). **Model-based analysis of postcranial osteology of marsupials from the Palaeocene of Itaboraí (Brazil), and the phylogenetics and biogeography of Metatheria**. Geodiversitas 23(2), 139–302 (concurrently published on the Geodiversitas site: *http://www.mnhn.fr/publication/geodiv/g01n2a1.pdf*)
199. (McFarlin, S. C, Warshaw J., Szalay F. S., and Bromage T. G.). **Microstructural features in mammalian bone as indicators of organismal life history: a survey of the order Primates**. Journal of Vertebrate Paleontology 21(Supplement 3), 79A.
200. **Problems with hominid fossil species taxa and the construction of taxograms**. Ludus Vitalis 9, 143–169.
201. (Szalay, F. S., and F. Schrenk). **An enigmatic new mammal (Dermoptera?) from the Messel Middle Eocene, Germany**. Kaupia 11, 153–164.
202. (Warshaw J., S.C. McFarlin, T. G. Bromage, C. J. Terranova, F. S. Szalay). **The effects of life history, locomotion and phylogeny on bone microstructural features in mammals**. Journal of Morphology 248(3), 298.

2002

203. (Warshaw J., T. G. Bromage, C. J. Terranova, F. S. Szalay). **Diversity in bone collagen fiber orientation patterns among primates and other mammals**. Journal of Vertebrate Paleontology 22(Supplement 3), 117A–118A.

2003

204. (Warshaw J., T. G. Bromage, C. J. Terranova, F.S. Szalay). **Primate bone microstructural variability: relationships to mechanical and life history adaptation**. American Journal of Physical Anthropology Supplement 36, 220.

2004

205. (Salton, J. A. and F. S. Szalay). **The tarsal complex of Afro-Malagasy Tenrecoidea: A search for phylogenetically meaningful characters**. Journal of Mammalian Evolution 11, 73–104.
206. **Review of** "Late Cretaceous and Cenozoic Mammals of North America: Biostratigraphy and Geochronology" (M. O. Woodburne, Ed.), 2004, Journal of Mammalian Evolution 11, 205–206.

2005

207. **Review of** "The Rise of Placental Mammals. Origins and Relationships of the Major Clades" (Rose, K. D., and J. D. Archibald, Eds.), 2005, Journal of Mammalian Evolution 12, 533–542.

2006

208. **Review of** "Macroevolution-Diversity, Disparity, Contingency: Essays in Honor of Stephen Jay Gould" (Vrba E., and N. Eldredge, Eds.), 2006, Journal of Mammalian Evolution 13, 165–166.
209. (Szalay, F. S., and E. J. Sargis). **Cretaceous therian tarsals and the Metatherian-Eutherian dichotomy**. Journal of Mammalian Evolution 13, 171–210.

2007

210. **Ancestral locomotor modes, placental mammals, and the origin of euprimates: Lessons from history**. In: M. J. Ravosa and M. Dagosto (Eds.). Primate Origins: Adaptations and Evolution. Springer, New York, pp. 457–487.
211. (Chester, S. G. B., Sargis, E. J., Szalay, F. S., Archibald, J. D., Averianov, A. O.) **Functional analysis of mammalian humeri from the Late Cretaceous of Uzbekistan**. Journal of Vertebrate Paleontology 27(Supplement), 58A.

In press

212. **Connections between modeling, vertebrate paleobiology, and systematics**. American Zoologist (Walter J. Bock Festschrift volume).

References (not including those listed in Szalay bibliography)

Amrine-Madsen, H., Scally, M., Westerman, M., Stanhope, M. J., Krajewski, C. and Springer, M. S., 2003. Nuclear gene sequences provide evidence for the monophyly of australidelphian marsupials. Molecular Phylogenetics and Evolution 28, 186–196.

Beard, K. C., 1987. *Jemezius*, a new omomyid primate from the early Eocene of northwestern New Mexico. Journal of Human Evolution 16, 457–468.

Bloch, J. I. and Boyer, D. M., 2002. Grasping primate origins. Science 298, 1606–1610.

Bloch, J. I., Silcox, M. T., Boyer, D. M. and Sargis, E. J., 2007. New Paleocene skeletons and the relationship of plesiadapiforms to crown-clade primates. Proceedings of the National Academy of Sciences USA 104, 1159–1164.

Bown, T. M. and Rose, K. D., 1991. Evolutionary relationships of a new genus and three new species of omomyid primates (Willwood Formation, Lower Eocene, Bighorn Basin, Wyoming). Journal of Human Evolution 20, 465–480.

de Muizon, C. and Cifelli, R., 2001. A new basal “didelphoid” (Marsupialia, Mammalia) from the early Paleocene of Tiupampa (Bolivia). Journal of Vertebrate Paleontology 21, 87–97.

Gregory, W. K., 1910. The orders of mammals. Bulletin of the American Museum of Natural History 27, 1–524.

Horovitz, I. and Sanchez-Villagra, M.R., 2003. A morphological analysis of marsupial mammal higher-level phylogenetic relationships. Cladistics 19, 181–212.

Luo, Z.-X., Ji, Q., Wible, J. R. and Yuan, C.-X., 2003. An early Cretaceous tribosphenic mammal and metatherian evolution. Science 302, 1934–1940.

McKenna, M. C., 1975. Toward a phylogenetic classification of the Mammalia. In: Luckett, W. P. and Szalay, F. S. (Eds.), Phylogeny of the Primates: A Multidisciplinary Approach. Plenum, New York, **pp**. 21–46.

Murphy, W. J., Eizirik, E., O’Brien, S. J., Madsen, O., Scally, M., Douady, C. J., Teeling, E., Ryder, O. A., Stanhope, M. J., De Jong, W. W. and Springer, M. S., 2001. Resolution of the early placental mammal radiation using Bayesian phylogenetics. Science 294, 2348–2351.

Rosenberger, A. L., Hartwig, W. C. and Wolff, R. G., 1991. *Szalatavus attricuspis*, an early platyrrhine primate. Folia Primatologica 56, 225–233.

Silcox, M. T., 2001. A phylogenetic analysis of Plesiadapiformes and their relationship to Euprimates and other archontans. Ph.D. dissertation, Johns Hopkins University, School of Medicine, Baltimore, MD.

Simpson, G. G., 1978. Variations and details of macroevolution. Paleobiology 4, 217–221.

Sloan, R. E., 1981. Systematics of Paleocene multituberculates from the San Juan Basin, New Mexico. In: Lucas, S. G., Rigby, J. K. and Kues, B. S. (Eds.), Advances in San Juan Basin Paleontology. University of New Mexico Press, Albuquerque, **pp**. 127–160.

Acknowledgments

We thank Rob Asher, Bert Covert, Darin Croft, John Fleagle, Dick Fox, Haviva Goldman, Terry Harrison, Luke Holbrook, Zhexi Luo, Greg McDonald, Mike Plavcan, James Rossie, Guillermo Rougier, Tony Tosi, Blaire Van Valkenburgh, Robert Walker, John Wible, and Steve Wroe for serving as outside reviewers of the chapters in this book, and especially Carl Terranova, Erik Seiffert, Bruce Shockey, and Mary Silcox for reviewing multiple manuscripts. We also thank the series editors, Eric Delson and Ross MacPhee, for all their help with this volume. Finally, we thank Frederick S. Szalay who served as our graduate advisor, and to whom we both owe a huge debt. Although it will be impossible to fully repay that debt, we consider this volume to be a small token of our appreciation for all the support, encouragement, and friendship he has provided.

List of Contributors

Federico Anaya
Facultad de Ingeniería Geológica
Universidad Autónoma "Tomás Frías"
Potosí, Bolivia

Christine Argot
Département Histoire de la Terre
Muséum national d'Histoire naturelle
8 rue Buffon
75005 Paris, France
argot@mnhn.fr

K. Christopher Beard
Section of Vertebrate Paleontology
Carnegie Museum of Natural History
4400 Forbes Avenue
Pittsburgh, PA 15213, USA
beardc@carnegiemnh.org

Lilian P. Bergqvist
Universidade Federal do Rio de Janeiro
Departamento de Geologia
Av. Athos da Silveira Ramos, 274
Rio de Janeiro/RJ
21941-916 Brasil
bergqvist@geologia.ufrj.br

Jonathan I. Bloch
Florida Museum of Natural History
University of Florida
P.O. Box 117800
Gainesville, FL 32611, USA
jbloch@flmnh.ufl.edu

Doug M. Boyer
Department of Anatomical Sciences
Stony Brook University
T8 040 Health Science Center
Stony Brook, NY 11733, USA
dboyer@ic.sunysb.edu

Richard L. Cifelli
Sam Noble Oklahoma Museum of Natural History
and Department of Zoology
University of Oklahoma
2401 Chautauqua Ave.
Norman, OK 73072, USA
rlc@ou.edu

Sébastien Couette
Ecole Pratique des Hautes Etudes
UMR 5143 Paléobiodiversité et
Paléoenvironnements
Case Courrier 38
Département d'Histoire de la Terre
Muséum national d'Histoire naturelle
8 rue Buffon
75005 Paris, France
couette@mnhn.fr

Marian Dagosto
Department of Cell and Molecular Biology
Feinberg School of Medicine
Northwestern University
303 E. Chicago Ave.
Chicago, IL 60611, USA
m-dagosto@northwestern.edu

Brian M. Davis
Sam Noble Oklahoma Museum of Natural History and
Department of Zoology
University of Oklahoma
2401 Chautauqua Ave.
Norman, OK 73072, USA
bmdavi@ou.edu

Eric Delson
Department of Anthropology, Lehman College/CUNY
New York Consortium in Evolutionary Primatology
(NYCEP)
Department of Vertebrate Paleontology

American Museum of Natural History
Central Park West & 79th Street
New York, NY 10024, USA
eric.delson@lehman.cuny.edu

Timothy F. Flannery
Division of Environment and Life Sciences
Macquarie University
Sydney, 2109, Australia

Stephen R. Frost
Department of Anthropology
University of Oregon
Eugene, OR 97403, USA
sfrost@uoregon.edu

Daniel. L. Gebo
Department of Anthropology
Northern Illinois University
DeKalb, IL 60115, USA
dgebo@niu.edu

Wayne R. Gerdtz
School of Ecology and Environment
Melbourne Campus
Deakin University
Burwood, Victoria
3125, Australia

Marc Godinot
Ecole Pratique des Hautes Etudes
UMR 5143 Paléobiodiversité et
Paléoenvironnements
Case Courrier 38
Département d'Histoire de la Terre
Muséum national d'Histoire naturelle
8 rue Buffon
75005 Paris, France
godinot@mnhn.fr

William E. H. Harcourt-Smith
Department of Vertebrate Paleontology
American Museum of Natural History
Central Park West & 79th Street
New York, NY 10024, USA
willhs@amnh.org

Russell Hogg
Department of Anthropology
The Graduate Center
City University of New York
365 Fifth Avenue
New York, NY 10016, USA
hogg@nycep.org
New York Consortium in Evolutionary
Primatology (NYCEP)
Hard Tissue Research Unit
New York University

Benjamin P. Kear
Department of Genetics
School of Molecular Sciences
La Trobe University
Melbourne, Victoria 3086
South Australian Museum
North Terrace, Adelaide
South Australia 5000, Australia
kear.ben@saugov.sa.gov.au

Zofia Kielan-Jaworowska
Instytut Paleobiologii PAN
ul. Twarda 51/55
PL-00-818 Warszawa, Poland
zkielan@twarda.pan.pl

Michael S. Y. Lee
South Australian Museum
North Terrace, Adelaide
South Australia 5000,
Australia

Wolfgang Maier
Lehrstuhl Spezielle Zoologie
Universität Tübingen
Auf der Morgenstell 28
D-72076 Tübingen, Germany
wolfgang.maier@uni-tuebingen.de

Xijun Ni
Department of Vertebrate Paleontology
American Museum of Natural History
Central Park West & 79th Street
New York, NY 10024, USA
nixj@amnh.org

Jay A. O'Sullivan
Department of Exercise Science and
Sport Studies
University of Tampa
401 West Kennedy Boulevard
Tampa, FL 33606, USA
josullivan@ut.edu

Tonya A. Penkrot
Marshall University
Department of Biological Sciences
1 John Marshall Drive
Huntington, WV 25755, USA
penkrot@marshall.edu

P. David Polly
Department of Geological Sciences
Indiana University
1001 East 10th Street
Bloomington, IN 47405, USA
pdpolly@indiana.edu

Tao Qi
Institute of Vertebrate Paleontology and Paleoanthropology
Chinese Academy of Sciences
Beijing, 100044, China

F. James Rohlf
Department of Ecology and Evolution
Stony Brook University
Stony Brook, NY 11794, USA
rohlf@life.bio.sunysb.edu

Kenneth D. Rose
Johns Hopkins University School of Medicine
Center for Functional Anatomy and Evolution
1830 E. Monument St.
Baltimore, MD 21205, USA
kdrose@jhmi.edu

Alfred L. Rosenberger
Department of Anthropology and Archaeology
Brooklyn College, CUNY
2900 Bedford Ave.
Brooklyn, NY 11210, USA
alfredr@brooklyn.cuny.edu
Department of Anthropology
The Graduate Center, CUNY
New York Consortium in Evolutionary
Primatology (NYCEP)
American Museum of Natural History/Mammalogy

Justine A. Salton
Program in Biology
Bard College
Annandale-on-Hudson, NY 12504, USA

Eric J. Sargis
Department of Anthropology
Yale University
P.O. Box 208277
New Haven, CT 06520, USA
Eric.Sargis@yale.edu
Division of Vertebrate Zoology
Peabody Museum of Natural History

Bruce J. Shockey
Biology Department
Manhattan College
Manhattan College Parkway
Riverdale, NY 10471, USA
bshockey@amnh.org
Department of Vertebrate Paleontology
American Museum of Natural History
Central Park West & 79th Street
New York, NY 10024, USA

Mary T. Silcox
University of Winnipeg
Department of Anthropology
515 Portage Ave.
Winnipeg, MB R3B 2E9, Canada
m.silcox@uwinnipeg.ca

Melissa Tallman
Department of Anthropology
The Graduate Center
City University of New York
365 Fifth Avenue
New York, NY 10016, USA
tallman@nycep.org
New York Consortium in Evolutionary
Primatology (NYCEP)

Carl J. Terranova
Department of Anatomy
Touro College of Osteopathic Medicine
230 West 125th St.
New York, NY 10027, USA
carl.terranova@touro.edu

Johanna Warshaw
Hard Tissue Research Unit
Departments of Biomaterials & Basic Sciences
Rm 817-S
New York University College of Dentistry
NYU Mail Code: 9448
345 East 24th Street
New York, NY 10010, USA
jw143@nyu.edu

David F. Wiley
Institute for Data Analysis and
Visualization
University of California
Davis, CA 95616, USA
wiley@cs.ucdavis.edu

Sai Man Wong
Department of Anthropology and Archaeology
Brooklyn College, CUNY
2900 Bedford Ave.
Brooklyn, NY 11210, USA
vmax137@yahoo.com

Shawn P. Zack
Marshall University
Department of Biological Sciences
1 John Marshall Drive
Huntington, WV 25755, USA
zack@marshall.edu

Contents

Section I
Non-primate Mammals

1. Earliest Evidence of Deltatheroida (Mammalia: Metatheria) from the Early Cretaceous of North America

Brian M. Davis*
Sam Noble Oklahoma Museum of Natural History
Department of Zoology
University of Oklahoma
2401 Chautauqua Ave.
Norman, OK 73072, USA
bmdavi@ou.edu

Richard L. Cifelli
Sam Noble Oklahoma Museum of Natural History
Department of Zoology
University of Oklahoma
2401 Chautauqua Ave.
Norman, OK 73072, USA
rlc@ou.edu

Zofia Kielan-Jaworowska
Instytut Paleobiologii PAN
ul. Twarda 51/55
PL-00-818 Warszawa
Poland
zkielan@twarda.pan.pl

1.1 Introduction

Deltatheroida are small therian mammals known only from the Cretaceous of Asia and North America. As fossils, they are represented mainly by isolated teeth and dentigerous jaws, though rostra, a petrosal, and the calcaneus, at least, have been described for the best known genus, Asiatic *Deltatheridium* (Rougier et al., 1998; Horovitz, 2000). Aside from two dubious forms: *Oxlestes* (Nessov, 1982) and *Khuduklestes* (Nessov et al., 1994), Deltatheroida are unambiguously represented by only four genera (*Deltatheridium, Deltatheroides, Deltatherus,* and *Sulestes*), all Asiatic in distribution and all placed in the family Deltatheridiidae (see Kielan-Jaworowska et al., 2004). The Asian record of Deltatheridiidae ranges from Coniacian to late Campanian. In North America, one genus, Aptian-Albian *Atokatheridium*, has been tentatively referred to Deltatheroida (Kielan-Jaworowska and Cifelli, 2001). Other records of the group on this continent consist of poorly represented, unnamed taxa from the Turonian (Cifelli, 1990a), late Campanian, and late Maastrichtian (Fox, 1974).

The dentition of deltatheroidans has beguiled mammalian systematists since the first specimens were found some 80 years ago, the main problem areas being molar structure and dental formula. It has long been generally agreed that the molars are characterized by a number of plesiomorphies: Gregory and

* Address for correspondence: bmdavi@ou.edu

E.J. Sargis and M. Dagosto (eds.), *Mammalian Evolutionary Morphology: A Tribute to Frederick S. Szalay*, 3–24.

Simpson (1926, p. 2), for example, described them as exhibiting a "pretritubercular stage of evolution" (in the sense of Osborn, 1907), and they lack certain apomorphies common to eutherians and metatherians (Cifelli, 1993a). Combined with this primitiveness (for example, the small protocone, broad stylar shelf, and weak conules on upper molars; small, poorly basined talonid, often with only hypoconid and hypoconulid, on lower molars), however, are certain specializations suggestive of carnivory (Butler, 1990a, b). Most significant in this regard is the hyperdevelopment of postvallum-prevallid shearing, as indicated by a salient, elongate postmetacrista on upper molars and enlarged paraconid-paracristid on lowers. This functional complex is associated with carnivory in living mammals and has been identified in various fossil forms: in addition to Deltatheroida, three groups of marsupials and as many as three groups of eutherians are characterized by the hypertrophied postvallum-prevallid shearing system (Muizon and Lange-Badré, 1997). Reduction of crushing and grinding function often accompanies hypertrophy of postvallum-prevallid shearing in molars of mammalian carnivores (MacIntyre, 1966; Muizon and Lange-Badré, 1997). This leaves open the door for interpretation of certain features of deltatheroidan molars (e.g., small protocone and small talonid) as correlates of carnivorous specialization, rather than plesiomorphies. The general consensus, however, seems to be the interpretation that deltatheroidans represent the first therians specialized for carnivory; and that otherwise, their molar structure is exceedingly primitive (e.g., Szalay, 1994). In any event, molar structure has proven to be of limited use in assessing broader relationships of Deltatheroida (however, Rougier et al. (2004) have identified several informative molar characteristics in deltatheroidans).

Interpretation of the postcanine dental formula in deltatheroidans has changed through the years, with significant implications for higher relationships of the group. Gregory and Simpson (1926) found the molar structure to be structurally antecedent to that of creodonts and certain insectivores, a view that attained wide acceptance (Matthew, 1928; Simpson, 1928, 1945). The first specimens to be described are poorly preserved; that of *Deltatheridium* preserved six upper and lower postcanine loci, and that of *Deltatheroides* preserved seven upper molar loci. Gregory and Simpson (1926) interpreted the specimens as preserving P/p1–3, M/m1–3 and P/p1–4, M/m1–3, respectively: that is, a eutherian (four premolars and three molars) or eutherian-derived pattern, rather than the count seen in marsupials (three premolars and four molars). This interpretation, based on poorly preserved specimens then available, was to lead mammalian systematists astray for nearly 50 years. Van Valen (1966) erected the order Deltatheridia to include creodonts and certain insectivores; and this view, or minor variants thereof, attained some general acceptance in the late 1960s and early 1970s (McKenna et al., 1971; Szalay and McKenna, 1971; McKenna, 1975). Based on new, more numerous, and better preserved specimens, Butler and Kielan-Jaworowska (1973; see also Kielan-Jaworowska, 1975) documented the presence of three premolars and four molars in *Deltatheroides* and the lower jaw of *Deltatheridium*; the existence of a fourth upper molar in *Deltatheridium* was later reported by Rougier et al. (1998). Despite this similarity to marsupials, deltatheroidans were for a time relegated to the taxonomic *Erebus* of "Theria of metatherian-eutherian grade" (Kielan-Jaworowska, 1975; Kielan-Jaworowska et al., 1979) or "tribotheres" (Butler, 1978; Clemens and Lillegraven, 1986). A metatherian relationship for Deltatheroida was first championed by Kielan-Jaworowska and colleagues (Kielan-Jaworowska and Nessov, 1990; Kielan-Jaworowska, 1992; Marshall and Kielan-Jaworowska, 1992), in part based on perceived similarities to Stagodontidae (North American, Cretaceous marsupials) and/ or Borhyaenoidea (South American, Cenozoic marsupials). More substantial support for this hypothesis has come from newly collected specimens of *Deltatheridium* from Mongolia, which show marsupial similarities in the pattern of tooth replacement, structure of the dentary, and aspects of cranial anatomy (Rougier et al., 1998). Nonetheless, the position of Deltatheroida as basal Metatheria remains precarious, differing even in some studies conducted by the same authors (e.g., Luo et al., 2002; Luo et al., 2003).

Though higher-level relationships of Deltatheroida are not directly relevant to the present paper (except, perhaps, in our conclusions as to molar count), we accept deltatheroidans as a stem group of Metatheria. This provides us with an excellent opportunity to link our chapter thematically with the purpose of this book: to honor Fred Szalay. Fred has worn many hats during his long, magnificently productive scientific career as a student of mammalian evolution. Lest physical anthropology attempt to lay proprietary claim on Fred Szalay, we point out that he is widely recognized for his seminal contributions on the evolutionary radiations of metatherian mammals. Szalay was the first to recognize fundamental, adaptively important differences in the ankle of metatherians and eutherians (Szalay, 1984), is the progenitor of a once-radical but now universally accepted hypothesis that South American microbiotheres are closely related to Australian marsupials (Szalay, 1982), and is the co-describer of the first Cretaceous marsupial from Asia (Trofimov and Szalay, 1994; Szalay and Trofimov, 1996). Szalay and Sargis (2001) reconstructed the early adaptive radiation of marsupials in South America based on form-function analysis, and tested hypotheses of marsupial relationships using the same data. In this context, it is also relevant to mention that Fred is author of a widely-cited book on the evolutionary history of marsupials (Szalay, 1994). We are pleased to offer this small contribution as a tribute to Fred Szalay, who has added so much to understanding of metatherian history.

1.1.1 Conventions and Abbreviations

We follow the general practice of abbreviating molars and premolars with the letters "M" and "P", respectively; teeth belonging to the lower dentition are indicated with a lower case letter. Right and left are abbreviated "R" and "L", respectively. Molar terminology follows that of Bown and Kraus (1979). Standards of measurement are illustrated in Figure 1.1, and measurements of all described specimens are listed in Table 1.1.

Institutional abbreviations: FMNH, Field Museum of Natural History, Chicago, Illinois; OMNH, Oklahoma Museum of Natural History, Norman, Oklahoma; SMP-SMU, Shuler Museum of Paleontology, Southern Methodist University, Dallas, Texas.

1.1.2 Historical Background

Of the many significant scientific advances made by the American Museum Central Asiatic Expeditions (Andrews, 1932), the discovery of Cretaceous mammals clearly ranks among the most groundbreaking. Mesozoic mammals had, of course, been known to science as early as the first half of the nineteenth century (Broderip, 1828); and both Cope (1882, 1892) and Marsh (1889a, b, 1892) described a number of Late Cretaceous taxa from western North America. These, however, were based on scant remains: jaws, or – more commonly – bits and pieces thereof. It thus came as a welcome surprise when skulls were reported from the Djadokhta Formation, in the Mongolian part of the Gobi Desert (Gregory and Simpson, 1926). Three of the five genera described by Gregory and Simpson were placed in the then new family Deltatheridiidae: *Deltatheridium*, known by two rostral parts of the skull and associated dentaries; *Deltatheroides*, known by a partial rostrum preserving partial crowns for the last four postcanine teeth; and *Hyotheridium*, represented by a snout with the upper and lower tooth rows interlocked. Of these, *Hyotheridium* is so poorly known as to be indeterminate (it may be a eutherian); we follow Kielan-Jaworowska et al. (2004) in considering the type and only species, *H. dobsoni*, as a *nomen dubium*. As noted above, Gregory and Simpson (1926) accorded Deltatheridiidae a basal position among Eutheria, an interpretation that was to remain unchallenged until new, more complete fossils were described in the 1970s.

The first record of deltatheroidans from North America is that of Fox (1974), who reported several *Deltatheroides*-like isolated teeth from Campanian and Maastrichtian horizons (units follow current stratigraphic nomenclature): an upper molar from the Scollard Formation, Alberta; a lower molar and a trigonid of another from the Lance Formation, Wyoming; and a talonid from the Dinosaur Park Formation, Alberta. Subsequently, Cifelli (1990a) described another fragmentary specimen (a lower molar trigonid) from the Smoky Hollow Member of the Straight Cliffs Formation (Turonian), Utah, referring the fossil to Deltatheridiidae, indet.

Returning to Asia, Kielan-Jaworowska (1975) had, in the meantime, described new material from the Gobi Desert, Mongolia, assembled by the Polish-Mongolian Palaeontological Expeditions. Recovered from both the Djadokhta and Baruungoyot formations, these fossils include five specimens (three rostra with dentaries, a maxilla, and a dentary) of *Deltatheridium* and a dentary assigned to *Deltatheroides*.

The geographic range of Deltatheroida was extended to middle Asia by Nessov (1985), who described *Sulestes karakshi* from the Bissekty Formation (Coniacian) of Uzbekistan. The holotype is a maxillary fragment with M1–2; Nessov (1987) later referred an isolated lower molar to the genus. In recognition of its distinctness from *Deltatheridium* and *Deltatheroides*, Nessov (1985) placed *Sulestes* in its own subfamily, Sulestinae. Kielan-Jaworowska and Nessov (1990) elaborated on the systematics of the group by removing *Deltatheroides* and some unnamed taxa to their own family, Deltatheroididae. Subsequent authors (McKenna and Bell, 1997; Kielan-Jaworowska et al., 2004; Rougier et al., 2004) have abandoned formal subdivision of Deltatheroida, recognizing (as we do herein) the single family Deltatheridiidae. A second deltatheroidan from the Bissekty Formation of Uzbekistan was initially described by Nessov (1993) as *Deltatheroides kizylkumensis* and later (Nessov, 1997) transferred to its own genus, *Deltatherus*. *D. kizylkumensis* is known by two lower molars and an edentulous fragment of a maxilla. A more recent record of a Mongolian deltatheroidan genus in the Cretaceous of middle Asia was provided by Averianov (1997), who named *Deltatheridium nessovi* from the Darbasa Formation (Campanian) of Kazakhstan. *D. nessovi* is known only by the labial part of an upper molar, perhaps M2.

Several significant fossils from Mongolia have been reported in recent years. Two new specimens of *Deltatheridium pretrituberculare*, represented by partial skulls with well-preserved upper and lower dentition and postcranial fragments, were collected in the Gobi Desert at the Ukhaa Tolgod locality, Nemegt Basin, by members of the Mongolian Academy of Sciences–American Museum of Natural History Expeditions. The most significant details of these specimens were published by Rougier et al. (1998) and Horovitz (2000). The nearby locality of Kholbot (Red Rum) yielded to the same field parties a maxilla of *Deltatheroides cretacicus*, including all four molars in a good state of preservation. As a result, this hitherto poorly understood taxon is incomparably better known (Rougier et al., 2004). For the sake of completeness, several other Asiatic taxa deserve passing mention: an undescribed specimen known as the "Gurlin Tsav skull", first thought to represent a deltatheroidan (Kielan-Jaworowska and Nessov, 1990) but now considered to be more closely related to stagodontid marsupials (Rougier et al., 1998; Rougier et al., 2004); and *Oxlestes* and *Khuduklestes*, each based on an isolated axis vertebra (see Nessov, 1982; and Nessov et al., 1994, respectively) and, for all intents and purposes, indeterminate (Kielan-Jaworowska et al., 2004).

Atokatheridium boreni was described by Kielan-Jaworowska and Cifelli (2001) on the basis of a single upper molar from the Antlers Formation of southern Oklahoma (a lower molar was regarded as probably representing the species but was not formally referred to *A. boreni*). This taxon, of possible reference to Deltatheroida as suggested by Kielan-Jaworowska and Cifelli (2001), is notable in its occurrence: *Atokatheridium* is of Aptian-Albian age, significantly older than the Asiatic taxa securely referred to the group. Tentative placement of *Atokatheridium* in Deltatheroida was subsequently adopted by Kielan-Jaworowska et al. (2004) and was provisionally supported by the preliminary cladistic analysis of Rougier et al. (2004). Herein we describe additional fossils of *Atokatheridium boreni*, together with those representing a new but allied species. These new specimens allow us to refer both taxa to Deltatheroida, family Deltatheridiidae, with some confidence, to present morphological comparisons among relevant genera, and to make preliminary faunal comparisons between the classic "Trinity therian" sites of Texas and the Antlers Formation of Oklahoma.

1.1.3 Geological Context

The specimens described herein were collected from the Antlers Formation in extreme southeast Atoka County, Oklahoma (Figure 1.1). The Antlers Formation is a terrigenous unit comprised of sandstones, together with variegated siltstones and mudstones that were deposited under deltaic, fluvial, and strandplain systems, not far from the paleocoastline (Hobday et al., 1981). In Oklahoma, the Antlers Formation crops out as a narrow band extending westward from the Arkansas border across the southeastern part of the state, turning southward into northcentral Texas. From there it extends southward and westward into central Texas, where its lateral equivalent, the Trinity Group, can be subdivided into three formations on the basis of an interposed marine unit, not present in northcentral Texas or Oklahoma. These three units are, in ascending order, the Twin Mountains, Glen Rose, and Paluxy formations; the Glen Rose being a nearshore limestone of marine origin (see detailed discussion of stratigraphy in Winkler et al., 1990), famous for its dinosaur trackways (Bird, 1985). Invertebrates from the Glen Rose Formation and marginal marine facies of the Twin Mountains Formation show the latter unit to be Aptian in age, and that the basal Albian lies near the bottom of the Glen Rose Formation. The marine Walnut Formation of the Fredericksburg Group, together with data from the Glen Rose

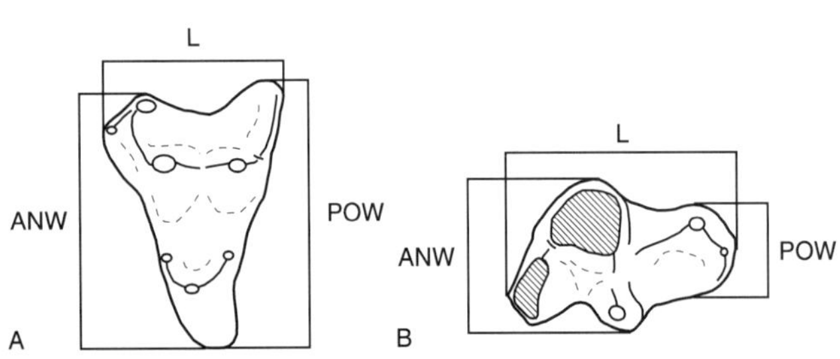

FIGURE 1.1. Standards of measurement for upper (A) and lower (B) molars. ANW, anterior width; L, length; POW, posterior width. Line drawings based on *Atokatheridium boreni*.

Formation, suggest that the Paluxy Formation correlates with the lower Albian (Jacobs and Winkler, 1998). Many sites in the Trinity Group of Texas and Oklahoma have yielded fossil vertebrates, mainly fishes and reptiles (e.g., Langston, 1974; Thurmond, 1974).

The Twin Mountains and Paluxy formations are lithologically similar, so that they cannot be distinguished northward and eastward of the pinchout of the Glen Rose Formation, where they are laterally represented by the Antlers Formation. As such, the undifferentiated Antlers Formation is simply regarded as being of Aptian-Albian age (e.g., Winkler et al., 1990; Jacobs and Winkler, 1998). Most of the published mammals from the Trinity Group of Texas are from sites north of the Glen Rose pinchout, and hence are placed in the Antlers Formation. Most notable among these sites are Greenwood Canyon, worked by Bryan Patterson and associates in the early 1950s (Patterson, 1951, 1955, 1956), and Butler Farm, worked by Bob Slaughter and associates in the 1960s (e.g., Slaughter, 1965, 1968a, b, 1969, 1971). Both of these sites are close to the top of the Antlers Formation, suggesting that they may lie within the younger part of the age range for the unit, perhaps around 108 Ma (Jacobs and Winkler, 1998, Figure 1.2).

The Antlers Formation thins northward and eastward into Oklahoma and it is estimated (Rennison, 1996) to be about 150 m thick in the vicinity of OMNH locality V706, which yielded the specimens reported herein. Correlation with parts of the Trinity Group in Texas, including sites that have yielded mammals there, is hampered by a number of factors, including lateral variability in lithology and lack of intercalated marine units. Based on data from a nearby well hole (Hart and Davis, 1981), OMNH V706 appears to lie near the local middle of the Antlers Formation (Cifelli et al., 1997; see Brinkman et al., 1998 for more complete discussion of stratigraphy, sedimentology, and age of the Antlers Formation at OMNH locality V706). This was corroborated by Rennison (1996) who, based on ratios of stable carbon isotopes, proposed two possible correlations of the lower to middle part of the Antlers Formation in Oklahoma: with the middle part of the Twin Mountains Formation and/or the lower to middle part of the Glen Rose Formation. Summarizing the limited and somewhat equivocal data now available, OMNH locality V706 (1) lies within the Antlers Formation of Oklahoma; (2) probably correlates with the upper Aptian to lowest Albian; and (3) appears to be older than the most productive mammal sites in the Antlers Formation of Texas, Greenwood Canyon and Butler Farm.

Given this possible difference in age of important mammal sites, together with some obvious faunal differences among vertebrate-bearing sites of the Antlers Formation and Trinity Group (see below) in general, we believe that it is no longer appropriate to recognize a collective, generalized "Trinity fauna." Meticulous studies by L. L. Jacobs, D. A. Winkler, and others at Southern Methodist University, Dallas, have

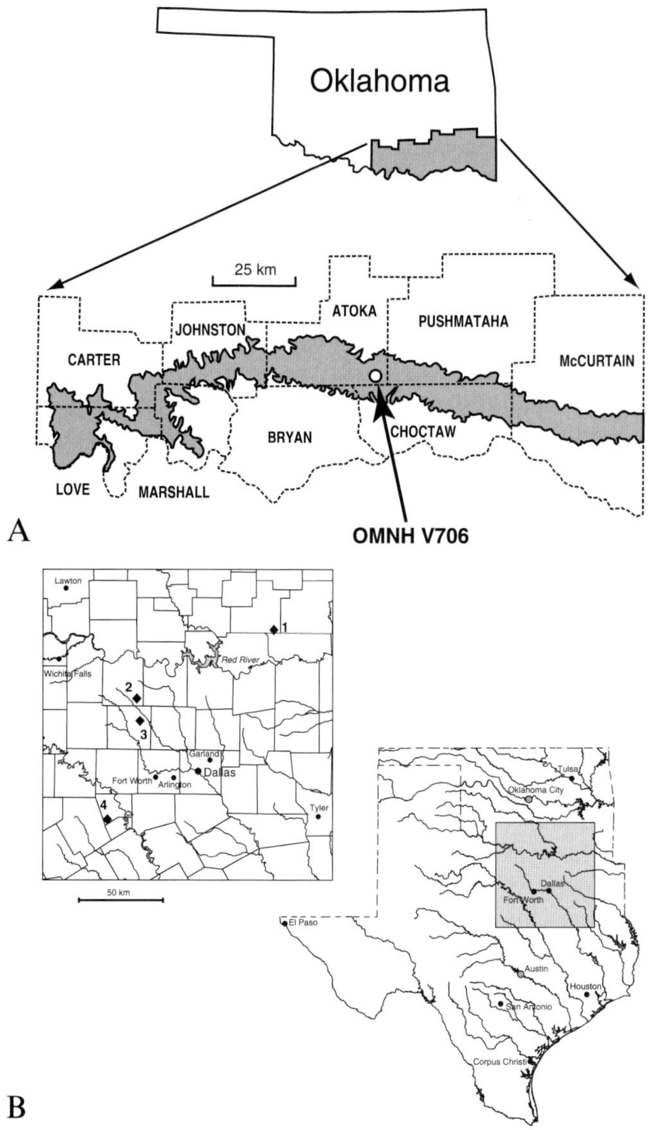

FIGURE 1.2. Early Cretaceous mammal localities, Trinity Group, Texas and Oklahoma. A, Map detailing outcrop of Antlers Formation (shaded) in southeastern Oklahoma. McLeod Honor Farm (OMNH microvertebrate locality V706) indicated by dot. B, Map detailing mammal-bearing microvertebrate localities from the Trinity Group (Aptian–Albian): 1, McLeod Honor Farm; 2, Greenwood Canyon; 3, Butler Farm (all Antlers Formation); 4, Paluxy Church (Twin Mountains Formation, late Aptian).

resulted in the discovery of many new sites in the Trinity and overlying groups, with recognition of important faunal changes within the sequence (e.g., Winkler et al., 1990; Jacobs and Winkler, 1998). To promote comparisons (both geographically and stratigraphically) and precision of usage, we herein introduce the term Tomato Hill local fauna in referring to the vertebrate assemblage from OMNH locality V706. A list of the 42 vertebrate taxa (including eight mammalian varieties) currently recognized from the Tomato Hill local fauna is given in Table 1.1.

TABLE 1.1. Measurements of described specimens. Numbers in brackets indicate estimates due to breakage. Standards of measurement illustrated in Figure 1.1.

	ID	Locus	L	ANW	POW
Oklatheridium szalayi					
	62411	LM1	1.4	[1.3]	[1.6]
	62410	LM2	1.5	[1.7]	[1.8]
	61180	LM2	[1.5]	[1.7]	[1.8]
	63727	RM3	[1.5]	[2.2]	[1.6]
	33945	Lm1	–	1.0	–
	33940	Rm2 or 3	–	1.0	–
cf. Oklatheridium sp.					
	33455	LMx	[1.6]	1.9	[2.1]
Atokatheridium boreni					
	61151	LM1	[0.9]	1.1	1.3
	61623	RM2	1.2	1.6	1.7
	63725	LM3	[1.5]	[2.0]	[2.0]
	61624	Lmx	1.3	0.8	0.5
	61181	Lmx	–	0.8	–

Tomato Hill is the local name for the immediate vicinity of OMNH locality V706, which lies on the flank of the first major terrace above the Muddy Boggy River and on the grounds of the Howard McLeod Corrections Center, operated by the Department of Corrections of the State of Oklahoma. As such, the fossil site was secure from unwanted collecting or other activities, until several years ago, when prison officials determined that landscaping of the area was needed. As a result, the site no longer exists, its former location now lying several meters above current ground level. Vertebrate fossils can occasionally be found in the nearby roadbed, up to 300 m from the former site.

Local exposure of the Antlers Formation at OMNH locality V706 consisted mainly of alternating gray-green and red claystones, together with localized, intermittent lenses of poorly consolidated sandstones and occasional horizons with small limonitic carbonate nodules, suggesting the presence of paleosols. There were two fossil horizons at OMNH V706, the upper of which represented a mass death assemblage, consisting of numerous, mostly articulated dinosaur skeletons. Most of these belong to the basal iguanodontian *Tenontosaurus tilletti* (see Ostrom, 1970; Forster, 1990), representing a wide variety of growth stages. At least one partial skeleton of the maniraptoran *Deinonychus antirrhopus* was also recovered from this horizon (Brinkman et al., 1998). Both of these species are otherwise known only from the Cloverly Formation, Wyoming and Montana (Ostrom 1969, 1970).

The lower fossil horizon at Tomato Hill, located immediately adjacent to and approximately 1.5 m below the dinosaur assemblage, lay in a dark gray, mottled mudstone with numerous localized, thin sandstone lenses, often bearing small mud clasts. Limonitic carbonate nodules were also abundant in this horizon. Fossils from this horizon, which include those described herein, consist mainly of microvertebrate remains, together with dinosaur and crocodile teeth, as well as small fragments of larger bone (e.g., turtle carapace). Preservation of the bone varies from

TABLE 1.2. Vertebrate fauna of the Tomato Hill Local Fauna (OMNH locality V706), Antlers Formation, Atoka County, Oklahoma. References are given in footnotes.

- Chondrichthyes
  - Hybodontiformes
    - Hybodontidae
      - *Hybodus butleri*[1]
      - ?*Hybodus* sp.[1]
    - Polyacrodontidae
      - *Lissodus anitae*[1]
- Osteichthyes
  - ?Semionitiformes
    - ?Semionotidae
      - gen. and sp. indet.[1]
  - ?Lepisosteiformes
    - ?Lepisosteidae
      - gen. and sp. indet.[1]
  - Pycnodontiformes
    - Pycnodontidae
      - ?*Palaeobalistum* sp.[1]
      - *Gyronchus dumblei*[1]
  - Amiiformes
    - ?Amiidae
      - gen. and sp. indet.[1]
  - Order and family indet.
    - gen. and sp. indet.[1]
- Lissamphibia
  - Allocaudata
    - Albanerpetontidae
      - *Albanerpeton arthridion*[2]
  - ?Caudata, family indet.
    - gen. and sp. indet.[1]
  - Anura, family indet.
    - gen. and spp. (2) indet.[1]
- Reptilia
  - Testudines
    - Family indet.
      - gen. and sp. indet.[1]
    - Pleurosternidae
      - *Naomichelys* sp.[1]
    - Glyptopsidae
      - ?*Glyptops* sp.[1]
  - Squamata
    - ?"Paramacellodidae"
      - *Atokasaurus metarsiodon*[3]
    - Teiidae
      - *Ptilotodon wilsoni*[3]
      - gen. and sp. indet.[3]
    - ?Scincomorpha
      - gen. and spp. indet. (2)[3]
    - ?Anguimorpha
      - gen. and sp. indet.[3]
  - Crocodylia
    - Bernissartiidae
      - *Bernissartia* sp.[1]
    - ?Atoposauridae
      - gen. and sp. indet.[1]
    - ?Goniopholididae
      - gen. and sp. indet.[1]
    - ?Pholidosauridae
      - gen. and sp. indet.[1]
  - Ornithopoda
    - Family *incertae sedis*
      - *Tenontosaurus tilletti*[5]
  - Sauropoda
    - Brachiosauridae
      - *Astrodon* sp.[1]
  - Theropoda
    - Carcharodontisauridae
      - *Acrocanthosaurusatokensis*[1]
    - Dromaeosauridae
      - *Deinonychus antirrhopus*[4]
  - ?Aves
    - Order and family indet.
      - gen. and sp. indet.[1]
- Mammalia
  - Eutriconodonta
    - Triconodontidae
      - *Astroconodon denisoni*[6]
  - Multituberculata
    - Family *incertae sedis*
      - ?*Paracimexomys crossi*[7]
      - gen. and spp. (2) indet.
  - "Stem Cladotheria"
    - Spalacotheriidae
      - ?*Spalacotheroides* sp.[8]
  - Boreosphenida, order uncertain
    - Holoclemensiidae
      - *Holoclemensia texana*[8]
    - Pappotheriidae
      - ?*Pappotherium* sp.[8]
    - Family uncertain
      - gen. and spp. (3) indet.[8]
  - Deltatheroida
    - Deltatheridiidae
      - *Atokatheridium boreni*[10]
      - *Oklatheridium szalayi*[9]

[1] Cifelli et al. (1997).
[2] Gardner (1999).
[3] Nydam and Cifelli (2002).
[4] Brinkman et al. (1998).
[5] Werning (2005).
[6] Turnbull and Cifelli (1999).
[7] Cifelli (1997).
[8] RLC, unpublished data.
[9] This study.
[10] Kielan-Jaworowska and Cifelli (2001).

excellent to abraded and rolled; some of the mammal teeth are lacking the enamel and have an etched appearance, suggesting that their owners had become meals and had passed through digestive tracts. The accumulation of vertebrate fossils at this horizon has been interpreted as lag formed within a fluvial overbank deposit, probably laid down in a localized depression on a floodplain (Cifelli, 1997).

1.2 Systematic Paleontology

Infraclass Metatheria Huxley, 1880
Cohort Deltatheroida Kielan-Jaworowska, 1982
Family Deltatheridiidae Gregory and Simpson, 1926
Included genera: *Deltatheridium* Gregory and Simpson, 1926, type genus; *Atokatheridium* Kielan-Jaworowska and Cifelli, 2001; *Deltatheroides* Gregory and Simpson, 1926; *Deltatherus* Nessov, 1997; *Oklatheridium* gen. nov.; *Sulestes* Nessov, 1985; and taxa left in open nomenclature (Fox, 1974; Cifelli, 1990a).
Distribution: See Kielan-Jaworowska et al. (2004).
Diagnosis: See Kielan-Jaworowska et al. (2004).
Comments: Additional characters used to define Deltatheroida can be found in Appendix 3 of Rougier et al. (2004). *Atokatheridium* (Kielan-Jaworowska and Cifelli, 2001) and *Oklatheridium* gen. nov. are referred to the Deltatheridiidae based primarily on the presence of hypertrophied shearing crests (postmetacrista and paracristid) and an enlarged paraconid, which are apomorphies relative to the condition in basal Boreosphenida. *Sulestes* has been demonstrated to be phylogenetically removed from the core of the family in recent analyses (Luo et al., 2003; Rougier et al., 2004), and is clearly derived in a separate direction from the rest of the Deltatheroida. As noted, however, we follow recent studies (Rougier et al., 1998, 2004; Kielan-Jaworowska et al., 2004) and include *Sulestes* in Deltatheridiidae, without subdividing the family. *Khudukulestes* (Nessov et al., 1994) and *Oxlestes* (Nessov, 1982), known only by isolated axis vertebrae, have been referred to the Deltatheroida based on their large size with respect to contemporaneous mammals; however, this element is unknown from and non-comparable to most other Mesozoic mammals, leaving no support for their inclusion in the cohort.
Type species: *Oklatheridium szalayi* sp. nov., type species by monotypy.
Etymology: *Okla-*, in reference to the state of Oklahoma, where specimens belonging to the taxon were discovered, and *–theridium*, from the Greek *theridion*, meaning small beast, a common suffix for Cretaceous mammals; *szalayi*, in honor of Frederick S. Szalay, for his invaluable contributions to our understanding of mammalian paleobiology, and especially for his work on the Metatheria.
Holotype: OMNH 62410, a LM2 lacking the protoconal region of the crown (Figure 1.3B).
Referred specimens: OMNH 62411, LM1; OMNH 61180, LM2; OMNH 63727, RM3; OMNH 33945, Lm1; OMNH

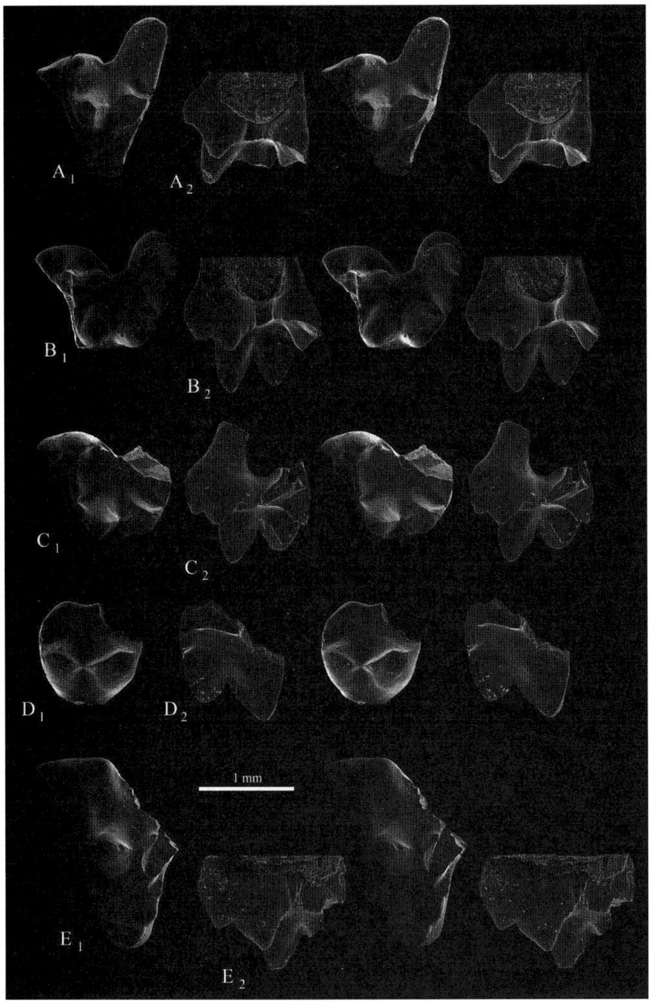

FIGURE 1.3. Upper molars of *Oklatheridium szalayi* gen. et sp. nov. (A–D) and *?Oklatheridium* sp. (E) OMNH locality V706, Antlers Formation (Aptian–Albian), Atoka County, Oklahoma. A, OMNH 62411, LM1, in occlusal (A_1) and labial (A_2) views; B, OMNH 62410 (holotype), LM2, in occlusal (B_1) and labial (B_2) views; C. OMNH 61180, LM2, in occlusal (C_1) and lingual (C_2) views; D, OMNH 63727, RM3, in occlusal (D_1) and lingual (D_2) views; E, OMNH 33455, LM2 or 3, in occlusal (E_1) and labial (E_2) views.

61643, Rm2 or 3; OMNH 33940, Rm2 or 3; OMNH 63728, Rm2 or 3; OMNH 63730, Lm2 or 3.

Distribution: OMNH locality V706, Antlers Formation (Aptian–Albian), Atoka County, Oklahoma.

Diagnosis: Moderate-sized deltatheroidan, larger than *Atokatheridium* but smaller than other taxa, and distinctive from all other deltatheroidans in having a deeper ectoflexus, larger and more anteriorly placed parastyle, stronger and more inflated stylocone (*Sulestes* might have had a stylocone of similar size, though breakage prevents direct comparison), taller, more labially oriented postmetacrista, and well developed conular cristae (all presumed apomorphies). Differs from all deltatheroidans except *Sulestes* in less height differential between the paracone and metacone, metacone somewhat broader than the paracone, portion of the preprotocrista labial to paraconule strong and broad, protoconal region anteroposteriorly expanded, and conules well developed (all presumed apomorphies). Differs specifically from *Sulestes* in greater width of the protoconal region (plesiomorphy), and lack of marginal cuspules on the metastylar lobe of the stylar shelf (polarity uncertain). Differs from all deltatheroidans except *Deltatheridium* in strong suppression of the metastylar lobe on M3. Differs from *Pappotherium* in stronger development of the metastylar lobe, less height differential and less divergence between metacone and paracone, stronger postmetacrista, and weaker developed postprotocrista terminating at the base of the metacone. Differs from *Holoclemensia* in less development of the parastylar lobe and greater development of the metastylar lobe, and absence of cusp "C" and other stylar cusps.

Description: Three upper molar loci are known for *Oklatheridium szalayi*. It is unknown whether or not this taxon possessed four molars, as is the case in *Deltatheridium* and *Deltatheroides*, but a reasonable case in the affirmative may be made based on other close morphological similarities between these deltatheroidans (see the Morphological Comparisons section of the Discussion). The upper dentition of *O. szalayi* is based on five isolated molars, all incomplete. Two specimens, OMNH 62411 and 62410 (Figures 1.3A, B; LM1 and LM2, respectively) are very similar in terms of wear patterns, relative morphology, preservation, and breakage and almost certainly belong to the same individual. The specimens were also found in relatively close association, though both are isolated teeth.

The M1, represented by one specimen (OMNH 62411; Figure 1.3A), is the most complete, lacking only the protocone. The parastylar lobe is significantly narrower than the metastylar lobe, giving the crown an asymmetrical outline. The stylocone is large, occupies the entire surface of the parastylar lobe, and is positioned directly labial to the paracone. The stylocone is roughly conical, though somewhat transversely compressed, and stands approximately one-half the height of the paracone (an exact comparison is impossible due to slight breakage at the apex of the paracone). The stylocone is connected via a weak crest to the parastyle, which is positioned lower on the crown and slightly more lingually. The parastyle is closely appressed to the stylocone, and situated at the terminal end of a moderately well developed preprotocrista. Both this crest and the parastyle are heavily worn, presumably due to occlusion with the protoconid of the opposing lower molar. There are no other stylar cusps present, though the labial margin is rimmed by a strong crest. The metastylar lobe is relatively narrow and runs obliquely to the long axis of the crown. The occlusal surface is occupied almost entirely by the slope of the postmetacrista, which is very tall and sharp. The ectoflexus is somewhat shallow due to the size difference between the parastylar and metastylar lobes. The paracone and metacone are conical, closely appressed, and roughly equal in size at their bases.

The cusps were presumably divergent (most of the metacone is missing due to breakage), since the paracone leans anteriorly. The bases of both cusps are equal in lingual extent. The apices of the paracone and stylocone are connected via a short, salient, and somewhat weakly developed preparacrista. The crest dips low in the middle, forming a weak notch. The centrocrista is straight and very weak. The postmetacrista is extremely well developed, with a deep carnassial notch present near the base of the metacone. The crest runs posterolabially. The protoconal region of the crown is small, comparable in width to that of *Atokatheridium*. This region is considerably better developed than in most other deltatheroidans (except *Sulestes*). The protocone itself is entirely missing. The conules are well developed, approximately equal in size, and possess weak internal cristae (the postparaconular crista is slightly the better developed of the two). The paraconule is positioned more labiad relative to the metaconule. The preprotocrista is continuous from the paraconule to the parastyle, creating a narrow shelf on the anterior margin of the crown. The postprotocrista extends past the metaconule only to the base of the metacone before terminating. The trigon basin is very small and restricted.

The M2 is represented by two specimens (OMNH 62410 (holotype) and 61180; Figure 1.3B, C, respectively). Both specimens are missing the protoconal region of the crown; OMNH 61180 additionally lacks the metastylar lobe and the tip of the metacone. From what is present, the M2 was larger than the M1. The parastylar lobe on M2 is much wider than on M1, though still not as wide as the metastylar lobe. The stylocone is similar in all relative dimensions to M1, though this cusp is much taller on OMNH 62410 than on OMNH 61180. The parastyle is positioned somewhat more labially than on M1. The ectoflexus is very deep on the type specimen, though the stylar shelf is rather broad centrally on OMNH 61180, indicating a shallower ectoflexus (breakage prevents determination of the actual depth on this specimen). The metastylar lobe is very broad and similar to that of M1 in all respects. There are no stylar cusps present posterior to the stylocone, though both specimens exhibit a small cuspule positioned on the posterior margin of the stylocone. The paracone and metacone are somewhat more transversely compressed than on M1. The paracone is taller than the metacone, but both cusps are approximately equally long in labial view. They share a significant portion of their bases and are somewhat divergent. The preparacrista is similar to that of M1, though the crest OMNH 61180 is significantly sharper and more deeply notched. The centrocrista is sharp and straight. The postmetacrista on M2 is very strong, sharp, and deeply notched, even more so than on M1. The crest runs much more directly labially than on M1. The preprotocrista is relatively narrow but still complete. Both specimens are broken labial to the conules, but OMNH 61180 shows evidence of a sharp crest running up the lingual surface of the paracone, likely representing an internal crista from the paraconule. This feature is absent on OMNH 62410, and shows a different orientation of the postparacrista from the condition on M1. The postprotocrista is similar to that of M1 in that it terminates at the base of the metacone.

The M3 is represented by one fragmentary specimen (OMNH 63727; Figure 1.3D), preserving only the paracone, metacone, and metastylar lobe. Based solely on the central portion of the crown, the M3 was larger still than the M2, falling in line with the typical deltatheroidan molar size progression of M1<M2<M3>?M4 (M4 is not known for this taxon, but it is reasonable to assume that the tooth at this locus would have been smaller than the M3). The metastylar lobe is strongly reduced relative to M1 and M2, consisting of a narrow, flat, gently rounded shelf. A slight concavity exists at the posterolabial corner of the metastylar lobe, which could have fit the parastylar lobe of a succeeding molar. This feature provides possible evidence for the presence of four molars in *Oklatheridium szalayi*. The parastylar lobe appears to have broken away at the deepest point of the ectoflexus, which was apparently very shallow. The paracone and metacone are closely appressed at their bases and strongly divergent, with the paracone significantly taller than the metacone, which is very short relative to the metacone on the other loci. Both cusps are somewhat transversely compressed, with nearly flat labial faces. The preparacrista is preserved from the apex of the paracone to its base, and is relatively strong and sharp. The centrocrista is straight and sharper than in the other loci. The postmetacrista, however, is very low and weak, though a small carnassial notch is still present at the base of the metacone. No trace of a postprotocrista is present on the base of the metacone, implying that it terminated more lingually, if it progressed past the metaconule.

The lower dention of *Oklatheridium szalayi* is based on six isolated molars, all of which preserve only the trigonid. These trigonids can be confidently referred to the upper molars based on expected size and morphology; it is also noteworthy that upper and lower molars referred to *O. szalayi* achieve the highest frequency of tribosphenic specimens in the collection from this locality. The trigonid is tall; though the talonid is missing, the trigonid cusps are much higher than the break that roughly indicated the position of the talonid. All three trigonid cusps are strong, with the protoconid being the tallest. The paraconid is taller and anteroposteriorly longer than the metaconid. As in other deltatheroidan taxa, this height difference appears to increase posteriorly through the molar series. *O. szalayi* differs from other deltatheroidans in having a more "closed" trigonid, with the bases of the paraconid and metaconid contacting each other. Both cusps support sharp crests with carnassial notches; however, the paracristid is much stronger than the protocristid, as would be expected in a dentition specialized for postvallum-prevallid shear. A well-developed wear facet is present on the anterior surface of the paracristid (facet 2 of Crompton, 1971). The lower molars are primitive in retaining well developed cusps e and f on the anterior surface of the trigonid. A short, strong precingulid runs nearly vertically, associated with cusp f, but is restricted to the anterior surface of the paraconid lingual to the paracristid notch (primitively, cusp f is placed more anterolabially on the molar, as is the case in the aegialodontid *Kielantherium*). Additionally, *O. szalayi* possesses a distal metacristid (see Section 1.3 for comments regarding the interpretation of this feature), though it appears to be variable in strength between specimens. This feature is shared by many early tribosphenic mammals, as a vertical continuation of the cristid obliqua from the talonid.

Though only trigonids are preserved, the morphology of the break where the talonid was connected does shed light on what the talonid would be expected to look like. The morphology of the break where the talonid was connected suggests the talonid was smaller than the trigonid (OMNH 33945 and 33940; Figure 1.4). Though its length cannot be assessed, it was likely narrower than the trigonid (though it could have been expanded posteriorly, in a "flexed" manner similar to that of *Kermackia*, cf. Butler, 1978: Figure 1.3K). All deltatheroidans possess a small talonid relative to the trigonid. However, the upper molars of *Oklatheridium szalayi* possess relatively well developed protocone and conules, so it would be expected that the talonid of this taxon would also be broader and better developed than is typical of Deltatheroida (*Sulestes*, which has the best developed talonid among previously known Deltatheroida, also has a strong protocone and conules on the upper molars. See Kielan-Jaworowska and Nessov, 1990: Figures 1.1–1.4). Confirmation of these speculations must, however, await discovery of more complete material.

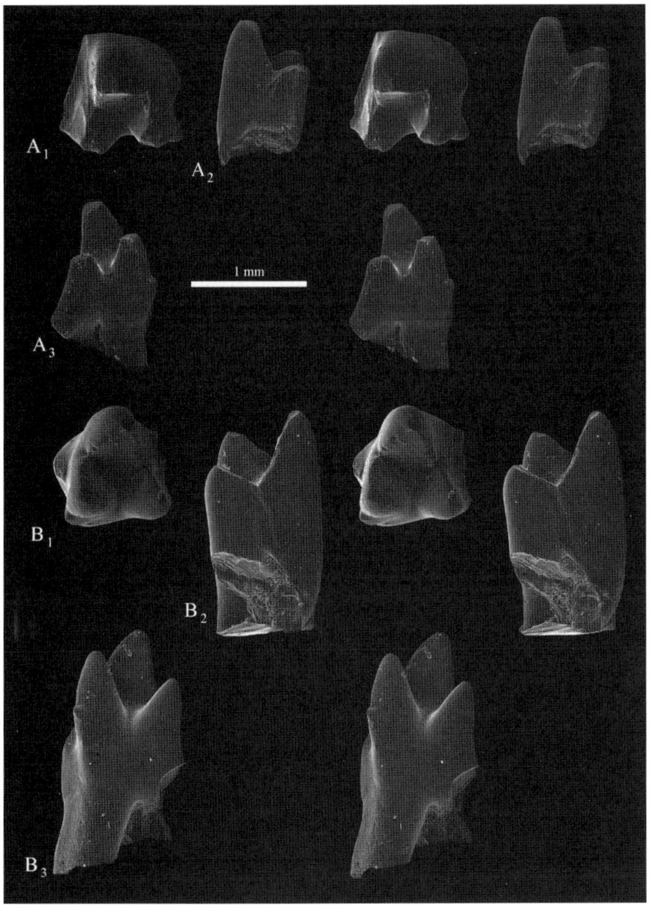

FIGURE 1.4. Lower molars of *Oklatheridium szalayi* gen. et sp. nov. OMNH locality V706, Antlers Formation (Aptian–Albian), Atoka County, Oklahoma. A, OMNH 33945, Lm1, in occlusal (A_1), posterior (A_2), and lingual (A_3) views; B, OMNH 33940, Rm2 or 3, in occlusal (B_1), posterior (B_2), and lingual (B_3) views.

Despite the lack of knowledge concerning talonid morphology, the lower molars of *Oklatheridium szalayi* compare favorably with those of deltatheroidan mammals. However, a number of features are common to other early tribosphenic mammals, prompting comparisons to non-deltatheroidan taxa. *Kielantherium* is similar in having a relatively taller paraconid than metaconid, but *O. szalayi* differs in having a lesser height differential between protoconid and paraconid, less separation of paraconid and metaconid (presumed apomorphies), and a more lingual placement of cusp f and the precingulid (polarity uncertain). Molars of *O. szalayi* differ from the stem boreosphenidan *Potamotelses* (Fox, 1975) in being generally higher-crowned (even on m1), in having a relatively taller paraconid and a transversely wider trigonid, and in retaining a stronger cusp e (presumed plesiomorphy). *O. szalayi* differs from all "Trinity therians" in the fact that the paraconid is substantially taller and more robust than the metaconid. However, it is similar to both *Pappotherium* and *Holoclemensia* in the degree of development of strong shearing crests on both the anterior and posterior edges of the trigonid.

The molar loci of *Oklatheridium szalayi* are defined on the basis of general morphological trends present in most primitive tribosphenic mammals (and specifically the resemblance of the lower molar specimens to equivalents in deltatheroidans where tooth locus can be established with certainty), as one moves posteriorly through the molar series. The m1 (represented by one specimen, OMNH 33945; Figure 1.4A) is smaller than the posterior molars. The tooth is also relatively lower crowned, with the protoconid slightly recumbent posteriorly. The posterior margin of the trigonid (most notably the posterolingual margin) slopes gently posteriorly down toward the talonid (or where the talonid would be in a complete molar). Without a dentary with associated teeth or at least a larger sample of isolated teeth that includes complete lower molars, it is unclear whether the remaining trigonids (OMNH 33940, 61643, 63728, and 63730; Figure 1.4B) represent the second or third molar locus. In *Deltatheridium*, the m2 is the largest molar, though this difference is not as clear in *Deltatheroides*. In both taxa, however, the occlusal outline of m1 is preserved on m2, while the m3 has a somewhat broader trigonid. In occlusal outline, the four trigonids designated m2 or m3 are all very similar to the specimen designated as m1, hence by analogy it is possible that they all represent m2. However, due to difficulties in differentiating between the second and third molar loci without a better sample, the specimens are assigned as m2 or 3. These trigonids are somewhat larger in all dimensions than m1. The crown is higher, and there is a greater size difference between the paraconid and metaconid. The two cusps are nearly subequal on m1, but on m2 or 3 the paraconid is noticeably taller and longer than the metaconid. One of these trigonids (OMNH 61643) was described briefly by Kielan-Jaworoska and Cifelli (2001) as "Family et gen. indet., sp. B".

Comments: Upper molars of *Oklatheridium szalayi* exhibit features typical of primitive boreosphenidans, such as a paracone taller than the metacone and a relatively small protocone. However, they are derived in a number of important features. In most early tribosphenic taxa, prevallum/postvallid shearing is dominant (Crompton, 1971; Clemens and Lillegraven, 1986). *O. szalayi*, as well as some other taxa (e.g., *Sulestes, Pappotherium*, and *Potamotelses*), possesses an enlarged shelf-like preprotocrista in addition to a strong preparacrista that would allow second-rank or *en echelon* shear along the anterior margin of the upper molar (Fox, 1975). However, *O. szalayi* departs from most other early tribosphenic mammals in also possessing an hypertrophied postmetacrista which, coupled with a tall, sharp paracristid on the lower molars, would have provided strong postvallum-prevallid shearing capability.

Referred specimen: OMNH 33455, RM2 or M3 (Figure 1.3E).

Distribution: OMNH locality V706, Antlers Formation (Aptian–Albian), Atoka County, Oklahoma.

Description: (from Cifelli, 1997, p. 10) "Judged by the relatively wide stylar shelf labial to the paracone and the elongate preparacrista, OMNH 33455 (Figure 1.3) appears to be a penultimate tooth, M2 or 3, depending on whether three or four molars were present in the dentition (see Fox, 1975 for discussion). The posterolabial corner of the tooth and the metacone are missing; the tip of the protocone is also broken (Figure 1.3A). Damage precludes some standard measurements; ANW is 1.96 mm; protocone width and length (as defined by Butler, 1990a) are 0.64 and 0.79, respectively. Stylar cusps A and B (terminology follows Clemens, 1979) are prominent, the latter being nearly as tall as the paracone. A well-marked preprotocrista extends labially from the protocone to stylar cusp A; this crest is uninterrupted in the region of the paracone (Figure 1.3A, B), such as in primitive marsupials and eutherians. By contrast, the postprotocrista extends only to the base of the metacone. Both conules are well developed and project slightly beyond the occlusal margin of the tooth. The paraconule is positioned about halfway between protocone and paracone and bears a small postparaconular crista that terminates at the base of the paracone; the metaconule is placed distinctly closer to the protocone and its internal crista is weak or lacking."

Comments: OMNH 33455 is similar to upper molars of *Oklatheridium szalayi* in terms of general outline and morphology (Figure 1.3), such as the shape and proportions of the stylocone and paracone, but it is distinct in a number of important ways. As mentioned above, the locus represented by OMNH 33455 cannot be confidently determined, primarily due to loss of the metastylar lobe, so direct comparisons with other specimens must be approached cautiously. For example, OMNH 33455 is larger than the type specimen of *O. szalayi* (OMNH 62410), an M2, though it appears to be smaller than the M3 (OMNH 63727) in some dimensions. But for present purposes, morphological similarities suggest the most appropriate comparisons are with the M2 of *O. szalayi* (OMNH 62410).

The cusps on OMNH 33455 are more robust than in *O. szalayi* (likely due to its larger size). The parastyle is better separated from the stylocone and positioned lower and more labially. The preparacrista is sharper and more distinct, and the preprotocrista is substantially wider and stronger in OMNH 33455. These differences leave some doubt as to the association of this specimen with *Oklatheridium*, but given the nature of the specimens in the Tomato Hill Local Fauna, it is most likely that the similarities between OMNH 33455 and *Oklatheridium* indicate the specimen represents a similar, related taxon. However, breakage of the metastylar lobe on OMNH 33455 prevents us from referring this specimen confidently or placing it elsewhere.

Holotype: OMNH 61623, RM2 (Figure 1.5B).

Newly referred specimens: OMNH 61151, LM1; OMNH 63725, LM3; OMNH 61624, Lmx; OMNH 61181, Lmx; OMNH 34905, Rmx.

Distribution: OMNH locality V706, Antlers Formation (Lower Cretaceous: Aptian–Albian), Atoka County, Oklahoma.

Revised diagnosis: Small deltatheroidan differing from all other deltatheroidans in smaller size, weaker stylocone, shallower ectoflexus, slightly narrower parastylar lobe, trend of increasing width of metastylar lobe posteriorly through molar series (excluding the unknown but hypothesized M4), greater height differential between the paracone and metacone, extremely weak development of conules, lack of conular cristae, transversely wider protoconal region, and a taller protocone. Differs from *Oklatheridium* and *Sulestes* in weaker conules. Differs from *Oklatheridium* in slightly narrower metastylar lobe and more posteriolabially oriented postmetacrista on M2, and in retention of a wide metastylar lobe on M3.

Description: Three upper molar loci are known with some confidence in *Atokatheridium boreni*. Each locus is represented by a single specimen; two are complete, but all three are rather worn (Figure 1.5). The M1 (OMNH 61151), despite being heavily worn, is confidently referred based on numerous general morphological similarities between it and the M2 (OMNH 61623). The M3 (OMNH 63725), however, is both broken and heavily worn or digested, making its referral somewhat more tentative.

The M1 (Figure 1.5A) is very small, though its original size is impossible to determine due to loss of nearly all the enamel. The parastylar lobe is very narrow and bears a small stylocone. The parastyle is small and closely approximated to the stylocone, though placed considerably lower on the crown.

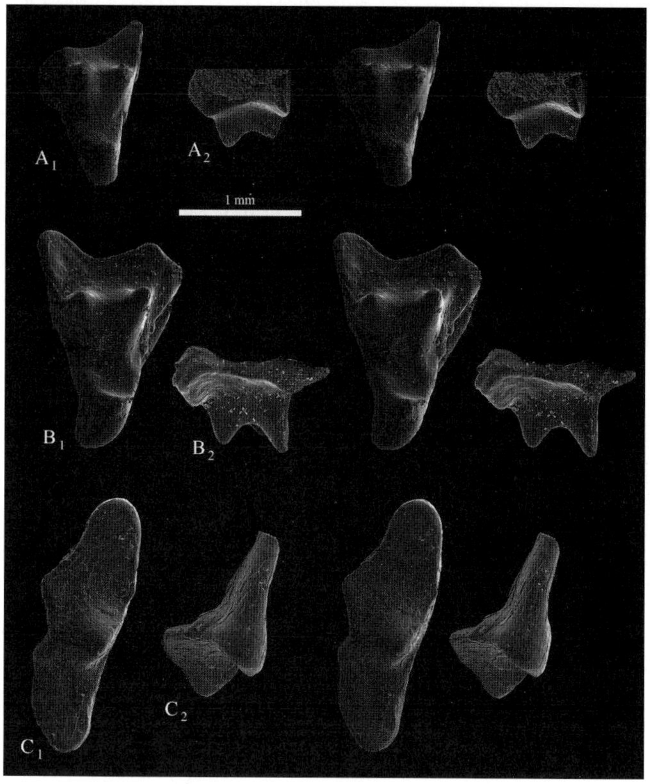

FIGURE 1.5. Upper molars of *Atokatheridium boreni* Kielan-Jaworowska and Cifelli, 2001. OMNH locality V706, Antlers Formation (Aptian–Albian), Atoka County, Oklahoma: A, OMNH 61151, LM1, in occlusal (A_1) and labial (A_2) views; B, OMNH 61623 (holotype), RM2, in occlusal (B_1) and labial (B_2) views; C, OMNH 63725, LM3, in occlusal (C_1) and labial (C_2) views.

The ectoflexus is nearly absent, despite the metastylar lobe being significantly wider than the parastylar lobe. No cusps or cuspules appear to have been present posterior to the stylocone, and no evidence of a cingulum along the labial edge of the stylar shelf is preserved. The paracone is taller than the metacone, though it cannot be determined how great the original difference was. The cusps are connate, share a large part of their bases, and are divergent. A short, low preparacrista runs from the apex of the paracone directly to the stylocone. The centrocrista appears to have been straight. The postmetacrista, though also worn, is still very tall and strong. The crest runs posterolabially from the apex of the metacone, and bears a strong carnassial notch near the base of the metacone. The entire surface of the metastylar lobe slopes anteriolabially from the postmetacrista. The protoconal region of the crown is short but transversely wide (relative to other deltatheroidans), and shallowly basined. Given the apparent amount of wear, the protocone is very tall and procumbent. No evidence of conules remains, though they were likely present since they occur on the type specimen. The preprotocrista runs uninterrupted to the parastyle and is not particularly well developed. The postmetacrista terminates at the base of the metacone.

The M2 (OMNH 61623; Figure 1.5B) is strongly similar to M1 in almost every respect, though its improved preservation provides more information (though the tooth is still somewhat abraded). The M2 is considerably larger than the M1, though enamel loss on the M1 makes direct size comparison difficult. The parastylar lobe is wider than on M1, though the stylocone is still small. The ectoflexus is slightly deeper, and a weak cingulum rims the labial margin of the crown. The metastylar lobe is wider than on M1. The paracone is significantly taller than the metacone, and both cusps are spire-like and strongly divergent. Conules are present as weak bulges along the pre- and postprotocristae, approximately equidistant between the bases of the paracone and metacone and the protocone.

The M3 (OMNH 63725; Figure 1.5C) is missing the paracone and parastylar lobe, and what remains of the molar is heavily abraded, with a melted appearance. The M3 is larger still than the M2, and much of the size difference is due to the presence of a significantly wider metastylar lobe on the M3. In both these respects, the relationships between the M2 and the presumed M3 of *Atokatheridium* are strikingly similar to those seen in *Deltatheroides cretacicus* (see Section 1.3.1 for further discussion). Though the entire parastylar lobe is missing, the ectoflexus was likely very deep, in sharp contrast with the preceding molars (however, a relatively narrow parastylar lobe would result in a shallower, short ectoflexus, less of a departure from the morphology of the M1 and M2). The metastylar lobe is very wide, nearly as wide as the portion of the crown from the metacone to protocone. No trace of a rimming cingulum or stylar cusps remains. The metacone is short and stout, and was closely appressed to the paracone. The postmetacrista is tall and deeply notched, running much more directly labiad than on M1 or M2. The protoconal region is very wide and short, but worn almost smooth. No evidence of conules is present, but their small size on the M2 makes it unlikely that they would be preserved on a tooth as worn as OMNH 63725. The protocone is heavily worn, so height cannot be determined. The preprotocrista is broken not far from the protocone, but the postprotocrista terminates in a similar spot as on the M2, posterolingual to the base of the metacone.

The lower dentition of *Atokatheridium boreni* is represented by one complete molar and two trigonids (Figure 1.6). The only lower molar specimen bearing a talonid was described by Kielan-Jaworoska and Cifelli (2001, p. 382) in the initial publication of *Atokatheridium boreni*, though it was referred to ?*A. boreni*: "OMNH 61624 is complete except for the tip of the protoconid and some loss of enamel fragments on the precingulid. The tooth is 1.28 mm long, trigonid width is 0.86, and the talonid width is 0.49 mm. The precingulid extends to the lingual margin of the tooth, forming a small, mesiolingual projection at the base of the paraconid. The paracristid is heavily worn; the protocristid and talonid also show wear, though major shearing surfaces (see Crompton, 1971) are clear and well developed. The paraconid and metaconid are well separated,

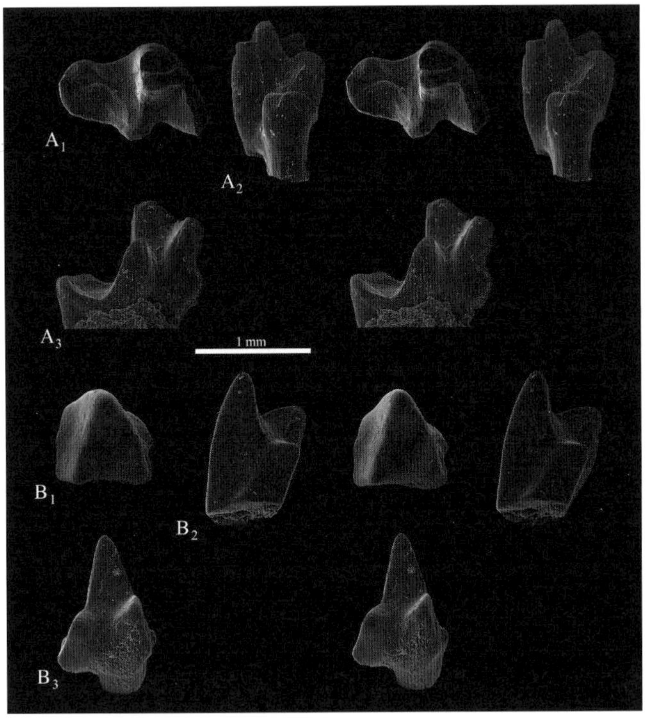

FIGURE 1.6. Lower molars of *Atokatheridium boreni* Kielan-Jaworowska and Cifelli, 2001. OMNH locality V706, Antlers Formation (Aptian–Albian), Atoka County, Oklahoma: A, OMNH 61624, Lmx, in occlusal (A_1), posterior (A_2), and lingual (A_3) views; B, OMNH 61181, Lmx, in occlusal (B_1), posterior (B_2), and lingual (B_3) views.

so that the trigonid angle is rather obtuse compared to that seen in *Pappotherium, Holoclemensia*, and early members of Marsupialia and Eutheria. The paraconid is much taller and more robust than the metaconid and appears to slant anteriorly, though this appearance may be an artifact of preservation. A distal metacristid (see Fox, 1975) extends distolabially from the apex of the metaconid. The talonid is much lower and narrower than the trigonid and has a very small, shallow basin that is open lingually. Two cusps, hypoconid and hypoconulid, are present; despite the presence of some wear on the rim of the talonid, it is clear no entoconid was ever present." This specimen is referred to the upper molars based on deltatheroidan characteristics, such as a paraconid much taller than the metaconid and a small, poorly-developed talonid; and to *Atokatheridium* specifically because of its small size (*Atokatheridium* is considerably smaller than *Oklatheridium*). In comparison with *Deltatheridium*, OMNH 61624 compares most favorably with the m3 based on the angle formed by the trigonid cusps, but the talonid is considerably better developed in *Atokatheridium*. This is plausible, since the protoconal region on the M3 is the widest of the molar series, and if the protocone was as tall on that tooth as it is on the M2, one would expect a more discernable talonid on the m3 compared to taxa such as *Deltatheridium*.

Both of the isolated trigonids (OMNH 34905 and OMNH 61181) compare very well with the trigonid of OMNH 61624 in both size and morphology. Both are complete, but virtually all the enamel is missing except for two small chips still attached to the anterior and posterior surfaces of OMNH 34905, so the full height of the trigonid cusps is still difficult to evaluate. OMNH 61181 likely represents the same locus as OMNH 61624. The protoconid is spire-like, and is by far the tallest cusp. The metaconid is strongly reduced in size, with the paraconid higher and slightly anteriorly projecting. The trigonid is open lingually. The paracristid is sharp and notched, though not as strongly as in *Oklatheridium*. A distal metacristid is present, running steeply posteriorly from the metaconid. OMNH 34905 is identical in morphology, but it preserves chips of enamel on the anterior base of the paraconid and the posterior surface of the protoconid.

Comments: With the description of *Oklatheridium szalayi*, the large majority of tribosphenic lower molar specimens from the Tomato Hill Local Fauna are partitioned into two morphological groups; this allows a more confident assignment of OMNH 61624 to *Atokatheridium boreni* than was made by Kielan-Jaworoska and Cifelli (2001).

1.3 Discussion

1.3.1 Morphological Comparisons

Both the age (Early Cretaceous) and generally plesiomorphic nature of the molars of *Oklatheridium* and *Atokatheridium* invite comparison with primitive boreosphenidan taxa (such as *Aegialodon, Kielantherium*, and *Potamotelses*). The lower molars of the Tomato Hill taxa resemble those of early boreosphenidans in a number of ways (Figure 1.7). Cuspule e is present, situated on the anterolingual margin of the paraconid (see Appendix 1 of Luo et al., 2002 for distribution of this character), and cuspule f is cuspate on most trigonids. A distal metacristid is present on the posterior aspect of the trigonid, running ventrolabially from the metaconid, and similarly developed as in the Trinity therians (see Patterson, 1956; Turnbull, 1971; Butler, 1978) but much weaker than in *Aegialodon* or *Kielantherium*. Our observations suggest that the distal metacristid is a true crest and not a wear feature, despite the fact that both the preparacrista and preprotocrista are expected to produce separate facets on the posterior wall of the trigonid (Crompton, 1971). In *Aegialodon*, the paraconid is taller than the metaconid, but the height difference is greater in the Tomato Hill taxa, as it is in other deltatheroidans. The talonid is shallow, open lingually, and poorly developed relative to the trigonid; only two cusps are present (hypoconid and hypoconulid). The trigonid/talonid proportions of *Atokatheridium* (a talonid is not known for *Oklatheridium*) are certainly primitive. The talonid is better developed than in *Aegialodon*, much

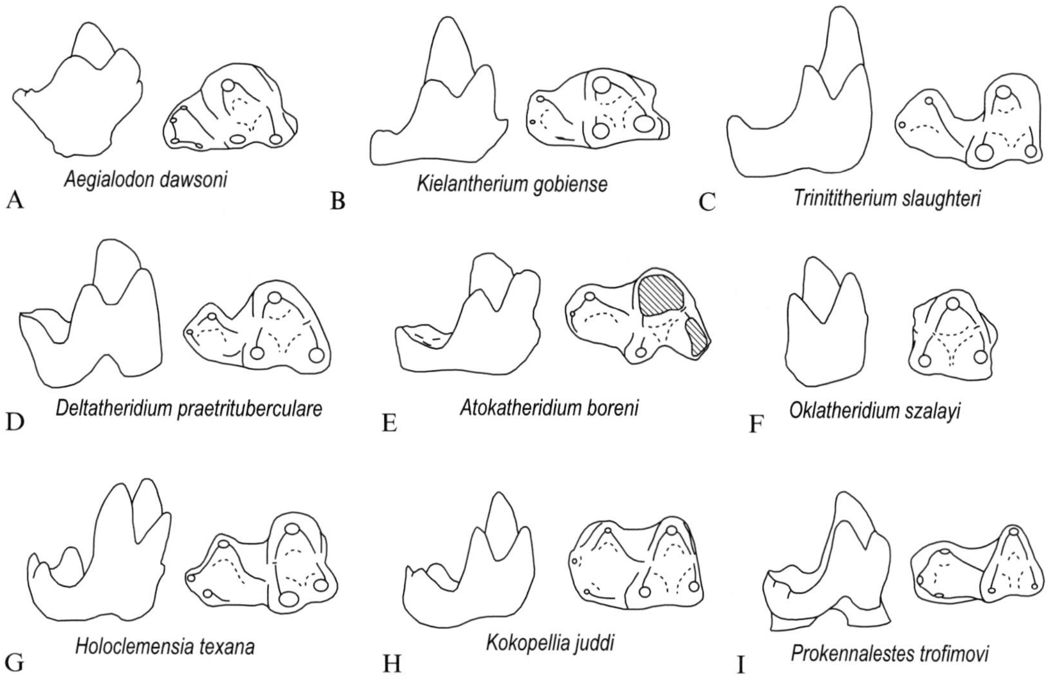

FIGURE 1.7. Lower molar comparisons (lingual and occlusal views): A, *Aegialodon dawsoni;* B, *Kielantherium gobiense* (reversed); C, *Trinititherium slaughteri,;* D, *Deltatheridium praetrituberculare;* E, *Atokatheridium boreni;* F, *Oklatheridium szalayi* (reversed); G, *Holoclemensia texana;* H, *Kokopellia juddi;* I, *Prokennalestes trofimovi.* (A modified from Kermack et al. 1965; B–I original.) Line drawings not to scale.

less developed than in *Potamotelses*, but of similar structural grade as *Kielantherium. Atokatheridium* is also similar to stem boreosphenidans in lacking a postcingulid.

The lower molars of the Tomato Hill taxa possess a few characters that serve to distance them somewhat from the primitive boreosphenidan condition. Most notable is the hypertrophy of the paraconid and paracristid. The paraconid is significantly taller than the metaconid (somewhat similar to *Kielantherium*), projecting somewhat anteriorly and supporting a very strong crest and carnassial notch (this shearing surface is particularly well developed on trigonids referred to *Oklatheridium*). Notably, the paraconid of both taxa lacks the distinct mesiolingual keel present in *Kokopellia* and more derived metatherians (Luo et al., 2002). Additionally, the trigonid basin is somewhat more closed in *Oklatheridium* compared to primitive taxa and *Atokatheridum*, due to a swelling of the bases of the paraconid and metaconid.

The upper molars of *Oklatheridium* and *Atokatheridium* provide strong support for a molar count of four, though only three loci are represented in each taxon. It should be noted, however, that in some early taxa the upper and lower molar counts are not equal (*Sinodelphys*, for example, has four upper molars but only three lower molars, see Luo et al., 2003). Though the ancestral boreosphenidan molar count is unknown, all stem taxa with a known (or at least surmised) dentition suggest four molars were present. The aegialodontid *Kielantherium gobiense*, initially described on the basis of a single lower molar (Dashzeveg, 1975), is known by a dentary preserving four molars (Dashzeveg and Kielan-Jaworowska, 1984). The "Trinity therians" *Holoclemensia* and *Pappotherium* were reconstructed by Butler (1978) as having four upper molars (but see Fox, 1975 for contrasting interpretation of *Pappotherium*). Additionally, the deltatheridiids *Deltatheroides cretacicus* and *Deltatheridium pretrituberculare* have been demonstrated to possess four molars (Gregory and Simpson, 1926; Rougier et al., 1998), though the ultimate upper molar is tiny and was assumed absent in early descriptions (Butler and Kielan-Jaworowska, 1973; Kielan-Jaworowska, 1975; Rougier et al., 1998). Though an ultimate molar is not known for either of the Tomato Hill taxa, one is assumed to have been present based on morphological similarities at the third molar locus between these taxa and the two aforementioned deltatheridiids. Additionally, the slight concavity on the posterolabial margin of the metastyle of the M3 of *Oklatheridium* might suggest an interlocking mechanism with the parastyle of an M4 (however, it should be noted that this feature is absent on the M3 of *Atokatheridium*). Figure 1.8 shows the molar series of both *Oklatheridium* and *Atokatheridium* with a hypothetical M4. The large metastylar lobe on M3 of *Atokatheridium* suggests its M4 (if present) was large compared with that of

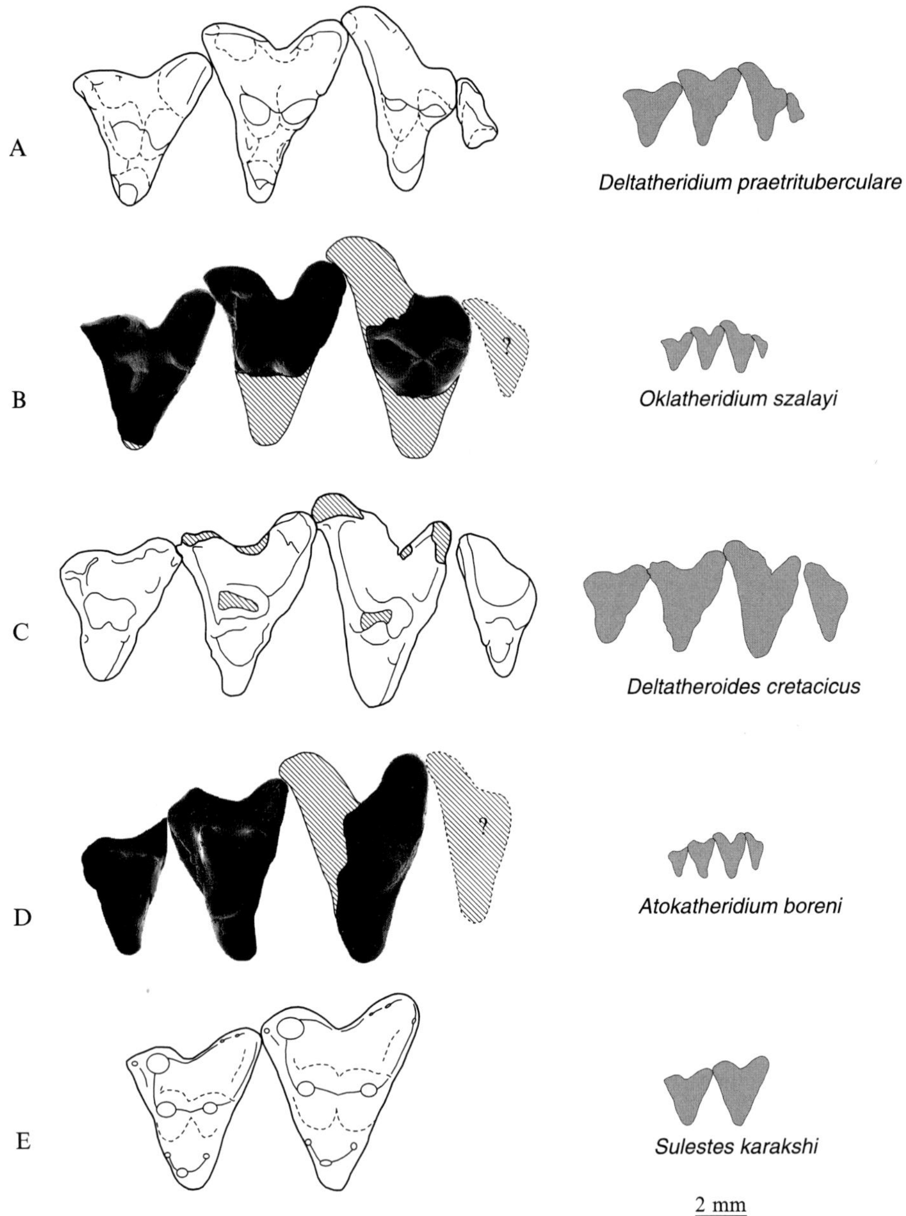

FIGURE 1.8. Upper molar series comparisons: A, *Deltatheridium praetuberculare;* B, *Oklatheridium szalayi;* C, *Deltatheroides cretacicus;* D, *Atokatheridium boreni;* E, *Sulestes karakshi.* (A and C from Rougier et al. 2004; B, D, and E original.) Line drawings resized for comparison; scale bar applies only to silhouettes.

other deltatheroidans (as is the case in *Deltatheroides*), where reduction of the posterolabial portion of M3 correlates with strong overall reduction of the fourth molar (as is the case in *Deltatheridium*).

The primitive condition for the protoconal region is well illustrated by the recently-described upper molar of the aegialodontid *Kielantherium* (Lopatin and Averianov, 2006, see our Figure 1.9A). The protocone is very small and situated on a short, narrow shelf, and the conules are lacking. The postprotocrista is very short, though the preprotocrista provided double-rank shearing on the anterior margin of the molar. The protoconal region differs significantly between the two Tomato Hill taxa. The protocone is tall and transversely wide in *Atokatheridium*, but the conules are virtually absent. In *Oklatheridium*, the entire width of the protoconal region is uncertain, but it was likely nearly as wide and longer still than in *Atokatheridium*. The conules in *Oklatheridium* are relatively large compared to those of stem boreosphenidans, deltatheroidans, or other contemporaneous taxa, providing another point of contrast. All considered, the protoconal region of *Atokatheridium* shows a blend of primitive and advanced features, while *Oklatheridium* is generally

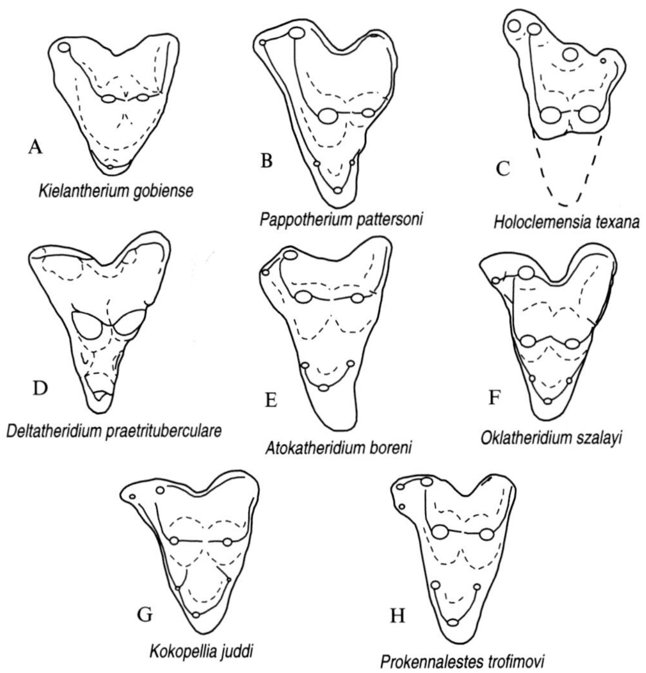

FIGURE 1.9. Upper molar comparisons: A, *Kielantherium gobiense* (reversed); B, *Pappotherium pattersoni* (reversed); C, *Holoclemensia texana* (reversed); D, *Deltatheridium praetrituberculare*; E, *Atokatheridium boreni* (reversed); F, *Oklatheridium szalayi* (lingual half reconstructed); G, *Kokopellia juddi*; H, *Prokennalestes trofimovi*. (A from Lopatin and Averianov 2006; B modified from Slaughter 1965; C modified from Slaughter 1968; D from Rougier et al. 2004; E–F original.) Line drawings not to scale.

more advanced. The protocone of both taxa is much better developed than the tiny, poorly developed protocone of *Kielantherium* and *Picopsis* (Fox, 1980), and is more similar to that of *Holoclemensia*.

Upper molars of the Tomato Hill taxa exhibit primitive morphology in a number of other features (Figure 1.9). The postprotocrista terminates just slightly around the base of the metacone; though the crest is somewhat better developed than in *Potamotelses*, it is equal to or less developed than in *Pappotherium*. The paracone and metacone also show some primitive features, such as the sharing of a significant portion of their bases and height. The cusps are connate at their bases and divergent, with the notch occupied by the centrocrista straight but shallow. The paracone is taller than the metacone, as in other early boreosphenidans, but the height difference is less in *Oklatheridium* (and thus more advanced). The parastylar lobe is generally primitive in the Tomato Hill taxa. The stylocone is large, with a much smaller and closely appressed parastyle situated basally on a small anterior projection of the shelf. It should be noted that this anterior projection in *Oklatheridium* is somewhat larger (advanced) than in *Atokatheridium*. Finally, the absence of any stylar cusps posterior to the stylocone can be interpreted as primitive, though as Clemens and Lillegraven (1986, p. 77), wisely observed, "evolution of stylar cusps in the region between the stylocone and metastylar corner of the crown poses vexatious questions." Posterior stylar cusps (with the exception of non-homologous cuspules) are absent in *Peramus*, most stem boreosphenidans, all deltatheroidans, and basal metatherians. However, a stylar cusp in the "D" position is present in the "tribotheres" *Pappotherium, Holoclemensia*, and *Comanchea*, and the basal eutherian *Paranyctoides*, suggesting some degree of homology, or raising the possibility of some functional importance of this portion of the stylar shelf.

There are a number of advanced features present in the upper molars of the Tomato Hill taxa, though the two taxa differ somewhat in many of these. The protocone is wide and well developed, especially in *Oklatheridium* (coupled with strong conules). The preprotocrista is moderately developed but continuous from the paraconule to the parastyle, a feature present in *Pappotherium* but absent in *Holoclemensia*. The metacone is relatively taller in *Oklatheridium* than in all other early taxa, approaching (though still shorter than) the paracone in height. Most notably, the metastylar lobe and postmetacrista are much more strongly developed than in any contemporaneous taxa.

The main features that serve to differentiate the Tomato Hill taxa from stem boreosphenidans such as *Aegialodon, Kielantherium*, and *Potamotelses* are those commonly associated with adaptations for carnivory. The primary shearing surfaces (in this case, the postmetacrista and paracristid) are hypertrophied. It should be noted that these features appear independently multiple times through evolution, for example, in the Late Cretaceous Stagodontidae as well as the South American "dog-like" marsupials, the Borhyaenidae (see Muizon and Lange-Badré, 1997). Additionally, the protocone is better developed than in some early boreosphenidans. Significance of the relatively large protocone is unclear; it appears that the protocone became enlarged independently in later metatherians, eutherians, and perhaps among the "Trinity therians". In the context of Early Cretaceous mammals, the Tomato Hill taxa are unique in possessing such an advanced suite of characters, such as a large protocone and specializations for carnivory, especially given their small body size.

The only tribosphenic mammals with both geographic and temporal proximity to the Tomato Hill taxa hail from stratigraphically younger horizons in the Trinity Group of northern Texas. The "Trinity therians", traditionally referred to as "Theria of metatherian-eutherian grade" (Patterson, 1956; Kielan-Jaworowska et al., 1979) because their affinities to the two higher groups of mammals are debatable, are a motley group of six described taxa that vary widely in size and morphology (Patterson, 1956; Slaughter, 1965, 1968a, b, 1971; Turnbull, 1971; Butler, 1978, 1990a; Jacobs et al., 1989). Direct comparison of lower molars between these taxa and the Tomato Hill taxa yields only a few similarities, mainly with the larger and more advanced *Pappotherium* (Figure 1.7). No "Trinity therian" has cusp proportions that approach the height differential between the paraconid and metaconid seen in either *Oklatheridium* or *Atokatheridium*. However, both *Pappotherium* and *Holoclemensia* have well developed shearing crests on the trigonid (due to the size of the metaconid, *Holoclemensia* shows emphasis on the protocristid as opposed to the paracristid). The talonid of *Atokatheridium* is much less developed than that of any described "Trinity therian", in that it is very small relative to the trigonid and it bears only two cusps (*Trinititherium* has been described as having an incipient entoconid (Butler, 1978), but no evidence of any entoconid is present in *Atokatheridium*). The talonid of *Atokatheridium* is very primitive, but the trigonids of the Tomato Hill taxa are advanced and clearly divergent from any forms seen among the "Trinity therians".

The differences among upper molars of the Tomato Hill taxa and the "Trinity therians" are more striking (Figure 1.9). *Pappotherium* is more generally primitive than either, with a narrower metastylar lobe, smaller metacone, and poorly developed protocone. *Holoclemensia*, on the other hand, is oddly derived. It differs from the Tomato Hill taxa in having a very large cusp in the mesostylar position (absent on all other contemporaneous taxa), a narrow metastylar lobe, a small stylocone, and a very small metacone. However, the protocone is about as well developed as in *Oklatheridium*.

The Tomato Hill taxa differ from basal Eutheria in a number of features (Figures 1.7, 1.9). The oldest North American eutherian, *Montanalestes keeblerorum* (Cifelli, 1999), from the Aptian–Albian Cloverly Formation, is derived in having a well developed, three-cusped talonid; it retains a vestige of a distal metacristid on the first molar only. Upper molars from early eutherians (such as *Murtoilestes*) differ from the Oklahoma taxa in having a smaller stylocone and a larger protocone and conules. Neither *Atokatheridium* nor *Oklatheridium* compares well with any basal eutherian.

There are also important differences between basal metatherians and the Tomato Hill taxa (Figures 1.7, 1.9). Disregarding *Sinodelphys szalayi* (Luo et al., 2003) (for which little occlusal morphology is known), *Kokopellia juddi* (Cifelli, 1993b; see also Cifelli and Muizon, 1997) is the oldest uncontested metatherian. It differs from the Tomato Hill taxa in having subequal paraconid and metaconid; a well developed, three-cusped talonid; some approximation of entoconid and hypoconulid; and presence of a postcingulid. Upper molars have a similarly developed protocone (though the conules are weaker in *Kokopellia* than in *Oklatheridium*), but differ in having a subequal paracone and metacone and subequal parastylar and metastylar lobes.

Comparison of *Oklatheridium* and *Atokatheridium* to *Kokopellia* is noteworthy in that the Deltatheroida have often been placed basally within Metatheria (Rougier et al. 1998, 2004; Kielan-Jaworowska et al. 2004; a trend followed in this paper). The inclusion of the Tomato Hill taxa in Deltatheroida indicates that the group was already morphologically diverse and distinctive by the end of the Early Cretaceous, in turn implying that diversification of Metatheria was already well under way by that time.

Atokatheridium and *Oklatheridium* compare favorably with other deltatheroidans. Lower molars (Figure 1.7) are very similar, sharing a tall paraconid and strong, sharp paracristid. The height difference between the paraconid and metaconid is the least in *Oklatheridium* and the greatest in *Atokatheridium* (the full height of the metaconid is unknown in *Sulestes* due to breakage). The trigonid is open lingually in *Atokatheridium*, as in *Deltatheridium* and *Sulestes*, though it is more closed in *Oklatheridium*. A distal metacristid is present in all deltatheroidans. The small, low talonid preserved in *Atokatheridium* is very similar to that of *Deltatheridium*. Both have a shallow, open talonid basin and only two cusps (hypoconid, hypoconulid) are present in the majority of specimens belonging to *Deltatheridium pretrituberculare*, as is the case for the only known lower molar of *Atokatheridium boreni*. The talonid in *Sulestes*, however, is somewhat stronger and possesses three equally developed cusps. The height difference between the trigonid and talonid is less in *Sulestes* than in other deltatheroidans, though it is likely that the only known lower molar referred to this genus is an m1, and in *Deltatheridium* the m1 has the relatively lowest trigonid.

The upper molars of the Tomato Hill taxa share a number of similarities with other deltatheroidans (Figures 1.8, 1.9), most notably retention of a large stylocone, emphasis on the postmetacrista, and a taller paracone that shares a significant portion of its base with the metacone. The general outline of the M2 of *Atokatheridium* is very similar to that of *Deltatheroides*. However, *Atokatheridium* has a significantly narrower stylar shelf and wider protoconal region, with a taller protocone. *Atokatheridium* differs similarly from *Deltatheridium*, but also has a shallower ectoflexus and shows less emphasis on prevallum shear, with a reduced preparacrista. Conversely, *Oklatheridium* has the deepest ectoflexus among deltatheroidans, as well as the greatest width disparity between the parastylar and metastylar lobes. The postmetacrista is very strongly developed, as in other deltatheroidans, but the crest is oriented more directly labiad. The metacone in *Oklatheridium* is lower than the paracone, but the height difference is relatively much less than in other taxa. The protocone and conules of *Oklatheridium* are much better developed than in any other deltatheroidan; though *Sulestes* has strong conules, the protoconal region is relatively narrow.

The third upper molar locus of each of the Tomato Hill taxa suggests the presence of a fourth molar, as noted earlier, but in each taxon the expected morphology of that fourth molar is quite distinct (Figure 1.8). The M3 referred to *Oklatheridium* shows strong reduction of the metastylar lobe, leaving only a rim labial to the metacone. This condition is seen in other taxa in which the fourth molar is heavily reduced, with the third molar assuming the morphology typical of the ultimate molar. The unique Late Cretaceous (Lancian) marsupial *Glasbius* exhibits this condition, with a tiny fourth molar present; this condition is also seen in *Deltatheridium*, where the presence of the fourth upper molar remained undocumented until exceptionally complete material became available (Rougier et al., 1998). The size of the M3 referred to *Oklatheridium* indicates that a succeeding molar was present; morphology of the M3 suggests that the last molar was strongly reduced. In this respect, the most appropriate model for restoration of the tooth row is *Deltatheridium* (Figure 1.8A, B). By contrast, the M3 assigned to *Atokatheridium* possesses a very wide metastylar lobe and a deep ectoflexus that is distinct from the anterior molars. Morphology indicates that an M4 was certainly present in this taxon as well. Though loss of the anterior half of the tooth makes direct comparison with the other loci impossible, the relative strength of the metastylar lobe suggests a much larger M4 than would have been present in *Oklatheridium*, perhaps more similar to the condition seen in *Deltatheroides* (Figure 1.8C, D; see Rougier et al., 2004).

1.3.2 Faunal Comparisons

Vertebrates of the Tomato Hill local fauna are listed in Table 1.1. Many of the 42 taxa known thus far are only identified to higher taxonomic level, reflecting incompleteness of the fossils, the fact that many groups remain to be studied, and the poor state of knowledge regarding Early Cretaceous terrestrial vertebrates in general. Among mammals, for example, only the triconodontid *Astroconodon* (see Turnbull and Cifelli, 1999) and the deltatheroidans (Kielan-Jaworowska and Cifelli, 2001, this study) have received treatment to date, with the remainder being currently under study. With these caveats, a few general comments on mammals of the Tomato Hill local fauna may be made.

The most obvious comparison lies with the two main mammal-yielding sites in the Trinity Group of Texas, both of which are in reasonably close geographic (and possibly stratigraphic) proximity (Figure 1.1): Greenwood Canyon (Montague County) and Butler Farm (approximately 20 km to the south, in Wise County). As noted under "Geological Context", above, both of these sites (which, like OMNH V706, no longer exist) lay within the local uppermost part of the Antlers Formation, and mammals collected from them are probably somewhat geologically younger than those of the Tomato Hill local fauna. A minimum of some eight mammalian varieties is known from Greenwood Canyon, which is reasonably close to the estimate for the Tomato Hill local fauna. Similarly, published reports suggest the presence of at least eight mammalian taxa at Butler Farm, though the actual number may be closer to six, if the synonymies suggested in Table 1.3 are verified by further study. The Greenwood Canyon and Butler Farm faunas are strikingly similar to each other, even at the species level: the triconodontid *Astroconodon denisoni* is present in both, as are the stem boreosphenidans *Kermackia texana, Pappotherium pattersoni*, and *Holoclemensia texana*. Indeed, if (as we suspect)

TABLE 1.3. Mammals from Greenwood Canyon (Montague County) and Butler Farm (Wise County), upper Antlers Formation, northern Texas. References and comments are given in footnotes.

Greenwood Canyon	Butler Farm
Eutriconodonta	Eutriconodonta
Triconodontidae	Triconodontidae
Astroconodon denisoni[1]	*Astroconodon denisoni*[1]
Multituberculata	Multituberculata
Family uncertain	Family uncertain
gen. and sp. indet. (2)[2]	gen. and sp. indet. (2)[2]
"Stem Cladotheria"	
Spalacotheriidae	
Spalacotheroides bridwelli[3]	
Boreosphenida, Order uncertain	Boreosphenida, Order uncertain
Family uncertain	Kermackiidae
gen. and sp. indet.[4,5]	*Kermackia texana*[7]
Kermackiidae	[*Trinititherium slaughteri*][8]
Kermackia texana[4]	Pappotheriidae
Pappotheriidae	*Pappotherium pattersoni*[9]
Pappotherium pattersoni[4]	[*Slaughteria eruptens*][10]
Holoclemensiidae[6]	Holoclemensiidae
Holoclemensia texana[4]	*Holoclemensia texana*[11]

[1] Patterson (1951), Slaughter (1969), Turnbull and Cifelli (1999).
[2] Krause et al. (1990).
[3] Patterson (1955, 1956).
[4] Butler (1978).
[5] Passing mention should be made of a partial edentulous dentary, FMNH PM 583, described as an unidentified therian by Patterson (1956, Figures 10, 11), and later designated as the holotype and only known specimen of *Adinodon pattersoni* by Hershkovitz (1995), who placed it in his marsupial family "Marmosidae". We follow Cifelli and Muizon (1997) in regarding this as a *nomen dubium* and, like Patterson, consider the specimen to represent an interesting but presently unidentifiable "therian of metatherian-eutherian grade" (i.e., stem boreosphenidan).
[6] Family erected by Aplin and Archer (1987), who followed Slaughter (e.g., 1968a, b, 1971) in regarding *Holoclemensia* to be marsupial. Butler (1978) placed the genus in the Pappotheriidae, within his suprafamilial group (infraclass) Tribotheria, a collocation of basal tribosphenic mammals. Most subsequent workers (e.g., Kielan-Jaworowska et al., 1979; McKenna and Bell, 1997; Kielan-Jaworowska et al., 2004; but see Fox, 1980; Luo et al., 2003) have followed Butler in excluding *Holoclemensia* from Metatheria, but a formal revision remains in the future, and for present purposes we provisionally recognize Aplin and Archer's monotypic family Holoclemensiidae.
[7] Slaughter (1971).
[8] We follow W. A. Clemens (cited in Butler, 1978: 11) in considering the holotype and only known specimen of *T. slaughteri*, a posterior lower molar (SMP-SMU 61728), to probably represent a positional variant of *Kermackia texana*, also known only by a lower molar, though we do not formally synonomize them.
[9] Slaughter (1965, 1971).
[10] The holotype and only known specimen of *Slaughteria eruptens*, a fragment of dentary bearing four teeth (SMP-SMU 61992), was originally described and illustrated by Slaughter (1971, pl. 9) as *Pappotherium pattersoni*. Butler (1978) erected the new genus and species *Slaughteria eruptens* for the specimen, partly on the strength of Slaughter's 1971: 137) observation (based on X-rays) that no unerupted teeth were present in the specimen. Slaughter (1971) considered the first molariform tooth of SMP-SMU 61992 to be a molarized last premolar (and therefore evidence of eutherian affinities), whereas Butler (1978) supposed it to be a first molar. Subsequent study using ultra high-resolution X-ray computed tomography shows that both interpretations were incorrect: the tooth in question is the last deciduous premolar (Kobayashi et al., 2002). We follow Slaughter (1971) and Kobayashi et al. (2002) in regarding the specimen as probably belonging to *Pappotherium pattersoni*, but do not formally sink *Slaughteria eruptens* into synonymy.
[11] Slaughter (1968a, b, 1971).

the boreosphenidans *Slaughteria eruptens* and *Trinititherium slaughteri* (both from Butler Farm) are junior subjective synonyms of *Pappotherium pattersoni* and *Kermackia texana*, respectively, then the two mammalian assemblages are identical, with one exception: the stem cladotherian *Spalacotheroides bridwelli*, which is known only from Greenwood Canyon. Whether or not this is a significant difference is difficult to judge, especially in light of the tiny, fragile nature of spalacotheriid fossils, and the small sample sizes in general. Based on published reports (Patterson, 1956; Slaughter, 1971; Turnbull, 1971; Butler, 1978), the most characteristic feature of the Greenwood Canyon–Butler Farm assemblages appears to be the great abundance of the stem boreosphenidans *Holoclemensia* and, to a lesser extent, *Pappotherium*.

By comparison, current data show that the mammalian assemblage of the Tomato Hill local fauna is quite different. The triconodontid *Astroconodon denisoni* is shared with Greenwood Canyon and Butler Farm. However, this species has a considerable stratigraphic range, as it is also known from Paluxy Church, one of the stratigraphically lowest vertebrate sites known in the Twin Mountains Formation (Winkler et al., 1990, p. 102). In a broader context, occurrence of the genus *Astroconodon* is widespread, both geographically and stratigraphically: it is also known from the?Aptian–Albian of the Cloverly Formation, Wyoming and Montana (Cifelli et al., 1998), and from near the Albian–Cenomanian (Early–Late Cretaceous) boundary in the Cedar Mountain Formation, Utah (Cifelli and Madsen, 1998).

Two other mammalian varieties possibly shared between the Tomato Hill local fauna and the Texas sites are the aforementioned *Spalacotheroides* (a stem cladotherian) and *Pappotherium* (a stem boreosphenidan), both of which are tentatively identified by rare, incomplete fossils from OMNH locality V706. Unfortunately, the holotype of *Spalacotheroides bridwelli* (the only species known for the genus), which consists of a dentary fragment bearing an incomplete molar (FMNH PM 933, see Patterson, 1955, Figure 145), is not particularly diagnostic, owing to recent discovery of a number of other spalacotheriids with similar lower molars (see review by Cifelli and Madsen, 1999). Identification of *Pappotherium* in the Tomato Hill local fauna is based on a heavily worn, incomplete upper molar, and should be regarded as tentative.

The most important difference between the mammalian assemblages from Butler Farm and Greenwood Canyon on one hand, and the Tomato Hill local fauna on the other, concerns the presence and relative abundance of boreosphenidans. Whereas the Butler farm and Greenwood Canyon faunas are dominated by *Holoclemensia* (especially) and *Pappotherium*, these are rare at Tomato Hill, where most of the boreosphenidan fossils are referable to the two deltatheroidans *Atokatheridium boreni* and *Oklatheridium szalayi*. Given the limited data at hand, the source(s) of these faunal differences (geological age, paleoecology, or both) cannot be identified at present.

1.3.3 Deltatheroida In Space and Time

1.3.3.1 North America

With the exception of a few fossils from the Neocomian Lakota Formation of South Dakota (Cifelli and Gordon, 2005), no mammals older than Aptian–Albian age are known from the Early Cretaceous of North America. Hence, *Atokatheridium* and *Oklatheridium* are, by default, the oldest deltatheroidans known from the continent. Among boreosphenidans, the Aptian–Albian record from North America otherwise consists of "tribotheres" (e.g., *Holoclemensia, Pappotherium, Kermackia*, see Butler, 1978 and comments above) and a single eutherian, *Montanalestes* (Cifelli, 1999). Beginning at the Early–Late Cretaceous boundary, North American assemblages became dominated by marsupials (Cifelli and Davis, 2003; Cifelli, 2004), with occasional "tribotheres" also present (e.g., Fox, 1972, 1976, 1980, 1982; Clemens and Lillegraven, 1986); eutherians did not reappear until the late Santonian or early Campanian (Fox, 1984), and did not diversify appreciably until the late Maastrichtian (e.g., Lillegraven, 1969). Following the Aptian–Albian, the next possible deltatheroidan is from the Turonian part of the Straight Cliffs Formation, southern Utah, represented by a relatively large but incomplete lower molar (Cifelli, 1990a). The only other specimens of deltatheroidans from the Cretaceous of North America were reported by Fox (1974), who identified an upper molar (Scollard Formation, late Maastrichtian, Alberta) and three lowers or parts thereof (Oldman Formation, late Campanian, Alberta; Lance Formation, late Maastrichtian, Wyoming) as cf. *Deltatheroides* sp. Rougier et al. (2004) considered the upper molar, at least, unidentifiable to genus, and we follow their judgment. In a strict consensus tree resulting from the analysis of Rougier et al. (2004), this upper molar came out as part of an unresolved tetrachotomy with *Atokatheridium, Deltatheridium*, and *Deltatheroides*; the four collectively forming a sister taxon to *Sulestes* within Deltatheroida.

1.3.3.2 Asia

The geologically oldest boreosphenidans of Asia come from the Barremian of Liaoning Province, China. The two best known are *Eomaia scansoria* and *Sinodelphys szalayi*, referred to Eutheria and Metatheria, respectively (Ji et al., 2002; Luo et al., 2003). Through the remainder of the Cretaceous, the boreosphenidan fauna of Asia was dominated by eutherians, and in this respect it differs markedly from North American assemblages for this time interval (e.g., Lillegraven, 1974; Cifelli, 2000; Cifelli and Davis, 2003).

The antiquity of Deltatheroida in Asia is debatable. Excluding *Kielantherium* (see above), which is from the ?Aptian–Albian "Höövör Beds", Mongolia (Dashzeveg and Kielan-Jaworowska, 1984), the next geologically oldest taxa that have been referred to Deltatheroida are of Cenomanian age. Both of these, *Oxlestes grandis* (from the Khodzhakul Formation, Uzbekistan, see Nessov, 1982; Nessov et al.,

1994) and *Khuduklestes bohlini* (Gansu Province, China, geological unit unknown see Nessov et al., 1994), are based on axis vertebrae and were referred to Deltatheroida on the basis of their relatively large size. Neither of these constitutes a verifiable record of the group, and we follow Rougier et al. (2004) in dismissing them from further consideration.

The geologically oldest, generally accepted records of deltatheroidans in Asia come from Coniacian strata in the Bissekty Formation at Dzharakuduk, Uzbekistan. *Sulestes* (represented by *S. karakshi* Nessov 1985 and *Sulestes* sp., see Kielan-Jaworowska and Nessov, 1990), known by a maxilla fragment with M1–2 and a referred lower molar, is relatively advanced, despite its geological age; as noted, Rougier et al. (2004) place it as sister taxon to remaining deltatheroidans. *Deltatherus kizylkumensis* was initially placed in *Deltatheroides* by Nessov (1993), but later transferred to its own genus (Nessov, 1997). The only informative specimens are two lower molars, one of which was illustrated by Kielan-Jaworowska et al. (2004, Figure 12.7C). We accept *Deltatherus* as a deltatheroidan and included it within the family Deltatheridiidae, but cannot comment further on its affinities. Finally, the geologically youngest and incomparably best known deltatheroidans come from beds of probable Campanian age in Asia. These are *Deltatheroides*, from the? early Campanian (and possibly younger strata) in Mongolia (Gregory and Simpson, 1926; Kielan-Jaworowska, 1975; Rougier et al., 2004); and *Deltatheridium*, known from the? early through? late Campanian of Mongolia (*D. pretrituberculare*, Gregory and Simpson, 1926; Kielan-Jaworowska, 1975; Rougier et al., 1998) and the Campanian of Kazakhstan (*D. nessovi*, see Averianov, 1997).

1.3.3.3 *Origin and Dispersal of Deltatheroida*

Ever since Deltatheroida were given ordinal status and recognized as being a monophyletic clade (Kielan-Jaworowska, 1982), they have been recognized as a mainly Asiatic group. Given the perceived similarity of *Deltatheroides*-like fossils described by Fox (1974) to the Mongolian form, presence of the group in North America could be reasonably explained by immigration from Asia, probably not long before first occurrence of relevant fossils. Though specimens from the Gobi Desert remain indisputably the most complete and abundant, the waters have become considerably murkier with the discoveries of the past 2 decades. Summing up evidence then available, Cifelli (2000) concluded that Deltatheroida dispersed twice between North America and Asia. A significant result of the analysis by Rougier et al. (2004) is that known distribution of the group may be explained by a single dispersal between the two continents.

Where did Deltatheroida originate? Data at hand are insufficient for anything more than speculation: the answer may be summarized as "source unknown". If pressed to speculate, however, we are inclined to favor a North American origin for deltatheroidans, as suggested by Rougier et al. (2004). This group of elegant little carnivores first appeared in North American Aptian–Albian, antedating their appearance in Asia. New data presented above also show that there was some morphological diversity, at least, among deltatheroidans in the Aptian–Albian of Oklahoma.

Though rare and poorly represented, Deltatheroida appear to have been continuously present in North America from the Aptian–Albian until nearly the end of the Cretaceous, and in Asia from the Coniacian through the ?late Campanian, at least. Hence, the minimum age constraint for dispersal to Asia is Coniacian. Occurrences of several other boreosphenidan groups are germane here, though they provide little in the way of clarification. Stratigraphic distributions suggest the following examples: marsupials (stem-based definition, see Kielan-Jaworowska et al., 2004) appeared in North America no later than the Albian–Cenomanian boundary (Cifelli, 2004) and appeared in Asian no later than Coniacian (Averianov and Kielan-Jaworowska, 1999); Ungulatomorpha were present in Asia by the Cenomanian or Turonian (Setoguchi et al., 1999), were diverse on the continent by the Coniacian (Nessov et al., 1998), and had appeared in North America by the late Campanian (Cifelli, 2000) or, perhaps, as early as Santonian (Nessov et al., 1998); and the ?nyctitheriid lipotyphlan *Paranyctoides* appears to have been present in Asia by the Coniacian (Nessov, 1993; Archibald and Averianov, 2001), whereas its first appearance in North America is late Santonian or early Campanian (Fox, 1984; Cifelli, 1990b). Though these distributions do not provide much in the way of constraints, they do suggest the working hypothesis that a mammalian dispersal event between North America and Asia may have occurred sometime between the Early–Late Cretaceous boundary and the Coniacian, and that Deltatheroida may have dispersed between the continents during this interval.

Acknowledgments. We would first like to thank Drs. Eric Sargis and Marian Dagosto for inviting us to contribute to this volume, and for their supreme patience during the preparation of this manuscript. We also thank Dr. Cynthia Gordon for her role in initial collection and curation of the described specimens, and Dr. Cyprian Kulicki for his assistance with the SEM at the Instytut Palaeobiologii, Polska Akademia Nauk, Warsaw. Funding for this project was provided in part by grants from the American Chemical Society, the National Science Foundation (RLC), and Instytut Paleobiologii (ZK-J).

References

Andrews, R. C., 1932. The New Conquest of Central Asia. American Museum of Natural History, New York, 678 pp.

Archibald, J. D., Averianov, A. O., 2001. *Paranyctoides* and allies from the Late Cretaceous of North America and Asia. Acta Palaeontologica Polonica 46, 533–551.

Averianov, A. O., 1997. New Late Cretaceous mammals of southern Kazakhstan. Acta Palaeontologica Polonica 46, 243–256.

Averianov, A. O., Kielan-Jaworowska, Z., 1999. Marsupials from the Late Cretaceous of Uzbekistan. Acta Palaeontologica Polonica 44, 71–81.

Bird, R. T., 1985. Bones for Barnum Brown: Adventures of a Fossil Hunter. Texas Christian University Press, Fort Worth, 225 pp.

Bown, T. M., Kraus, M. J., 1979. Origin of the tribosphenic molar and metatherian and eutherian dental formulae. In: Lillegraven, J. A., Kielan-Jaworowska, Z., Clemens, W. A. (Eds.), Mesozoic Mammals: The First Two-thirds of Mammalian History. University of California Press, Berkeley, CA, pp. 172–181.

Brinkman, D. L., R. L. Cifelli, Czaplewski, N. J., 1998. First occurrence of *Deinonychus antirrhopus* (Dinosauria: Theropoda) from the Antlers Formation (Lower Cretaceous: Aptian–Albian) of Oklahoma. Bulletin, Oklahoma Geological Survey 146, 1–26.

Broderip, W. J., 1828. Observations on the jaw of a fossil mammiferous animal found in the Stonesfield slate. Zoological Journal of London 3, 408–412.

Butler, P. M., 1978. A new interpretation of the mammalian teeth of tribosphenic pattern from the Albian of Texas. Breviora 446, 1–27.

Butler, P. M., 1990a. Early trends in the evolution of tribosphenic molars. Biological Reviews 65, 529–552.

Butler, P. M., 1990b. Tribosphenic molars in the Cretaceous. In: Smith, P., Tchernov, E. (Eds.), Structure, Function and Evolution of Teeth. Freund Publishing House, Tel Aviv, pp. 125–138.

Butler, P. M., Kielan-Jaworowska, Z., 1973. Is *Deltatheridium* a marsupial? Nature 245, 105–106.

Cifelli, R. L., 1990a. Cretaceous mammals of southern Utah. III. Therian mammals from the Turonian (early Late Cretaceous). Journal of Vertebrate Paleontology 10, 332–345.

Cifelli, R. L., 1990b. Cretaceous mammals of southern Utah. IV. Eutherian mammals from the Wahweap (Aquilan) and Kaiparowits (Judithian) formations. Journal of Vertebrate Paleontology 10, 346–360.

Cifelli, R. L., 1993a. Theria of metatherian-eutherian grade and the origin of marsupials. In: Szalay, F. S., Novacek, M. J., McKenna, M. C. (Eds.), Mammal Phylogeny, Volume 2 – Mesozoic Differentiation, Multituberculates, Monotremes, Early Therians, and Marsupials. Springer, New York, pp. 205–215.

Cifelli, R. L., 1993b. Early Cretaceous mammal from North America and the evolution of marsupial dental characters. Proceedings of the National Academy of Sciences USA 90, 9413–9416.

Cifelli, R. L., 1997. First notice on mesozoic mammals from Oklahoma. Oklahoma Geology Notes, Oklahoma Geological Survey 57, 4–17.

Cifelli, R. L., 1999. Tribosphenic mammal from the North American Early Cretaceous. Nature 401, 363–366.

Cifelli, R. L., 2000. Cretaceous mammals of Asia and North America. Journal of the Paleontological Society of Korea, Special Publication 4, 49–85.

Cifelli, R. L., 2004. Marsupial mammals from the Albian-Cenomanian (Early-Late Cretaceous) boundary, Utah. Bulletin of the American Museum of Natural History 285, 62–79.

Cifelli, R. L., Davis, B. M., 2003. Marsupial origins. Science 302, 1899–1900.

Cifelli, R. L., Gordon, C. L., 2005. First Neocomian (earliest Cretaceous) mammals from North America. Journal of Vertebrate Paleontology 25, Supplement to no. 3.

Cifelli, R. L., Madsen, S. K., 1998. Triconodont mammals from the medial Cretaceous of Utah. Journal of Vertebrate Paleontology 18, 403–411.

Cifelli, R. L., Madsen, S. K., 1999. Spalacotheriid symmetrodonts (Mammalia) from the medial Cretaceous (upper Albian or lower Cenomanian) Mussentuchit local fauna, Cedar Mountain Formation, Utah, USA. Geodiversitas 21, 167–214.

Cifelli, R. L., Muizon, C., de, 1997. Dentition and jaw of *Kokopellia juddi*, a primitive marsupial or near marsupial from the medial Cretaceous of Utah. Journal of Mammalian Evolution 4(4), 241–258.

Cifelli, R. L., J. D. Gardner, R. L. Nydam, Brinkman, D. L., 1997. Additions to the vertebrate fauna of the Antlers Formation (Lower Cretaceous), southeastern Oklahoma. Oklahoma Geology Notes, Oklahoma Geological Survey 57, 124–131.

Cifelli, R. L., Wible, J. R., Jenkins, F. A., Jr., 1998. Triconodont mammals from the Cloverly Formation (Lower Cretaceous), Montana and Wyoming. Journal of Vertebrate Paleontology 18, 237–241.

Clemens, W. A., 1979. Marsupialia. In: Lillegraven, J. A., Kielan-Jaworowska, Z., Clemens, W. A. (Eds.), Mesozoic Mammals: The First Two-thirds of Mammalian History. University of California Press, Berkeley, CA, pp. 192–220.

Clemens, W. A., Lillegraven, J. A., 1986. New Late Cretaceous, North American advanced therian mammals that fit neither the marsupial nor eutherian molds. Contributions to Geology, University of Wyoming, Special Paper 3, 55–85.

Cope, E. D., 1882. Mammalia in the Laramie Formation. American Naturalist 16, 830–831.

Cope, E. D., 1892. On a new genus of Mammalia from the Laramie Formation. American Naturalist 26, 758–762.

Crompton, A. W., 1971. The origin of the tribosphenic molar. In: Kermack, D. M., Kermack, K. A. (Eds.), Early Mammals. Zoological Journal of the Linnean Society, London, pp. 65–87.

Dashzeveg, D., 1975. New primitive therian from the Early Cretaceous of Mongolia. Nature 256, 402–403.

Dashzeveg, D., Kielan-Jaworowska, Z., 1984. The lower jaw of an aegialodontid mammal from the Early Cretaceous of Mongolia. Zoological Journal of the Linnean Society 82, 217–227.

Forster, C. A., 1990. The postcranial skeleton of the ornithopod dinosaur *Tenontosaurus tilletti*. Journal of Vertebrate Paleontology 10, 273–294.

Fox, R. C., 1972. A primitive therian mammal from the Upper Cretaceous of Alberta. Canadian Journal of Earth Sciences 9, 1479–1494.

Fox, R. C., 1974. *Deltatheroides*-like mammals from the Upper Cretaceous of North America. Nature 249, 392.

Fox, R. C., 1975. Molar structure and function in the Early Cretaceous mammal *Pappotherium*: evolutionary implications for Mesozoic Theria. Canadian Journal of Earth Sciences 12, 412–442.

Fox, R. C., 1976. Additions to the mammalian local fauna from the upper Milk River Formation (Upper Cretaceous), Alberta. Canadian Journal of Earth Sciences 13, 1105–1118.

Fox, R. C., 1980. *Picopsis pattersoni*, n. gen. and sp., an unusual therian from the Upper Cretaceous of Alberta, and the classification of primitive tribosphenic mammals. Canadian Journal of Earth Sciences 17, 1489–1498.

Fox, R. C., 1982. Evidence of new lineage of tribosphenic therians (Mammalia) from the Upper Cretaceous of Alberta, Canada. Géobios, Mémoire Spécial 6, 169–175.

Fox, R. C., 1984. *Paranyctoides maleficus* (new species), an early eutherian mammal from the Cretaceous of Alberta. Special Publication, Carnegie Museum of Natural History 9, 9–20.

Gardner, J. D., 1999. The amphibian *Albanerpeton arthridion* and the Aptian–Albian biogeography of albanerpetontids. Palaeontology 42, 529–544.

Gregory, W. K., Simpson, G. G., 1926. Cretaceous mammal skulls from Mongolia. American Museum Novitates 225, 1–20.

Hart, D. L., Davis, R. E., 1981. Geohydrology of the Antlers aquifer (Cretaceous), southeastern Oklahoma. Oklahoma Geological Survey Circular 81, 1–33.

Hershkovitz, P. 1995. The staggered marsupial third lower incisor: hallmark of cohort Didelphimorphia, and description of a new genus and species with staggered i3 from the Albian (Lower Cretaceous) of Texas. Bonner Zoologische Beiträge 45, 153–169.

Hobday, D. K., C. M. Woodruff, McBride, M. W., 1981. Paleotopographic and structural controls on non-marine sedimentation of the Lower Cretaceous Antlers Formation and correlatives, north Texas and southeastern Oklahoma. Society of Economic Paleontologists and Mineralogists, Special Publication 31, 71–87.

Horovitz, I., 2000. The tarsus of *Ukhaatherium nessovi* (Eutheria, Mammalia) from the Late Cretaceous of Mongolia: an appraisal of the evolution of the ankle in basal therians. Journal of Vertebrate Paleontology 20, 547–560.

Huxley, T. H., 1880. On the application of the laws of evolution to the arrangement of the Vertebrata and more particularly of the Mammalia. Proceedings of the Zoological Society of London 43, 649–662.

Jacobs, L. L., Winkler, D. A., 1998. Mammals, archosaurs, and the Early to Late Cretaceous transition in north-central Texas. In: Tomida, Y., Flynn, L. J., Jacobs, L. L. (Eds.), Advances in Vertebrate Paleontology and Geochronology. National Science Museum Monographs 14, Tokyo, pp. 253–280.

Jacobs, L. L., Winkler, D. A., Murry, P. A., 1989. Modern mammal origins: evolutionary grades in the Early Cretaceous of North America. Proceedings of the National Academy of Sciences USA 86, 4992–4995.

Ji, Q., Luo, Z., Wible, J. R., Zhang, J.-P., Georgi, J. A., 2002. The earliest known eutherian mammal. Nature 416, 816–822.

Kielan-Jaworowska, Z., 1975. Evolution of the therian mammals in the Late Cretaceous of Asia. Part I. Deltatheridiidae. Palaeontologia Polonica 33, 103–132.

Kielan-Jaworowska, Z., 1982. Marsupial-placental dichotomy and paleogeography of Cretaceous Theria. In: Gallitelli, E. M. (Ed.) Palaeontology, Essential of Historical Geology. S.T.E.M. Mucci, Modena, pp. 367–383.

Kielan-Jaworowska, Z., 1992. Interrelationships of Mesozoic mammals. Historical Biology 6, 185–202.

Kielan-Jaworowska, Z., Cifelli, R. L., 2001. Primitive boreosphenidan mammal (Deltatheroida) from the Early Cretaceous of Oklahoma. Acta Palaeontologica Polonica 46, 377–391.

Kielan-Jaworowska, Z., Nessov, L. A., 1990. On the metatherian nature of the Deltatheroida, a sister group of the Marsupialia. Lethaia 23, 1–10.

Kielan-Jaworowska, Z., Eaton, J. G., Bown, T. M., 1979. Theria of metatherian-eutherian grade. In: Lillegraven, J. A., Kielan-Jaworowska, Z., Clemens, W. A. (Eds.), Mesozoic Mammals: The First Two-thirds of Mammalian History. University of California Press, Berkeley, CA, pp. 182–191.

Kielan-Jaworowska, Z., Cifelli, R. L., Luo, Z.-X., 2004. Mammals from the Age of Dinosaurs: Structure, Relationships, and Paleobiology. Columbia University Press, New York, 630 pp.

Kobayashi, Y., Winkler, D. A., Jacobs, L. L., 2002. Origin of tooth replacement patterns in mammals: evidence from a 110-million-year-old fossil. Proceedings of the Royal Society, London, 269, 369–373.

Krause, D. W., Kielan-Jaworowska, Z., Turnbull, W. D. 1990. Early Cretaceous Multituberculata (Mammalia) from the Antlers Formation, Trinity Group, of southcentral Texas. Journal of Vertebrate Paleontology 10, Supplement to no. 3, 31A.

Langston, W., Jr., 1974. Nonmammalian Comanchean tetrapods. Geoscience and Man 8, 77–102.

Lillegraven, J. A., 1969. Latest Cretaceous mammals of upper part of Edmonton Formation of Alberta, Canada, and review of marsupial-placental dichotomy in mammalian evolution. University of Kansas Paleontological Contributions 50, 1–122.

Lillegraven, J. A., 1974. Biogeographical considerations of the marsupial-placental dichotomy. Annual Review of Ecology and Systematics 5, 263–283.

Lopatin, A. V., Averianov, A. O., 2006. An aegialodontid upper molar and the evolution of mammalian dentition. Science 313, 1092.

Luo, Z.-X., Kielan-Jaworowska, Z., Cifelli, R. L., 2002. In quest for a phylogeny of Mesozoic mammals. Acta Palaeontologica Polonica 47, 1–78.

Luo, Z.-X., Ji, Q., Wible, J. R., Yuan, C., 2003. An Early Cretaceous tribosphenic mammal and metatherian evolution. Science 302, 1934–1940.

MacIntyre, G. T., 1966. The Miacidae (Mammalia, Carnivora). Part 1. The systematics of *Ictidopappus* and *Protictis*. Bulletin of the American Museum of Natural History 131, 117–209.

Marsh, O. C., 1889a. Discovery of Cretaceous Mammalia. American Journal of Science 38, 81–92.

Marsh, O. C., 1889b. Discovery of Cretaceous Mammalia. Part II. American Journal of Science, ser. 3 38, 81–92.

Marsh, O. C., 1892. Discovery of Cretaceous Mammalia. Part III. American Journal of Science 43, 249–262.

Marshall, L. G., Kielan-Jaworowska, Z., 1992. Relationships of the dog-like marsupials, deltatheroidans and early tribosphenic mammals. Lethaia 25, 361–374.

Matthew, W. D., 1928. The evolution of mammals in the Eocene. Proceedings of the Zoological Society of London 1928, 947–985.

McKenna, M. C., 1975. Toward a phylogenetic classification of the Mammalia. In: Luckett, W. P., Szalay, F. S. (Eds.), Phylogeny of the Primates. Plenum, New York, pp. 21–46.

McKenna, M. C., Bell, S. K., 1997. Classification of Mammals above the Species Level. Columbia University Press, New York, 631 pp.

McKenna, M. C., Mellett, J. S., Szalay, F. S., 1971. Relationships of the Cretaceous mammal *Deltatheridium*. Journal of Paleontology 45, 441–442.

Muizon, C., de, Lange-Badré, B., 1997. Carnivorous dental adaptations in tribosphenic mammals and phylogenetic reconstruction. Lethaia 30, 353–366.

Nessov, L. A., 1982. Ancient mammals of the USSR (in Russian). Ezegodnik vsesoûznogo Paleontologiceskogo Obsestva 35, 228–243.

Nessov, L. A., 1985. Rare osteichthyans, terrestrial lizards and mammals from the estuaries and costal plain zones of the Cretaceous of the Kyzylkum Desert (in Russian). Ezegodnik vsesouznogo Paleontologiceskogo Obsestva 28, 199–219.

Nessov, L. A., 1987. Results of search and study of Cretaceous and early Paleogene mammals on the territory of SSSR (in Russian). Ezhegodnik Vsesojuznogo Paleontologiceskogo Obscestva 30, 119–218.

Nessov, L. A., 1993. New Mesozoic mammals of middle Asia and Kazakhstan, and comments about evolution of theriofaunas of Cretaceous coastal plains of ancient Asia (in Russian). Trudy Zoologiceskogo Instituta RAN 249, 105–133.

Nessov, L. A., 1997. Cretaceous Non-marine Vertebrates of Northern Eurasia (in Russian). University of Sankt Petersburg, Institute of Earth Crust, Sankt Petersburg, 218 pp.

Nessov, L. A., Sigogneau-Russell, D., Russell, D. E., 1994. A survey of Cretaceous tribosphenic mammals from middle Asia (Uzbekistan, Kazakhstan and Tajikistan), of their geological setting, age and faunal environment. Palaeovertebrata 23, 51–92.

Nessov, L. A., Archibald, J. D., Kielan-Jaworowska, Z., 1998. Ungulate-like mammals from the Late Cretaceous of Uzbekistan and a phylogenetic analysis of Ungulatomorpha. Bulletin of the Carnegie Museum of Natural History 34, 40–88.

Nydam, R. L., Cifelli, R. L., 2002. The first report of lizards from the Early Cretaceous of Oklahoma, Montana, and Wyoming. Journal of Vertebrate Paleontology 22, 286–298.

Osborn, H. F., 1907. Evolution of Mammalian Molar Teeth. MacMillan, New York, 250 pp.

Ostrom, J. H., 1969. Osteology of *Deinonychus antirrhopus*, an unusual theropod from the Lower Cretaceous of Montana. Peabody Museum of Natural History Bulletin 30, 1–165.

Ostrom, J. H., 1970. Stratigraphy and paleontology of the Cloverly Formation (Lower Cretaceous) of the Bighorn Basin area, Montana and Wyoming. Peabody Museum of Natural History Bulletin 35, 1–234.

Patterson, B., 1951. Early Cretaceous mammals from northern Texas. American Journal of Science 249, 31–46.

Patterson, B., 1955. A symmetrodont mammal from the Early Cretaceous of northern Texas. Fieldiana Zoology 37, 689–693.

Patterson, B., 1956. Early Cretaceous mammals and the evolution of mammalian molar teeth. Fieldiana Geology 13, 1–105.

Rennison, C. J., 1996. The Stable Carbon Isotope Record Derived from Mid-Cretaceous Terrestrial Plant Fossils from North-Central Texas. Southern Methodist University, Dallas, 120 pp.

Rougier, G. W., Wible, J. R., Novacek, M. J., 1998. Implications of *Deltatheridium* specimens for early marsupial history. Nature 396, 459–463.

Rougier, G. W., Wible, J. R., Novacek, M. J., 2004. New specimen of *Deltatheroides cretacicus* (Metatheria, Deltatheroida) from the Late Cretaceous of Mongolia. Bulletin of the Carnegie Museum of Natural History 36, 245–266.

Setoguchi, T., Tsubamoto, T., Hanamura, H., Hachiya, K., 1999. An early Late Cretaceous mammal from Japan, with reconsideration of the evolution of tribosphenic molars. Paleontological Research 3, 18–28.

Simpson, G. G., 1928. Affinities of the Mongolian Cretaceous insectivores. American Museum Novitates 330, 1–11.

Simpson, G. G., 1945. The principles of classification and a classification of mammals. Bulletin of the American Museum of Natural History 85, 1–350.

Slaughter, B. H., 1965. A therian from the Lower Cretaceous (Albian) of Texas. Postilla 93, 1–18.

Slaughter, B. H., 1968a. Earliest known eutherian mammals and the evolution of premolar occlusion. The Texas Journal of Science 20, 3–12.

Slaughter, B. H., 1968b. Earliest known marsupials. Science 162, 254–255.

Slaughter, B. H., 1969. *Astroconodon*, the Cretaceous triconodont. Journal of Mammalogy 50, 102–107.

Slaughter, B. H., 1971. Mid-Cretaceous (Albian) therians of the Butler Farm local fauna, Texas. Zoological Journal of the Linnean Society 50 (Supplement 1), 131–143.

Szalay, F. S., 1982. A new appraisal of marsupial phylogeny and classification. In: Archer, M. (Ed.) Carnivorous Marsupials. Royal Zoological Society of New South Wales, Sydney, pp. 621–640.

Szalay, F. S., 1984. Arboreality: is it homologous in metatherian and eutherian mammals? In: Hecht, M. K. (Ed.) Evolutionary Biology, Volume 6. Plenum, New York, pp. 215–258.

Szalay, F. S., 1994. Evolutionary History of the Marsupials and an Analysis of Osteological Characters. Cambridge University Press, Cambridge, 481 pp.

Szalay, F. S., McKenna, M. C., 1971. Beginning of the age of mammals in Asia: the late Paleocene Gashato fauna, Mongolia. Bulletin of the American Museum of Natural History 144, 269–318.

Szalay, F. S., Sargis, E. J., 2001. Model-based analysis of postcranial osteology of marsupials from the Palaeocene of Itaboraí (Brazil) and the phylogenetics and biogeography of Metatheria. Geodiversitas 23, 139–302.

Szalay, F. S., Trofimov, B. A., 1996. The Mongolian Late Cretaceous *Asiatherium*, and the early phylogeny and paleobiogeography of Metatheria. Journal of Vertebrate Paleontology 16, 474–509.

Thurmond, J. T., 1974. Lower vertebrate faunas of the Trinity Division in north-central Texas. Geoscience and Man 8, 103–129.

Trofimov, B. A., Szalay, F. S., 1994. New Cretaceous marsupial from Mongolia and the early radiation of the metatheria. Proceedings of the National Academy of Sciences USA 91, 12569–12573.

Turnbull, W. D., 1971. The Trinity therians: their bearing on evolution in marsupials and other therians. In: Dahlberg, D. D. (Ed.) Dental morphology and evolution. University of Chicago Press, Chicago, IL, pp. 151–179.

Turnbull, W. D., Cifelli, R. L., 1999. Triconodont mammals of the Aptian–Albian Trinity Group, Texas and Oklahoma. In: Mayhall, J. T., Heikkinen, T. (Eds.), Dental Morphology, '98. University of Oulu Press, Oulu, pp. 252–272.

Van Valen, L., 1966. Deltatheridia, a new order of mammals. Bulletin of the American Museum of Natural History 132, 1–126.

Werning, S., 2005. Long bone histology of *Tenontosaurus tilletti* Ostrom, 1970 (Early Cretaceous, North America), with comments on ontogeny. Unpublished M.S. thesis. 154 p. University of Oklahoma, Norman.

Winkler, D. A., Murry, P. A., Jacobs, L. L., 1990. Early Cretaceous (Comanchean) vertebrates of central Texas. Journal of Vertebrate Paleontology 10, 95–116.

2. Evolution of Hind Limb Proportions in Kangaroos (Marsupialia: Macropodoidea)

Benjamin P. Kear*
Department of Genetics
School of Molecular Sciences
La Trobe University
Melbourne, Victoria 3086
South Australian Museum
North Terrace
Adelaide, South Australia 5000
Australia

Michael S.Y. Lee
South Australian Museum
North Terrace
Adelaide, South Australia 5000
Australia

Wayne R. Gerdtz
School of Ecology and Environment
Melbourne Campus, Deakin University
Burwood, Victoria 3125
Australia

Timothy F. Flannery
Division of Environment and Life Sciences
Macquarie University
Sydney 2109
Australia

2.1 Introduction

Kangaroos (Macropodoidea: Marsupialia) are a characteristic group of Australo-New Guinean mammals that diversified during the geographic isolation of the Australian continent in the Cenozoic. They are first recorded in the Late Oligocene, although the clade diverged from other diprotodontians around 38 million years ago (mya; Westerman et al., 2002), with early forms perhaps resembling small arboreal 'phalangerids' (Flannery, 1982).

Living macropodoids vary widely in body size (<500 g in *Hypsiprymnodon* to >60 kg in larger species of *Macropus*), and show a high degree of ecological diversity. They include forms specialized for climbing (e.g., *Dendrolagus*), burrowing (e.g., *Bettongia leseur*), and occupation of closed rainforest/woodland (e.g., *Hypsiprymnodon, Setonix*) through to open temperate/tropical and/or arid zone grassland (e.g., *Macropus*). Despite this variability, the appendicular skeleton of macropodoids is remarkably conservative with all members of the group showing similar modifications (particularly in the long bones of the hind limb, tarsus, and pes) favoring a bipedal hopping gait. Windsor and Dagg (1971) standardized terminology for kangaroo locomotion designating 'slow pentapedal progression' as that involving synchronous use of the limbs and tail (present in all macropodoids and extensively used by species of *Dorcopsis*; Bourke, 1989), 'walking' as a gait involving asynchronous use of all limbs (confined to species of *Dendrolagus*; Windsor and Dagg, 1971), 'quadrupedal bounding' as movement employing synchronous use of all limbs (present in species of *Dendrolagus*, Windsor and Dagg, 1971; Flannery et al., 1996; and *H. moschatus*, Johnson and Strahan, 1982), and bipedal hopping characterized by synchronous use of the hind limbs only (used at high speeds by

* Address for correspondence: kear.ben@saugov.sa.gov.au

E.J. Sargis and M. Dagosto (eds.), *Mammalian Evolutionary Morphology: A Tribute to Frederick S. Szalay*, 25–35.

all Recent macropodoids except *H. moschatus*; Johnson and Strahan, 1982).

The macropodoid taxa studied in this paper (Appendix) can be placed within four major family-level clades (Balbaridae, Hypsiprymnodontidae, Potoroidae and Macropodidae; see Kear and Cooke, 2001) each of which exhibits a range of characteristic locomotor strategies. The first of these, Balbaridae, is an extinct Oligo-Miocene group of basal macropodoids that is thought to have used slow quadrupedal bounding as their primary gait (Cooke and Kear, 1999). The presence of an opposable first toe, together with a high degree of lateral flexibility in the foot and robust fore limbs, is also potentially indicative of climbing ability (Cooke and Kear, 1999). The record of confidently attributed balbarid appendicular elements is scant, and most inferences about locomotor behavior are drawn from a single near complete skeleton (representing a new species of *Nambaroo* from Riversleigh, northwestern Queensland; see Cooke and Kear, 1999). However, because of the close similarities with living hypsiprymnodontids, a potential analogue for the locomotor habits of extinct balbarids is available. Modern hypsiprymnodontids are represented by the single species *Hypsiprymnodon moschatus* (Hypsiprymnodontinae); a small plesiomorphic macropodoid currently restricted to the tropical rainforest areas of northeastern Australia (Johnson and Strahan, 1982). However, the fossil occurrences of hypsiprymnodontids are geographically widespread (as far as southeastern Australia) indicating a broader distribution during the mid-late Tertiary (Flannery and Archer, 1987; Flannery et al., 1992; Wroe, 1996). Locomotor behavior in extant hypsiprymnodontids (*H. moschatus*) is characterized by consistent use of quadrupedal bounding at both high and low speeds (Johnson and Strahan, 1982). This contrasts with most other living macropodoid groups (Macropodidae, Potoroidae), which predominantly employ both slow pentapedal locomotion during feeding and full bipedal hopping at higher speeds. However, some notable exceptions include the potoroos (*Potorous*: Potoroidae: Potoroinae), which, like hypsiprymnodontids, mainly use quadrupedal bounding (not incorporating the tail) at slower speeds (see Buchmann and Guiler, 1974), and tree kangaroos (*Dendrolagus*: Macropodidae: Macropodinae), which utilize an asynchronous walk when moving along branches and/or climbing (Windsor and Dagg, 1971). Some intriguing fossil taxa are also thought to have employed distinctive locomotor strategies. For example, Plio-Plesitocene sthenurines (Macropodidae), a group that includes some of the largest kangaroos (e.g., *Procoptodon* ~2.5 m high), have been interpreted as specialized high level browsers that habitually used bipedal hopping at the expense of quadrupedal and/or pentapedal gaits (Wells and Tedford, 1995).

Considerable work has been devoted to the structure and function of the limb skeleton in marsupials (e.g., Elftman, 1929; Jenkins, 1971; Van Valkenburgh, 1987; Szalay, 1994). A number of contributions have also discussed functional aspects in fossil taxa (e.g., Finch and Freedman, 1988; Munson, 1992; Muizon, 1998; Szalay and Sargis, 2001; Argot, 2001, 2002, 2003a, b, 2004). For macropodoids, research has focused largely on functional analysis of particular species or clades (e.g., Flannery, 1982; Bishop, 1997; Ride et al., 1997; Kear et al., 2001a, b), but as yet few studies have investigated broad-scale evolutionary trends in the group as a whole.

The purpose of this study is to examine the relationship between locomotor strategy and proportional changes in the proximal limb bones and metatarsals of a range of modern and extinct macropodoids using morphometric analyses. In addition, trends in hind limb evolution through time are investigated, firstly using only observations on extant taxa and inferring ancestral conditions on dated molecular phylogenies, and secondly by adding information from the fossil record. The results suggest that incorporating fossils can drastically change inferences about past diversity and evolutionary trends.

2.2 Materials and Methods

One hundred and eighty-six specimens belonging to 44 species of macropodoids (Appendix) were included together with a phalangerid (*Trichosurus vulpecula*), phascolarctid (*Phascolarctos cinereus*), and vombatid (*Vombatus ursinus*), which served as outgroups. Material was derived from collections of the South Australian Museum, Museum Victoria, Australian Museum, Queensland Museum, and University of New South Wales. Measurements for some fossil taxa were also derived from the literature; these include *Sthenurus tindalei* (Wells and Tedford, 1995), *Procoptodon goliah* (Tedford, 1967), *Protemnodon tumbuna* (Menzies and Ballard, 1994), and *Macropus mundjabus* (Flannery, 1980). All skeletal remains examined were from adults and only articulated or definitively associated fossil elements were used.

A set of three measurements for the maximum lengths of the long bones were taken for each specimen using digital calipers (where <150 mm) to the nearest 0.01 mm, or steel tape (where >150 mm) to the nearest 0.1 mm. Mean and standard deviation values for each species are reported in the Appendix.

(1) *Femur length* (FL) was the distance from the distal apex of the greater trochanter to the distal point of the femoral condyles.
(2) *Tibia length* (TL) was the distance between the proximal and distal articular surfaces of the tibia.
(3) *Metatarsal IV length* (MtL) was the distance between the proximal and distal articular surfaces of metatarsal IV.

Measurements were combined into two functional indices derived from the literature (see Howell, 1944; Hildebrand, 1985, 1988; Finch and Freedman, 1988; Garland and Janis, 1993; Christiansen, 2002). These represent indicators of primary locomotor habits.

(1) *Femoro-tibial index* (T/F = [TL/FL] × 100) is the tibia length divided by the femur length. It gives a measure of proportional change in the proximal limb elements.

(2) *Femoro-metatarsal index* (Mt/F = [MtL/FL] × 100) is the longest metatarsal (metatarsal IV in macropodoids) length divided by the femur length. This gives an indication of proportional change in the metapodials relative to the propodial part of the limb.

To test for potential correlations between limb bone lengths and inferred primary locomotor strategy, measurements were log-transformed and regression lines fitted to the data using standard least-squares. Tibia and metatarsal lengths were arbitrarily treated as the dependent variables. Regression analyses were carried out using *Prism* 4.0a, which also provided 95% confidence intervals (CI) for the slopes.

The relationships between limb bone length, locomotion, and phylogeny was examined using the functional index (T/F, Mt/F) values, which were averaged for each species (Appendix) and optimized onto dated phylogenies based on molecular and fossil information. The most common molecule sequenced for macropodoids is a ~2.5 kB region of mtDNA spanning 12S, valine tRNA, and 16S rRNA. Eighteen of the living species measured above have been sequenced for this gene. The other widely sequenced molecule, protamine P1, was not used, as it was available for fewer taxa and produced poorly resolved trees (see Westerman et al., 2002). Alignments followed Westerman et al. (2002). The arrangement for these 18 taxa found in the larger taxon set of Westerman et al. (2002) was used to infer branch lengths (analyses of the 18 taxa alone yielded a very similar topology). Branch lengths were inferred using PAUP (Swofford, 2000) and the optimal model selected by hierarchical likelihood ratio tests (Posada and Crandall, 1998), the GTRig model. As the chi-squared test detected no significant rate heterogeneity ($P > 0.05$), the molecular clock constraint was enforced to generate an ultrametric tree. This was calibrated to absolute time using the first calibration point used by Westerman et al. (2002); this is one of the most robust and precise calibration points for macropodoids. *Purtia* has traditionally been considered a primitive member of the potoroine (*Bettongia* + *Aepyprymus*) clade and occurs in late Oligocene deposits around 24 my old (see Case, 1984; Woodburne et al., 1993). Accordingly, we set the potoroine-macropodine split at 25 mya. The other fossil Westerman et al. (2002) used to date this divergence, the putative basal macropodine *Nambaroo*, has been reinterpreted as a basal macropodoid (Kear and Cooke, 2001) and thus can no longer be used to date this split. However, the ~23 my old (early Miocene) *Ganguroo* is a true basal macropodid, further supporting the interpretation that this split occurred at least 24 mya but not much earlier. Based on these two calibration fossils, the branch lengths of the ultrametric tree were scaled with *Mesquite* (Maddison and Maddison, 2003) so that the depth of the potoroine-macropodine split was 25; branch lengths throughout this rescaled tree thus represented millions of years. This tree is hereafter termed the 'extant tree'.

Fossil lineages were then added to the 'extant tree', to generate the 'full tree'. *Purtia* and *Gangaroo* were assumed to insert low on the potoroine and macropodine stem lineages (diverging at 24 mya), due to their plesiomorphic characteristics. Within sthenurines, the split between *Sthenurus tindalei* and *S. stirlingi* was set at 1.4 mya, the split between *S. andersoni* and the previous two species at 3.5 mya, and the split between *Procoptodon* and *Sthenurus* at 4.2 mya (see Prideaux, 2004). In the absence of more precise information, other fossil taxa are assumed to have diverged mid-way along the branch connecting relevant extant taxa. For example, *Nambaroo* is a sister taxon to all other macropodoids (*sensu* Balbaridae; Cooke and Kear, 1999; Kear and Cooke, 2001), and is assumed to have diverged mid-way along the stem leading from the outgroups to macropodoids. The three species of *Protemnodon* diverged along the stem leading to derived macropodines (Dawson, 2004), *Macropus mundjabus* diverged along the stem leading to *M. giganteus* (Flannery, 1980), and the sthenurine clade diverged along the stem leading to the *Wallabia-Macropus* clade (e.g., Szalay, 1994).

For each of the two trees (extant and full) and for each of the two traits (T/F and Mt/F), the values for each species (terminal branches) were used to infer ancestral conditions along internal branches (extinct ancestral lineages) with square-change parsimony (Huey and Bennett, 1987) in *Mesquite* (Maddison and Maddison, 2003). The other available option in *Mesquite*, linear parsimony, appears less reliable (Webster and Purvis, 2001) and was not employed. The trends through time were then examined by plotting the inferred values of lineages passing through each time slice. This was done for the extant and full trees to investigate the effects of adding fossil taxa on inferences of past diversity. The fossil taxa contribute the only direct observations for the past, and could also potentially change the inferred values for other (internal) branches, which are not directly observed.

2.3 Results

2.3.1 Regression Analyses of Hind Limb Bone Lengths

Both tibia (Figure 2.1A) and metatarsal IV lengths (Figure 2.1B) were found to scale differently with femur length in bipedal saltating, and in obligate quadrupedal diprotodontians (see Table 2.1). When tibia length is plotted against femur length, the regression slope is >1 in bipedal saltators, and not significantly different from 1 in quadrapeds. This implies strong positive allometry in tibia length in hopping forms, and corroborates the conclusions of others including Windsor and Dagg (1971), who noted that tibia length in particular increased in proportion to that of the femur in larger-bodied macropodoids. Furthermore, for the size ranges considered here, the hopping forms had consistently longer tibiae (relative to femur length) than did the quadrapeds. When metatarsal IV length is plotted against femur length, the regression slopes for both bipedal and quadrapedal taxa do not differ

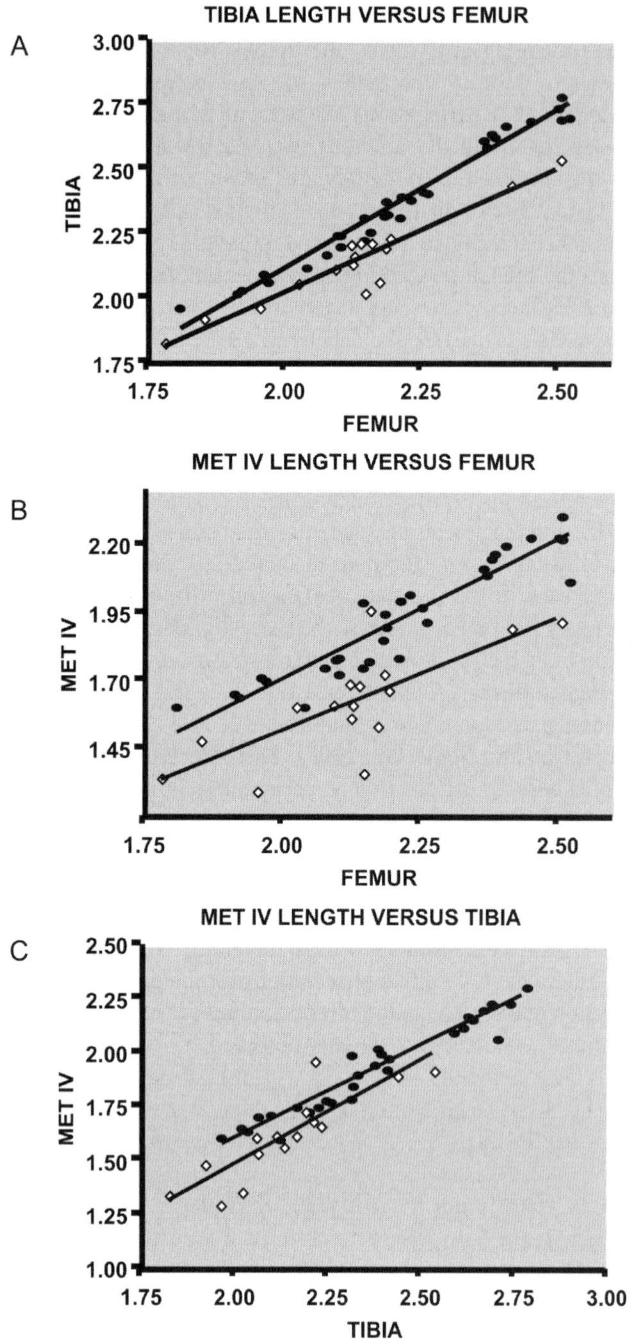

FIGURE 2.1. Regression plots of, A, log tibia versus log femur, B, log metatarsal IV versus log femur, and C, log metatarsal IV versus log tibia lengths, showing close fit of standard regression lines to the data for consistently bipedally saltating (•), and consistently quadrupedal (◊) macropodoid taxa.

from 1, suggesting that metatarsal IV scales approximately isometrically in both groups. However, as before, for the size ranges considered here, the hopping forms had consistently longer metatarsals (relative to femur length) than did the quadrupeds. These results suggest that high tibia or metatarsal IV lengths correlate with locomotor mode.

2.3.2 Comparison of Index Values In Macropodoids

2.3.2.1 Femoro-Tibial Index

Amongst mammals, high T/F values (around 100–200) are traditionally correlated with adaptations towards bipedal saltating gaits (Finch and Freedman, 1988). This trend is evident in the present sample, which indicates that kangaroos of all body sizes consistently have T/F values >100 (see Appendix). The ougroup taxa generally fall below this range (e.g., *Vombatus ursinus*, 74.8; *Phascolarctos cinereus*, 77.8), although the primarily arboreal phalangerid *Trichosurus vulpecula* (101.3) does exhibit tibia/femur proportions similar to some 'short-legged' kangaroo taxa (most notably tree kangaroos; e.g., *Dendrolagus bennettitanus*, 101.8). This is significant given that phalangerids are thought to be closely related to macropodoids (Aplin and Archer, 1987; Kear and Cooke, 2001), and indeed utilize a high speed quadrupedal bound similar to that of some more plesiomorphic kangaroos when moving about on the ground (Goldfinch and Molnar, 1978).

Within Macropodoidea, the lowest values occur in tree kangaroos (ranging from 101.1 in *Dendrolagus matschei* to 109.1 in *D. goodfellowi*) in which the tibia and femur show little differentiation in relative length. Similar T/F proportions also occur in some species of the consistently pentapedal (see Windsor and Dagg, 1971) *Dorcopsis* (*D. atrata*, 109.8), and in the New Guinean Plio-Pleistocene species of *Protemnodon* (105.8 in *P. hopei*, 107.8 in *P. tumbuna*). Murray (1991) suggested that *Protemnodon* might have been a 'low-geared' macropodid, favouring slower speed locomotor modes, and requiring considerable energy expenditure and distance to achieve high-speed saltation. The only other macropodoid to exhibit markedly low T/F values is *Setonix brachyurus* (108.2). Windsor and Dagg (1971) noted that this species also utilizes a quadrupedal bounding gait similar to that of *Dendrolagus*.

The plesiomorphic *Hypsiprymnodon moschatus* (110.6) has T/F values close to those of *Potorous tridactylus* (117.6) and the Riversleigh *Nambaroo* (Balbaridae) species (113.9). Both *H. moschatus* and species of *Potorus* are known to be habitual quadrupedal bounders (Buchmann and Guiler, 1974; Johnson and Strahan, 1982), and a similar locomotor strategy has been suggested for balbarids (Cooke and Kear, 1999).

T/F values are also comparable in Recent bipedal saltating potoroines (123.9 for *Aepyprymnus rufescens*, 126.9 for species of *Bettongia*) and the enigmatic Oligo-Miocene taxon *Purtia* (121.2). Whether this relationship reflects similar locomotor habits or ecology is unclear. Notably, however, *Purtia* has been considered an early potoroid (Case, 1984), although Kear and Cooke (2001) have alternatively suggested affinity with the extinct macropodid subfamily Bulungamayinae.

Some species of the small forest-dwelling *Dorcopsis* (*Dorcopsis* sp., 121.7; *D. luctosa* 119.1) and *Thylogale* (120.3

TABLE 2.1. 95% confidence intervals (significance level set at P < or = 0.05) derived from standard least squares regression analyses of log limb bone lengths for bipedally saltating, and consistently quadrupedal (including pentapedal locomotors; see Introduction for gait definitions) macropodoid and outgroup diprotodontian taxa.

	Consistent saltating bipeds		Consistent quadrupeds	
Regression analysis	95% CI	P	95% CI	P
Log tibia length versus log femur length	1.157 to 1.311	<0.0001	0.7681 to 1.144	<0.0001
Log metatarsal IV length versus log femur length	0.8832 to 1.185	<0.0001	0.4134 to 1.262	0.0008
Log metatarsal IV length versus log tibia length	0.7731 to 0.9420	<0.0001	0.6463 to 1.276	<0.0001

in *T. billiarderii*, 124.9 in *T. thetis*) have values comparable to those of potoroines. *Dorcopsis* in particular has an unusual locomotor mode, favouring slow pentapedal progression with the tail used as an arched prop rather than laying flat on the ground (Bourke, 1989). However, both *Dorcopsis* and *Thylogale* regularly use bipedal hopping at higher speeds (as in potoroines; see Bourke, 1989; Strahan, 1998), and it is probably this habitus that is reflected in the elongation of the tibia relative to the femur (a feature thought to be intimately linked to the use of more efficient high speed bipedal progression; Windsor and Dagg, 1971).

The majority of both fossil and Recent macropodids have high T/F values. Most modern macropodines have means >126 (e.g., 126.7 in *Petrogale penicillata* to 185.5 in *Macropus fuliginosus*), reflecting their consistent use of high-speed bipedal hopping. Interestingly, the bulungamayine *Ganguroo bilamina* also falls within this range (143.2), suggesting that some of the hind limb adaptations necessary for effective bipedal saltation may have already evolved within the macropodid clade by at least the early Miocene.

The Plio-Pleistocene species of *Protemnodon* (Macropodinae) show considerable disparity in their T/F values. For example, those of the large-bodied *P. anak* from southeastern Australia (152.4) are comparable to modern *Macropus* species (see Appendix), whereas the New Guinean *P. hopei* and *P. tumbuna* show values (see above) closer to those of tree kangaroos (*Dendrolagus*). Such high variability within a single genus is surprising, but is consistent with recent indications of taxonomic diversity within the clade (Dawson, 2004). In terms of locomotor behavior, this suggests that the species of *Protemnodon* were adapted for a number of primary gait types, ranging from quadrupedal bounding (*P. hopei* and *P. tumbuna*) to full bipedal hopping (*P. anak*). This may have occurred in response to differing habitat preferences between individual species, with some (e.g., *P. anak*) favoring more open woodland and grassland environments (typifying southern mainland Australia in the Late Pleistocene; Macphail, 1997), as opposed to closed dense forest conditions (i.e., New Guinea) in which quadrupedal progression enables easier directional changes when moving among obstacles covering the ground (Windsor and Dagg, 1971).

The highest T/F values within Macropodoidea occur within the extinct giant Plio-Pleistocene sthenurines (Macropodidae). These include species of *Sthenurus* (approximating some of the larger macropodines in size; Wells and Tedford, 1995), which has values (*S. tindalei*, 154.9; *S. stirlingi*, 172.4; *S. andersoni*, 178.1) comparable to *Macropus* (see Appendix), and *Procoptodon goliah*, which has the highest T/F values of any macropodoid tested (189.1). Wells and Tedford (1995) suggested that *Sthenurus* might have been a habitual bipedal hopper with little or no dependence on pentapedal or quadrupedal movement. Similarly, *Procoptodon* is thought to have been specialized for bipedal progression (Murray, 1991). Indeed, the extreme elongation of the hind limb bones in both *Sthenurus* and *Procoptodon* is likely to have conferred some selective advantage towards this habitus by increasing stride length (critical for bipedal hopping at larger body sizes; see Windsor and Dagg, 1971) and/or height when standing erect for browsing.

2.3.2.2 *Femoro-Metatarsal Index*

As with the T/F index, bipedal hopping mammals are known to show consistent Mt/F values ranging from around 40 to 60 (Howell, 1944; Finch and Freedman, 1988). This trend is also evident in the present sample, with most macropodoids scoring between 45 and 65 (see Appendix). Notably, however, some taxa, namely the habitually quadrupedal species of *Dendrolagus* (*D. matschei*, 27.3; *D. goodfellowi*, 29.6; *D. lumholtzi*, 33; *D. bennettitanus*, 34.7) and *Dorcopsis* (*D. atrata*, 29.6; *D. luctosa*, 34.9; *Dorcopsis* sp., 37.1), *H. moschatus* (30), *S. brachyurus* (38.1), the Oligo-Miocene *Purtia* (36.6), and the enigmatic species of *Protemnodon* (*P. tumbuna*, 26; *P. hopei*, 30; *P. anak*, 35.4), have considerably lower values. Despite this, these figures are still significantly higher than any of those for the outgroup taxa (*V. ursinus*, 15.8; *P. cinereus*, 21.2; *T. vulpecula*, 21.6), suggesting that marked metatarsal elongation may be a common feature shared by all macropodoids.

Most other Recent and fossil potoroids and macropodids in the present study have values that fall within the expected range for bipedal saltators (see Appendix). Taxa with significantly higher values include the larger-bodied species of *Macropus* (*M. giganteus*, 60.4; *M. parryi*, 62.3; *M. fuliginosus*, 63; *M. rufus*, 63.5) and the giant Late Pleistocene sthenurine *P. goliah* (65). Interestingly, the small 'wallaby-sized' (see Kear and Cooke, 2001) Oligo-Miocene bulungamayine *G. bilamina* (63) and balbarid *Nambaroo* (63.3) also have Mt/F values comparable to these large bipedal macropodids. This is unusual given that both these small-bodied taxa are thought to have utilized a considerable degree of quadrupedal movement in their primary gaits (Cooke and Kear, 1999; Kear et al., 2001a). Regardless of these conflicting locomotor strategies and body sizes, the presence of high Mt/F values in a number of independent macropodoid taxa is important because it indicates that elongate metatarsals evolved several times (perhaps in response to similar environmental constraints) in a number of kangaroo groups during the late Oligocene to Recent.

2.3.3 Trends Through Time

The reconstructed, least-squares values for T/F and Mt/F on all branches in the 'extant tree' are shown in Figure 2.2A. On each branch, the lower number (*in italics*) is the T/F value, and the upper number (in plain text) is the Mt/F value. The reconstructed values for each branch in the 'full tree' are shown in Figure 2.2B, using the same notation. The trends through time, namely the inferred values for all lineages existing at each given time slice, are also plotted. The trends through time for T/F are shown in Figure 2.3A (inferred using the 'extant tree') and in Figure 2.3B (inferred using the 'full tree'). Similarly, trends through time for Mt/F are shown in Figures 2.3C and 2.3D (extant and full trees, respectively).

2.4 Discussion

2.4.1 Limb Proportions Versus Locomotion and Ecology

Macropodoids have been a ubiquitous element of the Australian mammal fauna since at least the late Oligocene (Cooke and Kear, 1999). Their characteristic adaptation to bipedal hopping has led to a number of important modifications in the hind limb skeleton, particularly elongation of the femur, tibia, and metapodium. Windsor and Dagg (1971) examined the relationship between limb morphology and movement, recognizing that kangaroos employ a number of primary locomotor strategies (i.e., walking, slow pentapedal progression, quadrupedal bounding, and bipedal hopping), and that these vary considerably according to habitat preference. Importantly, they found no reliable correlation between bipedal hop pattern and relative lengths of the femur and tibia; however, tibia length was reported to increase over that of the femur in larger-bodied taxa. This is confirmed by the present analysis, which shows that T/F values do not differ greatly between bipedal saltators, although they do tend to reach a maximum in the largest species tested (e.g., *P. goliah*; see Figure 2.2). Conversely, Mt/F values fail to conform to a similar pattern, instead varying widely between body sizes, major clades, and in taxa through time (Figure 2.3C, 2.3D).

Although unable to discern between bipedal hop patterns, the T/F (and to a lesser degree Mt/F) values recorded here do seem to reflect major differences in primary gait, especially between quadrupedal taxa (which tend to have low scores) and habitual bipedal saltators. In addition, these figures correlate well with preferred habitat in extant taxa (where discernable); primarily quadrupedal macropodoids (e.g., *H. moschatus, S. brachyurus*, and species of *Dendrolagus*) generally occupying more densely vegetated environments (see Windsor and Dagg, 1971). However, this trend is less evident in the fossil taxa where low T/F forms such as the apparently quadrupedal New Guinean *Protemnodon tumbuna* and *P. hopei*, and the Riversleigh *Nambaroo* species, correlate with a range of interpreted paleohabitats including closed rainforest to alpine grassland (e.g., Flannery, 1992, 1994; Megirian, 1992; Hope et al., 1993; Archer et al., 1997; Archer et al., 2001; Guerin, 2004).

Some early fossil forms (e.g., the Oligo-Miocene *G. bilamina, Purtia* sp.) show high T/F and Mt/F values well within the range of modern bipedal saltators. This suggests that like today, ancient macropodoids probably exhibited a wide variety of locomotor strategies (including quadrupedal bounding to habitual bipedal saltation), and thus were probably able to occupy a similar spectrum of habitats. Indeed, such locomotor diversity, which appears to have been established in a range of taxa (e.g., quadrupedal balbarids and bipedal bulungamayines) by at least the early Miocene, might have facilitated the successful diversification of modern kangaroos (including macropodines and sthenurines, which replaced many of these earlier forms) into the mosaic of open forest and grassland environments that spread across the Australian continent after the onset of aridity in the Miocene-Pliocene (Megirian, 1992; Macphail, 1997; McGowran and Li, 2002).

2.4.2 Limb Proportions Versus Phylogeny

As comparison of Figures 2.2A and 2.2B clearly shows, incorporation of fossil taxa demonstrably changes the inferred T/F and Mt/F values for internal branches within macropodoids. This effect is more evident in Mt/F than T/F values. The results for Mt/F will therefore be discussed below; however a similar (but weaker) pattern is also found in the T/F results. Amongst currently living taxa, high (>60%) Mt/F values are restricted to three species of *Macropus* (*M parryi, M. giganteus, M. rufus*); consideration of extant taxa alone would thus suggest that this condition evolved fairly recently (<1 mya), with all early lineages having low inferred proportion values. However, some basal fossil macropodoids (*Nambaroo* and *Gangaroo*) have unexpectedly high Mt/F values. Addition of these fossils to the analysis (1) adds lineages with directly observed high Mt/F values to the basal parts of the tree, and (2) increases the inferred values of nearby internal

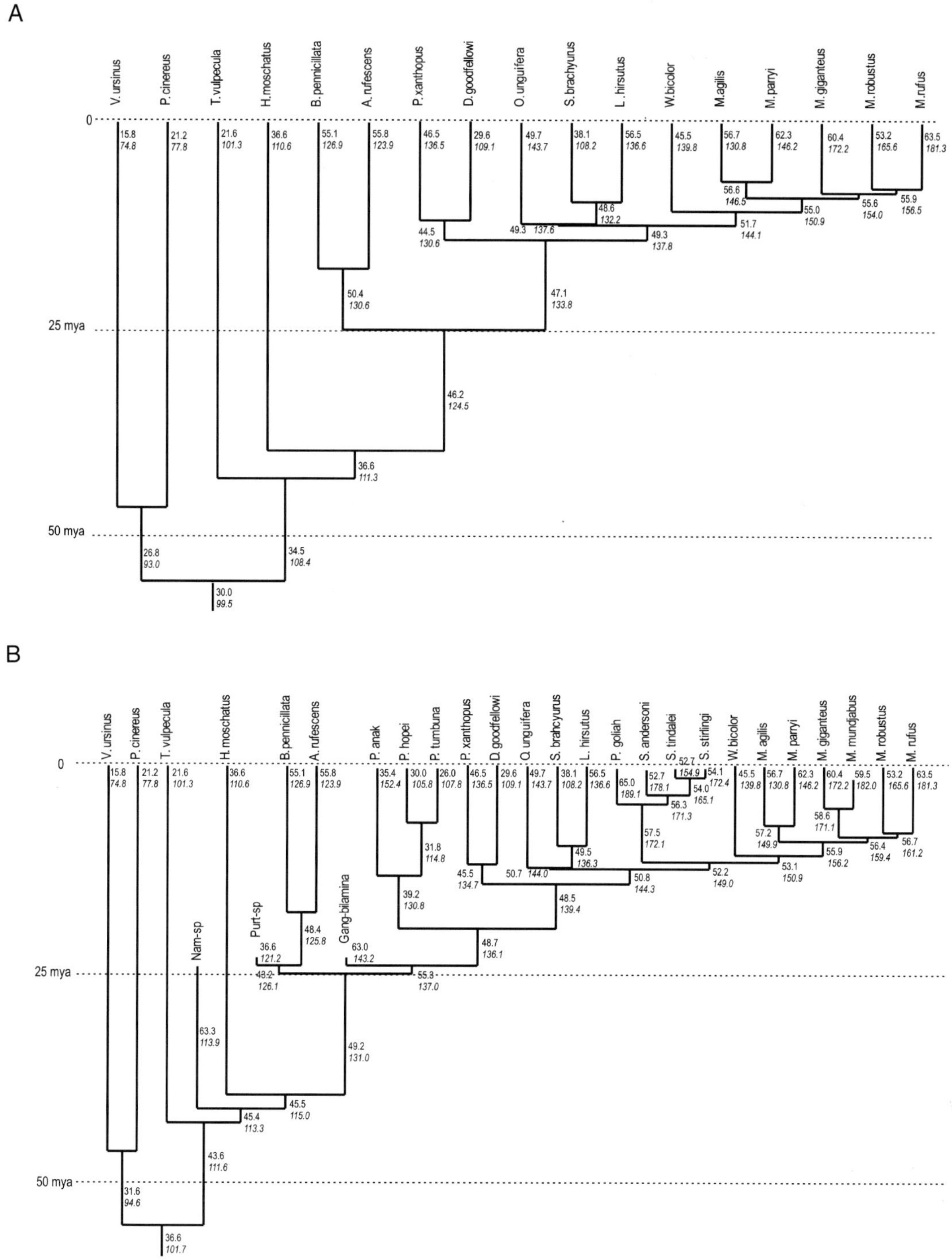

FIGURE 2.2. Macropodoid phylogenies showing reconstructed least-squares values for T/F and Mt/F. A, 'extant tree' (fossil taxa excluded), B, 'full tree' (fossil taxa included). On each branch, the lower number (in italics) is the T/F value, and the upper number (in plain text) is the Mt/F value.

branches. In the 'extant tree' (Figure 2.2A), the inferred value of the branch leading to crown-clade macropodoids (*Hypsiprymnodon* upwards) is 36.6. The inferred value for the same branch increases to 45.5 with the incorporation of the fossil taxa. When plotted through time, the differences in the trends implied by the extant and full phylogenies are striking (see Figures 2.3C, 2.3D). Consideration of extant species alone suggests that low Mt/F values (<37%) characterized all early macropodoid lineages, and that high Mt/F values (>40%) evolved only recently (around 25 mya) within

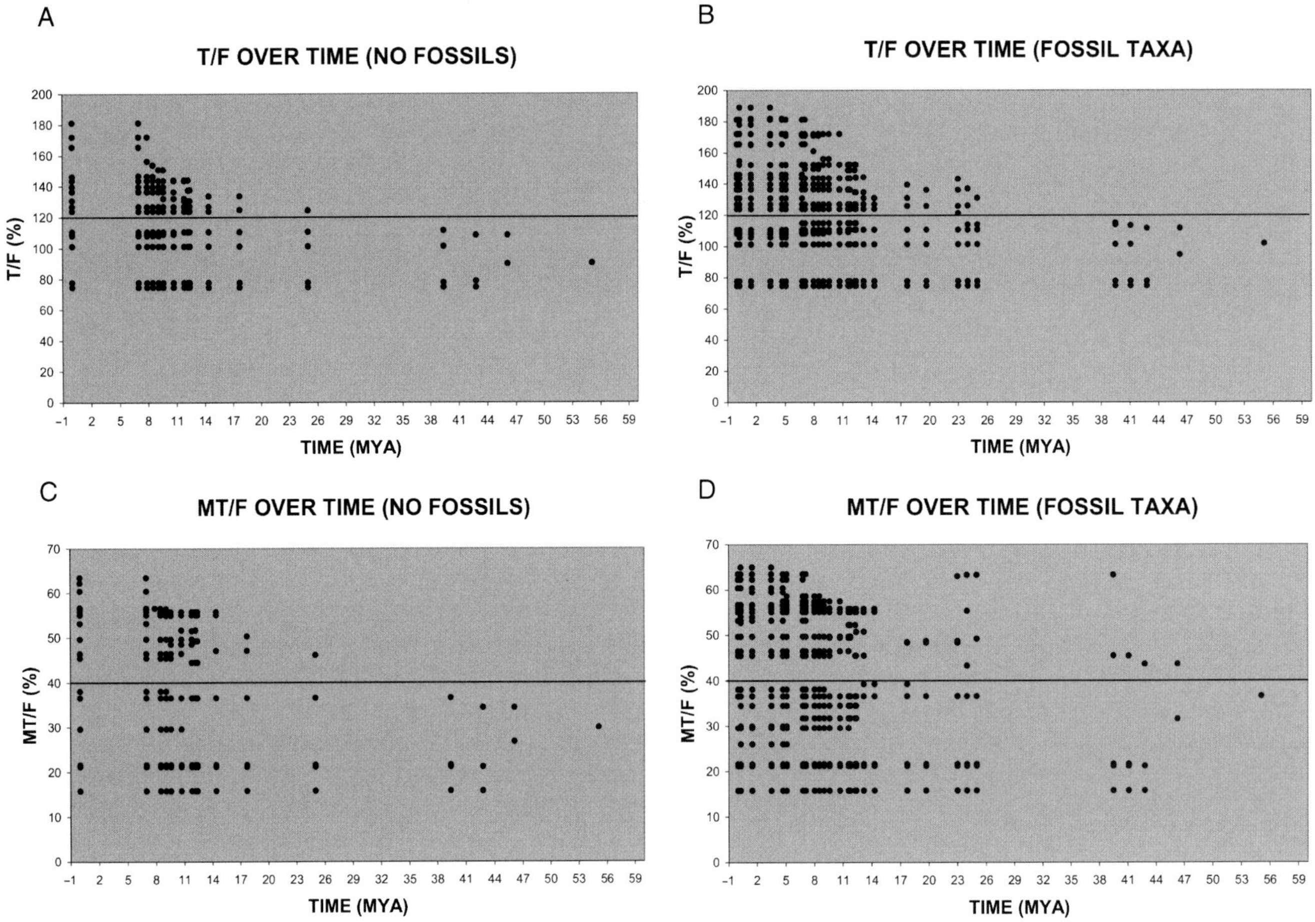

FIGURE 2.3. Plots of reconstructed least-squares values for T/F and Mt/F at each node versus time. A, T/F values inferred using 'extant tree'; B, 'full tree'. C, Mt/F values inferred using 'extant tree'; D, 'full tree'. Horizontal lines at T/F value 120 (A, B) and Mt/F value 40 (C, D) indicate minimum mean value correlating with habitual bipedal saltation.

macropodids (Figure 2.3C). The timing of this change, functionally related to bipedal saltation, is consistent with previous suggestions that the evolution of derived macropodoid locomotor strategies was driven by the onset of aridity and adaptation to increasingly open environments (e.g., Flannery, 1982). However, inclusion of fossil taxa changes the picture. These demonstrate that some early kangaroos had high Mt/F proportions, and indeed that the ancestors of macropodoids had values over 45. When plotted over time, there is no trend towards gradual increase of Mt/F values; rather macropodoids seem to have established a broad range of metapodial proportions very early in their evolutionary history, and then maintained them through to the present day (see Figure 2.3D). Thus, when the full evidence is considered, there is no clear correlation between the evolution of high Mt/F values and increasing aridity in the late Tertiary. Because living taxa with high Mt/F ratios (*Macropus*) represent only one of the many clades that evolved this feature, consideration of living taxa alone gives distorted interpretations; such problems will be manifest whenever groups that are ecological analogues replace each other over geological time.

Acknowledgments. S. Ingleby (Australian Museum), N. W. Longmore and D. Henry (Museum Victoria), N. Pledge and D. Stemmer (South Australian Museum), M. Archer and H. Godthelp (University of New South Wales) all generously assisted in making information and specimens available. La Trobe University, the Australian Research Council, South Australian Museum, Umoona Opal Mine and Museum, Outback at Isa, Origin Energy, The Advertiser, The Waterhouse Club, the Coober Pedy Tourism Association, Commercial and General Capital Ltd and Kenneth J. Herman Inc. gave financial support for this research.

Appendix

Hind limb bone measurements and index values of 47 diprotodontian marsupial taxa (mm)

	N	FL		TL		MtL		T/F		Mt/F	
Taxon		Mean	SD	Mean	SD	Mean	SD	Mean	SD	Mean	SD
Vombatus ursinus	8	9.2		6.2		2.7		74.8	1.8	15.8	0.6
Phascolarctos cinereus	7	9.5		6.4		1.8		77.8	0.8	21.2	0.8
Trichosurus vulpecula	5	14.3		10.2		2.3		101.3	5.9	21.6	1
Nambaroo sp.	1	–		–		–		113.9	–	63.3	–
Hypsiprymnodon moschatus	2	3.5		2.5		0.1		110.6	2.1	36.6	2.3
Potorous tridactylus	6	23.1		83.6	12.9	30.2	2.9	117.6	5.9	43	5.7
Bettongia lesuer	8	83.6	4.7	108.2	5.3	1.2		129.5	3.9	51.8	1.8
Bettongia penicillata	7	4.6		103.9	4.7	45.2	3.2	126.9	3.3	55.1	2.6
Aepiprymnus rufescens	10	15.3		20.3		50.25	10.5	123.9	9.5	55.8	12
Purtia sp.	1	109.4	–	–		–		121.2	–	36.6	–
Ganguroo bilamina	1	64.1	–	91.8	–	–		143.2	–	63	–
Protemnodon hopei	1	–		–		78.1	–	105.8	–	30	–
Protemnodon tumbuna	1	319	–	–		–		107.8	–	26	–
Protemnodon cf. anak	2	40.3		506.5	105.4	14.1		152.4	13.3	35.4	–
Dorcopsis sp.	3	132.7	8	6.5		3.4		121.7	2.3	37.1	0.7
Dorcopsis luctosa	2	137.9	11.7	163.9	3.7	0.8		119.1	7.4	34.9	2.4
Dorcopsis atrata	1	–		170.8	–	46	–	109.8	–	29.6	–
Dendrolagus goodfellowi	2	134.3	0.4	1		0.1		109.1	4.2	29.6	1.3
Dendrolagus matschei	1	–		135	–	36.5	–	101.1	–	27.3	–
Dendrolagus bennettitanus	1	153.2	–	–		53.2	–	101.8	–	34.7	–
Dendrolagus lumholtzi	2	124.5	11.7	129.9	8.6	41.1	3.3	104.5	2.9	33	0.5
Petrogale godmani	1	162	–	206.1	–	60.8	–	127.2	–	37.5	–
Petrogale xanthopus	3	152.1	6.9	207.5	8.2	70.7	3.8	136.5	1.6	46.5	3.4
Petrogale penicillata	4	143.5	5.5	181.8	6.6	59.3	2.4	126.7	1.5	41.1	0.6
Thylogale thetis	2	126.7	13.2	158.8	27.2	53.1	0.2	124.9	8.4	42.1	4.2
Thylogale billarderii	5	139.7	13.4	168	16	56.2	1.5	120.3	1.5	40.5	4.8
Onychogalea fraenata	10	124.7	19.4	176.4	8.2	60.5	3.5	143.7	17.1	49.7	7
Setonix brachyurus	3	105.9	6.1	114.6	6.7	40.3	1.9	108.2	1.6	38.1	0.4
Lagorchestes hirsutus	3	91.7	6.4	125.1	5.6	51.7	1.3	136.6	4	56.5	2.9
Wallabia bicolor	8	183	12.1	255.5	11.3	83.1	4.1	139.8	4.4	45.5	3.1
Macropus agilis	3	156.7	14	205.8	35.3	97.5	–	130.8	10.7	56.7	–
Macropus parryi	6	170.1	31.5	241	55	104.3	9.1	146.2	4.9	62.3	6.4
Macropus giganteus	7	242.2	39.4	420.7	91.5	147	28.8	172.2	56.3	60.4	6.2
Macropus fuliginosus	12	253.3	18.5	465.4	45.5	157.8	10.7	185.5	16.6	63	6
Macropus mundjabus	1	–		434	–	–		182	–	59.5	–
Macropus robustus	4	39.2		74.8		16.7		165.6	5.9	53.2	2.5
Macropus rufus	5	281	42.4	486	97.7	168.7	14.4	181.3	11.6	63.5	7.1
Macropus irma	1	–		–		–		154.7	–	57.9	–
Macropus dorsalis	6	17.5		213.3	16	79.5	4	142.8	5.5	52.9	3.5
Macropus parma	6	119.3	6.8	147.4	14.6	2.6		127.4	2	49.2	3
Macropus rufogriseus	8	18.5		21.3		13.3		144.6	9.9	54.3	5.7
Macropus eugenii	9	126.8	7.5	11.3		60.6	1.3	138.2	4.1	47.7	2.2
Macropus greyi	2	4.6		8.5		–		151	0.9	59.5	–
Procoptodon goliah	1	320	–	605	–	208	–	189.1	–	65	–
Sthenurus andersoni	1	–		–		130.5	–	178.1	–	56.6	–
Sthenurus tindalei	1	319	–	494	–	168.1	–	154.9	–	52.7	–
Sthenuus strirlingi	4	316.1	24.6	544.5	33.7	170.5	6.3	172.4	3.3	54.1	2.3

References

Aplin, K. P., Archer, M., 1987. Recent advances in marsupial systematics with a new syncretic classification. In: Archer, M. (Ed.), Possums and Opossums: Studies in Evolution, Volume 1. Surrey Beaty, Chipping Norton, New South Wales, pp. 15–72.

Archer, M., Hand, S. J., Godthelp, H., 2001. Australia's Lost World: Prehistoric Animals of Riversleigh. Indiana University Press, Bloomington, MN.

Archer, M., Hand, S. J., Godthelp, H., Creaser, P., 1997. Correlation of the Cainozoic sediments of the Riversleigh world heritage fossil property, Queensland, Australia. In: Aguilar, J. P., Legendre, S., Michaux,

J. (Eds.), Actes du Congres BiochroM'97. Ecole Pratique des Hautes estudes Institut de Montpellier, Montpellier, France, pp. 131–152.

Argot, C., 2001. Functional-adaptive anatomy of the forelimb in Didelphidae, and the palaeobiology of the Paleocene marsupials *Mayulestes ferox* and *Pucadelphys andinus*. Journal of Morphology 247, 51–79.

Argot, C., 2002. Functional-adaptive anatomy of the hindlimb anatomy of extinct marsupials, and the palaeobiology of the Paleocene marsupials *Mayulestes ferox* and *Pucadelphys andinus*. Journal of Morphology 253, 76–108.

Argot, C., 2003a. Postcranial functional adaptations in the South American Miocene borhyaenoids (Mammalia: Metatheria): *Cladosictis, Pseudonotictis* and *Sipalocyon*. Alcheringa 27, 303–356.

Argot, C., 2003b. Functional adaptations of the postcranial skeleton of two Miocene borhyaenoids (Mammalia: Metatheria), *Borhyaena* and *Prothylacinus* from South America. Palaeontology 46, 1213–1267.

Argot, C., 2004. Functional-adaptive features and palaeobiologic implications of the postcranial skeleton of the Late Miocene sabretooth borhyaenoid *Thylacosmilus atrox* (Metatheria). Alcheringa 28, 229–266.

Bishop, N., 1997. Functional anatomy of the macropodid pes. Proceedings of the Linnean Society, New South Wales 117, 17–50.

Bourke, D. W., 1989. Observations on the behaviour of the Grey Dorcopsis Wallaby *Dorcopsis luctosa* (Marsupialia: Macropodidae) in captivity. In: Grigg, G., Jarman, P., Hume, I. (Eds.), Kangaroos, Wallabies and Rat Kangaroos.Surrey Beatty, Sydney, pp. 633–640.

Buchmann, O. L. K., Guiler, E. R., 1974. Locomotion in the potoroo. Journal of Mammalogy 55, 203–206.

Case, J. A., 1984. A new genus of potoroinae (Marsupialia: Macropodidae) from the miocene Ngapakaldi local fauna, South Australia, and a definition of the potoroinae. Journal of Paleontology 58, 1074–1086.

Christiansen, P., 2002. Locomotion in terrestrial mammals: the influence of body mass, limb length and bone proportions on speed. Zoological Journal of the Linnean Society 136, 685–714.

Cooke, B. N., Kear, B. P., 1999. Evolution and diversity of kangaroos (Macropodoidea: Marsupialia). Australian Mammalogy 21, 27–29.

Dawson, L., 2004. A new fossil genus of forest wallaby (Marsupialia: Macropodinae) and a review of *Protemnodon* from eastern Australia and New Guinea. Alcheringa 28, 275–290.

Elftman, H. O., 1929. Functional adaptations of the pelvis in marsupials. Bulletin of the American Museum of Natural History 58, 189–232.

Finch, M. E., Freedman, L., 1988. Functional morphology of the limbs of *Thylacoleo carnifex* Owen (Thylacoleonidae: Marsupialia). Australian Journal of Zoology 36, 251–272.

Flannery, T. F., 1980. *Macropus mundjabus*, a new kangaroo (Marsupialia: Macropodidae) of uncertain age from Victoria, Australia. Australian Mammalogy 3, 35–51.

Flannery, T. F., 1982. Hindlimb structure and evolution in the kangaroos (Marsupialia: Macropodoidea). In: Rich, P. V., Thompson, E. M. (Eds.) The Vertebrate Fossil Record of Australasia. Monash University Press, Clayton, Australia, pp. 508–524.

Flannery, T. F., 1992. New pleistocene marsupials (Macropodidae, Diprotodontidae) from subalpine habitats in Irian Jaya. Alcheringa 16, 321–331.

Flannery, T. F., 1994. The fossil land mammal record of New Guinea: a review. Science in New Guinea 20, 39–48.

Flannery, T. F., Archer, M., 1987. *Hypsiprymnodon bartholomaii* (Potoroidae: Marsupialia), a new species from the Dwornamor local fauna and a reassessment of the phylogenetic position of *H. moschatus*. In: Archer, M. (Ed.), Possums and Opossums: Studies in Evolution. Surrey Beatty, Sydney, pp. 749–758.

Flannery, T. F., Martin, R., Szalay, A., 1996. Tree Kangaroos: A Curious Natural History. Reed Books, Melbourne, Australia.

Flannery, T. F., Rich, T. H., Turnbull, W. D., Lundelius, E. L. Jr., 1992. The macropodoidea of the early Pliocene Hamilton local fauna from Victoria, Australia. Fieldiana Geology 25, 1–37.

Garland, T. Jr, Janis, C. M., 1993. Does metatarsal/femur ratio predict the maximum funning speed in cursorial mammals? Journal of Zoology, London 229, 133–151.

Goldfinch, A. J., Molnar, R. E., 1978. Gait of the brush-tail possum (*Trichosurus vulpecula*). Australian Zoology 19, 277–289.

Guerin, G., 2004. Plant macrofossils associated with the Riversleigh macrofauna. Australian Biology 17, 55–62.

Hildebrand, M., 1985. Digging in quadrupeds. In: Hildebrand, M., Bramble, D. M., Leim, K. F., Wake, D. B. (Eds.), Functional Vertebrate Morphology. Belknap, Cambridge, MA, pp. 89–109.

Hildebrand, M., 1988. Analysis of Vertebrate Structure, 3rd ed. Wiley, New York.

Hope, G. S., Flannery, T. F., Boeardi, N., 1993. A preliminary report of changing quaternary mammal faunas in subalpine New Guinea. Quaternary Research 40, 117–126.

Howell, B. A., (1944). Speed in Animals. Their Specialisation for Running and Leaping. University of Chicago Press, Chicago.

Huey, R. B., Bennett, A. F., 1987. Phylogenetic studies of coadaptation: preferred temperatures versus optimal performance temperatures of lizards. Evolution 41, 1098–1115.

Jenkins, F. A. Jr., 1971. Limb posture and locomotion in the Virginia opossum (*Didelphis marsupialis*) and in other non-cursorial mammals. Journal of Zoology, London 165, 303–315.

Johnson, P. M., Strahan, R., 1982. A further description of the musky rat kangaroo, *Hypsiprymnodon moschatus* Ramsay, 1876 (Marsupialia: Potoroidae), with notes on its biology. Australian Zoology 21, 27–46.

Kear, B. P., Cooke, B. N., 2001. A review of macropodoid systematics with the inclusion of a new family. Memoirs of the Association of Australasian Palaeontologists 25, 83–101.

Kear, B. P. Archer, M., Flannery, T. F., 2001a. Postcranial morphology of *Ganguroo bilamina* Cooke, 1997 (Marsupialia: Macropodidae) from the middle Miocene of Riversleigh, northwestern Queensland. Memoirs of the Association of the Australasian Palaeontologists 25, 123–138.

Kear, B. P. Archer, M., Flannery, T. F., 2001b. Bulungamayine (Marsupialia: Macropodoidea) postcranial remains from the late Miocene of Riversleigh northwestern Queensland. Memoirs of the Association of the Australasian Palaeontologists 25, 103–122.

Macphail, M. K., 1997. Late Neogene climates in Australia: fossil pollen- and spore-based estimates in retrospect and prospect. Australian Journal of Botany 45, 425–464.

Maddison, W. P., Maddison, D. R., 2003. *Mesquite*: a modular system for evolutionary analysis. Version 1.0. http://mesquiteproject.org.

McGowran, B., Li, Q. Y., 2002. Sequence biostratigraphy and evolutionary palaeoecology: foraminifera in the Cenozoic Era. Memoirs of the Association of the Australasian Palaeontologists 27, 167–188.

Megirian, D., 1992. Interpretation of the Miocene Carl creek limestone, northwestern Queensland. The Beagle, Records of the Northern Territory of Museum of Arts and Science 9, 219–248.

Menzies, J. I., Ballard, C., 1994. Some new records of pleistocene megafauna from New Guinea. Science in New Guinea 20, 113–139.

Muizon, C. de., 1998. *Mayulestes ferox*, a borhyaenoid (Metatheria: Mammalia) from the early Paleocene of Bolivia. Phylogenetic and palaeobiologic implications. Geodiversitas 20, 19–142.

Munson, C. J., 1992. Postcranial morphology of *Ilaria* and *Ngapakaldia* (Vombatiformes: Marsupialia) and the phylogeny of the vombatiforms based on postcranial morphology. University of California Publications in Zoology 125, 1–99.

Murray, P. F., 1991. The Pleistocene megafauna of Australia. In: Vickers-Rich, P., Monaghan, J. M., Baird, R. F., Rich, T. H. (Eds.), Vertebrate Palaeontology of Australasia. Pioneer Design Studio, Monash University, Melbourne, Australia, pp. 1071–1164.

Posada, D., Crandall, K., 1998. Modeltest: testing the model of DNA substitution. Bioinformatics 14, 817–818.

Prideaux, G., 2004. Systematic and evolution of the sthenurine kangaroos. University of California Publications in Geological Sciences 146, 1–646.

Ride, W. D. L., Pridmore, P. A., Barwick, R. E., Wells, R. T., Heady, R. D., 1997. Towards a biology of *Propleopus oscillans* (Marsupialia: Propleopinae: Hypsiprymnodontidae). Proceedings of the Linnean Society, New South Wales 117, 243–328.

Strahan, R., 1998. The Mammals of Australia. Reed New Holland, Sydney, Australia.

Swofford, D. L., 2000. PAUP*. Phylogenetic Analysis Using Parsimony (*and Other Methods). Version 4. Sinauer, Sunderland, MA.

Szalay, F. S., 1994). Evolutionary History of the Marsupials and an Analysis of Osteological Characters. Cambridge University Press, New York.

Szalay, F. S., Sargis, E. J., 2001. Model-based analysis of postcranial osteology of marsupials from the Paleocene of Itaborai, Brazil, and the phylogenetics and biogeography of Metatheria. Geodiversitas 23, 139–302.

Tedford, R. H., 1967. The fossil Macropodidae from Lake Menindee, New South Wales. University of California Publications in Geological Sciences 64, 1–156.

Van Valkenburgh, B., 1987. Skeletal indicators of locomotor behaviour in living and extant carnivores. Journal of Vertebrate Paleontology 7, 162–182.

Webster, A. J., Purvis, A., 2001. Testing the accuracy of methods for reconstructing ancestral states of continuous characters. Proceedings of the Royal Society of London, Biological Sciences 269, 143–149.

Wells, R. T., Tedford, R. H., 1995. *Sthenurus* (Macropodidae: Marsupialia) from the Pleistocene of Lake Callabonna, South Australia. Bulletin of the American Museum of Natural History 225, 1–111.

Westerman, M., Burk, A., Amrine-Madsen, H., Prideaux, G., Case, J. A., Spinger, M. S., 2002. Molecular evidence for the last survivor of an ancient kangaroo lineage. Journal of Mammalian Evolution 9, 209–223.

Windsor, D. E., Dagg, A. I., 1971. Gaits in the Macropodinae (Marsupialia). Journal of Zoology, London 163, 165–175.

Woodburne, M. O., MacFadden, B. J., Case, J. A., Springer, M. S., Pledge, N., Power, J. D., Woodburne, J. M., Springer, K. B., 1993. Land mammal biostratigraphy of the Etadunna formation (late Oligocene) of South Australia. Journal of Vertebrate Paleontology 14, 483–515.

Wroe, S., 1996. An investigation of phylogeny in the giant extinct rat kangaroo *Ekaltadeta* (Propleopinae: Potoroidae: Marsupialia). Journal of Paleontology 70, 681–690.

3. Changing Views in Paleontology: The Story of a Giant (*Megatherium*, Xenarthra)

Christine Argot*
Département Histoire de la Terre
Muséum national d'Histoire naturelle
UMR 5143 Paléobiodiversité et paléoenvironnements
57 rue Cuvier, C.P. 38
75005 Paris
France
argot@mnhn.fr

3.1 Introduction

Until the eighteenth century, fossils were included in the legends composing the history of life on Earth, which included biblical myths like the Deluge. Knowledge of the history of life is still evolving, shared between numerous geological, biological, and ecological scenarios (e.g., what we know about function, use of ecological niches, and integration of organisms within communities), each one trying to explain a small part of the whole. The main interest of these scenarios seems to provide the opportunity to open the imagination and propose new hypotheses, more than explaining how events really occurred, a point definitely beyond what we can reach (Cohen, 1994). Paleontology is a historical science, which tries to make sense of scattered remains through the composition of a linear story that organizes facts through time and is plausible in the context of current knowledge. But several stories are plausible according to the data known, the supposed rhythms and modalities of evolution, the representations of time …, etc. Therefore, an evolution in the specific field of fossil reconstructions usually refers to an evolution in our representation of the remains of vanished organisms. Here is the story of the theories and interpretations that developed around a spectacular and now extinct animal.

The story of the discovery of the first skeleton of the giant ground sloth *Megatherium americanum* has been told by Cuvier (1812) and Simpson (1984). It is summarized here, in order to provide a basis for the present analysis. In 1788 the skeleton of a big mammal was discovered in northern Argentina, on the bank of the river Luján, near the city of the same name located 65 km west of Buenos Aires. The giant skeleton was discovered by a Dominican, Manuel Torres. The following year it was sent to Madrid and placed in the royal Cabinet of Natural History. One of the employees, Juan Bautista Bru, assembled and drew the skeleton and its various elements in five plates.

French anatomist Georges Cuvier determined the nature and systematic affinities of this mammal on the basis of Bru's drawings. He published the first paper on this subject in 1796 (the transcription of a lecture given previously at the French Academy of Sciences), and complemented this paper in 1804, including in the appendix an original description of the bones of the skeleton written by Bru. The 1804 paper is reproduced in full in Cuvier's famous book "Recherches sur les ossemens fossiles de quadrupèdes" (first edition published in 1812). The controversy concerning the priority of Cuvier's and Bru's descriptions of this specimen has been the subject of two reviews (Hoffstetter, 1959; López-Piñero, 1988) and will not be discussed here. Cuvier gave the taxon the name *Megatherium americanum*, i.e., "big beast of America," following the rules of Linnean nomenclature. *Megatherium* thus became the first fossil mammal to be identified with both generic and specific names.

Although the order Xenarthra is now known from an abundance of extinct species and could be considered as the symbolic group of a continent, South America, it was known only from living species during Cuvier's lifetime, except for *Megalonyx* (i.e., "great claw"), another giant sloth known from a few elements discovered in Virginia, USA, and described by Thomas Jefferson in 1797 (see Simpson, 1984, for a historical review). Only four xenarthran families are extant today, providing little indication of the past richness of the order. *Megatherium americanum* belongs to the

* Address for correspondence: argot@mnhn.fr

E.J. Sargis and M. Dagosto (eds.), *Mammalian Evolutionary Morphology: A Tribute to Frederick S. Szalay*, 37–50.

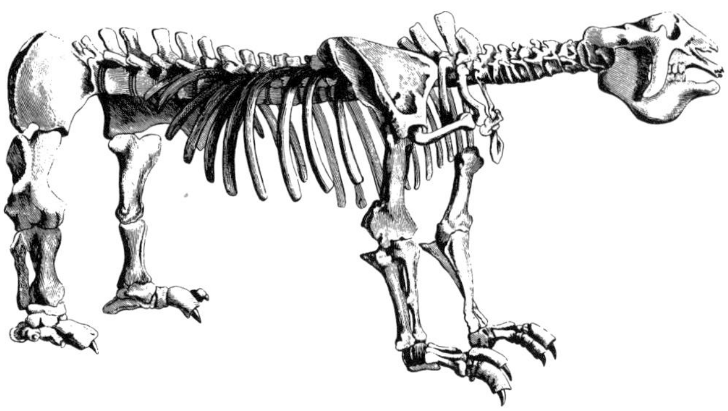

FIGURE 3.1. First representation of a nearly complete skeleton of *Megatherium americanum*, drawn and engraved by Juan Bautista Bru, and reproduced in the first edition of Cuvier's book "Recherches sur les ossemens fossiles" (1812). The head and body length is up to 4 m long, and the shoulder height is 2.25 m. Note that the tail is lacking, as well as parts of the pelvis and sternum. The stiffness of the limbs and axial skeleton is remarkable, the feet lie flat on the ground, and the head of the radius is placed distally. No life dynamics stand out in this drawing, which just reproduces a lifeless object.

family Megatheriidae that appeared during the Santacrucian (end of Early Miocene, i.e., about 17 million years ago) in Patagonia, and disappeared at the end of the Pleistocene. *Megatherium americanum*, known from the Argentinean plains and Bolivian altiplano, belongs to this last period, the "Lujanian" age (middle-late Pleistocene of South America, approximately 800,000–10,000 years BP; McKenna and Bell, 1997), an age named after the first place where *Megatherium* was found, Luján, and characterized by a distinctive "megafauna", i.e., a great number of very large vertebrate taxa.

At the end of the eighteenth century, the interest of paleontologists focused on the bones of quadrupeds, and these bones became the basis of many discussions concerning the problem of vanished species (Cohen, 1994). The Luján specimen is the first sub-complete skeleton of *Megatherium* (and the first fossil vertebrate) to have been assembled (Figure 3.1). As emphasized by Rudwick (1992), "in style, Bru's engraving belongs to a pictorial tradition as long as comparative anatomy itself: a strictly lateral profile drawing providing the most effective visual summary of almost any animal" (p. 32). The skeleton is nearly complete, big and spectacular, and once mounted and exhibited it became accessible to everybody. However, in contrast to dinosaurs and mammoths, *Megatherium* has never become the hero of novels, comics, movies or advertisements. A reason might be that it lived far away, mainly in Patagonia, whereas mammoths and dinosaurs were known from all over the world and during a longer period of time. Moreover, modern sloths, discovered by Europeans at the beginning of the sixteenth century, i.e., only two centuries before Cuvier's description of *Megatherium*, were probably not familiar to most people, except specialized naturalists. Lastly, there are no mythical animals similar to the giant sloth, in contrast to the similarity of medieval dragons to dinosaurs. However, *Megatherium* might have appealed to popular imagery, especially because of its strangeness and enormity. From these two points of view at least, it indeed stands out from the "norms" of the modern fauna, as it is bigger than the three largest living terrestrial animals (the elephant, rhinoceros, and hippopotamus) and as such, it allows both traveling through time and imagination, and illustrates the problems and questions met by paleontologists when they try to reconstruct vanished organisms.

3.2 Lessons from Anatomy

3.2.1 Cuvier's Anatomical Observations (1812)

Cuvier is the creator of four "embranchements" or morphological types, each one representing the basic configurations of animal functions (Eigen, 1997). Guiding Cuvier's conclusions was his belief in the "conditions of existence," the coordination of vital parts that made an animal's life possible in a given milieu. The analysis of these conditions led to the formulation of general laws; Cuvier showed in particular that the objects of nature are connected in predictable conformations, as is the organization of our knowledge about them (Eigen, 1997). In this context, the principle of correlation of forms probably served as a unifying principle of Cuvier's classifications.

The case of *Megatherium* illustrates Cuvier's method of working particularly well. He performed an osteological analysis of the skeleton, comparing it to the skeletons of living species that he had at hand, which were called "Edentata" at that time: extant sloths, giant anteaters, tamanduas, armadillos, as well as scaly anteaters (pangolins), aardvarks, and monotremes (especially echidnas). There were no temporal relationships between extinct and extant mammals in Cuvier's thought process. According to him, these faunas were fully distinct, without genealogical links, but it is useful to compare them in anatomical studies, especially when a vanished animal has, even at a different scale, all the details of organization that characterize an extant species. In this case, the consequences of this organization should be similar (Cuvier, 1812). This comparative process follows a method already used by a predecessor, another French anatomist, Daubenton. It is still followed nowadays, as it remains the key that allows us to resolve the identity of fossil bones. This thought process was regarded with mistrust by some of Cuvier's contemporaries; in their way of thinking, the behavioral and osteological peculiarities of modern sloths precluded any comparison with the giant discovered. However, the rational comparative process began to convince more and more people and we can say that it is from this time that anatomists are "seeking the truth in speaking bones."

Another interest of comparative anatomy outlined by Goethe at the end of the eighteenth century is that the science of morphology might provide insight into primordial forms (Steigerwald, 2002). Morphology was thought to be capable of providing objective knowledge of organisms by discerning the pure forms guiding their formation; these primordial forms were the "necessary" forms of organisms in which the specific forms realized by specific organisms are contained as possibilities (Steigerwald, 2002). Goethe also defended the comparison of all animals with every animal and of every animal with all animals *versus* the comparison of animals to human beings, which was traditional in anatomy until the eighteenth century (Steigerwald, 2002). Recalling this tradition, Richard Owen noted (1858) that no single bone would have better excused the common conclusion of the medieval anatomists concerning the nature of large fossil bones (i.e., that they were those of human giants) than the clavicle of *Megatherium*, since the largest claviculate mammal known at the time of its discovery was humans. Morphology became then and since a descriptive discipline that broadly consists of tracing topological correspondences in organs that are considered to be homologous (Camardi, 2001).

From his analysis, Cuvier established that *Megatherium* had the head and shoulder of a sloth, whereas its legs and feet showed a peculiar mix of characters belonging to both anteaters and armadillos (1812). Cuvier determined that *Megatherium* was a sloth primarily on the basis of skull morphology, and secondly on the dental formula (although the fossil had no canine, in contrast to the modern sloths he had at hand). He noted that the limbs are sub-equal in length (Figure 3.2), in contrast to modern sloths, which led him to deduce that *Megatherium* probably did not crawl on the ground (like modern sloths), nor run or jump, since the runners and especially jumpers usually have longer hind limbs than forelimbs. The presence of a clavicle did not suggest to him the possibility of human affinities, but rather that *Megatherium* probably used its hands to grasp or even to climb. Another behavioral reference from the development of the humeral crests is that the muscles attached there, which are useful in moving the hand and digits, were probably extremely well-developed. This, according to Cuvier, appears to be a clue concerning the important use of the hands by this animal. The development

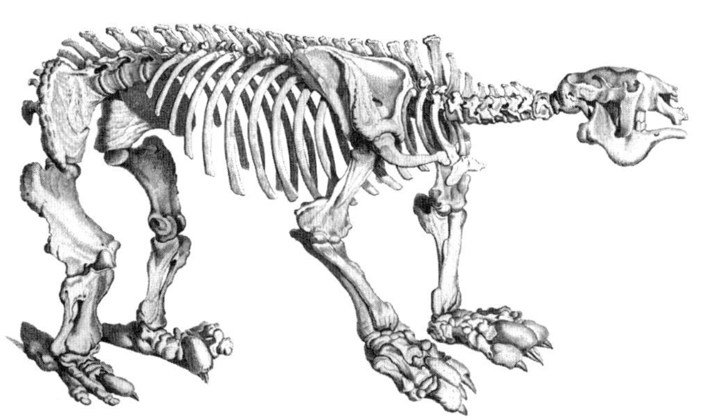

FIGURE 3.2. *Megatherium americanum* as it appears in the second edition of "Recherches sur les ossemens fossiles" (Cuvier, 1821–1824). Note that the animal looks less awkward and stiff in this drawing than in Bru's engraving, with more flexed limbs, and that the radius is reversed (i.e., the head is proximal).

of the anterior crest of the radius suggests powerful pronator-supinator muscles, and this feature, associated with a rounded and concave radial head rotating freely on the ulna, indicates a skillful hand. The pelvis is incomplete on the Spanish skeleton, and the iliac wings are the only remaining parts (Figures 3.1, 3.2). However, they allowed Cuvier to make one of the only paleoecological interpretations that appears in his description; the shape of the ilium suggested to him that the development of the intestines was like that of extant herbivorous species, which was consistent with the shape of the molars. The diet is therefore inferred within the context of the entire organism.

The tibia and fibula are fused to one another on both the right and left hind limb, and Cuvier noted this feature as being characteristic of this animal. This is still true today, *Megatherium americanum* being the only megatheriid to exhibit a full ankylosis of these two bones, which increases the width of the leg. Is this feature related to the weight that the leg has to support? This is not yet explained functionally. Finally, and still only from Bru's engravings, Cuvier described, on the inner side of the hind foot, a bone that he (rightly) interpreted as the fusion of the first two cuneiforms and the first two digits of the foot.

Problems may appear when we see something that does not correspond to what we know, or to what we expect. Cuvier's misinterpretation regarding the hand of *Megatherium* illustrates this point. In the first edition of his "Recherches sur les ossemens fossiles" (1812), Cuvier did not criticize the way Bru mounted the skeleton, but described linearly the three digits with enormous ungual phalanges, the clawless fifth digit, the vestigial pollex, and the unfused carpal bones (Figure 3.3). In the second edition of his book, his description appears to be influenced by the hand structure of anteaters and armadillos, and doubts concerning the interpretation of the hand of *Megatherium* appear: is it the pollex or fifth digit that is vestigial? Now considering the hand as a digging apparatus, Cuvier interpreted the fifth digit as the vestigial one, and the pollex as the clawless digit. This deduction appears "true" to him, as it better mimics the model chosen and follows a general rule indicating that when a reduction of digits occurs, the most external digit gets reduced first. In this work, Cuvier suggests that the ungual phalanges have not been attributed to the correct digits, especially because they do not show a regular decrease in size toward the most lateral digit.

Cuvier's general comments concerning the hind foot are disappointing because he did not seem to pay attention to the information provided by the shape of the bones, in sharp contrast to his usual habits. First, he did not say anything about the peculiarities (shape and position) of the astragalus. The position of this bone and its pivot-like structure both relate to the lateral rotation of the hind foot, which results in *Megatherium* walking on the calcaneum and fifth digit. Such rotation (Figure 3.3) is unique among mammals and is still unexplained. This rotation is thought to be related to the presence of an enormous claw on the third digit, which prevents the

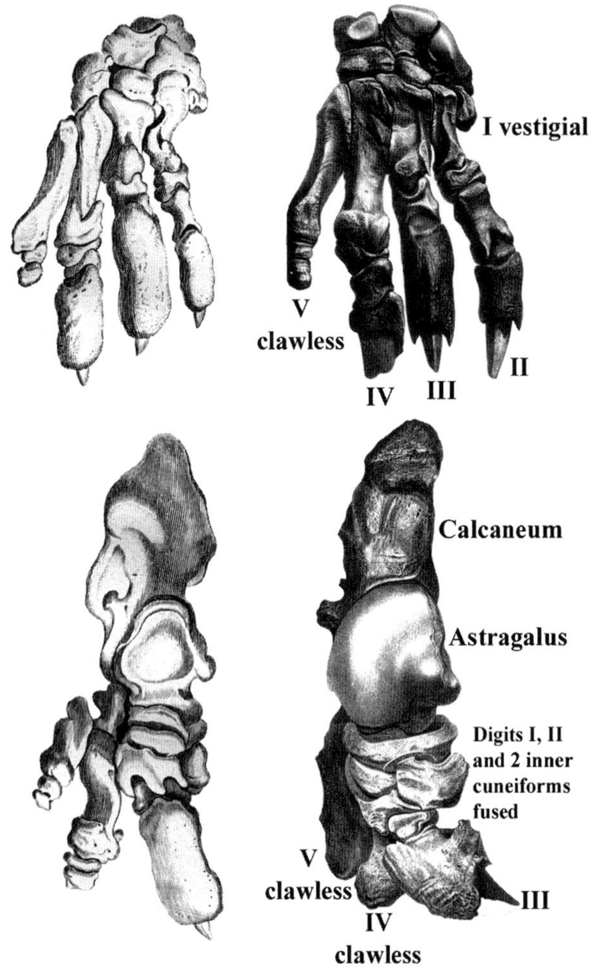

FIGURE 3.3. Extremities of *Megatherium americanum*. The right hand in dorsal view is above, the right foot in dorsal view is below. On the left, the elements as they appear in the second edition of "Recherches sur les ossemens fossiles" (Cuvier, 1821–1824), from Bru's engravings. On the right, as they appear in Owen's monographs (1858, 1860). Observe that the more precise drawings represented in Owen's monographs allow a better understanding of joint function, especially the tibio-astragalar one (see details in the text).

animal from putting its foot flat on the ground. Is this the only possible explanation? Could this pedolateral rotation not be the *consequence* of the presence of this claw, but the *cause* of the disappearance of the two most lateral claws? Why did two digits disappear and only one clawed digit remain? Could the pedolateral rotation be related to bipedalism in the giant sloths? It is a hypothesis that is difficult to test, as no other graviportal mammal developed bipedalism, or even occasional erect posture (see a more extensive discussion below). The pedolateral rotation is not represented in Bru's engravings, or in most representations of *Megatherium* (see below). Was this peculiarity difficult to show because it was unknown in modern mammals and not understood then? Cuvier was astonished by the fact that only three digits remained on the hind foot of *Megatherium*; all the clawed-mammals he knew (e.g., carnivores, rodents) have five clawed digits. But modern sloths also exhibit only three digits on the hind foot, which outlines, according to Cuvier, the similarity between the structure of the hind foot of *Megatherium* and its closest extant relatives. However, there is actually no similarity at all; in modern sloths, the remaining digits are II-III-IV, not III-IV-V, and there is no pedolateral rotation with its additional modifications. The tibioastragalar joint is nevertheless extremely mobile, which allows, for example, the positioning of the plantar sole in continuation with the leg, thereby facilitating suspensory behavior. The pedolateral rotation of giant ground sloths therefore brings into question the various adaptive potentialities present in the ancestral condition.

3.2.2 An Ancestral Constraint Kept By Modern Sloths: The Acromio-coracoid Bridge

Modern sloths are the highly specialized living representatives of Tardigrada, and use an original inverted suspensory behavior during their active periods. They are represented today by a half dozen species placed in two genera: *Choloepus* spp., the two-toed sloth, and *Bradypus* spp., the three-toed sloth. The peculiar suspensory mode of life characterizing these animals represents a remarkable case of convergent evolution, since many studies support the diphyletic origin of tree sloths: *Choloepus* being recognized as a megalonychid, whereas the affinities of *Bradypus* are unclear (Patterson and Pascual, 1968; Webb, 1985; Gaudin, 1995, 2004). Despite superficial resemblances, the two genera differ both in their anatomy (e.g., morphology of the skull; dentition; number of fore digits; number of cervical, dorsal, and caudal vertebrae; presence/absence of the entepicondylar foramen of the humerus; relative length of the forearms; development of volar pads; structure of the hairs) and physiology (e.g., behavior, habits, diet, quality of thermoregulation) (Goffart, 1971; Aiello, 1985; Webb, 1985; Gaudin, 2004). The split between the two extant sloth genera is ancient, dating back perhaps 40 My (Gaudin, 2004), and the appearance and evolution of their suspensory locomotion is not yet understood, mainly because of the lack of fossil tree sloths and transitional forms.

Modern sloths are so specialized that they appear unable to provide any useful information for better understanding the giant members of their group. However, even these highly specialized mammals exhibit – of course – inherited features. This inherited morphology, transmitted from ancestors to their more recent descendents represent genetic, developmental, and physical constraints that existed together with various adaptive traits related to a specific environment exploited by the group during its evolutionary history (see, for example, Bock and von Wahlert, 1965; Szalay, 1999; Szalay and Sargis, 2001, and references therein for a review of this point). Recently, some paleontologists have stressed the stability (or stasis) of a species through time (issue of punctuated equilibrium; see Gould and Eldredge, 1977), which allowed us to focus our attention on the stability of anatomical traits in the evolution of species (Camardi, 2001). An example of a

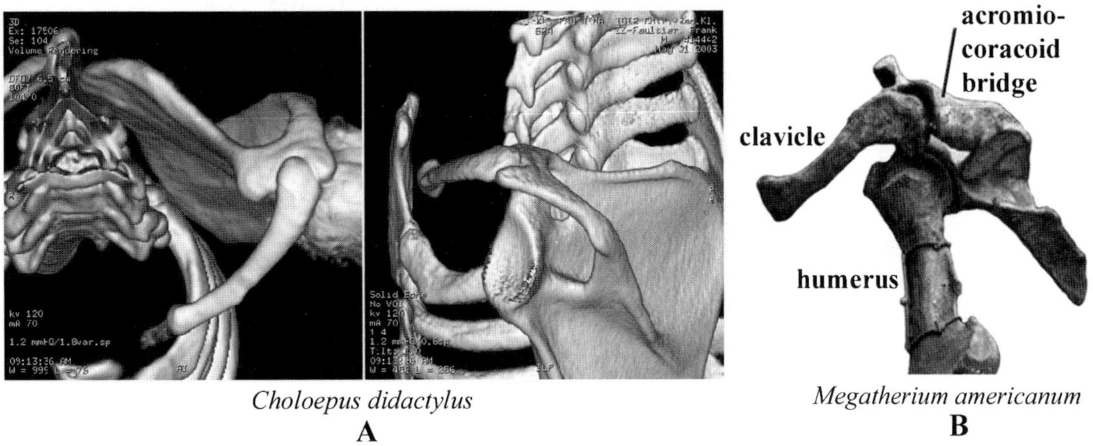

FIGURE 3.4. A, three-dimensional reconstruction of the shoulder joint of a frozen adult male specimen of *Choloepus didactylus*, from a series of 600 μm slices obtained by medical computer tomography. Note that the clavicle has no sternal articulation (only a ligamentous bond), whereas the acromial end is tightly articulated with a robust acromiocoracoid bridge. Compare with the anatomy of the shoulder joint of *Megatherium americanum* (B), which emphasizes the robustness of the acromiocoracoid bridge and the broad articular facet with the clavicle.

stable anatomical trait, whose appearance and function is not yet fully understood, is discussed below.

Sloths are characterized by a very unique shoulder joint, with the acromial end of the clavicle articulating with an acromiocoracoid bridge specific to this order (Figure 3.4). Furthermore, there is no sternal articulation, the clavicle being linked to the manubrium by a ligament. The lack of this medial articulation does not restrict the animal to maintaining a fixed distance between the manubrium and the acromion. Hence, the clavicle loses its usual role, which is to ensure that relative movement between these structures is arcuate (Jenkins, 1974). This condition increases the range of movement in the arm of sloths (Mendel, 1985), in contrast to fully aclaviculate mammals that exhibit linear shoulder excursions restricted to the parasagittal plane (Jenkins, 1974). In mammals that employ generalized patterns of quadrupedal, terrestrial locomotor movements, the weight-bearing forefoot is placed lateral to the sagittal plane through the shoulder joint, and the medially directed component of propulsive force is resisted by the clavicle acting as a strut and preventing medial displacement of the shoulder (Jenkins, 1974). In sloths subjected to "inversed gravity," the role of the clavicle as a strut is less important, which might explain the loss of the medial attachment, the sternal ligamentous attachment of the clavicle providing support only when it is placed under tension (i.e., when the animal is hanging down; see Mendel, 1985).

The bony bridge between the distal part of the acromion and the coracoid process characterizes the scapula of all sloths, regardless of the size, substrate preference, or geological age of the species. This bridge is present in *Megatherium americanum*, and it does not disappear with the specialized locomotion of modern hanging sloths. It is also present in the oldest sloth known from nearly complete skeletons, *Hapalops* sp., which is known from the end of the Early Miocene of Patagonia, Argentina (see Scott, 1903: plate XXX). The locomotion of this relatively small sloth is still unknown, although it has been interpreted as being semi-arboreal, a conclusion based on a very small set of characters (White, 1997). This means that the terrestrial specialization of giant ground sloths many millions of years later, and the appearance of a suspensory mode of life in the most recent sloths, with very peculiar anatomical constraints and environmental conditions, did not lead to the disappearance of this structure. What could be its function?

Modern sloths are characterized (among other features) by the development of the acromioclavicular head of the m. deltoideus, an abductor and protractor of the arm that wraps extensively around the shoulder joint. Some fibers fuse with the radial head of the m. biceps brachii and with the superficial head of the m. pectoralis major, which suggests an emphasis on powerful flexion of the arm and supination of the forearm (Mendel, 1985). The development of such a flexor and supinator unit is helpful both in climbing and digging taxa, which might explain the development of this acromiocoracoid bridge (and associated muscles) in both fossil sloths (some of them likely having been semi-arboreal, whereas others were powerful diggers; see White, 1997; Vizcaíno et al., 2001) and digging armadillos and glyptodonts. This peculiar feature therefore seems to reinforce the general statement made by Szalay and Schrenk (1998) that "postcranial similarities in the stem of both groups [Xenarthra and Palaeanodonta] were likely to be adaptations at least to digging (in a broad sense) from an unknown ancestry" (p. 171), although these authors acknowledge the uncertainties related to the detailed biological role of the postcranial attributes characterizing these groups. The bony bridge between the distal end of the acromion and the coracoid process might help buttress the distal end of the flared acromion, in order to resist the muscular tensile forces acting on it and in the absence of the strut function normally exerted by the clavicle (in modern sloths). However, this bridge exists regardless of the development of the clavicle. Moreover, xenarthrans are not the only mammals to have a mobile shoulder, useful for climbing or digging, and, therefore, the following question arises: why would such a bridge develop *only* in one group within this order? How can we understand an original structure when no other living mammal developed a similar structure by convergence?

Convergent adaptations are indeed sometimes used to stimulate the imagination, as their appearance in unrelated groups suggests similar relationships between structure and function (Ricqlès, 1991). In morphology, it seems that there are only a limited number of solutions to a given functional problem. However, each solution is achieved from distinct premises, which usually leads to different (although superficially similar) final structures. This led some authors to conclude that morphology remains confined to inquiries into the geometric processes of conservation of forms, the science of argumentation based on analogies enduring a definitive lack of explanatory power (Camardi, 2001). When analogs are lacking it becomes almost impossible to understand a structure, and despite the improvements of science, we are still lacking a theoretical, functional science of forms that would allow us to deduce the behavior of a fossil directly from its morphology, without using any guide represented by living organisms (Ricqlès, 1991). It is clear, then, that a morphological theory of organisms is still missing. What paleontologists would need is to be able to identify the "signature" left by different modes of life, this signature being the minimal complex of characters necessary to identify an adaptation, isolated within a larger mass of information.

3.2.3 Interpreting Fossil Remains

Animal reconstructions go back to Cuvier at the start of the nineteenth century (Rudwick, 1992). With Cuvier, the reconstitution of lost species was mainly ruled by the application of the principle of correlation of forms (1812) that mimics mathematical theorems but leads to some exaggerations; e.g., from the teeth, you could know not only the diet, but also the shape of the skull, the morphology of the limbs, the organization of the digestive apparatus, and even the orientation of the eyes, one part determining the whole. From this principle, Cuvier created a true myth: from one fragment it would be possible to recreate the entire animal, a rather holographic thought process. However, a fragmentary skeleton often does not allow us to build correct reconstructions, as the imagination is not constrained enough by the shape. This has been expressed especially by Armand de Ricqlès (1991) who outlined the importance of the information given by the different parts of the body. Ricqlès emphasized that this is the combination of various clues that is judged significant or not, a process that he called the "convergence of presumptions." The problem is that a collection of independent tests only gives the status of a probability to the solution reached.

We might indeed consider that the adaptations of a fossil examined stands in a field of probability, a field born with the first fragment discovered and that narrows with each new fragment discovered. With the entire skeleton, only one of the possibilities suggested by the initial fragment would materialize. However, since we shall never know all the biotic and abiotic conditions surrounding the fossil organism while it was living, the conclusion reached will always remain as a small cloud of probabilities (i.e., with a probability < 1). And the worlds reconstructed will then remain, of course, very ambiguous.

The major problem is still to represent something that we do not know and never saw alive, especially when the principle of correlation of forms cannot be applied as, for example, in modern sloths that exhibit in particular an ankle joint and enormous claws that do not allow the foot to lie flat on the ground, a neck that is too short and stiff to graze, unusual proportions of the limbs, and a peculiar orientation of the femur (because of the position and orientation of the acetabulum). Hence, these mammals appeared imperfect, weak, and vulnerable to Cuvier, who wrote that they probably escaped by a miracle from one of the cataclysms that destroyed previous, imperfect faunas (1821–24).

Anatomical reconstructions look easy to do, although they are usually far from being infallible. They include, for example, a rational process of identification of the elements discovered, the use of living species as models of reference, imagination, and practical skill. A reconstruction therefore reflects the techniques and ways of thinking from a specific time, and always stays at the frontier of art and science. However, too much fantasy would appear to be counterproductive if it is too far from the knowledge we possess. Changes in modern representations should therefore be attributed, at least in part, to the subsequent growth of scientific knowledge: "In a science such as paleontology … there is indeed an unmistakable element of progressive improvement in the development of knowledge" (Rudwick, 1992, p. 220).

However, Rudwick stressed an essential point: do the scenes from deep time simply represent in visual terms the process of scientific discovery (i.e., discovery and identification of bones, assemblage of a partial skeleton belonging to a particular individual, reconstruction of a complete skeleton sometimes based on the remains of many individuals, comparative analysis for reconstructing the soft parts, inferences about the habits and mode of life, and integration in a scene with coexisting individuals and specific ecological elements)? And, in the end, is there a correct manner in which to reconstruct fossils? Would a textual description be better than a visual one? Textual conclusions may be very ambiguous when paleontologists try to take into account all the information provided by the elements found. Alternatively, some visual reconstructions remain too close to the modern forms, hardly indicative of fossil forms unknown in modern faunas. However, it is clear that "clothing" a skeleton requests a deep knowledge and understanding of a living animal's body, the only relevant source of information concerning, for example, the organization and attachment of muscles. Bryant and Seymour (1990) outlined the limits of such muscular reconstructions, showing in particular that although most muscular attachments in Carnivora can be associated with osteological features, which provide information concerning the *position* of the attachments and the *orientation* of the muscle bundles, the inference of muscle *size* and *functional significance* from the same osteological features is problematic. It is clear that muscular reconstructions in extinct taxa are based on assumptions regarding the similarity of musculature in closely related taxa, which emphasizes the importance of phylogenetic relationships. In the case of *Megatherium*, we are unfortunate from this point of view, as it does not make sense to infer that the organization of the musculature of the giant ground sloths was similar to that of modern sloths. Reconstructions cannot, however, be restricted to muscular reconstructions, and a functional analysis also depends on the movements that can be inferred from typically well-preserved and informative skeletal parts, i.e., the shape of joint articulations. Only a small part of the broad articular facet of the humeral or femoral head seems to be involved in locomotor movements (M. Schmidt, personal communication, 2003), the rest of it allowing the animal to perform more extreme postural movements. It is therefore likely that more thorough analyses of articular shapes may help to infer movements and behaviors beyond basic locomotion.

3.3 Paleoecological Interpretations of *Megatherium* over One Century

3.3.1 Nineteenth Century: Cuvier, Pictet, And Owen

Cuvier hardly goes further than the pure description of the skeleton itself. According to his comments (1812), *Megatherium* appears to be a quadruped more or less similar to other large living quadrupeds, not as specialized as ruminants, and lacking any important specificity. Cuvier does not even mention the locomotor specialties of modern sloths, but he probably never had the opportunity to observe a modern sloth alive. This is one of the important points to keep in mind when considering reconstructions that have been made of extinct organisms, namely that the knowledge of those who do the reconstructions determines how they interpret the available elements.

Successive researchers went beyond Cuvier's description, such as the French geologist Pictet (1853–57) who, on the basis of more complete remains, discussed several hypotheses concerning the possible mode of life of *Megatherium*. Pictet based his interpretations on two main points: (1) the forelimbs were probably not restricted to locomotion, and (2) the tail could have played a role in locomotion or provided a strong support. His three main hypotheses are:

(1) *Megatherium* might have been a fossorial taxon because of the enormous claws and herbivorous diet, although the country would have been endangered by the galleries dug up by such an enormous creature. As described by Owen (1860), it would have been an earth whale! But another giant mammal, the mammoth, has also been considered as a giant mole in many Siberian legends (Cohen, 1994) because it was usually found with the tusks emerging from the ground. However, Pictet rejected this hypothesis, concluding that the orientation of the claws precluded the hand from being an efficient digging apparatus, although the hands may have been capable of grasping objects.
(2) Instead, *Megatherium* might have been arboreal because of some features shared with modern sloths, such as the pronation-supination capabilities of the forearm. However, the size of the body still represents the largest problem with this hypothesis, unless the trees associated with the Lujanian megafauna were much more robust than extant trees. Was the tail prehensile and used during climbing? It was not; Pictet observed that the articular facets of the caudal vertebrae indicate a tail that curved upward.
(3) Finally, *Megatherium* might have been terrestrial, and able to uproot trees to better handle them when eating their foliage. In 1860, Richard Owen, on the basis of more complete remains discovered in 1832 and 1837, recognized

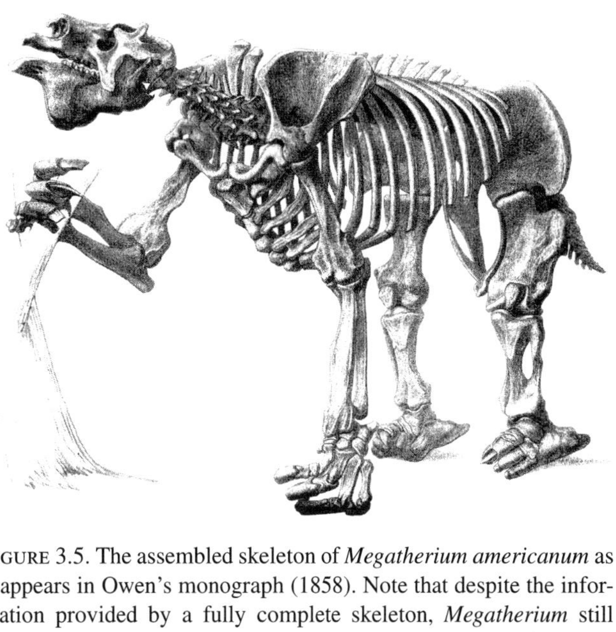

FIGURE 3.5. The assembled skeleton of *Megatherium americanum* as it appears in Owen's monograph (1858). Note that despite the information provided by a fully complete skeleton, *Megatherium* still appears as a typically quadrupedal mammal, the right hand grasping the branch being the only original feature separating it from any other ruminant. Also, note that despite the very precise drawings of the hands and feet that provide much information about the function of the wrist and ankle joints (reproduced in Figure 3.3), the extremities are drawn here in an implausible stance.

the triangular base represented by the posterior part of the body (pelvis, tail, and hind limbs); the tail and a major portion of the pelvis were unknown during Cuvier's lifetime. According to Owen, this would have represented a reliable support from which the animal was able to easily uproot trees, an activity made easier by the powerful clawed forelimbs.

Beyond the details, these interpretations outline how the shape of bones constrains interpretations, since the most complete elements known to Owen (in contrast to Cuvier) helped the English anatomist to imagine stances unknown in typical quadrupeds. Unfortunately, despite this original textual interpretation, *Megatherium* remained quadrupedal in the illustrations from Owen's monograph (Figure 3.5).

Concerning the hind limbs, an interesting observation made by Pictet (1853–57) is that the acetabula are directed ventrally, so that the femora are oriented vertically and are not oblique as is usually the case in typical quadrupeds. According to Pictet, this orientation would have led to an awkward quadrupedal gait, and is consistent with a semi-erect posture. However, the vertical orientation of the femora is also seen in very large mammals like elephants, although the knee joint of *Megatherium* suggests that the giant sloth, in contrast to elephants, was characterized by a flexed, abducted position of the hind limb (Toledo, 1996).

3.3.2 Two Additional Reconstructions: Hawkins And Riou

Despite his limited comments concerning the postures and paleoecology of *Megatherium*, Cuvier often used skeletal and life reconstructions of fossils during his lectures, and described the possible habits of these animals, putting them into a paleobiological context (Rudwick, 1999). However, body profiles, as represented in the second edition of Cuvier's book on fossil vertebrates (1821–24) are less informative than the textual reconstruction of their likely appearance and habits. From this time, however, skeletal reconstructions became the first stage in the reconstruction of complete prehistoric scenes (Rudwick, 1992). It is still the greatest goal of functional anatomy, namely to give life again to vanished organisms, surrounded by the environment in which they once lived.

In the middle of the nineteenth century, a series of life-sized three-dimensional reconstructions of some of the most spectacular fossils that geological research had revealed in the preceding decades was produced by the sculptor Benjamin Waterhouse Hawkins, with the scientific advice of the anatomist Richard Owen, for the Great Exhibition (the first major international event of that kind), which took place in London in 1851. These vanished worlds were then allowed to penetrate the consciousness of a large public (Rudwick, 1992) and from that time, artists and writers began to take another look at these elements, adding to the naked skeletons a look, stance, fur, color, and expression of feelings. Although these attributes come directly from their imagination, there is at least one common basis to the artistic and scientific reconstructions, which is drawing. Drawing represents a means to focus the eye on realities that must be the same for both artists and scientists, as illustrated by the multiple talents of various past scientists and philosophers. In the present context, it is noteworthy that Cuvier was a skillful artist (Rudwick, 1992), whereas Goethe would have liked to be able to express himself entirely in drawings (Steigerwald, 2002), a complementary form of expression for a poet.

The dinosaur series is certainly the most famous series known from this Great Exhibition, but Rudwick (1992) noted that B. W. Hawkins was commissioned by the Department of Science and Art to continue his educational work by drawing a whole set of paleontological scenes, thereby making his reconstructions available to those unable to visit them. One of the wall posters he created illustrates *Megatherium americanum* (Figure 3.6). Two individuals are represented, one quadruped and the other a biped, grasping a tree trunk. The small head contrasts with the huge belly and hindquarters. They seem to suffer from obesity and it is difficult to imagine how they could raise their bellies while moving. The individual on the right is represented with a small mane on the neck and shoulders. Its hind feet lie flat on the ground, the single huge claw pointing forward. By contrast, it rests on the lateral side of its left hand. The apparent lack of pedolateral rotation, in contrast to the "manolateral" rotation is particularly interesting. If *Megatherium* walked primarily quadrupedally, the hand would indeed probably exhibit features related to lateral rotation, as the hind feet do. Cuvier (1812), influenced more by the direct observations he made on living mammals than by the information provided by the Spanish skeleton that he never actually saw, did not interpret the reduction and loss of the claw on the fifth digit of the hand as an incipient specialization toward the lateral rotation of the hand. By contrast, Owen (Hawkins' scientific advisor) suggested a relationship between the loss of the fifth digit claw and the load from the weight of the body (1858). Recently, lateral contact of the hand with the ground has been suggested for several Andean Pleistocene megatheriids, in relation to a peculiar morphology of metacarpals IV and V (Pujos et al., 2002). The clawless fifth digit of *Megatherium americanum* would illustrate an incipient lateral load displacement, not achieved because of the lack of specialization of the hand, still involved in various movements. The reconstruction seems therefore to support this hypothesis.

In 1867 French physician Louis Figuier, involved in the popularization of science (see details in Rudwick, 1992), published some 'ideal landscapes of the ancient world,' painted by Edouard Riou, the illustrator for French writer Jules Verne, an occasional creator of vanished worlds. Among these landscapes, a view of South America during the Pleistocene was represented (Figure 3.7). In this landscape, two quite indistinct sloths are shown. The most familiar silhouette is a long-nosed, hairy mammal represented in a

FIGURE 3.6. A wall poster by B. W. Hawkins showing *Megatherium americanum*. This poster is reproduced in Rudwick (1992, p. 164).

FIGURE 3.7. A typical South American landscape, drawn by Edouard Riou and represented in a book describing the Earth before the Deluge, as understood during the second part of the nineteenth century. According to the author of the book, Louis Figuier, this scene groups four typical South American mammals of the 'Quaternary Epoch': *Glyptodon, Megatherium, Mylodon,* and the Mastodon. Number 1 probably represents *Mylodon* and number 2, *Megatherium*. Landscape shown in Figuier (1874, p. 401) – numbers added.

bipedal stance, grasping a tree trunk. However, although this posture can be considered as being narrowly associated with *Megatherium*, Rudwick (1992) identified this sloth as *Mylodon* (its posture following the skeletal reproduction of this genus represented in Pictet [1853–57: pl. VII, figure 5]), with *Megatherium* as a hairy, quadrupedal mass standing just beside it. As a large, clawed herbivore, it has been represented as being similar to a huge bear, lacking any distinctive, recognizable character except the long nose. The animal, reduced to an indistinct mass, suggests a reconstruction based only on the skull, and especially teeth. Tooth shape provides information about the diet of the animal, and tooth dimensions allow the estimation of body mass. Many regression equations have been developed with modern taxa that relate tooth dimensions (or mandibular length) to body weight (e.g., see Martin, 1990; Myers, 2001; Wroe et al., 2004). However, the inferences from the skull do not provide any detailed information concerning other aspects of the animal's behavior, including, for example, limb use in feeding, locomotion, preferred stances, differential use of the fore and hind limbs, and defense behavior. Riou's illustration outlines the problem of visual reconstructions that remain too close to modern forms. A reconstruction should always be based on a rigorous analysis of bones, something the artist occasionally forgets. A typical detail showing that the information provided by the skeleton has not been taken into account is that the hand of *Megatherium* has four distinct claws of equal size like that of a modern bear, even though it was well-known from the first discovery that this animal had only three clawed manual digits. In the same illustration, the mastodon looks like a modern elephant, whereas *Glyptodon* is quite similar to a large turtle. This close association of past faunas with modern ones suppresses what gives a fossil species an "identity," and illustrates that a fossil cannot differ from any present-day quadruped if the artist only tries to design it from modern natural evidence, without taking into account its peculiarities.

3.3.3 The Parisian Exhibition: Albert Gaudry's Choice

In 1854, the Museum of Natural History of Paris included the first skeleton of *Megatherium americanum* in the comparative anatomy exhibition, and associated it with the ruminants. It is quadrupedal like the modern herbivores and similar to the Madrid sloth specimen. The Parisian skeleton, collected in Tarija (Bolivia), includes only a few original elements of the head and limbs. The missing parts were reconstructed from the Spanish skeleton. The specimens chosen for this exhibition were supposed to illustrate Cuvier's theoretical ideas related to the Revolutions of the Globe (Derieux, 1998). This theory was an attempt to provide a coherent story about the disappearance of extinct species, supposing a discontinuous succession of lost worlds separated by terrifying cataclysms. The successive faunas were therefore totally distinct, since nobody had a precise idea about the time during which these extinct species lived. *Megatherium* was included to illustrate such vanished worlds, and how different fossils could be from extant species.

The skeleton that can be seen today in the exhibition of paleontology (Figure 3.8) is registered in the catalog of Comparative Anatomy in 1871 by Paul Gervais, professor of comparative anatomy since 1866. This specimen was previously part of the Great Exhibition that took place in Paris in 1867 (Derieux, 1998). It was mounted by Dr. Sénéchal, with bones collected in the pampas of the Santa-Fé province, Argentina. This skeleton has been included in the gallery of paleontology since its opening in 1898, and since then it has stood erect on its hind limbs, with its fore feet lying on a tree. The public presentation of this skeleton was influenced by the beliefs and personality of the designer of the gallery, Albert Gaudry, who was also an active paleontological researcher. Gaudry was an evolutionist and in this gallery he wished to consider each fossil as a member of the evolutionary chain, its age illustrating the evolutionary degree achieved: "if it is true that the geological strata are nothing but stages in the history of the development of beings, the knowledge of these stages of evolution will provide a precious aid for the determination of the ages of the earth" (translated by Simpson, 1984, p. 100). In contrast to Cuvier's catastrophism,

FIGURE 3.8. *Megatherium americanum*, as it appears today in the gallery of paleontology of the Museum of natural history of Paris.

Gaudry organized the gallery with the idea that a single world has evolved since the oldest ages, with the specimens exhibited illustrating the idea of progress and increased complexity as was typically the case at the time (Gould, 1995). This representation of a progressive evolution of living beings was probably born from the representations of historical periods illustrating the progress of humanity (Cohen, 1994). In this historical context, the Parisian, bipedal *Megatherium* is placed among other giants and spectacular fossil mammals, such as *Mastodon angustidens, Elephas meridionalis*, and *Glyptodon asper*, this group representing, according to Gaudry (1895), "l'apogée du monde."

However, Gaudry does not interpret the bipedalism of *Megatherium* based on anatomical clues, but mainly because he wished to surprise and impress the public: "This gigantic edentate probably had an odd bearing, like the living anteaters; it walked on the external side of its feet, flexing obliquely its phalanges in order to press the top side of its huge nails [sic] against the ground. This arrangement is favorable not to walk but to climb. Nobody will suppose that *Megatherium* climbed trees; what trees would have been able to carry such a heavy creature! But *it is natural to believe* that it often rested on its powerful hindquarters, and stood up against trees, grasping them with its forefeet in order to devour their fruits and foliage. *We thought that it would be curious* to represent our *Megatherium* erected like this on its hind legs, resting on a tree. Its mouth is 3.15 m above the ground; it could easily reach 3.50 m high.... We hope that our *Megatherium* will make a great impression in the future gallery of paleontology" (1895, p. 253; personal translation, italics added – see the construction of this *Megatherium* in Figure 3.9).

Bipedalism in this context has an aesthetic meaning rather than a scientific one. However, Gaudry also seemed to have the desire to represent this animal in a "natural" way, and the posture chosen indeed belongs to one of those described by Pictet (1853–57). Touching the visitor with the aesthetic, external appearance is supposed to improve the understanding of scientific knowledge linked to the object (Déotte, 1993); first you have to be touched before being deeply moved, whereas trying to understand would be the last step. However, the knowledge may sometimes add more darkness than light to the first impression captured by the visitor. The exhibition of objects in a particular context is supposed to lead the public to open passively to specific knowledge (Derieux, 1998). All exhibitions are intellectual constructions, with an irreducible distance between the shape of the object, the ever-growing knowledge of scientists, and the imagination of the public. Hence, once included in an exhibition, a fossil escapes partly from science and stands far from the "truth" that it is supposed to hold in its fragments. Moreover, in the specific case of paleontology, fossils are supposed to help the public's understanding of the history of Earth. Fossils represent particular shapes expressed during different epochs of the evolution of life. Although this shape is one of the only elements of identity available, fossils cannot simply be reduced to it, as they are indicative of past events and environments. It is in this framework that they can help researchers produce something like an evolutionary history of functions.

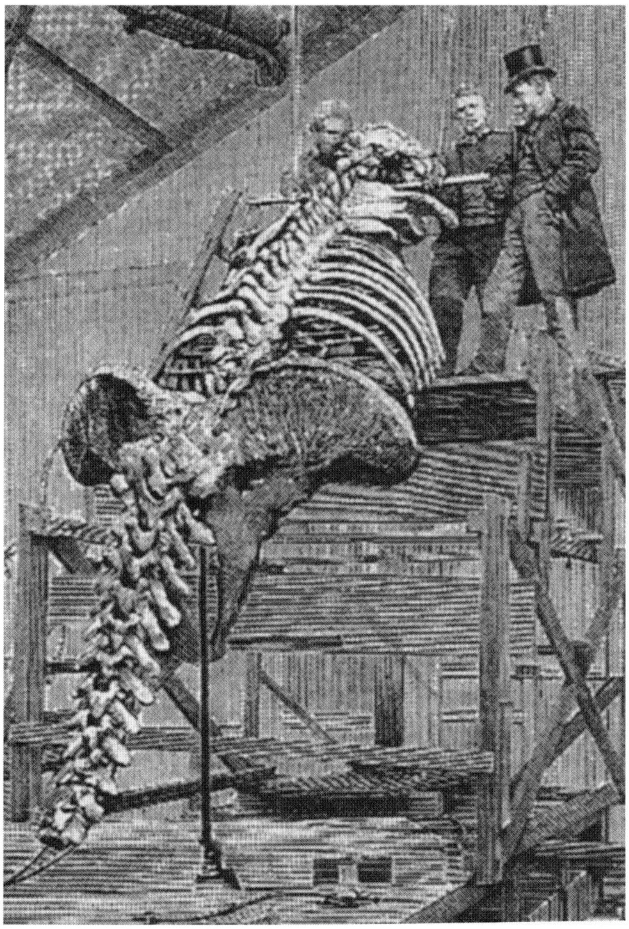

FIGURE 3.9. *Megatherium americanum* being assembled before the opening of the gallery of paleontology of the Museum of natural history of Paris. Figured in Glangeaud (1898).

3.3.4 Recent Information About The Locomotion Of *Megatherium*

Concerning the locomotion of these animals, in his famous book "Splendid Isolation," Simpson (1980) described megatheriids as such: "Large as they are, they surely were not even semi-arboreal, and with their very stout but relatively rather short hind legs, they probably could walk bipedally on occasion if not habitually" (p. 91). Simpson did not provide a reference for this speculative judgment, which seems to have been instinctive, or depending simply on common sense. According to new data, one point is in favor of at least occasional bipedalism: tracks of footprints, discovered in the Pliocene (Casamiquela, 1974) and late Pleistocene of Pehuen-Co (Aramayo and Bianco, 1996), both in the province of Buenos Aires. Although it is impossible to attribute these tracks specifically to *Megatherium americanum*, the hind foot prints indicate pedolateral rotation, with the claw of the third toe pointing toward the inner side of the foot. No tail print is observed, suggesting that the tail was held a few centimeters above the ground, playing a role in the equilibrium of the animal.

Based on these tracks, *Megatherium* could reach 5–6 km/h (1.4–1.67 m/s; Casinos, 1996, Table 3) when walking bipedally. Casinos (1996) also further investigated whether bipedalism was possible, considering some biomechanical characteristics of the skeleton like the moment of resistance of the vertebral column and the bending moment at breaking of the femur. To my knowledge, this is the only biomechanical study to have been performed on *Megatherium*. From this study, it appears that the skeleton was able to support the forces generated by bipedalism. When compared to average mammalian values, the humerus appears to be the longest and thinnest bone among the long bones of the skeleton, and the tibia the shortest and thickest one (Casinos, 1996). It has also been estimated that the body mass of a "typical specimen" like the one from Madrid was close to four tons. The function of peculiar features like the unique outline of the femur, which is very wide transversely, flattened anteroposteriorly, and twisted, although possibly related to bipedal walking on the lateral side of the foot and therefore to high lateral stresses exerted on the hind limb, is still unclear. As discussed by Casinos (1996, p. 95), "perhaps this bipedalism was the only one possible" for such a giant, with the particular historical constraints it had to deal with. In this case, is the specialization toward pedolateral rotation dependent on the weight of the animal? The fact that a pivot-like astragalus (suggesting pedolateral rotation) is present in a large variety of sloths of all sizes, including, for example, a small sloth from Salla, Bolivia (Late Oligocene) weighing less than 50 kg and having five functional digits (Pujos, 2002), precludes any absolute link between these two variables.

In the context of the order Xenarthra, one issue is to determine why bipedalism would have evolved in sloths. Was it to occasionally get some leaves from the tops of trees, or possibly for defense? *Megatherium* has indeed been described as an efficient stabber by Fariña and Blanco (1996), although, as outlined by these authors, a giant adult mammal with an adult body mass of four tons should not be particularly concerned about the possibility of being attacked by, for example, a saber-toothed *Smilodon*. An erect posture would have freed the forelimbs for roles other than locomotion, such as manipulating large prey (e.g., turning the glyptodonts of the Lujanian fauna upside down to reach their unprotected ventral region) or tearing branches out (Fariña and Blanco, 1996), which would explain the lack of specialized "manolateral" rotation.

The major problem faced when reconstructing the locomotion of extinct animals is that modern kinematics and dynamics data that model the locomotion of living forms through quantified parameters like angular variations or ground force reaction are of limited value when dealing with fossil elements. As outlined by Gould (1993), nothing is more complex than the integrated parameters of the form and behavior of a living organism, compared with what exists in the realm of human construction. The use of technology to render accurate and believable animals is therefore one of the greatest all-time challenges to human ingenuity (Gould, 1993). Today the relationship between behavior (function, *sensu lato*) and skeletal structures has not yet been achieved, so it remains difficult to improve the interpretation of fossils using the most modern tools. In this context, the fossil postcranial material housed in museums represents a great but not yet fully exploited source of information concerning extinct faunas. "The reconstruction of the deep past, although hailed as one of the "wonders" of science, still had to depend on far more traditional resources to be made comprehensible and persuasive to the general public" (Rudwick, 1992, p. 172). Paleontologists still work at this level.

3.4 Conclusion

I wished to show in this chapter how the "only true story" told by fossils is developed day after day, illustrating (and sometimes hiding) a complex network of scientific hypotheses based on very fragmentary evidence. It is astonishing to realize that *Megatherium americanum*, a famous fossil mammal used in many paleontological exhibitions, has never really been studied functionally. In fact, we still know almost nothing concerning, for example, the constraints and mechanisms that have led to the formation of the acromiocoracoid bridge or to the peculiar pedolateral rotation, if this rotation is related to the occasional bipedalism of a giant mammal, and why the order Xenarthra is the only one to have evolved both characters, which are unknown in modern faunas.

The reconstruction of fossils transports modern human beings back into a scene that no human beings actually witnessed (Rudwick, 1992), but such reconstructions are, and will always be, pure human constructions. Paleontologists, like any researchers working in the historical sciences, have to deal with events that occurred only once. They only have access to morphological characters and have to assume that their reconstruction hypotheses will never be tested with other types of evidence. Despite the absence of experimentations and life dynamics in the material at hand, reconstructions and models are built day after day by the well-thought-out interpretation of the observation of a static material (Babin, 1991). In this context, paleontology remains a dialog between imagination and materialization, fossils falling into the area of imagination as soon as they are discovered so that they do not simply remain cadavers. The work of paleontologists is to make sense of the fragments studied and to share the story obtained despite the fact that most of the constitutive elements of the identity of a fossil leave no trace. In this framework, paleontology does not relate especially to the "truth," whose access is now denied forever, but more with the knowledge that we have about the vanished worlds, with the ideas built around them. The most important point is to give coherence and sense to what is observed, keeping in mind the bases on which the conclusions have been made, both scientifically and historically; the story told has its own story! And, despite their imperfections, the reconstructions are in some ways the best visual tools we have with which to present the conclusions of specialists to a large public audience.

Acknowledgments. I thank particularly François Pujos for giving me the opportunity to get more familiar with sloths and for providing many references concerning *Megatherium*, and Cédric Crémière for his advice concerning the historical background of this chapter. I thank two reviewers, Greg McDonald and an anonymous one, for helpful comments that improved the manuscript. I also thank Eric Sargis and Marian Dagosto for inviting me to contribute to this book, doctor in medicine Alexander Petrovitch who performed the CT scan (Figure 3.4), and Philippe Taquet who provided much information concerning the historical publications on *Megatherium*, as well as the opportunity to consult the first two editions of Georges Cuvier's famous book. The photograph in

Figure 3.8 was taken by Denis Serrette and Philippe Loubry from the Museum of Natural History of Paris. This chapter was written in the context of postdoctoral research that took place in 2003–2004 at the Institut für spezielle Zoologie und Evolutionsbiologie of Jena (Germany), granted by the German Foundation Alexander von Humboldt.

References

Aiello, A., 1985. Sloth hair: unanswered questions. In: Montgomery, G. G. (Ed.), The Ecology and Evolution of Armadillos, Sloths, and Vermilinguas. Smithsonian Institution Press, Washington DC, pp. 213–218.

Aramayo, S. A., Bianco, T. M. de., 1996. Edad y nuevos hallazgos de icnitas de mamiferos y aves en el yacimiento paleoicnologico de Pehuen-Co (Pleistoceno tardio), provincia de Buenos Aires, Argentina. Asociación Paleontológica Argentina (Publicación Especial) 4, 47–57.

Babin, C., 1991. De la morphologie à la morphogenèse: un champ ouvert aux investigations paléontologiques. Geobios (Mémoire Spécial) 13, 9–12.

Bock, W. J., von Wahlert, G., 1965. Adaptation and the form-function complex. Evolution 19, 269–299.

Bryant, H. N., Seymour, K. L., 1990. Observations and comments on the reliability of muscle reconstruction in fossil vertebrates. Journal of Morphology 206, 109–117.

Camardi, G., 2001. Richard Owen, morphology and evolution. Journal of the History of Biology 34, 481–515.

Casamiquela, R. M., 1974. El bipedismo de los megaterioideos. Estudio de pisadas fósiles en la formación Rio Negro tipica. Ameghiniana 11, 249–282.

Casinos, A., 1996. Bipedalism and quadrupedalism in *Megatherium*: an attempt at biomechanical reconstruction. Lethaia 29, 87–96.

Cohen, C., 1994. Le destin du mammouth. Editions du Seuil, Paris, 350 pp.

Cuvier, G., 1796. Notice sur le squelette d'une très grande espèce de quadrupède inconnue jusqu'à présent, trouvé au Paraguay, et déposé au cabinet d'Histoire naturelle de Madrid. Magasin encyclopédique 1, 303–310.

Cuvier, G., 1804. Sur le *Megatherium*, autre animal de la famille des paresseux, mais de la taille du rhinocéros, dont un squelette fossile presque complet est conservé au cabinet royal d'Histoire naturelle à Madrid. Annales du Muséum national d'Histoire naturelle 5, 376–387.

Cuvier, G., 1812. Sur le *Megatherium*, autre animal de la famille des paresseux, mais de la taille du rhinocéros, dont un squelette fossile presque complet est conservé au cabinet royal d'histoire naturelle à Madrid. In: Recherches sur les ossemens fossiles de quadrupèdes: où l'on rétablit les caractères de plusieurs espèces d'animaux que les révolutions du globe paroissent avoir détruites. Deterville, Paris, France, Volume IV, part. IV, 19–43.

Cuvier, G., 1821–1824. Recherches sur les ossemens fossiles de quadrupèdes: où l'on rétablit les caractères de plusieurs animaux dont les révolutions du globe ont détruit les espèces, 2ème édition. G. Dufour et E. d'Ocagne, Paris, France, 7 vols.

Déotte, J. L., 1993. Le Musée, l'origine de l'esthétique. Paris: L'Harmattan. 443 pp.

Derieux, V., 1998. Le statut de l'objet paléontologique: du terrain aux collections. Mémoire de D.E.A. non publié, Muséum national d'Histoire naturelle, Paris, France. 75 pp.

Eigen, E. A., 1997. Overcoming first impressions: Georges Cuvier's types. Journal of the History of Biology 30, 179–209.

Fariña, R. A., Blanco, R. E., 1996. *Megatherium*, the stabber. Proceedings of the Royal Society of London, Series B 263, 1725–1729.

Figuier, L., 1874. La terre avant le déluge. Paris, Hachette et Cie, 7e édition.

Gaudin, T. J., 1995. The ear region of Edentates and the phylogeny of the Tardigrada (Mammalia, Xenarthra). Journal of Vertebrate Paleontology 15, 672–705.

Gaudin, T. J., 2004. Phylogenetic relationships among sloths (Mammalia, Xenarthra, Tardigrada): the craniodental evidence. Zoological Journal of the Linnean Society 140, 255–305.

Gaudry, A., 1895. Le nouveau montage du *Megatherium*. Bulletin du Muséum d'Histoire Naturelle 1, 252–253.

Glangeaud, P., 1898. Les nouvelles galeries du Muséum. La Nature 1297, 307–311.

Goffart, M., 1971. Function and Form in the Sloth. New York: Pergamon, 225 pp.

Gould, S. J., 1993. Dinomania. New York Review of Books 40(14), 51–56.

Gould, S. J., 1995. Evolution by walking. Dinosaur in a Haystack. Harmony Books, New York, pp. 248–259.

Gould, S. J., Eldredge, N., 1977. Punctuated equilibria: the tempo and mode of evolution reconsidered. Paleobiology 3, 115–151.

Hoffstetter, R., 1959. Les rôles respectifs de Bru, Cuvier et Garriga dans les premières études concernant *Megatherium*. Bulletin du Muséum national d'Histoire naturelle, Paris, 2ème série, 31, 536–545.

Jenkins, F. A. Jr., 1974. The movement of the shoulder in claviculate and aclaviculate mammals. Journal of Morphology 144, 71–84.

López-Piñero, J. M., 1988. Juan Bautista Bru (1740–1799) and the description of the genus *Megatherium*. Journal of the History of Biology 21, 147–163.

Martin, R. A., 1990. Estimating body mass and correlated variables in extinct mammals: travels in the fourth dimension. In: Damuth, J., MacFadden, B. J. (Eds.), Body Size in Mammalian Paleobiology: Estimation and Biological Implications. Cambridge University Press, Cambridge, pp. 49–68.

McKenna, M. C., Bell, S. K., 1997. Classification of Mammals above the Species Level. Columbia University Press, New York, 631 pp.

Mendel, F. C., 1985. Adaptations for suspensory behavior in the limbs of two-toed sloths. In: Montgomery, G. G. (Ed.), The Ecology and Evolution of Armadillos, Sloths, and Vermilinguas. Smithsonian Institution Press, Washington, DC, pp. 151–162.

Myers, T. J., 2001. Prediction of marsupial body mass. Australian Journal of Zoology 49, 99–118.

Owen, R., 1858. On the *Megatherium (Megatherium americanum*, Blumenbach). Part IV. Bones of the anterior extremities. Philosophical Transactions of the Royal Society of London, 261–278 + plates XVIII–XXII.

Owen, R., 1860. On the *Megatherium (Megatherium americanum*, Blumenbach). Part V. Bones of the posterior extremities. Philosophical Transactions of the Royal Society of London, 809–829 + plates XXXVII–XLI.

Patterson, B., Pascual, R., 1968. Evolution of mammals on southern continents. V. The fossil mammal fauna of South America. The Quarterly Review of Biology 43, 409–451.

Pictet, F. J., 1853–1857. Traité de Paléontologie: ou, histoire naturelle des animaux fossiles considérés dans leurs rapports zoologiques et géologiques, 2ème édition. JB. Baillière, Paris, France, 4 vols.

Pujos, F., 2002. Contribution à la connaissance des Tardigrades (Mammalia: Xenarthra) du Pléistocène péruvien: systématique, phylogénie, anatomie fonctionnelle et extinction. Ph.D. dissertation, Muséum national d'Histoire naturelle, Paris, France, 513 pp.

Pujos, F., Salas R., Mattos, J., 2002. Andean lineage of Pleistocene *Megatherium*: geographical implications. Journal of Vertebrate Paleontology 22, 97A.

Ricqlès, A. de, 1991. Les fossiles sont en forme: quelques aspects du problème des relations entre formes et fonctions en paléontologie. Geobios (Mémoire Spécial) 13, 127–134.

Rudwick, M. J. S., 1992. Scenes from Deep Time. University of Chicago Press, Chicago, IL/London, 280 pp.

Rudwick, M. J. S., 1999. A la recherche des monstres perdus. In: Les Cahiers de Science et Vie n°49, coll. 1000 ans de sciences, hors-série n°VII: XIXe siècle, la passion des mondes disparus, pp. 35–41.

Scott, W. B., 1903. Mammalia of the Santa Cruz beds: Edentata, Glyptodonta, and Gravigrada. In: Scott, W. B. (Ed.), Reports of the Princeton University Expedition to Patagonia, 1896–1899. Princeton University Press, Stuttgart, Volume V, part I, 2: 107–227 + plates XVII–XXXV.

Simpson, G. G., 1980. Splendid Isolation. Yale University Press, New Haven/London, 266 pp.

Simpson, G. G., 1984. Discoverers of the Lost World; an Account of Some of Those Who Brought Back to Life South American Mammals Long Buried in the Abyss of Time. Yale University Press, New Haven/London, 222 pp.

Steigerwald, J., 2002. Goethe's morphology: Urphänomene and aesthetic appraisal. Journal of the History of Biology 35, 291–328.

Szalay, F. S., 1999. Paleontology and macroevolution: on the theoretical conflict between an expanded synthesis and hierarchic punctuationism. In: Bromage, T. G., Schrenk, F. (Eds.), African Biogeography, Climate Change and Human Evolution. Oxford University Press, New York/Oxford, pp. 35–61.

Szalay, F. S., Schrenk, F., 1998. The middle Eocene *Eurotamandua* and a Darwinian phylogenetic analysis of "edentates". Kaupia 7, 97–186.

Szalay, F. S., Sargis, E. J., 2001. Model-based analysis of postcranial osteology of marsupials from the Palaeocene of Itaboraí (Brazil) and the phylogenetics and biogeography of Metatheria. Geodiversitas 23, 139–302.

Toledo, P. M. de., 1996. Locomotory patterns within the Pleistocene sloths. Ph.D. dissertation, University of Colorado, Department of Geological Sciences, 316 pp.

Vizcaíno, S. F., Zárate, M., Bargo, M. S., Dondas, A., 2001. Pleistocene burrows in the Mar del Plata area (Argentina) and their probable builders. Acta Palaeontologica Polonica 46, 289–301.

Webb, S. D., 1985. The interrelationships of tree sloths and ground sloths. In: Montgomery, G. G. (Ed.), The Ecology and Evolution of Armadillos, Sloths, and Vermilinguas. Smithsonian Institution Press, Washington, DC, pp. 105–112.

White, J. L., 1997. Locomotor adaptations in Miocene Xenarthrans. In: Kay, R. F., Madden, R. H., Cifelli, R. L., Flynn, J. J. (Eds.), Vertebrate Paleontology in the Neotropics. The Miocene fauna of La Venta, Colombia. Smithsonian Institution Press, Washington, DC, pp. 246–264.

Wroe, S., Argot, C., Dickman, C., 2004. On the rarity of big fierce carnivores and primacy of isolation and area: tracking large mammalian carnivore diversity on two isolated continents. Proceedings of the Royal Society of London, Series B 1544, 1203–1211.

4. Evolutionary Morphology of the Tenrecoidea (Mammalia) Forelimb Skeleton

Justine A. Salton
Program in Biology
Bard College
Annandale-on-Hudson, NY 12504, USA

Eric J. Sargis*
Department of Anthropology
Yale University
P.O. Box 208277
New Haven, CT 06520, USA
Eric.Sargis@yale.edu

4.1 Introduction

Functional morphology of the mammalian forelimb skeleton and the details of its joints have been explored and discussed in great depth relative to other postcranial regions, despite potential difficulties with interpreting the morphology of this region. The mammalian forelimb performs a variety of biological roles, including postural, locomotor, feeding, exploratory, grooming, and defense related behaviors. Detailed morphology might therefore reflect several overlapping functions and compromises between various demands. Much work has focused on primates, with a particular interest in climbing and rotational mechanics of the shoulder and elbow (e.g., Roberts, 1974; Roberts and Davidson, 1975; Fleagle and Simons, 1982; Rose, 1988, 1989; Harrison, 1989; Ciochon, 1993; Gebo and Sargis, 1994). Function-based analyses of mammalian diggers such as geomyids and vermilinguans focus on aspects of the shoulder, elbow, and wrist that correlate with digging and movement of soil (e.g., Campbell, 1939; Reed, 1951; Yalden, 1966; Taylor, 1978, 1985; Rose and Emry, 1983; Szalay and Schrenk, 1998; Stein, 2000). Studies of proportional differences and details of the shoulder and elbow joints in cursorial mammals have identified a suite of characteristics associated with lengthening the stride and stabilizing joints in the parasagittal plane for high-speed locomotion (e.g., Hopwood, 1947; Smith and Savage, 1956; Taylor, 1974; Hildebrand, 1995). There has been less published work on the functional morphology of aquatic mammals (but see Osburn 1903; Howell, 1970; Smith and Savage, 1956; Kerbis Peterhans and Patterson, 1995). This chapter is a comparative morphological study of the tenrecoid scapula, humerus, ulna, and radius, with particular emphasis on the shoulder and elbow joints. The following questions are addressed:

(1) Do aspects of the tenrecoid forelimb exhibit intergeneric variation that correlate with expected differences based on positional behavior in other mammalian locomotor specialists?
(2) Do taxon-specific features of the tenrecoid forelimb suggest phylogenetic affiliation among members of the tenrecoid subfamilies, such as those found in the hindlimb?
(3) Do *Solenodon, Petrodromus*, and/or *Echinosorex* share characteristics of the forelimb with tenrecoids that might be phylogenetically meaningful?

4.1.1 General Form and Variation of the Mammalian Scapula and Forelimb

Studies on mammalian forelimb form and function focus on a series of general skeletal characteristics that demonstrate considerable variability among taxa. The study of highly

* Address for correspondence: Eric.Sargis@yale.edu

E.J. Sargis and M. Dagosto (eds.), *Mammalian Evolutionary Morphology: A Tribute to Frederick S. Szalay*, 51–71.

variable aspects of form can result in differences in interpretations of what a particular aspect of form is, i.e., where it begins and ends, in addition to how it is defined. Particular characters of form are, therefore, briefly defined and discussed, especially those that are often identified and described in the literature discussed here. Some aspects of their variability are also illustrated (for this section, refer to Table 4.1 for proposed locomotor correlates of form). Functional and phylogenetic interpretation of these characters in relation to taxon differences are addressed further in the Results/Discussion section.

Overall scapular shape varies considerably among mammals. At one end of the spectrum of form there is a triangular scapula, as in humans, with an expanded vertebral (medial) border and the humeral articular surface at the apex. This is generally an effect of a relatively small supraspinous fossa and expanded infraspinous fossa. At the other end of the spectrum is a more rectangular form, usually the correlate of a more moderate vertebral border, a broader axillary (lateral) border towards the glenoid fossa, and a broader supraspinous fossa with a steeply inclined cranial (superior) border towards the glenoid fossa (Figure 4.1; see Argot, 2001, for scapular morphotypes in metatherians). Differences in form are attributed to various attachments of muscles that protract, retract, and rotate the scapula and humerus, stabilize the shoulder joint, and anchor the scapula, yet there

TABLE 4.1. Aspects of the mammalian forelimb with proposed relationship to locomotor behavior.

	Climber	Digger	Terrestrial/runner	Leaper	Swimmer
SCAPULA					
Scapula shape[3,8,13]	Short and broad	Elongated	Long and narrow		
Scapula shape[6]		Short	Long, narrow		Short
Scapular spine[6]		High and long	Present, not enlarged		Low
Supraspinous fossa[1]	Large		Less well-developed		
Supraspinous fossa[10]	Cranially expanded			Large	
Infraspinous fossa[1,10,11,22]	Broad		Narrow and deep		
Vertebral border[10,13,22]	Extended relative to length				
Acromion[3,6,8,10,13]	Large, angled cranially	Long, flaring	Not as large		
Coracoid process[8,10,13]	Long, caudally oriented	Stout, prominent	Short, medially oriented		
Glenoid fossa[3,8,15]	Wide	Elliptical	Tall and narrow		
HUMERUS					
Humerus shape[6,7,12,14,23]	Long, narrow	Robust, short, wide			
Humerus/radius length[16]	Long, narrow	Short	Long	Long	
Humeral head[3,8,10]	Hemispherical	Elliptical	Anteroposteriorly elongated		
Humeral head[17]	Large		Smaller		
Bicipital groove[3,11]	Clearly defined	Well-formed into tunnel	Not as well-formed		
Lesser tuberosity[8,10,13,17,22]	Low, small (but bigger than greater tuberosity)	Pronounced	Higher, larger		
Greater tuberosity[8,10,13,17,18,23]	Lower than head	Pronounced	Prominent, high		
Deltopectoral crest[8,10,19,22]	Large, distally extended	Prominent, distally extend.	Small, short		
Midshaft[2]		Wide			
Distal end of humerus[13]	Wide		Narrow		
Entepicondylar foramen[3]		Elongated			
Medial epicondyle[8,9,10,13,14,20,22]	Well-developed, long	Enlarged	Short		
Lateral epicondyle[8,9,10,14,22]	Well-extended	Enlarged			
Capitulum[4,13]	Spherical		Spindle-shaped		
Trochlea[10]	Developed anteriorly more than posteriorly		More concave posteriorly		
Trochlea[10,13]	Well-separated from capitulum		Continuous with capitulum		
Trochlea[13,19,20]	Mediolaterally wide, shallow		Mediolaterally narrow, deep		
Coronoid fossa[10]			Deep		
Olecranon fossa[10,21]	Shallow		Deep		
ULNA					
Ulnar length[2,8,11,12,14,16]	Long	Short, wide	Long		
Olecranon process[2,5,8,9,10,13,14,21,22]	Less prominent	Large	Prominent		
Olecranon process[5,13]	Curved anteriorly		Straight or curved posteriorly		
Trochlear notch (proximal lip)[3]		Long	Shorter		
Trochlear notch[10]			Deep		
RADIUS					
Radius shape[2,10,12,13,14,23]	Long, bowed	Short, wide			
Radial head[3,4,8,10,13]	Circular	Elliptical	Elliptical		

[1]Roberts and Davidson (1975); [2]Verma (1963); [3]Reed (1951); [4]Szalay and Dagosto (1980); [5]Van Valkenburgh (1987); [6]Smith and Savage (1956); [7]Yalden (1966); [8]Stein (2000); [9]Biknevicius (1993); [10]Argot (2001); [11] Taylor (1974); [12] Casinos et al. (1993); [13]Sargis (2002); [14] Grand and Barboza (2001); [15] Larson (1993); [16] Hildebrand (1995); [17] Rose (1989); [18] Heinrich and Rose (1997); [19] Gebo and Sargis (1994); [20] Szalay and Sargis (2001); [21] Ciochon (1993); [22] Rose and Emry (1983); [23] Hopwood (1947)

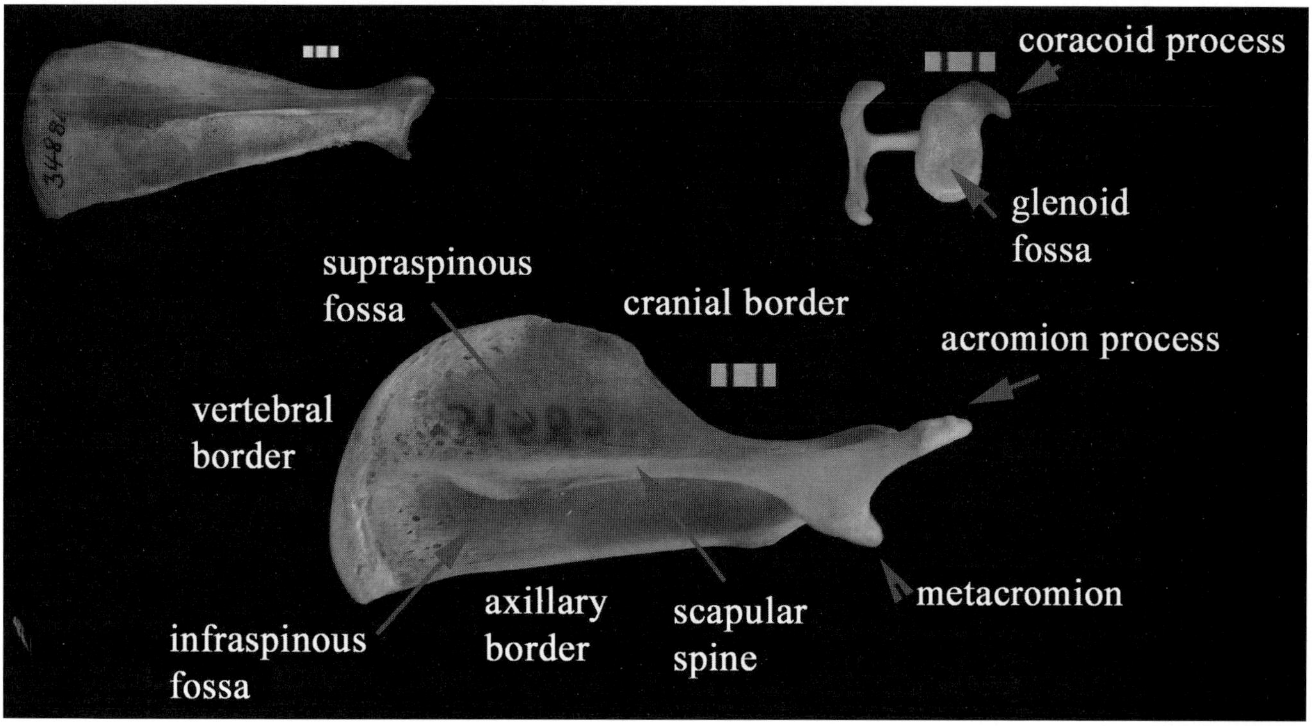

FIGURE 4.1. Right scapulae of *Potamogale* and *Echinops*. Lateral view of *Potamogale* (top left), demonstrating a greatly attenuated triangular form and no articular processes. Lateral view of *Echinops* (center), which is more rectangular, somewhat circular cranially, with distinct acromion, metacromion, and coracoid processes. Articular surface of scapula of same *Echinops* specimen (top right). Subdivisions on scale are 1.0 mm.

are also differences in the position of the scapula against the lateral ribcage resulting in a variety of muscle mass distributions. Scapular position is difficult to determine on a disarticulated skeleton, although it is undoubtedly strongly correlated with differences in scapular shape.

The scapular spine denotes the border between the supraspinous fossa and infraspinous fossa and the Mm. supraspinatus and infraspinatus. Some taxa develop a secondary spine, posterior and ventral to the primary spine, which is associated with an expanded M. teres major and scapular head of the M. triceps brachii (Taylor, 1978, Rose and Emry, 1983). A secondary spine located superior to the primary scapular spine, within the space of the supraspinous fossa, might be associated with an expanded M. rhomboideus or a laterally expanding M. subscapularis.

The acromion process of the scapular spine, when present as a process, is highly variable and can reach well beyond the humeral articulation (Figure 4.1). A metacromion process may or may not be present, hanging caudally from the acromion and extending back along the scapular spine. Development, presence, and absence of the acromion and metacromion are associated with protraction and lateral rotation of the humerus, as well as scapular stabilization. The coracoid is another highly variable feature of the scapula, and, when present, can extend proximally/ventrally and laterally to differing degrees. Its relative length is correlated with the M. coracobrachialis and associated with humeral adduction (Stein, 2000; Argot, 2001; Sargis, 2002). Finally, the shape and size of the glenoid fossa varies amongst mammalian locomotor specialists, presumably in correlation with a shoulder joint that facilitates multiaxial rotation vs. one that restricts movement to a particular plane (Figure 4.1).

Relative length and width of the humerus vary dramatically, from the relatively slender, elongated humerus of a brachiating primate (e.g., *Hylobates*) to a short and robust block-like humerus of a golden mole (Figure 4.2). Relative differences in length and width are generally ascribed to differences in functional mechanics of the musculoskeletal lever system; a relatively short humerus is related to increased force of the muscles originating on the scapula, and a longer humerus contributes to a longer stride for high-speed motion (at the expense of power). Yet the more distal forelimb bones also need to be considered relative to the humerus to interpret mechanical output. For example, in high-speed cursors, lengthened and narrow limbs are expected for long-strides with minimal resistance. Yet the humerus is often short and somewhat robust and the distal elements of the limb are long and thin because muscle mass of the limb is concentrated at the shoulder and proximal arm with long elastic tendons extending to the distal elements (see Hildebrand, 1995).

Shape and relative size of the humeral head varies with differences in rotational facilitation of the glenohumeral joint, yet characteristics of the head do not reliably or necessarily intuitively correlate with features of the glenoid fossa (see Taylor, 1974). The greater tuberosity, attachment site for the humeral retractor M. infraspinatus and protractor M.

supraspinatus, is generally interpreted in terms of its robusticity and proximodistal height above or below the humeral head. The lesser tuberosity is the primary attachment site for M. subscapularis, a medial rotator and adductor of the humerus, and is also discussed in terms of its length and robusticity. The position of the greater and lesser tuberosities might also be of functional relevance; more anteriorly positioned tuberosities result in increased, uninterrupted surface area along the proximal surface of the humeral head, and may be related to rotational facilitation (Figure 4.2). The bicipital groove (or tunnel in some cases), positioned anteriorly between the two tuberosities, transmits a tendon of the M. biceps brachii, and its development might be correlated with powerful forelimb flexion (Figure 4.2).

Muscles associated with the deltoids and pectorals attach at several sites along the anterior and lateral humerus, and are usually associated with characters designated as the deltopectoral crest (ridge, process, or eminence), deltoid tubercle (tuberosity), and/or pectoral process (Figure 4.2). Many mammalian taxa have a deltopectoral crest running down the anterior third of the humerus, with a deltoid tubercle towards the distal end of the crest, as in *Didelphis* (Taylor, 1978). In some forms, the deltoid musculature inserts on the lateral edge of the humerus where a deltoid tubercle is formed and the pectorals attach on the anterior surface, in which case the ridge is referred to as a pectoral ridge (e.g., in tamanduas, Taylor, 1978; Szalay and Schrenk, 1998). The deltoids often act as lateral rotators and abductors of the humerus, whereas the pectorals adduct and retract the humerus (Larson, 1993; Argot, 2001).

At the distal end of the humerus, the coronoid (ulnar) fossa marks the point at which the coronoid process of the ulna (ulnar distal trochlear crest of the semilunar or trochlear notch) rests when the forearm is completely flexed. When the forearm is extended, the ulnar proximal trochlear crest (olecranon beak) inserts into the olecranon fossa of the humerus. Deep or perforated coronoid and/or olecranon fossae are generally attributed to more extreme degrees of forearm flexion and extension, respectively.

The trochlea and capitulum of the distal humerus mark the articular surfaces with the ulna and radius, respectively (Figure 4.2). Differences in mediolateral widths of each sug-

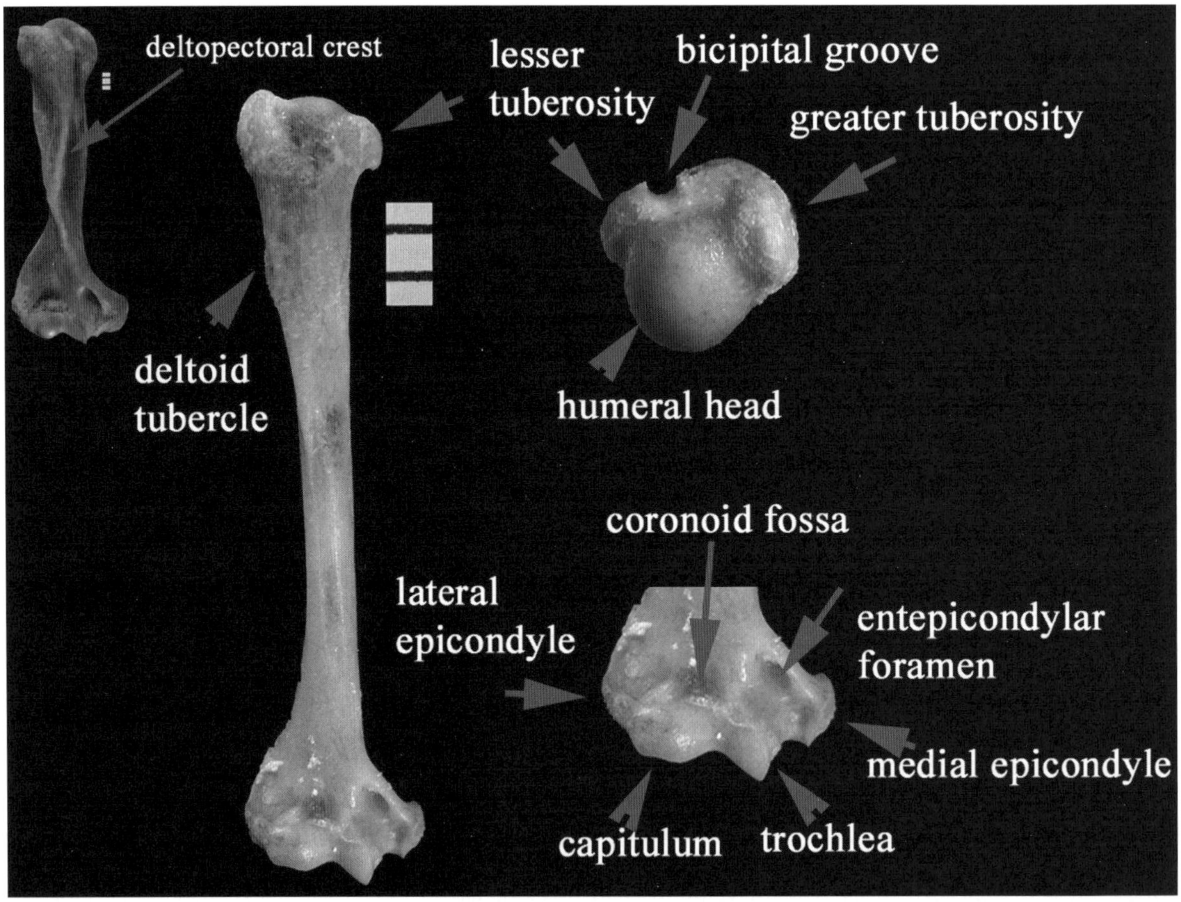

FIGURE 4.2. Right humeri of *Solenodon* and *Microgale*. Anterior view of *Solenodon* (top left), demonstrating a humerus with pronounced crests and processes, including a deltopectoral crest, and wider shaft. Anterior view of whole humerus (center left) and views of proximal humerus (upper right) and distal humerus (lower right) of *Microgale dobsoni*. Subdivisions on scale are 1.0 mm.

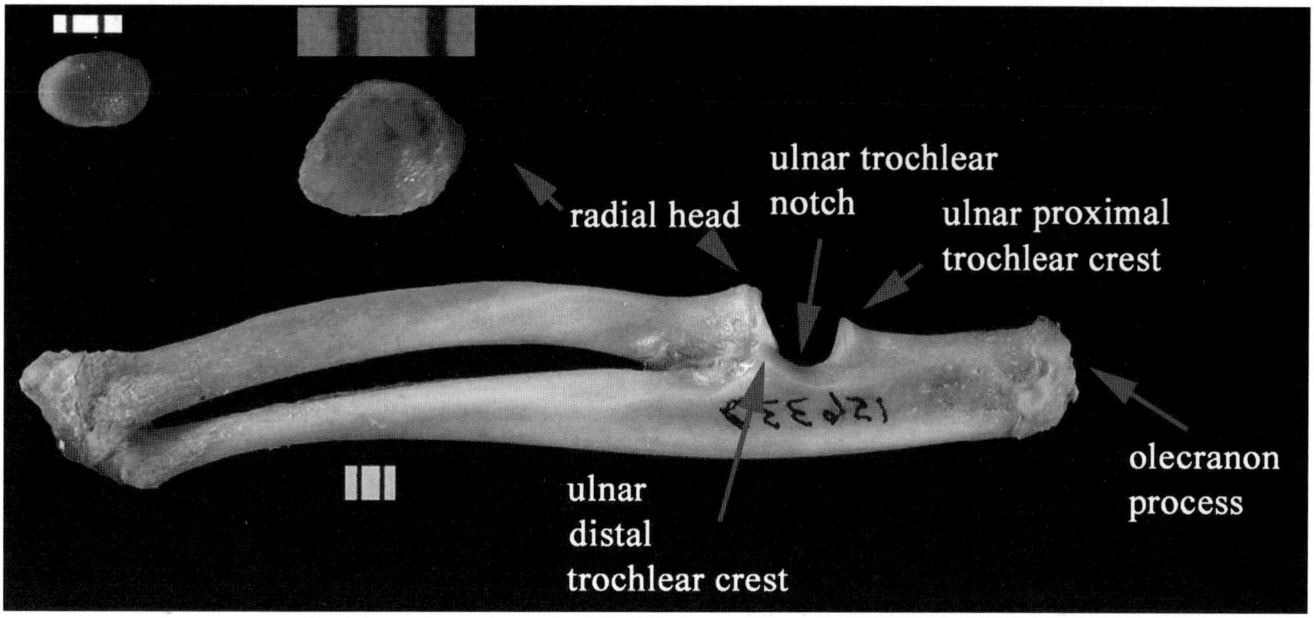

FIGURE 4.3. Right radii and ulna of *Setifer, Limnogale*, and *Tenrec*. Proximal view of *Setifer* radius (top left), illustrating mediolaterally elliptical radial head, proximal view of *Limnogale* radius (middle top), demonstrating a rounded head, and medial view of *Tenrec* antebrachium (bottom). Subdivisions on scale are 1.0 mm.

gest how much body weight is distributed on one side of the humerus relative to the other. Capitulum shape is correlated with movement of the radius on the humerus, and a more spherical shape is generally indicative of multiaxial movement, whereas a trochleated capitulum is correlated to varying degrees with fast flexion/extension of the ulna that requires lateral bracing.

The olecranon process of the ulna is the attachment site for the M. triceps brachii, which is the primary forearm extensor (Figure 4.3). The olecranon process is generally considered in terms of its robusticity and length relative to the rest of the ulna. Overall ulnar and radial proportions are commonly compared to humerus length to determine mechanical compromises between speed (a relatively longer forearm) and power (a relatively shortened forearm). Lastly, the articular surface of the radial head ranges from completely rounded to a mediolaterally-expanded ellipse, indicating greater degrees of mobility in the former and a more restricted lateral elbow joint in the latter (Figure 4.3). All of these characteristics are considered in tenrecoids below.

4.2 Materials and Methods

The scapula, humerus, radius, and ulna of 12 tenrecoids and 3 outgroups were studied and digitally photographed (Nikon Coolpix 995) in several standardized views. Skeletal specimens were examined at the American Museum of Natural History (AMNH), Field Museum of Natural History (FMNH), Harvard University Museum of Comparative Zoology (MCZ), and United States National Museum of Natural History (USNM). Two *Echinops* specimens were borrowed from H. Kuenzle's laboratory at the University of Munich (UMUN), Germany, and a *Hemicentetes* and *Tenrec* specimen were borrowed from the University of Darmstadt (DARM), Germany (see Salton, 2005, for specimen list).

Digital image files were written into TPSdig (Version 1.31, 2001, F.J. Rohlf), which allows for superimposition of landmarks (x,y coordinates) onto images and calibration of image scale from a millimeter ruler. Linear measurements were then calculated from specific coordinates (Salton, 2005). Measurements included those that incorporate features with proposed functional and/or phylogenetic significance (see Table 4.1). Precision of digital measurements was tested against fine-point caliper measurements from three complete specimens, and there were no significant differences ($P < 0.05$) between caliper and digital values.

The following tenrecoid species were studied: *Echinops telfairi, Setifer setosus, Hemicentetes semispinosus, Tenrec ecaudatus, Microgale cowani, M. dobsoni, M. talazaci, Oryzorictes tetradactylus* (or *O. hova*), *Limnogale mergulus, Geogale aurita*, and *Potamogale velox* (Table 4.2). The following species were included as outgroups for comparison with tenrecoids (orders according to Springer et al., 2004): *Solenodon paradoxus* (Eulipotyphla), *Petrodromus tetradactylus* (Macroscelidea), and *Echinosorex gymnurus* (Eulipotyphla). These taxa were chosen because tenrecoids have traditionally been included in Lipotyphla, but have more recently been allied with other African mammals in Afrotheria (Springer et al., 2004; for further discussion of outgroup choices see Salton and Szalay, 2004; Salton, 2005; Salton and Sargis, 2008).

TABLE 4.2. Taxonomy and primary locomotor behavior of study taxa.

Family/subfamily	Genus	Species	n	Locomotor Behavior
Tenrecidae/Tenrecinae	*Echinops*	*telfairi*	13	Arboreal/terrestrial
Tenrecidae/Tenrecinae	*Hemicentetes*	*semispinosus*	18	Terrestrial/fossorial
Tenrecidae/Tenrecinae	*Setifer*	*setosus*	19	Terrestrial
Tenrecidae/Tenrecinae	*Tenrec*	*ecaudatus*	14	Terrestrial
Tenrecidae/Oryzorictinae	*Limnogale*	*mergulus*	5	Aquatic/terrestrial
Tenrecidae/Oryzorictinae	*Microgale*	*cowani*	22	Terrestrial
Tenrecidae/Oryzorictinae	*Microgale*	*dobsoni*	21	Terrestrial
Tenrecidae/Oryzorictinae	*Microgale*	*talazaci*	13	Terrestrial
Tenrecidae/Oryzorictinae	*Oryzorictes*	*tetradactylus/hova*	35	Fossorial/terrestrial
Tenrecidae/Geogalinae	*Geogale*	*aurita*	4	Terrestrial
Potamogalidae	*Potamogale*	*velox*	3	Aquatic/terrestrial
Macroscelididae	*Petrodromus*	*tetradactylus*	3	Terrestrial/saltatory
Solenodontidae	*Solenodon*	*paradoxus*	10	Terrestrial/fossorial
Erinaceidae	*Echinosorex*	*gymnurus*	1	Terrestrial

TABLE 4.3. Indices.

APLI	Acromion Process Length Index = length of scapula from distal end to tip of acromion process/length of scapula to base of glenoid fossa
GFSI	Glenoid Fossa Shape Index = Glenoid fossa dorsoventral length/mediolateral width
HHSI	Humeral Head Shape Index = Humeral head length/width
HRLI	Humerus/Radius Length Index (Brachial Index) = Humerus length/radius length
HSI	Humerus Shape Index = Humerus width/length
MEWI	Medial Epicondyle Width Index = Medial epicondyle width/trochlear width (distal view)
OPLI	Olecranon Process Length Index = Olecranon process length/ulna length
RSI	Radius Shape Index = Radius depth/length
SSI	Scapula Shape Index = Scapula width/length
USI	Ulna Shape Index = Ulna depth/length

In order to control for size differences between species, linear measurements (see Salton, 2005) were transformed into ten indices (Table 4.3). Statistical analyses were performed using STATISTICA (Version 6.0, StatSoft Inc., Tulsa, OK). Indices were each compared between species using one-way ANOVA and the Tukey honest significant difference (HSD) post hoc test ($P < 0.05$). All ANOVA tables are in Salton (2005).

4.3 Results and Discussion

4.3.1 Scapula

There are no subfamily-level differences in the Scapular Shape Index (SSI) between the tenrecines and oryzorictines due to the considerable variation within Oryzorictinae and their overlapping ranges with Tenrecinae (Table 4.4). A narrow, elongated scapula is characteristic of some fossorial rodents and soricids (Reed, 1951; Stein, 2000), and this might be expected in *Hemicentetes*. Lengthening of the scapula is presumably correlated with a large and posteriorly displaced origin of the M. teres major and M. triceps brachii caput longum, which retract and rotate the shoulder and extend the forearm, respectively (Yalden, 1966; Taylor, 1978; Neveu and Gasc, 2002). *Microgale cowani* has a significantly longer and narrower scapula than *M. dobsoni* (Figure 4.4, Table 4.4; $P < 0.05$). This is consistent with a series of other postcranial traits that suggests *M. cowani* is more of a habitual digger than previously recorded. *Oryzorictes* has a narrow and elongate scapula (Figure 4.4), significantly more so than in any of the other study taxa (Table 4.4; $P < 0.05$), and similar in form to subterranean talpids.

The swimmers *Limnogale* and *Potamogale* do not have similar scapular morphology (Figure 4.4), yet they both have long and narrow scapulae relative to the other tenrecoids (except *Oryzorictes*), which suggests considerable retraction-based loading during aquatic propulsion. Although the supraspinous fossa is well-developed in leaping marsupials (Argot, 2001), the supraspinous fossa in the elephant shrew *Petrodromus* is not remarkable (Figure 4.4). Rather, its infraspinous fossa is deep and expanded at the caudal vertebral border, highlighting the importance of the M. teres major in powerful forelimb retraction. Unlike *Hemicentetes* and *Tenrec, Echinops* and *Setifer* have a relatively flat (vs. angled) axillary border and steeply rising cranial border, resulting in an enlarged, broad supraspinous fossa (Figure 4.4). The supraspinous fossa is large and cranially expanded in arboreal scandentians, primates, and xenarthrans (Roberts and Davidson, 1975; Gebo and Sargis, 1994; Monteiro and Abe, 1999; Sargis, 2002), related to an enlarged attachment area for the M. supraspinatus and its function as a scapular suspensor and forelimb protractor (Taylor, 1974; Taylor, 1978; Roberts and Davidson, 1975; Argot, 2001; Vasquez-Molinero et al., 2001).

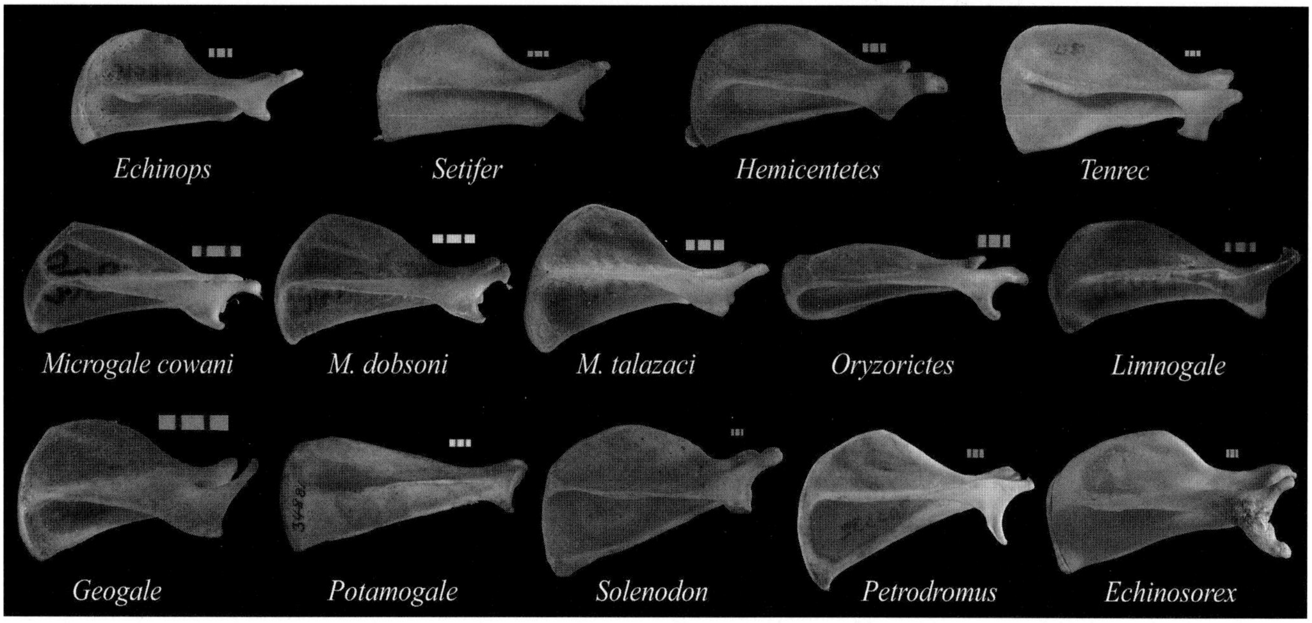

FIGURE 4.4. Lateral view of right scapulae scaled to length in tenrecines (top), oryzorictines (middle), two other tenrecoids (bottom left), and three outgroup taxa (bottom right). Note differences in the relative length and width of the whole scapula, angle of the axillary border, breadth of the vertebral border, depth of the supraspinous and infraspinous fossae, and shape of the acromion and metacromion processes. Subdivisions on scale are 1.0 mm.

Within the Tenrecinae, the acromion process reaches ventrally well beyond the glenoid fossa in *Echinops* and *Hemicentetes*, whereas that in *Tenrec* and *Setifer* is less ventrally extended (Figures 4.4, 4.5). Though there is considerable intraspecific variation for the Acromion Process Length Index (APLI), the trend of a longer acromion in a climber and a digger is consistent with data from rodents and marsupials (Stein, 2000; Argot, 2001). *Oryzorictes* has a significantly longer acromion process than *Setifer, Tenrec*, and *Microgale* (Figures 4.4, 4.5, Table 4.4; $P < 0.05$). *Limnogale* is unlike *Potamogale*, which lacks an acromion process almost altogether (Figures 4.4, 4.5; of the three available *Potamogale* scapulae, two had broken scapular spines, so $n = 1$ for this variable). The acromion process is the site of origin for M. deltoideus pars acromialis (Neveu and Gasc, 2002), which acts as a protractor and lateral rotator of the humerus. Although some forelimb diggers have large acromion processes, they are reduced in some, e.g., talpids and erinaceids (Reed, 1951; Verma, 1963). Length of the acromion in a digger might therefore be indicative of whether the animal is generating force from the shoulder musculature, as in *Oryzorictes* and xenarthrans (Smith and Savage, 1956), or more from the forearm, as in *Talpa* and erinaceids.

The metacromion is an attachment site for Mm. trapezius, atlantoscapularis, omotransversarius anterior, and for the extension of the deltoideus pars acromialis (Campbell, 1939; Neveu and Gasc, 2002), muscles involved with scapular stabilization and humeral rotation. A conspicuous arc of the metacromion is only present in the two most extreme locomotor specialists, *Oryzorictes* and *Petrodromus* (Figure 4.4), suggesting that the metacromion, when present, is a good indicator of heavy loading at the shoulder. *Geogale* has an unusual metacromion process, which does not project anteriorly in a characteristic "c" shape, but extends back (posteriorly along the long axis of the scapula) and forms a wide, thin sheet of bone confluent with the scapular spine (Figure 4.4). It is unclear which of the attached muscles is most influencing this form, but given the lack of rotational arm movement in *Geogale* (JAS pers. obs.), it is most likely a reflection of a strongly anchored scapula.

Although there are apparent differences in the shape of the glenoid fossa between taxa, the Glenoid Fossa Shape Index (GFSI) may not be a reliable variable because it is difficult to discern the limits of humeral head rotation against the fossa (see Taylor, 1974). Nonetheless, the glenoid fossa of tenrecoids appears to be generally dorsoventrally (anteroposteriorly if facing ventrally) narrow with some variation between taxa. Within Tenrecinae, the shape of the glenoid fossa in *Hemicentetes* is distinctive in its high, narrow, almost rectangular shape (Figure 4.6), which is consistent with glenoid fossa shape in other mammalian diggers (Reed, 1951; Stein, 2000). *Limnogale* has a significantly narrower glenoid fossa than in any other tenrecoid (Figure 4.6, Table 4.4; $P < 0.05$), yet the functional interpretation of this trait is unclear. Though it seems as if a narrow glenoid fossa would restrict motion to a single plane, the highly restricted shoulder joints of *Potamogale* and *Petrodromus* (based on their humeral morphology) have rounded glenoid fossae (Figure 4.6).

TABLE 4.4. Index summary statistics*

Taxon		SSI	APLI	GFSI	HSI	HRLI	HHSI	MEWI	USI	OPLI	RSI
Echinops telfairi	Mean	0.47	1.17	1.45	10.46	1.12	1.02	1.07	0.079	0.17	0.10
	SD	0.05	0.04	0.16	0.95	0.12	0.06	0.16	0.008	0.02	0.01
	n	13	10	13	13	12	13	12	11	11	12
Setifer setosus	Mean	0.50	1.07	1.48	11.01	1.09	1.10	0.95	0.083	0.16	0.11
	SD	0.04	0.28	0.11	0.82	0.05	0.08	0.12	0.006	0.01	0.01
	n	19	17	18	19	15	19	19	17	17	16
Hemicentetes semispinosus	Mean	0.44	1.20	1.60	**7.66**	1.11	**1.24**	**1.38**	**0.094**	**0.22**	**0.13**
	SD	0.02	0.02	0.15	0.45	0.11	0.10	0.21	0.009	0.01	0.01
	n	16	12	15	18	14	17	18	16	16	15
Tenrec ecaudatus	Mean	0.47	1.09	1.54	10.52	1.19	1.10	1.17	**0.091**	**0.23**	**0.13**
	SD	0.03	0.01	0.18	0.85	0.11	0.09	0.12	0.007	0.01	0.01
	n	11	10	11	12	8	12	12	13	14	9
Microgale cowani	Mean	0.41	1.15	1.52	13.06	0.97	1.05	1.14	0.069	0.16	0.08
	SD	0.04	0.02	0.13	0.78	0.08	0.10	0.14	0.005	0.01	0.02
	n	20	18	19	22	14	22	22	12	13	14
Microgale dobsoni	Mean	0.53	1.14	1.57	13.04	0.97	1.10	0.92	0.063	0.12	0.08
	SD	0.29	0.02	0.11	3.89	0.11	0.05	0.10	0.007	0.01	0.01
	n	21	21	18	21	10	20	21	10	10	10
Microgale talazaci	Mean	0.51	1.14	1.58	14.33	0.93	1.09	1.01	0.065	0.12	0.08
	SD	0.05	0.02	0.10	0.78	0.02	0.07	0.11	0.003	0.01	0.004
	n	13	10	11	12	8	12	12	8	8	8
Oryzorictes sp.	Mean	**0.31**	**1.26**	1.59	**7.83**	1.10	**1.40**	**1.49**	**0.094**	**0.27**	**0.15**
	SD	0.03	0.03	0.17	0.49	0.12	0.13	0.18	0.006	0.03	0.01
	n	35	31	30	34	7	34	34	10	10	7
Limnogale mergulus	Mean	0.36	1.21	**1.88**	12.34	0.94	1.08	1.03	0.077	0.17	0.10
	SD	0.02	0.08	0.33	0.58		0.11	0.05			
	n	5	3	5	5	1	5	5	1	1	1
Geogale aurita	Mean				13.59	1.09	1.00	0.83	0.069	0.13	0.09
	SD				1.88		0.08	0.08			
	n				4	1	4	4	1	1	1
Potamogale velox	Mean				14.75	**1.32**	0.96	**0.58**	**0.093**	0.19	**0.13**
	SD				1.36	0.05	0.03	0.17	0.006	0.01	0.01
	n				3	3	3	2	3	3	3
Solenodon paradoxus	Mean				**9.33**	1.09	**1.15**	**1.26**	**0.099**	0.19	**0.14**
	SD				0.95	0.03	0.08	0.10	0.008	0.02	0.01
	n				10	7	10	10	10	10	7
Petrodromus tetradactylus	Mean				13.74	**0.72**	0.89	**0.56**	**0.044**	0.11	0.05
	SD				0.82		0.04	0.004	0.002	0.01	0.003
	n				2	1	2	2	3	3	3
Echinosorex gymnurus					11.49	1.29	0.93	0.75	0.072	0.19	0.10
	n				1	1	1	1	1	1	1

* See Table 4.3 for index descriptions; values in bold are discussed in the text

4.3.2 Humerus

Overall shape of the humerus in terms of its length relative to width does not appear to be reliably correlated with positional behavior, except for the consistent finding of a relatively short, wide humerus correlated with digging (Smith and Savage, 1956; Yalden, 1966; Casinos et al., 1993; Hildebrand, 1995; Grand and Barboza, 2001; Luo and Wible, 2005). This is also the case with the taxa studied here; the humeri of *Hemicentetes, Oryzorictes*, and *Solenodon* are significantly wider at midshaft than those of the other study taxa (Figure 4.7, Table 4.4; $P < 0.05$). With the exception of *Oryzorictes*, the oryzorictines have longer, thinner humeri than the tenrecines (Figure 4.7). Despite other traits that correlate with digging in the *M. cowani* postcranium, its humeral shape as defined by the Humeral Shape Index (HSI) is within the range of the other *Microgale* species (Table 4.4).

There are no significant differences among tenrecines in humerus length relative to the radius (HRLI, or brachial index); all have a humerus that is slightly longer than the radius, although *Tenrec* has a slightly higher value than the others (Table 4.4). In oryzorictines, the humerus tends to be shorter than the radius, except in the digging *Oryzorictes*, in which the humerus is just longer than the radius, as in tenrecines (Table 4.4). The swimmer *Potamogale* and the saltatory *Petrodromus* represent two ends of a spectrum; *Potamogale* has an extremely long humerus relative to the radius, whereas *Petrodromus* has a low brachial index (Table 4.4). Lengthening of the distal limb elements has been well-correlated with the mechanics of higher-speed locomotion, whereas shortened distal limbs and short limbs in general are correlated with more powerful forelimb (and hind limb) thrust. *Petrodromus* most likely concentrates muscle mass at the proximal end

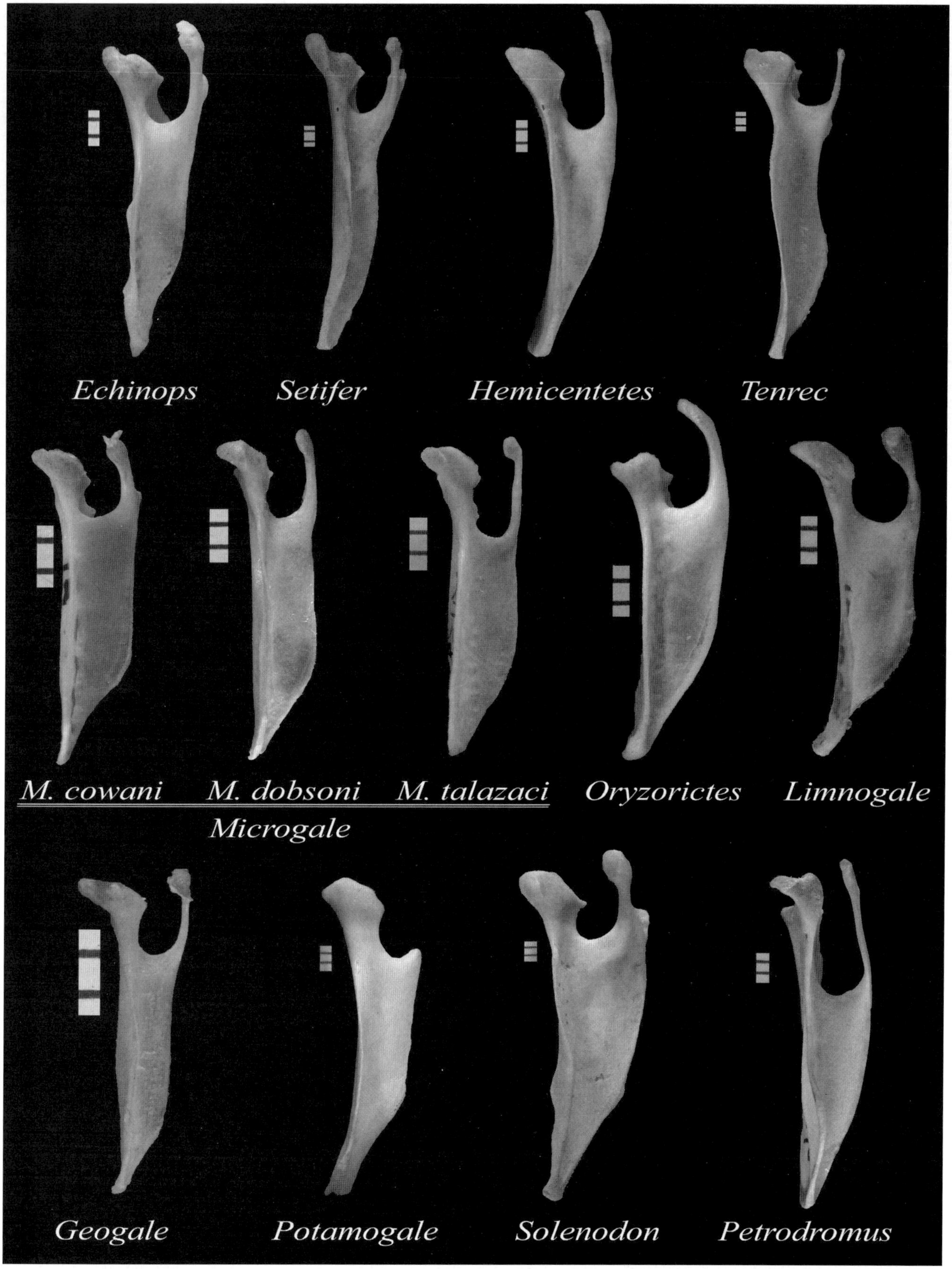

FIGURE 4.5. Dorsal view of right scapulae scaled to length in tenrecines (top), oryzorictines (middle), two other tenrecoids (bottom left), and two outgroup taxa (bottom right). Note length of the acromion process, which is longer in the diggers. Subdivisions on scale are 1.0 mm.

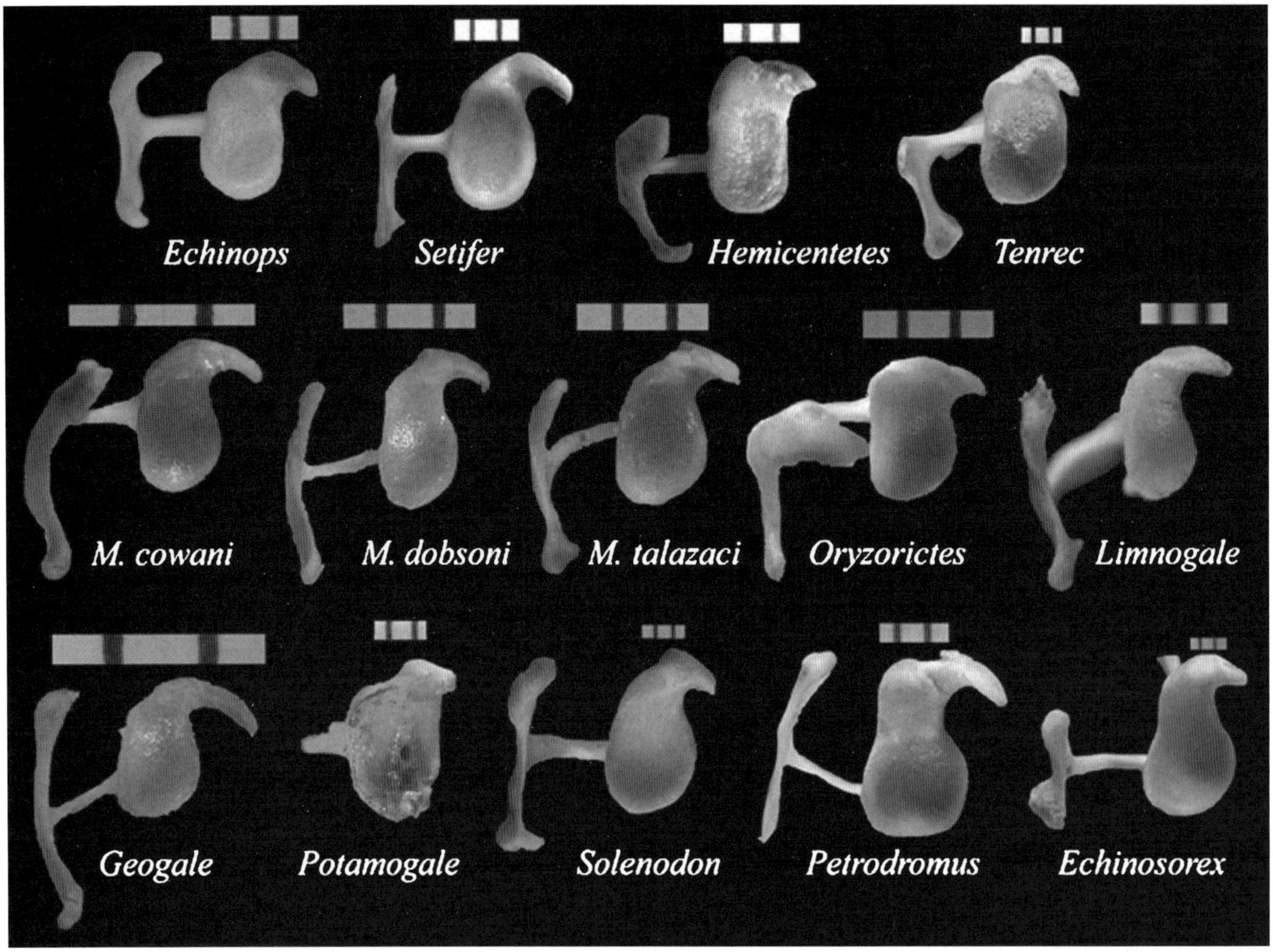

FIGURE 4.6. Articular surface of right scapulae scaled to height in tenrecines (top), oryzorictines (middle), two other tenrecoids (bottom left), and three outgroup taxa (bottom right). Note the shape of the glenoid fossa, ranging from rectangular in *Hemicentetes* to spherical in *Petrodromus*. Subdivisions on scale are 1.0 mm.

of the limb, as in other mammalian cursors, and effectively lengthens its stride with a long distal limb and long tendinous insertions (Hildebrand, 1995). Smith and Savage (1956) noted similarities in scapular form between aquatic mammals and fossorial mammals. The extreme shortening of the distal limb in *Potamogale* suggests that it uses its arms for some aquatic paddling, which, in terms of movement and direction of reactive force, is similar to digging in *Oryzorictes* (though differences in humeral shape reflect the lighter resistance of water vs. soil, and considerably less powerful elbow flexion/extension in *Potamogale*).

Humeral head shape (HHSI) varies with locomotor behavior in the Tenrecinae. The digging *Hemicentetes* has a significantly (anteroposteriorly) longer head than the other tenrecines (Table 4.4; $P < 0.05$), whereas the climber *Echinops* has a more rounded humeral head (Figure 4.8). This is consistent with data from arboreal primates and several small digging mammals, and reflects multiaxial rotational movement in the climbers and more restricted shoulder motion in the diggers (Reed, 1951; Stein, 2000; Argot, 2001). A comparison across all taxa demonstrates that the diggers *Hemicentetes, Oryzorictes*, and *Solenodon* share an elliptical articular surface of the humeral head vs. a more rounded head in the others (Figure 4.8, Table 4.4), and *Oryzorictes*, like *Hemicentetes*, has a significantly higher HHSI than the other tenrecoids (Table 4.4; $P < 0.05$).

A well-formed bicipital groove is likely correlated with the size of the tendon of the M. biceps brachii that passes through it, and may be indicative of powerful flexion associated with climbing (Taylor, 1974; Argot, 2001) or digging (Campbell, 1939; Reed, 1951). There is tremendous intraspecific variation in the formation of the bicipital groove. In several *Hemicentetes* and *Oryzorictes* specimens, the groove is completely closed to form a bicipital tunnel (Figure 4.9), characteristic of talpids (Barnosky, 1982), but this is not the norm for either tenrecoid genus. The presence of a well-formed bicipital groove or tunnel may be indicative of digging, yet the absence of this trait is not clear in terms of positional behavior.

Tenrec has a larger greater tuberosity (in terms of mediolateral width and anteroposterior length) than the other tenrecines (Figure 4.8; Salton, 2005), which might indicate

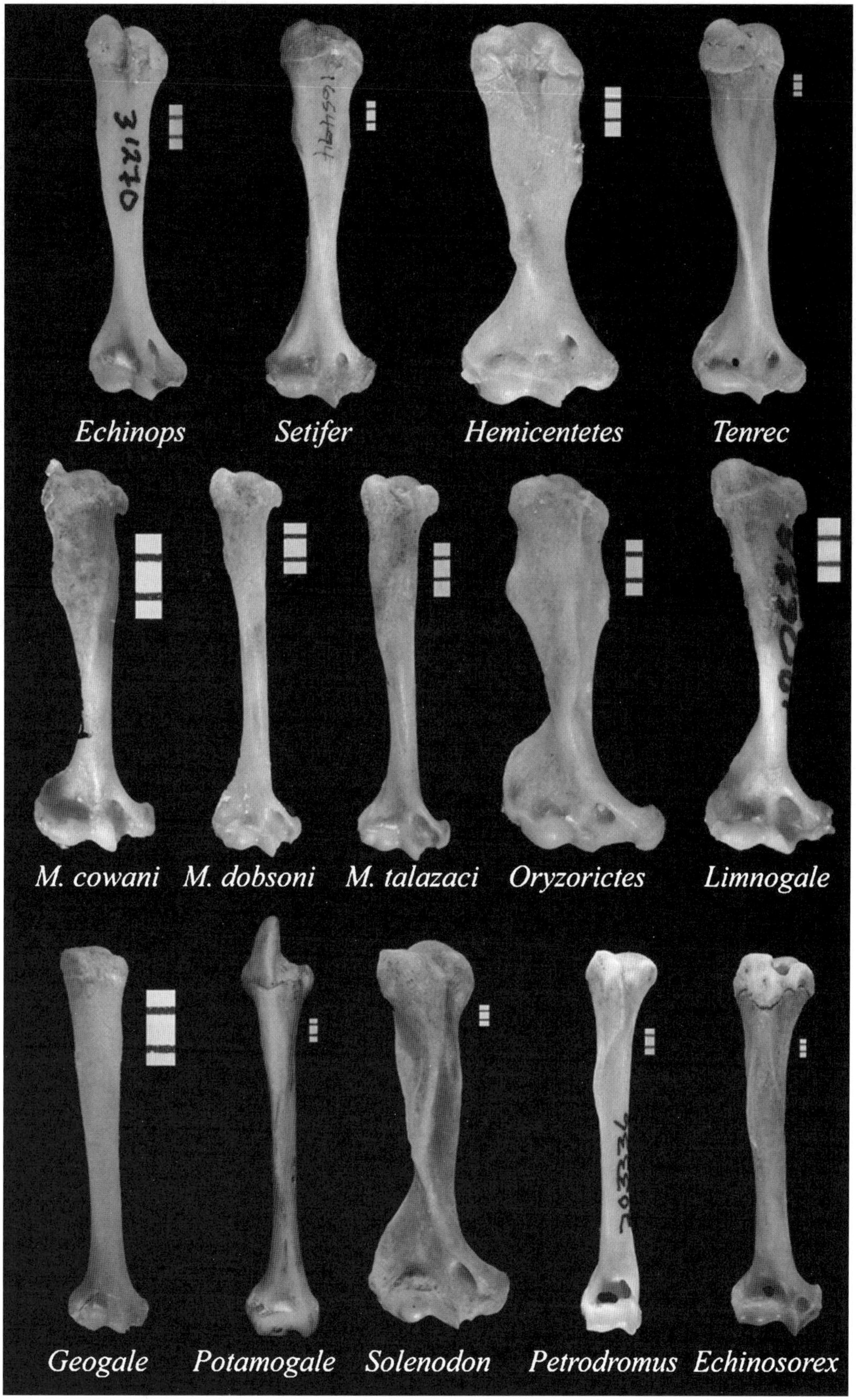

FIGURE 4.7. Anterior view of right humeri scaled to height in tenrecines (top), oryzorictines (middle), two other tenrecoids (bottom left), and three outgroup taxa (bottom right). Note differences in relative midshaft width, greater and lesser tuberosity height, distal humerus width, epicondyle widths, trochlea and capitulum shape, presence/absence of entepicondylar foramen and coronoid fossa, and deltoid tuberosity/deltopectoral crest shape. Subdivisions on scale are 1.0 mm.

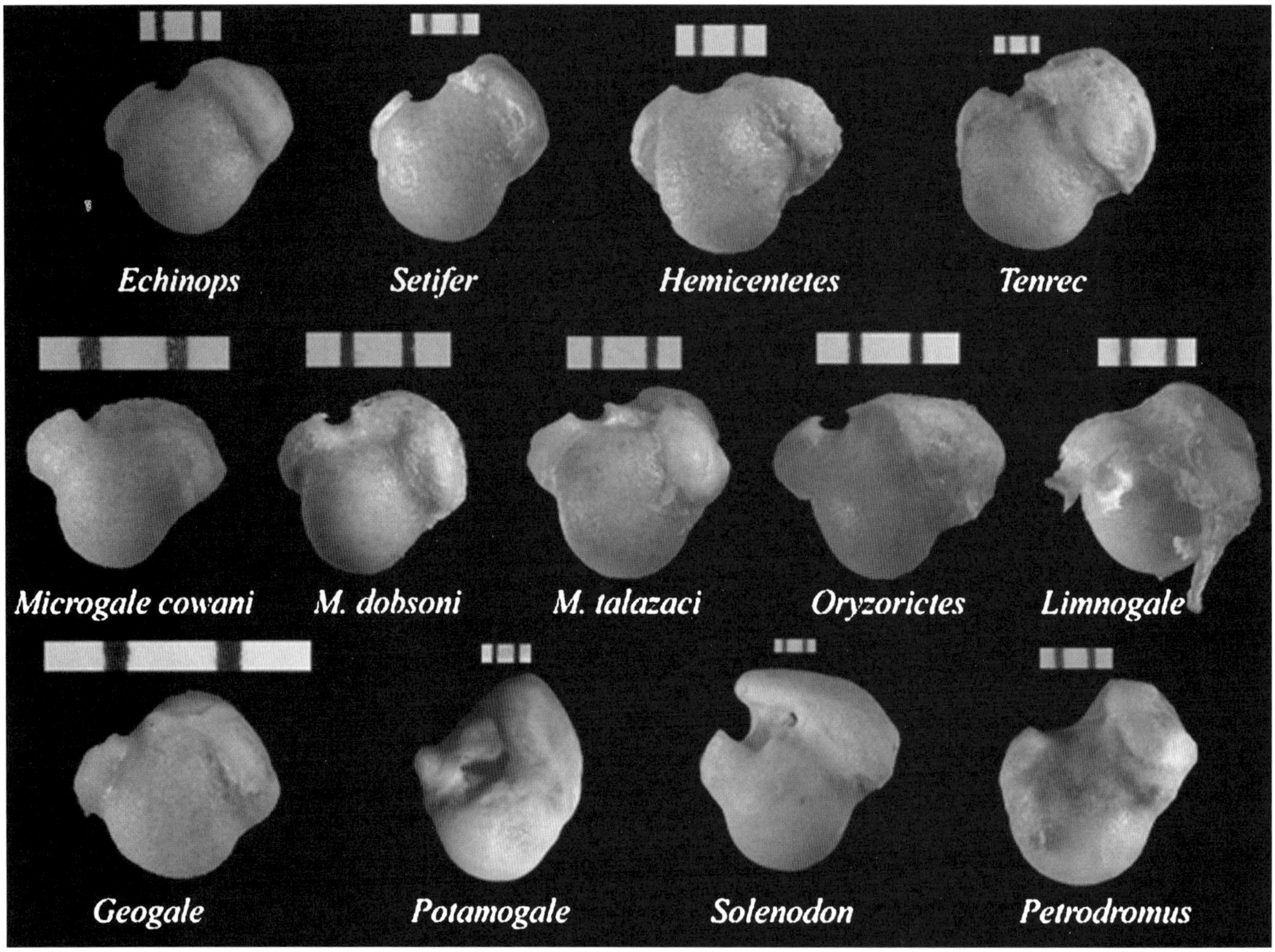

FIGURE 4.8. Proximal articular surfaces of the right humeri of tenrecines (top), oryzorictines (middle), two other tenrecoids (bottom left), and two outgroup taxa (bottom right). Note differences in greater and lesser tuberosity size and presence/absence of a bicipital groove. Subdivisions on scale are 1.0 mm.

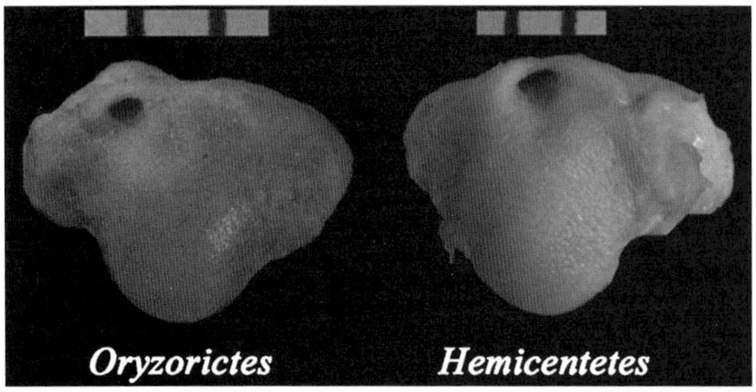

FIGURE 4.9. Proximal articular surfaces of right humeri in two digging tenrecoids, illustrating the formation of a complete bicipital tunnel. Most specimens from each of these two genera have a bicipital groove; only a few have a completely formed tunnel. Subdivisions on scale are 1.0 mm.

a more restricted shoulder joint with powerful parasagittal forelimb movement. *Hemicentetes* does not have a more pronounced greater tuberosity despite predictions based on other mammalian diggers, nor does *Echinops* have a smaller one relative to the others (see Table 4.1). The greater tuberosity of the humerus serves as an attachment site for M. infraspinatus, which retracts the humerus, and M. supraspinatus, which protracts the humerus. Both muscles serve to stabilize the shoulder joint, so the relative size of the greater tuberosity may correlate with restriction of shoulder mobility (Roberts and Davidson, 1975; Argot, 2001; Sargis, 2002). Among the oryzorictines, *Oryzorictes* has a larger greater tuberosity (Figure 4.8; Salton, 2005), suggesting a more powerful and restricted fore and aft stroke, and perhaps implying that *Oryzorictes* and *Hemicentetes* utilize different types of arm strokes when digging (Figure 4.8). The greater tuberosity in *Potamogale* is remarkable compared to that of the tenrecids and highly unusual for any mammal (Figures 4.7, 4.8), although it is somewhat similar to the condition found in microchiropteran bats. It extends proximally and anteriorly as a sharp process that is claw-like in shape. It reaches high beyond the proximal surface of the humeral head. Although different from bats in its shape and anterior position on the humerus, the greater tuberosity projection in *Potamogale* may serve as a protective lock against the scapula and help to prevent overextension of the forelimb during swimming.

The digger *Hemicentetes* has a broad lesser tuberosity relative to other tenrecines, and *Tenrec* has a smaller lesser tuberosity (Figure 4.8; Salton, 2005). The lesser tuberosity is the primary attachment site for M. subscapularis, a medial rotator and adductor of the arm (Taylor, 1978; Argot, 2001). One might therefore expect a larger lesser tuberosity in diggers (Rose and Emry, 1983; Stein, 2000), and perhaps a larger tuberosity (relative to the greater tuberosity) in climbers (Argot, 2001; Sargis, 2002). The small lesser tuberosity in *Tenrec* is consistent with its large greater tuberosity, suggesting a forelimb with very limited rotational mobility. There is a strong negative correlation between the size of the greater tuberosity and lesser tuberosity in tenrecines, as well as the other study taxa (−0.87 for tenrecines, −0.70 for all study taxa; Salton, 2005). Like *Tenrec, Potamogale* has a diminutive lesser tuberosity (Figure 4.8), which, when coupled with the large greater tuberosity, suggests an armstroke that is limited to one major directional plane (as in forward and backward paddling). The other semiaquatic taxon, *Limnogale*, has a relatively large lesser tuberosity (Figure 4.7), as in *Oryzorictes*, suggesting more rotational arm movement during swimming. This is consistent with tarsal and hind limb morphology, which suggests more varied limb movements in *Limnogale* compared to more restricted and powerful swimming strokes in *Potamogale* (Salton and Szalay, 2004; Salton, 2005).

Interpretation of the deltopectoral region of the humerus is difficult, due to the interplay between the attachment of the Mm. deltoideus and pectoralis musculature, which results in their varying functions as lateral rotators and abductors (deltoids) and adductors and retractors (pectorals). A large and/or expanded deltopectoral crest is found in arboreal marsupials (Argot, 2001), arboreal primates (Gebo and Sargis, 1994), and fossorial rodents (Rose and Emry, 1983; Stein, 2000). In tenrecines, there is little development of the deltopectoral crest or deltoid tubercle. There is a small deltoid tubercle on the proximal quarter of most *Echinops* and *Setifer* specimens, a moderate crest in *Hemicentetes* and *Tenrec*, and a moderate tubercle at the distal third of the *Hemicentetes* humerus (Figure 4.7). Among the oryzorictines, *Microgale cowani* and *Oryzorictes* have noticeable anterior pectoral crests and lateral deltoid tubercles, whereas *M. dobsoni, M. talazaci*, and *Limnogale* do not (Figure 4.7). The similarity between *M. cowani* and *Oryzorictes* is another indication (in addition to a lengthened scapula and other postcranial traits, see below) that *M. cowani* utilizes digging behavior more than the other two *Microgale* species.

Most of the variation in the width of the distal humerus is accounted for by the medial and lateral epicondyles. These structures serve as areas of origin for the wrist and digital flexors (medially) and extensors (laterally). They are therefore reliable indicators of flexion and extension of the hand, and are particularly well-developed in a taxonomic range of climbers and diggers (e.g., Rose and Emry, 1983; Biknevicius, 1993; Stein, 2000; Argot, 2001; Grand and Barboza, 2001; Sargis, 2002). Overall width of the distal humerus is particularly great in the diggers *Hemicentetes, Oryzorictes*, and *Solenodon* (Figure 4.7). *Potamogale* and *Petrodromus* have the narrowest distal humeri, reflecting less powerful wrist and digital flexion/extension.

There are significant differences among the tenrecines in medial epicondyle width (MEWI; Figures 4.7, 4.10, Table 4.4; $P < 0.05$), indicating varying development of the wrist and digital flexors. *Hemicentetes* has the widest medial epicondyle, which is consistent with data from other mammalian diggers (Biknevicius, 1993; Stein, 2000; Grand and Barboza, 2001), and the relatively wide medial epicondyle of *Tenrec* suggests that it utilizes some manual scratch digging that is not reflected at the shoulder joint. Of the oryzorictines, *Oryzorictes* has a significantly wider medial epicondyle than the others (Figures 4.7, 4.10, Table 4.4; $P < 0.05$), and *Microgale cowani* has a wider medial epicondyle than the other *Microgale* species (Figures 4.7, 4.10, Table 4.4). The medial epicondyle of *Solenodon* is wide, reflecting its digging behavior, whereas that of *Potamogale* and *Petrodromus* is narrow (Figures 4.7, 4.10, Table 4.4).

The entepicondylar foramen, which transmits the median nerve (Reed, 1951), is considered to be a primitive therian trait that has been lost in several mammalian taxa such as bats, catarrhine primates, and some treeshrews (e.g., Szalay and Dagosto, 1980; Ciochon, 1993; Simmons, 1994; Sargis, 2002). Interestingly, the presumably more basal of the tenrecoid taxa, *Geogale* and *Potamogale*, do not have an entepicondylar foramen, whereas there is a moderate entepicondylar foramen in all of the other tenrecoids (Figure 4.7). Sargis (2002) suggested that its absence in the tupaiid *Urogale* might be related to digging, yet the tenrecid diggers (and *Solenodon*) have large entepicondylar foramina. In the Tenrecoidea there is little intraspecific variability (i.e., a foramen is always present or absent in adults of a given species), and the entepicondylar foramen is retained in all the Malagasy tenrecoids except for *Geogale*. The loss of the entepicondylar foramen in *Potamogale* might have functional significance given the narrowing and specialization of its distal humerus, but this does not apply to *Geogale*, which has few specializations of the forelimb. Additionally, macroscelidids retain an entepicondylar foramen, despite the narrowing and specialization of their distal humeri. This is likely a trait that is easily lost in any particular taxon, and was perhaps lost relatively late in both the *Potamogale* and *Geogale* lineages.

Another highly variable, simple feature of the mammalian humerus is the perforation of the coronoid fossa through to the olecranon fossa. In tenrecoids, occasional perforation of the fossa occurs in the more terrestrial taxa, but, when present, this is an intraspecifically variable characteristic. None of the *Echinops* or *Hemicentetes* specimens had a perforated coronoid fossa, whereas 11% of *Setifer* specimens and 61% of *Tenrec* specimens had a complete perforation. Perforations were present in 10% of *Microgale cowani* and *M. dobsoni* humeri and 25% of *M. talazaci* specimens, but none were present in the humeri of *Oryzorictes, Geogale, Limnogale*, or *Potamogale*.

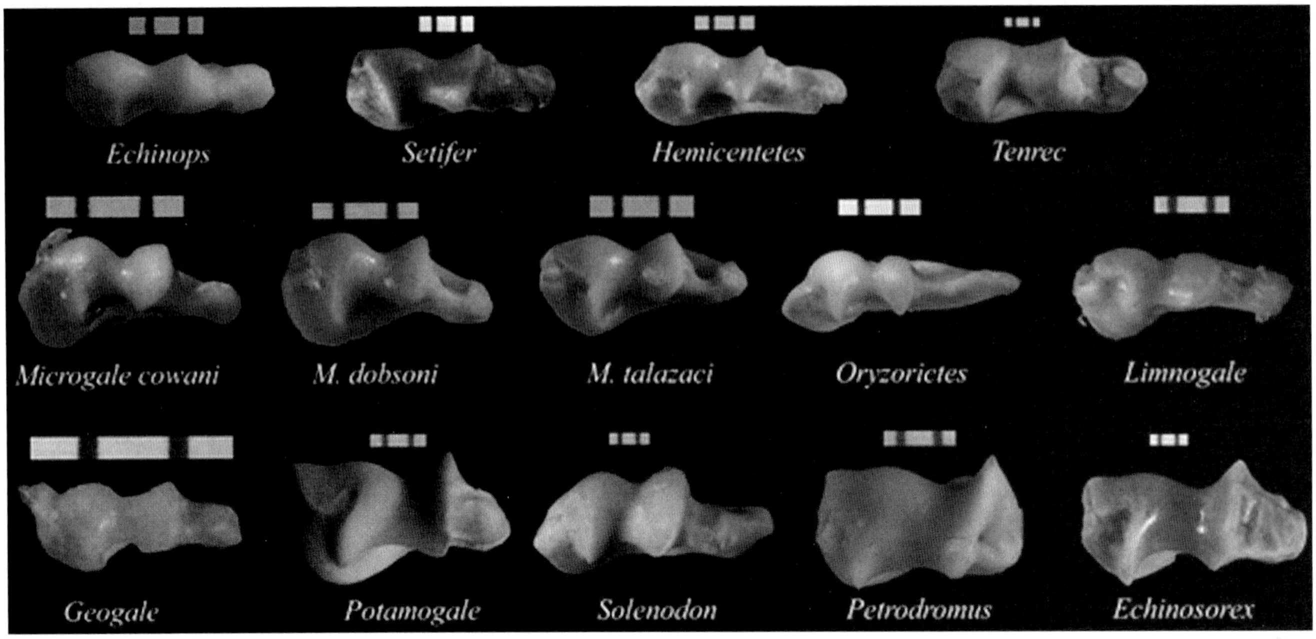

FIGURE 4.10. Distal surfaces of the right humeri in tenrecines (top), oryzorictines (middle), two other tenrecoids (bottom left), and three outgroup taxa (bottom right). Note differences in medial epicondyle mediolateral length and sharpness of trochlear and capitular edges. Subdivisions on scale are 1.0 mm.

The complete absence of this trait in a taxon is correlated with generally shallow coronoid and olecranon fossae, and its presence indicates deeper fossae. Similar to the bicipital groove, the absence of a perforated coronoid fossa is not particularly meaningful functionally, as all taxa include specimens with a non-perforated fossa. Yet the presence of this trait is likely correlated with great extension of the forearm during terrestrial locomotion (see Szalay and Sargis, 2001).

All Malagasy tenrecoids have a well-defined, slightly rounded capitulum, whereas that of *Potamogale* is mediolaterally lengthened and rectangular, with sharply defined medial and lateral borders (Figures 4.7, 4.11). A distally flattened (as opposed to rounded) capitulum with well-defined borders is characteristic of more terrestrial vs. arboreal carnivorans, primates, and scandentians (Szalay and Dagosto, 1980; Harrison, 1989; Rose, 1989; Gebo and Sargis, 1994; Sargis, 2002), and represents a restriction of radial rotation against the humerus. Aside from the flattened capitulum in *Potamogale*, the other tenrecoids have a rather uniformly rounded capitulum that does not seem to vary with locomotor behavior. The capitulum in all tenrecoids remains relatively large and plays a significant role in load-bearing at the elbow, as opposed to a more derived mammalian condition where the trochlea takes over more direct loads at the elbow, and the capitulum is reduced, playing a more important role in movement associated with radial rotation (Szalay and Dagosto, 1980). *Potamogale's* distal humerus suggests a highly stabilized forearm that does not allow for mediolateral excursion at the elbow. Its trochlea is mediolaterally narrow, medially bound by a steep incline, and laterally bound by the sharp rectangular edge of the capitulum (Figure 4.10). *Potamogale* and the elephant shrew *Petrodromus* share similar capitulum/trochlea articular form, yet other aspects of their distal humeri are distinct: *Petrodromus* has a wide entepicondylar foramen, complete perforation of the coronoid fossa, and a less extended medial epicondyle (Figures 4.7, 4.11).

Setifer has a deeper (proximodistally) trochlea than the other tenrecines (Figure 4.11; Salton, 2005), which is typical of a more terrestrial mammal (Szalay and Dagosto, 1980; Gebo and Sargis, 1994; Szalay and Sargis, 2001; Sargis, 2002) and represents extended surface area for ulnar articulation and medial restriction of that articulation. *Microgale talazaci* has a deeper trochlea than the other oryzorictines (Figure 4.11; Salton, 2005), yet other postcranial traits do not suggest that this species is more or less terrestrial than the others.

4.3.3 Ulna

The Ulna Shape Index (USI) is extremely variable among tenrecoid taxa and highly correlated with locomotor behavior (Figures 4.12, 4.13, Table 4.4). Other mammalian diggers exhibit relatively short, curved, and deep ulnae, whereas those of climbers tend to be relatively long and shallow (Verma, 1963; Taylor, 1974; Casinos et al., 1993; Hildebrand, 1995; Stein, 2000; Grand and Barboza, 2001). Of the tenrecines, *Hemicentetes* has the highest USI, and *Echinops* has the lowest (Figure 4.13, Table 4.4). The USI in *Setifer* is not significantly different from its sister taxon *Echinops*, and that of *Tenrec* is not significantly different from *Hemicentetes* (Table 4.4; $P < 0.05$), perhaps reflecting some climbing and digging, respectively, in these taxa. All

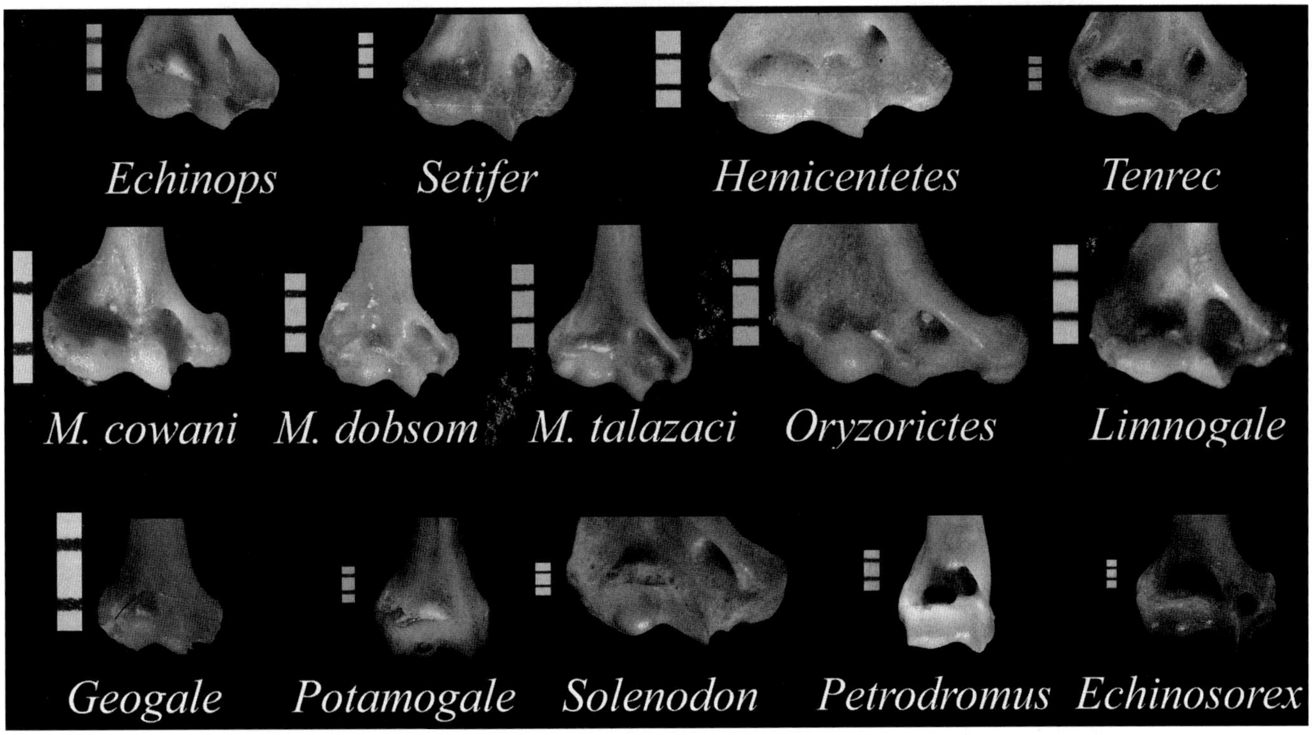

FIGURE 4.11. Anterior view of right distal humerus in tenrecines (top), oryzorictines (middle), two other tenrecoids (bottom left), and three outgroup taxa (bottom right). Note differences in the shape of the trochlea and capitulum, especially the convergence in form between *Potamogale* and *Petrodromus*. Subdivisions on scale are 1.0 mm.

three *Microgale* species have relatively long ulnae, though *M. cowani* has a slightly deeper ulna than the other two (Figure 4.13, Table 4.4). *Oryzorictes, Potamogale*, and *Solenodon* have the shortest, deepest ulnae (Figure 4.13), with similar USI values as *Hemicentetes* (Table 4.4), reflecting a decreased out-lever of the forearm for increased out-force against a resistant substrate (i.e., soil/water). Despite some other shared traits with *Potamogale* that are related to stabilizing articulations at potentially vulnerable joints, *Petrodromus* has an extremely long and shallow ulna (Figure 4.13, Table 4.4), which denotes its high-speed terrestrial mode of locomotion.

The olecranon process, the attachment site for the M. triceps brachii, has been well-correlated with locomotor behavior in arboreal, terrestrial, and fossorial mammals (Verma, 1963; Rose and Emry, 1983; Van Valkenburgh, 1987; Biknevicius, 1993; Ciochon, 1993; Stein, 2000; Argot, 2001; Grand and Barboza, 2001; Sargis, 2002). As the olecranon process length increases, triceps gains leverage for powerful ulnar extension against the humeral trochlea. Fossorial mammals have a particularly elongated olecranon process for digging, whereas that of climbers is less elongated, which allows for maximal elbow extension (Hildebrand, 1995).

As with ulnar shape, there are significant function-based differences among tenrecoids in the length of the olecranon process. Within Tenrecinae, both *Hemicentetes* and *Tenrec* have high Olecranon Process Length Index (OPLI) values, whereas *Echinops* and *Setifer* have low values, indicating a shorter process (Figure 4.13, Table 4.4). All three *Microgale* species have relatively short olecranon processes, yet that of *M. cowani* is significantly longer than the others (Figure 4.13, Table 4.4; $P < 0.05$), suggestive of some digging. The fossorial *Oryzorictes* has a significantly longer olecranon process than any of the other study taxa (Figures 4.12, 4.13, Table 4.4; $P < 0.05$). *Limnogale* and *Potamogale* both have long processes, with similar OPLI values as *M. cowani*. Surprisingly, *Solenodon* has a shorter olecranon process than the other diggers and *Tenrec*, although it is still of moderate size, in the range of *Potamogale* and *Echinosorex*. However, *Solenodon* is similar to *Oryzorictes* in the medial curvature of its olecranon process (Figure 4.12), which, like the wide medial epicondyle, is related to the origin of powerful wrist and digital flexors necessary for scratch digging (Hildebrand, 1985).

4.3.4 Radius

Results from the Radial Shape Index (RSI) are almost identical to those from the Ulna Shape Index: digging tenrecoids, *Potamogale*, and *Solenodon* all have relatively deep radii, whereas the climber and more terrestrial genera have longer, shallower radii (Figure 4.13, Table 4.4). This makes sense from the same general function-based perspective for the ulna; the deep forearm bones are related to powerful displacement of dirt and water during digging and swimming, respectively (see above). Radial form in the diggers *Hemicentetes, Oryzorictes*, and *Solenodon* is also distinct in its transition

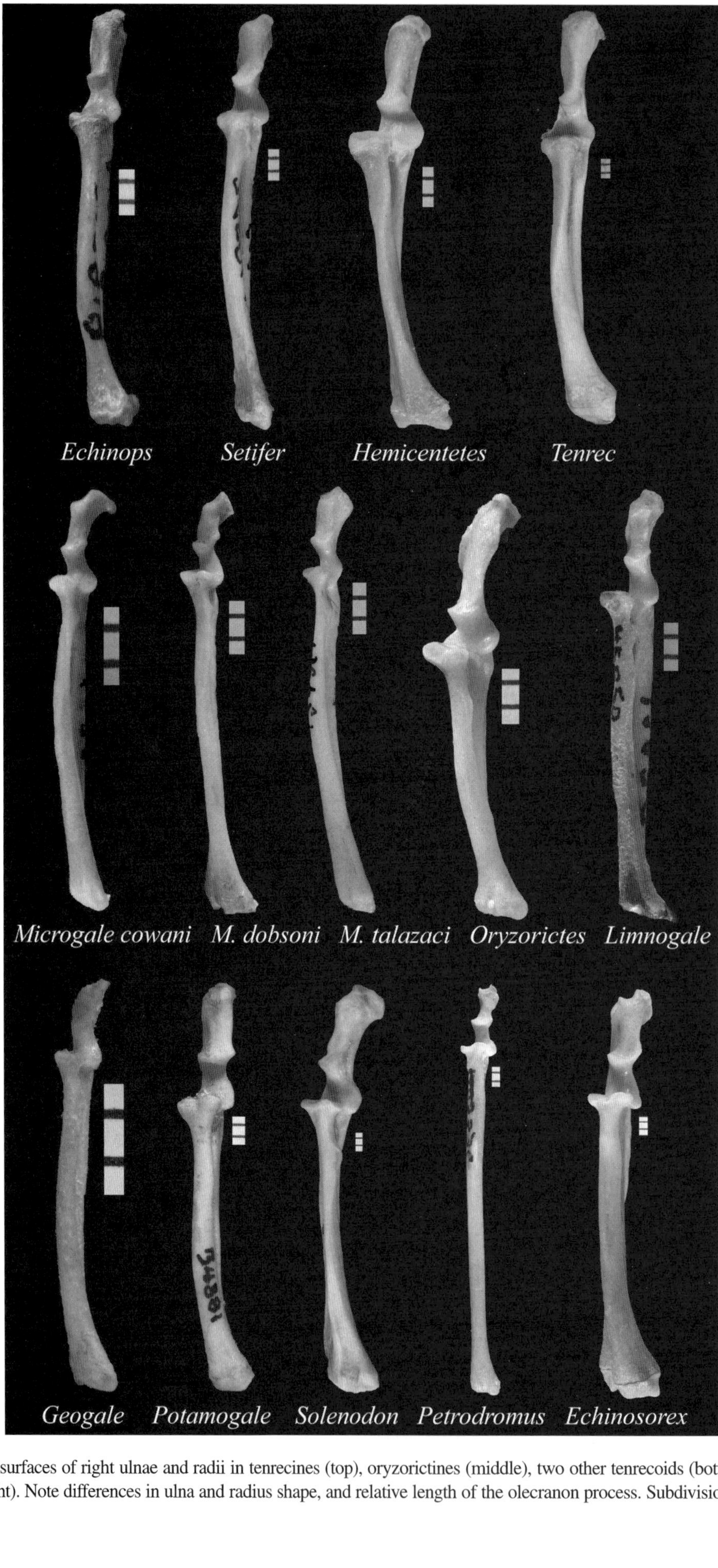

FIGURE 4.12. Anterior surfaces of right ulnae and radii in tenrecines (top), oryzorictines (middle), two other tenrecoids (bottom left), and three outgroup taxa (bottom right). Note differences in ulna and radius shape, and relative length of the olecranon process. Subdivisions on scale are 1.0 mm.

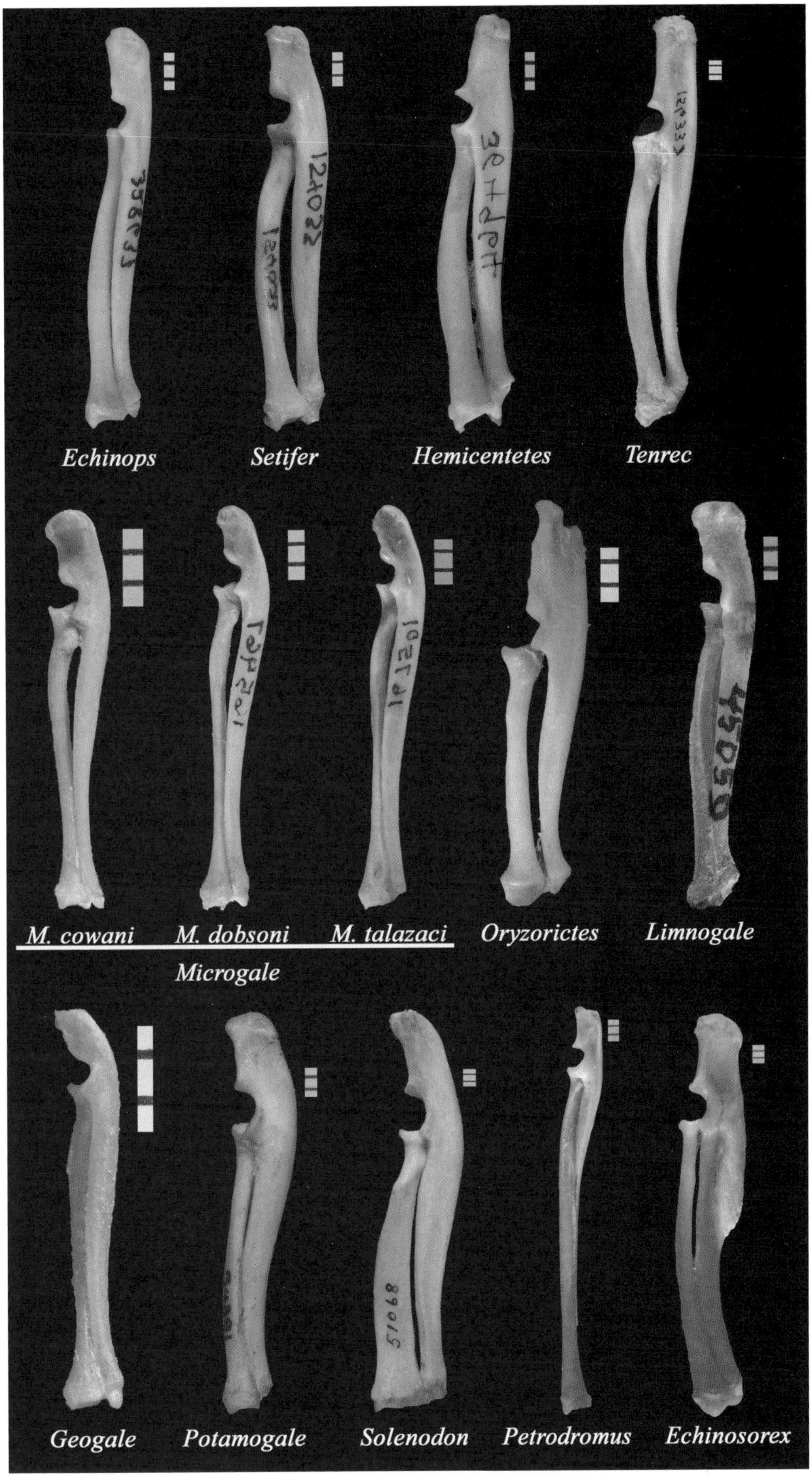

FIGURE 4.13. Medial view of right ulnae and radii in tenrecines (top), oryzorictines (middle), two other tenrecoids (bottom left), and three outgroup taxa (bottom right). Note differences in olecranon process length and shape, trochlear notch shape, and widening of the radius along the shaft. Subdivisions on scale are 1.0 mm.

from being relatively shallow at the proximal shaft to up to twice its proximal depth at the distal shaft. In contrast, the ulna, with its long and thick olecranon process, narrows at its distal end to the same degree as in the other taxa (Figure 4.13). The radius plays an important load-bearing role at the proximal wrist joint and limits rotational movement of the carpus, whereas the proximal ulna plays more of a load-bearing role at the elbow joint and the relatively small radial head does little to facilitate rotation. Although other studies have found a more elliptical radial head in diggers (Reed, 1951; Stein, 2000), the shape of the radial head is not more elliptical in *Oryzorictes* than in the other taxa, yet it is distinct in form (Figure 4.14). Two processes of the anterior radial head surface (which are present but small in some other taxa) serve to fold over and cup the capitulum, stabilizing the elbow joint along its anteroposterior axis (Figure 4.14). *Potamogale* and *Solenodon* have similar outgrowths of the proximal radial head, and *Potamogale* has an additional posterior notch capitulum (see *Potamogale* distal humerus, Figure 4.10). The radial head in the arboreal *Echinops* is more rounded than in the other three tenrecines, but not more so than in *Limnogale* or *Geogale*, which also have rounded radial head surfaces. *Petrodromus* represents the extreme in having an enormously mediolaterally widened radial head, which offers a large surface area for humeral articulation and restricts rotation.

4.4 Summary and Conclusions

4.4.1 *Echinops* and *Setifer*: Arboreal vs. Terrestrial Tenrecines

The tenrecines *Echinops* and *Setifer* offer a good model for investigating skeletal differences that have been strongly influenced by an arboreal habitat. These sister taxa are extremely difficult to distinguish with superficial characteristics, though *Setifer* has an additional molar in its dental formula and tends to have a greater average body mass. Postcranial regions other than the forelimb show several similarities between the two taxa that seem to be related to climbing behavior (see Salton and Szalay, 2004; Salton, 2005), which may indicate that their common ancestor was arboreal. Postcranial differences, especially in the tarsus, demonstrate convergences between *Echinops* and other mammalian climbers, and between *Setifer* and more terrestrial taxa (Salton and Szalay, 2004).

The forelimb of *Echinops* exhibits several differences from *Setifer* that are indicative of arboreal behavior in the former, including a longer acromion process; a slightly wider glenoid fossa; a rounder, larger humeral head; a mediolaterally wider medial epicondyle; a shallower, longer ulna; and a rounder articular surface of the radial head. Several features in common between the two taxa and not shared by *Hemicentetes* or

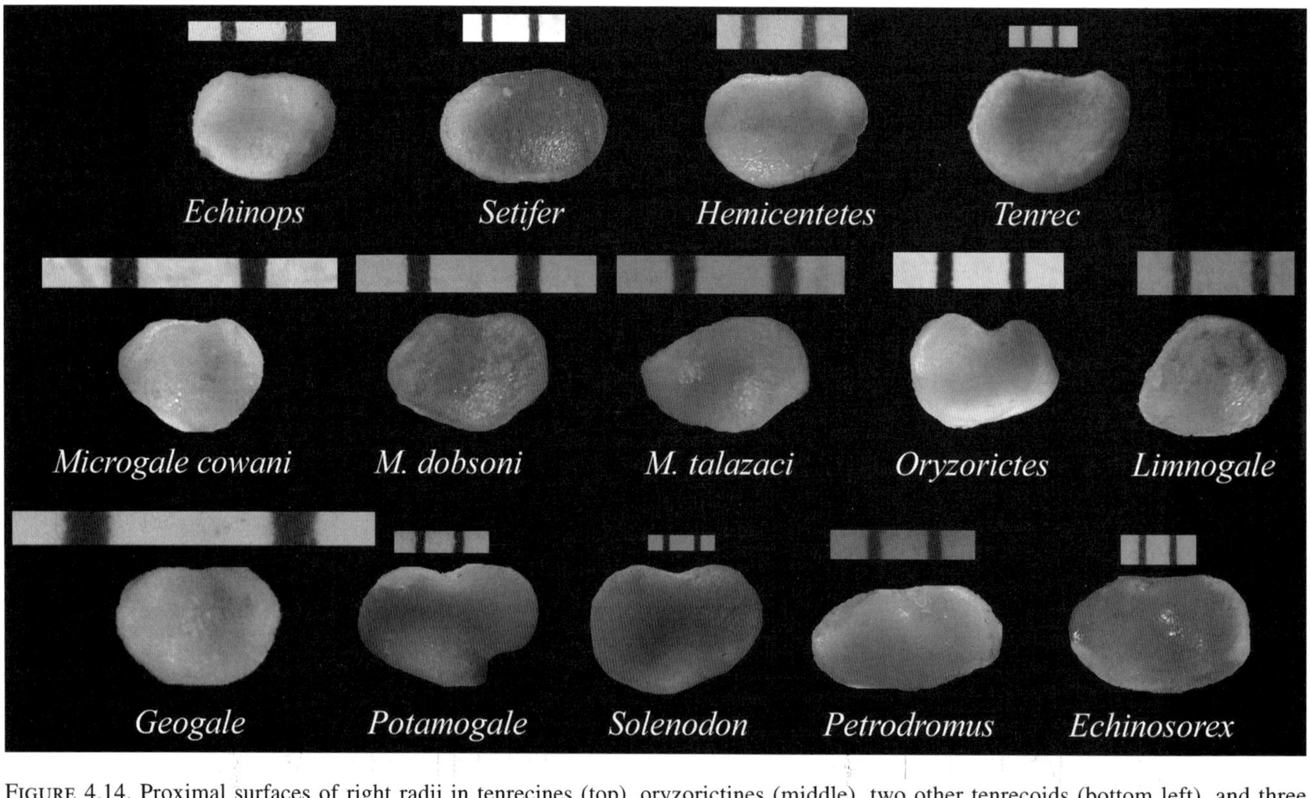

FIGURE 4.14. Proximal surfaces of right radii in tenrecines (top), oryzorictines (middle), two other tenrecoids (bottom left), and three outgroup taxa (bottom right); top is anterior, bottom is posterior. Note differences in radial head shape, ranging from rounded in *Limnogale* to elliptical in *Petrodromus*. Subdivisions on scale are 1.0 mm.

Tenrec include a large, rounded scapular surface area for the supraspinatus with a steep cranial border; rectangular infraspinous fossa; deltoid tubercle on the lateral edge of the humerus; short olecranon process; and long, shallow radius. These shared traits have been associated with climbing in other mammals (see Table 4.1), and other shared postcranial characteristics also tend to be characteristic of arboreal behavior (see Salton and Szalay, 2004; Salton, 2005; Salton and Sargis, 2008). It is possible that *Setifer* climbs more than reports would suggest. JAS observed *Setifer* in the field and found that although *Setifer* nests on the ground, it is able to climb when prodded to do so (as can *Tenrec*; see Eisenberg and Gould, 1970). The traits shared by *Setifer* and *Echinops* more likely reflect a common ancestor that was arboreal rather than being *Setifer*-like.

4.4.2 *Hemicentetes, Oryzorictes*, and *Solenodon*: Fossorial/Semi-fossorial

As already established by many other studies on fossorial mammals, digging behavior is strongly indicated in forelimb morphology. The semi-fossorial *Hemicentetes* and *Solenodon*, as well as the fossorial *Oryzorictes*, have a suite of characteristics that demonstrate extremely high loads incurred by the elbow and shoulder during digging. The *Tenrec* postcranium has some characteristics that suggest digging behavior as well, though not to the extent of the other three genera. Traits related to digging in these taxa include: an elongated, narrow scapula; short, wide humerus with a widened medial epicondyle; pronounced attachment sites for the deltoid and pectoral musculature; well-developed bicipital groove (sometimes forming a complete tunnel); proximodistally and anteroposteriorly elliptical humeral head; long (and sometimes medially curved) olecranon process; and short, deep ulna and radius.

4.4.3 *Limnogale* and *Potamogale*: Semi-aquatic

Other regions of the postcranium demonstrate some similarities between *Limnogale* and *Potamogale*, which are likely based on a shared semi-aquatic habitus. However, differences in the details of postcranial form do not point to a close common ancestry between *Limnogale* and *Potamogale*, and rather suggest that *Limnogale* is an oryzorictine, as supported by recent molecular data (Olson and Goodman, 2003). Morphology of the forelimb demonstrates very little similarity at all between *Limnogale* and *Potamogale*, despite their shared swimming behavior, and suggests that they use their arms in very different ways. *Limnogale* has a large acromion process (*Potamogale* has almost none), large lesser tuberosity, small greater tuberosity, relatively short humerus, wide distal humerus and medial epicondyle with an entepicondylar foramen, and rounded capitulum. *Limnogale* shares several of these features with the other oryzorictines, and it seems that its similarities to *Potamogale* are function-based convergences rather than synapomorphies. The differences between them emphasize the importance of forelimb stability and unilateral motion in the *Potamogale* forelimb, whereas *Limnogale* probably uses its arms for steering and changing direction (and perhaps more grooming and digging while on land), in addition to aquatic paddling. Unlike the other tenrecoids, *Potamogale* and *Geogale* have no entepicondylar foramen, which is an interesting observation given that *Potamogale* and *Geogale* are hypothesized to be basally divergent tenrecoid taxa (see Olson and Goodman, 2003).

4.4.4 *Microgale* spp.: Terrestrial/Fossorial?

The three *Microgale* species examined in this study are usually referred to as terrestrial, with some possible climbing in *M. talazaci* (based on foot and tail length; Eisenberg and Gould, 1970). This study confirmed a series of characteristics correlated with terrestrial running, and did not reveal any traits in *M. talazaci* that suggest climbing behavior. Rather, the forelimb (and other regions) of the *M. cowani* skeleton exhibits several features characteristic of a digger, such as a long, narrow scapula; large pectoral crest and deltoid tubercle; wide medial epicondyle; and short, deep ulna with a long olecranon process. Olson and Goodman's (2003) molecular phylogeny of *Microgale* does not place *M. cowani* anywhere near the root of the *Microgale* tree. This suggests that *M. cowani*, rather than being a close relative of *Oryzorictes*, may have convergently evolved a series of similar traits based on more frequent digging behavior than is recognized in the literature.

4.4.5 *Petrodromus*: Cursorial

The elephant shrew *Petrodromus* is the fastest running animal of all the taxa examined in this study, and the forelimb exhibits many traits (as does the hind limb) that reflect the importance of joint stabilization and restriction of movement to the parasagittal plane. Though its overall scapular shape is unremarkable, the metacromion process is long and narrow, and the glenoid fossa is spherical with a long overhanging coracoid process. The humerus is narrow and long with an enormous, perforated coronoid/olecranon fossa and a flat, spindle-shaped capitulum. The greater tuberosity is very robust and rises above the humeral head, and the medial and lateral epicondyles are almost nonexistent. The ulna is completely straight, long, and shallow, and has a very short olecranon process. The radius is also long and shallow, and the radial head is mediolaterally elliptical to an extreme. *Petrodromus* and *Potamogale* share a similar form of the humeral distal articular surface. But small-scale differences, such as the angle of the trochlea and shape of the capitular tail, strongly suggest that the similarities are convergent and based on the need for joint stabilization.

Unlike other regions of the postcranial skeleton (Salton and Szalay, 2004; Salton, 2005; Salton and Sargis, 2008), the forelimb offered little in terms of understanding phylogenetic relationships between taxa. Features of the shoulder and elbow joints, and the associated bones, are highly variable between particular genera and species and show few consistent subfamily-level differences. Forelimb form is highly dependent on species-specific behavior, and is not as constrained as, for example, aspects of the hind limb (Salton and Szalay, 2004; Salton, 2005).

As discussed in Salton and Szalay (2004) and Salton (2005), tenrecines and oryzorictines have a series of differences in the hind limb skeleton that are correlated with differences in basic posture. The tenrecine hind limb is more laterally rotated, and allows for much more general rotational movement than the oryzorictine hind limb. Oryzorictines have a parasagitally directed knee and foot and show much more constraint against mediolateral leg movement (Salton and Szalay, 2004). Both *Geogale* and *Potamogale* vary from the tenrecine and oryzorictine hind limb patterns, exhibiting some novel aspects of form, as well as some combination of tenrecine and oryzorictine traits. The forelimb, however, does not exhibit subfamily-level differences between tenrecines and oryzorictines that can be attributed to any basic differences in known positional behavior, or that may be attributed to phylogenetic inertia. Analysis of forelimb indices did not differentiate tenrecines from oryzorictines, unlike analyses of other postcranial regions. Aside from some similarities between the sister taxa *Echinops* and *Setifer* that appear to be phylogenetically important and unrelated to locomotor specialization, each tenrecoid genus exhibits function-based variation that often corresponds with hypotheses based on other mammalian locomotor specialists.

Acknowledgments. Thanks to Raphael Allison, Rochelle Buffenstein, John Wahlert, Johanna Warshaw, Link Olson, Lawrence Heaney, William Stanley, Judy Chupasko, Linda Gordon, Heinz Künzle, Angelos Tianarifidy Stila, Carl Terranova, Marian Dagosto, Stephen Chester, and the mammal collection staff at MCZ, FMNH, AMNH, and USNM. We especially thank our graduate advisor, Fred Szalay, for all his guidance, encouragement, and friendship over the years. This project was funded in part by the National Science Foundation, National Geographic Society, Association for Women in Science, Field Museum of Natural History, CUNY GC Mario Cappeloni Fellowship, and Scholarships Foundation.

References

Argot, C., 2001. Functional-adaptive anatomy of the forelimb in the Didelphidae, and the paleobiology of the Paleocene marsupials *Mayulestes ferox* and *Pucadelphys andinus*. Journal of Morphology 247, 51–79.

Barnosky, A. D., 1982. Locomotion in moles from the middle Tertiary of North America. Science 216, 183–185.

Biknevicius, A. R., 1993. Biomechanical scaling of limb bones and differential limb use in caviomorph rodents. Journal of Mammalogy 74, 95–107.

Casinos, A., Quintana, C., Viladiu, C., 1993. Allometry and adaptation in the long bones of a digging group of rodents. Zoological Journal of the Linnean Society 107, 107–115.

Campbell, B., 1939. The shoulder anatomy of the moles: a study in phylogeny and adaptation. American Journal of Anatomy 64, 1–39.

Ciochon, R. L., 1993. Evolution of the cercopithecoid forelimb: phylogenetic and functional implications from morphometric analyses. University of California Publications in Geological Sciences 138, 1–251.

Eisenberg, J. F., Gould, E., 1970. The Tenrecs: A Study in Mammalian Behavior and Evolution. Smithsonian Institution Press, Washington, DC.

Fleagle, J. G., Simons, E. L., 1982. The humerus of *Aegyptopithecus zeuxis*: a primitive anthropoid. American Journal of Physical Anthropology 59, 175–193.

Gebo, D. L., Sargis, E. J., 1994. Terrestrial adaptations in the postcranial skeletons of guenons. American Journal of Physical Anthropology 93, 341–371.

Grand, T. I., Barboza, P. S., 2001. Anatomy and development of the koala, *Phascolarctos cinereus*: an evolutionary perspective on the superfamily Vombatoidea. Anatomy and Embryology 203, 211–223.

Harrison, T., 1989. New postcranial remains of *Victoriapithecus* from the middle Miocene of Kenya. Journal of Human Evolution 18, 3–54.

Heinrich, R. E., Rose, K. D., 1997. Postcranial morphology and locomotor behaviour of two early Eocene miacoid carnivorans, *Vulpavus* and *Didymictis*. Palaeontology 40, 279–305.

Hildebrand, M., 1985. Digging of quadrupeds. In: Hildebrand, M., Bramble, D. M., Liem, K. F., Wake, D. B. (Eds.), Functional Vertebrate Morphology. Belknap, Cambridge, MA, pp. 89–109.

Hildebrand, M., 1995. Analysis of Vertebrate Structure, 4th ed. Wiley, New York.

Hopwood, A. T., 1947. Contributions to the study of some African mammals. III. Adaptations in the bones of the fore-limb of the lion, leopard and cheetah. Journal of the Linnean Society of London (Zoology) 41, 259–271.

Howell, A. B., 1970. Aquatic Mammals: Their Adaptation to Life in the Water. Dover, New York.

Kerbis Peterhans, J. C., Patterson, B. D., 1995. The Ethiopian water mouse *Nilopegamys* Osgood, with comments on semi-aquatic adaptations in African Muridae. Zoological Journal of the Linnean Society 113, 329–349.

Larson, S. G., 1993. Functional morphology of the shoulder in primates. In: Gebo, D. L. (Ed.), Postcranial Adaptation in Nonhuman Primates. Northern Illinois University Press, Dekalb, IL, pp. 45–69.

Luo, Z-X., Wible, J. R., 2005. A Late Jurassic digging mammal and early mammalian diversification. Science 308, 103–107.

Monteiro, L. R., Abe, A. S., 1999. Functional and historical determinants of shape in the scapula of xenarthran mammals: evolution of a complex morphological structure. Journal of Morphology 241, 251–263.

Neveu, P., Gasc, J-P., 2002. Lipotyphla limb myology comparison. Journal of Morphology 252, 183–201.

Olson, L. E., Goodman, S. M., 2003. Phylogeny and biogeography of tenrecs. In: Goodman, S. M., Benstead, J. P. (Eds.), The Natural History of Madagascar. University of Chicago Press, Chicago, pp. 1235–1242.

Osburn, R. C., 1903. Adaptation to aquatic, arboreal, fossorial and cursorial habits in mammals. I. Aquatic adaptations. American Naturalist 37, 651–665.

Reed, C. A., 1951. Locomotion and appendicular anatomy in three soricoid insectivores. American Midland Naturalist 45, 513–671.

Roberts, D., 1974. Structure and function of the primate scapula. In: Jenkins, F. A., Jr. (Ed.), Primate Locomotion. Academic Press, New York, pp. 171–200.

Roberts, D., Davidson, I., 1975. The lemur scapula. In: Tattersall, I., Sussman, R. W. (Eds.), Lemur Biology. Plenum, New York, pp. 125–147.

Rose, K. D., Emry, R. J., 1983. Extraordinary fossorial adaptations in the Oligocene palaeanodonts *Epoicotherium* and *Xenocranium* (Mammalia). Journal of Morphology 175, 33–56.

Rose, M. D., 1988. Another look at the anthropoid elbow. Journal of Human Evolution 17, 193–224.

Rose, M. D., 1989. New postcranial specimens of catarrhines from the middle Miocene Chinji formation, Pakistan: descriptions and a discussion of proximal humeral functional morphology in anthropoids. Journal of Human Evolution 18, 131–162.

Salton, J. A, 2005. Evolutionary morphology of the postcranial skeleton in Afro-Malagasy Tenrecoidea (Mammalia). Ph.D. dissertation, City University of New York.

Salton, J. A., Sargis, E. J., 2008. Evolutionary morphology of the Tenrecoidea (Mammalia) carpal complex. Biological Journal of the Linnean Society 93, 267–288.

Salton, J. A., Szalay, F. S., 2004. The tarsal complex of Afro-Malagasy Tenrecoidea: a search for phylogenetically meaningful characters. Journal of Mammalian Evolution 11, 73–104.

Sargis, E. J., 2002. Functional morphology of the forelimb of tupaiids (Mammalia, Scandentia) and its phylogenetic implications. Journal of Morphology 253, 10–42.

Simmons, N. B., 1994. The case for chiropteran monophyly. American Museum Novitates 3103, 1–54.

Smith, J. M., Savage, R. J. G., 1956. Some locomotory adaptations in mammals. Zoological Journal of the Linnean Society 42, 603–622.

Springer, M. S., Stanhope, M. J., Madsen, O., de Jong, W. W., 2004. Molecules consolidate the placental mammal tree. Trends in Ecology and Evolution 19, 430–438.

Stein, B., 2000. Morphology of subterranean rodents. In: Lacey, E. A., Patton, J. L., Cameron, G. N. (Eds.), Life Underground: The Biology of Subterranean Rodents. University of Chicago Press, Chicago, pp. 19–61.

Szalay, F. S., Dagosto, M., 1980. Locomotor adaptations as reflected on the humerus of Paleogene primates. Folia Primatologica 34, 1–45.

Szalay, F. S., Sargis, E. J., 2001. Model-based analysis of postcranial osteology of marsupials from the Palaeocene of Itaboraí (Brazil), and the phylogenetics and biogeography of Metatheria. Geodiversitas 23, 139–302.

Szalay, F. S., Schrenk, F., 1998. The middle Eocene *Eurotamandua* and a Darwinian phylogenetic analysis of "edentates." Kaupia 7, 97–186.

Taylor, M. E., 1974. The functional anatomy of the forelimb of some African Viverridae (Carnivora). Journal of Morphology 143, 307–336.

Taylor, B. K., 1978. The anatomy of the forelimb in the anteater (*Tamandua*) and its functional implications. Journal of Morphology 157, 347–368.

Taylor, B. K., 1985. Functional anatomy of the forelimb in vermilinguas (anteaters). In: Montgomery, G. G. (Ed.), The Evolution and Ecology of Armadillos, Sloths, and Vermilinguas. Smithsonian Institution Press, Washington, DC, pp. 163–171.

Van Valkenburgh, B., 1987. Skeletal indicators of locomotor behavior in living and extinct carnivores. Journal of Vertebrate Paleontology 7, 162–182.

Vasquez-Molinero, R., Martin, T., Fischer, M. S., Frey, R., 2001. Comparative anatomical investigations of the postcranial skeleton of *Henkelotherium guimarotae* Krebs, 1991 (Eupantotheria, Mammalia) and their implications for its locomotion. Mitteilungen aus dem Museum für Naturkunde zu Berlin – Zoologische Reihe 77, 207–216.

Verma, K., 1963. The appendicular skeleton of Indian hedgehogs. Mammalia 27, 564–580.

Yalden, D. W., 1966. The anatomy of mole locomotion. Journal of Zoology, London 149, 55–64.

5. Postcranial Morphology of *Apheliscus* and *Haplomylus* (Condylarthra, Apheliscidae): Evidence for a Paleocene Holarctic Origin of Macroscelidea

Tonya A. Penkrot
Marshall University
Department of Biological Sciences
1 John Marshall Drive
Huntington, WV 25755, USA
penkrot@marshall.edu

Shawn P. Zack
Marshall University
Department of Biological Sciences
1 John Marshall Drive
Huntington, WV 25755, USA
zack@marshall.edu

Kenneth D. Rose
The Johns Hopkins University School of Medicine
Center for Functional Anatomy and Evolution
1830 E. Monument St.
Baltimore, MD 21205, USA
kdrose@jhmi.edu

Jonathan I. Bloch
Florida Museum of Natural History
University of Florida
P.O. Box 117800
Gainesville, FL 32611, USA
jbloch@flmnh.ufl.edu

5.1 Introduction

Small-bodied eutherian mammals with bunodont teeth from the Paleocene and Eocene have long been the subjects of taxonomic contention, with regard to both the assignment of individual genera to supergeneric clades (such as Hyopsodontidae, Pentacodontidae, and Dormaalidae) and to the place of those larger groups within Eutheria. Taxa traditionally placed in the family Hyopsodontidae have proven particularly problematic from the first standpoint, the clade has become a wastebasket for small-bodied, bunodont taxa whose morphology suggests affinities with the basal ungulate order Condylarthra (in this study, Ungulata and ungulate refer to the traditional morphological concept of this group, minimally including Artiodactyla, Perissodactyla, Hyracoidea,

* Address for correspondence: penkrot@marshall.edu

E.J. Sargis and M. Dagosto (eds.), *Mammalian Evolutionary Morphology: A Tribute to Frederick S. Szalay*, 73–106.

Proboscidea, Sirenia, and their extinct relatives, particularly condylarths and the South American ungulate radiation (e.g., McKenna and Bell, 1997); for visual simplicity, we use condylarth and Condylarthra without quotes, although we recognize that this group is likely para- or polyphyletic).

At the root of the problem of hyopsodontid monophyly is that hyopsodontids have been united primarily by their small size, bunodont dentitions, and by a suite of vague and likely plesiomorphic dental characters, such as semimolariform P4/p4, distinct entoconids, unreduced M3/m3, and the presence of a variably developed hypocone (Simpson, 1937; Archibald, 1998). This lack of clearly diagnostic synapomorphies has led several authors to suggest that Hyopsodontidae is likely polyphyletic (Rigby, 1980; Cifelli, 1983; Archibald, 1998; Tabuce et al., 2001; Hooker and Dashzeveg, 2003; Zack et al., 2005a), particularly if the subfamily Mioclaeninae is included. Mioclaeninae has been recognized as a subfamily of Hyopsodontidae (e.g., Matthew, 1937; Simpson, 1937), but most recent works recognize Mioclaenidae as a distinct clade (Cifelli, 1985; McKenna and Bell, 1997; Archibald, 1998; Muizon and Cifelli, 2000).

Lack of support for the monophyly of Hyopsodontidae is particularly problematic, as hyopsodontids (used in the broad sense to include mioclaenids) have been identified as potential ancestors of a number of other ungulate clades, including extant artiodactyls (Simpson, 1937; Schaeffer, 1947), hyracoids (Godinot et al., 1996), and the extinct South American ungulate radiation (Cifelli, 1983; Muizon and Cifelli, 2000). It has also been suggested that members of the extant order Macroscelidea (elephant-shrews or sengis) evolved from hyopsodontids (Hartenberger, 1986; Simons et al., 1991; Butler, 1995; Tabuce et al., 2001).

Compounding the dual problems of the monophyly of Hyopsodontidae and of its potential relationships to later groups is the fact that most hyopsodontids are only known from dental remains. Until recently, only the type genus, *Hyopsodus*, has been known from substantial postcranial remains (Matthew, 1915; Gazin, 1968). Isolated proximal tarsal elements have also been ascribed to the European genus *Paschatherium* (Godinot et al., 1996), and it was these bones that prompted the suggestion of a relationship of hyopsodontids to hyracoids. Matthew (1918) described several fragmentary elements of the genus *Apheliscus*, including an ulna and partial humeri on the basis of a skeletal association (AMNH 15696), which also includes a badly crushed partial skull and both dentaries. Unfortunately, these postcranial elements were too poorly preserved to make meaningful functional or phylogenetic inferences.

The impetus for the present study was the identification of a new skeletal association of *Apheliscus chydaeus* (USNM 525597, collected by Dr. Thomas M. Bown; Willwood Formation, early Eocene, Bighorn Basin, Wyoming), including the partial upper and lower dentitions (at a light wear stage, indicating a young adult individual), a partial ilium, portions of all long bones (epiphyses unfused), nearly complete astragalus, calcaneum, and cuboid, a partial entocuneiform, partial metapodials, and an intermediate phalanx. In contrast to the specimen of *Apheliscus* described by Matthew (1918), the postcranial elements of USNM 525597 are reasonably well preserved, although initially encrusted in hematite – a condition that prevented recognition of the specimen's significance for over 15 years. This dental-postcranial association provides more complete elements of *Apheliscus*, as well as postcranial elements of a second North American Eocene hyopsodontid, *Haplomylus* (see Materials and Methods).

Apheliscus is a dentally distinctive small condylarth found in Paleocene and Eocene deposits of the greater Bighorn Basin, Wyoming (Matthew, 1918; Bown, 1979; Rose, 1981; Gingerich, 1994), as well as the Powder River and Washakie basins, Buckman Hollow, and Togwotee Pass of Wyoming, the Sand Wash Basin of Colorado, and the San Juan Basin of New Mexico (Cope, 1875; McKenna, 1960, 1980; Delson, 1971; Rose, 1981; Wilf et al., 1998). In the Bighorn Basin, where it is best represented, *Apheliscus* is a rare taxon in surface assemblages, but it is one of the most common taxa in early Eocene quarry samples, which are typically rich in small-bodied taxa. Affinities of *Apheliscus* have been enigmatic since the original description (Cope, 1874). In prior studies, the phylogenetic position of *Apheliscus* has been assessed based on dental morphology, the most distinctive feature of which is an expanded P4/p4 shearing complex, which has led to variable interpretations of the genus as a creodont (Cope, 1874, 1877), as an insectivoran in its own family Apheliscidae (Matthew, 1918), or as a pentacodontid pantolestan (Gazin, 1959). Most recent workers have favored affinities to Condylarthra, with specific affinities to either a restricted Hyopsodontidae (Van Valen, 1967; Rose, 1981) or to Mioclaenidae (McKenna, 1960; Archibald, 1998).

Eocene *Apheliscus*, and *Haplomylus*, have been known only from their cheek dentition. Cheek dentitions of *Haplomylus* are abundant in late Paleocene and early Eocene deposits of the greater Bighorn Basin and contemporaneous North American strata (Delson, 1971; Bown, 1979; Rose, 1981; Gingerich, 1994, Robinson and Williams, 1997) but no anterior dentitions, cranial, or postcranial remains have previously been attributed to the genus. Unlike *Apheliscus*, the hyopsodontid affinities of *Haplomylus* have been hypothesized since the first description of the genus by Matthew (1915). Aside from several studies describing stratigraphic variation within the lineage (Gingerich, 1976; Bown, 1979; Rose, 1981; Bown et al., 1994a; Robinson and Williams, 1997), very little attention has been devoted to the genus, particularly with regard to its phylogenetic position within the Hyopsodontidae. A notable exception to this is Simons et al. (1991), who hypothesized potential affinities between *Haplomylus* and the extant order Macroscelidea (elephant-shrews or sengis) based on several dental characters shared by *Haplomylus* and Eocene macroscelideans, including: a molariform p4 with an enlarged median paraconid and a bicuspid talonid; a well-developed paraconid on p2 and p3;

a postprotocrista on P4-M2 that is directed toward the hypocone rather than toward the metaconule; trigonids wider than talonids on m1 and m2; reduction of hypoconulids on m1 and m2; and, P4 transversely widened, with an expanded parastyle.

A recent preliminary description of the new postcranial material of *Apheliscus* and *Haplomylus* described in detail here, established the significance of these specimens for clarifying the phylogenetic integrity of hyopsodontids, and for understanding the relationships of these taxa to other groups (Zack et al., 2005a). This study confirmed the likely polyphyly of Hyopsodontidae and favored an origin of Macroscelidea from hyopsodontids allied to *Apheliscus* and *Haplomylus* (classified in the family Apheliscidae; Zack et al., 2005b). Zack et al. (2005b) include most taxa traditionally classified as hyopsodontids in Apheliscidae, while Hyopsodontidae includes only *Hyopsodus* and taxa usually assigned to Mioclaenidae. Zack et al. (2005a) informally referred to *Apheliscus* and *Haplomylus* as apheliscines. Apheliscidae and apheliscids replace this informal designation. The purpose of this paper is to present more complete descriptions and illustrations of this material, as well as more extensive comparisons to macroscelideans, *Hyopsodus*, and other contemporary taxa that show similar postcranial adaptations to apheliscids than published in Zack et al. (2005a).

5.2 Materials and Methods

A dentally associated partial skeleton of *Apheliscus chydaeus* (USNM 525597) was used to identify additional (and in many cases, better preserved) isolated postcranial elements of *Apheliscus* from the Bighorn Basin. Most of these additional elements come from quarry samples including Rose Quarry (D-1460Q) and Dorsey Creek Quarry (D-2035) where, in contrast to its scarcity in surface collections, *Apheliscus* tends to be relatively common.

Initial recognition of *Haplomylus* postcrania was based on the identification of isolated tarsals with morphologic similarities to *Apheliscus* from several Willwood localities- all low in the Willwood Formation- corresponding to the earlier part of the Wasatchian (Sandcouleean and Graybullian subages). The size and morphology of these tarsals suggested that they belonged to a close relative of *Apheliscus*. The newly identified tarsals are tentatively ascribed to *Haplomylus*, the only remaining Willwood hyopsodontid (traditional sense) whose postcrania was unknown. Supporting this supposition is the restriction of these tarsal morphs to the early Wasatchian and their abundance at these levels compared to other small tarsal material. This matches the dental record of *Haplomylus*, which is abundant in Sandcouleean and Graybullian faunas but absent from the later Wasatchian. Based on comparisons to *Apheliscus*, the size of these elements is also appropriate for *Haplomylus*.

Subsequently, abundant tarsals matching this morphology were identified in the 8abc limestone from the early Wasatchian (University of Michigan locality, SC-4; Wa-1) in the Clarks Fork Basin. Other postcranial material consistent with the morphologic pattern seen in the tarsals is also present in the 8abc limestone. The size and relative abundance of this postcranial material in the 8abc limestone is consistent with the size and abundance of *Haplomylus speirianus* dentitions known from this fauna. This material cannot be ascribed to any of the other taxa known from dentitions in the 8abc assemblage, as the size and morphology of these postcranials do not match those of other taxa represented dentally in that assemblage. Most taxa represented dentally in the 8abc assemblage, or their close relatives, are known elsewhere from skeletal associations, which do not match the morphology of the elements ascribed to *Haplomylus*. The few taxa whose postcrania are unknown, including *Plagiomene, Viverravus*, and *Niptomomys*, are not of appropriate size to go with these elements, and the first two are also very rare in the 8abc assemblage, in contrast to the relative abundance of *Haplomylus* postcranials. Thus, strong indirect evidence supports the reassociation of isolated postcranial elements from the Willwood Formation, particularly from the 8abc limestone, to *Haplomylus speirianus*. Nevertheless, until definitive associations of *Haplomylus* teeth and postcrania are found, attributions to elements to *Haplomylus* must remain somewhat tentative.

5.2.1 Institutional Abbreviations

AMNH, American Museum of Natural History, New York; **DMNH**, Denver Museum of Natural History, Denver, Colorado; **UCMP**, University of California Museum of Paleontology, Berkeley, California; **UM**, Museum of Paleontology, University of Michigan, Ann Arbor, Michigan; **USGS**, United States Geological Survey, Denver registry, Denver, Colorado; **USNM**, Department of Paleobiology (fossil specimens) or Mammalogy (modern specimens), United States National Museum of Natural History, Washington, DC; **YPM**, Yale Peabody Museum, Yale University, New Haven.

5.2.2 Locality Abbreviations

D, Bighorn Basin localities, USGS, Denver; **SC**, Clarks Fork Basin localities, Sand Coulee area, University of Michigan Museum of Paleontology, Ann Arbor; **W**, Bighorn Basin Localities, University of Wyoming, Laramie; **Y**, Bighorn Basin localities, YPM.

5.2.3 Principal Specimens Examined

Most fossil specimens examined in this study come from the southern Bighorn Basin, except for most *Haplomylus* postcrania and some leptictid postcrania examined, which comes from University of Michigan locality SC-4 (8abc limestone) in the Clarks Fork Basin. One specimen of *Hyopsodus*, USNM 23740, comes from the Bridger Basin. For southern Bighorn Basin specimens, locality abbreviations follow Bown et al. (1994b). For more detailed locality information, see Bown et al. (1994b), Bloch and Bowen (2001), and Silcox and Rose (2001).

5.2.3.1 *Apheliscus chydaeus*

W-22: USNM 525597, associated left P3-M3, right P3-4, M2-3, left p2-4, m2-3, left astragalus, calcaneum, cuboid, distal femur, proximal radius, proximal humerus, right humeral shaft, ulnar fragment, proximal tibia, fragments of both tibial shafts, vertebrae.

W-16A, Banjo Quarry: USNM 525594, left astragalus.

5.2.3.2 *Apheliscus sp. (Intermediate Between A. chydaeus and A. insidiosus)*

D-2037Q, McNeil Quarry: USNM 493819, unassociated left distal humerus, distal femur, and proximal tibia, right distal femur and calcaneum. Though unassociated, these could belong to a single individual, as there is no duplication of elements and all elements are of appropriate size to represent a single individual. Semiarticulated postcranial remains of other taxa are known from McNeil Quarry (personal observation) strengthening possibility that the *Apheliscus* postcrania from this quarry represent a single individual.

5.2.3.3 *Apheliscus insidiosus*

D-1350Q: USNM 488325, left proximal femur and proximal tibia, right distal femur, unassociated but possibly from a single individual based on degree of epiphysial fusion.

D-1460Q, Rose Quarry: USNM 493903, unassociated left proximal tibia and distal tibia-fibula, right distal humerus and cuboid; USNM 521789, right calcaneum; USNM 521790, right astragalus; USNM 521791 right astragalus; USNM 525593, unassociated right calcaneum, proximal femur, and distal humerus.

D-2035Q, Dorsey Quarry: USNM 488326, left femur; USNM 491971, left distal tibia-fibula (identified as a possible leptictid by Rose, 1999); USNM 495051, right tibia-fibula (identified as a possible leptictid by Rose, 1999); USNM 525591, unassociated left astragalus and calcaneum; USNM 525592, right astragalus; USNM 525646, associated right proximal tibia and fibula.

Bighorn Basin, locality unknown: AMNH 15696, associated crushed cranium with left P4-M3, right P4, M2-3, left mandible with p4-m3, right mandible with p4-m1, m3, left humeral shaft and proximal ulna, right distal humerus, pelvic and vertebral fragments.

5.2.3.4 *Haplomylus speirianus*

SC-4 (8abc limestone): USNM 513057-513062, proximal femora; USNM 513140, femur; USNM 513173-513175, distal femora; USNM 513239, distal tibia-fibula; USNM 513245-513247, proximal tibiae; USNM 513512, humerus; USNM 513555-513557, distal humeri; USNM 513632-513635, astragali; USNM 513655-513665, calcanei; USNM 513668, cuboid; USNM 513868, tibia-fibula.

D-1223: USNM 488321, left astragalus.

W-37: USNM 493902, right astragalus.

W-44: USNM 488327, two unassociated left calcanei.

W-46: USNM 488328, left calcaneum.

W-86: USNM 493901, left calcaneum; USNM 525595, unassociated left and right calcanei.

Anthill across from W-86: USNM 488329, right calcaneum.

Y-327: USNM 525596, right astragalus.

5.2.3.5 *Leptictidae*

We have examined the specimens listed by Rose (1999). Note that USNM 491971 and 495051, identified as possible leptictids in that work are here reidentified as *Apheliscus*. We have also examined newly identified isolated leptictid elements, all either *Prodiacodon* or *Palaeictops*, from D-1460Q (USNM 493931, USNM 493778, USNM 493761, uncataloged specimens), D-2035Q (uncataloged specimens), and SC-4 (USNM 513235, 513240, 513636, 513666, 513667, 513063).

5.2.3.6 *Macrocranion*

Isolated *Macrocranion* tarsals are abundant in almost all Willwood quarries and screenwash localities and have been identified based on specimens illustrated in Godinot et al. (1996). Fused distal tibia-fibulas from these localities were reassociated with *Macrocranion* based on their fit to *Macrocranion* proximal tarsals and by reference to *Macrocranion* skeletons from Messel, Germany, which indicate extensive distal fusion of the tibia and fibula in this genus (Storch, 1993, 1996). Most of the *Macrocranion* material examined for this study comes from three localities, D-1460Q, D-2037Q, and D-2018 (Castle Gardens) and remains uncataloged.

5.2.3.7 *Hyopsodus*

The primary specimens of *Hyopsodus* examined are three skeletons, USNM 23740, USNM 17980 (both *Hyopsodus paulus*, see Gazin, 1968), and an uncataloged specimen in the YPM from Y-332a that includes most of the posterior portion of the skeleton including both hind limbs with nearly complete, articulated pedes. Other specimens examined are isolated elements, including humeri (USGS 25179, USNM 493816, 493823, 521829, 521832, 521694), tibiae (USGS 4725, USNM 493816) astragali (USGS 25331, USNM 493782), and calcanei (USGS 4725, USNM 527532).

5.2.3.8 Modern Comparative Specimens Examined

Hypsiprymnodon moschatus: USNM 238443, 238444.
Rhynchocyon cirnei: USNM 537657.
Petrodromus tetradactylus: USNM 521009.
Elephantulus rufescens: USNM 283463.
Echinosorex gymnurus: USNM 448861.
Hemiechinus auritus: USNM 396508; KDR personal collection uncataloged.

Erinaceus europaeus: USNM 251764, 251765.
Dolichotis patagonum: USNM 175890.
Jaculus jaculus: USNM 308400, 477276.
Pedetes capensis: USNM 344334, 295258.
Dipodomys deserti deserti: USNM 034369, 034370.
Dipodomys ordi palmeri: USNM 05372.
Ochotona rufescens vizier: USNM 326747, 326748, 326750.
Ochotona alpina argentata: USNM 240727.
Sylvilagus sp.: KDR personal collection L1, L2.
Tupaia tana: USNM 574901.

5.3 Description of *Apheliscus* and *Haplomylus* Postcrania

5.3.1 Forelimb

The associated fore- and hindlimb fragments of USNM 525597 indicate that the forelimb of *Apheliscus* was relatively short in comparison to the hindlimb, although the lack of complete elements makes this observation impossible to quantify. Unassociated elements of *Haplomylus* also indicate a forelimb that is significantly shorter than the hindlimb. The complete humerus of *Haplomylus* from the 8abc limestone (USNM 513512) is shorter than the preserved length of the most complete femora (USNM 513058 and 513140) and tibia-fibula (USNM 513868) of *Haplomylus* from this locality, despite the fact that the lengths of the latter elements would certainly be even greater if they were complete. As such, it is reasonable to infer that the forelimbs of both apheliscid genera were reduced in comparison to the hindlimbs, although complete, associated elements of both genera will be required to confirm this.

It should be noted that in the following descriptive sections, unless a feature is explicitly ascribed to either *Apheliscus* or to *Haplomylus*, that the statement applies to both genera.

5.3.1.1 Humerus

The humeral head is ovoid with an articular surface that tapers posteriorly (Figure 5.1). The greater and lesser tuberosities are broad, flat, and about even with the level of the head. The deltopectoral crest is weakly developed, restricted to the proximal one-third of the humeral shaft. It is broad proximally, tapering to a sharp, slightly elevated crest distally. The shaft itself is gracile and long.

The distal end of the humerus is narrow, with only a moderately prominent medial epicondyle. There is an entepicondylar bar, forming a patent entepicondylar foramen. The supinator crest is weakly developed in *Apheliscus*, and virtually absent in *Haplomylus*. The olecranon and coronoid fossae are very deep in *Apheliscus*, leaving only a thin sheet of bone separating the two fossae, while in *Haplomylus* the olecranon fossa is perforate. The trochlea is sharp, and the capitulum is round to subovoid, with the capitulum of *Haplomylus* rounder and more prominent than that of *Apheliscus*. Features of the distal humerus (capitulum, trochlea, entepicondyle) are approximately aligned in the same transverse plane.

5.3.1.2 Radius

The apheliscid radius is known from a single proximal radius of *Apheliscus* associated with USNM 525597. In proximal view, the radial head is subrectangular; i.e., the facet for articulation with the ulna is flat, but the medial and lateral rims of the head are slightly rounded. The capitular eminence is high, giving the radial head a considerable amount of relief. The surface of the radial head is virtually perpendicular to the long axis of the radial shaft (Figure 5.1D, E).

5.3.2 Hind Limb

Although complete, associated femora and tibiae of a single individual are not known for either *Apheliscus* or *Haplomylus*, evidence from isolated elements suggests that the tibia was relatively elongate in comparison with the femur in both

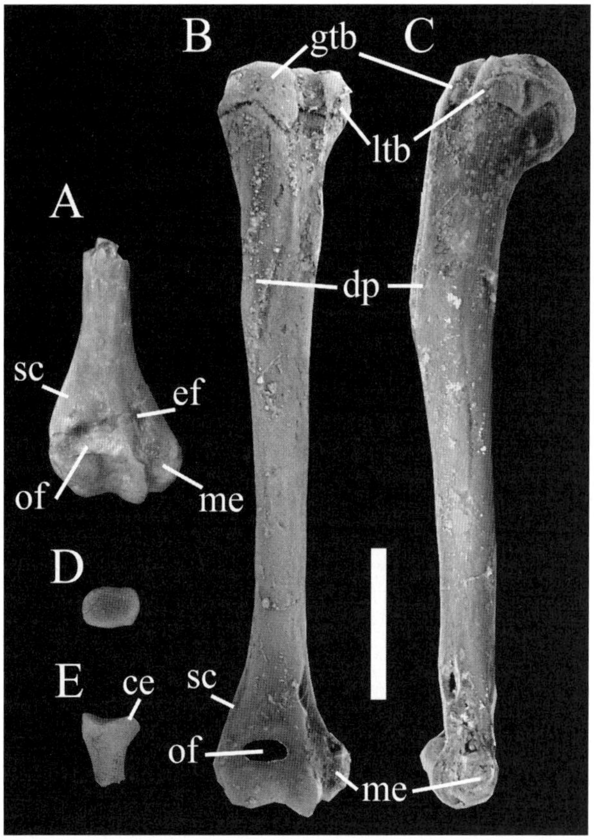

FIGURE 5.1. Apheliscid forelimb elements. A. Distal humerus of *Apheliscus* (USNM 493903) in anterior view. B, C. Anterior (B) and medial (C) views of the humerus of *Haplomylus* (USNM 513512). D, E. Proximal (D) and anterior (E) views of the proximal radius of *Apheliscus* (USNM 525597). Scale bar equals 5 mm. See appendix for anatomical abbreviations used in this and subsequent figures.

genera. The complete femur of *Apheliscus* (USNM 488326) is slightly shorter than the complete tibia-fibula (USNM 495051) from the same quarry, suggesting a crural index near one. Lengths of the hindlimb elements of *Haplomylus* can only be estimated as complete elements are not known, but the available material suggests that the crural index of this genus was higher than in *Apheliscus*. As in the case of relative forelimb size, this inference should be viewed cautiously until complete, associated elements become available.

5.3.2.1 Femur

The femur is gracile (more so in *Haplomylus* than in *Apheliscus*) (Figures 5.2, 5.3). The femoral head is small, subspherical to ovoid, and the articular surface does not extend onto the long neck. In *Apheliscus* the greater trochanter is slightly higher than the head, whereas in *Haplomylus* it projects distinctly above the head. It is also anteroposteriorly extended and mediolaterally compressed in *Haplomylus*. The trochanteric fossa for attachment of the pyriformis, gemelli, and obturator muscles is large and deep in both taxa (Figure 5.2B). The lesser trochanter is well developed and projects posteromedially (Figure 5.2C). There is a prominent, proximally located third trochanter on the lateral aspect of the femur (slightly better defined in *Haplomylus* than in *Apheliscus*); the proximal extent of this third trochanter is at the level of the distal termination of the lesser trochanter.

The distal femur is well preserved in *Apheliscus* (USNM 488326, 493819), and is deep and narrow (Figure 5.3B, C). The patellar groove is deep, narrow, long, and extends far proximally along the shaft of the femur, and the medial lip is higher than the lateral lip (Figure 5.3A, C). The articular surfaces of the condyles are anteroposteriorly elongate and anteriorly extensive. Distal femoral material of *Haplomylus* is limited to three poorly preserved epiphyses. These appear to conform to the morphology of *Apheliscus* in being deep and narrow, but otherwise, their poor preservation does not reveal much useful morphological information.

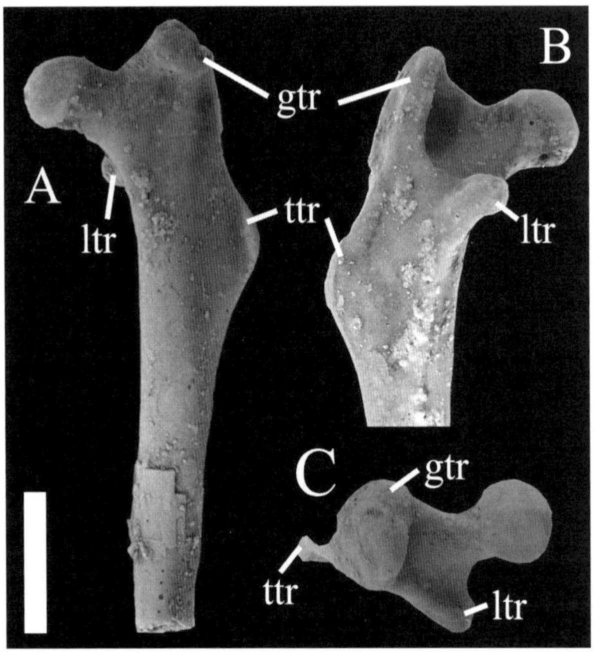

FIGURE 5.2. Proximal femur of *Haplomylus* (USNM 513057) in anterior (A), posterior (B), and proximal (C) views. Scale bar equals 5 mm.

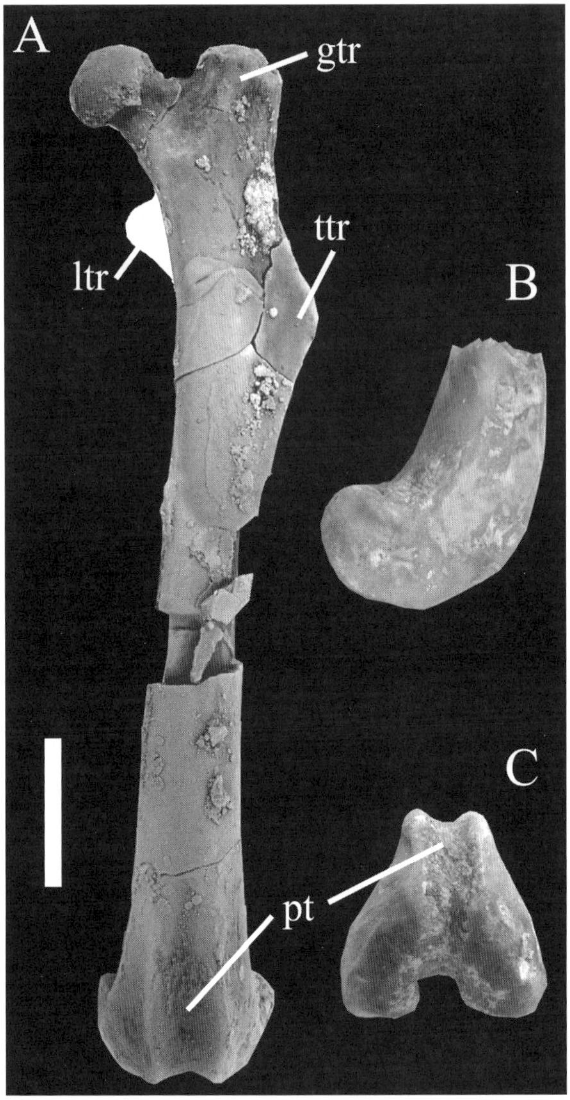

FIGURE 5.3. Femur of *Apheliscus*. A. Anterior view of whole femur (USNM 488326, lesser trochanter reconstructed from USNM 525593). B, C. Medial (B) and distal (C) views of distal femur (USNM 493819, reversed). Scale bar equals 5 mm.

5.3.2.2 Crus

The proximal apheliscid tibia, when viewed proximally, is roughly an equilateral triangle (or rather, heart-shaped) in outline (Figures 5.4B, 5.5B). The tibial condyles are nearly equal in area, and the lateral condyle is elevated slightly

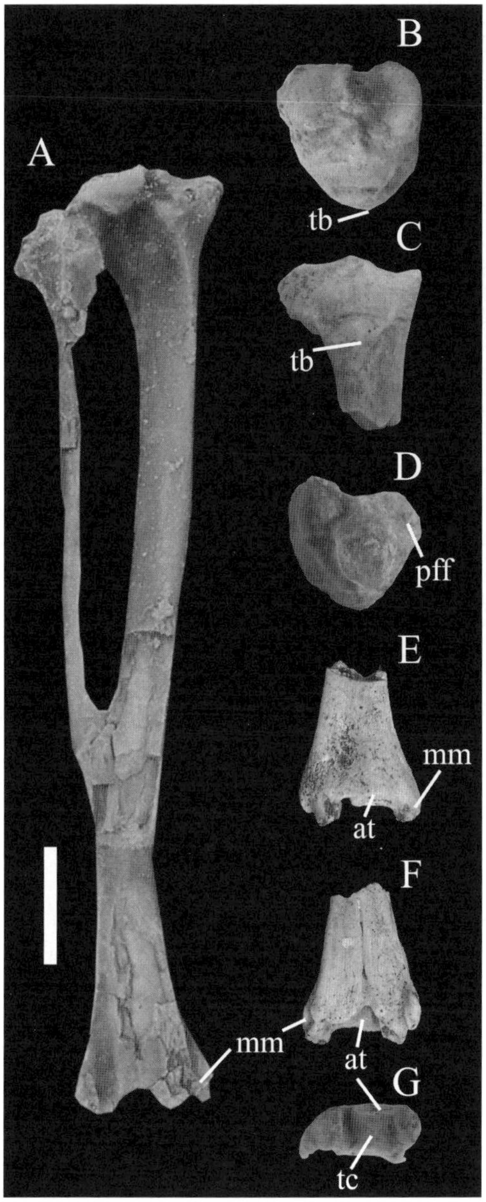

FIGURE 5.4. Tibia-fibula of *Apheliscus*. A. Posterior view of whole tibia-fibula (USNM 495051, reversed; anterior face is encased in matrix). B–D. Proximal tibia (USNM 493903, reversed) in proximal (B), anterior (C), and distal (D) views. E–G. Fused distal tibia-fibula (USNM 493903) in anterior (E), posterior (F), and distal (G) views. Scale bar equals 5 mm.

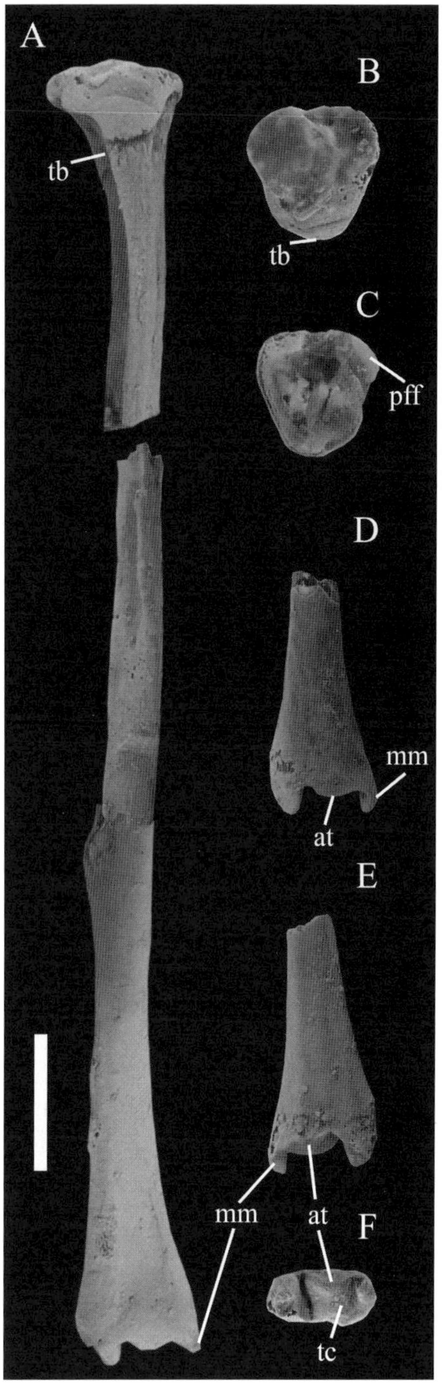

FIGURE 5.5. Tibia-fibula of *Haplomylus*. A, Composite whole tibia-fibula in anterior view (proximal: USNM 513245, reversed; distal: USNM 513868, reversed). Image of USNM 513868 has been digitally straightened to correct for midshaft breakage. B, C, Proximal tibia (USNM 513245, reversed) in proximal (B) and distal (C) views. D–F, Fused distal tibia-fibula (USNM 513239) in anterior (D), posterior (E), and distal (F) views. Scale bar equals 5 mm.

above the medial condyle (Figure 5.4C, 5.5A). The tibial tuberosity is narrow and projects anteriorly. The fibular facet on the proximal tibia is situated just posteroinferior to the lateral tibial condyle (Figures 5.4D, 5.5C). This facet appears to have a small extension onto the lateral surface in *Haplomylus*, but its absolute position on the proximal tibia is the same. There is a prominent and sharp cnemial crest that defines the medial margin of a well excavated lateral fossa for the tibialis anterior muscle. If size is removed as a distinguishing factor, the proximal tibiae of *Apheliscus* and *Haplomylus* are so similar that only the relative size and depth of the proximal fibular facet can reliably differentiate them.

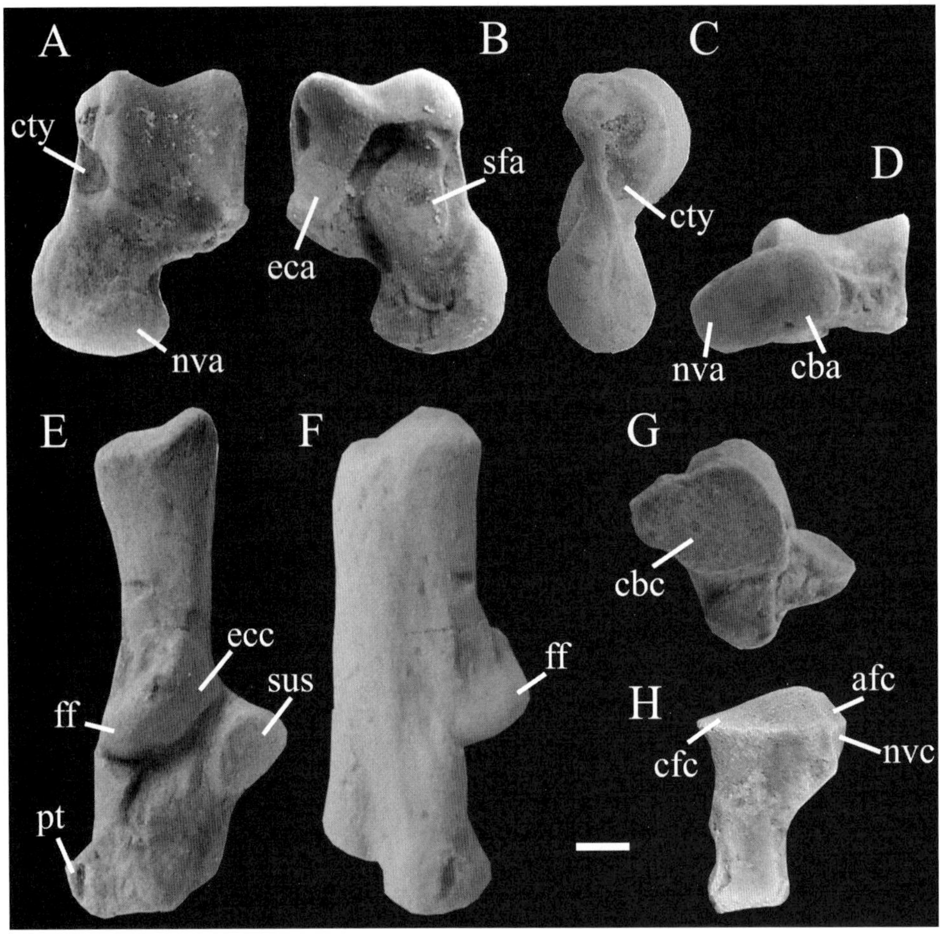

FIGURE 5.6. Tarsal elements of *Apheliscus*. A–D, Astragalus (USNM 521791, reversed) in dorsal (A), ventral (B), medial (C), and distal (D) views. E–G, Calcaneum (USNM 521789) in dorsal (E), lateral (F), and distal (G) views. H, Cuboid (USNM 493903) in dorsal view. Scale bar equals 1 mm.

The tibia is extensively fused to the fibula distally, for slightly more than one-third of the total tibial length in *Apheliscus* (Figure 5.4A). Fusion is more extensive in *Haplomylus*, although the full extent of fusion in this genus cannot be assessed due to a lack of complete elements (Figure 5.5A). Proximal to the point of fibular fusion, the tibial shaft is narrow. Distally, on the posterior surface of the fused distal tibia-fibula of *Apheliscus* there is a narrow cleft marking the line of fusion between the two elements, but there is no trace of the fusion on the posterior surface in *Haplomylus* (Figures 5.4E, F; 5.5D, E). There is a shallow depression at the point of fusion on the anterior surface in both taxa. The anterior surface of the distal tibia continues past the level of the distal articular surface to form an anterior tubercle.

The distal articular surface of the tibia is about twice as wide mediolaterally as anteroposteriorly and deeply excavated (particularly laterally between the distal ends of the tibia and fibula) to accommodate the relief on the astragalar trochlea (Figures 5.4G, 5.5F). In *Haplomylus*, the medial malleolus curves laterally to produce a hook-like structure (Figures 5.4E, F; 5.5D, E); that is, the medial malleolus is slightly torqued so that the articular surface for the astragalus faces slightly anteriorly rather than laterally. In *Apheliscus* the medial malleolus is less gracile (USNM 4930903), but it is not well enough preserved to determine the orientation of the articular surface for the astragalus. The fibular malleolus is laterally expanded, with four proximodistally-oriented ridges on the posterior surface. These processes form sulci for alignment of the peroneus tendons. The presence of these specialized malleoli promotes further stabilization of the crurotarsal joint for parasagittal motion. On the posteromedial surface of the distal tibia in *Apheliscus* there is also a deep sulcus for passage of the tibial flexor tendons, but this excavation is either absent or not preserved in the *Haplomylus* distal tibia.

The free, unfused proximal fibula of apheliscids is slender distally but widens proximally (USNM 495051, 491971).

5.3.2.3 Astragalus

The astragalar trochlea is moderately well grooved, and especially in *Apheliscus*, asymmetrical (Figures 5.6A, D; 5.7A, D). The medial rim of the trochlea is sharp, has a smaller

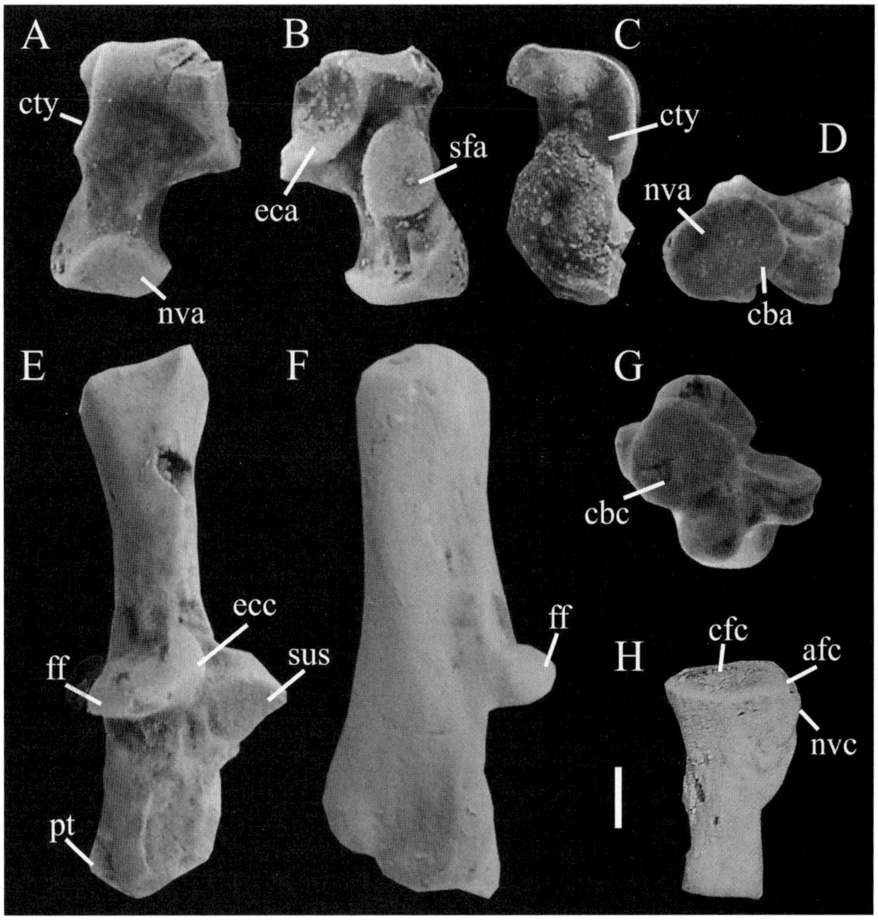

FIGURE 5.7. Tarsal elements of *Haplomylus*. A–D, Astragalus in dorsal (A, USNM 513632), ventral (B, USNM 513632), medial (C, USNM 493902, reversed), and distal (D, USNM 513632) views. E–G, Calcaneum (USNM 493901, reversed) in dorsal (E), lateral (F), and distal (G) views. H, Cuboid (USNM 513668) in dorsal view. Scale bar equals 1 mm.

radius of curvature, and is lower than the lateral rim. The articular surface of the trochlea extends to the posterior margin of the astragalar body, and there is no astragalar foramen. The astragalar body is expanded and excavated medial to the trochlea to form a distinct cotylar fossa for articulation with the medial malleolus of the tibia (Figures 5.6A, C; 5.7A, C; see also Figure 5.13). This cotylar fossa is enlarged in *Haplomylus* to include most of the medial surface of the astragalar body (Figures 5.6C, 5.7C). In contrast to the cotylar fossa in *Apheliscus*, which faces dorsomedially, the cotylar fossa in *Haplomylus* faces medially; i.e., the medial wall of the astragalus is more vertical in *Haplomylus*. The margin of the cotylar fossa is delimited by a bony rim that is most prominent anteromedially. Between the distal margin of the cotylar fossa and the medial extension of the articular surface of the head onto the neck there is a small, dorsomedially-oriented process in *Apheliscus*. The fibular facet, which is deeper dorsoventrally in *Apheliscus*, has a small ventrodistal process on its distal margin; this process is reduced in *Haplomylus*.

Ventrally, the ectal facet is broadest posteromedially, tapers anterolaterally, and is smoothly concave; the ectal facet is wider proximally in *Haplomylus* (Figures 5.6B, 5.7B). Its lateral margin is defined by the anteroventral termination of the fibular facet. The sustentacular facet is essentially circular, slightly convex, and does not meet the navicular facet distally. The astragalar neck of *Apheliscus* is moderately elongate and projects anteromedially away from the astragalar body; the neck may be relatively longer in *Haplomylus* but otherwise similar. The articular surface of the head itself is oblique, with considerable extensions ventrally and (particularly) dorsally along the long axis of the head. The boundary between the navicular and cuboid facets is particularly sharp in *Haplomylus*. A narrow cuboid facet is present in both taxa (Figures 5.6D, 5.7D; see also Figure 5.13). The navicular facet is expanded dorsoventrally to give the astragalar head a rhomboidal shape in distal view. Medially, the facet for the medial tarsal bone and/or spring ligament extends far proximally onto the astragalar neck. The contact between this facet and the navicular facet is gentle in *Apheliscus* and sharper in *Haplomylus*. On the ventral surface of the astragalar head, a small calcaneal facet, confluent with the navicular facet, is present laterally. The long axis of the head is oriented at approximately 45° to the mediolateral axis of the trochlea in dorsal view; that is, the lateral edge of the head is situated farther dorsally than the medial edge at an angle of roughly 45°.

5.3.2.4 Calcaneum

The calcaneal tuber is elongate, and is slightly longer relative to the distal length of the calcaneum in *Haplomylus* than in *Apheliscus* (Figures 5.6E, 5.7E; see also Figure 5.14). The distal calcaneum is not elongate. The ectal facet is unevenly convex, with a noticeable inflection at mid-length such that the posterior half faces almost medially, while the anterior half faces much more distally. The ectal facet in *Apheliscus* faces somewhat dorsally, but in *Haplomylus* the distal portion of the ectal facet faces distally. The proximal portion of the ectal facet does not extend far up onto the tuber. Lateral to the ectal facet is a prominent fibular facet (Figures 5.6F, 5.7F). Posteroventral to the fibular facet is a pit for articulation with the lateral malleolus.

The sustentacular facet is small, gently concave, and ovoid to subtriangular in outline (Figures 5.6E, 5.7E). In *Apheliscus* the sustentacular facet faces dorsally; in *Haplomylus* it faces slightly distally. In apheliscids the sustentacular and ectal facets are nearly in transverse alignment (vs. a sustentacular facet that is situated farther distally along the calcaneal body). At the ventrodistal margin of the calcaneum is a prominent plantar tubercle. Laterally, there is a small, distally situated peroneal tubercle. The cuboid facet is somewhat wider mediolaterally than dorsoventrally and is oriented oblique to the transverse plane (Figures 5.6G, 5.7G). In *Haplomylus* the cuboid facet is relatively smaller, narrower, and possibly more concave than in *Apheliscus*, approximating a sellar joint. Dorsal to the cuboid facet in both taxa is a small astragalar facet.

5.3.2.5 Cuboid

The cuboids of both *Apheliscus* (USNM 493903, 525597, Figure 5.6H) and *Haplomylus* (513668, Figure 5.7H) are gracile and elongate. In dorsal view, the cuboid is mediolaterally constricted distal to the ectocuneiform facet; this distal constriction is more obvious in *Apheliscus*, as the body of the cuboid in *Haplomylus* is overall relatively narrower mediolaterally. The calcaneal facet is gently helical, while the astragalar facet is oriented proximomedially. The navicular facet is flat and faces medially. The ectocuneiform facet is large, faces mediodistally, and includes a small continuation of the navicular facet dorsally. The distal (metatarsal) facet is subtriangular in *Apheliscus* (shortest side dorsally) except for a shallow notch on the medial side. In *Haplomylus*, the metatarsal facet is semilunar with a deeper medial notch. The long plantar tubercle is large, rhomboidal in ventral view, well separated from the calcaneal facet and metatarsal facets, and projects distinctly from the body of the cuboid.

5.4 Functional Interpretations

5.4.1 Forelimb Function

The functional signal implied by the apheliscid forelimb is consistent with specialized cursorial (or saltatorial) locomotion (Gambaryan, 1974; Berman, 1985). In fact, both modes of progression may have been used, as occurs in some extant small mammals (Berman, 1985; Emerson, 1985; Bramble, 1989; Fischer, 1994; Stein and Casinos, 1997). The proximal humerus provides some indication of rapid flexion and extension at the shoulder joint, as the deltopectoral crest is relatively low and proximally located. The oval shape and comparatively small articular area of the humeral head also imply a limited range of movement at the shoulder. The distal humerus suggests a substantial range of extension (e.g., deep coronoid and olecranon fossae), with simultaneous restriction to predominantly parasagittal motion (narrow distal end, sharp trochlea, rounded capitulum). The proximal radius of *Apheliscus* suggests a limited ability to supinate (flat ulnar facet, distinct capitular eminence), consistent with terrestrial habits (Taylor, 1974; Heinrich and Rose, 1997; Argot, 2001, 2003). In addition, the articular surface of the radial head is nearly perpendicular to the radial shaft, also indicative of terrestrial locomotion (see Heinrich and Rose, 1997). Finally, the small medial epicondyle and weak supinator crest indicate reduced attachment sites for forelimb muscles associated with complex movements of the forearm and manus (Taylor, 1974; Argot, 2001, 2003, 2004).

The relatively small size of the forelimb elements in comparison with hind limb elements suggests that much of the thrust in locomotion was generated by the hind limbs, which may favor a hopping or bounding gait over running (Berman, 1985; Offermans and de Vree, 1987). However, the forelimb is not as small as in bipedally hopping (ricochetal) mammals such as *Pedetes* or *Dipodoides*. Moreover, forelimb elements show modifications for rapid terrestrial locomotion (Heinrich and Rose, 1997; see also Gebo and Rose, 1993). In ricochetal forms, in which the forelimb does not participate in locomotion, the forelimb is more apt to be specialized for mobility to allow manipulation of objects, to dig, or to perform other tasks (see Offermans and de Vree, 1988).

The forelimbs of *Apheliscus* and *Haplomylus* differ somewhat in their degree of specialization for rapid terrestrial locomotion, with *Haplomylus* showing slightly greater specialization in the latter direction. The distal humerus of *Apheliscus* is slightly broader than that of *Haplomylus* due to a more prominent supinator crest and medial epicondyle. While the coronoid and olecranon fossae of *Apheliscus* are deep, there is no supratrochlear foramen, as is present in *Haplomylus*. The actual differences in the depths of the fossae, however, are minimal, suggesting that the range of extension at the elbow in *Apheliscus* was only slightly less than in *Haplomylus*.

5.4.2 Hind Limb Function

The morphology of the hind limb also indicates specialization toward rapid terrestrial locomotion in apheliscids. The high greater trochanter provides a long level arm for the deep gluteals. The height of the greater trochanter would also limit the potential range of thigh abduction (Taylor, 1976; Rose, 1999; Argot, 2002, 2003, 2004). The lesser trochanter projects posteromedially – a common configuration in specialized cursors and saltators (Taylor, 1976; Rose, 1999; Argot, 2002) – suggesting an emphasis on fore-aft flexion of the hip. Together, the morphology of the femoral trochanters is consistent with ability for rapid flexion and extension of the hip through the increase of lever arm lengths. The small size of the femoral head, its medial position on a distinct neck, and its restricted articular surface would further limit the range of mobility at the hip joint in favor of primarily parasagittal motion, while simultaneously maintaining the ability for rapid flexion and extension. The distal end of the femur is anteroposteriorly deep with anteriorly extensive articular surfaces, suggesting that apheliscids could extend the knee to a considerable degree and were not necessarily restricted to a habitually flexed position.

The relatively short femur compared to the tibia-fibula is similar to cursors, and is consistent with the pattern of lengthening distal limb elements to increase stride (Taylor et al., 1974; Hildebrand, 1985; Argot, 2004; see also Berman, 1985). The large, deeply excavated attachment site for m. tibialis anterior on the lateral side of the proximal tibia suggests powerful dorsiflexion of the ankle; this feature occurs in cursorial taxa (e.g., lagomorphs, macroscelideans), as well as fossorial (e.g., dasypodids) and semi-aquatic taxa (e.g., castorids, less so in *Lutra*). Distal fusion of the tibia and fibula is an adaptation to stresses on the distal crus, and, while present in a number of cursorial taxa (e.g., lagomorphs, artiodactyls, macroscelideans, many rodents), is also characteristic of diggers, some semiaquatic and aquatic mammals, as well as leapers (e.g., dasypodids, castorids, erinaceids, talpids, soricids, pinnipeds) (Barnett and Napier, 1953; Argot, 2002). The elongate, gracile nature of the apheliscid hind limb argues for cursorial habits, as cursors typically have long, gracile limbs to increase stride length while minimizing limb mass, while fossorial and semi-aquatic taxa tend to have much more robust and relatively shorter limb elements (Taylor et al., 1974; Hildebrand, 1985; Argot, 2004).

Both apheliscids possess deeply grooved astragali that restrict the range of motion at the crurotarsal joint to the parasagittal plane. The astragalus of *Haplomylus* in particular resembles those of modern lagomorphs, rodents, canids, and especially macroscelideans. The distal fusion of the tibia and fibula, in concert with the greater relief of the astragalar trochlea and more prominent development of the medial and lateral malleoli, allow the formation of a hinge joint at the crurotarsal articulation; that is, only parasagittal motion is permitted at the articulation of the fused tibia-fibula and the astragalus. The distinctive anterior tubercle on the distal tibia contacts the anterodistal-most extension of the articular surface of the trochlea when in articulation, perhaps preventing hyperflexion of the ankle or stabilizing the ankle when dorsiflexed. Although well developed in *Apheliscus*, the anterior tubercle on the distal tibia is even more prominent in *Haplomylus*.

The cotylar fossa, which contacts the medial malleolus, helps to stabilize the ankle joint, particularly in dorsiflexion, and also serves to help prevent hyperdorsiflexion. The latter is especially the case in the *Apheliscus*, in which the cotylar fossa faces dorsomedially. At the same time, continuation of the trochlear articular surface proximoventrally implies an increased range of plantarflexion at the ankle joint. Stabilization for parasagittal motion, enhancement of the range of plantarflexion at the crurotarsal joint, and adaptations to prevent hyper-dorsiflexion together strongly suggest specialized saltatorial or cursorial locomotion in apheliscids (see Taylor, 1976; Szalay, 1985; Argot, 2002, 2003, 2004).

The other tarsal joints suggest a general restriction of motion to a single plane of action. The astragalocalcaneal, astragalonavicular, naviculocuboid, and calcaneocuboid articulations are tight and permit very little or no motion, while the presence of an alternating tarsus supports the emphasis on tarsal stability. Most motion in the ankle took place at the crurotarsal joint, although some parasagittal motion at the transverse tarsal joint (possibly accompanied by slight rotation) may have also been possible. This restriction of movement at the transverse tarsal joint is particularly evident in *Haplomylus* in which many of the tarsal articular surfaces are flat or angular (e.g., the articular surfaces of the astragalar head). The more gently curved ectal facets of *Apheliscus* suggests that a small amount of motion may have been possible at the astragalocalcaneal joint in this taxon. Restriction of tarsal mobility mainly to parasagittal motion at the crurotarsal and transverse tarsal joints is consistent with cursorial specialization (Taylor, 1976; Szalay, 1985; Argot, 2002, 2003, 2004), while the tight articulations at all tarsal joints suggest further stabilization of the tarsus for rapid terrestrial locomotion (Taylor, 1976; Szalay, 1985; Argot, 2002).

Every character indicative of rapid locomotion in *Apheliscus* is also present in *Haplomylus*, but many suggest greater specialization for speed *Haplomylus*. This corroborates the signal apparent in apheliscid forelimbs, which also indicate somewhat greater specialization in *Haplomylus*. Both taxa, however, show clear specialization for rapid terrestrial locomotion.

5.5 Comparisons to Other Taxa

As described above, the postcranial morphology of *Apheliscus* and *Haplomylus* indicates that these two small-bodied forms were adapted for rapid terrestrial locomotion. Detailed comparisons to contemporary taxa showing similar adaptations, particularly those of similar body size, are warranted for two reasons. First, such comparisons will hopefully be of use in distinguishing isolated apheliscid elements from those of contemporary forms that are at least superficially similar in morphology. Second, comparisons will help evaluate whether these similarities represent evidence of recent common ancestry. In addition to comparisons with taxa that show postcranial similarities to apheliscids, comparisons should also be made to taxa that show sufficient dental similarities to *Apheliscus* and/or *Haplomylus* to have been postulated to share a recent common ancestry with one or both taxa.

Based on these criteria, the most relevant taxa (either by hypothesized phylogenetic affinity, functional similarity, or both) for detailed comparison to apheliscids include *Hyopsodus*, pentacodontids, Eocene leptictids, *Macrocranion*, and extant macroscelideans. On the basis of dental morphology, previous assessments of the relationships of *Apheliscus* and *Haplomylus* have suggested affinities to hyopsodontids in the broad sense (Matthew, 1915; Simpson, 1937; McKenna, 1960; McKenna and Bell, 1997) and, in the case of *Apheliscus*, pentacodontids (Gazin, 1959), making comparisons to both groups appropriate. Unfortunately, postcranial remains of pentacodontids have not been described, aside from an abstract reporting on new skeletons of the pentacodontid *Aphronorus* (Boyer and Bloch, 2003). This brief account indicates similarities to the basal palaeanodont *Escavadedon* (Boyer and Bloch, 2003) and generally suggests a robust, fossorially adapted animal, quite different from the taxa described here; however, detailed comparisons to pentacodontids must await full description of this material. In the case of hyopsodontids, *Hyopsodus* remains the only other hyopsodontid for which substantial postcrania have been described (Gazin, 1968), and is therefore the logical choice for comparisons between *Apheliscus, Haplomylus*, and other hyopsodontids.

Comparisons to leptictids and *Macrocranion* are warranted because the postcranial morphology of these two contemporaries of *Apheliscus* and *Haplomylus* suggest locomotor adaptations similar to those reconstructed for apheliscids (Storch, 1993, 1996; Godinot et al., 1996) and because dental and postcranial remains indicate relatively similar body sizes. While apheliscid postcranial elements are at least superficially similar to those of Eocene leptictids, their dentitions are quite different and, to our knowledge, a close relationship between these taxa has never been suggested (cf. the relative positions these taxa in McKenna and Bell, 1997). Unlike apheliscids, leptictids have comparatively sectorial teeth with characteristically tall trigonids, narrow talonids, transverse upper molars, and low hypocones (see Figure 5.8). This configuration is quite dissimilar from the bunodont teeth of apheliscids, which have lower trigonids, broader talonids, more quadrate upper molars, and stronger, more distinct hypocones. With no dental evidence for a close phylogenetic relationship, therefore, leptictids should serve as a useful example of postcranial convergence with apheliscids. On the other hand, several authors have used dental morphology to suggest potential phylogenetic affinities between some hyopsodontid condylarths, as traditionally defined, and some putative erinaceomorph insectivores, including *Macrocranion* (Russell, 1964; Russell et al., 1975; Rigby, 1980; Bown and Schankler, 1982) (see Figure 5.8), making comparisons to the latter taxon even more pertinent. Although clear differences do exist between the dentition of *Macrocranion* and those of apheliscids (e.g., the salient postmetacrista on *Macrocranion* upper molars), the dentitions of *Macrocranion* and apheliscids are both small, relatively bunodont, with low trigonids, inflated metaconids and protoconids, reduced paracristids, and distinct hypocones (Rigby, 1980). Finally, postcranial comparisons to modern macroscelideans are included to illustrate more clearly the numerous similarities to apheliscids, which, in combination with previously recognized dental similarities (Simons et al., 1991) strongly suggest a close phylogenetic relationship between these two taxa (Zack et al., 2005a). For comparison, Figure 5.8A illustrates the dentition of the Eocene macroscelidean *Chambius*.

Beyond the above taxa, detailed comparisons between the postcrania of apheliscids and those of other early Tertiary ungulates (artiodactyls, perissodactyls, arctocyonids, phenacodontids, mesonychids) will not be explored in detail here for several reasons. First, aside from the loose association of all being ungulates, no other close phylogenetic relationship has been hypothesized between apheliscids on one hand and any of these groups on the other, making direct comparisons less appropriate from a phylogenetic standpoint. Second, early Eocene representatives of these groups are all much larger than their contemporary apheliscids, making direct functional comparisons between these taxa and apheliscids less applicable.

Among early Tertiary ungulates, generalized features consistent with rapid terrestrial locomotion are found in artiodactyls (Franzen, 1981, 1988; Rose, 1982, 1985; Thewissen and Hussain, 1990; Erfurt, 2000), perissodactyls (Kitts, 1956; Rose, 1990, 1996), phenacodontids (Radinsky, 1966; Thewissen, 1990; Williamson and Lucas, 1992), and mesonychids (Zhou et al., 1992; O'Leary and Rose, 1995), unlike the arboreal or scansorial morphology of some other early ungulates (e.g., arctocyonids or periptychids) (Matthew, 1937; Russell, 1964; Rigby, 1981; Rose, 1987). These features include: a long, gracile humerus with a high greater tuberosity, and a low, proximally-restricted deltopectoral crest; a narrow distal humerus with a reduced medial epicondyle, prominent trochlea and capitulum, and deep (or perforate) olecranon/coronoid fossae; a proximal radius with a subrectangular head, and high capitular eminence; a long, gracile femur with a high greater trochanter, a small, spherical head, and a long neck; a narrow, deep distal femur with a long patellar trochlea; a narrow and deep tibial plateau; a sharp cnemial crest; tight articulation at the crurotarsal joint promoted by a deeply grooved distal tibia and astragalar trochlea; elongation and/ or

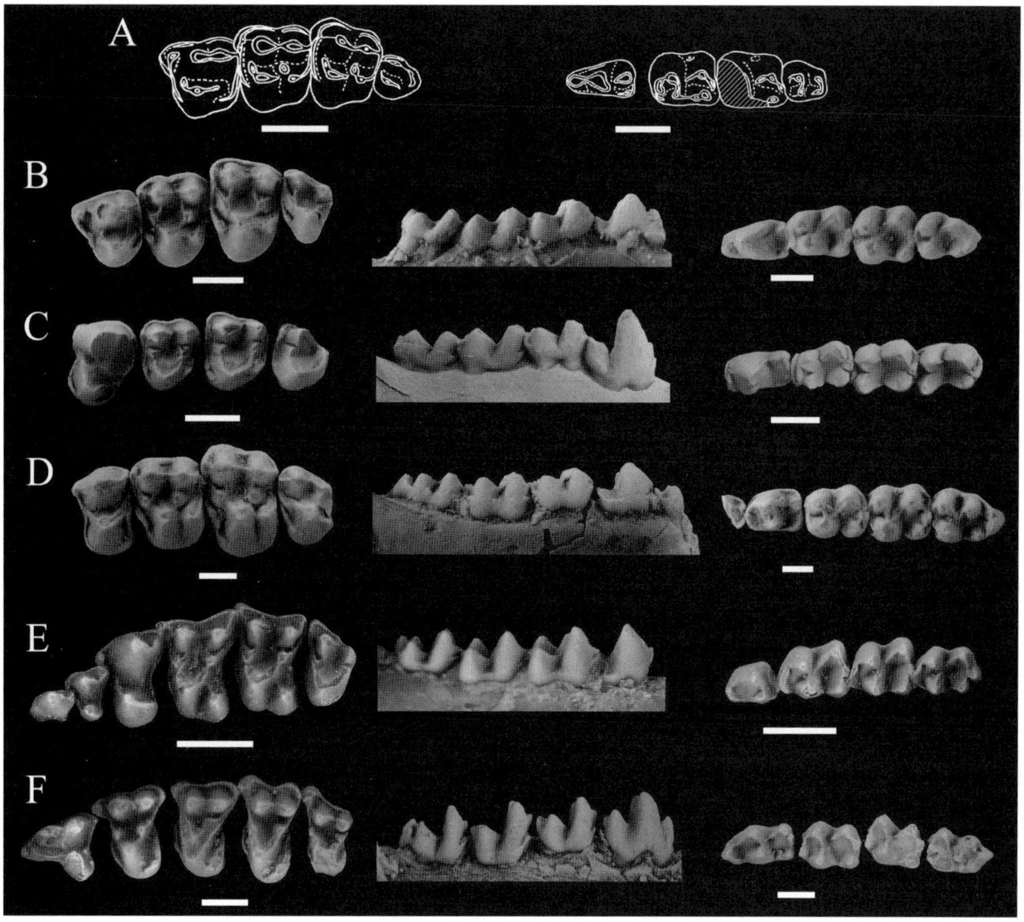

FIGURE 5.8. Comparison of apheliscid dentitions to other relevant taxa. Except for A, the left column illustrates the upper cheek dentition in occlusal view, the center column illustrates the lower cheek dentition in buccal view, and the right column illustrates the lower cheek dentition in occlusal view. In A, there is no buccal view of the lower dentition. Except where noted, illustrations are of left P4-M3 and right p4-m3. A, *Chambius* (Macroscelidea): modified from Hartenberger (1986). B, *Haplomylus*: USNM 493936 (uppers) and USNM 521645 (lowers, p2-3 present but not illustrated). C, *Apheliscus*: USNM 494896 (uppers) and USGS 12608 (lowers, reversed). D, *Hyopsodus*: USNM 521652 (uppers) and USNM 521661 (lowers, partial p3 included). E, *Macrocranion*: USNM 509582 (P2-M3, reversed) and USNM 495560 (lowers). F, *Palaeictops* (Leptictidae): USGS 9160 (P3-M3) and USGS 308 (lowers, reversed). All scale bars equal 2 mm.

compaction of the tarsal joints. These features, however, are found in most cursorial or saltatorial mammals (Taylor, 1974, 1976; Hildebrand, 1982; Gebo and Rose, 1993; Heinrich and Rose, 1997; Argot, 2002, 2003; Rose and Chinnery, 2004), and their presence does not argue persuasively for a close phylogenetic relationship.

In contrast to these vague similarities, there are numerous dissimilarities, particularly in details of tarsal morphology, which argue against a close phylogenetic affinity between apheliscids and these other taxa. With the exception of the derived phenacodontid *Meniscotherium* (Williamson and Lucas, 1992), none of these taxa has a cotylar fossa, and none shows other features such as extensive fusion of the tibia and fibula or elongation of the cuboid. In turn, apheliscids lack the distinctive features, particularly in the tarsus, that characterize artiodactyls (e.g., "double-pulley" astragalus: Schaeffer, 1947) or perissodactyls (e.g., short astragalar neck with a saddle shaped navicular facet: Radinsky, 1966). In the case of both mesonychids and phenacodontids, even some of the more generalized similarities are lacking in the earliest members of these groups (e.g., *Ankalagon* and *Tetraclaenodon*) (Matthew, 1937) suggesting independent development of cursorial ability. Taken in combination with well-documented differences in size and dental morphology (e.g., Cifelli, 1983 and Archibald, 1998), it seems safe to conclude that the general postcranial similarities shared by apheliscids, artiodactyls, perissodactyls, mesonychids, and phenacodontids reflect similar locomotor strategies, rather than suggesting a close phylogenetic relationship.

5.5.1 *Hyopsodus*

As supposed hyopsodontids, apheliscids are compared to *Hyopsodus* – the type genus of the family Hyopsodontidae

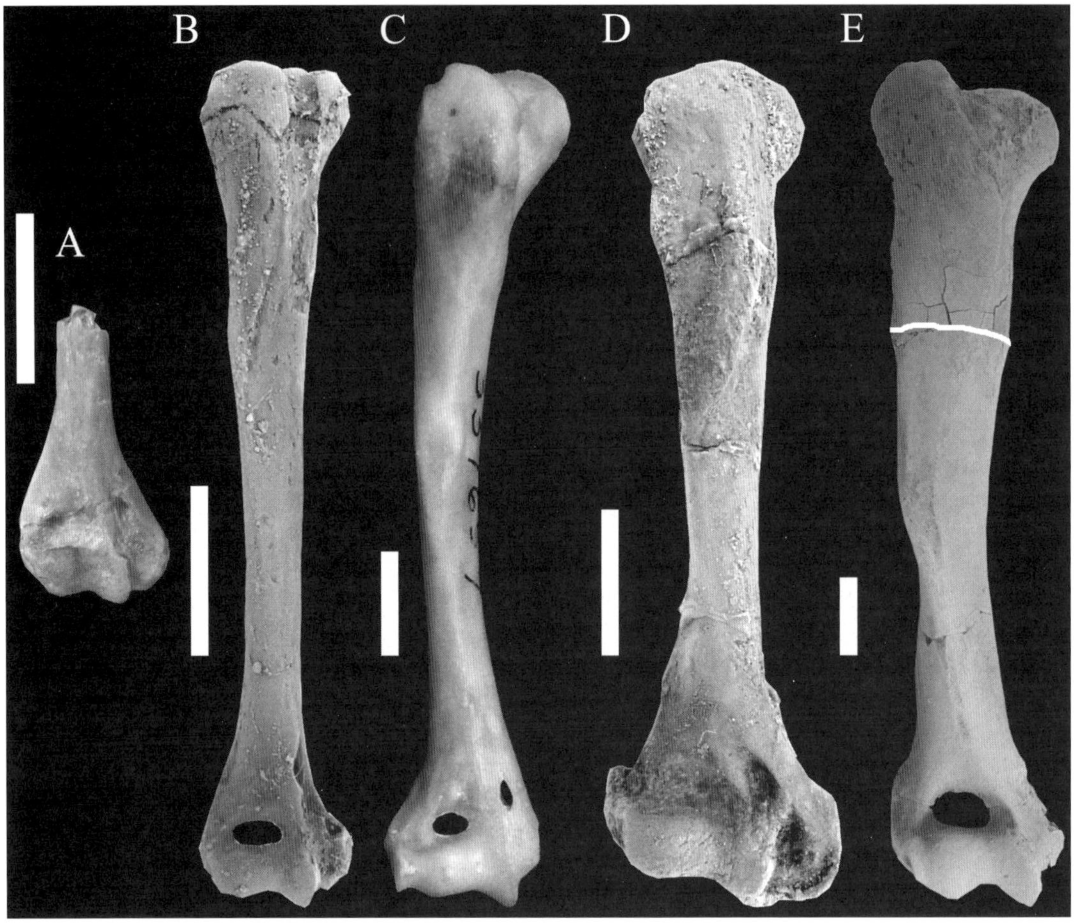

FIGURE 5.9. Comparison of apheliscid humeri to other taxa in anterior view. A, *Apheliscus* (USNM 493903). B, *Haplomylus* (USNM 513512). C, *Rhynchocyon* (USNM 537657). D, Eocene Leptictidae (UM 88105). E, *Hyopsodus* (proximal: USNM 25179; distal: USNM 493816, reversed). All scale bars equal 5 mm.

– to emphasize the substantial differences between these taxa. The postcranium of *Hyopsodus*, described and illustrated by Matthew (1915), Gazin (1968), and Godinot et al. (1996), is more distinct from that of apheliscids than the similar dentitions of the taxa would imply. Apheliscids and *Hyopsodus* do share a few noteworthy postcranial similarities that suggest some capacity for rapid terrestrial locomotion in *Hyopsodus*. The greater tuberosity on the humerus of *Hyopsodus* projects well above the humeral head (Figure 5.9E), as in apheliscids, although *Hyopsodus* shows even greater development of this feature (Gazin, 1968). This morphology would have increased the lever mechanical advantage at the shoulder, and is associated with cursorial (or at least terrestrial) progression. In the distal humerus, the capitulum is rounded and the trochlea is sharp in both the apheliscids and in *Hyopsodus*. A supratrochlear foramen is present in *Hyopsodus*, similar to *Haplomylus* but in contrast to *Apheliscus*, and all three taxa have a deep olecranon fossa indicating a comparable capacity for considerable extension and consistent with cursorial habits. On the proximal femur, the lesser trochanter of *Hyopsodus* is similar to apheliscids in pointing posteromedially (Gazin, 1968) as is typical of terrestrial (particularly cursorial) taxa. The distal end of the *Hyopsodus* femur is anteroposteriorly deep (compared to its width), as is the patellar trochlea. In both cases, the morphology of *Hyopsodus* is similar to, but not as strongly developed, as in apheliscids and is again consistent with incipient cursorial habits.

Contrasting with these limited similarities is a much more extensive suite of differences that suggest that the locomotor repertoire for *Hyopsodus* was very different from apheliscids. At a gross level, the long bones of *Hyopsodus* are much more robust than those of apheliscids, and a size disparity (seen in apheliscids) between forelimb and hind limb elements is not evident. This is true of both large (e.g., YPM uncat.) and small-bodied *Hyopsodus* (e.g., USNM 23740), the latter being similar in size to *Apheliscus* and *Haplomylus*.

On the humerus, the greater tuberosity of *Hyopsodus* is displaced anteriorly from the head (not the case in apheliscids), presumably allowing for greater abduction at the shoulder in *Hyopsodus*, resulting in a more mobile shoulder that is more typical of arboreal or scansorial than cursorial mammals. The deltopectoral crest of *Hyopsodus*, sharper

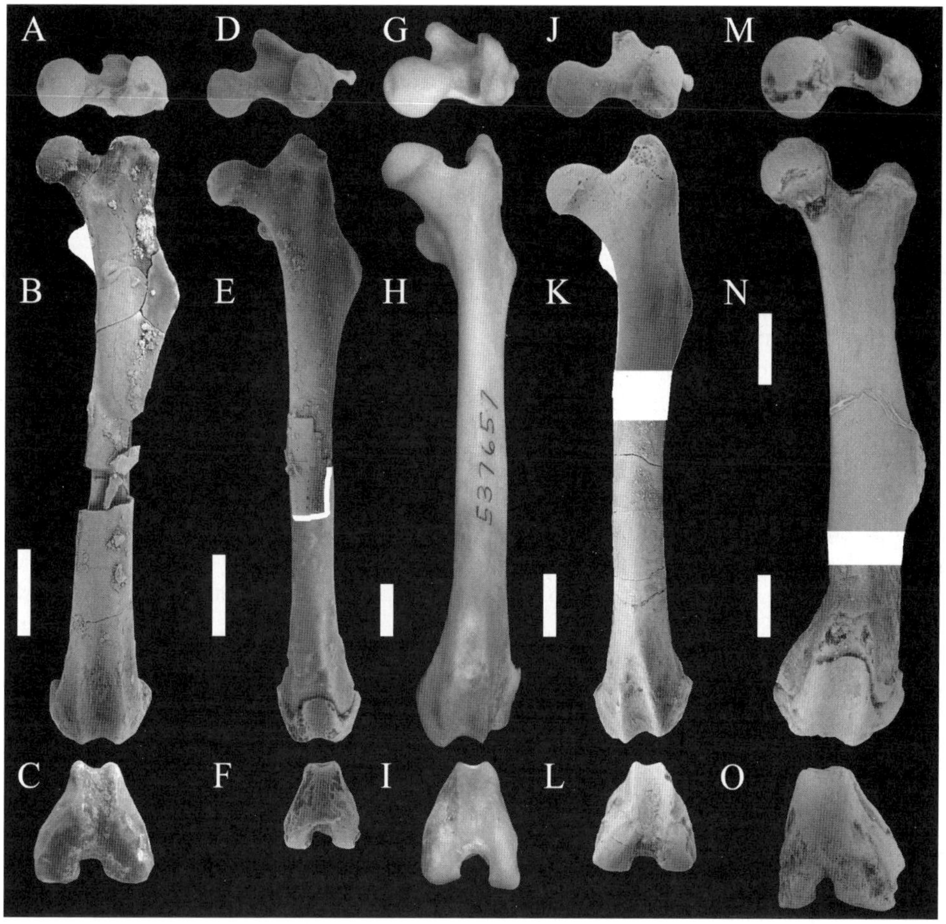

FIGURE 5.10. Comparison of apheliscid femora to other taxa. A–C. *Apheliscus*: A, proximal (USNM 488325); B, anterior (USNM 488326); C, distal (USNM 493819, reversed). D–F. *Haplomylus*: D, proximal (USNM 513057); E, anterior (proximal: USNM 513057; distal shaft: USNM 513140; distal epiphysis: USNM 513173); F, distal (USNM 513173). G-I. *Rhynchocyon* (USNM 537657): G, proximal; H, anterior; I, distal. J–L. Eocene Leptictidae: J, proximal (USNM 493932); K, anterior (proximal: USNM 493932; distal: UM 88105); L, distal (USNM 495152). M–O. *Hyopsodus*: M, proximal (USNM 17980, reversed); N, anterior (proximal: USNM 17980, reversed; distal: YPM uncat.); O, distal (YPM uncat.; same as in N). Portions in white denote reconstructions based on more complete but poorly preserved elements; all scale bars equal 5 mm.

and higher than in *Haplomylus*, extends distally past the midpoint of the shaft, and resembles small, slow-moving terrestrial taxa such as *Solenodon* and *Erinaceus* (Taylor, 1974; Rose, 1999). The supinator crest that is better developed than in apheliscids and the medial epicondyle is larger, indicating that the forearm and hand were better adapted for manipulation in *Hyopsodus*.

The femoral head of *Hyopsodus* is relatively larger than that of apheliscids (Figure 5.10M, O). The neck is short and almost vertically oriented in *Hyopsodus*, in contrast to the longer, medially oriented neck in apheliscids. Unlike apheliscids, the greater trochanter is lower than the head, and it is more widely separated from the head mediolaterally. The trochanteric fossa is shallower in *Hyopsodus* than in apheliscids, making the whole proximal femur appear anteroposteriorly compressed. The low greater trochanter and shallow trochanteric fossa indicate that the gluteal muscles and lateral rotator muscles in *Hyopsodus* likely had low mechanical advantage, with the possible exception of the superficial gluteal, which had a comparatively distal insertion on the third trochanter. The lateral displacement of the greater trochanter suggests that *Hyopsodus* could abduct its thighs to a considerable degree. These characters are generally consistent with fossorial or arboreal habits (see Rose et al., 1991). As alluded to above, the prominent, posterolaterally-projecting third trochanter of *Hyopsodus* is situated one-half to one-third of way down the shaft. The third trochanter in apheliscids is more proximally located (i.e., no more than one-quarter of the length from the proximal end). This suggests an emphasis on increasing the relative mechanical advantage of the superficial gluteal muscles in *Hyopsodus* in a pattern similar to some fossorial taxa (Dasypodidae, Tubulidentata, Pholidota, *Plesiorycteropus*: see MacPhee, 1994, Figures 5.33–5.35). While the distal femora of *Hyopsodus* and apheliscids are generally similar, the articular surfaces of the femoral condyles are posteriorly restricted in *Hyopsodus*, indicating a limited range of extension at the knee and a habitually flexed hind limb posture.

The tibia of *Hyopsodus* is shorter than the femur, resulting in a low crural index that would be atypical of a specialized cursor. The proximal tibia of *Hyopsodus* is mediolaterally broad, unlike the anteroposteriorly deeper proximal tibiae of apheliscids (Figure 5.11I, J), while the cnemial crest is lower and less sharp than in apheliscids, and the shaft of the tibia is slightly bowed medially. In a particularly striking contrast to apheliscids, the tibia and fibula of *Hyopsodus* are separate along their entire length. Finally, the distal articular surface of the tibia in *Hyopsodus* is flat and is inclined proximolaterally to distomedially, unlike the transverse, deeply grooved articular surface in apheliscids (Figure 5.12E). Overall, in contrast to the numerous specializations for running or leaping in apheliscids, the tibia of *Hyopsodus* is very generalized and is similar to, but more robust than, the crus of the scansorial *Tupaia tana*.

The astragalus of *Hyopsodus* is dissimilar from those of apheliscids in most respects. Major differences from apheliscids include: virtually flat astragalar trochlea; equivalent lengths of the medial and lateral rims of the trochlea; retention of astragalar foramen and restriction of the trochlear articular surface to the dorsodistal portion of the astragalar body; large but poorly defined fibular facet; reduced lateral process; absence of cotylar fossa; broader ectal facet; larger, flatter, and more circular sustentacular facet; the sustentacular facet merges into cuboid facet; short astragalar neck; subspherical astragalar head with long axis transverse; proximomedial extension of articular surface of head onto neck absent; relatively smaller cuboid facet (Figure 5.13).

As with the astragalus, *Hyopsodus* calcaneal morphology differs significantly from that of apheliscids. Differences include: generally shorter and more robust calcaneum, particularly the tuber; less convex ectal facet, facing more dorsally and lacking a sharp turn at midlength; long axis of ectal facet long axis oriented proximodistally; larger fibular facet that parallels the ectal facet and faces laterally rather

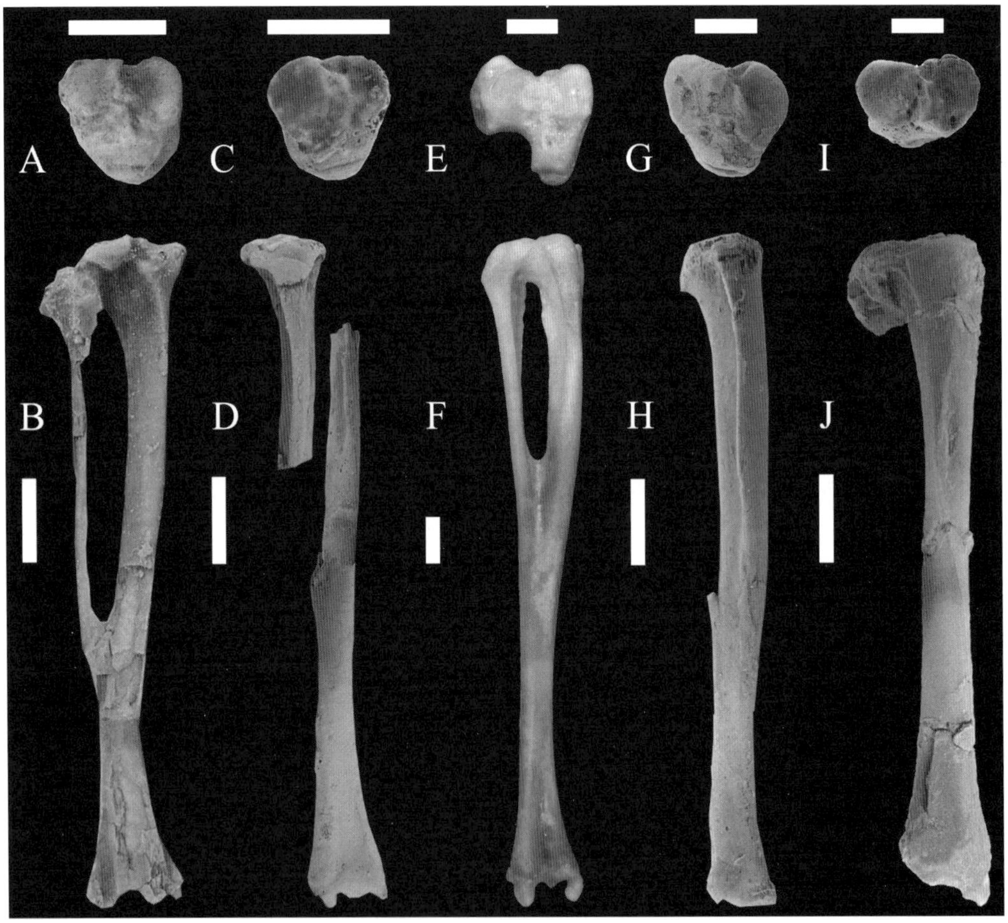

FIGURE 5.11. Comparison of apheliscid tibia-fibulae to other taxa. A, B. *Apheliscus*: A, proximal (USNM 493903, reversed); B, posterior (USNM 495051, reversed). C, D. *Haplomylus*: C, proximal (USNM 513245, reversed); D, anterior (proximal: 513245, reversed; distal: USNM 513868, reversed and digitally straightened to correct for midshaft breakage). E, F. *Rhynchocyon* (USNM 537657, reversed): E, proximal; F, anterior. G, H. Eocene Leptictidae: G, proximal (DMNH 29264); H, anterior (USNM 513235). I, J. *Hyopsodus*: I, proximal (USNM 493816); J, anterior (USNM 23740, reversed). All scale bars equal 5 mm.

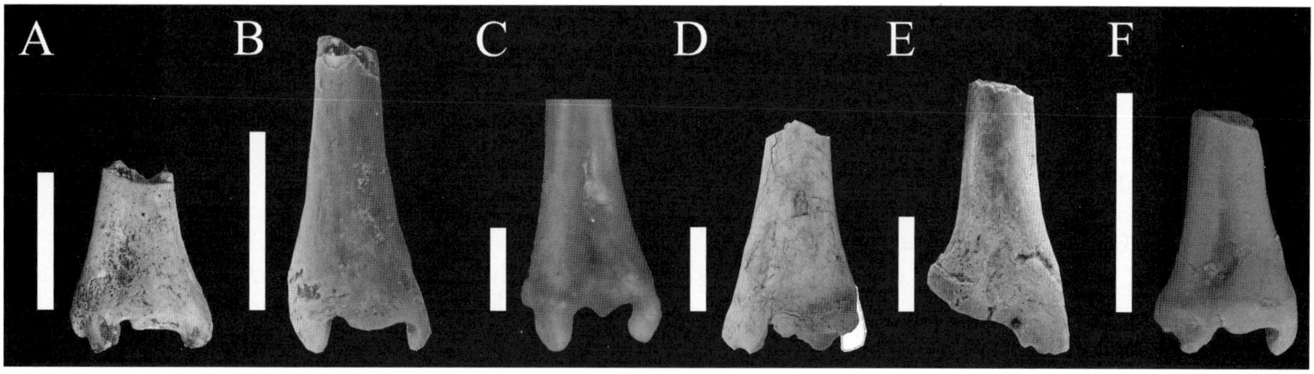

FIGURE 5.12. Comparison of apheliscid distal tibia-fibulae to other taxa. A, *Apheliscus* (USNM 493903). B, *Haplomylus* (USNM 513239). C, *Rhynchocyon* (USNM 537657, reversed). D, Eocene Leptictidae (USNM 493935; medial malleolus reconstructed from an uncatalogued specimen from Rose Quarry [D-1460]). E, *Hyopsodus* (USNM 493816, reversed). F, *Macrocranion* (UCMP uncat., reversed). All scale bars equal 5 mm.

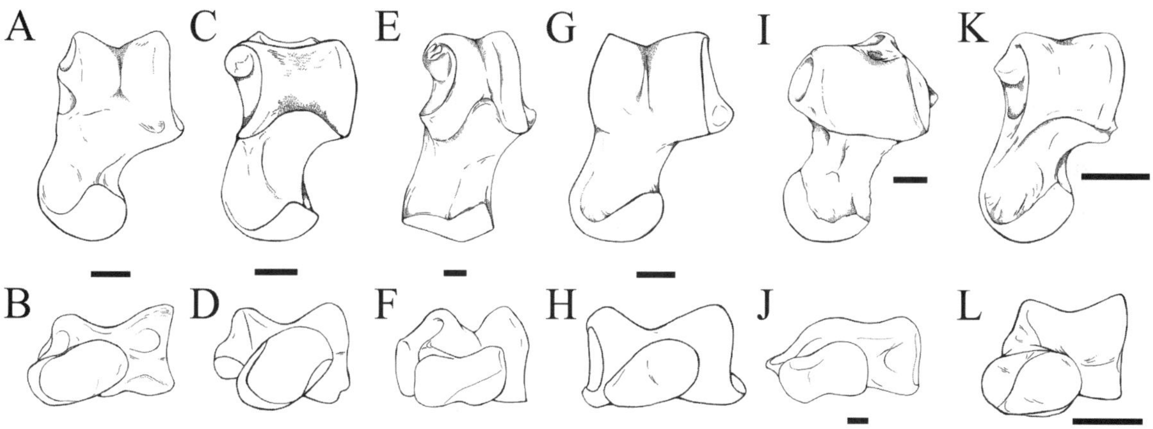

FIGURE 5.13. Comparison of apheliscid astragali to other taxa in dorsal and distal views. A, B. *Apheliscus* (USNM 521791, reversed): A, dorsal; B, distal. C, D. *Haplomylus*: C, dorsal (composite of USNM 493902 and 513632); D, distal (USNM 513632). E, F. *Rhynchocyon* (USNM 537657): E, dorsal; F, distal. G, H. *Prodiacodon* (USNM 513636, reversed: G, dorsal; H, distal. I, J. *Hyopsodus*: I, dorsal (USNM 25331, reversed; details from USNM 493782 and YPM unnumbered); J, distal (YPM unnumbered). K, L. *Macrocranion*: K, dorsal (UCMP uncataloged); L, distal (USNM 493780). All scale bars equal 1 mm.

than proximodorsally; larger, rounder sustentacular facet; distal extension of sustentacular facet on the medial aspect of the calcaneal body present; cuboid facet transversely narrower, circular, more concave, and oriented perpendicular to the long axis of the calcaneum; larger peroneal tubercle that is proximally expanded via a bony shelf on the lateral aspect of the calcaneum (Figure 5.14).

The apheliscid cuboid is relatively much more elongate and gracile than that of *Hyopsodus* (Figure 5.15D). As illustrated by Gazin (1968), the *Hyopsodus* cuboid is comparatively short and cylindrical. There is no distal constriction of the cuboid in *Hyopsodus*. The metatarsal facet in apheliscids is relatively much smaller than the proximal articular surface, unlike the condition in *Hyopsodus*, in which the proximal and distal ends of the cuboid are roughly comparable in size.

Overall, the tarsal morphology of *Hyopsodus* is consistent with substantial mobility at the tibiotarsal, astragalocalcaneal, and transverse tarsal joints, as opposed to the restriction of motion to the parasagittal plane in the ankles of apheliscids. The tibiotarsal joint of *Hyopsodus* shows features that would permit mediolateral rotation (flat distal tibia and astragalar trochlea) while limiting plantarflexion (posteriorly restricted trochlea). Features of the astragalocalcaneal articulation on *Hyopsodus* (i.e., large ectal and sustentacular facets, smoothly convex calcaneal ectal facet) also imply some freedom of movement and potential for inversion and eversion, while the rounded, transverse astragalar head suggests that motion at the transverse tarsal joint was not restricted to the parasagittal plane. In contrast, the apheliscid tarsus indicates greater restriction of possible motions to the parasagittal plane.

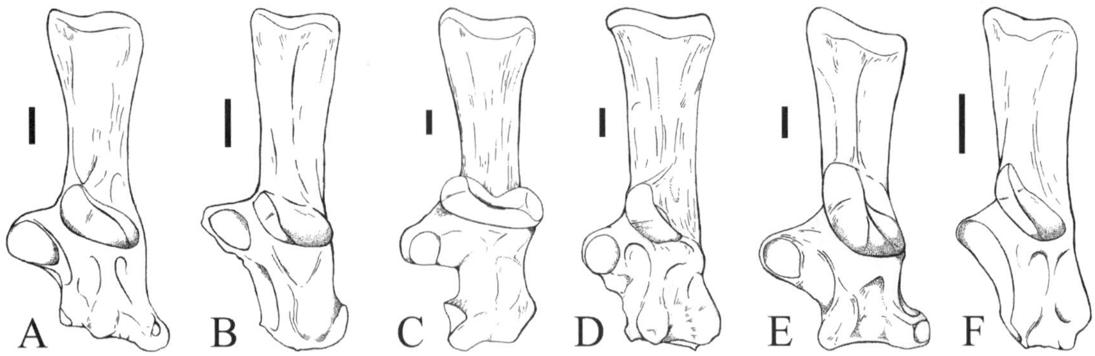

FIGURE 5.14. Comparison of apheliscid calcanei to other taxa in dorsal view. A, *Apheliscus* (USNM 521789, reversed). B, *Haplomylus* (USNM 513655, reversed). C, *Rhynchocyon* (USNM 537657). D, cf. *Prodiacodon* (USNM 493931; details from UM 88105). E, *Hyopsodus* (USNM 23740). F, *Macrocranion* (USNM 493780). All scale bars equal 1 mm.

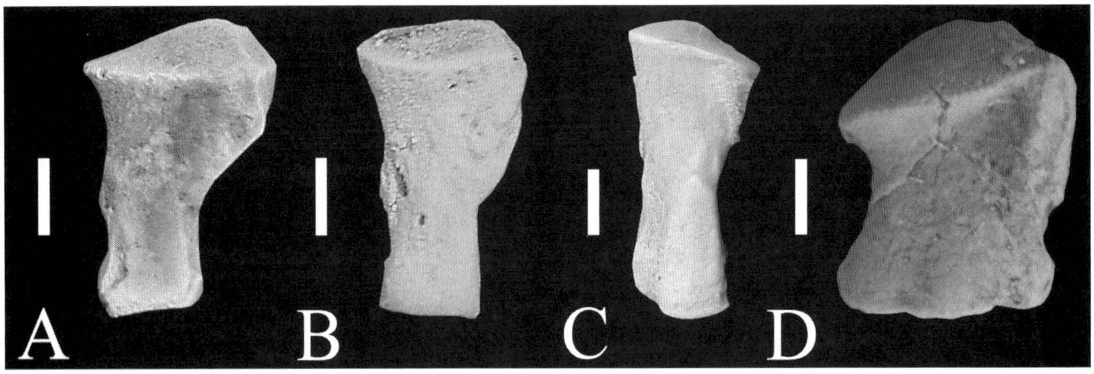

FIGURE 5.15. Comparison of apheliscid cuboids to other taxa in dorsal view. A, *Apheliscus* (USNM 493903). B, *Haplomylus* (USNM 513668). C, *Petrodromus* (USNM 521009, reversed). D, *Hyopsodus* (USNM 23740, reversed). All are in dorsal view; all scale bars equal 1 mm.

The postcrania of *Hyopsodus* overall suggest terrestrial or scansorial habits. Despite several apparent indicators of incipient cursorial ability (i.e., high humeral greater tuberosity, supratrochlear foramen, sharp humeral trochlea, deep femoral condyles, deep patellar trochlea), the majority of characters indicate that the postcranium of *Hyopsodus* was not well suited to running (at least, not efficient running). Although some of the postcranial features are contradictory, *Hyopsodus* may best be characterized as a generalized, small terrestrial mammal, perhaps capable of occasional bursts of speed, with some additional propensity to both climb and dig (see also Gazin, 1968). Moreover, in addition to the differences in postcranial function between *Hyopsodus* and apheliscids, no distinctive shared features in the postcranium suggest a close phylogenetic relationship between these two taxa. The few characters in *Hyopsodus* consistent with incipient cursoriality may be most appropriately viewed as convergent on apheliscid morphology.

5.5.2 Eocene Leptictidae

While the dentition is not suggestive of a close phylogenetic relationship between apheliscids and leptictids, here we show that these two families have superficially similar postcrania suggesting that they share similar locomotor repertoires.

Apheliscids and Eocene leptictids share a number of functionally significant characters, particularly in the long bones of the hind limb, which suggest similar locomotor habits. The femora of both taxa have similarly positioned and robust proximal attachment sites for hip flexor-extensor muscles (i.e., gluteals, lateral rotators, iliopsoas). The femoral shaft is long and gracile, and the distal end is narrow and anteroposteriorly deep. The patellar groove is well defined and proximally extensive. The proximal tibia is correspondingly deep. Together, the morphology of elements contributing to the knee suggests the ability for rapid and full extension of the knee. The tibia and fibula are longer than the femur in the case of both apheliscids and leptictids (Rose, 1999). These overall patterns in the hind limb are very similar to those of extant artiodactyls, perissodactyls, lagomorphs, some carnivorans, and macroscelideans, and are strongly suggestive of cursoriality.

Apheliscid and leptictid humeri are short compared to hindlimb elements. In both taxa, the short humerus suggests that the hindlimbs provided most of the thrust during rapid locomotion. Beyond this basic similarity, the humeri of leptictids are quite different from those of apheliscids, indicating differences in forelimb use. Humeri of leptictids are more robust than in apheliscids (Figure 5.9D). The greater and lesser tuberosities are lower in leptictids than in *Haplomylus*, but attachment sites for rotator cuff muscles are better developed in leptictids. The deltopectoral crest in leptictids is broader (especially proximally), has a flatter anterior surface, and is more distally projecting in leptictids. The site of insertion of teres major is also more evident in leptictids. Leptictid humeri have a much larger medial entepicondyle (Rose, 1999), which protrudes distomedially beyond the level of the trochlea. The supinator crest is much stronger, the trochlea is not as sharp, and the capitulum is wider in leptictids. The radial fossa in leptictids is never perforate, and the olecranon fossa is shallow. Finally, the capitulum, trochlea, and entepicondylar bar lie in the same mediolateral plane in the apheliscids, the medial structures (entepicondyle, trochlea) project farther distally than lateral structures (capitulum, lateral epicondyle) in leptictids.

The disparity in leptictid and apheliscid forelimbs probably reflects differences in substrate use. Rose (1999) noted that leptictid humeri were particularly similar to those of burrowing extant lipotyphlans in being short and robust, with a long, well-developed deltopectoral crest, low greater and lesser tuberosities, a prominent supinator crest, a broad distal articular surface, and a well-developed medial epicondyle. He concluded that leptictids were likely habitual forelimb diggers that used their hind limbs to brace themselves while burrowing. Apheliscid forelimbs do not necessarily suggest fossorial habits, nor are they entirely inconsistent with such behavior. Apheliscid humeri are, however, primarily adapted for speed and parasagittal mobility, and are more consistent with running than with digging behavior (Rose, 1990). Therefore it is likely that, unlike leptictids, the forelimb participated in rapid locomotion in apheliscids, suggesting that both *Apheliscus* and *Haplomylus* were quadrupedal cursors to a degree.

Femora of leptictids are more similar to those of apheliscids than are the forelimb elements. The proximal femur of leptictids can be distinguished from that of apheliscids primarily by the presence of a slightly larger femoral head with a dorsolateral extension of the articular surface onto the neck in leptictids, suggesting greater capacity for hindlimb abduction (Rose, 1999) (Figure 5.10J–L). The greater trochanter is perhaps slightly higher in leptictids than in *Haplomylus*, and noticeably higher than in *Apheliscus*, but otherwise the arrangement of the trochanters, shape of the greater trochanter, and depth of the trochanteric fossa are similar in leptictids and apheliscids. Differences in distal femoral morphology between leptictids and apheliscids are subtler. Aside from size (apheliscids tend to be smaller than leptictids from comparable stratigraphic levels), the most notable difference is that apheliscids have a more proximally extensive patellar groove.

Eocene leptictids share with apheliscids extensive fusion of the distal tibia and fibula (absent in Paleocene *Prodiacodon*: Matthew, 1918), and the elements are quite similar, but again there are differences in the details. The leptictid proximal tibia is deeper anteroposteriorly while the tibial tuberosity is broader and less anteriorly projecting than those of the apheliscids. The proximal fibular facet in leptictids is relatively smaller than that of apheliscids and is situated distolateral to the lateral tibial plateau. The proximal one-third of the tibial shaft is slightly more mediolaterally compressed in apheliscids, and in leptictids the cnemial crest tends to be lower than in apheliscids (Figure 5.11G, H).

Leptictids have a moderately developed posterior process on the distal tibia (lacking in apheliscids), while apheliscids possess an anterior tubercle (lacking in leptictids) (Figure 5.12D). Both features appear to be mechanisms to stabilize the tibioastragalar joint: in apheliscids, contact of the anterior tubercle with the astragalus would prevent hyper-dorsiflexion and rotation (in dorsiflexion), while in leptictids the posterior process would prevent hyper-plantarflexion and rotation of the pes (in plantarflexion). The configuration of grooves on the posterior surface of the malleoli for passage of muscle tendons is similar between leptictids and apheliscids.

The differences between leptictids and apheliscids are most notable in the tarsus, particularly the astragalus. There are some general cursorial similarities such as the presence of a well-grooved trochlea, lack of an astragalar foramen, and presence of an elongate astragalar neck, but there are several differences. As in apheliscids, the lateral rim of the trochlea is higher than the medial rim in leptictids, but the asymmetry in leptictids is much less pronounced. The ectal facet is narrower and more concave in apheliscids than in leptictids. The sustentacular facet is well separated from the navicular facet in apheliscids, but in leptictids these two facets are nearly confluent distomedially (Figure 5.13).

Significantly, leptictids lack the cotylar fossa present in apheliscids, and instead possess a lateral process off the astragalar body with a well-developed fossa for articulation of the lateral malleolus (Rose, 1999). These features may represent two different solutions for stabilizing the tibioastragalar joint during rapid locomotion, with each depression functioning as a guiding pivot for its respective malleolus. The cuboid facet is relatively larger and more sharply offset on apheliscid astragali than in leptictids.

Calcaneal features shared by apheliscids and leptictids include a moderately elongate calcaneal tuber, a small sustentacular facet that faces dorsodistally, and a reduced peroneal tubercle, but again there are more differences than similarities.

The ectal facet in leptictid calcanei faces dorsally, is essentially oriented proximodistally, and is smoothly convex; the ectal facets in apheliscids are angled at approximately the midpoint of the facet (Figure 5.14). Unlike apheliscids, leptictids lack a fibular facet on the calcaneum. The sustentaculum in leptictids is slightly more distal, overlapping less with the ectal facet in transverse alignment than in apheliscids. The cuboid facet in both groups are slightly concave, but the leptictid facet is subcircular and more nearly perpendicular to the long axis of the calcaneum in leptictids, while it is ovoid and oblique to the long axis of the calcaneum in apheliscids. The peroneal tubercle is slightly more proximally extensive in leptictids.

Overall similarities in the hindlimb indicate that *Apheliscus* and leptictids were probably comparable in cursorial ability, while *Haplomylus* was potentially more specialized for cursoriality than either of these taxa. Despite the differences just discussed, among their contemporaries, apheliscids most strongly resemble leptictids in postcranial morphology. The details of postcranial and dental morphology, however, indicate that this resemblance is convergent.

5.5.3 *Macrocranion*

It has been suggested that some small-bodied condylarths are closely related to certain putative erinaceomorph insectivores (e.g., *Macrocranion*: Russell, 1964; Russell et al., 1975; Rigby, 1980; Bown and Schankler, 1982). In fact, here we report several striking postcranial similarities between the apheliscids and *Macrocranion* from North America. Pending a full review of North American *Macrocranion* postcrania being undertaken by T. Penkrot and S. Zack (Penkrot et al., 2004 and in preparation), we restrict our comparisons to the proximal tarsals and distal tibia-fibula.

The distal tibia-fibula of *Macrocranion* is so similar to that of *Haplomylus* that size is the best distinguishing feature (the distal tibia-fibula of *Macrocranion* is about one-third the size of the *Haplomylus* tibia-fibula at comparable stratigraphic levels). Specific similarities shared by *Macrocranion* and apheliscids include complete distal fusion of the tibia and fibula, deep excavation of the distal articular surface of the tibia-fibula, presence of a hook-like, laterally recurved medial malleolus and presence of a prominent anterior tubercle on the distal tibia that is virtually identical to that of *Haplomylus* (Figure 5.12F).

Macrocranion and apheliscids share a deeply grooved astragalar trochlea; the trochlea of *Macrocranion* is particularly similar to that of *Haplomylus* (Figure 5.13). Both taxa lack an astragalar foramen, show a similar degree of posterior extension of the articular surface of the trochlea, and the lateral rim of the trochlea has a greater radius of curvature than the medial rim. Of particular significance, *Macrocranion* possesses a well-developed cotylar fossa on the medial aspect of the astragalar body that is virtually identical to that of *Haplomylus*. Ventrally, the ectal facet is similar in overall shape and orientation to that of *Haplomylus*, particularly in being more concave than in *Apheliscus*. The sustentacular facet of *Macrocranion* is similar to that of *Haplomylus* in being shifted medially on the posterior aspect of the astragalar neck. In both *Macrocranion* and apheliscids, the astragalar neck is comparably elongate and the articular surface of the astragalar head extends onto the neck. The navicular and cuboid facets on the astragalar head are similarly proportioned to those of apheliscids, although they are less well delineated from one another in *Macrocranion*.

Both apheliscids and *Macrocranion* have an elongate calcaneal tuber and a comparatively short calcaneal body (Figure 5.14). *Macrocranion* also has a well-developed fibular facet comparable in size and orientation to the fibular facet in *Apheliscus*. The cuboid facet in *Macrocranion* is slightly concave and angled proximomedially to distolaterally, similar to the morphology of the cuboid facet in apheliscids. Finally, the peroneal tubercle on the lateral aspect of the *Macrocranion* calcaneum is comparably distally situated but smaller than in either apheliscid.

Along with these similarities, there are also notable differences between *Macrocranion* and apheliscids, particularly at the subastragalar joint. Unlike the apheliscids, the sustentacular facet of the *Macrocranion* astragalus is large, anteroposteriorly elongate, and confluent with the articular surface of the head. Similarly, there is a distal extension of the calcaneal sustentaculum in *Macrocranion* that appears as a continuous, distolaterally tapering web of bone contacting the medial margin of the cuboid facet. The articular surface of the sustentacular facet in *Macrocranion* continues distolaterally along this distal extension of the sustentaculum. Apheliscids lack these distal extensions of the sustentaculum and facet. The calcaneal ectal facet of the *Macrocranion* calcaneum arcs smoothly instead of turning sharply at the midpoint as in apheliscids. This configuration of facets in *Macrocranion* is superficially similar to the expanded sustentacular facet present in artiodactyl astragali and implies an increased range of parasagittal motion of the navicular/cuboid on the astragalar head and of the astragalus on the calcaneum (Schaeffer, 1947).

Other tarsal differences between *Macrocranion* and apheliscids include a narrower astragalar body in *Macrocranion*, particularly proximally where the medial margin of the trochlea is angled proximolaterally to mediodistally. Additionally, in *Macrocranion* the facet for the median tarsal bone or spring ligament of the astragalus abruptly turns proximally at the base of its proximomedial extension onto the neck. In the apheliscids the articular surface of the head curves proximomedially in a smoother arc onto the neck.

The many morphologic similarities in the distal tibia-fibula and tarsus shared by *Macrocranion* and the apheliscids suggest that the resemblances may be more than simple convergence. Minimally, the numerous resemblances between *Macrocranion* and apheliscids indicate that *Macrocranion* was specialized for rapid terrestrial locomotion, a hypothesis that has already been suggested in reference to the specimens from the Grube Messel, Germany (Storch, 1993, 1996). The presence of both an anterior tubercle on the distal tibia-fibula and a cotylar fossa on the astragalus are notable. Moreover, several details of the distal tibia-fibula and tarsal morphology are virtually indistinguishable from those of *Haplomylus*. No other known taxa from the Paleocene or Eocene of North America are adapted for cursoriality using this same suite of postcranial characters. The fact that *Macrocranion* is also specialized for rapid locomotion in the "apheliscid" fashion may imply relatively recent shared common ancestry. The blurred line between some hyopsodontid (i.e., apheliscid) condylarths and erinaceomorph insectivores has been noted by previous authors (Gingerich, 1983; Bown and Schankler, 1982; Novacek et al., 1988). The two groups have been distinguished by a relatively small number of dental characters, such as the strength of the M1 postmetacrista or the relative sizes of m1-3, with the caveat that relationships are likely to change as both groups become better known and more taxa and morphologic features can be compared. With regard to the latter point, despite the well-preserved postcrania known for *Macrocranion*, data on these elements have seldom been used to help assess the phylogenetic position of the genus. With more comprehensive phylogenetic studies, the relationship between erinaceomorph insectivores and small-bodied condylarths may be better understood.

5.5.4 Recent Macroscelidea

Postcranial comparisons to modern macroscelideans more clearly illustrate the numerous similarities to apheliscids, strongly suggesting a close phylogenetic relationship between these two taxa (Zack et al., 2005a). In most features relevant to this study, extant macroscelideans are morphologically uniform and, unless otherwise noted, the following comparisons apply to *Rhynchocyon, Petrodromus*, and *Elephantulus*.

The postcrania of apheliscids resemble those of macroscelideans in features that indicate similarities in locomotor repertoire. The similarities to macroscelideans, particularly in features that are not widespread among mammals, also suggest a close phylogenetic relationship between apheliscids and living macroscelideans. In both macroscelideans and apheliscids the forelimb elements are shorter and more gracile than the hindlimb elements. In the proximal humerus, apheliscids and macroscelideans share an ovoid humeral head with a posteriorly restricted articular surface, moderately tall greater tuberosities, and well-developed lesser tuberosities (Figure 5.9C). The greater tuberosity tends to be relatively taller in macroscelideans, but otherwise macroscelidean proximal humeri are very similar to those of apheliscids. The height of the tuberosities indicates some restriction of motion at the shoulder to the parasagittal plane and increased lever arm for rotator cuff muscles (in use during running), although not to the degree seen in cursorial mammals in which the forelimb provides significant thrust (e.g., artiodactyls or canids: Gambaryan, 1974). *Haplomylus* and macroscelideans also share a gracile humeral shaft and a deltopectoral crest that is proximally broad, but tapers distally to its termination in the proximal half of the humerus, one-fourth (i.e., *Haplomylus*) to one-half (i.e., *Rhynchocyon*) the distance along the shaft. The deltopectoral crest is relatively low in both taxa except for a slight anterior projection at its distal termination.

In the distal humerus, both taxa have weak supinator crests, with apheliscids possessing a relatively stronger crest. Extant macroscelideans have a reduced medial epicondyle relative to apheliscids, but both taxa retain an entepicondylar foramen, a primitive feature that is lost in many cursors (e.g., Artiodactyla: Rose, 1985). The radial and olecranon fossae are deep in both apheliscids and macroscelideans, permitting a substantial range of flexion and extension at the elbow. In *Haplomylus* and macroscelideans, the fossae are deep enough to produce a supratrochlear foramen, in contrast to *Apheliscus*. In both apheliscids and macroscelideans, the trochlea projects sharply distally, and the capitulum is ovoid. The proximal radius of *Apheliscus* is similar to *Petrodromus* and *Elephantulus* in being ovoid and in having a flat ulnar facet. *Rhynchocyon* has a much broader proximal radius, but in all taxa, little or no capacity for supination is evident. Taken together, the elbows of apheliscids exhibit adaptations for rapid locomotion similarly to those of macroscelideans (Taylor, 1974; Rose, 1999).

Apheliscids and macroscelideans also show similar adaptations for rapid locomotion at the hip joint. The femoral head in both taxa is small and subspherical, with an elongate neck (Figure 5.10G–I). *Rhynchocyon* differs from the remaining taxa in having an extension of the dorsal margin of the head onto the neck (Rose, 1999). The greater trochanter is high and well developed in both taxa approaching (i.e., *Apheliscus*) or exceeding (i.e., *Haplomylus* and macroscelideans) the height of the femoral head, while the lesser trochanter is strong and points posteromedially. This development of the trochanters provides long lever arms for the gluteals and iliopsoas. In both apheliscids and macroscelideans, the trochanteric fossa is deep and the third trochanter is strong, bladelike, and proximally located. These features are not unique to either apheliscids or macroscelideans, as they are present in many cursorial mammals (Taylor, 1976; Heinrich and Rose, 1997; Rose, 1999; Rose and Chinnery, 2004).

The knee joint in both apheliscids and macroscelideans also strongly suggests similarly specialized cursorial locomotion. The distal femur is deep in both taxa, with anteriorly extensive condylar articular surfaces and a deep, proximally extensive patellar groove. Together, these features suggest both a large range of extension possible at the knee (articular surfaces of femoral condyles), as well as an increased moment arm for the quadriceps muscles (deep distal femur, long and well-defined patellar groove). While most of these features are frequently found among cursorial mammals (Taylor, 1976; Hildebrand, 1982; Rose, 1999; Argot, 2002, 2003), the proximal extent of the patellar groove is relatively unusual.

Apheliscid and macroscelidean tibiae share a prominent, sharp cnemial crest, and a narrow, projecting tibial tuberosity (Figure 5.11E, F). In both macroscelideans and apheliscids, the medial and lateral tibial condyles are anteroposteriorly long and approximately equal in area, and the proximal fibular facet is located posterodistal to the lateral tibial plateau. In *Apheliscus*, the proximal fibula is large and has a small facet that extends the articular surface of the lateral tibial condyle. The presence of an expanded proximal fibula is unusual among cursorial mammals, which tend to reduce the proximal fibula, while the extension of the articular surface of the lateral tibial condyle onto the fibula is unusual among eutherian mammals generally (Meng et al., 2003), and was only found in macroscelideans among modern eutherians examined. Both features are well developed in macroscelideans, in which the proximal fibula is large and articulates with the distal femur when the knee is flexed. While the proximal fibula of *Haplomylus* remains unknown, it shares with macroscelideans an extension of the tibial facet for the proximal fibula onto the lateral rim of the tibial plateau, a feature lacking in *Apheliscus*. Macroscelideans have a fused tibia and fibula at the proximal fibular facet, whereas the two bones meet proximally in a synovial joint in apheliscids.

In both apheliscids and macroscelideans, the tibia is as long as or longer than the femur, although the precise crural indices of apheliscids have yet to be determined, pending identification of complete proximal and distal elements of a single individual. In both groups, the shafts of the tibia and fibula are gracile and fused distally for a significant portion of their lengths. Fusion of the tibia-fibula begins more proximally in macroscelideans at approximately one-third of the way from the proximal end. In apheliscids, fusion starts at roughly midshaft (*Haplomylus*) or at about one-third of the way from the distal end of the tibia (*Apheliscus*). This high degree of tibia-fibula fusion in combination with gracile and elongate shafts is typical of leaping mammals, including both terrestrial taxa like macroscelideans and some arboreal leaping primates (Barnett and Napier, 1953; Hildebrand, 1982; Rose, 1999). Thus, the morphology shared by apheliscids and macroscelideans, although not exclusively restricted to saltators or cursors, does reinforce the argument of locomotor analogy between these two taxa.

Similarities in distal tibia-fibula morphology shared by apheliscids and macroscelideans are striking. In both groups, the distal surface of the tibia-fibula is deeply grooved to accommodate the astragalar trochlea, the medial malleolus is elongate and recurved laterally, and there is a strong anterior tubercle (Figure 5.12C). Neither taxon has a posterior process, in contrast to rodents, lagomorphs, and leptictids, among other mammals (Szalay, 1985; Rose, 1999). The only difference of note is that the lateral malleolus is relatively shorter in macroscelideans than in apheliscids.

Matching the morphology of the distal tibia, the astragalar trochlea is asymmetric and deeply grooved in both macroscelideans and apheliscids (Figure 5.13). Both taxa lack an astragalar foramen, have a posteroventrally extensive articular surface on the trochlea, and have a medial trochlear rim with a smaller radius of curvature than the lateral rim. Both medial and lateral rims are sharply defined. Modern macroscelideans have a well-developed cotylar fossa on the medial aspect of the astragalar body, which articulates with the hook-like medial malleolus. The prominent malleoli interlock with the astragalar trochlea to permit only flexion and extension at the crurotarsal joint.

The distribution of cotylar fossae among fossil and extant Eutheria is quite limited (e.g., Hyracoidea, Cercopithecoidea), and its presence in both macroscelideans and apheliscids therefore suggests a close phylogenetic relationship. The similarity in cotylar fossa morphology between modern macroscelideans and *Haplomylus* is particularly close. The cotylar fossae of *Apheliscus* and macroscelideans are less immediately similar, due to the greater difference in relative size and orientation in the cotylar fossae between *Apheliscus* and macroscelideans (versus between *Haplomylus* and macroscelideans). As a result of the configuration of the cotylar fossa in *Apheliscus*, there would have been a relatively limited period of contact between the cotylar fossa and the medial malleolus in *Apheliscus*. In *Apheliscus*, the cotylar fossa appears to have served as a stop against extreme dorsiflexion, rather than as a guiding pivot between the medial malleolus and astragalus as in both *Haplomylus* and macroscelideans.

On the calcaneum, apheliscids and macroscelideans share a well-defined fibular facet posterolateral to the ectal facet (Figure 5.14). The morphology of this fibular facet is unusual: its semilunar shape, proximoventral orientation, and sharply defined margins allow it to function as an extension of the trochlear articular surface, providing a pivot for the lateral malleolus.

In both apheliscids and macroscelideans, the astragalocalcaneal articulation is tight, and little or no movement is possible at the subastragalar joint. The calcaneal ectal facet changes orientation at approximately its midpoint in apheliscids such that approximately half of the facet faces medially, while the remainder faces distally. Macroscelideans show the same change in orientation, but a larger proportion of the facet faces distally, with only a minor portion facing medially. The change in orientation is particularly sharp in *Haplomylus* and macroscelideans, effectively preventing movement between the astragalus and calcaneum. The sustentacular facets of the astragalus and calcaneum are small and not proximodistally elongate, again limiting potential movements between these bones. Unlike in apheliscids where the sustentaculum projects medially from the calcaneal body, in macroscelideans it juts mediodistally. Matching this on the astragalus, the macroscelidean sustentacular facet is set entirely on the medial aspect of the astragalar neck, and, unlike apheliscids, is confluent with the articular surface of the head distally. In *Haplomylus* the facet is shifted medially on the astragalar neck, and faces ventromedially, while it is more central in position and ventral in orientation in *Apheliscus*.

The proximal tarsals are generally elongate in both apheliscids and macroscelideans, particularly with an elongate calcaneal tuber, gracile body, and a moderately long astragalar neck. Both apheliscids and macroscelideans have a cuboid facet on the astragalar head. The delineation between the navicular and cuboid facets is defined in *Haplomylus* by an abrupt change in orientation of the articular surface; this distinction is more marked in macroscelideans. The navicular and cuboid facets are less well delineated in *Apheliscus*. Apheliscids and macroscelideans share an oblique orientation of the navicular facet, although this facet is dorsoventrally flattened in macroscelideans and slightly convex in apheliscids. Apheliscids retain a facet on the medial side of the astragalar neck, possibly for the medial tarsal bone or spring ligament; a similar facet is strongly reduced or absent in macroscelideans. On the distal calcaneum, modern macroscelideans lack a distinct peroneal tubercle, whereas apheliscids retain a small tubercle. The cuboid facet in macroscelideans is roughly quadrate, and faces almost directly distally, whereas in apheliscids the cuboid facet is either subcircular (*Haplomylus*) or ovoid (*Apheliscus*) and faces mediodistally. The cuboid facet in macroscelideans is more deeply concave than in apheliscids, but in both groups the facet is concave dorsoventrally and not mediolaterally, restricting transverse movements.

On the proximal cuboid of macroscelideans, the astragalar and calcaneal facets can be distinguished by a subtle change in the orientation of the articular surface (Figure 5.15C). Otherwise, the proximal articular surface of the cuboid is faintly convex, and faces proximolaterally and dorsally. The proximal articular surface in apheliscid cuboids is more dorsoventrally convex than in macroscelideans, with a more abrupt change in orientation of the articular surface at the margin between the astragalar and calcaneal facets, matching the more angled cuboid facet of the calcaneum. Although an apheliscid navicular has yet to be identified, the presence of a astragalocuboid contact and the shapes of the astragalar navicular facet, the calcaneal cuboid facet, and the cuboid calcaneal facet imply that non-parasagittal motion at this joint was limited or absent. Our manipulations of elements indicate that little or no movement is possible at the transverse tarsal joint in modern macroscelideans (T. Penkrot, personal observation), but with no apheliscid navicular available, limited parasagittal motion at the transverse tarsal joint of apheliscids is possible.

The cuboids of apheliscids and macroscelideans are long and gracile, with a mediolateral constriction immediately distal to the ectocuneiform facet (more pronounced in apheliscids). Whereas many modern cursorial mammals (lagomorphs, some artiodactyls, some perissodactyls, some carnivores) truncate and/ or fuse the cuboid, navicular, and cuneiforms (Hildebrand, 1982), macroscelideans and apheliscids are unusual among cursors in lengthening the cuboid. The configuration of facets on the cuboid suggests tight articulation with adjacent tarsal elements, with little or no movement possible between distal tarsal bones (T. Penkrot, personal observation). Distally, the cuboids are mediolaterally constricted just distal to the ectocuneiform facet. The navicular facet is narrow, as is the ectocuneiform facet. There is a marked change in orientation of the articular surface between the navicular and ectocuneiform facets, particularly in *Apheliscus* due to the relatively broader proximal cuboid in this genus. The offset between the navicular and ectocuneiform facets is less in *Haplomylus*, and further reduced in macroscelideans. Just proximal to the metatarsal facet in *Haplomylus* there is a small subcircular facet on the posteromedial margin of the cuboid body, a facet that was not apparent in the cuboid of *Apheliscus*. In macroscelideans this small facet articulates with the distal part of the ectocuneiform. The metatarsal facet in macroscelideans is crescentic, and in apheliscids the metatarsal facet is semilunar, with either a shallow (*Apheliscus*) or a deep (*Haplomylus*) medial notch. The metatarsal facet is flat. The presence in all of these taxa of a small, flat metatarsal facet on the distal cuboid argues against any significant ability for abduction of the lateral metatarsals.

The postcrania of both apheliscids and macroscelideans are strongly indicative of specialized rapid terrestrial locomotion. The close overall similarity in the postcrania between apheliscids and macroscelideans, and the means by which these taxa approach cursoriality, imply phylogenetic affinity.

5.6 Phylogenetic Position of *Apheliscus* and *Haplomylus*

The postcranial features described above suggest that *Apheliscus* and *Haplomylus* are more closely related to extant Macroscelidea than to any other living group of mammals. This hypothesis invites a rigorous cladistically based test that incorporates other living and extinct taxa that may be closely related to Macroscelidea. The higher-level systematic position of macroscelideans is contentious, making it difficult to determine what other taxa are appropriately included in such a test. Recent workers have entertained three different hypotheses for the position of Macroscelidea within Eutheria.

The first hypothesis, which links Macroscelidea to Rodentia, Lagomorpha, and their extinct relatives traces its origin to McKenna's (1975) efforts to resolve the interrelationships of eutherian mammals based on shared derived characters. McKenna (1975) allied Macroscelidea with Lagomorpha and the Asian Paleogene family Anagalidae. Szalay (1977) provided the first support for this relationship in his groundbreaking analysis of the phylogenetic significance of eutherian tarsal morphology, in which he documented derived similarities in the tarsus of macroscelideans, lagomorphs, anagalids, and additional Asian Cretaceous and Paleogene taxa (Pseudictopidae, Eurymylidae, Zalambdalestidae).

Subsequent investigations of the interrelationships of eutherian orders based on morphology have supported a relatively close phylogenetic relationship between macroscelideans and lagomorphs, although most studies have linked

Lagomorpha most closely with Rodentia in a clade termed Glires (Novacek, 1982, 1986; Novacek and Wyss, 1986; Novacek et al., 1988; Shoshani and McKenna, 1998; Asher et al., 2003, 2005b). The combined grouping of Macroscelidea with Glires and the Asian Paleogene taxa mentioned above has been called Anagalida (Novacek, 1986; McKenna and Bell, 1997), the name used by McKenna (1975) for a macroscelidean-lagomorph clade. While tarsal morphology was instrumental in the initial recognition of Anagalida, recent diagnoses of the clade have relied primarily on dental, cranio-mandibular, and embryologic characters (Novacek, 1982, 1986). Szalay (1985) reevaluated the similarities in the tarsals of lagomorphs and macroscelideans and concluded that they probably represent convergence. Szalay noted that significant differences exist in the crurotarsal morphology of macroscelideans and lagomorphs, including the presence/absence of a cuboid facet on the astragalus, size and orientation of the fibular facet or the calcaneum, and the form of the tibioastragalar joint. Based on this evidence, he concluded that the similarities shared by the two groups represent convergent adaptation to similar locomotor strategies, and not evidence of a close phylogenetic relationship.

A second potential phylogenetic position for Macroscelidea emerged from Hartenberger's (1986) description of the first Eocene macroscelidean, *Chambius kasserinensis*. Hartenberger (1986) was impressed by similarities between the dentitions of *Chambius* and louisinine hyopsodontid condylarths such as *Louisina* and *Microhyus* (see Figure 5.8). Based on this material, he hypothesized that macroscelideans are ungulates, derived from the same radiation that produced artiodactyls and perissodactyls. Subsequent descriptions of additional material of Eocene and Oligocene macroscelideans have confirmed that the earliest known African macroscelideans have dentitions strikingly similar to those of early ungulates (Simons et al., 1991; Tabuce et al., 2001). Simons et al. (1991) suggested that *Haplomylus*, rather than Louisininae, is the condylarth sister taxon to Macroscelidea, while Tabuce et al. (2001) placed the louisinine *Microhyus* in this role. There is general agreement, however, that the dentitions of Eocene and Oligocene macroscelideans support derivation of Macroscelidea from a basal ungulate stock (Simons et al., 1991; Butler, 1995; Tabuce et al., 2001; Holroyd and Mussell, 2005). This hypothesis has yet to be corroborated by non-dental morphology, however, partly because the cranial and postcranial morphology of early African macroscelideans and potential "condylarth" relatives has been almost unknown prior to the present work.

A radically different view of the position of Macroscelidea within Eutheria has emerged recently from molecular phylogenetic studies. An early study based on eye lens protein sequences suggested that macroscelideans might be linked with the ungulate orders Hyracoidea, Proboscidea, Sirenia, and Tubulidentata (Jong et al., 1993). As molecular studies began to sample Eutheria more extensively, a novel clade (Afrotheria) including these taxa, as well as the lipotyphlan families Tenrecidae and Chrysochloridae was recognized (Stanhope et al., 1998; Madsen et al., 2001; Murphy et al., 2001). Afrotheria has become one of the best-supported and most unexpected results of molecular phylogenetic studies of eutherians. The linkage of several ungulate orders with some insectivores to the exclusion of other ungulates stands in particular contrast to all morphology-based studies of eutherian higher-level phylogeny. In fact, subsequent morphological studies have found little if any support for Afrotheria (Asher, 1999; Whidden, 2002). Despite its lack of morphological support, the persistent recognition of Afrotheria in molecular studies suggests that this hypothesis should still be given serious consideration by morphologists.

In summary, there are three viable hypotheses of the superordinal phylogenetic position of Macroscelidea, each based on a different approach to reconstructing eutherian phylogeny. Studies with a broad morphologic base that sample across Eutheria at the ordinal or subordinal level (Novacek, 1986; Novacek and Wyss, 1986; Novacek et al., 1988; Shoshani and McKenna, 1998; Asher et al., 2003) support Anagalida based on craniodental and embryologic characteristics. In contrast, studies based on dental morphology but sampling more densely and at lower taxonomic levels continue to support a relationship of macroscelideans to a broadly conceived Ungulata, and more specifically to what are now considered apheliscid condylarths (Hartenberger, 1986; Simons et al., 1991; Butler, 1995; Tabuce et al., 2001). Finally, molecular evidence favors the inclusion of Macroscelidea in Afrotheria (Stanhope et al., 1998; Madsen et al., 2001; Murphy et al., 2001). Thus, reconstruction of the affinities of Macroscelidea is largely dependent on the source of data employed.

To evaluate these varied potential phylogenetic positions of Macroscelidea, in the initial presentation of this material (Zack et al., 2005a), we tested the hypothesized link between apheliscids and macroscelideans using three character-taxon matrices, each with a different taxonomic focus. In that study, *Apheliscus* and *Haplomylus* were incorporated into a character-taxon matrix (Meng et al., 2003) that sampled all well-known anagalidan clades. *Apheliscus* and *Haplomylus* were also added to a character-taxon matrix (Asher et al., 2003), which samples all extant afrothere clades, along with many other living and extinct eutherians. Because, in this context, our primary interest was in testing the affinities of apheliscids to macroscelideans against a broad sample of afrotheres, we analyzed the matrix of Asher et al. (2003), with only living afrotheres and fossil taxa with a potential relationship to Afrotheria included. Finally, a new character-taxon matrix was constructed, including *Apheliscus, Haplomylus*, macroscelideans, and a diversity of taxa placed by morphologic studies in Ungulata. Characters in the latter matrix were coded from the cheek dentition and tarsus. All three analyses also included *Hyopsodus*, providing an opportunity to test the hypothesis that this taxon is not closely related to *Apheliscus* and *Haplomylus*. A full account of the new matrix, modifications to existing matrices, and the methods used to analyze

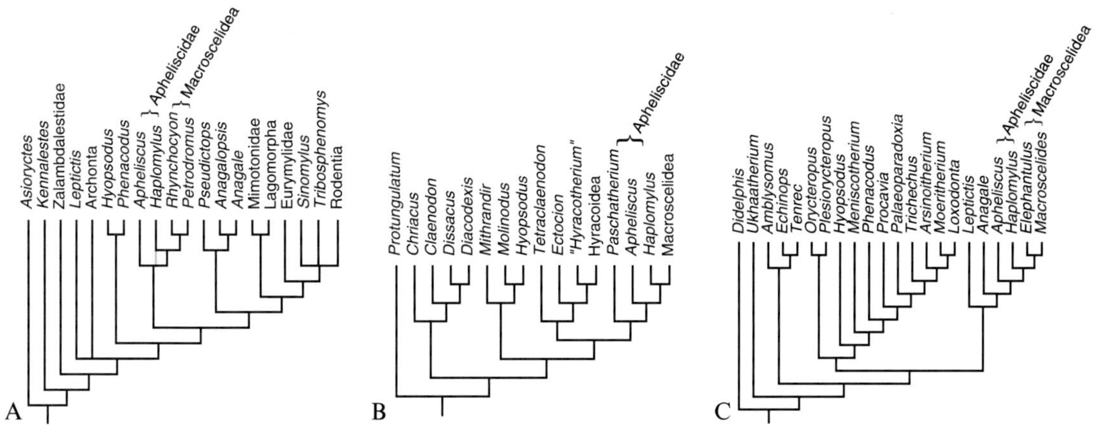

FIGURE 5.16. Phylogenetic relationships of apheliscids (modified from Zack et al., 2005a). A, Strict consensus of 16 most parsimonious trees from reanalysis of the Meng et al. (2003) character-taxon matrix with some characters ordered (black lines only). The gray line indicates the position of *Haplomylus* when all characters are unordered. For visual simplicity, Zalambdalestidae, Archonta, Mimotonidae, Lagomorpha, Eurymylidae, and Rodentia have been collapsed into single terminal taxa. B, Single most parsimonious tree from analysis of basal ungulates with all characters unordered or with some characters ordered. C, Single most parsimonious tree derived from reanalysis of the Asher et al. (2003) character-taxon matrix with all characters unordered and only afrotheres and their potential extinct relatives included.

all three matrices can be found in the online supplementary material to Zack et al. (2005a).

One reviewer of an earlier version of this paper suggested that the similarities shared by *Macrocranion* and apheliscids might indicate that apheliscids have erinaceomorph, rather than macroscelidean affinities. To evaluate this hypothesis, it was suggested that we reanalyze Asher et al.'s (2003) matrix with all taxa included and use a molecular scaffold to constrain the phylogenetic positions of extant forms (Springer et al., 2001). To further evaluate the phylogenetic position of *Apheliscus* and *Haplomylus* we have followed this suggestion, with a slight modification. Rather than use Asher et al.'s (2003) original sample, we have used the sample from Asher et al. (2005a), which includes two additional taxa (*Centetodon* and *Solenodon*) and lower diversity of some clades (e.g., Euprimates). This modified matrix has already been analyzed using a molecular scaffold, making it easier to evaluate the effect of the inclusion of *Apheliscus* and *Haplomylus*. Following Asher et al. (2005a), we used the molecular topology presented in Roca et al. (2004) to constrain the phylogeny of extant taxa. The three changes made to the matrix by Asher et al. (2005a) were also made for the present analysis. The new analysis was performed using PAUP*4.0b10 (Swofford, 1999), with the Roca et al. tree used as a backbone constraint. One thousand heuristic replicates were performed.

Analysis of the three original matrices placed *Apheliscus* and *Haplomylus* on the macroscelidean stem (Figure 5.16) and generally supported a sister-group relationship between *Haplomylus* and African macroscelideans. Because the characters and taxa included in the three matrices differ, the specific characters supporting apheliscid-macroscelidean ties also differed between the three analyses. However, in all cases, postcranial characters provided significant character support for an apheliscid-macroscelidean link. The molecular scaffold analysis produced five most parsimonious trees (L: 1224; CI: 0.27; RI: 0.47), the consensus of which differs from that presented by Asher et al. (2005a: figure 5.9b) only in the addition of *Apheliscus* and *Haplomylus* as the monophyletic sister taxon to Macroscelidea. The latter result indicates that apheliscids are not closely related to Erinaceomorpha and the postcranial similarities shared by *Macrocranion* and apheliscids are either convergent or indicate that *Macrocranion* is not an erinaceomorph.

All three of the original analyses resolved *Hyopsodus* as phylogenetically distant from *Apheliscus, Haplomylus*, and Macroscelidea, supporting separation of these taxa at the familial level. In contrast, the ungulate analysis resolved the louisinine *Paschatherium*, another form traditionally included in Hyopsodontidae, as phylogenetically close to *Apheliscus, Haplomylus*, and Macroscelidea, but basal to the latter taxa. This finding agrees with another study by the senior authors (Zack et al., 2005b) that analyzed the interrelationships of taxa traditionally placed in Hyopsodontidae and Mioclaenidae based on an extensive sample of both groups. That study recognizes a dichotomy between *Hyopsodus* and mioclaenids, on the one hand, and other hyopsodontids (including *Apheliscus, Haplomylus*, and *Paschatherium*). This result has been formalized by once again incorporating Mioclaenidae in Hyopsodontidae, while placing most other hyopsodontids in a resurrected Apheliscidae.

Apheliscids and hyopsodontids can be distinguished by several dental features, most notably the structure of the molar trigonids (Zack et al., 2005b). In addition, several consistent differences in astragalar morphology separate the two families. *Paschatherium, Apheliscus*, and *Haplomylus* share the derived presence of a well-developed cotylar fossa, a deeply grooved astragalus, and loss of the astragalar foramen. These features are also present in recently described tarsals of a second louisinine, *Microhyus* (Tabuce et al., 2006). In contrast, *Choeroclaenus* (Schaeffer, 1947), South American kollpaniine hyopsodontids (Muizon et al., 1998), and *Hyopsodus* lack a cotylar fossa, retain an astragalar foramen and ungrooved trochlea, and have a derived medial expansion of the astragalar body.

While there is strong support for linking apheliscids with macroscelideans, the higher-level phylogenetic position of Macroscelidea remains unclear. Apheliscids can be of some help in assessing which of the three hypotheses outlined above is most likely. As the paraphyletic stem group for Macroscelidea, apheliscids clarify the sequence of morphologic changes leading to macroscelideans, particularly with regard to postcranial morphology.

Overall, the postcranium of apheliscids does not provide particularly strong support for anagalidan affinities. Although representatives of several anagalidan clades show evidence for cursorial locomotion (Sulimski, 1968; Bleefeld and McKenna, 1985; Szalay, 1985; Meng et al., 2003), only lagomorphs and mimotonids, both members of Glires, show the degree of specialization seen in apheliscids and macroscelideans. In fact, *Haplomylus* and *Apheliscus* support Szalay's (1985) conclusion that the crurotarsal features shared by extant macroscelideans and lagomorphs were developed by parallel evolution and do not represent evidence of common ancestry. Unlike lagomorphs, but as in macroscelideans, apheliscids have a well-developed cotylar fossa, a large astragalar cuboid facet, and an elongate cuboid. Neither macroscelideans nor apheliscids have the proximally extensive calcaneal ectal facet or enlarged distal calcaneoastragalar facet seen in lagomorphs. Macroscelideans and lagomorphs are similar in lacking a peroneal tubercle, having a reduced facet for the medial tarsal or spring ligament, and having a sharp bend in the ectal facet. The morphology of apheliscids indicates that these similarities developed in parallel in macroscelideans and lagomorphs, as apheliscids retain a peroneal tubercle and a proximally extensive medial tarsal/spring ligament facet, while *Apheliscus* lacks a sharp bend in the ectal facet. Absence of a fibular facet in mimotonids (Szalay, 1985; Meng et al., 2004), a likely sister taxon of lagomorphs, may also suggest convergent development of a large fibular facet in macroscelideans and lagomorphs, although loss of the fibular facet in mimotonids would be equally parsimonious. Apheliscids and macroscelideans lack a posterior process on the distal tibia, a critical feature shared by all Glires. On the other hand no gliran has the equally distinctive combination of a cotylar fossa and enlarged medial malleolus shared by apheliscids and macroscelideans.

Taken together, the identification of apheliscids as stem macroscelideans combined with the discovery of basal members of Glires indicates that the crurotarsal similarities present in modern lagomorphs and macroscelideans probably represent convergent adaptation for similar modes of locomotion, rather than evidence of shared common ancestry. In contrast, many of the features in which modern macroscelideans and lagomorphs differ appear to represent more fundamental distinctions that were already established early in the histories of both groups. Consequently, similarities in the crura and tarsals of macroscelideans and lagomorphs do not provide convincing evidence for including Macroscelidea within Anagalida. Although the linkage of Apheliscidae to Macroscelidea weakens the hypothesized inclusion of macroscelideans in Anagalida, it cannot yet discount any potential connection between macroscelideans and anagalidans. Much of the recent support for Anagalida in higher-level phylogenetic studies has come from embryology and, particularly, cranial morphology, neither of which can be meaningfully assessed in apheliscids.

The postcranial morphology of apheliscids is more consistent with affinities to some ungulates. While apheliscid tarsals bear no particular similarity to the morphotypic ungulate tarsus, as represented by taxa such as *Protungulatum* and *Mithrandir* (=*Gillisonchus*) (Szalay and Decker, 1974; Rigby, 1981), other ungulates, including perissodactyls, artiodactyls, hyracoids, and mesonychid and phenacodontid condylarths, show many of the derived features present in apheliscids, such as a deeply grooved astragalar trochlea, oblique astragalar head, and an enlarged fibular facet. However, most of these similarities are common among cursorial mammals and their shared presence in apheliscids, macroscelideans, and other ungulates does not provide particularly compelling evidence for a close phylogenetic relationship. In most cases, more distinctive similarities are lacking.

A more compelling character complex is the combination of a cotylar fossa and expanded medial malleolus. In addition to apheliscids and macroscelideans, this complex occurs in several other putative ungulates, including hyracoids, proboscideans, tubulidentates, *Meniscotherium*, and *Plesiorycteropus*. This morphology otherwise has a very restricted distribution within Eutheria, occurring elsewhere in some catarrhine primates (MacPhee, 1994) and the tillodont *Esthonyx* (T. Penkrot and S. Zack, personal observation), as well as in macropodid marsupials (Szalay, 1994). The cotylar fossa/medial malleolus complex occurs in taxa that show a variety of locomotor repertoires ranging from semi-arboreal (extant Hyracoidea) to cursorial (Macroscelidea, *Antilohyrax*) to graviportal (basal Proboscidea) to fossorial (Tubulidentata). This broad functional distribution indicates that convergence due to similar locomotor habits is unlikely to be the reason for the recurrence of this complex. As such, the cotylar fossa/expanded medial malleolus offers a potential synapomorphy linking apheliscids and macroscelideans to a suite of putative ungulates.

There are hints that other features may be supportive of such a relationship. For instance, the basal hyracoid *Antilohyrax* shows a degree of fusion of the tibia and fibula similar to that seen in apheliscids and macroscelideans (Rasmussen and Simons, 2000), a condition that is otherwise rare among early mammals with an ungulate dentition. At present, however, the limited record of early Paleogene paenungulates and tubulidentates, combined with the morphologic diversity of known members of these taxa obscures other potential synapomorphies. This one character complex is not itself sufficient to overturn the strong morphologic support for other groupings such as Altungulata (Perissodactyla plus Paenungulata; e.g., Thewissen and Domning, 1992), but it suggests that more similarities between African ungulates will be revealed when more of the early history of these groups is known.

While the cotylar fossa/expanded medial malleolus suggests that macroscelidean affinities may lie with paenungulates and tubulidentates, it does not resolve whether macroscelideans are ungulates or afrotheres, because both of these groups are also members of Afrotheria in molecular studies. Similarly, *Meniscotherium* and *Plesiorycteropus* have been allied with afrotheres in a recent phylogenetic analysis combining morphological and molecular data (Asher et al., 2003). The cotylar fossa complex does not, however, represent a clear morphological synapomorphy of Afrotheria, as it is lacking in chrysochlorids and tenrecids, with the exception of the tenrecid *Potamogale* (Salton and Szalay, 2004), but it does suggest a link between the more ungulate-like afrotheres (potentially including *Meniscotherium* and *Plesiorycteropus*), which also share an herbivorously adapted dentition, in contrast to the insectivorous dentitions of tenrecids and chrysochlorids. Alternatively, the cotylar fossa could have been present primitively in afrotheres and lost in most tenrecoids, a possibility that has been suggested in the case of dental features (Robinson and Seiffert, 2004). Either way, dental and tarsal morphology indicates that a close relationship between apheliscids, macroscelideans, paenungulates, and *Meniscotherium* is not unreasonable. Tubulidentates and *Plesiorycteropus* may also be related to this group, but in the absence of meaningful dental evidence, such a relationship is more tentative. Placement of this clade in either Ungulata or Afrotheria must await resolution of the larger conflict between morphological and molecular data in reconstructing the higher-level phylogeny of Eutheria.

5.7 Macroscelidean Biogeography

The finding that *Apheliscus* and *Haplomylus* are stem macroscelideans has significant implications for the biogeography of Macroscelidea. *Apheliscus* and *Haplomylus* represent the first taxa from outside of Africa strongly linked to macroscelideans. The sister taxon relationship between *Haplomylus* and African Macroscelidea (excluding *Apheliscus*) further suggests a North American origin for the order. The late Paleocene first appearance of *Phenacodaptes* (Archibald, 1998), widely considered the sister taxon or ancestor of *Apheliscus* (Gazin, 1959; Rose, 1981), also represents a downward extension of the temporal range of the macroscelidean stem of almost 10 million years.

The full biogeographic implications of these findings are largely dependent on the true higher-level affinities of macroscelideans. If macroscelideans are members of Ungulata, as traditionally defined, a North American origin for the order would not be surprising, given that North America was home to a diverse Paleocene radiation of basal ungulates placed in the order "Condylarthra." Moreover, early Paleocene North American taxa such as *Protungulatum* are widely viewed as plesiomorphic enough to have given rise to all remaining ungulates (Sloan and Van Valen, 1965; Van Valen, 1978; Cifelli, 1983; Archibald, 1998).

If macroscelideans are, instead, anagalidans, the present data simply replace one biogeographic conundrum with another. With the notable exception of Macroscelidea, all major anagalidan clades make their first appearances in Asia. The Asian Paleocene played host to a diversity of anagalidans, mirroring the diversity of Paleocene ungulates in North America. Both groups occupied a number of small- to medium-sized herbivorous and omnivorous niches on their respective continents. Regardless of whether they originated in North America or Africa, current evidence places the origin of Macroscelidea outside of Asia. Therefore, if macroscelideans are anagalidans, dispersal from Asia must still be invoked to explain their presence outside that continent, and the lack of stem macroscelideans in Asia remains a problem.

The implications of macroscelidean affinities for apheliscids are most dramatic if the afrothere hypothesis proves to be correct. As indicated by the name of the group, Africa has been considered the center of origin of Afrotheria, based on the dominantly African distributions and first appearances of most afrothere groups (Figure 5.17). In fact, prior to the present work, Macroscelidea was one of three afrothere clades (the others being Tenrecidae and Chrysochloridae) restricted to Africa, and one of four (the above three plus Hyracoidea) with unambiguously African first appearances (Butler, 1984; Mahboubi et al., 1986; Gheerbrant et al., 2003). Of the three remaining afrothere clades, the first appearance of Sirenia is unambiguously non-African (Savage et al., 1994). Sirenians are aquatic and generally marine, though, which facilitates intercontinental dispersal. While the oldest definitive proboscideans come from the early Eocene of Africa (Gheerbrant et al., 1996, 2002, 2003), anthracobunids, a poorly known group sometimes allied with proboscideans (Wells and Gingerich, 1983), appear at the same time in Indo-Pakistan. The oldest definitive tubulidentate is *Myorycteropus* from the early Miocene of Africa (MacInnes, 1956; Patterson, 1975). A possible tubulidentate, *Leptomanis*, is known from the late Eocene or early Oligocene of Europe (Thewissen, 1985), but this material may represent a pholidotan (Storch, 1978).

The presence of Paleocene macroscelidean relatives in North America provides the strongest evidence to date for a non-African origin of an afrothere group. The Paleocene first appearance of apheliscids would make them the oldest known afrotheres by approximately 5 million years, and would mark the first record of afrotheres in the Paleocene. In conjunction with the weaker evidence that sirenians, proboscideans, and tubulidentates had non-African origins, this weakens the

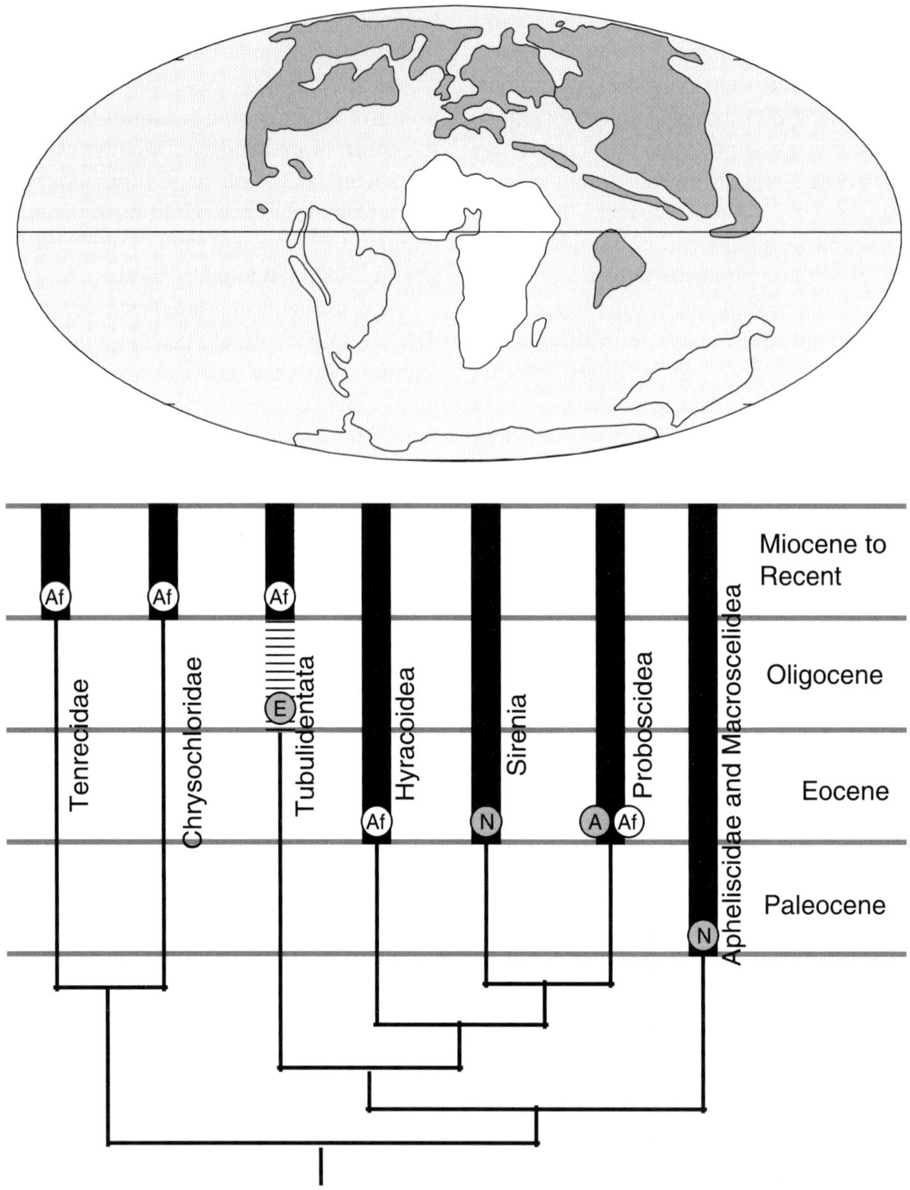

FIGURE 5.17. Biogeography of the first appearances of extant afrothere clades. Cladogram simplified from Figure 15C. Several extinct clades have been excluded. However, we note that, with the exception of *Plesiorycteropus*, all of the excluded groups have Holarctic first appearances. Heavy lines indicate the temporal ranges of afrothere clades. Dots indicate first appearances on either Africa (white) or Holarctic continents (gray). Letters within dots correspond to specific continents: A, Asia; Af, Africa; E, Europe; N, North America. The hatched line for Tubulidentata represents a controversial record from the Oligocene of Europe. The pair of dots for Proboscidea represent undoubted early Eocene records from Africa and a disputed record from Indo-Pakistan. (Map modified from Scotese, 2001.)

evidence for an African origin of Afrotheria. The possibility of an African origin for afrotheres cannot be dismissed, but the recognition of the earliest potential afrotheres in the Paleocene of North America suggests that Holarctic origins are equally likely. More data are clearly needed, particularly from the Paleocene of Europe, which would provide one likely route between North America and Africa. There is evidence for Paleocene and earlier Eocene faunal exchange between Europe and both North America (Russell, 1964; Hooker, 1994; Hooker and Dashzeveg, 2003) and Africa (Gheerbrant and Russell, 1989; Tabuce et al., 2005). Of particular interest is the recent report of forms with similarities to two European apheliscid genera, *Paschatherium* and *Microhyus*, in the middle Eocene of Africa (Tabuce et al., 2005). While the latter genera are tarsally somewhat divergent from *Apheliscus, Haplomylus*, and African macroscelideans (Godinot et al., 1996; Tabuce et al., 2006), their presence in Africa indicates that apheliscids were capable of intercontinental dispersal and highlights the gaps in our knowledge of the history of these taxa.

5.8 Conclusions

Apheliscid postcranial morphology strongly indicates specialized cursorial (or possibly saltatorial) locomotion. Several features indicative of cursoriality in *Apheliscus* are present in a more extreme form in *Haplomylus*. Together with the dentition, which indicates particular similarities between *Haplomylus* and macroscelideans (Simons et al., 1991), the morphology of the previously unknown apheliscid postcranium strongly supports a close phylogenetic relationship between apheliscid condylarths and Macroscelidea. Multiple shared derived characters unite apheliscids with macroscelideans to the exclusion of other taxa, including the presence of an anterior tubercle on the distal tibia, a recurved, hook-like medial malleolus, and a well-defined cotylar fossa on the astragalus. The apheliscid genera *Apheliscus* and *Haplomylus* represent two steps in a continuum of increasing specialization for rapid locomotion when compared with extant macroscelideans, which show further elongation of the crus and more substantial adaptations to restrict mobility at the elbow.

Comparison of apheliscid postcranium to that of *Hyopsodus* illustrates fundamentally different functions in the two taxa. *Hyopsodus* was likely capable of generalized terrestrial locomotion, mixing traits typically associated with scansorial, cursorial, and fossorial mammals, while apheliscids were likely capable of cursoriality. Cladistic analyses presented here and elsewhere do not support a close phylogenetic relationship between apheliscids and *Hyopsodus* (Rigby, 1980; Tabuce et al., 2001; Zack et al., 2005a, b).

There are superficial similarities between apheliscid postcrania and those of contemporary leptictids. Both taxa appear to be adapted for hindlimb-propelled, rapid terrestrial locomotion, although leptictid forelimb morphology suggests somewhat more habitual fossorial behavior than in apheliscids. Despite the superficial resemblance between leptictids and apheliscids, detailed comparisons of postcranial elements and dentition do not support a close relationship between these taxa, implying that the similarities are convergent.

Morphologic similarities in the distal tibia-fibula and tarsus of apheliscids and the erinaceomorph "insectivore" *Macrocranion* are also intriguing. A combination of features, including an anterior tubercle on the distal tibia, a recurved medial malleolus, a well-defined cotylar fossa on the medial aspect of the astragalar body, and several calcaneal characters, is virtually identical in configuration to the tibiotarsal joint in *Haplomylus*. Further study of the postcranium of *Macrocranion* is needed to resolve its relationship to apheliscids, but these similarities indicate that the boundary between small-bodied condylarths and erinaceomorph insectivores deserves renewed investigation.

Our study of apheliscid postcrania raises several issues relevant to future work. First, the fact that the oldest known macroscelidean relatives are now recognized from the Paleocene and early Eocene of western North America indicates that the earliest stages of macroscelidean evolution may have taken place outside of Africa. This poses a biogeographic problem, but we hope that future discoveries from North America, Europe, and Africa will fill these gaps in the sengi fossil record. Second, as mentioned above in the discussion of *Macrocranion* postcrania, the interrelationships of apheliscids, other small-bodied condylarths, and erinaceomorph insectivores requires further study. Addressing this issue may help to resolve the biogeographic questions noted above by (potentially) identifying additional macroscelidean relatives among putative erinaceomorph insectivore taxa, including forms present in Europe such as *Macrocranion, Adunator*, and *Adapisorex*. The identification macroscelidean relatives among European erinaceomorphs, in combination with the existing European record of louisinine apheliscids would provide a clearer link between North American apheliscids of the Paleocene and Eocene and the more recent African macroscelideans, although it would imply some homoplasy in either the dentition or the tarsus. Third, while the new material does not fully resolve the controversy surrounding the higher-level systematic position of Macroscelidea, it does weaken the support for a relationship to Anagalida, while strengthening the case for a relationship to African ungulates, within either Ungulata or Afrotheria. Finally, the discovery of fossil sengis in the Paleocene and Eocene of western North America, as the oldest known representatives of any potential afrotherian group, has implications for the biogeography of Afrotheria. That the earliest known afrothere taxa now occur outside of Africa does not support the hypothesis that afrotheres originated in Africa very early in eutherian history, but instead suggests a post-Cretaceous, Holarctic origin.

Acknowledgments. We thank T. M. Bown, for discovery of the original *Apheliscus* skeletal association; D. Diveley, J. Meng (AMNH), R. Emry, L. Gordon, H. Kafka, J. Mead, R. Purdy (USNM), S. G. Strait (Marshall University, Department of Biological Sciences), and P. Houde (New Mexico State University) for access to specimens; R. Ayoob, L. Fet, B. Sang, N. Smith, C. Williamson (Marshall University), and P. Holman for collection of many of the *Macrocranion* specimens from Castle Gardens; D. M. Boyer, J. C. Mussell, and J. A. Salton for helpful discussions; C. B. Ruff, M. F. Teaford, and D. B. Weishampel (JHU) for access to equipment and facilities; R. J. Asher for making his updated matrix available; M. Habib (JHU) for original illustrations (Figures 5.13 and 5.14); K. C. Beard, P. A. Holroyd, Z.-X. Luo, and M. T. Silcox for encouragement; and two anonymous reviewers whose comments on an earlier version of this manuscript have substantially improved it. Thanks to M. Dagosto and E. J. Sargis for the opportunity to contribute to this volume. We thank F. S. Szalay for his many contributions to mammalian paleontology, particularly his studies of functional morphology and the significance of postcranial morphology to mammalian phylogenetics. Finally, we would also like to thank the many people who have helped collect, prepare, identify, and sort specimens from the Willwood quarries throughout the years, for their work in particular made this project possible. Support of fieldwork leading to these discoveries has been provided by the National Geographic Society, the US National Science Foundation (grants BSR-8215099, BSR-8500732, DEB-8918755, IBN-9419776, and EAR-0000941), and the Bureau of Land Management.

Appendix

Abbreviations for Figures

afc	astragalar facet (cuboid)
at	anterior tubercle
cba	cuboid facet (astragalus)
cbc	cuboid facet (calcaneum)
ce	capitular eminence
cfc	calcaneal facet (cuboid)
cty	cotylar fossa
dp	deltopectoral crest
eca	ectal facet (astragalus)
ecc	ectal facet (calcaneum)
ef	entepicondylar foramen
ff	fibular facet
gtb	greater tuberosity
gtr	greater trochanter
ltb	lesser tuberosity
ltr	lesser trochanter
me	medial epicondyle
mm	medial (tibial) malleolus
nva	navicular facet (astragalus)
nvc	navicular facet (cuboid)
of	olecranon fossa/ supratrochlear foramen
pff	proximal fibular facet
pt	patellar trochlea
pt	peroneal tubercle
sc	supinator crest
sfa	sustentacular facet (astragalus)
sus	sustentaculum/ calcaneal sustentacular facet
tb	cnemial crest (tibial crest)
tc	tibial crest (eminence of distal articular surface)
tro	astragalar trochlea
ttr	third trochanter

References

Archibald, J. D., 1998. Archaic ungulates ("Condylarthra"). In: Janis, C. M., Scott, K. M., Jacobs, L. L. (Eds.), Evolution of Tertiary Mammals of North America. Cambridge University Press, Cambridge, pp. 292–328.

Argot, C., 2001. Functional-adaptive anatomy of the forelimb in the Didelphidae, and the paleobiology of the Paleocene marsupials *Mayulestes ferox* and *Pucadelphys andinus*. Journal of Morphology 247, 51–79.

Argot, C., 2002. Functional-adaptive analysis of the hindlimb anatomy of extant marsupials and the paleobiology of the Paleocene marsupials *Mayulestes ferox* and *Pucadelphys andinus*. Journal of Morphology 253, 76–108.

Argot, C., 2003. Postcranial functional adaptations in the South American Miocene borhyaenoids (Mammalia, Metatheria): *Cladosictis, Pseudonotictis* and *Sipalocyon*. Alcheringa 27, 303–356.

Argot, C., 2004. Functional-adaptive features and palaeobiologic implications of the postcranial skeleton of the late Miocene sabretooth borhyaenoid *Thylacosmilus atrox* (Metatheria). Alcheringa 28, 229–266.

Asher, R. J., 1999. A morphological basis for assessing the phylogeny of the "Tenrecoidea" (Mammmalia, Lipotyphla). Cladistics 15, 231–252.

Asher, R. J., Novacek, M. J., Geisler, J. H., 2003. Relationships of endemic African mammals and their fossil relatives based on morphological and molecular evidence. Journal of Mammalian Evolution 10, 131–194.

Asher, R. J., Emry, R. J., McKenna, M. C., 2005a. New material of *Centetodon* (Mammalia, Lipotyphla) and the importance of (missing) DNA sequences in systematic paleontology. Journal of Vertebrate Paleontology 25, 911–923.

Asher, R. J., Meng, J., Wible, J. R., McKenna, M. C., Rougier, G. W., Dashzeveg, D., Novacek, M. J., 2005b. Stem Lagomorpha and the antiquity of Glires. Science 307, 1091–1094.

Barnett, C. H., Napier, J. R., 1953. The rotatory mobility of the fibula in eutherian mammals. Journal of Anatomy 87, 11–21.

Berman, S. L., 1985. Convergent evolution in the hindlimb of bipedal rodents. Zeitschrift fuer Zoologische Systematik und Evolutionsforschung 23, 59–77.

Bleefeld, A. R., McKenna, M. C., 1985. Skeletal integrity of *Mimolagus rodens* (Lagomorpha, Mammalia). American Museum Novitates 2806, 1–5.

Bloch, J. I., Bowen, G. J. (2001). Paleocene-eocene microvertebrates in freshwater limestones of the Willwood formation, Clarks Fork Basin, Wyoming. In: Gunnell G. F. (Ed.), Eocene Biodiversity: Unusual Occurrences and Rarely Sampled Habitats. Kluwer/ Plenum, New York, pp. 95–129.

Bown, T. M., 1979. Geology and mammalian paleontology of the Sand Creek facies, lower Willwood Formation (lower Eocene), Washakie County, Wyoming. Memoirs of the Geololgical Survey of Wyoming 2, 1–151.

Bown, T. M., Schankler, D. M., 1982. A review of the Proteutheria and Insectivora of the Willwood Formation (lower Eocene), Bighorn Basin, Wyoming. Geololgical Survey Bulletin 1523, 1–79, 10 pls.

Bown, T. M., Holroyd, P. A., Rose, K. D., 1994a. Mammal extinctions, body size, and paleotemperature. Proceedings of the National Academy of Sciences of the United States of America 91, 10403–10406.

Bown, T. M., Rose, K. D., Simons, E. L., Wing, S. L., 1994b. Distribution and stratigraphic correlation of Upper Paleocene and Lower Eocene fossil mammal and plant localities of the Fort Union, Willwood, and Tatman Formations, southern Bighorn Basin, Wyoming. USGS Professional Papers 1540, 1–103, 2 pls.

Boyer, D. M., Bloch, J. I., 2003. Comparative anatomy of the pentacodontid *Aphronorus orieli* (Mammalia: Pantolesta) from the paleocene of the Western Crazy Mountains Basin, Montana. Journal of Vertebrate Paleontology 23, 36A.

Bramble, D. M., 1989. Cranial specialization and locomotor habit in the Lagomorpha. American Zoologist 29, 303–317.

Butler, P. M., 1984. Macroscelidea, Insectivora, and Chiroptera from the Miocene of East Africa. Palaeovertebrata 14, 117–200.

Butler, P. M., 1995. Fossil Macroscelidea. Mammal Review 25, 3–14.

Cifelli, R. L., 1983. The origin and affinities of the South American Condylarthra and early Tertiary Litopterna (Mammalia). American Museum Novitates 2772, 1–49.

Cifelli, R. L., 1985. Biostratigraphy of the Casamayoran, Early Eocene, of Patagonia. American Museum Novitates 2820, 1–26.

Cope, E. D., 1874. Report upon vertebrate fossils discovered in New Mexico, with description of new species. Annual Report Chief Engineers, US Army Appendix FF, 589–606.

Cope, E. D., 1875. Systematic catalogue of Vertebrata of the Eocene of New Mexico, collected in 1874. Geographic Explorations and Surveys west of the 100th meridian, G. M. Wheeler, Corps of Engineers, US Army, Washington, 4, 37–282.

Cope, E. D., 1877. Report upon the extinct Vertebrata obtained in New Mexico by parties of the expedition of 1874. In: Wheeler, G. M. (Ed.), Chapter XII: Fossils of the Eocene Period, from Geological Surveys west of the 100th Meridian. Washington: Corps of Engineers, US Army, pp. 37–282.

Delson, E., 1971. Fossil mammals of the early Wasatchian Powder River local fauna, Eocene of northeast Wyoming. Bulletin of the American Museum of Natural History 146, 305–364.

Emerson, S. B., 1985. Jumping and leaping. In: Bramble, D. M., Hildebrand, M., Liem, K. F., Wake, D. B. (Eds.), Functional Vertebrate Morphology. Belknap Press of Harvard University Press, Cambridge, MA, pp. 58–72.

Erfurt, J., 2000. Rekonstruktion des Skelettes und der Biologie von *Anthracobunodon weigelti* (Artiodactyla, Mammalia) aus dem Eozän des Geiseltales. Hallesches Jahrbuch für Geowissenschaften 12, 57–141.

Fischer, M. S., 1994. Crouched posture and high fulcrum, a principle in the locomotion of small mammals: the example of the rock hyrax (*Procavia capensis*) (Mammalia: Hyracoidea). Journal of Human Evolution 26, 501–524.

Franzen, J. L., 1981. Das erste Skelett eines Dichobuniden (Mammalia, Artiodactyla), geborgen aus mitteleozänen Ölschiefern der "Grube Messell" bie Darmstadt (Deutschland, S-Hessen). Senckenbergiana Lethaea 61, 299–353.

Franzen, J. L., 1988. Skeletons of *Aumelasia* (Mammalia, Artiodactyla, Dichobunidae) from Messel (M. Eocene, W. Germany). Courier Forschungsinstitut Senckenberg 107, 309–321.

Gambaryan, P. P., 1974. How Mammals Run. Wiley, New York.

Gazin, C. L., 1959. Early Tertiary *Apheliscus* and *Phenacodaptes* as pantolestid insectivores. Smithsonian Miscellaneous Collections 139, 1–7, pls 1–2.

Gazin, C. L., 1968. A study of the Eocene condylarthran mammal *Hyopsodus*. Smithsonian Miscellaneous Collections 153, 1–90, 13 pls.

Gebo, D. L., Rose, K. D., 1993. Skeletal morphology and locomotor adaptation in *Prolimnocyon atavus*, an early Eocene hyaenodontid creodont. Journal of Vertebrate Paleontology 13, 125–144.

Gheerbrant, E., Russell, D. E., 1989. Presence of the genus *Afrodon* [Mammalia, Lipotyphla (?), Adapisoriculidae] in Europe; new data for the problem of trans-Tethyan relations between Africa and Europe around the K/T boundary. Palaeogeography, Palaeoclimatology, Palaeoecology 76, 1–15.

Gheerbrant, E., Sudre, J., Capetta, H., 1996. A Palaeocene proboscidean from Morocco. Nature 383, 68–71.

Gheerbrant, E., Sudre, J., Capetta, H., Iarochene, M., Amaghzaz, M., Bouya, B., 2002. A new large mammal from the Ypresian of Morocco: evidence of surprising diversity of early proboscideans. Acta Palaeontologica Polonica 47, 493–506.

Gheerbrant, E., Sudre, J., Capetta, H., Mourer-Chauviré, C., Bourdon, E., Iarochene, M., Amaghzaz, M., Bouya, B., 2003. Les localités à mammifères des carrières de Grand Daoui, bassin des Ouled Abdoun, Maroc, Yprésien: premier état des lieux. Bulletin de la Société Géologique de France 174, 279–293.

Gingerich, P. D., 1976. Paleontology and phylogeny: patterns of evolution at the species level in early Tertiary mammals. American Journal of Science 276, 1–28.

Gingerich, P. D., 1983. New Adapisoricidae, Pentacodontidae, and Hyopsodontidae (Mammalia, Insectivora and Condylarthra) from the late Paleocene of Wyoming and Colorado. Contributions from the Museum of Paleontology University of Michigan 26, 227–255.

Gingerich, P. D., 1994. New species of *Apheliscus, Haplomylus*, and *Hyopsodus* (Mammalia, Condylarthra) from the late Paleocene of southern Montana and early Eocene of northwestern Wyoming. Contributions from the Museum of Paleontology, University of Michigan, Ann Arbor, MI 29, 119–134.

Godinot, M., Smith, T., Smith, R., 1996. Mode de vie et affinites de *Paschatherium* (Condylarthra, Hyopsodontidae) d'apres ses os du tarse. Palaeovertebrata 25, 225–242.

Hartenberger, J.-L., 1986. Hypothèse paléontologique sur l'origine des Macroscelidea (Mammalia). Comptes Rendus de l'Académie des Sciences, Serie II 302, 247–249.

Heinrich, R. E., Rose, K. D., 1997. Postcranial morphology and locomotor behaviour of two early Eocene miacoid carnivorans, *Vulpavus* and *Didymictis*. Palaeontology 40, 279–305.

Hildebrand, M., 1982. Analysis of Vertebrate Structure. Wiley, Toronto.

Hildebrand, M., 1985. Walking and running. In: Hildebrand, M., Bramble, D. M., Liem, K. F., Wake, D. B. (Eds.), Functional Vertebrate Morphology. Belknap, Cambridge.

Holroyd, P. A., Mussell, J. C., 2005. Macroscelidea and Tubulidentata. In: Rose, K. D., Archibald, J. D. (Eds.), The Rise of Placental Mammals, Origins and Relationships of the Major Extant Clades. Johns Hopkins University Press, Baltimore, MD, pp. 71–83.

Hooker, J. J., 1994. The beginning of the equoid radiation. Zoological Journal of the Linnean Society 112, 29–63.

Hooker, J. J., Dashzeveg, D., 2003. Evidence for direct mammalian faunal interchange between Europe and Asia near the Paleocene-Eocene boundary. Geological Society of America Special Paper 369, 479–500.

Jong, W. W. de, Leunissen, J. A. M., Wistow, G. J., 1993. Eye lens crystallins and the phylogeny of placental orders: evidence for a macroscelidid-paenungulate clade? In: Szalay, F. S., Novacek, M. J., McKenna, M. C. (Eds.), Mammal Phylogeny. Volume 2, Placentals. Springer, New York, pp. 5–12.

Kitts, D. B., 1956. American *Hyracotherium* (Perissodactyla, Equidae). Bulletin of the American Museum of Natural History 110, 1–60.

MacInnes, D. G., 1956. Fossil Tubulidentata from East Africa. Fossil Mammals of Africa. Bulletin of the British Museum (Natural History) 10, 1–38.

MacPhee, R. D. E., 1994. Morphology, adaptations, and relationships of *Plesiorycteropus*, and a diagnosis of a new order of eutherian mammals. Bulletin of the American Museum of Natural History 220, 1–214.

Madsen, O., Scally, M., Douady, C. J., Kao, D., DeBry, R. W., Adkins, R., Amrine, H. M., Stanhope, M. J., de Jong, W. W., Springer, M. S., 2001. Parallel adaptive radiations in two major clades of placental mammals. Nature 409, 610–614.

Mahboubi, M., Ameur, R., Crochet, J.-Y., Jaeger, J.-J., 1986. El Kohol (Saharan Atlas, Algeria): a new Eocene mammal locality in northwestern Africa. Palaeontographica Abteilung A 192, 15–49.

Matthew, W. D., 1915. A revision of the lower Eocene Wasatch and Wind River faunas: Part II. Order Condylarthra, Family Hyopsodontidae. Bulletin of the American Museum of Natural History 34, 311–328.

Matthew, W. D., 1918. A revision of the lower Eocene Wasatch and Wind River faunas (Part V.– Insectivora (continued), Glires, Edentata). Bulletin of the American Museum of Natural History 34, 565–657.

Matthew, W. D., 1937. Paleocene faunas of the San Juan Basin, New Mexico. Transactions of the American Philosophical Society 30, 1–510.

McKenna, M. C., 1960. Fossil Mammalia from the early Wasatchian four mile fauna, Eocene of northwest Colorado. University of California Publications in Geological Sciences 37, 1–130.

McKenna, M. C., 1975. Toward a phylogenetic classification of the Mammalia. In: Luckett, W. P., Szalay, F. S. (Eds.), Phylogeny of the Primates, a Multidisciplinary Approach. Plenum, New York, pp. 21–46.

McKenna, M. C., 1980. Late Cretaceous and Early Tertiary vertebrate paleontological reconnaissance, Togwotee Pass area, northwestern Wyoming. In: Jacobs, L. L. (Ed.), Aspects of Vertebrate History: Essays in Honor of Edwin Harris Colbert. Museum of Northern Arizona Press, Flagstaff, pp. 321–343.

McKenna, M. C., Bell, S. K., 1997. Classification of Mammals above the Species Level. Columbia University Press, New York.

Meng, J., Hu, Y., Li, C., 2003. The osteology of *Rhombomylus* (Mammalia, Glires): implications for phylogeny and evolution of Glires. Bulletin of the American Museum of Natural History 275, 1–247.

Meng, J., Bowen, G. J., Ye, J., Koch, P. L., Ting, S., Li, Q., Jin, X., 2004. *Gomphos elkema* (Glires, Mammalia) from the Erlian Basin: evidence for the early Tertiary Bumbanian Land Mammal Age in Nei-Mongol, China. American Museum Novitates 3425, 1–24.

Muizon, C. de, Cifelli, R. L., 2000. The "condylarths" (archaic Ungulata, Mammalia) from the early Palaeocene of Tiupampa (Bolivia): implications on the origin of the South American ungulates. Geodiversitas 22, 47–150.

Muizon, C. de, Cifelli, R. L., Bergqvist, L. P., 1998. Eutherian tarsals from the early Paleocene of Bolivia. Journal of Vertebrate Paleontology 18, 655–663.

Murphy, W. J., Eizirik, E., Johnson, W. E., Zhang, Y. P., Ryder, O. A., O'Brien, S. J., 2001. Molecular phylogenetics and the origins of placental mammals. Nature 409, 614–618.

Novacek, M. J., 1982. Information for molecular studies from anatomical and fossil evidence on higher eutherian phylogeny. In: Goodman, M. (Ed.), Macromolecular Sequences in Systematic and Evolutionary Biology. Plenum, New York, pp. 3–41.

Novacek, M. J., 1986. The skull of leptictid insectivorans and the higher-level classification of eutherian mammals. Bulletin of the American Museum of Natural History 183, 1–111.

Novacek, M. J., Wyss, A., 1986. Higher-level relationships of the recent eutherian orders: morphological evidence. Cladistics 2, 257–287.

Novacek, M. J., Wyss, A., McKenna, M. C., 1988. The major groups of eutherian mammals. In: Benton, M. J. (Ed.), The Phylogeny and Classification of the Tetrapods, Volume 2: Mammals. Clarendon, Oxford, pp. 31–71.

O'Leary, M. A., Rose, K. D., 1995. Postcranial skeleton of the early Eocene mesonychid *Pachyaena* (Mammalia: Mesonychia). Journal of Vertebrate Paleontology 15, 401–430.

Offermans, M., de Vree, F., 1987. Locomotion patterns in the springhare *Pedetes capensis* (Rodentia; Pedetidae). Revue de zoologie africaine 101, 271–274.

Offermans, M., de Vree, F., 1988. Appendicular morphology of the saltatorial rodent *Pedetes capensis* (Forster, 1778). Annales de la Societe royale zoologique de Belgique 118, 211–228.

Patterson, B., 1975. The fossil aardvarks (Mammalia: Tubulidentata). Bulletin of the Museum of Comparative Zoology 147, 185–237.

Penkrot, T. A., Zack, S. P., Strait, S. G., 2004. New postcrania of *Macrocranion* (Eutheria: Amphilemuridae) from the early Eocene, Bighorn Basin, WY. Journal of Vertebrate Paleontology 24, 101A.

Radinsky, L. B., 1966. The adaptive radiation of the phenacodontid condylarths and the origin of the Perissodactyla. Evolution 20, 408–417.

Rasmussen, D. T., Simons, E. L., 2000. Ecomorphological diversity among Paleogene hyracoids (Mammalia): a new cursorial browser from the Fayum, Egypt. Journal of Vertebrate Paleontology 20, 167–176.

Rigby, J. K., 1980. Swain Quarry of the Fort Union Formation, middle Paleocene (Torrejonian), Carbon County, Wyoming: geologic setting and mammalian fauna. Evolutionary Monographs 3, 1–178.

Rigby, J. K., 1981. A skeleton of *Gillisonchus gillianus* (Mammalia; Condylarthra) from the early Paleocene (Puercan) Ojo Alamo Sandstone, San Juan Basin, New Mexico, with comments on the local stratigraphy of Betonnie Tsosie Wash. In: Lucas, S., Rigby, J. K., Jr., Kues, B. (Eds.), Advances in San Juan Basin Paleontology. University of New Mexico Press, Albuquerque, pp. 89–126.

Robinson, P., Williams, B. A., 1997. Species diversity, tooth size, and shape of *Haplomylus* (Condylarthra, Hyopsodontidae) from the Powder River Basin, northeastern Wyoming. Contributions to Geology, University of Wyoming, 31, 59–78.

Robinson, T. J., Seiffert, E. R., 2004. Afrotherian origins and interrelationships: new views and future prospects. Current Topics in Developmental Biology 63, 37–60.

Roca, A. L., Bar-Gal, G. K., Eizirik, E., Helgen, K. M., Maria, R., Springer, M. S., O'Brien, S. J., Murphy, W. J., 2004. Mesozoic origin for West Indian insectivores. Nature 429, 649–651.

Rose, K. D., 1981. The Clarkforkian land-mammal age and mammalian faunal composition across the Paleocene-Eocene boundary. University of Michigan, Papers in Paleontology 26, 1–189, pls. 1–4.

Rose, K. D., 1982. Skeleton of *Diacodexis*, oldest known artiodactyl. Science 216, 621–623.

Rose, K. D., 1985. Comparative osteology of North American dichobunid artiodactyls. Journal of Paleontology 59, 1203–1226.

Rose, K. D., 1987. Climbing adaptations in the early Eocene mammal *Chriacus* and the origin of Artiodactyla. Science 236, 314–316.

Rose, K. D., 1990. Postcranial skeletal remains and adaptations in early Eocene mammals from the Willwood Formation, Bighorn Basin, Wyoming. GSA Special Paper 243, 107–133.

Rose, K. D., 1996. Skeleton of early Eocene *Homogalax* and the origin of Perissodactyla. Palaeovertebrata 25, 243–260.

Rose, K. D., 1999. Postcranial skeleton of Eocene Leptictidae (Mammalia), and its implications for behavior and relationships. Journal of Vertebrate Paleontology 19, 355–372.

Rose, K. D., Chinnery, B. J., 2004. The postcranial skeleton of early Eocene rodents. Bulletin of the Carnegie Museum of Natural History 36, 211–244.

Rose, K. D., Krishtalka, L., Stucky, R. K., 1991. Revision of the Wind River faunas, early Eocene of central Wyoming. Part 11. Palaeanodonta (Mammalia). Annals of the Carnegie Museum 60, 63–82.

Russell, D. E., 1964. Les mammifères paléocènes d'Europe. Memoires du Museum National d'Histoire Naturelle 13, 1–324, pls. 1–16.

Russell, D. E., Louis, P., Savage, D. E., 1975. Les Adapisoricidae de l'Eocene Inferieur de France. Reevaluation des formes considerees affines. Bulletin du Museum National d'Histoire Naturelle, Section C, Paris 327, 129–194, 7 pls.

Salton, J. A., Szalay, F. S., 2004. The tarsal complex of afro-malagasy Tenrecoidea: a search for phylogenetically meaningful characters. Journal of Mammalian Evolution 11, 73–104.

Savage, R. J. G., Domning, D. P., Thewissen, J. G. M., 1994. Fossil Sirenia of the West Atlantic and Carribean region. V. The most primitive known sirenian, *Prorastomus sirenoides* Owen, 1855. Journal of Vertebrate Paleontology 14, 427–449.

Schaeffer, B., 1947. Notes on the origin and function of the artiodactyl tarsus. American Museum Novitates 1356, 1–24.

Scotese, C. R., 2001. Digital Paleogeographic Map Archive on CD-ROM, PALEOMAP Project. Arlington, Texas.

Shoshani, J., McKenna, M. C., 1998. Higher taxonomic relationships among extant mammals based on morphology, with selected comparisons of results from molecular data. Molecular Phylogenetics and Evolution 9, 572–584.

Silcox, M. T., Rose, K. D., 2001. Unusual vertebrate microfaunas from the Willwood Formation, early Eocene of the Bighorn Basin, Wyoming. In: Gunnell, G. F. (Ed.), Eocene Biodiversity: Unusual Occurrences and Rarely Sampled Habitats. Kluwer/Plenum, New York, pp. 131–164.

Simons, E. L., Holroyd, P. A., Bown, T. M., 1991. Early tertiary elephant-shrews from Egypt and the origin of the Macroscelidea. Proceedings of the National Academy of Sciences of the United States of America 88, 9734–9737.

Simpson, G. G., 1937. The Fort Union of the Crazy Mountain field, Montana, and its mammalian faunas. Bulletin of the United States National Museum 169, 1–287.

Sloan, R. E., Van Valen, L. M., 1965. Cretaceous mammals from Montana. Science 148, 220–227.

Springer, M. S., Teeling, E. C., Madsen, O., Stanhope, M. J., de Jong, W. W., 2001. Integrated fossil and molecular data reconstruct bat echolocation. Proceedings of the National Academy of Sciences of the United States of America 98, 6241–6246.

Stanhope, M. J., Waddell, V. G., Madsen, O., Jong, W. W. d., Hedges, S. B., Cleven, G. C., Kao, D., Springer, M. S., 1998. Molecular evidence for multiple origins of Insectivora and for a new order of endemic African insectivore mammals. Proceedings of the National Academy of Sciences of the United States of America 95, 9967–9972.

Stein, B. R., Casinos, A., 1997. What is a cursorial mammal? Journal of Zoology, London 242, 185–192.

Storch, G., 1978. *Eomanis waldi*, ein Schuppentier aus dem Mittel-Eozän der "Grube Messel" bei Darmstadt (Mammalia: Pholidota). Senckenbergiana Lethaea 59, 503–529.

Storch, G., 1993. Morphologie und Paläobiologie von *Macrocranion tenerum*, einem Erinaceomorphen aus dem Mittel-Eozän von Messel bei Darmstadt (Mammalia, Lipotyphla). Senckenbergiana Lethaea 73, 61–81.

Storch, G., 1996. Paleobiology of Messel erinaceomorphs. Palaeovertebrata 25, 215–224.

Sulimski, A., 1968. Paleocene genus *Pseudictops* Matthew, Granger, and Simpson, 1929 (Mammalia) and its revision. Acta Palaeontologica Polonica 19, 101–129, pls. 10–14.

Swofford, D. L., 1999. PAUP*. Phylogenetic Analysis Using Parsimony (*and Other Methods), Version 4. Sinauer, Sunderland, MA.

Szalay, F. S., 1977. Phylogenetic relationships and a classification of the eutherian Mammalia. In: Hecht, M. K., Goody, P. C., Hecht, B. M. (Eds.), Major Patterns in Vertebrate Evolution. Plenum, New York, pp. 317–374.

Szalay, F. S., 1985. Rodent and lagomorph morphotype adaptations, origins, and relationships: some postcranial attributes analyzed. In: Luckett, W. P., Haretneberger, J.-L. (Eds.), Evolutionary Relationships among Rodents – a Multidisciplinary Analysis. Plenum, New York, pp. 83–132.

Szalay, F. S., 1994. Evolutionary History of the Marsupials and an Analysis of Osteological Characters. Cambridge University Press, Cambridge.

Szalay, F. S., Decker, R. L., 1974. Origins, evolution, and function of the tarsus in late cretaceous eutheria and paleocene primates. In: Jenkins, F. A., Jr. (Ed.), Primate Locomotion. Academic Press, New York, pp. 223–259.

Tabuce, R., Coiffait, B., Coiffait, P. E., Mahboubi, M., Jaeger, J.-J., 2001. A new genus of Macroscelidea (Mammalia) from the Eocene of Algeria; a possible origin for elephant-shrews. Journal of Vertebrate Paleontology 21, 535–546.

Tabuce, R., Adnet, S., Capetta, H., Noubhani, A., Quillevere, F., 2005. Aznag (Ouarzazate basin, Morocco), a new African middle Eocene (Lutetian) vertebrate-bearing locality with selachians and mammals. Bulletin de la Société Géologique de France 176, 381–400.

Tabuce, R., Antunes, M. T., Smith, R., Smith, T., 2006. Dental and tarsal morphology of the European Paleocene/Eocene "condylarth" mammal *Microhyus*. Acta Palaeontologica Polonica 51, 37–52.

Taylor, C. R., Shkolnik, A., Dmi'el, R., Baharav, D., Borut, A., 1974. Running in cheetahs, gazelles, and goats: energy cost and limb configuration. American Journal of Physiology 227, 848–850.

Taylor, M. E., 1974. The functional anatomy of the forelimb of some African Viverridae (Carnivora). Journal of Morphology 143, 307–336.

Taylor, M. E., 1976. The functional anatomy of the hindlimb of some African Viverridae (Carnivora). Journal of Morphology 148, 227–254.

Thewissen, J. G. M., 1985. Cephalic evidence for the affinities of Tubulidentata. Mammalia 49, 257–284.

Thewissen, J. G. M., 1990. Evolution of Paleocene and Eocene Phenacodontidae (Mammalia, Condylarthra). University of Michigan Papers in Paleontology 29, 1–107.

Thewissen, J. G. M., Domning, D. P., 1992. The role of phenacodontids in the origin of the modern orders of ungulate mammals. Journal of Vertebrate Paleontology 12, 494–504.

Thewissen, J. G. M., Hussain, S. T., 1990. Postcranial osteology of the most primitive artiodactyl *Diacodexis pakistanensis* (Dichobunidae). Anatomia, Histologia, Embryologia 19, 37–48.

Van Valen, L. M., 1967. New Paleocene insectivores and insectivore classification. Bulletin of the American Museum of Natural History 135, 217–284.

Van Valen, L. M., 1978. The beginning of the age of mammals. Evolutionary Theory 4, 45–80.

Wells, N. A., Gingerich, P. D., 1983. Review of Eocene Anthracobunidae (Mammalia, Proboscidea) with a new genus and species, *Jozaria palustris*, from the Kuldana Formation of Kohat (Pakistan). Contributions from the Museum of Paleontology, University of Michigan Ann Arbor, MI 26, 117–139.

Whidden, H. P., 2002. Extrinsic snout musculature in Afrotheria and Lipotyphla. Journal of Mammalian Evolution 9, 161–183.

Wilf, P., Beard, K. C., Davies-Vollum, K. S., Norejko, J. W., 1998. Portrait of a late Paleocene (early Clarkforkian) terrestrial ecosystem: big multi quarry and associated strata, Washakie Basin, southwestern Wyoming. Palaios 13, 514–532.

Williamson, T. E., Lucas, S. G. (1992). *Meniscotherium* (Mammalia, "Condylarthra") from the Paleocene-Eocene of western North America. New Mexico Museum of Natural History and Science Bulletin 1, 30–34.

Zack, S. P., Penkrot, T. A., Bloch, J. I., Rose, K. D. 2005a. Affinities of "hyopsodontids" to elephant shrews and a Holarctic origin of Afrotheria. Nature 434, 497–501.

Zack, S. P., Penkrot, T. A., Krause, D. W., Maas, M. C., 2005b. A new apheliscine "condylarth" mammal from the late Paleocene of Montana and Alberta and the phylogeny of "hyopsodontids." Acta Palaeontologica Polonica 50, 809–830.

Zhou, X., Sanders, W. J., Gingerich, P. D., 1992. Functional and behavioral implications of vertebral structure in *Pachyaena ossifraga* (Mammalia, Mesonychia). Contributions from the Museum of Paleontology, The University of Michigan, Ann Arbor, MI 28, 289–319.

6. Postcranial Skeleton of the Upper Paleocene (Itaboraian) "Condylarthra" (Mammalia) of Itaboraí Basin, Brazil

Lilian P. Bergqvist*
Universidade Federal do Rio de Janeiro
Departamento de Geologia
Av. Athos da Silveira Ramos, 274
Rio de Janeiro/RJ, Brazil, 21941-916
bergqvist@geologia.ufrj.br

6.1 Introduction

The Itaboraí Basin, in the state of Rio de Janeiro, is one of the smallest depositional basins in Brazil and is the only one to have yielded a terrestrial fauna of late Paleocene age. For 50 years its limestone was commercially exploited, allowing the recovery of a great amount of fossil vertebrates, predominantly from fissure fill deposits (Sequence S2 *sensu* Medeiros and Bergqvist, 1999), of Itaboraian age. Among the vertebrates, the fossil mammals are the most abundant, with marsupials being the most diverse but ungulates the most abundant. Fossil edentates are very rare. Since the end of the 1980s, the Itaboraí basin has been completely flooded and no further fossil collecting has been possible (Figure 6.1).

Among the ungulates, the "condylarths" comprise the second least abundant group next to the Xenungulata. The relationships of most "condylarths" are very uncertain and they probably represent a paraphyletic assemblage. Some authors have advocated abandonment of the concept of Condylarthra altogether. Herein I follow the concept of "Condylarthra" as advocated by Archibald (2005), and the recommendation of Prothero et al. (1988) to add quotation marks to the term "condylarth" in order to emphasize its paraphyly, as already done by Muizon and Cifelli (2000).

The first "condylarth" fossils discovered in the Itaboraí basin were assigned by Paula-Couto (1949) to *Didolodus* Ameghino, 1897. A few years later (1952), the same author recognized the presence of four species in the basin, then known as *Ernestokokenia protocenica, Ernestokokenia parayirunhor* (renamed as *Paulacoutoia protocenica* and *Miguelsoria parayirunhor*, respectively, by Cifelli, 1983a), *Lamegoia conodonta* (placed in the Didolodontidae), and *Asmithwoodwardia scotti*, originally considered to be the first South American hyopsodontid but later placed in the Didolodontidade by Paula-Couto (1978). The first two species were differentiated mainly on their lower molars, while the larger size and the presence of a protocone-hypocone crest characterized *Lamegoia*, the least common "condylarth" at Itaboraí.

Lamegoia conodonta is the largest "condylarth" at Itaboraí and approximates the size of a wolf; only twelve isolated teeth are currently recognized for this species. *Paulacoutoia protocenica* is the smallest of the three "condylarth" species of Itaboraí, but the most abundant. Its size is similar to the coati, *Nasua nasua. Victorlemoinea prototypica* is only a little smaller than *L. conodonta*, but much more abundant than the former. A review of all dental specimens of "Condylartha" suggests the presence of two new species of Didolodontidae, not yet described. The length and width (in mm) of m2 of the Itaboraí "condylarths" and other ungulates are provided in Table 6.1. As for all other mammalian species of the Itaboraí basin, except for *Carodnia vieirai* Paula-Couto, 1952, the "condylarth" species were established exclusively on their dental features. Although postcranial bones are almost as abundant as fossil teeth, they were found dissociated and mixed. These remained unstudied until the 1980s, when Cifelli (1983b) undertook the first attempt to assign some isolated postcranials (tarsals) to the Itaboraí species.

Cifelli's (1983b) work resulted in important changes to the current taxonomy of the Itaboraí ungulates, mainly to the "Condylarthra" and Litopterna. Using four different methods, Cifelli (1983b) assigned the isolated foot bones to the Itaboraí ungulate species. His association showed that certain taxa with typical "condylarth" dental morphology presented tarsal bones

* Address for correspondence: bergqvist@geologia.ufrj.br

E.J. Sargis and M. Dagosto (eds.), *Mammalian Evolutionary Morphology: A Tribute to Frederick S. Szalay,* 107–133.

FIGURE 6.1. Itaboraí basin. A, 1957; B, 2003.

TABLE 6.1. Length and width of the m2 of Itaboraí ungulates.

	Length		Width		
Taxa	(min–max)	(mean)	(min–max)	(mean)	Number of specimens
"Condylarthra"					
Paulacoutoia protocenica	7.8–9.7	8.2	6.3–7.7	6.9	8
Lamegoia conodonta	14.8	14.8	12.6	12.6	1
Species indet 1	7.0	7.0	6.4	6.4	1
Species indet 2	10.5	10.5	9.0	9.0	1
Victorlemoinea prototypica	14.5	14.5	9.0	9.0	1
Litopterna					
Asmithwoodwardia scotti	3.3–3.9	3.6	2.6–3.0	2.8	5
Protolipterna ellipsodontoides	3.8–5.5	4.3	3.0–4.2	3.5	75
Miguelsoria parayirunhor	4.3–6.2	5.2	3.4–5.2	4.2	40
Paranisolambda prodromus	8.7–9.4	9.1	5.4–6.4	6.0	12
Notoungulata[a]					
Itaboraitherium atavum	5.1	5.1	3.1	3.1	1
Camargomendesia pristina	4.8–5.6	5.2	3.8	4.7	12
Colbertia magellanica	5.9–8.0	6.9	4.1–6.0	5.0	17
Astrapotheria[a]					
Tetragonostylops apthomasi	10.2–11.9	11.1	6.7–8.0	7.3	5
Xenungulata					
Carodnia vieirai	28.8–33.5	31.2	27.2–28.2	27.7	2

[a] Measurements taken from Cifelli (1983a)

with derived litoptern features, while some primitive tarsals were assigned to a species placed in the order Litopterna due to its derived dental features. He then placed *Miguelsoria parayirunhor* in the order Litopterna and transferred *Victorlemoinea prototypica* to the order "Condylarthra". The placement of *V. prototypica* within "Condylarthra" was not widely accepted (e.g., Bond *et al.*, 1995; McKenna and Bell, 1997; Soria, 2001).

This work is the first of a series that will present the results of a new proposal for reassociation of other postcranial bones (besides ankle bones) to the Itaboraí ungulate species, originally part of my doctoral dissertation (Bergqvist, 1996). Although some may question this study, as the postcranials are not in direct association with teeth, I followed the steps of Dr. Frederick Szalay, who was one of the pioneers to study mammalian isolated postcranials and showed their ultimate importance. In this chapter I describe, illustrate, and comment on all bones assignable to the various "condylarth" species from Itaboraí.

6.2 Abbreviations

CV, coefficient of variation; AMNH, American Museum of Natural History, New York, USA; DGM, Divisão de Geologia e Mineralogia of Departamento Nacional da Produção Mineral,

Rio de Janeiro, Brazil; MCN-PV, Museu de Ciências Naturais – Paleontologia de Vertebrados, Porto Alegre, Brazil; MCT, Museu de Ciências da Terra (continuing DGM collection), Rio de Janeiro, Brazil, UM, University of Michigan, Museum of Paleontology, Michigan, USA; USGS, United States Geological Survey; USNM, United States National Museum, Washington D.C., USA.

6.3 Materials and Methods

The fossil bones of Itaboraí basin are, in general, well-preserved, with little or no abrasion, making the study of even delicate bone structures straightforward. Most of the long bones are broken at the diaphysis, preserving just one end. All were recovered from fissure fill deposits and bear different colors (white, cream, orange, brown). Diogenes Campos and Llewelin Price recovered, in 1968, all the brownish fossils, fortunately a large and important sample from a single fissure in the northeastern side of the basin, named because of its importance as the "1968 Fissure". The other colors come from fissures worked in 1948 and 1949 (northern and southeastern parts of the basin, respectively; Figure 6.2). Many of the fossils lack precise collecting information, and exact provenance cannot be established.

Bones selected for the analysis are those that include at least one end preserved. All long bones (except the fibula), metacarpals, pelvis, tarsals, metatarsals, and ungual phalanges were selected for reassociation. Vertebrae, ribs, and non-ungual phalanges were excluded from this work, as their morphology and intraspecific variation are poorly known. The fossils examined for each taxon are listed in the Appendix.

Several studies based on recent mammals showed that there is a high correlation between different body measurements and body mass (see Damuth and MacFadden, 1990). Such studies have also shown a high correlation between body mass and dental measurements (e.g., Gingerich *et al.*, 1982; Fortelius, 1990).

The use of m1/M1 or m2/M2 as independent variables in the prediction of body mass became more widely accepted after the work of Gingerich (1974), who observed in different mammal species that these teeth are the least variable in size. Damuth (1990) and Fortelius (1990), studying extant ungulates, concluded that the length of the series p4-m3/P4-M3 presents a higher coefficient of correlation ($r = 0.967$) with body mass than do measurements of isolated teeth. This variable, however, has little application in many fossil species, due to fact that complete series are not always (or even usually) preserved. Fortelius (1990) also observed that the width of a tooth is more related to diet than to the size of an animal, suggesting that prediction of body mass should be based on length measurements. Damuth (1990) indicated that the area of a tooth should not be used in correlation, as some lineages show decrease in molar width through time. He showed that the length of a tooth presents a higher coefficient of correlation than the area.

Cifelli (1983a) was a pioneer in the use of logarithmic linear regression in early Paleogene species, and the methodology employed here follows his proposal, which was based on three independent methods: morphology, relative size and abundance, and, in some cases, "fit". Due to the large temporal and morphological hiatus between Paleocene and recent ungulates, new regression equations were established in this chapter based on archaic ungulates for which the skeleton and teeth are known by association. Sixteen species of North American "condylarths" and two litopterns were selected for this purpose, not only for their availability, but also because they represent a similar level of dental and pedal organization to the species in question (Table 6.2). As taxa with similar teeth tend to have a similar diet, a relationship should exist between tooth and body size (Fortelius, 1990).

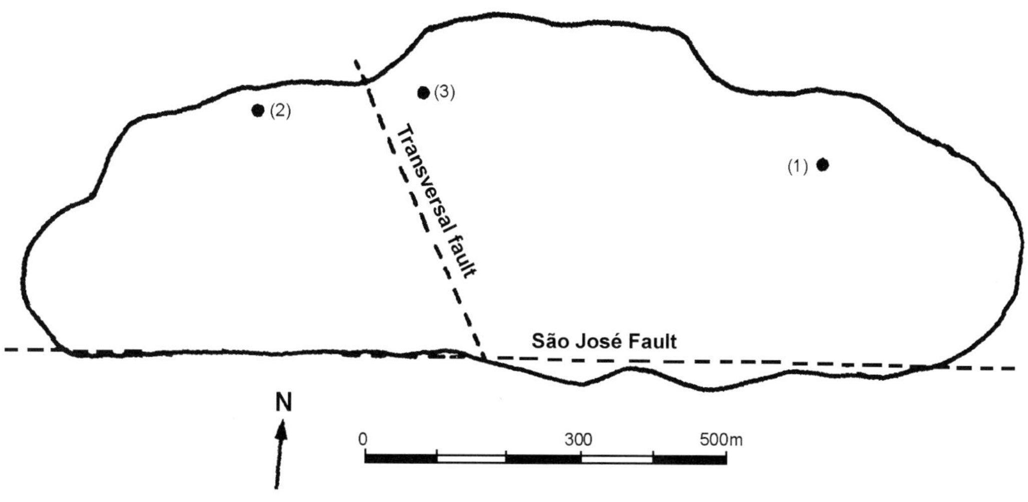

Figure 6.2. Outline of the São José de Itaboraí basin, showing the location of the fissures worked in 1948 (1); 1949 (2) and 1968 (3).

TABLE 6.2. North American "condylarths" and South American litopterns with m2/M2 and skeleton directly associated used in the regressions. The measurements under M2 and m2 were used for establishing the regression lines – the upper refers to the length and the lower to the width. AS = astragalus; CA = calcaneum; EC = ectocuneiform; FE = femur; ID = collection number; HU = humerus; MC = metacarpal; MT = metatarsal; NA = navicular; RA = radius; SP = species; TI = tibia; UL = ulna.

ID	SP	M2	m2	UL	RA	HU	MC	FE	TI	CA	AS	NA	EC	MT
AMNH 4378	*Phenacodus wortmani*	8.6	8.6	X		X		X	X	X	X	X	X	
		10.2	7.4											
AMNH 15262	*Phenacodus wortmani*	11.7	11.8	X										
		13.6	10.9											
AMNH 15283	*Phenacodus wortmani*	12.0	-			X								
		5.2	-											
AMNH 4370	*Phenacodus wortmani*	12.4	13.1						X					
		15.8	12.0											
UM 64179	*Copecion brachypternium*	-	7.0	X				X	X					
		-	6.0											
AMNH 16343	*Loxolophus hyattianus*	5.7	6.2	X				X	X	X				
		7.2	4.7											
AMNH48699	*Chriacus* sp.	7.5	7.1	X										
		8.5	4.4											
USGS 2353	*Chriacus* sp.	7.4	7.8	X			X		X	X	X	X	X	X
		9.1	4.7											
USGS 48006	*Chriacus* sp.	-	6.2		X	X		X	X					
		-	4.7											
AMNH 3115	*Chriacus pelluidens*	-	6.3					X						
		-	4.5											
AMNH 17384	*Thryptacodon australis*	-	6.6	X		X								
		-	5.1											
AMNH 16542	*Arctocyon ferox*	10.9	11.7	X		X		X	X					
		12.4	9.4											
AMNH 27601	*Carsioptycus coarctatus*	8.4	-		X	X			X					
		11.2	-											
AMNH 16517	*Carsioptycus coarctatus*	-	9.1			X								
		-	8.3											
AMNH 3636	*Periptycus rhabdodon*	10.9	10.3	X	X	X					X		X	
		11.6	8.3											
AMNH 3637	*Periptycus rhabdodon*	9.8	9.7					X						
		11.5	8.8											
AMNH 16500	*Ectoconus* sp.	-	12.3	X	X	X	X	X	X				X	
		-	9.4											
AMNH 48002	*Meniscotherium chamense*	9.5	8.3	X	X	X	X	X	X					
		10.3	5.8											
USNM 22675	*Meniscotherium chamense*	-	8.5	X				X	X		X			
		-	6.3											
USNM 22918	*Meniscotherium chamense*	-	9.7			X								
		-	11.1											
USNM 19555	*Meniscotherium robustum*	-	9.9	X	X	X		X	X	X	X	X		
		-	7.3											
USNM 23740	*Hyopsodus paulus*	4.0	-			X	X						X	
		5.7	-											
USNM 17980	*Hyopsodus paulus*	4.1	4.2	X				X	X					
		5.8	3.5											
AMNH 14654	*Hyopsudus walcottianus*	-	6.6							X	X			
		-	5.1											
AMNH 9270	*Diadiaphorus majusculus*	19.0	-	X					X	X	X	X	X	
		24.4	-											
PU 15799	*Diadiaphorus majusculus*	23.7	20.5					X						
		23.1	11.9											
IGM 183544	*Megadolodus molariformis*	-	17.8							X	X			
		-	16.6											

The logarithmic linear regressions were established using the equation Y = a + bX, where Y and X are the dependent (postcranial measurement) and independent (tooth measurement) variables, respectively, and the parameters a (y intercept) and b (slope) were calculated mathematically using the software SYSTAT, version 5.03. Although the length of the m2/M2 is better correlated with body mass than the area in extant mammals (Damuth, 1990), several tests accomplished within "condylarths" in this work indicated that, in most cases, the area of the tooth presented a higher coefficient of correlation.

Besides these measurements, all postcranial bones of "condylarths" and Paleogene South American ungulates were studied directly or from published descriptions. Although many species available for comparison belong to more recent and presumably derived groups, the contrast with primitive "condylarths" permitted the identification of apomorphies particular to each order, facilitating a preliminary assignment of specimens to major taxonomic groups.

The specimens were arranged in groups (morphotypes) based on similarity. One hundred and eight measurements were taken and scatterplot graphs were generated using the software SYSTAT 5.03. Some of these graphs are figured in Bergqvist (1996). Groupings based on measurements were then compared to the morphotypes, to check if there was an agreement between morphology and size. The same 108 measurements were tested in relation to the area and length of the second lower molar, to check, for each bone, which bone measurement presented the highest correlation with tooth length or area. Using the software SYSTAT, regression equations and lines were constructed and the ones with a coefficient of correlation higher than 0.900 were used in the reassociations proposed here. All measurements were log transformed to remove the size effect on variability (Gingerich, 1974). The square root of the area was calculated, to allow comparison between values from one variable (distance between two points) and two variables (area).

The dental remains of each Itaboraí species, where feasible, were counted, and the minimal individual number (MIN) was calculated (Table 6.3). Most of the dental material consists of isolated cheek teeth, and it was not always possible to sort

TABLE 6.3. Total specimen number and minimal individual number (MIN) of the ungulate species of Itaboraí Basin. A, number of specimens sorted; B, minimal individual number; C, fossil used for calculation of the MIN.; #, species represented by different teeth.

	1948/49	1949	1953	1961	1967	1968	No Date[a]	MCN-PV	Total
Paulacoutoia protocenica	A = 32	0	0	0	0	0	18	7	57
	B = 5	0	0	0	0	0	3	4	
	C = left p4	0	0	0	0	0	Right M2	Right M2	
Lamegoia conodonta	A = 6	0	0	0	0	0	5	1	12
	B = 1	0	0	0	0	0	1	1	
	C = #	0	0	0	0	0	#	Left dp4	
Victorlemoinea protoptypica	A = 19	1	0	0	0	0	3	1	24
	B = 3	1	0	0	0	0	1	1	
	C = left dp4	Left dentary	0	0	0	0	#	Left Dp3	
Asmithwoodwardia scotti	A = 0	1	1	0	0	0	2	1	5
	B = 0	1	1	0	0	0	1	1	
	C = 0	Skull	Dentary?	0	0	0	#	Left dentary	
Miguelsoria parayirunhor	A = 8	7	6	0	10	12	59	15	117
	B = 4	3	3	0	3	2	12	4	
	C = left dentary	Left dentary	Left dentary	0	Left m2	Right m3	Left m2	Left m1	
Protolipterna ellipsodontoides	A = 0	0	0	0	0	575	0	0	575
	B = 0	0	0	0	0	50	0	0	
	C = 0	0	0	0	0	M_1E	0	0	
Paranisolambda prodromus	A = 7	7	0	1	1	54	19	14	103
	B = 4	2	0	1	1	8	3	4	
	C = left m2	Left M1	0	M	Right m1	Left dP3	Right M2	Left m2	
Colbertia magellanica[b]	A = 122	23	?	?	?	?	?	***	145
	B = 18	7	?	?	?	?	?		
	C = right m2	Right M2							
Camargomendesia pristina[b]	A = 0	2	?	?	?	26	?	***	28
	B = 0	1	?	?	?	16	?		
	C = 0	#	?	?	?	Right dentary	?		
Itaboraitherium atavum[b]	A = 2	4	6	?	?	?	?	***	12
	B = 1	1	?	?	?	?	?		
	C = #	#	?	?	?	?	?		
Tetragonostylops apthomasi[b]	A = 343	16	?	?	?	25	?	27	411
	B = 25	6	?	?	?	?	?	3	
	C = left m3	Left m3	?	?	?	?	?	Left dp4	
Carodnia vieirai[b]	A = 13	4	?	?	?	0	?	0	17
	B = 2	2	?	?	?	0	?	0	
	C = right m3	#	?	?	?	0	?	0	

[a] It also includes the specimens of the orders "Condylarthra" and Litopterna from the AMNH collection.
[b] Only the cataloged specimens were counted, with the exception of the dentaries of *C. pristina* collected in 1968.
? = Specimens not catalogued or classified
*** = Specimens not classified

some of the closely similar forms (i.e., among the notoungulates), as many deciduous teeth are present and the locus of certain isolated cheek teeth is difficult to determine.

In most cases, the bones were reassociated by size, whereby groups (morphotypes) were assigned to the species in which the dependent variable (bone measurement) was closest to the mean actual value of the specimens of a group. When the specimens of a group presented synapomorphies with more derived species, the morphology was first used for the assignment of this group to one of the orders present in Itaboraí Basin. Postcranial bones of 48 species of fossil "Condylarthra" and South American ungulates were studied for this purpose. The regression was then used to assign elements to species based on size. In cases where the coefficient of correlation of m2/M2 was lower than 0.900 (minimum value considered here for the regression) the bones were tentatively associated on the basis of coloration (only for the brownish ones), relative abundance, and direct articulation. For direct articulation, as the goal was articulation of specimens of a single species, not specimens of the same individual, the criteria used for defining the "most appropriate" articulation were less rigorous.

For the humerus, femur, tibia and astragalus, logarithmic linear regressions were proposed. For the humerus, the measurement that presented the highest coefficient of correlation with the length of m2 ($r = 0.925$) was body width above entepicondylar foramen (BWAEF; Figure 6.3), and the resultant equation was:

$$\text{Log(BWAEF)} = -0.654 + 1.614\ \text{Log(m2 length)}.$$

The femur presented several measurements with a high coefficient of correlation with the area of the m2 ($r > 0.950$). The body area above lesser trochanter (BAALT) (Figure 6.3), though presenting a slightly lower coefficient of correlation ($r = 0.966$) than the area of the body between lesser trochanter and third trochanter, was selected for this purpose as it was based on more taxa (11). The regression equation was:

$$\text{Log(BAALT)} = 0.093 + 1.302\ \text{Log(m2 area)}$$

For the tibia it was necessary to establish two different equations, as some of the bones preserved only the proximal half, while others only the distal half of the body. For the first case, the width of the medial face (WMF) (Figure 6.3) presented $r = 0.989$ with the area of m2, and the resultant equation was:

$$\text{Log(WMF)} = -0.665 + 1.830\ \text{Log(m2 area)}$$

For the distal half of the tibia, the coefficient of correlation of the area of the distal end (ADE) (Figure 6.3) with the area of m2 was $r = 0.963$, and the equation:

$$\text{Log(ADE)} = -0.147 + 1.433\ \text{Log(m2 area)}$$

As proposed by Cifelli (1983a), the total length of the astragalus presented the highest coefficient of correlation with the area of m2 ($r = 0.930$). However, to test the association proposed by Cifelli (1983a), I used the maximum length of the trochlea (MLT) (Figure 6.3), which had a slightly lower coefficient of correlation ($r = 0.921$) with the area of m2. The resulting equation was:

$$\text{Log(MLT)} = 0.283 + 1.345\ \text{Log(m2 area)}$$

The proposed assignment of the remaining bones (except metapodials) was based on the "expected morphology" and direct articulation. For metapodials, associations were based exclusively on morphology. For all other bones considered here, the relative abundance of postcranial and dental specimens was considered. The relative frequency of the postcranial and dental specimens was first considered in the associations proposed for the fossils collected from the "1968 fissure" (the brownish ones).

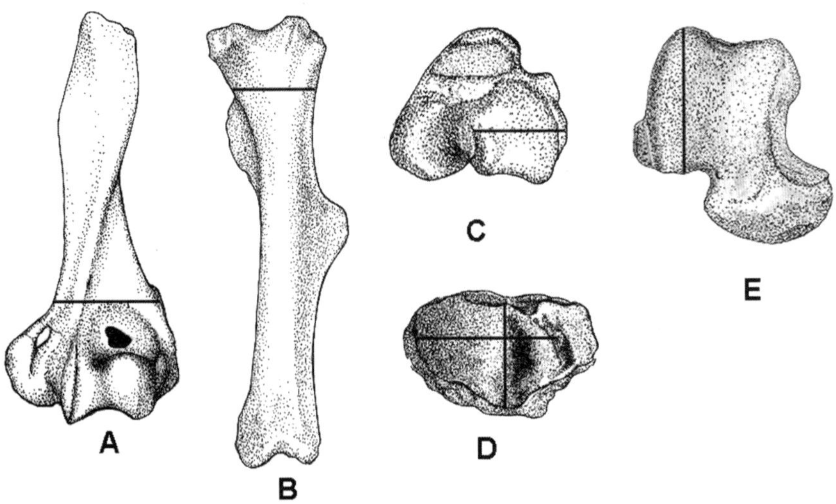

FIGURE 6.3. Measurements taken for establishing the regression lines. A, body width above entepicondylar foramen of humerus, B, body width above lesser trochanter of femur, C, width of the medial face of tibia, D, area of the distal end of tibia, E, maximum length of the trochlea of the astragalus.

The "expected morphology" of the bones studied here, of each Itaboraí order, was conceived after the direct observation of all "condylarths" and South American ungulates (of the same orders present at Itaboraí) with bones and teeth in direct association. Through this comparative observation it was possible to realize the morphological features typical of each order. The phylogenetic analysis undertaken by Bergqvist (1996) showed the apomorphic ones. In most features, the "condylarths" bear the most primitive morphology.

All humeri recovered at Itaboraí are of primitive morphology, retaining an entepicondyle and entepicondylar foramen. This condition prevented reassociation based on morphology. Size and relative abundance proved to be the only available criteria for associating humeri.

The low coefficient of correlation for most ulnar measurements led to an alternative method of reassociation based on direct articulation with the humerus. Observations on the humeroulnar articulation in extant mammals have shown a series of variations in the morphology of this joint, each of which was taken into account for the proper association of these bones:

(1) The orientation of the medial crest of the trochlea is directly related to the orientation of the ulnar coronoid process: if the medial crest is weak and more obliquely placed (in relation to the main axis of the humerus), the coronoid process projects more anteriorly than distally, contrary to when the medial crest is more vertical and projected
(2) The direction of the posterior borders of the trochlea is directly correlated to the placement of the semilunar notch in relation to the ulnar body: laterally oriented borders imply an oblique position of the semilunar notch, whereas a more longitudinal orientation is associated with more vertical borders
(3) The symmetry and parallelism of posterior borders of the trochlea are related to the shape of the anconeus process: a more developed upper projection of the lateral border of this process implies higher asymmetry and feeble parallelism
(4) Variation in the anteroposterior length of the trochlea (at the level of the constriction between the medial crest and the capitulum) is related to the length of the notch: a slender trochlea articulates with a notch that is proportionally short (in relation to the total length of the ulna) and deep

The morphology of the femur is variable among "condylarths", and features such as head shape and placement of the third and lesser trochanters may vary within a family. However, the femur of notoungulates (with the exception of Homalodotheriidae) presents little divergence from the femur of Arctocyonidae (i.e., greater trochanter and head at about the same level, shaft almost straight, lesser and third trochanter at the same level), preventing reassociation based exclusively on morphology. However, most of the femur measurements presented a high coefficient of correlation with the area of the second lower molar.

Morphology of the tibia varies dramatically among South American ungulates of the Paleogene and Neogene, and this variation is observed among the Itaboraí fossils. This great morphological variation facilitated the sorting of the tibiae into morphotypes. As some tibiae preserved the proximal half, while others preserved the distal half, the reassociation of this bone required two separate regressions. The association of the distal halves was also supported by direct articulation with the astragalus.

Generally, the astragalus of each mammalian order has a distinctive morphology, and its value in mammalian taxonomy has long been recognized (at least since Matthew, 1909). In some orders, the astragalus of early forms presents the derived features of the order, while the teeth remain primitive (Schaeffer, 1947; Cifelli, 1983a). Since the Paleocene, most of the derived features of each South American ungulate order are already present in the astragali, which made the association of the astragalus the most confident among the bones studied.

Besides having an important functional role, the calcaneum retains basic characteristics at various taxonomic levels, being an important taxonomic tool (Stain, 1959). Like the astragali, the calcanea of early forms also bear the derived features of the order. Both astragali and calcanea were first reassociated to one of the Itaboraí species by Cifelli (1983a); he used direct articulation with the astragalus to reach his conclusions. Several attempts to establish a regression line were undertaken, but the coefficient of correlation with both area and length of m2 was low, except for the total length of the calcaneum, which, however, was established for only six species. Thus, I employed the same methodology used by Cifelli (1983a).

No regression line was necessary for the association of the navicular and ectocuneiform, once it could be confidently suggested based on morphology, relative abundance, and direct articulation with the astragalus and navicular, respectively. Unfortunately, only a few fossil ungulate ectocuneiforms have been studied, and most of the descriptions are superficial. However, the available information shows that its morphology is more constant among ungulates than is that of the navicular.

6.4 Systematic Paleontology

Order "Condylarthra" Cope, 1881
Family Didolodontidae Scott, 1913
Paulacoutoia Cifelli, 1983b
Paulacoutoia protocenica Paula-Couto, 1952
Lamegoia Paula-Couto, 1952
Lamegoia conodonta Paula-Couto, 1952
Family Sparnotheriodontidae Soria, 1980
Victorlemoinea Ameghino, 1901
Victorlemoinea prototypica Paula-Couto, 1952

6.4.1 *Paulacoutoia protocenica*

6.4.1.1 *Humerus*

The assignment of these bones to *P. protocenica* was tentatively made only by size, as the morphology of the humerus is very conservative in most Paleogene taxa.

The three specimens retain only the distal part (Figure 6.4), which is anteroposteriorly compressed. The deltoid tuberosity is not preserved, but based on the specimen MCN-PV 1702, it seems to be placed on the distal half of the shaft, as in *Hyopsodus* Leidy, 1870 (Figure 6.5). The distal end is broad. The supinator crest is long, convex, weakly developed, and distally thick. The radial and olecranon fossae are deep and perforated, forming a broad supratrochlear foramen. Perforation of the olecranon fossa allows a greater arc of movement for the antebrachium, and the olecranon may pass into it when the antebrachium is fully extended (Taylor, 1974). This feature is often present in cursorial mammals, though it can be sporadically present in mammals with other locomotor habits (O'Leary and Rose, 1995). The existing bone between the olecranon and radial fossae is very thin. The medial epicondyle is longer than wide, as in *Meniscotherium* Cope, 1874, but more prominent than in this taxon. Proximodistally short but longer entepicondyles are present in more cursorial forms, such as *Phenacodus* Cope, 1873 (Rose, 1990) and *Pachyaena* Cope, 1874 (O'Leary and Rose, 1995). Above it sits a well-developed entepicondylar foramen. The lateral epicondyle is much smaller than the medial. The medial border of the trochlea is steeply inclined and projects strongly downward, forming a prominent sharp crest. The trochlea is confluent with the capitulum, which is transversely broader than the trochlea and less convex mediolaterally than proximodistally. According to Taylor (1974), a more angular trochlea and capitulum limits movement to the anteroposterior plane.

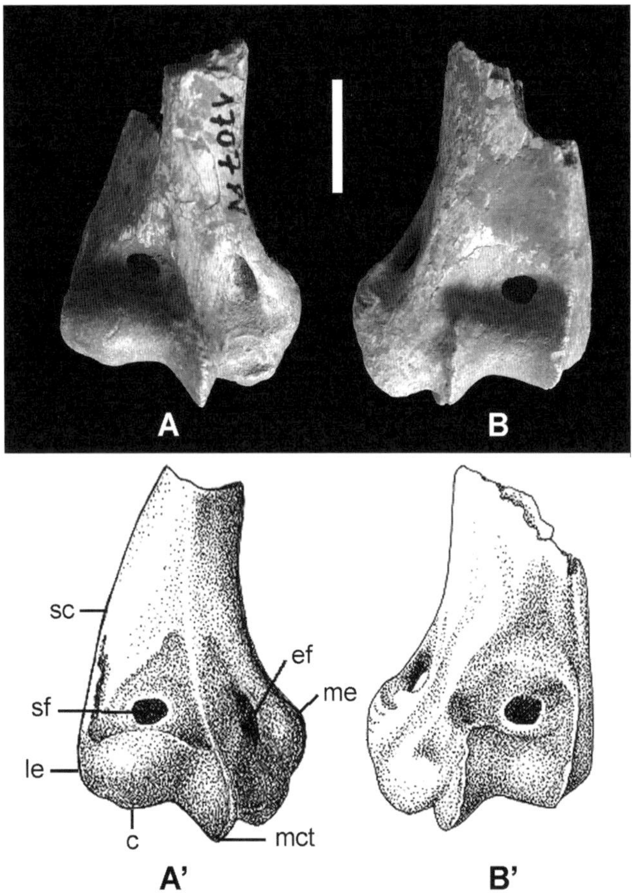

FIGURE 6.4. Right humerus of *Paulacoutoia protocenica*. **A,** anterior and **B,** posterior views of MCN-PV 1711. Outline drawings of MCN-PV 1711 in **A'** and **B'** in the same views. **c,** capitulum; **ef,** entepidondylar foramen; **le,** lateral epicondyle, **mct,** medial crest of trochlea; **me,** medial epicondyle; **sc** supinator crest; **sf,** supratrochlear foramen. Scale bar: 1 cm.

6.4.1.2 *Tibia*

The bone assigned to *P. protocenica* is also the appropriate size for the proterotheriid *Paranisolambda prodromus* (Paula-Couto, 1952). The order Litopterna has a derived and consistent skeletal morphology, especially in the hind feet (Bergqvist, 2005), present since the most primitive forms of the Paleocene (Bergqvist, 1996). On the other hand, the tibia of "condylarths" (with the exception of Phenacodontidae that are incipiently cursorial), in particular *Loxolophus* Cope, 1885, retains most of the primitive features of Eutheria (see Dagosto, 1985).

The tibia assigned to *P. protocenica* resembles that of "condylarths". The bone is almost complete, except for the absence of both epiphyses, suggesting that it probably belonged to a juvenile (Figure 6.6). In general, it resembles the tibia of *Meniscotherium chamense* Cope, 1874 (Figure 6.7). The shaft is slightly convex anteriorly, deeper than wide, and roughly triangular in proximal cross-section. The three faces are of different lengths, the medial being more extended anteroposteriorly than the lateral, so the cnemial crest is more laterally placed and (barely) visible in posterior view, a feature very common among "condylarths" (e.g., *Chriacus* Cope, 1883; *Ectoconus* Cope, 1884; *Hyopsodus*). The shaft becomes gradually thinner distally and more or less oval in cross section. At the distal end it strongly widens again. The cnemial crest is moderately prominent and extends halfway down the shaft before becoming indistinct. The popliteal notch is deep, but no popliteal crests are discernible. The interosseous crest is blunt but well-defined all the way down the shaft. No fibular articular surfaces are preserved.

6.4.1.3 *Astragalus*

The same two specimens assigned by Cifelli (1983a) to *P. protocenica*, based on the total length as the dependent variable, were assigned to this taxon here using the maximum length of the trochlea. However, the specimens were

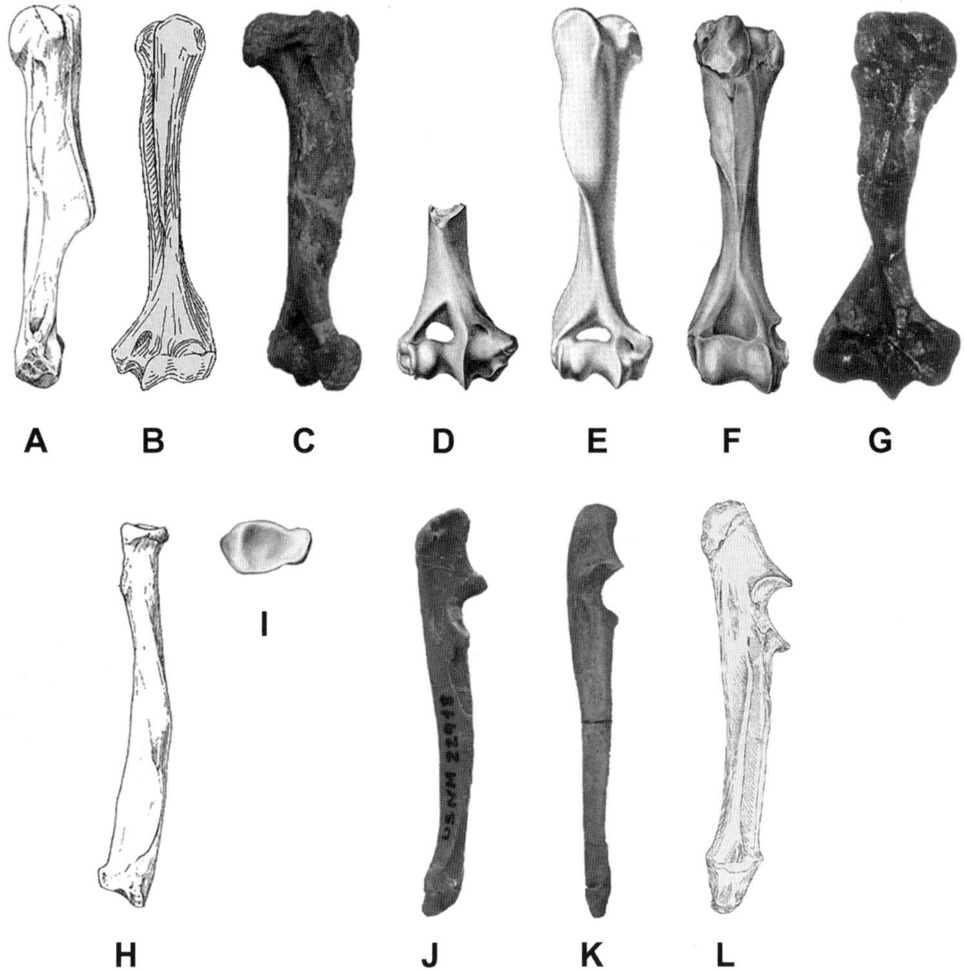

FIGURE 6.5. Forelimb bones of some North American mammals used for comparison with Itaboraí "condylarths". Left humerus of **A,** *Arctocyon primaevus*, in medial view (from Russell, 1964), **B,** *Chriachus* sp., in anterior view (USGS 2353; from O'Leary and Rose, 1995), **C,** *Phenacodus primaevus*, in medial view (AMNH 15262); right humerus of **D,** *Hyopsodus paulus*, in anterior view (USNM 23740; from Gazin, 1965), **E,** *Meniscotherium chamense*, in anterior view (USNM 19555; from Gazin, 1968), **F,** *Pachyaena gigantea*, in anterior view (USNM 14915; from O'Leary and Rose, 1995), **G,** *Periptychus rhabdodon*, in anterior view (AMNH 837); left radius of **H,** *Arctocyon primaevus*, in anterior view (from Russell, 1964), **I,** *Phenacodus trilobatus, in proximal view* (USGS 7146; from O'Leary and Rose, 1995); left ulna of **J,** *Meniscotherium chamense*, in medial view (USNM 22435; from Gazin, 1968), **K,** *Phenacodus wortmani*, in medial view (AMNH 4378); **L,** right ulna of *Pachyaena gigantea*, in lateral view (USNM 14915; from O'Leary and Rose, 1995). Not to scale.

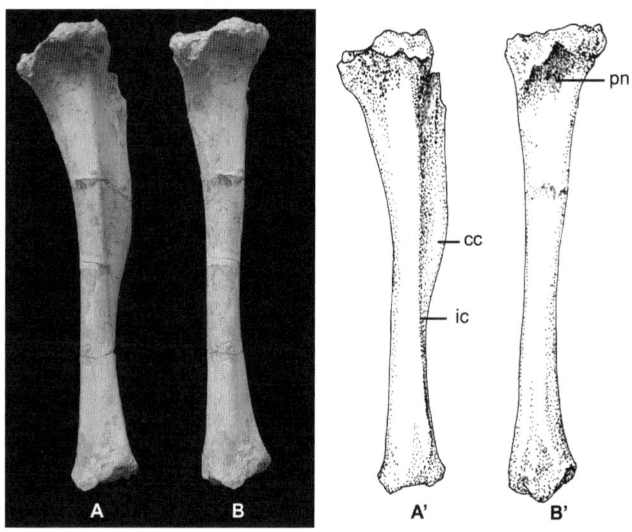

FIGURE 6.6. Right tibia of *Paulacoutoia protocenica*. **A,** posterior and **B,** posterolateral views of DGM 345M. Outline drawings of tibia in **A'** and **B'** in the same views. **cc,** cnemial crest; **ic,** interosseous crest; **pn,** popliteal notch. Scale bar = 1 cm.

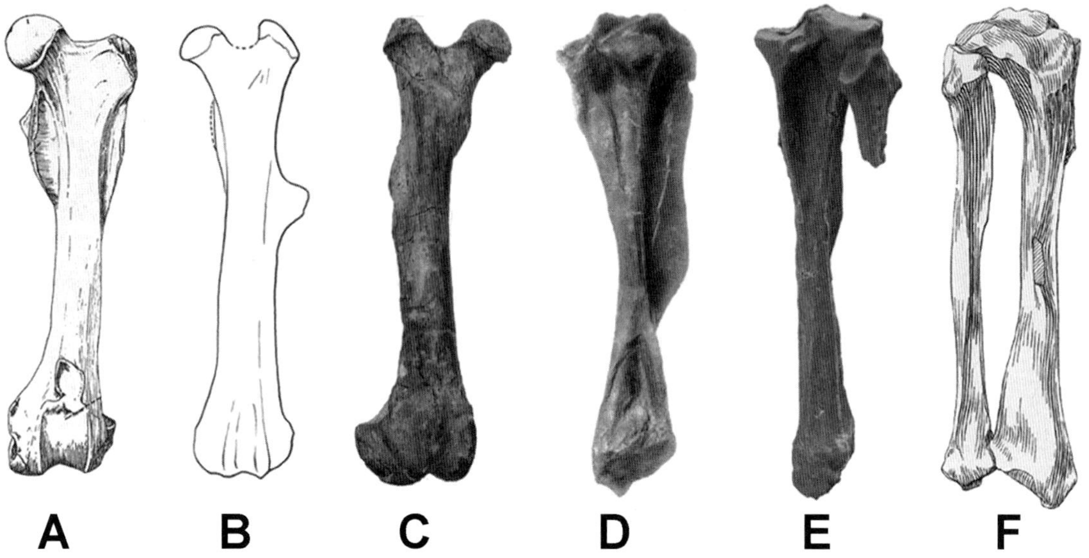

FIGURE 6.7. Hind limb bones of some North American mammals used for comparison with Itaboraí "condylarths". Left femur of **A,** *Arctocyon primaevus*, in anterior view (from Russell, 1964), **B,** Copecyon brachypternus, in anterior view (UM 64179; from Thewissen, 1990); **C,** right femur of *Phenacodus wortmani*, in anterior view (AMNH 4378); right tibia of **D,** *Ectoconus majusculus*, in posterolateral view (AMNH 16500; from Matthews, 1937), **E,** *Meniscotherium chamense*, in posterior view (USNM 17917), **F,** right tibia and fibula of *Periptychus rhabdodon*, in anterior view (AMNH 17075; from Matthew, 1937). Not to scale.

also of appropriate size for *Paranisolambda prodromus*. As mentioned before, the litopterns have a very constant skeletal morphology (Bergqvist, 2005), and an astragalus with typical litoptern morphology would be expected for this species. Moreover, *P. prodromus* is much more abundant in the basin than *P. protocenica*, and more litoptern-like astragali would be expected to be found in the basin, as was shown by Bergqvist (1996).

These astragali were described by Cifelli (1983a) under the name *Ernestokokenia protocenica*. The specimen provisionally numbered as LE 443 in his paper has now received the collection number MCT 1388M. Cifelli's descriptions of the astragalus (and calcaenum) were transcribed here with minor changes, but with new information added.

The adult astragalus of *P. protocenica* is approximately 20% smaller than the specimen USNM 17917 of *Meniscotherium chamense*. The astragalar body is relatively deep and bears a moderately grooved tibial trochlea (Figure 6.8), intermediate between *Phenacodus intermedius* Granger, 1915 and *M. chamense* (Figure 6.9). The medial and lateral crests are moderately sharp, with about the same anteroposterior length, but the lateral exhibits a larger arc than the medial, although not as much as in *M. chamense*. The astragalar foramen persists in a small and presumably juvenile specimen (DGM 1388M), but is filled with cancellous bone and virtually obliterated in a larger specimen. The medial malleolar facet extends far anteriorly onto the neck and nearly to the head of the astragalus, where it curves abruptly medially. This distinctive structure is synapomorphic for Didolodontidae (Cifelli, 1993; Bergqvist, 1996). An extensive fibular facet, terminating anteroinferiorly in a well-developed fibular shelf, covers the lateral wall of the astragalar body, which is also vertical.

The neck is medially offset from the trochlea, forming an angle of approximately 30° with respect to the anteroposterior axis of the trochlea. This is another feature that is clearly distinct between Itaboraian forms and the more primitive astragali of Tiupampan "condylarths" (Muizon et al., 1998). The head is somewhat narrower transversely, and bears a navicular facet that is obliquely oriented in a very similar way to *M. chamense* (Williamson and Lucas, 1992). The shape of the head and trochlea (moderately grooved) are suggestive of little cursorial abilities. However, Van Valkenburgh (1987) concluded that astragalar trochlea depth is a character that better reflects heritage than behavior. Medial collateral ligament and cuboid facets are absent. A small supplementary facet, continuous with that for the navicular but not with the sustentacular, is present in the specimen MCT 1388-M, and in life contacted the dorsolateral neck of the calcaneum, as articulation of the two tarsals demonstrates. The sustentacular facet is expanded distally and may or may not be continuous with the navicular facet. The interarticular sulcus is deep and the ectal facet is triangular and moderately concave, and approximates *Arctocyon* de Blainville, 1841 in size and orientation. Posteriorly, the groove for the digital flexor tendons is well-marked and somewhat offset from that of the posteroinferior margin of the tibial trochlea.

According to Wang (1993), the astragalus is one of the most important hind limb elements in the transformation

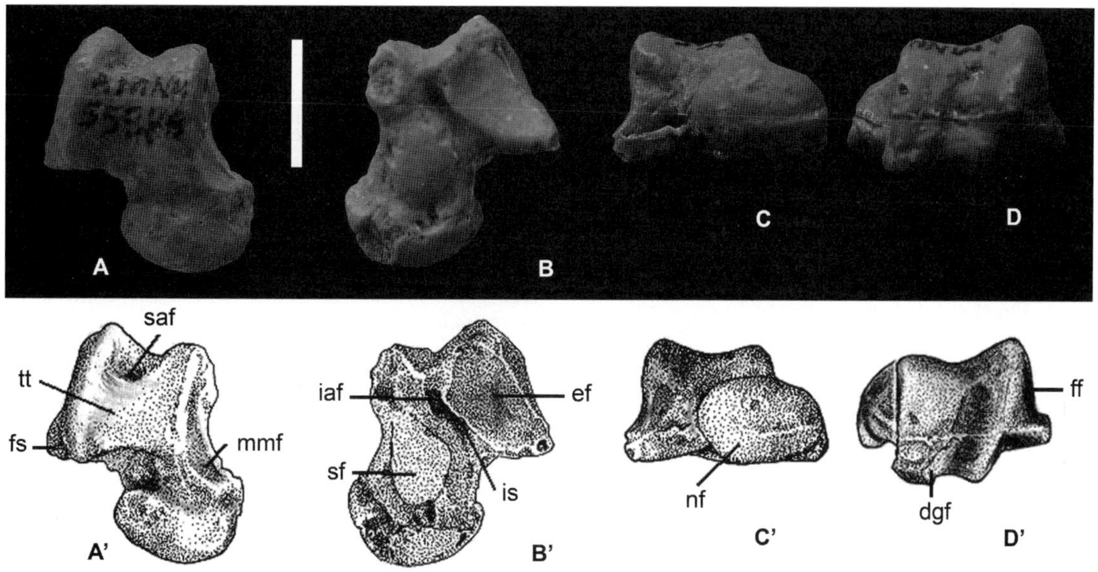

FIGURE 6.8. Right astragalus of *Paulacoutoia protocenica*. **A,** proximal; **B,** plantar; **C,** distal, and **D,** posterior views of AMNH 55388. Outline drawings of astragalus in **A', B', C'** and **D'** in the same views. **dfg,** groove for the deep digital flexor tendon(s); **ef,** ectal facet; **fs,** fibular shelf; **ff,** fibular facet; **iaf,** inferior astragalar foramen; **is,** interarticular sulcus; **mmf,** medial maleolar facet; **nf,** navicular facet; **saf,** superior astragalar foramen; **sf,** sustentacular facet; **tt,** tibial trochlea. Scale bar: 1 cm.

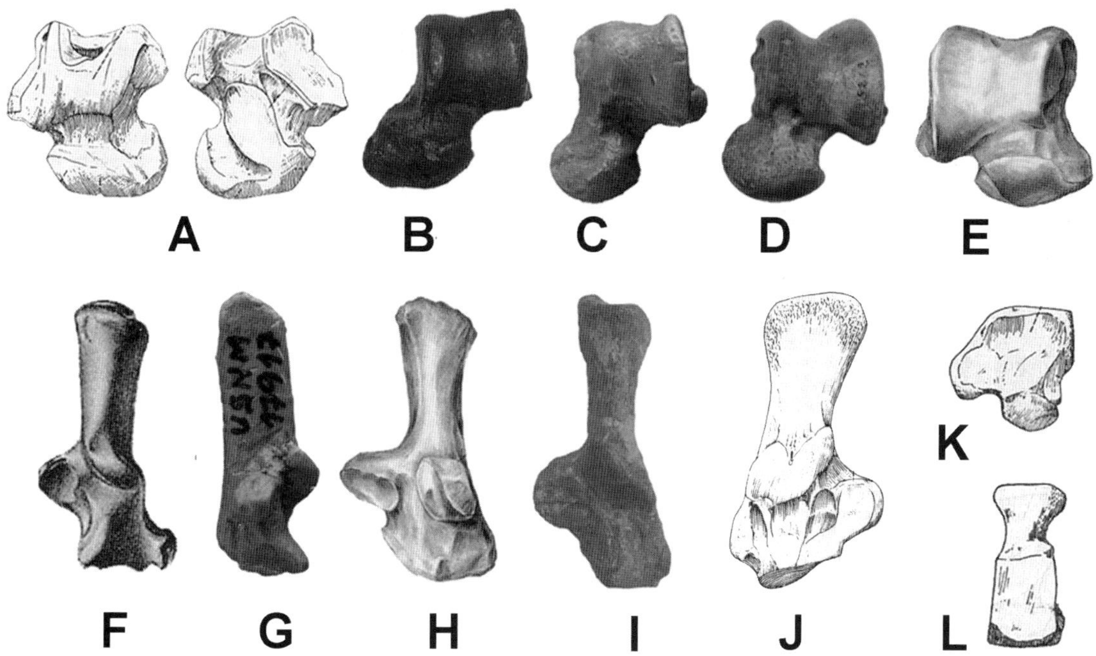

FIGURE 6.9. Tarsal bones of some North American mammals used for comparison with Itaboraí "condylarths". Right astragalus of **A,** *Arctocyon primaevus*, in proximal and plantar views (from Russell, 1964), Left astragalus of **B,** *Didolodus* sp. (AMNH 117457), in proximal view, **C,** *Meniscotherium chamense* (USNM 17917), in proximal view, **D,** *Phenacodus primaevus* (AMNH 15262), in proximal view; **E,** right astragalus of *Pachyaena gigantea* (AMNH 15228), in proximal view; left calcaneum of **F,** *Hyopsodus paulus*, in anterior view (USNM 23740; from Gazin, 1965), **G,** *Meniscotherium chamense*, in medial view (USNM 17917); **H,** *Pachyaena gigantea*, in anterior view (AMNH 2959), **I,** *Phenacodus primaevus*, in anterior view (AMNH 15262); **J,** right calcaneum of *Arctocyon primaevus* in anterior view (from Russell, 1964); **K** and **L,** right navicular and ectocuneiform of *Arctocyon primaevus* (from Russell, 1964). Not to scale.

from plantigrade to digitigrade, and the main difference between both postures is the plantar extension between the tibia and the astragalus. A large astragalar foramen between the trochlea and the plantar tendinal groove prevents rotation beyond the foramen, restricting dorso-plantar extension. Although in *P. protocenica* there is a marked depression for the superior astragalar foramen, it seems not to have been functional. Moreover, the plantar tendinal groove has a different depth but the same orientation of the trochlea. On the sides of the depression, a polished surface indicates that the trochlea continued posteriorly to the depression. This feature is suggestive of, at least, a digitigrade posture in *P. protocenica*. Carrano (1997) observed that the crests of the trochlea differ in size, being distinctly asymmetrical in plantigrade animals, as is the case in this species. However, in plantigrade taxa, the crests of the trochlea are wider and shallower than in digitigrade taxa, which is different from what is present in *P. protocenica*.

6.4.1.4 Calcaneum

This bone (Figure 6.10) resembles closely that of *Meniscotherium chamense* (e.g., USNM 17917; Figure 6.9), mainly in the very salient dorsally projected beak at its anterior extremity. This feature is absent in other North American "condylarths", and also in all Paleocene "condylarths" of Tiupampa, Bolivia, (Muizon et al., 1998). The proportionally short neck and developed fibular tubercle are primitive features also present in North American "condylarths" (except in *Copecion* Gingerich, 1989).

It exhibits a robust tuber calcis that enlarges at its posterior half, being slightly deeper than thick. The tuberosity bears a medial process, well-developed and separated from the lateral portion by a shallow and oblique sulcus. Plantolaterally, a shallow groove may have received the flexor digitorum superficialis muscle.

The ectal protuberance on the dorsal surface of the calcaneum is prominent and is situated farther posteriorly on the robust body, but less so than in *M. chamense*. The articular

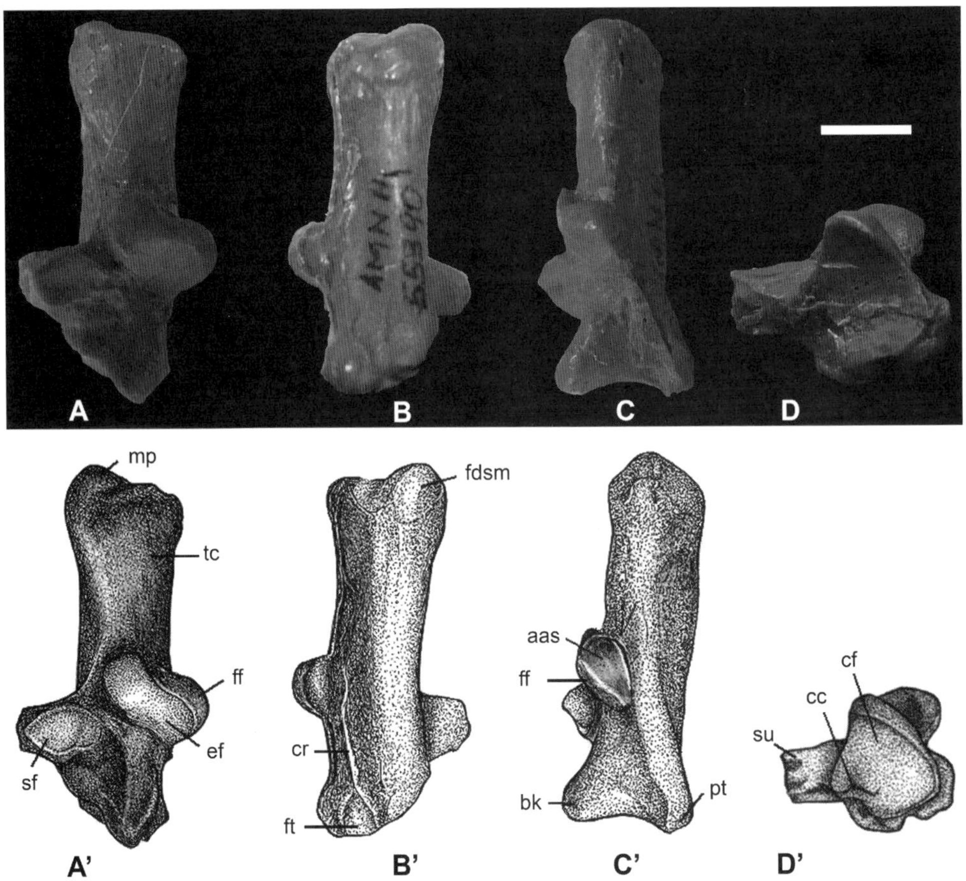

FIGURE 6.10. Left calcaneum of *Paulacoutoia protocenica*. **A,** anterodorsal; **B,** posteroplantar; **C,** lateral, and **D,** distal views of AMNH 55390. Outline drawings of calcaneum in **A', B', C'** and **D'** in the same views. **aas,** accessory articular surface; **bk,** beak; **cc,** invagination for the calcaneum-cuboid ligament; **cf,** cuboid facet; **cr,** crest; **ef,** ectal facet; **fdsm,** groove for the flexor digitorum superficialis muscle; **ff,** fibular facet; **ft,** peroneal (fibular) tubercle; **mp,** medial process; **pt,** plantar tubercle; **sf,** sustentacular facet; **su,** sustentaculum; **tc,** tuber calcis. Scale bar: 1 cm.

surface for the fibula is well-developed and strongly convex anteroposteriorly but relatively flat transversely; an extensive articular surface is also developed on the lateral surface of the protuberance. The calcaneal ectal facet is broad and obliquely oriented with respect to the tuber. The sustentaculum is thick and much broader transversely than proximodistally. A moderately transversely broad sustentaculum occurs in arctocyonids and may be primitive (O'Leary and Rose, 1995). It bears a rather small, ovoid (with longer transverse axis), and slightly concave facet, which does not attain contact with the cuboid facet at the distal end of the calcaneum.

The most conspicuous feature of the bone is the verysalient, dorsally projecting beak at the anterior extremity (ahead of the anterior level of the ectal facet). It is proportionately more projecting than in *M. chamense*, a larger species than *P. protocenica*. The dorsal surface of this prominence is rugose, and probably gave origin to one of the heads of the extensor digitorium brevis muscle and the anterior astragalocalcaneal ligament. A narrow articular surface for the lateral side of the astragalar head descends from the apex of the prominence down to its medial surface. A similar facet, named the distal astragalar facet by O'Leary and Rose (1995), is present in *Pachyaena gigantea* Osborn and Wortman, 1892 and *Mesonyx* Cope, 1872 (Figure 6.9). The cuboid facet is dorsoventrally elongate and somewhat concave in that direction. Its major axis, following the curvature of concavity, is oblique and bears a medial invagination for the calcaneum-cuboid ligament. Inferior to the cuboid facet, the tubercle for attachment of the calcaneotarsal and tarsal fibrocartilage ligaments is modestly developed. On the distolateral surface of the calcaneum the peroneal tubercle is robust but not large, and rugose, suggesting that the lateral collateral ligaments and quadratus plantae muscle are well-developed. The crest extending proximally from it terminates at the base of the ectal prominence.

6.4.2 *Lamegoia conodonta*

6.4.2.1 *Humerus*

This bone is much larger than but similar to that of *P. protocenica* (Figure 6.11). In some ways (shape of the trochlea, projection and size of the entepicondyle), it resembles that of *Periptychus* Cope, 1881. The proximal end is not preserved, but the shaft is broken above the deltoid tuberosity, which is pronounced but ends smoothly. The deltoid tuberosity is less pronounced than in more primitive arboreal or scansorial arctocyonids (such as *Chriacus, Anacodon*, and *Arctocyon*), but more so than in more cursorial forms like *Phenacodus* (Figure 6.5). The deltoid crest is more prominent than the pectoral one. Proximally, the shaft is transversely compressed, but distally it becomes anteroposteriorly compressed. The supinator crest is almost completely broken, but from the remaining portion, it appears to have been long (extending up to the deltoid tuberosity) but weak. The radial and olecranon fossae are as in *P. protocenica*. The supratrochlear foramen has an irregular outline, making it unclear whether it is a natural feature or was artificially (taphonomically) made. The medial epicondyle is almost twice as long as it is wide, as in more cursorial forms like *Phenacodus*, and the entepicondylar foramen is very large. The medial crest of the trochlea is less prominent and sharp than in *P. protocenica*; its distal border is almost at the same level as the capitulum, which is more rounded and longer proximodistally than transversely. Proximolateral to the capitulum there is a shallow groove and a lateral crest, which is less prominent than the medial one. A similar crest is developed, to a greater or lesser degree, in arboreal taxa such as *Chriacus* and *Oxyaena* Cope, 1874, in terrestrial forms such as *Ursus*, and in cursorial taxa like *Pachyaena gigantea, Hyracotherium* Owen, 1840, and *Diacodexis* Cope, 1882 (Rose, 1990; Figure 6.5).

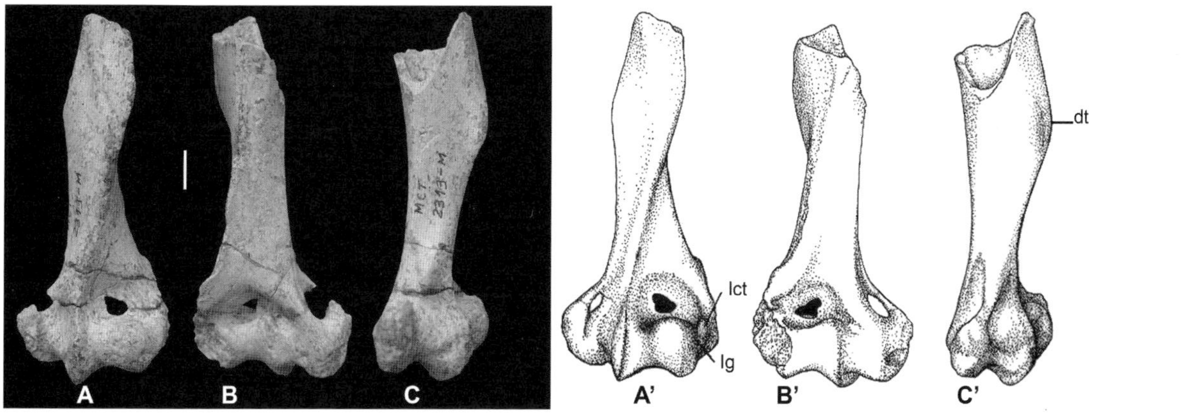

FIGURE 6.11. Left humerus of *Lamegoia conodonta*. **A,** anterior; **B,** posterior, and **C,** medial views of MCT 2313M. Outline drawings of humerus in **A', B'** and **C'** in the same views. **dt,** deltoid tuberosity; **lct,** lateral crest of trochlea; **lg,** lateral groove. Scale bar: 1 cm.

6.4.2.2 *Ulna*

The ulnae assigned to *L. conodonta* are the largest recovered in the basin (with the exception of *Carodnia vieirai*, the tapir-sized ungulate of Itaboraí). The largest specimen (MCT 1833M) articulates so perfectly with the humerus assigned to this species that they may have belonged to the same individual. This is supported by the same coloration and fossilization type in both specimens.

The specimen MCT 2326M is quite complete, lacking only the proximal and distal epiphyses (Figure 6.12). It is a strong element, showing no tendency toward reduction or fusion with the radius. The entire bone is mediolaterally compressed and deep, especially at the semilunar notch. It is almost straight, in both lateral and anterior profile, being somewhat convex at the level of the semilunar notch and slightly concave distally, as in *Phenacodus wortmani* (Cope, 1880; Rose, 1990; Figure 6.5). This might suggest that *L. conodonta* had incipient cursorial capability. The preserved part of specimen MCT 1833M is slightly more convex posteriorly than the figured specimen, but it seems to have belonged to an older (with olecranon epiphysis fused) and rather larger individual. Its lateral and medial surfaces are slightly convex at the olecranon and concave on the rest of the shaft – shallowly on the medial surface and deeply on the lateral one. The absence of a longitudinal crest on the medial surface is a feature present only in these specimens. Among the comparative sample it is only observed in some "condylarths", such as *Arctocyon*, *Hyopsodus*, and *Ectoconus*, as well as notoungulates (except for *Thomashuxleya* Ameghino, 1901; Figure 6.5).

The olecranon is prominent and long (longer than the semilunar notch and ~25% of the total preserved ulnar length, excluding the epiphysis), providing a long lever arm for the elbow extensor muscles, triceps brachii and anconeous. It is transversely compressed and bends slightly medially. As in *Pachyaena gigantea*, the semilunar (trochlear) notch outlines a wide oval in lateral view, being more tightly curved at the proximal than at the distal end. In anterior view, it is saddle-shaped and oblique to the proximodistal axis of the shaft. Forms in which the semilunar notch is more open and set out from the axis of the ulna have greater freedom of supination (Taylor, 1974). The anconeus and coronoid processes are only slightly prominent and equally projected anteriorly, but well-projecting medially and laterally. The radial notch is somewhat worn, but seems to have been smoothly concave and facing anterolaterally. Thus the proximal part of the radius sits anterolateral to the ulna, as in non-cursorial mammals.

The fossa for the brachioradialis muscle is best defined in specimen DGM 1833M. It is deep and proximodistally elongated, as in *Meniscotherium chamense*, indicating strong capability for flexion of the antebrachium. The deep and well-defined anterolateral fossa provides a broad origin for the abductor pollicis longus muscle. It occupies almost

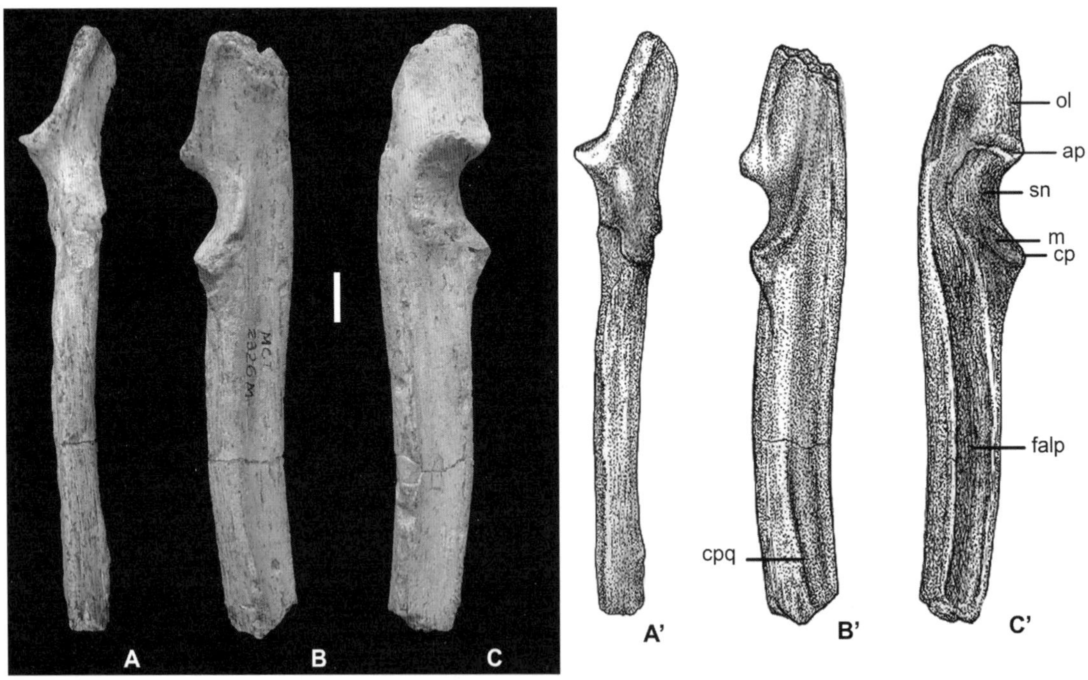

FIGURE 6.12. Right ulna of *Lamegoia conodonta*. **A,** anterior; **B,** medial, and **C,** lateral views of MCT 2326M. Outline drawings of ulna in **A', B'** and **C'** in the same views. **ap,** anconeus process; **cp,** coronoid process; **cpq,** crest for the pronator quadratus muscle; **fapl,** fossa for the adutor pollicis longus muscle; **ol,** olecranon; **rn,** radial notch; **sn,** semilunar notch. Scale bar: 1 cm.

completely the lateral face of the body and may be a primitive eutherian feature (O’Leary and Rose, 1995). It is bound anteromedially by a more or less sharp interosseous crest, and posterolaterally by a salient rounded crest that projects laterally from the shaft, as in *Pachyaena* (O’Leary and Rose, 1995). This anterolateral fossa may suggest the presence of a well-developed digit I (still unknown for this species), although this is not true for *Pachyaena*, in which a vestigial pollex is present (O’Leary and Rose, 1995). In *Pachyaena*, the abductor pollicis longus might have inserted on the second metacarpal, as in *Sus* and *Tapirus* (Getty, 1975). Distally, the shaft enlarges laterally and has a sharp crest where the distal portion of the pronator quadratus muscle probably inserted. The distal articular facet was probably located completely on the distal epiphysis, which is not preserved on any of the specimens.

6.4.2.3 *Femur*

A nearly complete femur is known for *Lamegoia*, missing only the epiphyses, which indicates that it belonged to a juvenile individual (Figure 6.13). It also suffered postmortem deterioration, as shown by erosion and the presence of several fractures. Proximally, the bone is much wider than deep, but it is only slightly wider distally. It is slightly bent medially. Proximally, the dorsal surface of the shaft is almost flat, bearing a discrete concavity between the head and the greater trochanter. Distally, on the same surface, an elliptical depression above the trochlea appears to be the result of postmortem damage or deformation.

The femoral head and the epiphysis of the greater trochanter are missing, but from what is preserved, the latter probably either projected slightly above the head or was even with it. The neck is short and slightly directed anteriorly. The trochanteric fossa is deep, but a trochanteric crest is lacking. The lesser trochanter is a narrow but long wing on the medial edge of the shaft, directed medially as in Arctocyonidae and *Pachyaena* (Figure 6.7). Unlike most “condylarths”, the lesser trochanter does not have a triangular shape, but is rounded in outline, suggestive of the condition in *Phenacodus wortmani* and *Copecion brachypternus* (Cope, 1882) (Thewissen, 1990). Taylor (1976) associated large size and medial position of the lesser trochanter in Viverridae with greater climbing ability. Although this is incompatible with other lines of evidence for locomotion in *L. conodonta*, these features may indicate higher capacity for outward rotation of the femur and capability for locomotion on a variety of substrates. The third trochanter is distal to the lesser trochanter, being placed about halfway down the length of the shaft. It is well-developed and its shape also resembles that of *C. brachypternus*, although it is less salient. Gazin (1968) indicated that this position provides considerable leverage to the gluteus superficialis muscle for abducting the limb and flexing the hip joint. Howell (1944), however, attributed little adaptive but great phylogenetic significance to this condition. The greater development would be a reflection of large musculature, rather than a disproportionately powerful superficial gluteus muscle.

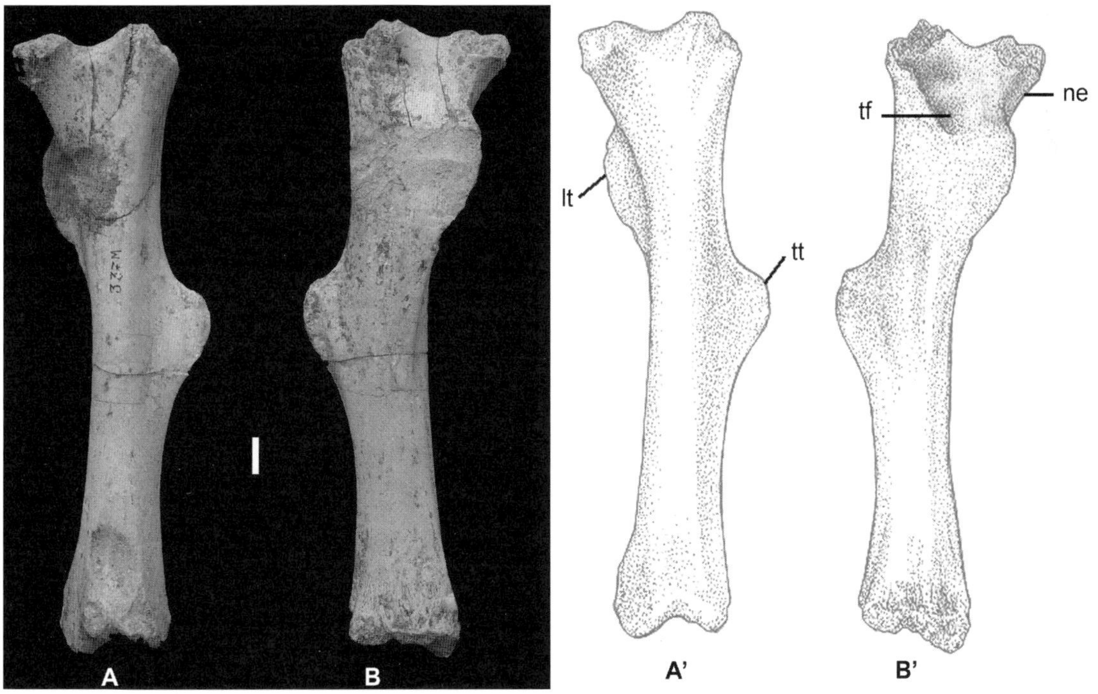

FIGURE 6.13. Left femur of *Lamegoia conodonta*. **A,** anterior and **B,** posterior views of DGM 337M. Outline drawings of femur in **A**’ and **B**’ in the same views. **lt,** lesser trochanter; **ne,** neck; **tf,** trochanteric fossa; **tt,** third trochanter. Scale bar: 1 cm.

6.4.2.4 *Astragalus*

The first description of the astragalus of *L. conodonta* (Cifelli, 1983a) was based on a very poorly preserved specimen (DGM 940M). A complete and well-preserved specimen was found later and a detailed description is presented here.

The astragalus of *L. conodonta* is very similar to that of *Paulacoutoia protocenica*, but 50% larger (Figure 6.14). It is also similar to an unassociated specimen of *? Didolodus* (AMNH 11457) from the Casamayoran of Chubut (Figure 6.9). The body has a moderately deep tibial trochlea with well-defined and rounded crests, the medial crest forming a shorter arc than the lateral. Compared to *P. protocenica*, the crests are more equally developed, while in this species the lateral crest is noticeably more pronounced than the medial one. The astragalar channel is reduced, with tiny openings. The superior foramen is more posterioly placed and not seen in dorsal view, differing from *P. protocenica, Pachyaena*, and *Meniscotherium*, in which the superior astragalar foramen is larger and placed more anteriorly (Figure 6.9). Posteroplantar to the foramen, the groove for the deep digital flexor tendons is deeply depressed but almost indistinct from the trochlea. The medial wall of the body is vertical; the facet for the medial malleolus of the tibia is well-developed and extends anteriorly onto the neck, where it flares sharply medially, more than it does in *P. protocenica*. The tubercle for the medial collateral ligament is proportionately as pronounced as in *P. protocenica*. The lateral face is also vertical and has a fibular shelf.

The sustentacular facet has an elliptical outline and does not attain contact anteriorly with the navicular facet or posteriorly with the groove for the digital flexor tendons. The ectal facet is similar to that of *P. protocenica*, but the head is distinct from this taxon in being transversely elongated and almost quadrangular. The cuboid facet is lacking and the facet for the medial collateral ligaments is small.

The size and position of the superior astragalar foramen, together with the shape of the plantar tendinal groove, are suggestive of a capability for plantar extension between the tibia and astragalus, compared to *P. protocenica*. As in this species, *Lamegoia conodonta* was, at least, semidigitigrade.

6.4.2.5 *Calcaneum*

This bone was first described by Cifelli (1983a) on a partially broken specimen of a juvenile. Discovery of an almost complete specimen belonging to an adult provides new information on its morphology.

The calcaneum of *L. conodonta* is larger than that of *P. protocenica*, but otherwise in many features they are similar (Figure 6.15). The heel (tuber) is relatively longer than in *P. protocenica*, representing 56% of the total length of the bone (compared to 51% in *P. protocenica*). The tuberosity is convex (with no medial process) and has a longitudinal groove on its plantar side for attachment of the flexor digitorum superficialis muscle.

The calcaneal ectal facet bears a similar orientation and placement to that of *P. protocenica*, but is somewhat less convex and transversely larger. The fibular facet is smaller in *L. conodonta*, and is visible only posteriorly. The area of the (possible) accessory facet is eroded. The depression for the short part of the fibular collateral ligament is anteroposteriorly long and deep. The sustentaculum is thick and has a relatively small, rounded facet, which does not extend to the cuboid articulation, as in *P. protocenica*. The dorsal surface of the calcaneal neck, as in *P. protocenica*, is elongated into a conspicuous beak (a supplementary facet for the astragalus is present on its medial border, but is very

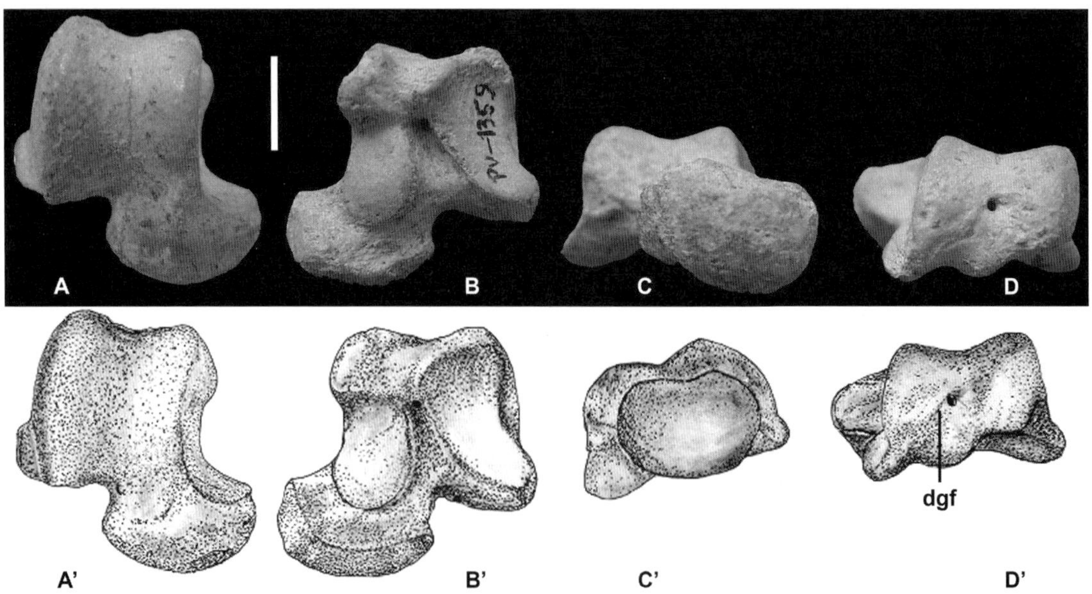

FIGURE 6.14. Right astragalus of *Lamegoia conodonta*. **A,** anterior; **B,** posterior; **C,** distal, and **D,** proximal views of MCN-PV 1359M. Outline drawings of astragalus in **A', B', C'** and **D'** in the same views. **dgf,** groove for the deep digital flexor tendon(s). Scale bar: 1 cm.

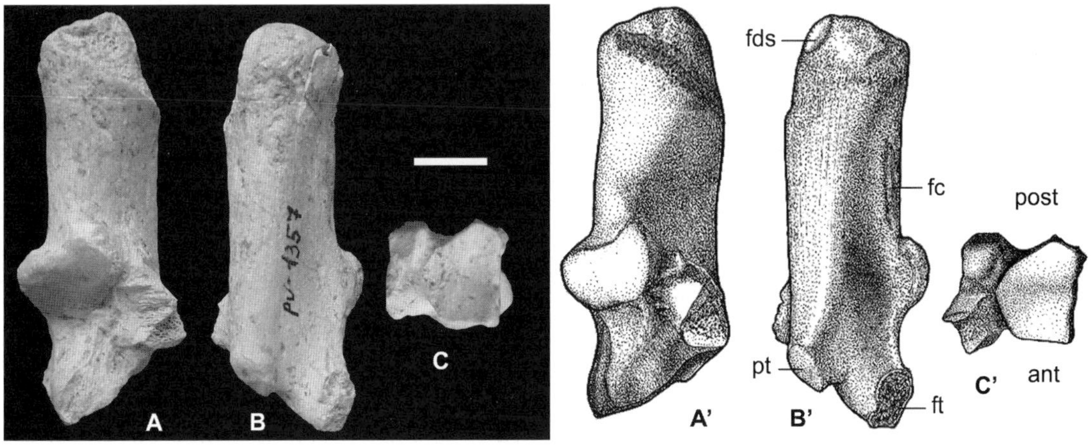

FIGURE 6.15. Right calcaneum of *Lamegoia conodonta*. **A,** mediodorsal; **B,** lateral; **C,** distal of MCN-PV 1271. Outline drawings of astragalus in **A', B'** and **C'** in the same views. **fcl,** depression for fibular collateral ligament; **fds,** groove for the flexor digitorum superficialis muscle; **ft,** peroneal (fibular) tubercle; **pt,** plantar tubercle. Scale bar: 1 cm.

reduced); the cuboid facet is as in *P. protocenica*. Inferiorly, the protuberance for attachment of calcaneotarsal and tarsal fibrocartilage ligaments is proportionally more developed than in *P. protocenica*. The peroneal tubercle, on the distolateral corner of the calcaneum, is more robust and better developed than in *P. protocenica*. Its great development is suggestive of a stronger quadratus plantae muscle and the capability for strong plantar flexion of the pes. The peroneal tubercle extends posterosuperiorly to join the base of the fibular side of the ectal protuberance.

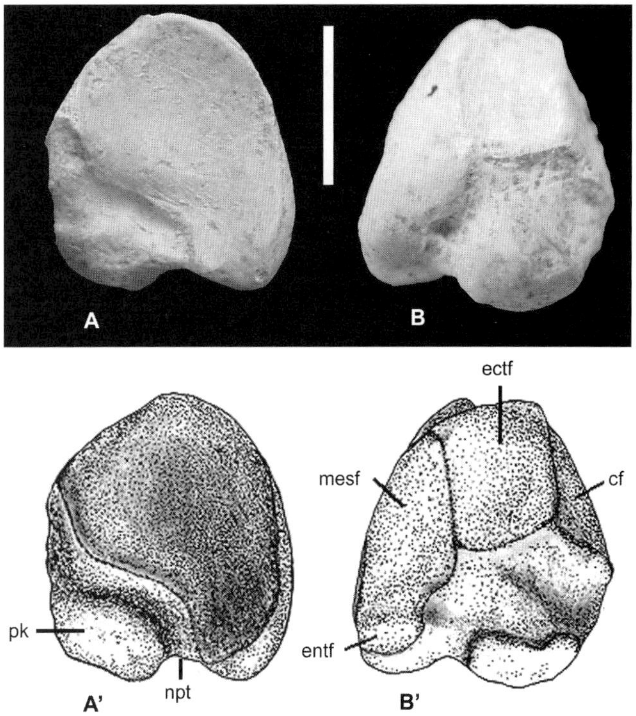

FIGURE 6.16. Left navicular of *Lamegoia conodonta*. **A,** proximal and **B,** distal views of MCN-PV 1827. Outline drawings of navicular in **A'** and **B'** in the same views. **cf,** cuboid facet; **ectf,** ectocuneiform facet; **entf,** entocuneiform facet; **mesf,** mesocuneiform facet; **npt,** notch for the tendon of the posterior tibialis; **pk,** plantar knob. Scale bar: 1 cm.

6.4.2.6 Navicular

Seven navicular morphotypes are present among ungulate fossils from the Itaboraí basin, but they can be grouped into four distinct clusters. The first is characterized by being elongate proximodistally, having a proximal articular surface with a deep dorsoplantar concavity and a prominent plantar process with a notch lateral to it. In the second group, there is a proximal projection of the medioplantar angle, the plantar process is short, the cuboid facet is large, and a notch is absent. The main features of the third group are: short proximodistally, proximal articular surface shallow, articular facet for ectocuneiform quadrangular in outline, and short plantar process with a notch medial to it. The last group is characterized by: proximal articular surface very shallow, prominent plantar process, and cuneiform facets facing distally and equally developed. The morphology of the first cluster is very similar to that of litopterns; the second cluster is very close to Typotheria (Notoungulata), being distinct in minor details only; some features of the third cluster are observed among "condylarths" (such as *Phenacodus, Tetraclaeonodon, Arctocyon, Meniscotherium,* and *Hyopsodus*), but its general morphology is very similar to the navicular of *Arctocyon primaevus* (Russell, 1964; Figure 6.9). The navicular of the last cluster is very large and certainly belonged to *Carodnia vieirai*.

Besides being very similar to the navicular of *Arctocyon primaevus* Blainville, 1841 (sensu Russell, 1964), the navicular assigned to *Lamegoia conodonta* bears the same color and pattern of fossilization of the astragalus and calcaneum. It is short proximodistally and has a subtriangular outline in proximal view, with a medioplantar angle that is prominent and projecting proximally (Figure 6.16). The astragalar facet is moderately deep. The plantar surface is notched for the tendon of the tibialis posterior muscle, which inserts on a moderately developed plantar knob, lateral to the notch. The cuboid facet is flat, pentagonal, and extends along the entire lateral surface; it is contiguous with the squared facet for the ectocuneiform. The facet for the mesocuneiform is the largest, extending partially onto the medial surface. It is continuous with the entocuneiform facet, which is the shortest and the only one that is convex, and it is partially directed plantarly.

6.4.2.7 *Ectocuneiform*

The shape of the navicular facet and the transverse expansion of the plantar hook are uncommon features among fossil ungulates, but both are present in the ectocuneiform of *Arctocyon primaevus* (see Russell, 1964; Figure 6.9), suggesting that they may be primitive for ungulates. Their perfect articulation with the navicular of *L. conodonta*, together with the same color and type of fossilization of the remaining fossils assigned to this species, suggests reference to this species. Assuming this referral is correct, the ectocuneiform of *Lamegoia* appears to be little modified from the primitive condition.

Both ectocuneiforms are perfectly preserved (Figure 6.17). The body is almost as large as it is long, presenting in dorsal view a nearly square shape, concave proximodistally, and rugose for ligamentous attachment. The navicular facet is quadrangular in outline and almost flat and contiguous with the mesocuneiform facet, which is dorsoplantarly concave and elongated. This facet is also continuous distally with the dorsal facet for metatarsal II, as well as with the plantar facet for the same metatarsal in specimen MCN-PV 1760. Gazin (1965) observed a similar pattern of facet variation in the ectocuneiform of *Meniscotherium*. The dorsal facet for metatarsal II is medially projecting at its distal end; the plantar facet is flat, and when isolated, has an elliptical outline. Both facets are contiguous with the distal "T" shaped and dorsoplantarly concave facet for metatarsal III. The cuboid facet, on its lateral side, is flat and square or rectangular. The plantar surface has a prominent, transversely expanded, and rugose tuberosity for attachment of the tendons of the tibialis posterior and flexor hallucis brevis muscles, as well as ligaments.

6.4.3 *Victorlemoinea prototypica*

6.4.3.1 *Humerus*

As in the other two species, only the distal part of the humerus of *V. prototypica* is known (Figure 6.18). The four

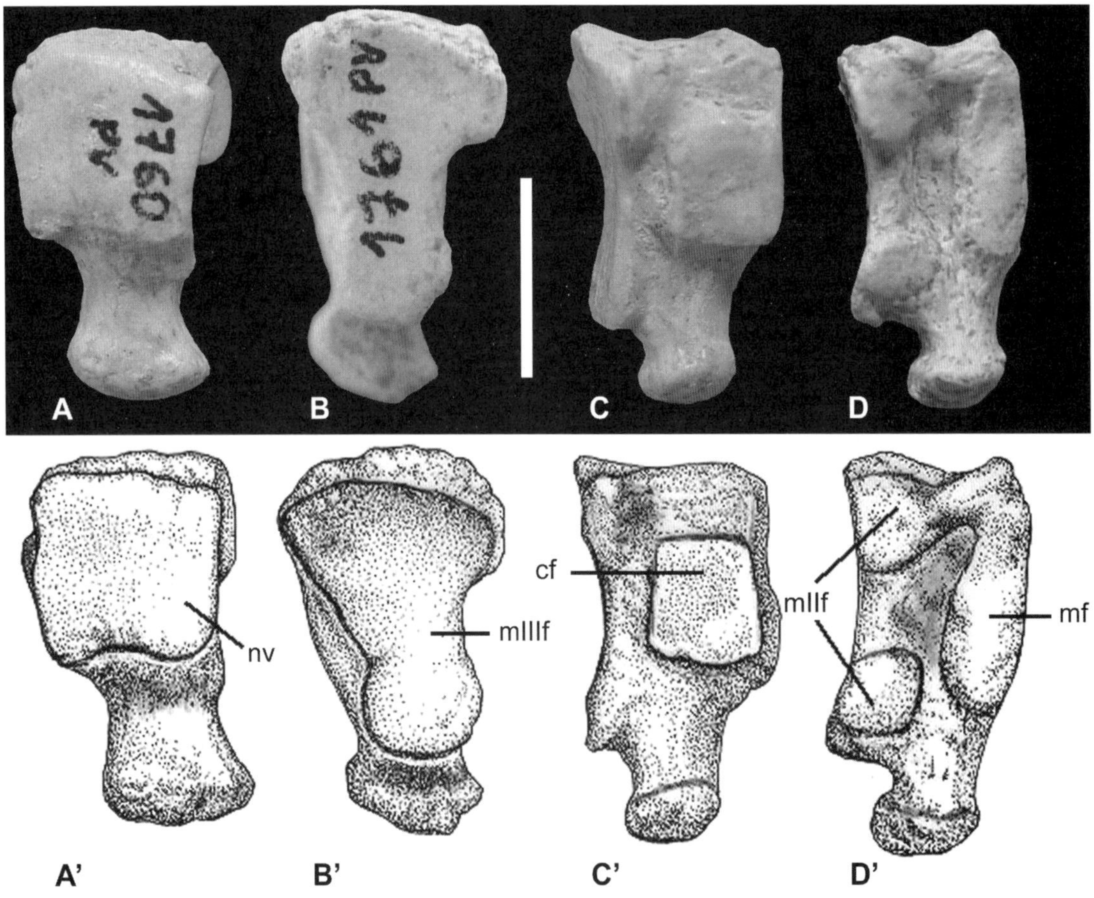

FIGURE 6.17. Left (**A, C**) and right (**B, D**) ectocuneiform of *Lamegoia conodonta*. **A,** proximal; **B,** distal; **C,** lateral and **D,** medial views of MCN-PV 1760 and 1761 (respectively). Outline drawings of ectocuneiform in **A', B', C'** and **D'** in the same views. **cf,** cuboid facet; **mIIf,** metatarsal II facet; **mIIIf,** metatarsal III facet; **mf,** mesocuneiform facet; **nv,** navicular facet. Scale bar: 1 cm.

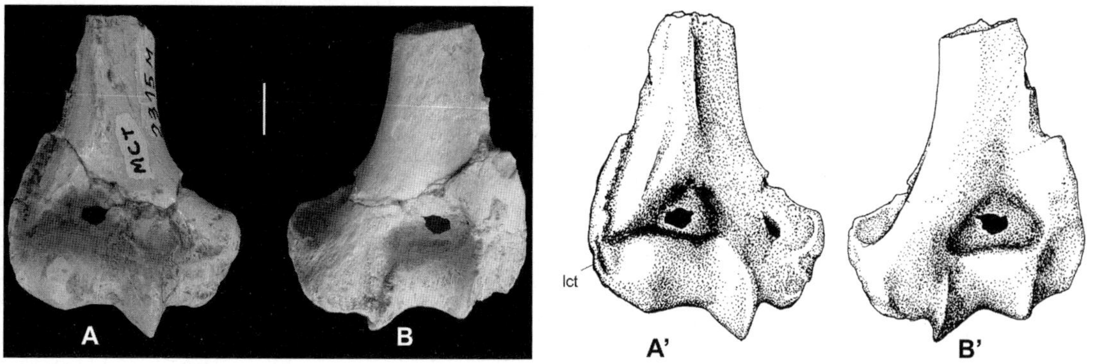

FIGURE 6.18. Right humerus of *Victorlemoinea prototypica*. **A,** anterior and **B,** posterior views of MCT 2315M. Outline drawings of humerus in **A'** and **B'** in the same views. **lct,** lateral crest of trochlea. Scale bar: 1 cm.

specimens grouped here present small variations in the development of the supinator crest and in the diameter of the supratrochlear foramen, but they are all of appropriate size for this species, and for this reason they are grouped together. These humeri are slightly smaller than the one assigned to *L. conodonta*.

From the preserved portion, the supinator crest is more prominent and the medial crest of the trochlea more developed than in *Lamegoia* and *Paulacoutoia*, but not as sharp as in the latter taxon. The distal part of the shaft has a triangular cross-section, indicating the presence of a long deltoid crest with a prominent deltoid tuberosity. The supratrochlear fossa is as in the previous species. As in *L. conodonta*, the trochlea has a lateral crest, but the lateral groove is less marked.

6.4.3.2 *Radius*

The reassociation of the radius was based on direct articulation between the head of this bone and the capitulum of the humerus. The correlation between both measurements is observed in extant mammal species (Bergqvist, 1996). In all mammals, the radial head is slightly larger than the capitulum.

Two complete radii of different individuals are preserved. The larger one (Figure 6.19) belongs to a juvenile, whereas the shorter one belongs to an adult. They are moderately robust, especially distally. Both proximal and distal ends are transversely expanded relative to the dimensions of the shaft. Overall, the radial head bears general resemblance to those of "condylarths" (such as *Phenacodus, Carsioptychus, Hyopsodus*, and *Meniscotherium*). It has an elliptical outline, almost twice as wide as in anteroposterior diameter. The humeral surface is dominated by a central depression for the capitulum, flanked by a steeply inclined medial trochlear surface. The proximal ulnar facet is gently convex, transversely elongate, and extends across almost two-thirds of the head. The shape of the head and the moderate development of the capitular eminence suggest some restriction of supination capability.

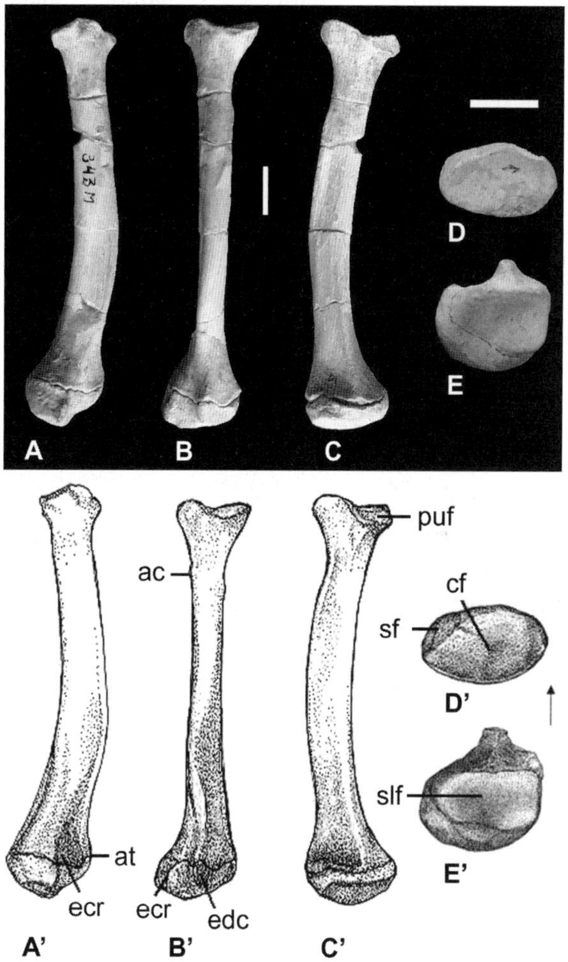

FIGURE 6.19. Right radius of *Victorlemoinea prototypica*. **A,** lateral; **B,** anterolateral; **C,** posteromedial; **D,** proximal, and **E,** distal views of DGM 343M. Outline drawings of radius in A', B', C', D', and E' in the same views. **ac,** anterolateral crest; **at,** anterior tubercle; **cf,** capitulum facet; **ecr,** groove for the extensor carpi radialis muscle; **edc,** groove for the extensor digitorum communis tendon; **ic,** interosseous crest; **puf,** proximal ulnar facet; **sf,** supplementar facet for humerus; **slf,** scapho-lunate facet. Arrow indicates anterior surface of proximal and distal ends. Scale bars: 1 cm.

A small, supplementary, crescent-shaped facet is present on the anterolateral corner of the proximal ulnar facet on specimen DGM 343M. This facet is uncommon in ungulates, but is seen in *Homalodotherium* Flower, 1873 (a toxodontian notoungulate) and *Pachyaena* (O'Leary and Rose, 1995). A similar facet is also present in the Oligocene *Patriomanis* Emry, 1970. Scott (1930) proposed that this facet in *Homalodotherium* is for articulation with a sesamoid, as it is in recent manids, in which a small sesamoid is present in the tendon of origin of the supinator brevis muscle (Emry, 1970). However, in *Pachyaena* (and, possibly, also in *Victorlemoinea prototypica*) it articulated with a matching surface on the humerus (lateral trochlea) when the elbow was flexed (O'Leary and Rose, 1995).

The shaft curves gently posteromedially. Taylor (1974) concluded that in Viverridae the curvature of the shaft is related to plantigrade posture and the ability to supinate the manus. The shaft is anteroposterioly compressed proximally, transversely compressed at mid-shaft, and gradually thickens at its distal end, becoming nearly round in cross-section. The bicipital tuberosity is weakly expressed on specimen DGM 348-M and barely discernible on DGM 343-M. This suggests that in this animal the flexion of the antebrachium was not powerful, as the biceps brachii muscle was poorly developed. The interosseous crest is sharper on its distal half. An anterolateral crest, probably for attachment of the supinator muscle, is present on the proximal half. It cuts obliquely across the anterior surface, and distally it runs medial to the groove for the passage of the tendons of the extensor carpi radialis muscle, which is bound laterally by a prominent anterior tubercle. A more lateral, shorter, and weakly defined groove probably housed the extensor digitorum communis tendon.

The distal ulnar articular surface is posterolateral in position, wide, short, and confluent with the carpal articular surface, which bears a unique facet for both the scaphoid and the lunate. This primitive condition is seen in *Arctocyon primaevus*, a generalist "condylarth" (Russell, 1964). The distal facet of the radius is subtriangular, larger and concave on the portion for the lunate, and flat to smoothly convex medioposteriorly. The radial styloid process is small, approximating the size seen among arctocyonids.

6.4.3.3 *Ulna*

The ungulate ulnae collected in the Itaboraí basin can be separated into three major groups based on the posterior border of the shaft. Most of the specimens have a slightly or markedly convex posterior border, as is observed in almost all "condylarths", typotherians, and primitive toxodonts. Some have a posteriorly concave shaft, which is typical of derived litopterns, Xenungulata, Pyrotheria, some toxodonts, and *Phenacodus*. A straight posterior border is present in few specimens, and is also observed in the ulnae of astrapotherians and homalodotheriid toxodonts.

The ulna of *Victorlemoinea* is a robust element (Figure 6.20). In lateral profile it is slightly concave posteriorly only on the distal half, the proximal half being flat to smoothly convex.

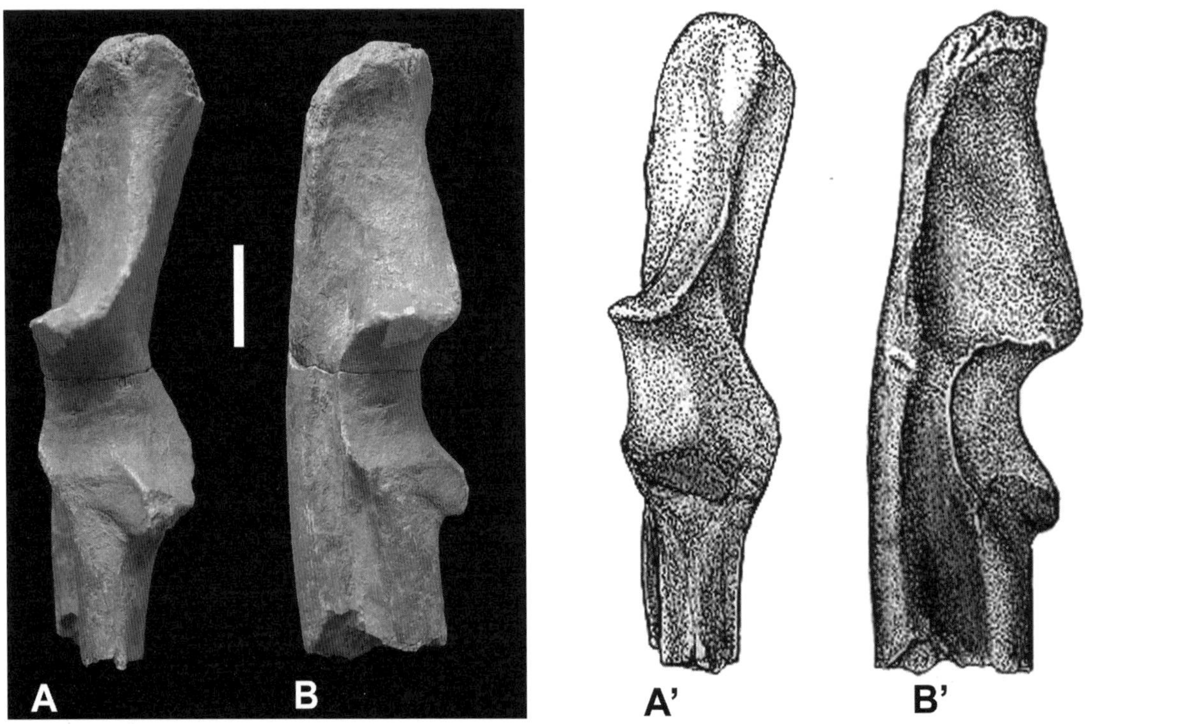

FIGURE 6.20. Right ulna of *Victorlemoinea prototypica*. **A**, lateral; **B**, medial view of MCN-PV 1718. Outline drawings of ulna in **A'** and **B'** in the same views. Scale bar: 1 cm.

The shaft is transversely compressed, though not as deep as in *L. conodonta*.

The olecranon process is prominent and relatively long (longer than the semilunar notch and 25–26% of the total preserved ulnar length), differing from *L. conodonta* in being much thicker posterioly than anteriorly, and in having expanded and rough epiphyses. In this way, it is comparable to *Pachyaena*. In lateral profile, the semilunar notch resembles that of *Pachyena gigantea* and *L. conodonta*. The articular surface is like that of *L. conodonta*, except that the coronoid process projects farther over the shaft in *V. prototypica*. The radial facet is flat, narrow, elongated, and faces anterolaterally, as it does in non-cursorial mammals. The fossa for the tendon of the brachialis and biceps brachii muscles is shallow and less marked than in *L. conodonta*.

The anterolateral fossa is conspicuously different from that of *L. conodonta*. It is eye-shaped, not completely occupying the anteroposterior extension of the lateral surface, and terminates proximal to the distal end. Whereas its depth in *L. conodonta* is almost equal throughout, in *V. prototypica* it is notably deeper in its mid-portion. As in *Lamegoia*, the anterolateral fossa originates proximally at the midpoint of the semilunar notch. The fossa is bounded anteromedially by a more or less rounded interosseous crest, and posterolaterally by a salient sharp crest.

Distally the shaft enlarges and has a sharp lateral crest, where the distal portion of the pronator quadratus muscle probably inserted. The distal articular facet was probably completely located on the distal epiphysis, which is not preserved in any of the specimens.

6.4.4 "Condylarthra" Indet

A tibia, very similar to that of *Arctocyon ferox* (1833), and several phalanges, close in morphology to *Phenacodus, Tetraclaenodon*, and *Meniscotherium*, could not be confidently assigned to any of the Itaboraí "condylarth" species (see Appendix). They were described by Bergqvist (1996), but are not included here, except for the astragali and calcanea, some of which were previously assigned to *Victorlemoinea prototypica* by Cifelli (1983a).

6.4.4.1 *Astragalus*

These astragali approximate the size of the astragalus of *Lamegoia conodonta*, but they differ from it in the absence of a deep tibial trochlea with well-defined crests (Figure 6.21). They exhibit some morphological variation, and many different groupings could be proposed depending on the feature or group of features used to distinguish them. This appears to comprise a single but variable group.

The astragalar body is robust and of comparable depth to that of *Arctocyon* (Figure 6.9). The tibial trochlea is short and very shallow, reducing the extent of movement of the proximal astragalar articulation; it is interrupted posteriorly by an unreduced superior astragalar foramen and is bordered by sharp lateral and rounded medial crests. Posteroplantarly, the groove for the digital flexor tendons is broad, shallow or deep depending on the specimen, and directed posteromedially. Posterior to the superior astragalar foramen, and directed posterolaterally, a sulcus extends to the posterior border of the ectal facet. This sulcus probably protected

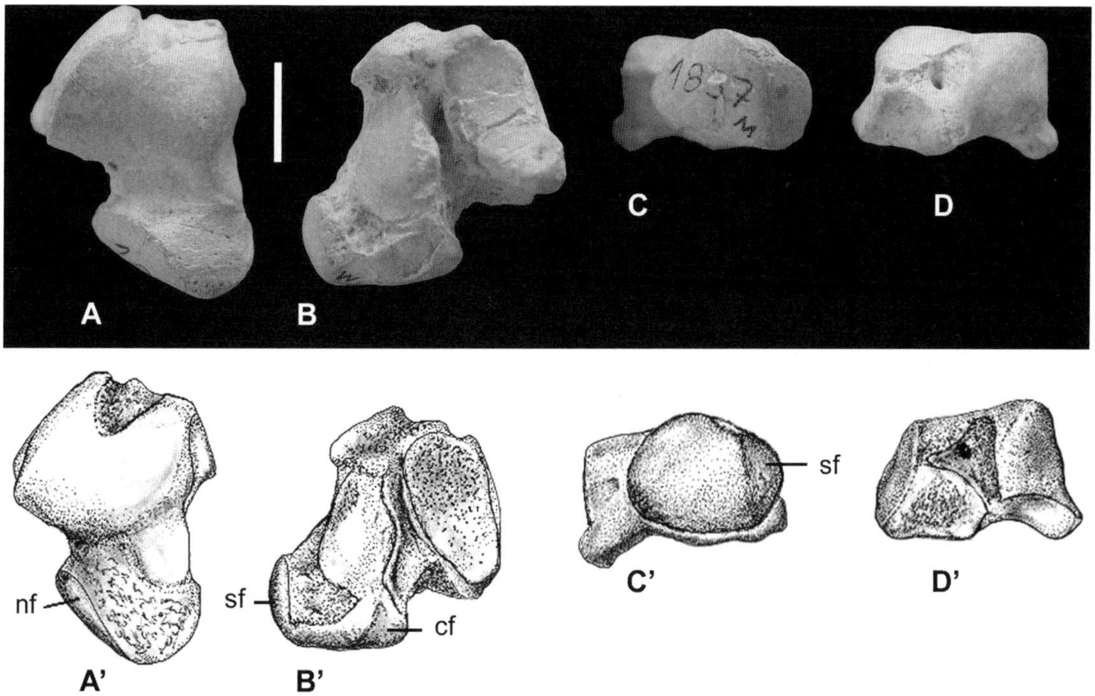

FIGURE 6.21. Left astragalus of "Condylarthra" indet. **A,** dorsal; **B,** plantar; **C,** distal; **D,** proximal of MCT 1837M. Outline drawings of astragalus in **A', B', C'** and **D'** in the same views. **cf,** cuboid facet; **nf,** navicular facet; **sf,** sesamoid facet (in the medial collateral ligament). Scale bar: 1 cm.

the veins and nerves that run through the astragalar channel (Cifelli, 1983a).

The medial wall is relatively vertical, and the facet for the medial malleolus of the tibia is broad and extends well down the neck (nearly to the head in juvenile specimens) of the astragalus. The tubercle for the medial collateral ligament is weak. The fibular facet on the lateral side of the astragalus is extensive and vertical, and a prominent fibular shelf is developed anteroinferiorly.

The neck is moderately long and oblique in relation to the trochlea. The head is transversely narrow but deep, both medially and laterally. In some specimens it is relatively large in relation to the trochlea, while in others it is relatively small. The head has three facets. The medial facet is very broad, dorsoventrally and proximodistally convex, and meets the navicular facet at a relatively high angle. It probably articulated with a sesamoid in the medial collateral ligament. The navicular facet, restricted to the middle portion of the head, is oriented much more vertically with respect to the astragalar body in most of the specimens, compared to that of *L. conodonta*. The cuboid facet, occupying the inferolateral portion of the head, is broadly continuous posteroinferiorly with the sustentacular facet. Part of the cuboid facet was for contact with an accessory facet on the medial side of the calcaneal neck, but articulation of appropriately sized astragali and calcanea indicates that astragalocuboid contact was also well-developed (as in an alternating tarsus). The sustentacular facet is anteroposteriorly elongate and, in some specimens, achieves very broad contact with the navicular facet, while in others only with the cuboid facet. The astragalar ectal facet is somewhat transversely broader than in *Arctocyon*, but narrower than in *Lamegoia*. The ectal and sustentacular facets are separated by a deep interarticular sulcus. The groove for the digital flexor tendons at the posteroinferior margin of the astragalus is broad and sharply distinct.

6.4.4.2 Calcaneum

The specimens assigned to "Condylarthra" indet. by Bergqvist (1996) were separated into two different groups on the basis of the sustentacular shelf and sustentacular facet morphology. Morph 1 (DGM 890M, MCT 2575M, and MCN-PV 1271) presents a deep sustentacular shelf and the sustentacular facet extends anteriorly to achieve broad contact with the calcaneocuboid facet (Figure 6.22). In Morph 2 (MCT 2576 and MCN-PV 1268) the sustentacular shelf is shallow and the sustentacular facet is isolated, large, and rounded.

The specimens of Morph 1 are slightly shorter than those assigned to *Lamegoia*, but more robust, and bear general resemblance to *P. protocenica*. They exhibit extensive biometric variation due to the juvenile condition of the smaller specimen (MCN-PV 1271). They also show some degree of morphological variation.

The tuberosity has two parallel and oblique sulci, dividing it into three parts. The ectal facet is less obliquely located than in *L. conodonta* and *P. protocenica*, and the fibular facet is transversely broad and well-developed. The depression for the fibular collateral ligament is more marked and the peroneal tubercle is rather enlarged and rugose, with an incipient shelf, suggesting a well-developed quadratus plantae muscle and lateral collateral ligaments.

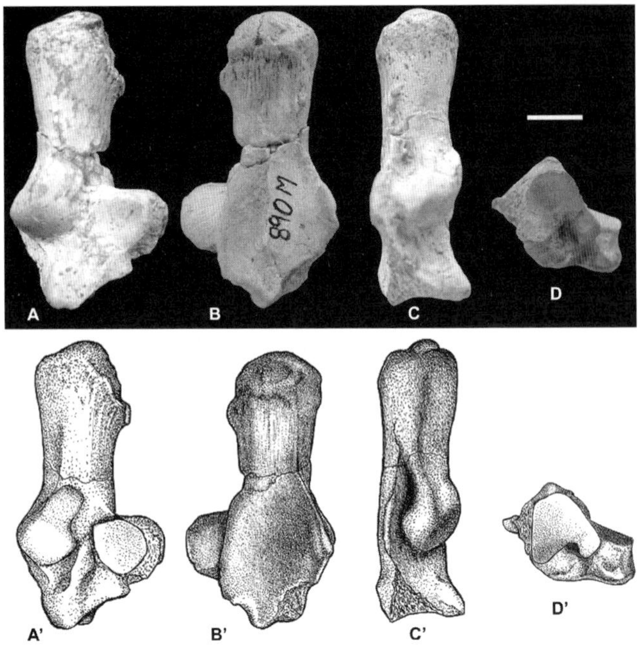

Figure 6.22. Right calcaneum of "Condylarthra" indet. **A,** dorsomedial; **B,** plantar; **C,** lateral, and **D,** distal views of DGM 890M. Outline drawings of calcaneum in **A', B', C',** and **D'** in the same views. Scale bar: 1 cm.

As in *P. protocenica*, the anterodorsal portion of the calcaneal neck is elongated dorsally into a distinct "beak" (very pronounced in DGM 890M); the cuboid facet, extending up the distal face of this eminence, is also very similar to that of *P. protocenica*, except that it extends medially, forming an inverted V-shape (excluding MCN-PV 1271, in which it is continuous). Inferior to the cuboid facet, the tuberosity for attachment of the calcaneotarsal and tarsal fibrocartilage ligaments is modestly developed.

The specimens of Morph 2 are quite similar to *P. protocenica*, but differ from this species in the greater lateral projection of the sustentaculum, larger sustentacular facet, and weaker development of the fibular facet. The cuboid facet, though partially eroded, is almost flat, and the plantar tubercle better developed than in *P. protocenica*.

6.5 Discussion

Since *Victorlemoinea prototypica* was proposed (Paula-Couto, 1952), it has been regarded as a litoptern based on its lophodont tooth morphology. Cifelli (1983b) observed that the connecting crest between the hypocone and metaconule on the upper molars could be regarded as reminiscent of the specialization in Macraucheniidae. However, by the Deseadan/Colhuehuapian ages, the Macraucheniidae were far more primitive than *Victorlemoinea*, leading Cifelli (1983b) to the conclusion that *Victorlemoinea*'s tooth morphology was aberrantly specialized and convergent with macrauchenids. He then referred it, together with *Phoradiadus* and *Sparnotheriodon*, to the family Sparnotheriodontidae, and placed this family in the order "Condylarthra" based on the morphology of the ankle bones he referred to *V. prototypica* (Cifelli, 1983b, 1993). Its assignment, as mentioned above, was not accepted by most subsequent authors, who kept the family Sparnotheriodontidae in the order Litopterna. Soria (2001) was the first and sole author to undertake a phylogenetic revision of the Sparnotheriodontidae, though he did not present a cladogram or list the synapomorphies that supported the placement of Sparnotheriodontidae within Litopterna.

Cifelli (1983a) associated, with some doubt, a group of seven relatively large tarsal bones to *V. prototypica*, noting that there were no litoptern tarsals large enough to belong to this species among the fossils recovered from the Itaboraí basin. In spite of great morphologic and size variation, he assigned the entire group of tarsals to *V. prototypica*.

Eighteen astragali (including the sample studied by Cifelli, 1983a) with the same morphology as those assigned to *V. prototypica* by Cifelli (1983a) were studied here. In spite of their great size (CV = 10.4) and morphological variation (see Table XXXI of Bergqvist, 1996), they were grouped together because they are visibly distinct from the other morphotypes proposed by Bergqvist (1996), but similar in all basic and important features, as mentioned in the description. However, in contrast to Cifelli (1983a), this group was not assigned here to *V. prototypica*, as a revision of all Itaboraí "condylarth" and litoptern teeth (in progress) has shown the existence of two new species similar in size to *V. prototypica*. One of them is clearly bunodont and represents a new form of Didolodontidae, and the other, as already observed by Soria (2001), represents a new sparnotheriodontid, more primitive than *V. prototypica*. This large group of astragali may include specimens belonging to these three large species of the Itaboraí basin, but the assignment of some of them to any one of these species is impractical.

Although no astragali or calcanea were assigned here to *V. prototypica*, it does not weaken Cifelli's (1983a, b) proposition that the ankle bones of this species are primitive and "Condylarthra-like", and that *V. prototypica* is not a litoptern. Cifelli (1993a) and Bergqvist (1996, 1997, 2005) have shown that several synapomorphies of the order Litopterna are present in the postcranial skeleton, mainly in the tarsal bones. The discovery of associated cranial and postcranial bones of Neogene fossils such as *Megadolodus molariformis* McKenna, 1956 (Cifelli and Villaroel, 1997) and *Prothoatherium colombianus* (Hoffstetter and Soria, 1986) (Cifelli and Guerrero-Diaz, 1989), with primitive bunodont teeth and derived litoptern postcranial features, support the derived postcranial morphology of litopterns, even when they retain primitive bunodont dental morphology. The opposite situation is observed among North American "condylarths". The Paleocene-Eocene genus *Meniscotherium* combines an entirely selenolophodont dentition, similar in several ways to that of *Victorlemoinea*, with generalized postcrania and primitive tarsal morphology (Williamson and Lucas, 1992). So, *V. prototypica* is not the first or sole "condylarth" to exhibit a derived dentition in combination with a primitive skeleton. Furthermore, the humerus, radius, and ulna assigned to *V. prototypica* do not exhibit the derived features shared by litoptern species, such as the absence of the entepicondyle and entepicondylar foramen of the humerus, markedly distinct lunate and scaphoid facets of the radius, and ulnar olecranon longer than the semilunar notch (Bergqvist, 1996, 1998). All of these features are related to cursorial abilities developed by Litopterna since the origin of the lineage, as seen in *Protolipterna ellipsodontoides*, for instance.

Another line of evidence supporting the placement of *V. prototypica* among "condylarths" is the entire absence among the fossils recovered from the Itaboraí basin of large enough tarsals with litoptern morphology to be assigned to this species (Bergqvist, 1996, 1998). It is possible, but less probable, that no tarsal bones of *V. prototypica* were preserved. Bergqvist (1996) showed that astragali and calcanea comprise more than 50% of the fossil postcrania studied in her dissertation, and Almeida (2005) also observed that astragali and calcanea are the most abundant fossil bones recovered from the "1968 Fissure" (with the exception of vertebrae, phalanges, and metapodials, which are more numerous in the skeleton than ankle bones). This, in my view, weakens the possibility that no tarsals of *V. prototypica* are present in the sample. In this case, some of the larger and more primitive astragali of Itaboraí might belong to this species, and on the basis of the synapomorphies of the Litopterna proposed by Cifelli (1993) and Bergqvist (1996, 1997, 2005), *V. prototypica* could not be placed in this group.

Soria (2001) is correct when he points out that the relative abundance and size of the m2 of *Paulacoutoia protocenica* and *Paranisolambda prodromus* are similar (see Cifelli, 1983a). Bergqvist (1996) reached the same conclusion based on more specimens. Soria (2001) then suggested that it would be possible to assign primitive tarsals (named "phenacodontoid" by him) to *P. prodromus* and derived tarsals (named "litopternoid") to *P. protocenica*. He continues to state that this would solve the problem of Sparnotheriodontidae (both *V. prototypica* and *P. prodromus* would have primitive phenacondontoid tarsals), but in the same paper he recognized that it would represent a false solution.

Since Matthew (1909), it has been recognized that the astragalus is of great importance in the study of mammalian affinities, as each mammalian order has a peculiar and distinct astragalus, with relatively consistent morphology within the group. The minor variations in proportions that might be seen are related to adaptive divergences. Stain (1959) also observed that, even though function undoubtedly plays an important role in shaping the calcaneum, the basic pattern of the bone is fixed genetically and retains a similar shape in closely related species. Bergqvist (1996) undertook a detailed study of most "Condylarthra", Litopterna, Notoungulata, Astrapotheria, and Xenungulata known by associated dentitions and skeletons, from every Paleogene and Neogene age. She noticed that since the Paleocene, the astragalus of these South American ungulates exhibits some of the derived features observed in later taxa from the order to which they belong.

For Soria (2001), the most reasonable postcranial association proposed by Cifelli (1983b) was that of *Protolipterna ellipsodontoides* Cifelli, 1983a, as all (not the majority, as stated by Soria on page 18) bones and teeth came from the single fissure in which all fossils have brown color. Considering the dental morphology of *P. ellipsodontoides* and the association of a "litopternoid" astragalus to *Asmithwoodwardia subtrigona* Ameghino, 1901 by Ameghino (1904), and the fact that both species have a very primitive didolodontid dental morphology, Soria (2001) concluded that the "litopternoid" tarsal bone was typical of Didolodontidae.

Although no didolodontid tarsal bone is known by association, it is not unreasonable to expect (challenging Soria, 2001) that they would have the "phenacodontoid" morphology, as the family Didolodontidae is placed among "condylarths". The tarsals assigned to *Lamegoia* and *Paulacoutoia* are not only of the appropriate size for these taxa, but also exhibit the primitive general morphology observed in North American "condylarths" such as *Meniscotherium* and *Arctocyon*. Soria (2001) also suggested that since they are few in number, they might have belonged to species unknown from dental morphology. Teeth are the hardest part of the skeleton and they form a prominent fraction of the mammal remains from paleontological and archaeological sites (Hillson, 1996; Bergqvist, 2003). Thousands of teeth were recovered in the Itaboraí basin and it is quite unlikely that postcranial remains of a species would be preserved without any segment of its dentition.

Among South American ungulates, the tarsals of notoungulates are also somewhat primitive in morphology, and, in some ways, similar to those of "condylarths". Soria (2001) proposed that the tarsals assigned by Cifelli (1983a) to *P. protocenica*, *V. prototypica*, and *L. conodonta* may have belonged to a notoungulate. However, a phylogenetic analysis based on postcranial features undertaken by Bergqvist (1996) revealed three tarsal synapomorphies of the order Notoungulata: constriction of the astragalar neck (exclusive to notoungulates), a well-developed medial collateral ligament tubercle on the astragalus (only moderately developed in *Hyopsodus*), and a calcaneal tuberosity with grooves following different directions. These features are clearly present in the tarsal bones assigned to notoungulates by Cifelli (1983a) and Bergqvist (1996). Moreover, the notoungulate species of the Itaboraí basin have smaller teeth than the "condylarth" taxa (Bergqvist, 1996), and at least one species (*Colbertia magellanica* Price and Paula-Couto, 1950) is much more abundant than any of the "condylarth" species. Tarsals appropriate in size and morphology for *C. magellanica* are readily identified in the collections, and, as might be expected, they are also much more abundant than are those referred to the "condylarths".

Most of the bones recovered from the Itaboraí basin are incomplete and some appear to have belonged to juveniles (due to the absence of epiphyses), making functional interpretation problematic. However, some interpretations can be proposed.

The postcrania assigned to the Itaboraí "condylarths" are primitive in general morphology, although a few features of the fore- and hind limbs are indicative of slight cursorial capabilities, mainly in the family Didolodontidae. The bones assigned to *Lamegoia conodonta* and *Paulacoutoia protocenica* suggest that the Itaboraí Didolodontidae were generalized mammals with few cursorial abilities, indicated mainly by the morphology of the astragalus. In both species the trochleae have relatively sharp crests, restricting the mediolateral mobility of the ankle joint; in addition, the posterior displacement and reduction of the superior astragalar foramen increases the range of orthal movement. The remaining astragali, among which some certainly belonged to the sparnotheriodontid *Victorlemoinea prototypica*, are less specialized than the astragali of Didolodontidae, in retaining a large astragalar foramen that would have restricted anteroposterior rotation of the ankle (Wang, 1993), and in having flat and short trochleae, which would have allowed relatively greater mediolateral mobility. They are also more primitive than those referred to Didolodontidae in the presence of astragalus-cuboid contact (Cifelli, 1983a, 1993).

The forelimb bones of the two families are primitive in general morphology, although a few features of the humerus (inclination of the medial crest of the trochlea; Taylor, 1974), radius (wide, oval radial head; Rose, 1990), and ulna (relative length of the olecranon; O'Leary and Rose, 1995) are also suggestive of slight cursorial capabilities. However, the distal ends of the humeri show no sign of reduction, all bearing a well-developed entepicondyle with a large entepicondylar foramen, and the distal articulation of the radius of *Victorlemoinea prototypica* lacks divided articular facets for the scaphoid and lunate.

O'Leary and Rose (1995) showed that, depending on the regression used, the body weight estimation of *Pachyaena gigantea* varied from 129 to 396 kg. Hence, no body weight estimations were proposed here due to the great inconsistency presented by fossil samples with no extant relatives.

6.6 Conclusions

Among the isolated bones recovered in the Itaboraí basin some could be confidently reassociated to species of "Condylarthra" established on the basis of dental morphology.

The reassociations proposed by Cifelli (1983b) and here are supported by abundance, color, size, and morphology. Although for some authors these reassociations are questionable, they appear to be more parsimonious than other propositions (e.g., Soria, 2001) that assume the total absence of teeth or tarsals of some species. Considering the abundance of Itaboraí fossils and the size and shape of the basin, absence of these elements seems totally implausible.

Although no astragali or calcanea were assigned to the Sparnotheriodontidae, there is a large sample of these bones currently lacking specific assignments. The morphology of these ankle bones is the most primitive among Itaboraí "condylarth" tarsals and is suggestive of generalized terrestrial habits. Based on bone morphology, the Itaboraí Didolodontidae are generalized terrestrial mammals, with few cursorial specializations.

Acknowledgments. I thank the following people for giving me assistance and allowing me to examine fossil specimens in their care: Dr. Sérgio Alex K. Azevedo of the Museu Nacional, Rio de Janeiro; Dr. Diogenes de Almeida Campos of the Departamento Nacional da Produção Mineral, Rio de Janeiro; Dr. Jorge Ferigolo of the Fundação Zoobotânica do Rio Grande do Sul, Porto Alegre; Dr. Marcelo Reguero of the Museo de La Plata, La Plata; Dr. José Bonaparte of the Museo Argentino de Ciencias Naturales, Buenos Aires; Dr. Richard Cifelli of the San Noble Oklahoma Museum of Natural History, Norman; Dr. Malcolm McKenna of the American Museum of Natural History, New York; Dr. Robert Emry of the National Museum of Natural History, Washington; Dr. Elisabeth Vrba of the Peabody Museum of Natural History, New Haven; Dr. John Flynn of the Field Museum of Natural History, Chicago; Dr. Leigh Van Valen, of the University of Chicago, Chicago; Dr. Philip Gingerich of the University of Michigan, Ann Arbor; Dr. Kenneth Rose of the John Hopkins University, Baltimore. I am also grateful to Dr. Leonardo Avilla, of Universidade do Rio de Janeiro, for providing me with digital photos of *Phenacodus*.

I am in debt to Drs. Richard Cifelli and Jorge Ferigolo, my doctorate advisors, for their revision, comments, and suggestions to my dissertation, and also to the former for reviewing a previous version of this manuscript. A special thanks to Dr. Fred Szalay for inspiring and encouraging me to expand my dissertation to include a study of the postcranial bones of all Itaboraí ungulates.

This study was made possible largely through grants from the Coordenação de Aperfeiçoamento de Pessoal de Nivel Superior (CAPES), Brasília; Theodore Roosevelt Memorial Fund, New York; Field Museum of Natural History, Chicago; Smithsonian Institution, Washington. This paper is a contribution to Instituto Virtual de Paleontologia/FAPERJ.

Appendix

Fossils Examined for Each Taxon

Paulacoutoia protocenica: DGM 345M – right tibia lacking epiphysis; 1388M – left astragalus (juvenile); 1854M – distal third part of right humerus; MCN-PV 1702 – distal part of right humerus; 1707 – distal part of right humerus; AMNH 55388 – right astragalus; 55390 – left calcaneum.

Lamegoia conodonta: DGM 337M – left femur; 940M – right astragalus; MCT 1833M – proximal half of left ulna lacking olecranon epiphysis; 2313M – distal half of right humerus; 2326M – incomplete right ulna lacking olecranon epiphysis; MCN-PV 1357 – right calcaneum; 1359 – right astragalus; 1760 – left ectocuneiform; 1761 – right ectocuneiform; 1826 – right navicular; 1827 – left navicular; AMNH 55389 – left calcaneum.

Victorlemoinea prototypica: DGM 340M – right ulna lacking epiphysis; 341M – right ulna lacking epiphysis; 343M – left radius; 889M – distal part of left humerus; MCT 2314M – distal third of left humerus; 2315M -; 2327M – distal part of left humerus; 2423M – incomplete right ulna; 2431M – proximal part of right radius; 2433M – proximal part of left radius; MCN-PV 1718 – proximal part of right ulna lacking olecranon epiphysis.

"Condylarthra" indet.: DGM 890M – right calcaneum; MCT 1837M – left astragalus; 1861M – ungual phalanx; 1908M – left third metatarsal; 1919M – ungual phalanx; 1940M – ungual phalanx; 2316M – right astragalus; 2317M – left astragalus; 2318M – right astragalus; 2319M – right astragalus; 2321M – left astragalus; 2453M – right astragalus; 2460M – right astragalus; 2575M – left calcaneum; 2576M – left calcaneum lacking tuberosity; 2604M – incomplete right calcaneum; 1268 – left calcaneum lacking tuberosity; MCN-PV 1271 – right calcaneum; 1336 – left astragalus; 1360 – left astragalus; 1361 – left astragalus; 1362 – left astragalus; 1364 – left astragalus; 1365 – right astragalus; 1366 – right astragalus; 1367 – right astragalus; 1716 – proximal part of left tibia; 1721 – right third metacarpal; 1757 – ungual phalanx; AMNH 55383 – left astragalus; 55391 – right astragalus; 55392 – left astragalus; 55393 – right astragalus; 55397 – left calcaneum.

References

Almeida, E. B., 2005. Aspectos tafonômicos dos fósseis de mamíferos procedentes da Bacia de São José de Itaboraí, RJ – Paleoceno superior. Master thesis, Universidade Federal do Rio Grande do Sul.

Archibald, D., 2005. Archaic ungulates ("Condylarthra"). In: Janis, C. M., Soctt, K. M., Jacobs, L. L. (Eds.), Evolution of Tertiary Mammals of North America. Cambridge University Press, Cambridge, pp. 292–331.

Bergqvist, L. P., 1996. Reassociação do pós-crânio às espécies de ungulados da bacia de S. J. de Itaboraí (Paleoceno), Estado do Rio de Janeiro, e filogenia dos "Condylarthra" e ungulados sul-americanos com base no pós-crânio. Ph.D. dissertation, Universidade Federal do Rio Grande do Sul.

Bergqvist, L. P., 1997. New Postcranial Apomorphies Shared by Primitive and Advanced Taxa of the Order Litopterna. Boletim de Resumos do XV Congresso Brasileiro de Paleontologia, São Pedro, p. 131.

Bergqvist, L. P., 1998. Postcranial bones reassociated to Condylarthra species of Itaboraí basin (Late Paleocene, Brazil). Journal of Vertebrate Paleontology 18(3), 27A.

Bergqvist, L. P., 2003. The role of teeth in mammal history. Brazilian Journal of Oral Science 2(6), 249–257.

Bergqvist, L. P., 2005. Postcranial synapomorphies supporting the monophyly of the Order Litopterna. Boletim de Resumos do II Congresso Latino-Americano de Paleontologia de Vertebrados, Rio de Janeiro, pp. 48–49.

Bond, M., Carlini, A. A., Goin, F. J., Legarreta, L., Ortiz-Jaureguizar, E., Pascual, R., Uliana, M. A., 1995. Episodes in South American land mammal evolution and sedimentation: testing their apparent concurrence in a Paleocene succession from Central Patagonia. Actas del VI Congreso Argentino de Paleontología y Bioestratigrafia, Trelew, pp. 47–58.

Carrano, M. T., 1997. Morphological indicators of foot posture in mammals: a statistical and biomechanical analysis. Zoological Journal of the Linnean Society 121, 77–104.

Cifelli, R. L., 1983a. Eutherian tarsals from the late Paleocene of Brazil. American Museum Novitates 2761, 1–31.

Cifelli, R. L., 1983b. The origin and affinities of the South American "Condylarthra" and early Tertiary Litopterna (Mammalia). American Museum Novitates 2772, 1–49.

Cifelli, R. L., 1993. The phylogeny of the native South American ungulates. In: Szalay, F.S., Novacek, M. J., McKenna, M. C. (Eds.), Mammal Phylogeny – Placentals. Springer, New York, pp. 195–216.

Cifelli, R. L., Guerrero-Diaz, J., 1989. New remains of *Prothoatherium colombianus* (Litopterna, Mammalia) from the Miocene of Colombia. Journal of Vertebrate Paleontology 9(2), 222–231.

Cifelli, R. L., Villaroel, C., 1997. Paleobiology and affinities of *Megadolodus*. In: Kay, R. F., Madden, R. H., Cifelli, R. L., Flynn, J. J. (Eds.), Vertebrate Paleontology in the Neotropics. Smithsonian Institution Press, Washington, DC, pp. 265–288.

Damuth, J., 1990. Problems in estimating body masses of archaic ungulates using dental measurements. In: Damuth, J., MacFadden, B. J. (Eds.), Body Size in Mammalian Paleobiology. Cambridge University Press, Cambridge, pp. 229–253.

Damuth, J., MacFadden, B. J., 1990. Body Size in Mammalian Paleobiology. Cambridge University Press, Cambridge, 397 pp.

Emry, R. S., 1970. A North American Oligocene pangolin and other additions to the Pholidota. Bulletin of the American Museum of Natural History 142(6), 459–510.

Fortelius, M., 1990. Problems with using teeth to estimate body sizes of extinct mammals. In: Damuth, J., MacFadden, B. J. (Eds.), Body Size in Mammalian Paleobiology. Cambridge University Press, Cambridge, pp. 207–228.

Gazin, C. L., 1965. A study of the early Tertiary condylarthran mammal *Meniscotherium*. Smithsonian Miscellaneous Collections 149(2), 1–98.

Gazin, C. L., 1968. A study of the Eocene condylarthran mammal *Hyopsodus*. Smithsonian Miscellaneous Collections 153(4), 1–90.

Getty, R., 1975. The Anatomy of the Domestic Animals. W.B. Saunders, Philadelphia, 2,095 pp (2 volumes).

Gingerich, P. D., 1974. Size variability of the teeth in living mammals and the diagnosis of closely related sympatric fossil species. Journal of Paleontology 48(5), 895–903.

Gingerich, P. D., Smith, B. H., Rosenberg, K., 1982. Allometric scaling in the dentition of primates and prediction of body weight from tooth size in fossils. American Journal of Physical Anthropology 58, 81–100.

Hillson, S., 1996. Teeth. Cambridge University Press, Cambridge, 376 pp.

Matthew, W. D., 1909. The Carnivora and Insectivora of the Bridger Basin, middle Eocene. Part IV. Memoirs Bulletin of the American Museum of Natural History 9(6), 289–567.

Matthew, W. D., 1937. Paleocene faunas of the San Juan Basin, New Mexico. Transaction of the American Philosophical Society, new series, 30, 1–372.

Medeiros, R. A., Bergqvist, L. P., 1999. Paleocene of the São José de Itaboraí Basin, Rio de Janeiro, Brazil: lithostratigraphy and biostratigraphy. Acta Geologica Leopoldensia 22(48), 3–22.

McKenna, M. C., Bell, S. K. 1997. Classification of Mammals above the Species Level. Columbia University Press, New York, 963 pp.

Muizon, C., Cifelli, R. L., Bergqvist, L. P., 1998. Eutheran tarsals from the early Paleocene of Bolivia. Journal of Vertebrate Paleontology 18(3), 655–663.

Muizon, C., Cifelli, R. L., 2000. The "condylarths" (archaic Ungulata, Mammalia) from the early Paleocene of Tiupampa (Bolivia): implications on the origin of the South American ungulates. Geodiversitas 22(1), 47–150.

O'Leary, M. A., Rose, K. D., 1995. Postcranial skeleton of the early Eocene mesonychid *Pachyaena* (Mammalia: Mesonychia). Journal of Vertebrate Paleontology 15(2): 401–430.

Paula-Couto, C., 1949. Novas observações sobre paleontologia e geologia do depósito calcário de São José de Itaboraí. Notas Preliminares e Estudos, Divisão de Geologia e Mineralogia 49, 1–13.

Paula-Couto, C., 1952. Fossil mammals from the beginning of the Cenozoic in Brazil. Condylarthra, Litopterna, Xenungulata and Astrapotheria. Bulletin of the American Museum of Natural History 99(6), 359–394.

Paula-Couto, C., 1978. Ungulados fósseis do Riochiquense de Itaboraí, RJ, Brasil. II – Condylarthra e Litopterna. Anais da Academia Brasileira de Ciências 50(2), 209–218.

Prothero, D. R., Manning, E. M., Fischer, M., 1988. The phylogeny of the ungulates. In: Benton, M. J. (Ed.), The Phylogeny and Classification of the Tetrapods, Volume 2: Mammals. Systematics Association Special Volume 35b, Clarendon, Oxford, pp. 201–234.

Rose, K. D., 1990. Postcranial skeleton remains and adaptation in early Eocene mammals from the Willwood Formation, Bighorn Basin, Wyoming. In: Bown, T. W., Rose, K. D. (Eds.), Dawn of the Age of Mammals in the Northern Part of the Rocky Mountain Interior, North America: Boulder, Colorado. Geological Society of America, Special Paper, 243, 107–133.

Russell, D. E., 1964. Les mammifères Paléocènes d'Europe. Memoires du Museum National D' Histoire Naturelle, Series C, 13, 1–324.

Schaeffer, B., 1947. Notes on the origin and function of the artiodactyl tarsus. American Museum Novitates 1356, 1–24.

Scott, W. B., 1930. A partial skeleton of *Homalodotherium* from the Santa Cruz beds of Patagonia. Publications of the Field Museum of Natural History, Geological Memoirs 1(1), 7–34.

Soria, M. F., 2001. Los Proterotheriidae (Litopterna, Mammalia) sistemática, origem y filogenia. Monografias del Museu Argentino de Ciencias Naturales 1, 1–167.

Stain, H. J., 1959. Use of the calcaneum in studies of taxonomy and food habits. Journal of Mammalogy 40(3), 392–401.

Taylor, M. E., 1974. The functional anatomy of the forelimb of some African Viverridae (Carnivora). Journal of Morphology 143, 307–336.

Taylor, M. E., 1976. The functional anatomy of the hindlimb of some African Viverridae (Carnivora). Journal of Morphology 148, 227–254.

Thewissen, J. G. M., 1990. Evolution of Paleocene and Eocene Phenacodontidae (Mammalia, Condylarthra). University of Michigan Papers on Paleontology 29, 1–107.

Van Valkenburgh, B., 1987. Skeletal indicators of locomotor behavior in living and extinct carnivores. Journal of Vertebrate Paleontology 7(2), 162–182.

Wang, X., 1993. Transformation from plantigrady to digitigrady: functional morphology of locomotion in *Hesperocyon* (Canidae: Carnivora). American Museum Novitates 3069, 1–23.

Williamson, T. E., Lucas, S. G., 1992. *Meniscotherium* (Mammalia, "Condylarthra") from the Paleocene-Eocene of Western North America. New Mexico Museum of Natural History and Science Bulletin 1, 1–75.

7. Postcranial Osteology of Mammals from Salla, Bolivia (Late Oligocene): Form, Function, and Phylogenetic Implications

Bruce J. Shockey*
Biology Department
Manhattan College
Manhattan College Parkway
Riverdale, NY 10471
& Department of Vertebrate Paleontology
American Museum of Natural History
Central Park West & 79th Street
New York, NY 10024, USA
bshockey@amnh.org

Federico Anaya
Facultad de Ingeniería Geológica
Universidad Autónoma "Tomás Frías"
Potosí, Bolivia

7.1 Introduction

7.1.1 General Overview

South America was a remote island continent throughout the greatest part of the Cenozoic. Such a "splendid isolation" (sensu Simpson, 1980) drove natural experiments in the organic evolution of terrestrial faunas on a continental scale. Thus, the fossil record of Cenozoic South America documents distinctive faunas, peculiar to that "lost" continent. These land mammal faunas were initially composed of primarily marsupials, xenarthrans, and native ungulates ("Stratum I" of Simpson, 1980). Somehow, in the mid-Tertiary, rodents and primates immigrated to South America (defining Simpson's Stratum II). Then, in the late Tertiary, South America's "splendid isolation" ended with the invasion of numerous North American land mammals upon the formation of the Panamanian land bridge (Stratum III: Simpson, 1980; see Stehli and Webb, 1985 for an overview of this "Great American Biotic Interchange"). Now, all the native ungulate orders are extinct, as are the glyptodont and pampathere xenarthrans. Even the once spectacular diversity of sloths has been reduced to just a couple of genera of small, arboreal folivores.

For its species richness and early appearances of derived and immigrant taxa, the Deseadan South American Land Mammal "age" (SALMA, late Oligocene) is of considerable interest (Patterson and Pascual, 1972). It is characterized by numerous derived native South American ungulates of four orders, the first evidence of sloth diversity, some of the earliest records of rodents in South America, and the earliest record of primates on that continent (Ameghino, 1895, 1897; Gaudry, 1906; Loomis, 1914; Patterson and Pascual, 1972; Hoffstetter, 1969; MacFadden et al., 1985).

7.1.2 Historical Background

Carlos Ameghino discovered the classic Deseadan localities during expeditions to Patagonia from 1893 to 1896 (see Simpson, 1984). These localities range from Chubut down into Santa Cruz provinces of Argentina and include Cabeza

* Address for correspondence: bshockey@amnh.org

E.J. Sargis and M. Dagosto (eds.), *Mammalian Evolutionary Morphology: A Tribute to Frederick S. Szalay*, 135–157.

Blanca, the Gran Barranca, and La Flecha, the later exposures lying near the Río Deseado, the inspiration for the name of the age (Gaudry, 1906). Also located in Patagonia, is the Deseadan "Scarritt Pocket", discovered by Simpson and colleagues in their 1933–34 expedition to Patagonia (Marshall et al., 1984).

The first important Deseadan locality discovered outside of Patagonia was Salla (Hoffstetter, 1968). Other extra-Patagonian Deseadan localities include Taubaté of Brazil (Soria and Alvarenga, 1989), Fray Bentos Formation in Uruguay with exposures also in northern Argentina (Mones and Urbilla, 1978), and the newly discovered localities of Moquegua, Peru (Shockey et al., 2006). Salla is the best sampled of these extra-Patagonian localities, with collections held at MNHN-Paris, MNHN-Bol, PU, UF, and, now, UATF (see Methods section regarding abbreviations used).

The Salla Beds take their name from a mapmaker's misspelling of the Aymaran village of Sahalla, Bolivia. Bolivian geologist G. Bejarano discovered fossils near Sahalla in 1962 (first announced by Baird et al., 1966). Leonardo Branisa, also a Bolivian geologist, and Robert Hoffstetter of Paris (Hoffstetter, 1968) accomplished further fieldwork that provided material for numerous publications regarding the geological setting and the fauna. The discovery of the primate *Branisella boliviana* Hoffstetter, 1969 sparked interest in Salla that was followed up with works on other faunal members such as the rodents (Hoffstetter and Lavocat, 1970; Hoffstetter, 1976; Lavocat, 1976; Patterson and Wood, 1982), marsupials (Patterson and Marshall, 1978; Villarroel and Marshall, 1982; Hoffstetter and Petter, 1983; Wolff, 1984b; Sanchez-Villagra and Kay, 1997), a suspected "condylarth" and the

TABLE 7.1. Faunal list of mammals of the Salla Beds, late Oligocene (Deseadan SALMA).

- Cohort Marsupiala Illiger, 1811
  - Order Sparassodonta Ameghino, 1894
    - Family Borhyaenidae Ameghino, 1894
      - *Fredszalaya hunteri* gen. et sp. nov.
      - *Pharsophorus lacerans* Ameghino, 1897
      - *Notogale mitis* (Ameghino, 1897)
      - *Sallacyon hoffstetteri* Villarroel and Marshall, 1982
        - (= *Adinogale sallensis* Hoffstetter and Petter, 1983)
      - Unnamed genus
      - *Paraborhyaena boliviana* Hoffstetter and Petter, 1983
  - Order Paucituberculata Ameghino, 1894
    - Family Caenolestidae Trouessart, 1898
      - *Evolestes hadrommatos* Goin et al. 2007
      - *Palaeothentes boliviensis* Patterson and Marshall, 1978
    - Family Argyrolagidae Ameghino, 1894
      - *Proargyrolagus bolivianus* Wolff, 1984
- Cohort Placentalia Owen, 1837
  - Order Cingulata Illiger, 1811
    - Family Dasypodidae Gray, 1821
      - Euphractini Wing, 1923
      - Eutatini Bordas, 1933
    - Family Peltephilidae Ameghino, 1894
      - Unnamed genus, cf. *Peltephilus* sp.
    - Family Glyptodontidae Gray, 1869
      - Glyptatelinae Castellanos, 1932
    - Family Palaeopeltidae Ameghino, 1895
  - Order Pilosa Flower, 1883
    - Suborder Folivora DeSulc et al., 2001
      - *Pseudoglyptodon sallaensis* Engelmann, 1987
    - Family Mylodontidae Gill, 1872
      - Unnamed genus, Shockey and Anaya, in preparation
      - Unnamed small "orophodontids" (2 spp.) Pujos and de Iulius, 2007.
    - Family Megalonychidae Gervais, 1855
      - Unnamed small species Pujos and de Iulius, 2007
  - Order Primates Linnaeus, 1758
    - Family Incerta Cedis

(continued)

TABLE 7.1. (continued)

      - *Branisella boliviana* Hoffstetter, 1969
        - (=*Szalatavus attricuspis* Rosenberger et al., 1991)
  - Order Rodentia Bowdich, 1821
    - Family Agoutidae Gray, 1821
      - *Incamys bolivianus* Hoffstetter and Lavocat, 1970
      - *Branisamys luribayensis* Hoffstetter and Lavocat, 1970
      - *Cephalomys bolivanus* Lavocat, 1976
    - Family Octodontidae Waterhouse, 1839
      - *Migraveramus beatus* Patterson and Wood, 1982
      - *Sallamys pascuali* Hoffstetter and Lavocat, 1970
  - Order Pyrotheria
    - Family Pyrotheriidae
      - *Pyrotherium macfaddeni* Shockey and Anaya, 2004
      - *P. romeroi* Ameghino, 1889
      - unnamed gen. and sp.
  - Order Astrapotheria
      - genus indeterminate
  - Order Litopterna Ameghino, 1889
    - Family Proterotheriidae Ameghino, 1887
      - Genus and species indeterminate
      - *Salladolodus deuterotherioides* Sora and Hoffstetter, 1983
        - (regarded as a didolodontid by some students)
    - Family Macraucheniidae Gervais, 1855
      - *Coniopternium primitivum* Cifelli and Soria, 1983a
    - Family Adianthidae Ameghino, 1891
      - *Thadaniuus hoffstetteri* Cifelli and Soria, 1983b
      - *Tricoelodus boliviensis* Cifelli and Soria, 1983b
  - Order Notoungulata Roth, 1903
    - Interatheriidae Ameghino, 1887
      - Two unnamed species (See Hitz, 1997)
  - Suborder Typotheria Zittel, 1892
    - Archaeohyracidae Ameghino, 1897
      - *Archaeohyrax* Ameghino, 1897
      - *Protarchaeohyrax* Reguero et al., 2003
    - Mesotheriidae Alston, 1876
      - *Trachytherus alloxus* Billet et al., 2008
    - Hegetotheriidae Ameghino, 1894
      - *Prohegetotherium schiaffinoi* (Kraglievich, 1932)
      - *Sallatherium altiplanense* Reguero and Cerdeño, 2005
  - Suborder Toxodonta Owen, 1853
    - Family Leontiniidae Ameghino, 1895
      - *Anayatherium ekecoa* Shockey, 2005
      - *Anayatherium fortis* Shockey, 2005
    - Family Notohippidae Ameghino, 1894
      - *Eurygenium pacegnum* Shockey, 1997
      - *Pascualihippus boliviensis* Shockey, 1997
      - *Rhynchippus* cf. *R. brasiliensis* Soria and Alvarenga, 1989
    - Family Toxodontidae Owen, 1845
      - *Proadinotherium saltoni* sp. nov.

litopterns (Soria and Hoffstetter, 1983; Cifelli and Soria, 1983a,b). More recently, one or both of us have described some of the other ungulates, including notohippid, leontiniid, and mesotheriid notoungulates (Shockey, 1997; Shockey, 2005; and Shockey et al., 2007, respectively), the postcranials of the litopterns (Shockey, 1999), and the common pyrothere of Salla (Shockey and Anaya, 2004). Reguero and Cerdeño (2005) have described the hegetotheriid notoungulates. As the oldest known primate of South America, *Branisella* continues to be a subject of intensive study (e.g., Wolff, 1984a; Rosenberger et al., 1991; Takai and Anaya, 1996; Kay et al., 2002).

7.1.3 Goals of Paper

The purpose of this present work is to summarize and update the state of knowledge regarding the fauna of Salla, especially in regard to postcranial skeletal form and function. This will be accomplished by general descriptions of selected taxa with summaries of their comparative and functional anatomy. For both practical and principled reasons, special attention is given to tarsal elements. On the practical side, tarsals elements are less fragile than other elements and are thus often preserved. But also, as complex working elements of the hind limb, they provide both functional and phylogenetic information (Szalay, 1985). That is, hind limbs are more exclusively devoted to locomotion than the forelimbs, which may be involved in (and adapted for) other functions, such as gathering food, modifying the environment for housing, or for grooming (Szalay, 1985, 1994). Thus, tarsal elements are more likely to provide pure information regarding locomotion than elements of the manus or other components of the forelimb. Also, since the tarsals function as integrated parts of the hind limb they are not especially phenotypically plastic since alteration of one element may change its role in relation to the remaining members of the functional complex.

A secondary goal is to document, in a single work, a summary of the fauna, including an updated faunal list (Table 7.1) and figures of some instructive specimens. This will be accomplished by a general description of selected specimens, followed by discussions regarding their comparative and functional anatomy. Phylogenetic implications will also be noted.

7.2 Materials and Methods

7.2.1 Fossil Material and Abbreviations

Fossils of Salla examined for our studies are housed at the following institutions (with abbreviations used in the text): Vertebrate Paleontology Division of the Florida Museum of Natural History, University of Florida (FLMNH, with UF indicating FLMNH specimens); the Princeton University collection in the Yale Peabody Museum, Yale University (PU); the Museúm National d'Histoire Naturelle, Paris (MNHN-Paris); Museo Nacional de Historia Natural, La Paz, Bolivia (MNHN-Bol); and in the developing collection of the Facultad de Ingeniería Geológica, Universidad Autónoma "Tomás Frías," Potosí, Bolivia (UATF).

Other abbreviations include SALMA, South American Land Mammal "age"; Ma, millions of years before present; I, C, P, M represent upper incisors, canines, premolars, and molars (lower case letters for the respective lower teeth); to emphasize the lack of understanding of tooth homologies of xenarthrans compared to other mammals, we designate Cf for upper caniniformes and Mf for upper molariformes (lower case letters for the respective lower teeth); Mt, metatarsal; Mc, metacarpal.

7.2.2 Categories of Locomotion and Confidence in Functional Interpretations

To summarize general functions of taxa examined, we use a modified version of the locomotor categories of Argot (2003) and define general feeding categories (Table 7.2).

We infer function via one or (preferably) more of the following independent methods:

(1) Morphology is consistent with paradigm (sensu Rudwick, 1964) for hypothetical function
(2) Morphology is consistent with that of extant modern analogs of known function
(3) Taxon is bracketed by taxa of known function (sensu Witmer, 1995)
(4) External (non-morphological) physical evidence for function (e.g., diet known via stomach content or associated coprolites; putative digger found in fossil burrow)

As usual for faunal overviews, we will provide the best functional hypothesis for the taxa discussed (see Section 7.4.1). However, since the quality of evidence varies, we wish to provide information regarding the confidence one may have in the hypotheses. For example, functional interpretations based upon complete skeletons will likely be closer to the true function than those based upon fragmentary material. Also, interpretations based upon two converging lines of evidence will be superior to those derived from a single principle or observation (or pseudoreplicates, like multiple observations of interdependent phenomena). The following means will be used to communicate our level of confidence in the hypotheses, from low levels of confidence (α) to high confidence in robust hypotheses (δ):

Alpha (α) *level hypothesis*: Hypothesis *plausible* from only one line of evidence or the hypothesis has a higher level of confidence ($>\beta$), but it is in conflict with plausible and competing hypothesis.

Beta (β) *level hypothesis*: Hypothesis *probable* from one line of evidence or, α-level hypothesis that is compatible

TABLE 7.2. Locomotor and feeding categories (locomotor categories adapted from Argot, 2003).

Locomotor type	
Arboreal	Rarely on ground; typically forages and shelters in trees; usually exhibits particular specializations for climbing.
Scansorial	Adept climber that also forages on the ground during a considerable proportion of its time.
Subcursorial	May never climb, displays incipient adaptations for running.
Cursorial	Never climbs, displays marked adaptations for running.
Fossorial	Adept digger, forages and/or shelters below ground, shows marked adaptations for digging.
Graviportal	Massive, never climbs, may move rapidly for brief periods, but typically moves slowly.
Feeding categories	
Carnivory	Meat is most significant source of calories, dental adaptations (or other oral adaptations [e.g., shearing beak]) for shearing meat
Omnivory	Meat an important source of calories, but also relies largely upon arthropods, plants or fungi for nutrition. Usually has unspecialized dentition, but may display incipient adaptations for shearing meat.
Herbivore, unspecialized	General classification for animal that gets nearly all of its nutrition from plant material. Adaptations to crush and/or slice plant material, such as lophs on occlusal surfaces of cheek teeth. Mandibular condyle usually high above tooth row in order to facilitate simultaneous occlusion of all grinding teeth.
Browser	Herbivore that consumes a variety of nutrient rich foods, such as new leafy growth, buds, fruits, and seeds. Adaptations for selective feeding may include narrow muzzle, proboscis, or dexterous manus.
Grazer	Herbivore that receives nearly all of its calories from grasses. Adaptations for grazing include high crowned cheek teeth, broad muzzle.

with a second, independently derived α-level hypothesis or, internal independent confirmation via evidence of same general function in two systems (e.g., grazing deduced via hypsodonty *and* broad muzzle; digging via specialized forelimb and specialized pelvis).

Gamma (γ) *level hypothesis*: β-level hypothesis consistent with another independently derived and congruent β-level hypothesis.

Delta (δ) *level hypothesis*: Unusually compelling evidence, which may include direct physical evidence of function (e.g., method d, above), or morphological feature is a direct result of function (see macraucheniid example in text), or three compatible and independent β-level hypotheses.

7.3 Mammal Fauna of Salla

7.3.1 General Mammalian Fauna

A revised faunal list is provided in Table 7.1. It conservatively records 47 species of mammals, including many of the natives of Simpson's Stratum I: marsupials (six sparassodont and three paucituberculate species), xenarthrans (five species of cingulates and five species of sloths) and four orders of extinct, endemic, South American ungulates: Pyrotheria (three species), Astrapotheria (one indeterminate species), Litopterna (five species), and Notoungulata (13 species). Immigrant (Stratum II) taxa include rodents (five species) and the primate *Branisella boliviana* (=*Szalatavus attricuspis*).

For the sake of convenience and for stability of nomenclature, we generally follow the higher level classifications of McKenna and Bell, 1997. Such may not reflect the actual phylogenies of taxa, but the systematics of most of the above family level groups discussed are not so confidently known that they may be regarded as dogma.

7.3.2 Systematic Paleontology

SUPERCOHORT THERIA PARKER AND HASWELL, 1897
COHORT MARSUPIALIA ILLIGER, 1811
MAGNORDER AMERIDELPHIA SZALAY, 1982
ORDER SPARASSODONTA AMEGHINO, 1894a
FAMILY BORHYAENIDAE AMEGHINO, 1894a
FREDSZALAYA GEN. NOV.

Material –UF 172501 (Holotype), partial skull with partial upper dentition (right P3-M1, roots of right C – P2, and left M2-4) and associated vertebral and costal fragments and the left calcaneum.

Locality – The holotype (Figures 7.1, 7.2 and Appendix) was collected 15 m below the El Planimiento by Roger Portell, Gary Morgan, and Bruce J. MacFadden in 1986.

Type species –*Fredszalaya hunteri* sp. nov.

Etymology – To honor Fred Szalay, especially for his contributions to our understanding of marsupial evolution.

Diagnosis – Same as for species.

FREDSZALAYA HUNTERI sp. nov.

Etymology – In reference to Dr. Szalay's longtime affiliation with Hunter College, City University of New York, and to suggest the predacious nature of the animal.

Diagnosis – Medium size borhyaenid with short muzzle and broad posterior palate, short but distinctive molar protocones, weak parastyle, but well developed stylar cusp B, with stylar shelf, especially on M1-2, well developed carnassial postmet-

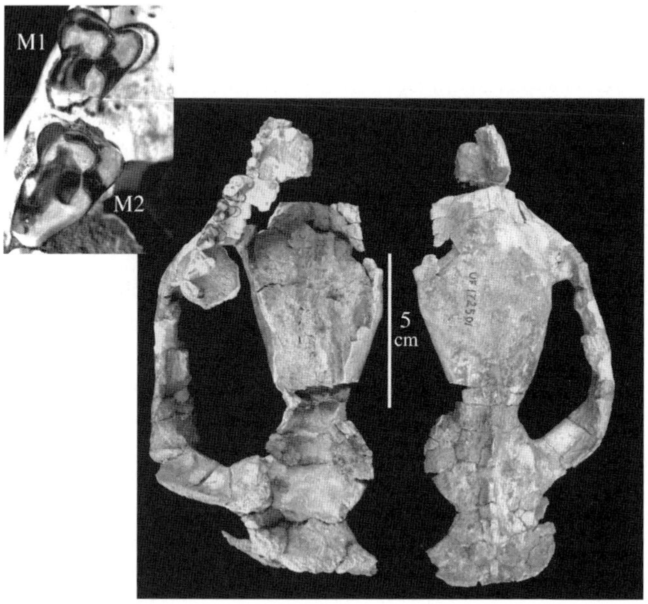

FIGURE 7.1. Holotype of *Fredszalaya hunteri* gen. et sp. nov. (UF 172501). Views of the skull are ventral (left) and dorsal (right). The insert at upper left is occlusal view of the right M1–2.

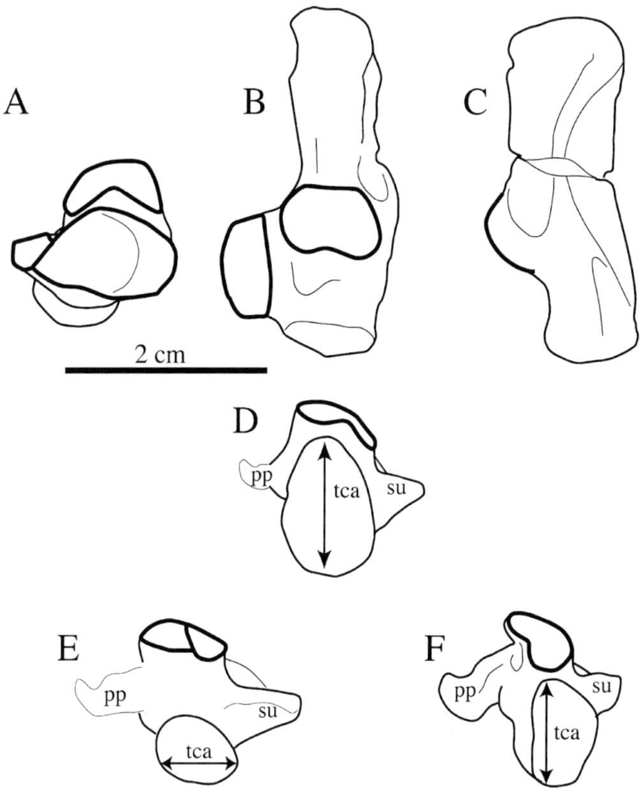

FIGURE 7.2. Calcaneum (left) of the holotype of *Fredszalaya hunteri* gen. et sp. nov. (A–D) compared to the posterior view of the morphotype of a terrestrial marsupial calcaneum (E) and that of the grasping morphotype (F). Views of the calcaneum of *Fredszalaya hunteri* (UF 172501) are A, distal; B, dorsal; C, lateral; and D, posterior. E and F are adapted from Szalay, 1994: Figure 6.16. Abbreviations are pp, peroneal process; su, sustentacular process; and tca, tuber calcani.

acrista, M4 short, lacking metacone, but nearly as wide as M3 with well developed, elongated, blade-like preparacrista. Foramen ovale absent.

Smaller than species of *Pharsophorous* with relatively wider posterior palate, upper molars with smaller stylar cusp A and larger stylar cusp B, more conspicuous stylar shelf; M4 wider with greater development of protocone and with carnassial preparacrista.

Larger than *Notogale mitis* with greater development of upper molar protocones, especially that of M4.

Larger than *Sallacyon hoffstetteri* Villarroel and Marshall, 1982, also differing by the presence of well-developed stylar cusp B of M1-2, better developed protocone of M4, M4 relatively and absolutely wider, zygomatic arch more rectangular.

Differs from *Prothylacynus patagonicus* by it smaller size, shorter muzzle with crowded premolars, better develop molar protocones, protocone retained in a larger M4 and absence of foramen ovale.

Similar to *Borhyaena* spp. by the absence of the foramen ovale on the alisphenoid, but differs by its smaller size, relatively and absolutely shorter muzzle, relatively larger protocones and stylar shelves of upper molars.

Description – The skull is grossly similar to that of the Santacrucian *Prothylacynus patagonicus* Ameghino, 1891 (see Sinclair, 1906; Marshall, 1979). The cranial vault is quite flat and the zygomatic arches are exceedingly wide, suggestive of massive jaw muscles. The incisors are not preserved and only parts of the roots of the canines remain. These indicate that the canines were fairly robust.

The P1 is broken, but both roots remain and indicate that the tooth was set obliquely to the tooth row, with the anterior root being more labial than the posterior. Crowns are missing from P2-3, but enough of these teeth are present to indicate that the P2 is larger than P1 and that P3 is the largest premolar, having a slightly larger anteroposterior dimension than the M1.

The protocone of M1 is basined and sits low on the crown. The paracone is much higher and it is slightly worn such that it grades into a stylar cusp (stylar cusp B of many workers) that lies adjacent to its labial surface. A small stylar shelf is present. The metacone is conspicuously higher and more robust than the paracone. The post metacrista is broken on the M1, but is preserved on M2 and the left M3. This forms the distinctive, oblique shearing blade of the upper molars. The protocones, stylar cusps, and stylar shelves are progressively smaller on M2-3. M4 is smaller than the proceeding molars, lacking a metacone, but the protocone is retained and better developed than that of the M3. A well-developed crista connects the paracone to the parastyle, forming a shearing blade.

The zygomatic arches and glenoid fossae are preserved. The glenoid is oval and deep, being buttressed posteriorly by bony ridge nearly the entire width of the fossa and anteriorly by the terminal end of the jugal. Unlike *Prothylacynus*, but like *Borhyaena* (see Sinclair, 1906), there is no foramen ovale.

Given its distinctive morphology and close association to the skull in the field, there is no reasonable doubt that the associated calcaneum (Figure 7.2) is that of the *Fredszalaya*. It is similar to that of *Borhyaena* (see Argot, 2003), differing by its more transverse (rather than strongly oblique) distal border (as seen in dorsal view) and its shorter sustentaculum, resulting in a narrower lower ankle articulation. It has no fibular facet, which appeared also to be lacking in *Borhyaena*, but damage made its absence uncertain (Argot, 2003). The calcaneal tuber of *Fredszalaya* is narrow and deep. The cuboid facet is concave and circular, suggesting the socket of a ball-and-socket calcaneocuboid articulation that would have allowed significant rotation of the foot. There is no proximal component of the cuboid facet as seen in didelphids, but the cuboid facet is somewhat medially placed such that nearly half of the facet is sculpted into the body of the short sustentacular process. The lateral portion of the distal calcaneal surface is not excavated, but has a slight convexity that suggests that the cuboidal surface of the calcaneocuboid joint was not just a simple "ball". The distolateral area where the peroneal process would have been (if present) is broken.

The narrow transverse ankle joint and narrow but deep tuber calci are typical of grasping metatheres (Figure 7.2; Szalay, 1994 text and Figure 6.16) and unlike that expected of most borhyaenids, which are typically regarded as being terrestrial animals (Sinclair, 1906; Marshall, 1978; Szalay, 1994). A relevant exception is provided by Argot (2003), who documented a variety of skeletal features of *Prothylacynus* that are likely related to adaptations for climbing (see her discussion and, especially, her Figure 21).

Biology of Fredszalaya hunteri – The shearing nature of the elongated metacristae of M1-M3 and elongated blade-like preparacrista of the M4 predicts meat-eating habits for *Fredszalaya*. This prediction, derived from the principle of shearing as a means of cutting meat, is supported by the observation of carnivorous habits in extant marsupials having similar shearing crests, as well as historical observations of the recently extinct carnivorous *Thylacinus* (Tasmanian "wolf"). Carnivory has even been confirmed in an extinct, Tertiary borhyaenid (*Lycopsis logirostrus* of the middle Miocene, La Venta) by the presence of rodent remains within the body cavity of the borhyaenid (Marshall, 1977:p. 641). *Fredszalaya*, however, is less extreme in its meat cutting adaptations than some other borhyaenids. For example, the stylar shelves and protocones are retained, whereas these are significantly reduced in *Borhyaena*. Thus, it is likely that *Fredszalaya* was not a meat specialist, but foraged on other food items, much like the morphologically similar *Prothylacynus* (Marshall, 1978; Argot, 2003).

As noted above, climbing abilities are suggested by the morphology of the calcaneum. Such a hypothesis, based upon a single element, must be regarded as tentative.

Phylogeny of Fredszalaya hunteri – Although the skull of *Fredszalaya* is superficially more similar to that of *Prothylacynus* than that of *Borhyaena*, we regard the absence (loss) of the foramen ovale to represent a synapomorphy uniting *Fredszalaya* with *Borhyaena*. Similarities of form between *Fredszalaya* and *Prothylacynus* would represent the plesiomorphic condition for the lineage leading to *Borhyaena*.

PROBORHYAENIDAE AMEGHINO, 1897

PARABORHYAENA BOLIVIANA HOFFSTETTER AND PETTER, 1983

Comments – Proborhyaenids were large to huge carnivorous marsupials, with the skull of *Proborhyaena gigantea* reaching two feet in length (Marshall, 1978). The last records of these giant terrestrial carnivores were in the Deseadan, with two named genera: *Proborhyaena* and *Paraborhyaena*, the later described from Salla (Hoffstetter and Petter, 1983).

In their summary of the Salla fauna, MacFadden et al. (1985) noted that all proborhyaenids remains lacked precise stratigraphic data. The discovery of a jaw (UATF-V-000129) during our recent expedition to Salla by colleagues Darin Croft and Rodolfo Salas at the base of Unit 3 at Pasto Grande provides the first stratigraphic control for this huge carnivorous taxon.

Work is in progress on this specimen, but for now we document its stratigraphic position at the very base of Unit 3. This places it in a normal paleomagnetic horizon that, according to the "best fit" hypothesis of Kay et al. (1998), would be Chron 10n.2n. This would indicate an age of about 28.6 Ma. We also provide the observation that the animal had but one pair of large, blunt lower incisors. This was previously unknown for *Paraborhyaena* (Hoffstetter and Petter, 1983; Babot et al., 2002) and it represents the most derived condition of incisors known for proborhyaenids (*Proborhyaena* has two pairs and the romantically named *Callistoe* Babot et al., 2002 has three pairs).

No postcranials are known from *Paraborhyaena*. Aside from various fragments tentatively referred to *Proborhyaena* (see Marshall, 1978), the only postcranial remains of any proborhyaenid recovered are those of *Callistoe vincei*, currently under study by Judith Babot (see Babot et al., 2002). So, for now little can be deduced regarding locomotion for proborhyaenids in general, but in regards to *Paraborhyaena* we may be content with commenting that it was one of the largest carnivorous marsupials.

ORDER PAUCITUBURCULATA AMEGHINO, 1894

ARGYROLAGIDAE AMEGHINO, 1904

PROARGYROLAGUS BOLIVIANA WOLFF, 1984

Comments – Postcranials of *Proargyrolagus* are yet to be known. However, Sanchez-Villagra and Kay (1997) methodically considered cranial characters to establish a general hypothesis regarding its feeding habits. Based on its small size (~100 g, which made folivory most unlikely), unrooted lower incisors, and the unusually high crowned cheek teeth, they regarded *Proargyrolagus* as being an herbivore that gathered food items, like seeds, at ground level. They also noted the well-developed nasal region (similar to that seen in water-conserving rodents of arid regions) that suggests that *Proargyrolagus* could have tolerated dry environments.

Based upon the remarkable convergence between the postcranial skeletons of Plio-Pleistocene argyrolagids (*Argyrolagus* and *Microtragulus*) with those of desert dwelling heteromyid kangaroo rats and dipodid jerboas, Simpson (1970) proposed that argyrolagids were also specialized for bipedal, ricocheted locomotion. Applying such a hypothesis to *Proargyrolagus*, however, would represent an extrapolation from the data, not an interpolation, since none of the proposed phylogenies of argyrolagids implies that *Proargyrolagus* is bracketed by *Argyrolagus* and *Microtragulus* (see Sánchez-Villargra and Kay, 1997).

COHORT PLACENTALIA OWEN, 1837
MAGNORDER XENARTHRA COPE, 1889
ORDER CINGULATA ILLIGER, 1881
PELTEPHILIDAE AMEGHINO, 1894
CF. *PELTEPHILUS* SP.

Comments – For their dermal horns, cranial shield, and slicing anterior teeth, peltephilid armadillos have inspired much curiosity. This family of armored Xenarthrans is best known from the early middle Miocene Santacrucian SALMA, however, until now, little more than isolated osteoderms had been known from the Deseadan.

The few peltephilid specimens of Salla are variable in terms of size, tooth number, robustness of mandible, and fusion or absence thereof of the mandibular symphysis. This variation suggests that more than one species is present at Salla, but we are unable to rule out within species variables such as ontogenetic changes or sexual dimorphism at this time. Further study is indicated and is being undertaken.

The peltephilid specimens of Salla (Figure 7.3) are grossly similar to the well-known Santacrucian *Peltephilus* Ameghino, 1887 (see Scott, 1903). Similarities include seven teeth of the mandible (though two specimens of Salla have eight), hoof-like ungual phalanges of the pes, and the presence of horn-like cranial osteoderms, including an anterior pair (preserved in a MNHN-Paris specimen), as predicted by Ameghino (1894). Some differences between the Salla and Santacrucian animals are significant, with the Salla peltephilids having generally plesiomorphic characters. For example, whereas other known peltephilids have fused mandibular symphyses, both peltephilid mandibular specimens in the UF collection have unfused symphyses. These may represent immature individuals, as a larger specimen in the PU collection (PU 21143) does have a fused symphysis. Also, whereas other known peltephilids have but seven teeth in the mandible, two of the specimens of Salla (UF 93587 and PU 21143) have an additional small tooth anterior to the seven that appear to be homologous to the seven of UF 93586 (this anterior tooth is designated as "mf 0" in the Appendix).

Much of the pes is preserved in UF 93515 (Figure 7.3). Mt II, III, and IV are subequal in size, suggesting a similar form to that of *Peltephilus strepens* which was functionally tridactyl having much reduced Mt I and V (Mt I and V are missing on the Salla specimen). Like *P. strepens*, the Salla peltephilid has concavities of the distal metatarsals, though not as pronounced as in the Santacrucian example. Also, the peltephilid of Salla has hoof-like ungual phalanges, even blunter than those of the Santacrucian animal (compare the hoof in our Figure 7.3 with that of Scott, 1903: plate 16.14).

Ameghino's interpretation of the biology of peltephilids was sensational. He wrote of them as, "ferocious and meat eating, like a tiger and armed with horns like a rhinoceros – one's imagination could not conjure anything more lively (Ameghino, 1934:317: translated from Spanish by BJS)." Modern interpretations are less imaginative. Vizcaíno and Fariña (1997) dispute this traditional "killer armadillo" reconstruction. Citing evidence from the skull and forelimbs of the Santacrucian *P. ferox*, they interpret it as being fossorial and herbivorous.

Although there is insufficient postcranial material of the Salla peltephilid to obtain any metatarsal index, we note that the metatarsals are not long, but proportionally similar to those reported for Santacrucian peltephilids (Scott, 1903) and to those of our comparative sample of Florida "road kill" *Dasypus* specimens. Neither the Santacrucian taxa nor the Salla peltephilid show any specializations for which they could be regarded as cursorial.

Based upon the cranial morphology of *Peltephilus*, Vizcaíno and Fariña (1997) argued against meat eating in peltephilids; they noted that the teeth were too slender to resist struggling prey and that the apparent location of the main bite force was at the anterior jaw rather than half way between the tempromandibular joint and anterior grasping teeth (as predicted from mechanical models [e.g., Greaves, 1995] and observed in extant carnivores). Their arguments for herbivory included the wide zygomatic arch, which strongly suggests lateral jaw movements, and their high crowned teeth (though the relevance of the later for xenarthrans may be questioned). Their favored hypothesis was that *Peltephilus* ate subterranean plant material.

Though the mandibular symphysis is unfused in most of our sample of peltephilids of Salla, there is still evidence that significant forces were applied at the anterior region of the jaw. That is, the anterior region is the deepest and thickest part of the mandible. So, the greatest forces (and the equal opposing forces) associated with the bite appear to have occurred at that anterior region, just as in *Peltephilus* (Vizcaíno and Fariña, 1997). Whether these forces involved vegetable matter or animal, we offer no opinion, but subterranean feeding seems well-suited for an animal with a specialized anterior bite, since, at initial contact, the anterior snout would usually be the only part of the animal in contact with the food item in an underground environment (the rest of the animal would be separated from its food by soil).

SUPERFAMILY GLYPTODONTOIDEA GRAY, 1869

Comments – Glyptodonts are not common at Salla, but some osteoderms are in the various collections of Salla. These are

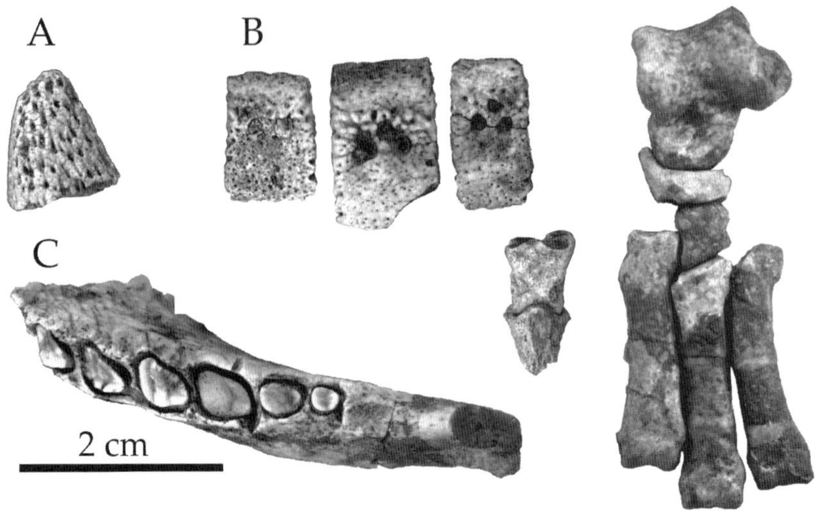

FIGURE 7.3. Peltephilidae. A, UF 93551, nasal horn; B, UF 93515, associated osteoderms, partial pes with medial and distal phylanges; C, UF 93586, left mandibular ramus.

largely unstudied, but Freddie Carlini (personal communication) identified two species from two families (Glyptodontidae and Palaeopeltidae) during visits to the UF collection.

Shockey (2001) described an isolated distal femur that he referred to the Glyptodontoidea. It was quite distinctive, having exceedingly asymmetric trochlear ridges, with the medial being much higher and nearly conical in shape, and a patellar trochlea having a sinuous path such that the patella must have rotated transversely during knee flexion-extension. This rotation of the patella likely resulted in differential tension on the crus, which in turn would have caused rotary movement of the crus in addition to the flexion-extension.

Evidence of this complex knee extension is also implied by the complex ball-and-socket medial knee articulation coupled with the sliding lateral articulation of various glyptodonts and sloths. The biomechanical consequences of this joint are poorly understood, but the near ubiquity of the ball-n-socket/sliding knee joint among sloths and glyptodonts is curious.

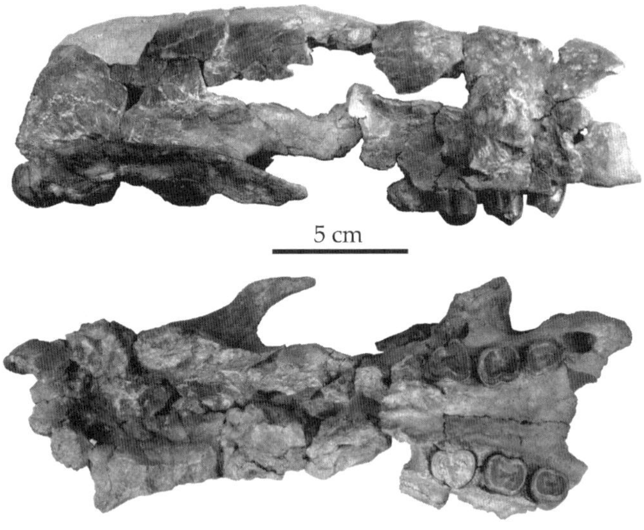

FIGURE 7.4. Mylodontidae. Lateral and ventral views of cranium, UATF-V-000127.

ORDER PILOSA FLOWER, 1883
SUBORDER FOLIVORA DELSUC ET AL., 2001
MYLODONTIDAE GILL, 1872
UNNAMED GENUS (Shockey and Anaya, in preparation)

Comments – Along with UATF geology student, Luis Lopez, we recovered a fairly complete skull (UATF-V-000127, Figure 7.4) of an unnamed genus of mylodontid sloth during our visit to Salla in January of 2003. It came from Unit 4 (the "Principle Guide Zone") at Calaboza Pata, Salla.

It is distinguished by its broad muzzle, large external nares, oval to sub-figure-eight molariform occlusal surfaces, and teeth composed of relatively equal amounts of vasodentin, orthodentin, and cement. It differs from other Deseadan sloths by being smaller than species of *Octodontotherium* and *Orophodon*, and its distinctive tooth histology of nearly equal proportions of the three tissues. The orthodentine does not appear to have been much harder than the other tissues, since the teeth wore quite smoothly, without the orthodentine forming a palpable ridge, as is typical for sloths.

The broad muzzle of UATF-V-000127 suggests that the animal ate grass, since many modern grazers have broad muzzles that help them acquire much grass (a poor quality food) with a single bite (Gwynne and Bell, 1968; Solounias et al., 1993). Mylodontids have frequently been interpreted as being grazers (McDonald, 1997 and references therein), an interpretation confirmed in *Mylodon darwini* via the contents of its coprolites (Moore, 1978).

Based upon the broad muzzle, we suspect that this mylodont of Salla also was a grazer. The only caveat at this time is that we are curious in regard to the functional significance of the enlarged external nares and the consequences of its development. Did the large nose have some critical function that had selective value (with the broad muzzle being merely a consequence of developing the tall and broad external nares)? Of course, the two could have a complementary adaptive-functional history: e.g., the nose as an adaptation for water conservation in arid environments and the broad muzzle for eating about the only commonly found vegetation available in such environments (grass).

ORDER PYROTHERIA AMEGHINO, 1895
FAMILY PYROTHERIIDAE AMEGHINO, 1889
PYROTHERIUM AMEGHINO, 1888
PYROTHERIUM MACFADDENI SHOCKEY AND ANAYA, 2004

Comments – Various postcranial remains of *Pyrotherium macfaddeni* are in the collections of MNHN-Paris, MNHN-Bol, PU, and UF. These illustrate the extreme graviportal nature of the beast. These collections also contain various tarsal elements that illustrate the curious foot of the animal (see Shockey and Anaya, 2004). *Pyrotherium* is plantigrade and almost uniquely has a reversed form of the calcaneoastragalar articulation (i.e., the ectal facet of the calcaneum is concave and sustentacular facet is convex). It also has well-developed fibular-calcaneal and cuboastragalar articulation. As far as we know, this form is only seen in the embrithropod *Arsinoitherium* and is quite unlike the tarsus of any other known South American native ungulate.

We are confident that this graviportal beast was a slow, terrestrial herbivore. The huge surface area of its teeth indicates that mechanical digestion was important for the animal and may suggest that chemical digestion was not as efficient as that of other ungulates.

ORDER LITOPTERNA AMEGHINO, 1889
FAMILY PROTEROTHERIIDAE AMEGHINO, 1887
SUBFAMILY PROTEROTHERIINAE AMEGHINO, 1885
GENUS INDETERMINATE

Comments – The presence of proterotheriids at Salla was confirmed with the discovery of a functionally monodactyl partial pes referable to the group (Shockey, 1999; see also Figure 7.5). The enlarged Mt III was broken such that it was not possible to estimate the length of this element. The length of the Mt III of the Salla proterotheriid remains unknown, but a left Mt III, bulk cataloged (AMNH 14153) with several Loomis notoungulate specimens from Cabeza Blanca, is complete (see Figure 7.5b). AMNH 14153 is 81.5 mm long and has a proximal width of 16.9 mm. Since this Patagonian specimen has a distinctive dorsal component of the distal keel, a feature absent in the Salla specimen, it is probably a species distinct from that of Salla and apparently is more derived towards cursorial habits. Neither specimen was associated with dental remains, so they can only be noted as protertheriine proterotheriids.

Although not as advanced in regard to running abilities as the Loomis specimen, the Salla proterotheriid can confidently be classified in the general category of a cursor. The more

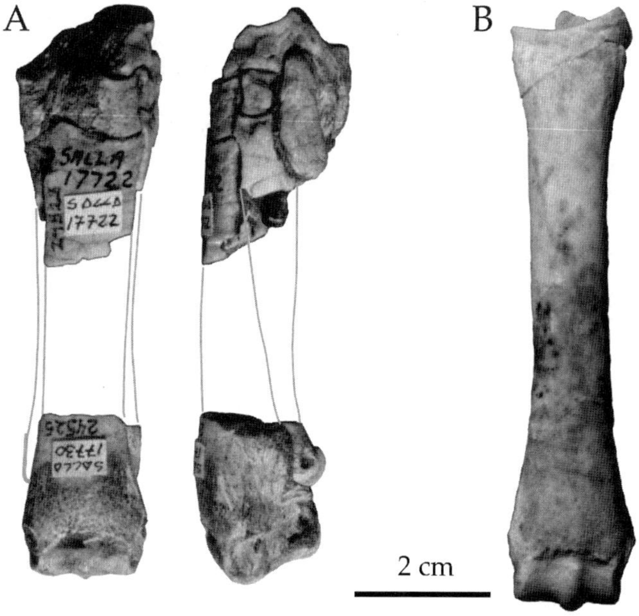

FIGURE 7.5. Proterotheriid pes. A, partial right pes of the proterotheriid of Salla, composite of the distal tarsals and proximal Mt II-IV (PU 24528) and distal metapodials (PU 24525). (Reconstructed as if from a single individual.) B, left Mt III from Cabeza Blanca (AMNH 14153).

specific reconstruction as it being a forest dwelling running animal similar to dasyproctid rodents and Old World, Recent forest ungulates, like duikers (Shockey, 1999 and references within), should be regarded as tentative.

FAMILY MACRAUCHENIIDAE GERVAIS, 1855
CONIOPTERNIUM AMEGHINO, 1895

Comments – With a little doubt, Cifelli and Soria (1983a) referred the Salla macraucheniid (?*Coniopternium primitivum*) to *Coniopternium* Ameghino, 1895. We are content to drop the query and suggest that the slight angle of the calcaneal tuber of the Salla macraucheniid(s) (Figure 7.6) and that of *Coniopternium andium* is a homologous, derived character that can be regarded as a synapomorphy for *Coniopterium* spp.

Due to various skeletal modifications restricting movement to the parasagittal plane (deep trochlea of astragalus and humerus and transversely elongated astragalonavicular joint), the non-supinating antebrachium, and elongated metatarsals, Shockey (1999) inferred that the macraucheniids of Salla were adapted for a cursorial mode of locomotion (Figures 7.5 and 7.6). Additionally, he proposed that the deep suprapatellar fossa ("patellar pit") of the femur served as a pit "into which the patella *could* have slid (Shockey, 1999:p. 385, emphasis not in original)," thus serving as a passive stay: a knee lock structurally different from that of modern horses (see Hermanson and MacFadden, 1996; Shockey, 2001). At that time, no macraucheniid patella had been recovered from Salla, so Shockey constructed a model patella that indeed "locked" and resisted movement when cloth "ligaments" attached to it were pulled.

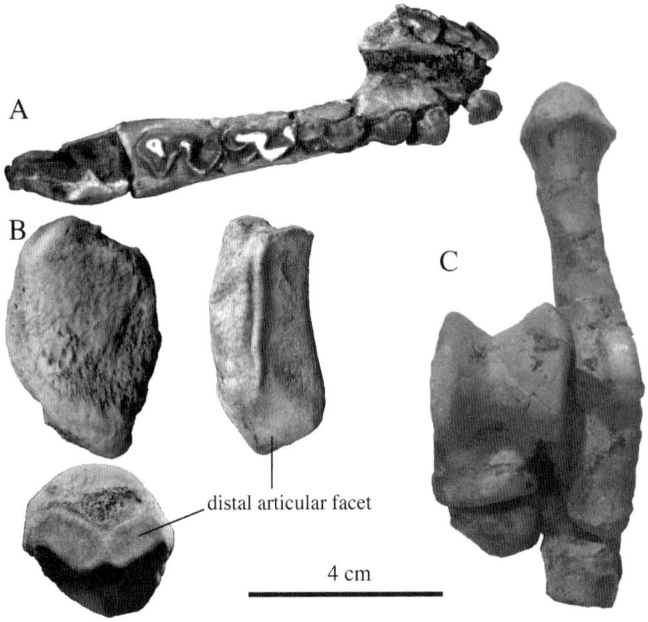

FIGURE 7.6. *Coniopternium* cf. *C. primitivum*. A, UF 172122, mandible of juvenile with left di2-dc and right di2, di3-dp4, with m1 still in crypt; B, UATF-V-000133, patella (views clockwise from upper left: dorsal, lateral, distal); C, UF 172426, left calcaneum, astragalus, navicular, and cuboid.

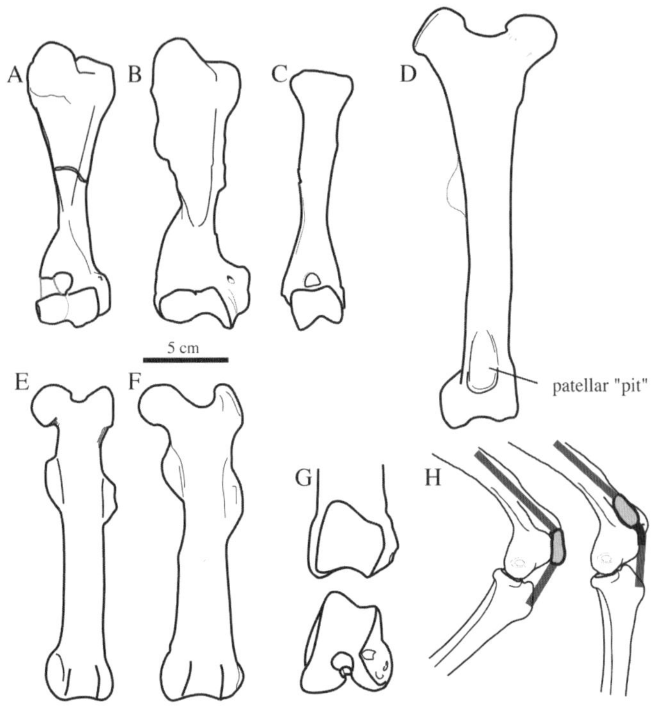

FIGURE 7.7. Ungulate limb bones. Cranial views of humeri (A, *Eurygenium pacegnum* Bol-V-00364; B, *Trachytherus alloxus*, UF 91933; C, *Coniopternium* sp., UF 149207) and femora (D, *Coniopternium* sp. MNHN-Bol-V-004502; E, *Eurygenium pacegnum pacegnum* Bol-V-00364; F, *Trachytherus alloxus*, UF 90960; and G, *Proadinotherium*, cf. *P. saltoni* sp. nov. (anterior view also provided) MNHN-Paris [uncataloged]), and H, a functional reconstruction of the knee lock of *Coniopternium* sp. show as flexed (left) and hyper-extended and locked (right). (H modified from Shockey, 1999.)

As originally proposed, the suggestion of a macraucheniid knee lock was an untested hypothesis. The model patella merely illustrated what *could* have occurred, but said nothing as to what the animal actually did in life. Fortunately, during our expedition to Salla in January 2004, we recovered patellae associated with two partial skeletons of macraucheniids. These patellae unambiguously show distal articular surfaces (Figure 7.6), in addition to the typical facet for articulation with the patellar groove during knee flexion-extension. The only plausible explanation for the distal articulation of the patella is that it contacted the distal surface of the patellar pit of the femur into which its shape conforms (reconstructed in Figure 7.7).

In response to the incorrect and over-used dogma of "form follows function", investigators are now quick to note non-functional explanations for morphology (e.g., exaptations, phylogenetic inertia, multiple functions for a single form; see Ross et al., 2002 for summary). Articular facets, however, do provide information regarding the relative position of bones. Indeed, in the case of the *Coniopternium* patella, the form (presence of distal articular facets) is a direct result of function (patellar "locking" in the suprapatellar fossa). Thus, we can say with considerable confidence that the knee of *Coniopternium* did indeed hyperextend during the life of the animal such that the patella locked into the patellar pit. Such an adaptation would have allowed the animal to stand for considerable amounts of time without expending much energy.

ORDER NOTOUNGUALTA ROTH, 1903
SUBORDER TYPOTHERIA ZITTEL, 1892
FAMILY MESOTHERIIDAE ALSTON, 1876
TRACHYTHERUS ALLOXUS BILLET ET AL., 2008

Comments – Trachytherus is a sheep sized notoungulate with distinctive gliriform incisors and a robust postcranial skeleton (see Figures 7.7 and 7.8). Individuals vary considerably, but there is no discontinuity in the size or other characters to suggest the presence of either two species at Salla or even sexual dimorphism of the one present.

In an unpublished master's thesis, Heidy Sydow (1988) described two partial skeletons of *Trachytherus* and concluded that this Deseadan mesothere was a "scratch digger", sensu Hildebrand, 1985. Her hypothesis was supported by our functional analysis (with D. Croft) of mesotheres, which included *Trachytherus*, as well as the mesotheriines, *Plesiotypotherium* and *Mesotherium* (Shockey et al., 2007). In terms of limb proportions and development of specializations associated with strength of the forelimb, *Trachytherus, Plesiotypotherium* and *Mesotherium* compared most favorably to extant scratch diggers, like wombats

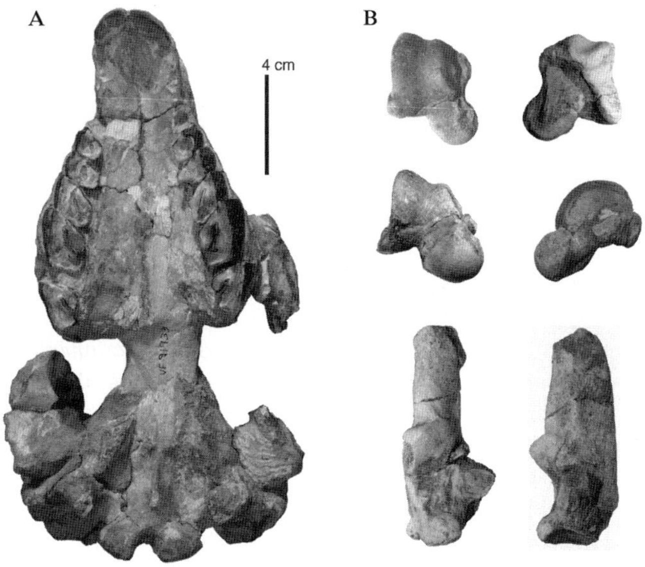

FIGURE 7.8. *Trachytherus alloxus*. A, UF 91933 skull (palatal view); B, UF 172437, astragalus (views clockwise upper left, dorsal, plantar, medial, distal); and UF 172514, calcaneum in dorsal view (left) and lateral view (right).

(*Vombatus, Lasiorhinus*), badgers (*Taxidea*), and, especially, aardvarks (*Orycteropus*).

Adaptations for digging of the forelimb of *Trachytherus* include the well developed and distally placed crests for the deltoid and pectoralis muscles, enlarged medial epicondylar process of the humerus (Figure 7.7b), enlarged and medially curved olecranon, enlarged pisiform, and fissured ungual phalanges. The manus is pentadactyl, with just modest reduction of the first digit, a condition nearly identical to the Pleistocene *Mesotherium* (see Shockey et al., 2007).

The lower limb is similar to that described for *Eurygenium* (Figure 7.7), except that there is less calcaneofibular contact in *Trachytherus*.

The astragalus is remarkable for the conspicuous asymmetry of the trochlear ridges (lateral being much higher than the medial crest), a constricted neck, which is fairly long in some specimens, but always longer than those of the notohippids of Salla. The head is subspherical, forming the ball of a ball-and-socket joint with the navicular. The lateral and medial walls of the astragalus are oblique with distinctive lateral and medial processes such that the plantar surface is broader than the dorsal trochlea. A well-developed groove for the digital flexor is separate from the trochlea.

The calcaneum has a small fibular facet that is obliquely oriented on the dorsal prominence. The ectal facet is convex and broad. The lateral calcaneal border has a groove for the tendon of the peroneus longus and one specimen (UF 172514) preserves a small peroneal process this is not directly adjacent to the distal region of the peroneal groove. The apex of the tuber is rugose and lacks a distinctive groove for the Achilles tendon. The cuboid facet is teardrop shaped, slightly concave, and obliquely oriented and appeared to have allowed a sliding articulation with the cuboid; thus, along with the ball-and-socket articulation of the astragalonavicular joint, would have permitted some rotation of the pes.

Although no articulated pes of *Trachytherus* of Salla is available for study, much of the pes is preserved in a specimen of *T. spegazzinianus* recently collected from the Deseadan of Moquegua, Peru (Shockey et al., in preparation). This illustrates that the pes of *Trachytherus* was pentadactyl. Other than its larger size and retention of the hallux, the form and relative position of the elements are similar to those of the hegetothere pes described below.

We regard the scratch digging hypothesis for *Trachytherus* as being robust, since the postcranial morphology of the animal is consistent with a digging paradigm and with the morphology of known, extant scratch diggers as noted above (see also Shockey et al., 2007). Also, other mesotheres (*Plesiotypotherium* and *Mesotherium*) have fossorial adaptations. Indeed, the mesotheriine mesotheres were more derived in this respect as they evolved ossified reinforcement of their pelvis, like that seen in some extant diggers (Hildebrand, 1985; Shockey et al., 2007).

FAMILY HEGETOTHERIIDAE AMEGHINO, 1894

Comments – Reguero and Cerdeño (2005) demonstrate the presence of two species of hegetotheres at Salla, *Prohegetotherium schiaffinoi* (Kraglievich, 1932) and *Sallatherium altiplanense* Reguero and Cerdeño, 2005. They noted, but did not describe, the femur and a partial pes (Reguero and Cerdeño, 2005: Figure 7.5d) associated with cranial material of *P. schiaffinoi*. We provide here brief descriptions of these elements.

UF 172445 is a damaged skull of *Prohegetotherium schiaffinoi* associated with fragmentary postcranials and a nearly complete left femur (missing the greater trochanter) and a distal right femur. These distal femora have long, narrow, but deep, petallar grooves, suggestive of running animals (Rose, 1999).

The partial pes (UF 172502, Figure 7.9 and Appendix) was found in close association with cranial material of two individuals of *Prohegetotherium* (UF 91661 and 91662), but it is unknown to which, if either, it pertains. This pes is very similar to that of the Santacrucian *Hegetotherium mirabile*, described and figured by Sinclair (1909:Figure 7.4a and plate 6.19). Like that of *Hegetotherium*, the *Prohegetotherim* pes is tetradactyl and has a small fibular facet of the calcaneum. A distal tibia-fibula, into which the astragalus perfectly fits, is solidly fused, indicating that the *Prohegetotherium* also had a fused distal crus. Also, as in *Hegetotherium*, the Mt II is shorter than the Mts III and IV, having a more proximal articulation with the tarsals, overlapping part of the ectocuneiform. The Mt V was not preserved, but an impression in the matrix on the lateral side of Mt IV unambiguously indicates that it was present and smaller than the Mt IV.

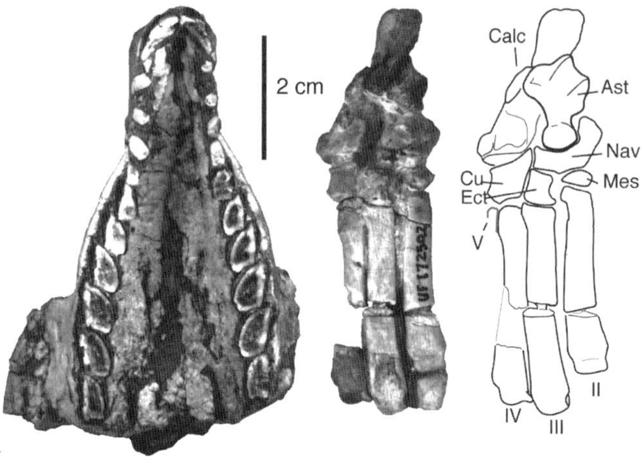

FIGURE 7.9. Hegetotheriids of Salla. *Sallatherium altiplanense* (Reguero and Cerdeño, 2005), Holotype, UF 91621, partial skull (palatal view); *Prohegetotherium schiaffinoi* (Kraglievich, 1932), UF 172502, partial pes in dorsal view (left) and dorsal view in outline (right). (See also Reguero and Cerdeño, 2005.)

The pes of *Prohegetotherium* is similar to that of *Trachytherus* (noted above), but differs by its smaller size, lack of the first digit, fused distal tibia-fibula, lesser asymmetry of the astragalar trochlea, and its relatively and absolutely smaller fibular facet. Otherwise, the hind feet of these animals are remarkably similar in their general appearance, the relative lengths of the their metatarsals to one another, and the form by which Mt II overlaps the ectocuneiform and Mt III, and Mt III slightly overlapping Mt II. The pes of *Prohegetotherium* is quite distinct from that of the interatheriid as indicated in the description and comments below.

ARCHAEOHYRACIIDAE AMEGHINO, 1897
Comments – The archaeohyracids of Salla are under study by Marcelo Reguero and his colleagues. They recognize two genera at Salla, *Archaeohyrax* and *Protarchaeohyrax*, which are represented by numerous teeth, jaws, and cranial material in all noted collections of Salla. Postcrania, however, have never been reported.

We note the presence of two astragali (UF 17069 [left] and UF 17089 [right]), possibly of the same individual, found in close association with fragmentary dental remains of adult and juvenile archaeohyracids. The astragli are very similar to those known for *Prohegetotherium*, but are larger (see Appendix) and have a greater asymmetry of trochlear ridges. Also the groove for the digital flexor is further removed from the astragalar trochlea, sitting upon a distinctive process. Such would have provided greater leverage for digital flexion, offering a modest clue regarding archaeohyracid locomotion compared to that of the hegetothere.

SUBORDER INCERTAE SEDIS
FAMILY INTERATHERIIDAE AMEGHINO, 1887
UNNAMED TAXA, HITZ, 1997
Comments – We deviate from McKenna and Bell (1997) and nearly all authorities, by not classifying interatheriids within the Typotheria. We leave their higher ordered classification unresolved (see Section 7.4.3).

Hitz (1997) described (but did not name) two distinct and otherwise unknown interatheriid taxa of Salla in his doctoral dissertation (formal naming is a work in progress). He also described a nearly complete skeleton of the smaller taxon. This skeleton, however, lacked foot bones, so the pes has been unknown for any Deseadan interatheriid. However, several UF and MNHN-Bol tarsal specimens have been found in association with interatheriid teeth. These are so similar to the tarsal form seen in Santacrucian interatheriids (*Interatherium, Protypotherium*) that there can be little doubt that they are from interatheres. The description below is based upon UATF-V-000132, a partial left pes, and UF 173247, a right proximal tarsus (calcaneal tuber missing) found articulated and fused by matrix (see Figure 7.10 and Appendix).

The proximal tarsus of the interatheres of Salla is so distinct that it cannot be confused with that of similar sized typotheres of Salla. The most conspicuous feature of the calcaneum is its well-developed, rabbit-like fibular facet. This appears as a semicircular, dorsal process in lateral view. The articular surface is proximodistally straight, covering the strongly convex surface of the protuberance that supports it. This is quite distinct from the obliquely oriented and weaker fibular facets of the calcani of *Trachytherus* and *Prohegetotherium* (Figures 7.7–7.10). The ectal facet is also distinguished by its inclined, more vertical orientation, such that the calcaneoastragular contact may be characterized as side-to-side rather than overlapping.

The astragalus has a well-defined, fairly deep trochlea. The lateral and medial sides are vertical and parallel to one another, rather than oblique as seen in *Trachytherus, Prohegetotherium* and the archaeohyracid (below). It lacks the astragalar peroneal process that inserts between the distal fibula and the calcaneum in *Trachytherus* and *Prohegetotherium.* The neck is relatively longer and much more conspicuous than that of the notohippids noted below, and the head is subspherical, forming the ball of the ball-and-socket joint with the navicular. The ectal facet is convex and has a nearly vertical orientation to meet the steeply inclined ectal facet of the calcaneum. The fibular facet is vertical and lacks the peroneal process that is present in typotheres (e.g., *Trachytherus*, see Figure 7.8b) and basal notoungulates, such as *Colbertia* (see Cifelli, 1983).

The pes of the Salla interatheriid appears to be tetradactyl, but functionally tridactyl, with Mt I being absent and Mt V being reduced in size compared to Mt II–IV. The articulation of the Mt III and IV with the distal tarsals lies in about the same plane, similar to that of the Santacrucian *Protypotherium australe* and *Interatherium robustum*, but unlike that of *P. attenuatum* in which the articulation between the cuboid and Mt IV and V appear more distal than the navicular/Mt III joint.

Elements of the pes of UATF-V-000132 are about the size as the homologous elements reported for the Santacrucian interatheriid *Protypotherium attenuatum* (Sinclair, 1909: p. 46), but smaller than *P. australe* (p. 39), and quite a bit

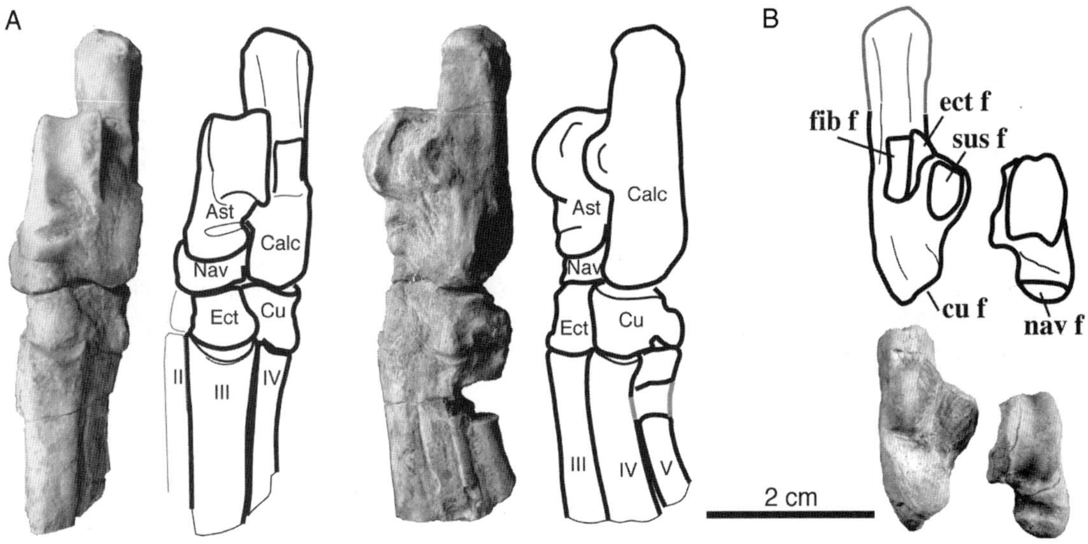

FIGURE 7.10. Intertheriidae. **A**, UATF–V-000132, partial left pes, in frontal and lateral views; **B**, UF 173247, proximal tarsus: calcaneum (left) and astraqalus (right) (Shown as photos and in outline).

larger than *Interatherium robustum* (p. 63). The foot of the Salla interatheriid differs from those of *Protypotherium* spp. by having a narrower cuboid and by the compact manner in which Mt III and IV of the Salla specimen interface with one another.

The form of the proximal tarsus of the Salla interatheriid and the Santacrucian interatheres is distinct from those of known typotheres. These differences include parallel sides of the astragalus, lacking the astragalar peroneal process, the robust, dorso-ventrally oriented fibular articulation of the calcaneum, and the more transverse articulation between the two proximal tarsal elements (strongly inclined ectal facets). This form is more similar to that of the notohippid tarsals described below (see Figure 7.11), that noted by Chaffee (1952) for *Rhynchippus pumilus*, and of the tarsus of the early toxodontids *Adinotherium* and *Nesodon* (Scott, 1912).

SUBORDER TOXODONTIA OWEN, 1853
FAMILY NOTOHIPPIDAE AMEGHINO, 1894
cf., *PASCUALIHIPPUS BOLIVIENSIS* SHOCKEY, 1997

Comments – The partial pes described below (Figure 7.12) was found at the type locality of *Pascualihippus boliviensis* (Unit II of Pasto Grande) in association with lower molar fragments of a notohippid. These teeth are of a size similar to teeth of *Pascualihippus* and *Eurygenium pacegnum*, but are not referable to the latter taxon due to the presence of an entolophid fossetid, a feature lacking in *E. pacegnum*. The tarsus described below is similar to those of early toxodontids (e.g., *Adinotherium*), so it is possible that it may be that of the toxodont, *Proadinotherium*, and the association with the notohippid teeth is merely a coincidence. However, we note the phylogenetic analysis of Toxodontia by Shockey (1997) in which *Pascualihippus* was shown as being the sister taxon to toxodontids;

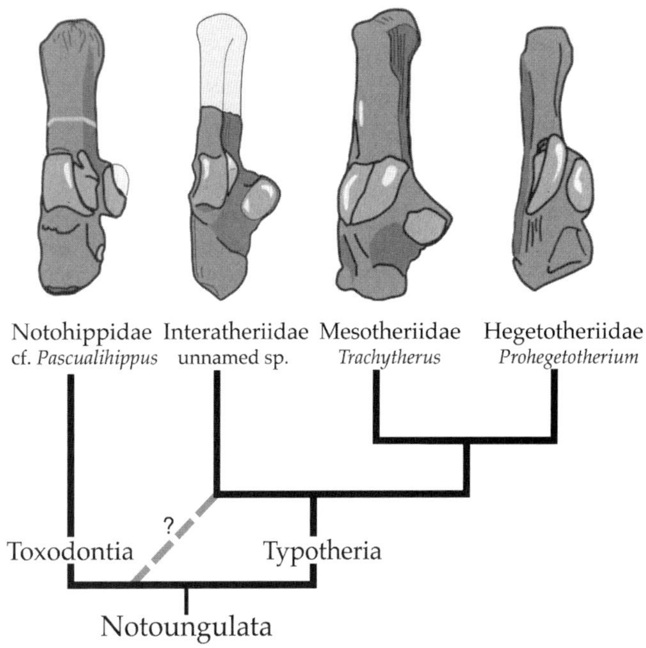

FIGURE 7.11. Comparative calcani in phylogenetic context. Bold lines indicate hypothesis from phylogenetic analysis of Cifelli (1993) and dashed line ambiguously placed to suggest alternative hypotheses as interatheriids, Typotheria and Toxodontia as unresolved tritomy, or interatheriids sister taxa to Toxodontia.

thus such a similar tarsus is a reasonable probability for *Pascualihippus*.

The calcaneum of UF 172410 has a fairly robust tuber, and is rectangular in dorsal view. The fibular facet is large and has proximodistal orientation, like that of the interathere noted above, but unlike that of *Trachytherus* and *Prohegetotherium*,

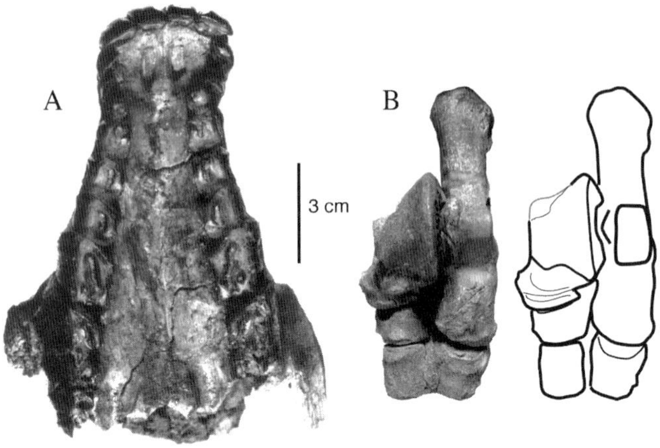

FIGURE 7.12. *Pascualihippus boliviensis.* A, Holotype MNHN-Bol-V-003642, ventral view of palate; B, referred left tarsus (UF 172410), calcaneum, astragalus, navicular, cuboid, and ectocuneiform, dorsal views (photo and in outline).

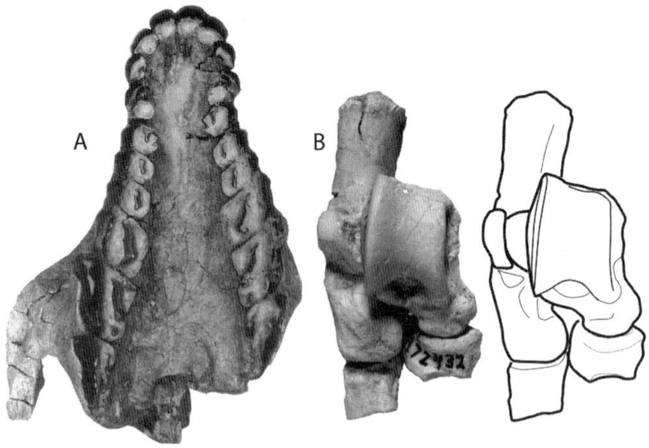

FIGURE 7.13. *Eurygenium pacegnum.* A, palatal view of holotype (MNHN-Bol-V-003643); and B, dorsal view of tarsus (UF 172432) shown as photo and in outline.

both of which have obliquely oriented fibular facets. Also, the ectal facet is vertically oriented. A small, but well-developed oval facet is present on the medial side of the calcaneum, which articulates with a similar facet of the navicular, clearly indicating the "reverse alternating tarsus" of Cifelli (1993) where tarsal alternation occurs by way of robust calcaneonavicular contact, rather than the more familiar cuboastragular alternating tarsus. On the navicular, just distal to the facet for calcaneal articulation, is a smaller facet for cuboid articulation.

The astragalus has a short neck and a modestly well-developed, somewhat asymmetric trochlea. The astragalar neck is much shorter than those of *Trachytherus, Prohegetotherium*, and the interatheriids noted above. The head is broader than deep, much less spherical than the notoungulate specimens described previously, indicating limited mobility, except in the parasagittal plane.

EURYGENIUM PACEGNUM SHOCKEY, 1997

Comments – Eurygenium pacegnum is the most completely known notohippid, being represented by a nearly complete skeleton (Shockey, 1997). One of us (Shockey, 1997) provided a brief description of this skeleton, including a limited account of the poorly preserved pes. The strength of the forelimb and relatively low Mt/femur ratio was noted, suggesting that the animal was not adapted for speed as originally suggested for notohippids (e.g., Loomis, 1914).

Since the tarsus of the skeleton of *Eurygenium* (MNHN-Bol-V-003643) was poorly preserved (thus not figured in Shockey, 1997), we provide a figure of another specimen (UF 172432: Figure 7.13), a partial pes, similar to that of cf. *Pascualihippus*, but smaller and lacking the distinctive navicular facet on the calcaneum. A faint facet is seen on the navicular for cuboid articulation, suggesting that *Eurygenium* had some, perhaps transient, articulation with the calcaneum, but not the strong "reverse alternating tarsus" of *Pascualihippus*.

Data from the *Eurygenium* skeleton was recently included in a multivariate analysis that included extant species of known function (Shockey et al., 2007). Like *Trachytherus, Eurygenium* shared morphometric space among the larger bodied fossorial taxa. But it also tracked closely with the semi aquatic capybara (*Hydrochoerus*). This was a consequence of similar body size and limb proportions with capybara and suggests the hypothesis that *Eurygenium* was a competent swimmer. The discriminant function analysis of this study classified *Eurygenium* with the extant "generalists".

Caution should be used regarding the swimming hypothesis, since it was only empirically derived, not generated from any *a priori* principles. Thus, a semiaquatic hypothesis should be considered as being unsupported, though there may not be any particular evidence against it. So, we continue to regard *Eurygenium* as having general terrestrial adaptations and suggest that it was capable of digging and swimming. Support for the digging hypothesis, independent of the aforementioned multivariate analysis, is found in the cleft ungual phalanges. Such digits are frequently found in extant diggers (see Hildebrand, 1985; Shockey et al., 2007).

FAMILY TOXODONTIDAE OWEN, 1845

PROADINOTHERIUM AMEGHINO, 1895

PROADINOTHERIUM SALTONI, SP. NOV.

Holotype – UF 149222 (Figure 7.14) damaged, but reasonably complete mandible containing the complete dentition.

Locality – The holotype comes from Unit 3 of Pasto Grande of the Salla Beds.

Diagnosis – Relatively small toxodontid having rooted incisors, no diastema between c and p1, lower premolars without fossettids, molar entolophid transverse with fossetid. Differs from *Proadinotherium leptognathum* by its smaller size (linear dimension about 80% those of *P. leptognathum*), the lack of diastema, and presence of enamel on both the external and internal surfaces of the incisors (internal enamel absent in adults of *P. leptognathum*).

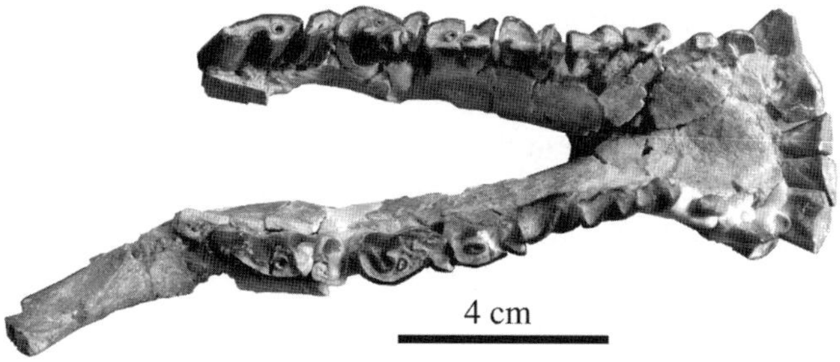

FIGURE 7.14. *Proadinotherium saltoni* sp. nov. Mandible (holotype, UF 149222) in occlusal view.

Etymology – In memory of Justine Salton, with special regard to the field season that she worked with us at Salla.

Description – Plant roots grew through the holotype, breaking it to bits, but these fragments have been reconstructed to give a reasonably good indication of the jaw morphology (Figure 7.14, dental metrics given in Appendix). The lower incisive battery is similar to that of *Adinotherium* (see Scott, 1912: Plate XVII: Figure 7.10), but the incisors of *P. saltoni* have shorter crowns. These incisors are enlarged, spatulate, and have long, but closed roots; those of the i3s extend posteriorly to the level below the p2. The lower canine is much smaller than the incisors and may be described as "incisiform", with the qualification that it looks nothing like the incisors of this animal.

No diastema occurs between the c and p1 as in the *P. leptognathum* specimen figured by Loomis (1914: Figure 81). The p1 is smaller than the canine and is little more than a peg-like structure, ovoid in occlusal view with a tiny fossettid in the middle.

The p2 is shaped like a double crescent and lacks a fossettid. None of the lower premolars of the holotype, or those of the more heavily worn referred specimen, UF 149223, has any fossettids, which is indicative of the deep, broad ectoflexid (premolars of other known nesodontine toxodontids form fossettids early in wear [see Scott, 1912: plate XVIII, Figures 7.4 and 7.5]).

The m1 is moderately worn, having formed a trigonid-talonid fossettid. It and the other lower molars have an entolophid fossettid. The m2 is less worn and demonstrates the generalized form of advanced toxodontids (sensu Cifelli, 1993) a 7/9 morphology (sensu Shockey et al., 2004), where the trigonid has the form of the number seven and the talonid a number nine, the entolophid representing the upper part of the nine, pierced by the entolophid fossettid.

No postcranials have been found associated with *P. saltoni* teeth at Salla, but we refer a distal femur of the MNHN-Paris collection to this taxon (Figure 7.7). This referral is based upon the distinctive, enlarged medial trochlear ridge, similar to that of other toxodontids and unlike the femora of similar-sized taxa of Salla (e.g., *Eurygenium, Trachytherus*; the femur of *Pascualihippus* is unknown).

The medial trochlear ridge of the Salla toxodontid is considerably higher than the lateral ridge of the patellar trochlea, but is not bulbous, as in *Toxodon* or *Hoffstetterius*, which have been shown to function just like knee locks of horses (Shockey, 2001). The functional significance of the less modified, but enlarged MTR of *Proadinotherium* is unknown, but it may have served to prevent medial dislocation of the patella or to prevent lateral movements of the lower leg (Shockey, 2001). Whatever the function, it has been shown that having an enlarged MTR, at least in bovids, is correlated with living in open habitats (Kappelmann, 1988). This morphology likely, and incidentally, served as a "preadaptation" for the knee locks seen in later toxodontids. These have been documented to the late Miocene (Shockey, 2001), but our observations of the distal femur of cf. *Pericotoxodon* from the La Venta Beds in the MNHN-Paris collection extents the toxodontid knee lock back to middle Miocene (about 13 Ma). This is roughly the same time that knee locks evolved in horses (Hermanson and MacFadden, 1996).

7.4 Discussion

7.4.1 Biology of Form

The species richness of the Salla fauna is mirrored by the diversity of morphological adaptations; some mammals show specializations for running, digging, and climbing (uncommon), whereas others were less specialized and perhaps more versatile in their locomotor abilities.

7.4.1.1 Marsupials

Given the great diversity of primary consumers at Salla, it not surprising to see a fair diversity of carnivorous marsupials. With our description of *Fredszalaya hunteri* and the work in progress on a small dog-like marsupial, the species richness of sparassodonts now includes six taxa. The huge *Paraborhyaena*, was presumably terrestrial (α-level hypothesis), but the calcaneum of *Fredszalaya* suggests some climbing abilities (α-level hypothesis), perhaps like the morphologically similar *Prothylacynus* (Argot, 2002).

7.4.1.2 Xenarthrans

Though instructive postcranials of dasypodids are lacking (e.g., complete ulnae, metatarsals), one might presume typical dasypodid digging and insectivorous habits for these animals (α-level hypotheses) until there is evidence to the contrary. Likewise, we presume fossorial locomotion (α-level hypothesis) for the peltephilid based upon the indirect (extrapolated) link of phylogenetic relationship to *Peltephilus* and the more fully developed hypothesis of digging for that taxon offered by Vizcaíno and Fariña (1997). The postcranial evidence presented here, though incomplete, is at least consistent with a digging hypothesis. Like *Peltephilus*, the Salla peltephilid appears have focused forces for biting at the anterior region of its jaw, though its mandibular symphysis was not fused in most known specimens. A strong anterior bite is unusual for carnivorous or herbivorous mammals, or any kind of mammal for that matter. The closest analog among mammals we think of is the anterior cutting teeth of fruit eating and blood drinking bats. For now, we are content to continue wondering about the strange peltephilids.

Like most of the notoungulates summarized below, the mylodontid sloth of Salla has high crowned (indeed, evergrowing) teeth. Additionally, it had a broad muzzle. The hypselodonty and broad muzzle each is suggestive of grazing in this most ancient of mylodontids (β-level hypothesis).

7.4.1.3 Pyrotheria

The relatively huge head and grossly robust postcranial skeleton of *Pyrotherium macfaddeni* leaves no real doubt about its unspecialized, terrestrial locomotion (γ-level hypothesis). The tusks and probable proboscis provided the animal a means to probe around and manipulate its environment to find and obtain food items. Its relatively low crowned teeth makes it unlikely that it was a grazer, but in absolute terms the crowns were fairly high, so it probably could have consumed some grasses. Despite that caveat, it is probably best classified as a browser (β-level hypothesis).

7.4.1.4 Litopterna

The proterotheriid and macraucheniid litopterns show marked adaptations for cursorial habits (β-level hypotheses). The macraucheniids are regarded as open habitat cursors (α-level hypothesis) and the small proterotheriid has been compared to forest dwelling cursorial rodents (e.g., dasyproctids) and artiodactyls (e.g., duikers and tragulids) (Shockey, 1999) (α-level hypothesis). (No postcranials are known for the adianthid litopterns.) The narrow muzzles (known in the macraucheniid and assumed for the proterotheriid) and low crowned dentition suggest browsing (β-level hypothesis).

7.4.1.5 Notoungulata

The smaller notoungulates (interatheriids, hegetotheriids, and archaeohyracids) appear to show modest developments associated with quick locomotion, such that they may be regarded as cursorial (γ-level hypothesis for interatheres and hegetotheres, but only α-level hypothesis for the poorly known archaeohyracids). There is no compelling evidence at Salla for saltatory habits in these small notoungulates, but given their size and general cursorial adaptations such should be considered (α-level hypothesis). Suggestive evidence for saltatory activity includes the fused tibia-fibula at the upper ankle joint in the interatheriid (Hitz, 1997) and *Prohegetotherium*. This is unknown for the archaeohyracid, but we note the greater mechanical advantage at the astragalus for the digital flexors.

The body of evidence strongly supports the hypothesis of fossorial habits (δ-level hypothesis) for the mesothere *Trachytherus* (for details see Shockey et al., 2007). Modern wombats, with their digging abilities, high crowned cheek teeth, and gliriform incisors, serve as a model for *Trachytherus* biology, though due to the extreme convergence of the postcranial skeletons of mesotheres with those of aardvarks (*Orycteropus*) the precise method of digging in *Trachytherus* is probably more similar to that of *Orycteropus* (see Shockey et al., 2007).

The notohippid *Eurygenium* is regarded as subcursorial generalist (β-level hypothesis) as we note only modest adaptations for speed in its skeleton. Weaker evidence suggests that this generalist was capable of swimming and digging (α-level hypotheses). Postcranial elements of *Pascualihippus* and the toxodontid *Proadinotherium* are poorly known and thus are only suggestive of terrestrial, subcursorial habits (α-level hypothesis). No postcranials of the leontiniid *Anayatherium* are known, so we defer any comments on its locomotion, but we note that its narrow muzzle and mesodont cheek teeth imply browsing (β-level hypothesis).

Deseadan faunas typically contain numerous high crowned notoungulates. Remarkably for the late Oligocene, the notoungulate taxa almost exclusively have high crown, or hypsodont, dentitions (11/13 species). The only notoungulates at Salla lacking hypsodont teeth are the leontiniids. Leontiniids at Salla are exceedingly rare, but even they have moderately high crowned, mesodont teeth (Shockey, 2005).

Compared to contemporaneous faunas throughout the world, Deseadan faunas have a remarkably large number of herbivores having high crowned teeth (Patterson and Pascual, 1972; Flynn and Wyss, 1998; MacFadden, 2000). The phenomenon of this "precocious hypsodonty" began in South America around 35–30 Ma, about 15 million years earlier than in North America (MacFadden, 2000). For some, this precocious hypsodonty has implied the early spreading of grasslands in South America (Stebbins, 1981; MacFadden, 2000). Pascual and Ortiz Jaureguizar (1990) provided an alternative explanation for this precocious hypsodonty, suggesting that it was a response to tooth wear caused by volcanic grit that dusted the plants during that time. In more general terms, Janis (1995 and references therein) demonstrated that any dust or grit covering low lying plants in open country may contribute to the evolution of high crowned teeth.

The robust evidenced for *Trachytherus* being a digging specialist (Sydow, 1988; Shockey et al., 2007), along with suggestive evidence of fossorial habits in other Tertiary notoungulates (e.g., homalodotheres [Scott, 1930], *Scarrittia* [Chaffee, 1952], toxodontids [Hildebrand, 1985]), adds another dimension to the problem of precocious hypsodont of notoungulates. Did hypsodonty evolve in response to eating subterranean foods covered with abrasive grit?

To have confidence in the grazing hypothesis, there must be some independent evidence for grazing in notoungulates. Stable isotopic studies have no utility in this context, since the global carbon shift did not occur until much later (MacFadden, 2000). That is, prior to the global carbon shift,

C4 photosynthesis was insignificant; so even grasses would have been predominantly C3.

Microwear or other dental abrasion analyses are wanting, but they have the potential to be illuminating. A major challenge for investigators beginning this work will be to develop techniques that can discriminate between wear caused by phytoliths of grass versus non-biogenic abrasives, such as volcanic grit, dust, or dirt.

Meanwhile, we must use morphological characters independent of hypsodonty to test grazing hypotheses for these hypsodont herbivores. One such morphological feature that appears to be independent of hypsodonty, but associated with grazing is muzzle width (Gwynne and Bell, 1968; Solounias et al., 1993). Shockey (1997) noted the broad muzzle of *Pascualihippus* and argued that it, in the context of the animal's hypsodont dentition, indicated that grass was a significant portion of the animal's diet (indeed, the generic name, in addition to honoring Rosendo Pascual, means "grazing horse") (γ-level hypothesis). The toxodontid *Proadinotherium saltoni* also had a broad cropping dentition and hypsodont teeth, providing two lines of evidence suggestive of grazing (γ-level hypothesis).

The smaller notoungulates (interatheres, archaeohyracids, hegetotheres) have hypsodont-to-hypselodont dentitions, suggestive of grazing, but they have narrow muzzles, suggestive of more selective feeding. Also, due to their small body size with the implied high mass specific metabolic rate, it is unlikely that they could have obtained the quality of nutrients they would have required from grasses alone. Thus, we tentatively characterize them as being selective feeders, foraging upon richer food items (seeds, fruits, tubers and protein and calorie-rich new-growth, leafy material) at ground level (α-level hypothesis).

7.4.1.6 Immigrant Taxa

Rodents of Salla are numerous and diverse but nothing is known of their postcranial skeletons, other than that we have referred a couple of proximal tarsals to the order, but not to any particular genus. We offer no hypotheses regarding their locomotion or feeding ecology at this time.

The impulse to assume that the monkey *Branisella* was arboreal is tempered by the evidence provided by Kay et al. (2002). Based upon the relatively high crowned teeth and the low primate diversity at Salla, they interpret *Branisella* as being more terrestrial than other platyrrhines. This evidence, along with some details of the sediments, were used to suggest that Salla was dry and not forested, an interpretation consistent with the interpretation of MacFadden, 1990, but at odds with the results of the body size distribution analysis of Croft (2001; discussion below).

7.4.2 Paleoecology of Salla

There is a near consensus that the habitat of Salla was fairly open and dry, with much grass and scattered patches of brush and trees. Such would explain the numerous hypsodont taxa, which include nearly all of the numerous notoungulates, as well as the marsupial *Proargyrolagus*, and (relative to other primates) *Branisella*. However, we note the small body size of most of the hypsodont notoungulates, animals that may have had difficulty getting enough nutrients to fuel their presumably high mass specific metabolic rates if they were feeding on grass. Of the larger hypsodont notoungulates, *Trachytherus* had a narrow muzzle, suggestive of specialized feeding, and it was almost certainly a digger. Their high crowned teeth may have just served to protect the animal against the rapid tooth wear caused be eating dirt-covered vegetation, rather than phytolith-filled grasses. Of the larger notoungulates, only *Pascualihippus* and *Proadinotherim* show independent evidence of grazing – their broad muzzle, a character also seen in the hypsodont/hypselodont mylodontid sloth.

Croft (2001) showed that the distribution of body size of the herbivores of Salla was similar to body size distributions seen in extant, forested habitats but distinct from the patterns of modern arid regions. He was conscious that his findings were in conflict with the prevailing hypotheses and suggested that the slope of his cenogram could have been artificially flattened since it included taxa from all horizons. That is, there was not a single Salla fauna, but a dynamic fauna that changed over time.

Given that the Salla Beds contain up to 600 m of sediments, deposited over a time span of about 3 million years (MacFadden et al., 1985), it is probably inappropriate to discuss the paleoecology of Salla as if it were a single phenomenon. Clearly, the depositional environment changed over time and work in progress suggests that there were some changes in the fauna, though many taxa are found at all horizons. In the mean time, caveats must accompany any comments about the environment of Salla.

7.4.3 Phylogenetic Considerations

7.4.3.1 Xenarthrans

The occurrence of the complex (medial ball-and-socket/lateral sliding) knee joint in Tertiary glyptodonts and sloths (Shockey, 1999; Salas et al., 2005) presents an interesting problem. This morphology is absent in armadillos and pampatheres. The current and essentially universally accepted dichotomous model of xenarthran phylogeny (armored cingulates vs. hairy pilosans) is incompatible with any suggestion that the complex knee articulation of cingulated glyptodonts and pilosan sloths is homologous. Thus, one might assume that the complex, rotary knee joint of sloths and glyptodonts evolved independently. With much curiosity and some discomfort, we include the complex knee joint as a homoplasy among sloths and glyptodonts; along with other such similar distinctive structures, including their fused mandibular symphyses with mandibular spout, deep mandibles, short and wide nasals, descending process of the jugal, and the tri-lobe ("carved tooth") nature of glyptodonts and the most primitive sloths known, *Pseudoglyptodon sallaensis* of Salla and *P. chilensis*, from the early Oligocene Tinguiririca fauna (McKenna et al., 2006).

7.4.3.2 Notoungulata

The interatheriid tarsus reveals problems for systematists. Although the interatheriids had been almost universally considered to be nested within the Typotheria (e.g., Simpson, 1945; Cifelli, 1993; but see Reguero, 1999 for another interpretation), they have several derived tarsal traits that occur in the "advanced Toxodontia" (sensu Cifelli, 1993; = leontiniids, notohippids, and toxodontids), but not Typotheria. These tarsal characters include the well-developed calcaneofibular articulation, which is dorsoventrally oriented (rather than oblique), absence of the astragalar peroneal process, nearly vertical lateral and medial walls of the astragalar body, and the steeply inclined orientation of the articulation of the calcaneum and astragalus at their ectal (lateral) contact. Characters of the tarsus were not included in the phylogenetic analysis of Reguero (1999), thus, our finding here provide independent support for the exclusion of the Interatheriidae from Typotheria.

These shared characteristics of the interatheriid and "advanced Toxodontia" tarsus is significant. If they are indeed homologous, it will have a profound effect on our interpretations of the inter-familial relationships of notoungulates. It would suggest that interatheriids are more closely related to the "advanced Toxodontia", which would have even broader implications, since this Toxodontia-interatheriid form is quite similar to the tarsus of the Arctostylopida (Cifelli et al., 1989; Missiaen et al., 2006; and discussion below). If, instead, this suite of characters represents a homoplasy, then it evolved independently evolved at least three times (Arctostylopida, "advanced Toxodontia", and interatheriids [and similar to the lagomorph morphology]), and it likely suggests a tarsal form required when near equal forces are transmitted through the fibula-calcaneum and the tibia-astragalus.

The similarities of the Toxodontia-interatheriid form to that of the Arctostylopida are of interest. The Arctostylopida are a small group of ungulates, mostly from Asia but represented in North America by *Artostylops* (see Cifelli and Schaff, 1999 for a review of the Arctostylopida). Initially, arctostylopids were regarded as being notoungulates (e.g., Matthew, 1915; Simpson, 1945), but later students of the groups generally regarded the dental similarities between Artostylopida and notoungulates to represent homoplasies (Cifelli et al., 1989; Cifelli and Schaff, 1999; Missiaen et al., 2006), largely based upon differences in the tarsi of notoungulates and arctostylopids.

Previous comparisons of the arctostylopid tarsus to that of notoungulates have noted significant differences (Cifelli et al., 1989; Missiaen et al., 2006). However, these authors used the putative primitive notoungulate form (e.g., *Colbertia*) to compare notoungulates with the actostylopid, *Paleostylops* (= *Gashatostylops* of Cifelli et al., 1989). Cifelli et al. (1989) at least noted the presence of the arctostylopid-like calcaneofibular joint of "advanced Toxodontia", but reasoned that this would have evolved too late (perhaps not until the Oligocene) to have any relevance on the arctostylopid-notoungulate question. However, they were not aware of the same form occurring in interatheriids. This form is unknown prior the Oligocene, but if it is homologous with that of the "advanced Toxodontia", then its origins must have occurred much earlier. Even if the interatheriid form is not homologous with that of the Toxodontia, the record of the interatheriids goes back to the Riochican (Cifelli, 1993; generally regarded as Paleocene). If these early Tertiary notopithicine interatheriids had a similar form, then it would be most relevant in regard to the arctostylopid-notoungulate problem. Unfortunately, however the tarsus of these Paleocene-Eocene notopithicine interatheriids remains unknown. Indeed, prior to this report, the oldest record of any interatheriid tarsus was Santacrucian, early-middle Miocene [Sinclair, 1909]).

Bloch (1999) reported on the discovery of a partial skeleton of the North American arctostylopid, *Arctostylops*. He noted several tarsal characters that he regarded as putative synapomorphies to unite arctostylopids with notoungulates. These included astragalus with tibial protuberance and elongated, constricted neck with oblique dorsal ridge, and calcaneum with proximally positioned sustentaculum). Since he had compared the tarsus of *Arctostylops* with the primitive notoungulate morphotype, he listed the "steeply inclined ectal facet" of arctostylopids as being different from the form of notoungulates. The steeply inclined ectal facet of the Toxodontia-interatheriid morphotype negates that difference and strengthens Bloch's argument for artostylopid-notoungulate affinities. We also add the well-developed, anteroposteriorly oriented fibular facet of the calcaneum, the lack of astragalar peroneal process, and the nearly vertical medial and lateral walls of the astragalar body to Block's list of putative synapomorphies. Further study of the skeleton of *Arctostylops* and the postcrania of notoungulates should help resolve this interesting notoungulate-arctostylopid problem.

7.4.4 Concluding Remarks

Regarding Salla, Simpson (1984:p. 214) wrote: "A needed monograph, including all the species present in the various collections, has not yet been published." Such a work is still wanting, but we hope that our efforts here will serve in the interim to better document the varied and curious fauna of Salla.

This overview of the Salla fauna includes a review of previous works of Salla and a report of some of our new findings. The new includes *Fredszalaya hunteri* gen. et. sp. nov.; stratigraphic context for one of the last surviving proborhyaenines (*Paraborhyaena*) with the observation that it had a single pair of incisors; the first description of the peltephilid of Salla (with the documentation of the unfused mandibular symphysis); the documentation of one of the oldest mylodontid skulls; the significant elongation of Mt

III of a Deseadan proterotheriid; the macraucheniid patella that provides an example of form resulting from function (essentially proving that Deseadan macraucheniids had knee locks); the analysis of interatheriid tarsals (casting doubt upon its classification as a typothere and suggesting a closer relationship with Toxodontia and raising the problematic possibility of a notoungulate-arctostylopid relationship); the first descriptions of the tarsals of *Pascualihippus, Eurygenium, Prohegetotherium*, and the unnamed interatheriids of Salla; as well as the distal femur of *Proadinotherium* along with the description of the new species of toxodontid, *P. saltoni*.

In addition to being a review and to documenting new data, we like to think of this work as being a preview of works to come. We have noted some of our works in progress and those of some of our colleagues. Perhaps most importantly, we should note work in the Tinguiririican faunas (early Oligocene), which has great potential for resolving some of the current conflicts regarding some of the phylogenetic hypotheses.

We are conscious of the fact that we have raised more questions than illuminated answers (e.g., the complexities regarding peltephilid biology, phylogenies of interatheres and arctostylopids, and the function of notoungulate hypsodonty) and hope to see continued works regarding these varied and interesting problems. We have deferred (i.e., avoided) paleoecological conclusions and note work in progress by D. Croft and ourselves.

A saying in Bolivia goes, "*Chancho limpio nunca engorda* (the clean pig never fattens)." The relevance here is that some answers still lie in the field, at Salla and at other Tertiary localities of South America, and that we and other investigators need to continue to soil ourselves with Tertiary sediments in order to resolve some of these little mysteries of life on that ancient, "lost" continent.

Acknowledgments. Recent fieldwork at Salla has been sponsored by the National Geographic Society (7507–03 to BJS). Previous field seasons have been sponsored by various NSF grants to B. J. MacFadden, BJS, and to Kay, Madden, and MacFadden. We thank the many people who assisted us in our museum studies. These include Mary Ann Turner, Lyndon Murray, and Daniel Brinkman of the Yale Peabody Museum; Christian de Muizon of the Museúm National d'Histoire Naturelle, Paris; Denny Dively and Jin Meng of the American Museum of Natural History; and Bruce J. MacFadden and Richard Hulbert of the Florida Museum of Natural History. Alfredo Carlini provided identifications for most of the cingulate specimens. Darin Croft, Jay O'Sullivan, and Eric Sargis provided critical reviews of this manuscript for which we are thankful. We are especially grateful to the administration of the Universidad Autónoma "Tomás Frías," Potosí, Bolivia, especially Pedro Lopez, for helping revive our studies in Bolivia.

Appendix

Metric data (mm) of proximal tarsals and other elements of taxa discussed in text

Proximal tarsals

Taxon	Specimen ID	Astragalus Length	Trochlear Width	Calcaneum Length
Marsupialia				
Fredszalaya hunteri	UF 172501	–	–	33.2
Xenarthra				
cf. *Peltephilus* sp.	UF 93515	15.3	12.7	–
Pyrotheria				
Pyrotherium macfaddeni	UF 172765	–	–	150
Litopterna				
Coniopternium sp.	UF 172426	38.4	25.5	88.5
Notoungulata				
Trachytherus alloxus	UF 172437	35	15.4	–
Trachytherus alloxus	MNHN-Bol-F-94-k	37.5	14.7	–
Trachytherus alloxus	UF 90960	34.6	14.8	58.8
Trachytherus alloxus	UF 172514	–	–	66.7
Prohegetotherium schiaffinoi	UF 172502	15.1	7.1	67.3
Prohegetotherium schiaffinoi	UF 172445	15.9	7.2	–
Archaeohyracid	UF 176069	20.0	8.5	–
Archaeohyracid	UF 176089	19.8	8.6	–
Interatheriid, large sp.	UATF-V-000132	17.5	7.8	29.2
Interatheriid, large sp.	UF 173247	17.7	7.5	–
Interatheriid, small sp.	UF 172970	–	–	20.8

(continued)

Appendix (continued)

Taxon	Specimen ID	Astragalus Length	Trochlear Width	Calcaneum Length
Eurygenium pacegnum	UF 172432	32.7	16.0	59.6
Eurygenium pacegnum	Bol-V-003644	–	–	54.0
Eurygenium pacegnum	Bol-V-004077	26.3	15.0	–
cf *Pascualihippus boliviensis*	UF 172410	39.6		17.9 74.5

Other elements

Taxon	Specimen ID	Element	L	W
Marsupialia				
Fredszalaya hunteri[a]	UF 172501	C-nuchal crest	162	–
		C – M4	34	–
		P3	8.8	6.4
		M1	(7.5)	6.5
		M2	9.5	7.6
		M3	10.3	9.1
		M4	4.8	8.8
Xenarthra				
Peltephilid (unnamed taxon)	UF 93586	mf1–mf7	32.5	
	UF 93587	mf “0” – mf7	34.0	
	PU 21143	mf “0” – mf7	40.5	
	UF 93515	Mt II	23.7	6.3 (proximal)
		Mt III	25.1	6.1 (prox.)
		Mt IV	24.0	6.0 (prox.)
Mylodontid (unnamed genus)	UATF-V-127	Cf-Mf4	67.1	–
		Mf1	13.2	10.1
		Mf2	14.0	11.6
		Mf3	19.3	13.0
		Mf4	(≈10)	(≈10)
Litopterna				
Coniopternium spp.	UF 149207	Humerus	158	40 (condyles)
	MNHN-Bol-V-004502	Femur	238	48 (condyles)
	MNHN-Bol-(no #)	Fémur	265	–
	UF 172425	Mt III	118	–
Notoungulata				
Trachytherus alloxus	UF 91933	Humerus	185	66
	UF 90960	Humerus	146	55
		Mt III	60	–
		Femur	179	–
Prohegetotherium schiaffinoi	UF 172502	Mt II	28.8	6.0
		Mt III	32.9	6.1
		Mt IV	(32)	5.5
	UF 172445	Femur	91.3	20.2
Interatheriid, large sp.	UATF-V-132	Mt III	–	7.5 (proximal)
		Mt IV	–	7.0 (prox.)
		Mt V	–	3.5 (prox.)
Eurygenium pacegnum	MNHN-Bol-V-3644	Humerus	156	–
		radius	126	–
		Mc II	52.1	–
		Mc III	52.1	–
		Mc IV	48.5	–
		Mc V	(30)	–
		Femur	175	–
		Tibia	164	–
		Mt III	44.5	–
		Mt IV	44.3	–
		Mt V	42.5	–

(continued)

Appendix (continued)

Taxon	Specimen ID	Element	L	W
Proadinotherium saltoni[a]	UF 149222	i1	8.4	4.8
		i2	11.8	4.8
		i3	(13)	4.6
		c	6.9	2.5
		p1	5.5	3.5
		p2	8.5	4.3
		p3	9.5	5.5
		p4	10.8	6.5
		m1	12.6	6.9
		m2	16.9	6.8
		m3	23.0	6.9

[a] Holotype

References

Ameghino, F., 1891. Nuevos restos mamíferos fósiles descubiertos por Carlos Ameghino en el eoceno inferior de Patagonia austral. Revista Argentina Historia Natural 1, 289–328.

Ameghino, F., 1895. Premiére contribution la connaissance de la fauna mammalogique de couches à *Pyrotherium*. Boletin Instituto Geográfico Argentino 15, 603–660.

Ameghino, F., 1897. Mammifères crétacés de l'Argentine. Deuxième contribution à la connaissance de la fauna mammalogique de couches à *Pyrotherium*. Boletin Instituto Geográfico Argentino 18, 406–521.

Ameghino, F., 1904. Nuevas especies de mamíferos, cretáceos y terciarios de la República Argentina. Anales de Sociedad de Ciencias de Argentina, Buenos Aires 56, 193–208.

Ameghino, F., 1934 (originally published in 1910). Geología, paleogeografía, paleontología y anthropología de la República de Argentina. Obras Completas 18, 1–317.

Argot, C., 2003. Functional adaptations of the postcranial skeleton of two Miocene borhyaenoids (Mammalia, Metatheria), *Borhyaena* and *Prothylacynus*, from South America. Palaeontology 46, 1213–1267.

Baird, D., Woodburne, M., Lawrence, A., 1966. *Pyrotherium* and other mammals from Bolivia. Society of Vertebrate Paleontology News Bulletin 77, 18.

Babot, J., Powell, J. E., Muizon, C. de., 2002. *Callistoe vincei*, a new Proborhyaenidae (Borhyaenoidea, Metatheria, Mammalia) from the early Eocene of Argentina. Geobios 35, 615–629.

Billet, G., Muizon, C. de, Mamani, B. 2008. Late Oligocene mesotheriids (Mammalia, Notoungulata) from Salla and Lacayani (Bolivia): implications for basal mesotheriid phylogeny and distribution. Zoological Journal of the Linnean Society 152, 153–200.

Bloch, J. I., 1999. Partial skeleton of *Arctostylops* from the Paleocene of Wyoming: arctostylopid-notoungulate relationship revisited. Journal of Vertebrate Paleontology 19 (Supplement), 32a.

Chaffee, R. G., 1952. The Deseadan vertebrate fauna of the Scarritt Pocket, Patagonia. Bulletin of the American Museum of Natural History 98, 509–562.

Cifelli, R. L., 1983. Eutherian tarsals from the late Paleocene of Brazil. American Museum Novitates 2761, 1–31.

Cifelli, R. L., 1993. The phylogeny of native South American ungulates. In: Szalay, F. S., Novacek, M. J., McKenna, M. C. (Eds.), Mammal Phylogeny, Volume 2: Placentals. Springer, New York.

Cifelli, R., Schaff, C. R., 1999. Arctosylopida. In: Janis, J. M., Scott, K. M., Jacobs, L. L. (Eds.), Evolution of Tertiary mammals of North America, Volume 1: Terrestrial Carnivores, Ungulates, and Ungulatelike Mammals. Cambridge University Press, Cambridge, pp. 332–336.

Cifelli, R., Soria, M. 1983a. Notes on Deseadan Macraucheniidae. Ameghinana 20, 141–153.

Cifelli, R., Soria, M. 1983b. Systematics of the Adianthidae (Litopterna, Mammalia). Novitates 2771, 1–25.

Cifelli, R., Schaff, C. R., McKenna, M. C., 1989. The relationships of the Arctostylopida (Mammalia): new data and interpretation. Bulletin of the Museum of Comparative Zoology 152, 1–44.

Croft, D. A., 2001. Changing environments in South America as indicated by mammalian body size distributions (cenograms). Diversity and Distributions 7, 271–278.

Delsuc, F., Catzeflis, F. M., Stanhope, M. J., Douzery, E. J. P., 2001. The evolution of armadillos, anteaters, and sloths depicted by nuclear and mitochondrial phylogenies: implications for the status of the enigmatic fossil *Eurotamandua*. Proceedings of the Royal Society of London B 268, 1605–1615.

Flynn, J. J., Wyss, A. R., 1998. Recent advances in South American mammalian paleontology. Trends in Ecology and Evolution 13, 449–454.

Gaudry, A., 1906. Fossiles de patagonie, etude sur un portion du monde Antarctique. Annales de Paleóntologue 2, 101–143.

Goin, F. J., Sanchez-Villagra, M. R., Abello, A., and Kay, R. F. 2007. A new generalized paucituberculatan marsupial from the Oligocene of Bolivia and the origin of "shrew-like" opossums. Palaeontology 50, 1267–1276.

Greaves, W. S., 1995. Functional predictions from theoretical models of the skull and jaw in reptiles and mammals. In: Thomason, J. J. (Ed.), Functional Morphology in Vertebrate Paleontology. Cambridge University Press, Cambridge, pp. 99–115.

Gwynne, M. D., Bell, R. H. V., 1968. Selection of vegetation components by grazing ungulates in the Serengeti National Park. Nature 220, 390–393.

Hermanson, J., MacFadden, B. J., 1996. Evolutionary and functional morphology of the knee in fossil and extant horses (Equidae). Journal of Vertebrate Paleontology 16, 349–357.

Hildebrand, M., 1985. Digging in quadrupeds. In: Hildebrand, M., Bramble, D. M., Liem, K. F., Wake, D. B. (Eds.), Functional Vertebrate Morphology. Belknap, Cambridge, MA/London, pp. 89–109.

Hitz, R., 1997. Contributions to South American mammalian paleontology: new interathres (Notoungulata) from Chile and Bolivia, typothere phylogeny, and paleosols from the late Oligocene Salla Beds. Ph.D. dissertation, University of California, Santa Barbara.

Hoffstetter, R., 1968. Un gisement de mammifère Déséadiens (Oligocène Inférieur) en Bolivie. Comptes rendus des séances de l'Académie des Sciences 267D, 1095–1097.

Hoffstetter, R., 1969. Un primate de l'Oligocène Inférieur sud-américain: *Branisella boliviana* gen. et sp. nov. Comptes rendus des séances de l'Académie des Sciences 269, 434–437.

Hoffstetter, R., 1976. Rongeurs caviomorphes de l'Oligocène de Bolivie. Paleovertebrata 7, 1–14.

Hoffstetter, R., Lavocat, R., 1970. Découverte dans le Déséadien de Bolivie de genres pentalophodontes appuyant les affinités africaines des Rongeurs Caviomorphes. Comptes rendus des séances de l'Académie des Sciences 273, 2215–2218.

Hoffstetter, R., Petter, G. 1983. *Paraborhyaena boliviana* et *Andinogale sallensis*, deux Marsupiaux (Borhyaenidae) nouveaux du Désédien (Oligocéne Infériur) de Salla (Bolivie). Comptes rendus des séances de l'Académie des Sciences 296, 205–208.

Janis, C., 1995. Correlations between craniodental morphology and feeding behavior in ungulates: reciprocal illumination between living and fossil taxa. In: Thomason, J. J. (Ed.), Functional morphology in vertebrate paleontology, Cambridge University Press, Cambridge, pp. 76–98.

Kappelmann, J., 1988. Morphology and locomotor adaptations of the bovid femur in relation to habitat. Journal of Morphology 198, 119–130.

Kay, R. F., MacFadden, B. J., Madden, R., Sandeman, H., Anaya, F., 1998. Revised age of the Salla beds, Bolivia, and its bearing on the age of the Deseadan South American Land Mammal "Age." Journal of Vertebrate Paleontology 18, 189–199.

Kay, R. F., Williams, B. A., Anaya, F., 2002. The adaptations of *Branisella boliviana*, the earliest South American monkey. In: Plavcan, J. M., Kay, R. F., Jungers, W. L., van Schaik, C. P. (Eds.), Reconstructing behavior in the primate fossil record. Kluwer/Plenum, New York, pp. 339–370.

Kraglievich, L., 1932. Nuevos apuntes para la geología y paleontología uruguayas. Anales del Museo de Historia Natural de Montevideo 3, 1–65.

Lavocat, R., 1976. Rongerus caviomorphes de l'Oligocène de Bolivie. II. Rongeurs de Bassin Déséadien de Salla-Luribay. Palaeovertebrata 7, 15–90.

Loomis, F., 1914. The Deseado Formation of Patagonia. Rumford Press, Concord, NH.

MacFadden, B. J., 1990. Chronology of Cenozoic primate localities in South America. Journal of Human Evolution 19, 7–22.

MacFadden, B. J., 2000. Origin and evolution of the grazing guild in Cenozoic New World terrestrial mammals. In: Sues, H.-D. (Ed.), Evolution of Herbivory in Terrestrial Vertebrates: Perspectives from the Fossil Record. Cambridge University Press, Cambridge, pp. 223–244.

MacFadden, B. J., Campbell, K. E., Cifelli, R. L., Siles, O., Johnson, N. M., Maeser, C. W., Zeitler, P. K., 1985. Magnetic polarity stratigraphy and mammalian fauna of the Deseadan (Late Oligocene-Early Miocene) Salla beds of northern Bolivia. Journal of Geology 93(3), 223–250.

McDonald, H. G., 1997. Xenarthrans: pilosans. In: Kay, R. F., Madden, R. H., Cifelli, R., Flynn, J. J. (Eds.), Vertebrate Paleontology in the Neotropics. The Miocene Fauna of La Venta, Colombia, pp. 233–245.

McKenna, M. C., Bell, S. K., 1997. Classification of Mammals above the Species Level. Columbia University Press, New York.

McKenna, M. C., Wyss, A., Flynn, J. J., 2006. Paleogene pseudoglyptodont xenarthrans from central Chile and Argentine Patagonia. American Museum Novitates 3536, 1–18.

Marshall, L. G., 1977. A new species of *Lycopsis* (Borhyaenidae; Marsupialia) from the La Venta fauna (late Miocene) of Colombia, South America. Journal of Paleontology 51, 633–642.

Marshall, L. G., 1978. Evolution of the Borhyaenidae, extinct South American predaceous marsupials. University of California Publications in Geological Sciences 117, 1–89.

Marshall, L. G., 1979. Review of the prothylacyninae, an extinct subfamily of South American "dog-like" marsupials. Fieldiana, Geology New Series 3, 1–50.

Marshall, L. G., Cifelli, R. L., Drake, R. E., Curtis, G. H., 1984. Vertebrate paleontology, geology, and geochronology of the Tapera de Lopez and Scarritt Pocket, Chubut Province, Argentina. Journal of Paleontology 60, 920–951.

Matthew, W. D., 1915. A revision of the lower Eocene Wasatch and Wind River faunas, Part IV: Entelonychia, primates, insectivora. Bulletin of the American Museum of Natural History 34, 429–483.

Missiaen, P., Smith, T., Guo, D.-Y., Bloch, J. I., Gingerich, P. D., 2006. Asian gliriform origin for arctostylopid mammals. Naturwissenschaften 93, 407–411.

Mone, A., Urbilla, M., 1978. La edad Deseadense (Oligoceno Inferior) de la Formación Fray Bentos y su contenido paleontológico, con especial referencia a la presencia de *Proborhyaena* cf. *gigantea* Ameghino (Marsupialia:Borhyaenidae) en el Uruguay. Nota preliminary. Comunicaciones Paleontológicas del Museo de Historia Natural de Montevideo 1, 151–158.

Moore, D. M., 1978. Post-glacial vegetation in the South Patagonian territory of the giant ground sloth, *Mylodon*. Botanical Journal of the Linneaen Society 77, 177–202.

Pascual, R., Ortiz Jaureguizar, O. E., 1990. Evolving climates and mammal faunas in Cenozoic South America. Journal of Human Evolution 19, 23–60.

Patterson, B., Marshall, L., 1978. The Deseadan, early Oligocene, Marsupialia of South America. Fieldiana Geology 42, 37–100.

Patterson, B., Pascual, R., 1972. The fossil mammal fauna of South America. In: Keast, A., Erk, F. C., Glass, B. (Eds.), Evolution, Mammals, and Southern Continents. State University of New York Press, Albany, NY, pp. 247–309.

Patterson, B., Wood, A., 1982. Rodents from the Deseadan Oligocene of Bolivia and the relationships of the Caviomorpha. Bulletin of the Museum of Comparative Zoology 149, 371–543.

Pujos, F., De Iuliis, G., 2007. Late Oligocene Megatherioidea fauna (Mammalia: Xenarthra) from Sall-Luribay (Bolivia): new data on basal sloth radiation and Cingulata-Tardigrada split. Journal of Vertebrate Paleontology 27(1), 132–144.

Reguero, M., Cerdeño, E., 2005. New late Oligocene Hegetotheriidae (Mammalia, Notoungulata) from Salla, Bolivia. Journal of Vertebrate Paleontology 25(3), 674–684.

Reguero, M., Croft, D., Flynn, J. J., Wyss, A. R., 2003. Small archaeohyracids (Typotheria, Notoungulata) from Chubut Province, Argentina, and central Chile: implications for trans-Andean temporal correlation. Fieldiana (Geology) New Series 48, 1–17.

Reguero, M. A., 1999. El problema de las relaciones sistemáticas y filogenéticas de los Typotheria y Hegetotheria (Mammalia, Notoungulata): análisis de los taxones de Patagonia de la Edad-mamífero Deseadense (Oligoceno). Ph.D. dissertation. Universidad de Buenos Aires.

Rose, K. E., 1999. Postcranial skeleton of Eocene Lepticidae (Mammalia), and its implications for behavior and relationships. Journal of Vertebrate Paleontology 19, 355–372.

Rosenberger, A. L., Hartwig, W. C., Wolff, R. G., 1991. *Szalatavus attricuspis*, an early platyrrhine primate. Folia Primatolology 56, 225–233.

Ross, C. F., Lockwood, C. A., Fleagle, J. G., Jungers, W. L. 2002. Adaptation and behavior in the primate fossil record. In: Plavcan, J. M., Kay, R. F., Jungers, W. L., van Schaik, C. P. (Eds.), Reconstructing Behavior in the Primate Fossil Record. Kluwer/Plenum, New York, pp. 1–41.

Rudwick, M. J. S., 1964. The inference of function from structure in fossils. British Journal of Philosophy of Science 15, 27–40.

Salas, R., Pujos, F., de Muizon, C., 2005. Ossified meniscus and cyamofabella in some fossil sloths: a morpho-functional interpretation. Geobios 38, 389–394.

Sanchez-Villagra, M., Kay, R. F. 1997. A skull of *Proargyrolagus*, the oldest argyrolagid (late Oligocene Salla Beds, Bolivia), with brief comments concerning it paleobiology. Journal of Vertebrate Paleontology 17, 717–724.

Scott, W. B., 1903–1904. Mammalia of the Santa Cruz beds. I. Edentata. Reports of the Princeton University Expeditions to Patagonia, 1896–1899, Princeton and Stuttgard 5, 1–364.

Scott, W. B., 1912. Toxodonta of the Santa Cruz Beds. Reports of the Princeton University Expeditions to Patagonia, 1896–1899, Princeton and Stuttgard 6, 111–300.

Scott, W. M., 1930. A partial skeleton of *Homalodontotherium* from the Santa Cruz Beds of Patagonia. Memoire Field Museum of Natural History, Geology 1, 1–34.

Shockey, B. J., 1997. Two new notoungulates (Family Notohippidae) from the Salla Beds of Bolivia (Deseadan: Late Oligocene): Systematics and functional morphology. Journal of Vertebrate Paleontology 17, 584–599.

Shockey, B. J., 1999. Postcranial osteology and functional morphology of the Litopterna of Salla, Bolivia (late Oligocene). Journal of Vertebrate Paleontology 19, 383–390.

Shockey, B. J., 2001. Specialized knee joints in some extinct, endemic, South American herbivores. Acta Palaeontologica Polonica 46, 277–288.

Shockey, B. J., 2005. New leontiniids (Class Mammalia, Order Notoungulata) from the Salla Beds of Bolivia (Deseadan, late Oligocene). Bulletin of the Florida Museum of Natural History 45, 249–260.

Shockey, B. J., Anaya, F., 2004. *Pyrotherium macfaddeni*, sp. nov. (late Oligocene, Bolivia) and the pedal morphology of pyrotheres. Journal of Vertebrate Paleontology 24, 481–488.

Shockey, B. J., Hitz, R., Bond, M., 2004. Paleogene notoungulates from the Amazon Basin of Peru. Natural History Museum of Los Angeles County, Science Series 40, 61–69.

Shockey, B. J., Salas, R., Quispe, R., Flores, A., Sargis, E. J., Acosta, J., Pino, A., Jarica, N., Urbina, M., 2006. Discovery of Deseadan fossils in the Upper Moquegua Formation (late Oligcene-?early Miocene) of southern Perú. Journal of Vertebrate Paleontology 26, 205–208.

Shockey, B. J., Croft, D. A., Anaya, F., 2007. Analysis of function in absence of extant functional homologues: a case study of mesotheriid notoungulates. Paleobiology 33, 227–247.

Simpson, G. G., 1945. The principles of classification and a classification of mammals. Bulletin of the American Museum of Natural History 85, 1–350.

Simpson, G. G., 1970. The argyrolagidae, extinct South American marsupials. Bulletin of the Museum of Comparative Zoology 139, 1–86.

Simpson, G. G., 1980. Splendid Isolation: the Curious History of South American Mammals. Yale University Press, New Haven, CT.

Simpson, G. G., 1984. Discovers of the Lost World: an Account of Some of Those Who Brought Back to Life South American Mammals Long Buried in the Abyss of Time. Yale University Press, New Haven, CT.

Sinclair, E. J., 1906. Mammalia of the Santa Cruz beds: Marsupialia. Reports of the Princeton University Expeditions to Patagonia 4(3), 333–460.

Sinclair, E. J., 1909. Typotheria of the Santa Cruz Beds. Reports of the Princeton University Expeditions to Patagonia, 1896–1899, Princeton and Stuttgard 6, 1–110.

Solounias, N., Teaford, M., Walker, A., 1993. Interpreting the diets of extinct ruminants: the case of a non-browsing giraffid. Paleobiology 14, 287–300.

Soria, M. F., Alvarenga, H., 1989. Nuevos restos de mamíferos de la Cuenca de Taubaté, Estado de São Paulo, Brazil. Anais Académia Brasileira de Ciências 61, 157–175.

Soria, M. F., Hoffstetter, R., 1983. Présence d'un Condylarthre (*Salladolodus deuterotheriodes* gen. et sp. nov.) dan le Déséadien de Salla, Bolivie. Comptes rendus des séances de l'Académie des Sciences 297, 549–552.

Stehli, F. G., Webb, S. D., 1985. The Great American Biotic Interchange. Plenum, New York.

Stebbins, G. L., 1981. Coevolution of grasses and herbivores. Annals of the Missouri Botanical Garden 68, 75–86.

Sydow, H. K., 1988. Postcranial skeleton of *Trachytherus* (Mammalia, Notoungulata) with an evaluation of dentition. Masters thesis, Department of Geology, University of Florida, Gainesville, FL.

Szalay, F. S., 1985. Rodent and lagomorph morphotype adaptations, origins and relationships: some postcranial attributes analyzed. In: Luckett, W. P., Hartenbergber, J.-L. (Eds.), Evolutionary Relationships among Rodents – a Multidisciplinary Analysis. Plenum, New York.

Szalay, F. S., 1994. Evolutionary History of the Marsupials and an Analysis of Osteological Characters. Cambridge University Press, Cambridge.

Takai, M., Anaya, F., 1996. New specimens of the oldest fossil platyrrhine, *Branisella boliviana* from Salla, Bolivia. American Journal of Physical Anthropology 99, 301–318.

Villarroel, C., Marshall, L. G., 1982. Geology of the Deseadan (early Oligocene) age Estratos Salla in the Salla-Luribay Basin, Bolivia, with description of new Marsupialia. Geobios, Mémoire Spécial 6, 201–211.

Vizcaíno, S. F., Fariña, R. A., 1997. Diet and locomotion of the armadillo *Peltephilus*: a new view. Lethaia 30, 70–86.

Witmer, L. M., 1995. The extant phylogenetic bracket and the importance of reconstructing soft tissues in fossils. In: Thomason, J. J. (Ed.), Functional Morphology in Vertebrate Paleontology, Cambridge University Press, New York, pp. 19–33.

Wolff, R. G., 1984a. New specimens of the primate *Branisella boliviana* from the early Oligocene of Salla, Bolivia. Journal of Vertebrate Paleontology 4, 570–574.

Wolff, R. G., 1984b. New early Oligocene Argyrolagidae (Mammalia, Marsupialia) from Salla, Bolivia. Journal of Vertebrate Paleontology 4, 108–113.

8. Evolution of the Proximal Third Phalanx in Oligocene-Miocene Equids, and the Utility of Phalangeal Indices in Phylogeny Reconstruction

Jay A. O'Sullivan*
Department of Exercise Science and Sport Studies
University of Tampa
401 West Kennedy Boulevard
Tampa, FL 33606, USA
josullivan@ut.edu

8.1 Introduction

The late Oligocene – early Miocene of Florida contain *Miohippus, Archaeohippus, Anchitherium*, and *Parahippus*, equid genera that possess and define many of the character state transitions that occurred between advanced anchitheriine and primitive equine horses. Although much previous research regarding *Archaeohippus* has emphasized its uniqueness, the genus is equally interesting for those characters that suggest its affinities to other taxa.

The affinities of *Archaeohippus* are obscured in part by a complicated taxonomic history. Specimens of this small, brachydont Miocene horse were first mentioned in publication by Cope (1886) from the early Barstovian Mascall Fauna of Oregon. Cope named the species *ultimus*, and assigned it to *Anchitherium*, a genus of large, tridactyl Miocene horses with brachydont teeth. Osborn (1910) placed both *Anchitherium* and *Archaeohippus* in the grossly paraphyletic subfamily "Anchitheriinae," what I will refer to as "Anchitheriinae" *sensu lato* (ASL). Osborn's (1910) formulation of this subfamily also included *Mesohippus, Miohippus, Parahippus*, and *Hypohippus*, as well as the European palaeothere *Anchilophus*. The inclusion of the palaeothere renders this concept of the subfamily polyphyletic. In a more recent review of Osborn's grouping (MacFadden, 1992), ASL is defined as those horses with fully molarized P2-M3 that lack the dental characters that define the subfamily Equinae (Hulbert, 1989; Hulbert and MacFadden, 1991). This gradistic concept is MacFadden's (1992, 1998) paraphyletic "Anchitheriinae," derived from Osborn's (1910) definition and used by many museum collections today. Thus, ASL includes the late Eocene-Oligocene genus *Mesohippus*, its Oligocene-Miocene descendent *Miohippus*, and at least eight genera derived from one or more species of *Miohippus* (and perhaps *Mesohippus*). These taxa comprise the Arikareean anchithere radiation (AAR) of the New and Old Worlds, which begins in the late Oligocene (early Arikareean) and ends in the middle Miocene (late Clarendonian). The eight taxa in the AAR are *Archaeohippus, Desmatippus sensu* MacFadden (1998; *Anchippus sensu* Albright, 1998, 1999), and *Parahippus*, as well as the "Anchitheriinae" *sensu stricto* (ASS) of MacFadden (1992): *Anchitherium, Kalobatipus, Sinohippus, Megahippus*, and *Hypohippus*. All members of the AAR are united by possession of a connection between the metaloph and the ectoloph (Evander, 1989). This connection is absent in *Mesohippus* and only occasionally present in *Miohippus* (both members of ASL). All members of ASS are united by "greatly increased tooth crown area and estimated body size, relatively well developed cingula, and loss of ribs between styles on cheek teeth" (MacFadden, 1992, p. 101), as well as a mesentocuneiform facet on MTIII (Osborn, 1918). Generally, they are further distinguished from other members of the AAR by the possession of robust, strongly divergent lateral digits on the manus and pes. Thus, these groups, arranged from most to least inclusive, are ASL > AAR > ASS. Of these, only ASS is possibly holophyletic.

Although the systematics of the AAR is poorly understood, it has a sizeable fossil record. This record indicates an increase in morphological diversity unparalleled in the earlier evolutionary history of horses (Webb et al., 1995). Prior to the AAR, horse evolution in North America was much more conservative. Seminal work on the subject (Osborn, 1918; Matthew, 1924; Stirton, 1940) interprets this horse phylogeny as a series of gradistic genera, each genus distinguished from its ancestor by increased molar complexity and a slight increase in body size. This gradistic reconstruction of the

* Address for correspondence: josullivan@ut.edu

E.J. Sargis and M. Dagosto (eds.), *Mammalian Evolutionary Morphology: A Tribute to Frederick S. Szalay*, 159–165.

evolutionary history of early horses may be due in part to very gradual evolution, which produced only subtle morphological differences between different species. However, it may also reflect the perspectives of horse systematists. Much of the systematics of equids from the middle Miocene and younger is based on distinct characters in the complex occlusal surfaces of their molars, characters that are not always present in earlier equids. Another factor that explains the lack of resolution in the early part of the phylogeny of horses is the paucity of species-level cladistic studies on pre-Miocene taxa other than *Hyracotherium*. A recent revision of *Hyracotherium* (Froehlich, 1999) demonstrated that the traditional definition of this taxon includes a variety of primitive equids and other perissodactyls. Modern revisions of such taxa as *Epihippus, Orohippus, Mesohippus*, and *Miohippus* may similarly reveal more complex relationships than previously envisioned. Even considering this possibility, overall known equid morphological diversity was relatively low until the AAR was fully underway in the late Arikareean.

Certain lineages in the AAR demonstrate early phases of the trends in limb and tooth evolution that characterize the later radiation of advanced equids in the middle Miocene. The adaptive radiation of equine horses during the middle Miocene of North America is a well-studied macroevolutionary phenomenon that resulted in at least 11 late Miocene clades (Webb and Hulbert, 1986; MacFadden and Hulbert, 1988; Hulbert and MacFadden, 1991; Hulbert, 1993; for a review, see MacFadden, 1992). Morphological trends characteristic of this radiation include reduction of the side toes, elongation of distal limb elements, and increase in tooth crown height and occlusal complexity. These trends have been interpreted as adaptations to life in open country and a diet that included grasses (Marsh, 1879; Simpson, 1951; Janis, 1976; Behrensmeyer et al., 1992; Janis et al., 1994). These evolutionary trends can be traced back to the AAR in members of the genus *Parahippus*. Primitive members of this genus, sometimes assigned to the genus *Anchippus* (*sensu* Albright, 1998) or *Desmatippus* (*sensu* MacFadden, 1998), are dentally little more derived than advanced species of *Miohippus*. The most derived species of *Parahippus*, such as the Hemingfordian *P. leonensis*, possessed cheek teeth that were incipiently hypsodont and usually covered with cement. Its feet were tridactyl, but the lateral digits were reduced in length and thickness and held close to the middle digit, such that it was probably functionally monodactyl under normal locomotor conditions (Sondaar, 1968). These and other derived characters led Hulbert and MacFadden (1991) to identify *P. leonensis* as the nearest sister group of the middle Miocene adaptive radiation of equines.

The clade of large-bodied horses designated Anchitheriinae *sensu stricto* (ASS) is characterized by a suite of morphological trends that differ fundamentally from those that led to the advanced grazing horses (MacFadden, 1992, Figure 5.15, node 3). These include an increase in body size without an increase in relative crown height or occlusal complexity of the molars (MacFadden, 1992), and perhaps an even more functionally tridactyl foot than that seen in many species of *Miohippus*. The lateral metapodials and phalanges are very robustly built and the lateral metapodials are not firmly appressed to the medial metapodial (Sondaar, 1968). Whereas the morphology of *P. leonensis* suggests that it may have been an early inhabitant of the first North American savannas (Hulbert and MacFadden, 1991), the morphology of members of the ASS (brachydont teeth and splayed digits) reflects a continuation of the forest-dwelling ecology of earlier equids (Sondaar, 1968).

Archaeohippus is perhaps the most enigmatic genus in the AAR and shows an interesting mosaic of primitive and derived features. Among anchitheres, it possesses a unique facial morphology, including a long pre-orbital region of the skull with a deeply pocketed malar fossa confluent with a deep lacrimal fossa. It possessed primitively brachydont teeth, but with slightly more occlusal complexity than that seen in the ASS. However, its pedal adaptations are as advanced as those of *Parahippus* (Matthew, 1932; Sondaar, 1968), with strongly reduced lateral metapodials entirely attached by ligaments to the medial metapodial. In addition, *Archaeohippus* has been cited as an example of phyletic dwarfism (MacFadden, 1987, 1998). At approximately 20 kg (Janis et al., 1994), the estimated body weight of *Archaeohippus* is about half that of most species of *Miohippus*, the common equid of the late Oligocene. The ecology of *Archaeohippus* must have bridged that of the more ecologically distinct members of the AAR. Its primitively brachydont teeth indicate a diet of browse, like that of *Anchitherium*, whereas its limb morphology suggests an affinity for open country, like *Parahippus*.

As stated above, in the original description of the type species *Archaeohippus ultimus*, Cope (1886) assigned material from Cottonwood Creek, Oregon, to the genus *Anchitherium*. In his description of fossils from the same locality from the Mascall Fauna of Oregon, Gidley (1906) erected a new genus, *Archaeohippus*, to distinguish this small brachydont horse from the anchitheres *sensu stricto*. The next named species, the somewhat larger and younger *Archaeohippus mourningi* (Merriam, 1913), was originally assigned to *Parahippus. Archaeohippus penultimus* was described from the Sheep Creek of Nebraska by Matthew (1924). Hay (1924) described both *Miohippus blackbergi* and *Parahippus minutalis* from the Garvin Gully Local Fauna of Texas. Simpson (1932) described *A. nanus* from the Thomas Farm Local Fauna of Florida.

Matthew (1932) recognized the derived nature of the pes and manus shared by *Archaeohippus* and *Parahippus* and the facial fossa shared by *Archaeohippus* and *Parahippus pristinus*, and suggested that *Archaeohippus* was a subgenus of *Parahippus*. In an excellent synthesis, Bode (1933) rediagnosed the species *Archaeohippus ultimus, Archaeohippus penultimus*, and *Archaeohippus mourningi*, and defended the generic status of *Archaeohippus*. Schlaikjer (1935, 1937) considered *Archaeohippus blackbergi* (=*Archaeohippus minutalis*) to be a dwarf *Parahippus*, as did White (1942). White's (1942) justification was that individuals in the Thomas Farm population of *A. blackbergi* variably possess advanced dental characters such as a crochet, additional plications, and a hypostyle that connects to the ectoloph and metaloph to close the postfossette. White (1942, p. 19) noted that the patterns of variation of the dentitions of other species of *Archaeohippus* do not display these advanced characters, but are "simple and stable." Bode (1933) and Downs (1956) also noted that these characters were rare and weak when present in populations of *A. mourningi, A. penultimus*, and *A. ultimus*. Rather than accept the possibility that reduced variation in later species might involve the loss of advanced dental characters, Schlaikjer (1935, 1937) and White (1942) concluded that the other species of *Archaeohippus* were convergent with *A. blackbergi*. White (1942) considered "*Parahippus*" *blackbergi* to be an intermediate between *Miohippus* and more advanced *Parahippus*. White (1942) also identified several teeth from Thomas Farm lacking a metaloph connected to the ectoloph as belonging to *Miohippus*. However, this character is variable within individual dentitions and cannot be considered diagnostic when found in isolated teeth (Forsten, 1975).

Downs (1956) compared *A. blackbergi* with the western species and returned *A. blackbergi* to *Archaeohippus*. For the next two decades the debate was dropped, to resurface briefly in the work of Forsten (1975), who agreed with Downs (1956) that *A. blackbergi* is the correct name for the species of tiny horse found in both Texas and Florida during the Hemingfordian. More recently, Storer and Bryant (1993) identified *A. stenolophus* (Lambe, 1905) from the early Hemingfordian of Saskatchewan.

8.2 Abbreviations and Conventions Used in this Study

ADP = *Archaeohippus/Desmatippus/Parahippus* clade; ASL = "Anchitheriinae" *sensu lato*; ASS = Anchitheriinae *sensu stricto*; AAR = Arikareean Anchithere Radiation; FAM = Frick American Mammals, American Museum of Natural History; LSUMG = Louisiana State University Museum of Geoscience; Ma = Mega anna (millions of years ago), MCIII = metacarpal III; MTIII = metatarsal III; MCZ = Museum of Comparative Zoology, Harvard University; PPIIIL/MW = Proximal Phalanx III Length vs. Midshaft Width index; UF = University of Florida.

8.3 Specimens Used in this Study

Anchitherium clarencei: UF 175395, UF 58782, UF 47570, Thomas Farm, FL
cf. *Anchitherium*: UF uncatalogued, La Camelia Mine, FL
Anchippus texanus: LSUMG V-2258, LSUMG V-2549
Archaeohippus blackbergi: 101 phalanges, 37 uncatalogued UF, 64 lot catalogued as UF V-6414, Thomas Farm, FL
Archaeohippus mannulus: UF 160784, Curlew Creek, FL
Archaeohippus penultimus: FAM 71650, Thomson Quarry Sheep Creek, NE
Hypohippus wardi: uncatalogued FAM
Mesohippus bairdi: 3 phalanges lot catalogued as MCZ 20475, White River Badlands, SD
Mesohippus sp.: UF 200610, Toadstool Park, NE; UF 191530, Turkey Foot East High, NE; UF 191842, Horse Hill High NE; UF 208155, Suzan's Cat Site, NE; UF 208165, Sagebrush Flats, NE; UF 207944, Horse Hill New, NE; UF 207642, Sagebrush Flats 1, NE; UF 207923, Twin Buttes, NE; UF 207124, Horse Hill Low, NE; UF 201879, Twin Buttes, NE; UF 203240, Sagebrush Flats 1, NE; UF 209585, Twin Buttes, NE; UF 209566, Sand Creek Flats North, NE; UF 209584, Sagebrush Flats 2, NE
Miohippus intermedius: AMNH 1196 (cast), Protoceras Beds, White River, SD
Miohippus sp.: UF 200375, Turkey Foot, NE; UF 16872, I-75, FL; UF 163794, UF 178933, UF 178934, Brooksville 2, FL; UF/FGS V 3442, Franklin Phosphate, FL
Parahippus leonensis: UF 188515, UF 188711, UF 188022, UF 188418, UF 188021, UF 188776, UF 188497, UF 188020, UF 192872, UF 192325, UF 190381, UF 192873, UF 192621, UF 192620, UF 190361, UF 186430, UF 186431, UF 187542, UF 187715, UF 187716, UF 185568, UF 185890, UF 195591, UF 195004, UF 193194, UF 195001, UF 193030, UF 195059, UF 195003, UF 192975, UF 195002, UF 193031, Thomas Farm, FL
Parahippus pawniensis: FAM 71705, Elder Ranch, Dawes County, NE

8.4 Discussion

A phylogenetic analysis (O'Sullivan, 2002; in preparation) of 21 ASL equids and 62 characters in PAUP 4.04b4a for MacIntosh produced 106 shortest trees 190 steps long. A strict consensus tree (Figure 8.1) supports the monophyly of a clade that includes *Archaeohippus* and *Parahippus*, and excludes the ASS. The analysis included the Proximal Phalanx III Length vs. Midshaft Width index (PPIIIL/MW; see Table 8.1). A character analysis performed on MacClade 4.0 demonstrates that this index and several supporting dental characters define a clade including *Archaeohippus, Parahippus*, and primitive parahippines included in the genus *Desmatippus* (the ADP clade). All taxa within this clade possess derived elongate phalanges (Figure 8.2), and have a PPIIIL/MW index of 2.0 or greater.

The elongation of the proximal third phalanx is one of the most significant morphological developments in the complex of character transformations that signify the evolutionary transition among tridactyl equids from the digitigrade "pad-foot" to the unguligrade "springfoot", the acknowledged precursor to the monodactyl state found in modern *Equus* (Camp and Smith, 1942; Sondaar, 1968; Hussain, 1975;

FIGURE 8.1. Strict consensus tree of anchithere *sensu lato* relationships without the constraint of an outgroup. To the right of designated clades are silhouettes depicting proportions of proximal third phalanges of the types of *Mesohippus protoeulophus* (above; AMNH 524a) and *Archaeohippus mannulus* (below; UF 160784). The ASS clade is supported by derived dental character states including: (1) protoconule submerged in protoloph, and (2) metaconule submerged in metaloph. The ADP clade is supported by derived dental character states including (1) hypostyle tall and (2) hypostyle connected to the metaloph.

Thomason, 1986). The shared derived character complexes of the manus and pes of the springfoot equids was recognized by Matthew (1932). While the padfoot equids (in the form of the ASS clade) successfully radiated throughout Europe and Asia during the Miocene, they were less common in North America during this same period. Presumably, their autecology was better suited to the persistent forested biomes of the Old World than to savanna-dominated North America. The padfoot ASS clade equids went extinct during the middle Miocene as the springfoot equids of the ADP clade were experiencing unprecedented taxonomic diversity (see MacFadden, 1992 for an overview).

The dental characters that support the ADP clade pertain to trends in the development of the hypostyle and the protocone, and are rather subtle. The pedal adaptations, on the other hand, are easily evaluated with the PPIIIL/MW index. A phalanx with an index ≥ 2 came from a springfoot ADP equid, while a phalanx with an index <2 came from a padfoot equid, either a member of the ASS clade or a more plesiomorphic equid such as *Mesohippus*. Thus, a simple index of two linear measurements from one of the most common skeletal elements in the equid fossil record is a powerful indicator of one of the most significant ecomorphological transitions in the evolution of the Equidae, and of the evolution of the ancestors of the subfamily Equinae.

TABLE 8.1. Measurements and indices from phalanges of ASL equids included in this study.

	Proximal phalanx III		Proximal phalanx III	
Taxon	n	Length	Midshaft ML width	PPIIIL/MW
Anchippus texanus	1	24.6	12.1	2.03
Anchitherium clarencei	4	38.5	24.3	1.58
Archaeohippus blackbergi	101	25.1	10.4	2.41
Archaeohippus mannulus	1	16.8	7.5	2.24
Archaeohippus penultimus	1	27.2	12	2.27
Hypohippus wardi	1	40	25	1.60
Miohippus intermedius	1	21.6	12.1	1.79
Parahippus leonensis	32	31.9	14.6	2.18
Parahippus pawniensis	1	35	15.5	2.26
Mesohippus sp.	19	15.9	10.3	1.54
Miohippus sp. *(I-75)*	1	20.9	13	1.61
Brooksville 2	3	21.5	12.7	1.70
Franklin phosphate	1	27	15.7	1.72

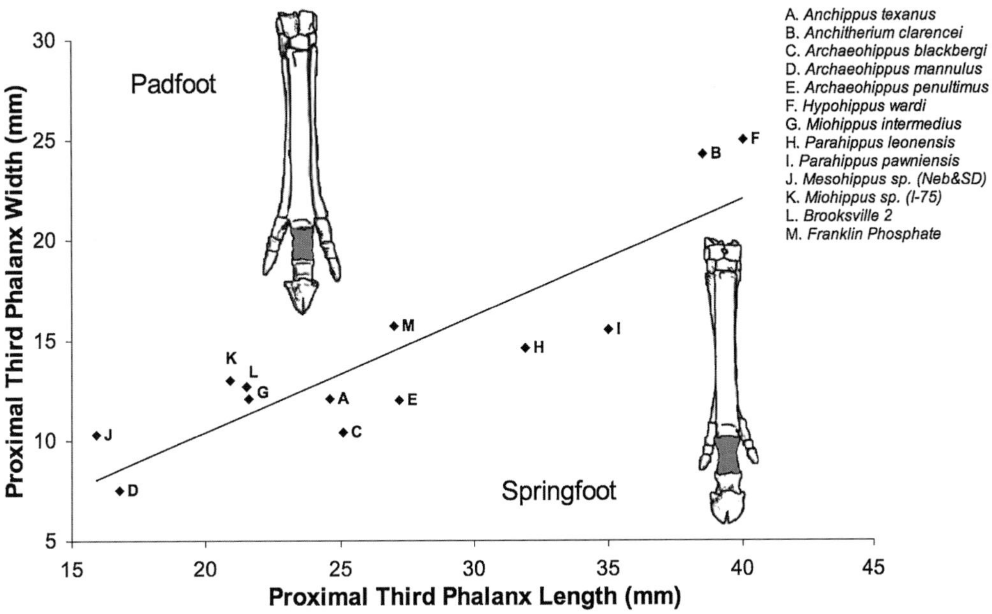

FIGURE 8.2. Plot of PPIIIL/MW index for ASL equids included in this study. The trend line differentiates the broad phalanx of the padfoot equid from the narrow waisted phalanx of the springfoot equid. The *Mesohippus* (padfoot) and *Merychippus* (springfoot) manual skeletons are redrawn from Simpson (1951).

8.4.1 Using Phalangeal Indices As a Diagnostic

The fossil record of Florida indicates that the peninsula has at times been an island, and perhaps at other times an archipelago (White, 1942; Frailey, 1980; Huddleston, 1993). The strong selectional imperatives of island endemism have been cited as influencing body size evolution in Florida equids (Frailey, 1980), possibly resulting in the extreme size reduction seen in the earliest known species of *Archaeohippus, A. mannulus* (O'Sullivan, 2003) from the Gulf Coast of Florida. In the phylogenetic analysis cited above, one species of *Parahippus, P. pawniensis*, usually nests within the *Archaeohippus* clade, not with the other parahippines. Therefore, *Archaeohippus* likely shares a common ancestor with *P. pawniensis*, a larger, fairly advanced equid, and is therefore a dwarfed parahippine itself.

Several other tiny fossil equids have been recovered from late Oligocene and early Miocene sediments in Florida. As absolute dating techniques are usually not available for these early terrestrial sites in Florida, the exact chronological relationships of these sites are unknown.

These unresolved chronological and geographical relationships could be very complex. For instance, if a small peninsular Florida during the Oligocene was subsequently subdivided into numerous islands, a single taxon of equids may have diverged simultaneously through this single vicariance event into multiple dwarf taxa. Conversely, minor cycles of sea level increase/decrease could have placed a sequence of different taxa on peninsular Florida, then isolated and dwarfed each. Rising sea levels might drive some of these dwarf taxa to extinction, while decreasing sea levels might permit reattachment to the mainland and emigration of island taxa (as is probably the case with *Archaeohippus*).

The above speculation on sea level change serves but one purpose in this discussion—to establish that there is no *a priori* reason to assume that small equids in Oligocene-Miocene Florida are closely related. While some may be assignable to *Archaeohippus*, instances of convergent body size evolution are also likely. When proximal phalanges are available for analysis, application of the PPIIIL/MW index can be used to assess the ADP affinities of these specimens.

The small equid from Brooksville 2 described by Hayes (2000) is within the size range of *A. blackbergi*. However, it possesses some plesiomorphic character states that exclude it from that genus. Among these is a PPIIIL/MW index of 1.70. Undescribed fossils from the Cowhouse Slough site of Hillsborough County (Albright, 1998) include elements from a small equid that cannot be included in the ADP due to its broad proximal phalanx III and metapodial. The Buda Local Fauna (Frailey, 1979) does not include a proximal third phalanx, but does include a fairly large, primitive proximal lateral phalanx with a *Miohippus* aspect to it. The Franklin Phosphate Local Fauna (Simpson, 1930) has several teeth assigned to *P. leonensis*. However, some plesiomorphic dental characteristics, plus a PPIIIL/MW index of 1.72, make this assignment unlikely. The PPIIIL/MW index in ADP equids is 2.0 or greater; thus, this equid is more likely an advanced *Miohippus*. A phalanx from the I-75 site (Patton, 1969) near Gainesville has an index of 1.6, and is therefore probably correctly identified as *Miohippus* sp., as it is definitely not an ADP equid.

The utility of this index is immediately apparent. Dental indices pertaining to crown height evolution are not reliable for Oligocene-early Miocene taxa. The dietary trends these indices pertain to were not yet established among equids. Except for the ASS equids, most AAR equids were probably mixed feeders (MacFadden, 1997, 2004), with specialist grazers evolving in the middle Miocene. In contrast, the pedal adaptations found in the ADP clade were already evident by the late Oligocene. Thus, the PPIIIL/MW index can distinguish advanced springfoot ADP equids from plesiomorphic padfoot equids among the taxa recovered from some of the earliest terrestrial fossil sites in Florida. Variation in this index in Oligocene-early Miocene equid taxa from the Florida Gulf Coast suggests the possibility that multiple, convergent dwarf taxa inhabited Florida. Thus, a single index, viewed as a proxy for a suite of pedal adaptations that occurred at the base of the equine radiation, provides improved resolution of phylogenetic relationships within the Equidae.

Acknowledgments. I would like to thank the editors of this Festschrift for inviting me to submit this contribution. Dr. Frederick S. Szalay was my M.A. advisor in the Department of Anthropology at Hunter College. I think it is fair to say that I never thought harder than in those days of trying to keep up with Fred. The observations contained in this present study were made while I was engaged in doctoral research at the University of Florida. While there I studied equids—not one of Fred's groups—but his influence on my thought processes was inescapable. It is no accident that I looked to the feet when the dentition proved uncooperative. This study also reflects two additional influences of Professor Szalay, for it was he who introduced me to the use of indices and acronyms (see O'Sullivan [1995] for gratuitous use of both). I would also like to thank Dr. Bruce J. Shockey and an anonymous reviewer for their many helpful comments.

References

Albright, L. B., III, 1998. The Arikareean Land Mammal Age in Texas and Florida: southern extension of Great Plains faunas and Gulf Coastal Plain endemism. In: Terry, D. O., Jr., LaGarry, H. E., Hunt, R. M., Jr. (Eds.), Depositional Environments, Lithostratigraphy, and Biostratigraphy of the White River and Arikaree Groups (Late Eocene to Early Miocene, North America). Geological Society of America Special Paper 325, Boulder, Colorado, pp. 167–183.

Albright, L. B., III, 1999. Ungulates of the Toledo Bend Local Fauna (late Arikareean, early Miocene), Texas coastal plain. Bulletin of the Florida Museum of Natural History 42, 1–80.

Behrensmeyer, A. K., Damuth, J. D., DiMichele, W. A., Potts, R., Wing, S. L., 1992. Terrestrial Ecosystems Through Time: Evolutionary Paleoecology of Terrestrial Plants and Animals. University of Chicago Press, Chicago, IL.

Bode, F. D., 1933. Anchitheriine horses from the *Merychippus* zone of the North Coalinga district, California. Carnegie Institution of Washington 440, 43–58.

Camp, C. L., Smith, N., 1942. Phylogeny and functions of the digital ligaments of the horse. University of California Memoirs 13, 69–124.

Cope, E. D., 1886. On two new species of three-toed horses from the upper Miocene, with notes on the fauna of the *Ticholeptus* beds. Proceedings of the American Philosophical Society 23, 357–361.

Downs, T., 1956. The Mascall fauna from the Miocene of Oregon. University of California Publications in Geological Science 31, 199–354.

Evander, R. L., 1989. Phylogeny of the Equidae. In: Prothero, D. R., Schoch, R. M. (Eds.), The Evolution of Perissodactyls. Oxford University Press, New York, pp. 109–127.

Forsten, A., 1975. The fossil horses of the Texas Gulf Coastal Plain: A revision. Texas Memorial Museum, Pearce-Sellards-Series 22, 1–87.

Frailey, D., 1979. The large mammals of the Buda Local Fauna (Arikareean: Alachua County, Florida). Bulletin of the Florida State Museum, Biological Sciences 24, 123–173.

Frailey, D., 1980. The beginning of the age of mammals in Florida. The Plaster Jacket 33, 4–13.

Froehlich, D. J., 1999. Phylogenetic systematics of basal perissodactyls. Journal of Vertebrate Paleontology 19, 140–159.

Gidley, J. W., 1906. A new genus of horse from the Mascall beds, with notes on the small collections of equine teeth in the University of California. Bulletin of the American Museum of Natural History 22, 385–388.

Hay, O. P., 1924. Description of some fossil vertebrates from the upper Miocene of Texas. Proceedings of the Biological Society of Washington 37, 1–20.

Hayes, F. G., 2000. The Brooksville 2 Local Fauna (Arikareean, latest Oligocene): Hernando County, Florida. Bulletin of the Florida Museum of Natural History 43, 1–47.

Huddleston, P. F., 1993. Revision of the lithostratigraphic units of the coastal plain of Georgia – the Oligocene. Georgia Geologic Survey Bulletin 105, 1–152.

Hulbert, R. C., Jr., 1989. Phylogenetic interrelationships and evolution of North American late Neogene Equidae. In: Prothero, D. R., Schoch, R. M. (Eds.), The Evolution of Perissodactyls. Oxford University Press, New York, pp. 176–196.

Hulbert, R. C., Jr., 1993. Taxonomic evolution in North American Neogene horses (subfamily Equinae): the rise and fall of an adaptive radiation. Paleobiology 19, 216–234.

Hulbert, R. C., Jr., MacFadden, B. J., 1991. Morphological transformation and cladogenesis at the base of the adaptive radiation of Miocene hypsodont horses. American Museum Novitates 3000, 1–61.

Hussain, S. T., 1975. Evolutionary and functional anatomy of the pelvic limb in fossil and recent Equidae (Perissodactyla, Mammalia). Anatomy, Histology, Embryology 4, 179–222.

Janis, C. M., 1976. The evolutionary strategy of the Equidae and the origins of rumen and caecal digestion. Evolution 30, 757–774.

Janis, C. M., Gordon, I. J., Illius, A. W., 1994. Modelling equid/ruminant competition in the fossil record. Historical Biology 8, 15–29.

Lambe, L. M., 1905. Fossil horses from the Oligocene of the Cypress Hills, Assiniboia. Transactions, Royal Society of Canada, Series 2, 11, 43–52.

MacFadden, B. J., 1987. Fossil horses from "*Eohippus*" (*Hyracotherium*) to *Equus*: scaling, Cope's Law, and the evolution of body size. Paleobiology 12, 355–369.

MacFadden, B. J., 1992. Fossil Horses: Systematics, Paleobiology, and Evolution of the Family Equidae. Cambridge University Press, Cambridge.

MacFadden, B. J., 1997. Origin and evolution of the grazing guild in New World terrestrial mammals. Trends in Ecology and Evolution 12, 182–186.

MacFadden, B. J., 1998. Equidae. In: Janis, C. M., Scott, K. M., Jacobs, L. L. (Eds.), Evolution of Tertiary Mammals of North America, Volume One: Terrestrial Carnivores, Ungulates, and Ungulatelike Mammals. Cambridge University Press, New York, pp. 537–559.

MacFadden, B. J., 2004. Fossil horses – evidence for evolution. Science 307, 1728–1730.

MacFadden, B. J., Hulbert, R. C., Jr., 1988. Explosive speciation at the base of the adaptive radiation of Miocene grazing horses. Nature 336, 466–468.

Marsh, O. C., 1879. Polydactyle horses, recent and extinct. American Journal of Science 17, 499–505.

Matthew, W. D., 1924. Third contribution to the Snake Creek fauna. Bulletin of the American Museum of Natural History 50, 59–210.

Matthew, W. D., 1932. New fossil mammals from the Snake Creek quarries. American Museum Novitates 540, 1–8.

Merriam, J. C., 1913. New anchitheriine horses from the Great Basin area. University of California Publications, Bulletin of the Department of Geology 7, 419–434.

Osborn, H. F., 1910. The Age of Mammals in Europe, Asia, and North America. Macmillan, New York.

Osborn, H. F., 1918. Equidae of the Oligocene, Miocene, and Pliocene of North America. Iconographic type revision. Memoirs of the American Museum of Natural History, New Series 2, 1–326.

O'Sullivan, J. A., 1995. Skeletal indicators of arboreal and terrestrial positional behavior in didelphid marsupials. M.A. thesis, Hunter College, City University of New York, New York.

O'Sullivan, J. A., 2002. Paleobiology of *Archaeohippus* (Mammalia; Equidae), a three-toed horse from the Oligocene-Miocene of North America. Ph.D. dissertation, University of Florida, Gainesville, FL.

O'Sullivan, J. A., 2003. A new species of *Archaeohippus* (Mammalia, Equidae) from the Arikareean of central Florida. Journal of Vertebrate Paleontology 23, 877–885.

Patton, T. H., 1969. An Oligocene land vertebrate fauna from Florida. Journal of Paleontology 43, 543–546.

Schlaikjer, E. M., 1935. Contributions to the stratigraphy and paleontology of the Goshen Hole area, Wyoming IV: new vertebrates and the stratigraphy of the Oligocene and early Miocene. Bulletin of the Museum of Comparative Zoology 76, 97–189.

Schlaikjer, E. M., 1937. A study of *Parahippus wyomingensis* and a discussion of the phylogeny of the genus. Bulletin of the Museum of Comparative Zoology 80, 255–280.

Simpson, G. G., 1930. Tertiary land mammals of Florida. Bulletin of the American Museum of Natural History 59, 149–211.

Simpson, G. G., 1932. Miocene land mammals from Florida. Florida Geological Survey Bulletin 10, 7–41.

Simpson, G. G., 1951. Horses. Oxford University Press, New York.

Sondaar, P. Y., 1968. The osteology of the manus of fossil and recent Equidae, with special reference to phylogeny and function. Verhandelingen der Koninklijke Nederlandse Akademie van Wetenschappen, Afd. Natuurkunde eerste reeks 25, 1–76.

Stirton, R. A., 1940. Phylogeny of North American Equidae. University of California Publication, Bulletin of the Department of Geological Science 25, 165–198.

Storer, J. E., Bryant, H. N., 1993. Biostratigraphy of the Cypress Hills Formation (Eocene to Miocene), Saskatchewan: equid types (Mammalia; Perissodactyla) and associated faunal assemblages. Journal of Paleontology 67, 660–669.

Thomason, J. J., 1986. The functional morphology of the manus in the tridactyl equids *Merychippus* and *Mesohippus*: paleontological inferences from neontological models. Journal of Vertebrate Paleontology 6, 143–161.

Webb, S. D., Hulbert, R. C., Jr., 1986. Systematics and evolution of *Pseudhipparion* (Mammalia, Equidae) from the Late Neogene of the Gulf Coastal Plain and Great Plains. In: Flanagan, K. M., Lillegraven, J. A. (Eds.), Vertebrates, Phylogeny, and Philosophy. Contributions in Geology, University of Wyoming, Special Paper 3, Laramie, pp. 237–285.

Webb, S. D., Hulbert, R. C., Jr., Lambert, W. D., 1995. Climatic implications of large-herbivore distributions in the Miocene of North America. In: Vrba, E. S., Denton, G. H., Partridge, T. C., Burckle, L. H. (Eds.), Paleoclimate and Evolution with Emphasis on Human Origins. Yale University Press, New Haven, CT, pp. 91–108.

White, T. E., 1942. The lower Miocene mammal fauna of Florida. Bulletin of the Museum of Comparative Zoology 92, 1–49.

9. Adaptive Zones and the Pinniped Ankle: A Three-Dimensional Quantitative Analysis of Carnivoran Tarsal Evolution

P. David Polly*
Department of Geological Sciences
Indiana University
1001 East 10th Street
Bloomington, IN 47405, USA
pdpolly@indiana.edu

9.1 Introduction

History repeats the old conceits, the glib replies, the same defeats
Elvis Costello (1982), Beyond Belief

Bones are functional. Stated so abruptly, this observation is a truism, but its significance depends on the context in which it is made. In an individual animal, bones support loads, resist muscular contractions, and facilitate bodily movements. Bone form both constrains, and is shaped by, force and motion. In an environmental context, a bone's form is compatible with its owner's size and habits and is, thus, related indirectly to habitat and environment, although any particular bone (or, more properly, musculoskeletal configuration) can cope in diverse environments, and any substrate can be traversed by animals with different skeletal forms. Form and function are inseparable at the level of joint movements (Bock and von Wahlert, 1965), but they are only loosely correlated at the level of ecology, specifically locomotor types and habitats. The coarseness of the correlation between form and ecology come from the temporal lag of phylogenetic adaptation and the many-to-many relationship between form and habitat. Even though ecophenotypic plasticity allows bones to be modified during an individual's lifetime, bone form is largely heritable and evolutionary change requires generations of selective genetic and epigenetic reorganization (Cock, 1966; Grüneberg, 1967; Thorpe, 1981).

Some have argued that functional adaptation is incompatible with phylogenetic reconstruction. Assertions as to whether skeletal variation is primarily phylogenetic (Acero et al., 2005) or functional (Nadal-Roberts and Collard, 2005) color the entire canon of systematic literature, but the nature of the conflict is murky – functional adaptation is a phylogenetic process and the phylogenetic transformation of bone form does not occur outside a functional context. On one hand, all adaptive specializations, even those shared by different clades, arise phylogenetically, but on the other, no bone character is functionally neutral. The question, then, is not whether skeletal characters are functional – they are – but to what extent adaptation masks phylogenetic history, how the convergences can be recognized, whether adaptation impedes phylogeny reconstruction, and how the interplay between form, function, and phylogeny can be better understood. These questions are the main subject of this paper.

In this paper, function and phylogeny were analyzed using a new geometric morphometric technique that quantitatively represents the entire three-dimensional surface of the bones. This method was used to associate variation in the two bones, including the size and curvature of occluding joint facets, with locomotor type, stance, number of digits, and body mass. Principal components analysis was used to describe the major axes of variation in the two bones, and multivariate analysis of variance was used to test functional categories for significance. Correlated transformations in the interlocking surfaces of the two bones were also explored using two-block partial least squares. Phylogenetic components of variation were assessed by mapping the three-dimensional shape of the bones onto a cladogram and projecting the results back into the principal component morphospace to visualize the patterns of homoplasy. Rates of morphological evolution in the several clades were calculated from the mapped shapes. Homoplasy was also quantitatively assessed by measuring the scaling coefficient between evolutionary divergence and time since common ancestry. The final aim of this paper was to develop criteria for assessing the whether functional adaptation is likely to confound phylogenetic signal in a dataset for the taxa being considered. A quantitative redefinition of Simpson's adaptive zones was employed to assess the effect of adaptive convergence on phylogenetic divergence, and determine the circumstances in which associated homoplasy is likely to confound phylogeny reconstruction.

* Address for correspondence: pdpolly@indiana.edu

E.J. Sargis and M. Dagosto (eds.), *Mammalian Evolutionary Morphology: A Tribute to Frederick S. Szalay*, 167–196.

9.1.1 Adaptive Zones, Function, and Phylogeny

Adaptive zones are arguably the key to interpreting the relationship of form, function, and phylogeny. Simpson (1944, 1953) introduced the concept of adaptive zones for environmental spaces that accommodate several evolving lineages. Species inside a zone have latitude for phenotypic evolution and speciation, but those that evolve too near the zone's margins are normally weeded out by selection. Simpson hypothesized that new zones are colonized when rapid bursts of 'quantum' evolution propel a species across the selectively disadvantageous space between zones. Any lineage that escapes one zone for another will do so by evolving a new phenotype compatible with its new environment; those that do not escape will remain constrained within the zone's range of phenotypes. The adaptive zone underpinned Simpson's (1945) taxonomic concept of a phenotypically coherent, paraphyletic group, and rejection of that concept by cladists in the 1970s conflated adaptive zones and plesiomorphic characters (Rosen, 1974). During the same period, quantitative geneticists developed sophisticated mathematical models of adaptive peaks, adaptive landscapes, and trait covariances that are conceptually related to adaptive zones, but whose focus was not taxonomic (Lande, 1976, 1986; Kirkpatrick, 1982; Wake et al., 1983; Wright, 1988; Cheverud, 1996; Schluter, 1996; MacLeod, 2002; Polly, 2004, 2005; Salazar-Ciudad and Jernvall, 2004; Zelditch et al., 2004). Arnold et al. (2001) summarized much of the latter literature. In this paper, I will explore adaptive zones from the conceptual advantage of the quantitative genetics models, but apply them to problems of phylogeny.

For the purposes of this study, I operationally redefine adaptive zone as a bounded range of phenotypes that can be linked to a recognizable functional roles. Phenotypes are emphasized because they can be measured more objectively than the seemingly endless number of potentially constraining environmental variables like the ones suggested by Simpson (1944, 1953); testing selected a set of phenotypes for association with specific environmental variables is more feasible than testing a block of environmental variables for association with all phenotypes. Adaptive zone boundaries are recognized not as gaps in phenotype distribution – these, too, are difficult to objectively measure – but as margins of phenotypic space that are not normally crossed by phylogenetic trajectories and outside which the phenotype is incompatible with the range of functional parameters associated with the zone. Those lineages which do cross the boundaries must do so in association with changes in both their functional ecology and their phenotype if the zone is to be considered an adaptive one. Within the zone boundaries, we can expect that, given enough phylogenetic time, clades will have crisscrossed the phenotypic space as they converged on the finite number of functionally compatible morphologies. Thus, to recognize an adaptive zone using this operational definition: (1) the zone must be occupied, (2) the zone's occupants must have diversified throughout the zone, and (3) the phylogeny of the occupants must be at least partly known. Unoccupied or recently occupied zones may exist, but they cannot be unambiguously identified from the viewpoint adopted here.

The implication for phylogeny reconstruction is that long term occupation of an adaptive zone will increase the likelihood of convergent evolution, with phenotypic reticulation within the zone obscuring the recoverable phylogenetic history. Clades that escape from the adaptive zone will be difficult to associate with their closest relatives within it. Phylogenetic reconstruction will not be adversely affected if a clade is not evolving within the confines of an adaptive zone or if the zone is recently occupied and has not been fully explored. In those cases, phenotypic convergence will be no more common than expected by chance and phylogeny will be recoverable.

9.1.2 Fissipeds and Pinnipeds

Adaptive zones and phylogeny are explored in the Carnivora. This placental mammal order is diverse today, has a plentiful, well-studied fossil record, has a well-understood phylogeny, has evolved many locomotor types, and has been subject to numerous locomotor functional studies. Living carnivorans are divided phylogenetically into two major clades, Feliformia (or Aeluroidea), containing felids, hyaenids, viverrids, and herpestids, and Caniformia, containing canids (Cynoidea), ursids, mustelids, and procyonids (Arctoidea), and pinnipeds. Functionally, Pinnipedia (seals, sea lions, and walruses) are often distinguished from their more terrestrial kin, known collectively as Fissipedia. Some classification schemes, including Simpson's (1945), awarded pinnipeds a coordinate taxonomic rank to other carnivorans, even though the close relationship of pinnipeds to arctoid caniforms has never been seriously questioned.

Pinnipeds are specialized in many ways for their marine lifestyle (Howell, 1930; Bininda-Emonds and Gittleman, 2000), but most important for this paper is their derived locomotor morphology (Figure 9.1). Pinnipeds use their limbs for propulsion in the water – sea lions (Otariidae) use their fore limbs, seals (Phocidae) their hind limbs, and walruses (Odobenidae) both. The digits of both limbs are modified from the terrestrial condition into long flippers that trail fully extended behind the body while swimming. Sea lions and walruses are capable of dorsiflexing the hind foot into a plantigrade position, but seals are prevented from doing so by modifications of the tarsals and the attached tendons and ligaments (Howell, 1929, 1930). The proximal tarsal bones – calcaneum and astragalus – of pinnipeds thus differ in form and function from those of fissiped mammals. A further aim is to compare pinniped and fissiped tarsals in their functional and phylogenetic context, assessing evidence for a fissiped adaptive zone and a higher rate of tarsal evolution in the lineage leading to pinnipeds.

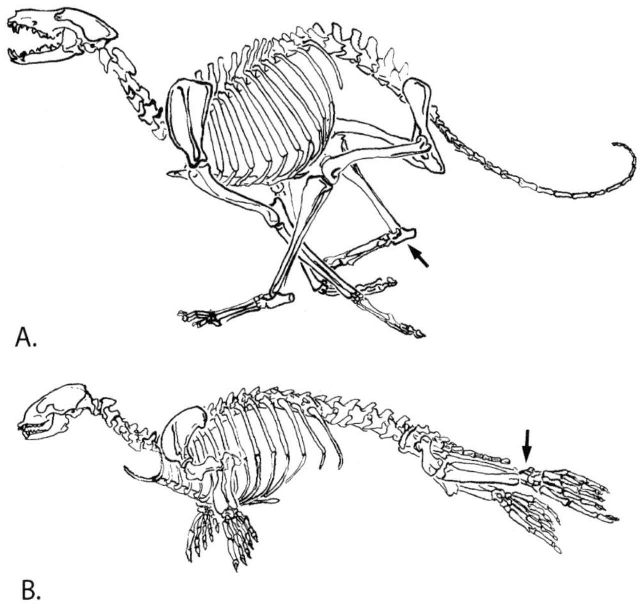

FIGURE 9.1. Skeleton of *Canis familiaris*, the domestic dog (A) and *Phoca vitulina*, the Harbour seal (B). (After Gregory, 1951.) Arrows point to the calcaneum and astragalus.

The diverse morphologies of living carnivorans are the result of approximately 40 million years of evolution. The earliest members of Carnivora, the viverravids, are older than that, coming from the late Paleocene of North America (Fox and Youzwyshyn, 1994; Polly, 1997); however, it is likely that these animals lie outside the crown group (Gingerich and Winkler, 1985; Wesley-Hunt and Flynn, 2005) making the age of the last common ancestor of the living species about 40 Ma (Megannum, or million years ago; Flynn, 1996).

9.1.3 Szalayian Analysis

The issues explored in this paper are not new. In paleontology, Szalay especially has written on the role of function in phylogeny reconstruction (Szalay, 1977a,b, 1981, 1994, 2000; Szalay and Bock, 1991; Szalay and Schrenk, 1998; Salton and Szalay, 2004). Building on the work of Bock (1965; Bock and Von Wahlert, 1965), Szalay advocated that functional analysis is required for assessing phylogenetic transformations. All phylogenetic hypotheses are, explicitly or implicitly, statements about character transformation. One of Szalay's points was that transitions in multiple characters must be functionally compatible with one another. Phylogenetic algorithms that treat characters as independent may produce biologically impossible transitions if functional integration was not considered at the stage of character definition. The use of character complexes and the rejection of automated tree optimization algorithms are major planks of the Szalay platform. For Szalay, the character complex is a rich, multidimensional source of phylogenetic data whose states can be compared among taxa, whose transformations can be tested for functional compatibility, and whose form has biological meaning, especially for reconstructing the lifestyles of incompletely preserved fossil taxa. The masticatory system, especially occluding cheek teeth, and the limbs, particularly interlocking tarsal bones, are conducive to Szalayian analysis because of their physical integration, their direct relevance to an animal's lifestyle, and their common preservation as fossils. An additional goal of this paper is to quantify the Szalayian analysis of tarsal evolution using geometric morphometrics. These techniques are ideal for quantitatively representing three-dimensional morphology, assessing correlations, testing associations with extrinsic functional and ecological data, and building trees (Rohlf and Slice, 1990; Bookstein, 1991; MacLeod and Rose, 1993; Dryden and Mardia, 1998; MacLeod, 1999; Rohlf and Corti, 2000; Polly, 2003a,b). Specifically, I will analyze the major patterns of covariation within the carnivore calcaneoastragalus complex to extract components associated with locomotor types, stance, number of digits, body mass, and the covariation of occluding surfaces of the two bones. I will explore the structural transformations of the complex in the context of current understanding of the phylogeny to assess the conditions under which quantitative phylogenetic reconstruction will be accurate.

For this paper, I developed a method for the three-dimensional geometric analysis of bone surfaces. As currently practiced, geometric morphometrics is limited to using a few landmark points or outline curves to represent a complex morphological structure, but functional bone characters are best studied as parts of complete three-dimensional structures. The technique used here is fundamentally the same as the geometric analysis of landmarks and outlines, but it uses as its data the complete surface of an object rather than the limited representation derived from points or curves. An advantage of this approach is that joint surfaces and bony processes are fully incorporated in the analysis and depicted in the results, making it possible to assess the quantitative results of statistical manipulations with the same visual criteria that would be used for a physical bone. A second advantage, and perhaps the more important one, is that the analysis can be applied to any homologous bone, regardless of its derived evolutionary transformations. Standard measurement or landmark morphometrics can only be applied when each bone has precisely the same component structures. If, for example, a bony process is present in only a few taxa, it cannot normally be incorporated into a quantitative analysis of all taxa. By analyzing the complete surface of homologous bones, the presence or absence of individual structures does not impede quantitative analysis so long as the bones can be placed in comparable orientations.

9.1.4 Tarsal Morphology and Function

The calcaneum and astragalus (or talus) will be analyzed in this paper (Figure 9.2). These two bones are the largest of the tarsals, or ankle bones, and lie distal to the tibia and fibula and proximal to the rest of the foot. The upper ankle joint (UAJ) lies proximal to them, formed by the large,

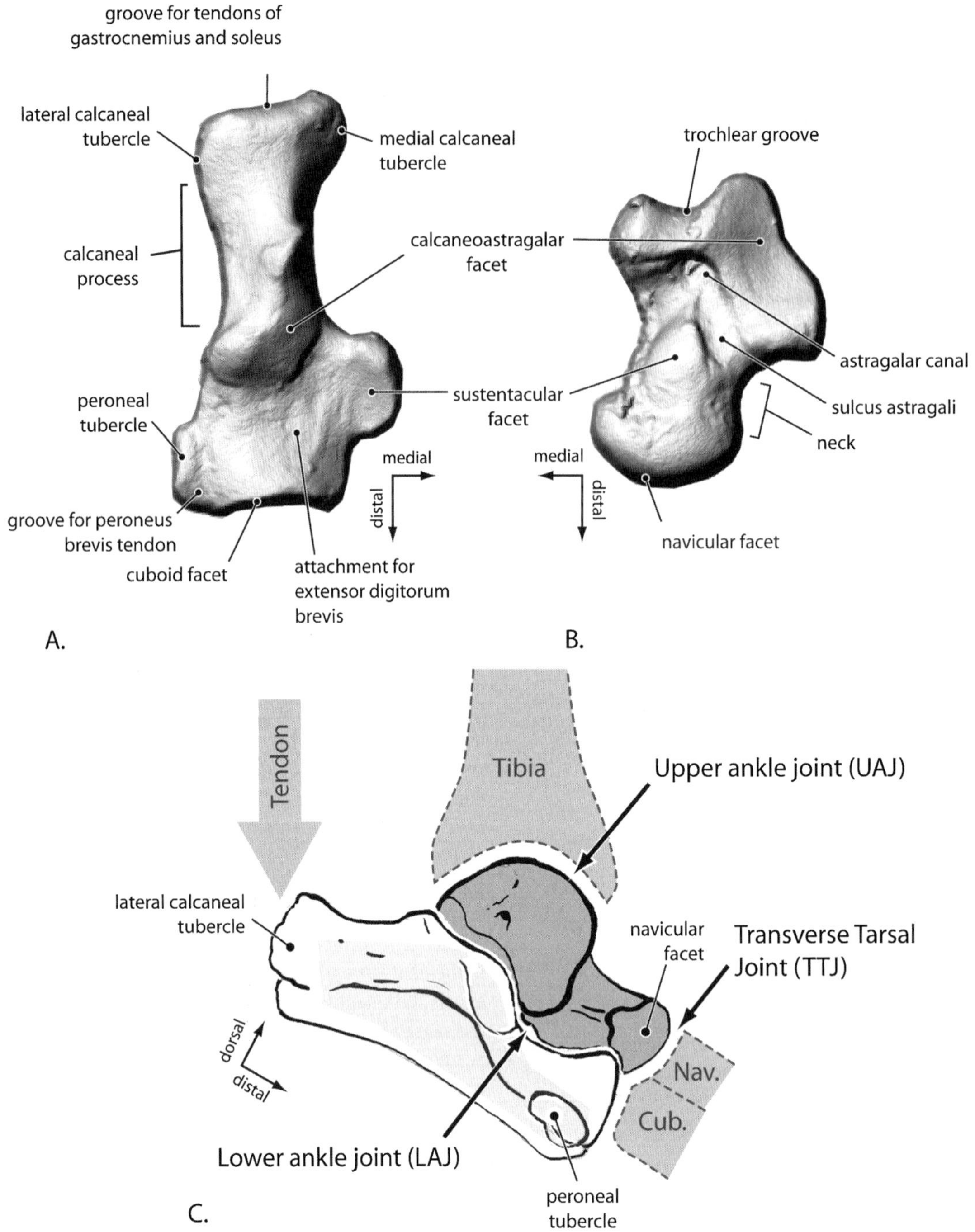

FIGURE 9.2. Morphology of the right calcaneum (A, dorsal view) and astragalus (B, plantar view) of the Badger, *Meles meles*. The bones fit together by flipping the astragalus over to the left so that its calcanealastragalar and sustentacular facets occlude against those with the same name on the calcaneum (C). The sustentacular and calcaneoastragalar facets of the two bones are the primary contact surfaces of the lower ankle joint.

dorsal trochlea of the astragalus and the concave, contoured surface on the distal ends of the tibia and fibula. In carnivorans, movement at this joint is associated with dorsiflexion and plantarflexion of the foot. Plantarflexion, an important propulsive movement on land or in water, is powered by the large gastrocnemius and soleus muscles, which insert at the proximal end of the calcaneal process. The astragalar trochlea forms the fulcrum of the joint, and the calcaneal process the moment arm of effort. The lower ankle joint (LAJ) lies between the astragalus and calcaneum and moves along two pairs of occluding facets. The calcaneoastragalar facets lie on the medial margin of the ventral side of the astragalus body and the medial side of the dorsal calcaneum. These are usually long and proximodistally oriented on both bones, with the astragalar facet concave and the calcaneal one convex. The sustentacular facets lie on the ventral side of the astragalus neck and on the dorsal side of the sustentacular process on the medial side of the distal calcaneum. These are usually subcircular in outline, with the astragalar facet convex and the calcaneal facet concave. Movement at the LAJ varies considerable among carnivorans. In terrestrial digitigrade taxa the LAJ may be tightly interlocked, allowing only small movements, but in arboreal taxa considerable movement at the LAJ may be associated with foot inversion. The transverse tarsal joint (TTJ) lies between the astragalus head and distal calcaneum on the one hand, and the proximal faces of the navicular and cuboid bones on the other. Movement at the TTJ also varies in Carnivora, but includes foot inversion and eversion. Other functionally important features of the tarsals include the medial and lateral calcaneal tubercles, which are sites of origination for flexors of the digits and sites of insertion for the plantarflexors of the ankle. The peroneal tubercle has grooves on the dorsal and plantar sides for the peroneus brevis and longus respectively and functions in flexion and inversion-eversion of the foot. The proportional lengths of the calcaneal process, the distal calcaneum (anterior to the calcaneoastragalar facet), and the astragalar neck are related to the lever advantages of dorsiflexion and plantarflexion. The breadth of the distal calcaneum, especially as contributed by the peroneal tubercle and sustentacular process, are partly related to lever advantages of foot inversion and eversion. Taylor (1989) reviewed locomotor morphology in Carnivora. Important original studies of carnivoran hind limb functional morphology include Howell (1929), Hildebrand (1954), Taylor (1970, 1976, 1988), Howard (1973), Jenkins and Camazine (1977), Goslow and Van de Graff (1982), Jenkins and McClearn (1984), Van Valkenburgh (1985), and Evans (1993). General references on functional morphology of the hindlimb in mammals include Howell (1944), Clevedon-Brown and Yalden (1973), Gambaryan (1974), Szalay (1977a, 1994), Lewis (1989), and Alexander (2003).

9.2 Materials and Methods

9.2.1 Materials

Twelve species were chosen to represent carnivoran phylogenetic and locomotor diversity (Table 9.1). All family-level groups except Ursidae were included, with representation balanced across the three major divisions Aeluroidea, Canoidea, and Arctoidea. *Phoca groenlandica*, the Harp seal, represented the pinnipeds. Seals have the most highly derived locomotor morphology of the pinnipeds (King, 1966; Wyss, 1988) and thus maximize differences between the pinniped and fissiped morphologies, giving the best chance to detect a high rate of phenotypic divergence.

Associated data were taken from published literature and personal observations. Average body mass was calculated from data compiled by Silva and Downing (1995), except for *Bassaricyon* (Eisenberg, 1989), and the domestic Greyhound (Kennel Club, 1998). Locomotor types follow Van Valkenburgh (1985) and Taylor (1976, 1989). Terrestrial animals spend most of their time on the ground (e.g., dogs and hyenas); scansorial animals spend considerable time on the ground, but are also good climbers (e.g., most felids); arboreal animals spend most of their time in trees (e.g., olingos, red pandas); natatorial animals spend time in both the water and on land (e.g., otters); and aquatic animals spend most time in water and are only capable of awkward locomotion on land (e.g., seals, sealions). Stance refers to the position of the heel during normal locomotion (Clevedon Brown and Yalden, 1973; Gambaryan, 1974; Gonyea, 1976; Hildebrand, 1980). Plantigrade animals walk with their heels touching the ground (e.g., red pandas); semidigitigrade animals often keep their heels elevated during locomotion (e.g., many mustelids); and digitigrade animals always have their heels elevated during normal locomotion, using the metatarsus as an additional limb segment (e.g., dogs, felids). The combination of stance and locomotor type distinguishes some common categories, such as ambulatory from cursorial. Pinnipeds do not fall into normal stance categories and so have been classified as

TABLE 9.1. Carnivoran species and associated data used in this study.

Species	Common name	Family	Body mass (kg)	Stance	Digits	Locomotor type
Ailurus fulgens	Red panda	Ailuridae	5.1	Plantigrade	5	Arboreal
Bassaricyon gabbii	Bushy-tailed olingo	Procyonidae	1.2	Plantigrade	5	Arboreal
Canis familiaris	Dog (Greyhound)	Canidae	29.0	Digitigrade	4	Terrestrial
Crocuta Crocuta	Spotted hyaena	Hyaenidae	63.9	Digitigrade	4	Terrestrial
Felis catus	Domestic cat	Felidae	3.7	Digitigrade	4	Scansorial
Leptailurus serval	Serval	Felidae	10.6	Digitigrade	4	Terrestrial
Lutra lutra	European otter	Mustelidae	7.4	Semidigitigrade	5	Natatorial
Lynx rufus	Bobcat	Felidae	9.6	Digitigrade	4	Scansorial
Meles meles	Badger	Mustelidae	10.7	Semidigitigrade	5	Semifossorial
Mustela putorius	Polecat	Mustelidae	1.0	Semidigitigrade	5	Terrestrial
Paradoxurus hermaphroditus	Palm civet	Viverridae	3.1	Semidigitigrade	5	Arboreal
Phoca groenlandica	Harp seal	Phocidae	167.0	Specialized	5	Aquatic

"specialized". Number of toes on the hind foot was recorded as four or five.

9.2.2 Scanning and Post-processing

The calcaneum and astragalus of each species were scanned in three dimensions. The calcaneum was scanned in dorsal view. Most of the functional features of the calcaneum are on the dorsal side, including the sustentacular facet, the astragalocalcaneal facet, and many muscle and tendon attachments. The astragalus was scanned in plantar (ventral) view. The plantar surface of the astragalus contains the structures that directly interface with the calcaneum. The dorsal side of the astragalus also contains important functional features, such as the trochlea, which were not analyzed.

Scans were made with a Roland PICZA PIX-4 pin scanner. This instrument records the Cartesian x y z coordinates of an object's surface by translating it in the x dimension below a carriage that moves along the y dimension. The carriage drops a pin to touch the object, recording the z coordinate. The density of x y point coordinates can be set between 0.05 and 5 mm; the density of z depends on the vertical relief of the object, with a higher density recorded on flatter surfaces. Several recent studies used such pin scanners (e.g., Eguchi et al., 2004).

Resolution of the tarsal scans was set according to the size of the bone. The smallest species, *Mustela putorius*, was scanned at 0.05 mm x y density, which produced a point grid of about 100 × 55 points; large species, such as *Canis familiaris*, were scanned at a lower density of 0.40 mm to produce a similar size point grid. Fine detail, including bone texture, was visible at these resolutions.

The lower z margins of the scans were standardized because variation in scan depth would influence the apparent variation in shape. Data were standardized by truncating each calcaneum scan just below the level of the sustentacular process and each astragalar scan just below the neck margin. Because the spacing of points in the z dimension depends on the slope of the surface, the lower z margin was irregular. To prevent the irregularity from influencing the shape comparisons, the margin was evened by dropping a series of new points to a z value of 0.0 directly below the original margin points.

9.2.3 Three-dimensional "Fishnet" Surface Points

Each bone was characterized for morphometric analysis with points interpolated across the surface. Quantitative comparisons of shape require that each surface be represented by an equal number of points. Different scan resolutions and bone sizes produce point grids that cannot be compared without interpolating an equal number of regularly space points on each surface.

Interpolation was a two step process. First an equal number of points were interpolated on each row of the original scan coordinates from proximal to distal. The original scan coordinates formed an $n_j \times j$ matrix, where n_j was the number of points in row j, and j was the number of rows along the proximodistal axis. Each n_i row was replaced with m interpolated points to produce an $m \times j$ matrix of points. The algorithm used for standard eigenshape analysis was used to interpolate each row (Lohmann, 1983; Lohmann and Schweitzer, 1990; MacLeod, 1999; MacLeod and Rose, 1993). Then for each m, a k number of evenly spaced points were interpolated to produce an $m \times k$ matrix of points.

The resulting interpolated surface can be likened to an elastic fishnet stocking stretched around the object (Figure 9.3). Each node of the stocking fabric represents a point on the surface. Before fitting, the nodes are equally spaced, but on the object they are stretched to fit the contours of the surface beneath them.

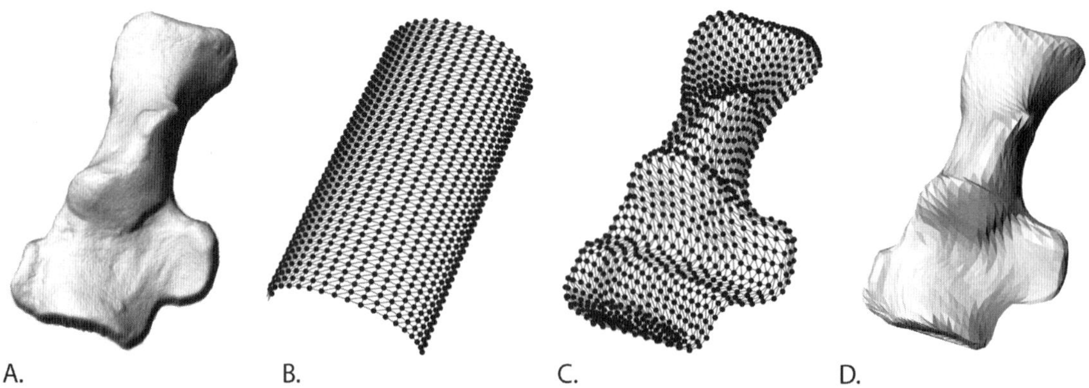

FIGURE 9.3. A, Calcaneum of *Meles meles*, three-dimensional rendering of original scan. B, three-dimensional "fishnet" point grid before fitting. C, 25 × 50 point fishnet grid fit to the calcaneum of *Meles*. D, Raster rendering of the fishnet grid to show quality of resolution. Surface lines are an artefact of the interpoint mesh representation used to make the rendering.

A fishnet grid of 25 × 50 points was placed on each calcaneum, with the first point at the distolateral corner of the bone and the last at the proximomedial corner. A 25 × 30 grid was placed on each astragalus in the same orientation. Because of computational limitations, the number of points was reduced to 20 × 50 and 20 × 30 respectively for two-block partial least squares analysis.

9.2.4 Ordination and Shape Modeling

Bone surfaces were aligned to a common size and orientation using Procrustes analysis. Each set of fishnet surfaces was translated, rotated, and rescaled to unit size using the generalized least squares Procrustes algorithm described by Rohlf (1990) and Rohlf and Slice (1990). An orthogonal projection into shape tangent space was applied using the algorithm of Rohlf and Corti (2000). To create a standard orientation for quantitative analysis, the aligned shapes were rotated to their three-dimensional principal components.

The bones were ordinated in shape space using principal components analysis (PCA; Dryden and Mardia, 1998). The consensus, or mean shape, was subtracted from each set of aligned fishnet surfaces. This step centers the shape space on the mean shape. A covariance matrix was calculated from the resulting residuals. PC vectors were calculated from the singular value decomposition of the covariance matrix:

$$\mathbf{SVD[P]=USV^T} \tag{9.1}$$

where **P** is the covariance matrix of the Procrustes residuals, **U** is the matrix of PC vector weightings, **S** is the matrix of singular values, and **V** is the transpose of **U** (when **P** is square, symmetric, positive definite covariance matrix).

The computationally limiting factor for three-dimensional surface analysis was the size of the covariance matrix. Matrices for three-dimensional surfaces are large because of the number are the number rows in each direction of the fishnet and 3 is the number of dimensions of each point. For the 25 × 50 calcaneum fishnet, the covariance matrix had 14,062,500 cells, requiring more than 112 Mb of memory. The efficiency of the analysis can be improved by doing the SVD on the covariance matrix of the objects rather than variables. In this case, the 12 taxa require a very small matrix of 12 × 12, or 144 cells. The eigenvectors and eigenvalues for the variable matrix can be back-calculated from those of the object matrix.

The coordinates of each bone in the PC shape space are called scores. The scores were used as shape variables for statistical analysis and tree building. Each bone has a score on every PC axis. The consensus shape has a score of 0.0 on each axis because it lies at the center of the shape space. Scores were calculated as $\mathbf{RU^T}$, where **R** is the matrix of surface residuals and $\mathbf{U^T}$ is the transpose of **U**.

Every point in the PC shape space corresponds to a different three-dimensional surface, regardless of whether a real bone lies there or not. Shapes at any particular point can be modeled by multiplying the value of the position on the PC axis by the corresponding vectors and adding the shape consensus to the result. Modeling can be done in the full multidimensional space or at positions along particular vectors or subsets of vectors. The former is useful for representing the shape of a particular taxon or locomotor category, whereas the latter is useful for illustrating the range of shapes associated with a particular PC or PLS axis. The estimated shape is:

$$\hat{X} = PU^T \tag{9.2}$$

where $\hat{X}$ is the modeled shape, P is the position being modeled, and U^T is the transpose of the vectors being modeled. $\hat{X}$ is a vector of three-dimensional fishnet points. These can be represented as points with or without a connecting mesh (Figure 9.3c) or as a shadowed surface rendering of the surface defined by the points (Figure 9.3d). The shadowed renderings in Figures 9.3, 9.7–9.9, 9.11, 9.12, and 9.16 were created by importing $\hat{X}$ into Rhinoceros 3.0, a three-dimensional vector graphics program.

9.2.5 Facet Size and Shape

The size and shape of joint facets were analyzed using the original scan data. The sustentacular and calcaneoastragalar facet surfaces were extracted from both the calcaneum and astragalus of each taxon (Figure 9.4a) and the margin of each synovial capsule was traced. Facet area was calculated in mm^2.

Facet curvature was measured as the curvature of a sphere fit to the surface of the facet. To do this, the surface area was first converted to $x\ y\ z$ coordinates (Figure 9.4b) and rotated to its principal components with the concave side up. The rotation provided a common, horizontal orientation for comparison. The center of the facet was estimated by calculating the centroid (or arithmetic average) of the surface coordinates. The centroid of a three-dimensional shape, such as these facets, does not necessarily lie on the surface itself, so the center of the facet was estimated as the projection of the centroid onto the surface along the z axis. The rate of curvature was represented as the coefficient of curvature of a sphere fit to the surface and passing through the center point. The fit was achieved by subtracting the coordinates of the center point from the surface points and fitting the function $\hat{Z} = bx^2$ (Figure 9.4c). The coefficient b was used as the measure of curvature. For perfectly flat facets, height on the z axis does not increase away from the center and $b = 0$; for curved facets, height on the z axis increases away from the center and $b > 0$. Because the facets were compared concave side up regardless of whether the surface on the bone was concave or convex, all coefficients of curvature were positive.

The fit of occluding facets was measured as the ratio of the larger or most curved of the occluding facets over the smaller or less curved. The index of fit for area was:

$$Index_{area} = \frac{area_{\max}}{area_{\min}}, \tag{9.3}$$

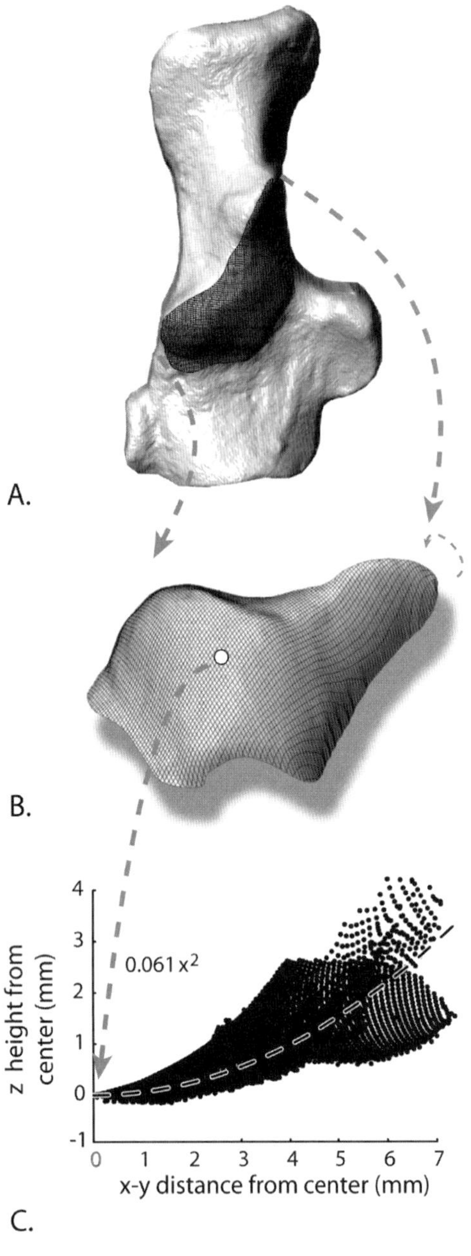

FIGURE 9.4. The measurement of facet curvature. A, The boundaries of the synovial capsule are traced on the surface of the bone. B, The facet is extracted and its centroid found (white circle). C, Each surface point is plotted by its *x y* distance and *z* height from the centroid when the facet is in a standard orientation. Curvature is measured by fitting the equation $y = a\,x^2$ to these transformed data. The coefficient *a* is larger when the surface is more curved.

where $area_{max}$ is the larger of the two occluding facets and $area_{min}$ is the smaller. The index for curvature was the same as Equation 9.3, but with *b* substituted for *area*. The index equals 1 when the occluding facets are the same size (or curvature) and increases with increasing discrepancy in size (or curvature). No data on errors estimating the area or curvature of the facets are available, but differences less than 0.3 are unlikely to be significant.

9.2.6 Two-block Partial Least Squares

The evolution of a form-function complex, such as the occluding faces of the astragalus and calcaneum, is an integrated process (Bock and von Wahlert, 1965; Szalay and Bock, 1991; Szalay, 2000). In order for the form of one bone to change, the form of the other must change in a compatible fashion, especially those parts that contact one another. One method for extracting correlated variation in the astragalus and calcaneum is two-block partial least squares analysis (2B-PLS). 2B-PLS extracts correlated shape variation as a series of orthogonal vector pairs (Sampson et al., 1989; Rohlf and Corti, 2000). In this study, each vector pair describes correlated variation in the plantar astragalus and dorsal calcaneum respectively. Different pairs are mathematically independent of one another such that change along one vector could, in principle, occur independently of that on another while still maintaining functional integration between the shapes of the two bones. Pairs of 2B-PLS vectors are ordered by the proportion of covariation they explain, with PLS 1 explaining the largest portion.

2B-PLS was performed on the matrix of covariances of the astragalus and calcaneum surface coordinate. Because of computational limitations mentioned above, smaller fishnet grids were used to represent each bone than for other analyses. Each bone was Procrustes superimposed, rotated to its principal components, and projected into orthogonal tangent space. Procrustes residuals were calculated by subtracting the consensus configuration. The matrix of covariances between astragalus and calcaneum residuals was calculated. 2B-PLS vectors and singular values were calculated by SVD of the covariance matrix following Rohlf and Corti (2000). 2B-PLS scores were calculated by projecting the shapes onto the vectors as described above. Shape variation on the 2B-PLS axes was modeled as described above.

9.2.7 Maximum-likelihood Trees

Tree diagrams are useful for depicting the similarities between morphometric shapes. Trees can be compared to phylogenies based on independent evidence, to locomotor categories, and to other biological groupings to explore possible causes of shape similarity. Many tree algorithms are available, but one of the most appropriate is maximum-likelihood (ML) method for quantitative traits because it does not assume equal rates of change and it treats independent aspects of shape as separate characters (Felsenstein, 1973, 1981, 1988; Polly, 2003a, b; Caumul and Polly, 2005). The Brownian motion model of evolution used in most ML algorithms appears to be appropriate for multivariate morphological shape (Polly, 2004; see discussion below).

PCA scores were used as the data for tree building. The scores satisfy the requirements for ML because they are uncorrelated and continuously distributed. Scores were standardized to a mean of zero and variance equal to the proportion explained by the corresponding vector. Standardization to unit variance, which is the usual procedure (Felsenstein, 1973) puts undesirable weight on less important vectors that contain little information about shape. The CONTML module of PHYLIP (Felsenstein, 1993) was used for tree building. Input order of the taxa was randomized and three global rearrangements were made.

9.2.8 Reconstruction of Ancestral Morphologies

Morphological shape scores can be reconstructed at the nodes of a phylogenetic tree when branch lengths and the shape of all terminal taxa are known using any one of the several available methods (Grafen, 1989; McArdle and Rodrigo, 1994; Martins and Hansen, 1997; Garland et al., 1999; Garland and Ives, 2000; Polly, 2001; Rohlf, 2001). The estimated ancestral shape can be modeled from the reconstructed scores using the modeling procedure described above. Importantly, the phylogenetic tree can be projected into the morphospace by projecting the node scores into the space and connecting the branches. Such a projection provides a visual means of assessing the history of phylogenetic occupation of the morphospace.

The generalized linear model (GLM) method of estimating ancestral states (Martins and Hansen, 1997) was used here to reconstruct node scores. This method requires two matrices derived from the phylogenetic tree. The phylogenetic error variance matrix, **var[Y]**, is an $n \times n$ matrix, where n is the number of terminal taxa. The diagonal contains the length from the taxon to the base of the tree and the off diagonal elements contain the length from the last common ancestor of two taxa to the bottom of the tree. The node error variance matrix, **var[A, Y]**, is an $n \times m$ matrix, where n is the number of terminal taxa and m is the number of nodes in the tree. Each element contains the length of shared history between the node and each terminal taxon. The scores at the base of the tree were estimated as:

$$M_G = (J' \operatorname{var}[Y]^{-1} J)^{-1} J' \operatorname{var}[Y]^{-1} Y, \quad (9.4)$$

where $\mathbf{M_G}$ is the estimated score values at the base of the tree, **J** is a unit vector of length n, and **Y** is the matrix of scores of the terminal taxa. The other node values are then calculated from the residuals of the scores from the node reconstruction as:

$$\hat{A} = \operatorname{var}[A,Y] \operatorname{var}[Y]^{-1} Y' + M_G, \quad (9.5)$$

where $\hat{\mathbf{A}}$ are the estimated node scores and **Y'** is the matrix of residual scores after the subtraction of $\mathbf{M_G}$.

Ancestral shapes were reconstructed on a phylogenetic tree derived from Flynn (1996). Some points about carnivoran phylogeny are controversial. The broad agreement 15 years ago on relationships among the major family-level groups (e.g., Flynn et al., 1988; Wayne et al., 1989; Wozencraft, 1989; Wyss and Flynn, 1993; Wolsan, 1993, but see Hunt and Tedford, 1993) has dissolved with subsequent studies (Lento et al., 1995; Slattery and O'Brien, 1995; Ledge and Árnason, 1996a, b; Flynn, 1996; Werdelin, 1996; Wang, 1997; Flynn and Nedbal, 1998; Janis et al., 1998; Koepfli and Wayne, 1998, 2003; Flynn et al., 2000; Veron and Heard, 2000; Gaubert and Veron, 2003; Sato et al., 2003; Sato et al., 2004; Davis et al., 2004; Veron et al., 2004; Yu et al., 2004; Flynn and Wesley-Hunt, 2005; Flynn et al., 2005; Wesley-Hunt and Flynn, 2005; Yu and Zhang, 2005). Only the dichotomy between Feliformia and Caniformia remains completely uncontroversial, at least with respect to the taxa included here. I did not use the consensus "super-tree" of Carnivora (Bininda-Emonds et al., 1999) because of the well-known pitfalls affecting consensus trees that are based on different, overlapping datasets (Miyamoto, 1985; Kluge, 1989).

The tree adopted for ancestral reconstruction is shown in Figure 9.5. The choice of this tree did not affect ancestral node reconstructions because the controversial nodes, such as the phylogenetic placement of the Red panda, *Ailurus*, are closely spaced in geological time. The three feliform groups – Hyaenidae, Viverridae, and Felidae – diverged around 22 Ma, regardless of which of the three are most closely related. The shared history of the two most closely related was brief relative to the history shared by all three. The duration of shared history exerts more influence on ancestral reconstruction than the ordering of closely spaced branching events (Martins and Hansen, 1997; Polly, 2001). Reversing the order of branching among *Paradoxurus, Crocuta*, and the felids would not change the reconstruction at those nodes (because of that, I have grouped the nodes together as Node 1). Similarly, Canidae, Phocidae (and other pinnipeds), *Ailurus*, Procyonidae, and Mustelidae shared a last common ancestor around 36 Ma. The lineages leading to the three mustelids – *Lutra, Mustela*, and *Meles* – branched in quick succession around 24 Ma, and these have been grouped as Node 4 for purposes of ancestral reconstruction. The divergence times are the younger of Flynn's (1996) two estimates, which are the ones supported by recent palaeontological studies (Wesley-Hunt and Flynn, 2005).

Ancestral node reconstructions have large confidence intervals (Martins and Hansen, 1997; Garland and Ives, 2000; Polly, 2001). The reconstructed shapes are those that are the most likely given the morphology of the terminal taxa and the topology of the tree; however, many other shapes are both biologically and statistically plausible. The range of statistically likely shapes can be described by a 95% confidence spheroid in shape space with the optimal reconstruction at its centre. At the deep nodes of the tree, these spheroids will encompass the range of variation found in the terminal taxa, something that should be remembered when comparing the reconstructions to real fossil taxa. No attempt was made to model the range of shapes falling within the node confidence spheroids, though it is easily done (Martins and Hansen, 1997), because a very large number of reconstructions would be required to give even a hint of the morphological range encompassed by a confidence spheroid in the 11-dimensional shape space.

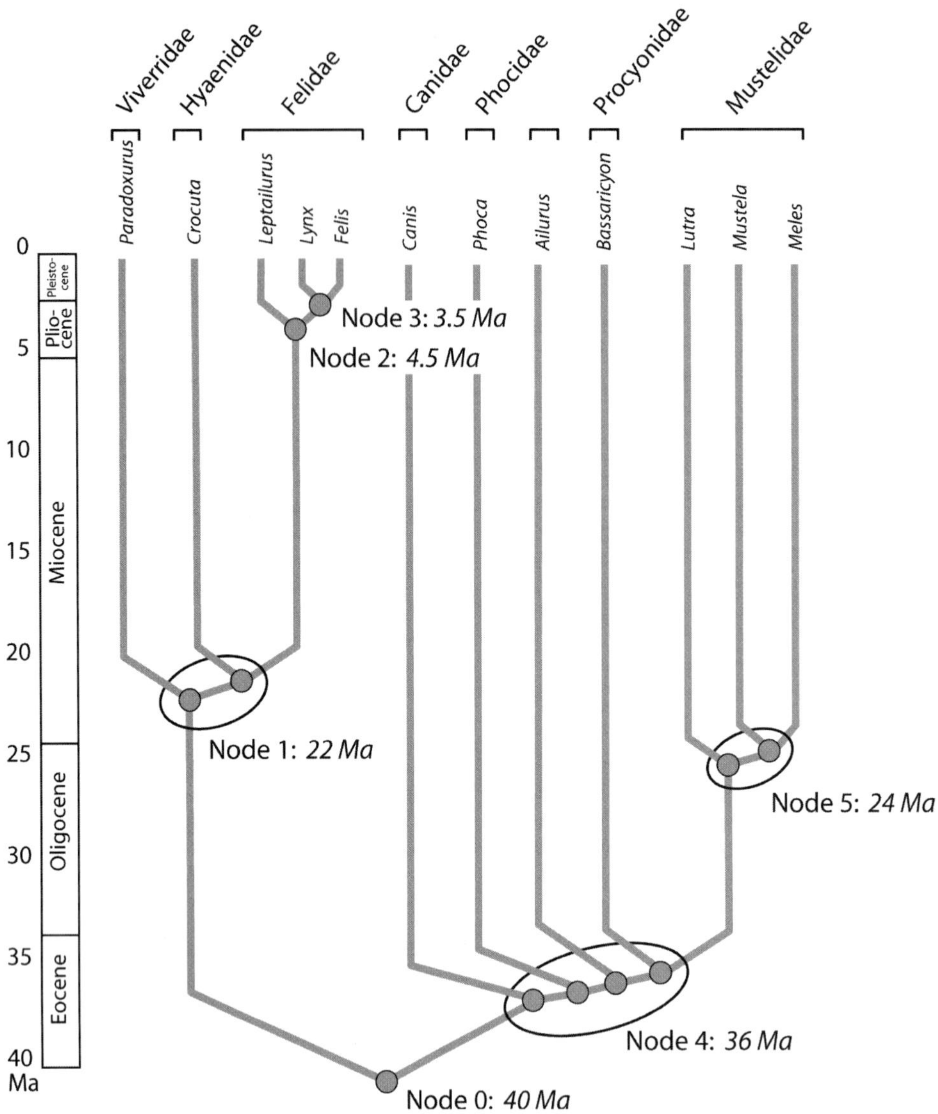

FIGURE 9.5. Phylogenetic tree of the twelve carnivoran species. The topology and branch lengths are based on Flynn (1996). Nodes whose estimated divergence times are equal were combined for purposes of ancestral shape reconstruction.

9.2.9 Rates And Mode of Tarsal Evolution

Rate of morphological evolution was quantified as Procrustes distance over time. Procrustes distance is the sum of distances between corresponding fishnet points when two bones are in optimal superimposition:

$$D = \sqrt{\sum_{i=1}^{m\times k}(p_{1i} - p_{2i})^2} \tag{9.6}$$

where p_{1i} and p_{2i} are points i on bones 1 and 2, and $m \times k$ is the total number of points.

Rates along individual tree branches were calculated by dividing the squared distance between the end shapes by the length of the branch. Shapes at terminal branches are the interpolated point grids from the original scans and the shapes at the nodes are the ancestral reconstructions. Squared distances were used because morphological variance increases linearly with time when there is a Brownian motion mode of evolution (Felsenstein, 1988; Polly, 2004), making D^2 rates less biased than ordinary ones. Rate estimation depends on the accuracy of node reconstructions. The optimal reconstructions used here minimize rates across the tree. Alternative reconstructions increase some branch rates at the expense of others, but also increase the average rate across all branches. For tree-specific rates, branch length was measured in generations, which are the natural units of evolutionary processes (Gingerich, 1993, 2001; Polly, 2001, 2002). Lengths were converted from millions of years to generations using a mean carnivoran generation length – 3.8 years – calculated from data compiled by Eisenberg (1981).

The pattern of shape divergence over time was used to estimate the mode of evolution. By mode, I mean the pattern of divergence resulting from long-term directional, stabilizing, or randomly varying selection (Polly, 2004). Long-term directional selection continually pushes all species in the same direction, causing correlated evolution among lineages. An example would be the general increase in size expected under Cope's Rule (Stanley, 1973; Alroy, 1998; Polly, 1998; Van Valkenburgh et al., 2004). Stabilizing selection prevents species from evolving away from some optimum, and is variously called stabilizing selection (Schmalhausen, 1949), centripetal selection (Simpson, 1953), adaptive peak model (Lande, 1976; Felsenstein, 1988), and Ornstein-Uhlenbeck process (Felsenstein, 1988; Martins and Hansen, 1997). Stasis – the complete absence of evolutionary change – is an extreme example. Randomly fluctuating selection changes direction and magnitude, usually as species adapt to new and changing environments. This is a typical Brownian motion process (Lande, 1986; Felsenstein, 1988; Martins and Hansen, 1997) and may include components of directional or stabilizing selection at particular times. The evolutionary pattern produced by randomly fluctuating selection is identical to that produced by neutral genetic drift except in the rate of divergence. Drift is slower and it its rate is a function of population size. I subsume drift under randomly fluctuating selection because the data required to distinguish them are unavailable. Evolutionary modes may be distinguished by the scaling relationship of shape divergence to time since common ancestry (Bookstein et al., 1978; Gingerich, 1993; Polly, 2004; Pie and Weitz, 2005). Directional selection causes a constantly increasing, linear divergence. Randomly fluctuating selection causes both divergence and convergence such that shape diverges on average with the square root of time. Stabilizing selection limits divergence so there is no relation with time since common ancestry after a point. That point depends on the strength of the stabilizing selection, the rate of evolution, and the maximum time of common ancestry of the clade.

Mode was estimated by fitting the equation:

$$y = x^{a} \tag{9.7}$$

where y is shape difference between two taxa, x is the time elapsed since their last common ancestor, and a is a coefficient ranging from 1 to 0. A coefficient of 1 corresponds to directional selection where divergence increases linearly with time. A coefficient of 0.5 corresponds to random selection where divergence increases with the square root of time. A coefficient of 0 corresponds to stabilizing selection where divergence does not increase with time. Intermediate values of a correspond to random selection with a predominance of stabilizing or directional selection. To find a, Equation 9.6 was fit to the data using values ranging from 0.1 to 1 at 0.1 intervals. The value that that minimized the residual variance was chosen (Butler and King, 2004).

Two measures of time since common ancestry were used. Palaeontological estimates of the age of the last common ancestor of each species pair were taken from the tree in Figure 9.5 (see above). Cytochrome *b* sequence distance was used as a proxy measure (Brown et al., 1979; Springer, 1997). The advantage of cyt *b* is that it is measured in each species independently, so an error (including atypical mutational history) in one species does not affect all of the pairs. Error in a palaeontological node age affects all species connected through that node. The disadvantage of cyt *b* is that it is only a proxy for time since common ancestry and has its own set of errors (Graur and Martin, 2004). One such error is saturation, or the effect of "multiple hits", which causes divergences older than 15–20 million years to be underestimated (Nei, 1987). Cyt *b* divergence was measured as the Kimura 2-parameter sequence distance, which weights transition and transversion mutations differently and which is corrected for the effects of saturation (Kimura, 1980). Eleven sequences were taken from GenBank. Cyt *b* was not available for *Leptailurus*. Clustal X, version 1.8 (Thompson et al., 1997) was used to align the sequences and calculate distances.

9.2.10 Adaptive Landscape Contours

The fissiped adaptive landscape contours in Figure 9.16 were calculated from the fissiped scores on the first two PCs. Ellipses were centered on the mean fissiped shape and encompass 5 and 95 percentiles. The outer ellipse is equivalent to a 95% confidence interval. In more than two dimensions, the adaptive landscape is a multidimensional spheroid; the ellipses shown in Figure 9.16 are the projection of that spheroid onto the first two PCs.

9.2.11 Association of Tarsal Shape and Locomotor Factors

Association of tarsal shape with locomotor mode, stance, toe number, body mass, and the facet indices was tested using regression and one-way analysis of variance (ANOVA). Multivariate analysis of variance (MANOVA) was used to test the entire three-dimensional shape against categorical factors; univariate ANOVA was used to test individual PCs. Significance was determined with a multivariate F, or Wilks' Lambda test.

9.3 Results and Discussion

9.3.1 The Bones

Twenty-four bones were scanned, two each from the twelve species (Figure 9.6). The most obvious variation in the calcaneum was in the size of the peroneal tubercle, which was largest in *Ailurus* and *Lutra* and nearly absent in *Crocuta* and *Phoca*. The position and size of the sustentacular process also varied from a proximal position in the three felids to a distal position in *Paradoxurus* and *Bassaricyon*. The shape of the calcaneoastragalar facet was proximodistally short and curved in the felids, *Crocuta*, and *Canis*, but long and less curved

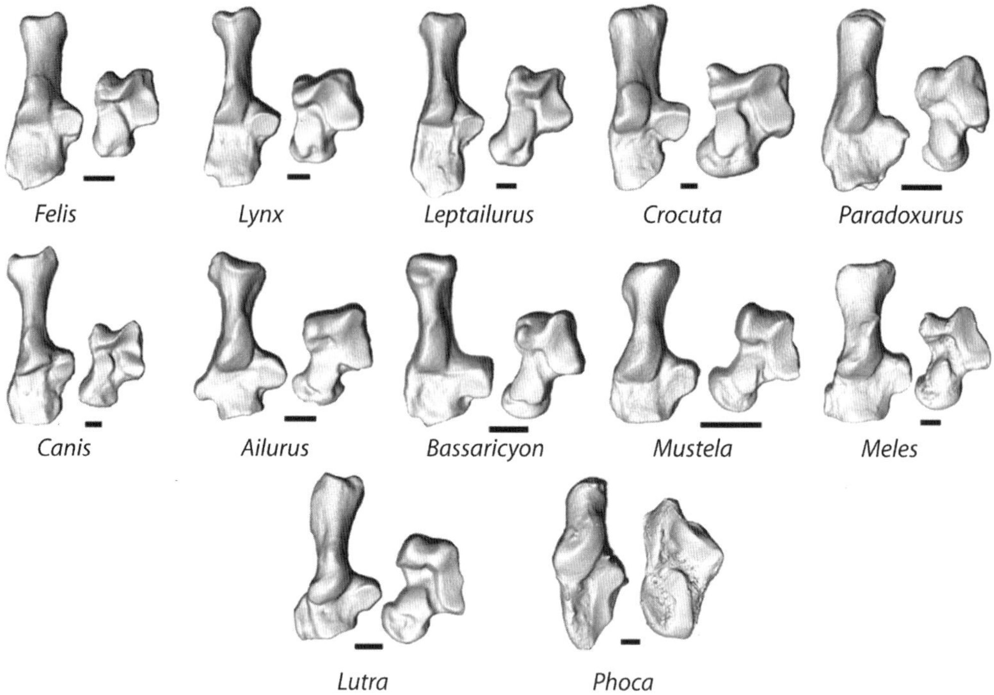

FIGURE 9.6. Three-dimensional scans of the right calcaneum (dorsal view) and astragalus (plantar view) from 12 carnivorans. Orientations are the same as in Figure 9.2. Scale bars equal 5 mm.

in other taxa. The length of the calcaneal tubercle was proportionally long in *Lutra*, but shorter in the felids. The bony processes of the tubercle also varied, especially in their degree of lateral and medial asymmetry. On the astragalus, the length of the neck, the curvature of the calcaneoastragalar facet, and the shape of the body were the most visibly variable features.

The tarsals of *Phoca* were notably different from the fissiped taxa. *Phoca's* calcaneum was diamond-shaped with a pointed distal end. The calcaneal process was short and without lateral tubercles for attachment of the superficial digital flexor. The groove at the end of the calcaneal process for the attachment of the semitendinosus and gastrocnemius was absent. In seals, these tendons probably insert parallel to the long shaft of the bone instead of at a vertical angle (Howell, 1929). The astragalus was blocky without noticeable neck or trochlear groove. A bony process extended from the proximal end of the astragalus, paralleling the calcaneal tuber. This bony process and the tendon for the flexor hallucis longus, which passes along a groove on the process, prevent the foot from being brought into a plantigrade position (Howell, 1929). The calcaneoastragalar and sustentacular facets were long, especially the latter (King, 1966; Wyss, 1998). In fissipeds, the sustentacular facet is always an ovate basin.

9.3.2 Principal Component Analysis

Principal component analysis (PCA) was used to construct a phenotypic morphospace for the bones and to extract major axes of correlation. The results of the PCAs for calcaneum and astragalus are shown in Figures 9.7 and 9.8. Plots of the bones in the first three PC dimensions are shown in the top panels. Each taxon is represented by a shape model calculated from scores on the first three axes. Models of shape variation along the individual PC axes are shown at the bottom of the figures. Each axis is illustrated by five models showing shape variation along the axis at quartile points. The right and leftmost models represent the shape at the positive and negative extremes of variation, and the middle model represents the average, or consensus shape. The consensus models of the three PCs are identical because all axes are centered on the same mean sample shape. Differences among models along a particular PC represent correlated variation, but differences between series represent independent variation. The eigenvalues and percent variance explained by each PC are reported at right.

The shape models in Figures 9.7a and 9.8a represent real bones reconstructed from scores on the first three PCs, but the quartile series in Figures 9.7b–d and 9.8b–d are hypothetical bones constructed from only one PC each. The shape of the *Phoca* calcaneum illustrates this. In Figure 9.7a, the *Phoca* model can be understood as the visual summation of the models in Figure 9.7b–d that correspond to *Phoca's* position on each respective axis. The *Phoca* model can be mentally reconstructed from the PC shape models by combining those at position 0.075 on PC 1, position –0.12 on PC 2, and position 0.09 on PC 3. Even though the models in Figure 9.7a are realistic enough for visual identification, they differ from the

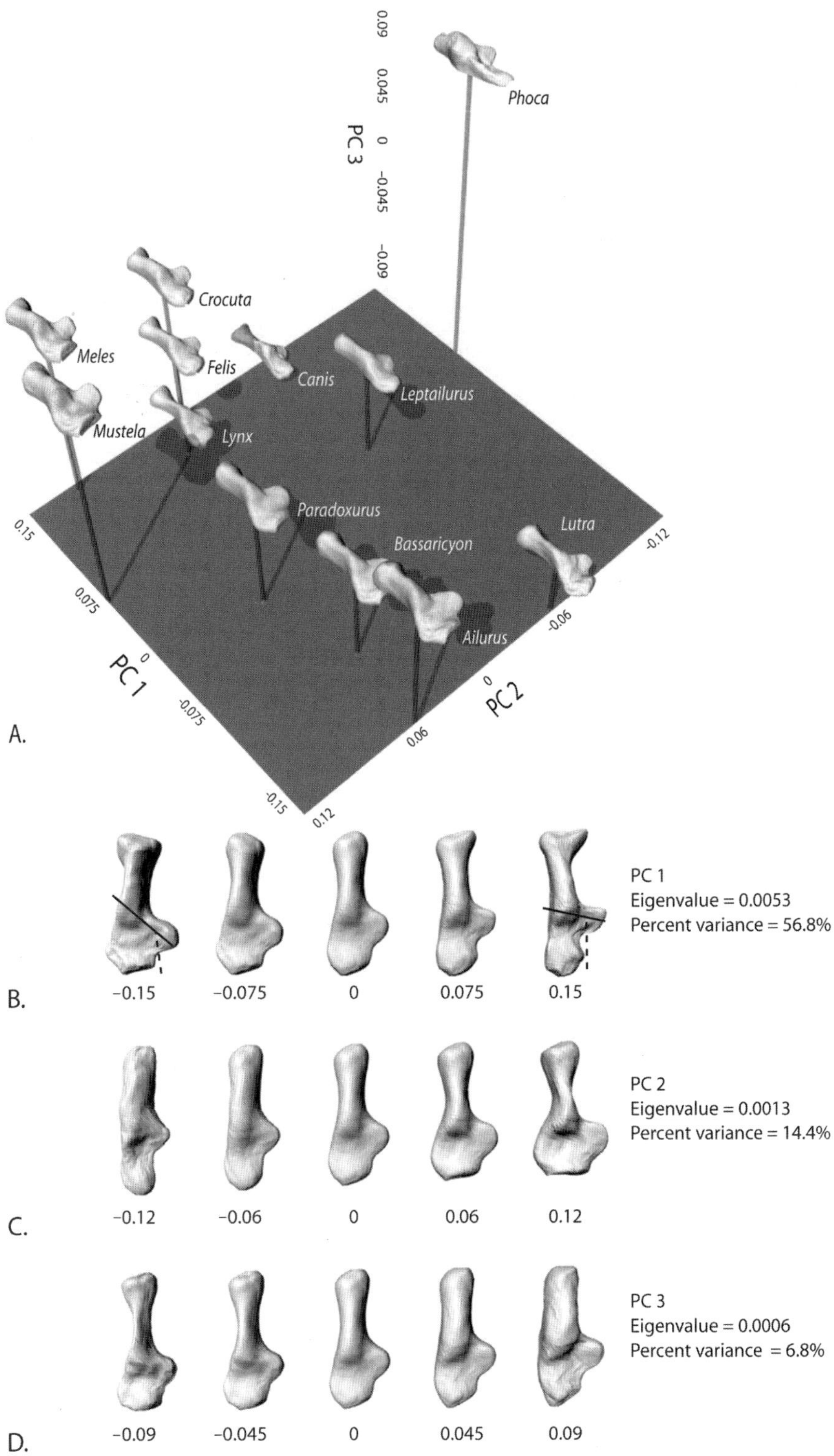

FIGURE 9.7. Principal components analysis of calcaneal shape. A, Ordination on the first three PCs. Taxa are represented by shape models constructed from their scores on all three axes. B, Shape models showing variation along PC 1. The position of each model is indicated by the number below it. An important part of variation on PC 1 is the orientation of the two facets (unbroken line) and the astragalar neck (broken lines), the curvature of the calcaneoastragalar facet, development of the peroneal tubercle, dominance of the lateral and medial calcaneal tubercles, and angle of the cuboid facet relative to the long axis of the bone. C, Shape models along PC 2. Important variation is width of the distal calcaneum, proximodistal position of the sustentacular process, and development of the calcaneal tubercle as a whole. The negative end of the axis is dominated by *Phoca*. D, Shape models along PC 3. Important variation includes shape of the calcaneoastragalar facet, development of the peroneal tubercle, and thickness of the calcaneal process. The positive end is dominated by *Phoca*.

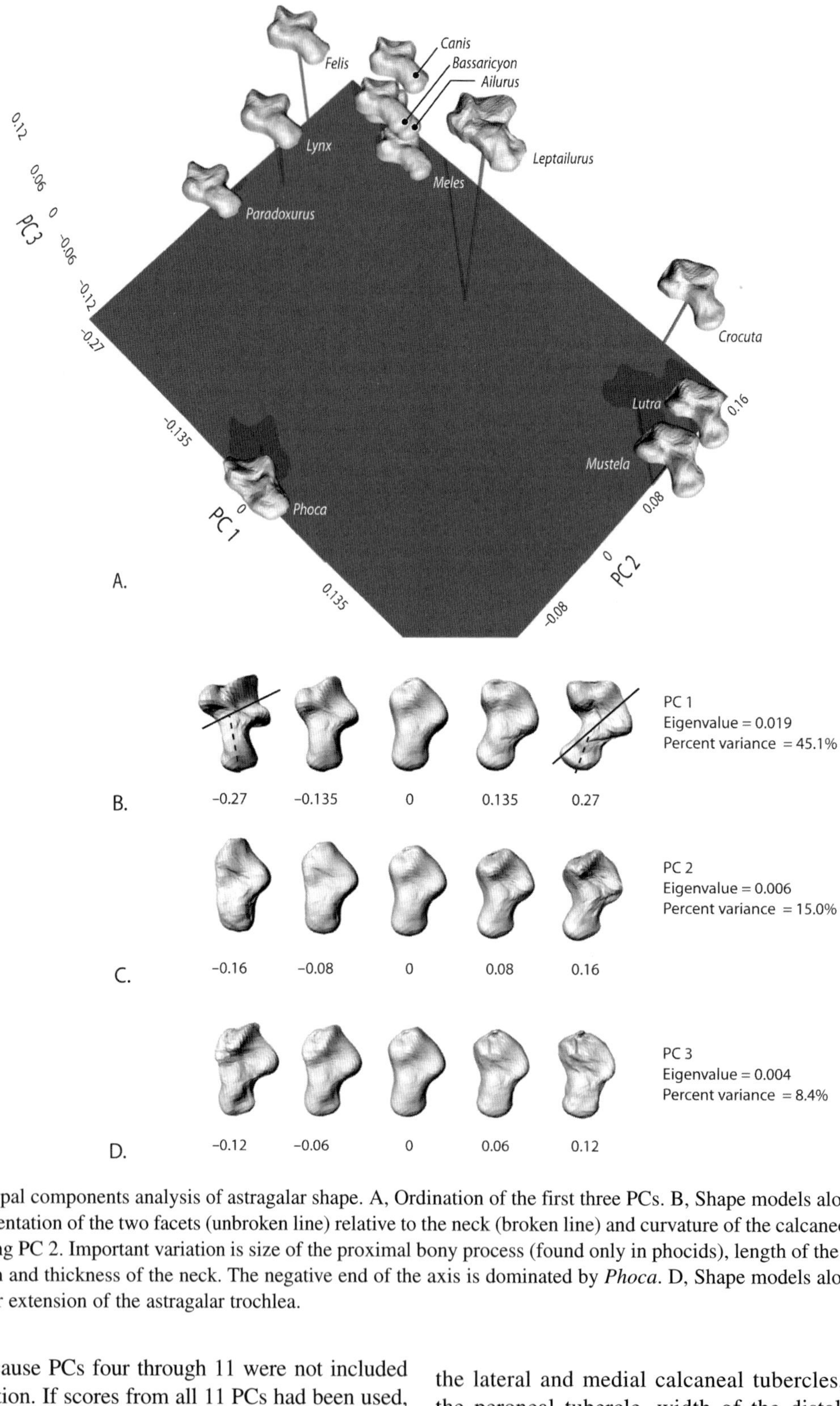

FIGURE 9.8. Principal components analysis of astragalar shape. A, Ordination of the first three PCs. B, Shape models along PC 1. Dominant variation is the orientation of the two facets (unbroken line) relative to the neck (broken line) and curvature of the calcaneoastragalar facet. C, Shape models along PC 2. Important variation is size of the proximal bony process (found only in phocids), length of the sustentacular facet, and the orientation and thickness of the neck. The negative end of the axis is dominated by *Phoca*. D, Shape models along PC 3. Important variation is plantar extension of the astragalar trochlea.

actual bones because PCs four through 11 were not included in their construction. If scores from all 11 PCs had been used, then the models would be indistinguishable from the original scans after interpolation.

9.3.2.1 Calcaneum

PC 1. The first principal component separated species by the orientation of the sustentacular facet, the dominance of the lateral and medial calcaneal tubercles, development of the peroneal tubercle, width of the distal calcaneum, and angle of the cuboid facet relative to the long axis of the bone. The negative end of the axis – dominated by *Lutra, Ailurus*, and *Bassaricyon* – combined a distally positioned, dorsally facing sustentacular facet; a broadly curved calcaneoastragalar facet; a large lateral calcaneal tubercle; a large peroneal tubercle; a broad distal calcaneum; and a medially angled cuboid facet. This morphology is associated with a mobile, plantigrade, five-digit foot. The narrowly acute angle between the line connecting the two facets and the main axis of the bone and the broad curvature of the calcaneoastragalar facet are functionally associated with the ability to rotate the calcaneum under the astragalus for foot inversion (Szalay and Decker, 1974; Jenkins and McClearn, 1984). The strong peroneal process is probably associated with abduction of the fifth digit and eversion of the foot, because the tendons of the abductor digiti quinti and peroneus brevis pass under the process (Greene, 1935; Szalay, 1977a; Evans, 1993). The length of the calcaneal process gives these species a mechanical advantage for plantar flexion, a motion important in both swimming and climbing.

The positive end of the axis – dominated by *Crocuta, Canis, Felis, Lynx* – combined a proximally positioned, distally facing sustentacular facet, a narrowly curved calcaneoastragalar facet, a long, narrow distal calcaneum, a short peroneal process, a distally angled cuboid facet, a deep groove for the gastrocnemius tendon, and strong development of the medial calcaneal tubercle. This morphology is associated with a parasagitally constrained, digitigrade, four-digit foot. The proximal position of the sustentaculum and the narrowly curved calcaneoastragalar facet restrict movement between the calcaneum and astragalus and help transmit weight straight down the shaft of the foot (Howell, 1944). The distally directed cuboid facet also transmits weight down the center axis of the foot. The proportionally long distal end of the calcaneum lengthens the moment arm of effort, emphasizing speed over strength during plantar flexion. *Meles* and *Mustela*, which do not have digitigrade, four-toed feet, also lay at the positive end of PC 1, but many of the features associated with that part of the axis are counteracted by the contribution of PC 2.

PC 1 is functionally associated with mobility in the LAJ and TTJ. This can be seen by following the axes of rotation of the upper ankle, lower ankle, and transverse tarsal joints through the series of models in Figures 9.7b and 9.8b. The upper ankle joint moves by rotation of the dorsal astragalar trochlea against the tibia and fibula, roughly perpendicular to the hashed line. The lower ankle joint moves by rotation of the calcaneum under the astragalus along the axis represented by the solid line. The transverse tarsal joint moves by rotation of the cuboid and navicular against the facets on the distal end of the calcaneum and astragalus. Angles of the axes are functionally correlated because they are related both to the flexibility of the ankle and the type of stance (Schaeffer, 1947; Szalay and Decker, 1974; Decker and Szalay, 1974). Species at the negative end have greater mobility in the LAJ and TTJ and species at the positive end have less.

PC 2. The second principal component separated species on the shape of the calcaneal tuber, the width and shape of the distal calcaneum, and the position of the sustentaculum. The positive end of PC 2 – dominated by *Mustela* and *Meles* – paralleled the negative end of PC 1, but with a distally directed cuboid facet, a more narrowly curved calcaneoastragalar facet, and a more strongly developed medial calcaneal tubercle. This morphology is associated with a semidigitigrade, five-digit foot. The narrow curvature of the calcaneoastragalar facet and the distally directed cuboid facet transmit weight associated with an upright stance.

The negative end of PC 2 – dominated by *Phoca* – combined a narrow calcaneal tubercle, narrow distal calcaneum, and a pointed, proximally positioned sustentaculum. At its extreme end, the axis is associated with the aquatic specializations found in pinnipeds, but towards the center it is associated with the more extreme cursorial specializations of *Canis* and *Leptailurus*, including a long, narrow calcaneum whose lever mechanics are optimized for speed over strength in plantarflexion.

PC 3. The third principal component separated species on the thickness of the calcaneal process and the orientation of the calcaneoastragalar and cuboid facets. The positive end is dominated by *Phoca* and provides the oblique angle of its calcaneoastragalar facet in combination with a proximally placed sustentaculum (an oblique angle is also found at the negative end of PC 1, but there it is associated with a distal sustentaculum). PC 1 also contributes further to the pointed distal end of *Phoca's* calcaneum. The negative end of PC 3 is dominated by *Canis* and *Lynx*, which have a narrow calcaneal process shaft with a wider tuber at the end.

9.3.2.2 *Astragalus*

PC 1. The first astragalar principal component separated species by the angle of the neck relative to the axis connecting the sustentacular and calcaneoastragalar facets, and by the development of the medial trochlear ridge. The positive end of the axis – dominated by *Mustela, Lutra*, and *Crocuta* – combined a narrow angle between neck and the axis connecting the two facets with a proximally prominent medial trochlear ridge. While part of this morphology parallels the positive end of the calcaneal PC 1, the taxa found at the positive end of the astragalar axis share no special locomotor or phylogenetic similarity.

The negative end – dominated by *Felis, Lynx*, and *Paradoxurus* – combined a deep but rounded calcaneoastragalar facet, a perpendicular angle between the neck and the axis connecting the sustentacular and calcaneoastragalar facets, and a trochlea dominated by the lateral ridge. The animals at the negative end are the scansoral and arboreal feliform species.

PC 2. The second principal component is dominated by *Phoca* and separated species on the curvature of the calcaneoastragalar facet, the length of the sustentacular facet, and the thickness of the neck.

PC 3. The third principal component is described a contrast between *Leptailurus* and *Meles*, separating species based on the curvature of the calcaneoastragalar facet and the shape of the trochlea. The positive end of the axis represents a twisted calcaneoastragalar facet, which faces laterally at the proximal end and ventrally at the distal end. This is combined with a dominant lateral trochlear ridge with sharp plantar relief. *Leptailurus, Crocuta, Canis, Felis*, and *Lynx*, all digitigrade species, lie towards the positive end of the axis.

The negative end of the axis represents calcaneoastragalar facet that faces ventrally along its entire length. The entire plantar surface of the astragalus has uniform, low relief. The arboreal and semifossorial species lie at the negative end.

9.3.3 Body Mass and Tarsal Shape

Body mass (Log_e) and multidimensional calcaneum shape were not significantly related (MANOVA: Wilk's Lambda = 0.114, $F_{10,1} = 0.777$, $p = 0.716$), but body mass was significantly related to PC 2 individually (ANOVA: $F_{1,10} = 11.43$, $p = 0.007$). PC 2, which describes the width of the tuber, distal calcaneum, and position of the sustentacular facet, was dominated by *Phoca*, the largest species in the analysis. When *Phoca* was excluded, none of the individual calcaneum PCs is significantly related to body mass. Body mass was not related to astragalar shape, neither multidimensionally (MANOVA: Wilk's Lambda = 0.041, $F_{10,1} = 2.341$, $p = 0.472$) nor to individual PCs with or without *Phoca*.

9.3.4 Locomotion and Tarsal Shape

Calcaneum shape differed among locomotor types (Figure 9.9a). Semifossorial and natatorial calcanea were similar in the size and shape of their peroneal process and the position of their sustentaculum, but differed in the proportional length of the calcaneal process. Arboreal, scansorial, terrestrial and aquatic mean shapes were unique. Not only was shape significantly associated with locomotor type (Wilk's Lambda = 0.000, $F_{30,6} = 12.12$, $p = 0.002$), but 64.8% of the total shape variation could be explained by it. Locomotor diversity was most closely associated with PC 1, which was the only axis that individually had a statistically significantly relation to locomotion ($F_{5,6} = 6.98$, $p = 0.017$). PC 1 described the position of the sustentaculum, the curvature of the calcaneoastragalar facet, the size of the calcaneal and peroneal tubercles, the width of the distal calcaneum, and the angle of the cuboid facet, features classically linked with locomotion because they influence the mobility, especially the rotation, of the ankle (Szalay, 1977a; Jenkins and McClearn, 1984). Locomotor mode explained 85% of the variation on PC 1.

Calcaneum shape was less influenced by stance, though averages of each of the three stance types were visibly different (Figure 9.9b). Digitigrade species shared a sharply convex calcaneoastragalar facet, a small peroneal process, and a proximally positioned sustentaculum. Plantigrade species shared a rounded calcaneoastragalar facet, a long peroneal process, and a larger, distally positioned sustentaculum. Semidigitigrade species were intermediate. As a whole, the association between stance and shape was not statistically significant (Wilk's Lambda = 0.000, $F_{24,8} = 5.50$, $p = 0.07$), even though 46% of the total shape variation was explained by stance. Correlation of stance with PC 2 was significant, however ($F_{3,8} = 4.21$, $p = 0.046$). The second PC describes correlations in the width of the distal calcaneum, the width of the tuber, and the position of the sustentaculum. Narrower distal calcanea result from a narrower peroneal process and cuboid facet, both associated with reduction in the number of digits, which itself is associated with digitigrady. The sustentaculum is also more posteriorly positioned in digitigrade species.

Calcaneum shape was also only marginally influenced by the number of toes (Figure 9.9c). The mean calcaneal shape of five-digit species had a rounded calcaneoastragalar facet, an open, distally positioned sustentaculum, an angled cuboid facet, a medium-sized peroneal process, and a relatively short calcaneal process. Four digit species had a more sharply curved calcaneoastragalar facet, a more posteriorly positioned sustentaculum, a narrower, more sharply defined peroneal process, and a longer calcaneal process. The difference was not significant (Wilk's Lambda = 0.012, $F_{10,1} = 8.34$, $p = 0.26$), though digit number was significantly related to PC 1 by itself ($F_{1,10} = 6.49$, $p = 0.029$). Digit number and locomotor type are themselves correlated, with four-digit species dominating the scansorial and terrestrial types.

Astragalus shape was associated with locomotor type (Figure 9.9d). Arboreal astragali were characterized by long necks, large sustentacular facets, and open calcaneoastragalar facets. Scansorial species had shorter necked astragali. The sustentacular facets of terrestrial species were smaller, and their calcaneoastragalar facets more sharply concave. Semifossorial taxa had blockier with a more angled trochlea, while natatorial astragali (represented only by *Lutra*) had a very long calcaneoastragalar facet compared to the neck. Aquatic astragali (represented only by *Phoca*) had a very thick neck, large sustentacular and calcaneoastragalar facets, and no ventral extension of the trochlea. Differences among locomotor types were significant (Wilk's Lambda = 0.000, $F_{30,6} = 7.75$, $p = 0.008$), with PC 5 having the strongest association ($F_{5,6} = 6.37$, $p = 0.022$); 62.3% of shape variation was explained by locomotor type.

Stance had a significant effect on astragalus shape (Figure 9.9e). Digitigrade species had sharply curved calcaneoastragalar facets, semidigitigrade taxa had more open ones, and plantigrade species had long, narrow necks. The difference among stance categories was significant (Wilk's Lambda = 0.000, $F_{24,8} = 20.78$, $p = 0.01$), with PC 2 having the strongest association ($F_{3,8} = 6.91$, $p = 0.013$).

Astragalus shape was not significantly related to digit number (Figure 9.9f). Five digit species had a flatter calcaneoastragalar facet and a narrower distal trochlea, while four digit species had a blockier body and anteriorly directed neck. The differences were not significant (Wilk's Lambda = 0.003, $F_{10,1} = 30.12$, $p = 0.141$), though digit number was significantly related to PC 3 by itself ($F_{1,10} = 11.42$, $p = 0.007$).

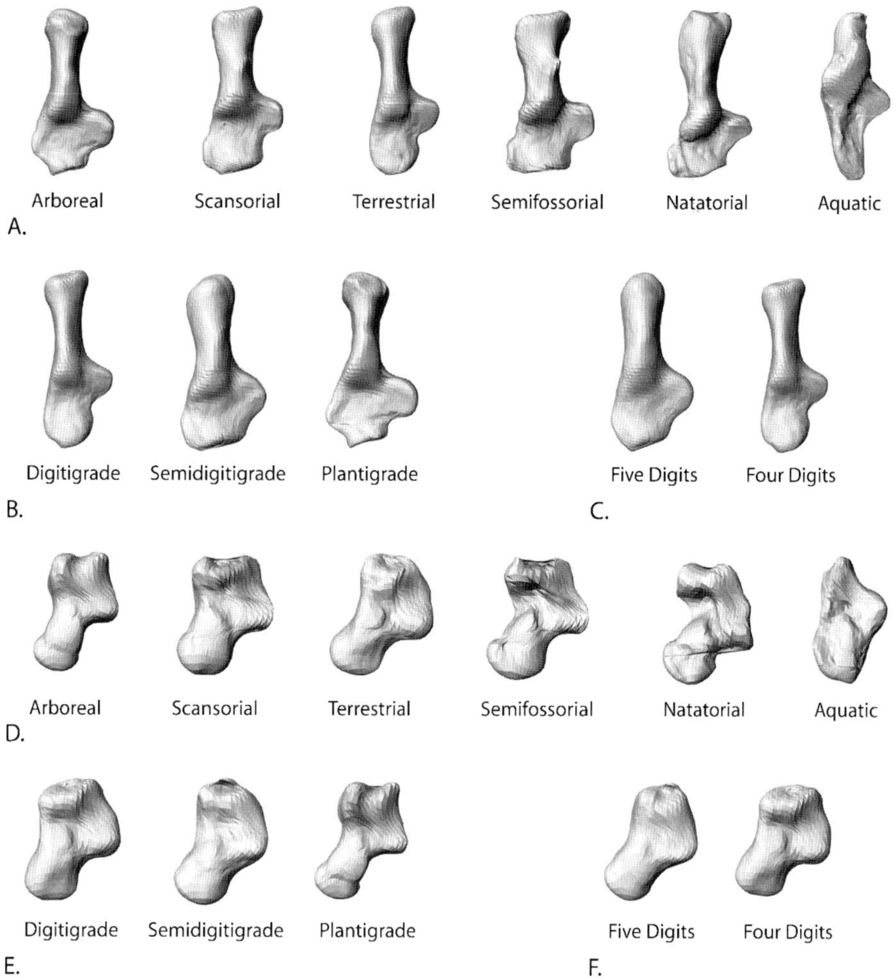

FIGURE 9.9. Mean tarsal shape for different locomotor categories. A, Mean calcaneum shape for the six locomotor types. Differences among types are significant ($p < 0.01$). B, Mean calcaneum shape for the three stance types. Differences are not significant ($p = 0.07$; but see text). C, Mean calcaneum shape for four and five digit species. Differences are not significant ($p = 0.26$; but see text). D, Mean astragalar shape for the six locomotor types. Differences are significant ($p < 0.01$). E, Mean astragalar shape for the three stance types. Differences are significant ($p < 0.01$). F, Mean astragalar shape for four and five digit species. Differences are not significant ($p = 0.14$).

9.3.5 Locomotion and Facets

Occluding facets are expected to have different areas and curvatures. Joints with greater mobility should have greater disparity in size and curvature than those with less (MacConaill, 1946a, b, c; Szalay and Drawhorn, 1980; Szalay, 1994). Surprisingly, neither facet area nor curvature was correlated with locomotor type in these carnivores (Table 9.2, Figure 9.10).

Locomotor type and sustentacular facet area index – the proportion of the larger to smaller facet – were not significantly correlated ($F_{5,6} = 0.82$, $p = 0.58$). *Lutra, Bassaricyon*, and *Ailurus* had the biggest difference in facet size, while *Crocuta, Paradoxurus*, and *Phoca* had the smallest. In all species except *Phoca*, the calcaneum had the largest of the two sustentacular facets, probably because in fissipeds the concave shape of the lower facet supports weight and constrains translation of the astragalar neck (Figure 9.10a).

The size of upper and lower calcaneoastragalar facets was also unrelated to locomotor type (Figure 9.10b). The statistical relationship was not significant ($F_{5,6} = 2.41$, $p = 0.16$) and there was no obvious phylogenetic relationship. *Canis, Felis, Crocuta*, and *Phoca* had facets that were the most equal in size, and *Lutra* had the ones with the biggest difference. About half the taxa had a larger calcaneal facet, while the remaining had a larger astragalar one.

The sustentacular facet was more strongly curved on the astragalus than on the calcaneum in all species (Figure 9.10c). *Bassaricyon, Leptailurus*, and *Lynx* had curvatures that were most similar, while *Paradoxurus, Meles* and *Crocuta* had the biggest difference. In *Paradoxurus* the curvature coefficient on the astragalus was especially large, 3.4 times greater than on the calcaneum. This value does not seem to be in error (the measurements and calculations were rechecked several times), but it does not correspond to any known locomotor peculiarity

TABLE 9.2. Area and curvature of the sustentacular and calcaneoastragalar facets on the astragalus and calcaneum. The index is the ratio of the larger over the smaller value for each facet (compare with plots in Figure 9.10).

	Area (mm^2)						Curvature					
	Calcaneum		Astragalus		Index		Calcaneum		Astragalus		Index	
Species	Sust.	Calc-astr.	Sust.	Calc-astr.	Sust.	Calc-astr.	Sust.	Calc-astr.	Sust.	Calc-astr.	Sust.	Calc-astr.
Ailurus	2.59	3.77	3.24	3.62	1.25	1.04	0.040	0.092	0.066	0.065	1.64	1.41
Bassaricyon	2.37	3.11	3.07	3.32	1.30	1.07	0.043	0.088	0.050	0.074	1.16	1.19
Canis	3.81	5.25	4.07	5.25	1.07	1.00	0.025	0.086	0.035	0.098	1.40	1.14
Crocuta	4.53	5.25	4.64	5.31	1.02	1.01	0.018	0.077	0.032	0.055	1.78	1.40
Felis	2.84	3.80	3.35	3.76	1.18	1.01	0.039	0.149	0.052	0.119	1.33	1.25
Leptailurus	3.29	4.25	3.92	4.57	1.19	1.08	0.034	0.119	0.041	0.105	1.21	1.13
Lutra	2.81	4.45	3.61	3.86	1.28	1.15	0.048	0.139	0.067	0.070	1.40	1.99
Lynx	3.43	4.39	4.03	4.54	1.17	1.03	0.029	0.081	0.036	0.079	1.24	1.03
Meles	2.90	4.53	3.16	4.32	1.09	1.05	0.037	0.061	0.069	0.044	1.86	1.39
Mustela	1.50	2.85	1.90	2.63	1.26	1.09	0.069	0.201	0.090	0.169	1.30	1.19
Paradoxurus	2.81	3.47	3.05	3.32	1.09	1.04	0.035	0.093	0.119	0.056	3.36	1.67
Phoca	4.71	4.89	4.46	4.99	1.06	1.02	0.018	0.047	0.028	0.043	1.56	1.10

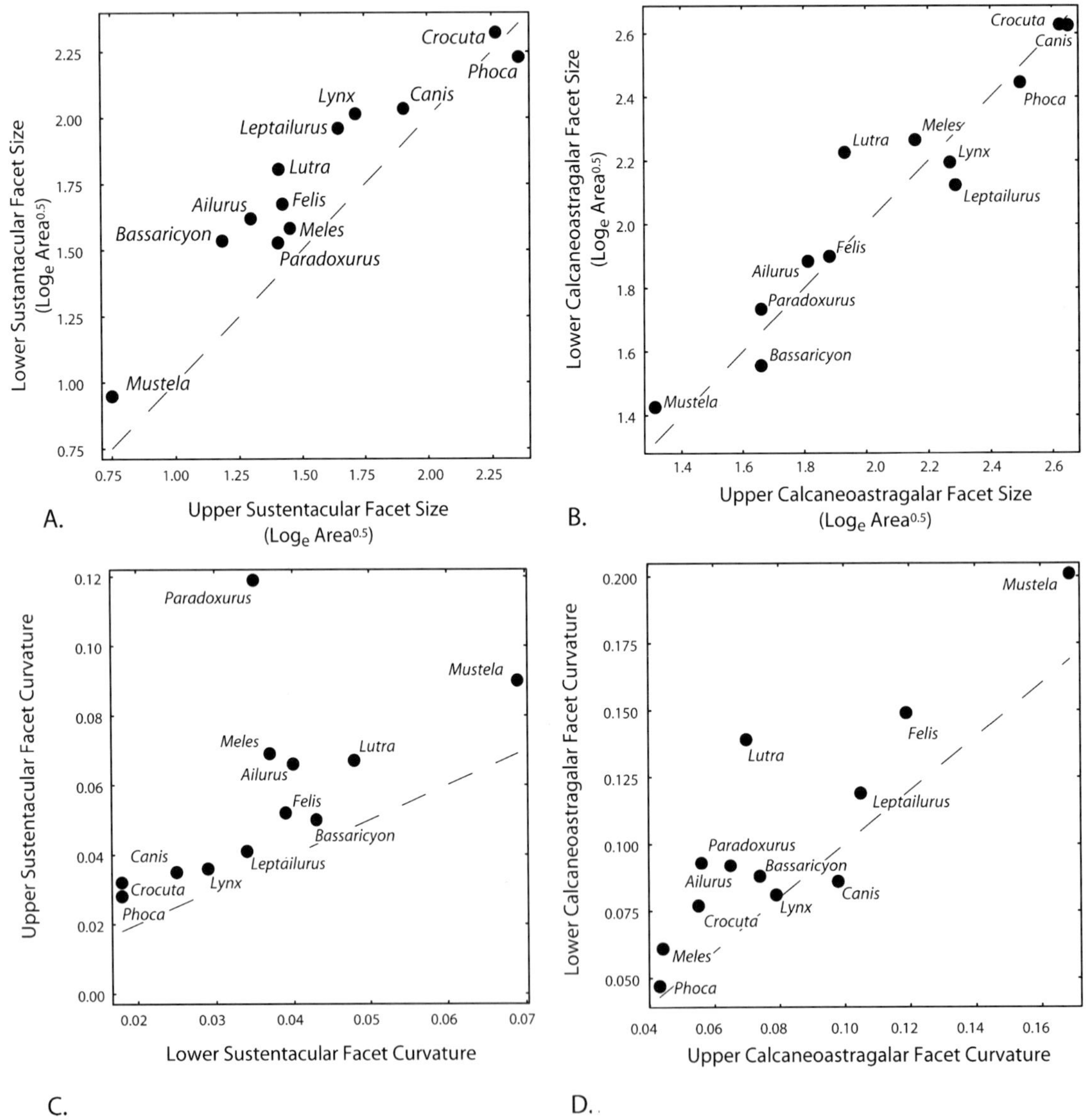

FIGURE 9.10. Area and curvature of occluding astragalar and calcaneal facets. A, Upper (astragalar) and lower (calcaneal) sustentacular facet size. B, Upper and lower calcaneoastragalar facet size. C, Upper and lower sustentacular facet curvature. D, Upper and lower calcaneoastragalar facet curvature. In each plot, the broken line shows a 1 to 1 relationship. The area plots use the natural logged square root of the measurements reported in Table 9.2. The curvature plots use the same values as Table 9.2.

of the animal. The large value may result from non-linearity in the curvature coefficient that could be compensated for by transformation before the calculation of ratios, but data were not available to assess this possibility. The *Paradoxurus* specimen was also from a relatively young animal, which could also explain the large difference in curvature between the two facets. There was no significant statistical relationship between curvature and locomotor mode ($F_{5,6} = 0.44$, $p = 0.80$).

Curvature in the calcaneoastragalar facets was greater on the calcaneum in all taxa except *Canis* (Figure 9.10c). *Lutra, Paradoxurus*, and *Ailurus* had the biggest difference in curvature, while *Lynx, Phoca*, and *Leptailurus* had the smallest. There was no statistical relationship between curvature and locomotor type for this facet either, though the association was stronger than for the sustentacular one ($F_{5,6} = 4.08$, $p = 0.06$).

These results suggest that mobility is constrained by something other than differences in curvature between occluding facets. The overall curvature of the joint, as opposed to the difference between its facets, is unquestionably related to mobility. The cartilaginous joint surfaces, which were not measured in this study, may also introduce differences in occluding facet curvature.

9.3.6 Body Mass and Facets

The area of all four facets was significantly related to body mass (Sustentacular: Calc, $F_{1,10} = 109.23$, $p < 0.01$; Astr, $F_{1,10} = 34.84$, $p < 0.01$. Calcaneoastragalar: Calc, $F_{1,10} = 49.59$, $p < 0.01$; Astr, $F_{1,10} = 49.78$, $p < 0.01$). Larger species had bigger facets. The index of proportional size was significantly related to body mass for the sustentacular facets ($F_{1,10} = 12.72$, $p < 0.01$). Larger species had sustentacular facets of similar size, while smaller species had proportionally larger sustentacular facets on the calcaneum. Body size was not related to the index of proportional size for the calcaneoastragalar facets ($F_{1,10} = 2.26$, $p = 0.16$).

Curvature was related to body mass for only some facets. The relation was significant for both sustentacular facets (Calc, $F_{1,10} = 25.56$, $p < 0.01$; Astr, $F_{1,10} = 7.36$, $p = 0.02$) and for the calcaneal calcaneoastragalar facet ($F_{1,10} = 7.32$, $p = 0.02$), but not for the astragalar calcaneoastragalar facet ($F_{1,10} = 3.99$, $p = 0.07$). All facets were less curved in larger species. The index of curvature was not related to body mass for either facet.

9.3.7 Correlation Between Calcaneum and Astragalus Shape

The calcaneum and astragalus are functionally integrated, especially at the synovial joint contacts. During the course of evolution, the position, size and shape of the facets must change together or the joint will not work. The PCAs identified major axes of variation, but those results cannot reveal anything about correlations between the bones. 2B-PLS was used to identify correlated variation in the calcaneum and astragalus. This method is similar to PCA in that it extracts a series of axes explaining shape variation but, unlike PCA, 2B-PLS finds axes of correlation between the two shapes. The results of a 2B-PLS analysis of calcaneum and astragalus shape are shown in Figure 9.11. Each PLS axis explains a certain percentage of the total covariation between the two bones, with the first axis explaining the most. Each axis has two vectors, one for each bone. Shape models along the two vectors illustrate variation correlated between the bones. Correlation in real species is usually not 100% and a perfect correlation would be represented on each plot by a diagonal line with a slope of one. Real data are scattered around that line, indicating residual variation in one or both of the bones that cannot be explained by the PLS axis. Each axis therefore has a correlation coefficient (*R*) that describes the strength of the correlation on that axis.

PLS 1 described correlation between the angles of the transverse tarsal joint and the lower ankle joint (Figure 9.11a). This axis explained 24.5% of the covariation between the two bones. On the calcaneum, the correlated variation included the orientation of the sustentacular facet, the angle of the calcaneoastragalar facet, the depth of the groove for the peroneus brevis, and the angle of the cuboid facet relative to the long axis of the bone. These features were correlated on the astragalus with the angle of the calcaneoastragalar facet to the neck, the blockiness of the body, and the relative size of the proximal trochlear ridges. The correlation coefficient (*R*) for the two bones on PLS 1 was 0.79. Note that neither PLS vector was exactly like any PC in Figure 9.7 or 9.8. The calcaneum PLS 1 was most similar to calcaneum PC 3, but contained some aspects of PC 1. Astragalus PLS 1 was a combination of PCs 1 and 2. The reason that PLS and PC axes were not identical was that part of the variation in either bone is not correlated with variation in the other. This bone-specific variation does not appear on any of the PLS axes, but it may contribute heavily to PC axes.

PLS 2 described the proximodistal position of the sustentacular facet (Figure 9.11b). PLS 2 explained 18.4% of the covariation. On the calcaneum this was manifested in the position of the sustentacular process and the width of the distal calcaneum and on the astragalus it was manifested in the length and angle of the neck. The calcaneum PLS vector closely resembled calcaneum PC 2, but the astragalus PLS vector was not really like any of its PC axes. Correlation between the two bones on PLS 2 was 0.90. PLS 2 has a strong functional component. With arboreal *Ailurus* and *Bassaricyon* at one end and aquatic *Phoca* at the other, the axis forms a transect from arboreal through natatorial, semiplantigrade terrestrial, scansorial, digitigrade cursorial, to aquatic.

PLS 3 and 4 explained 14.8% and 10.4% respectively, with correlation coefficients of $R = 0.77$ and $R = 0.83$. Readers can see for themselves the shape associated with each in Figure 9.11c and 9.11d.

9.3.8 Evolution of Tarsal Morphology

The fossil record yields the only direct data about morphologies and locomotor specializations in the past, but phylogenetic

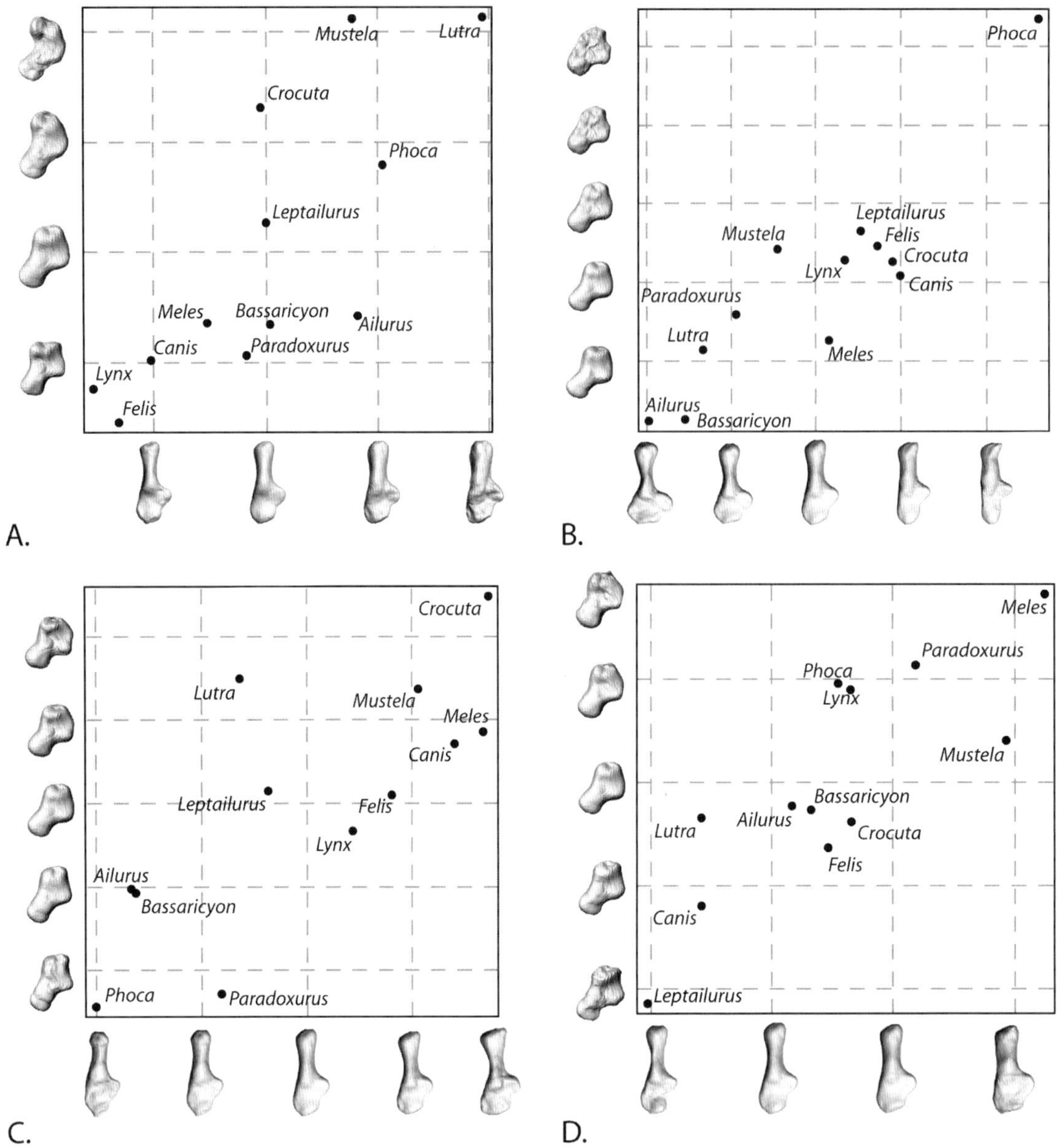

FIGURE 9.11. Correlation of calcaneum and astragalus shape 2B-PLS results. A, Axis 1. B, Axis 2. C, Axis 3. D, Axis 4. Each PLS axis has two vectors so that each graph has the calcaneum as its abscissa and the astragalus as its ordinate. The shape transformation along the vectors is shown as a series of models.

comparative analysis of living taxa can be used to make indirect inferences, inferences which can later be tested against palaeontological evidence. In this section, ancestral reconstructions of tarsal shape are presented, ancestral functional conditions are estimated by quantitative comparison of the reconstructions to the models in Figure 9.9, and the existence of an adaptive zone is tested (1) by projecting the phylogenetic tree back into the morphospace to look for boundaries and convergence; (2) by testing shape divergence relative to recency of common ancestry for saturation; and (3) by comparing terminal outliers to branching pattern and to ecological changes. A tree is also constructed from tarsal shape for comparison to the established phylogenetic tree.

Ancestral reconstructions and locomotor interpretations are shown in Figure 9.12. Only one reconstruction is shown for the multiple branching events at Nodes 1, 4, and 5 because those splits were spaced too closely in time to be separated paleontologically (Flynn, 1996). Node reconstructions are identical if the branching events are not separated in time. The shape at Node 3 is not shown because it was not visibly different from Node 2. The reconstructions were based on the 11 PC axes for each bone. Shapes were also reconstructed from the PLS axes, which maximize the correlation between the two bones, but they were not different than the ones shown here and will not be considered further. The reconstructions are the most parsimonious shapes for

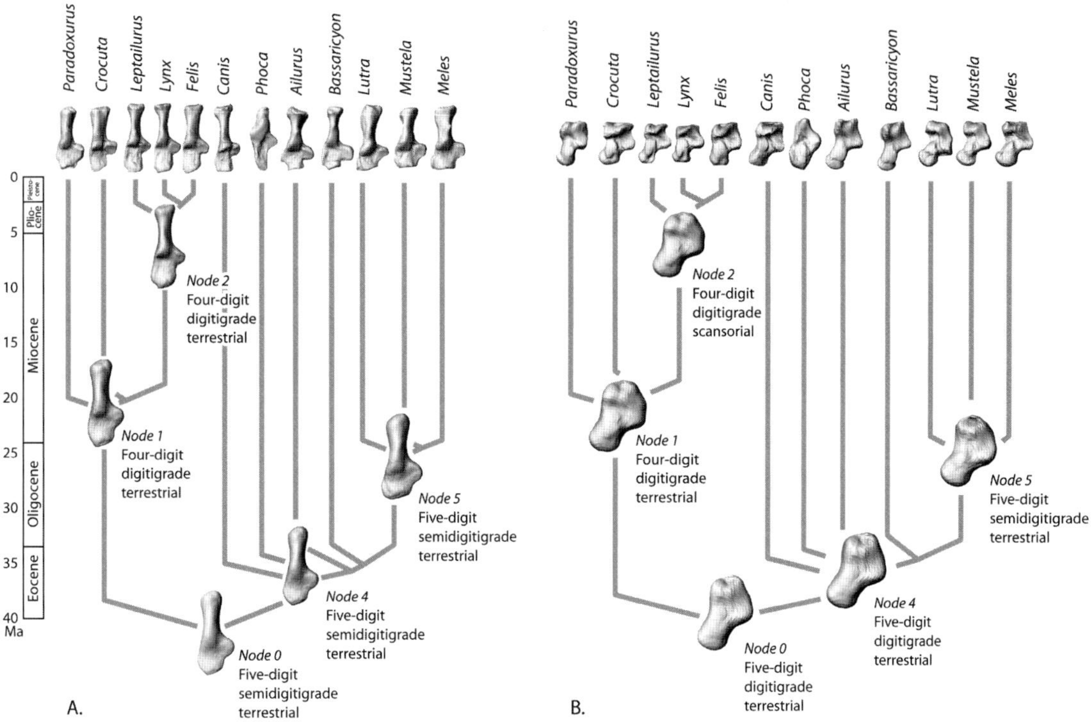

FIGURE 9.12. Ancestral node reconstructions based on the tree in Figure 9.5. A, Calcaneum. B, Astragalus. Locomotor descriptions are from the best match between the reconstructions and the models in Figure 9.9. Tip shapes are shown for reference. Node 3 is not shown because of its similarity to Node 2.

the tarsals using a likelihood model that assumes that the tree and its branch lengths are correct, assumes a Brownian motion model of evolution (which may not be the case because evolution in carnivoran tarsals may be constrained by on Ornstein-Uhlenbeck process, as discussed below), assumes that the shape of each species is accurately represented, and assumes that these taxa are representative of the true diversity of the group. Even though the reconstructions presented here are the most likely given the data, many other ancestral shapes may be nearly as likely (Martins and Hansen, 1997; Garland and Ives, 2000; Polly, 2001). The relative support for other hypotheses, such as independently observed fossil morphologies, can, in principle, be judged using log-likelihood ratios, but such tests were not attempted here because they are tangential to the main purpose of the paper.

The calcaneum and astragalus reconstructions at the base of the tree (Node 0) and at the base of the caniform clade (Node 4) were similar, in part because of the short temporal interval between them. Both shapes were interpreted as belonging to five-digit terrestrial species by finding the closest match among the locomotor models in Figure 9.9. The interpretation of stance differed in the two bones, with the calcaneum most closely matching the semidigitigrade mean and the astragalus matching the digitigrade one. The match is closer in the calcaneum, lending support to the semidigitigrade hypothesis, but differences among stance categories were more significant in the astragalus, lending contradictory support to the digitigrade hypothesis. These reconstructions are consistent with two previous hypotheses of a terrestrial ancestry for carnivorans, one based on eutherian mammals in general (Szalay, 1977a, 1984) and the other on the early carnivoran *Didymictis* (Heinrich and Rose, 1997). The ancestral shape of Mustelidae (Node 5) had a longer peroneal process on the calcaneum and a more trapezoidal body shape on the astragalus. The locomotor reconstruction was five digit, semidigitigrade terrestrial for both calcaneum and astragalus. The reconstruction at the base of the feliform clade (Node 1) was similar to the earliest carnivoran ancestor, but with a shorter distal calcaneum and blockier astragalus body. Both bones suggested a four digit, digitigrade, terrestrial ancestor. If this is correct, then the five digit condition of some viverrids and herpestids may be an evolutionary reversal, a scenario already feasible because of the variation in digit number among these groups and disagreements about their phylogeny (Taylor, 1970, 1976, 1988; Veron and Heard, 2000; Gaubert and Veron, 2003). The felid ancestor was the most visibly different, with a dainty peroneal process and a proximally placed sustentaculum. Both bones suggested a four digit, digitigrade stance, but the calcaneum suggested a terrestrial locomotor mode while the astragalus a scansorial one, an ambiguity contributed by the specializations of *Leptailurus*.

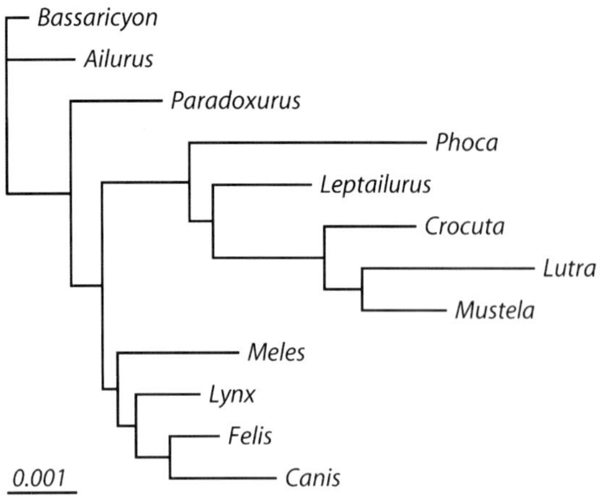

FIGURE 9.13. A maximum likelihood tree constructed from calcaneum and astragalus shape. The groupings are influenced by low-level phylogenetic relationship and locomotor type, but do not clearly indicate either.

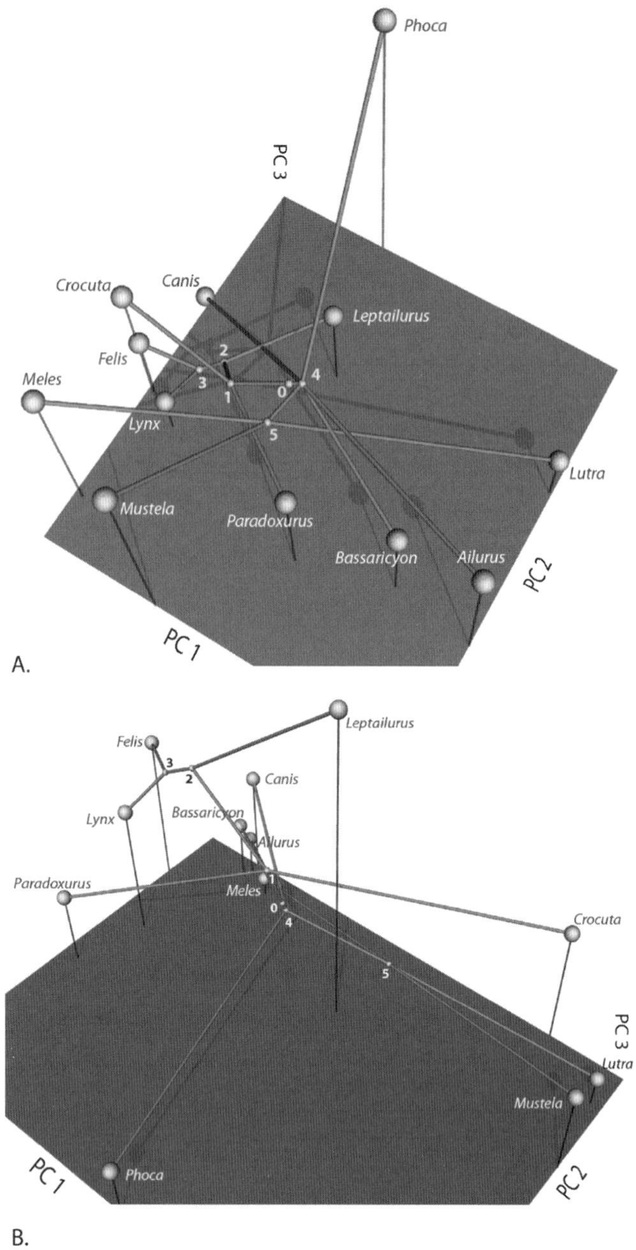

FIGURE 9.14. Tree projection into morphospace. A, Calcaneum. B, Astragalus. Node reconstructions from Figure 9.12 were projected into the PC spaces (Figures 9.7 and 9.8) and the branches connected. Node numbering follows Figure 9.5. Tarsal morphology crisscrossed the morphospace, erasing phylogenetic history as various locomotor types re-evolved in different clades. The branch leading to *Phoca* set off in a new direction not otherwise traversed.

The similarity among the reconstructed nodes and the diversity among the tips suggest considerable homoplasy in shape. I tested this first by building a maximum-likelihood tree from combined calcaneum and astragalus shape scores (Figure 9.13). This tree had a log likelihood of 600.1. Comparison with Figures 9.7 and 9.8 shows that the topology is influenced mostly by the astragalus. *Lutra, Mustela*, and *Crocuta*, which are united in the tree, are closely clustered in the astragalus PC plot, but not so in the calcaneum one. The groupings in the tree are not phylogenetic, but neither do they appear to be functional; they are a compromise between the two, and between the conflicting signals from the two bones. The lack of clear phylogenetic signal has two causes: (1) considerable shape convergence among the fissipeds has arisen because of their long history within a constrained adaptive zone, and (2) *Phoca*, which has escaped from that adaptive zone, has an especially different shape branch, preventing it from being sensibly grouped with the fissipeds.

The extent of the homoplasy, the outlines of the fissiped adaptive zone, and the position of *Phoca* outside the adaptive zone are clear in Figure 9.14. The diagrams were made by projecting the ancestral node reconstructions into the shape space and connecting the branches of the tree. Terminal taxa are represented by large, labeled balls and nodes by small numbered ones. The phylogenetic pattern in the shape space is a tangled mess. Descendants of Node 5, for example, have colonized all extremes of both the calcaneum and astragalus shape space, and the same is true for Node 4. The immediate ancestors of *Paradoxurus* and *Bassaricyon* evolved in different directions from node 0, but then the terminal lineages moved in parallel, ending up with similar forms. Despite the chaos, the fissipeds were constrained within a limited area of the space, while *Phoca* followed a trajectory out of the tangle into a shape region of its own. These features fulfill the main criteria proposed above for the recognition of an adaptive zone: the zone is occupied, the zone has phenotypic limits that are associated with functional differences (the functional associations described above for the first three PC axes), phylogenetic history suggests that evolution beyond the boundaries was rare, and the only lineage to travel outside the boundaries evolved into a radically different functional context.

The interplay between rate of shape evolution, the boundaries of the adaptive zone, and recency of common ancestry are important for understanding homoplasy and phylogenetic signal. For most of these taxa, the rate was rapid enough and common ancestry ancient enough for the ancestral condition not to be a good predictor of terminal shape. In other words, unrelated terminal taxa are likely to be more similar to one another than they are to their immediate ancestors. Node 5 is a good example. Descendants of that node have moved to three extremes in both shape spaces in only 24 million years. The rate of evolution is fast enough that any point in the fissiped adaptive zone can be reached from any other in much less time than the groups have been evolving. This situation makes convergent evolution highly probable. Only the felid clade has not diffused to the far corners. All three species occupy a relatively small proportion of both shape spaces. *Felis* and *Lynx* are closer to one another than they are to the more distantly related *Leptailurus* and subtending nodes 2 and 3 occupy phylogenetically logical positions. This is a recent clade, sharing its last common ancestor only 4.5 million years ago, a time too short to reach the boundaries of the adaptive zone. Note, however, that *Leptailurus* has moved nearly halfway across the space. Based on these observations, we can guess that good phylogenetic signal could be extracted when the last common ancestor is no more than 3–5 million years older than the taxa in question. Phylogenetic analysis based on tarsal shape would have to be confined to closely related taxa or have enough fossil taxa at intermediate ages to ensure that the longest branches are no more than a few million years long. This restriction only exists because of the boundaries of the adaptive zone. If evolution could continue indefinitely in any morphological direction, then homoplasy would be less likely because lineages would not have to double back in the morphospace. As long as the rate of evolution is slow enough, the adaptive boundaries broad enough, and the branch lengths short enough, phylogenetic reconstruction from morphological shape is possible.

Rates on the tree varied, but fell within an order of magnitude of one another. Rates were calculated by measuring shape change along each branch and divided by branch length scaled to generations. Branch lengths and rates along terminal branches are reported in Table 9.3. The highest rate was 2.5×10^{-8} Procrustes units/gen, measured in the astragalus of *Leptailurus*.

TABLE 9.3. Branch lengths and rates of shape change from terminal taxa to their immediately ancestral node. Branch lengths are in Procrustes units^2 and rates in Procrustes units^2 per generation.

	Calcaneum		Astragalus	
Species	Branch length	Rate	Branch length	Rate
Ailurus	0.021	2.2×10^{-9}	0.018	1.9×10^{-9}
Bassaricyon	0.014	1.5×10^{-9}	0.017	1.8×10^{-9}
Canis	0.019	2.0×10^{-9}	0.026	2.8×10^{-9}
Crocuta	0.008	1.4×10^{-9}	0.048	8.3×10^{-9}
Felis	0.003	3.3×10^{-9}	0.015	1.6×10^{-8}
Leptailurus	0.007	6.0×10^{-9}	0.029	2.5×10^{-8}
Lutra	0.023	3.7×10^{-9}	0.032	5.0×10^{-9}
Lynx	0.004	4.3×10^{-9}	0.009	1.0×10^{-8}
Meles	0.009	1.4×10^{-9}	0.068	1.1×10^{-8}
Mustela	0.012	1.9×10^{-9}	0.031	5.0×10^{-9}
Paradoxurus	0.008	1.3×10^{-9}	0.030	5.2×10^{-9}
Phoca	0.026	2.7×10^{-9}	0.050	5.2×10^{-9}

The lowest rate was 1.3×10^{-9} in the calcaneum of *Paradoxurus*. The highest calcaneum rate was 6.0×10^{-9} in *Leptailurus*, and the lowest astragalus rate was 1.8×10^{-9} in *Bassaricyon*. Some of the highest rates were in the shortest branches, especially the felid ones. This distribution could be related to the scaling phenomenon of rate to interval, where shorter intervals yield higher rates all else being equal (Gingerich, 1993), but some of the highest rates (e.g., the rate of astragalus change in *Meles*) were associated with long branches.

The rates in *Phoca* were intermediate. Its calcaneum rate of 2.7×10^{-9} fifth highest, and its astragalus rate of 5.2×10^{-9} was sixth highest. Despite *Phoca* having traveled far beyond the boundaries of the fissiped adaptive zone, there is no evidence that it did so at a higher rate, contrary to the expectation of Simpson (1944, 1953) that crossing adaptive zone boundaries will be accompanied by an exceptionally high rate of change. The remarkable, derived morphology of *Phoca's* tarsals was not produced by evolving faster, but simply by evolving in a different direction than other fissiped taxa, following trajectories with different covariances, a situation in keeping with Hecht's (1965) view that seemingly rapid rates may result from morphologic reorganization. The choice of *Phoca* among the extant pinnipeds should not influence the conclusion drawn here, however. Because phocids have a more derived tarsal morphology than otariids or odobenids, the estimated rate of change will be maximized so that quantum evolution, if it were present, should be easily detectable using *Phoca*. But the rate calculated here was based on the amount of change between an extant species and a reconstructed ancestor. If fossil taxa were included, one might find a localized region of the pinniped clade which did experience rapid evolution.

The constraints on phylogenetic divergence can be measured a second way using the scaling of morphological divergence to time since common ancestry (Figure 9.15). The scaling relationship indicates whether the phenotype is evolving freely, whether it is constrained in how much it can change, or whether it is being pushed uniformly in a particular direction (Gingerich, 1993; Hansen and Martins, 1996; Polly, 2001, 2004; Roopnarine, 2001). When evolution is unconstrained and unbiased (i.e., Brownian motion), phenotypic divergence will increase with the square-root of time (it diverges linearly when measured in variance units); when the phenotype changes constantly in a single direction, such as under long-term directional selection, divergence will scale linearly with time; and when divergence is constrained, such as within an adaptive zone, it will reach a plateau with respect to time. These modes of evolution can be measured by regression fitting. For untransformed data, a function of divergence to time near $x^{0.5}$ will describe unconstrained evolution, a function near $x^{0.0}$ will describe constrained evolution, and a function of $x^{1.0}$ will describe directional evolution. Figure 9.15 shows four plots of divergence and time since common ancestry. In the first two, time was measured from the fossil record; in the second two, it was measured as mitochondrial cytochrome *b* genetic distance. A series of functions were fit to each plot with powers of *x* varying between 0.1 and 1 (broken lines). The line that minimized the residual error

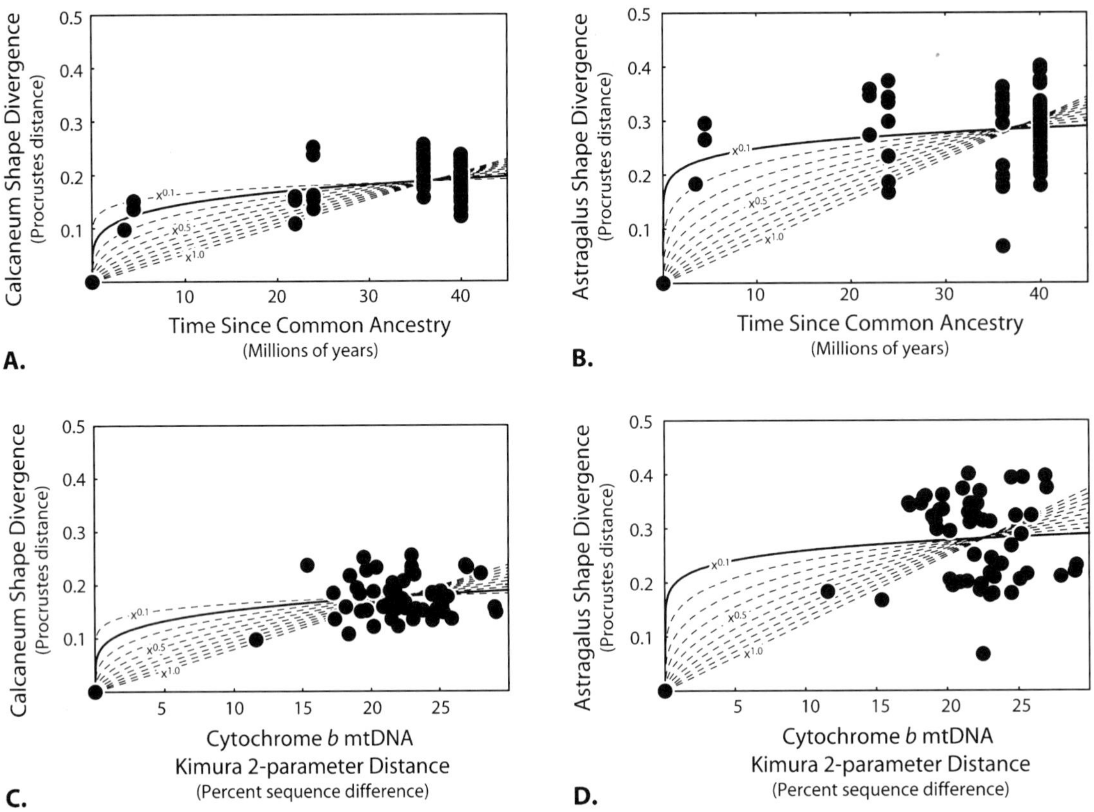

FIGURE 9.15. Mode of evolution in tarsal morphology inferred by fitting functions of x. A, Calcaneum divergence as a function of time since common ancestry. B, Astragalus divergences as a function of time since common ancestry. C, Calcaneum divergences as a function of genetic distance. D, Astragalus divergence as a function of genetic distance. Ten functions were fit (broken curves) to find the best (solid curve). The best fit for all four was $x^{0.1}$ or $x^{0.2}$, values consistent with the broad constraints characteristic of adaptive zones.

was chosen as the best hypothesis of evolutionary mode. For the calcaneum, the best fit was obtained by $x^{0.2}$ and for the astragalus $x^{0.1}$ (solid lines). Both coefficients indicate that divergence was highly constrained, a finding consistent with the phenotypes being trapped within an adaptive zone.

From the perspective of clades evolving within a constrained phenotypic space, Rosen's (1974) characterization of an adaptive zone as a collection of taxa sharing primitive features is nonsensical. All of the taxa here, fissiped and pinniped, have derived morphologies, easily seen in the difference between the shapes of terminal taxa and their subtending nodes. Indeed, the branch distances reported in Table 9.3 demonstrate that *Phoca's* calcaneum is not much more derived than *Ailurus*'. The fissiped taxa cannot be said to be united by symplesiomorphy, because they do not share anything in common that they do not also share with *Phoca*. Rather, fissiped taxa are constrained by the adaptive zone boundaries so that their morphology suffers from considerable convergence (convergence is a better term than homoplasy, because the latter implies identical character states, which do not easily occur in continuously variable quantitative data). Phylogenetically convergent similarity is not shared ancestral similarity. Consequently, the criteria proposed here for recognizing an adaptive zone rather belatedly answer Rosen's (1974) challenge that no evolutionist has been able to formulate adaptive zone concepts into a "recognizable methodology".

9.4 Conclusions

The main functional findings of this study are not new. The significance of the position of the sustentaculum, the length of the peroneal process, the shape of facets, and the length of the astragalus neck for carnivore ankle mobility has been long recognized. Rather, this study contributed a quantitative assessment of the three-dimensional form of these features, allowing direct statistical analysis. This study found, for example, that the long held opinion that intertarsal mobility is enhanced by larger and flatter facets on one of the two bones was not supported in Carnivora. This study also generated new hypotheses about carnivoran tarsal evolution: the reconstructions based on modern taxa predict that the ancestral carnivoran had tarsals adapted for a semidigitigrade terrestrial locomotion and the quantification of the rate of shape change on the phylogenetic tree suggested that pinniped tarsals no faster than in fissiped clades, even though pinniped morphology appears to be more derived.

The more far reaching significance of the paper concerns the relationship between form, function, and phylogeny.

Function and phylogeny are not mutually exclusive explanations for morphological variation. Skeletal morphology is clearly functional and therefore susceptible to convergent selection, a process which confuses phylogenetic interpretation. The carnivoran tarsals in this study are a good example, because their shape is statistically associated with locomotor type, stance, and number of digits and because the regions of shape space associated with these functional categories had been converged upon by unrelated clades. Yet, evolutionary transformations, even convergent ones, occur through phylogenetic divergence, meaning that variance among the most closely related taxa will have a strong phylogenetic component. In this study, young clades, such as the felids, were not affected by convergence even though deeper clades were. Phylogenetic history leaves an imprint on evolving morphology – the issue for systematics is to determine how the interaction between common ancestry and adaptive convergence affect the dominance of phylogeny and function in morphometric data from a given set of taxa.

Simpson's adaptive zones provide a suitable analogy for understanding the tradeoffs between common ancestry and functional adaptation. If an adaptive zone is considered to be a region of morphospace whose axes are defined by functionally correlated variation and within whose space evolution is constrained, then the concept can be used to assess the conditions under which phylogeny should be reconstructable. Fissiped tarsals fell within such a region whose boundaries were not normally crossed during the evolution of the group (Figure 9.16). The functional, terrestrial nature of the boundaries of the morphospace can be clearly seen in the morphologies that

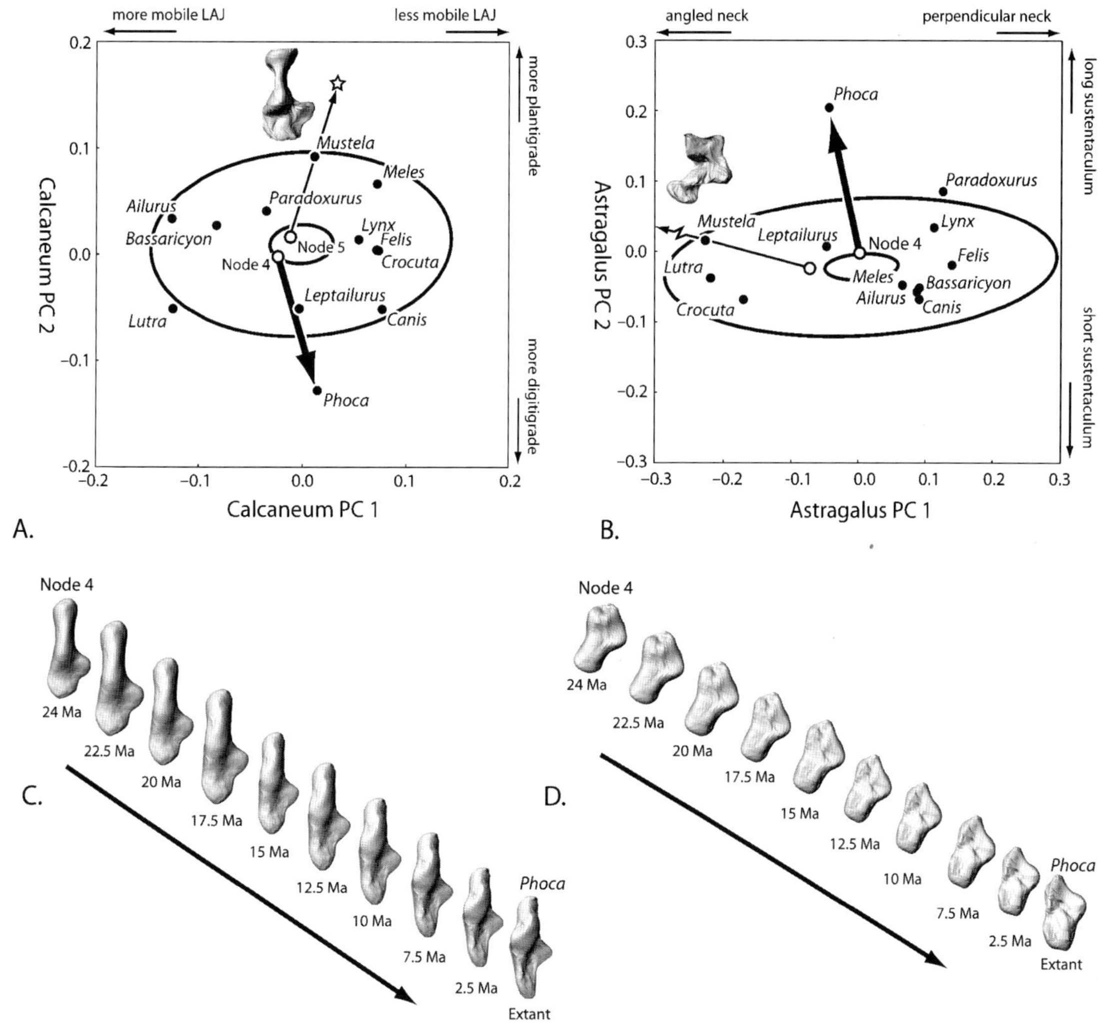

FIGURE 9.16. Escape from the Fissiped adaptive zone. A, Calcaneum adaptive zone projected onto the first two PCs. Ellipses characterizing the adaptive zone are the 5th and 95th percentiles of fissiped shape variation. The heavy arrow shows the branch leading from Node 4 to *Phoca*, which is modeled in C. The light arrow shows a hypothetical trajectory from Node 5 to a point outside the adaptive zone (star) with *Mustela* as its midpoint. The shape at the hypothetical point is shown. B, Astragalus adaptive zone with the same conventions. Note the hypothetical point beyond *Mustela* lies outside the graph area. C, Calcaneum shape models along the branch leading from Node 4 to *Phoca*, the most probable trajectory for the escape. D, Astragalus shape models for the same branch. The trajectories are shown as dark arrows in A and B.

lie outside them. For example, when the phylogenetic branch leading to *Mustela* is projected outside the zone boundary, the resulting morphology has a combination of features that are functionally incompatible with terrestrial locomotion. That morphology has a broad distal end, but no peroneal tubercle; its sustentacular and calcaneoastragalar facets are oriented like a digitigrade terrestrial species, even though these normally have narrow distal calcanea; and its astragalar neck is thin and sharply angled. The branch leading to *Phoca* also traces morphologies that are incompatible with terrestrial locomotion – something known from both visual inspection of the bones and knowledge of movement in living seals. The elongation of the calcaneum facets combined with the enlargement of the astragalus neck and sustentacular facet correspond to a side-to-side movement in the LAJ that undesirable for stable movements on a firm substrate.

If prevalent, adaptive zones like the fissiped one may place identifiable constraints on morphological phylogeny reconstruction. The adaptive zone identified here acts as a loosely constrained Ornstein-Uhlenbeck process, which by erasing deep phylogenetic history as different lineages explore the same morphospace again and again, violates the Brownian motion assumptions required for effective phylogeny reconstruction (Felsenstein, 1988, 2002). The departure of *Phoca* from the adaptive zone further confounds the phylogenetic picture by creating a long-branch that does not easily link to taxa within the zone.

While these findings sound pessimistic in regards to the reconstruction of phylogeny from tarsal morphology, echoing Nadal-Roberts and Collard (2005), such reconstruction is possible in the right circumstances. The limits on reconstruction depend on the size of the adaptive zone, the rate of evolution, and the recency of common ancestry. In the fissiped zone, rate and zone size dictate that the best reconstructions will be with taxa who shared a common ancestor within the previous 5–15 million years, as shown by the position of the felid tarsals and the age of their common ancestor. Thus, we can expect good phylogenetic signal, for example, from living taxa that have radiated since the end of the Miocene, as well as for fossil taxa that radiated within the Early Oligocene. Relationships among taxa whose last common ancestor is older than 5–15 million years could, in principle, be reconstructed if fossils of intermediate ages were strategically included in the analysis, an idea that invites further investigation. The limit for phylogeny reconstruction of 5–15 million years is unlikely to hold across taxa or character complexes, though it is compatible with findings from teeth, mandibles, and skulls in rodents (Caumul and Polly, 2005). Sweeping statements about the lack of phylogenetic signal in morphometric data cannot be expected to hold at all levels of analysis; rather the nature of functional and phylogenetic components of morphological variance need to be scientifically assessed.

Acknowledgments. Thanks to Eric Sargis and Marian Dagosto for organizing this volume and allowing my contribution. Annalisa Berta, Andrea Cardini, Alistair Evans, Joseph Felsenstein, Jason Head, Paula Jenkins, Jukka Jernvall, Lauren Morris, Marcelo Sánchez-Villagra, Fred Szalay, Heidi Schutz, Rebecca Spang, Mark Uhen, Blaire Van Valkenburgh, Vera Weisbecker, and two anonymous reviewers provided information, discussed issues, or made comments that contributed to this paper. Norman MacLeod deserves special acknowledgment for lengthy discussions that culminated in the method of placing points on three-dimensional surfaces, discussions without which the 3D surface analysis would not have been possible. Steve Le Comber, and, by extension, George Gamow, made a creative contribution to the measure of joint surface curvature. Andy Currant (NHM) and Haidee Price-Thomas (QMUL) provided access to specimens in their care. Thank you Fred Szalay for always being challenging, controversial, interesting, inspiring, and, more recently, a friend.

References

Acero, A., Tavera, J. J., Reyes, J., 2005. Systematics of the genus *Bagre* (Siluriformes: Ariidae): a morphometric approach. Cybium 29, 127–133.

Alexander, R. M., 2003. Principles of Animal Locomotion. Princeton University Press, Princeton, NJ.

Alroy, J., 1998. Cope's rule and the dynamics of body mass evolution in North American fossil mammals. Science 280, 731–734.

Arnold, S. J., Pfrender, M. E., Jones, A. G., 2001. The adaptive landscape as a conceptual bridge between micro- and macroevolution. Genetica 112–113, 9–32.

Bininda-Emonds, O. R. P., Gittleman, J. L., 2000. Are pinnipeds functionally different from fissiped carnivores? The importance of phylogenetic comparative analyses. Evolution 54, 1011–1023.

Bininda-Emonds, O. R. P., Gittleman, J. L., Purvis, A., 1999. Building large trees by combining phylogenetic information: a complete phylogeny of the Carnivora (Mammalia). Biological Reviews of the Cambridge Philosophical Society 74, 143–175.

Bock, W. J., 1965. The role of adaptive mechanisms in the origin of higher levels of organization. Systematic Zoology 14, 272–287.

Bock, W. J., von Wahlert, G., 1965. Adaptation and the form-function complex. Evolution 19, 269–299.

Bookstein, F. L., 1991. Morphometric Tools for Landmark Data. Cambridge University Press, Cambridge.

Bookstein, F. L., Gingerich, P. D., Kluge, A. G., 1978. Hierarchical linear modeling of the tempo and mode of evolution. Paleobiology 4, 120–134.

Brown, W. M., George, M. Jr., Wilson, A. C., 1979. Rapid evolution of animal mitochondrial DNA. Proceedings of the National Academy of Sciences USA 76, 1967–1971.

Butler, M. A., King, A. A., 2004. Phylogenetic comparative analysis: a modeling approach for adaptive evolution. American Naturalist 164, 683–695.

Caumul, R., Polly, P. D. 2005. Comparative phylogenetic and environmental components of morphological variation: skull, mandible and molar shape in marmots (*Marmota*, Rodentia). Evolution 59.

Cheverud, J. M., 1996. Developmental integration and the evolution of pleiotropy. American Zoologist 36, 44–50.
Clevedon Brown, J., Yalden, D. W., 1973. The description of mammals -2. Limbs and locomotion of terrestrial mammals. Mammal Review 3, 107–135.
Cock, A. G., 1966. Genetical aspects of metrical growth and form in animals. Quarterly Review of Biology 41, 131–190.
Davis, C. S., Delisle, I., Stirling, I., Siniff, D. B., Strobeck, C., 2004. A phylogeny of the extant Phocidae inferred from complete mitochondrial DNA coding regions. Molecular Phylogenetics and Evolution 33, 363–377.
Decker, R. L., Szalay, F. S., 1974. Origins and function of the pes in the Eocene Adapidae (Lemuriformes, Primates). In: Jenkins, F. A., Jr. (Ed.), Primate Locomotion. Academic Press, New York, pp. 261–291.
Dryden, I. L., Mardia, K. V., 1998. Statistical Analysis of Shape. Wiley, New York.
Eguchi, S., Townsend, G. C., Richards, L. C., Hughes, T., Kasai, K., 2004. Genetic contribution to dental arch size variation in Australian twins. Archives of Oral Biology 49, 1015–1024.
Eisenberg, J. F., 1981. The Mammalian Radiations: An Analysis of Trends in Evolution, Adaptation, and Behavior. Chicago University Press, Chicago, IL.
Eisenberg, J. F., 1989. Mammals of the Neotropics: the Northern Neotropics. University of Chicago Press, Chicago, IL.
Evans, H. E., 1993. Miller's Anatomy of the Dog, 3rd Ed. W. B. Saunders, Philadelphia.
Felsenstein, J., 1973. Maximum-likelihood estimation of evolutionary trees from continuous characters. American Journal of Human Genetics 25, 471–492.
Felsenstein, J., 1981. Evolutionary trees from gene frequencies and quantitative characters: finding maximum likelihood estimates. Evolution 35, 1229–1242.
Felsenstein, J., 1988. Phylogenies and quantitative characters. Annual Review of Ecology and Systematics 19, 445–471.
Felsenstein, J., 1993. PHYLIP Phylogeny Inference Package version 3.5c. Distributed by the author. Department of Genetics, University of Washington, Seattle.
Felsenstein, J., 2002. Quantitative characters, phylogenies, and morphometrics. In: MacLeod, N., Forey, P. (Eds.), Morphology, Shape, and Phylogenetics. Taylor & Francis, London, pp. 27–44.
Flynn, J. J., 1996. Carnivoran phylogeny and rates of evolution: morphological, taxic, and molecular. In: Gittleman, J. L. (Ed.), Carnivore Behavior, Ecology, and Evolution, Cornell University Press, Ithaca, NY, pp. 542–581.
Flynn, J. J., Nedbal, M. A., 1998. Phylogeny of the Carnivora (Mammalia): congruence vs. incompatibility among multiple data sets. Molecular Phylogenetics and Evolution 9, 414–426.
Flynn, J. J., Wesley-Hunt, G. D., 2005. Carnivora. In: Archibald, D., Rose, K. (Eds.), Origin, Timing, and Relationships of the Major Clades of Extant Placental Mammals. Johns Hopkins University Press, Baltimore, MD, pp. 175–198.
Flynn, J. J., Neff, N. A., Tedford, R. H., 1988. Phylogeny of the Carnivora. In: Benton, M. J. (Ed.), The Phylogeny and Classification of Tetrapods, Volume 2. Clarendon, Oxford, pp. 73–116.
Flynn, J. J., Nedbal, M. A., Dragoo, J. W., Honeycutt, R. L., 2000. Whence the red panda? Molecular Phylogenetics and Evolution 17, 190–199.
Flynn, J. J., Finarelli, J. A., Zehr, S., Hsu, J., Nedbal, M. A., 2005. Molecular phylogeny of the Carnivora Mammalia: assessing the impact of increased sampling on resolving enigmatic relationships. Systematic Biology 54, 317–337.
Fox, R. C., Youzwyshyn, G. P., 1994. New primitive carnivorans Mammalia from the Paleocene of Western Canada, and their bearing on relationships of the order. Journal of Vertebrate Paleontology 14, 382–404.
Gambaryan, P. P., 1974. How Mammals Run: Anatomical Adaptations. Wiley, New York.
Garland, T. Jr., Ives, A. R., 2000. Using the past to predict the present: confidence intervals for regression equations in phylogenetic comparative methods. American Naturalist 155, 346–364.
Garland, T. Jr., Midford, P. E., Ives, A. R., 1999. An introduction to phylogenetically based statistical methods, with a new method for confidence intervals on ancestral values. American Zoologist 39, 374–388.
Gaubert, P., Veron, G., 2003. Exhaustive sample set among Viverridae reveals the sister-group of felids: the linsangs as a case of extreme morphological convergence within Feliformia. Proceedings of the Royal Society of London B 270, 2523–2530.
Gingerich, P. D., 1993. Quantification and comparison of evolutionary rates. American Journal of Science 293-A, 453–478.
Gingerich, P. D., 2001. Rates of evolution on the time scale of the evolutionary process. Genetica 112–113, 127–144.
Gingerich, P. D., Winkler, D. A., 1985. Systematics of Paleocene Viverravidae (Mammalia, Carnivora) in the Bighorn Basin and Clark's Fork Basin, Wyoming. Contributions from the Museum of Paleontology, University of Michigan 27, 87–128.
Grafen, A., 1989. The phylogenetic regression. Philosophical Transactions of the Royal Society of London B 326, 119–137.
Graur, D., Martin, W., 2004. Reading the entrails of chickens: molecular timescales of evolution and the illusion of precision. Trends in Genetics 20, 80–86.
Gonyea, W. J., 1976. Adaptive differences in the body proportions of large felids. Acta Anatomica 96, 81–96.
Goslow, G. E., Van de Graff, K., 1982. Hindlimb joint angle changes and action of the primary extensor muscles during posture and locomotion in the Striped skunk *Mephitis mephitis*. Journal of Zoology (London) 1982, 405–419.
Greene, E. C., 1935. Anatomy of the rat. Transactions of the American Philosophical Society 27, 1–370.
Gregory, W. K., 1951. Evolution Emerging. Macmillan, New York.
Grüneberg, H., 1967. The Pathology of Development: A Study of Inherited Skeletal Disorders in Animals. Blackwell, Oxford.
Hansen, T. F., Martins, E. P., 1996. Translating between microevolutionary process and macroevolutionary patterns: the correlation structure of interspecific data. Evolution 50, 1404–1417.
Hecht, M. K., 1965. The role of natural selection and evolutionary rates in the origin of higher levels of organization. Systematic Zoology 14, 301–317.
Heinrich, R. E., Rose, K. D., 1997. Postcranial morphology and locomotor behaviour of two early Eocene miacoid carnivorans, *Vulpavus* and *Didymictis*. Palaeontology 40, 279–305.
Hildebrand, M., 1954. Comparative morphology of the body skeleton in recent Canidae. University of California Publications in Zoology 52, 399–470.
Hildebrand, M., 1980. The adaptive significance of tetrapod gait selection. American Zoologist 20, 255–267.

Howard, L. D., 1973. Muscular anatomy of the hind limb of the otter *Enhydra lutris*. Proceedings of the California Academy of Sciences 40, 335–416.

Howell, A. B., 1929. Contribution to the comparative anatomy of the eared and earless seals genera *Zalophus* and *Phoca*. Proceedings of USNM 73, 1–142.

Howell, A. B., 1930. Aquatic Mammals: Their Adaptations to Life in the Water. Charles Thomas, Springfield, IL.

Howell, A. B., 1944. Speed in Animals. University of Chicago Press, Chicago, IL.

Hunt, R. M., Tedford, R. A., 1993. Phylogenetic relationships within the aeluroid carnivora and implications of their temporal and geographic distribution. In: Szalay, F. S., Novacek, M. J., McKenna (Eds.), Mammalian Phylogeny: Placentals. Springer, New York, pp. 53–73.

Janis, C. M., Baskin, J. A., Berta, A., Flynn, J. J., Gunnell, G. F., Hunt, R. M., Martin, L. D., Munthe, K., 1998. Carnivorous mammals. In: Janis, C. M., Scott, K. M., Jacobs, L. J. (Eds.), Tertiary Mammals of North America. Cambridge University Press, Cambridge, pp. 73–90.

Jenkins, F. A., Camazine, S. M., 1977. Hip structure and locomotion in ambulatory and cursorial carnivores. Journal of Zoology (London) 181, 351–370.

Jenkins, F. A., McClearn, D., 1984. Mechanisms of hind foot reversal in climbing mammals. Journal of Morphology 182, 197–219.

Kennel Club, 1998. Illustrated Breed Standards: the Official Guide to Registered Breeds. Ebury, London.

Kimura, M., 1980. A simple method for estimating evolutionary rates of base substitutions through comparative studies of nucleotide-sequences. Journal of Molecular Evolution 16, 111–120.

King, J. E., 1966. Relationships of the hooded and elephant seals (Genera *Cystophora* and *Mirounga*). Journal of Zoology (London) 148, 385–398.

Kirkpatrick, M., 1982. Quantum evolution and punctuated equilibria in continuous genetic characters. American Naturalist 119, 833–848.

Kluge, A. G., 1989. A concern for evidence and a phylogenetic hypothesis of relationships among *Epicrates* (Boidae, Serpentes). Systematic Zoology 38, 7–25.

Koepfli, K.-P., Wayne, R. K., 1998. Phylogenetic relationships of otters Carnivora: Mustelidae based on mitochondrial cytochrome b sequences. Journal of Zoology 246, 401–416.

Koepfli, K.-P., Wayne, R. K., 2003. Type-1 STS markers are more informative than cytochrome b in phylogenetic reconstruction of the Mustelidae (Mammalia: Carnivora). Systematic Biology 52, 571–593.

Lande, R., 1976. Natural selection and random genetic drift in phenotypic evolution. Evolution 30, 314–344.

Lande, R., 1986. The dynamics of peak shifts and the pattern of morphological evolution. Paleobiology 12, 343–354.

Ledge, C., Árnason, Ú., 1996a. Phylogenetic analyses of complete cytochrome b genes of the Order Carnivora with particular emphasis on the Canifornia. Journal of Molecular Evolution 42, 135–144.

Ledge, C., Árnason, Ú., 1996b. Phylogenetic relationships within caniform carnivores based on analyses of the mitochondrial 12S rRNA gene. Journal of Molecular Evolution 43, 641–649.

Lento, G. M., Hickson, R. E., Chambers, G. K., Penny, D., 1995. Use of spectral analysis to test hypotheses on the origin of pinnipeds. Molecular Biology and Evolution 12, 28–52.

Lewis, O. J., 1989. Functional Morphology of the Evolving Hand and Foot. Clarendon, Oxford.

Lohmann, G. P., 1983. Eigenshape analysis of microfossils: a general morphometric method for describing changes in shape. Mathematical Geology 15, 659–672.

Lohmann, G. P., Schweitzer, P. N., 1990. On eigenshape analysis. The University of Michigan Museum of Zoology, Special Publication 2, 145–166.

MacConnaill, M. A., 1946a. Studies in the mechanics of synovial joints. I. Fundamental principles and diadochal movements. Irish Journal of Medical Science 246, 190–199.

MacConnaill, M. A., 1946b. Studies in the mechanics of synovial joints. II. Displacements on articular surfaces and the significance of saddle joints. Irish Journal Medical Science 247, 223–235.

MacConnaill, M. A., 1946c. Studies in the mechanics of synovial joints. III. Hinge-joints and the nature of intra-articular displacements. Irish Journal of Medical Science 249, 620–626.

MacLeod, N., 1999. Generalizing and extending the eigenshape method of shape space visualization and analysis. Paleobiology 25, 107–138.

MacLeod, N., 2002. Testing evolutionary hypotheses with adaptive landscapes: use of random phylogenetic-morphological simulation studies. Mathematical Geology 6, 45–55.

MacLeod, N., Rose, K. D., 1993. Inferring locomotor behavior in paleogene mammals via eigenshape analysis. American Journal of Science 293-A, 300–355.

Martins, E. P., Hansen, T. F., 1997. Phylogenies and the comparative method: a general approach to incorporating phylogenetic information into the analysis of interspecific data. American Naturalist 149, 646–667.

McArdle, B., Rodrigo, A. G., 1994. Estimating the ancestral states of a continuous-valued character using squared-change parsimony: an analytical solution. Systematic Biology 43, 573–578.

Miyamoto, M. M., 1985. Consensus cladograms and general classifications. Cladistics 1, 186–189.

Nadal-Roberts, M., Collard, M., 2005. The impact of methodological choices on assessments of the reliability of fossil primate phylogenetic hypotheses. Folia Primatologica 76, 207–221.

Nei, M., 1987. Molecular Evolutionary Genetics. Columbia University Press, New York.

Pie, M. R., Weitz, J. S., 2005. A null model of morphospace occupation. American Naturalist 166, E1–E13.

Polly, P. D., 1997. Ancestry and species definition in paleontology: a stratocladistic analysis of Viverravidae (Carnivora, Mammalia) from Wyoming. Contributions from the Museum of Paleontology, University of Michigan 30, 1–53.

Polly, P. D., 1998. Cope's Rule. Science 282, 50–51.

Polly, P. D., 2001. Paleontology and the comparative method: ancestral node reconstructions versus observed node values. American Naturalist 157, 596–609.

Polly, P. D., 2002. Phylogenetic tests for differences in shape and the importance of divergence times: Eldredge's enigma explored. In: MacLeod, N., Forey, P. (Eds.), Morphology, Shape, and Phylogenetics. Taylor & Francis, London, pp. 220–246.

Polly, P. D., 2003a. Paleophylogeography: the tempo of geographic differentiation in marmots (*Marmota*). Journal of Mammalogy 84, 369–384.

Polly, P. D., 2003b. Paleophylogeography of Sorex araneus: molar shape as a morphological marker for fossil shrews. Mammalia 68, 233–243.

Polly, P. D., 2004. On the simulation of the evolution of morphological shape: multivariate shape under selection and drift. Palaeontologia Electronica 7.2.7A, 28pp. http://palaeo-electronica.org/paleo/2004_2/evo/issue2_04.htm

Polly, P. D., 2005. Development, geography, and sample size in P matrix evolution: molar-shape change in island populations of *Sorex araneus*. Evolution and Development 7, 29–41.

Rohlf, F. J., 1990. Rotational fit Procrustes methods. In: Rohlf, F. J., Bookstein, F. L. (Eds.), Proceedings of the Michigan Morphometrics Workshop. The University of Michigan Museum of Zoology, Special Publication 2, 227–236.

Rohlf, F. J., 2001. Comparative methods for the analysis of continuous variables: geometric interpretations. Evolution 55, 2143–2160.

Rohlf, F. J., Corti, M., 2000. Use of two-block partial least-squares to study covariation in shape. Systematic Biology 49, 740–753.

Rohlf, F. J., Slice, D., 1990. Extensions of the Procrustes method for the optimal superimposition of landmarks. Systematic Zoology 39, 40–49.

Roopnarine, P. D., 2001. The description and classification of evolutionary mode: a computational approach. Paleobiology 27, 446–465.

Rosen, D. E., 1974. Cladism or gradism? A reply to Ernst Mayr Systematic Zoology 23, 446–451.

Salton, J. A., Szalay, F. S., 2004. The tarsal complex of Afro-malagasy Tenrecoidea: a search for phylogenetically meaningful characters. Journal of Mammalian Evolution 11, 73–104.

Salazar-Ciudad, I., Jernvall, J., 2004. How different types of pattern formation mechanisms affect the evolution of form and development. Evolution and Development 6, 6–16.

Sampson, P. D., Streissguth, A. P., Barr, H. M., Bookstein, F. L., 1989. Neurobehavioral effects of prenatal alcohol: Part II. Partial least squares analysis. Neurotoxicology and Teratology 11, 477–491.

Sato, J. J., Hosoda, T., Wolsan, M., Tsuchiya, K., Yamamoto, Y., Suzuki, H., 2003. Phylogenetic relationship and divergence times among mustelids Mammalia: Carnivora based on nucleotide sequences of the nuclear interphotoreceptor retinoid binding protein and mitochondrial cytochrome b genes. Zoological Science 20, 243–264.

Sato, J. J., Hosoda, T., Wolsan, M., Suzuki, H., 2004. Molecular phylogeny of *Arctoides* (Mammalia: Carnivora) with emphasis on phylogenetic and taxonomic positions of the ferret-badgers and skunks. Zoological Science 21, 111–118.

Schaeffer, B., 1947. Notes on the origin and function of the artiodactyl tarsus. American Museum Novitates 1356, 1–24.

Schluter, D., 1996. Adaptive radiation along genetic lines of least resistance. Evolution 50, 1766–1774.

Schmalhausen, I. I., 1949. Factors of Evolution: The Theory of Stabilizing Selection. Translated by T. Dobzhansky. Blakiston, Philadelphia.

Silva, M., Downing, J. A., 1995. The CRC Handbook of Mammalian Body Masses. CRC, Boca Raton, FL.

Simpson, G. G., 1944. Tempo and Mode in Evolution. Columbia University Press, New York.

Simpson, G. G., 1945. The principles of classification and a classification of mammals. Bulletin of the AMNH 85, 1–350.

Simpson, G. G., 1953. The Major Features of Evolution. Columbia University Press, New York.

Slattery, J. P., O'Brien, S. J., 1995. Molecular phylogeny of the red panda *Ailurus fulgens*. Heredity 86, 413–422.

Springer, M. S., 1997. Molecular clocks and the timing of placental and marsupial radiations in relation to the Cretaceous-Tertiary boundary. Journal of Mammalian Evolution 4, 285–302.

Stanley, S. M., 1973. Explanation for Cope's rule. Evolution 27, 1–26.

Szalay, F. S., 1977a. Phylogenetic relationships and a classification of the eutherian Mammalia. In: Hecht, M. K., Goody, P. C., Hecht, B. M. (Eds.), Major Patterns in Vertebrate Evolution. Plenum, New York, pp. 315–374.

Szalay, F. S., 1977b. Ancestors, descendants, sister groups, and testing of phylogenetic hypotheses. Systematic Zoology 26, 12–18.

Szalay, F. S., 1981. Functional analysis and the practice of the phylogenetic method as reflected by some mammalian studies. American Zoologist 21, 37–45.

Szalay, F. S., 1984. Arboreality: is it homologous in Metatherian and Eutherian mammals? In: Hecht, M. K., Wallace, B., Prance, G. T. (Eds.), Evolutionary Biology, Volume 18. Plenum, New York.

Szalay, F. S., 1994. Evolutionary History of the Marsupials and an Analysis of Osteological Characters. Cambridge University Press, Cambridge.

Szalay, F. S., 2000. Function and adaptation in paleontology and phylogenetics: Why do we omit Darwin? Palaeontologia Electronica 3.2.2, 25 pp. 372KB. http://palaeo-electronica.org/2000_2/darwin/issue2_00.htm

Szalay, F. S., Bock, W. J., 1991. Evolutionary theory and systematics: relationships between process and patterns. Zeitschrift für Zoologische Systematik und Evolutionsforschung 29, 1–39.

Szalay, F. S., Decker, R. L., 1974. Origins, evolution, and function of the tarsus in Late Cretaceous Eutheria and Paleocene Primates. In: Jenkins, F. A., Jr. (Ed.), Primate Locomotion. Academic Press: New York, pp. 223–259.

Szalay, F. S., Drawhorn, G., 1980. Evolution and diversification of the Archona in an arboreal milieu. In: Luckett, W. P. (Ed.), Comparative Biology and Evolutionary Relationships of Tree Shrews. Plenum, New York, pp. 133–169.

Szalay, F. S., Schrenk, F., 1998. The middle Eocene Eurotamandua and a Darwinian phylogenetic analysis of "edentates." Kaupia 7, 97–186.

Taylor, M. E., 1970. Locomotion in some East African viverrids. Journal of Mammals 51, 42–51.

Taylor, M. E., 1976. The functional anatomy of the hindlimb of some African Viverridae (Carnivora). Journal of Morphology 148, 227–254.

Taylor, M. E., 1988. Foot structure and phylogeny in the Viverridae (Carnivora). Journal of Zoology (London) 216, 131–139.

Taylor, M. E., 1989. Locomotor adaptations. In: Gittleman, J. L. (Ed.), Carnivore Behavior, Ecology, and Evolution.. Cornell University Press, Ithaca, NY, pp. 382–409.

Thompson, J. D., Gibson, T. J., Plewniak, F., Jeanmougin, F., Higgins, D. G., 1997. The ClustalX windows interface: flexible strategies for multiple sequence alignment aided by quality analysis tools. Nucleic Acids Research 24, 4876–4882.

Thorpe, R. S., 1981. The morphometrics of the mouse: a review. In: Berry, R. J. (Ed.), Biology of the House Mouse. Zoological Society of London, London, pp. 85–125.

Van Valkenburgh, B., 1985. Locomotor diversity within past and present guilds of large predatory mammals. Journal of Vertebrate Paleontology 11, 406–428.

Van Valkenburgh, B., Wang, X. M., Damuth, J., 2004. Cope's rule, hypercarnivory, and extinction in North American canids. Science 306, 101–104.

Veron, G., Heard, S., 2000. Molecular systematics of the Asiatic Viverridae Carnivora inferred from mitochondrial cytochrome b sequence analysis. Journal of Zoological Systematics and Evolutionary Research 38, 209–217.

Veron, G., Colyn, M., Dunham, A. E., Taylor, P., Gaubert, P., 2004. Molecular systematics and evolution of sociality in mongooses Herpestidae, Carnivora. Molecular Phylogenetics and Evolution 30, 582–598.

Wake, D. B., Roth, G., Wake, M. H., 1983. On the problem of stasis in organismal evolution. Journal of Theoretical Biology 101, 211–224.

Wang, X., 1997. New cranial material of *Simocyon* from China and its implications for phylogenetic relationships to the Red panda Ailurus. Journal of Vertebrate Paleontology 17, 184–198.

Wayne, R. K., Benveniste, R. E., Janczewski, D. N., O'Brien, S. J., 1989. Molecular and biochemical evolution of the Carnivora. In: Gittleman, J. L. (Ed.), Carnivore Behavior, Ecology, and Evolution, Volume 1. Comstock Cornell, Ithaca, NY, pp. 465–495.

Werdelin, L., 1996. Carnivoran ecomorphology: a phylogenetic perspective. In: Gittleman, J. L. (Ed.), Carnivore Behavior, Ecology, and Evolution, Volume 2. Comstock Cornell, Ithaca, NY, pp. 582–624.

Wesley-Hunt, G. D., Flynn, J. J., 2005. Phylogeny of the carnivora: basal relationships among the carnivoramorphans and assessment of the position of "Miacoidea" relative to crown-clade Carnivora. Journal of Systematic Palaeontology 3, 1–28.

Wolsan, M., 1993. Phylogeny and classification of early European Mustelida (Mammalia: Carnivora). Acta Theriologica 38, 345–384.

Wozencraft, W. C., 1989. The phylogeny of the recent Carnivora. In: Gittleman, J. L. (Ed.), Carnivore Behavior, Ecology, and Evolution, Volume 1. Comstock Cornell, Ithaca, NY, pp. 495–535.

Wright, S., 1988. Surfaces of selective value revisited. American Naturalist 131, 115–123.

Wyss, A. R., 1988. On "retrogression" in the evolution of the Phocinae and phylogenetic affinities of the monk seals. American Museum Novitates 2924, 1–38.

Wyss, A. R., Flynn, J. J., 1993. A phylogenetic analysis and definition of the Carnivora. In: Szalay, F. S., Novacek, M. J., McKenna, M. C. (Eds.), Mammal Phylogeny: Placentals. Springer, New York, pp. 32–52.

Yu, L., Zhang, Y.-P., 2005. Phylogenetic studies of pantherine cats (Felidae) based on multiple genes, with novel application of nuclear β-fibrinogen intron 7 to carnivores. Molecular Phylogenetics and Evolution 35, 483–495.

Yu, L., Li, Q.-W., Ryder, O. A., Zhang, Y.-P., 2004. Phylogenetic relationships within mammalian order Carnivora indicated by sequences of two nuclear DNA genes. Molecular Phylogenetics and Evolution 33, 694–705.

Zelditch, M. L., Ludrigan, B. L., Garland, T., 2004. Developmental regulation of skull morphology. I. Ontogenetic dynamics of variance. Evolution and Development 6, 194–206.

Section II
Primates

10. The Biogeographic Origins of Primates and Euprimates: East, West, North, or South of Eden?

Mary T. Silcox*
University of Winnipeg
Department of Anthropology
515 Portage Ave.
Winnipeg, MB R3B 2E9, Canada
m.silcox@uwinnipeg.ca

10.1 Introduction

The place of origin of Primates is a subject that has received surprisingly little treatment in the literature. Part of the difficulty in considering this question is terminological. Until recently, the order Primates had usually been considered to include both euprimates *sensu* Hoffstetter, 1977, which are clearly related to modern members of the order, and the more archaic "plesiadapiforms" (e.g., Simpson, 1945; Hoffstetter, 1977; Szalay and Delson, 1979; MacPhee et al., 1983; Szalay et al., 1987; note that "plesiadapiforms" is placed in quotation marks to signify that it is likely a non-monophyletic group; Gunnell, 1989; Silcox, 2001). Since the very influential publication of two papers back-to-back in *Nature* in 1990 (Beard, 1990; Kay et al., 1990), which suggested paromomyid "plesiadapiforms" were more closely related to dermopterans than to euprimates, it has become common practice for authors to exclude "plesiadapiforms" from the order Primates (e.g., Beard, 1998a; Hartwig, 2002; Tavaré et al., 2002; Soligo and Martin, 2006). In this case the taxon name Primates was equivalent in meaning to Euprimates (see also Martin, 1968, 1986; Cartmill, 1972, 1974; Wible and Covert, 1987).

Not all authors concur, however, with this more restricted definition of Primates, or with the interpretation of the data upon which this re-classification was based. McKenna and Bell (1997), for example, included all "plesiadapiforms" with euprimates and modern dermopterans in an expanded order Primates. Several workers (e.g., Krause, 1991; Sargis, 2000, 2002; Bloch and Silcox, 2001, 2007; Silcox, 2001, 2003; Bloch and Boyer, 2002, 2003) have posed questions about the validity and strength of the data supporting the supposed tie between "plesiadapiforms" and dermopterans, leading to the suggestion (Silcox, 2001, 2007; Bloch et al., 2007) that all "plesiadapiforms" be returned to membership in an order Primates that does not include dermopterans. For this paper the taxon name Primates will refer to Euprimates + "plesiadapiforms"

This paper seeks to consider, therefore, two distinct but interrelated questions: the place of origin of Primates, and that of Euprimates. The first of these two questions is probably of greater relevance to those interested in the general patterns of mammalian evolution, in that it marks the point at which primates became a lineage distinct from other mammalian orders. The latter question may be of greater relevance to primate specialists, in that it is at this evolutionary transition that characteristic primate features such as convergent orbits and postcranial traits for leaping appear to have arisen (Silcox et al., 2007), although the process by which these features were acquired is not currently documented in the fossil record.

Part of the puzzle surrounding the place of origin of Euprimates stems from the abruptness with which they appear in the fossil record of both Western North America and Western Europe at the beginning of the Eocene (Gingerich, 1986, 1989, 1993, 2006). An abrupt appearance at approximately the same time is also seen in the records of a number of other groups, including the modern orders Perissodactyla and Artiodactyla, coincident with a major climatic shift (Gingerich, 2006). In light of the suddenness of this faunal shift, and the lack of transitional stages available, an origin from somewhere other than these intensely sampled regions may seem likely. Some authors have nonetheless supported the idea of an American origin for Euprimates, with the transitional stages from a "plesiadapiform" ancestor to a euprimate occurring in non-sampled southern areas, and being associated with Paleocene climatic changes (Sloan, 1969; Gingerich, 1976; Schiebout, 1979). This idea has fallen out of favor in recent years due to the

* Address for correspondence: m.silcox@uwinnipeg.ca

E.J. Sargis and M. Dagosto (eds.), *Mammalian Evolutionary Morphology: A Tribute to Frederick S. Szalay*, 199–232.

lack of fossils that provide a link between euprimate and non-euprimate groups from the increasingly well-sampled North American fossil record (Gingerich, 1989, 1993; Beard, 1998a). It is worth noting, however, that while the North American fossil record is excellent for some time periods and regions, it is very strongly geotemporally patterned, so that substantial parts of the continent remain unsampled for critical time periods.

Other authors have suggested African (e.g., Tattersall, 1982; Gingerich, 1986, 1989, 1990; Aiello, 1993) and Asian (Szalay and Li, 1986; Hoffstetter, 1988; Beard, 1998a, b; Beard and Dawson, 1999) origins, in part relying implicitly on the less well-sampled fossil records of these areas to make the lack of a series of transitional fossils between primate and non-primate fossils seem more explicable. A euprimate is known from the late Paleocene of Africa (Sigé et al., 1990; Gheerbrant et al., 1998) and geological work suggests euprimates, artiodactyls, and perissodactyls appeared in Asia at or before the Paleocene/Eocene boundary, possibly earlier than they occur in North American and Europe (Bowen et al., 2002), which may also support an origin from one of these areas. Other source locations for the characteristic early Eocene fauna that have been discussed include northern North America (e.g., arctic regions; Hickey et al., 1983; but see Kent et al., 1984), the Indian subcontinent (Krause and Maas, 1990), and less well-sampled regions of Europe (Godinot, 1981).

Beard (1998a: p. 23; Figure 10.1) published an analysis that reconstructed the origin of Primates (meaning Euprimates) as being "unequivocally" or even "unambiguously" (Beard, 1998a: p. 23; Beard, 1998b: p. 27; Beard and Dawson, 1999: p. 699) Asian, which is the strongest statement made in recent years on the place of origin of the group. This conclusion formed part of a general hypothesis about mammalian evolution, in which Asia was seen as a major source area for the evolution of groups found subsequently in North America and Europe. Support for this "East of Eden" hypothesis came from similar analyses of a series of mammalian groups, all of which were reconstructed as dispersing after the major evolutionary events of the groups occurred in Asia. This hypothesis was used to suggest that two Asian Land Mammal Ages (ALMA), the Gashatan and Bumbanian, were older than had been previously supposed (Beard, 1998a, b; Beard and Dawson, 1999). The current paper will consider the East of Eden hypothesis as it applies to Primates, and assess whether or not the evidence currently known supports an unequivocal origin for Primates and/or Euprimates from Asia.

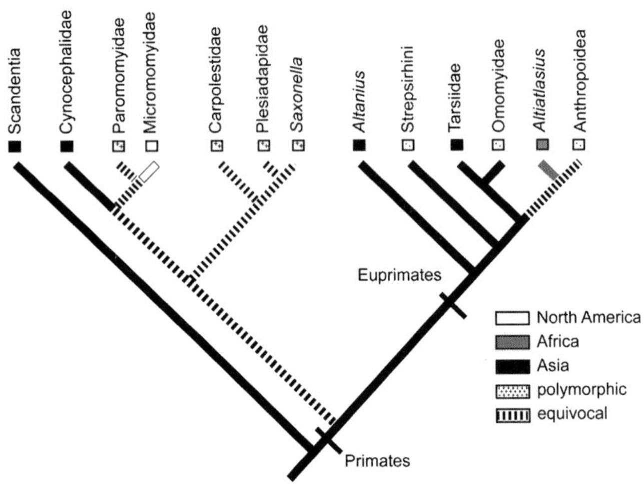

FIGURE 10.1. Hypothesis of relationships and biogeographic reconstruction redrawn from Beard (1998a: Figure 13). Pattern coding of branches was re-done to match Figure 10.3, and taxonomic names were revised in some cases to conform to terminology used in this paper (i.e., Cynocephalidae was used in place of Galeopithecidae [see Stafford and Szalay, 2000], and Euprimates in lieu of Primates *sensu* Beard, 1998a). As discussed in the text, this diagram was not based on an explicit data matrix. Under this hypothesis of relationships both Primates (sensu McKenna and Bell, 1997) and Euprimates originated in Asia.

10.2 Issues in Biogeographic Reconstruction

Studies of biogeography have been a major application of cladistics for many years. In particular, many techniques have been suggested that involve using or creating "area cladograms", showing patterns of branching for land masses based on geology or on common occurrences of taxa (e.g., Brooks and McLennan, 1991; da Silva and Oren, 1996). These area cladograms can be used to make comparisons to a cladogram of organisms, in order to elucidate patterns of dispersal among taxa (e.g., Brooks and McLennan, 1991; Raxworthy et al., 2002; but see Siddall, 2005).

Some authors have noted, however, that this approach can become very complicated over a long time scale. Particularly, barriers to dispersal may open and close, leading to a non-hierarchical pattern of area relationships that require complicated new techniques to be accommodated (Ronquist, 1997). In a much simpler approach to the problem, Beard (1998a; see also Ronquist, 1997) did not attempt to form any type of area cladogram. Instead, he simply optimized a character based on the geographic distributions of the included groups onto a hypothesized set of relationships. Using the same technique that has been applied to studying character distributions on a tree, this method produces a "decision" about whether there is a single most parsimonious solution to the ancestral place of origin (=an unequivocal optimization), or more than one equally parsimonious solution (=an equivocal optimization).

The parsimony-based determination of an ancestral character state or place of origin is influenced by a number of factors. First and foremost is the tree topology upon which the character is optimized. The algorithm used in optimizing characters is particularly sensitive to the sister group to the clade in question (Maddison et al., 1984). For example, if the sister group is only known from area "a", it will not be possible to find a place of origin for the ingroup as being unequivocally anything other than "a" (Maddison et al., 1984).

Members of the ingroup near the base of the tree also have a significant influence on the determination of the ancestral state or place of origin of the ingroup. Both of these factors imply that primitive taxa, and the pattern of relationships at or near the base of the tree, are particularly crucial to the parsimony decision. Related to this is the issue of taxon sampling. If a particular taxon is left out of a tree, its character state, or geographic distribution, will be unable to influence the parsimony decision. Another factor to consider in performing an optimization is the underlying resolving algorithm used. Two primary methods are available. The first, generally referred to as ACCTRAN, favors early acquisition of character states, which may be followed by reversals. The second, DELTRAN, favors late acquisition of character states, so that parallelisms tend to appear more frequently than reversals (Swofford and Maddison, 1987). These different approaches only apply when there are multiple equally parsimonious ways in which to map a character onto a tree. MacClade (Maddison and Maddison, 1992), the computer program most frequently used to perform character optimizations (and that used by Beard, 1998a), offers the option of showing all the equally parsimonious resolutions for each node, so that optimizations that differ only based on whether ACCTRAN or DELTRAN is used are shown as being equivocal.

In light of this discussion it is worth looking at Beard's arguments to consider the support for the tree topology he used for Primates, the breadth of his taxon sampling, and the optimization technique applied.

10.2.1 Tree Topology

In terms of the tree topology used by Beard (1998a), no data matrix was published to corroborate the branching pattern illustrated (see Figure 10.2a), and it is clear from the accompanying discussion that this tree has been pieced together from various sources rather than calculated in a cladistic analysis. Since the validity of the conclusions about the biogeographical implications of the tree depend on its topology, it seems important to assess how well-supported this pattern of relationships actually is. The part of the tree that shows the relationships between the included "plesiadapiforms" and dermopterans derives from a combination of the analyses of Beard and Wang (1995), which deals only with the Plesiadapoidea (see Figure 10.2b) and the conclusions (not supported by a cladistic analysis) given in Beard (1993b; see Figure 10.2c). The link between modern dermopterans ("Galeopithecidae"; note that Stafford and Szalay (2000) have argued convincingly that Cynocephalidae is the more appropriate name for the family including all living dermopterans) and paromomyids also receives some support from Beard and MacPhee's (1994; see Figure 10.2d) analysis, although only when an *a posteriori* reweighting scheme was applied to their dataset.

Beard cites several sources in support of the relationships within euprimates (Beard and MacPhee, 1994; Beard et al., 1996; Beard, 1998c; see Figure 10.2d–10.2f), although none of these studies include *Altiatlasius* (see below), and only one of these (Gingerich et al., 1991) includes *Altanius*. The phylogenetic position of *Altanius* has been a source of debate since Rose and Krause (1984) questioned whether this taxon was even a euprimate, identifying a number of similarities to carpolestid "plesiadapiforms". Gingerich et al. (1991) assembled a dataset of dental characters for *Altanius, Cantius, Teilhardina*, and *Elphidotarsius* to test this hypothesis. Gingerich et al. then performed two analyses. They first produced an unrooted network that was then rooted at the "midpoint of greatest patristic difference" (p. 644). The result was a tree that did indeed separate *Elphidotarsius* from the other taxa (see Figure 10.2g). The reason for this, however, is the large number of autapomorphous character states for *Elphidotarsius*. As such, this analysis is actually a phenetic, not a cladistic, analysis, whose result may simply reflect a great deal of divergent evolution in *Elphidotarsius*, rather than a close relationship between *Altanius* and the euprimates *Cantius* and *Teilhardina*. Since Rose and Krause (1984) never suggested that *Altanius* need be as aberrantly derived as any definitive carpolestid, this analysis is not an adequate test of their hypothesis.

Gingerich et al. (1991) also ran a cladistic analysis rooted with an outgroup. However, the outgroup they chose was *Elphidotarsius*, based on its greater age relative to the other sampled taxa. Without including an additional outgroup it is not possible to resolve the basal node of this tree, so that the result of this analysis is most accurately portrayed with an unresolved relationship between *Altanius* and *Elphidotarsius* (see Figure 10.2h). Assuming that *Elphidotarsius* is an appropriate outgroup to this analysis (which is debatable), the only meaningful conclusion that comes from this result is that *Teilhardina* and *Cantius* are more closely related to one another than either is to *Altanius*. In sum, this analysis is not actually informative about the euprimate status of *Altanius*, removing the support for this node on Beard's (1998a) tree. As the basal-most euprimate in Beard's topology (see Figures 10.1, 10.2a), the systematic position of *Altanius* is very influential to the character optimization of the biogeographic character at the euprimate node.

Beard (1998a) does not explicitly reference any support for the relationship he portrays between Anthropoidea and *Altiatlasius*. This node is presumably based on the "Cladogramme interprétif" published by Sigé et al. (1990). This diagram was not based, however, on an explicit cladistic analysis, and conflicts with the taxonomic conclusion reached by these authors in the same publication, which placed *Altiatlasius* in the Omomyidae. As discussed below, several subsequent analyses have included *Altanius* and/or *Altiatlasius*; however, since they were published after Beard (1998a), they are irrelevant to the issue of the support he had for his tree.

Another taxon that had a crucial influence over Beard's optimization is Scandentia. In support of Scandentia's position on his tree, Beard (1998a) cites a series of molecular studies (Cronin and Sarich, 1980; Adkins and Honeycutt, 1991, 1993; Ammerman and Hillis, 1992; Bailey et al., 1992;

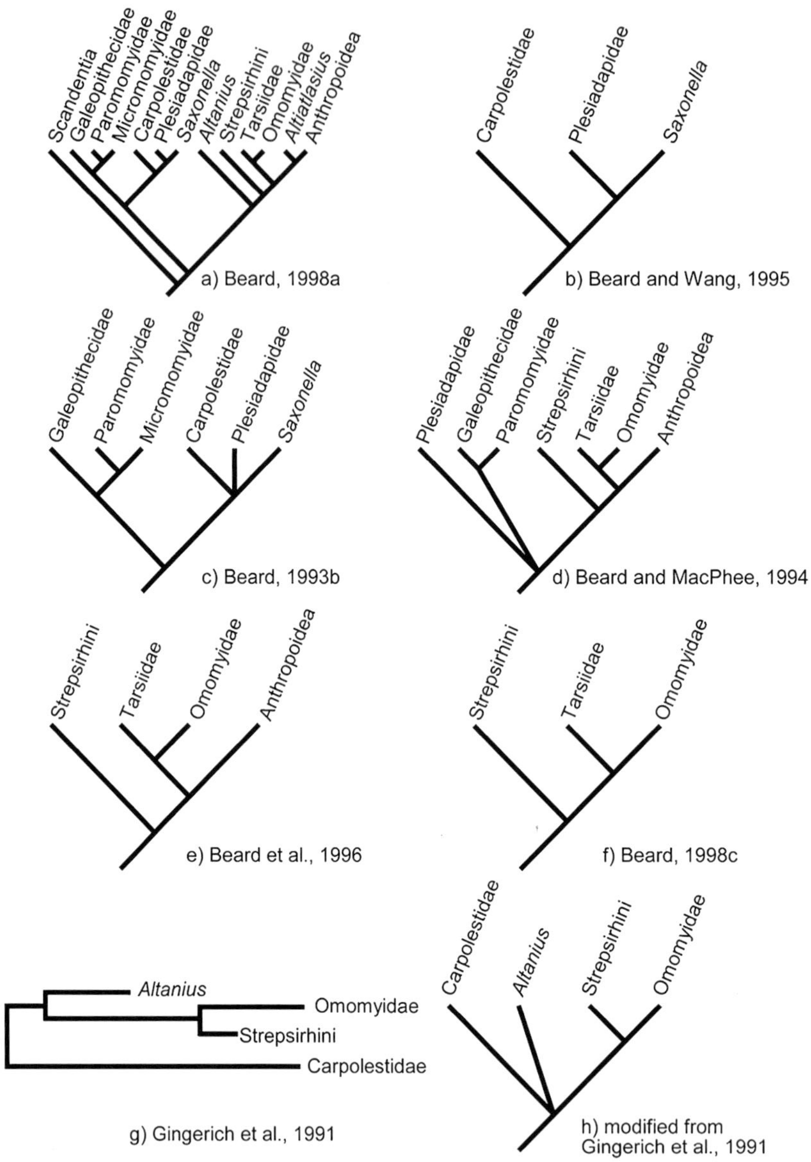

FIGURE 10.2. Phylogenetic hypotheses from sources cited by Beard (1998a) in support of the pattern of relationships he portrays among primates and their near kin. This figure demonstrates that it is not possible to build the hypothesis of relationships given by Beard (1998a; part a) from the various sources that he cites. In particular, none of these sources include *Altiatlasius*, and none of these analyses are capable of resolving the position of *Altanius* with respect to Carpolestidae.

(a) Hypothesis of relationships from Beard 1998a. (Redrawn from Beard, 1998a: Figure 13.)
(b) Pattern of relationships from Beard and Wang (1995) for taxa included by Beard (1998a). (Redrawn from Beard and Wang, 1995: Figure 14).
(c) Pattern of relationships implied by the taxonomic hypothesis given by Beard (1993b); note that this was not produced in a cladistic analysis and contradicts the results of the cladistic analysis in Beard (1993a).
(d) Results of a cladistic analysis of cranial characteristics by Beard and MacPhee (1994). In order to achieve this result, the authors had to apply an *a posteriori* re-weighting scheme; the results prior to the use of this procedure were almost entirely unresolved. (Redrawn for taxa included by Beard [1998a] from Beard and MacPhee [1994: Figure 7]).
(e) Pattern of relationships implied by the discussion in Beard et al. (1996); this hypothesis was not supported by a cladistic analysis.
(f) Results of a cladistic analysis of mostly dental and gnathic characters from Beard (1998c: Figure 10.4); redrawn for the taxa included by Beard (1998a).
(g) Pattern of relationships given by Gingerich et al. (1991: Figure 10.2), redrawn with taxonomic names used by Beard (1998c).
(h) A more accurate representation of the results of Gingerich et al.'s (1991) analysis. Since these authors did not include any additional outgroups, this analysis is incapable of resolving the basal node. As such, the relationships of *Altanius* and Carpolestidae are more correctly depicted as unresolved.

Stanhope et al., 1993), as well as the "long-standing hypothesis of a close relationship between tree shrews and primates" (p. 23), as discussed in Luckett (1980) and Novacek et al. (1988). This relationship was not found by Beard (1993a), however, and of the molecular papers he cited, most found a Scandentia-Dermoptera clade that excluded Primates in at least some of their results (i.e., Adkins and Honeycutt, 1991, 1993; Ammerman and Hillis, 1992; Stanhope et al., 1993). A Scandentia-Dermoptera clade (Sundatheria Olson, Sargis, and Martin, 2005) is also a common feature of more recent studies of both molecules and morphology (e.g., Liu and Miyamoto, 1999; Sargis, 2000, 2002, 2004, 2007; Liu et al., 2001; Madsen et al., 2001; Murphy et al., 2001a, b; Van Den Bussche and Hoofer, 2004).

There remain, therefore, a number of serious questions about the tree topology used by Beard (1998a) to support an unequivocally Asian origin for euprimates. Although his technique of combining the results of various datasets to produce a composite "best guess" might seem a logical way to deal with the magnitude of this problem, this approach has some serious drawbacks. One of the benefits of cladistic analysis is that it is explicit about the bases upon which the conclusion was reached. Also, in cladistic studies it is relatively easy to add additional evidence to the dataset to modify, or falsify a pattern of relationships. By relying on judgment rather than an explicit data matrix, Beard's tree becomes impossible to falsify or improve with new characters or new taxa. Because he has simply chosen to emphasize some lines of evidence over others, he could equally validly choose to discount any additional evidence that should be discovered. Stevens and Heesy (2000) made various changes to Beard's tree and found either an Asian or an equivocal (African or Asian) origin for Primates (=Euprimates of this paper), implying that using a different judgment about the pattern of relationships can substantially alter the key result.

There has been a movement recently in cladistic studies to favor total evidence analysis of data, so that all the available information from different systems is allowed to "compete" in a single simultaneous analysis (Kluge, 1989). Although generally applied to combining different data types (i.e., molecular and morphological), these same arguments apply to including all the relevant taxa in a simultaneous analysis, since it can be possible for synergistic combinations of characters to suggest branching patterns not found from a series of analyses including only parts of the total list of relevant taxa. Also, small subsets of a total dataset, when analyzed separately, can produce inaccurate results that are not found in a combined analysis (Hillis, 1998). Relationships between members of a large sample of taxa that incorporates primitive forms can actually be easier to reconstruct correctly than relationships between a smaller subset of those taxa (Hillis, 1996). The reason for this is the well-known phenomenon of long branch attraction, in which convergent evolution over the course of evolution in a set of lineages whose separation points are distant in time may lead to the noise from homoplasy overwhelming the evolutionary signal, with the result that it is impossible to find the correct tree (Felsenstein, 1978; Hendy and Penny, 1989). Simulation studies have found that adding taxa that can break those long branches (i.e., primitive forms; see below) can lead to a greater probability of finding the true tree (Hillis, 1996; Purvis and Quicke, 1997; Graybeal, 1998; Wiens, 2005). For these reasons, basing phylogenetic conclusions on an actual cladistic analysis that includes all the relevant taxa is likely to be a better approach than piecing together results from disparate sources that include only subsets of the relevant forms. In particular, all the most primitive taxa that have the greatest potential to break long branches should be included.

10.2.2 Taxon Sampling

There are a number of groups that were excluded from Beard's (1998a) tree that could, potentially, have a serious impact on the results of the optimization. Many authors focusing on morphological data (Gregory, 1910; Novacek, 1986, 1990, 1991, 1994; Novacek and Wyss, 1986; Wible and Novacek, 1988; Thewissen and Babcock, 1991, 1993; Simmons, 1994, 1995; Szalay and Lucas, 1993, 1996; Simmons and Quinn, 1994; Simmons and Geisler, 1998) have supported a relationship between dermopterans and chiropterans, called Volitantia. Molecular evidence for this grouping is lacking, with most studies placing chiropterans in a closer relationship to carnivores and ungulates than to other supposed archontans (e.g., Pumo et al., 1998; Miyamoto et al., 2000; Murphy et al., 2001b). The potential impact of a close relationship between chiropterans and other archontans is nonetheless worth assessing, particularly in light of Beard's (1993a) own finding of a closer relationship between chiropterans and primates than either shared with scandentians.

Three other taxonomic groupings that may be influential to this question are the extinct fossil groups Mixodectidae, Plagiomenidae, and Nyctitheriidae. Dental characteristics have suggested a tie between plagiomenids and dermopterans (Matthew, 1918; Rose, 1973; Rose and Simons, 1977; MacPhee et al., 1989), and between mixodectids and plagiomenids (Gunnell, 1989). Postcranials for *Mixodectes* do appear to show some similarities to other archontans (Szalay and Lucas, 1996), and Mixodectidae grouped with archontans in Silcox's (2001) analysis, although not as a close relative to Dermoptera or Plagiomenidae. The only published cranial evidence for Plagiomenidae is a poorly preserved skull of *Plagiomene* cf. *P. accola* (UM 65145; MacPhee et al., 1989). This specimen has been interpreted as being inconsistent with a dermopteran affinity for the group (MacPhee et al., 1989), although the basis for this is simply the highly derived nature of the specimen, rather than characters that show a clear tie to some other clade. While these derived features may make *Plagiomene* an unlikely direct ancestor for modern dermopterans, plagiomenids may still be more closely related to flying lemurs than to any other group as stem dermopterans.

Indeed, in an analysis that included both these cranial features, and dental traits, Silcox (2001) found plagiomenids to be the sister taxon to modern dermopterans. This is potentially significant to considerations of primate (and archontan) biogeography since plagiomenids are North American in distribution, and are known from very high latitudes in the Eocene (Ellesmere Island; West and Dawson, 1977). Hooker (2001) published the first ankle elements of the "insectivoran" family Nyctitheriidae, which he interpreted as demonstrating characteristic archontan features. Nyctitheriids are known from North America, Asia, and Europe (McKenna and Bell, 1997).

Finally, Beard's (1998a) tree only included 5 of the 12 families of "plesiadapiforms" that are currently recognized (Silcox, 2001; Tabuce et al., 2004). The exclusion of two of these families, Toliapinidae and Azibiidae, is understandable in light of the fact that they were not named until after 1998 (Hooker et al., 1999; Tabuce et al., 2004). Of the other five families excluded, however, three ("Palaechthonidae", Microsyopidae, and Purgatoriidae) include very primitive forms that may be of relevance to primate or euprimate origins (as Beard himself suggested for Microsyopidae; Beard, 1991). *Purgatorius* is of particular importance as the oldest known primate. This is relevant not only because of the importance to the optimization algorithm of basal forms, but because primitive taxa can be key in "breaking" long branches so that the correct phylogeny can be reconstructed (see above and Gauthier et al., 1988; Hillis, 1996, 1998; Purvis and Quicke, 1997; Graybeal, 1998; Wiens, 2005).

Following this discussion, it is clear that endeavoring to consider more adequately the questions of primate and euprimate origins requires including a broader range of archontans, as well as sampling more completely the "plesiadapiform" radiation, than did Beard (1998a).

10.2.3 Optimization Algorithm

Reanalyzing the dataset and tree topology suggested by Beard (1998a) makes it clear that he used the MacClade option of showing conflicting optimizations under ACCTRAN vs. DELTRAN as being equivocal. With these data under DELTRAN, both the ancestors of the plesiadapiform-dermopteran clade and the anthropoid-*Altiatlasius* clade are reconstructed as being Asian, while with ACCTRAN the plesiadapiform-dermopteran clade is reconstructed as having originated in North America, while the anthropoid-*Altiatlasius* clade is optimized as having originated in Africa. Both of these nodes are shown as equivocal on Beard's tree (Figure 10.1). Although applying a different optimizing algorithm does not influence the major conclusions of Beard's analysis, the influence of the resolving algorithm used needs to be considered in any further applications of this technique.

10.3 More Recent Attempts at Reconstructing Primate and Euprimate Biogeographic Origins Using Optimization

Silcox (2001) performed a cladistic analysis that dealt with many of the shortcomings of Beard's (1998a) optimization argument for the biogeographic origin of Euprimates. Her analysis was based on a dataset of 181 morphological characters including dental, cranial, and postcranial traits (see Appendices I and II). These characters were scored for representatives of all 11 "plesiadapiform" families known at the time (i.e., not including Azibiidae; Tabuce et al., 2004), primitive euprimates including *Altanius* and *Altiatlasius*, and representative chiropterans, dermopterans, scandentians, mixodectids, plagiomenids, and outgroups. Her results failed to uphold Beard's hypotheses of relationships, and instead found "plesiadapiforms" to be stem primates, leading to her suggestion that they should be returned to that order (Silcox, 2001, 2007). She applied a biogeographic character to the results of her analysis (Figure 10.3). Unlike Beard, her optimization did not support an unequivocal resolution at the Euprimates node. Rather, an origin for this group from Asia, Africa, North America, or Europe was found to be possible. The supported resolution varied under ACCTRAN and DELTRAN algorithms. Under DELTRAN, North America was the unequivocally resolved place of origin for Euprimates, whereas under ACCTRAN African or Asian origins were preferred. For the Primates node (including "plesiadapiforms" but not dermopterans; no comparable node was present on Beard's tree), an origin in North America was supported unequivocally, independent of the optimizing algorithm.

Bloch et al., (2007) performed a cladistic analysis on a dataset that was updated from that used by Silcox (2001), including information from new "plesiadapiform" skeletons, and from the postcranium of *Ptilocercus lowii*, but with a more restricted range of taxa. These authors also performed a biogeographic analysis, although they made a significant change from the approach adopted by Beard (1998a) and Silcox (2001) in considering two different types of biogeographic character. The first included the total distributions of the groups in question, which was also the approach taken by Beard (1998a) and Silcox (2001). The results of this analysis were basically similar to those of Silcox (2001), except that Europe was not found as a possible place of origin of Euprimates. This is simply a product of the exclusion of the poorly sampled toliapinids from Bloch et al.'s dataset. The second was based on a more optimistic view of the fossil record. Bloch et al. (2007) coded groups based on the location of their first occurrences (scoring the character as multi-state if the place of first occurrence was equivocal). In this case the optimization of the biogeographic character was unequivocally North American for both Euprimates and Primates.

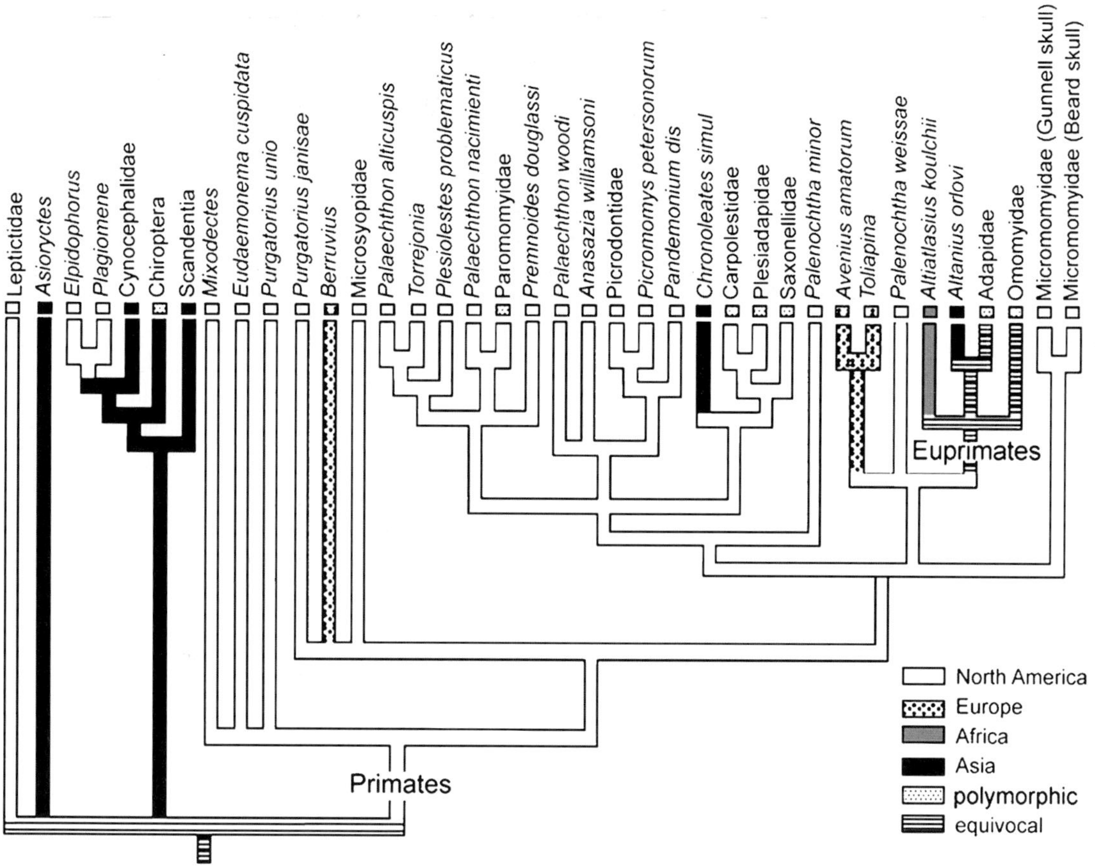

FIGURE 10.3. The Adams consensus tree for the combined analysis of 181 dental, cranial, and postcranial characters, with a biogeographic character optimized on the cladogram, from Silcox (2001). (Redrawn from Silox, 2001: Figure 7.7.) Note that the place of origin of Primates is unequivocally North American, and the place of origin of Euprimates is equivocal.

These results highlight the important point that when a more inclusive set of taxa and characters are included, Beard's results are not upheld. As a result, viewing an Asian origin of Primates or Euprimates as "unequivocal" or "unambiguous" is clearly an optimistic view at best.

However, one must ask whether or not these more recent analyses can be used to argue convincingly for an origin for either or both of these groups from North America (or from anywhere). Because of the lack of clear precursors for Euprimates in the relatively well-sampled North American record, a North American origin for this group seems implausible to some workers. This highlights an important point about using parsimony decisions in forming arguments about issues of biogeography. Just because an optimization decision is "unequivocal" in terms of parsimony does not imply that it is "unequivocal" in a larger sense. While these types of optimization analyses can be interesting for looking at patterns in the data, they will always be influenced by factors of sampling. In particular, it is unlikely that the samples available from different geographic areas will ever be comparably large or equally good. For this reason, I would argue that such analyses should form only the first step in an examination of the data relevant to the biogeographic origins of major groups. Just as hypotheses of relationships stemming from cladistic analyses should be accompanied by a critical assessment of the character data on which they are based, so too should cladistic biogeographic analyses incorporate a critical consideration of the data driving the results. For this reason, the taxa central to the issue of the biogeographic origins of primates and euprimates are assessed individually and in detail below.

10.4 Fossil Evidence Relevant to the Issue of Primate and Euprimate Biogeographic Origins

The purpose of the following discussion is to assess the morphological evidence for an origin of Primates and Euprimates from the most likely continental landmasses. The discussion is structured for convenience using current geopolitical boundaries – for example, North America and Europe are considered separately, even though they were connected at various points in the past, and the Indian subcontinent is considered with Asia even though its connection to that continent is geologically quite recent. The intention here is not to suggest a specific biogeographic scenario for how and when various lineages dispersed from one area to another. The timing of these events is too poorly resolved at present for it to be possible to integrate these data with a particular geographic reconstruction. For example, some authors consider the origin of Primates to have occurred near the Cretaceous-Tertiary boundary (Springer et al., 2003, 2004; Silcox et al., 2005), while others advocate an origin in "deep time", substantially before the Cretaceous-Tertiary boundary (e.g., Miller et al., 2005; Soligo and Martin, 2006).

10.4.1 North America

As noted above, North America has been discounted in recent years as a likely place of origin for Primates or Euprimates on the basis of the lack of intermediates in the generally well-sampled North American fossil record. North America also lacks any evidence of crown-clade scandentians or dermopterans, the extant groups most likely to be sister taxa to Primates (Beard, 2004). However, it must be noted that many of the most likely *fossil* sister taxa to Primates are known from North America, including Plagiomenidae, Mixodectidae, Nyctitheriidae (see discussion above) and Apatemyidae (Silcox et al., 2005). And the fossil record for early Primates includes more primitive taxa than known from anywhere else in the world. In particular, there is no primate known from anywhere in the world that is as old or as primitive as *Purgatorius*, which consistently appears at the very base of the order in cladistic analyses (e.g., Hooker et al., 1999; Silcox, 2001; Ni et al., 2004; Tabuce et al., 2004; Seiffert et al., 2005; Bloch et al., 2007). The oldest possible specimen of *Purgatorius* may be Cretaceous in age (Van Valen and Sloan, 1965; Van Valen, 1994; but see Lofgren, 1995; Clemens, 2004). Unfortunately, the potential Cretaceous record of *Purgatorius* consists of only one damaged lower molar, which is not a terribly diagnostic tooth. It can be difficult, for example, to differentiate some "condylarths" from primitive primates from only the lower molars. Interestingly, the next oldest material for *Purgatorius*, comes from fairly far north – the Ravenscrag Formation of Saskatchewan (Johnston and Fox, 1984). This collection includes a very diagnostic tooth (an upper molar). The northern location of this primitive form raises an interesting point about the fossil record of North America. Although there are a lot of fairly primitive primate fossils known from the continent, the record is highly patterned geographically, making an origin from poorly sampled areas to the north (or south, or east) still worth considering.

In addition to *Purgatorius*, almost all of the other species near the base of the primate tree come from North America. This includes all the known "palaechthonids" (with the possible exception of *Asioplesiadapis*; see below), most of the microsyopids, all of the micromomyids (the potential Asian micromomyid discussed by Tong and Wang, 1998 probably does not belong to that family based on my observations of the relevant specimen), and the most primitive members of most of the other "plesiadapiform" families. Put simply, the best evidence for the origin of Primates comes from North America, and there is no compelling reason to dismiss its importance.

In terms of Euprimates, there are very primitive representatives of this group known from North America, including an adapid and an omomyid from the earliest Eocene (Gingerich, 1986, 1989, 1993), and many of the most likely euprimate sister taxa (i.e., "plesiadapiforms") also come from this continent. Discussions of the origins of this group should at least acknowledge these data. It is true, nonetheless, that the appearance of Euprimates at the start of the Eocene is abrupt, suggesting an origin elsewhere (Gingerich, 1986, 1989, 1993). For this reason, while a North American origin for Primates seems quite plausible, an origin on this continent (at least from the west) for Euprimates may be less likely.

10.4.2 Europe

Like North America, Europe is rarely considered as a likely place of origin for Primates or Euprimates, in part because of the combination of a relatively good (although geographically patterned) fossil record, and a lack of intermediates. However, there are a few primitive primates (*Berruvius*) and euprimates (*Donrusellia, Teilhardina*) from Europe that need to be considered in any discussions of this issue. *Berruvius*, from the late Paleocene, may be the most primitive microsyopid known (Silcox, 2001); microsyopids are among the most primitive primates. Also, two of the four potential fossil sister taxa to Primates or Archonta discussed above have European representatives (Apatemyidae, Nyctitheriidae). Nevertheless, most of the other "plesiadapiforms" known from Europe are fairly well-nested in their families (e.g., *Platychoerops, Chiromyoides, Arcius, Plesiadapis*), or superfamilies (e.g., *Saxonella*). The lack of primitive primates from Europe makes a European origin for the order seem unlikely.

The opposite is the case for the early euprimate families Adapidae and Omomyidae, however. European *Donrussellia* is generally held to be the most primitive adapid (Godinot, 1978, 1992, 1998; Rose and Bown, 1991; Ni et al., 2005; but see Gingerich et al., 1991), and *Teilhardina belgica* may be even more primitive an omomyid than North American members of that genus, although it has some features that may be derived relative to the Asian *Teilhardina asiatica* (Ni et al., 2004). Both of these genera date from the earliest Eocene in Europe, at or near the same time that the earliest euprimates appear in North America and Asia (Godinot et al., 1978; Godinot, 1998). In light of these considerations, it seems surprising that a European origin for Euprimates is so routinely dismissed (e.g., it was not even considered by Ni et al., 2005, in spite of possible support for this idea in their own results; see below). One obvious reason for this is the lack of any potential sister taxa for Euprimates in the European record. With the possible exception of *Berruvius*, all known European "plesiadapiforms" are too well nested within their families to be considered basal primates and bear derived features (e.g., the oddly enlarged p3 of *Saxonella*) that make them unlikely direct euprimate ancestors. However, Silcox's (2001) analysis supported a sister group relationship between Toliapinidae *sensu stricto* (including only *Toliapina* and *Avenius*) and Euprimates. The support for this node comes in part from similarities between these taxa and the basal euprimate *Altiatlasius* (see below), including small size and bunodont cusps. *Toliapina* also shares a feature with adapids that is not seen in any other taxa in Silcox's analysis: a buccally positioned hypoconulid on the molars. Although the support for this node is admittedly extremely limited, this possible link is tantalizingly suggestive of a potential role for Europe in euprimate origins.

10.4.3 Africa

The early Tertiary fossil record of Africa is very limited. Nonetheless, the oldest likely euprimate, *Altiatlasius koulchii*, hails from this continent. In their original description, Sigé et al. (1990) placed the species in the Omomyidae, noting similarities to *Omomys* and *Chumashius* in particular. However, as Hooker et al. (1999) note, these forms are derived members of omomyid clades that do not exhibit similar traits otherwise, weakening support for an omomyid attribution. Hooker et al. (1999) favored a placement for *Altiatlasius* in their "plesiadapiform" family Toliapinidae, and presented a cladistic analysis of dental traits in support of this view. The characters in the diagnosis for Toliapinidae are not diagnostic, however, when viewed in the broader context of the range of variation seen in "plesiadapiforms" and primitive euprimates. For example, having lower molars with a metaconid that is lower than the protoconid is a fairly common configuration, with most "plesiadapiform" families exhibiting intrafamilial variability in this trait (Silcox, 2001), suggesting that it has relatively low phylogenetic valence. Another of their supposed diagnostic toliapinid traits, "m1-2 with hypoconulid median" (Hooker et al., 1999: p. 378) is characteristic of most "plesiadapiform" families, and is actually not true of *Toliapina*, which has buccally shifted hypoconulids (Silcox, 2001).

The cladistic analysis that Hooker et al. (1999) used to bolster their hypothesis of toliapinid *sensu lato* monophyly (i.e., including *Berruvius* and *Altiatlasius*) includes a rather limited sample of taxa. In particular, the only euprimate they include is *Teilhardina*, which forms a very unlikely clade with *Paromomys* on their tree. This odd position for *Teilhardina* influences the internal optimizations of characters, since *Paromomys* becomes the functional outgroup to Euprimates. This makes it not at all surprising that *Altiatlasius* did not fall out with *Teilhardina*, as it does not look much like *Paromomys*, and therefore does not approach the morphology that euprimates are reconstructed (likely incorrectly) as having evolved from in this analysis. Since Hooker et al. (1999) published their results, two more broadly sampled analyses (Silcox, 2001; Tabuce et al., 2004) have failed to uphold a "plesiadapiform" position for *Altiatlasius*, finding instead that it is a euprimate near the base of that group.

Altiatlasius does share some superficial features with *Toliapina* and *Avenius* – it is a relatively small taxon, which nonetheless has somewhat bunodont molars. The combination of small size and rounded cusps is not that common – bunodonty is a trait that usually accompanies larger size in primates, in association with a shift from a more insectivorous to a more frugivorous diet (Kay and Covert, 1984; Godinot and Mahboubi, 1992). One of the other places this combination turns up is in some early anthropoids, including the African anthropoid *Algeripithecus* (Godinot and Mahboubi, 1992). Godinot (1994) performed a detailed character analysis of the small, bunodont, African anthropoids including *Algeripithecus*, and concluded that *Altiatlasius* shared with this group some important similarities, such as a continuous lingual cingulum, which might suggest that *Altiatlasius* is in fact a very primitive anthropoid. This idea is actually not wholly inconsistent with Sigé et al. (1990), since their "Cladogramme interprétif" positions *Altiatlasius* as possibly near "simiiformes" (=anthropoids). Similarities between *Altiatlasius* and *Eosimias*, including the absence of a postprotocingulum and presence of a complete lingual cingulum (Beard, 2004; Beard and Wang, 2004), add to the support for a possible anthropoid attribution for the former.

It is worth noting that the position suggested by Godinot (1994) for *Altiatlasius* is completely consistent with the cladograms of both Silcox (2001) and Tabuce et al. (2004). Neither of these studies included any anthropoids, and both supported the basal position of *Altiatlasius* in Euprimates, which is where Godinot portrays anthropoids as branching off. This depends on a view of anthropoid origins that is controversial, in that it portrays the group as having a very ancient origin, and as being equally distantly related to both Adapidae and Omomyidae. This viewpoint is at odds, for example, with Kay et al.'s (2004) results, which position omomyids as haplorhines, closely related to anthropoids and tarsiers. Seiffert et al. (2005) supported a basal anthropoid position for *Altiatlasius*, as part of a monophyletic Haplorhini including omomyids, so that anthropoids again did not appear as a basal branch from Euprimates. This analysis included only three "plesiadapiforms" (*Purgatorius unio, Plesiadapis tricuspidens*, and *Plesiolestes problematicus*), however, which would impact the features reconstructed as primitive for Euprimates. Kay et al.'s (2004) and Seiffert et al.'s (2005) results are not consistent with Silcox's (2001) and Tabuce et al.'s (2004) cladograms; to be consistent, *Altiatlasius* should have grouped with the omomyids included by these authors. However, both Silcox (2001) and Tabuce et al. (2004) had samples of euprimates that were too limited to provide a convincing test of alternative placements of this taxon. What is needed to confirm *Altiatlasius*' position, and to assess its potential importance to the broader issue of anthropoid origins, is an analysis that includes *Altiatlasius*, "plesiadapiforms", and anthropoids, along with a representative sample of other fossil euprimate taxa (e.g., omomyids, adapids, eosimiids, etc.).

In any case, it now seems clear that *Altiatlasius* is a euprimate. Whether it is a basal euprimate that is an anthropoid, a basal euprimate that is not, or a basal anthropoid nested within Haplorhini, is still a subject to be resolved. Regardless, the Paleocene age of this taxon (Gheerbrant et al., 1998) implies it is extremely relevant to discussions of early euprimate evolution, making Africa a likely candidate for the location of origin for this group.

Since it appears unlikely that *Altiatlasius* is a "plesiadapiform", the only potential stem primates from Africa are the Eocene-age azibiids of Algeria. Tabuce et al. (2004) recently argued that *Azibius trerki* and *Dralestes hammadaensis* are "plesiadapiforms" closely related to carpolestids (see also Sudre, 1975, 1979; Schwartz, 1986), rather than "condylarths", macroscelideans, or adapids, as had been previously argued (Szalay, 1975; Gingerich, 1976; Holroyd and Simons, 1991; Hartenberger et al., 1997). Primate features that these taxa exhibit include a postprotocingulum on the upper molars and on one specimen of P4. The most notable similarities to carpolestids are in the enlarged, exodaenodont p4, and in the form of the m1, which has an elongate trigonid that continues the shearing surface from the last premolar. However, the form of the p4 in *Azibius* is rather different from that of carpolestids. In carpolestids the trigonid is elongated through displacement of the trigonid cusps into a line (Rose, 1975), while the talonid is very short. In spite of the worn condition of the only known azibiid p4, in this tooth there is a remnant of the hypoflexid visible, and it is clear that the talonid is nearly as long as the trigonid. This suggests that in *Azibius* both the talonid and the trigonid lengthened to produce the large p4, a pattern that is fundamentally different from, and thus possibly non-homologous to, that of carpolestids. Also, the azibiids exhibit a rather confusing combination of derived and primitive traits relative to carpolestids. Unlike all plesiadapoids (except *Pandemonium* if it is a plesiadapoid), P4 in *Dralestes* does not have a centroconule. However, *Dralestes* has an m1 with cusps that are more linearly arranged than in some carpolestids (e.g., *Elphidotarsius florencae*, in which the trigonid cusps are only "slightly splayed"; Rose, 1975: p. 16), and unlike in carpolestids, m2 also has a similarly modified trigonid, extending the shearing surface further distally.

Tabuce et al. (2004) acknowledge the existence of such character conflicts, and the fact that the plesiadapoid status of the azibiids is supported in part by characters that also occur in some euprimates (e.g., the central position of the protocone on the upper molars). Although the oddly elongate molar trigonids of azibiids is not matched in any primitive euprimates, it is worth noting that a stepped postvallid on m1 (i.e., a pronounced offset between the metaconid and protoconid, with a cristid obliqua that does not extends as a "step" up the metaconid; Silcox et al., 2001), which could be viewed as a precursor morphology to the trigonid morphology of azibiids, is present in a number of euprimates, including *Altanius*. Therefore, it is not inconceivable that azibiids are euprimates with adaptations to produce a shearing complex convergent on that of carpolestids, in spite of the results of Tabuce et al.'s cladistic analysis.

However, even if azibiids are euprimates and not "plesiadapiforms", they are unlikely to be adapids. Godinot (1992: p. 237) has identified certain traits as key adapid synapomorphies: the absence of a postmetaconule crista on the upper molars, the presence of a postprotocrista running from protocone to metacone, "a cristid obliqua directed towards the summit of the metaconid", and a buccally offset hypoconulid on the lower molars. *Dralestes* lacks a postmetaconule crista and has a postprotocrista running from protocone to metacone, as in adapids. However, in *Dralestes* the metaconule is entirely missing, so the absence of the postmetaconule crista may be explicable as a consequence of the lack of this cusp. Since this cusp is typically not missing in adapids, this would suggest the independent evolution of this trait. *Dralestes* also has a cristid obliqua that is directed towards the summit of the metaconid; however, in this case it forms part of the modifications to this tooth to make it a more effective "tool" for shearing, which contrasts with the situation in adapids, and again suggests that these similarities may not be homologous. The rather weak hypoconulid on all known molars makes it difficult to assess the last trait listed above; however, a wear facet on m2 of *Azibius* likely represents the hypoconulid, and is not buccally offset. For this reason, an adapid attribution seems unlikely.

A third possibility needs to at least be acknowledged: that azibiids are "none of the above". That is, they could be an otherwise unknown endemic group of African mammals, not particularly closely related to primates. In light of the poor quality of the early Tertiary record in Africa, there are probably a number of endemic groups that are poorly sampled and/or unrecognized, and it may be that the azibiids are one of them. The molecular evidence supporting Afrotheria, an endemic radiation of mammals in Africa (Stanhope et al., 1998; but see Zack et al., 2005 for a contrary view), would suggest a high degree of endemism in early African mammalian evolution, which one might predict would produce a number of uniquely derived endemic groups that are unrelated to extant forms. The upper molars of *Dralestes* look very odd for a primate, even though they do possess a postprotocingulum. Features that would not be expected in a plesiadapoid include the large parastyle, the total lack of a metaconule or any conule wings, the small trigon basin, and the extremely long lingual slope to the tooth, so that the cusp tip of the protocone is positioned very far buccally (approximately 1/3 the width of the tooth). Although some of these features occur in particular euprimate taxa (e.g., the very long lingual slope in the upper molars is found in the omomyids *Strigorhysis* and *Gazinius* and a decrease in the size of the metaconule is common in anthropoids.; Godinot, 1994), their combination produces very odd looking upper molars for a euprimate. While I am not advocating this viewpoint, it is a possibility that must be acknowledged.

If the azibiids *are* primates, and even if they are "plesiadapiforms", they are too derived to be considered as either basal primates or primitive euprimate ancestors. As such they do not directly support an African origin for these groups. However, what they do indicate is the possible presence of a poorly sampled primate radiation in Africa. An important message that comes from the discovery of the new material of *Dralestes* is that there is still much of interest to learn from the late Mesozoic and early Tertiary of Africa, which may fundamentally change our views about early primate evolution.

10.4.4 Asia

In contrast to the poor early Tertiary African record, the Asian record of early primates has seen a massive improvement in the last 11 years, beginning with Beard and Wang's (1995) paper describing the first Asian "plesiadapiforms". There have been eight species of potential "plesiadapiforms" from Asia mentioned in the literature, which have been described as coming from four distinct families (contra Smith et al., 2004). These include four species described as carpolestids (*Chronolestes simul, Carpocristes oriens, Parvocristes oligocollis*, and *Subengius mengi*; Beard and Wang, 1995; Thewissen et al., 2001; Smith et al., 2004), two species described as plesiadapids (*Asioplesiadapis youngi* and *Jattadectes mamikheli*; Thewissen et al., 2001; Fu et al., 2002), and a paromomyid and supposed micromomyid from the Wutu fauna, mentioned by Tong and Wang (1998). Some authors (e.g., Rose and Krause, 1984; Godinot, 1994) consider *Altanius orlovi* to be a "plesiadapiform"; however, in light of what is now known of the anatomy of this taxon, and of the "plesiadapiform" fauna from Asia, this now seems unlikely (see below).

Of the eight above-listed potential "plesiadapiforms", the two mentioned by Tong and Wang (1998) have not been published in enough detail yet to make an assessment of their relationships possible (although, as noted above, my preliminary observations of the supposed micromomyid suggest that it does not belong in that family). Two of the other species' phylogenetic positions are fairly clear-cut, at least in broad terms. *Subengius mengi*, the oldest known Asian "plesiadapiform" (late Paleocene, Erlian Basin, Inner Mongolia, China; Smith et al., 2004; Beard et al., 2005), is a relatively primitive carpolestid. The presence of only four apical cuspules on the p4 of this taxon places it with the paraphyletic North American genus *Elphidotarsius* near the base of the family (Silcox et al., 2001). Indeed, it is unclear why this form merits a generic distinction from *Elphidotarsius*, apart from its Asian location. The relatively derived P4 of *S. mengi* (bearing both a well developed pericone and hypocone) suggests that this species is more derived than the North American carpolestids *Elphidotarius florencae* and *E. wightoni*, which lack these cusps. This taxon is interestingly similar to *E. wightoni* in having a relatively low p4 blade, lower even than in *E. florencae*, suggesting a possible special relationship between these taxa. Clarification of the precise position of *S. mengi* will have to await its inclusion in a cladistic analysis (currently being undertaken by K.C. Beard; Beard et al., 2005).

Carpocristes oriens (early Eocene or late Paleocene Wutu Formation, China; Beard and Wang, 1995) on the other hand, is a derived carpolestid, with features such as seven apical cusps on the p4 suggesting a substantially more advanced position in the carpolestid radiation than occupied by *Subengius mengi*. Phylogenetic hypotheses position this taxon in a sister group relationship with the North American *Carpodaptes hobackensis* (Beard and Wang, 1995; Bloch et al., 2001). Interestingly, this suggests that multiple dispersals of carpolestids occurred between North America and Asia in the early Tertiary (Beard et al., 2005; Beard, 2006). A possible alternative is that the similarities between *Carpocristes oriens* and advanced North American carpolestids are the product of rampant parallelism – although this prospect is extremely unparsimonious based on present evidence it will be interesting to see if future finds in Asia make the notion more plausible.

For the other four Asian "plesiadapiforms" there are reasons to query the systematic position suggested by their original describers. For *Chronolestes simul*, there has already been a debate in the literature about its phylogenetic position (Beard and Wang, 1995; Silcox et al., 2001; Fox, 2002). This debate is of import because the significance of *Chronolestes* to euprimate and primate origins depends on its phylogenetic position. If, as Beard and Wang (1995) originally suggested, it is a basal carpolestid, then its relevance to these questions is limited, since it would be a member of a family that is presumably well nested within Plesiadapoidea (but see McKenna and Bell, 1997; Bloch and Boyer, 2002). However, if Silcox et al. (2001) and Fox (2002) are correct, then this taxon is a basal plesiadapoid. Recent reconstructions of "plesiadapiform" relationships (Bloch et al., 2007) suggest that Plesiadapoidea is the sister taxon to Euprimates. As such, basal plesiadapoids are very relevant to the issue of euprimate origins. In this case, the Asian location of *Chronolestes* would provide some support for an Asian origin for Euprimates.

While I stand by my conclusion that *Chronolestes simul* is a basal plesiadapoid (Silcox, 2001; Silcox et al., 2001; see also Fox, 2002), not a carpolestid, it must be admitted that either phylogenetic position mandates some considerable homoplasy in the evolution of this species. The absence of features found not only in carpolestids, but also in their likely sister taxa the plesiadapids (i.e., multicuspate upper central incisor, conule on P3, no cristid obliqua on p4, margoconid on i1 [present only in *E. wightoni*, but reconstructed as primitive for Plesiadapidae + Carpolestidae; Silcox et al., 2001]) provides fairly convincing evidence against a carpolestid affinity for this taxon. However, there are a few carpolestid-like traits of this species (e.g., exodaenodont p4 with a metaconid) which would be unexpected in a basal plesiadapoid. In either case *Chronolestes* must have a very long ghost lineage – perhaps as that ghost lineage is filled in the issue of its relationships will be settled to everyone's satisfaction.

Thewissen et al. (2001) published material from two supposed "plesiadapiform" species from the early Eocene of Pakistan, including an additional purported carpolestid, *Parvocristes oligocollis* (Kuldana Formation, northern Pakistan). The hypodigm that they published for this specimen includes H-GSP 92164, which they identify as a right I1, and H-GSP 92163 (the holotype), which they consider to be a right p4. These specimens were not found in direct association, although they apparently came out of the same screen-washing sample. Neither of these specimens are clearly carpolestid in affinities. H-GSP 92164 looks more like an anterior upper or lower premolar of a non-plesiadapiform (perhaps an insectivoran?) than an upper incisor of a carpolestid. The supposed p4 simply does not look like a carpolestid p4. It is much too low-crowned (indeed, it is unclear why these authors consider this a plagiaulacoid tooth), and the cuspules that make up the tooth are too well individuated (note that the published illustrations [Thewissen et al., 2001: Figures 10.2c, d] of this specimen are misleading on this point). A much more likely attribution for the latter tooth is to Dermoptera. *Galeopterus variegatus*, in particular, has several odd, multicuspate teeth (Stafford and Szalay, 2000). Although H-GSP 92163 is not a precise match to any of the teeth in this modern species, it is more similar in gestalt to the distal incisors and canines of *G. variegatus* than it is to the p4 of carpolestids. In any case, neither specimen referred to *Parvocristes* is diagnostically carpolestid.

The other species Thewissen et al. (2001) describe as a "plesiadapiform" is *Jattadectes mamikheli*, from the Mami Khel Formation of northern Pakistan, which they consider to be a plesiadapid. Unfortunately the only known specimen of *Jattadectes*, a maxilla with a fragment of M2 and a fairly complete M3 (H-GSP 97203), is not diagnostic of the family. Most of the distinctive features of this specimen (e.g., a centrally positioned protocone, a well developed postprotocingulum) are also found in many euprimates. The characteristic that Thewissen et al. (2001) focus on as indicating that this specimen pertains to a "plesiadapiform", rather than a euprimate, is the relatively large M3, which they claim is larger than the M2. They suggest that an M3 larger than the M2 is a feature found in "plesiadapiforms" but not Eocene euprimates.

In light of the damage to M2, it is unclear what the precise size relationships are between these teeth – although M3 does appear to be longer lingually than M2, it is unclear how these teeth would compare in buccal length or overall width. Moreover, it is not consistently true that the M3 is as large as or larger than the M2 in "plesiadapiforms", and it is also not consistently true that M3 is reduced in Eocene euprimates. For example, M3 is quite similar in size to M2 in the adapid *Pronycticebus gaudryi*, and M3 is substantially smaller than M2 in *Carpodaptes hazelae*. Even in the primitive plesiadapid *Pronothodectes matthewi* the M3 is somewhat reduced distally relative to M2. Therefore, there are no clear features of H-GSP-97203 that indicate an attribution to Plesiadapidae, or to "plesiadapiforms" more generally, and it seems equally likely that this specimen pertains to a euprimate.

Another potential Asian plesiadapid is *Asioplesiadapis youngi* Fu et al., 2002, which is from the early Eocene of the Wutu Formation, China. The authors who named this taxon argued that it belongs in the Plesiadapidae on the basis of its reduced lower dental formula (1-0-3-3) and the presence of a margoconid and margocristid on the i1. However, dental reduction is a common trend in "plesiadapiforms", and one carpolestid (*Elphidotarsius wightoni*) is also known to have a margoconid and margocristid on i1 (Fox, 1984b). *Asioplesiadapis* also bears a number of features that would be surprising in a plesiadapid, such as no m3 hypoconulid lobe (although the hypoconulid is enlarged over the condition of m1-2, it is very narrow and does not form a distinct lobe that approaches the talonid in width, as is seen in plesiadapids), a basined p4 talonid with a strong cristid obliqua and hypoconid, and no stepped postvallid on m1. Fu et al. (2002) acknowledge that it is lacking typical plesiadapid traits, but consider this to be a product of the primitive nature of this taxon. The first problem with this notion is the apparently derived dental formula of *Asioplesiadapis*. The lower dental formula of *Pronothodectes* is 2.1.3.3. (Gingerich, 1976; Fox, 1990). If the reduced dental formula of *Asioplesiadapis* is a meaningful similarity linking it to Plesiadapidae, then that would suggest it should be nested within *Pronothodectes*. This would mean that it should have all the traits expected in a plesiadapid – including a lobate m3 hypoconulid, an unbasined p4 with no cristid obliqua and weak/absent hypoconid, and a stepped postvallid on m1. There is a clear inconsistency in "explaining away" the absence of these plesiadapid traits as being related to the primitive nature of the taxon, while considering the reduced dental formula to be a plesiadapid similarity.

Second, it is fairly well established that Plesiadapidae forms part of a superfamily, Plesiadapoidea, with Carpolestidae, Saxonellidae, and *Chronolestes simul* (Beard and Wang, 1995; Silcox et al., 2001); Paromomyidae and/or *Pandemonium dis* are also sometimes included (e.g., Maas et al., 1988; Van Valen, 1994; Silcox et al., 2001) although Silcox (2001; see also Bloch et al., 2007) failed to support these hypotheses in her cladistic analysis. It has also been argued by Silcox et al. (2001) that Plesiadapidae shares a sister group relationship with Carpolestidae, and that *Chronolestes simul* is the sister taxon to a clade including Carpolestidae, Saxonellidae, and Plesiadapidae. Traits shared by all plesiadapoids, missing in *Asioplesiadapis*, include a lobate m3 hypoconulid and an unbasined p4 talonid. Carpolestidae, Saxonellidae, and Plesiadapidae additionally lack a cristid obliqua on p4, present in *Asioplesiadapis*. And Carpolestidae and Plesiadapidae share a stepped m1 postvallid (the optimization of this trait is equivocal since it is present in *Chronolestes* but absent in saxonellids), which is also missing in *Asioplesiadapis*. All of these features would be expected in any primitive plesiadapid, since they are reconstructed as primitive to more inclusive groupings that include this family. Their absence in *Asioplesiadapis* indicates that, if this taxon is to be considered a plesiadapid, an enormous amount of homoplasy must be invoked.

So what is *Asioplesiadapis* if it is not a plesiadapid? One possibility that must be considered is that it is not a "plesiadapiform" at all. Another early Tertiary fossil group that has an enlarged first incisor that sometimes bears a margoconid is the Apatemyidae. However, other aspects of the dentition of *Asioplesiadapis* make its attribution to Apatemyidae unlikely. Most apatemyids share derived characteristics of the postcanine dentition that are missing in *Asioplesiadapis*, including an anterolabial expansion of m1, which may produce a fourth trigonid cusp, a distinct curved profile to the lingual aspect of the molars resulting from a lack of an invagination at the talonid notch, a lack of distinct molar talonid cusps so the talonid is ringed by a fairly continuous crest, and a reduced p4. *Jepsenella*, the most primitive known apatemyid, is perhaps more similar to *Asioplesiadapis* in lacking many of these traits. However, compared to this taxon *Asioplesiadapis* is more like a "plesiadapiform" in having talonids that are similar in width to the trigonids, lower molar trigonids, and an m3 hypoconulid that is enlarged over the condition in m1-2. Therefore, *Asioplesiadapis* lacks derived features shared by most apatemyids, and exhibits derived features missing in the primitive members of the group, making its attribution to Apatemyidae unlikely.

Another possibility is that *Asioplesiadapis* is a basal plesiadapoid. Although a margoconid and margocristid are missing from known saxonellids and from *Chronolestes simul*, their presence in both plesiadapids and one carpolestid makes it possible (although not most parsimonious) that this is actually a primitive plesiadapoid trait. However, *Asioplesiadapis* lacks other traits that would be expected in any plesiadapoid. Most notable of these is the lobate m3 hypoconulid, which is a trait not only ubiquitous amongst plesiadapoids, but also common amongst the likely sister taxa to Plesiadapoidea (i.e., Paromomyoidea, Silcox, 2001; *Pandemonium dis*, Van Valen, 1994; Euprimates, Bloch et al., 2007).

As Fu et al. (2002) indicate, *Asioplesiadapis* is missing the traits characteristic of most other "plesiadapiform" families, including, for example, the enlarged m1 of picrodontids and the enlarged p3 of saxonellids. They failed to make explicit comparisons to three "plesiadapiform" families known in 2002: Toliapinidae, Picromomyidae and Micromomyidae. A relationship with toliapinids seems very unlikely in light of the Asian location of this species, and the relatively sharp-cusped molars of *Asioplesiadapis* (toliapinid molars are lower-crowned). Picromomyids are characterized by very small size and an oddly enlarged p4, both features missing in *Asioplesiadapis*, which makes a relationship between these taxa seem improbable. The possibility of a relationship with Micromomyidae is worth considering, in light of the similarly small m3 hypoconulid and the shared presence of rather crestiform paracristids in the most primitive known micromomyid, *Micromomys fremdi* (Fox, 1984a). However, even the otherwise primitive *M. fremdi* has the distinctive micromomyid p4 with an enlarged trigonid that is exodaenodont and a talonid that is much shorter than the trigonid, features lacking in *Asioplesiadapis*. As such, the similarities between these taxa seem likely to be primitive.

This leaves the two most primitive families of "plesiadapiforms", Purgatoriidae and "Palaechthonidae", as the most likely taxonomic placements for *Asioplesiadapis*. Fu et al. (2002) made extensive comparisons between *Purgatorius* and *Asioplesiadapis* in order to demonstrate the contrasts between these forms. However, in some of their characterizations of the anatomy of *Purgatorius* they are imprecise. In particular, in their Table 2 they list *Purgatorius* as having a p4 paraconid and a "developed" lower molar paraconid. As Buckley (1997) noted, the p4 paraconid is variable in its development in *Purgatorius unio* (="*titusi*"; see discussion in Silcox, 2001). In terms of the second characteristic, the form of the paraconid is actually very similar in its rather crestiform shape in *Purgatorius janisae* and *Asioplesiadapis*, making these more similar than the characterization of their form in Fu et al.'s table would suggest. As Fu et al. (2002) note, the reduced dental formula of *Asioplesiadapis* would be unexpected for a purgatoriid. However, this trait may be explicable by the presumably long ghost lineage of this taxon, stretching (at least) the length of the Paleocene.

The other possibility worth considering is that *Asioplesiadapis* is some kind of "palaechthonid". It is true that all "palaechthonids" known from the i1 lack a margoconid on this tooth. However, the form of the i1 is only known for a few "palaechthonids" and since the family is unlikely to be monophyletic (Silcox, 2001), its morphology in one "palaechthonid" cannot be assumed to be representative of its form in other members of the group. Most of the other characteristic features of *Asioplesiadapis* can be found in some "palaechthonids". For example, Fu et al. (2002) report that *Asioplesiadapis* has weak lower molar mesoconids. Mesoconids are common features of "palaechthonids", occurring in at least some specimens of most species. A non-lobate hypoconulid that is nonetheless enlarged relative to that cusp on m1-2 is present in some "palaechthonids", including *Palenochtha minor*, which is also similar to *Asioplesiadapis* in having trigonids that are less mesially inclined than is typical for a "plesiadapiform". While most "palaechthonids" (including *P. minor*) have a more premolariform p4 than *Asioplesiadapis*, with a shorter talonid, the proportions of this tooth are more comparable to the Asian form in *Palenochtha weissae*.

Although *Asioplesiadapis youngi* does not fit easily into any particular genus of "palaechthonid", it is conceivable that this species is a member of this family. Of course, since "Palaechthonidae" is a non-monophyletic assemblage of fairly primitive "plesiadapiforms" (Silcox, 2001), this does not tell us much about its potential wider relationships. However, the general impression produced by the anatomy of *Asioplesiadapis* is that it is a relatively primitive form, with a few derived oddities, such as the margoconid and reduced dental formula. This means that *Asioplesiadapis* could be a very important taxon for arguments in favour of an Asian origin for Primates.

Rose and Krause (1984; see also Godinot, 1994) suggested possible "plesiadapiform" affinities for *Altanius orlovi*, a diminutive primate from the early Eocene of Mongolia, based on similarities to carpolestids including exodaenodont molars and a stepped postvallid on p4. Since they made this suggestion, the upper dentition of *Altanius orlovi* has been described (Gingerich et al., 1991), demonstrating that this form lacks a conule on P4; this feature is characteristic not only of all carpolestids, but of all clear plesiadapoids (with *Pandemonium* being the only possible exception). The discoveries in Asia of clear carpolestids, and of the enigmatic *Chronolestes*, which do not resemble *Altanius* in most aspects of morphology, have also made a carpolestid affinity for *Altanius* seem less likely. If *Chronolestes* is a basal plesiadapoid, as Silcox et al. (2001) and Fox (2002) suggest, then the presence of distinctive plesiadapoid traits in this taxon (e.g., enlarged i1, conule on p4) that are missing in *Altanius* makes a basal carpolestid position very unlikely for *Altanius*, since these traits would be expected in any member of the superfamily.

Although, as noted above, Gingerich et al.'s (1991) attempt to assess the position of *Altanius* was flawed by the lack of an additional outgroup, several more comprehensive analyses performed since then (e.g., Silcox, 2001; Seiffert et al., 2005; Bloch et al., 2007) have supported a position for this taxon near the base of Euprimates, with a special relationship with adapids appearing possible. Ni et al. (2004, 2005) found a "plesiadapiform" position for *Altanius* in their analysis; however, their poor sampling of "plesiadapiforms" and adapids makes this result questionable. In any case, *Altanius orlovi* offers one of the best pieces of potential evidence in favor of an Asian origin for Euprimates.

Two additional species known from China have been suggested as candidates for some of the oldest and most primitive euprimates: *Decoredon anhuiensis* (middle Paleocene, Upper Member, Wanghudun Formation, Qianshan Basin) and *Petrolemur brevirostre* (late Paleocene, Datangxu Member, Nongshan Formation, Nanxiong Basin; Tong, 1979). In both cases the primate status of these species has been questioned (e.g., Russell and Gingerich, 1980; Szalay, 1982 [but see Szalay and Li, 1986, Szalay et al., 1986]; Rose and Krause, 1984; Sigé et al., 1990; Beard and Wang, 1991; Gingerich et al., 1991; Rose and Bown, 1991; Rose et al., 1994; Rose, 1995; Godinot, 1998). For both of these taxa the preserved material that has been published is inadequate to either confirm, or definitively dismiss, euprimate attributions.

The material grouped by Szalay and Li (1986) as *Decoredon anhuiensis* was originally published by Xu (1976) as coming from two species (*Diacronus anhuiensis* and *Decoredon elongatus*) of Anagalidae; he later (Xu, 1977) reclassified *D. elongatus* as a hyopsodontid "condylarth". Van Valen (1978) was the first to recognize the possible primate status of *D. elongatus*. The material included in this genus by Szalay and Li (1986) includes a fragment of a right maxilla with P2-M3 (IVPP 4271), and associated right and left dentaries each bearing p4-m3 (IVPP 4281). These specimens are quite damaged – for example, m1 and m2 are not well preserved in either dentary. The association between the upper and lower teeth has been questioned, with Rose et al. (1994: p. 8) indicating that they "occlude poorly" (see also Sigé et al., 1990). There are certainly features of both the upper and lower dentitions that would be surprising for a euprimate (see also discussions in Rose et al., 1994; Sigé et al., 1990). In particular, the very long, narrow talonid of m3 would be unexpected in a primitive euprimate. Increases to the length of the m3 talonid are typically a product of an expansion of the hypoconulid in primates, rather than a lengthening of the talonid basin proper. The upper molars are extremely transverse for Primates, even compared to *Purgatorius* (Sigé et al., 1990). One characteristic that would be very convincing as evidence of primate (although not necessarily euprimate) affinities would be the presence of a postprotocingulum (=nannopithex fold). Szalay et al. (1986) indicated that this feature was present on the upper molars. However, Szalay and Li's (1986: p. 362, emphasis added) description of this area is more equivocal: "The so called lingual slope of the protocone, and the area immediately distal to it *suggesting* the protocone fold (hypocone, nannopithex fold etc.) is extensive…" My assessment of published figures (Szalay and Li, 1986: Figure 10.1) and casts of this taxon indicates that the upper molars are not well enough preserved lingually to determine whether or not a postprotocingulum was originally present. In all, the primate, and especially the euprimate, status of this form remains to be confirmed. As Gingerich et al. (1991) note, an allocation to "Condylarthra" is also a likely possibility for this material.

McKenna and Bell (1997) classified *Petrolemur* as an oxyclaenid (which they exclude from "Condylarthra", but which most workers would place in that group), and Godinot (1998) suggested it could be a primate, an artiodactyl, or a "condylarthran". Szalay (1982) made a fairly convincing argument that *Petrolemur* was not a primate, but might instead be a dichobunid artiodactyl. Only 4 years later, however, Szalay and Li (1986: p. 396) referred to this form as an "undoubted primate…astutely recognized by Tong…"! The basis for this changed opinion was apparently a new specimen, alluded to in an abstract by Szalay et al. (1986). However, to my knowledge this new specimen has never been described in publication. The specimen that has been published is a damaged maxilla with P3-M3 (IVPP V5298); the buccal aspects of P4-M3 are missing. Although showing general similarities to adapids, this specimen exhibits some features that suggest it may not belong to a primate. In particular, the lingual aspect of P4 is very narrow, with no hint of a postprotocingulum. Although it is true that some adapids that approach *Petrolemur* in morphology lack a clear postprotocingulum on the molars, a remnant of this crest is still clearly present on P4. Rose et al. (1994) also suggest that its age is inconsistent with an adapid or omomyid attribution. While not definitive evidence against a euprimate affinity, certainly the current evidence is inadequate to use this taxon to draw broad-reaching evolutionary or biogeographic conclusions. Until additional material is discovered and/or published, this taxon will remain an "irritating enigma" (Godinot, 1998: p. 243).

There are somewhat later occurring, but less irritating, forms from Asia that are undoubtedly fairly primitive euprimates (e.g., *Kohatius coppensi, Asiomomys changbaicus, Panobius afridi*). In most cases these taxa are clearly referable to Omomyidae or Adapidae, making them no more convincing of Asian origins for Euprimates as the North American or European adapids and omomyids are for origins for this group from those continents. The recent discovery of an exquisitely preserved skull of *Teilhardina asiatica* adds an extremely primitive omomyid to the picture, perhaps strengthening suggestions of an Asian origin for Euprimates (Ni et al., 2004, 2005). However, as Bloch and Silcox (2006) note, this form is several nodes removed from the base of Euprimates, so that it cannot be used as a proxy for the common ancestor of this group. In Ni et al.'s (2004) tree, *Teilhardina asiatica* and *T. belgica* are sister taxa at the base of Haplorhini, and *Donrussellia* is positioned at the base of Strepsirhini. Since basal forms in each of the two major euprimate clades are European in origin on this tree, whereas only one basal haplorhine is Asian, this result could be interpreted as more supportive of a European origin for Euprimates than an Asian one. What's more, if one adds a biogeographic character to this matrix, a North American origin for Euprimates actually comes out as most parsimonious! The limited sampling of adapids (e.g., the absence of any species of *Cantius*) and "plesiadapiforms" in this analysis makes over-interpreting its biogeographic significance unwise, but nonetheless the point is clear that, while generally supportive of an Asian origin for Euprimates, *Teilhardina asiatica* is hardly a "smoking gun".

A point emphasized by Beard (2004) in recent arguments in favor of an Asian origin for Primates is the presence of all potential living sister taxa (Scandentia, Dermoptera) to Primates in Asia. This perspective is rather ironic coming from a researcher who spent the first part of his career (Beard, 1989, 1990, 1993a,b, 1998a) arguing that North American paromomyids were dermopterans, an opinion that he has never reversed in the literature. Nonetheless, this is a point worth considering. It is true that there is no record of crown-clade scandentians or dermopterans outside of Asia. However, as noted above, potential fossil relatives of dermopterans (plagiomenids) are known from North America, and other likely fossil archontan groups come from North America and Europe. As such, this argument is also not definitive.

10.5 Conclusions

It is entirely possible that both Primates and Euprimates originated in Asia. Indeed, there are a number of reasons to consider this notion likely, such as the Asian location of the living sister taxa to Primates, the presence of the relatively primitive primate *Asioplesiadapis youngi* and euprimates *Altanius orlovi* and *Teilhardina asiatica* on that continent, and the growing record of Asian stem primates ("plesiadapiforms"). However, the following considerations make it clear that to consider the place of origin for these clades as "unequivocally" or "unambiguously" determined is to fail to account for all of the evidence currently known:

- No continent preserves evidence of any primate as old or as primitive as *Purgatorius* except for North America. Most of the other very primitive primates also come from North America, including "palaechthonids", micromomyids, most microsyopids, etc., suggesting that the earliest phases of evolution of the group occurred on that continent.
- Most of the potential *fossil* sister taxa to Primates and/or Archonta (e.g., Nyctitheriidae, Apatemyidae, Plagiomenidae, Mixodectidae) are almost exclusively North American and European in location (nyctitheriids are also known from Asia).
- Although North America and Europe are well-sampled compared to Africa and Asia, that sampling is highly patterned in terms of both time and space, leaving large areas (e.g., northern, southern, and eastern North America) rather poorly known for the critical time periods (e.g., late Cretaceous-early Eocene).
- Even well-sampled areas in North America and Europe continue to produce surprising discoveries, such as the existence of *Picromomys petersonorum* (one of the smallest primates known, which was not published until 1996; Rose and Bown, 1996; Silcox et al., 2002), the presence of a nail on a divergent hallux in a "plesiadapiform" (Bloch and Boyer, 2002), and the presence of a possible euprimate sister taxon (Toliapinidae) in Europe (Hooker et al., 1999; Silcox, 2001). It is unwise to over-interpret negative evidence.
- The European record includes the most primitive adapid (*Donrussellia*), one of the most primitive omomyids (*Teilhardina belgica*), and a potential sister taxon to Euprimates (Toliapinidae).
- The oldest euprimate currently published is *Altiatlasius koulchii* from Africa. Although this form is isolated in time and space, in light of the poor sampling of Africa in the Paleogene and late Mesozoic, such negative evidence is not a reason to consider this form to be isolated evolutionarily. It may be part of a more extensive African early primate radiation.

- The azibiids suggest that there may be much still to learn about the early primate fossil record of Africa.

Although some authors may prefer an air of certainty to a cautious consideration of all the available data, this has potentially disastrous consequences to attempts to flesh out our understanding of early primate and euprimate evolution. It is clear that Asia should not be the only focus for fieldwork seeking to understand these events. Rather, continued work in North America, Europe, and Africa has the potential to fundamentally change our views about the history of primate and euprimate origins.

Acknowledgments. I thank E. J. Sargis and M. Dagosto for the opportunity to contribute to this Festschrift. This paper benefited from discussions with D. M. Boyer on *Asioplesiadapis*, and from reviews by J. I. Bloch and an anonymous reviewer. The author would like to thank the following for providing access to specimens, casts or photographs of the taxa discussed here: M. Godinot, J. J. Hooker, X. Ni, B. Sigé, K. D. Rose, R. Tabuce, P. Tassy, J. G. M. Thewissen, Y. -S. Tong. This research was supported by NSERC, NSF, the Paleobiological Fund, Wenner-Gren, and Sigma Xi.

Appendix I

Combined cranial, postcranial, and dental datasets at the family level from Silcox, 2001

	d1	d2	d3	d4	d5	d6	d7	d8	d9	d10	d11	d12	d13	d14	d15	d16	d17	d18	d19	d20	d21	d22	d23	d24	d25	d26	d27	d28
Leptictidae	0	0	1	0	0	0	0	0	0	0	0	0	0	0	0	0	0	1	0	0	3	0	0	0	0&1	1	0	2
Asioryctes nemegetensis	0	0	0	0	0	1	0	0	0	0	0	1	0	0	0	0	0	1	0	1	2	0	0	0	1	0	0	1
Purgatorius unio	?	?	?	?	?	?	?	?	?	?	?	?	?	?	?	?	?	?	0	0	0	0	0	0	0&1	0	0	1
Purgatorius janisae	0	0	?	?	1	1	?	?	1	2	0	?	0	0	0	?	?	1	0	0	0	0	0	0	0	0	0	0
Palaechthon alticuspis	1	0	?	?	1	1	?	?	?	?	1	?	1	1	0	?	0	1	0	0	0	0&1	0	0	0&1	0&1	0	1
Palaechthon woodi	?	?	?	?	?	?	?	?	?	?	?	?	?	?	?	?	?	?	0	0	0	0	0	0	0	0	0	1
Palaechthon nacimienti	0	0	?	?	1	1	?	?	?	?	0	0	1	1	0	1	0	1	0	0	0	0	0	0	?	?	0	?
Talpohenach torrejonius	?	?	?	?	?	?	?	?	?	?	?	0	?	?	?	1	0	1	?	?	?	?	?	?	?	?	?	?
Palenochtha weissae	?	?	?	?	1	1	?	?	?	?	0	?	0	1	0	?	?	?	0	0	0	0	0	0	0	1	0	1
Palenochtha minor	1	1	?	?	1	1	1	0	?	?	0	?	1	1	0	?	?	?	0	0	0	0	0	0	1	0	0	1
Plesiolestes problematicus	1	0	?	?	1	1	2	0	?	?	0	0	1	1	0	1	0	1	1	0	3	1	0	1	0&1	1	0	1
Anasazia williamsoni	?	?	?	?	1	1	?	?	?	?	0	?	0	1	0	?	?	?	?	0	0	0	0	0	0	0	0	1
Torrejonia	1	0	?	?	1	1	?	?	?	?	0	?	1	2	0	?	?	?	0	0	0&3	1	1	0	0&1	0&1	0	1
Premnoides douglassi	?	?	?	?	1	1	?	?	?	?	?	?	?	1	0	?	?	?	1	0	0	0	0	0	1	0	0	1
Berruvius	?	0	?	?	1	1	1	0	?	?	1	?	?	0/1	0	?	?	?	0	0	0&2	0	0	0	1	1	0	0
Avenius amatorum	?	?	?	?	?	?	?	?	?	?	?	?	?	?	?	?	?	?	0	0	1	0	0	0	1	0	0	2
Toliapina	?	?	?	?	?	?	?	?	?	?	?	?	?	?	?	?	?	?	0	0	0	0	0	0	0	1	0	2
Microsyopidae	1	1	?	?	1	1	3	0	?	?	1	1	1	1	0	1	0	0	0	0	0&1	0	0	0	1	1	0	0
Paromomyidae	1	0	?	?	1	1	?	?	?	?	0&1	?	1	0	0	1	0	1	0	0	0	0	0	0	1	0	0	1
Picrodontidae	1	0	?	?	1	1	2	0	?	?	2	2	1	1	1	1	0	0	3	1	1	2	1	0	2	0	0	2
Picromomys petersonorum	1	1	?	?	1	1	1	0	?	?	2	?	1	1	1	?	?	?	2	2	1	1	1	0	3	?	0	2
Micromomyidae (Gunnell skull)	1	1	1	0	1	1	1	0	1	2	1	1	1	0	0	1	0	1	2	0	0&1	0	0	0	3	0	0	2
Micromomyidae (Beard skull)	1	1	1	0	1	1	1	0	1	2	1	1	1	0	0	1	0	1	2	0	0&1	0	0	0	3	0	0	2
Chronolestes simul	1	0	0	0	1	1	1	0	1	1	1	2	1	1	1	0	0	1	2	1	2	0	0	0	1	1	0	1
Carpolestidae	0	0	0	0	1	1	2	1	1	3	0	0	1	1	0	0	1	1	2	1	2	2	0	0	0	1	1	1
Pandemonium dis	?	?	?	?	?	?	?	?	1	1	?	?	?	?	0	?	?	0	3	0	0	0	0	0	2	0	0	1
Plesiadapidae	1	0	?	?	1	1	2	1	1	3	1	0	1	1	0	1	0	1	3	1	1	2	0	0	2	0	0	2
Saxonellidae	1	1	?	?	1	1	1	0	1	3	2	?	1	2	0	1	0	1	0&3	1	0&2	2	1	0	1	0	0	1&2
Mixodectes	1	2	1	0	1	1	?	?	?	?	2	0	1	1	0	?	?	0&1	0&1	3	1	0	0	0	0	0	0	1
Eudaemonema cuspidata	1	2	?	?	0	0	?	?	?	?	1	?	0	1	0	?	?	?	1	0	3	0	0	0	0	1	0	1
Elpidophorus	?	0	0	0	1	1	4	0	?	?	?	?	?	1	0	?	3	1	0&1	0	3	0&1	0	0	0	1	0	1&2
Plagiomene	0	0	0	0	1	1	4	0	?	?	0	?	0	1	0	0	3	1	1	0	3	1	0	0	0	1	0	2
Cynocephalidae	0	2	0	0	0	1	4	0	0	0	0	1	1	0&2	0	1	2	1	1	0	3	1	0&1	0	0	1	0	2
Chiroptera	0	0	0	0	0	0	0	0	0	0	0	1	1	2	0	1	0	1	0	1	1	2	1	0	0	0	0	1
Altiatlasius koulchii	?	?	?	?	?	?	?	?	?	?	?	?	?	?	0	?	?	?	?	?	?	?	?	?	?	?	?	?
Altanius orlovi	1	0	?	?	0	0	?	?	?	?	0	0	0	0	0	0	0	1	0	0	0	0	0	0	0	1	0	1
Omomyidae	1	0	?	?	0	0	?	?	?	?	0	0	0&1	1	0	1	1	1	0	0	0&1	0	0	0	0&1	1	0	0
Adapidae	1	0	1	0	0	0	0	0	0	0	0	0	0	0	0	0	0	1	0	0&1	1	0&1	0&1	0	0&1	1	0	1
Scandentia	0	2	0	0	0	1	0	0	1	0	0	0&1	1	1	0&1	1	0	1	3	1	1	0&2	0	0	0	1	0	2

(continued)

Appendix I (continued)

	d29	d30	d31	d32	d33	d34	d35	d36	d37	d38	d39	d40	d41	d42	d43	d44	d45	d46	d47	d48	d49	d50	d51	d52	d53
Leptictidae	0	0	1	2	0	3	0	0	1	0	0	0	0	0	0	0	0	0	2	0	0	1	0	0	0
Asioryctes nemegetensis	0	0	1	0	0	0	0	0	0	1	1	1	0	0	0	0	0	0	2	0	0	0	0	0	0
Purgatorius unio	0	0	?	?	?	?	?	?	?	?	?	?	?	?	?	?	?	?	?	?	?	?	?	?	?
Purgatorius janisae	0	0	1	0	0	0	0	0	1	0	0	0	1	0	0	0	0	0	2	0	0	0	1	0	0
Palaechthon alticuspis	0	0	0	0	0	2	0	0	0	0	0	0	1	0	0	0	0	0	0	1	0	0	1	0	1
Palaechthon woodi	0	0	?	0	0	?	?	?	?	?	1	?	?	?	?	?	?	?	?	?	?	?	?	0	?
Palaechthon nacimienti	0	0	0	0	0	0	0	0	1	0	0	0	1	0	0	0	0	0	0	1	1	2	1	0	0
Talpohenach torrejonius	?	?	1	0	0	2	0	0	1	0	0	0	0	0	0	0	0	0	2	1	0	2	1	0	1
Palenochtha weissae	0	0	?	?	?	?	?	?	?	?	?	?	?	?	?	?	?	?	?	?	?	?	?	?	?
Palenochtha minor	0	0	?	?	?	0	0	?	?	?	0	0	0	0	0	?	0	0	?	0	?	0	1	0	1
Plesiolestes problematicus	0	0	0	0	0	2	0	0	1	0	0	0	0	0	0	0	0	0	2	1	0	0	1	0	1
Anasazia williamsoni	0	?	?	?	?	?	?	?	?	?	?	?	?	?	?	?	?	?	?	?	?	?	?	?	?
Torrejonia	0	0	?	?	?	2	0	?	?	?	0	0	?	0	0	?	0	0	?	1	?	?	1	0	1
Premnoides douglassi	0	0	?	?	?	?	?	?	?	?	?	?	?	?	?	?	?	?	?	?	?	?	?	?	?
Berruvius	0	0	?	?	?	2	0	?	?	?	0	0	?	0	0	?	0	0	?	1	?	0	0	0	1
Avenius amatorum	0	0	?	?	?	?	?	?	?	?	?	?	?	?	?	?	?	?	?	?	?	?	?	?	?
Toliapina	0	0	?	?	?	0	0	?	?	?	0	0	?	1	0	?	0	0	?	1	?	2	0	0	0
Microsyopidae	0	0	0	0	0	0	0	0	0	0	1	0	1	0	0	0	1	0	0	0	1	0	1	0	1
Paromomyidae	0	0	0	0	0	0	0	0	0	0	0&1	0	1	0&1	0	0	0	0	0	1	0	1&2	1	0	1
Picrodontidae	0	2	?	?	?	0	0	?	?	?	1	0	?	1	0	?	0	0	?	1	?	?	0	1	1
Picromomys petersonorum	0	0	?	?	?	?	?	?	?	?	?	?	?	?	?	?	?	?	?	?	?	?	?	?	?
Micromomyidae (Gunnell skull)	2	0	0	0	0	0	0	0	0	1	0	0	1	0	0	0	0	0	2	0	0	0	0	0	0
Micromomyidae (Beard skull)	2	0	0	0	0	0	0	0	0	1	0	0	1	0	0	0	0	0	2	0	0	0	0	0	0
Chronolestes simul	1	0	1	0	0	1	1	0	0	0	1	0	1	1	0	0	1	0	2	1	1	0	1	1	1
Carpolestidae	2	1	1	1	1	1	1	0	1	0	0	1	1	0	0	0	0	0	1	1	1	1	1	1	1
Pandemonium dis	0	0	1	0	0	0	0	0	0	0	1	0	1	0	0	0	0	0	0	1	0	?	1	0	1
Plesiadapidae	0	0	1	1	0	1	0	0	1	0	0	0	0&1	0&1	0	0	0	0	1	1	0	2	1	0	1
Saxonellidae	0	0	1	1	1	1	1	0	1	0	0	0	0	0&1	0	0	0	0	2	1	2	1	1	1	1
Mixodectes	0	0	0&1	0	0	0	0	0	0	0	1	0	1	1	0	0	0	0	0	0	1	1	1	0	1
Eudaemonema cuspidata	1	1	1	?	?	2	0	0	?	?	0	0	?	0	0	0	0	0	2	1	?	?	0	0	0
Elpidophorus	1	1	1	0	0	3	0	0	1	0	0	0	0	0	0	0	1	0	?	1	?	0	0	0	0
Plagiomene	1	2	1	2	0	3	0	0	1	1	0	1	0	0	0	0	0	0	1	1	0	0	0	0	0
Cynocephalidae	0	0	0	0	0	3	0	0	1	0	0	0	0&1	1	0	0	0	0	2	1	?	?	0	0	0
Chiroptera	0	0	0	0	0	0	0	0	0	0	1	0	0	0	0	0	0	0	2	0	0	0	0	1	1
Altiatlasius koulchii	?	0	?	?	?	?	?	?	?	?	?	?	?	?	?	?	?	?	?	?	?	?	?	?	?
Altanius orlovi	1	?	1	0	0	0	0	0	0	0	1	?	0	?	?	0	0	0	?	?	?	?	?	1	1
Omomyidae	0	0	1	0	0	0	0	0	0	0	1	0	1	1	0	0	0	0	0	0	0	2	1	0	1
Adapidae	0	0	1	0	0	0	0	0	0	1	1	0	0	0	0	0	0	0	0	1	0	0	1	0	1
Scandentia	0	0	0&1	0	0	0	0	0	0	0	1	0	1	1	0	0	0&1	0	0&2	0	?	?	1	0	1

(continued)

Appendix I (continued)

	d79	d80	d81	d82	d83	d84	d85	d86	d87	d88	d89	d90	d91	d92	d93	d94	d95	d96	d97	sk1	sk2	sk3	sk4	sk5	sk6
Leptictidae	0	0	0	1	0	0	0	1	0	0	0	0	0	0	0	0	0	0	1	0	1	0	1	0	0
Asioryctes nemegetensis	0	0	0	0	0	0	0	0	0	0	0	1	2	0	0	0	0	0	0	0	0	0	0	0	0
Purgatorius unio	0	0	0	1	1	0	0&1	0	0	0	0	0	0	0	0	0	0	0	0	?	?	?	?	?	?
Purgatorius janisae	0	0	0	0	1	0	0&1	0	0	0	0	1	0	0	0	0	0	0	0	?	?	?	?	?	?
Palaechthon alticuspis	0	0	0	1	1	1	2	1	0	0	0	0	0	0&1	0&1	0	1	?	0	?	?	?	?	?	?
Palaechthon woodi	0	0	0	?	?	1	2	1	0	0	0	0	0	0	0	0	0	0	0	?	?	?	?	?	?
Palaechthon nacimienti	0	0	0	1	1	1	2	1	0	0	0	0	0	0	0	0	0	0	0	?	?	?	?	?	?
Talpohenach torrejonius	?	?	0	0	?	1	2	1	0	1	0	0	0	1	1	0	1	0	0	?	?	?	?	?	?
Palenochtha weissae	?	0	0	?	?	1	?	1	0	?	0	0	0	0	0	0	?	?	?	?	?	?	?	?	?
Palenochtha minor	0	0	0	1	1	1	2	1	0	0	0	0	0	0	0	0	1	0	0	?	?	?	?	?	?
Plesiolestes problematicus	1	0	0	0	1	1	2	1	0	0	0	0	0	0	0	0	0	0	0	?	?	?	?	?	?
Anasazia williamsoni	?	?	?	?	?	0	2	1	0	0	0	0	0	0	0	0	?	0	0	?	?	?	?	?	?
Torrejonia	0	0	1	?	1	1	2	1	0	0&1	0	0	0	0	0	0	0&1	0	0	?	?	?	?	?	?
Premnoides douglassi	1	0	0	?	1	?	?	?	0	?	?	?	?	?	?	?	?	?	?	?	?	?	?	?	?
Berruvius	0	0	0	1	0	1	1	1	0	0	1	0	0	0	0	0	1	0	0	?	?	?	?	?	?
Avenius amatorum	0	0	1	1	1	1	0	1	0	0	0	0	2	0	0	0	1	0	1	?	?	?	?	?	?
Toliapina	0	0	0&1	1	1	1	0	1	0	0	0	0	1	0	0	0	1	0	1	?	?	?	?	?	?
Microsyopidae	0	0	0	1	0	1	1	1	0	0	1	0	0	0	0	0	1	0	1	0	?	0	?	0	0
Paromomyidae	2	0	0	1	1	1	2	1	1	0	0	0	0	1	0	0	1	0	0	0	2	1	1	0&2	1
Picrodontidae	0	0	0	1	0	0	2	?	2	1	0	0	0	0	0	0	0	0	0	?	?	?	?	?	?
Picromomys petersonorum	1	0	?	?	1	?	?	?	?	?	?	?	0	?	?	?	?	?	?	?	?	?	?	?	?
Micromomyidae (Gunnell skull)	0	0	0	0	1	1	0	1	0	0	0	0	0	0	0	0	0	0	0	?	?	?	?	0	0
Micromomyidae (Beard skull)	0	0	0	0	1	1	0	1	0	0	0	0	0	0	0	0	0	0	0	?	?	?	?	2	?
Chronolestes simul	0	1	0	0	1	1	2	1	0	0	0	1	0	0	0	0	1	0	1	?	?	?	?	?	?
Carpolestidae	0	1	0	0	1	1	2	1	0	0	1	1	0	0	0	0	1	0	1	1	?	1	?	0	0
Pandemonium dis	0	0	0	?	1	0	2	1	0	0	0	0	0	0	0	0	0	0	0	?	?	?	?	?	?
Plesiadapidae	1	1	1	0	1	1	2	1	0	0	0	1	0&1	1	0&1	2	1	0	?	1	?	1	1	2	0
Saxonellidae	0	0	0	0	1	1	2	1	0	0	0	1	0	0&1	0&1	0	1	0	1	?	?	?	?	?	?
Mixodectes	0	0	0	0	1	0	0	0	0	0	1	0	0	0	0	3	0	1	0	?	?	?	?	?	?
Eudaemonema cuspidata	1	0	0	0	0	0	0	0	0	0	1	1	0	0	0	3	0	1	0	?	?	?	?	?	?
Elpidophorus	?	0	0	0	1	0	0	0	0	1	1	0	0	0	0	3	1	1	0	?	?	?	?	?	?
Plagiomene	0	0	0	1	1	0	0	0	0	1	0	0	0	1	0	3	1	1	0	1	0/2	1	?	1	0
Cynocephalidae	0	0	0	0	0	0	0	0	0	2	0	1	2	0	0	3	1	0	0	0	2	1	0	2	0
Chiroptera	0	0	0	0	0	0	0	?	2	2	0	0	1	0	0	3	1	0	0	0	3	0	0	0	0
Altiatlasius koulchii	0	0	1	?	?	0	0	1	0	0	0	1	0&1	0&1	0	0	1	0	1	?	?	?	?	?	?
Altanius orlovi	0	1	0	1	1	1	1	0	0	1	1	1	0	0	0	0	1	0	1	?	?	?	?	?	?
Omomyidae	0	0&1	0	1	1	1	1	1	0	0	0	1	0	0	0	0	1	0	1	1	0	0	1	1	1
Adapidae	0	1	0&1	1	1	1	1	0	0	0	0&1&2	0	0&1	1	0&1	0	1	0	1	1	0	0	1	1	1
Scandentia	0	0	0	1	0	0	0	?	2	2	1	0	1&2	0&1	0	1&3	0&1	0&1	0&1	0	1&2	0	1	1	0

(continued)

Appendix I (continued)

	sk7	sk8	sk9	sk10	sk11	sk12	sk13	sk14	sk15	sk16	sk17	sk18	sk19	sk20	sk21	sk22	sk23	sk24	sk25	sk26	sk27	sk28	sk29
Leptictidae	0	0	0	0	0	0	0	0	0	0	0	1	0	1	0	1	0	0	0	0	0	0	0
Asioryctes nemegetensis	0	0	0	0	0	1	0	0	0	1	0	0	0	0	?	0	0	0	2	0	?	?	0
Purgatorius unio	?	?	?	?	?	?	?	?	?	?	?	?	?	?	?	?	?	?	?	?	?	?	?
Purgatorius janisae	?	?	?	?	?	?	?	?	?	?	?	?	?	?	?	?	?	?	?	?	?	?	?
Palaechthon alticuspis	?	?	?	?	?	?	?	?	?	?	?	?	?	?	?	?	?	?	?	?	?	?	?
Palaechthon woodi	?	?	?	?	?	?	?	?	?	?	?	?	?	?	?	?	?	?	?	?	?	?	?
Palaechthon nacimienti	?	0	?	0	1	?	?	0	0	0	?	0	0	0	?	1	0	?	2	?	?	?	?
Talpohenach torrejonius	?	?	?	?	?	?	?	?	?	?	?	?	?	?	?	?	?	?	?	?	?	?	?
Palenochtha weissae	?	?	?	?	?	?	?	?	?	?	?	?	?	?	?	?	?	?	?	?	?	?	?
Palenochtha minor	?	?	?	?	?	?	?	?	?	?	?	?	?	?	?	?	?	?	?	?	?	?	?
Plesiolestes problematicus	?	?	?	?	?	?	?	?	?	?	?	?	?	?	?	?	?	?	?	?	?	?	?
Anasazia williamsoni	?	?	?	?	?	?	?	?	?	?	?	?	?	?	?	?	?	?	?	?	?	?	?
Torrejonia	?	?	?	?	?	?	?	?	?	?	?	?	?	?	?	?	?	?	?	?	?	?	?
Premnoides douglassi	?	?	?	?	?	?	?	?	?	?	?	?	?	?	?	?	?	?	?	?	?	?	?
Berruvius	?	?	?	?	?	?	?	?	?	?	?	?	?	?	?	?	?	?	?	?	?	?	?
Avenius amatorum	?	?	?	?	?	?	?	?	?	?	?	?	?	?	?	?	?	?	?	?	?	?	?
Toliapina	?	?	?	?	?	?	?	?	?	?	?	?	?	?	?	?	?	?	?	?	?	?	?
Microsyopidae	1	0	0	0	1	0	0	0	0	0	?	0	0	0	0	0	?	?	1	0	0	0	0
Paromomyidae	1	1	0	0	0	0	0	2	0	0	1	0	0	0	?	1	0	0	0	0	2	0	1
Picrodontidae	?	?	?	?	?	?	?	?	?	?	?	?	?	?	?	?	?	?	?	?	?	?	?
Picromomys petersonorum	?	?	?	?	?	?	?	?	?	?	?	?	?	?	?	?	?	?	?	?	?	?	?
Micromomyidae (Gunnell skull)	?	?	?	?	?	?	0	?	?	?	?	?	?	?	?	?	?	?	?	?	3	0	0
Micromomyidae (Beard skull)	?	?	?	?	?	?	0	?	?	?	?	?	?	?	?	?	?	?	?	?	2	0	0
Chronolestes simul	?	?	?	?	?	?	?	?	?	?	?	?	?	?	?	?	?	?	?	?	?	?	?
Carpolestidae	1	1	0	0	0	0	0	0	0	0	0	0	1	0	1	1	0	1	2	0	2	0	0
Pandemonium dis	?	?	?	?	?	?	?	?	?	?	?	?	?	?	?	?	?	?	?	?	?	?	?
Plesiadapidae	1	1	0	0	0	0	0	0	0	0	0	0	1	0	1	0	0	1	1	0	2	0	0
Saxonellidae	?	?	?	?	?	?	?	?	?	?	?	?	?	?	?	?	?	?	?	?	?	?	?
Mixodectes	?	?	?	?	?	?	?	?	?	?	?	?	?	?	?	?	?	?	?	?	?	?	?
Eudaemonema cuspidata	?	?	?	?	?	?	?	?	?	?	?	?	?	?	?	?	?	?	?	?	?	?	?
Elpidophorus	?	?	?	?	?	?	?	?	?	?	?	?	?	?	?	?	?	?	?	?	?	?	?
Plagiomene	1	?	?	0	?	0	0	2	0	0	?	0	?	0	?	?	?	?	?	0	?	?	0
Cynocephalidae	1	0	0	0	1	0	1	1	0	1	0	1	0	1	0	0	1	0	0	1	0	1	0
Chiroptera	1	0	0	0	1 & 2	0	0	2	1	0	0	1	1	0	0	0	2	0	1	0	0	1	0
Altiatlasius koulchii	?	?	?	?	?	?	?	?	?	?	?	?	?	?	?	?	?	?	?	?	?	?	?
Altanius orlovi	?	?	?	?	?	?	?	?	?	?	?	?	?	?	?	?	?	?	?	?	?	?	?
Omomyidae	1	1	1	1	2	1	0	0	0	0	0	0	1	0	?	1	2	?	1	0	1	0	0
Adapidae	1	1	0	1	2	0	0	0 & 1	0	0	0	0	0	0&1	1	1	2	?	1	0	1	0	0
Scandentia	1	0	0	0	2	1	1	0	0	0	0	0 & 1	1	0&1	0	0&1	1&2	0&1	2	0	0&1	0	2

(continued)

Appendix I (continued)

	sk30	ps1	ps2	ps3	ps4	ps5	ps6	ps7	ps8	ps9	ps10	ps11	ps12	ps13	ps14	ps15	ps16	ps17	ps18	ps19	ps20	ps21	ps22	ps23
Leptictidae	0	0	0	0	0	0	0	0	0	0	1	0	0	0	0	0	0	0	0	0	0	0	?	1
Asioryctes nemegetensis	0	?	?	?	?	?	?	?	?	?	?	?	?	?	?	?	0	0	?	0	0	0	0	?
Purgatorius unio	?	?	?	?	?	?	?	?	?	?	?	?	?	?	?	?	?	?	?	?	?	?	?	?
Purgatorius janisae	?	?	?	?	?	?	?	?	?	?	?	?	?	?	?	?	?	?	?	?	?	?	?	?
Palaechthon alticuspis	?	?	?	?	?	?	?	?	?	?	?	?	?	?	?	?	?	?	?	?	?	?	?	?
Palaechthon woodi	?	?	?	?	?	?	?	?	?	?	?	?	?	?	?	?	?	?	?	?	?	?	?	?
Palaechthon nacimienti	?	?	?	?	?	?	?	?	?	?	?	?	?	?	?	?	?	?	?	?	?	?	?	?
Talpohenach torrejonius	?	?	?	?	?	?	?	?	?	?	?	?	?	?	?	?	?	?	?	?	?	?	?	?
Palenochtha weissae	?	?	?	?	?	?	?	?	?	?	?	?	?	?	?	?	?	?	?	?	?	?	?	?
Palenochtha minor	?	?	?	?	?	?	?	?	?	?	?	?	?	?	?	?	?	?	?	?	?	?	?	?
Plesiolestes problematicus	?	?	?	?	?	?	?	?	?	?	?	?	?	?	?	?	?	?	?	?	?	?	?	?
Anasazia williamsoni	?	?	?	?	?	?	?	?	?	?	?	?	?	?	?	?	?	?	?	?	?	?	?	?
Torrejonia	?	?	?	?	?	?	?	?	?	?	?	?	?	?	?	?	?	?	?	?	?	?	?	?
Premnoides douglassi	?	?	?	?	?	?	?	?	?	?	?	?	?	?	?	?	?	?	?	?	?	?	?	?
Berruvius	?	?	?	?	?	?	?	?	?	?	?	?	?	?	?	?	?	?	?	?	?	?	?	?
Avenius amatorum	?	?	?	?	?	?	?	?	?	?	?	?	?	?	?	?	?	?	?	?	?	?	?	?
Toliapina	?	?	?	?	?	?	?	?	?	?	?	?	?	?	?	?	?	?	?	?	?	?	?	?
Microsyopidae	0	0	1	0	1	0	?	0	0	2	?	1	?	1	1	1	?	?	?	?	?	?	?	?
Paromomyidae	1	0	1	0	1	0	1	0	0	2	1	0	0	1	1	1	0	1	0	0	1	?	1	1
Picrodontidae	?	?	?	?	?	?	?	?	?	?	?	?	?	?	?	?	?	?	?	?	?	?	?	?
Picromomys petersonorum	?	?	?	?	?	?	?	?	?	?	?	?	?	?	?	?	?	?	?	?	?	?	?	?
Micromomyidae (Gunnell skull)	0	0	1	0	1	0	0	0	0	2	1	0	0	1	1	1	0	0	0	1	?	?	?	1
Micromomyidae (Beard skull)	1	0	1	0	1	0	0	0	0	2	1	0	0	1	1	1	0	0	0	1	?	?	?	1
Chronolestes simul	?	?	?	?	?	?	?	?	?	?	?	?	?	?	?	?	?	?	?	?	?	?	?	?
Carpolestidae	1	1	1	0	1	0	0	0	0	2	1	0	?	?	?	?	?	?	?	?	?	?	?	?
Pandemonium dis	?	?	?	?	?	?	?	?	?	?	?	?	?	?	?	?	?	?	?	?	?	?	?	?
Plesiadapidae	1	1 & 2	1	0	1	0	0	0	0	1	1	0	0	1	1	1	0	0	0	0	1	0	0	0
Saxonellidae	?	?	?	?	?	?	?	?	?	?	?	?	?	?	?	?	?	?	?	?	?	?	?	?
Mixodectes	?	?	?	?	?	?	0	?	0	2	?	0	0	1	1	0	?	?	0	?	?	?	?	?
Eudaemonema cuspidata	?	?	?	?	?	?	?	?	?	?	?	?	?	?	?	?	?	?	?	?	?	?	?	?
Elpidophorus	?	?	?	?	?	?	?	?	?	?	?	?	?	?	?	?	?	?	?	?	?	?	?	?
Plagiomene	?	?	?	?	?	?	?	?	?	?	?	?	?	?	?	?	?	?	?	?	?	?	?	?
Cynocephalidae	0	0	1	1	0	0	0	1	1	1	1	0	1	0	1	1	1	1	1	2	1	1	1	1
Chiroptera	0	1	1	1	0	1	1	1	1	0	0	0	1	1	1	0	1	1	1	2	0	1	2	1
Altiatlasius koulchii	?	?	?	?	?	?	?	?	?	?	?	?	?	?	?	?	?	?	?	?	?	?	?	?
Altanius orlovi	?	?	?	?	?	?	?	?	?	?	?	?	?	?	?	?	?	?	?	?	?	?	?	?
Omomyidae	1	0	1	0	1	0	0	0	0	2	1	1	0	1	1	1	?	?	?	?	?	?	?	?
Adapidae	1	0	1	0	1	0	0	0	0	2	1	1	0	1	1	1	0	0	0	1	0	0	0	0
Scandentia	0	0	0	0	0	0 & 1	0	0	1	0	0	0	0	0	0	0	0	0	0	0	0	1	0	0

(continued)

Appendix I (continued)

	ps24	ps25	ps26	ps27	ps28	ps29	ps30	ps31	ps32	ps33	ps34	ps35	ps36	ps37	ps38	ps39	ps40	ps41	ps42	ps43	ps44	ps45
Leptictidae	0	0	0	1	1	1	0	0	1	0	0	0	?	0	1	1	0	0	0	1	0	0
Asioryctes nemegetensis	?	?	0	0	0	0	0	0	0	?	?	?	?	?	?	?	?	?	?	?	?	?
Purgatorius unio	?	?	?	?	?	?	?	?	?	?	?	?	?	?	?	?	?	?	?	?	?	?
Purgatorius janisae	?	?	?	?	?	?	?	?	?	?	?	?	?	?	?	?	?	?	?	?	?	?
Palaechthon alticuspis	?	?	?	?	?	?	?	?	?	?	?	?	?	?	?	?	?	?	?	?	?	?
Palaechthon woodi	?	?	?	?	?	?	?	?	?	?	?	?	?	?	?	?	?	?	?	?	?	?
Palaechthon nacimienti	?	?	?	?	?	?	?	?	?	?	?	?	?	?	?	?	?	?	?	?	?	?
Talpohenach torrejonius	?	?	?	?	?	?	?	?	?	?	?	?	?	?	?	?	?	?	?	?	?	?
Palenochtha weissae	?	?	?	?	?	?	?	?	?	?	?	?	?	?	?	?	?	?	?	?	?	?
Palenochtha minor	?	?	?	?	?	?	?	?	?	?	?	?	?	?	?	?	?	?	?	?	?	?
Plesiolestes problematicus	?	?	?	?	?	?	?	?	?	?	?	?	?	?	?	?	?	?	?	?	?	?
Anasazia williamsoni	?	?	?	?	?	?	?	?	?	?	?	?	?	?	?	?	?	?	?	?	?	?
Torrejonia	?	?	?	?	?	?	?	?	?	?	?	?	?	?	?	?	?	?	?	?	?	?
Premnoides douglassi	?	?	?	?	?	?	?	?	?	?	?	?	?	?	?	?	?	?	?	?	?	?
Berruvius	?	?	?	?	?	?	?	?	?	?	?	?	?	?	?	?	?	?	?	?	?	?
Avenius amatorum	?	?	?	?	?	?	?	?	?	?	?	?	?	?	?	?	?	?	?	?	?	?
Toliapina	?	?	?	?	?	?	?	?	?	?	?	?	?	?	?	?	?	?	?	?	?	?
Microsyopidae	?	?	?	?	?	?	?	?	?	?	?	?	?	?	?	?	?	?	?	?	?	?
Paromomyidae	1	1	1	0	1	0	1	1	0 & 1	1	1	1	1	0	0	1	1	1	1	0	0	0&1
Picrodontidae	?	?	?	?	?	?	?	?	?	?	?	?	?	?	?	?	?	?	?	?	?	?
Picromomys petersonorum	?	?	?	?	?	?	?	?	?	?	?	?	?	?	?	?	?	?	?	?	?	?
Micromomyidae (Gunnell skull)	?	?	1	0	0	0	1	1	0	?	1	1	0	0	0	0	1	1	1	0	1	0
Micromomyidae (Beard skull)	?	?	1	0	0	0	1	1	0	?	1	1	0	0	0	0	1	1	1	0	1	0
Chronolestes simul	?	?	?	?	?	?	?	?	?	?	?	?	?	?	?	?	?	?	?	?	?	?
Carpolestidae	?	?	?	?	?	?	?	?	?	?	?	?	?	?	?	?	?	?	?	?	?	?
Pandemonium dis	?	?	?	?	?	?	?	?	?	?	?	?	?	?	?	?	?	?	?	?	?	?
Plesiadapidae	1	0	1	0	1	0	1	1	0	1	0&1	1	0	0	0	1	1	0	1	0	0	0
Saxonellidae	?	?	?	?	?	?	?	?	?	?	?	?	?	?	?	?	?	?	?	?	?	?
Mixodectes	1	?	0	0	0	?	1	1	0	0	?	?	1	?	?	?	?	?	?	?	?	?
Eudaemonema cuspidata	?	?	?	?	?	?	?	?	?	?	?	?	?	?	?	?	?	?	?	?	?	?
Elpidophorus	?	?	?	?	?	?	?	?	?	?	?	?	?	?	?	?	?	?	?	?	?	?
Plagiomene	?	?	?	?	?	?	?	?	?	?	?	?	?	?	?	?	?	?	?	?	?	?
Cynocephalidae	1	1	1	2	0	0	1	1	1	1	1	1	1	1	1	1	1	1	1	0	1	0
Chiroptera	0&1	1	2	2	0	0	0	0	1	1	1	1	2	1	1	2	1	0	1	0	1	1
Altiatlasius koulchii	?	?	?	?	?	?	?	?	?	?	?	?	?	?	?	?	?	?	?	?	?	?
Altanius orlovi	?	?	?	?	?	?	?	?	?	?	?	?	?	?	?	?	?	?	?	?	?	?
Omomyidae	?	?	0	1	0	1	0	1	2	?	?	?	?	0	1	0	1	1	0	1	0	0
Adapidae	2	0	1	1	0	1	0	1	2	1	1	1	0	0	0&1	0	1	1	0	1	0	0
Scandentia	0	0	0	0&2	0	0&1	0	0	0&1	0	0	0	1	0	1	1	0&1	0	0	0	0	0

(continued)

Appendix I (continued)

	ps46	ps47	ps48	ps49	ps50	ps51	ps52	ps53	ps54
Leptictidae	0	2	0	0	0	0	2	0	0
Asioryctes nemegetensis	?	0	?	1	?	?	?	?	?
Purgatorius unio	?	?	?	?	?	?	?	?	?
Purgatorius janisae	?	?	?	?	?	?	?	?	?
Palaechthon alticuspis	?	?	?	?	?	?	?	?	?
Palaechthon woodi	?	?	?	?	?	?	?	?	?
Palaechthon nacimienti	?	?	?	?	?	?	?	?	?
Talpohenach torrejonius	?	?	?	?	?	?	?	?	?
Palenochtha weissae	?	?	?	?	?	?	?	?	?
Palenochtha minor	?	?	?	?	?	?	?	?	?
Plesiolestes problematicus	?	?	?	?	?	?	?	?	?
Anasazia williamsoni	?	?	?	?	?	?	?	?	?
Torrejonia	?	?	?	?	?	?	?	?	?
Premnoides douglassi	?	?	?	?	?	?	?	?	?
Berruvius	?	?	?	?	?	?	?	?	?
Avenius amatorum	?	?	?	?	?	?	?	?	?
Toliapina	?	?	?	?	?	?	?	?	?
Microsyopidae	?	?	0	?	0	0	?	?	?
Paromomyidae	0	1	0	1	0	0	0	?	?
Picrodontidae	?	?	?	?	?	?	?	?	?
Picromomys petersonorum	?	?	?	?	?	?	?	?	?
Micromomyidae (Gunnell skull)	0	?	?	?	?	?	?	?	?
Micromomyidae (Beard skull)	0	?	?	?	?	?	?	?	?
Chronolestes simul	?	?	?	?	?	?	?	?	?
Carpolestidae	?	?	?	?	?	?	?	?	?
Pandemonium dis	?	?	?	?	?	?	?	?	?
Plesiadapidae	0	1	0	1	0	0	0	?	0
Saxonellidae	?	?	?	?	?	?	?	?	?
Mixodectes	?	?	?	?	?	?	?	?	?
Eudaemonema cuspidata	?	?	?	?	?	?	?	?	?
Elpidophorus	?	?	?	?	?	?	?	?	?
Plagiomene	?	?	?	?	?	?	?	?	?
Cynocephalidae	0	1	1	1	1	1	0	0	1
Chiroptera	0	0	1	1	1	1	1	0	1
Altiatlasius koulchii	?	?	?	?	?	?	?	?	?
Altanius orlovi	?	?	?	?	?	?	?	?	?
Omomyidae	1	0	?	0	?	?	2	0	?
Adapidae	1	1	0	0	0	0	2	1	0
Scandentia	0	0	0	0	0	0	0	0	0

Note that data was originally collected at the species level for most taxa. See Silcox (2001) for more details.

Appendix II

Brief character descriptions of dental, cranial, and postcranial characters scored by Silcox (2001)

	Incisors	
d1	Presence of I_3	0 = present, small; 1 = absent
d2	Presence of I_2	0 = present, small; 1 = absent; 2 = present, also enlarged and/or larger than I_1
d3	Presence of I^3	0 = present; 1 = absent
d4	Presence of I^2	0 = present; 1 = absent
d5	Size of I_1	0 = comparable to other incisors (or premolars if I_{2-3} are lost); 1 = hypertrophied
d6	Orientation of I_1	0 = essentially vertical; 1 = procumbent-horizontal
d7	Form of I_1	0 = simple, not laterally compressed; 1 = laterally compressed with no broad, flattened dorsal surface; 2 = as 1, with a flattened dorsal surface; 3 = as 2, but rotated medially; 4 = multicuspate and "comb-like"
d8	Presence of a margoconid on I_1	0 = absent; 1 = present
d9	Enlargement of I^1	0 = not markedly enlarged relative to other incisors or premolars (if I^{2-3} are missing); 1 = markedly enlarged
d10	Morphology of I^1	0 = no accessory cuspules; 1 = posterocone present but no apical cuspules; 2 = posterocone and small cuspules developed around the tip, no strong apical division; 3 = strong apical division into an anterocone and laterocone in addition to the presence of a posterocone
	Canines	
d11	Relative size of the lower canine	0 = larger than adjacent teeth; 1 = smaller than adjacent teeth; 2 = absent
d12	Number of upper canine roots	0 = single rooted; 1 = double rooted; 2 = absent
	Premolars	
d13	Presence of P_1	0 = present; 1 = absent
d14	Number of alveoli for P_2	0 = 2 alveoli; 1 = 1 alveolus; 2 = tooth absent
d15	Number of P_3 roots	0 = 2 roots; 1 = 1 root; 2 = tooth absent
d16	Presence of P^1	0 = present; 1 = absent
d17	Number of roots for P^2	0 = double rooted; 1 = single rooted; 2 = absent; 3 = triple rooted
d18	Number of roots for P^3	0 = double rooted; 1 = triple rooted
d19	Mesiodistal length of P_4	0 = P_4 somewhat smaller than M_1; 1 = subequal; 2 = $P_4 >> M_1$; 3 = $M_1 >> P_4$
d20	Morphology of the P_4 talonid	0 = basined; 1 = not basined, reduced; 2 = not basined, expanded; 3 = not basined, flap-like
d21	Number of P_4 talonid cusps	0 = 2 well defined; 1 = all poorly defined; 2 = 1 solo distinct cusp; 3 = 3 well defined
d22	Position of the P_4 cristid obliqua	0 = joins postvallid near midline of tooth or more lingually; 1 = joins postvallid near buccal margin of trigonid; 2 = absent
d23	Morphology of the hypoflexid on P_4	0 = distinct, deep; 1 = not distinct, shallow
d24	Presence of a mesoconid on P_4	0 = absent; 1 = present (at least in some specimens)
d25	P_4 paraconid presence	0 = paraconid distinct, cuspate; 1 = cusp indistinct but paracristid present, not markedly elongate; 2 = paraconid and paracristid absent or weak; 3 = paracristid elongate with or without a distinct cusp
d26	Presence of a P_4 metaconid	0 = absent; 1 = present
d27	Addition of apical cusps to P_4 beyond the paraconid and metaconid	0 = none; 1 = one neomorphic cusp (4 apical cusps in total); 2 = two neomorphic cusps (five apical cusps in total); 3 = three or more neomorphic cusps (six or more apical cusps in total)
d28	Morphology of the crest between the protoconid and cristid obliqua on P_4	0 = crest not continuous from postvallid to cristid obliqua; 1 = crest present and continuous; 2 = no comparable crest on postvallid
d29	Presence of exodaenodont lobes on P_4	0 = tooth not exodaenodont at all; 1 = little lobe development; 2 = distal lobe developed more than mesial; 3 = both distal and mesial lobes strongly developed
d30	Presence of exodaenodont lobes on M_1	0 = tooth not exodaenodont at all; 1 = little lobe development; 2 = distal lobe developed more than mesial; 3 = both distal and mesial lobes strongly developed
d31	Presence of a protocone on P^3	0 = absent; 1 = present
d32	Presence of conules on P^3	0 = absent; 1 = 1 present; 2 = 2 present
d33	Presence of a preparaconule crista on P^3	0 = absent; 1 = present

(continued)

Appendix II (continued)

d34	Presence of conules on P^4	0 = absent; 1 = large conule present; 2 = small paraconule present; 3 = both conules strong
d35	Presence of a intermediate ridge on P^4	0 = absent; 1 = present
d36	Presence of an indentation on P^3 postcingulum	0 = absent; 1 = present (with or without a cuspule)
d37	Presence of a P^3 metacone	0 = absent; 1 = present
d38	Presence of a P^3 metastyle	0 = absent; 1 = present
d39	Presence of a P^4 metacone	0 = present; 1 = absent
d40	Presence of a P^4 metastyle	0 = absent; 1 = present
d41	Presence of a P^3 parastyle	0 = present; 1 = absent
d42	Presence of a P^4 parastyle	0 = present; 1 = absent
d43	Presence of an additional "preparaconal" cusp on P^4	0 = absent; 1 = present between parastyle and paracone
d44	Presence of a hypocone on P^3	0 = absent; 1 = present
d45	Presence of a hypocone on P^4	0 = totally absent; 1 = present; at least incipiently
d46	Presence of a pericone on P^4	0 = totally absent; 1 = present; at least incipiently
d47	Shape of P^3 (buccal length/lingual length)	0 = < 1.7; 1 = 1.8–2.0; 2 = > 2.0
d48	Shape of P^4 (buccal length/lingual length)	0 = > 1.8; 1 = < 1.8
d49	Relative size of P^3 compared to P^4	0 = 0.6–1.2; 1 = < 0.6; 2 = > 1.2 based on (ln(buccal lengthxwidth)P^3/ln(buccal lengthxwidth)P^4)
d50	Relative size of P^4 compared to M^1	0 = 0.90–0.98; 1 = > 0.98; 2 = < 0.9 based on (ln(buccal lengthxwidth) P^4/ln(buccal lengthxwidth)M^1)
d51	Presence of a postprotocingulum on P^4	0 = absent; 1 = present
d52	Presence of a preprotocrista on P^4	0 = present; 1 = absent
d53	Presence of a postprotocrista on P^4	0 = present; 1 = absent
d54	Morphology of the P^4 parastylar lobe	0 = large, projecting; 1 = smaller and not projecting
d55	Shape of the P^4 protocone lobe	0 = shorter mesiodistally than wide; 1 = equally long and wide
d56	Acuteness of P^4 cusps	0 = acute; 1 = bulbous
	Molars	
d57	Distinctiveness of the molar hypoflexids	0 = distinct, invaginated; 1 = not distinct
d58	Molar trigonid height on M_{1-2}	0 = tall relative to talonid; 1 = of a similar height to talonid
d59	Molars high-crowned	0 = high crowned (index value > 0.79); 1 = moderate (index value 0.60–0.78); 2 = low crowned (index value < 0.60)
d60	Presence of buccal cingulids on lower molars	0 = anterobuccal "precingulid"only; 1 = separate anterior and posterior cingulids; 2 = continuous buccal cingulid; 3 = absent, no buccal cingulid or "precingulid"
d61	Length of molar trigonids	0 = trigonids become less mesiodistally compressed from M_{1-3}; 1 = no change; 2 = trigonids become more mesiodistally compressed from M_{1-3}
d62	Height of the lower molar paraconids	0 = lower than metaconid; 1 = subequal to or taller than the metaconid
d63	Distinctiveness of the paraconid on M_1	0 = indistinct from paracristid; 1 = distinct from paracristid
d64	Presence of paraconids on M_2	0 = comparably distinct to the paraconid on M_1; 1 = less distinct than the paraconid on M_1
d65	Presence of lower molar mesoconids	0 = weak-absent; 1 = present (if variably) on some molars
d66	Molar enamel roughness	0 = smooth; 1 = crenulated; 2 = papillated
d67	Relative positions of the M_{1-2} hypoconulid and entoconid	0 = unpaired; 1 = paired; 2 = twinned
d68	Position of M_{1-2} hypoconulid relative to the central axis of the tooth	0 = hypoconulid centrally placed or shifted lingually towards the entoconid; 1 = hypoconulid shifted bucally
d69	Presence of an entoconid notch (between the entoconid and hypoconulid) on M_{1-2}	0 = absent; 1 = present
d70	Presence of a hypoconulid notch (between the hypoconulid and hypoconid) on M_{1-2}	0 = absent-weak; 1 = present
d71	Presence of mesial inflection of the molar trigonids	0 = absent; 1 = weak; 2 = pronounced
d72	Basal breadth of the M_{1-2} trigonids	0 = not swollen at the base; 1 = swollen basally
d73	Breadth of the M_{1-2} talonids near the apices of the cusps	0 = narrower than trigonid; 1 = wider than trigonid
d74	Breadth of the M_3 talonids	0 = much narrower than trigonid; 1 = similar in breadth to trigonid or wider
d75	Size of the M_3 hypoconulid	0 = similar to that cusp on M_{1-2}; 1 = larger than on M_{1-2} but not developed into a lobe; 2 = developed into a lobe
d76	Relative height of the molar talonid cusps	0 = hypoconid taller than entoconid; 1 = entoconid taller
d77	Relative height of the protoconid and metaconid on M_1	0 = protoconid higher; 1 = subequal; 2 = metaconid higher
d78	Presence of curving molar paracristids	0 = absent; 1 = present

(continued)

Appendix II (continued)

d79	Morphology of the protoconid-metaconid notch on the lower molars	0 = strong and sharp; 1 = more rounded; 2 = fold of enamel bridges notch
d80	Presence of a stepped postvallid on M_1	0 = absent; 1 = present
d81	Acuteness of upper and/or lower molar cusps	0 = relatively acute; 1 = blunter
d82	Relative size of M^3	0 = > 0.9; 1 = < 0.9 based on (ln(buccal LxW)M^3/ln (buccal LxW)M_1)
d83	Relative length of M_3	0 = $M_3 < M_2$; 1 = $M_3 > M_2$
d84	Morphology of the molar stylar shelf	0 = broad; 1 = narrow (buccal cingulum only) or absent
d85	Presence of a postprotocingulum on M^1 or M^2	0 = absent; 1 = weak; 2 = pronounced
d86	Position of the conules on M^{1-2}	0 = central or closer to protocone than to paracone/metacone; 1 = apressed to paracone/metacone
d87	Presence of conules on M^{1-2}	0 = both conules present; 1 = metaconule absent; 2 = both absent
d88	Relative sizes of metacone and paracone on M^{1-2}	0 = paracone > metacone; 1 = cusps are subequal; 2 = metacone > paracone
d89	Presence of a hypocone on M^{1-2}	0 = absent; 1 = present (true hypocone, coming off the cingulum); 2 = present (pseudohypocone, budding off the protocone)
d90	Position of the M^{1-2} protocone	0 = skewed mesiobuccally; 1 = central on the tooth
d91	Presence of a precingulum on M^{1-2}	0 = present, doesn't connect to postcingulum; 1 = present, connects to postcingulum in at least some specimens; 2 = absent
d92	Continuity of post- and metacingula	0 = not continuous; 1 = continuous
d93	Separateness of pre- and paracingula on M^{1-2}	0 = not continuous; 1 = continuous; 2 = no paracingulum
d94	Morphology of the centrocrista on the upper molars	0 = moderate; 1 = strong and straight; 2 = absent or v. weak; 3 = strong and v-shaped
d95	Morphology of parastylar lobe on M^1	0 = projecting beyond the plane of the mesiolingual corner of the tooth; 1 = not projecting
d96	Presence of one or more mesostyle(s) on the upper molars	0 = absent; 1 = one or more present
d97	Orientation of the preparacrista on M^{1-2}	0 = angled buccally; 1 = straight
	Cranium	
sk1	Structure of the auditory bulla	0 = membranous, or bony but non-petrosal in origin; 1 = no suture separating bulla from petrosal (and/or no developmental evidence for additional elements)
sk2	Relations of the entotympanic	0 = no entotympanic is present; 1 = entotympanic contacts the petrosal medially; 2 = entotympanic contacts the basioccipital medially; 3 = no medial contact
sk3	Form of the External Auditory Meatus (EAM)	0 = not expanded into a tube; 1 = expanded into a bony tube
sk4	Presence of a subtympanic recess (between the tympanic ring and the bulla)	0 = subtympanic recess absent and the ectotympanic does not include a distinct ring-like element; 1 = subtympanic recess present and the ectotympanic includes a ring-like element separated by an annular bridge, membrane or gap between it and the bulla
sk5	Presence of branches of the internal carotid artery (ICA)	0 = grooves for at least the promontorial branch, no tubes; 1 = tubes present for one or both arteries; 2 = ICA absent
sk6	Posterior carotid foramen position (or the position of the entry of the internal carotid artery and/or nerves into the middle ear)	0 = posteromedial; 1 = posterolateral
sk7	Presence of the subsquamosal foramen	0 = present, large; 1 = very small or absent
sk8	Width of the central stem and relative size of the hypotympanic sinus	0 = broad (hypotympanic sinus restricted); 1 = narrow (hypotympanic sinus expansive)
sk9	Choanae shape	0 = broad; 1 = very narrow and "peaked"
sk10	Relative length of snout	0 = long; 1 = reduced
sk11	Presence of a postorbital bar	0 = absent; 1 = postorbital process of the frontal is present, but doesn't meet zygomatic; 2 = complete postorbital bar present
sk12	Presence of a mastoid process	0 = strong tubercle or inflation in the mastoid region; 1 = no strong tubercle or inflation in the mastoid region
sk13	Number of posterior lacerate foramina	0 = single; 1 = dual
sk14	Position of the caudal midsagittal margin of the palate	0 = near M^3/; 1 = well rostral to M^3/; 2 = caudal to M^3/
sk15	Presence of a posterior palatine torus	0 = present; 1 = absent
sk16	Number of pterygoid plates	0 = 2 plates present; 1 = 1 plate present
sk17	Presence of a foramen in the lateral pterygoid plate	0 = absent; 1 = present
sk18	Presence of the supraorbital foramen	0 = absent; 1 = present
sk19	Shape of the nasals and width of the suture with the frontal	0 = nasals flare laterally at the caudal extent (wide contact with frontal); 1 = do not flare (narrow suture with frontal)
sk20	Diameter of the infraorbital foramen	0 = large; 1 = small
sk21	Contact between lacrimal and palatine in the orbit	0 = present; 1 = obscured by maxillofrontal contact
sk22	Presence of a lacrimal tubercle	0 = absent; 1 = present

(continued)

Appendix II (continued)

sk23	Size of the optic foramen	0 = small; 1 = moderate; 2 = large
sk24	Presence of the foramen rotundum	0 = absent; 1 = present
sk25	Position of the lacrimal foramen	0 = in orbit; 1 = on face; 2 = on orbital rim
sk26	Presence of the postglenoid foramen	0 = present; 1 = absent
sk27	Presence of a shielded cochlear window	0 = absent; 1 = rostrally shielded by an arterial tube; 2 = caudally shielded; 3 = ventrally shielded
sk28	Orientation of the cochlear window	0 = directed posterolaterally; 1 = directed posteriorly
sk29	Presence of a longitudinal septum	0 = absent, septa that are present do not extend onto the promontorium; 1 = long septum, presumably petrosal in origin, extending rostro-caudally onto the promontorium; 2 = septa are present in the middle ear cavity, but are formed by the entotympanic
sk30	"Fattened" area on medial promontorium	0 = absent; 1 = present
	Humerus	
ps1	Size of the greater tuberosity	0 = small, not projecting above the head; 1 = intermediate; 2 = prominent and even with, or above head
ps2	Robusticity of lesser tuberosity	0 = gracile; 1 = protrudes medially away from humeral shaft
ps3	Deltopectoral crest length	0 = more than 33% total length of the bone; 1 = less than 33% total length of humerus
ps4	Position of the deltopectoral crest	0 = anterior; 1 = lateral
ps5	Presence of entepicondylar foramen	0 = present; 1 = absent
ps6	Presence of a radial and/or olecranon fossa	0 = distinct; 1 = indistinct or absent
ps7	Medial epicondyle (=entepicondyle) breadth	0 = prominent (30% or more of total distal breadth); 1 = reduced (less than 30% of total distal breadth or generally reduced)
ps8	Morphology of the supinator crest	0 = broad and well developed; 1 = reduced or absent
ps9	Capitulum shape	0 = spindle shaped; 1 = subspheroidal; 2 = ball-like
ps10	Morphology of the attachement for M. teres major	0 = no distinct attachment; 1 = present as a distinct protrusion on the crest leading down from the lesser tuberosity
ps11	Humeral trochlea morphology	0 = only a medial edge; 1 = both medial and lateral edges are present, trochlea and capitulum are well separated by a distinct gap
	Radius	
ps12	Distinctness of the bicipital tuberosity	0 = clearly distinct from the rest of the shaft; 1 = absent
ps13	Shape of radial head	0 = ovoid (ratio of mediolateral breadth/anteroposterior breadth is greater than 1.26); 1 = round (ratio of mediolateral breadth/anteroposterior breadth is less than 1.26)
ps14	Degree of excavation of the radial head	0 = central fossa flat; 1 = central fossa deeply excavated
ps15	Morphology of the lateral lip on the radial head	0 = broad but limited to the lateral side; 1 = narrow but more extensive
ps16	Presence of a ridge on the distal end of the anterior radius	0 = absent; 1 = present
	Ulna	
ps17	Form of the ulnocarpal articulation	0 = mediolaterally and dorsopalmarly extensive, ulnocarpal articulation occurs in a transverse plane; 1 = limited to radial and palmar aspects of distal ulna, ulnocarpal articulation lies in a proximodistal plane
ps18	Length of the olecranon process	0 = similar in length to the height of the semilunar notch (ratio of olecranon process length/semilunar notch height greater than 0.8); 1 = very reduced (ratio of olecranon process length/semilunar notch height less than 0.75)
ps19	Nature of the distal radioulnar articulation	0 = syndesmosis, no ulnar head; 1 = synovial, with ulnar head; 2 = fused
	Manus and Pes	
ps20	Arrangement of the scaphoid and lunate	0 = arranged approximately ulnoradially (=mediolaterally); 1 = arranged more proximodistally
ps21	Condition of the scaphoid and lunate	0 = unfused; 1 = fused
ps22	Shape of triquetrum, and nature of the articulation between the pisiform and triquetrum	0 = triquetrum is squared in dorsal view due to an unreduced articulation with an unreduced pisiform 1 = triquetrum is more triangular in dorsal view as a result of a reduced articulation with a reduced pisiform; 2 = pisiform absent
ps23	Relative length and proportions of the intermediate phalanges	0 = short and fat (maximum elongation index less than 7); 1 = long and thin (maximum elongation index more than 7)
ps24	Shape of the distal phalanges	0 = laterally compressed, moderately high; 1 = very high and laterally compressed; 2 = dorsoventrally flattened and mediolaterally wide
ps25	Shape of the proximal phalanges	0 = no dorsopalmar reinforcement (broadly oval in cross-section); 1 = with dorsopalmar reinforcement (triangular in cross-section)

(continued)

Appendix II (continued)

ps26	Position of the groove for M. flexor fibularis on the astragalus	0 = midline; 1 = shifted laterally; 2 = absent
ps27	Morphology of the astragalar body	0 = shallowly grooved; 1 = narrow and more deeply grooved; 2 = not grooved at all, medial and lateral guiding ridges absent
ps28	Relative height of the borders of the astragalar trochlea	0 = subequal in height; 1 = lateral much higher
ps29	Relative length of astragalar neck	0 = astragalar neck is less than 30% of total length of the bone; 1 = astragalar neck is more than 30% of the bone's total length
ps30	Presence of a secondary articulation between the posterior side of the sustentaculum tali and the astragalus	0 = absent; 1 = present
ps31	Shape of the calcaneocuboid articulation	0 = distal calcaneus is flat; 1 = concave pit on the calcaneus
ps32	Proportions of the calcaneus	0 = distal end of calcaneus is short, tuber calcaneus is not enlarged; 1 = distal end of calcaneus not elongate, tuber calcaneus is more than 40% of total length; 2 = distal end of calcaneus is elongate.
ps33	Form of distal facet on the entocuneiform	0 = bears a strong plantodistal process; 1 = plantodistal process reduced or absent
	Pelvis	
ps34	Acetabular shape	0 = circular (ratio of craniocaudal length/dorsoventral breadth less than 1.1); 1 = craniocaudally elongate in lateral view (ratio of craniocaudal length/ dorsoventral breadth more than 1.1)
ps35	Pattern of acetabular bony buttressing	0 = even around the entire rim; 1 = markedly more emphasized cranially
ps36	Morphology of the ischial spine	0 = not expanded or hooklike; 1 = pronounced, hooklike; 2 = absent completely
	Femur	
ps37	Height of the greater trochanter	0 = taller or comparable in proximal extent to femoral head; 1 = markedly shorter than femoral head
ps38	Size of the lesser trochanter	0 = enlarged, extends medially beyond the level of the head; 1 = not enlarged, not so extensive medially
ps39	Position of the third trochanter	0 = same level as the lesser trochanter; 1 = distal to lesser trochanter; 2 = absent
ps40	Position of the fovea capitis femoris	0 = centrally placed with respect to the anteroposterior axis of the femoral head; 1 = posterior to anteroposterior midline of femoral head
ps41	Area for insertion of M. quadratus femoris	0 = limited; 1 = extensive (enlarged flattened triangular area on posterior femoral shaft between the greater and lesser trochanters)
ps42	Shape of the patellar groove	0 = narrow mediolaterally; 1 = wide mediolaterally
ps43	Depth of the distal femur	0 = shallow or moderately deep; 1 = very deep (anteroposteriorly)
ps44	Trochanteric fossa depth	0 = deep; 1 = shallow or absent
ps45	Femoral neck length	0 = long (ratio of neck length/head diameter > 0.16); 1 = short (ratio of neck length/head diameter > 0.16)
ps46	Relative height of lateral femoral patellar ridge	0 = lateral and medial ridges subequal; 1 = lateral ridge much taller (more anteriorly projecting) than medial
	Tibia	
ps47	Nature of the distal tibiofibular joint	0 = syndesmosis; 1 = synovial; 2 = fused
ps48	Strength of the tibial tuberosity or cnemial crest	0 = robust tibial tuberosity or strong cnemial crest; 1 = no well demarcated tibial tuberosity or cnemial crest
ps49	Morphology of the crurotarsal joint	0 = medial malleolus is long, joint is less mobile; 1 = medial malleolus is short, joint is more mobile
ps50	Presence of a prominent intercondylar eminence on the proximal tibia	0 = present; 1 = absent
ps51	Proximal extent of the condyles on the tibial plateau	0 = lateral condyle projects further proximally than the medial condyle 1 = medial and lateral project to a similar extent proximally;
	Limb proportions	
ps52	Humerofemoral proportions	0 = humerofemoral index (HFI = [humerus length/femoral length]*100) between 70 and 150; 1 = HFI > 150; 2 = HFI < 70
ps53	Crural index	0 = crural index (CI = [tibia length/femoral length]*100) greater than 90; 1 = CI < 90
ps54	Brachial index	0 = brachial index (BI = [radius length/humerus length]*100) less than 120; 1 = BI > 120

See Silcox (2001) for more detailed character descriptions.

References

Adkins, R. M., Honeycutt, R. L., 1991. Molecular phylogeny of the superorder Archonta. Proceedings of the National Academy of Science, USA 88, 10317–10321.

Adkins, R. M., Honeycutt, R. L., 1993. A molecular examination of archontan and chiropteran monophyly. In: MacPhee, R. D. E. (Ed.), Primates and their Relatives in Phylogenetic Perspective. Plenum, New York, pp. 227–249.

Aiello, L. C., 1993. The origin of the new world monkeys. In: George, W., Lavocat, R. (Eds.), The Africa-South America Connection. Clarendon, Oxford, pp. 100–118.

Ammerman, L. K., Hillis, D. M., 1992. A molecular test of bat relationships: monophyly or diphyly? Systematic Biology 41, 222–232.

Bailey, W. J., Slightom, J. L., Goodman, M., 1992. Rejection of the "flying primate" hypothesis by phylogenetic evidence from the ε-globin gene. Science 256, 86–89.

Beard, K. C.,1989. Postcranial anatomy, locomotor adaptations, and paleoecology of Early Cenozoic Plesiadapidae, Paromomyidae, and Micromomyidae (Eutheria, Dermoptera). Ph.D. dissertation, Johns Hopkins University School of Medicine, Baltimore, MD.

Beard, K. C., 1990. Gliding behavior and palaeoecology of the alleged primate family Paromomyidae (Mammalia, Dermoptera). Nature 345, 340–341.

Beard, K. C., 1991. Postcranial fossils of the archaic primate family Microsyopidae. American Journal of Physical Anthropology 34(Supplement 12), 48–49.

Beard, K. C., 1993a. Phylogenetic systematics of the Primatomorpha, with special reference to Dermoptera. In: Szalay, F. S., Novacek, M. J., McKenna, M. C. (Eds.), Mammal Phylogeny: Placentals. Springer, New York, pp. 129–150.

Beard, K. C., 1993b. Origin and evolution of gliding in early Cenozoic Dermoptera (Mammalia, Primatomorpha). In: MacPhee, R. D. E. (Ed.), Primates and their Relatives in Phylogenetic Perspective. Plenum, New York, pp. 63–90.

Beard, K. C., 1998a. East of Eden: Asia as an important center of taxonomic origination in mammalian evolution. In: Beard, K. C., Dawson, M. R. (Eds.), Dawn of the Age of Mammals in Asia. Bulletin of the Carnegie Museum of Natural History 34, pp. 5–39.

Beard, K. C., 1998b. Biostratigraphy and paleobiogeography of Asian land mammals near the Paleocene-Eocene boundary: a phylogenetic approach. Strata 9, 25–28.

Beard, K. C., 1998c. A new genus of Tarsiidae (Mammalia: Primates) from the middle Eocene of Shanxi Province, China, with notes on the historical biogeography of tarsiers. In: Beard, K. C., Dawson, M. R. (Eds.), Dawn of the Age of Mammals in Asia. Bulletin of the Carnegie Museum of Natural History 34, pp. 260–277.

Beard, K. C., 2004. The Hunt for the Dawn Monkey: Unearthing the Origins of Monkeys, Apes, and Humans. University of California Press, Los Angeles.

Beard, K. C., 2006. Iterative dispersal across Beringia by early Cenozoic primates. American Journal of Physical Anthropology 129(Supplement to 42), 62–63.

Beard, K. C., Dawson, M. R., 1999. Intercontinental dispersal of Holarctic land mammals near the Paleocene/Eocene boundary: paleogeographic, paleoclimatic and biostratigraphic implications. Bulletin de la Société Géologique de France 170, 697–706.

Beard, K. C., MacPhee, R. D. E., 1994. Cranial anatomy of *Shoshonius* and the antiquity of Anthropoidea. In: Fleagle, J. G., Kay, R. F. (Eds.), Anthropoid Origins. Plenum, New York, pp. 55–97.

Beard, K. C., Wang, J., 1991. Phylogenetic and biogeographic significance of the tarsiiform primate *Asiomomys changbaicus* from the Eocene of Jilin Province, People's Republic of China. American Journal of Physical Anthropology 85, 159–166.

Beard, K. C., Wang, J., 1995. The first Asian plesiadapoids (Mammalia: Primatomorpha). Annals of the Carnegie Museum 64, 1–33.

Beard, K. C., Wang, J., 2004. The eosimiid primates (Anthropoidea) of the Heti formation, Yuanqu Basin, Shanxi and Henan Provinces, People's Republic of China. Journal of Human Evolution 46, 401–432.

Beard, K. C., Tong, Y., Dawson, M. R., Wang, J., Huang, X., 1996. Earliest complete dentition of an anthropoid primate from the late middle Eocene of Shanxi Province, China. Science 272, 82–85.

Beard, K. C., Ni, X., Wang, Y., Gebo, D., Meng, J., 2005. Phylogenetic position and biogeographic significance of *Subengius mengi* (Mammalia, Carpolestidae), the oldest Asian plesiadapiform. Journal of Vertebrate Paleontology 25(Supplement to 3), 35A.

Bloch, J. I., Boyer, D. M., 2002. Grasping primate origins. Science 298, 1606–1610.

Bloch, J. I., Boyer, D. M., 2003. Response to comment on "Grasping primate origins". Science 300, 741c.

Bloch, J. I., Silcox, M. T., 2001. New basicrania of Paleocene-Eocene *Ignacius*: re-evaluation of the plesiadapiform-dermopteran link. American Journal of Physical Anthropology 116, 184–198.

Bloch, J. I., Silcox, M. T., 2006. Cranial anatomy of Paleocene plesiadapiform *Carpolestes simpsoni* (Mammalia, Primates) using ultra high-resolution x-ray computed tomography, and the relationships of plesiadapiforms to Euprimates. Journal of Human Evolution 50, 1–35.

Bloch, J. I., Fisher, D. C., Rose, K. D., Gingerich, P. D., 2001. Stratocladistic analysis of Paleocene Carpolestidae (Mammalia, Plesiadapiformes) with description of a new late Tiffanian genus. Journal of Vertebrate Paleontology 21, 119–131.

Bloch, J. I., Silcox, M. T., Boyer, D. M., Sargis, E. J., 2007 New Paleocene skeletons and the relationship of plesiadapiforms to crown-clade primates. Proceedings of the National Academy of Science 104(4); 1159–1164.

Bowen, G. J., Clyde, W. C., Koch, P. L., Ting, S., Alroy, J., Tsubamoto, T., Wang, Y., Wang, Y., 2002. Mammalian dispersal at the Paleocene/Eocene boundary. Science 295, 2062–2065.

Brooks, D. R., McLennan, D. A., 1991. Phylogeny, Ecology, and Behavior: A Research Program in Comparative Biology. University of Chicago Press, Chicago.

Buckley, G. A., 1997. A new species of *Purgatorius* (Mammalia: Primatomorpha) from the lower Paleocene Bear formation, Crazy Mountains Basin, south-central Montana. Journal of Paleontology 71, 149–155.

Cartmill, M., 1972. Arboreal adaptations and the origin of the order Primates. In: Tuttle, R. (Ed.), The Functional and Evolutionary Biology of Primates. Aldine-Atherton, Chicago, pp. 97–122.

Cartmill, M., 1974. Rethinking primate origins. Science 184, 436–443.

Clemens, W. A., 2004. *Purgatorius* (Plesiadapiformes, Primates, Mammalia), a Paleocene immigrant into northeastern Montana: stratigraphic occurrences and incisor proportions. In: Dawson, M. R., Lillegraven, J. A. (Eds.), Fanfare for an Uncommon Paleontologist: Papers in Honor of Malcolm C. McKenna. Bulletin of the Carnegie Museum of Natural History 36, 3–13.

Cronin, J. E., Sarich, V. M., 1980. Tupaiid and Archonta phylogeny: the macromolecular evidence. In: Luckett, W. P. (Ed.), Comparative Biology and Evolution Relationships of Tree Shrews. Plenum, New York, pp. 293–312.

da Silva, J. M. C., Oren, D. C., 1996. Application of parsimony analysis of endemicity in Amazonian biogeography: an example with primates. Biological Journal of the Linnean Society 59, 427–437.

Felsenstein, J., 1978. Cases in which parsimony or compatibility methods will be positively misleading. Systematic Zoology 27, 401–410.

Fox, R. C., 1984a. The dentition and relationships of the Paleocene primate *Micromomys* Szalay, with description of a new species. Canadian Journal of Earth Sciences 21, 1262–1267.

Fox, R. C., 1984b. A new species of the Paleocene primate *Elphidotarsius* Gidley: its stratigraphic position and evolutionary relationships. Canadian Journal of Earth Sciences 21, 1268–1277.

Fox, R. C., 1990. *Pronothodectes gaoi* n. sp. from the late Paleocene of Alberta, Canada, and the early evolution of the Plesiadapidae. Journal of Paleontology 64, 637–647.

Fox, R. C., 2002. The dentition and relationships of *Carpodaptes cygneus* (Russell) (Carpolestidae, Plesiadapiformes, Mammalia), from the late Paleocene of Alberta, Canada. Journal of Paleontology 76, 864–881.

Fu, J. -F., Wang, J. -W., Tong, Y. -S., 2002. The new discovery of the Plesiadapiformes from the early Eocene of Wutu Basin, Shandong Province. Vertebrata PalAsiatica 40, 219–227.

Gauthier, J., Kluge, A. G., Rowe, T., 1988. Amniote phylogeny and the importance of fossils. Cladistics 4, 105–209.

Gheerbrant, E., Sudre, J., Sen, S., Abrial, C., Marandat, B., Sigé, B., Vianey-Liaud, M., 1998. Nouvelles données sur les mammifères du Thanetien et de l'Ypresien du Bassin d'Ouarzazate (Maroc) et leur contexte stratigraphique. Palaeovertebrata 27, 155–202.

Gingerich, P. D., 1976. Cranial anatomy and evolution of early Tertiary Plesiadapidae (Mammalia, Primates). University of Michigan Papers on Paleontology 15, 1–141.

Gingerich, P. D., 1986. Early Eocene *Cantius torresi*–oldest primate of modern aspect from North America. Nature 320, 319–321.

Gingerich, P. D., 1989. New earliest Wasatchian mammalian fauna from the Eocene of northwestern Wyoming: composition and diversity in a rarely sampled high-floodplain assemblage. University of Michigan Papers on Paleontology 28, 1–97.

Gingerich, P. D., 1990. African dawn for primates. Nature 346, 411.

Gingerich, P. D., 1993. Early Eocene *Teilhardina brandti*: oldest omomyid primate from North America. Contributions from the Museum of Paleontology, University of Michigan 28, 321–326.

Gingerich, P. D., 2006. Environment and evolution through the Paleocene-Eocene thermal maximum. Trends in Ecology and Evolution 21, 246–253.

Gingerich, P. D., Dashzeveg, D., Russell, D. E., 1991. Dentition and systematic relationships of *Altanius orlovi* (Mammalia, Primates) from the early Eocene of Mongolia. Geobios 24, 637–646.

Godinot, M., 1978. Un nouvel Adapidé (primate) de l'Éocène inférieur de Provence. Comptes rendus de l'Académie des sciences (sér. D) 286, 1869–1872.

Godinot, M., 1981. Les mammifères de Rians (Éocène inférieur, Provence). Palaeovertebrata 10-II, 43–126.

Godinot, M., 1992. Apport à la systématique de quatre genres d'Adapiformes (Primates, Éocène). Comptes rendus de l'Académie des sciences (sér II) 314, 237–242.

Godinot, M., 1994. Early North African primates and their significance for the origin of Simiiformes (=Anthropoidea). In: Fleagle, J. G., Kay, R. F. (Eds.), Anthropoid Origins. Plenum, New York, pp. 235–295.

Godinot, M., 1998. A summary of adapiform systematics and phylogeny. Folia Primatologica 69, 218–249.

Godinot, M., Mahboubi, M., 1992. Earliest known simian primate found in Algeria. Nature 357, 324–326.

Godinot, M., Broin, F. de., Buffetaut, E., Rage, J. -C., Russell, D. 1978. Dormaal: une des plus ancienes faunes éocènes d'Europe. Comptes rendus de l'Académie des sciences (sér. D) 287, 1273–1276.

Graybeal, A., 1998. Is it better to add taxa or characters to a difficult phylogenetic problem? Systematic Biology 47, 9–17.

Gregory, W. K., 1910. The orders of mammals. Bulletin of the American Museum of Natural History 27, 1–524.

Gunnell, G. F., 1989. Evolutionary history of Microsyopoidea (Mammalia, ?Primates) and the relationship between Plesiadapiformes and Primates. University of Michigan Papers on Paleontology 27, 1–157.

Hartenberger, J. -L., Crochet, J. -Y., Martinez, C., Feist, M., Godinot, M., Mannai Tayech, B., Marandat, B., Sigé, B., 1997. Le gisement de mammifères de Chambi (Eocène, Tunisie centrale) dans son contexte géologique. Apport à la connaissance de l'évolution des mammifères en Afrique. In: Aguilar, J. -P., Legendre, S., Michaux, J. (Eds.), Proceedings of the BiochroM'97 Symposium. Mémoires et travaux de l'Institut de Montpellier, pp. 263–274.

Hartwig, W. C. (Ed.), 2002. The Primate Fossil Record. Cambridge University Press, Cambridge.

Hendy, M. D., Penny, D., 1989. A framework for the quantitative study of evolutionary trees. Systematic Zoology 38, 297–309.

Hickey, L. J., West, R. M., Dawson, M. R., Choi, D. K., 1983. Arctic terrestrial biota: paleomagnetic evidence of age disparity with mid-northern latitudes during the late Cretaceous and early Tertiary. Science 221, 1153–1156.

Hillis, D. M., 1996. Inferring complex phylogenies. Nature 383, 130–131.

Hillis, D. M., 1998. Taxonomic sampling, phylogenetic accuracy, and investigator bias. Systematic Biology 47, 3–8.

Hoffstetter, R., 1977. Phylogénie des primates. Bulletins et mémoires de la Société d'anthropologie de Paris t.4, série XIII, 327–346.

Hoffstetter, R., 1988. Origine et évolution des primates non humains du nouveau monde. In: Marois, M. (Ed.), L'Évolution dans sa Réalité et ses Diverses Modalités. Masson, Paris, pp. 133–170.

Holroyd, P., Simons, E. L., 1991. The phyletic relationships of *Azibius*. American Journal of Physical Anthropology 12, 94.

Hooker, J. J., 2001. Tarsals of the extinct insectivoran family Nyctitheriidae (Mammalia): evidence for archontan relationships. Zoological Journal of the Linnean Society 132, 501–529.

Hooker, J. J., Russell, D. E., Phélizon, A., 1999. A new family of Plesiadapiformes (Mammalia) from the Old World lower Paleogene. Palaeontology 42, 377–407.

Johnston, P. A., Fox, R. C., 1984. Paleocene and late Cretaceous mammals from Saskatchewan, Canada. Palaeontographica Abt. A 186, 163–222.

Kay, R. F., Covert, H. H. 1984. Anatomy and behavior of extinct primates. In: Chivers, D. J., Wood, B. A., Bilsborough, A. (Eds.), Food Acquisition and Processing in Primates. Plenum, New York, pp. 467–508.

Kay, R. F., Thorington R. W., Jr., Houde, P., 1990. Eocene plesiadapiform shows affinities with flying lemurs not primates. Nature 345, 342–344.

Kay, R. F., Williams, B. A., Ross, C., Takai, M., Shigehara, N., 2004. Anthropoid origins: a phylogenetic analysis. In: Ross, C. F., Kay, R. F. (Eds.), Anthropoid Origins: New Visions. Plenum, New York, pp. 91–155.

Kent, D. V., McKenna, M. C., Opdyke, N. D., Flynn, J. J., MacFadden, B. J., 1984. Arctic biostratigraphic heterochroneity. Science 224, 173–174.

Kluge, A. G., 1989. A concern for evidence and a phylogenetic hypothesis of relationships among *Epicrates* (Boidae, Serpentes). Systematic Zoology 38, 7–25.

Krause, D. W., 1991.Were paromomyids gliders? Maybe, maybe not. Journal of Human Evolution 21, 177–188.

Krause, D. W., Maas, M. C., 1990.The biogeographic origins of late Paleocene-early Eocene mammalian immigrants to the western interior of North America. In: Bown, T. M., Rose, K. D. (Eds.), Dawn of the Age of Mammals in the Northern Part of the Rocky Mountain Interior, North America. Geological Society of America, Special Paper 243, Boulder CO, pp. 71–105.

Liu, F. -G. R., Miyamoto, M. M., 1999. Phylogenetic assessment of molecular and morphological data for eutherian mammals. Systematic Biology 48, 54–64.

Liu, F. -G. R., Miyamoto, M. M., Freire, N. P., Ong, P. Q., Tennant, M. R., Young, T. S., Gugel, K. F., 2001. Molecular and morphological supertrees for eutherian (placental) mammals. Science 291, 1786–1789.

Lofgren, D. L., 1995. The Bug Creek problem and the Cretaceous-Tertiary boundary at McGuire Creek, Montana. University of California Publications in Geological Sciences 140, 1–185.

Luckett, W. P., 1980. The suggested evolutionary relationships and classification of tree shrews. In: Luckett, W. P. (Ed.), Comparative Biology and Evolutionary Relationships of Tree Shrews. Plenum, New York, pp. 3–31.

Maas, M. C., Krause, D. W., Strait, S. G., 1988. The decline and extinction of Plesiadapiformes (Mammalia: ?Primates) in North America: displacement or replacement? Paleobiology 14, 410–431.

MacPhee, R. D. E, Cartmill, M., Gingerich, P. D., 1983. New Paleogene primate basicrania and the definition of the order Primates. Nature 301, 509–511.

MacPhee, R. D. E., Cartmill, M., Rose, K. D., 1989. Craniodental morphology and relationships of the supposed Eocene dermopteran *Plagiomene* (Mammalia). Journal of Vertebrate Paleontology 9, 329–349.

Maddison, W. P., Donoghue, M. J., Maddison, D. R., 1984. Outgroup analysis and parsimony. Systematic Zoology 33, 83–103.

Maddison, W. P., Maddison, D. R., 1992. MacClade 3, Program and Documentation. Sinauer, Sunderland, MA.

Madsen, O., Scally, M., Douady, C. J., Kao, D. J., DeBry, R. W., Adkins, R. M., Amrine, H. M., Stanhope, M. J., de Jong, W. W., Springer, M. S., 2001. Parallel adaptive radiations in two major clades of placental mammals. Nature 409, 610–614.

Martin, R. D., 1968. Towards a new definition of Primates. Man 3, 377–401.

Martin, R. D., 1986. Primates: a definition. In: Wood, B., Martin, L., Andrews, P. (Eds.), Major Topics in Primate and Human Evolution. Cambridge University Press, Cambridge, pp. 1–31.

Matthew, W. D., 1918. A revision of lower Eocene Wasatch and Wind River faunas. Part 5. Insectivora (continued), Glires, Edentata. Bulletin of the American Museum of Natural History 38, 565–657.

McKenna, M. C., Bell, S. K., 1997. Classification of Mammals Above the Species Level. Columbia University Press, New York.

Miller, E. R., Gunnell, G. F., Martin, R. D., 2005. Deep time and the search for anthropoid origins. Yearbook of Physical Anthropology 8, 60–95.

Miyamoto, M. M., Porter, C. A., Goodman, M., 2000. *c-Myc* gene sequences and the phylogeny of bats and other eutherian mammals. Systematic Biology 49, 501–514.

Murphy, W. J., Eizirik, E., Johnson, W. E., Zhang, Y. P., Ryder, O. A., O'Brien, S. J., 2001a. Molecular phylogenetics and the origins of placental mammals. Nature 409, 614–618.

Murphy, M. J., Eizirik, E., O'Brien, S. J., Madsen, O., Scally, M., Douady, C. J., Teeling, E., Ryder, O. A., Stanhope, M. J., de Jong, W. W., Springer, M. S., 2001b. Resolution of the early placental mammal radiation using Bayesian phylogenetics. Science 294, 2348–2351.

Ni, X., Wang, Y., Hu, Y., Li, C., 2004. A euprimate skull from the early Eocene of China. Nature 427, 65–68.

Ni, X., Hu, Y., Wang, Y., Li, C., 2005. A clue to the Asian origin of euprimates. Anthropological Science 113, 3–9.

Novacek, M. J., 1986. The skull of leptictid insectivorans and the higher-level classification of eutherian mammals. Bulletin of the American Museum of Natural History 183, 1–111.

Novacek, M. J., 1990. Morphology, paleontology, and the higher clades of mammals. In: Genoways, H. H. (Ed.), Current Mammalogy, Volume 2. Plenum, New York, pp. 507–543.

Novacek, M. J., 1991. Aspects of morphology of the cochlea in microchiropteran bats: an investigation of character transformation. Bulletin of the American Museum of Natural History 206, 84–100.

Novacek, M. J., 1994. Morphological and molecular inroads to phylogeny. In: Grande, L., Rieppel, O. (Eds.), Interpreting the Hierarchy of Nature: From Systematic Patterns to Evolutionary Process Theories. Academic, New York, pp. 85–131.

Novacek, M. J., Wyss, A. R., 1986. Higher-level relationships of the recent eutherian orders: morphological evidence. Cladistics 2, 257–287.

Novacek, M. J., Wyss, A. R., McKenna, M. C., 1988. The major groups of eutherian mammals. In: Benton, M. J. (Ed.), The Phylogeny and Classification of the Tetrapods, Volume 2: Mammals. Clarendon, Oxford, pp. 31–71.

Olson, L. E., Sargis, E. J., Martin, R. D., 2005. Intraordinal phylogenetics of treeshrews (mammalia: scandentia) based on evidence from the mitochondrial 12S rRNA gene. Molecular Phylogenetics and Evolution 35, 656–673.

Pumo, D. E., Finamore, P. S., Franek, W. R., Phillips, C. J., Tarzami, S., Balzarano, D., 1998. Complete mitochondrial genome of a neotropical fruit bat, *Artibeus jamaicensis* and a new hypothesis of the relationships of bats to other eutherian mammals. Journal of Molecular Evolution 47, 709–717.

Purvis, A., Quicke, D. L. J., 1997. Building phylogenies: are the big easy? Trends in Ecology and Evolution 12, 49–50.

Raxworthy, C. J., Forstner, M. R. J., Nussbaum R. A., 2002. Chameleon radiation by oceanic dispersal. Nature 415, 784–787.

Ronquist, F., 1997. Dispersal-vicariance analysis: a new approach to the quantification of historical biogeography. Systematic Biology 46, 195–203.

Rose, K. D., 1973. The mandibular dentition of *Plagiomene* (Dermoptera, Plagiomenidae). Breviora 411, 1–17.

Rose, K. D., 1975.The Carpolestidae: early Tertiary primates from North America. Bulletin of the Museum of Comparative Zoology147, 1–74.

Rose, K. D., 1995. The earliest primates. Evolutionary Anthropology 3, 159–173.

Rose, K. D., Bown, T. M., 1991. Additional fossil evidence on the differentiation of the earliest euprimates. Proceedings of the National Academy of Sciences, USA 88, 98–101.

Rose, K. D., Bown, T. M., 1996. A new plesiadapiform (Mammalia: Plesiadapiformes) from the early Eocene of the Bighorn Basin, Wyoming. Annals of the Carnegie Museum 65, 305–321.

Rose, K. D., Krause, D. W., 1984. Affinities of the primate *Altanius* from the early Tertiary of Mongolia. Journal of Mammalogy 65, 721–726.

Rose, K. D., Simons, E. L., 1977. Dental function in the Plagiomenidae: origin and relationships of the mammalian order Dermoptera. Contributions from the Museum of Paleontology, University of Michigan 24, 221–236.

Rose, K. D., Godinot, M., Bown, T. M., 1994. The early radiation of Euprimates and the initial diversification of Omomyidae. In: Fleagle, J. G., Kay, R. F. (Eds.), Anthropoid Origins. Plenum, New York, pp. 1–28.

Russell, D. E., Gingerich, P. D., 1980. Un nouveau primate omomyide dans l'Éocène du Pakistan. Comptes rendus de l'Académie des sciences 291, 621–624.

Sargis, E. J., 2000. The functional morphology of the postcranium of *Ptilocercus* and tupaiines (Scandentia, Tupaiidae): implications for the relationships of primates and other archontan mammals. Ph.D. dissertation, City University of New York.

Sargis, E. J., 2002. The postcranial morphology of *Ptilocercus lowii* (Scandentia, Tupaiidae): an analysis of primatomorphan and volitantian characters. Journal of Mammalian Evolution 9, 137–160.

Sargis, E. J., 2004. New views on tree shrews: the role of tupaiids in primate supraordinal relationships. Evolutionary Anthropology 13, 56–66.

Sargis, E. J., 2007. The postcranial morphology of *Ptilocercus lowii* (Scandentia, Tupaiidae) and its implications for primate supraordinal relationships. In: Ravosa, M. J., Dagosto, M. (Eds.), Primate Origins: Adaptations and Evolution. Springer, New York, pp. 51–82.

Schiebout, J. A., 1979. An overview of the terrestrial early Tertiary of southern North America–fossil sites and paleopedology. Tulane Studies in Geology and Paleontology 15, 75–93.

Schwartz, J. H., 1986. Primate systematics and a classification of the order. In: Swindler, D. R., Erwin, J. (Eds.), Systematics, Evolution, and Anatomy: Comparative Primate Biology. A. R. Liss, New York, pp. 1–41.

Seiffert, E. R., Simons, E. L., Clyde, W. C., Rossie, J. B., Attia, Y., Bown, T. M., Chatrath, P., Mathison, M. E., 2005. Basal anthropoids from Egypt and the antiquity of Africa's higher primate radiation. Science 310, 300–304.

Siddall, M. E., 2005. Bracing for another decade of deception: the promise of secondary Brooks parsimony analysis. Cladistics 21, 90–99.

Sigé, B., Jaeger, J. -J., Sudre, J., Vianey-Liaud, M., 1990. *Altiatlasius koulchii* n.gen et sp., primate omomyidé du paléocène supérieur du Maroc, et les origines des euprimates. Palaeontographica Abteilung A 212, 1–24.

Silcox, M. T., 2001. A phylogenetic analysis of Plesiadapiformes and their relationship to Euprimates and other archontans. Ph.D. dissertation, Johns Hopkins University, School of Medicine, Baltimore, MD.

Silcox, M. T., 2003. New discoveries on the middle ear anatomy of *Ignacius graybullianus* (Paromomyidae, Primates) from ultra high resolution x-ray computed tomography. Journal of Human Evolution 44, 73–86.

Silcox, M. T., 2007. Primate taxonomy, plesiadapiforms, and approaches to primate origins. In: Ravosa, M. J., Dagosto, M. (Eds.), Primate Origins: Adaptations and Evolution. Springer, New York, pp. 143–178.

Silcox, M. T., Krause, D. W., Maas, M. C., Fox, R. C., 2001. New specimens of *Elphidotarsius russelli* (Mammalia, ?Primates, Carpolestidae) and a revision of plesiadapoid relationships. Journal of Vertebrate Paleontology 21, 131–152.

Silcox, M. T., Rose K. D., Walsh, S., 2002. New specimens of picromomyids (Plesiadapiformes, ?Primates) with description of a new species of *Alveojunctus*. Annals of the Carnegie Museum 71, 1–11.

Silcox, M. T., Bloch, J. I., Sargis, E. J., Boyer, D. M., 2005. Euarchonta. In: Rose, K. D., Archibald, J. D. (Eds.), The Rise of Placental Mammals: Origins and Relationships of the Major Extant Clades. Johns Hopkins University Press, Baltimore, MD, pp. 127–144.

Silcox, M. T., Sargis, E. J., Bloch, J. I., Boyer, D. M. 2007. Primate origins and supraordinal relationships: morphological evidence. In: Henke, W., Tattersall, I. (Eds.), Handbook of Paleoanthropology Volume 2: Primate Evolution and Human Origins. Springer, Berlin.

Simmons, N. B., 1994. The case for chiropteran monophyly. American Museum Novitates 3103, 1–54.

Simmons, N. B., 1995. Bat relationships and the origin of flight. Symposia of the Zoological Society of London 67, 27–43.

Simmons, N. B., Geisler, G. H., 1998. Phylogenetic relationships of *Icaronycteris, Archaeonycteris, Hassianycteris*, and *Palaeochiropteryx* to extant bat lineages, with comments on the evolution of echolocation and foraging strategies in Microchiroptera. Bulletin of the American Museum of Natural History 235, 1–182.

Simmons, N. B., Quinn, T. H., 1994. Evolution of the digital tendon locking mechanism in bats and dermopterans: a phylogenetic perspective. Journal of Mammalian Evolution 2, 231–254.

Simpson, G. G., 1945. The principles of classification and a classification of mammals. Bulletin of the American Museum of Natural History 85, 1–350.

Sloan, R. E., 1969. Cretaceous and Paleocene terrestrial communities of western North America. North American Paleontological Convention, Proceedings, Part E, 427–453.

Smith, T., Van Itterbeeck, J., Missiaen, P., 2004. Oldest plesiadapiform (Mammalia, Proprimates) from Asia and its palaeobiogeographical implications for faunal interchange with North America. Comptes rendus de l'Académie des sciences, Palevol 3, 43–52.

Soligo, C., Martin, R. D., 2006. Adaptive origins of primates revisited. Journal of Human Evolution 50, 414–430.

Springer, M. S., Murphy, W. J., Eizirik, E., O'Brien, S. J., 2003. Placental mammal diversification and the Cretaceous–Tertiary boundary. Proceedings of the National Academy of Sciences, USA 100, 1056–1061.

Springer, M. S., Stanhope, M. J., Madsen, O., de Jong, W. W., 2004. Molecules consolidate the placental mammal tree. Trends in Ecology and Evolution 19, 430–438.

Stafford, B. J., Szalay, F. S., 2000. Craniodental functional morphology and taxonomy of dermopterans. Journal of Mammalogy 81, 360–385.

Stanhope, M. J., Bailey, W. J., Czelusniak, J., Goodman, M., Nickerson, J. St. J., Sgouros, J. G., Singer, G. A. M., Kleinschmidt, T. K., 1993. A molecular view of primate supraordinal relationships from the analysis of both nucleotide and amino acid sequences. In: MacPhee, R. D. E. (Ed.), Primates and their Relatives in Phylogenetic Perspective. Plenum, New York, pp. 251–292.

Stanhope, M. J., Waddell, V. G., Madsen, O., de Jong, W. W., Hedges, S. B., Cleven, G. C., Kao, D., Springer, M. S., 1998. Molecular evidence for multiple origins of the Insectivora and for a new order of endemic African mammals. Proceedings of the National Academy of Sciences, USA 95, 9967–9972.

Stevens, N. J., Heesy, C. P., 2000. Biogeographic origins of Primate higher taxa. Journal of Vertebrate Paleontology 20(Supplement to 3), 71A.

Sudre, J., 1975. Un prosimien du Paléogène ancien du Sahara nord-occidental: *Azibius trerki* n.g. n.sp. Comptes rendus de l'Académie des sciences 280, 1539–1542.

Sudre, J., 1979. Nouveaux mammifères éocènes du Sahara Occidental. Palaeovertebrata 9, 83–115.

Swofford, D. L., Maddison, W. P., 1987. Reconstructing ancestral character states under Wagner parsimony. Mathematical Biosciences 87, 199–229.

Szalay, F. S., 1975. Phylogeny, adaptations, and dispersal of the tarsiiform primates. In: Luckett, W. P., Szalay, F. S. (Eds.), A Phylogeny of the Primates: A Multidisciplinary Approach. Plenum, New York, pp. 357–404.

Szalay, F. S., 1982. A critique of some recently proposed Paleogene primate taxa and suggested relationships. Folia Primatologica 37, 153–162.

Szalay, F. S., Delson, E., 1979. Evolutionary History of the Primates. Academic, New York.

Szalay, F. S., Li, C. -K., 1986. Middle Paleocene euprimate from southern China and the distribution of Primates in the Paleogene. Journal of Human Evolution 15, 387–397.

Szalay, F. S., Lucas, S. G., 1993. Cranioskeletal morphology of archontans, and diagnoses of Chiroptera, Volitantia, and Archonta. In: MacPhee, R. D. E. (Ed.), Primates and their Relatives in Phylogenetic Perspective. Plenum, New York, pp. 187–226.

Szalay, F. S., Lucas, S. G., 1996. The postcranial morphology of Paleocene *Chriacus* and *Mixodectes* and the phylogenetic relationships of archontan mammals. New Mexico Museum of Natural History and Science Bulletin 7, 1–47.

Szalay, F. S., Li, C. K., Wang, B. Y., 1986. Middle Paleocene omomyid primate from Anhui Province, China: *Decoredon anhuiensis* (Xu, 1976), new combination Szalay and Li, and the significance of *Petrolemur*. American Journal of Physical Anthropology 69, 269.

Szalay, F. S., Rosenberger, A. L., Dagosto, M., 1987. Diagnosis and differentiation of the order Primates. Yearbook of Physical Anthropology 30, 75–105.

Tavaré, S., Marshall, C. R., Will, O., Soligo, C., Martin, R. D., 2002. Using the fossil record to estimate the age of the last common ancestor of extant primates. Nature 416, 726–729.

Tabuce, R., Mahboubi, M., Tafforeau, P., Sudre, J., 2004. Discovery of a highly-specialized plesiadapiform primate in the early-middle Eocene of northwestern Africa. Journal of Human Evolution 47, 305–321.

Tattersall, I., 1982. The Primates of Madagascar. Columbia University Press, New York.

Thewissen, J. G. M., Babcock, S. K., 1991. Distinctive cranial and cervical innervation of wing muscles: new evidence for bat monophyly. Science 251, 934–936.

Thewissen, J. G. M., Babcock, S. K., 1993. The implications of the propatagial muscles of flying and gliding mammals for archontan systematics. In: MacPhee, R. D. E. (Ed.), Primates and their Relatives in Phylogenetic Perspective. Plenum, New York, pp. 91–109.

Thewissen, J. G. M., Williams, E. M., Hussain, S. T., 2001. Eocene mammal faunas from northern Indo-Pakistan. Journal of Vertebrate Paleontology 21, 347–366.

Tong, Y. -S., 1979. A late Paleocene primate from S. China. Vertebrata PalAsiatica 17, 65–70.

Tong, Y., Wang, J., 1998. A preliminary report on the early Eocene mammals of the Wutu fauna, Shandong Province, China. In: Beard, K. C., Dawson, M. R. (Eds.), Dawn of the Age of Mammals in Asia. Bulletin of the Carnegie Museum of Natural History 34, 186–193.

Van Den Bussche, R. A., Hoofer, S. R., 2004. Phylogenetic relationships among recent chiropteran families and the importance of choosing appropriate out-group taxa. Journal of Mammalogy 85, 321–330.

Van Valen, L. M., 1978. The beginning of the age of mammals. Evolutionary Theory 4, 45–80.

Van Valen, L. M., 1994. The origin of the plesiadapid primates and the nature of *Purgatorius*. Evolutionary Monographs 15, 1–79.

Van Valen, L. M., Sloan, R. E., 1965. The earliest primates. Science 150, 743–745.

West, R. M., Dawson, M. R., 1977. Mammals from the Palaeogene of the Eureka Sound Formation: Ellesmere Island, Arctic Canada. Géobios special memoir 1, 107–124.

Wible, J. R., Covert, H. H., 1987. Primates: cladistic diagnosis and relationships. Journal of Human Evolution 16, 1–22.

Wible, J. R., Novacek, M. J., 1988. Cranial evidence for the monophyletic origin of bats. American Museum Novitates 2911, 1–19.

Wiens, J. J., 2005. Can incomplete taxa rescue phylogenetic analyses from long-branch attraction? Systematic Biology 54, 731–742.

Xu, Q., 1976. New materials of old Anagalidae from the Paleocene of Anhui (B). Vertebrata PalAsiatica 14, 242–251.

Xu, Q., 1977. New materials of old Ungulata from the Paleocene of Qianshan Basin, Anhui. Vertebrata PalAsiatica 15, 119–125.

Zack, S. P., Penkrot, T. A., Bloch, J. I., Rose, K. D. 2005, Affinities of 'hyopsodontids' to elephant shrews and a Holarctic origin of Afrotheria. Nature 434, 497–501.

11. Evaluating the Mitten-Gliding Hypothesis for Paromomyidae and Micromomyidae (Mammalia, "Plesiadapiformes") Using Comparative Functional Morphology of New Paleogene Skeletons

Doug M. Boyer*
Department of Anatomical Sciences
Stony Brook University
T8 040 Health Science Center
Stony Brook, NY 11794-8081, USA
dboyer@ic.sunysb.edu

Jonathan I. Bloch
Florida Museum of Natural History
University of Florida
P.O. Box 117800
Gainesville, FL 32611, USA
jbloch@flmnh.ufl.edu

11.1 Introduction

11.1.1 Plesiadapiformes: Historical Perspective

Teeth of primate-like mammals from the Paleogene ("plesiadapiforms") have been known for at least 130 years (Gervais, 1877). These fossil taxa are generally recognized as being closely related, but not monophyletic (e.g., Gingerich, 1976; Szalay et al., 1987; Beard, 1993a; Silcox, 2001; Bloch and Boyer, 2002a; Bloch et al., 2007) and we maintain that view here. Thus "plesiadapiforms" are referred to with quotation marks throughout the text to reflect that status. Vertebrate paleontologists have struggled to understand the nature of the phylogenetic relationship of "plesiadapiforms" to the extant and extinct members of crown group Primates [= Euprimates (Hoffstetter, 1977)], since *Plesiadapis* was first described by Gervais in 1877 (e.g., Lemoine, 1887; Stehlin, 1916; Teilhard-de-Chardin, 1922; Gidley, 1923; Simpson, 1935; Russell, 1959; Simons, 1972; Szalay, 1973; Gingerich, 1975, 1976; Szalay et al., 1975; MacPhee et al., 1983; Gunnell, 1989; Kay et al., 1990; Beard, 1993a; Silcox, 2001; Bloch and Boyer, 2002a). The strongest support for a close relationship between "plesiadapiforms" and Euprimates, specifically, has historically come from the excellent fossil record of teeth known for "plesiadapiforms," first in Europe and then in North America (e.g., Gidley, 1923). Cranial and postcranial fossils were relatively rare and fragmentary initially, such that the first researchers were unable to evaluate the presence of non-dental euprimate features in "plesiadapiforms." As non-dental fossils of "plesiadapiforms" were recovered there was some disagreement as to whether they suggested treeshrew or euprimate affinities. A humerus from the San Juan Basin (at the time attributed to *Nothodectes*) was figured by Gregory (1920) and interpreted to fit the "tupaioid" pattern. Later, a crushed skull and additional postcranial material associated with that specimen (and now referred to *Nannodectes gidleyi* Gingerich) was interpreted by Simpson (1935) as being similar to both lemurs and treeshrews. An implication of this acknowledged similarity was that *Nannodectes* spent time in the trees, as expected for the early forebears of

* Address for correspondence: dboyer@ic.sunysb.edu

E.J. Sargis and M. Dagosto (eds.), *Mammalian Evolutionary Morphology: A Tribute to Frederick S. Szalay,* 233–284.

the euprimate clade. However, Simpson (1935) discounted the similarities to treeshrews as being primitive (plesiomorphic), while he emphasized perceived shared-derived (synapomorphic) characters with lemurs as supporting a relationship with Euprimates. At the same time, he rejected any special relationship to *Daubentonia*, an idea that had been seriously considered based on the shared presence of procumbent incisors (Stehlin, 1916; Teilhard-de-Chardin, 1922). Instead, Simpson (1935) attributed this similarity to convergence. He interpreted the procumbent incisors of *Nannodectes* as feeding specializations for a way of life that likely differed in significant respects from that of the euprimates to which he had compared it (*Lemur* and *Notharctus*). He also interpreted the differences between *Nannodectes* and euprimates in other parts of the skeleton to reflect differing ecological specializations. These differences indicated to Simpson that *Nannodectes* could not be the direct ancestor to later occurring euprimates. Description of two skulls of *Plesiadapis tricuspidens*, from Cernay and Berru, France, appeared to support a euprimate relationship in certain characteristics. These included a bony auditory bulla continuous with the petrosal bone, and a thin, ring-like intrabullar component to the ectotympanic bone (Russell, 1959, 1964; Szalay, 1971; Gingerich, 1976; Szalay et al., 1987). Additional cranial material also supported a "plesiadapiform"-euprimate link. Szalay (1972b) described *Phenacolemur jepseni*, and found that it too had a petrosal bulla (although, see MacPhee et al., 1983; Bloch and Silcox, 2001), and additionally, that it had a large posterior carotid canal for the internal carotid artery. He concluded that it would have had a transpromontorial bony tube for this vessel, as in many early Eocene adapoid and omomyoid euprimates. Additional evidence for an arboreal lifestyle began to accumulate with fragmentary postcranials (e.g., Szalay and Decker, 1974).

Even as these specimens were described, providing limited evidence of euprimate features in some "plesiadapiforms," the primate status of the group was being even more seriously questioned by a few influential researchers. Cartmill (1972) provided a definition of Primates, requiring members of the group to possess the full suite of features interpreted by him to reflect the ancestral primate innovation of nocturnal, arboreal, visual predation. These included a postorbital bar with forward facing orbits, petrosal bulla, and a divergent, opposable hallux and pollex bearing a flattened nail (Cartmill, 1972). Martin (1972, 1979) and Cartmill (1972, 1974) noted that plesiadapiforms lacked most of these key "adaptive" features. Szalay (1975) objected to this particular adaptive definition of Primates. He maintained that the more general adaptive similarities (arboreality), and clear evidence for a close phylogenetic relationship required that meaningful definitions of Primates include "plesiadapiforms." Gingerich (1976) supported a primate relationship for "plesiadapiforms," but determined that *Plesiadapis* was likely predominantly terrestrial (but see below), further bolstering Martin and Cartmill's case for removing "plesiadapiforms" from Primates. Later, the existing cranial evidence was challenged by MacPhee et al. (1983) who noted that all known "plesiadapiform" skulls (1) lacked substantial carotid supply to the forebrain, or even any remnant of the artery (contrary to Szalay, 1972b); and (2) could not be shown to have a petrosally-derived bulla without developmental evidence. Furthermore, MacPhee and Cartmill (1986) argued that extant chinchillids among other rodents (Rodentia; *Lagostomus maximus*) have an ear convergent on that of *Plesiadapis tricuspidens*, including apparent continuity of the bulla and petrosal, and an intrabullar ring-like component to the ectotympanic bone. As a result, the consensus in the field became that "plesiadapiforms" were both adaptively different and phylogenetically disparate (Martin, 1986) from Euprimates. To what group(s) "plesiadapiforms" could be related became the most pressing question surrounding them.

Throughout this "dark" period in understanding, Szalay, his students and collaborators (e.g., Szalay, 1975; Szalay et al., 1975; Szalay and Dagosto, 1980; Szalay et al., 1987; Szalay and Dagosto, 1988) continued to support a euprimate relationship for "plesiadapiforms" based on cranial and postcranial anatomy. They developed adaptive scenarios leading from claw-climbing "plesiadapiforms" (or stem-primates) to grasp-leaping euprimates (e.g., Szalay and Dagosto, 1980).

Given the "lost-sibling" status the anthropological community had assigned to "plesiadapiforms," it is not surprising that new affinities were soon suggested and embraced. A skull of a paromomyid, by far the best-preserved skull ever recovered for any "plesiadapiform" group, was described and analyzed by Kay et al. (1990, 1992). Their cladistic analyses of 33 cranial characters supported a special relationship with Dermoptera (flying lemurs – also known as colugos), a strange order of gliding mammals from Southeast Asia, represented today by two species of a single family (*Cynocephalus volans* and *Galeopterus variegatus*) (Stafford and Szalay, 2000). Among extant mammals, dermopterans have been considered close relatives of euprimates, treeshrews, and bats (e.g., Gregory, 1910). In an independant study of postcranial material, then known for "plesiadapiforms," Beard (1990, 1993a, b) added substantial support to the hypothesis of dermopteran relationships. Furthermore, Beard concluded that certain paromomyid and micromomyid plesiadapiforms were "mitten-gliders" like modern dermopterans. These studies were influential, leading to the widespread acceptance of the phylogenetic hypothesis with some (or all) "plesiadapiforms" classified as dermopterans (e.g., McKenna and Bell, 1997). Certain scientists have remained skeptical of both the cranial support (Wible and Martin, 1993) and postcranial interpretations (Krause, 1991; Szalay and Lucas, 1996; Sargis, 2002c).

Now, over a decade after this view was presented and so convincingly argued, new evidence coming from (1) discovery of new fossils, and (2) new methods for accessing morphological information contained within fossils [e.g., Ultra High Resolution X-Ray Computed Tomography (UhrCT)] suggests that some aspects of original interpretations were in fact *not* wrong, and that some "plesiadapiforms" possess more euprimate-like features than imagined (except possibly by F.S. Szalay), making their primate status harder to dismiss on both adaptive and phylogenetic grounds. Specifically, in the cranium, an additional "plesiadapiform" (beyond *Plesiadapis tricuspidens*) is now interpreted to be euprimate-like in having a bulla continuous with the petrosal (*Carpolestes simpsoni*; Bloch and Silcox, 2006). Further analysis of the same *Ignacius* skull studied by Kay and others (1990, 1992) with UhrCT showed it to have a bony tube enclosing the internal carotid system (Silcox, 2003). Finally, new carpolestid specimens clearly had an unreduced internal carotid artery with promontory and stapedial branches, based on grooves crossing the promontorium (Bloch and Silcox, 2006). It has also been demonstrated that carpolestids are unique among known "plesiadapiforms" in having a foot with a divergent, opposable hallux tipped with a nail like that of euprimates (Bloch and Boyer, 2002a).

It is notable that, following descriptions of this new material, the original proponents of the "plesiadapiform"-dermopteran link have referred to "plesiadapiforms" as likely "stem-primates" (Kay, 2003: 840), and "primates" (Beard, 2006). It seems reasonable to assume that these changes in view are due largely to the fact that previous characterizations of "plesiadapiforms" could not consider the extent to which the morphologies of *Plesiadapis* (and *Nannodectes*) and paromomyids were different from those of other "plesiadapiforms" because the relevant fossil material was not available. Given the dental diversity of "plesiadapiforms," with remains known for over 120 species classified into 11 or 12 families from the Paleocene and Eocene of North America, Europe, Asia, and possibly Africa (Silcox, 2001), it would not be surprising if there existed a commensurate skeletal diversity yet to be discovered. Szalay (1972a, p.18) recognized this and commented on the need for a taxonomically broader sample of "plesiadapiform" postcranial skeletons, writing:

> "It may be that once postcranial elements of the Paleocene primate radiation become more common, *Plesiadapis* might become recognized as a relatively more aberrant form than the majority of early primates."

In fact, the last 15 years of paleontological research, has validated this prediction (Beard, 1989, 1990, 1993a, b; Bloch and Boyer, 2002a, 2007; Bloch et al., 2007). Various *Plesiadapis* species are currently understood to be generalized arborealists with specializations for vertical clinging and climbing (Gunnell and Gingerich, 1987; Beard, 1989, 1991; Gunnell, 1989; Youlatos and Godinot, 2004). Beard's (1989, 1990, 1993a, b) studies of postcranial elements attributed to paromomyid and micromomyid "plesiadapiforms," led him to conclude that they were mitten-gliders. Recent work on a new carpolestid plesiadapiform skeleton indicated that these animals were different from plesiadapids, paromomyids, and micromomyids in exhibiting capabilities for strong pedal grasping in a manner similar to euprimates (Bloch and Boyer, 2002a; Sargis et al., 2007). While the mitten-gliding hypothesis for paromomyids and/or micromomyids is seriously questioned here and elsewhere (Krause, 1991; Runestad and Ruff, 1995; Szalay and Lucas, 1996; Stafford and Thorington, 1998; Hamrick et al., 1999; Bloch et al., 2001a, 2002b, 2007; Bloch and Silcox, 2001; Boyer et al., 2001; Bloch and Boyer, 2002b, 2007; Boyer and Bloch, 2002a, b), it is still notable that micromomyids, paromomyids and carpolestids are postcranially distinct from each other and plesiadapids. Moreover, it appears that there is diversity even within Plesiadapidae (Boyer et al., 2004; Bloch and Boyer, 2007).

Even with these new finds, it is likely that we are barely scratching the surface of the full skeletal diversity that existed in either "plesiadapiforms" or basal euprimates. As more and better fossils are recovered, they are bound to influence and change our views of phylogenetic relationships. Thus, whether "plesiadapiforms" share a special relationship with dermopterans, euprimates, or neither will be debated far into the foreseeable future. We do not address that subject in this paper. Szalay has emphasized what is generally recognized by paleobiologists, that:

> "...one can often corroborate far better the adaptive strategies of fossil species than their phylogenetic affinities" (Szalay and Sargis, 2001: p.153).

While we do not claim to evaluate "adaptations," exclusively, we *do* evaluate functional morphology to address a more immediately testable hypothesis regarding inferred postional behaviors of some "plesiadapiforms." Simply, we ask the question: ***Do new skeletons of micromomyid and paromomyid "plesiadapiforms" support the hypothesis of "mitten-gliding" proposed by Beard (1989, 1990, 1991, 1993b)?*** We acknowledge that, while we certainly think we can shed some light on this particular subject, the answers to this question will not resolve the phylogeny of the group. Instead, we sincerely hope that this study will be of some interest to those of you interested in the paleoecology of Paleogene mammals, and who "just want to know."

11.1.2 Conceptual Framework

We use comparative functional morphology to address the question presented above within the conceptual framework (as we understand it) carefully outlined by Szalay for analyzing fossil taxa (e.g., Szalay and Drawhorn, 1980; Szalay and Dagosto, 1988; Szalay and Sargis, 2001). Szalay has derived justification for form-function relationships from John Stuart Mill (1872) and referred extensively to the writings of Bock (1977, 1981, 1988, 1991, 1993, 1999; Szalay and Bock, 1991) and Bock and von Wahlert (1965) to define his approach to functional-adaptive analyses. He championed the tenet that elucidating aspects of functional morphology and phylogeny are not separate endeavors, and one should not precede the other in reconstructing the evolutionary history of a group of animals (e.g., Szalay, 1977, 1981; Szalay and Lucas, 1996; Szalay, 1998, 2000; Szalay and Sargis, 2001). He contended that they are "temporally looped" (Szalay and Sargis, 2001, p. 152).

He emphasized that cladistic analyses lack the power to reveal adaptive trends (Szalay, 1981; Szalay and Bock, 1991; Szalay, 2000). On the other hand, he stressed that biological roles and adaptive significance of skeletal features in fossils cannot be inferred without a thorough appreciation of the evolutionary history of the clade to which the fossil under study belongs. He continually stressed the fact that organisms are comprised of features related to heritage and habitus (e.g., Szalay, 1981; Szalay and Sargis, 2001), those related to habitus being of most value for reconstructing biological roles, yet ineluctable without the context of the features of heritage (those reflecting previous biological roles of ancestral lineages). While he recognized the existence of "heritage" features, he did not believe in "phylogenetic baggage", or non-functional skeletal features. He saw natural selection's effect on morphology as powerful and capable of keeping pace with environmental change in most situations (Szalay and Sargis, 2001) such that the morphology of an organism should, at the very least, not contradict its ability to engage in its observed postional behaviors, if not support those observations (based on mechanical predictions or comparative observations). He took the historic observation of striking convergence among distantly phylogenetically separated mammalian groups as evidence for his view of the interaction between evolving populations, environmental change and natural selection. Ultimately, he endorsed the comparative method, in which observation of similar osteological forms among a diversity of mammals also characterized by similar functions (*sensu* Plotnick and Baumiller, 2000) or faculties, (*sensu* Bock and von Wahlert, 1965) establishes a form-function relationship that can be used to infer the presence of function and faculty in extinct forms (what he referred to as a "model-based approach" Szalay and Sargis, 2001).

He spurned the superficially similar practice of associating morphological features with particular biological roles without consideration of their adaptive (versus exaptive) status (Szalay, 1981), and emphasized that different characters and character complexes had different relative importance for reconstructing certain biological roles (Szalay, 1981). For instance, it had been suggested that dental and cranial features suggesting faunivory in the "plesiadapiform" *Palaechthon* also implied that this animal was a terrestrialist who spent its life "nosing" its way through the leaf litter (Kay and Cartmill, 1977). But Szalay (1981) saw craniodental features as basically irrelevant to the question of substrate preference in this case. Evidence from the postcranium, indicating arboreality in other "plesiadapiforms" suggested more strongly to him that *Palaechthon* was likely arboreal as well (Szalay, 1981).

In this study, we focus on evaluating positional (locomotor and postural) behaviors of paromomyids and micromomyids. Ultimately, we speculate about biological roles (Bock and von Wahlert, 1965) of inferred positional behaviors, but only after also considering what craniodental morphology suggests about food preferences and feeding behaviors.

A given positional behavior is characterized by various static and dynamic functions (*sensu* Plotnick and Baumiller, 2000) or faculties (*sensu* Bock and von Wahlert, 1965). Functions often have osteological correlates identifiable through mechanical modeling and the comparative method. However, identification of a correlate to a particular function in a fossil does not typically go very far towards elucidating positional behaviors because most functions are required by more than one positional behavior. For instance, taxa that utilize two very different positional behaviors, "quadrumanus suspension" and "terrestrial quadrupedal cursoriality" are characterized by elbow joints that share a static function: the capacity for full extension. But this is not to say these two groups are identical in their functional requirements: for example, unlike suspensory taxa, cursors are also characterized by the capacity for fast, sagittal extension of the femur, a dynamic function.

Thus, a hypothesis of positional behavior can be tested effectively by examining the skeleton for the presence of osteological correlates to a *suite* of functions that characterize the proposed behavior. The lack of osteological correlates of a function rules out its presence in a fossil. Likewise, positional behaviors that require those functions will also be refuted. From the example above, if, in a fossil taxon of interest, the elbow joint had features suggesting that it could be fully extended and the femur lacked osteological correlates of fast powerful extension, we could fairly confidently refute use of cursorial behaviors, but not suspensory ones.

Thus, summarizing various functional features in a fossil generates a limiting hypothesis regarding its positional behaviors. That is, some positional behaviors will be refuted while others will remain plausible, if not specifically supported. The discovery of additional functional attributes not predicted for the un-refuted positional behaviors does not then rule out those behaviors as well, unless such attributes somehow compromise one of the other required functions. Such "extra" features likely correspond to "adaptive" traits of an organism and thus may help elucidate the biological roles of these un-refuted positional behaviors (Szalay, 1981). For instance, if the fossil taxon under consideration had selenodont teeth indicating folivory and postcranial features indicating suspensory postures, we might speculate that the biological role of suspensory postures was to allow access to leaves on terminal branches.

Finally, the functional demands and morphological correlates of some positional behaviors may be poorly understood, as in the case of "mitten-gliding," the subject of focus in this study. However, such positional behaviors may be associated with others, the morphological indications of which are better understood [in this case quadrumanus suspensory behaviors are associated with mitten-gliding (Pocock, 1926; Beard, 1989)]. The presence of suspensory features in a fossil taxon would not positively support mitten-gliding, but the lack of them would definitely refute it.

11.1.3 Main Objectives and Summary of Findings to be Demonstrated

As indicated above, our main objective is to test Beard's (1989, 1990, 1991, 1993b) hypothesis of mitten-gliding for paromomyid and micromomyid "plesiadapiforms." While others have carried out studies with the same goal (Krause, 1991; Runestad and Ruff, 1995; Hamrick et al., 1999), this is the first such study that applies new postcranial material with documented craniodental associations (see below). The only extant mammal currently referred to as a "mitten-glider" is the dermopteran *Cynocephalus*. Unlike other gliding mammals (but similar to bats), dermopterans possess an inter-digital patagium, which gives their digits the appearance of being united, roughly similar in form to the hand of a person wearing a mitten. Beard (1993b) argued that this "mitten" has a critical role in adjusting the aerodynamic properties of the animal during gliding. Specifically, the mitten's most critical function was said to be its affect on billowing of the patagium. Thus, Beard suggested that the unique termination point for the patagium in *Cynocephalus*, compared to other gliders, had influenced the morphology of the phalanges by placing unusual mechanical demands on them. When it is not gliding, *Cynocephalus* utilizes quadrumanus suspension to forage for leaves or to rest, and bimanual suspension while urinating or defecating (Wharton, 1950). Beard further suggested that the mechanical demands of these suspensory resting/foraging postures had also influenced the phalangeal morphology, resulting in a functional complex of features relating to mitten-gliding and suspension that is unique to dermopterans among extant mammals. As Beard explained:

"Some of the unusual features of the manus of extant *Cynocephalus* are clearly functionally related to the mode of attachment of the patagium to the manus in this genus" (Beard, 1989, p.441)

"...some of the unusual features of the hand skeleton of this genus also appear to reflect its emphasis on suspensory postures and locomotion" (Beard, 1989, p.442)

"...the total morphology of the hands of *Cynocephalus* is an anatomical compromise that in several ways is unique among extant mammals." (Beard, 1989, p.442).

While Beard's explanation is plausible and accepted by us in general, Runestad and Ruff (1995) demonstrated that *Cynocephalus* is also similar to other gliding mammals (regardless of attachment point of the patagium) in having elongate limb bones. Stafford (1999) supported this conclusion and observed more points of similarity between dermopterans and other gliding mammals. In this study, we present evidence that agrees with the observations of these authors, showing that *Cynocephalus* and other gliding eutherian mammals are morphologically similar in many respects. Thus, we would modify Beard's functional explanation of dermopteran morphology by adding that much of the skeleton has additionally been influenced by the mechanical demands for gliding, despite differences in patagial attachment points.

Beard (1989, 1990, 1993b) observed that a number of otherwise rare morphological features of dermopterans potentially unite them with fossil paromomyids and micromomyids. He argued that these shared features reflected the shared presence of the dermopteran functional complex.

The new paromomyid skeletons presented here indicate that these mammals were drastically different from *Cynocephalus* in their positional behavior. Specifically, we demonstrate that (1) they are distinctly unlike living dermopterans in lacking all osteological correlates specific to gliding with an interdigital patagium or associated quadrupedal suspension (Pocock, 1926; Beard, 1993b), and (2) they lack more general osteological correlates to gliding found in *Cynocephalus*, as well as gliding rodents and in some cases gliding marsupials. Likewise, new skeletons of micromomyids lack mitten-gliding features and features unambiguously suggestive of dermopteran-like quadrupedal suspension. While micromomyids are similar to dermopterans and gliding sciurids in some other respects, many of these features are also found in the extant, primitive treeshrew *Ptilocercus lowii* (Sargis, 2002a, b, c), primitive euprimates, and taxa likely to represent euarchontan outgroups (Hooker, 2001; Meng et al., 2004; Rose and Chinnery 2004). Thus, in the context of still other features that appear inconsistent with a capacity for gliding, we are unable to entertain the possibility that micromomyids were gliders of any sort.

11.1.4 Institutional Abbreviations

AMNH	American Museum of Natural History, New York
CR	Cernay-lès-Reims (for MNHN specimens from that locality)
MNHN	Muséum Nationale d'Histoire Naturelle, Paris
UA	University of Alberta Laboratory or Vertebrate Paleontology, Edmonton
UCMP	University of California at Berkeley, Museum of Paleontology, Berkeley
UF	University of Florida Museum of Natural History, Gainesville
UKMNH	University of Kansas Museum of Natural History, Lawrence
UM	University of Michigan Museum of Paleontology, Ann Arbor (specimen reference)
UMMP	University of Michigan Museum of Paleontology, Ann Arbor (institution reference)
UMMZ	University of Michigan Museum of Zoology, Ann Arbor
USNM	United States National Museum Department of Paleobiology (Smithsonian Institution), Washington, DC
UW	University of Wyoming, Laramie
YPM	Peabody Museum of Natural History, Yale University, New Haven
YPM-PU	Princeton University collections at the Peabody Museum of Natural History, Princeton

11.2 Materials

In this study we observed and measured a number of fossil and extant mammalian osteological specimens, mainly from the UMMP and UMMZ collections respectively, however, other specimens from institutions given in the abbreviations were also examined. New fossil material focused on here includes *Ignacius clarkforkensis* UM 108210, 82606, *Acidomomys hebeticus* UM

108207 [all from UM locality SC-62 (1335 m above Cretaceous-Tertiary boundary in the Clarks Fork Basin): Clarkforkian (Cf)-2, North American Land Mammal Age (NALMA)], *Dryomomys szalayi* UM 41870 [from UM locality SC-327 (1420 m), Cf-3], and a currently uncatalogued cf. *Tinimomys graybulliensis* specimen from a Wasatchian-1 aged locality in the Clarks Fork Basin. Detailed locality information for these specimens is archived with the UMMP. Additionally, Appendix I provides a list of all known paromomyid and micromomyid postcranial material, some of which is also focused on here. Specifically, we also incorporate information from dentally-associated remains of *Phenacolemur praecox* (UM 66440, 86352), *Phenacolemur jepseni* (USGS 17847), and *Tinimomys graybulliensis* (UM 85176); and some isolated remains attributed to *Phenacolemur simonsi* (USNM 442260, 62) and *Ignacius graybullianus* (USNM 442259) (Beard, 1989). Fossil specimens used for comparative purposes include plesiadapiforms (*Carpolestes simpsoni* UM 101963, *Nannodectes gidleyi* AMNH 17379, *Nannodectes intermedius* USNM 442229, *Plesiadapis cookei* UM 87990), euprimates (*Smilodectes mcgrewi* UM 95526, *Omomys carteri* UM 14134, *Omomys sp.* UM 98604), and dentally-associated postcranial material from an uncatalogued nyctitheriid insectivoran from UM locality SC-327.

Appendix II, Table 1 is a comprehensive list of all specimens measured, observed or figured in the course of this study, including those listed above. All quantitative analyses and all figures with more than four specimens are represented by a column in this table. If a specimen was included in a given analysis or figure, this is indicated by an "x" in the row corresponding to that specimen in the appropriate column.

11.3 Methods

11.3.1 Documentation Of Association in New Specimens

Fossils were extracted from limestone nodules of the Clarks fork basin using acid reduction. The protocol by which this process was carried out and that by which dental, cranial and postcranial associations were documented is described in Bloch and Boyer (2001, 2007).

11.3.2 Functional Analysis: Evaluation of the Gliding Hypothesis

We carried out functional analyses using the comparative method (Bock, 1977; Szalay and Sargis, 2001) and with analytical and statistical techniques including regression, ANOVA, students t-tests and principal components analysis (Sokal and Rohlf, 1997). Morphology was quantified and compared using indices in many cases (e.g., intermembral index). For statistical comparisons, such indices were logged, because this transformation tended to make the distributions of these data normal. We used the software PAST and SPSS 11.0 to carry out statistical analyses, for which the accepted level of significance was 95% or better.

Other comparisons were made based on qualitative morphology or with too few specimens to assign a level of statistical significance to our observations. However, in such cases, the differences between gliding and/or suspensory taxa and those taxa that lack these locomotor repertoires were substantial.

Throughout the text we provide tables that list the key functional characters evaluated in each region of the skeleton. The state of each character is coded for a number of different generalized "locomotor/behavior categories" of modern mammals in each table. Character states that are shared by different locomotor/behavioral groups are inferred to reflect common functions among those groups (if not shared behaviors). With this method we draw conclusions such as the following: elongate intermediate phalanges and restricted mobility at the proximal interphalangeal joint in *Cynocephalus* are likely correlates of suspensory behaviors, not mitten gliding, because the two other taxa that share such features suspend themselves from their hands and/or feet, but are not mitten-gliders (see Table 11.4).

11.4 Results and Discussion

11.4.1 Documentation of Association in New Specimens

11.4.1.1 Acidomomys hebeticus

UM 108207. We began preparation of the cranium of *Acidomomys hebeticus* from limestone in 1999 (Bloch and Boyer, 2001: Figures 11.8, 11.11). The preparation took roughly a year and revealed other individuals of *Acidomomys hebeticus*, as well as craniodentally associated postcrania for several other taxa, including *Carpolestes simpsoni* (UM 101963; Bloch and Boyer, 2002a). Initially, it was difficult to identify postcrania for *A. hebeticus* because of its proximity with several animals of similar size (Figures 11.1 and 11.2). Fortunately, parts of the *A. hebeticus* skeleton were in semi-articulation and associated with cranial remains (Figure 11.1A). Having a portion of the skeleton articulated and associated allowed for identification of form, ontogenetic stage and preservational quality of the postcranials. With this information we were able to identify other less well-associated postcranials. *A. hebeticus* is represented by at least three individuals in the accumulation (Bloch et al., 2002a). All are juveniles in various stages of erupting their adult dentitions (UM 108206-8). UM 108207 is represented by a skull and dentaries; distal, intermediate and proximal phalanges of the hand and foot; metacarpals I and V; right and left scaphoids; astragalus; right radius, ulna and distal humeral fragment. Of these elements, the radius, ulna, scaphoids, metacarpals, four proximal phalanges, and three intermediate phalanges are confidently attributed to a single individual (Figures 11.1 and 11.2). Other phalanges, not associated with forelimb

TABLE 11.1. Measurements of "plesiadapiform" proximal phalanges, proximal phalanx length (PPL); intermediate phalanges, intermediate phalanx length (IPL) and mid-shaft area (MSA); and metapodials, metapodial length (MPL) are given in this table. Bold lettering indicates that the measurements are analyzed or plotted in one of the figures. Columns titled "Fig" give the figure number showing a particular specimen with its original skeletal associations, as well as the label or location of that specimen in the figure. Only paromomyids and the micromomyid have information recorded in these columns. For the micromomyid all specimens are shown in Figure 11.6c. The side of the body to which the element was attributed is also given (left-L or right-R). Raw measurements include precisely measured length on pristine specimens, approximately (~) measured length on badly damaged specimens, or estimated {est} length of incomplete specimens. Part of the identification criteria for *Acidomomys hebeticus* was the juvenile status of the elements attributed to it. As such, most measurements do not include proximal end epiphyses. The proximal phalanx epiphyses add from 0.51–0.68 mm to the proximodistal length of these elements. The intermediate phalanx proximal epiphyses are shorter ranging from 0.4–0.51 mm. If a measurement on a particular specimen was not taken, because it was not necessary (or not possible due to breakage), this is indicated by "nm" (not measured). The "autopod" column is coded to indicate whether a specimen is attributed to hand (1) or foot (2). "Ray" gives the digit ray to which an element belongs. In most cases, associations and morphology are not distinctive enough to give a position, and a "?" is put here. First digits and metapodials are exceptions. In Figure 11.8, intermediate phalanx lengths are compared to proximal phalanx lengths to evaluate the probable interphalangeal proportions of these specimens. Intermediate-proximal phalanx pairs used in that analysis occupy single rows in the table.

Catalogue number	Family	Taxon	Figure	Autopod	Ray	PPL	Figure	Autopod	Ray	IPL	MSA	Figure	Autopod	Ray	MPL
UM 101963	**Carpolestidae**	***Carpolestes simpsoni***		**1**	**3**	**6.61**		**1**	**3**	**4.60**	**0.42**		1	3	
7.18															
UM 101963	**Carpolestidae**	***Carpolestes simpsoni***		**1**	**2**	**6.60**		**1**	**2**	**5.05**	**0.41**				
UM 101963	Carpolestidae	*Carpolestes simpsoni*		2	5	5.82		2	5	3.88	nm		2	3	7.40
UM 101963	Carpolestidae	*Carpolestes simpsoni*		2	3	5.62		2	3	3.80	nm				
UM 101963	Carpolestidae	*Carpolestes simpsoni*		2	4	6.26		2	4	4.00	nm				
UM 101963	Carpolestidae	*Carpolestes simpsoni*		2	2	4.85		2	2	3.24	nm				
UM 41870	**Micromomyidae**	***Dryomomys szalayi***	**R (Fig. 6)**	**1**	**?**	**2.91**	**L (Fig. 6)**	**1**	**?**	**2.56**	**0.10**	**m-l (Fig. 6)**	**1**	**3**	**3.73**
UM 41870	**Micromomyidae**	***Dryomomys szalayi***	**L (Fig. 6)**	**1**	**?**	**3.03**	**L (Fig. 6)**	**1**	**?**	**2.70**	**0.09**				
UM 41870	**Micromomyidae**	***Dryomomys szalayi***	L (Fig. 6)	**1**	**?**	**3.07**	**L (Fig. 6)**	**1**	**?**	**2.72**	**0.09**				
UM 41870	**Micromomyidae**	***Dryomomys szalayi***	**R (Fig. 6)**	**1**	**?**	**2.97**	**L (Fig. 6)**	**1**	**?**	**2.61**	**0.09**				
UM 41870	**Micromomyidae**	***Dryomomys szalayi***	**R (Fig. 6)**	**1**	**?**	**2.88**	**R (Fig. 6)**	**1**	**?**	**2.53**	**0.11**				
UM 41870	**Micromomyidae**	***Dryomomys szalayi***	**L (Fig. 6)**	**1**	**?**	**3.07**	**R (Fig. 6)**	**1**	**?**	**2.74**	**0.09**				
UM 41870	**Micromomyidae**	***Dryomomys szalayi***	**R (Fig. 6)**	**1**	**?**	**3.10**	**R (Fig. 6)**	**1**	**?**	**2.78**	**0.09**				
UM 41870	**Micromomyidae**	***Dryomomys szalayi***	L (Fig. 6)	1	1	2.84	R (Fig. 6)	1	?	nm	nm				
UM 41870	Micromomyidae	*Dryomomys szalayi*	R (Fig. 6)	2	?	3.17	R (Fig. 6)	2	?	2.79	nm	p-l (Fig. 6)	2	3	5.09
UM 41870	Micromomyidae	*Dryomomys szalayi*	R (Fig. 6)	2	?	3.18	R (Fig. 6)	2	?	2.82	nm				
UM 41870	Micromomyidae	*Dryomomys szalayi*	R (Fig. 6)	2	?	3.11	R (Fig. 6)	2	?	nm	nm				
UM 41870	Micromomyidae	*Dryomomys szalayi*	R (Fig. 6)	2	?	3.00	R (Fig. 6)	2	?	nm	nm				
UM 41870	Micromomyidae	*Dryomomys szalayi*	L (Fig. 6)	2	?	2.97	L (Fig. 6)	2	?	2.73	nm				
UM 41870	Micromomyidae	*Dryomomys szalayi*	L (Fig. 6)	2	1	2.70	L (Fig. 6)	2	?	nm	nm				
UM 41870	Micromomyidae	*Dryomomys szalayi*	R (Fig. 6)	2	1	2.70	L (Fig. 6)	2	?	nm	nm				
UM 108207	**Paromomyidae**	***Acidomomys hebeticus***	**12 (Fig. 1)**	**1**	**?**	{est} **6.27**	**1 (Fig. 1)**	**1**	**?**	**4.82**	**0.43**				
UM 108207	**Paromomyidae**	***Acidomomys hebeticus***	**6 (Fig. 1)**	**1**	**?**	{est} **6.75**	**2 (Fig. 1)**	**1**	**?**	**~6.12**	**nm**				
UM 108207	**Paromomyidae**	***Acidomomys hebeticus***	**7 (Fig. 1)**	**1**	**?**	**5.90**	**3 (Fig. 1)**	**1**	**?**	**~4.80**	**0.52**				
UM 108207	**Paromomyidae**	***Acidomomys hebeticus***	**5 (Fig. 1)**	**1**	**?**	{est} **6.3**	**11 (Fig. 1)**	**1**	**4**	**5.50**	**0.44**				
UM 108207	**Paromomyidae**	***Acidomomys hebeticus***	**9 (Fig. 1)**	**1**	**?**	{est} **6.3**	**13 (Fig. 1)**	**1**	**5**	**5.30**	**0.43**				
UM 108207	Paromomyidae	*Acidomomys hebeticus*	8 (Fig. 1)	1	1	5.5									
UM 108207	Paromomyidae	*Acidomomys hebeticus*	15 (Fig. 2)	2	?	6.69	16 (Fig. 2)	2	2	6.51	nm				
UM 108207	Paromomyidae	*Acidomomys hebeticus*	17 (Fig. 2)	2	?	{est} 7.3	18 (Fig. 2)	2	4	~6.3	nm				
UM 108207	Paromomyidae	*Acidomomys hebeticus*	19 (Fig. 2)	2	1	5.21	20 (Fig. 2)	2	3	6.48	nm				
UM 108207	Paromomyidae	*Acidomomys hebeticus*	20 (Fig. 2)	2	?	6.84	21 (Fig. 2)	2	5	6.5	nm				
UM 108207	Paromomyidae	*Acidomomys hebeticus*	22 (Fig. 2)	2	?	6.77									
UM 108210	Paromomyidae	*Ignacius clarkforkensis*	5 (Figs. 3, 5)	1	?	10.30	1 (Figs. 3, 5)	1	?	nm		Figs. 3, 5	1	3	{est} 10.30

(continued)

Table 11.1. (continued)

Catalogue number	Family	Taxon	Figure	Autopod	Ray	PPL	Figure	Autopod	Ray	IPL	MSA	Figure	Autopod	Ray	MPL
UM 108210	**Paromomyidae**	***Ignacius clarkforkensis***	**4 (Figs. 3, 5)**	**1**	**?**	**10.80**	**2 (Figs. 3, 5)**	**1**	**?**	**~7.95**	**0.77**				
UM 108210	**Paromomyidae**	***Ignacius clarkforkensis***	**8 (Figs. 3, 5)**	**1**	**?**	**10.70**	**7 (Figs. 3, 5)**	**1**	**?**	**7.80**	**0.78**				
UM 108210	Paromomyidae	*Ignacius clarkforkensis*	6 (Figs. 3, 5)	1	5	8.30	3 (Figs. 3, 5)	1	?	nm					
UM 82606	Paromomyidae	*Ignacius clarkforkensis*	Figs. 3, 5									Figs. 3, 5	2	2	15.3
UM 87990	**Plesiadapidae**	***Plesiadapis cookei***		**1**	**?**	**15.22**		**1**	**?**	**12.11**	4.37		1	3	23.54
UM 87990	**Plesiadapidae**	***Plesiadapis cookei***		**1**	**?**	**12.70**		**1**	**?**	**10.30**	3.55				
UM 87990	Plesiadapidae	*Plesiadapis cookei*		1	?	15.17									
UM 87990	Plesiadapidae	*Plesiadapis cookei*		2	?	17.27		2	?	13.39	nm		2	3	31.10
UM 87990	Plesiadapidae	*Plesiadapis cookei*		2	?	17.27		2	?	13.39	nm		2	3	31.10
UM 87990	Plesiadapidae	*Plesiadapis cookei*		2	?	17.64		2	?	13.80	nm				
USNM 442229	Plesiadapidae	*Nannodectes intermedius*		1	1	6.95							1	3	11.50
USNM 442229	Plesiadapidae	*Nannodectes intermedius*		1	1	6.95									
USNM 442229	**Plesiadapidae**	***Nannodectes intermedius***		**1**	**?**	**8.59**		**1**	**?**	**6.72**	**1.22**				
USNM 442229	Plesiadapidae	*Nannodectes intermedius*		1	?	8.09									
USNM 442229	Plesiadapidae	*Nannodectes intermedius*		1	?	7.63									
USNM 442229	Plesiadapidae	*Nannodectes intermedius*		2	?	9.11		2	?	7.18	nm				
USNM 442229	**Plesiadapidae**	***Nannodectes intermedius***						**2**	**?**	**7.10**	**0.95**				
USNM 442229	Plesiadapidae	*Nannodectes intermedius*						2	?	7.39	nm				
AMNH 17379	**Plesiadapidae**	***Nannodectes gidleyi***		**1**	**?**	**9.17**		**1**	**?**	**6.88**	**0.91**		**1**	**3**	**12.02**
AMNH 17379	Plesiadapidae	*Nannodectes gidleyi*		1	?	8.63									
AMNH 17379	**Plesiadapidae**	***Nannodectes gidleyi***						**2**	**?**	**8.26**	**1.03**				
AMNH 17379	Plesiadapidae	***Nannodectes gidleyi***						**2**	**?**	**7.19**	**0.84**				

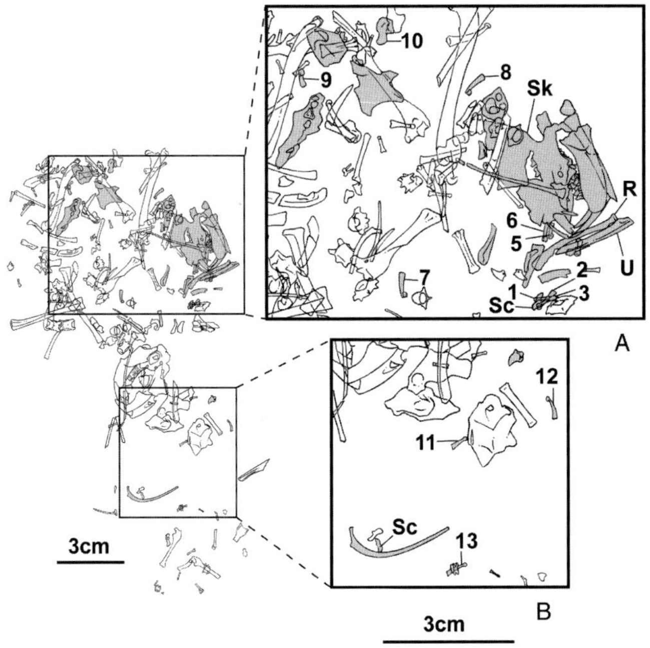

FIGURE 11.1. Documentation of association of *Acidomomys hebeticus* skeleton. Map of bones in upper layer of limestone "Block M," from the Clarks Fork Basin locality SC-62. Elements identified as *Acidomomys hebeticus* (UM 108207) are depicted in gray. A, Enlarged view to show articulated radius (R) and ulna (U) in association with proximal phalanges 5 and 6, intermediate phalanges 1–3 and a scaphoid (Sc), below the crushed skull (Sk). *A. hebeticus* was distinguished from other animals in the accumulation on the basis of associations, size, age and taphonomic state. *A. hebeticus* is a juvenile with porous bone, while the other taxa are adults. Bones not attributed to *A. hebeticus* are attributed to two or more individual paramyid rodents. Lengths and length estimates of phalanges are reported in Table 11.1. Scale = 3 cm.

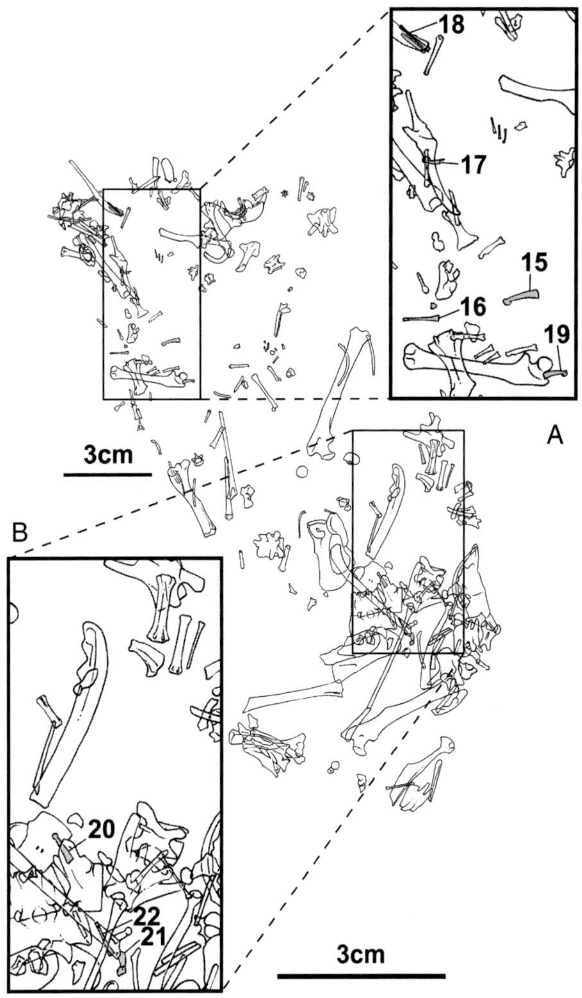

FIGURE 11.2. Documentation of association of *Acidomomys hebeticus* skeleton. Map of bones in lower layer of limestone "Block M" (see Figure 11.1). Elements identified as *Acidomomys hebeticus* (UM 108207) are depicted in gray. Bones not attributed to *A. hebeticus* belong to a large rodent, an erinaceomorph insectivore, a carpolestid plesiadapiform, or a small marsupial. Lengths and length estimates of phalanges are reported in Table 11.1. Scale = 3 cm.

elements and with consistently different proportions than those that were associated with the forelimb (although still clearly belonging to *A. hebeticus*), are interpreted as pedal phalanges (Figures 11.1 and 11.2). The pedal elements are longer than the manual ones. Table 11.1 gives measurements of phalanges attributed to UM 108207, depicted and labeled in Figures 11.1 and 11.2.

11.4.1.2 Ignacius clarkforkensis

UM 108210. (Holotype; Bloch et al., 2007) Associated dentitions and postcranium were preserved in their original positions in a freshwater limestone from University of Michigan locality SC-62 (Figure 11.3A).

UM 82606. This specimen is also from SC-62. While relative locations of UM 108210 to UM 82606 are undocumented, these specimens cannot represent the same individual, because each preserves an astragalus from the same side of the body (right side). In fact UM 82606 lacks association with teeth (Figure 11.3B), and was identified as belonging to *Ignacius clarkforkensis* based on (1) size and form of the astragalus (Figure 11.4) and calcaneum, which are nearly identical to those of UM 108210, and (2) by the similarity in hind limb morphology to that of previously described paromomyids (Beard, 1989).

Analysis of this specimen combined with UM 108210 allowed estimates of interlimb and inter body-segment proportions. We justify the use of these specimens as a composite based on the similar astragalus size and the fact that different long bones from each specimen yield overlapping body mass estimates. Together, these two specimens allow analysis of nearly the entire morphology and proportions of a composite individual of *Ignacius clarkforkensis* (Figure 11.5).

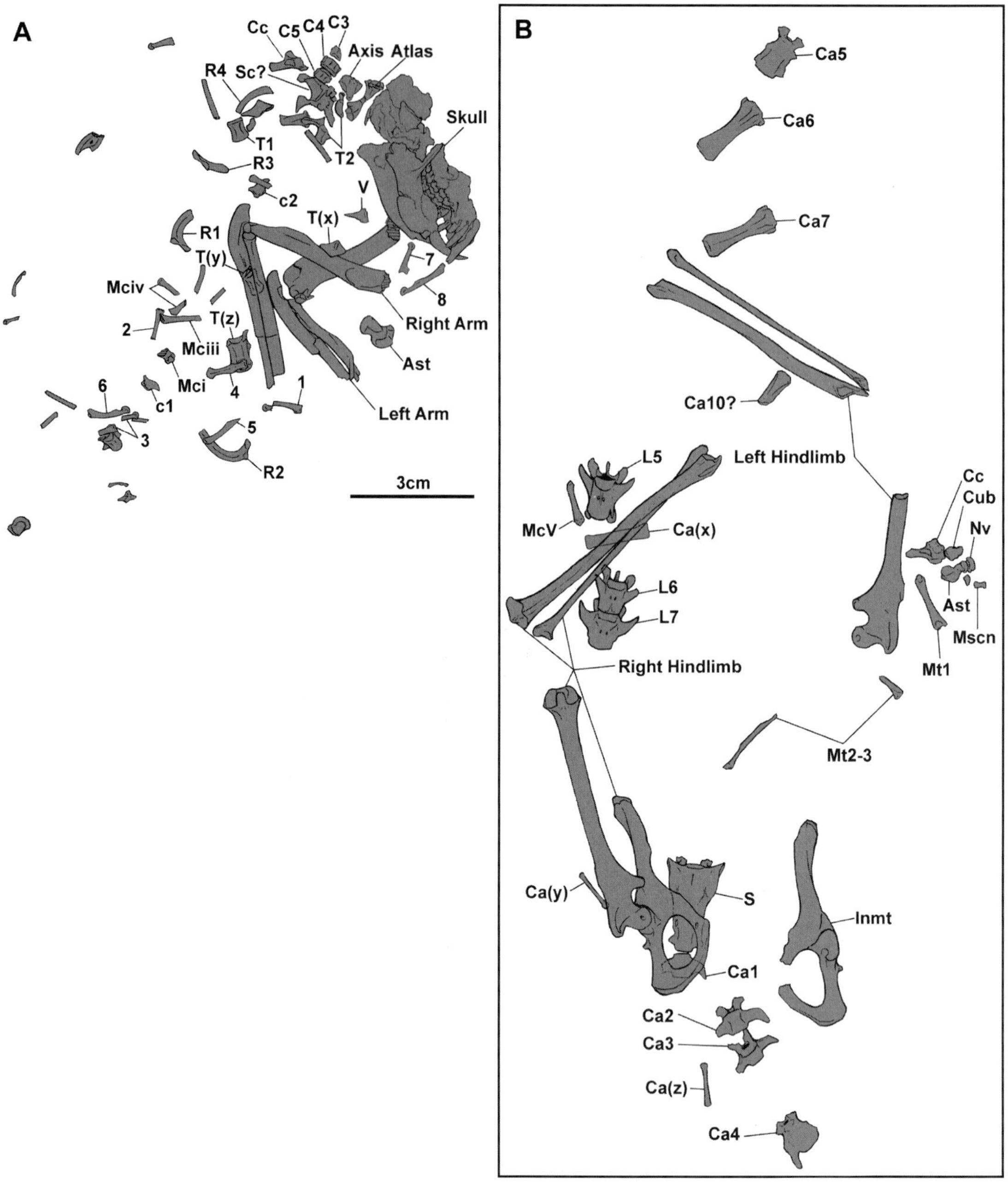

FIGURE 11.3. Documentation of association of *Ignacius clarkforkensis* skeletons. A, Map of semi-articulated skeleton of *Ignacius clarkforkensis* (UM 108210) prepared from a limestone from SC-62. The skull has been described by Bloch and Silcox (2001). B, Map of second semi-articulated skeleton of *Ignacius clarkforkensis* (UM 82606) prepared from a limestone from SC-62. See Figure 11.4 for a key to abbreviated labels on these diagrams. Scale = 3 cm.

11.4.1.3 *Dryomomys szalayi*

UM 41870. (Holotype; Bloch et al., 2007). In late 2000, we recognized the semi-articulated remains of a micromomyid plesiadapiform, preserved in a richly fossiliferous block of limestone collected by University of Michigan field crews in 1982 from locality SC-327 of the Clarks Fork Basin, Wyoming (Figure 11.6). Through methods described in Bloch and Boyer (2001) and, with regard to this particular specimen (Bloch and Boyer, 2007), we extracted the bones from their calcite tomb, while preserving critical information on the original position of each bone (Figure 11.6B). The specimen is the most complete and best articulated "plesiadapiform" yet recovered, and is represented by much of a skull (including auditory region and perfectly preserved premaxillae) and dentaries (with all tooth positions represented); cervical, thoracic, and caudal vertebrae (but no lumbars or sacrum);

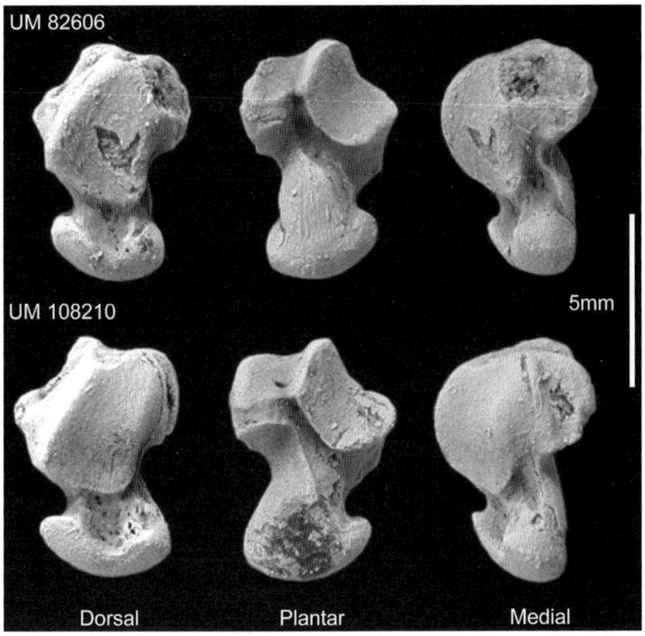

FIGURE 11.4. Right astragali of UM 82606 (top) and UM 108210 (bottom). While these elements differ in subtle respects, they clearly represent the same taxon, *Ignacius clarkforkensis* Scale = 5 mm.

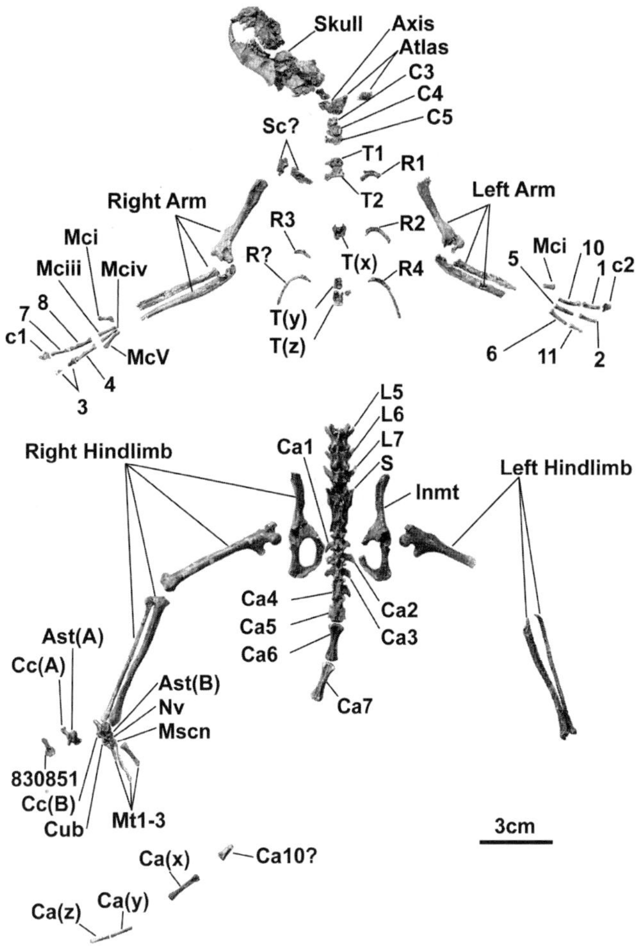

FIGURE 11.5. Composite layout of *Ignacius clarkforkensis* (UM 108210 and UM 82606) in rough anatomical position. Abbreviations: Ast – astragalus; C# – cervical vertebra; c# – claw; Ca# – caudal vertebra; Cc – calcaneum; Cub – cuboid; Inmt – innominate; L# – lumbar vertebra; Mc – metacarpal; Mscn – mesocuneiform; Mt – metatarsal; Nv – navicular; S – sacrum; Sc? – scapula?; T# – thoracic vertebra; R# – rib; "4" and "8" – left proximal phalanges; "6" and "10" – left proximal phalanges; "1", "2", "3", "7", and "11" – right intermediate phalanges. The right "Mci", "10", and "11" were recovered in the screen during preparation, thus their exact positions in the deposit is unknown. Scale = 3 cm.

several ribs; manubrium sternae; both scapulae; both humeri and radii; the left ulna; both scaphoids, trapezoids and a triquetrum; metacarpals of both hands (left hand preserved with the metacarpals in almost perfect articulation; Figure 11.6B); many proximal, intermediate and distal phalanges of the hands and feet; a left femur; the distal ends of the left tibia and fibula; both astragali, calcanea, naviculars, and mesocuneiforms; right ectocuneiform and cuboid; left entocuneiform; and metatarsals.

Bloch et al. (2003) explored the adaptive morphology of the euarchontan morphotype by comparing micromomyid morphology (reconstructed from this specimen and the one depicted in Figure 11.7) with that of *Ptilocercus lowii* and a possible stem-euarchontan, *Leptacodon* (Nyctitheriidae; see Hooker, 2001). They found a surprising degree of similarity between the micromomyid and *Ptilocercus* in morphology associated with committed arboreality in *Ptilocercus*. Thus, their results supported the hypothesis that the ancestral euarchontan was a committed arborealist (Szalay and Decker, 1974; Szalay and Drawhorn, 1980; Sargis, 2001b, 2002c). Furthermore, the retention of scansorial features in *Leptacodon*, considered to be a euarchontan outgroup by them, suggested an adaptive shift to committed arboreality at the base of the euarchontan radiation (Szalay and Drawhorn, 1980).

Another micromomyid was recently recognized from UM Locality SC-26 (Wa-1, early Eocene) and prepared beginning in 2002 (Figure 11.7). Although no teeth have been recovered, postcranial morphology and stratigraphic position make it attributable to *Tinimomys graybulliensis*. The limestone was recovered in the early 1980s by Dr. Peter Houde during Smithsonian sponsored field expeditions in the Clarks Fork Basin. The specimen is semi-articulated (Figure 11.7). It includes vertebrae from all anatomical regions of the spine; a left radius; many ribs; both innominates; parts of both femora; both tibiae and fibulae; the right wrist (scaphoid, capitate, lunate, hamate, triquetrum and centrale); the right hand (metacarpals and proximal, intermediate and distal phalanges); the left foot (distal tarsal row); and the right foot (navicular and third metatarsal).

The total lengths of the tibiae and radii can be measured in this Wasatchian specimen, and the total lengths of the femur, radius and humerus are measurable in the Clarkforkian specimen (UM 41870). Thus, by scaling one skeleton or the other

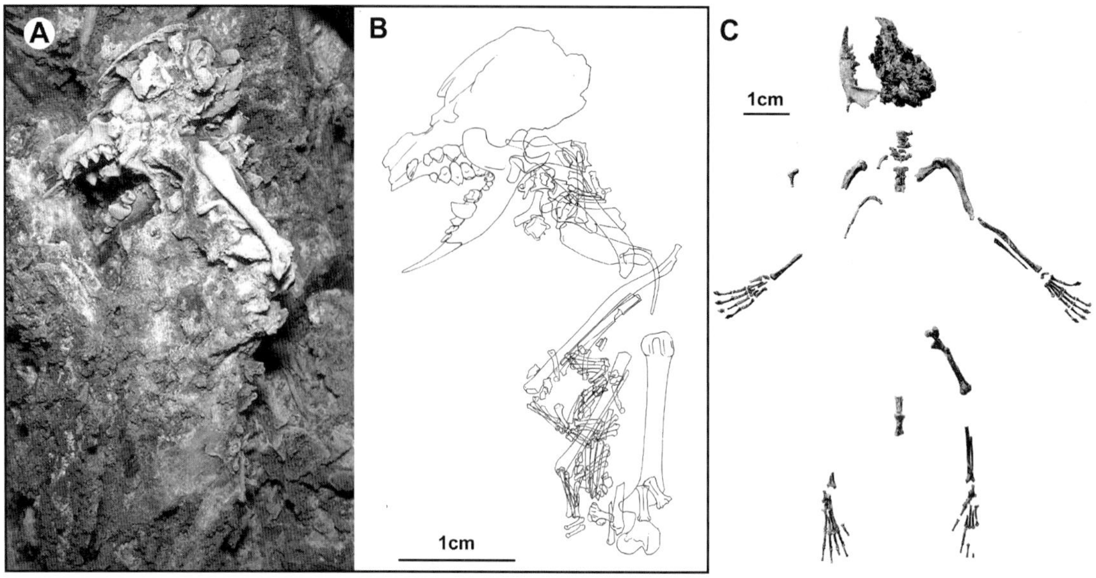

FIGURE 11.6. Documentation of association of micromomyid skeleton, *Dryomomys szalayi* (UM 41870). A, skull and skeleton partially prepared from fossiliferous limestone, University of Michigan Locality SC-327, late Clarkforkian (Cf-3) North American Land Mammal Age. B, Composite map of skull and skeleton with numbers on bones. C, Skull and skeleton in anatomical position. Note that C was made before all of the bones were prepared from the limestone and that not all bones depicted in B are in C. See Bloch and Boyer (2007) for (1) a similar figure that differs in providing numbers on all bones to show the exact correspondence between elements in B and C, and (2) a more detailed discussion of the preparation of this skeleton. Scale in A and B = 1 cm; Scale in C = 1 cm.

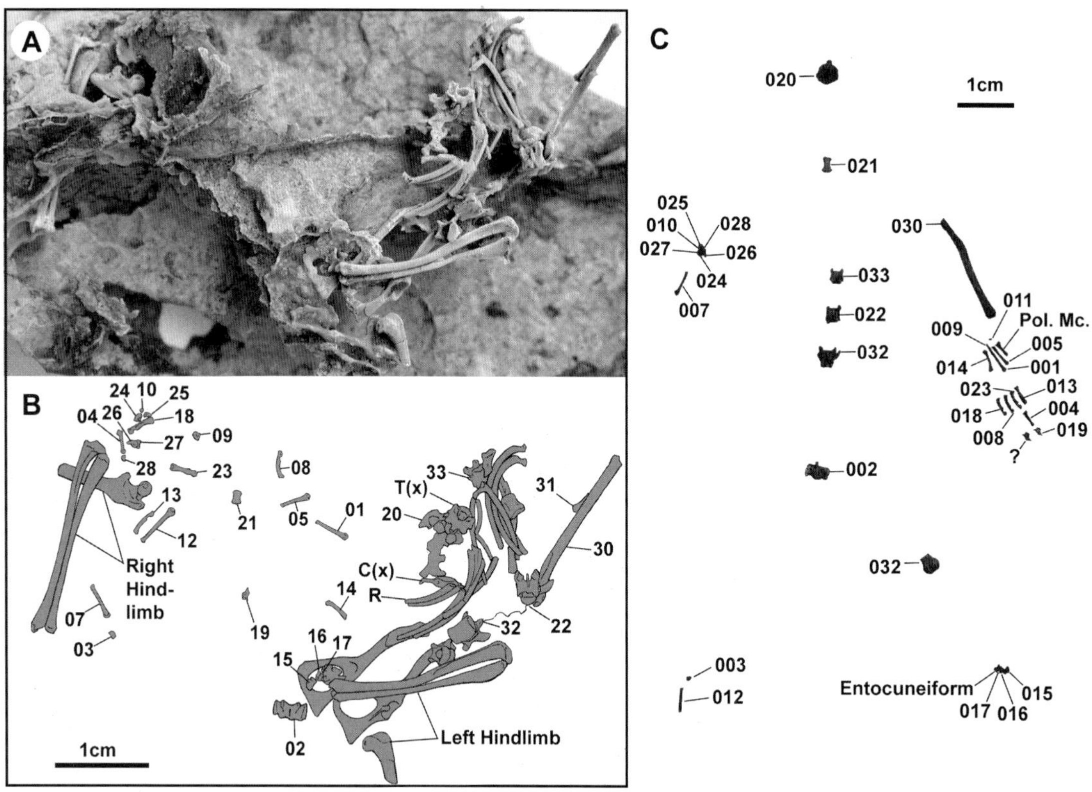

FIGURE 11.7. Documentation of association of micromomyid skeleton, cf. *Tinimomys graybulliensis* from UM locality SC-026 (currently uncatalogued). A, Photograph of a partly prepared limestone yielding a Wasatchian-aged micromomyid skeleton. B, Map of distribution of the bones in the limestone. Although it lacks craniodental remains, this specimen does preserve many of the same elements as that in Figure 11.6, so we can confidently identify it to the familial level, given its age, we tentatively refer to *Tinimomys graybulliensis*. The specimen includes complete tibiae, fibulae, innominates, and a complete radius. C, Some of the fully removed and prepared bones from this specimen in rough anatomical position. Note that most of the right wrist, the left hand, many vertebrae, and the left distal tarsal row are preserved. Scale = 1 cm.

so that radius lengths are equal between the two, the relative lengths of all limb segments of a hypothetical, composite micromomyid can be estimated.

11.4.2 Functional Analysis: Evaluation of the Gliding Hypothesis

11.4.2.1 Features Originally Marshaled to Support Mitten-Gliding

Beard (1990, 1993b) used similarity in morphological and proportional features of the digits of paromomyids, micromomyids and *Cynocephalus* to provide evidence for mitten-gliding in the fossil taxa. This morphology and behavior was inferred to have been present in the common ancestor of the three groups (Beard, 1993a, b). The following is a list of features discussed by Beard (1993b) in his latest comprehensive treatment of the issue, in sections titled "Anatomical Evidence for Gliding in Paromomyids" and "Anatomical Evidence for Gliding in Micromomyids." He argued that at least dermopterans and paromomyids, and in some cases micromomyids exhibited these features to the exclusion of most other mammals. While he did not argue that all of these features were immediate adaptations to the act of mitten-gliding, it was implied that they were at least consistent with and possibly expected in a mitten gliding mammal. Features hypothesized to be adaptations for mitten-gliding, specifically, are listed in bold. All listed features are addressed in functional analyses presented below:

Intermediate Phalanges

1. **Longer than proximal phalanges, as in *Cynocephalus***
2. **High elongation index, as in *Cynocephalus*.**
3. Proximal ends deeper (dorsopalmarly) than wide (mediolaterally), as in *Cynocephalus*.
4. Shafts straight, without dorsal recurvature at distal end, as in *Cynocephalus*

Proximal Phalanges

5. **Shorter than intermediate phalanges, as in *Cynocephalus*.**
6. Morphology of corresponding joint surfaces of the proximal interphalangeal (pip) joint limits extension and flexion, as in *Cynocephalus*.

Beard (1989) argued that additional features suggest a *Cynocephalus*-like habitus for the both paromomyids and micromomyids. We address these as well.

11.4.2.2 Morphological Evidence of Positional Behavior in Paromomyidae and Micromomyidae

Intermediate and Proximal Phalanges. The original evidence for mitten-gliding in paromomyids was based on the observation that their intermediate phalanges were similar to those of dermopterans in being longer than the proximal phalanges (Beard, 1990), extraordinarily gracile, and dorsoventrally deep at the proximal end (Beard, 1993b). In a paper titled "Were paromomyids gliders? Maybe, maybe not," Krause (1991) warned that these observations were based on unassociated postcranial elements that were not necessarily from single individuals or even the same species, bringing into question the true proportional relationship between the proximal and intermediate phalanges of paromomyids. In fact, the new specimens of *Ignacius* (UM 108210) and *Acidomomys* (UM 108207) presented here (Table 11.1), have exactly the opposite phalangeal proportions as those proposed by Beard (1990), with the intermediate phalanges shorter than the proximal phalanges. Interestingly, in the context of better-established associations and a larger sample size, regression analysis (Figure 11.8A) shows that paromomyids and other "plesiadapiforms" have intermediate phalanx to proximal phalanx proportions comparable to a sample of non-gliding extant eutherian mammals. While some extant taxa in the sample, as well as some "plesiadapiforms," fall slightly above the upper confidence interval for this regression, the only substantial outlier is *Cynocephalus*. In fact, it is the only taxon in the plots of Figure 11.8 with intermediate phalanges that are actually longer than the proximal phalanges (above the x = y line). The fact that the pedal digits of some fruit bats and all digits of clawed suspensory sloths share this feature with *Cynocephalus*, suggest that it may actually reflect suspensory postures (Table 11.4).

Analysis of this relationship in another way (Figure 11.8B) also shows that both paromomyids (as a group) and micromomyids differ substantially from dermopterans. Specifically, Model I ANOVA of natural log interphalangeal ratios showed significant variance (at $P < 0.05$) among "terrestrial," "arboreal," "other plesiadapiform," "paromomyid," "micromomyid," and "*Cynocephalus*" groups (Figure 11.8B). Unlike in the regression, in this analysis our arboreal group included gliding squirrels because the average ratio for these taxa was not actually higher than that of non-gliding tree squirrels (i.e., *Sciurus* has a mean ratio of 0.80, while *Glaucomys*, a glider, has 0.76). Subsequent comparisons of these groups using t-tests (Table 11.2) showed *Cynocephalus* to be significantly higher than the other two extant behavioral groups, which did not differ from one another. Micromomyids, paromomyids and other "plesiadapiforms" were significantly lower than *Cynocephalus*, but higher than extant terrestrialists. Micromomyids were indistinguishable from paromomyids, but had a higher ratio than other "plesiadapiforms" and arborealists. On the other hand, paromomyids were not distinguishable from other "plesiadapiforms" (including micromomyids) but did have a higher average ratio than extant arborealists. Finally, non-micromomyid and non-paromomyid "plesiadapiforms" were indistinguishable from extant arborealists. While the average ratio for paromomyids is slightly higher than for other plesiadapiforms and significantly higher than for extant arborealists, careful inspection shows that these differences are probably artifactual. Adult and juvenile paromomyid specimens have different indices. *Ignacius* has

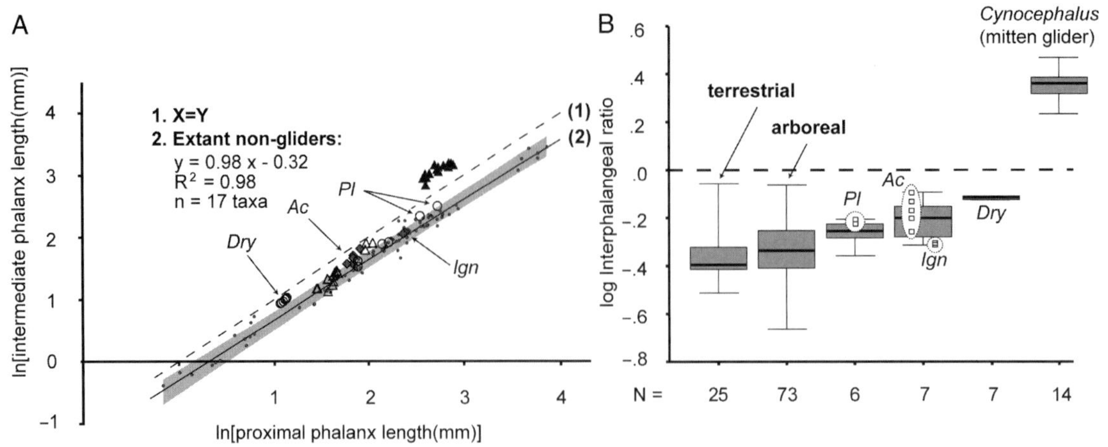

FIGURE 11.8. A, Reduced Major Axis (RMA) regression of intermediate phalanx on proximal phalanx length (solid line; 2) for select non-gliding mammals (small gray circles). Light gray area encompasses the 95% confidence limits for the relationship (y = 0.98 x – 0.32, R^2 = 0.98, n = 18 taxa). The slope is indistinguishable from that expected for an isometric relationship between the two variables (1). Mitten-gliding dermopterans (closed black triangles) have relatively much longer intermediate phalanges than do paromomyids (large gray diamonds), other plesiadapiforms (unfilled circles), and "mittenless" gliders (unfilled triangles). B, Box plots of logged interphalangeal ratios for a sample (same as for A) extant mammals and plesiadapiforms. Boxes encompass 50% of data points with medians depicted as a horizontal line within them. Whiskers encompass all data. Numbers below boxes represent the sample size for data points comprising it. *Ac = Acidomomys hebeticus, Ign = Ignacius clarkforkensis, Dry = Dryomomys szalayi, Pl = Plesiadapis cookei*. Dashed lines in A and B represent position of proximal and intermediate phalanges of equal length. See Table 11.1 for measurements of fossil specimens. See Table 11.2 for group means, t-values, and p-values of t-tests. See Appendix II, Table 1 for all specimens included in these plots.

TABLE 11.2. t-values and p-values of independent t-tests of natural log interphalangeal ratios of various extant behavioral and fossil taxonomic groups. In the right hand column, the group mean is given below its name. Significant p-values are in bold. In the top row, the number in parentheses after the group name is sample size for the group. Blank cells represent comparisons made elsewhere in the table. See Figure 11.8B for a plot of the samples analyzed here. See Appendix II, Table 1 for the specimens included.

	Ln(interphalangeal ratio) (Figure 11.8B)					
Group/mean	Terrestrial (25)	Arboreal (73)	Plesiadapiforms (6)	Paromomyidae (9)	Micromomyidae (4)	*Cynocephalus* (14)
Terrestrial	t = 0					
–0.35	p = 1					
Arboreal	t = 0.630	t = 0				
–0.33	p = 0.0530	p = 1				
Plesiadapiforms	t = 1.644	t = 1.454	t = 0			
–0.27	p = 0.111	p = **0.150**	p = 1			
Paromomyidae	t = 2.790	t = 2.761	t = 1.318	t = 0		
–0.21	p = **0.010**	p = **0.007**	p = 0.214	p = 1		
Micromomyidae	t = 9.405	t = 15.80	t = 6.533	t = 2.913	t = 0	
–0.12	p = **0.000**	p = **0.000**	p = **0.001**	p = **0.026**	p = 1	
Cynocephalus	t = 24.11	t = 32.40	t = 21.57	t = 17.76	t = 28.69	t = 0
0.35	p = **0.000**	p = **0.000**	p = **0.000**	p = **0.000**	p = **0.000**	p = 1

an average ratio of 0.73, while the average for *Acidomomys* is 0.84, with some *Acidomomys* as high as 0.91. The high value for *Acidomomys* cannot be interpreted with too much confidence because it is a juvenile. The fact that the adult *Ignacius* has lower values suggests that an older *Acidomomys* would also be lower. The average for *Ignacius* is actually lower than that reconstructed for at least two species of *Plesiadapis* (*P. cookei* = 0.81: see Figure 11.8b, *P. tricuspidens* ~ 0.82 based on MNHN R 5341 and 5305: see Godinot and Beard, 1991) a large bodied "plesiadapiform."

The finding that paromomyids do not really differ from other "plesiadapiforms" in the interphalangeal ratio is inconsistent with predictions of the mitten-gliding hypothesis. While the micromomyid *Dryomomys* is higher than all other "groups" in the analysis, it certainly does not share a position with *Cynocephalus*. Furthermore, both arboreal and terrestrial groups include taxa with average ratios that are actually as high as those of the micromomyid: if these had been analyzed separately they would have been indistinguishable from the micromomyid. Specifically, terrestrial *Gerbillus* has a mean interphalangeal ratio identical to that of *Dryomomys* (0.89). Anomalurids and some of the digit rays of the arboreal sciurids in the sample also share this value. A qualitative comparison of typical manual digit rays of a callitrichine euprimate, *Cynocephalus volans, Ignacius* and *Dryomomys* shows that the former extant taxon makes a better morphological analogue for both fossils, even though its claws evolved from an ancestor with euprimate-like nails (Figure 11.9).

Responding to critical questioning about the validity of conclusions drawn from comparisons using interphalangeal proportions of unassociated fossils (Krause, 1991), Beard (1993b) devised an index for evaluating elongation and gracility of intermediate phalanges that does not refer to the proximal elements. Based on this elongation index (shaft length divided by the square-root of the cross-sectional area at the mid-shaft) he again concluded that paromomyids, micromomyids and *Cynocephalus* were uniquely similar. Runestad and Ruff (1995), who rejected Beard's gliding interpretation of paromomyids based on evidence from the long bones, showed that the intermediate phalangeal length to square root cross-sectional area ratios for paromomyids (9–10) are more similar to those of non-gliding euprimates *Avahi* (8), *Tarsius* (8.5) and *Microcebus* (9) than to dermopterans (15–17). Hamrick et al. (1999) analyzed elongation in a different way (principal components and discriminant function analyses) and determined that dermopterans are unique among all the taxa included in their study (including paromomyids) in having extreme elongation. They concluded that paromomyids were probably not gliders. They did, however, conclude that dermopteran and paromomyid intermediate phalanges are similar to each other and those of bats in having a dorsoventrally deep proximal end. Hamrick et al. (1999) concluded that this was not a mitten-gliding feature, but potentially a dermopteran or volitantian synapomorphy, based on their use of galagonid and tupaiid morphology as reflective of the primitive archontan state. They did not consider this feature to reflect vertical-clinging with claws on large diameter supports.

We present results of analyses that support the findings of both Runestad and Ruff (1995) (Figure 11.10) and, in some respects, Hamrick et al. (1999) (Figures 11.11 and 11.12). Further, we use regression analysis to address a potential problem with all previous analyses of "elongation": that they have not been performed in an allometric context (Figure 11.14).

Analysis of the elongation index using a sample increased beyond that used by Beard (1993b) or Runestad and Ruff (1995) shows unambiguously (Figure 11.10) that dermopterans have significantly more elongate intermediate phalanges than either paromomyids or micromomyids. More specifically, Model I ANOVA of elongation indices revealed significant variance (at $P < 0.05$) among the "terrestrial," "arboreal," "other plesiadapiform," "paromomyid," "micromomyid," "gliding squirrel," and "*Cynocephalus*" groups. In this case gliding squirrels were kept as a separate group because their average elongation indices are substantially higher than those of non-gliding sciurids (gliding squirrels range from 9.3–11.4, *Sciurus* has a mean of 7.4). Comparison of these groups using t-tests showed

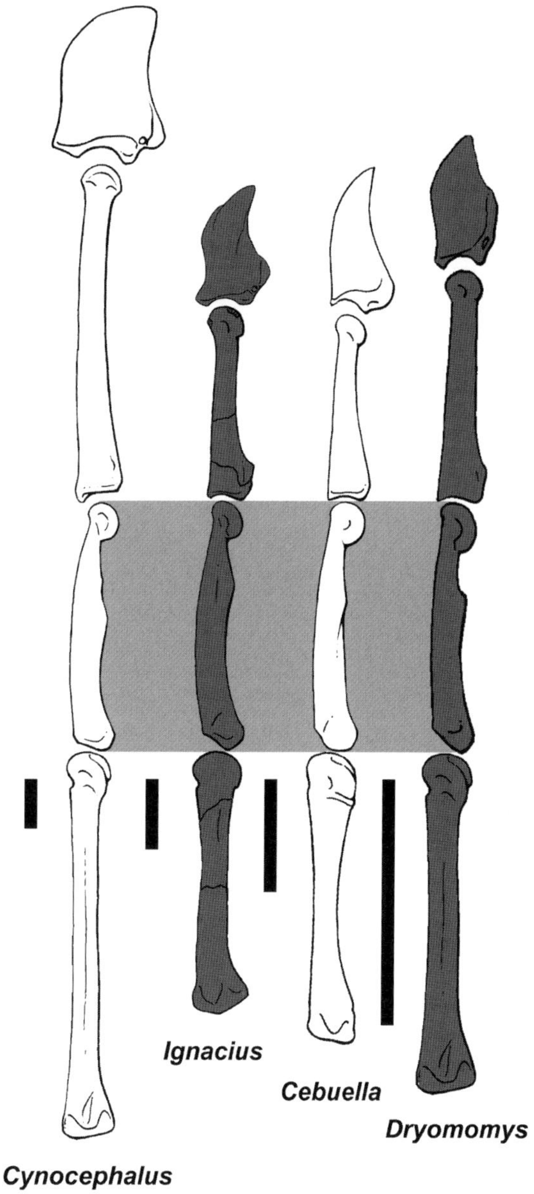

FIGURE 11.9. Comparison of phalangeal proportions in the manual third digit ray, normalized to the length of the proximal phalanx, for a dermopteran *Cynocephalus* (USNM 56530), a paromomyid, *Ignacius clarkforkensis* (UM 108210), a callitrichine euprimate *Cebuella* (UM 160146), and a micromomyid, *Dryomomys szalayi* (UM 41870). Rows from bottom-to-top are: metacarpal III, proximal phalanx, intermediate phalanx, and distal phalanx, respectively. Scales = 3 mm.

TABLE 11.3. t-values and p-values of independent t-tests of natural log elongation indices of various extant behavioral and fossil taxonomic groups. In the right hand column, the group mean is given below its name. Significant p-values are in bold. In the top row, the number in parentheses after the group name is sample size for the group. Blank cells represent comparisons made elsewhere in the table. See Figure 11.10 for a plot of the samples analyzed here. See Appendix II, Table 1 for specimens included.

	Ln(elongation Index) (Figure 11.10)						
Group/mean	Terrestrial (37)	Plesiadapiforms (9)	Arboreal (62)	Paromomyidae (6)	Micromomyidae (7)	Gliding squirrel (20)	*Cynocephalus* (18)
Terrestrial	t = 0						
1.59	p = 1						
Plesiadapiforms	t = 4.904	t = 0					
1.93	p = **0.000**	p = 1					
Arboreal	t = 9.022	t = 1.105	t = 0				
2.02	p = **0.000**	p = 0.273	p = 1				
Paromomyidae	t = 5.952	t = 2.060	t = 0.508	t = 0			
2.08	p = **0.000**	p = 0.060	p = 0.613	p = 1			
Micromomyidae	t = 14.16	t = 3.861	t = 1.435	t = 1.650	t = 0		
2.16	p = **0.000**	p = **0.002**	p = 0.156	p = 0.127	p = 1		
Gliding squirrel	t = 17.63	t = 7.498	t = 4.586	t = 4.221	t = 2.980	t = 0	
2.29	p = **0.000**	p = **0.000**	p = **0.000**	p = **0.000**	p = **0.006**	p = 1	
Cynocephalus	t = 20.39	t = 12.28	t = 13.33	t = 8.900	t = 11.52	t = 8.261	t = 0
2.63	p = **0.000**	p = **0.000**	p = **0.000**	p = **0.000**	p = **0.000**	p = **0.000**	p = 1

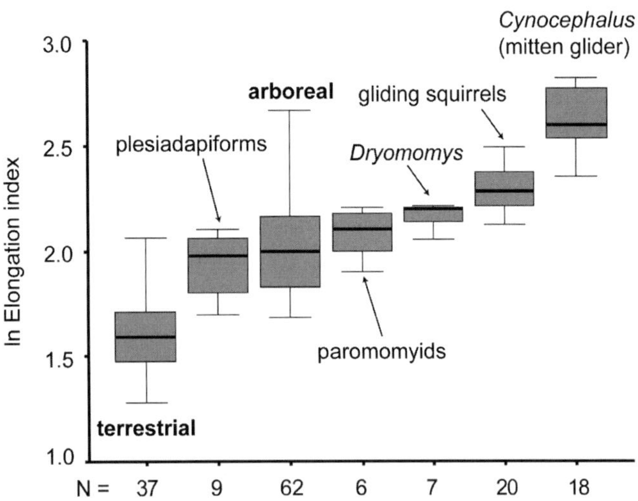

FIGURE 11.10. Elongation index of intermediate phalanx. Box plots of logged elongation indices for a sample of extant mammals and "plesiadapiforms". Boxes encompass 50% of data points with medians depicted as a horizontal line within them. Whiskers encompass all data. Numbers below boxes represent the sample size for data points comprising it. See Table 11.1 for a list of fossil specimens and measurements used in this analysis. See Table 11.3 for group means, t-values, and p-values of t-tests. See Appendix II, Table 1 for specimens included.

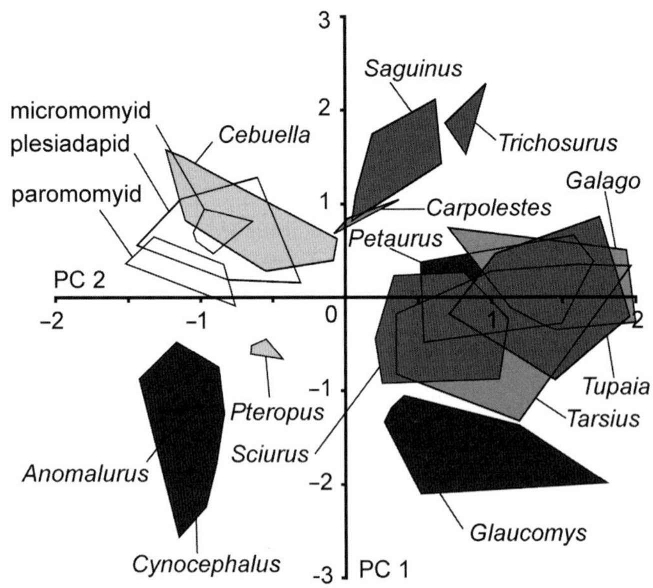

FIGURE 11.11. Principal Components (PC) Analysis (PCA) of manual intermediate phalangeal morphology. Data are derived from nine different measurements taken on 147 manual intermediate phalanges representing 18 genera and 22 species. Taxa are color-coded by functional group: darkest gray = gliders; lightest gray = vertical clinging arborealists; dark gray = clawed, pronograde quadrupeds; light gray = non-clawed, specialized grasping arborealists; and open = fossil. We controlled for body size by running the analysis on variables that were the logged ratio of the value of each raw measurement to the value of the geometric mean of all measurements on each specimen. Much of the comparative extant sample was provided by M. W. Hamrick and was originally used in Hamrick et al. (1999). Eight "plesiadapiform" species are plotted including the following: *Carpolestes simpsoni, Nannodectes intermedius, N. gidleyi, Plesiadapis cookei, Acidomomys hebeticus, Ignacius clarkforkensis*, and *Dryomomys szalayi*. For descriptions and illustrations of measurements taken see Hamrick et al. (1999). PC 1 represents 33.7% of the variance in the dataset, PC

FIGURE 11.11. (continued) 2 represents 27.1% and PC 3 represents 12.9%. PC 1 is most strongly correlated to increasing mediolateral breadth of the distal trochlea (articular surface), decreasing dorsopalmar depth of the proximal articular surface, and decreasing total length. Thus, narrow, deep, long phalanges have low PC 1 scores. PC 2 is most strongly correlated to increasing shaft and proximal end dimensions, decreasing breadth of dorsal margin of distal trochlea and decreasing length. Thus, phalanges that are relatively elongate with gracile shafts and ends have low PC 2 scores. See Appendix II, Table 1 for a list of specimens.

Cynocephalus to be significantly higher than the others (Table 11.3). Gliding squirrels were significantly higher than more generalized arborealists, which were higher than terrestrialists. All "plesiadapiforms," including paromomyids and micromomyids, were significantly higher than the terrestrial group, yet significantly lower than gliding squirrels and *Cynocephalus*. Paromomyids and the arboreal group were not distinguishable from one another or any of the other "plesiadapiform" groups. However, the micromomyid was found to be higher than non-paromomyid "plesiadapiforms".

These results strongly contradict either "mitten" or "mittenless" gliding in either paromomyids or micromomyids.

A principal components (PC) analysis (PCA) of manual intermediate phalanx shape based on data from Hamrick et al. (1999), but with an increased sample of extant mammals and fossil "plesiadapiforms", agrees with Hamrick et al.'s results in showing that *Cynocephalus* and *Anomalurus* are separated from most other extant mammals in the sample (including primates, treeshrews, rodents and marsupials) by having mediolaterally narrow shafts and trochleae (low PC 1 scores). However, restricting the analysis to manual elements *only* and adding callitrichine euprimates and other plesiadapiforms (including micromomyid UM 41870) shows the following: (1) dermopteran manual intermediate phalanges are unique among all extant and extinct taxa (except *Anomalurus*) in being extremely elongate (low PC 1 and PC 2 scores), and (2) the added taxa have manual intermediate phalanx shapes similar to those of paromomyids (UM 108210 and 108207) and dermopterans in being mediolaterally narrow and dorsoventrally deep (low PC 1 scores), but are separated from dermopterans in being relatively shorter (higher PC 2 scores) (Figure 11.11). Looking at this result from a functional perspective, we note that low PC 1 scores characterize extant taxa that use their digits for clinging and/or climbing on large diameter vertical supports (*Anomalurus*, the callitrichine euprimate (*Cebuella pygmaea*), *Pteropus* and *Cynocephalus*), while higher PC 1 scores characterize arboreal primates that predominately grasp small diameter supports, or terrestrial taxa that do not subject their phalanges to tensile forces, but load them in compression, instead. Based on these results, we agree with Hamrick et al. (1999), that narrow trochleae and shafts in intermediate phalanges are strongly linked to frequent exposure to tensile loads and sagittal bending moments associated with vertical clinging on large diameter supports and/or suspending with claws.

This comprehensive PCA of overall intermediate phalanx shape is not entirely sufficient to evaluate Beard's (1993b) suggestion that similarity in the shape of the intermediate phalanx proximal end, specifically, links paromomyids and dermopterans, or Hamrick et al.'s suggestion that this morphology, being additionally found in bats, is a character supporting Volitantia (now typically recognized as polyphyletic; e.g., Murphy et al., 2001b). Thus, we note that in making the foregoing suggestion, Beard did not mention the fact that he had illustrated another

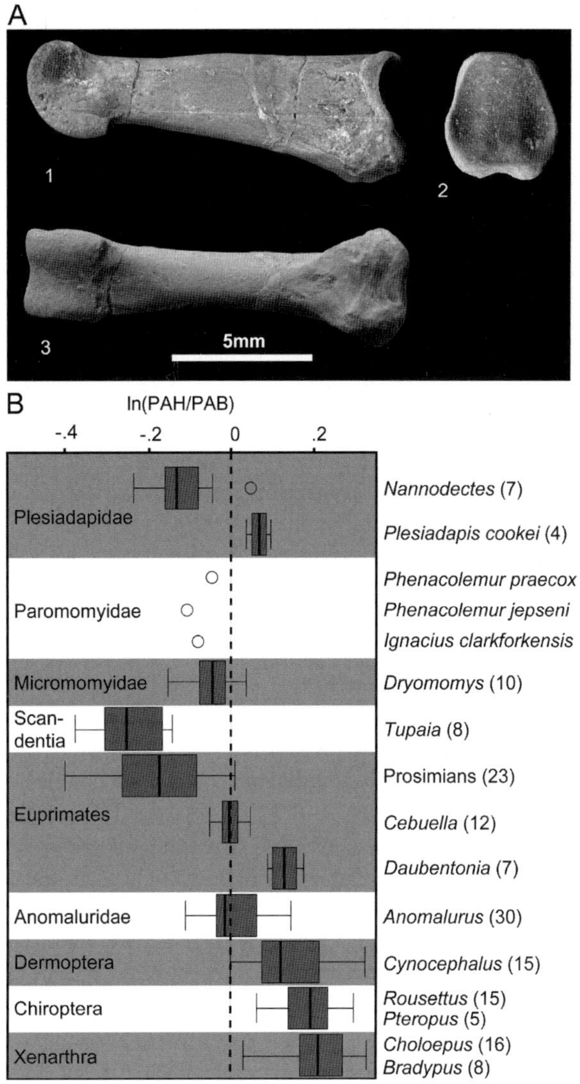

FIGURE 11.12. A, *Plesiadapis cookei* (UM 87990) intermediate phalanx. 1, lateral; 2, proximal; and 3, volar view. *P. cookei* has a proximal articular surface that is dorsoventrally taller than it is mediolaterally wide. The shaft is straight, or slightly dorsally convex (depending on how the central axis is defined) with no dorsal re-curvature at the distal end. Contrary to what this specimen shows, these features (and others) were said to be uniquely shared by paromomyids and dermopterans among "plesiadapiforms," primates, treeshrews and most other mammals (Beard, 1993b). Scale = 5 mm. B, Plot of natural log ratio of dorsoventral height of the proximal articular surface (PAH) to mediolateral breadth of surface (PAB) of intermediate phalanges. Alternating gray and white bars demarcate higher-level taxa. Contrary to previous claims paromomyids, dermopterans and bats are not uniquely characterized by a high value of this ratio (plotting right of the dashed line). *Plesiadapis* (as shown qualitatively in A), *Daubentonia*, and sloths plot in this realm, but paromomyids do not. Intermediate phalanges of *Acidomomys hebeticus* were not included here because none were preserved with their proximal epiphyses solidly attached. Sample sizes are given after taxon names in parentheses. See Appendix II, Table 1 for specimens included.

"plesiadapiform" *Plesiadapis tricuspidens* (MNHN R 5341: Godinot and Beard, 1991: Figures 11.1F, 11.2) such that it appeared to have this morphology. Given that another species of *Plesiadapis, P. cookei* (UM 87990) is also shown (Figure 11.12A) to have an intermediate phalanx with a dorsoventrally deep proximal end, it seems likely that at least this genus of "plesiadapiform" shares this feature with paromomyids and dermopterans. Importantly, the fact that more primitive plesiadapids lack this morphology (Figure 11.12B; Beard, 1993b), suggests that it is adaptive in *Plesiadapis* and does not reflect the probable close relationship between paromomyids and plesiadapids. *Plesiadapis* has never been seriously regarded as a glider by those who have studied it from a functional perspective (including Beard), throwing into further doubt the possibility that dorsoventral depth of the intermediate phalanx proximal end, has a special association with a lifestyle that includes mitten-gliding. Hamrick et al. (1999) never evaluated this feature from a univariate quantitative perspective. They appraised it qualitatively, as did Beard (1993b). However, we plot the ratio of the dorsoventral height of the proximal articular surface (PAH) to its mediolateral breadth (PAB) for a broader sample of mammals which shows (Figure 11.12B) that (1) dermopterans and paromomyids are not necessarily similar to one another, as paromomyids actually have a lower ratio (even those specimens available to Beard); and (2) the condition of having a dorsoventrally deep proximal end is distributed like other features linking together extant taxa that put tensile loads on their digits (i.e., it is present in bats, dermopterans, anomalurids, callitrichines, *Daubentonia* and sloths). Therefore, we interpret this feature to also reflect vertical clinging and/or suspending with claws. It may be that the extremely high ratio of bats, *Cynocephalus*, sloths and possibly even *Plesiadapis* can be specifically linked to use of suspensory behaviors.

Other paromomyids and *Dryomomys* have lower ratios suggesting against frequent quadrumanus suspension. Other features of the intermediate phalanges of paromomyids and micromomyids also suggest against this behavior: Compared to sloths and the pedal intermediate phalanges of bats, they have distal trochleae that face more distally (less ventrally). Compared to sloths, bats and *Cynocephalus* they have tubercles for annular ligaments that are more distinctly flaring, and more distally positioned (Figure 11.13). A ventrally facing trochlea appears to indicate a habitually ventriflexed distal interphalangeal (dip) joint, and reduction of tubercles for annular ligaments appears to reflect a habitually extended pip joint in the phalanges of bats (Simmons and Quinn, 1994) and sloths (Mendel, 1985). While dermopterans are noted for their inability to completely extend the pip joint (Pocock, 1926; Beard, 1993b), they are similar to other clawed-suspensory animals in also being incapable of tight flexion at these joints (Mendel, 1985; Simmons and Quinn, 1994). The inability for tight flexion as a reflection of suspensory behavior is a salient point. Micromomyids and paromomyids have a joint that indicates a capacity for relatively tight flexion in

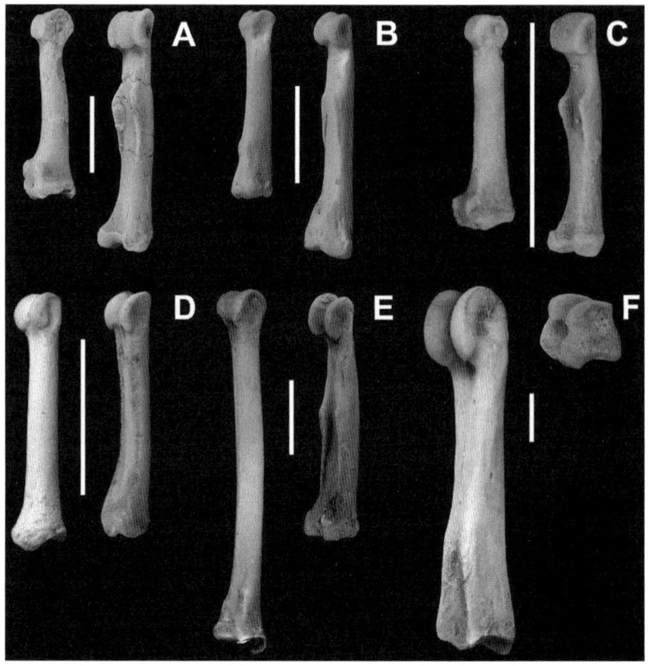

FIGURE 11.13. Intermediate and proximal phalanges of various taxa. A, *Ignacius clarkforkensis*; B, *Cebuella pygmaea*; C, *Dryomomys szalayi*; D, *Pteropus pumillio*; E, *Cynocephalus volans*; F, *Choloepus hoffmani*. Ventrolateral view of intermediate phalanx (on the left) and proximal phalanx (on the right). Manual elements are represented except in the case of *Pteropus*, for which toe bones are shown. Functionally, its toes are more comparable to the fingers of the other taxa shown, than are its own fingers, which are modified as wings. Phalanges are normalized to length of the proximal phalanx to show variation in intermediate phalanx length, except for F, which is normalized to the length of intermediate phalanx in E. Note that suspensory taxa D–F differ from fossils (A and C) and vertical clinger and climber (B) in having intermediate phalanges that are relatively longer, with more ventrally facing distal articular surfaces and less distinct flexor sheath tubercles. The proximal phalanges of the suspensory taxa have more deeply trochleated distal articular surfaces. See Appendix II, Table 1 for specimen numbers. Scale bars = 3 mm.

so much as they have morphology that is similar to callitrichines that evidently habitually tightly flex the pip joint of their digits because they frequently locomote on small diameter supports even though they forage predominately on large diameter vertical supports (Youlatos, 1999). This morphology is illustrated in Figure 11.13.

Finally, regression analysis using a reduced major axis (RMA) method (Figure 11.14) shows a pattern consistent with previous modes of analysis. "Plesiadapiforms," including paromomyids and micromomyids, have intermediate phalanges that are not significantly different from those of euprimates in their length (y-axis) to midshaft diameter (x-axis) proportions, whereas dermopterans and gliding squirrels differ in being more elongate. More specifically, gliders plot outside the confidence interval of a regression generated using non-gliding eutherian mammals. In contrast, all

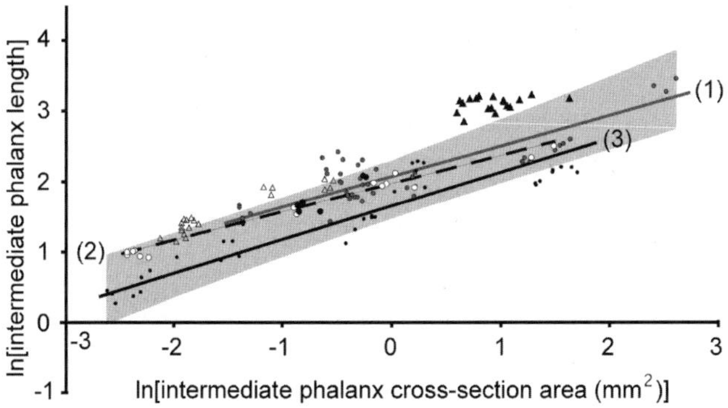

FIGURE 11.14. Intermediate phalanx elongation. Intermediate phalanx length vs. cross-section at midshaft – Solid lines are RMA regressions of the natural log of intermediate phalanx length on the natural log of the mid-shaft area for a sample of non-gliding primates [small gray circles (1) with gray line], "plesiadapiforms" [large white and black circles (2) with dashed line], and other extant non-gliding mammals [small black circles (3) with black line]. A composite "non-gliding mammals regression" is illustrated by a shaded gray area that encompasses the 95% confidence limits for the isometric relationship ($y = 0.54\,x + 1.88$, $R^2 = 0.83$). Gliders, including dermopterans (black, filled-in triangles) and flying squirrels (open triangles), have more elongate intermediate phalanges than non-gliders, generally falling outside of the 95% confidence limits. Among non-gliders, euprimates (1) and plesiadapiforms (2) have more elongate intermediate phalanges than do other non-gliding mammals (3). Note that paromomyids (large black circles) lack the elongation characteristic of mitten-gliding dermopterans and are in the range of other plesiadapiforms (2). Equations for the regressions are the following: (1) $y = 0.43\,x + 2.07$, $R^2 = 0.71$, n = 8 taxa; (2) $y = 0.40\,x + 1.94$, $R^2 = 0.98$, n = 6 taxa; and (3) $y = 0.48x + 1.645$, $R^2 = 0.94$, n = 10 taxa. See Table 11.1 for list of fossil specimens and measurements used in this analysis. See Appendix II, Table 1 for all specimens included.

plesiadapiforms plot within the confidence interval and have intermediate phalanges that fall on the same regression line as those of euprimates. Thus, the relatively high elongation indices of *Dryomomys* can be most easily explained as a consequence of extending a more general scaling relationship between intermediate phalanx length and cross-sectional area to a very small size.

Beard (1993b) stated that both *Cynocephalus* and paromomyids had straight intermediate phalanx shafts that lacked dorsal recurvature at the distal end. He associated these features with the presence of elongated volar pads in *Cynocephalus*. However, his own figures seem to contradict this observation. Beard (1993b: Figure 11.5) illustrates a phalanx attributed to *Ignacius graybullianus* that actually has slight dorsal recurvature, similar to the pronograde creatures with which he contrasted it. Furthermore, *Cynocephalus* actually possesses intermediate phalanx shafts that are dorsally convex, not perfectly straight (Figure 11.9 and 11.13E). The intermediate phalanx of *Ignacius* (UM 108210) has notable dorsal recurvature at the distal end of the shaft (Figure 11.13A). *Dryomomys* possesses straight-to-convex shafts, but this trait is not otherwise unique to *Cynocephalus*. Notably, it is also found in *P. tricuspidens* (Godinot and Beard, 1991), *P. cookei* (Figure 11.12A) and many euprimates (Stern et al., 1995).

As suggested by Beard (1989), and demonstrated by Hamrick et al. (1999), we also found that different gliders are not consistently distinguishable from non-gliders with regard to the morphology of the proximal phalanges (Table 11.5). For example, in gliding anomalurids (Hamrick et al., 1999) the proximal phalanges have shafts that are straight and relatively short, are deep dorsoventrally, and have flexor sheath ridges that are extended proximally and flare ventrally. On the other hand, the gliding squirrel *Glaucomys* has proximal phalanges that are slender, elongate and have dorsally curved shafts. *Cynocephalus* has proximal phalanx morphology extremely similar to that of anomalurids (Hamrick et al, 1999). This last fact had not yet been documented when Beard explained the "unusual" morphology of the proximal phalanges of *Cynocephalus* as relating to other idiosyncrasies of a mitten-glider in the following way:

> "because of elongation of its intermediate phalanges, the proximal phalanges are subjected to relatively higher bending moments in the anteroposterior plane during suspensory postures involving the hands than would otherwise be the case if the intermediate phalanges were more typical (i.e., shorter) in length." (Beard, 1989:p.453)

In *Ignacius* and the micromomyid, but not in *Acidomomys*, the proximal phalanges have a triangular cross-sectional shape and extended flexor sheath ridges somewhat similar to those in *Cynocephalus* (the difference between different paromomyids is very likely an ontogenetic artifact, as noted previously) (Figure 11.13). It seems likely that Beard's explanation for such morphology is correct with regard to the resistance of bending moments during postures in which the digits are loaded in tension (e.g., suspensory postures). However, such stresses appear to select for dermopteran-like proximal phalanges even if the intermediate phalanges are relatively short: Anomalurids, paromomyids and micromomyids lack intermediate phalanges that are "greatly elongated" relative to the proximal phalanges (Figure 11.8) but have proximal phalanx morphology similar to that of *Cynocephalus* in most respects. Furthermore, a number of euprimates have flexor sheath ridges similar to those of *Cynocephalus*, anomalurids, paromomyids, and micromomyids; including spider monkeys, gibbons, and some callitrichines, but (again) do not have elongated intermediate phalanges. Thus, the proximal phalanx morphology characterizing *Cynocephalus*, appears broadly associated with antipronograde behaviors (not just suspension) including clinging to large vertical supports with claws and/or with strong grasping as suggested by Hamrick et al. (1999).

Metapodials. Metacarpal proportions are distinctive for gliders that include the metacarpus in their patagium (Beard, 1993b) (Table 11.6). In *Cynocephalus volans* and *Petaurus breviceps*, metacarpal V is longer than metacarpal III and it is also longer than any metatarsal [in *C. volans* the ratio of metacarpal V to

TABLE 11.4. Comparative morphology of intermediate phalanges of fossil "plesiadapiforms" and extant arboreal mammals. The following explanation applies to this and (in most respects) the remaining tables: Columns represent different "positional behavior groups". Rows represent morphological features of the intermediate phalanges (or other skeletal elements). Gray shading in a box indicates that a feature is present in the corresponding behavior group. Letter codes are sometimes included in gray-shaded boxes to specify the functional significance of the feature. Functions include mitten-gliding (mg), suspension (s), gliding (g), vertical clinging (vc), grasping-clinging (g-c), and pronograde postures (p). Extant taxa used to represent behavioral groups in this table include: (1) *Cynocephalus volans* (mitten-glider); (2) *Choloepus hoffmanni* and *Pteropus pumilio* (non-gliding, clawed suspensory mammals); (3) *Cebuella pygmaea* (non-gliding vertical clinger and climber); (4) *Glaucomys volans* and *G. sabrinus* (rodent gliders); (5) *Petaurus breviceps* (marsupial glider); and (6) *Sciurus niger* and *S. carolinensis* (non-gliding clawed scansorialists). We tried to determine the functional significance of the coded features by (1) noting which behavioral groups exhibited a particular feature, (2) noting what functions were shared by those same behavioral groups, and (3) linking the shared features and shared functions. In some cases we were unable to identify a function distributed among behavior groups in the same way as a particular feature. We took this to mean that the feature had different functions in different behavioral groups. An example of how these tables can be used to determine a feature's functional significance is given in the methods section of the main text. Fossil taxa can be linked to extant behavioral groups by noting with which of those groups they share the most features. *Ignacius*, the paromomid, and the micromomyid have a suite of features most similar to non-gliding vertical clingers and climbers and non-gliding clawed scansorialists. They lack features that uniquely characterize mitten-gliders, as well as those that characterize mitten-gliders and quadrupedal suspensory taxa, together. They exhibit features lacking in rodent and marsupial gliders. Thus, the intermediate phalanges suggest these taxa were non-gliding vertical clingers and climbers in life. For this and the remaining tables, codings are based on observations of specimens listed in Appendix II, Table 1.

Intermediate Phalanges	Extant mammalian behavioral groups						Fossil plesiadapiforms	
	Mitten glider	Non-gliding clawed suspensory mammal	Non-gliding vertical clingers and climbers	Rodent gliders	Marsupial gliders	Non-gliding clawed scansorialist	micromomyid	paromomyid
Extremely gracile (high elongation index)	g			g				
Longer than proximal phalanges	s	s						
Distinct, distally positioned tubercles for annular ligaments			g-c			g-c	g-c	g-c
Deeply trochleated proximal articular surface	s	s						
Proximal articular surface relatively deep dorsopalmarly	vc	vc	vc				vc	vc
Ventrally facing distal articular surface		s						
Trochleated distal articular surfaces		s						
Dorsally recurved shaft at distal end					p	p		p

metacarpal III is 1.03 (Figure 11.15); that of metacarpal V to metatarsal III is 1.16 (Figure 11.16)]. In contrast to that of *C. volans*, metacarpal V of *Ignacius* and *Dryomomys* appears to have been shorter than metacarpal III-IV (Figure 11.15).

Furthermore, the metacarpals are roughly two-thirds the length of the second metatarsal (the only one for which the full length is preserved) in *Ignacius* (the ratio of metacarpal V to metatarsal II is 0.67 in UM 82606). The same is true for *Dryomomys* (the ratio of metacarpal III to metatarsal III is 0.73; Figure 11.16).

Not only are the metacarpals shorter than the metatarsals in the micromomyid, but the manual digits are shorter than the pedal digits (Figure 11.16). This also seems to have been true for *Acidomomys* UM 108207 (see above): phalanges attributed to its hands are shorter than those attributed to its feet (Table

TABLE 11.5. Comparative morphology of proximal phalanges. Non-gliding suspensory mammals include *Hylobates* and *Ateles*, as well as those listed for the suspensory group in Table 11.4. Functional categories include suspension (s), and vertical clinging (vc)

Proximal Phalanges	Extant mammalian behavioral groups						Fossil plesiadapiforms	
	Mitten glider	Non-gliding suspensory mammal	Non-gliding vertical clingers and climbers	Rodent gliders	Marsupial gliders	Non-gliding clawed scansorialist	micromomyid	paromomyid
Deeply trochleated distal articular surface	s	s						
Extensive, ventrally projecting flexor sheath ridges	vc	vc	vc				vc	vc
Dorsally convex shaft								

TABLE 11.6. Comparative morphology of metapodials. Functional categories include gliding (g). The distribution of "**Metacarpals longer than Metatarsals**" has an ambiguous functional significance. For gliders that include the metacarpus in the patagium (*Cynocephalus* and *Petaurus*), it likely serves to increase patagial area.

Metapodials	Extant mammalian behavioral groups						Fossil plesiadapiforms	
	Mitten glider	Non-gliding clawed suspensory mammal	Non-gliding vertical clingers and climbers	Rodent gliders	Marsupial gliders	Non-gliding clawed scansorialist	micromomyid	paromomyid
Metacarpal V ≥ length of Metacarpal IV	g				g			
Metacarpals longer than Metatarsals								

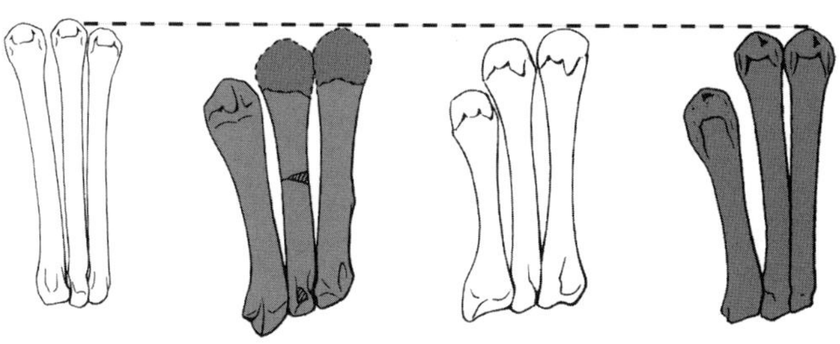

FIGURE 11.15. Comparison of relative lengths of metacarpals. Right metacarpals V-III (left to right) in palmar view. *Cynocephalus volans* (USNM 56530) *Ignacius clarkforkensis* (metacarpal V – UM 82606; metacarpal IV-III – UM 108210); *Cebuella pygmaea* (UM 160146); *Dryomomys szalayi* (UM 41870). Elements are standardized to the length of the third metacarpal. Note that *Cynocephalus volans* is unique in having a fourth and fifth metacarpal that extend distally beyond the third metacarpal.

11.1). This is an important point because the gliding hypothesis predicts an elongate manus relative to the pes (Beard, 1993b).

Finally, even though the manual proximal phalanges are shorter than the pedal elements in paromomyids and micromomyids, they are long relative to the metacarpals (Figure 11.9). This likely indicates effective grasping, as in euprimates (e.g., *Tarsius*) and some marsupials (e.g., *Caluromys*) (Lemelin, 1999; Hamrick, 2001; Bloch and Boyer, 2002).

Forelimb. The ability of a mammal to glide and/or use suspensory postures appears to be reflected in the relative lengths and morphology of its limb bones (Thorington and Heaney, 1981; Runestad and Ruff, 1995; Thorington et al., 2005). *Ignacius clarkforkensis* UM 108210 and *Dryomomys szalayi* UM 41870 preserve the first known articulated to semi-articulated forelimbs for their respective families (Figure 11.17, Figure 11.7c), allowing us to assess positional behavior by looking at relative lengths of elements for the first time.

With regard to the radius, Thorington and Heaney (1981) found that all gliding squirrels, regardless of body size, exhibit elongation of this element relative to the humerus. Furthermore, Runestad and Ruff (1995) examined the scaling relationship between radius length and humerus length in non-gliding and gliding mammals. They were able to generalize Thorington and Heaney's conclusion, by finding that gliders with patagia terminating on the wrist or digits (including *Cynocephalus*) have significantly longer radii relative to their humeri (higher brachial index), than non-gliders (Figure 11.18). We further note that suspensory taxa are also characterized by a high brachial index (Godfrey, 1988). Thus, whether a high brachial index is reflective of gliding or suspension in *Cynocephalus volans* (brachial index = 116) it should be present in other mitten-gliders as well. In this context, it is important to note that while a high brachial index is probably required for animals that exhibit these behaviors, it often present without gliding and suspensory behaviors in other extant euarchontan mammals. Specifically, the arboreal treeshrew *Ptilocercus lowii* and the vertically clinging and leaping euprimate *Tarsius* have elongate radii, with brachial indices of 107 and 127, respectively.

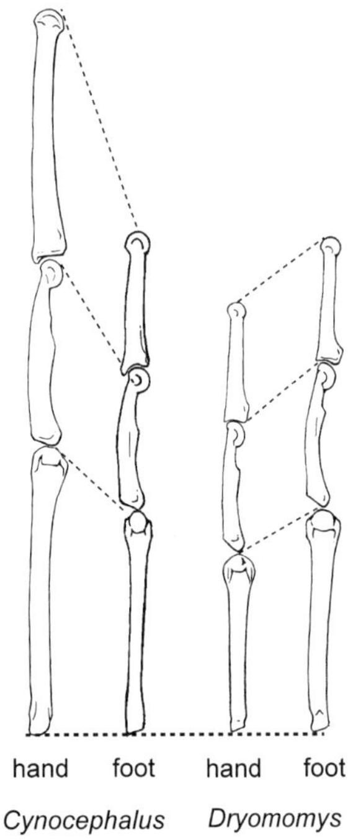

FIGURE 11.16. Comparison of relative lengths of metacarpals and metatarsals of a dermopteran and a micromomyid. Third digit rays of hands and feet of *Cynocephalus* (USNM 56530 – left) and *Dryomomys szalayi* (UM 41870 – right). Digit rays are standardized to the lengths of the metatarsals. Note that manual elements are longer than pedal elements in *Cynocephalus*, while the reverse is true of the micromomyid.

While distal radius morphology is not generally distinctive for gliders as a group, *C. volans*, sloths (e.g., *Choloepus*), and gibbons share a number of features in this region, which appear to reflect their use of suspensory behaviors. Specifically, the carpal articular surface of the radius is deeply cupped, faces palmarly and ulnarly, and is marked by a prominent ridge on its dorsal margin (Figure 11.19). Again, however, these traits are also present in *Ptilocercus* (Figure 11.20A), which is not committed to using suspensory postures, and is probably best characterized more generally, as a committed arborealist (e.g., Sargis, 2001a).

Morphological traits of the forearm that may relate to gliding in squirrels (Sciuridae: Pteromyini) include an ulna that has a relatively short olecranon process (Thorington et al., 2005); a deep trochlear notch (Figure 11.18); and a shaft that

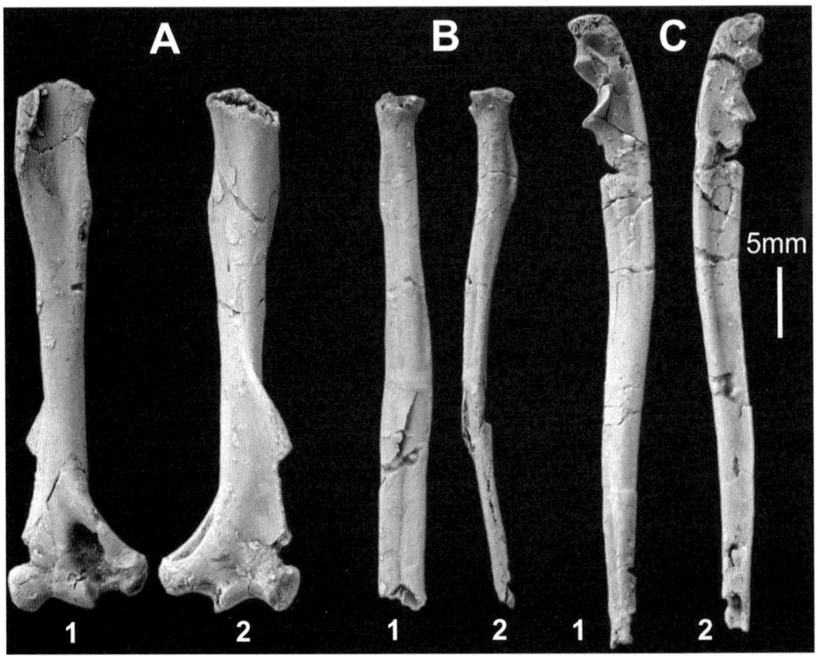

FIGURE 11.17. Forelimb elements of *Ignacius* (UM 108210). A, right humerus in (1) anterior and (2) posterior views. B, right radius in (1) posterior and (2) lateral views. C, right ulna in (1) medial, and (2) lateral views. Note that the proximal-most part of humerus is not preserved, nor are the distal tips of the radius and ulna. Scale bar = 5 mm.

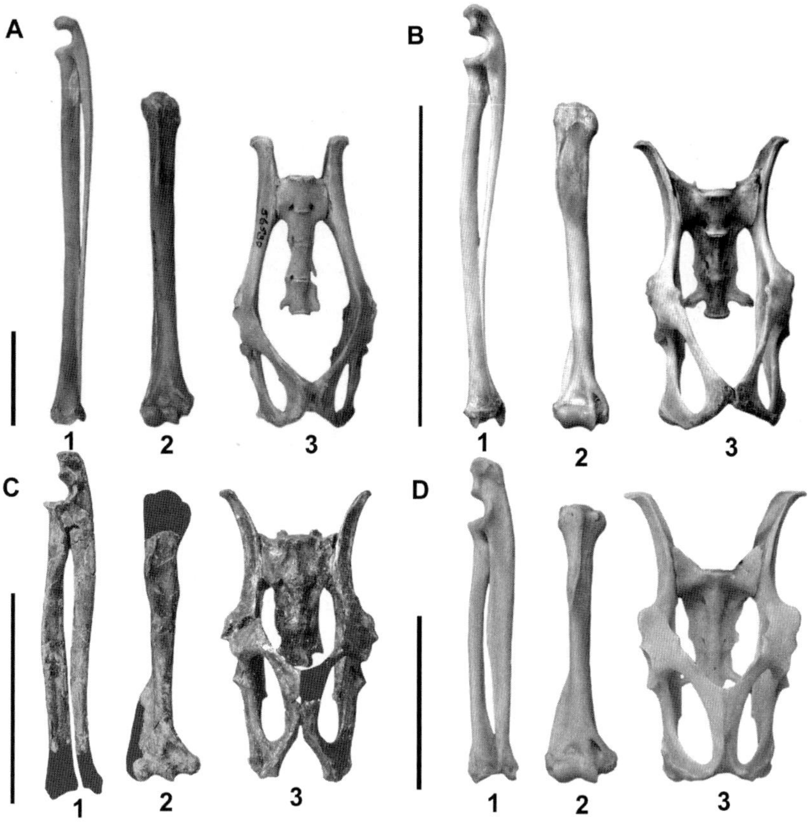

FIGURE 11.18. Proportions of select elements among different gliding and non-gliding taxa. For each lettered specimen, there are three numbered elements, or sets of elements: (1) right ulna and radius in medial view, (2) right humerus in anterior view, (3) articulated sacrum and innominates in ventral view. Elements are standardized to innominate length. A, *Cynocephalus volans* (USNM 56530); B, *Glaucomys sp*. (UMMZ 168356 – flying squirrel); C, *Ignacius clarkforkensis* (UM 108210 and UM 82606); D, *Sciurus niger* (UMMZ 3959 – gray squirrel). Gliding features present in A and B were most likely independently-evolved. Some of these features include a deep trochlear notch on the ulna; an ulnar shaft that is distally reduced and fused to radius; a radius that is substantially longer than the humerus; a humerus that has a proximally-restricted, anteriorly-oriented deltopectoral crest, a mediolaterally narrow distal end, a large capitular area on its distal articular surface, and a total length greater than that of the innominate; an innominate with narrow iliac crests, a long ilium relative to ischium, and a craniocaudally short caudally situated pubic symphysis. *Ignacius* (C) shares more features in common with *Sciurus* (D) than it does with either of the two gliding taxa. Scale bar = 3 cm.

is distally reduced (or synostosed to the radius), lacks ridges for the pronator quadratus muscle, and lacks a longitudinal groove along its lateral surface for the extensor carpi ulnaris muscle (Thorington et al., 2005). Additionally, gliding squirrels are said to have a distal radius with a large tubercle separating the first and second extensor compartments of the wrist, and a shaft with a more circular cross-section than non-gliders (Thorington et al., 2005). Like gliders, suspensory taxa are also typically characterized by a short olecranon process (Godfrey, 1988). Furthermore, Mendel (1979) related a reduced distal ulna in sloths (*Choloepus*) and gibbons to suspensory behaviors. We again note that many of these features are found in *Cynocephalus* (Figures 11.18, 11.19, Table 11.7), and should be present in a mitten-glider whether they reflect gliding or suspensory behaviors.

With regard to the arm, or humerus specifically, gliders have been shown to have a shaft that is relatively gracile (Runestad and Ruff, 1995), a distal end that is mediolaterally narrow (Thorington et al., 2005), and a deltopectoral crest (DPC) that is proximally restricted (Runestad and Ruff, 1995; Thorington et al., 2005) and anteriorly (versus laterally) oriented (Thorington et al., 2005). The first three features probably reflect elongation of the shaft relative to body mass, which functions to increase the surface area of the patagium (Beard, 1993b; Runestad and Ruff, 1995). The anterior orientation of the DPC in suspensory and gliding taxa may allow the arms to be abducted and flexed to a greater degree, while still maintaining mechanical efficiency of the attaching muscles.

In summary, there are many forelimb traits that characterize *C. volans*, suspensory taxa, and eutherian gliders, reflecting the functional demands they have in common. Thus the fossil forms considered here should also exhibit such features if we are to entertain a gliding hypothesis for them. However, in most cases these traits are not *exclusive* to gliders and their presence would not be sufficient to infer

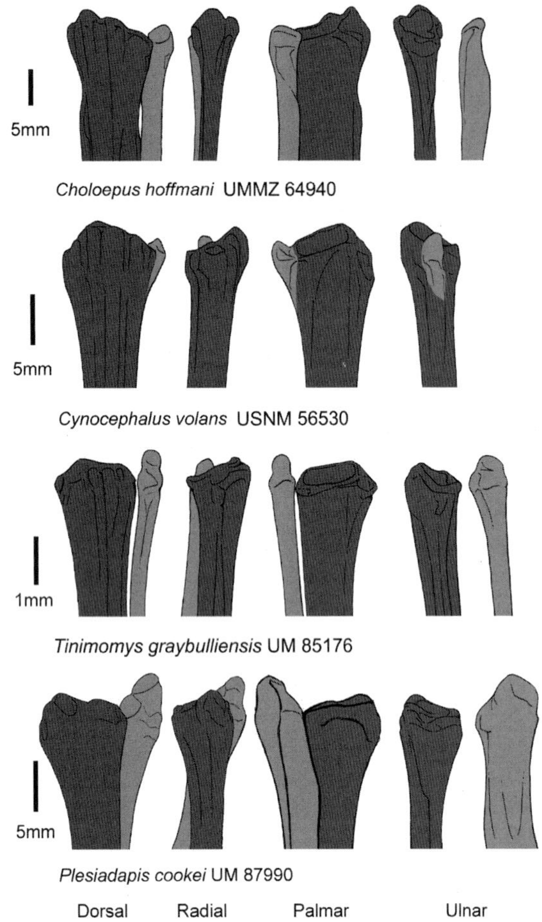

FIGURE 11.19. Right distal radii (dark gray shading) and ulnae (light gray shading) of select taxa (see labels in figure). *Choloepus* and *Cynocephalus* exhibit suspensory features shared by the micromomyid, but not by the larger plesiadapid. Note that *Plesiadapis* differs from the other three taxa in having a distal radius articular surface that is not as concave and is more distally oriented (compare elements in "palmar" and "ulnar" views), and an ulna that is relatively much larger. Scales on *Choloepus, Cynocephalus*, and *Plesiadapis* = 5 mm. Scale on *Tinimomys* = 1 mm.

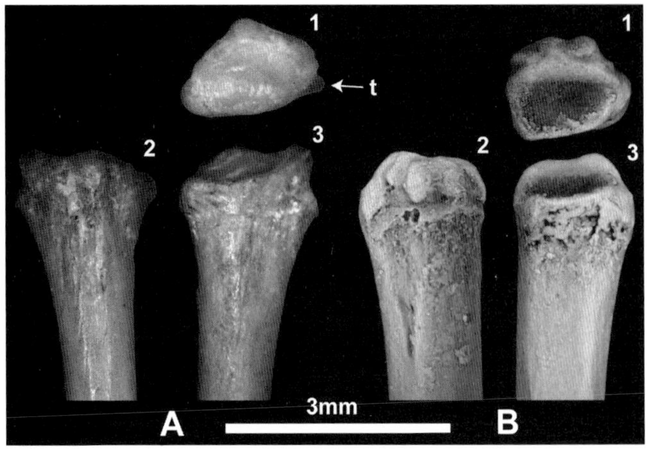

FIGURE 11.20. Comparison of distal radii of *Ptilocercus lowii* and micromomyid. (1) Distal, (2) ventral, and (3) dorsal views of the radii of A, primitive euarchontan mammal, *Ptilocercus lowii* (YPM 10179), and (B) a cf. *Tinimomys graybulliensis* (un-numbered, semi-articulated specimen from SC-26). Note that they are similar in having a ventrally oriented distal articular surface for the proximal carpal row that is also deeply cupped and marked by a prominent ridge projecting from the extensor surface. Note also in A1, a prominent tubercle (t) in *P. lowii*. This may be homologous to that discussed by Thorington et al. (2005) for sciurids, which separates the 1st from the 2nd extensor compartments. If so prominence of this feature is clearly not an indicator of gliding in euarchontans, as it is in squirrels. Scale = 3 mm.

gliding in a fossil. Given the similarities of *Ptilocercus* and *Tarsius* to sloths, gibbons and *Cynocephalus*, such a finding *could*, however, be confidently used to infer anti-pronograde behaviors in a fossil.

We began our evaluation of the paromomyid and micromomyid forelimb material by estimating the lengths of the forelimb bones of *Ignacius* (Figures 11.17 and 11.18). We were able to make well-constrained estimates of total shaft length for the humerus of *Ignacius* based on comparison to isolated elements of other paromomyids (Beard, 1989). Specifically, proportional scaling based on a humerus attributed to *Phenacolemur simonsi* USNM 442260 and one attributed to *Ignacius graybullianus* USNM 442259 allowed estimation of humerus length in UM 108210 (~42.02 mm). Radius length regressed on radial head area for a sample (n = 39) of extant archontans and rodents (gliding taxa included) yielded a high correlation ($r^2 = 0.953$) and allowed an estimate of the total shaft length of the radius of *Ignacius* UM 108210 (42.2 ± 1.4 mm). Reconstructions based on these estimates are shown in Figure 11.18. Our results indicate that the radius length of *Ignacius* was between 0.97 and 1.04 times the humerus length. This is identical to the proportion seen in the large-bodied, late-Paleocene plesiadapiform, *Plesiadapis cookei* (Brachial index = 101). On the other hand, the brachial index of the micromomyid *Dryomomys* is quite high (117).

Although the distal radius is not preserved for new specimens of paromomyids, one has been described previously for Eocene *Phenacolemur simonsi* (USNM 442262 – see Beard, 1989). It is unlike that of *Cynocephalus*, suspensory taxa or *Ptilocercus*, but is fairly similar to those of plesiadapids such as *P. cookei* in having a shallower, distally oriented, distal articular surface. On the other hand, *Dryomomys* is most similar to *Cynocephalus* and *Ptilocercus* in having a deeply cupped distal articular surface that faces palmarly and ulnarly, and has a prominent ridge on its dorsal margin (Figures 11.19, 11.20A; Table 11.7).

With regard to other forearm features, *Ignacius* and *Dryomomys* differ from gliding squirrels and suspensory taxa in having an ulna with a relatively long anteriorly inflected olecranon process, which would seem to limit full-extension of the forearm (Figures 11.17 and 11.18). They differ from gliders specifically, in having a shallow trochlear notch. *Ignacius* differs further from gliders in the morphology of its ulna by having a ridge for the pronator quadratus muscle and a longitudinal groove along its lateral surface for the extensor carpi ulnaris muscle (Figures 11.17 and 11.18). Unlike both suspensory taxa and gliders, its distal ulna is fairly robust. (Figures 11.17, 11.18, Table 11.7). On the other hand, *Dryomomys* and *Tinimomys* are characterized by a reduced distal ulna that lacks ridges for pronator quadratus (see Figures 11.6C and 11.19). In the arm, both *Ignacius* and known micromomyids are unlike gliders in having a laterally-oriented deltopectoral crest (e.g., Beard, 1993a; Sargis, 2002c)

TABLE 11.7. Comparative morphology of forelimb. Functional categories include antipronograde (ap) suspension (s), gliding (g), forearm extension (fe) and hand ventriflexion (hv). The forelimb shows a number of features that appear to be related to gliding, one of which is shared by mitten-gliders, marsupial gliders and rodent gliders. The distribution of the presence of a deep trochlear notch on the ulna suggests a function held in common between gliders and "non-clawed suspensory taxa." This similarity is probably related to the common need of both groups to have full arm extension (ae) at the elbow joint. The fact that clawed suspensory taxa tend to lack this feature suggests that in the mitten-glider, *Cynocephalus volans*, which has claws, the trochlear notch depth reflects its gliding behaviors. The forelimb of *C. volans* exhibits a number of features only explainable by its use of suspensory behaviors, given that it shares these features with suspensory taxa only. A deeply cupped radius with a dorsal ridge seems to be related to a habitually ventriflexed hand as used in suspensory postures (Figure 11.19). In the case of the arboreal treeshrew (*Ptilocercus*), it may reflect the use of under-branch clinging (Figure 11.20). The last feature, "long forearm relative to arm," is only lacking in the clawed vertical clinger and climber, *Cebuella pygmaea*, among extant taxa in this table. This is probably due to the fact that *C. pygmaea* has an arboreal quadrupedal component to its locomotor repertoire (Youlatos, 1999; see Jouffroy et al., 1973, for a study on the functional significance of the distribution of euprimate forelimb proportions).

Forelimb	Extant mammalian behavioral groups							Fossil plesiadapiforms	
	Mitten glider	Rodent gliders	Non-gliding suspensory mammal	Non-gliding clawed suspensory mammal	Marsupial gliders	Clawed committed arborealist	Non-gliding clawed vertical clingers and climbers	micromomyid	paromomyid
Ulna distally reduced or synostosed	ap	ap	ap	ap		ap		ap	
Narrow distal humerus	g	g							
Radial shaft narrow and sub-isometric in x-sec dimensions	g	g							
Proximally restricted deltopectoral crest on humerus	g	g			g				
Anteriorly projecting deltopectoral crest	g	g							
Deep trochlear notch of ulna	fe	fe	fe						
Short olecranon process of ulna	fe	fe	fe	fe					
Base of capitulum most inferior surface on humerus	s			s					
Deeply cupped, ventrolateral facing articular surface of radius	hv		hv	hv		hv		hv	
Antebrachium much longer than humerus	ap	ap	ap	ap	ap	ap		ap	

that extends almost as far distally as it does in non-gliding sciurids, and a mediolaterally broad distal end (Figure 11.18). Runestad and Ruff (1995) previously showed paromomyids to differ further from gliders in having humeri with relatively robust shafts, using a regression analysis where the dependent variable was humerus length and the independent variable was humeral TA (cross-sectional area at midshaft, calculated using the formula for the area of an ellipse). The humerus of UM 108210 has a cross-sectional area and reconstructed length (TA = 11.03 mm^2, L ~ 42.02 mm) that put it closer to the regression line describing the scaling of these dimensions in non-gliding mammals (Runestad and Ruff, 1995: Figure 11.3a) than that describing gliders; however, this result is clearly not independent of Runestad and Ruff's because the humeri they analyzed were the same as those used to generate a length estimate for UM 108210 (see above). Micromomyids are outside the range of the data used to generate Runestad and Ruff's regressions. We note, however, that the dimensions of the humerus of UM 41870 (TA = 1.34 mm^2, L = 14.58 mm), put it below both glider and non-glider lines of Runestad and Ruff (1995), indicating that it is fairly robust, unlike the humeri of extant gliding mammals.

In contrast to the features of suspensory taxa and gliders, the majority of traits of both paromomyids and micromomyids suggest habitually flexed forearms and adducted arms, characteristic of taxa that locomote using pronograde and/or orthograde postures (Bloch and Boyer, 2007). Additionally, a spherical capitulum and mediolaterally broad distal humerus in *Ignacius* and micromomyids (Beard, 1989) indicate axial mobility of the forearm, similar to that of arboreal euprimates that frequently incorporate manual grasping into locomotor and foraging activities. Features shared by micromomyids, *Cynocephalus*, gliders, suspensory taxa and *Ptilocercus* to the exclusion of *Ignacius*, may represent retentions from an ancestor shared by the micromomyid, *Cynocephalus*, and *Ptilocercus*, and/or frequent use of under-branch clinging and anti-pronograde behaviors by micromomyids (Bloch et al., 2003; Bloch and Boyer, 2007).

Axial Skeleton. The fact that different gliders have different means of locomotion when they are not gliding [e.g., *Cynocephalus volans* is suspensory, whereas *Glaucomys* locomotes above branches using an asymmetrical bounding gait (Thorington and Heaney, 1981)] results in vertebral columns that lack gliding-specific characteristics with regard to morphology or intrinsic proportions. However, all

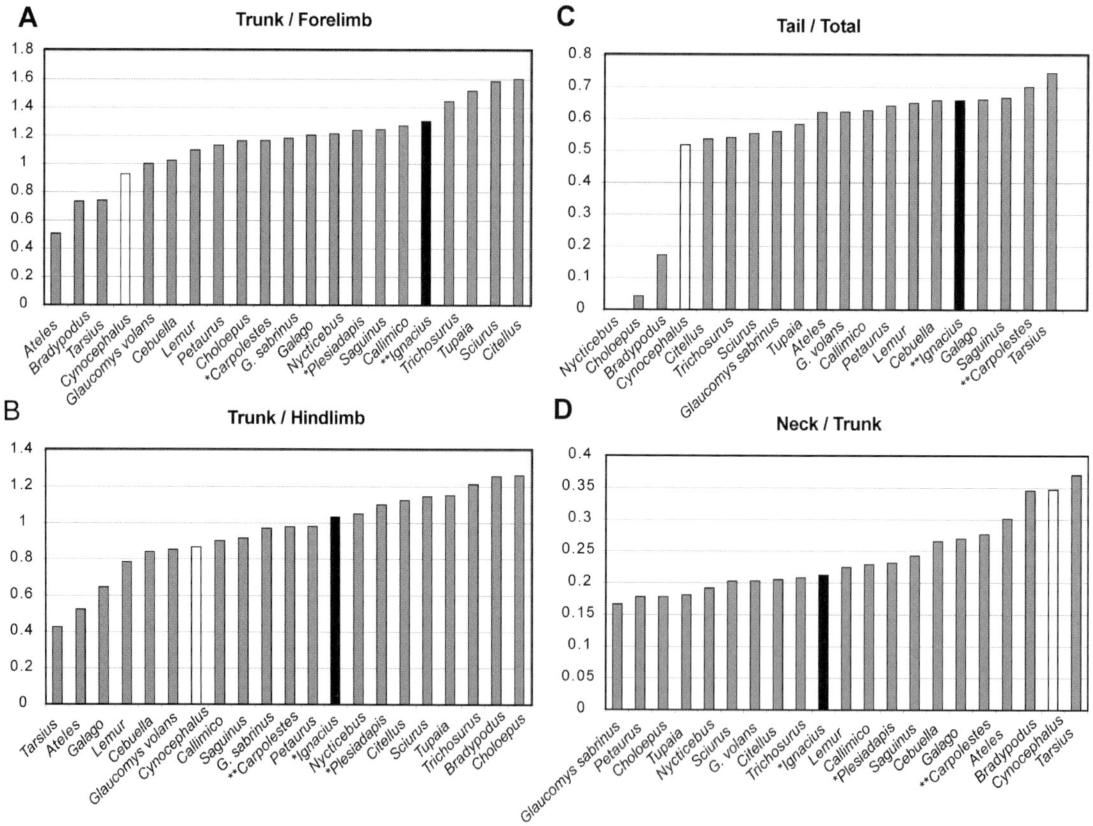

FIGURE 11.21. Selected trunk indices comparing *Ignacius* and *Cynocephalus*. Vertebral column and limb indices for a subset of the comparative sample of extant taxa used in this study are presented with individuals placed in order of increasing index to aid in identification of functional trends. *Ignacius clarkforkensis* (UM 108210 and UM 82606) is represented by a black bar in each plot, while *Cynocephalus volans* is represented by an unfilled bar. *Ignacius clarkforkensis* has indices that, in general, are separated from those of gliders and suspensory taxa by intermediate index values of agile arborealists and scansorialists. One asterisk by a taxon name indicates one parameter in the index has been estimated. Two asterisks by a taxon name indicate that both parameters were estimated. See Appendix II, Table 1 for specimens included.

extant gliding mammals have a short trunk (thoracic, lumbar and sacral regions) relative to the length of their limbs (Figures 11.21A, B), which actually reflects elongated limbs and is probably related to increasing the surface area for the patagium as mentioned previously (Thorington and Heaney, 1981; Beard, 1993b). Furthermore, gliding squirrels have been shown to possess relatively shorter tails than non-gliders (Thorington and Heaney, 1981). *C. volans* was also noted as having a short tail by Pocock (1926). Our data show its tail to be similar to that of non-primates in relative length, whereas quadrupedal arboreal euprimates (not including slow climbing lorisines) typically have longer tails (Figure 11.21C; see also Stafford, 1999). The short tail of *C. volans* appears to be a result of a reduced number (~20) of caudal vertebrae that are relatively small.

New specimens of *Ignacius* preserve many vertebrae, which allow estimates of neck, trunk, and tail length for the first time (Figures 11.21–11.23). In UM 108210, parts of six of its seven cervical vertebrae are preserved. For our calculations, we assumed it had 12 thoracic vertebrae (four preserved in UM 108210), and 7 lumbars (three preserved in UM 82606). It has three sacral vertebrae. We estimate there were

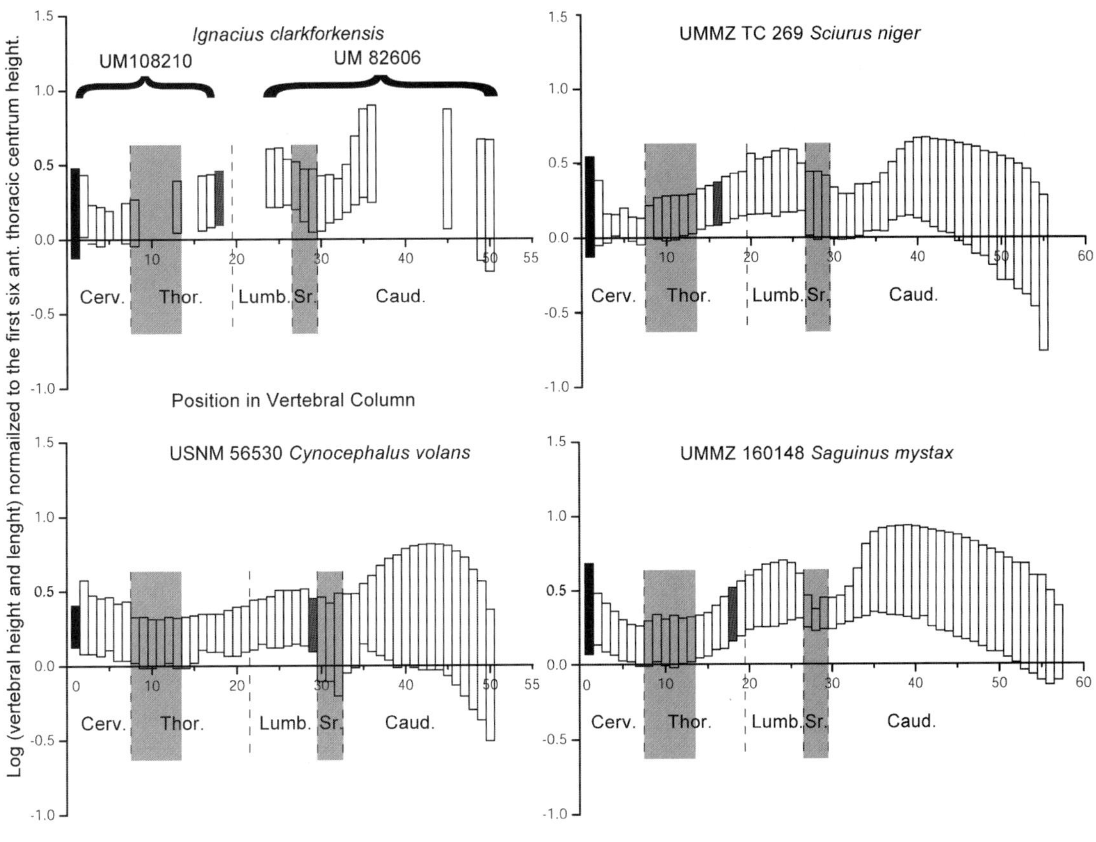

FIGURE 11.22. Comparison of vertebral profiles of *Ignacius* and *Cynocephalus*. Comparison of vertebral proportions following the method of Gingerich (1998). Y-axis depicts the logged value of two separate measurements, (1) vertebral body height and (2) vertebral body length: both are normalized to the average height of the first six anterior thoracic centra. Thus, if the boundary of a bar (vertebral body measurement) is positive, it is greater than the average height of the first six thoracics, whereas if it is negative it is lower. In white bars, the upper boundary represents the length (craniocaudal) of the body, while the base of the bar represents the height (dorsoventral). For black bars, the reverse is true (note that in all taxa depicted, it is only the atlas that has such proportions). Thus, the shorter the bar, the closer the vertebral body is to being square in lateral view. The gray areas depict those vertebrae most closely associated with pectoral (anterior thoracics) and pelvic (sacral) girdles. The gray bar represents the anticlinal vertebra, the boundary between vertebrae with caudally projecting spinous processes (vertebral positions to the left of the gray bar) and cranially projecting ones (vertebral positions to the right of the gray bar). In *Ignacius clarkforkensis* (UM 108210 and UM 82606) the neck is short, trunk vertebrae increase in height and length posteriorly, the sacrum is robust and the tail is long and robust. Such features suggest a relatively posteriorly shifted center of mass of the axial skeleton. Arborealists, *Sciurus* and *Saguinus*, depicted on the right have proportions similar to each other and *Ignacius*. *Cynocephalus volans*, on the other hand, exhibits a different pattern of vertebral proportions. In *C. volans*, the neck is long, trunk vertebrae remain roughly constant in size throughout the column, the sacrum is gracile, and the tail is shorter and more slender. Such features suggest a more anteriorly positioned center of gravity. We interpret proportional features in the paromomyid to be reflective of hind limb dominance in forward locomotion, while those of *C. volans* reflect a need for maneuverability while gliding, and equal emphasis on the fore and hindlimbs in suspensory locomotion. See Appendix II, Table 2 for measurements used to construct these plots.

TABLE 11.8. Comparative morphology of vertebral column. Functional categories include suspension (s), gliding (g), and bound galloping (bg). Note that long limbs relative to the trunk is a feature shared among all types of gliders. This feature may function to increase the area of the patagium in these taxa relative to their body mass. Features shared by *Cynocephalus volans* and suspensory taxa reflect vertebral columns that ventriflex with moderate, equivalent angular deviations among all intervertebral joints, producing a symmetrical arch. Taxa with such a configuration to the spine do not use pronograde bound-galloping behaviors in which extensive, powerful flexion and extension of the column is required (Slijper, 1946; Gambaryan, 1974). In bound-gallopers the back is rigid between most sets of vertebrae and there are just a few positions [e.g., T11–T12 and sacrolumbar joint in a treeshrew (see Jenkins, 1974)] where most of the flexion occurs.

Vertebral Column	Extant mammalian behavioral groups						Fossil plesiadapiforms	
	Mitten glider	Non-gliding clawed suspensory mammal	Non-gliding vertical clingers and climbers	Rodent gliders	Marsupial gliders	Non-gliding clawed scansorialist	micromomyid	paromomyid
Short trunk relative to limbs	g			g	g		?	
Long neck relative to trunk	s	s					?	
Caudally oriented lumbar spinous processes	s	s						
Craniocaudally deep lumbar spinous processes	s	s						
Short, laterally projecting transverse processes	s	s						
Four or more sacral vertebrae	s	s						
Ribs craniocaudally broad	s	s						
Longest sacral spinous process is 2nd or 3rd sacral vertebra						bg	?	bg
Short tail relative to trunk and neck								

~26 caudal vertebrae (10 preserved in UM 82606), which is the median from counts in 14 other arboreal and scansorial mammals from the UMMZ collection (including primates, rodents, treeshrews and marsupials) that have caudal vertebrae morphologies similar to those preserved in *Ignacius*. The length of the neck was obtained by adding the mean length of the third through sixth cervical bodies (representing the probable length of the seventh cervical body) to the sum of the lengths of the first through 6th cervical bodies. We obtained estimates of the total length of the trunk by estimating lengths of missing vertebrae. This was done by extrapolating trends of change in length along the column from preserved vertebral bodies into regions of the column for which vertebrae were not preserved (see Figure 11.22 for a graphic representation of the dimensions of vertebrae preserved for *Ignacius*). The same method was used to estimate tail length. The estimate of tail length is poorly constrained due to the greater variability in vertebral number in this region among mammals (e.g., Shapiro, 1993). Even though these methods are far more likely to underestimate total trunk length than to overestimate it, our calculations indicate that *Ignacius* had a longer trunk relative to its limbs than any mammalian glider in our sample (Figure 11.21A, B). Shorter limbs in *Ignacius* make it more similar to scansorial mammals such as tree squirrels and tupaiid treeshrews. Furthermore, the estimated tail length in *Ignacius* is relatively greater than that for gliders including *C. volans*, but is in the range exhibited by euprimates (Figure 11.21C).

In contrast to the vertebral columns of gliders, the trunks of suspensory taxa are not distinctive in their length relative to that of their limbs. They are, however, distinctive in morphological and proportional features (Table 11.8), differing from agile arboreal primates and scansorial rodents in having a thoracic region with an increased number of vertebrae and a lumbar region with fewer elements comprising it (Sargis, 2001a; Shapiro and Simons, 2002). Furthermore, their vertebral centra are short and roughly the same size throughout the trunk (as illustrated by *C. volans*, Figure 11.22). In the lumbar region, the transverse processes are reduced and oriented laterally, the spinous processes are wide craniocaudally and oriented caudally, the vertebral body articulations are oriented perpendicular to the body's long axis, and the zygapophyseal articulations are oriented nearly perpendicular to the sagittal plane or are revolute (Shapiro, 1993; Sargis, 2001a). In some cases, the sacrum is shallow and elongate, and includes extra caudal vertebrae, beyond the standard count of three seen in many agile arborealists and scansorialists. Finally, the tail tends to be reduced as dramatically illustrated by sloths, lorises and hominoids.

C. volans exhibits characteristic features of both gliders and suspensory taxa, while also exhibiting some features that do not typically characterize either of these groups. These unique features include a craniocaudally wide atlas (Sargis, 2001a) and a long neck relative to the trunk [Figures 11.21D and 11.22; (Pocock, 1926)]. In contrast, the morphology and proportions of the vertebrae of *Ignacius* suggest agility, as well as an emphasis on the hindlimb in forward propulsion. *Ignacius* is comparable to primates and squirrels in having a narrow atlas and a short neck relative to the trunk (Figures 11.21 and 11.22).

The lumbar vertebrae of *Ignacius* differ from those of *C. volans* in having narrow, cranially angled spinous processes (Figures 11.23 and 11.24). Furthermore, long, cranioventrally oriented lumbar transverse processes that extend below the level of the centrum in *Ignacius* provide a dorsoventrally deep trough for the erector spinae muscles

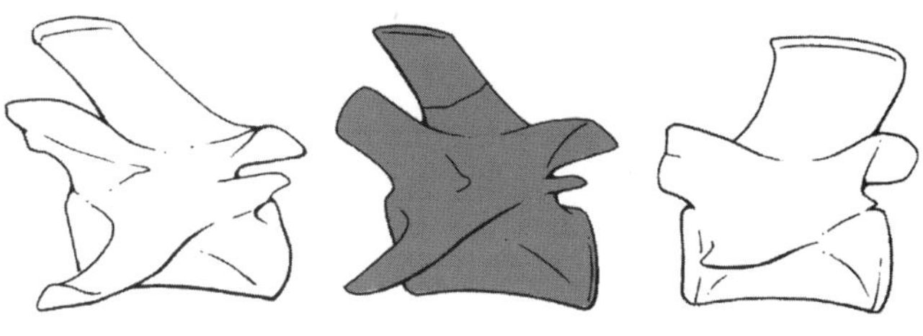

FIGURE 11.23. Comparison of antepenultimate lumbar vertebra of *Ignacius* and *Cynocephalus*. Antepenultimate lumbar vertebrae in lateral view of A, *Saguinus mystax* (UMMZ 160148), B, *Ignacius clarkforkensis* UM 82606 and C, *Cynocephalus volans* (USNM 56530). A and B have longer, more ventrally canted transverse processes, more cranially extended zygapophyses, and narrower, more cranially angled spinous processes than *C. volans*. The suite of features characterizing A and B reflects use of pronograde postures with a habitually ventriflexed back, in which a large range of powerful flexion and extension in the lumbus is possible compared to the condition of *C. volans*.

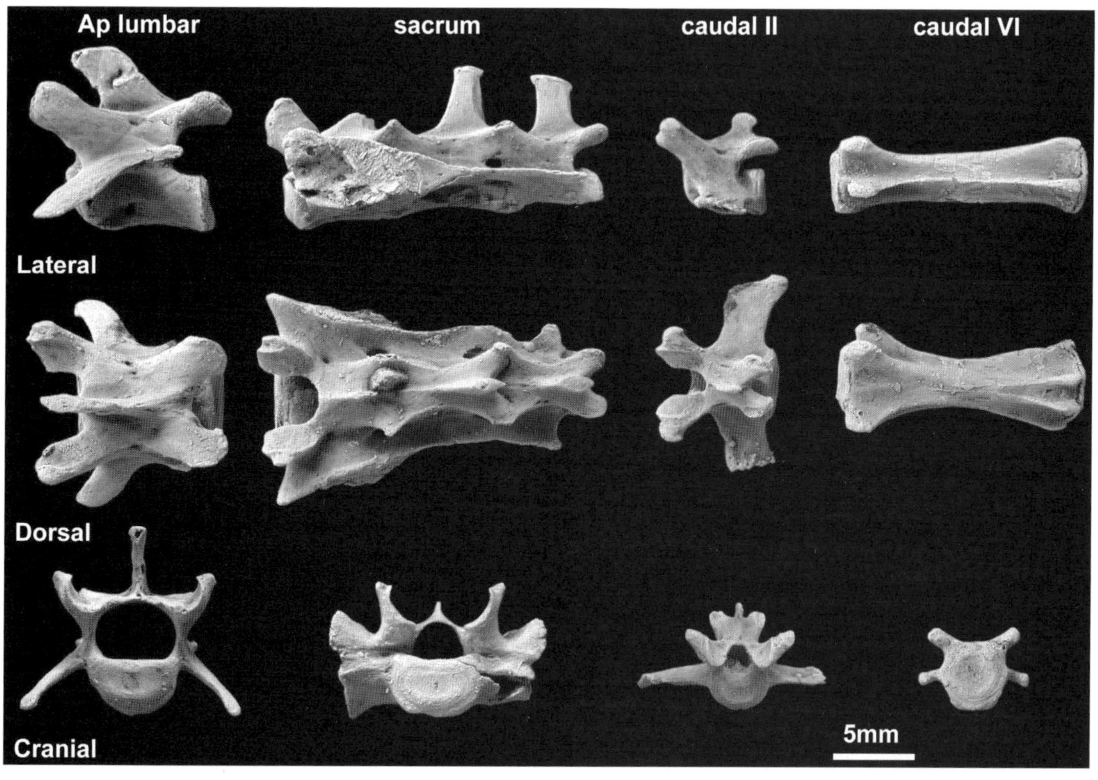

FIGURE 11.24. Vertebral elements from different regions of *Ignacius*. Representative vertebrae of *Ignacius clarkforkensis* (UM 82606). Note the cranioventrally oriented transverse processes in the antepenultimate (Ap) lumbar vertebra; the lack of a well-developed spinous process on the first sacral vertebra (which is unbroken) and large vertebral canal in the sacrum; the large vertebral canal in caudal II; and the robusticity and length of caudal VI. These features are not expected for an animal predominately using suspensory or gliding behaviors. Scale bar = 5 mm.

that extend the back (Benton, 1967; Shapiro, 1995; Sargis, 2001a).

We undertook two regression analyses to illustrate functional trends in lumbar vertebral morphology. These included (1) regression of body size on the dorsoventral depth of the trough formed by transverse processes and anterior zygapophyses of the posterior three lumbar vertebrae (Figure 11.25A), and (2) regression of lumbar spinous process shape (Figure 11.25B). These analyses successfully differentiated taxa that use asymmetrical bounding gaits from those that do not and allow us to comment on the locomotion of *Ignacius*. In the first analysis (Figure 11.25A), body size was represented by femoral cross-sectional area at mid-shaft (TA: x-axis) (see Runestad and Ruff, 1995). "Depth" (y-axis) was represented by the dorsoventral distance between the tip of the mammillary process on the prezygapophysis (dorsal) and the tip of the transverse process (ventral). The dorsoventral dimensions of the erector spinae and the dorsoventral distance of their attachments from the vertebral body should at least partly determine the strength of these muscles and the leverage they have on the intervertebral joints in the sagittal plane. These dimensions are captured by our measurements. We found the measurements from bounding taxa and non-bounding taxa to follow different regression lines. Both lines show slight positive allometry (Because TA [x-axis] is an area and the y-value is a length, isometry would be represented by a slope of 0.5 in log-log space). Taxa that leap and bound have deeper vertebrae (high y-value) for a given femur area, probably because they incorporate forceful flexion and extension of the back into their gaits and require better leverage and more force out of their erector spinae muscles than taxa that do not. *Ignacius* (large black circle in Figure 11.25A) falls in with bounding taxa. *Cynocephalus* (black triangle), as well as some other plesiadapiforms, plots with non-bounders. In the second regression (Figure 11.25B), lumbar spinous process axial (Ax) length (x-axis) is plotted against its craniocaudal (CC) length (y-axis). Again, bounders are separated from non-bounders. In this case, the bounder regression line is lower because the spinous processes of the posterior three lumber vertebrae of most bounders are smaller in their craniocaudal dimensions relative to their axial (dorsoventral) length compared to those of non-bounders. Narrower spinous processes result in more sagittally mobile backs, required for a bounding gait, which utilizes substantial flexion and extension of the vertebral column to increase the stride length. Again, *Ignacius* clearly falls with bounding taxa in having very narrow lumbar spinous processes, while *Cynocephalus* (black triangle) falls with the other group. Interestingly, other plesiadapiforms plot with *Ignacius* in this feature.

Finally, the posterior lumbar vertebrae are larger and more elongate than the thoracic vertebrae (Figure 11.22) in *Ignacius* and bounding taxa, compared to those of suspensory taxa. Together, these features provide strong evidence for utilization of bounding and leaping that incorporates sagittal flexion and extension of the trunk into the gait (Slijper, 1946; Jenkins, 1974; Shapiro and Simons, 2002). Unlike suspensory taxa, the paromomyid sacrum has only three vertebrae (Figure 11.24), which form a large vertebral canal. *Cynocephalus* has three to four sacrals, with the first one or two caudal vertebrae incorporated by fusion of the bodies and transverse processes into the sacrum in some individuals (Figure 11.18), making it similar to suspensory taxa. Furthermore, in *Ignacius* the spinous process on the first sacral vertebra is short (Figure

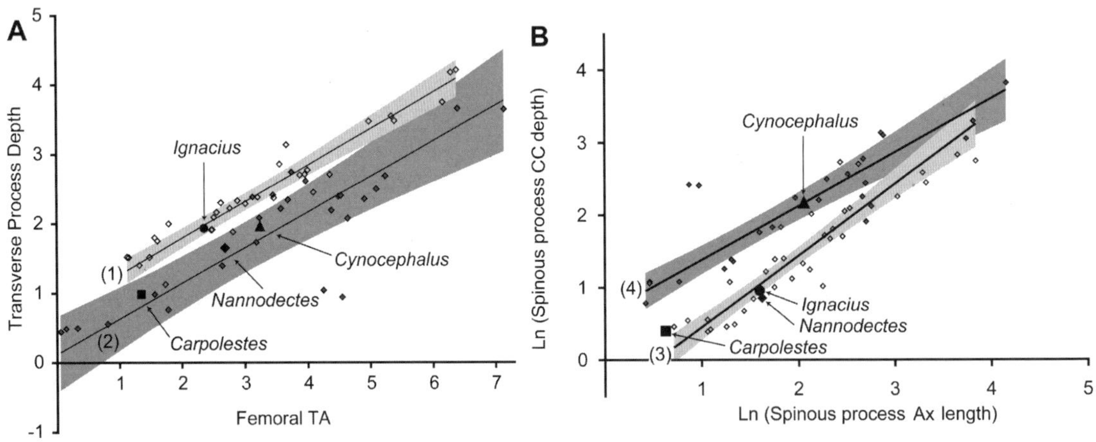

FIGURE 11.25. Regression of lumbar vertebrae against body mass proxies. A, Natural log femoral TA (transverse area) vs average depth of the posterior three lumbar vertebrae in bounding (open diamonds) and non-bounding/ambulatory (gray diamonds) taxa.. See Appendix II, Table 1 for specific taxa included in this analysis. (1) Bounder line: y = 0.524x + 0.74, R2 = 0.93. (2) Non bounder line: y = 0.512x + 0.12 R2 = 0.73. Sample includes 37 taxa. B, Natural log average spinous process length vs. depth for the posterior 3 lumbar vertebrae of bounding (open diamonds) and non-bounding/ambulatory (gray diamonds) taxa. (3) Bounder line: y = 0.91x – 0.45 R2 = 0.92. (4) Non-bounder line: y = 0.79 x + 0.6 R2 = 0.77. Sample includes 37 taxa. See Appendix II, Table 1 for specimens included in this analysis. Specimens marked with a "1" in the table were considered "bounders" for this analysis. Those marked with "0" were considered "non-bounders."

11.24). Such a configuration is similar to that in hindlimb-propelled taxa in which a large degree of flexibility at the lumbosacral joint is required, including scansorial tupaiid treeshrews (Jenkins, 1974) and squirrels. Not only does the spinous process of the first sacral vertebra not impede extension at this joint, but the supraspinous ligament, which spans two vertebrae instead of one, permits more mobility than it does when separated more distinctly into two segments (Gambaryan, 1974).

Morphology of the rib cage of both *Ignacius* and the *Dryomomys* suggests against the suspensory slow-climbing and bridging used by *C. volans*, suspensory taxa and *Ptilocercus lowii* (Sargis, 2001a) (Table 11.8). Whereas the ribs of suspensory or bridging taxa are craniocaudally broad, *Ignacius* and the *Dryomomys* have narrow ribs like those of scansorial tree squirrels and tupaiid treeshrews (Sargis, 2001a). Broad ribs may serve to increase rigidity of the thorax (Jenkins, 1970; Sargis, 2001a) and/or result in a more powerful forelimb by providing larger areas of attachment for muscles of the abdomen and shoulder girdle (e.g., serratus, pectoralis and obliquus abdominus muscles).

Innominate. The innominate of paromomyids is unlike that of either extant eutherian gliders or suspensory taxa. In these living forms, the innominate is distinctive in having a relatively long, narrow ilium (Walker, 1974) with a long post-auricular shaft; a short ischium; and a narrow pubic symphysis (Figures 11.18, 11.26; Table 11.9). Furthermore, in suspensory taxa the acetabulae are dorsolaterally oriented.

TABLE 11.9. Comparative morphology of innominate. Functional categories include suspension (s), gliding (g), bound-galloping (bg), and vertical clinging (vc). In the far left column (feature names), the first three features are separated from others by a gray background and a thicker line. This is to indicate that they are different states of the same feature. The only feature that seems to relate to suspensory behaviors is dorsolateral orientation of the acetabulum, although *C. volans* does not exhibit it. A ventrolaterally oriented acetabulum is found only in the most terrestrially-adapted rodents, among extant behavior groups in this table. The fourth and fifth features are present in taxa that hold the thigh in a flexed, abducted position while utilizing orthograde postures on vertical supports (Jenkins and Camazine, 1977; Beard, 1991). In these taxa, a large range of flexion-extension is sacrificed for mobility in abduction-adduction.

Innominate	Extant mammalian functional groups							Fossil plesiadapiforms	
	Mitten glider	Non-gliding clawed suspensory mammal	Non-gliding vertical clingers and climbers	Rodent gliders	Clawed committed arborealist	Marsupial gliders	Non-gliding clawed scansorialist	micromomyid	paromomyid
Acetabulum faces dorsolaterally		s							
Acetabulum faces laterally									
Acetabulum faces ventrolaterally							bg		
Craniocaudally elliptical acetabulum	vc	vc	vc		vc	vc		vc	vc
* Cranially buttressed acetabulum	vc	vc	vc		vc	vc		vc	vc
** Long ilium relative to ischium									
Narrow ilium		***							
Caudally positioned, narrow pubic symphysis									

[a]This feature is better referred to as "caudally-reduced" rather than "cranially buttressed," as it is seems to be the reduction of the caudal part of the acetabular lip that affects the apparent cranial buttressing. Such reduction also results in a shallower acetabulum and more mobile hip joint in these taxa.

[b]This feature does not seem to distinguish sloths or large bodied suspensory euprimates from vertical clinging and leaping euprimates, like *Tarsius*. Given that it is also seen in primitive "plesiadapiforms" (i.e., the micromomyid exhibits it, but *Ignacius* does not), there may be some phylogenetic valence to it. In fact, retention of primitive eutherian (or therian) morphology is a likely explanation for all gray features without functional codes in this table as they are present in *Ukhaatherium* (Horovitz, 2003).

[c]Neither sloths nor suspensory euprimates have a narrow ilium, but bats and the slow climbing euprimate, *Nycticebus*, do.

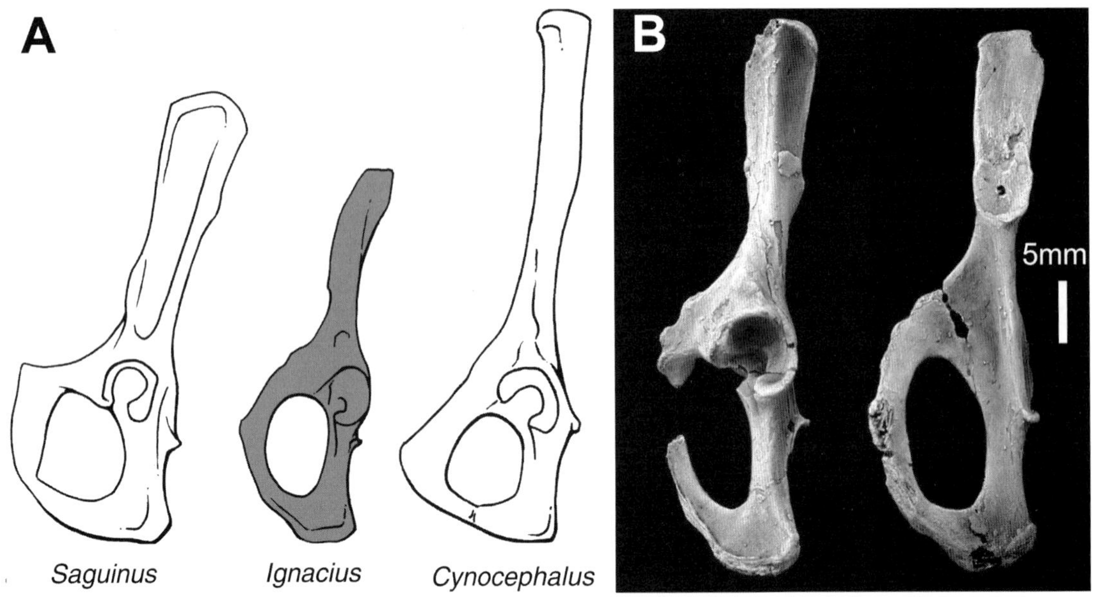

FIGURE 11.26. A, Left innominates in lateral view: *Saguinus mystax* (UMMZ 160148); *Ignacius clarkforkensis* (UM 82606); *Cynocephalus volans* (USNM 56530). Elements are standardized to ischium length. B, Left innominate in lateral view (on left) and medial view (on right) *Ignacius clarkforkensis* (UM 82606). Note the relatively shorter and more flaring ilia, the longer and/or more superiorly extended pubic symphysis, the larger ischial tuberosity and the inferiorly positioned ischial spines in both *Ignacius* and *Saguinus*. Scale bar in B = 5 mm.

Overall, both gliders and suspensory taxa have an elongate, gracile innominate with limited areas for attachment of muscles associated with hip flexion and extension during hindlimb-powered locomotion.

A gracile innominate with a relatively small pubic symphysis does not seem well-suited to resist strains experienced in pronograde postures that result from transmission of the animal's weight through this element. While these features characterize gliders, they are not necessarily exclusive to gliders, as was the case for many of the forelimb features. Specifically, didelphid marsupials, *Ptilocercus lowii* (Sargis, 2002a, b, c), and the proteutherian, *Ukhaatherium* (Horovitz, 2003) are characterized by narrow ilia. The latter two taxa are further characterized by a craniocaudally narrow, caudally positioned pubic symphysis, and a short ischium.

In general, *Ignacius* differs from gliding and suspensory taxa in having a robust innominate, with large areas of attachment for flexors and extensors of the hip (Figure 11.26). In this way, it is similar to other taxa that use an asymmetrical, hindlimb-propelled, bounding gait (e.g., tree and ground squirrels, and tupaiids). Specifically, *Ignacius* has a short ilium relative to its ischium, as well as a flaring dorsal iliac blade (visible in Figure 11.26B), allowing room for attachment of thigh extensors (gluteal muscles) and on its anterior aspect, thigh flexors (iliacus muscle). Its long ischium and large ischial tuberosity provide a long lever arm for other thigh extensors (hamstring muscles). There is a robust anterior inferior iliac spine for the origin of the rectus femoris muscle, a flexor of the hip joint and extensor of the knee joint. The extensive pubic symphysis provides stability and a large area of attachment for the adductor muscles. Finally, the acetabulum in *Ignacius* is more laterally directed than it is in suspensory taxa, indicating that habitual postures and stress orientations were different. In summary, the innominate of *Ignacius* is different from that of gliding and suspensory mammals in nearly every functionally salient respect identified by us.

Micromomyids actually appear quite similar to *Cynocephalus volans* and *Ptilocercus lowii* in the morphology of their innominate, having narrow ilia, a long post-auricular segment of the blade, a short ischium and a craniocaudally narrow, distally-positioned pubic symphysis (Figure 11.7A, B). However, given the distribution of these features outside of gliders and suspensory taxa, it is tenuous to argue that they offer support for gliding or suspension in this group. It is more likely that these reflect antipronograde behaviors more generally and/or are primitive euarchontan features (Sargis, 2002a), as we suggested for some of the forelimb features.

Hind limb. Paromomyids and micromomyids lack hind limb features that generally characterize gliders and suspensory taxa. With regard to interlimb proportions Thorington and Heaney (1981) demonstrated that the intermembral index is higher in gliding rodents (generally greater than 80) than non-gliders. The index of suspensory taxa is also high, generally over 100 (Godfrey, 1988) (Figure 11.27). *Cynocephalus* has an index of 93. Thus, we expect a similarly high index in fossil mitten-gliders, whether it reflects gliding or suspension. However, we note that among clawed euarchontans and marsupials, many generalized arborealists and vertical

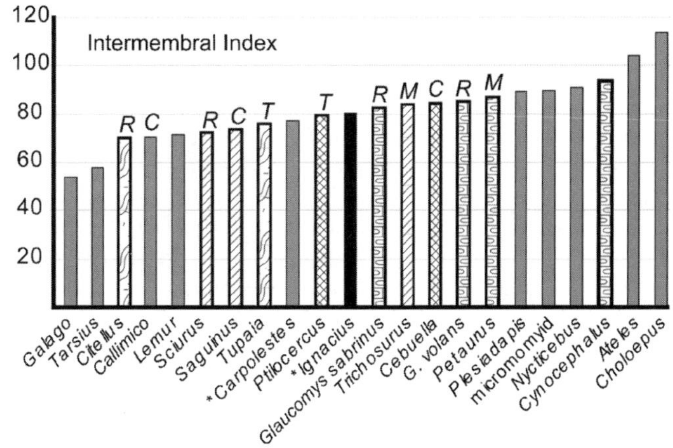

FIGURE 11.27. Intermembral indices of select taxa, presented in order of increasing index to aid in identification of functional trends. Taxa from a given higher-level clade are designated by a unique letter above the bars representing them. "R" = Rodentia; "C" = Callitrichinae; "T" = [treeshrew] Scandentia; "M" = marsupial. *Ignacius* is in black to highlight its position. An asterisk in front of a taxon name indicates that one of the parameters of the index was estimated. Note that in each clade of mammals the most terrestrial members (wavy lines) have the lowest intermembral indices, the more arboreal ones (diagonal lines) and those that spend time on large diameter vertical supports (cross-hatched) have successively higher indices; the gliders (closely packed horizontal lines) and suspensory taxa have the highest. For instance, among Callitrichinae, *Saguinus* is more pronograde and scansorial than *Cebuella* (e.g., Youlatos, 1999). Furthermore, *Callimico goeldii* has the shortest intermembral index of any callitrichine and is not known to utilize large diameter vertical supports or the exudate resources procurable there. Note that the ground squirrel *Citellus* has shorter forelimbs than the tree squirrel *Sciurus*. Both have shorter forelimbs than the gliding squirrels *Glaucomys sabrinus* and *volans*. This trend holds for marsupials as well, with the locomotor generalist *Trichosurus* exhibiting shorter forelimbs than the glider, *Petaurus*. Behavioral overlap occurs in the region of high indices, such that a high intermembral index is not evidence of gliding by itself; however, there seems to be a lower limit to the intermembral indices exhibited by gliders. That limit appears to be somewhere around 80 [although out of eight species of flying squirrels, one (*Eoglaucomys*) has an index below 80 (Thorington and Heaney, 1981)]. Thus, although a strong case for gliding in a fossil taxon cannot be made on the basis of a high index alone, a strong argument against it can be made on the basis of a low index. Almost no extant gliders have an index as low as that of *Ignacius*. See Appendix II, Table 1 for included specimens. If more than one individual is marked per taxon, the intermembral index shown represents an average of those individuals.

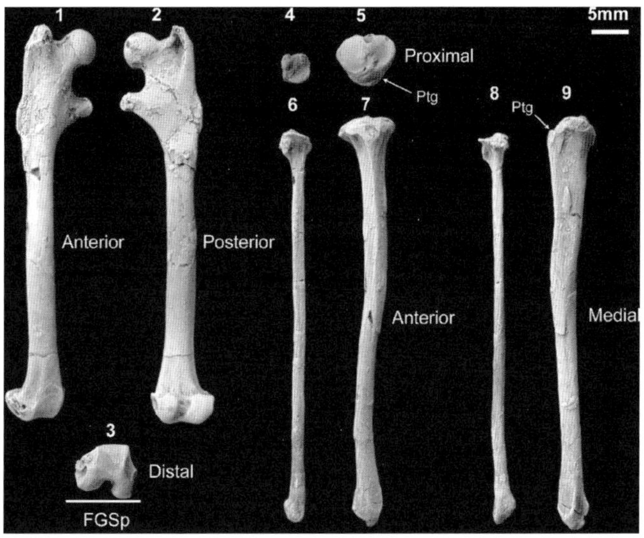

FIGURE 11.28. Illustration of hindlimb elements of UM 82606. Right femur of *Ignacius* in (1) anterior, (2) posterior and (3) distal views. View (3) is oriented so that the greater trochanter and fovea capitis femoris form a horizontal line (FGSp) on the page. Note that the condyles face posterolaterally and would have facilitated postures in which the feet were widely spaced (abducted). The right tibia of *Ignacius* is depicted in (5) proximal, (7) anterior and (9) medial views. The right fibula is shown in (4) proximal, (6) anterior, and (8) medial views. The broad shelf on the proximal fibula (4), oriented perpendicular to the shaft axis, gives it mobility with respect to the tibia. On the distal tibia (7), lateral inclination of the astragalar facet presumably accommodates asymmetry of margins of tibial facets on the astragalus when the foot is dorsiflexed at the crurotarsal joint. On (9), note that the patellar tendon groove (Ptg) is located distal to the tibial plateau. FGSp – Plane defined by Fovea capitis femoris-Greater trochanter-Shaft. Scale = 5 mm.

TABLE 11.10. Comparative morphology of hindlimb. Functional categories include mitten-gliding (mg), suspension (s), gliding (g), bound-galloping (bg), pronograde (p), vertical clinging (vc). The second and third features are considered to characterize eutherian suspensory taxa, as these states are rare among eutherians otherwise, especially primitive ones. These features are more common in phalangerid marsupials, and likely relate to function differently. For example they are present in *Petaurus*, but do not seem to reflect suspensory behaviors in it, because it is a glider and agile pronograde arborealist. For the feature "**Proximal insertion of patellar tendon on tibia,**" the present state signifies a tendon that inserts at the level of the tibial plateau, while the absent state signifies a tendon that inserts distal to the plateau. Paromomyids, micromomyids, sciurids and *Petaurus* have a groove positioned distal to the plateau by a proportionally similar distance. In vertically clinging and leaping euprimates, the groove is much more distally positioned. Taxa that exhibit the present state for the feature "**Proximal tibial shaft lacks cnemial crest**" include *Nycticebus, C. volans*, and *Choloepus*, but not hylobatids. Euprimates and tree squirrels have the most prominently developed crests, among those with the absent state for this feature. The lack of features 17 through 20 probably reflects more active, agile pronograde locomotion in rodent gliders and scansorialists.

Hindlimb	Extant mammalian functional groups							Fossil plesiadapiforms	
	Mitten glider	Non-gliding suspensory mammal	Non-gliding vertical clingers and climbers	Rodent gliders	Clawed committed arborealist	Marsupial gliders	Non-gliding clawed scansorialist (rodent)	micromomyid	paromomyid
Proximal fibula, reduced or fused to tibia	*s	*s							
Short femoral neck	s	s							
Femur lacking third trochanter	s	s							
Femoral condyles broadest distally	s	s							
Concave lateral tibial condyle	s	**s							
Tibial condyles perpendicular to tibial shaft	s	***s							
Proximal insertion of patellar tendon on tibia	s	***s							
Proximal tibial shaft lacks cnemial crest	ap	***ap		ap					
Shallow to absent fovea capitis femoris	s	s							
Distally facing femoral articular surface for tibia	s	s							
Reduced greater trochanter	s	s							
Superiorly extended greater trochanter			p	p			p		p
Reduced lesser trochanter	s	s							
lesser trochanter distally positioned and medially projecting			vc		vc			vc	vc
Femoral condyles face posteromedially	s	s							
Femoral condyles face posterolaterally			vc		vc			vc	vc

(continued)

TABLE 11.10. (continued)

Straight femoral shaft									
Distally restricted, shallow patellar groove									
Shallowly-grooved distal tibial facet									
Shallow femoral condyles (AP)									

[a]A reduced proximal fibula, specifically, only characterizes bats and *Cynocephalus volans*, among taxa that utilize suspension. Sloths fuse their proximal fibulae, as do arboreal porcupines. The common functional trend may be the reduction in number of independent bones that reach the knee. We therefore recognize these morphologically different states as functionally equivalent.

[b]The feature **"Concave lateral tibial condyle"** is not exhibited by the sloth, but characterizes the gibbon and other suspensory taxa. Its absence in the sloth is taken as a primitive retention.

[c]These features do not characterize *Hylobates*. Most features of the hindlimb that otherwise reflect suspensory postures are absent from gibbons, probably because they mainly use bimanual suspension which does not involve the hindlimbs.

clingers and climbers have an intermembral index between 80 and 100, such that this ratio does not distinguish them from gliders (Sussman and Kinzey, 1984; Godfrey, 1988; Fleagle, 1999; Sargis, 2002b) (Figure 11.27).

Ignacius appears to have an intermembral index of ~80, which is comparable to that of most callitrichines except *Cebuella*, for which it is ~84 (Fleagle, 1999). Other plesiadapiforms have indices that range between ~77 (for *Carpolestes*) and 89 (for *Plesiadapis cookei*) and 93 (for micromomyids). Such proportions argue strongly against use of gliding or suspensory behaviors by *Ignacius*, because its index is below the typical ranges of extant mammals that use such behaviors.

In addition to other findings reviewed by us above, Runestad and Ruff (1995) also found that gliders are characterized by a gracile femur and tibia. They evaluated gracility in hind limb by regressing cross-sectional area on length for each element, as they did for the forelimb elements. Plotting dimensions of known paromomyid and micromomyid material suggested against gliding for the former; however, they did not interpret the position of micromomyids, which have limbs that are beyond the limits of their regression because they are too small, as for the forelimb regression. Out of curiosity we plotted some euprimate postcranial material with Runestad and Ruff's data and found that *Saguinus mystax* (UMMZ 160148) plotted very close to the glider line for both femur and tibia dimensions, while *Smilodectes mcgrewi* (UM 95526), plotted with gliders for femur dimensions. This further supports Runestad and Ruff's (1995) warning that many non-gliders also have gracile limbs such that even this feature cannot be used on its own to infer gliding in a fossil. Nonetheless, we evaluated the gracility of the hind limb elements of *Ignacius*, UM 82606 (Figure 11.28), using the regressions of Runestad and Ruff (1995). Its femur plots midway between the glider and non-glider regressions (length = 53 mm, TA = 10.59 mm). However, the tibia has proportions that put it substantially closer to their non-glider regression line (length = 55 mm, TA = 7.70 mm). We take these results as consistent with those from our other analyses that suggest against a gliding habitus.

As for the forelimb, Thorington et al. (2005) noted a number of morphological features of the hind limb distinguishing arboreal sciurids from gliding sciurids, including a lesser trochanter that may extend medial to the femoral head; and a third trochanter that is more pronounced and positioned distal to the lesser trochanter in non-gliders. Gliding squirrels were proposed to also differ from non-gliding squirrels in having a rod-like tibia, a characteristically long distal tibia-fibula articulation and a sharply grooved tibial articular surface on the astragalus (Thorington et al., 2005). While some of these features may be useful indicators of gliding in euarchontans, as with the forelimb, others have a distribution among extant members showing that they do not necessarily reflect gliding (Table 11.10). Specifically, a proximally positioned third trochanter does not always indicate gliding because it characterizes many euprimates (e.g., Dagosto et al., 1999). Furthermore, the last two features of the leg and ankle cited by Thorington et al. (2005) to characterize gliding squirrels, while generally lacking in extant euarchontans and fossil "plesiadapiforms," are present in proposed euarchontan outgroups, Nyctitheriidae (Hooker, 2001) and basal gliroids (Murphy et al., 2001b; Meng et al, 2004; Rose and Chinnery, 2004). These outgroups have been reconstructed as more scansorial than most "plesiadapiforms" (Hooker, 2001; Bloch et al., 2003; Rose and Chinnery, 2004), and the ways in which they differ from "plesiadapiforms" in the distal crus and ankle likely represent a less axially mobile crurotarsal joint, corresponding to more frequent use of scansorial locomotion (Szalay 1984).

Both *Ignacius* and micromomyids differ from rodent gliders in having a femur with a lesser trochanter that extends medial to the femoral head, and a tibia with a prominent cnemial crest. *Ignacius* and other paromomyids differ further in having a femur with a distinct third trochanter that arises from the shaft distal to the lesser trochanter (Figure 11.28), a short distal tibia-fibula articulation, and a shallowly grooved tibial articular surface of the astragalus. On the other hand, micromomyids appear to have a third trochanter that arises lateral to the lesser trochanter (Figure 11.7B; Beard, 1993a: Figure 10.10), a longer distal tibia-fibula articulation (Figure 11.7A, B) and a more sharply grooved tibial articular surface on the astragalus, suggesting they retain the primitive condition for Euarchonta (Szalay and Drawhorn, 1980; Bloch et al., 2003; Bloch and Boyer, 2007).

Suspensory taxa and *Cynocephalus* exhibit features associated with habitually extended limbs that bear tensile (not compressive) loads, frequent use of small diameter supports, and infrequent reliance on the hind limbs in powerful propulsion. The hind limbs of *Ignacius* and micromomyids differ in having features that suggest habitual flexion at the hip and knee joints, use of large diameter supports instead of small diameter ones, and the capacity for powerful flexion and extension of the hind limb.

Features that suggest habitual flexion at the hip and knee of paromomyids and micromomyids are seen in the femoral head, and the femoral condyles and tibia, respectively. First, the articular surface of the femoral head extends onto the posterolateral part of the neck (Figure 11.28), which gives it a somewhat oval shape. Thus, the closest packed articulation between the head and elliptical acetabulum is achieved when the femur is in a flexed, abducted position (Jenkins and Camazine, 1977; Beard, 1991), as shown in Figure 11.32.

Second, in paromomyids and micromomyids, buttressing of the medial margin of the patellar groove relative to its lateral margin also reflects use of postures in which the thigh was flexed and abducted, and the knee joints were flexed. In such a posture, medial buttressing of the patellar groove would help prevent medial and ventral dislocation of the patella. Such dislocation would otherwise tend to occur in this posture, because the line of action of the quadriceps muscles (predominantly rectus femoris), which runs from the patellar groove on the tibia to the anterior inferior iliac spine on the innominate, is located medial and ventral to the anteroproximal aspect of the patellar groove of the femur, where the patella sits.

Third, the femoral condyles of *Ignacius* and the micromomyid are anteriorly restricted, posteriorly extensive, and broadest at their posteroproximal margin. These features result in a knee that is more stable when tightly flexed, because the tibia articulates with the posteroproximal part of the femoral condyles where they are broadest. Thus forces transmitted through the knee in a flexed position would be distributed over a greater area of articulation than positions in which the knee was extended. In contrast, suspensory taxa that utilize postures

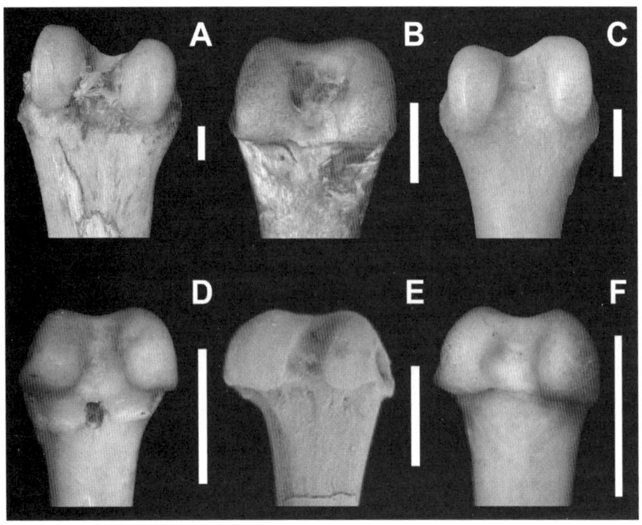

FIGURE 11.29. Comparison of femoral condyles of *Ignacius* to pronograde and antipronograde mammals. Distal femora in posterior view with the distal end pointing up. A, *Choloepus hoffmani* (left); B, *Cynocephalus volans* (right); C, *Nycticebus coucang* (right); D, *Tupaia glis* (right); E, *Ignacius clarkforkensis* (right); F, *Cebuella pygmaea* (right). Outer margins of the condyles on A–C are parallel to one another, reflecting the capacity for extreme knee extension (used during under-branch suspension), and postures that require variable degrees of knee flexion. The condyles of D–F appear wedge-shaped in this view because the outer margins of the condyles converge distally. These taxa use scansorial locomotion, resting postures in which the knees are flexed, and infrequently suspend below branches with extended limbs. See Appendix II, Table 1 for specimen numbers. Scale = 5 mm.

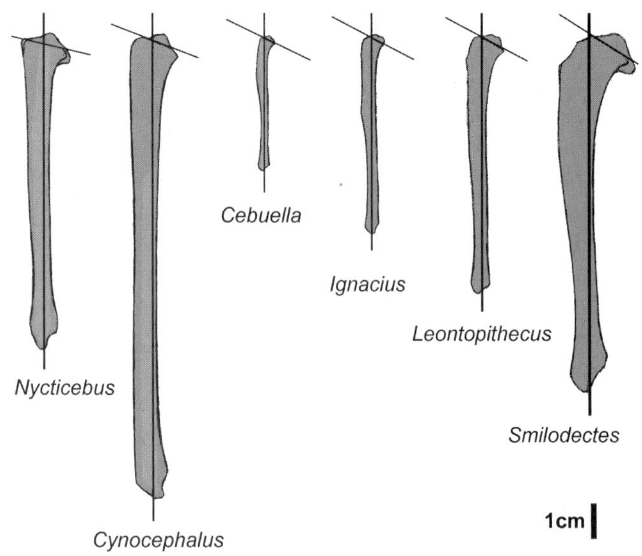

FIGURE 11.30. Right tibiae in medial view showing angle formed between shaft and medial facet of tibial plateau. From left to right, taxa depicted (and the angle formed) are the following: *Nycticebus coucang* (73); *Cynocephalus volans* (69); *Cebuella pygmaea* (65); *Ignacius clarkforkensis* (64); *Leontopithecus sp.* (62); *Smilodectes mcgrewi* (60). *Nycticebus* and *Cynocephalus*, which frequently use extended limb postures, have medial facets at more of an angle to the shaft (they approach perpendicular). Taxa that are more pronograde, or interpreted to have been leapers, have medial facets at less of an angle to the shaft (closer to parallel). See Appendix II, Table 1 for specimen numbers. Scale bar = 1 cm.

with varying degrees of knee flexure (including full extension) tend to have more anteriorly extensive condyles that are equal in mediolateral breadth along their circumference.

Looking at a posterior view of the distal femur (Figure 11.29), it is seen that the paromomyid and micromomyid condition is reflected by condyles that appear wedge-shaped because their outer margins "converge" distally. Such a form also characterizes other scansorial mammals, including some treeshrews and callitrichines. In contrast, Gebo (1989) noted that lorisines were characterized by "parallel" condyles and differed from galagos in this regard. We agree with this assessment by Gebo, and note that sloths and *Cynocephalus* also exhibit "parallel" condyles.

Yet another feature that suggests a flexed knee joint is the angle formed between the tibial shaft and the posteriorly tilted medial tibial condyle (Figure 11.30). *Ignacius* and the *Dryomomys* have more acute angles than *Cynocephalus* and *Nycticebus*. That is, in the former taxa, the medial facet faces more posteriorly than proximally, such that shear forces experienced across the knee joint when flexed, would be reduced.

Features indicating a wide foot stance, and thus large diameter support use in *Ignacius* include femoral condyles that face laterally with respect to the femoral neck and shaft (Figure 11.28, view 3), and tibial condyles that are oriented slightly laterally relative to the tibial shaft (the micromomyid specimens could not be evaluated for these features). As a result of these features a flush articulation between the posterior aspect of the femoral condyles and tibial plateau results in a laterally projecting tibial shaft.

Finally, paromomyid and micromomyid hind limbs appear to be suited for powerful flexion and extension relative to suspensory taxa. Features indicating this include a greater trochanter that extends superiorly above the femoral head in *Ignacius*, although not to the degree exhibited by tupaiid treeshrews (Sargis, 2002a); and a lesser trochanter that is distally positioned and medially extended in both *Ignacius* and the micromomyid (Figure 11.28). An extended greater trochanter increases the leverage that the gluteal muscles attaching to it have in thigh extension (e.g., Rose, 1999). The medial extension of the lesser trochanter allows the femur to remain somewhat abducted even when the iliopsoas muscles are fully contracted and the femur is fully flexed. Furthermore, the distal position of the lesser trochanter gives the iliopsoas muscles a long lever arm that would have made them effective at holding the thigh in flexed positions and capable of easily flexing the thigh during vertical climbing (Rose, 1987). Such prominent trochanters flanking the femoral head, while providing leverage for muscles, consequently also reduce mobility at the hip joint, which further suggests against suspensory behaviors. Unlike the fossil taxa, *C. volans* and tree sloths both have a femoral neck with a central axis nearly in line with that of the femoral shaft, and greatly reduced greater and lesser trochanters (White, 1993). The former two features have been directly related to hip joint mobility and suspensory behaviors in both primates and xenarthrans (White, 1993).

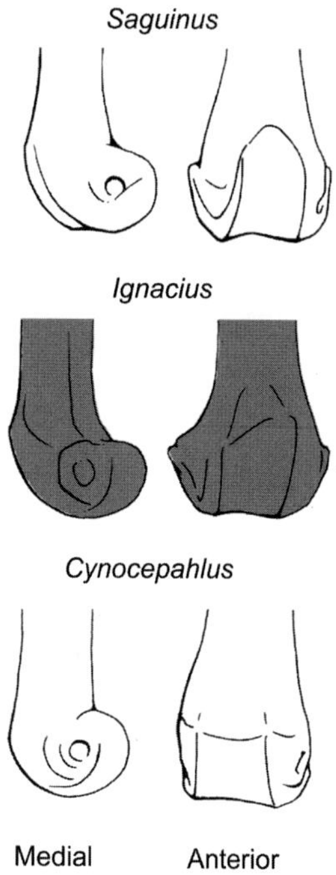

FIGURE 11.31. Distal femora showing patellar morphology. From top to bottom is *Saguinus mystax* (UMMZ 160148), *Ignacius clarkforkensis* (UM 82606), and *Cynocephalus volans* (USNM 56530) in medial and anterior view. Note that despite being an agile arborealist the distal femur in *Saguinus* is nearly identical to that in *Ignacius*. Both of these taxa differ only slightly from the condition in *C. volans* with respect to patellar groove morphology. However, other features discussed and figured in this chapter differentiate the fossil and *Saguinus* from *C. volans*, showing the former two taxa to be capable of pronograde postures and to be more agile than the latter. Images are standardized to distal femur mediolateral breadth.

Finally, features of the tibiae also suggest hind limbs used for quick forward propulsion in orthograde or pronograde postures in both the paromomyid and micromomyid. First, the tibiae are relatively long, with crural indices of the paromomyid and micromomyid being 104 and 127, respectively. Furthermore the tibiae of the paromomyid and micromomyid have a distally positioned groove for the patellar tendon (Figure 11.28: view 5 and 9), a prominent cnemial crest (as mentioned above) that provides room for extensors of the toes and foot, and deep popliteal fossa that provides room for pedal and digital flexor muscles (Figure 11.28: view 9). *C. volans*, suspensory, and slow climbing taxa have a more proximal attachment of the patellar tendon, and a reduced cnemial crest and popliteal fossa (consequently, the tibia is "rodlike").

We note that this view of paromomyid and micromomyid hind limbs as suggesting agile arboreality is slightly different than that previously supported (e.g., Beard, 1989; 1991). One feature frequently cited by Beard (1989, 1990,1991) as limiting "plesiadapiforms" in their agility is a proximodistally short patellar groove (Figure 11.31). It is characteristic of all known plesiadapiforms (e.g., Beard, 1993a). However, we note that callitrichine euprimates, many agile marsupials (e.g., *Petaurus breviceps*) and *Ptilocercus lowii* (e.g., Sargis, 2002b) have patellar grooves similar in shape to those of known "plesiadapiforms." While animals with narrow, proximally extended grooves, like tupaiid treeshrews and prosimian euprimates, are usually extremely agile and capable of acrobatic behaviors, the lack of this morphology in "plesiadapiforms" does not necessarily require a loris-like speed of progression, or severe limits on agility.

11.5 Summary and Conclusions

In summary, the features forwarded by Beard (1990, 1993b) as uniquely characterizing *Cynocephalus* and paromomyids are either not present in paromomyids or are present in other non-mitten-gliding taxa. Specifically: (1) Paromomyids and micromomyids do *not* have intermediate phalanges that are longer than the proximal phalanges. Furthermore, this proportional relationship is likely to reflect suspensory behaviors and thus would not be definitive evidence of mitten-gliding in the fossil taxa even if it had been substantiated. (2) Elongation of the intermediate phalanges does *not* group paromomyids and micromomyids with *Cynocephalus* to the exclusion of most arborealists including euprimates and other "plesiadapiforms." (3) Relatively great dorsopalmar depth of the proximal end and shaft of the intermediate phalanx does *not* uniquely group *Cynocephalus* with paromomyids and micromomyids, among "plesiadapiforms." It additionally characterizes *Plesiadapis*. Furthermore, callitrichine euprimates, distinctive for their use of vertical clinging and climbing, are actually more similar to "plesiadapiforms" than to typical euprimates in this regard. (4) An intermediate phalanx with a straight shaft that lacks dorsal recurvature of the distal end, does *not* characterize known paromomyids or unite them with *Cynocephalus*, although it does characterize micromomyids. However, it also characterizes *Plesiadapis*, which is traditionally thought to be a non-glider (Gingerich, 1976; Gunnell and Gingerich, 1987; Beard, 1989; Youlatos and Godinot, 2004; Bloch and Boyer, 2007), and at least some extant callitrichines.

These findings refute the gliding hypothesis as proposed by Beard (1993b) inasmuch as they contradict observations he made in the sections "Anatomical Evidence for Gliding in Paromomyids" and "Anatomical Evidence for Gliding in Micromomyids" which indicated that dermopterans, paromomyids, and in some cases micromomyids were uniquely similar. Given the nature of Beard's evidence we considered some of it to be invalidated by the previously unmentioned fact that large plesiadapids are also morphologically similar to dermopterans and paromomyids. While we (and others who have studied plesiadapid functional morphology) see gliding in plesiadapids as extremely unlikely, we feel it is important to acknowledge that if evidence were ever marshalled for gliding in *Plesiadapis*, from other (non-phalanx) regions of the skeleton, then Beard's observations regarding the functional significance of the phalangeal morphology would regain some validity. However, even in this unlikely scenario, our conclusions regarding paromomyids and micromomyids would probably still stand because analysis of the rest of the skeleton appears to further refute the possibility of *Cynocephalus*-like mitten-gliding, associated suspensory behaviors and squirrel-like gliding (although the functional significance of micromomyid morphology is admittedly ambiguous in several regions). Specifically: (1) The phalangeal morphology of paromomyids and micromomyids does not reflect functions characterizing suspensory behaviors, but suggests the ability for tight flexion of the digits, as is characteristic of euprimate-like grasping. (2) The finger-toe proportions, and the metapodial proportions of paromomyids and micromomyids do not reflect those expected for gliders incorporating the manus into the patagium. Instead, these features are more generalized and fit the pattern of many scansorial mammals (i.e., the fingers and metacarpals are short compared to the toes and metatarsals). (3) Features of eutherian gliders are absent from the forelimb of paromomyids, as are features usually associated with suspensory behaviors. Micromomyids possess some features that may reflect functions required in suspension and gliding, but are not exclusively associated with those behaviors. (4) The axial skeleton of paromomyids lacks features of *Cynocephalus*, more generalized gliders, and suspensory taxa. Specifically, it lacks the long neck and morphological traits of the lumbar vertebrae characterizing *Cynocephalus* and suspensory taxa, as well as the relatively short trunk and tail characterizing gliding taxa. Instead, the vertebrae are similar to those of extant sciurids and tupaiids that do not glide, but *do* utilize bounding gaits in which extensive sagittal flexion and extension of the trunk increases the stride length. (5) The ribs of both paromomyids and micromomyids are unlike those of *Cynocephalus*, sloths and *Ptilocercus*. This suggests against suspension or slow-climbing and bridging behaviors. (6) The innominate of gliders and suspensory taxa is quite distinctive from that of bound-galloping scansorialists or active arborealists. The innominate morphology of *Ignacius* is clearly similar to that of bound-gallopers. The morphology of micromomyids actually appears more similar to that of *Cynocephalus*, but it is also similar to that of some euprimates, *Ptilocercus lowii* and the basal eutherian mammal, *Ukhaatherium*. Thus, although the innominate characters strongly suggest against gliding and suspension in *Ignacius*, their functional significance is not clear, and the reverse cannot be argued for the case of micromomyids, especially in the face of many other characters that clearly suggest against gliding behaviors in micromomyids. (7) The proportion of the forelimb to hind limb (intermembral index) is distinctively different from that for specialized suspensory taxa. While not substantially different from some gliding taxa, it is not exclusively similar to any gliding taxa for either *Ignacius* or the micromomyid. Neither are the intermembral indices in these fossils outside the range of other known "plesiadapiforms," or extant, pronograde to orthograde arborealists. (8) The

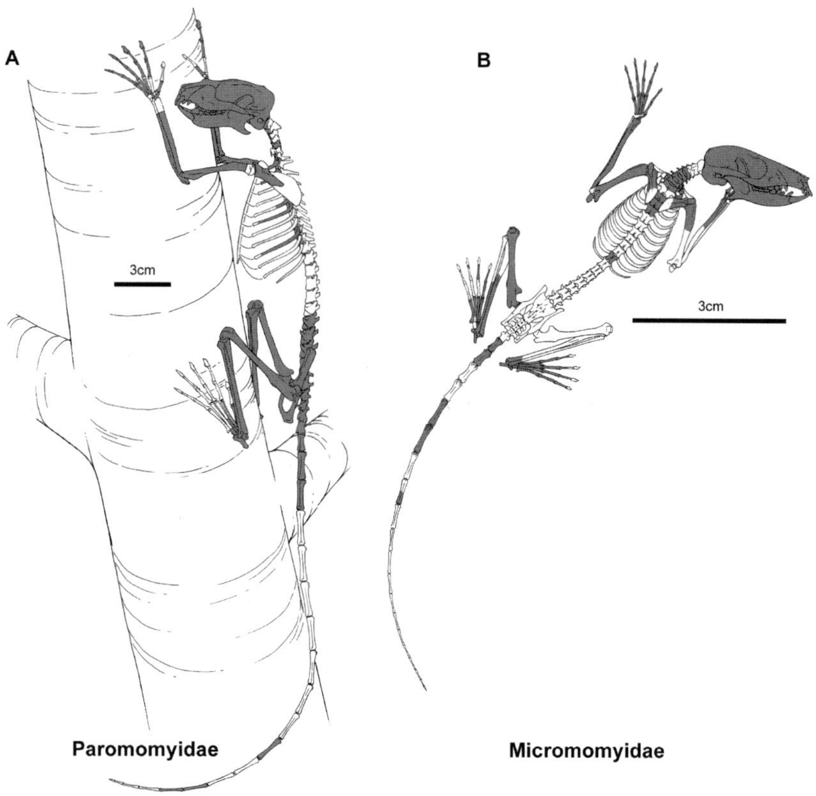

FIGURE 11.32. Reconstruction of paromomyid and micromomyid "plesiadapiforms" in typical postures. Reconstructions are based off of allavailable material for these animals, but skeletons are shaded to show bones present in *Ignacius clarkforkensis* (UM 108210 and UM 82606) and *Dryomomys szalayi* (UM 41870). Scales = 3 cm.

femur and tibia of paromomyids and micromomyids indicate habitually flexed, abducted thighs, flexed knee joints, a wide foot stance, and the capacity for relatively powerful extension and flexion of the thigh and leg, contrary to the expectations for suspensory taxa. Paromomyids also lack features of femur and tibia that characterize gliding squirrels and distinguish them from arboreal ones. Although the micromomyid has some features seen in gliding squirrels (if not in *Cynocephalus*), these features also characterize other non-gliding euarchontans (euprimates) and more basal, non-gliding taxa (nyctitheriids).

In summary, functional features expected for a mitten-glider are essentially absent from new paromomyid and micromomyid skeletons in every region analyzed. Therefore, the mitten-gliding hypothesis is confidently rejected for Paleogene paromomyids and micromomyids. Furthermore, habitual use of suspensory behaviors and more generalized gliding are also rejected for both paromomyids and micromomyids. The postcranial proportions and morphology of paromomyids are strikingly similar to those of callitrichine euprimates in features that signify the use of vertical clinging postures (Figure 11.32A) and more generalized agile arboreal activity (Bloch and Boyer, 2007). Micromomyids likely differed from paromomyids and callitrichines in being slightly less agile, and in utilizing antipronograde under-branch clinging postures more frequently (Figure 11.32B). Given these inferred behavioral differences, the arboreal treeshrew *Ptilocercus lowii* may make a slightly more appropriate model for the micromomyid (Bloch et al., 2003; Bloch and Boyer 2007; Bloch et al., 2007).

Given the fact that at least paromomyid dentitions suggest exudate-eating (Gingerich, 1974: but see Godinot, 1984), it is not outlandish to speculate that similarities between the functional skeletal and dental attributes of paromomyids and callitrichines reflect the presence of the same biological role. In other words, features reflecting vertical clinging and climbing in paromomyids and micromomyids may be part of an adaptive complex allowing access to exudates. This characterization matches paromomyids and micromomyids with Beard's (1991) view of the "Primatomorpha" morphotype. Thus, if vertical clinging in euarchontans can always be associated with the biorole of specialized exudativory, it is possible that evolutionary origins of Euprimates and Dermoptera were marked by a dietary shift to eating exudates. However, if this is true, Scandentia should also be primitively characterized by an exudate-eating phase, because recent phylogenetic analyses are overwhelmingly consistent in recovering it as the sister-group of Dermoptera (e.g., Liu and Miyamoto, 1999; Liu et al., 2001; Madsen et al., 2001; Murphy et al., 2001a, b; Sargis, 2004, 2007; Van Den Bussche and Hoofer, 2004 Bloch and Silcox, 2006; Bloch et al., 2007). This is to say that the common ancestor of dermopterans and euprimates also represents the common ancestor of Euarchonta (Sargis, 2002c; Silcox et al., 2005), which includes primates, dermopterans, and scandentians, but excludes chiropterans.

There are several problems with extending the exudate-eating, vertical clinging hypothesis to the base of a Euarchonta: (1) The basal euarchontan was likely characterized by features suggesting improved grasping abilities (i.e., long fingers relative to metacarpals, and incipiently divergent and mobile first digits (Sargis et al., 2005; Sargis et al., 2007, Bloch et al., 2007), not necessarily expected for an animal specialized for clinging to large-diameter supports, exclusively. (2) The dentitions of basal euarchontans and the earliest "plesiadapiforms," while consistent with de-emphasis on faunivory (Szalay, 1968), provide little evidence for initial specializations towards exudativory [i.e., teeth of early members of most "plesiadapiform" groups retain high shearing quotients (Kay and Cartmill, 1977; Biknevicius, 1986) and have relatively smaller central incisors (e.g., Gingerich, 1976; Bloch et al., 2001b)]. (3) Robust, procumbent incisors are also employed to access other types of angiosperm products, such as nuts, specifically (i.e., as in rodents and *Daubentonia*).

Given these caveats and new fossil data that have accumulated in the last 15 years, it now seems more likely that the dietary shift marking the origin of Euarchonta was one towards inclusion of more angiosperm products generally. Such a dietary shift also requires movement into an arboreal setting (Szalay, 1968; Sussman and Raven, 1978; Szalay, 1981). The *heritage* feature of *claws*, characterizing all euarchontans, probably necessitated a claw-climbing phase to this shift (Haines, 1958). Acquisition of claw-climbing, callitrichine-like arboreality (Szalay and Dagosto, 1988) would have opened an adaptive route towards specialization on exudates, which was apparently followed in paromomyid and possibly micromomyid evolution (Figure 11.32).

However, the innovation of an incipiently developed, divergent, opposable grasping hallux in basal Euarchonta also opened the way towards specialized manual and pedal grasping seen in carpolestids (Bloch and Boyer, 2002a) and Euprimates. Other taxa, such as plesiadapids (*Plesiadapis* and *Platychoerops*) evolved large body sizes [an order of magnitude greater than other plesiadapoid or paromomyoid "plesiadapiforms" (Gingerich and Gunnell, 2005)] that allowed utilization of foliage as a protein source, as suggested further by possible suspensory innovations in the postcranium and more selenodont dentitions in late occurring members (Gingerich, 1976; Bloch and Boyer, 2007). It would not be very surprising if evidence for gliding in this arboreal radiation of "plesiadapiforms" were to be eventually found, given that it is likely that much of its skeletal diversity still remains undocumented. However, as we have shown, evidence for such gliding remains elusive. Given the findings presented here and in light of current phylogenetic reconstructions (Bloch and Silcox, 2006; Bloch et al., 2007), if gliding is eventually demonstrated for some "plesiadapiform" group it is unlikely that it will also be demonstrably homologous to dermopteran mitten-gliding.

Acknowledgments. The studies reported on here would not have been possible without access to the facilities and resources of the University of Michigan Fossil Preparation Lab, granted to us by P. D. Gingerich, D. C. Fisher and W. J. Sanders. Furthermore, the University of Michigan Museum of Zoology provided critical access to comparative specimens. For access to specimens specifically, as well as other resources we thank P. D. Gingerich, P. Houde, and P. Meyers. For access to important datasets and other resources, as well as insightful discussion, we thank M. W. Hamrick, W. J. Sanders, and G. F. Gunnell. For insightful discussion and critical reviews of previous versions of this manuscript we thank W. J. Sanders, M. T. Silcox, E. J. Sargis, and K. C. Beard. For help with methodological issues, we thank F. S. Szalay, K. D. Rose, J. Trapani, M. Godinot, and R. W. Sussman. Funding was provided by an NSF grant to P. D. Gingerich, G. F. Gunnell, and J. I. Bloch; an NSF grant to D. W. Kravse, J. I. Bloch and D. M. Boyer (10333138), an NSF doctoral dissertion improvement grant to D. M. Boyer (BCS-0622544) and a University of Michigan undergraduate research grant to D. M. Boyer.

Appendix I

Postcranial specimens of Paromomyidae and Micromomyidae

Family Paromomyidae

Acidomomys hebeticus

UM 108207 Partial skeleton with associated skull and dentaries. Illustrated in Bloch and Boyer (2001) and Bloch et al. (2002a), and in Figures 11.1 and 11.2.

Ignacius graybullianus

USNM 442233 Damaged left humerus lacking its distal end. Specimen prepared out of the 8ABC Limestone from UM locality SC-4 (Wa-1, early Eocene) in the Clarks Fork Basin. Specimen briefly described, but not illustrated, by Beard (1989).

USNM 442259 Nearly complete left humerus. Specimen prepared out of the 8ABC Limestone from UM locality SC-4 (Wa-1, early Eocene) in the Clarks Fork Basin. Specimen described and illustrated by Beard (1989, Figure 50).

USNM 442232 Right proximal radius. Specimen prepared out of the 8ABC Limestone from UM locality SC-4 (Wa-1, early Eocene) in the Clarks Fork Basin. Specimen described and illustrated by Beard (1989, Figure 51).

USNM 442256 Proximal phalanx. Specimen prepared out of the 8ABC Limestone from UM locality SC-4 (Wa-1, early

Eocene) in the Clarks Fork Basin. Specimen illustrated (but not described) by Beard (1989, Figure 52). Identified as *I. graybullianus* based on size and similarity to dentally associated specimens of *Phenacolemur* (Beard, 1989). Identified as belonging to the manus (rather than the pes) based on elongation and "robusticity" (Beard, 1989; see Hamrick et al., 1999, for a different interpretation).

USNM 442253 Intermediate phalanx. Specimen prepared out of the 8ABC Limestone from UM locality SC-4 (Wa-1, early Eocene) in the Clarks Fork Basin. Specimen illustrated (but not described) by Beard (1989, Figure 53). Identified as *I. graybullianus* based on similarity to dentally associated specimens of *Phenacolemur* (Beard, 1989). Identified as belonging to the manus (rather than the pes) based on "robusticity" (Beard, 1989; see Hamrick et al., 1999, for a different interpretation).

USNM 442255 Intermediate phalanx. Specimen prepared out of the 8ABC Limestone from UM locality SC-4 (Wa-1, early Eocene) in the Clarks Fork Basin. Specimen discussed and illustrated by Beard (1989, Figure 53). Identified as possibly belonging to the pes based on lack of elongation (Beard, 1989; see Hamrick et al., 1999, for a different interpretation).

USNM 442284 Pelvic fragment. Specimen prepared out of the 8ABC Limestone from UM locality SC-4 (Wa-1, early Eocene) in the Clarks Fork Basin. Specimen mentioned, but not described or illustrated, by Beard (1989).

USNM 442285 Pelvic fragment. Specimen prepared out of the 8ABC Limestone from UM locality SC-4 (Wa-1, early Eocene) in the Clarks Fork Basin. Specimen mentioned, but not described or illustrated, by Beard (1989).

USNM 442286 Pelvic fragment. Specimen prepared out of the 8ABC Limestone from UM locality SC-4 (Wa-1, early Eocene) in the Clarks Fork Basin. Specimen described and illustrated by Beard (1989, Figure 35).

USNM 442245 Femur missing the femoral head. Specimen prepared out of the 8ABC Limestone from UM locality SC-4 (Wa-1, early Eocene) in the Clarks Fork Basin. Specimen described and illustrated by Beard (1989, Figure 54).

USNM 442246 Proximal femur. Specimen prepared out of the 8ABC Limestone from UM locality SC-4 (Wa-1, early Eocene) in the Clarks Fork Basin. Specimen described and illustrated by Beard (1989, Figure 54).

USNM 442234 Astragalus. Specimen prepared out of the 8ABC Limestone from UM locality SC-4 (Wa-1, early Eocene) in the Clarks Fork Basin. Specimen mentioned, but not illustrated, by Beard (1989).

USNM 442235 Right astragalus. Specimen prepared out of the 8ABC Limestone from UM locality SC-4 (Wa-1, early Eocene) in the Clarks Fork Basin. Specimen described and illustrated by Beard (1989, Figure 55).

USNM 442239 Calcaneum. Specimen prepared out of the 8ABC Limestone from UM locality SC-4 (Wa-1, early Eocene) in the Clarks Fork Basin. Specimen described, but not illustrated, by Beard (1989).

USNM 442240 Calcaneum. Specimen prepared out of the 8ABC Limestone from UM locality SC-4 (Wa-1, early Eocene) in the Clarks Fork Basin. Specimen described, but not illustrated, by Beard (1989).

USNM 442241 Calcaneum. Specimen prepared out of the 8ABC Limestone from UM locality SC-4 (Wa-1, early Eocene) in the Clarks Fork Basin. Specimen described, but not illustrated, by Beard (1989).

Ignacius clarkforkensis

UM 39893 Femur. Specimen prepared out of a Limestone from UM locality SC-117 (Cf-2, late Paleocene) in the Clarks Fork Basin. Specimen listed, but not described or illustrated, by Bloch (2001).

UM 39897 Astragalus. Specimen prepared out of a Limestone from UM locality SC-117 (Cf-2, late Paleocene) in the Clarks Fork Basin. Specimen listed, but not described or illustrated, by Bloch (2001).

UM 39926 Scaphoid. Specimen prepared out of a Limestone from UM locality SC-117 (Cf-2, late Paleocene) in the Clarks Fork Basin. Specimen listed, but not described or illustrated, by Bloch (2001).

UM 108210 Represented by skull and dentaries; distal, intermediate and proximal phalanges of the manus; metacarpals I, III and IV of the right hand and I of the left hand; left and right radius, ulna and humerus; cervical vertebrae I-V, VII; thoracic vertebrae I and three others; parts of the scapula; right astragalus; right and left calcanea. Specimens were retrieved from the same limestone lens as *A. hebeticus* but from a different area within it. See following section Figures 11.3 and 11.4 for documented association.

UM 82606 From the same locality, but from a separate nodule, as UM 108210. It includes metacarpal V; the last three lumbar vertebrae; sacrum; caudal vertebrae I-VII, and three others; left and right innominates; left and right femora, tibiae and fibulae; right astragalus, calcaneum, cuboid, navicular, ectocuneiform, mesocuneiform, and metatarsals I-III. See following section Figures 11.3 and 11.4 on documenting association.

Phenacolemur praecox

UM 37497 Femur. Specimen prepared out of a Limestone from UM locality SC-210 (Wa-1, early Eocene) in the Clarks Fork Basin. Specimen listed, but not described or illustrated, by Bloch (2001).

UM 86440 Dentally associated partial skeleton. Collected from UM locality SC-46 (Wa-2, early Eocene) of the Clarks Fork Basin and thought to be part of the same individual as UM 86352 (Beard, 1989, 1990). Discussed and described by Beard (1989). Partially illustrated by Beard (1989, Figures 32, 33, 41).

UM 86352 Dentally associated partial skeleton. Collected from UM locality SC-46 (Wa-2, early Eocene) of the Clarks Fork Basin and thought to be part of the same individual as UM 86440 (Beard, 1989, 1990). Discussed and described by Beard (1989). Partially illustrated by Beard (1989, Figures 31, 32, 43, 46).

Phenacolemur jepseni

USGS 17847 Dentally associated partial skeleton. Collected from USGS locality D-1651 (Wa-7, early Eocene) of the Bighorn Basin. Discussed and described by Beard (1989). Partially illustrated by Beard (1989, Figures 23, 24, 32, 33, 48). Also mentioned by Beard (1990).

Phenacolemur simonsi

The following specimens were partially discussed and figured by Beard (1989, 1990; also listed in Bloch, 2001). All are isolated elements prepared out of the 8ABC Limestone from UM locality SC-4 (Wa-1, early Eocene) in the Clarks Fork Basin: USNM 442230, right tibia; USNM 442231, tibia; USNM 442236, right astragalus; USNM 442237, astragalus; USNM 442238, right calcaneum; USNM 442242, proximal femur; USNM 442243, proximal femur; USNM 442244, proximal femur; USNM 442247, proximal phalanx, missing distal end; USNM 442248, proximal phalanx; USNM 442249, proximal phalanx; USNM 442250, intermediate phalanx; USNM 442251, intermediate phalanx; USNM 442252, intermediate phalanx; USNM 442254, intermediate phalanx; USNM 442257, left hallucal metatarsal; USNM 442258, left pollical metacarpal; USNM 442260, complete left humerus; USNM 442261, nearly complete left radius; USNM 442262, right proximal radius; USNM 442263, left radius lacking distal end; USNM 442264, fragment of left radial shaft; USNM 442265, right proximal ulna; USNM 442266, humerus; USNM 442267, left proximal ulna; USNM 442268, humerus; USNM 442287, right pelvic fragment.

Family Micromomyidae

Chalicomomys antelucanus

USNM 442282 Distal left humerus. Specimen prepared out of the 8ABC Limestone from UM locality SC-4 (Wa-1, early Eocene) in the Clarks Fork Basin. Described by Beard (1989).

USNM 442283 Nearly complete right humerus. Specimen prepared out of the 8ABC Limestone from UM locality SC-4 (Wa-1, early Eocene) in the Clarks Fork Basin. Described by Beard (1989).

Tinimomys graybulliensis

USNM 461201 Partial skeleton and skull from UM locality SC-26 (Wa-1, early Eocene). Illustrated in limestone (not fully prepared) by Beard (1993b, Figure 11.6). Briefly discussed, but largely undescribed, by Beard (1993b).

USNM 461202 Associated forelimb elements from UM locality SC-26 (Wa-1, early Eocene). Mentioned, but not described or illustrated, by Beard (1993b).

UM 85176 Distal radius and ulna from UM locality SC-327 (Cf-3, late Paleocene), discussed by Beard (1989). Associated with a petrosal fragment, mandible, and two isolated upper teeth, described and illustrated by Gunnell (1989, Figure 11.28).

The following specimens were partially discussed and figured by Beard (1989; also listed in Bloch, 2001). All are isolated elements prepared out of the 8ABC Limestone from UM locality SC-4 (Wa-1, early Eocene) in the Clarks Fork Basin:

USNM 442269, left proximal humerus; USNM 442270, distal humerus; USNM 442271, distal humerus; USNM 442272, distal humerus; USNM 442273, distal humerus; USNM 442274, distal humerus; USNM 442275, left proximal radius; USNM 442276, nearly complete right ulna; USNM 442277, right pelvic fragment; USNM 442278, right pelvic fragment; USNM 442279, left femur; USNM 442280, left femur; USNM 442281, right femur.

Dryomomys szalayi

UM 41870 Partial skeleton and skull from UM locality SC-327 (Cf-3, late Paleocene). Skull and jaws illustrated in Bloch (2001). Briefly discussed in Bloch (2001) and Bloch et al. (2003). See Figure 11.6 section for documented association.

Appendix II

Extant specimens used in comparative analyses

TABLE 1. This table gives all specimens measured or observed in detail in undertaking this research. Each figure depicting more than four specimens is included as a column. An "x" is placed in the row representing each specimen included in a particular figure. If the specimen has a letter label in the figure, that letter indicates its presence, rather than an "x". Some specimens are represented by more than one datum in the phalangeal analyses (if we measured more than one of their digits/phalanges). For such specimens, the number of digit rays representing them is given in the column titled "n phalanges."

Taxon	*Museum #*	N phalanges	Figure 11.8	Figure 11.10	Figure 11.11	Figure 11.12	Figure 11.13	Figure 11.14	Figure 11.21	Figure 11.25	Figure 11.27	Figure 11.29	Figure 11.30
Artiodactyla													
Bison bison	UMMZ 114227	na								0			
Ovis aries	UMMZ 123400	na								0			
Ovis aries	UMMZ 125084	na								0			
Sus scrofa	UMMZ 167210	na								0			
Carnivora													
Canis lupus	UMMZ 38960	na											
Canis lupus	UMMZ 103506	na								1			
Canis lupus	UMMZ 115800	na								1			
Panthera tigris	UMMZ 160141	na								1			
Lynx rufus	UMMZ 157258	na								1			
Panthera pardus	UMMZ 102451	na								1			
Paradoxurus hermaphroditus	UMMZ 157016	na								1			
Paradoxurus hermaphroditus	UMMZ 157019	na								1			
Potos flavus modestus	UMMZ 107906	na								0			
Potos flavus modestus	UMMZ 158610	na								0			
Procyon lotor	UMMZ 65802	na								1			
Procyon lotor hirtus	UMMZ 168355	na								1			
Ursus sp	UMMZ 103537	na								1			
Chiroptera													
Myotis	UMR 1504	na											
Pteropus pumilio	UMMZ 57081	4			x	x	D						
Pteropus pumilio	YPM 133	1				x							
Rousettus aegyptiacus	YPM 5086	5				x							
Rousettus aegyptiacus	YPM 5087	5				x							
Rousettus aegyptiacus	YPM 5088	5				x							
Dermoptera													
Galeopterus variegatus	Hamrick et al. (1999)	4	x	x	x	x		x					
Cynocephalus volans	Hamrick et al. (1999)	4	x	x	x	x		x					
Cynocephalus volans	Hamrick et al. (1999)	4	x	x	x	x		x					
Cynocephalus volans	Hamrick et al. (1999)	2	x	x	x	x		x					
Cynocephalus volans	USNM 56530	4		x	x	x	E	x	x	x	x		
Cynocephalus volans	UF 5969											B	
Lagomorpha													
Lepus crawshayi	UMMZ 123226	na								1			
Lepus sp	UMMZ x	na								1			
Lipotyphla													
Sorex sp.	UMMZ 157176	4	x	x				x					
Sorex sp.	UMMZ 165481	2	x	x				x					

(continued)

TABLE 1. (continued)

Taxon	*Museum #*	N phalanges	Figure 11.8	Figure 11.10	Figure 11.11	Figure 11.12	Figure 11.13	Figure 11.14	Figure 11.21	Figure 11.25	Figure 11.27	Figure 11.29	Figure 11.30
Marsupalia													
Didelphis virginianus	RD154	na								0			
Petaurus breviceps	Hamrick et al. (1999)	4			x								
Petaurus breviceps	Hamrick et al. (1999)	4			x								
Petaurus breviceps	Hamrick et al. (1999)	1			x								
Petaurus breviceps	UMMZ 158619	na								1			
Petaurus breviceps	UMMZ 160143	na							x	1	x		
Phalanger sp	UMMZ 159321	na								0			
Trichosurus vulpecula	UMMZ 157192	2	x		x				x	0	x		
Primates													
Alouatta caraya	UMMZ 12489	na											
Alouatta palliata aequatorialus	UMMZ 77300	na								0			
Alouatta palliata mexicana	UMMZ 116300	na											
Alouatta palliata pigra	UMMZ 63503	na								0			
Alouatta pigra	UMMZ 63511	na											
Ateles geoffroyi	UMMZ 63171	na							x		x		
Ateles geoffroyi	UMMZ 63165	na									x		
Callimico goeldii	UMMZ 160149	na							x		x		
Callithrix jacchus	UMMZ 91119	na											
Cebuella pygmaea	UMMZ 160146	4	x	x	x	x		x	x	1	x		x
Cebuella pygmaea	SBU NC-01	8		x	x	x	B	x	x		x	F	
Cebus apella paraguayanus	UMMZ 126129	na								1			
Cebus apella paraguayanus	UMMZ 126131	na								1			
Cebus sp.	UMMZ 126130	4	x	x				x					
Cercopithecus sp.	UMMZ 107905	4		x				x					
Daubentonia madagascariensis	AMNH 185643	7											
Eulemur fulvus	UMMZ 160910	na				x			x	1	x		
Galago senegalensis	Hamrick et al. (1999)	2	x	x	x	x		x					
Galago senegalensis	Hamrick et al. (1999)	2	x	x	x	x		x					
Galago senegalensis	Hamrick et al. (1999)	2	x	x	x	x		x					
Galago senegalensis	Hamrick et al. (1999)	3	x	x	x	x		x					
Galago senegalensis	UMMZ 113349	na									x		
Galago senegalensis	UMMZ 113350	na									x		
Galago senegalensis	UMMZ 113351	na							x	1	x		
Galago senegalensis	UMMZ 113352	na											
Galagoides demidovii	SBU PGa-02	na											
Hylobates hoolock	UMMZ 160909	3	x	x				x					
Hylobates hoolock	UMMZ 160908	3	x							0			
Leontopithecus rosallia	UMMZ 97464	na								1			x
Nycticebus coucang	UMMZ 113356	4	x							0	x		x
Nycticebus coucang	UMMZ 113353	na								0	x		
Nycticebus coucang	UMMZ 113354	na								0			
Nycticebus coucang	UMMZ 113355	na									x		
Nycticebus coucang	SBU PNc-01	na										C	
Otolemur crassicaudatus	UMMZ 113345	na											
Otolemur crassicaudatus	UMMZ 113347	na											
Otolemur crassicaudatus	UMMZ 113344	na											
Saguinus mystax	UMMZ 160148	4	x	x	x			x	x	1	x		

Saguinus oedipus	UMMZ 157195	na							1	x
Saguinus oedipus	UMMZ 156437	na								
Saguinus oedipus	UMMZ 160147	na								x
Saguinus sp.	SBU NSg06	8		x	x		x			
Saimiri sciureus	UMMZ 113357	na							1	
Saimiri sciureus	UMMZ 46415	3	x							
Saimiri sciureus	UMMZ 113358	na								
Saimiri sciureus	UMMZ 120304	na								
Tarsius bancanus	Hamrick et al. (1999)	3	x	x	x	x	x			
Tarsius bancanus	Hamrick et al. (1999)	4	x	x	x	x	x			
Tarsius spectrum	Hamrick et al. (1999)	4	x	x	x	x	x			
Tarsius syrichta	Hamrick et al. (1999)	3	x	x	x	x	x			
Tarsius syrichta	UMMZ 95741	na						x		x
Tarsius syrichta	UMR 139	na						x	1	x
Varecia variegata	UMMZ 172669	na								
Rodentia										
Anomalurus beecrofti	Hamrick et al. (1999)	3	x	x	x	x	x			
Anomalurus beecrofti	AMNH coll.	4		x		x	x			
Anomalurus derbianus beecroftii	AMNH 52100	4				x				
Anomalurus derbianus beecroftii	AMNH 52101	8				x				
Anomalurus derbianus	AMNH 241147	3				x				
Anomalurus derbianus nicrensis	AMNH 89389	3				x				
Anomalurus sp.	AMNH 5101	6				x				
Calomyscus sp.	UMMZ 167176	4	x	x			x			
Cavia porcellus	UMMZ 86465	na							1	
Cavia porcellus	UMMZ 88422	na							1	
Erethizon dorsatum bruneri	UMMZ 39970	na							0	
Erethizon dorsatum dorsatum	UMMZ 53682	na							0	
Erethizon sp.	UMMZ 116343	4		x			x			
Gerbillus sp.	UMMZ 117669	3	x	x			x			
Glaucomys sabrinus	UMMZ 113398	na						x		x
Glaucomys volans	Hamrick et al. (1999)	4	x	x	x		x			
Glaucomys volans	Hamrick et al. (1999)	2	x	x	x		x			
Glaucomys volans	Hamrick et al. (1999)	3	x	x	x		x			
Glaucomys volans	Hamrick et al. (1999)	4	x	x	x		x			
Glaucomys volans	UMMZ 168356	na						x		x
Marmota monax	UMMZ 166225	4		x			x			
Marmota monax	UMMZ 170527	na							0	
Mesocricetus auratus	UMMZ 86440	na							0	
Mesocricetus auratus	UMMZ 121017	na							0	
Phodopus sungorus	UMMZ 168149	na							0	
Phodopus sungorus	UMMZ 168150	na							0	
Rattus sp.	SBU MRD-28	na								
Sciurus carolinensis	SBU MRd10	na						x		x
Sciurus niger	UMMZ TC 269	3	x	x	x		x	x		
Sciurus niger	Hamrick et al. (1999)	2	x	x	x		x			
Sciurus niger	Hamrick et al. (1999)	2	x	x	x		x			
Sciurus niger	UMMZ 3959	na							1	
Sciurus niger	UMMZ 5195	na							1	
Citellus mexicanus	UMR 1716	na						x		x
Citellus mexicanus	UMMZ 126129	na							1	
Synaptomys cooperi cooperi	UMMZ 146833	na							0	
Synaptomys cooperi cooperi	UMMZ 160087	na							0	

(continued)

TABLE 1. (continued)

Taxon	*Museum #*	N phalanges	Figure 11.8	Figure 11.10	Figure 11.11	Figure 11.12	Figure 11.13	Figure 11.14	Figure 11.21	Figure 11.25	Figure 11.27	Figure 11.29	Figure 11.30
Tamias striatus rafescens	UMMZ 165187	na								1			
Tamias striatus rafescens	UMMZ 165188	4	x	x				x		1			
Scandentia													
Ptilocercus lowii	BMNH[a]	na									x		
Ptilocercus lowii	YPM 10179	na											
Tupaia glis	UMMZ 118389	na							x	1	x		
Tupaia glis	SBU collection	na										D	
Tupaia tana	Hamrick et al. (1999)	1	x	x	x	x		x					
Tupaia tana	Hamrick et al. (1999)	2	x	x	x	x		x					
Tupaia tana	Hamrick et al. (1999)	2	x	x	x	x		x					
Tupaia tana	Hamrick et al. (1999)	3	x	x	x	x		x					
Xenarthra													
Bradypus tridactylus	UMMZ 64943	na							x				
Bradypus tridactylus	SBU MEd-01	3				x							
Bradypus tridactylus	SBU MEd-02	3				x							
Choloepus hoffmanni	UMMZ 64944	na								0			
Choloepus hoffmanni	UMMZ 64940	na					F		x	0	x		
Choloepus hoffmanni	AMNH 15960	na										A	
Choloepus hoffmanni	SBU MEd-03	3				x							
Choloepus hoffmanni	YPM 268	8				x							
Choloepus hoffmanni	YPM 282	6				x							
Fossil specimens													
Carpolestes simpsoni	UM 101963	3	2	2	x			2	x	x	x		
Dryomomys szalayi	UM 41870	7	x	x		x	C	x	x		x		
Tinimomys graybulliensis	SC-026 specimen	na											
Tinimomys graybulliensis	UM 85176	na											
Acidomomyds hebeticus	UM 108207	5	x	4	4			4					
Acidomomyds hebeticus	UM 108208	na											
Acidomomyds hebeticus	UM 108209	na											
Ignacius clarkforkensis	UM 108210	2	x	x	1	x	A	x	x		x		
Ignacius clarkforkensis	UM 82606	na							x	x	x	E	x
Ignacius graybullianus	USNM 442259	na											
Phenacolemur praecox	UM 66440	1				x							
Phenacolemur praecox	UM 86352	4				x							
Phenacolemur jepseni	USGS 17847[b]	1				x							
Phenacolemur simonsi	USNM 442260[b]	na											
Phenacolemur simonsi	USNM 442262	na											
Nannodectes gidleyi	AMNH 17379	3	1	x	x	x		x					
Nannodectes intermedius	USNM 442229	4	1	2	x	x		2		x			
Plesiadapis cookei	UM 87990	4	2	2	x	x		2	x		x		
Plesiadapidis tricuspidens	MNHN CR 448	na											
Plesiadapidis tricuspidens	MNHN CR 450	na											
Plesiadapidis tricuspidens	MNHN CR 546	na											
Plesiadapidis tricuspidens	MNHN CR 550	na											
Plesiadapidis tricuspidens	MNHN CR 611	na											
Notharctus tenebrosus	UM 99999	na											
Omomys sp.	UM 98604	na											
Omomys carteri	UM 14134	na											
Smilodectes mcgrewi	UM 95526	na											x

[a]Measurements taken on photograph provided by Le Gros Clark (1927)

TABLE 2. This table gives raw measurements (mm) for vertebrae of the four specimens in Figure 11.22.

Ignacius clarkforkensis			*Saguinus mystax* UMMZ 160148			*Scuirus niger* UMMZ TC 269			*Cynocephalus volans*		
Vert level	Height	Length	Vert level	Height	Length	Vert level	Height	Length	Vert level	Height	Length
C1	6.0	1.5	C1	10.3	2.5	C1	10.3	2.2	C1	9.7	5.1
C2	2.1	5.4	C2	2.9	6.5	C2	2.6	7.1	C2	4.6	14.3
C3	1.9	3.4	C3	2.6	5.6	C3	2.7	4.2	C3	4.6	11.3
C4	1.8	3.3	C4	2.5	4.9	C4	3.0	4.2	C4	4.4	10.9
C5	2.0	3.1	C5	2.3	4.3	C5	2.9	4.6	C5	4.9	11.0
C6	-	-	C6	2.3	4.0	C6	2.8	4.0	C6	4.1	10.0
C7	1.9	3.5	C7	2.1	3.9	C7	2.6	3.9	C7	4.1	10.4
T1	1.8	3.7	T1	2.1	4.2	T1	2.9	4.8	T1	4.0	8.1
T2	-	-	T2	2.3	4.7	T2	2.9	5.4	T2	3.7	8.1
T3	-	-	T3	2.1	4.3	T3	2.8	5.6	T3	3.7	7.9
T4(?)	2.2	4.9	T4	2.1	4.6	T4	2.8	5.6	T4	3.9	7.9
T5	-	-	T5	2.2	4.4	T5	3.0	5.6	T5	4.0	8.1
T6	-	-	T6	2.2	4.5	T6	3.1	5.7	T6	3.7	8.0
T7(?)	2.3	5.3	T7	2.4	4.6	T7	3.3	6.2	T7	3.7	8.1
T8	-	-	T8	2.4	4.8	T8	3.5	6.5	T8	4.0	8.3
T9	-	-	T9	2.7	5.3	T9*	3.5	6.9	T9	4.9	8.5
T10(?)	2.4	5.4	T10*	2.7	6.1	T10	3.7	7.5	T10	4.7	8.5
T11*	-	-	T11	3.1	7.1	T11	3.9	7.9	T11	4.7	8.5
T12	-	-	T12	3.3	7.8	T12	4.1	8.1	T12	4.4	9.0
									T13	4.4	9.5
									T14	4.7	9.6
L1	-	-	L1	3.7	8.5	L1	4.2	10.7	L1	4.9	10.6
L2	-	-	L2	3.8	9.3	L2	4.2	9.8	L2	5.3	10.7
L3	-	-	L3	3.8	10.1	L3	4.2	10.1	L3	5.0	11.1
L4	-	-	L4	3.8	10.3	L4	4.0	11.1	L4	5.0	12.3
L5	3.3	8.1	L5	4.0	10.7	L5	4.3	11.5	L5	5.2	12.3
L6	3.3	8.2	L6	4.3	10.3	L6	4.3	11.4	L6	5.3	12.3
L7	3.4	6.9	L7	4.4	8.8	L7	4.4	9.2	L7	5.2	12.5
									L8*	4.7	10.8
S1	3.2	6.6	S1	3.8	6.2	S1	3.0	8.1	S1	3.0	11.0
S2	2.6	5.9	S2	3.6	5.1	S2	2.2	8.1	S2	3.0	10.0
S3	2.2	5.9	S3	3.7	6.0	S3	2.9	7.6	S3	2.4	11.7
Ca1	2.3	5.3	Ca1	3.7	6.0	Ca1	2.8	6.3	Ca1	3.4	11.6
Ca2	2.6	5.4	Ca2	4.0	6.2	Ca2	2.8	5.7	Ca2	3.7	11.6
Ca3	2.7	5.1	Ca3	4.1	7.1	Ca3	2.9	5.7	Ca3	5.3	13.5
Ca4	3.0	6.3	Ca4	4.4	9.4	Ca4	3.0	6.6	Ca4	4.7	15.4
Ca5	3.4	9.8	Ca5	4.5	13.8	Ca5	3.2	6.6	Ca5	4.4	17.9
Ca6	3.8	14.9	Ca6	4.8	16.6	Ca6	3.2	6.8	Ca6	4.1	20.0
Ca7	3.5	15.8	Ca7	4.7	17.6	Ca7	3.2	7.9	Ca7	3.7	21.1
Ca8	-	-	Ca8	4.6	18.0	Ca8	3.8	9.4	Ca8	3.9	23.2

(continued)

TABLE 2. (continued)

Ca9	2.3	14.8	Ca9	4.6	18.2	Ca9	4.0	11.3	Ca9	3.7	24.3
Ca10	-	-	Ca10	4.4	18.3	Ca10	4.1	12.7	Ca10	3.6	24.8
Ca11	-	-	Ca11	4.3	18.1	Ca11	3.9	13.6	Ca11	3.6	25.1
Ca12	1.4	9.3	Ca12	4.5	17.6	Ca12	3.7	13.7	Ca12	3.6	24.9
Ca13	1.2	9.2	Ca13	3.8	17.1	Ca13	3.4	13.3	Ca13	3.3	24.4
-	-	-	Ca14	3.6	16.7	Ca14	3.2	13.1	Ca14	2.8	22.3
-	-	-	Ca15	3.5	16.3	Ca15	3.1	12.9	Ca15	2.4	20.3
-	-	-	Ca16	3.3	15.9	Ca16	2.7	12.5	Ca16	1.9	16.8
-	-	-	Ca17	3.1	15.0	Ca17	2.6	12.1	Ca17	1.6	14.0
-	-	-	Ca18	3.0	14.3	Ca18	2.3	11.8	Ca18	1.2	9.0
-	-	-	Ca19	2.8	13.4	Ca19	2.2	11.0			
-	-	-	Ca20	2.6	12.7	Ca20	1.9	10.7			
-	-	-	Ca21	2.6	12.0	Ca21	1.5	10.2			
-	-	-	Ca22	2.4	11.1	Ca22	1.5	9.7			
-	-	-	Ca23	2.2	10.2	Ca23	1.3	9.1			
-	-	-	Ca24	2.0	10.2	Ca24	1.2	8.4			
-	-	-	Ca25	1.8	8.5	Ca25	1.0	6.7			
-	-	-	Ca26	1.7	8.5	Ca26	0.5	5.6			
-	-	-	Ca27	1.6	6.7						
-	-	-	Ca28	1.7	5.3						

References

Beard, K. C., 1989. Postcranial anatomy, Locomotor Adaptations, and Palaeoecology of Early Cenozoic Plesiadapidae, Paromomyidae, and Micromomyidae (Eutheria, Dermoptera). Ph.D. dissertation, Johns Hopkins University, Baltimore, MD.

Beard, K. C., 1990. Gliding behavior and palaeoecology of the alleged primate family Paromomyidae (Mammalia, Dermoptera). Nature 345, 340–341.

Beard, K. C., 1991. Vertical posture and climbing in the morphotype of Primatomorpha: implications for locomotor evolution in primate history. In: Coppens, Y., Senut, B. (Eds.), Origine(s) de la Bipedie chez les Hominides. CNRS, Paris, pp. 79–87.

Beard, K. C., 1993a. Phylogenetic systematics of the Primatomorpha, with special reference to Dermoptera. In: Szalay, F., Novacek, M., McKenna, M. (Eds.), Mammal Phylogeny: Placentals. Springer, New York, pp. 129–150.

Beard, K. C., 1993b. Origin and evolution of gliding in early cenozic dermoptera (Mammalia, Primatomorpha). In: MacPhee, R. (Ed.), Primates and their Relatives in Phylogenetic Perspective. Plenum, New York, pp. 63–90.

Beard, K. C., 2006. Iterative dispersal aross Beringia by early Cenozoic primates. American Journal of Physical Anthropology 129, 62A.

Benton, R., 1967. Morphological evidence for adaptations within the epaxial region of the primates. In: Vagtborg, H. (Ed.). The Baboon in Medical Research. University of Texas Press, Austin, pp. 201–216.

Biknevicius, A. R., 1986. Dental function and diet in the Carpolestidae (Primates: Plesiadapiformes). American Journal of Physical Anthropology 71, 157–171.

Bloch, J. I. 2001. Mammalian paleontology of freshwater limestones from the Paleocene-Eocene of the Clarks Fork Basin, Wyoming, Ph. D. dissertation, University of Michigan, Ann Arbor, MI.

Bloch, J.I., and Boyer, D. M. 2001. Taphonomy of small mammals in freshwater limestones from the Paleocene of the Clarks Fork Basin. In: Gingerich, P. D. (Ed.). Paleocene-Eocene Stratigraphy and Biotic Change in the Bighorn and Clarks Fork Basin, Wyoming, University of Michigan Papers on Paleontology (33) pp. 185–198.

Bloch, J. I., Boyer, D. M., 2002a. Grasping primate origins. Science 298, 1606–1610.

Bloch, J.I., Boyer, D.M., 2002b. Phalangeal morphology of Paleocene plesiadapiforms (Mammalia, ?Primates): evaluation of the gliding hypothesis and the first evidence for grasping in a possible primate ancestor. Geological Society of America Abstracts with Programs 34, 13A.

Bloch, J. I., Boyer, D. M., 2007. New skeletons of Paleocene-Eocene Plesiadapiformes: a diversity of arboreal positional behaviors in early primates. In: Ravosa, M., Dagosto, M. (Eds.). Primate Origins: Adaptations and Evolution. Springer, New York, pp. 535–581.

Bloch, J.I., Boyer, D.M., Gingerich, P.D., 2001a. Positional behavior of late Paleocene *Carpolestes simpsoni* (Mammalia, ?Primates). Journal of Vertebrate Paleontology 21, 34A.

Bloch, J. I., Boyer, D. M., Gingerich, P. D., Gunnell, G. F., 2002a. New primitive paromomyid from the Clarkforkian of Wyoming and dental eruption in Plesiadapiformes. Journal of Vertebrate Paleontology 22, 366–379.

Bloch, J. I., Boyer, D. M., Houde, P., 2003. New skeletons of Paleocene-Eocene micromomyids (Mammalia, Primates): functional morphology and implications for euarchontan relationships. Journal of Vertebrate Paleontology 23, 35A.

Bloch, J. I., Fisher, D. C., Rose, K. D., Gingerich, P. D., 2001b. Stratocladistic analysis of Paleocene Carpolestidae (Mammalia, Plesiadapiformes) with description of a new late Tiffanian genus. Journal of Vertebrate Paleontology 21, 119–131.

Bloch, J. I., Silcox, M. T., 2001. New basicrania of Paleocene-Eocene *Ignacius*: re-evaluation of the plesiadapiform-dermopteran link. American Journal of Physical Anthropology 116, 184–198.

Bloch, J. I., Silcox, M. T., 2006. Cranial anatomy of the Paleocene plesiadapiform *Carpolestes simpsoni* (Mammalia, Primates) using ultra high-resolution X-ray computed tomography, and the relationships of plesiadapiforms to Euprimates. Journal of Human Evolution 50, 1–35.

Bloch, J. I., Silcox, M. T., Boyer, D. M., Sargis, E. J., 2007. New Paleocene skeletons and the relationship of plesiadapiforms to crown-clade primates. Proceedings of the National Academy of Sciences USA 104, 1159–1164.

Bloch, J. I., Silcox, M. T., Sargis, E. J., Boyer, D. M., 2002b. Origin and relationships of Archonta (Mammalia, Eutheria): Re-evaluation of Eudermoptera and Primatomorpha. Journal of Vertebrate Paleontology 22, 37A.

Bock, W. J., 1977. Adaptation and the comparative method. In: Hecht, M., Goody, P., Hecht, B. (Eds.), Major Patterns in Vertebrate Evolution. Plenum, New York, pp. 57–82.

Bock, W. J., 1981. Functional-adaptive analysis in evolutionary classification. American Zoologist 21, 5–20.

Bock, W. J., 1988. The nature of explanations in morphology. American Zoologist 28, 205–215.

Bock, W. J., 1991. Levels of complexity of organismic organization. In: Valvassori, R. (Ed.), Form and Function in Zoology. Modena, Mucchi, pp. 181–212.

Bock, W. J., 1993. Selection and fitness: definitions and uses: 1859 and now. Proceedings of the Zoological Society of Calcutta, Haldane Commemorative Volume, pp. 7–26.

Bock, W. J., 1999. Functional and evolutionary explanations in morphology. Netherlands Journal of Zoology 49, 45–65.

Bock, W. J., von Wahlert, G., 1965. Adaptation and the form-function complex. Evolution 19, 269–299.

Boyer, D. M., Bloch, J. I., 2002a. Structural correlates of positional behavior in vertebral columns of Paleocene small mammals. Journal of Vertebrate Paleontology 22, 38A.

Boyer, D. M., Bloch, J. I., 2002b. Bootstrap comparisons of vertebral morphology of Paleocene Plesiadapiforms Mammalia, ?Primates): functional implications. Geological Society of America Abstracts with Programs 34, 13A.

Boyer, D. M., Bloch, J. I., Gingerich, P. D., 2001. New skeletons of Paleocene paromomyids Mammalia, ?Primates): were they mitten gliders? Journal of Vertebrate Paleontology 21, 35A.

Boyer, D. M., Bloch, J. I., Silcox, M. T., Gingerich, P. D., 2004. New observations on anatomy of *Nannodectes* (Mammalia, Primates) from the Paleocene of Montana and Colorado. Journal of Vertebrate Paleontology 24, 40A.

Cartmill, M., 1972. Arboreal adaptations and the origin of the Order Primates. In: Tuttle, R. (Ed.), The Functional and Evolutionary Biology of Primates. Aldine-Atherton, Chicago, IL, pp. 97–122.

Cartmill, M., 1974. Pads and claws in arboreal locomotion. In: Jenkins, F. (Ed.), Primate Locomotion. Plenum, New York, pp. 45–83.

Dagosto, M., Gebo, D. L., Beard, K. C., 1999. Revision of the Wind River faunas, Early Eocene of central Wyoming. Part 14. Postcranium of *Shoshonius cooperi* (Mammalia: Primates). Annals of the Carnegie Museum 68, 175–211.

Fleagle, J. G., 1999. Primate Adaptation and Evolution. Academic, New York.

Gambaryan, P. P., 1974. How Mammals Run. Wiley, New York.

Gebo, D. L., 1989. Postcranial adaptation and evolution in Lorisidae. Primates 30, 347–267.

Gervais, M. P., 1877. Enumeration de quelques ossements d'animaux vertebres, recueillis aux environs de Reims par M. Lemoine. Journal de Zoologie 6, 74–79.

Gidley, J. W., 1923. Paleocene primates of the Fort Union, with discussion of relationships of Eocene primates. Proceedings of the United States National Museum 63, 1–38.

Gingerich, P. D., 1974. Function of pointed premolars in *Phenacolemur* and other mammals. Journal of Dental Research 53, 497.

Gingerich, P. D., 1975. Systematic position of *Plesiadapis*. Nature 253, 111–113.

Gingerich, P. D., 1976. Cranial anatomy and evolution of Early Tertiary Plesiadapidae (Mammalia, Primates). University of Michigan Papers on Paleontology 15, 1–141.

Gingerich, P. D., Gunnell, G. F., 2005. Brain of *Plesiadapis cookei* (Mammalia, Propriamtes): surface morphology and encephalization compared to those of Primates and Dermoptera. University of Michigan Museum of Paleontology Contributions 31, 185–195.

Godfrey, L. R., 1988. Adaptive diversification of Malagasy strepsirrhines. Journal of Human Evolution 17, 93–184.

Godinot, M., 1984. Un nouveau de Paromomyidae (Primates) de l'Eocene Inferieur d'Europe. Folia Primatologica 43, 84–96.

Godinot, M., Beard, K. C., 1991. Fossil primate hands: a review and an evolutionary inquiry emphasizing early forms. Journal of Human Evolution 6, 307–354.

Gregory, W. K., 1910. The orders of mammals. Bulletin of the American Museum of Natural History 27, 1–524.

Gregory, W. K., 1920. On the structure and relations of *Notharctus*, an American Eocene primate. Memoirs of the American Museum of Natural History 3, 45–243.

Gunnell, G. F., 1989. Evolutionary history of Microsyopoidea (Mammalia, ?Primates) and the relationship between Plesiadapiformes and primates. University of Michigan, Papers on Paleontology 27, 1–157.

Gunnell, G. F., and Gingerich, P. D., 1987. Skull and partial skeleton of *Plesiadapis cookei* from the Clark Fork Basin, Wyoming. American Journal of Physical Anthropology 72, 206–206.

Haines, R. W., 1958. Arboreal or terrestrial ancestry of placental mammals. Quarterly Review of Biology 33, 1–23.

Hamrick, M. W., 2001. Primate origins: evolutionary change in digit ray patterning and segmentation. Journal of Human Evolution 40, 339–351.

Hamrick, M. W., Rosenman, B. A., Brush, J. A., 1999. Phalangeal morphology of the Paromomyidae (?Primates, Plesiadapiformes): The evidence for gliding behavior reconsidered. American Journal of Physical Anthropology 109, 397–413.

Hoffstetter, R., 1977. Phylogenie des primates. Confrontation des resultats obtenus par les diverses voies d'approche du probleme. Bulletin et mémories de la Société d'anthropologie de Paris 4, 327–346.

Hooker, J. J., 2001, Tarsals of the extinct insectivoran family Nyctitheriidae (Mammalia): evidence for archontan relationships. Zoological Journal of the Linnean Society 132, 501–529.

Horovitz, I., 2003, Postcranial skeleton of *Ukhaatherium nessovi* (Eutheria, Mammalia) from the late Cretaceous of Mongolia. Journal of Vertebrate Paleontology 23, 857–868.

Jenkins, F. A. Jr., 1970. Anatomy and function of expanded ribs in certain edentates and primates. Journal of Mammalogy 51, 288–301.

Jenkins, F. A. Jr., 1974. Tree shrew locomotion and the origins of primate arborealism. In: Jenkins, F. A. (Eds.), Primate Locomotion. Academic, New York, pp. 85–115.

Jenkins, F. A. Jr., Camazine, S. M., 1977. Hip structure and locomotion in ambulatory and cursorial carnivores. Journal of Zoology, London 181, 351–370.

Kay, R. F., 2003. Review of the Primate Fossil Record, edited by W. Hartwig. American Journal of Human Biology 15, 839–840.

Kay, R. F., Cartmill, M., 1977. Cranial morphology and adaptations of *Palaechthon nacimienti* and other Paromomyidae (Plesiadapoidea,? Primates), with a description of a new genus and species. Journal of Human Evolution 6, 19–35.

Kay, R. F., Thorington, R. W. Jr., Houde, P., 1990. Eocene plesiadapiform shows affinities with flying lemurs not primates. Nature 345, 342–344.

Kay, R. F., Thewissen, J. G. M., Yoder, A. D., 1992. Cranial anatomy of *Ignacius graybullianus* and the affinities of the Plesiadapiformes. American Journal of Physical Anthropology 89, 477–498.

Krause, D. W., 1991. Were paromomyids gliders? Maybe, maybe not. Journal of Human Evolution 21, 177–188.

Lemelin, P., 1999. Morphological correlates of substrate use in didelphid marsupials: implications for primate origins. Journal of Zoology 247, 165–175.

Lemoine, V., 1887. Sur genre *Plesiadapis*, mammifere fossile de le'eocene inferieur des environs de Reims. Comptes-rendus des Seances de I'Académie des Sciences, Paris. for 1887, 190–193.

Liu, F.-G. R., Miyamoto, M. M., 1999. Phylogenetic assessment of molecular and morphological data for eutherian mammals. Systematic Biology 48, 54–64.

Liu, F.-G. R., Miyamoto, M. M., Freire, N. P., Ong, P. Q., Tennant, M. R., Young, T. S., Gugel, K. F., 2001. Molecular and morphological supertrees for eutherian (placental) mammals. Science 291, 1786–1789.

MacPhee, R. D. E., Cartmill, M., 1986. Basicranial structures and primate systematics. In: Swindler, D., Erwin, J. (Eds.), Comparative Primate Biology, Volume 1. Alan R. Liss, New York, pp. 219–275.

MacPhee, R. D. E., Cartmill, M., Gingerich, P. D., 1983. New Palaeogene Primate Basicrania and the definition of the Order Primates. Nature 301, 509–511.

Madsen, O., Scally, M., Douady, C. J., Kao, D. J., DeBry, R. W., Adkins, R. M., Amrine, H. M., Stanhope, M. J., Jong, W. W. d., Springer, M. S., 2001. Parallel adaptive radiations in two major clades of placental mammals. Nature 409, 610–614.

Martin, R. D., 1972. Adaptive radiation and behaviour of the Malagasy lemurs. Philosophical Transactions of the Royal Society of London 264, 295–352.

Martin, R. D., 1979. Phylogenetic aspects of prosimian behavior. In: Doyle, G., Martin, R. (Eds.), The Study of Prosimian Behavior. Academic, New York, pp. 45–77.

Martin, R. D., 1986. Primates: a definition. In: Wood, B., Martin, L., Andrews, P. (Eds.), Major Topics in Primate and Human Evolution. Cambridge University Press, Cambridge, pp. 1–31.

McKenna, M. C., Bell, S. K., 1997. Classification of Mammals above the Species Level. Columbia University Press, New York.

Mendel, F. C., 1979. The wrist joint of two-toed sloths and its relevance to brachiating adaptations in the Hominoidea. Journal of Morphology 162, 413–424.

Mendel, F. C., 1985. Adaptations for suspensory behavior in the limbs of two-toed sloths. In: Montgomery, G. G. (Ed.), The Evolution and Ecology of Armadillos, Sloths, and Vermilinguas. Smithsonian Institute, Washington, DC, pp. 151–162.

Meng, J., Hu, Y., Li, C., 2004 The osteology of *Rhombomylus*: implications for the phylogeny and evolution of Glires. Bulletin of the American Museum of Natural History 275, 1–247.

Mill, J. S., 1872. A System of Logic: Ratiocinative and Inductive. Being a Connected View of the Principles of Evidence and the Methods of Scientific Investigation. Logman, London.

Murphy, W. J., Eizirik, E., Johnson, W. E., Zhang, Y. P., Ryderk, O. A., O'Brien, S. J., 2001a. Molecular phylogenetics and the origins of placental mammals. Nature 409, 614–618.

Murphy, W. J., Eizirik, E., O'Brien, S. J., Madsen, O., Scally, M., Douady, C. J., Teeling, E., Ryder, O. A., Stanhope, M. J., de Jong, W. W., Springer, M. S., 2001b. Resolution of the early placental mammal radiation using Bayesian phylogenetics. Science 294, 2348–2351.

Plotnick, R. E., Baumiller, T. K., 2000. Invention by evolution: functional analysis in paleobiology. Paleobiology supplement to 26, 305–323.

Pocock, R. I., 1926. The external characters of the flying lemur (*Galeopterus temminckii*). Proceedings of the Zoological Society of London 29, 429–441.

Rose, K. D., 1987. Climbing adaptations in the early Eocene mammal *Chriacus* and the origin of Artiodactyla. Science 236, 314–316.

Rose, K. D., 1999. Postcranial skeleton of the Eocene Leptictidae (Mammalia), and its implications for behavior and relationships. Journal of Vertebrate Paleontology 15, 401–430.

Rose, K. D., Chinnery, B. J., 2004. The postcranial skeleton of early Eocene rodents. Bulletin of the Carnegie Museum of Natural History 36, 211–243.

Runestad, J. A., Ruff, C. B., 1995. Structure adaptations for gliding in mammals with implications for locomotor behavior in paromomyids. American Journal of Physical Anthropology 98, 101–119.

Russell, D. E., 1959. Le crane de *Plesiadapis*. Bulletin de la Société Géologique de France 1, 312–314.

Russell, D. E., 1964. Les Mammiferes Paléocènes D'Europe. Mémoires du Muséum national d'histoire naturelle. sér. c 13, 1–324.

Sargis, E. J., 2001a. A preliminary qualitative analysis of the axial skeleton of tupaiids (Mammalia, Scandentia): functional morphology and phylogenetic implications. Journal of Zoology 253, 473–483.

Sargis, E. J., 2001b. The grasping behaviour, locomotion and substrate use of the tree shrews *Tupaia minor* and *T. tana* (Mammalia, Scandentia). Journal of Zoology 253, 485–490.

Sargis, E. J., 2002a. Functional morphology of the hindlimb of tupaiids (Mammalia, Scandentia) and its phylogenetic implications. Journal of Morphology 254, 149–185.

Sargis, E. J., 2002b. Functional morphology of the forelimb of tupaiids (Mammalia, Scandentia) and its phylogenetic implications. Journal of Morphology 253, 10–42.

Sargis, E. J., 2002c. The postcranial morphology of *Ptilocercus lowii* (Scandentia, Tupaiidae): An analysis of Primatomorphan and volitantian characters. Journal of Mammalian Evolution 9, 137–160.

Sargis, E. J., 2004. New views on tree shrews: the role of tupaiids in primate supraordinal relationships. Evolutionary Anthropology 13, 56–66.

Sargis, E. J., 2007. The postcranial morphology of *Ptilocercus lowii* (Scandentia, Tupaiidae) and its implications for primate supraordinal relationships. In: Ravosa, M., Dagosto, M. (Eds.), Primate Origins and Adaptations. Springer, New York, pp. 51–82.

Sargis, E. J., Bloch, J. I., Boyer, D. M., Silcox, M. T., 2005. Evolution of grasping in Euarchonta. Journal of Vertebrate Paleontology 25, 109A.

Sargis, E. J., Boyer, D. B., Bloch, J. I., Silcox, M. T. 2007. Evolution of pedal grasping in primates. Journal of Human Evolution 53, 103–107.

Shapiro, L. J., 1993. Functional morphology of the vertebral column in primates. In: Gebo, D. L. (Eds.), Postcranial Adaptation in Nonhuman Primates. Northern Illinois Press, DeKalb, pp. 121–149.

Shapiro, L. J., 1995. Functional morphology of indriid lumbar vertebrae. American Journal of Physical Anthropology 98, 323–342.

Shapiro, L. J., Simons, C. V. M., 2002. Functional aspects of strepsirrhine lumbar vertebral bodies and spinous processes. Journal of Human Evolution 42, 753–783.

Silcox, M. T., 2001. A Phylogenetic Analysis of Plesiadapiformes and Their Relationship to Euprimates and Other Archontans. Ph.D. dissertation, Johns Hopkins University, Baltimore, MD.

Silcox, M. T., 2003. New discoveries on the middle ear anatomy of *Ignacius graybullianus* (Paromomyidae, Primates) from ultra high resolution X-ray computed tomography. Journal of Human Evolution 44, 73–86.

Silcox, M. T., Bloch J. I., Boyer, D. M., Sargis E. J., 2005. Euarchonta (Dermoptera, Scandentia, Primates). In: Rose, K. D., Archibald, J. D. (Eds.), Rise of Placental Mammals. Johns Hopkins University Press, Baltimore, MD, pp. 127–144.

Simmons, N. B., Quinn, T. H., 1994. Evolution of the digital tendon locking mechanism in bats and dermopterans: a phylogenetic perspective. Journal of Mammalian Evolution 2, 231–254.

Simons, E. L., 1972. Primate Evolution: An Introduction to Man's Place in Nature. Macmillan, New York.

Simpson, G. G., 1935. The Tiffany fauna, upper Paleocene. II.-Structure and relationships of *Plesiadapis*. American Museum Novitates 816, 1–30.

Slijper, E. J., 1946. Comparative biologic-anatomical investigations on the vertebral column and spinal musculature of mammals. Verhandlungen des Koninklijke Nederlandse Akademie van Wetenschappen Tweede Sectie 42, 1–128.

Stafford, B. J., 1999. Taxonomy and Ecological Morphology of the Flying Lemurs (Dermoptera, Cynocephalidae). Ph.D. dissertation, City University of New York.

Sokal, R. R., Rohlf, J. F., 1997. Biometry, the Principles and Practice of Statistics in Biological Research. W. H. Freeman, New York.

Stafford, B. J., Szalay, F. S., 2000. Craniodental functional morphology and taxonomy of dermopterans. Journal of Mammalogy 81, 360–385.

Stafford, B. J., Thorington, R. W. J., 1998. Carpal development and morphology in archontan mammals. Journal of Morphology 235, 135–155.

Stehlin, H. G., 1916. Die Saugetiere des Schweizerischen Eocaens. 7 Teil, 2 Halfte. Schweizerische paläontologische Abhandlungen 41, 1299–1552.

Stern Jr., J. T., Jungers, W. L., Susman, R. L., 1995. Quantifying phalangeal curvature: an empirical comparison of alternative methods. American Journal of Physical Anthropology 97, 1–10.

Sussman, R. W., Raven, P. H., 1978. Pollination by lemurs and marsupials: an archaic coevolutionary system. Science 200, 731–736.

Sussman, R. W., Kinzey, W. G., 1984. The ecological role of the Callitrichidae: a review. American Journal of Physical Anthropology 64, 419–449.

Szalay, F. S., 1968. Beginnings of primates. Evolution 22, 19–36.

Szalay, F. S., 1971. Cranium of the late Palaeocene primate *Plesiadapis tricuspidens*. Nature 230, 324–325.

Szalay, F. S., 1972a. Paleobiology of the earliest primates. In: Tuttle, R. (Ed.), The Functional and Evolutionary Biology of the Primates. Aldine-Atherton, Chicago, IL, pp. 3–35.

Szalay, F. S., 1972b. Cranial morphology of early Tertiary *Phenacolemur* and its bearing on primate phylogeny. American Journal of Physical Anthropology 36, 59–76.

Szalay, F. S., 1973. New Paleocene Primates and a diagnosis of the new suborder Paromomyiformes. Folia primatologica 19, 73–87.

Szalay, F. S., 1975. Where to draw the primate-nonprimate boundary. Folia Primatologica 23, 158–163.

Szalay, F. S., 1977. Ancestors, descendants, sister groups and testing of phylogenetic hypotheses. Systematic Zoology 26, 12–18.

Szalay, F. S., 1981. Phylogeny and the problem of adaptive significance – the case of the earliest primates. Folia Primatologica 36, 157–182.

Szalay, F. S., 1984. Arboreality: is it homologous in metatherian and eutherian mammals? Evolutionary Biology 18, 215–258.

Szalay, F. S., 1998. The middle Eocene *Eurotamandua* and a Darwinian phylogenetic analysis of "edentates". Kaupia, Darmstader Beitrage zur Naturgeschichte 7, 97–186.

Szalay, F. S., 2000. Function and adaptation in paleontology: why omit Darwin? Palaeontologica electronica. 3, 366 KB; http://palaeo-electronica.org/paleo/2000_2/darwin/issue_2_00.htm.

Szalay, F. S., Bock, W. J., 1991. Evolutionary-theory and systematics – Relationships between process and patterns. Zeitschrift für Zoologische Systematik und Evolutionsforschung 29, 1–39.

Szalay, F. S., Dagosto, M., 1980. Locomotor adaptations as reflected on the humerus of Paleogene Primates. Folia Primatologica 34, 1–45.

Szalay, F. S., Dagosto, M., 1988. Evolution of hallucial grasping in the primates. Journal of Human Evolution 17, 1–33.

Szalay, F. S., Decker, R. L., 1974. Origins, evolution, and function of the tarsus in Late Cretaceous Eutheria and Paleocene Primates. In: Jenkins, F. (Ed.), Primate Locomotion. Academic, New York, pp. 223–259.

Szalay, F. S., Drawhorn, G., 1980. Evolution and diversification of the Archonta in an arboreal milieu. In: Luckett, W. P. (Ed.), Comparative Biology and Evolutionary Relationships of Tree Shrews. Plenum, New York, pp. 133–169.

Szalay, F. S., Lucas, S. C., 1996. The postcranial morphology of Paleocene *Chriacus* and *Mixodectes* and the phylogenetic relationships of archontan mammals. Bulletin of the New Mexico Museum of Natural History and Science 7, 1–47.

Szalay, F. S., Rosenberger, A. L., Dagosto, M., 1987. Diagnosis and differentiation of the order Primates. Yearbook of Physical Anthropology 30, 75–105.

Szalay, F. S., Sargis, E. J., 2001. Model-based analysis of postcranial osteology of marsupials from the Palaeocene of Itaboraí (Brazil) and the phylogenetics and biogeography of Metatheria. Geodiversitas 23, 139–302.

Szalay, F. S., Tattersall, I., Decker, R. L., 1975. Phylogenetic relationships of *Plesiadapis* – postcranial evidence. Contributions to Primatology 5, 136–166.

Teilhard-de-Chardin, P., 1922. Les mammiferes de l'Eocene inferieur francais et leurs gisements. Annales de Paléontologie 11, 9–116.

Thorington, R. W., Heaney, L. R., 1981. Body proportions and gliding adaptations of flying squirrels (Petauristinae). Journal of Mammalogy 62, 101–114.

Thorington, R. W. J., Schennum, C. E., Pappas, L. A., Pitassy, D., 2005. The difficulties of identifying flying squirrels (Sciuridae: Pteromyini) in the fossil record. Journal of Vertebrate Paleontology 25, 950–961.

Van Den Bussche, R. A., Hoofer, S. R., 2004. Phylogenetic relationships among recent chiropteran families and the importance of choosing appropriate out-group taxa. Journal of Mammalogy 85, 321–330.

Walker, A. C., 1974. Locomotor adaptations in past and present prosimian primates. In: Jenkins, F. (Ed.), Primate Locomotion. Academic, New York, pp. 349–381.

Wharton, C. H., 1950. Notes on the life history of the flying lemur. Journal of Mammalogy 31, 269–273.

White, J. L., 1993. Indicators of locomotor habits in xenarthrans: Evidence of locomotor heterogeneity among fossil sloths. Journal of Vertebrate Paleontology 13, 230–242.

Wible, J. R., Martin, J. R., 1993. Ontogeny of the tympanic floor and roof in archontans. In: MacPhee, R. (Ed.), Primates and their Relatives in Phylogenetic Perspective. Plenum, New York, pp. 111–148.

Youlatos, D., 1999. Positional behavior in *Cebuella pygmaea* in Yasuni National Park, Ecuador. Primates 40, 543–550.

Youlatos, D., Godinot, M., 2004. Locomotor adaptations of *Plesiadapis tricuspidens* and *Plesiadapis n. sp.* (Mammalia, Plesiadapiformes) as reflected in selected parts of the postcranium. Journal of Anthropological Science 82, 103–118.

12. Morphological Diversity in the Skulls of Large Adapines (Primates, Adapiformes) and Its Systematic Implications

Marc Godinot*, Sébastien Couette
Ecole Pratique des Hautes Etudes
UMR 5143 Paléobiodiversité et Paléoenvironnements
Case Courrier 38, Département d'Histoire de la Terre
Muséum National d'Histoire Naturelle
8 rue Buffon, 75005 Paris
France
godinot@mnhn.fr, couette@mnhn.fr

12.1 Introduction

The European Eocene adapiforms include two subfamilies, the Cercamoniinae, present in the early and middle Eocene, and the Adapinae, present in the late Eocene (Franzen, 1994; Godinot, 1998; Fleagle, 1999; Gebo, 2002). The large adapine species has a robust upper canine and other characters. It was named *Adapis magnus* by Filhol (1874), and was later placed in the genus *Leptadapis* by Gervais (1876). However, Gervais' choice was seldom followed by subsequent authors. Stehlin (1912) and Depéret (1917), for example, retained *Leptadapis* as a subgenus of *Adapis*. For Stehlin, a fossil species was equivalent to a living genus; he used the name *Adapis magnus* in his text and figures, and *Leptadapis magnus* in his final stratigraphic chart of the genus *Adapis* (p. 1280). A single genus, *Adapis*, is used by Genet-Varcin (1963), Simons (1972), and Gingerich (1977, 1981). Most recent authors (Godinot, 1998; Fleagle, 1999; Gebo, 2002) kept the genus *Leptadapis*, following Szalay and Delson (1979). Recent work by Lanèque emphasized the systematic complexity reflected by the skulls of *Adapis* sensu Szalay and Delson (1979) (Lanèque, 1992a, b, 1993), and also showed a marked heterogeneity in the larger adapine skulls (*Leptadapis*) (Lanèque, 1993). The radiation of these adapines is very complex (Godinot, 1998); there is an overlap in size between the large *Adapis* and the small *Leptadapis*. Furthermore, the dentally peculiar *Cryptadapis* (Godinot, 1984) lies in this zone of overlap. Deciphering the systematics and the phylogeny of the adapines is a long-term task.

One difficulty for adapine studies is that the systematics of Stehlin (1912) is based on skulls in the old Quercy collections, and these have no biochronological context. Even when specimen information includes the name of a village, these names are not sufficient because several fissure-fillings were exploited in most of these villages. It is impossible to know from which fissure the specimens were collected. Work started around 1960 by colleagues from the universities of Montpellier, Poitiers and Paris led to the recovery of many precisely located faunas, which have been placed in a good biochronological framework (Crochet et al., 1981; Remy et al., 1987; BiochroM'97, 1997). However, a difficulty remains because the newly collected material, from pocket remains or exploitation residue, is very poor in comparison with the old collections. This is especially true for the primates. The new collections include fragmentary primate material: jaws from a few localities, but more often isolated teeth. In contrast, the systematics of Stehlin (1912) was based on cranial characters, and the types he erected were crania without associated mandibles. It is very difficult to associate the new dental remains to the species defined by Stehlin. This work is nevertheless in progress, and it appears probable now that many *Adapis* species from the old collections belong to a late Eocene radiation of this genus (MP 19 reference-level in the European Paleogene Mammalian

* Address for correspondence: godinot@mnhn.fr

E.J. Sargis and M. Dagosto (eds.), *Mammalian Evolutionary Morphology: A Tribute to Frederick S. Szalay*, 285–313.

Scale, Schmidt-Kittler et al., 1987). By contrast, several species of *Leptadapis* come from earlier localities, MP 17-18, Priabonian (Remy et al., 1987; Godinot, 1998). Here we will concentrate on the skulls of large size present in the old Quercy collections, and only briefly mention specimens found in one stratified locality.

The skulls of large size in the old Quercy collections have not been well treated in the literature. This probably goes back to Grandidier (1905) who identified a beautiful cranium and associated mandible as "*Leptadapis magnus*" (Figure 1), whereas this specimen was in fact quite different from the type specimen of *L. magnus* described by Filhol (1874). When he described specimens from the collections of Montauban, Munich, and Basel, Stehlin (1912) recognized differences between two pairs of specimens (Figure 1). He discussed the idea that these pairs might possibly be males and females of one species, however, because the size differences between their canines were very small, insignificant ("geringfügig"), he could not definitely endorse this hypothesis. Without a clear conclusion, Stehlin did not propose a systematic distinction between them. He kept the question open and left all the specimens in "*Adapis magnus*". However, Stehlin saw clear morphological and size differences between the Montauban 3 cranium and the others, and he named a new variety for this specimen: "*Adapis magnus* var. *Leenhardti*". For Stehlin, a "variety" was equivalent to an extant "species". We do not know why Stehlin did not come to Paris to include the Paris collection in his study. Did he have only a short visit to Montauban? In his work, the figures are "reconstructions", for the most part very reliable, showing in white the reconstructed parts and in grey the actual fossil; this way, it is possible to identify the specimens he used. However, sometimes his drawings compensate for specimen deformations, and they can be inaccurate for some details. Some specimens also may have been partly damaged or lost, rendering their identification difficult.

Gingerich (1977) started a systematic revision of European adapiforms. He interpreted differences between two groups of small-sized *Adapis* specimens as due to sexual dimorphism within a single species *A. parisiensis*. He suggested that differences between two groups of large-sized specimens reflect sexual dimorphism among the large *"Adapis" magnus*. He also considered the Quercy *A. parisiensis* as a descendant of *"A." magnus* from the same region (Gingerich, 1977, 1981; Gingerich and Martin, 1981), in contrast with Stehlin's view of two distinct lineages (Stehlin, 1912, p 1280). Gingerich's suggestion was seldom followed. Szalay and Delson (1979) continued to see only one species, *L. magnus*, in the large Quercy skulls. In her study of orbital characters, Lanèque (1993) showed quite convincingly that two groups of *Leptadapis* could be distinguished based on interorbital breadth, however she did not pursue the systematic implications of this finding. On a stratophenetic diagram of biochronologically situated dental assemblages, Godinot (1998) found that two lineages could be distinguished, which might correspond to the groups delineated by Lanèque.

Several factors played a role in preventing an easy systematic study of these fossils, including the fact that they are spread between several distant institutions, and they differ in their preserved parts. The specimens are large enough that direct comparison of their dentitions under a binocular microscope is awkward or impossible. At superficial examination, they seem to have a relatively similar dental pattern, however, a detailed examination of these skulls reveals differences in dental characters, some of which have to be meaningful. In this paper, we first briefly describe the eight best-preserved skulls of large adapines. We then explain why we propose to distinguish two major groups (genera). We review each group and suggest how many species should be distinguished. We comment on the phylogeny of these groups. Lastly we propose a first morphometric approach to studying this material.

12.2 Material: The Best Preserved Large Adapine Skulls (Table 12.1)

The most complete specimen is a cranium, QU 10870, with associated lower jaws (QU 10871, Paris Museum, MNHN). There is little doubt that this is the specimen figured by Grandidier (1905) as "*A. magnus*", "coll. Filhol, Muséum de Paris" (idem, p 141). It possesses basically the same parts. The associated jaw is similarly preserved: breakage of the angular process of the left side (Figure 12.1B), same part missing anteriorly with a gap on the left side and the same teeth preserved, M/3 to P/2 on both sides (idem, Figure 4). This skull remains one of the best preserved Eocene primate skulls ever found. Grandidier's figures, labelled *"A." magnus* despite differences between it and the type specimen, were really a bad starting point for later work. Several of these figures were reproduced by Gregory (1920) and one by Piveteau (1957). This probably largely explains the failure of subsequent authors to recognize the groups distinguished below.

There are also some differences between the skull and the figures of Grandidier, revealing that those are not entirely accurate, and/or that the specimen possibly suffered some damage since 1905. The profile view of the left side (Figure 12.1B) shows that the sagittal crest was complete, whereas it is now broken (posterodorsal part missing); the zygomatic arch is shaped differently, the gap linked to a fissure in the anterior part being exaggerated in the figure. On the ventral aspect of the cranium, the same teeth are present on the actual specimen and in the figure (Figure 12.1E), the two canines showing breakage at a similar level, higher on the left than on the right canine. The left P3/ and M3/ are better preserved according to the figure than on the actual specimen. Were they

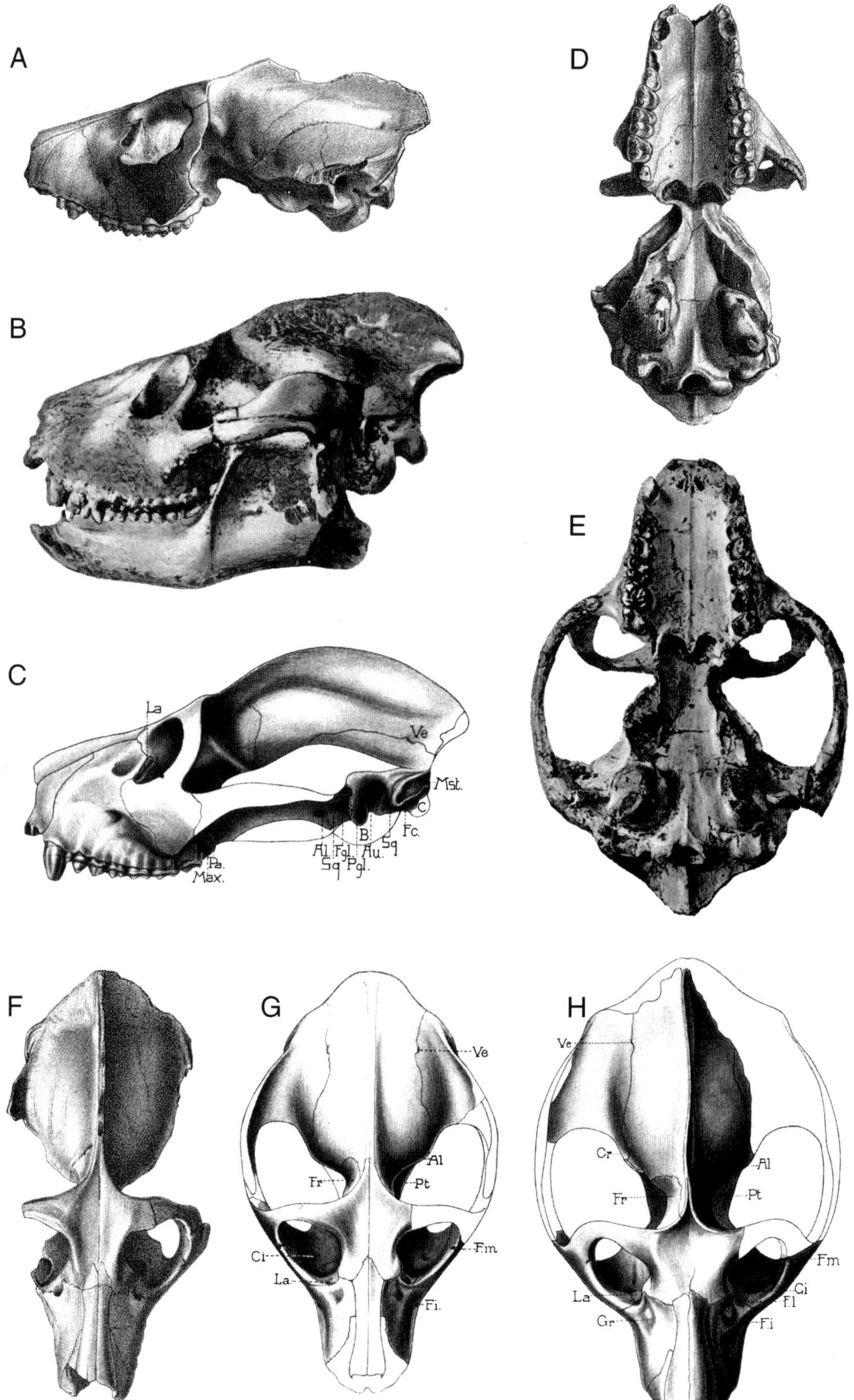

FIGURE 12.1. Illustrations of crania all ascribed to "*Adapis magnus*" by the authors who described them. A, D, F, the type specimen of *Leptadapis magnus*, QU 11002, three of the four drawings published by Filhol (1874), all inverted left/right on Filhol's engravings, now showing their natural side. A, left lateral view; D, ventral view; F, dorsal view (muzzle toward the top in Filhol, inverted here for comparison with Stehlin's figures). B, E, the skull QU 10870-10871 as figured by Grandidier (1905). B, lateral view of cranium (QU 10870) and mandible (QU 10871); E, ventral view of cranium. C, G, H, three illustrations from Stehlin (1912); C and H are reconstructions of MaPhQ 210 in lateral (C) and dorsal (H) views; G is the reconstruction of MaPhQ 211 in dorsal view. Not to scale, but G and H preserve their relative scale as present in Stehlin's illustrations. Differences in interorbital and muzzle breadth can be observed between F-G (*L. magnus*) and H (*Magnadapis fredi* n. gen. n. sp. in this paper), and differences in facial and muzzle height can be observed between A (*Leptadapis*) and B–C (two *Magnadapis* species).

TABLE 12.1. Listing of the specimens used in this study, indicating their preserved parts, aspects of their crests and canines, and their systematic allocation in this paper. Abbreviations for institutions are: MNHN, Muséum National d'Histoire Naturelle, Paris; MU, Montpellier University; NMB, Naturhistorisches Museum Basel; YPM, Yale Peabody Museum.

Specimen number	Preserved parts	Sagittal crest, nuchal projection, canine	Systematic allocation
MNHN QU 11002	Cranium lacking right postorbital bar and both zygomatic arches	High sagittal crest, strong nuchal projection, canine alveolus	Type of *Leptadapis magnus*
Montauban MaPhQ 211	Cranium somewhat distorted lacking left postorbital bar and both zygomatic arches	Broken probably high sagittal crest, strong nuchal projection, partial canine alveoli	*L.*aff. *magnus*
MU ACQ 209	Cranium with incomplete right zygomatic arch	Moderate sagittal crest and nuchal projection, partial left canine	Type of *L. filholi*
YPM PU 11481	Cranium with incomplete postorbital bars and zygomatic arches (reconstructed with plaster)	Very low sagittal crest, unknown nuchal projection, partial canine alveoli	Neotype of *L. leenhardti*
MNHN QU 10875	Cranium almost complete, lacking only right zygomatic arch	Low sagittal crest, weak nuchal projection, complete slender right canine	Type of *Magnadapis quercyi*
Leuven PLV 6	Muzzle with slightly deformed left orbit	Complete right canine	*M.*aff. *quercyi*
MNHN QU 10870-71	Complete cranium and mandible	High sagittal crest, strong nuchal projection, partial canines	Type of *M. intermedius*
MNHN QU 11035-36	Crushed cranium with complete left zygomatic arch and posterior extremity, and mandible	Posteriorly very high sagittal crest, strong nuchal projection, partial large canine alveolus	*M.*aff. *intermedius*
MU ACQ 214	Muzzle slightly crushed, without complete orbit	Complete large and robust left canine	*M. intermedius*
Montauban MaPhQ 210	Partial cranium, lacking zygomatic arches and posteroventral part	Very high sagittal crest, strong nuchal projection, almost complete large right canine	Type of *M. fredi*
NMB St.H. 1634	Crushed muzzle, palate	Almost complete partly deformed canines	*M. fredi*
MNHN QU 10872	Cranium lacking postorbital bars and zygomatic arches	Very low sagittal crest, moderate nuchal projection, broken bases of both canines	Type of *M. laurenceae*

damaged since, or restored on the figure? On this figure, the right P1/ should be more visible, and the right M2/ appears quite inaccurate in its rendering. Other aspects of the ventral view are also surprising. Small fissures at the same place on the zygomatic arches confirm that the specimen really is the same. However, a marked exaggeration of the concavity lying behind the left zygomatic arch suggests that a piece of bone is missing there, whereas it is present on the specimen: this demonstrates that the figure was modified, either by a poorly done cutting around the photograph, or through painting of the negative. A confirmation of the bad rendering of the outline is given by the interior of the left orbital aperture, which shows on the figure a long indentation not present on the specimen. Clearly, there was a heavy and partially inaccurate retouching of Grandidier's photographs, and not only on the outlines: the two occipital condyles are beautiful on the specimen, and poorly rendered on the same figure. Two other differences between the ventral view (Figure 12.1E) and the actual specimen must be taken with caution: on the figure, the back of the palate shows a marked posterior spine at its midline, and an osseous continuity between the right posterior end of the palate and the right median pterygoid lamina. These parts might have been broken since 1905, however they also might have been restored on the figure, the palatal spine by comparison with the type skull of *L. magnus*, on which it is conspicuous, and the continuous pterygoid plate by comparison with the crushed skull QU 11035, on which it shows a similar trajectory (different on other skulls, e.g., QU 10875, see below). Because no other skull pertaining to the same group (see below) shows such a palatal spine, and because the right pterygoid region of the figure suggests an inaccurate (asymmetric) external pterygoid plate, it seems that this figure also presents heavy and inaccurate retouching. The left bulla presents a slight crushing of its lateral wall, however the right one is perfectly preserved. Excessive retouching probably explains a more important and embarrassing aspect of Figure 2 in Grandidier (1905), in which the orbit is clearly too small, exaggerating the similarity between this specimen and *Adapis* (compare Figures 12.1B and 12.8C). The left postorbital bar shows a slight deformation, hence orbital characters should be measured on the right side. The right view gives a perfect outline of the orbit and the very peculiar zygomatic arch of a large adapine. A good photograph of this same specimen in dorsal view is given in Lanèque (1993, Figure 10a; in this paper the legend for Figure 10 is below Figure 9, and vice versa).

Another exquisitely preserved cranium is QU 10875, also from the Paris collection. This specimen appears in three excellent photographs by Genet-Varcin (1963), two photographs on one plate by Saban (1963), in one lateral view by Simons (1972), and in one dorsal view by Lanèque (1993, Figure 9b – legend below Figure 10). The two postorbital bars are even thinner than on QU 10870, and the left orbit shows a slight distortion, visible in anterior view; its right orbit seems undistorted. The right zygomatic arch is missing. The pterygoid laminae are almost complete. The teeth are quite well preserved; the small P1/ are lost, the right canine is complete, pointed and robust, showing the vertical grooves typical of these fossils. On the whole, this cranium including its teeth, resembles quite closely QU 10870, however there are also differences, which will be described below.

Another specimen preserving both postorbital bars is ACQ 209 from the University of Montpellier collection (UM). This specimen has probably never been previously figured (Figure 12.2). Its postorbital bars are thin and intact. Part of the right zygomatic arch is missing, however the left one is complete. There is slight deformation of the posterior part, visible in ventral view: the basioccipital appears pushed ventrally away from the bullar walls and above the more anterior ventral floor (probably the sphenoid). The bullae have experienced some deformation: the left one is less completely distorted than the right one. The pterygoid laminae are almost complete. A thick and short posterior palatal spine emerges at the confluence of two arcuate crests which border the palate posteriorly. On the dorsal side, the bone of the anterior part of the braincase is eroded on both sides, however, the sagittal crest, which is around 8 mm tall at its highest, is quasi intact, being one of the best preserved in all these skulls. It is very regularly round and its posterior part projects approximately 1 cm beyond the posterior border of the foramen magnum.

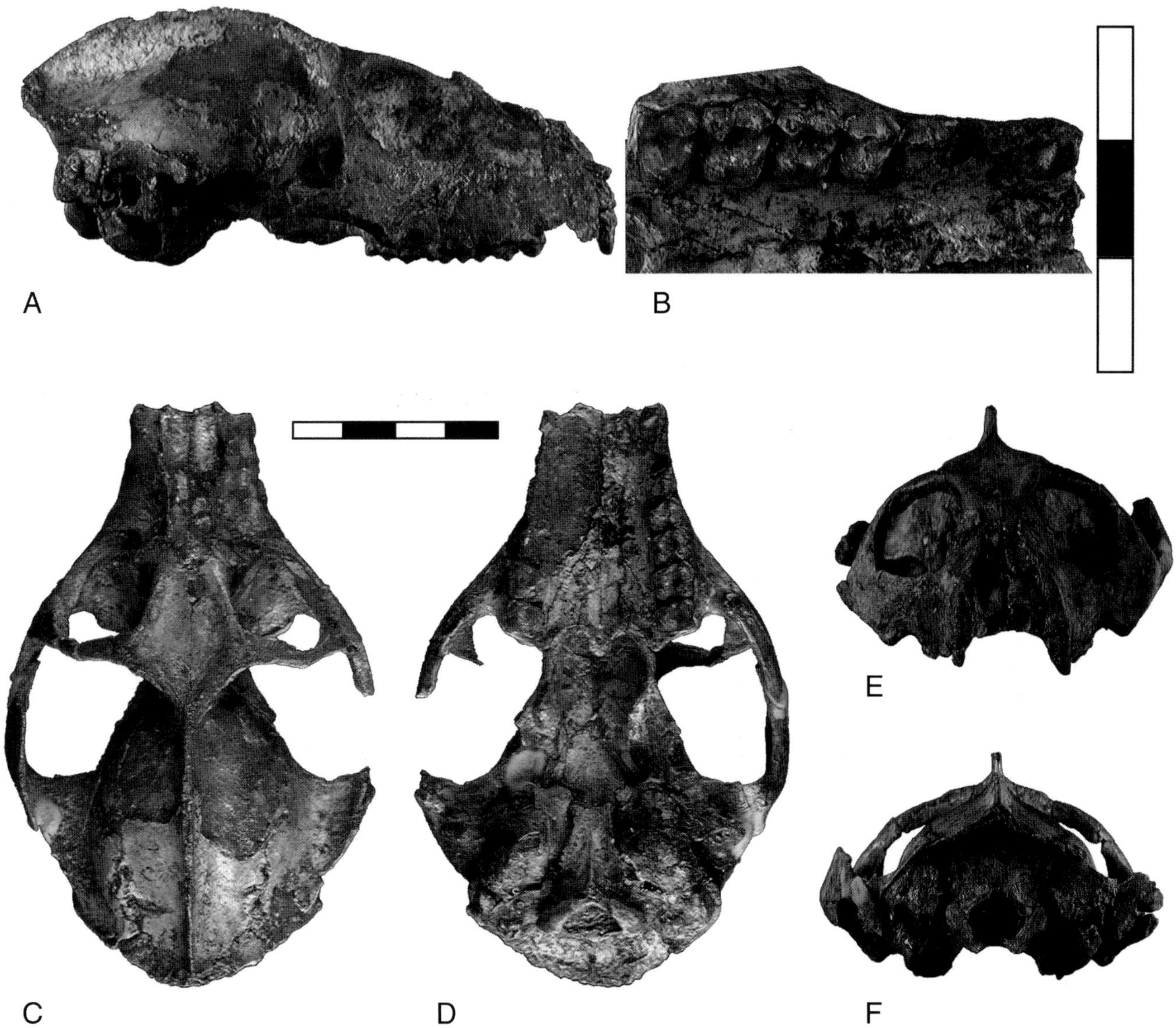

FIGURE 12.2. The cranium MU ACQ 209, type specimen of *Leptadapis filholi* n. sp., in lateral (A), dorsal (C), ventral (D), anterior (E), and posterior (F) views, all at the same scale (bar is 4 cm). The right part of its palate (B) is at a larger scale (bar is 3 cm).

The anterior part of the cranium is less well preserved. The right side of the muzzle is eroded (Figure 12.2). The anterior parts of the nasals are crushed against the palate. The premaxillae, the posterior parts of the nasals, and a small part of the frontal are missing. However, the left maxilla seems undistorted, allowing an estimate of the height of the muzzle without the nasals (the nasals make a thin straight ribbon in profile view on the similar *magnus*-type skull). The right dentition is relatively well preserved, from P3/ to M3/. The canine is moderate in size, partly eroded, and its outline in occlusal view is very rounded, almost as broad as it is long; P1/ is there and is relatively large; P2/ is missing. This specimen is very important because it is the only representative of the *Leptadapis* group as delineated below that has an intact zygomatic arch and that has the upper canine and the P1/ (see Figures 12.7, 12.9). Over many years of study, one of us (MG) wondered if this cranium could have been the one chosen by Stehlin (1912) as the type of his "variety *Leenhardti*". Some of the Montauban *Adapis* skulls were in a protestant college of theology and were later transferred to the Montpellier University collection, and this cranium resembles more closely Stehlin's figures than does any other present in the Montpellier or Montauban collections. Similarities include its small size, relative completeness, preservation up to the canine anteriorly and no further. However, there are also many discrepancies between the figures of Stehlin and this specimen, including a size discrepancy, this specimen being larger than that in Stehlin's figures (the scale is very accurate for the figure of the Montauban 2 specimen). This specimen preserves the left P4/, P3/ and P1/, which are absent in Stehlin's figure 297 [the right M1/ and P4/ of the figure could have been restored from those of the other side, and the right zygomatic likewise completed from the other side, as Stehlin's figures are "reconstructions"]. On the dorsal view (idem, Figure 296), the posterior extremity of the cranium is narrower and more salient than on the specimen, the posterior part of the temporal fossa is narrower and much more arcuate than on the specimen, and the orbits are more anteroposteriorly elongated. Although there are a few problems with some of Stehlin's reconstructions, these are very minor compared to the differences outlined above. There is no doubt for us that the type of the "variety *Leenhardti*", despite being similar to ACQ 209 in its general state of preservation, was actually not this specimen.

The last specimen that apparently preserves both orbits is YPM PU 11481, housed in the Yale University collection (Figure 12.3). This specimen comes from the former Princeton collection and may never have been figured. This specimen is the smallest of those studied here, and as such it is the closest to the type of the "variety *Leenhardti*". Its state of preservation looks superficially similar in dorsal view, only the nasals and premaxillae missing. On the ventral view, the same teeth are preserved, P4/ to M3/ on the right side, M1-3/ on the left side. The right bulla, open on the Yale specimen, could have been reconstructed based on the intact left bulla. The two canines, present on Stehlin's figure, might have been lost since.

Stehlin described the specimen as heavily deformed. This is not apparent on his reconstruction, and is not true of the Yale specimen. Furthermore, it is difficult to explain why Stehlin did not mention the Princeton specimen had he access to it. The specimen was, according to Yale records, purchased in the early to mid-1890s by Professor W. B. Scott from A. Rossignol in Caylus (W. Joyce, personal communication 2006). Many institutions have specimens labelled "Rossignol collection". This date refutes the hypothesis of a later purchase of the Montauban specimen, seen by Stehlin in or not long before 1912. A closer look at the Yale specimen confirms that it is different: it is actually heavily restored with plaster! Without the plaster, it would lack the right orbit, both zygomatic arches and its right and middle posterior extremity, as can be seen on CT scans of the specimen (Figure 12.3). The left orbit, almost complete, is anteroposteriorly shorter on the Yale specimen than on Stehlin's figure. In ventral view, if the canines had been lost since the drawing of the specimen, there would be breakages or big holes, which is not the case. Also, based on alveolar morphology, P2/ was single-rooted in the "variety *Leenhardti*" type specimen, whereas it was two-rooted on the Yale specimen. On the whole, although the Yale specimen is the closest to the type of Stehlin's variety in size and preserved teeth, for both historical and morphological reasons it cannot be the actual type. Further similarities in the outline in dorsal and ventral views suggest that the artist who did the plaster reconstruction probably used Stehlin's figures of the *Leenhardti* type as a model for making the reconstruction. This is why the plaster reconstruction looks realistic in dorsal and ventral views. However, he had no figure of a lateral view because Stehlin gave none, and he produced a very inaccurate reconstruction of the zygomatic arches in profile view: they are much too low and incorrectly proportioned for an adapine. The Yale specimen is still an important one, having an almost complete left orbit, a well preserved brain case and a very low sagittal crest. On its ventral side, the pterygoid laminae are complete anteriorly but incomplete posteriorly. The left bulla is slightly crushed, and the right one is open, showing the promontory. The teeth are little worn, suggesting that it was a young individual. The left M2-3/ are very well preserved, and the right M2-3/ are slightly damaged. The right P4/ is lacking two small chips which do not prevent the study of its morphology. On the whole, this specimen really could pertain to the same species as the "variety *Leenhardti*" of Stehlin, because it is by far the closest to its type specimen. The actual type is, as far as we know, lost. Therefore, we designate below YPM PU 11481 as a neotype for *L. leenhardti* Stehlin.

The type specimen of *Leptadapis magnus*, MNHN QU 11002, was described and illustrated by Filhol (1874). A photograph in dorsal view is given by Lanèque (1993, Figure 10b). It has a complete left orbit. Its postorbital bar is broader than that of any of the preceding specimens (Figures 12.1, 12.4, 12.6). Despite its missing parts, it is not distorted. The top of its muzzle is intact, and the left nasal appears to be complete. The left orbit is undistorted.

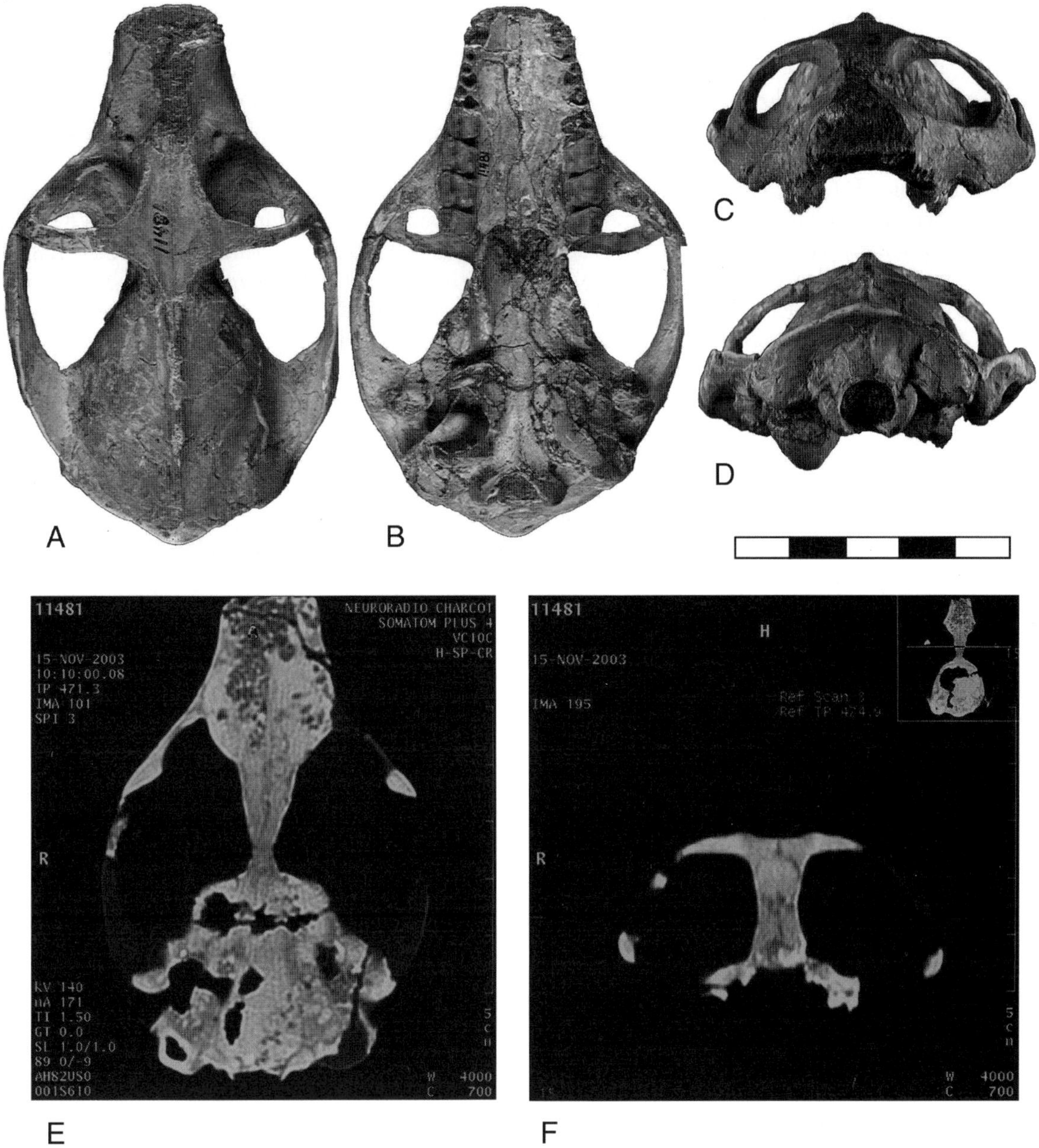

FIGURE 12.3. Cranium of YPM PU 11481, neotype of *Leptadapis leenhardti*, in dorsal (A), ventral (B), anterior (C), and posterior (D) views; all at the same scale (bar is 5 cm). Two scanner images of the same skull, in horizontal (E) and frontal (F) planes. The scanner images reveal parts reconstructed with plaster: zygomatic arches and right posterior extremity (E) and parts of the two postorbital bars (F). One can also see in D that the posterior extremity of the sagittal crest and the left nuchal crest are reconstructed with plaster.

A limited part of the high sagittal crest is missing, allowing a relatively accurate estimate of its shape. The posterior extremity of the cranium projects far behind the posterior rim of the foramen magnum (around 11 mm). The bullae are moderately distorted. The pterygoid laminae are quite well preserved posteriorly, but are incomplete anteriorly. This way, they do not prevent the observation in dorsal view of the marked postorbital narrowing of the skull, which is extreme on this specimen (Figure 12.4).

The two other relatively well preserved skulls are those from Montauban which were quite accurately described by Stehlin (1912) as Montauban 1 (now MaPhQ 210) and Montauban 2 (MaPhQ 211). However, Stehlin's figures were again reconstructions (Figure 12.1). These were on the whole accurate, and honest in the sense that they usually show in simple white outline the missing parts which were reconstructed. However, there are some differences, corresponding to actual specimen deformations which were compensated for in the drawings, and which are not shown in white and can be misleading. We mention some of them below and underline the need to study the original specimens and to use real photographs.

The right postorbital bar of MaPhQ 210 is deformed, whereas it is reconstructed undistorted on Stehlin's drawings. On the dorsal view, the posterior part of the right zygomatic arch appears more salient laterally and has a less arcuate outline on the specimen than on the figure, suggesting that the zygomatic arches might well have been more extended laterally than on the reconstruction (idem, Fig. 276). A comparison with the other very robust skull, QU 10870, shows that the flat area on the top of the posterior root of the zygomatic arch is anteroposteriorly more extended and mediolaterally flatter (less dorsally curved in posterior view) on MaPhQ 210 than on the other, again suggesting extremely laterally salient zygomatic arches, exceeding those of Stehlin's reconstruction. On the profile view, Stehlin showed the left side even though the right one is the one that preserves the orbit. The reconstruction shows the postorbital bar as having the same breadth along its entire length whereas the specimen narrows ventrally along its dorsal half. The reconstruction probably exaggerates slightly the height of the orbit (difficult to estimate due to deformation) and, in any case, misrepresents its ventral border. At the same time, it probably underestimates the vertical distance between the base of the orbit and the base of the zygoma. Stehlin, or his artist, seems to have reshaped the orbit with a very inclined anteroventral border, whereas that part does not appear to be deformed on the specimen, being more horizontal anterior to the anterior extremity of the zygoma. This partly diminished the extreme robusticity of the anterior zygomatic root. On this profile view (idem, Fig. 277), the shading suggests that the sagittal crest diminished in height posteriorly, whereas on the actual specimen the sagittal crest increases posteriorly, coming close to 2 cm in height (see Figure 12.8). This specimen has the largest sagittal crest of all. The ventral view shows longer portions of the zygomatic arches anteriorly on both sides: could some pieces have been broken since? On the contrary, posteriorly the zygomatic root is actually longer on the right side than represented in grey on Stehlin's figure, suggesting that these reconstructions should not be trusted. In the open right bulla, the drawing shows the tympanic ring, which is not present today.

The other Montauban cranium, MaPhQ 211 (Montauban 2 of Stehlin), is smaller and more deformed. Its right orbit is complete; some breakage makes it appear slightly smaller than in reality, however on the whole it is little distorted. There are fissures in the maxillae, nasals, and between the frontals. A part of the right premaxilla is still there, with a small part of the rim of the nasal aperture. The cranium was broken behind the orbits, however the left pterygoid lamina is relatively well preserved (incomplete further posteriorly). The right bulla is open, broken laterally, and shows the promontorium; the left bulla shows parts of its medial wall; some matrix remain inside the bullae. The braincase is not much distorted, except for its posterior extremity, which is pushed toward the right. Despite this displacement, one can see the extent of the nuchal projection (well marked, around 1 cm), and estimate the height of the sagittal crest, which must have been well developed. The two zygomatic arches are missing. The figures of Stehlin for this specimen (Figure 275 [12.1G] for the dorsal view, and 280 for the ventral view) show some evidence of restoration, and also make us wonder about possible damage to the specimen. The drawing in dorsal view evidently compensated for the displacement of the posterior extremity, therefore the outline is uncertain; it also modified the outline of the orbit. It shows a much more complete right zygomatic arch. The ventral view shows the same teeth, alveoli, anterior breakage and missing half of the right M3/. It shows the right pterygoid wing in white, i.e., reconstructed. It shows again a right zygomatic arch much more complete than on the actual specimen which suggests that breakage occurred since. The two bullae are drawn intact; they were probably less completely prepared at Stehlin's time because normally when the promontorium was visible it was drawn. However, some reconstruction was probably also added. The preserved parts look slightly different (the sutures medial to the bullae are very simplified on the drawing).

Another skull used in our study and never described until now is QU 10872 from the Paris collection (see Figures 12.8, 12.10). This specimen requires further cleaning. It is relatively small and not as well preserved as the others because it has no complete orbit. However, the upper part and the lower part of the right orbit are intact, allowing an estimate of orbit size. The orbit appears quite small in comparison with other specimens of similar size, e.g., ACQ 209. The zygomatic arches are missing and the left orbital region is damaged. However, the right side of the muzzle is well preserved and the dentition is relatively complete (P2/-M3/ and broken base of canine on both sides, and broken base of I1/ and I2/ on the left side). Posteriorly, the ventral side is poorly preserved, without pterygoids or bullae. The braincase on the whole is not distorted, however there is a slight global deformation of the specimen, which has, in dorsal view, its posterior part slightly pushed toward the left side relative to the anterior part. On the dorsal side, this specimen has one of the two lowest sagittal crests (with two missing chips) of any large adapine; only the Yale specimen has a lower one (the sagittal crest is less than 5 mm high at its maximum, difficult to measure precisely because the junction with the braincase is curved). This small sagittal and the nuchal crests nevertheless produce a nuchal projection of around 8 mm.

Five other specimens will be mentioned. The very crushed skull QU 11035 from the Paris collection is from a very large individual. The muzzle and the anterior part of the braincase are missing. However, the left zygomatic arch is very complete. The posterior part of the skull is well preserved and is very low (Figure 12.6). The large bullae are intact, and are the best preserved of any adapine. This specimen also has an associated mandible. The mandible is fused at the symphysis and bears a right series that includes a small P/1, broken canine and alveoli for the two incisors. The left posterior part of the mandible is also well preserved.

Another crushed and less complete specimen from the Basel collection comes from Euzet-les-Bains, having thus a stratigraphic provenance. The species from Euzet-les-Bains (or Saint-Hippolyte-de-Caton) was described by Depéret (1917), who figured two palates, mandibles and a fragment of maxilla with the base of the orbit in profile (idem, Pl. 25). The crushed muzzle of Basel, NMB St.H.1634 (Figure 12.10), clearly shows the broad interorbital distance and the breadth of the muzzle, which are important characters for us. Depéret (1917) ascribed the species to *Adapis (Leptadapis) magnus*; he considered that a subgenus was enough to distinguish the large species. He basically followed Stehlin (1912) concerning adapine systematics.

Another specimen of interest is a muzzle with part of the right orbit, UM ACQ 214, with beautiful teeth on the right side, complete canine and base of I2/ on the left side. This specimen has the largest intact upper canine of any Quercy cranium. The palate has a dorsal curvature, exaggerated posteriorly due to crushing. The anterior part of the muzzle is only slightly distorted, the premaxillae being almost complete and the right side of the nasal aperture being almost undistorted (nasal slightly pushed under the premaxilla). The anterolateral base of the right orbit is intact, and the height of the anterior root of the right zygomatic arch can be estimated. Large parts of the frontals are also preserved, showing the beginning of the right postorbital bar as it narrows laterally, and also the large interorbital breadth. However, some crushing between the frontals and the muzzle, close to the maxillary suture, pushed the frontal part ventrally, artificially diminishing orbital height in anterior and profile views.

Another specimen is an incomplete muzzle with left postorbital bar and zygomatic root, PLV 6 from Leuven University, Belgium. We mention it because it has a beautiful right canine. However, the specimen needs further cleaning and restoration, and it will be described more completely in a future study of dental material.

The last specimen which should be included in our series in the future is the partial cranium and associated mandibles which are housed at the University Geological Museum in Moscow. The specimen was figured by Pavlova (1910, p. 166). It is part of a collection brought back by Kovalewsky, and was a gift from Gaudry (T. Kouznetsova, personal communication 1995). The specimen resembles QU 10870 and QU 11035 in shape and preservation. It will be interesting in the future to compare these crania. We wonder if some of them might come from the same field collection, possibly made in one locality, and would as such be important for documenting skull variability in one putative species of large adapine.

12.3 First Group of Generic Value: *Leptadapis*

In her study of orbital characters in adapines, Lanèque (1993) showed that *Leptadapis* skulls exhibit variation in interorbital breadth greatly exceeding the variation found in living species. Scaling of this character (idem, Figure 8) clearly delineated two groups [her point 29 is an estimate, as ACQ 214 is a muzzle and has no associated condylobasal length]. Lanèque did not pursue the systematic implications of her study. We agree with her two groups, and we emphasize that other differences between them exist. These groups demand systematic recognition.

The first group includes four specimens: QU 11002, ACQ 209, MaPhQ 211, YPM PU 11481. These specimens have a smaller interorbital breadth (overlapping variations in living *Colobus* species in Lanèque's diagram). The dorsomedial rim of the orbit has a subrectilinear part directed slightly anteromedially and slightly extended on the muzzle. As a result, the frontal depression is bordered by two anteriorly converging rims, and is concave between the two orbits. In the other group the frontal depression is less extended anteriorly (Figures 12.2, 12.4). Specimens in the first group also have a much narrower muzzle (Figure 12.4). The palate is narrower, and the dental rows converge more anteriorly, with a canine alveolus protruding medially (ACQ 209 differs from the others in having a slightly broader muzzle and palate).

Several dental characters confirm this grouping: on all these specimens, there is a strong size contrast between a larger and especially longer M2/ and a smaller, narrower M3/; the P4/ has a more triangular outline, with a narrow lingual part. Only one of these specimens, ACQ 209, has an upper canine, which is partially worn, and a P1/. The P1/ seems larger than on specimens of the other group, and the canine seems to be smaller. This canine also seems simpler than on specimens of the other group, having less accentuated vertical grooves. In this group of specimens, only ACQ 209 shows a complete zygomatic arch, on its left side. This group includes QU 11002, which is the type specimen of *Leptadapis magnus*. Thus they must belong to the genus *Leptadapis*.

12.3.1 Comparison Between MaPhQ 211 and QU 11002

MaPhQ 211 is very similar to the type. Differences between them include the breadth of the postorbital bar, which is narrow on MaPhQ 211 whereas it is broader on the type (Figure 12.6). Possibly linked to this is the height of the anterior part of the zygomatic arch, below the orbit, which is greater on the type specimen than on MaPhQ 211. The ventral border of the nuchal projection is more horizontal on the type, but more inclined on MaPhQ 211. However, this part is deformed on the last specimen, making this difference of dubious value. The sagittal crests cannot be compared along most of their extent, however, in their anterior part, one can see clearly that the crest is higher on the type specimen than on the other. To properly observe this difference, it is necessary to look from the posterior side, in an antero- and slightly lateral direction; the distance between the border of the braincase and the top of the dorsal rim of the frontals is higher on the type specimen. In general the type specimen appears somewhat more robust than the other. This could be intraspecific variation, possibly

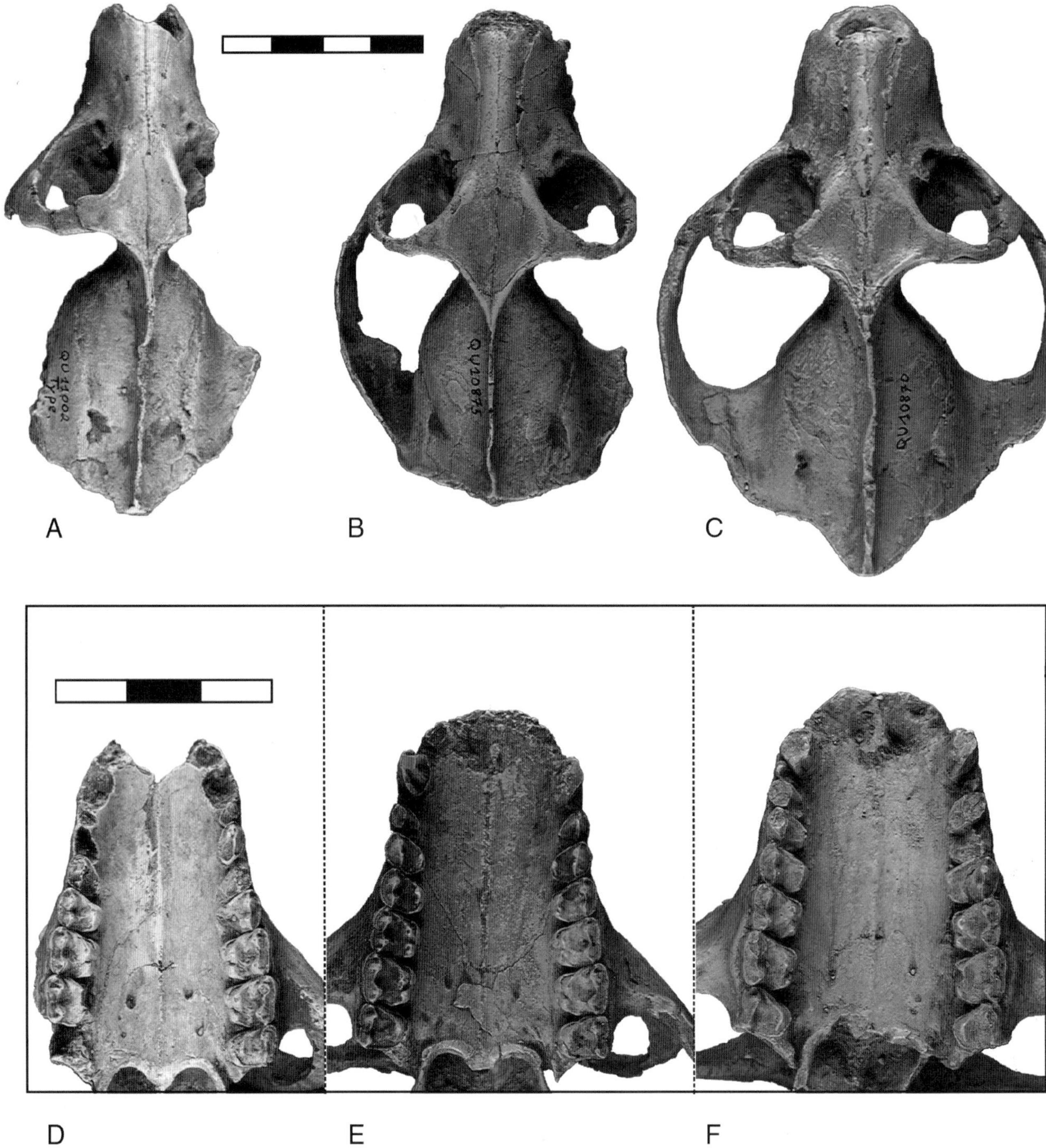

FIGURE 12.4. Crania of *Leptadapis* (A) and two *Magnadapis* species (B, C) in dorsal views (A–C) at the same scale (bar is 4cm), and ventral views of their palates (D–F) at another scale (bar is 3cm). QU 11002 is the type specimen of *L. magnus* (A, D); QU 10875 is the type specimen of *M. quercyi* n. gen. n. sp. (B, E), and QU 10870 is the type specimen of *M. intermedius* n. gen. n. sp. (C, F). Note differences in interorbital breadth, medial orbital rims and muzzle breadth between the two genera in A–C, and differences in palate breadth and anterior outline in D–F.

linked to age or sexual dimorphism; it could also reflect a small phylogenetic difference. Differences can be seen on the teeth. On MaPhQ 211, M1-2/ have a continuous lingual cingulum, whereas it is interrupted on the type specimen. M1/ on MaPhQ 211 appears more square, its posterior half is wider in comparison with the anterior half relative to the type specimen. On the type specimen, the crista obliqua seems slightly better developed. P4/ on the type specimen has a more triangular outline, and it has a distinct crista obliqua; on MaPhQ 211, P4/ is slightly broader lingually and there is no crista obliqua; the posterior cingulum is also isolated from the protocone at its lingual extremity. This extremity is slightly thickened on the right P4/, as if there were an incipient hypocone. On the right P4/, it seems that the cingulum was continuous on the lingual side, not very clearly seen because the tooth is eroded, whereas on the left side the lingual cingulum is clearly interrupted. The link between the protocone and the posterior cingulum is more continuous on the type specimen. P3/ on MaPhQ 211 is narrower in its lingual part, with a cingulum and no protocone cuspule, whereas on the type specimen the lingual part is slightly broader and there is a small recognizable protocone, with a short labial slope. These differences may slightly exceed intraspecific variability, however, they are not strong. The M2/ has a very similar outline. Hypocone size, which decreases from M1/ to M2/ to M3/ (tiny thickening of the cingulum on the type specimen; nothing on the other), appears very similar on both specimens. The differences could indicate just a very small distance within a single lineage. The continuous lingual cingulum on the molars and almost continuous on P4/, as well as the loss of the crista obliqua on P4/, could indicate a more derived stage for MaPhQ 211 than for the type specimen of *L. magnus*. In this case, the P3/ would be in the process of simplification, losing its protocone. However, at this moment, we refrain from naming a new species on such small differences without a better understanding of dental variability in *Leptadapis* species. The differences in cranial superstructures described above could also have evolutionary significance, and in that case would not reflect sexual dimorphism. We need to learn more about dental variability to better interpret these cranial differences. MaPhQ 211 is left in open nomenclature as *L.* aff *magnus*.

12.3.2 Comparison Between YPM-PU 11481 and QU 11002

The teeth of the Yale skull, YPM PU 11481, are similar to those of the *L. magnus* type specimen, QU 11002. M1/ is narrower in its posterior half, with a lingual flexus between the protocone and the hypocone, as on the type. There are still some differences: the M2/ of the Yale specimen has a complete lingual cingulum and a slightly smaller hypocone than does the type. Its M3/ is peculiar too: it is transversely short (a peculiarity of the "variety *Leenhardti*" type specimen noted by Stehlin) and is narrower in its lingual part. The P4/ of the Yale specimen has a triangular outline, slightly narrower lingually than the P4/ of QU 11002, without a postprotocrista, and with an anterior cingulum. On this P4/, the paracone is markedly higher than the metacone, whereas the two cusps are more similar in height on QU 11002 (with variation: the contrast is higher on the left than on the right side of the type; however, neither side has a metacone as reduced as on the P4/ of the Yale specimen). There are also differences in the alveoli of the anterior teeth. The Yale specimen has two alveoli for P2/, the posterior one is larger than the anterior. The alveolus for P1/ is small. On QU 11002, there is only one alveolus for P2/; it is broader posteriorly and is slightly larger. On the whole dental differences between the Yale specimen and the *L. magnus*-type pertain to P4/ and M3/. They tend to confirm a specific distinction between them, however their polarity is not straightforward.

Concerning the cranium, the Yale specimen is smaller than QU 11002, and they differ markedly in proportions. In ventral view, they differ little in palate size and tooth size (this is consistent with intraspecific variability). The palate is longer on the type; its posterior arcuate rim is at the level of the posterior margin of M3/. The rim is further anterior, clearly between the two M3/, on the Yale specimen. This could be due to the Yale individual being a younger at death one. The distance between palate and bullae is only slightly larger on the type than on the other. On the Yale skull, the pterygoid plates are interrupted anterior to the bullae. The right lateral plate is partly deformed, and breakage of the posterior part of these plates can be suspected. On the right side at the base of the pterygoid plate a smooth rounded surface shows the natural original interruption of the plate corresponding to the relatively large foramen pterygospinosum. On the left side, the small bony ridge which is an extension of the plate, possibly broken, would in any case have surrounded the foramen. Low and straight relief on the bulla might well indicate breakage of the pterygoid plate joining the bulla as in all other adapine skulls (this should be checked against the original). The apparent interruption between the pterygoid plate and the bulla on both sides, which would have been very unusual, is probably an artifact.

Size differences between the Yale specimen and QU 11002 are more marked posteriorly. The space between the bullae and the articular condyles is clearly larger on the type than on the Yale specimen. Adding these differences in length, the type appears markedly longer than the Yale specimen, whereas they differ little in breadth. This length difference is accentuated in dorsal view due to the strong posterior projection of the sagittal and nuchal crests on the type. This is absent on the Yale specimen. The braincase is clearly closer to the anterior part of the skull and the postorbital narrowing is slightly less expressed on the Yale specimen than on the type. The braincases of the two specimens seem to be similar in size. A big difference between them is the very strong development of the sagittal and nuchal crests, projecting further posteriorly on the type (more than 1 cm beyond the posterior margin of the foramen magnum) than on the Yale one (reconstructed with plaster in this part). The Yale individual has a very low sagittal crest. Even if this crest is slightly worn, one can see very clearly in the anterior part of the braincase the two frontal lines converging posteriorly just on the top of the braincase. This demonstrates that the crest is very low on the Yale individual. By contrast, on the type, the frontal lines join almost 1 cm above the anterior part of the braincase (the smallest distance between braincase and dorsal rim of frontals, along a dorso-anteriorly inclined line, is 8 mm). Hence there is a very strong difference between the two specimens in the development of the sagittal crest and the distance between the braincase and the frontal plane, which are both very reduced on the Yale specimen. Can such huge differences be explained by growth, sexual dimorphism, or both? The canine alveoli of the Yale specimen are too poorly preserved to be used in assessing differences in canine size. In lateral view, the zygomatic arches of the Yale specimen are disturbing because they are poorly restored in plaster. However, the anterior part of the zygomatic arch is intact: it is more gracile on the Yale individual. The anterior view clearly confirms that the zygomatic root, below the orbit, is higher on the type than on the Yale specimen, making the cranium look higher in profile view (Figure 12.6).

The Yale individual has a definitive P4/ and M3/, and only slight wear of the major crests on M1/ and M2/. It is probably a young adult, and certainly not a juvenile. The *L. magnus* type skull is an adult, however its teeth are little worn, so that it cannot be a very old individual. These individuals could not have been very different in age at death. Consequently, most of their morphological differences cannot be explained by growth. Might some of these differences be due to sexual dimorphism? If all the differences in size and cranial superstructures between these specimens were due to sexual dimorphism, this dimorphism would be extreme for a primate of that size, and one would expect the putative male, the type, to have enormous canines. Such is not the case based on its canine alveoli. More importantly, the distance between the braincase and the frontal rims reveals a marked difference in skull structure and not simply a difference in growth as could be expected between males and females. We acknowledge that a degree of sexual dimorphism in *Leptadapis* would explain a small part of the differences in crest development. However, on the whole, we think that the sum of the differences in size, cranial structure and cranial superstructures goes much beyond intraspecific variability, and must have an evolutionary significance. The two specimens very probably pertain to two different, closely related, species. As the Yale specimen is in its major characters close to *L. leenhardti*, and we do not want to excessively complicate the nomenclature of this group, we designate the Yale cranium as neotype for *L. leenhardti*.

12.3.3 Comparison Between MU ACQ 209 and MNHN QU 11002

The Montpellier cranium ACQ 209 differs from the *L. magnus* type specimen QU 11002 in several ways. The weaker development of its sagittal and nuchal crests could be due to a difference in age or sex, however both appear to be young adults and there is no indication from the canine alveolus of the type that the canine was larger than that of ACQ 209. The type has a peculiar broad postorbital bar; ACQ 209 has a narrower one. The anterior zygomatic root is slightly higher on the type; both roots present an anterior ventral spine almost at the level of the dental row. However, other differences appear more significant. In anterior view, the zygomatic arches project farther laterally on ACQ 209; this appears linked to the larger orbits of ACQ 209, which are more circular in anterior view, higher and more anteroposteriorly elongated in profile view (Figure 12.6). The infraorbital foramen is slightly more anterior on ACQ 209 in profile view. The most dramatic differences between them are in the height of the muzzle. The muzzle is higher anteriorly on ACQ 209 than on the type. Its interorbital breadth is slightly greater, and in ventral view its palate is markedly broader. In dorsal view, the braincase of ACQ 209 is further anterior. This is linked to its smaller sagittal crest. The anterior part of the braincase reaches less than 5 mm below the rims of the frontal depression. This is a clear difference in the lateral view of the two crania (Figure 12.6). The anterior slope of the muzzle is more inclined in relation to the tooth row in QU 11002, which appears to have a lower anterior muzzle. On the whole, differences in cranial superstructures and cranial elongation cannot be related to sexual dimorphism, because there is no evidence of canine size difference, and other marked differences exist in muzzle height and palate breadth. We think that such differences probably exceed intraspecific variability, however we admittedly have little reference in comparable skull morphologies.

There are differences in the teeth of the two specimens. On ACQ 209, M1/ is especially broad in its lingual part, due to the unusually large size of its hypocone lobe in occlusal view, and it has a more voluminous hypocone than does QU 11002 (Figure 12.5). The posterior extension of the hypocone is also present on M2/, which also shows a more salient and extended crest posterior to the protocone. The tooth has an anteroposteriorly elongated outline, which is derived for the group. The M3/ differs in some details. The lingual cingulum is complete on the type, which has a very weak hypocone (slight swelling of the cingulum), whereas the cingulum is interrupted on ACQ 209, which has a distinct cingular hypocone linked to the posterior crest of the protocone. This renders the M3/ of ACQ 209 broader lingually than on the type. The P4/ of the two specimens are similar for several characters, including a distinct crista obliqua. Both have a posterior crest descending from the protocone, joining the posterior cingulum on the type, but interrupted before the cingulum, which is more extended lingually, on ACQ 209. However, on ACQ 209, the P4/ metacone is especially small, being only a thin cusp on the crest descending from the high paracone (Figure 12.5). Both cusps are more equal in size on the type. Another difference between them is that P4/ is transversely less extended and lingually broader on ACQ 209. The P3/ does not show a clear difference. In general, the differences on the molars and P4/ are probably significant. This dental evidence reinforces our conclusion concerning cranial characters, and we conclude that ACQ 209 is a species of *Leptadapis* different from *L. magnus* and *L. leenhardti*.

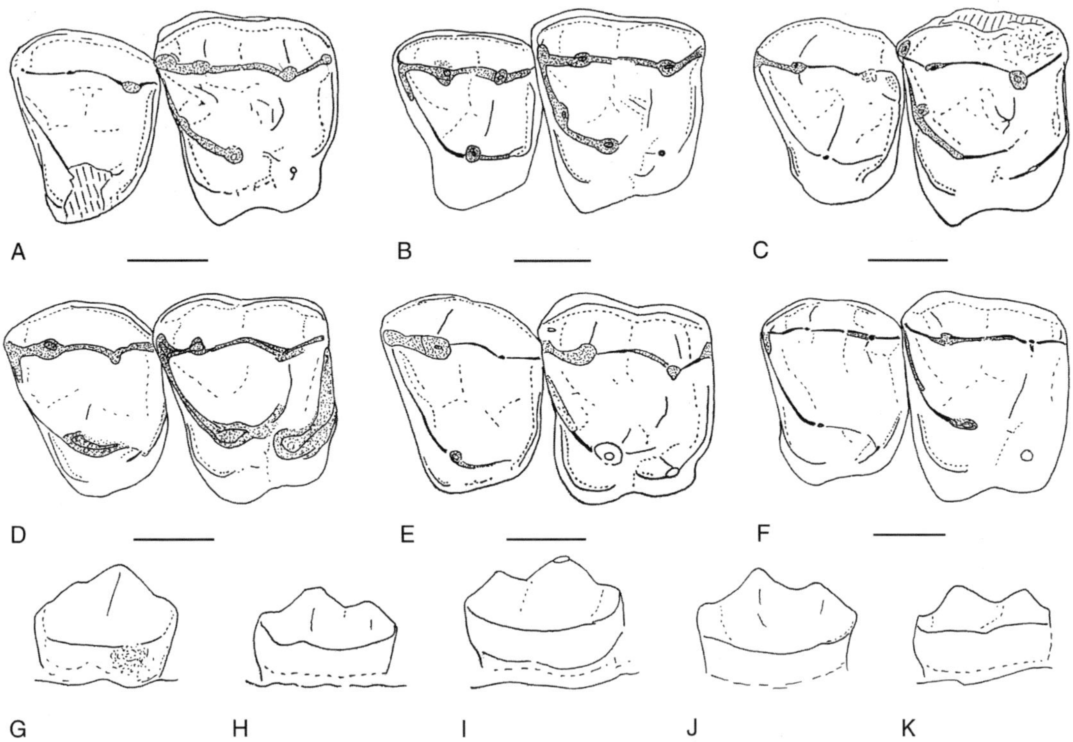

FIGURE 12.5. Schematic drawings of the occlusal views of P4/ and M1/ (A–F), and of the labial views of P4/ (G–K), in selected large adapines. Not to scale; all M1/ were drawn at the same maximal transversal breadth, and P4/ vary accordingly (scale bar is 3 mm). A is YPM PU 11481, neotype of *Leptadapis leenhardti*; B and H are MNHN QU 11002, type specimen of *L. magnus*; C and G are MU ACQ 209, type specimen of *Leptadapis filholi* n. sp.; D and I are MNHN QU 10875, the type specimen of *M. quercyi*, n. gen. n. sp.; E and J are MNHN QU 10870, the type specimen of *M. intermedius*, n. gen. n. sp.; F and K are MaPhQ 210, the type specimen of *M. fredi* n. gen. n. sp. Note that the M1/ vary in outline from markedly asymmetrical (B) to more subquadrate (E), and their hypocones vary from small and crested (B, E) to larger and round (F), or very large and moderately crested (C); there are also variations of the lingual cingulum. The P4/ are generally smaller and lingually narrower in *Leptadapis* species (A–C), and two have a well-developed crista obliqua (B and C); P4/ are generally larger and lingually broader in *Magnadapis* species (D–F); increasing lingual breadth can be observed from A and B to E and F (note incipient hypocone in F). Molarization of P4/ on its labial side can be seen from G (dominating paracone, metacone barely isolated), to K, the most molarized, with a lower paracone and well isolated metacone, through intermediates (H–J). Several specimens are left-right inverted for comparison in A–F, but not in G–K.

Is ACQ 209 also different from other previously named species? The maxilla of *L. assolicus* (Richard, 1940) bears molars which are incomplete but well enough preserved to show important differences from all the large adapines studied here. The P4/ is large compared to M1/. It is simple, very narrow lingually and unlike all the P4/ of the other large adapines. M1/ is difficult to analyze because pieces of enamel are missing. M2/ has a large and high hypocone, relatively close to the summit of the protocone. This morphology recalls that of *Cryptadapis tertius* and clearly differs from all the other large adapines analyzed here. The species *assolicus* is not a *Leptadapis* as defined here, and neither does it pertain to the group defined below.

It was suggested in a stratodimensional diagram that the species *A. stintoni*, proposed as a descendant of *L. magnus* by Gingerich (1977) and having a size similar to *assolicus*, could be a synonym of the latter (Godinot, 1998: Figure 5). However, a re-examination of the type material confirmed that it is a large species of *Adapis*. It is not a *Leptadapis* species nor is it a *Cryptadapis* species.

Lastly, the M3/ of ACQ 209 also differs from the M3/ described by Crusafont-Pairo (1967) as *"Arisella" capellae*, and later transferred to *Leptadapis capellae* by Szalay and Delson (1979). The M3/ of *L. (?) capellae* has a marked narrowing of its posterior half, a large paraconule, no lingual cingulum and no enlargement of the posterior cingulum corresponding to the location of a hypocone. These differences clearly exceed intraspecific variation of the M3/, and ACQ 209 cannot belong to the species *L. (?) capellae*. Hence ACQ 209 is a new species, which requires a new name. Our study of crania and teeth leads us to formally recognize three species in the genus *Leptadapis*, including a new species.

12.3.4 Systematics of the *Leptadapis* Species

Adapinae Trouessart 1879

Leptadapis Gervais, 1876

Type-species: *Leptadapis magnus* (Filhol, 1874)

Included species: *L. leenhardti* (Stehlin, 1912), *L. filholi*, n. sp.

Diagnosis: large adapines with crania possessing a relatively narrow interorbital breadth, narrow muzzle, salient posterior palatal spine; muzzle generally lower than in *Magnadapis* species; M1/ and M2/ with a well-developed hypocone, M3/ with a postprotocrista and a well-developed posterior cingulum; P4/ usually narrow in its lingual half; canines smaller than in *Magnadapis* species; no diastema between upper canine and I2/.

Leptadapis magnus (Filhol, 1874)

Type specimen: cranium MNHN QU 11002, Muséum National d'Histoire Naturelle, Paris (Figures 12.1, 12.4–12.7).

Horizon and locality: unknown; old collections, Quercy region, south France.

Emended diagnosis: *Leptadapis* with large sagittal and nuchal crests, marked distance between the braincase and the frontal plane, producing a relatively elongated skull.

Referred specimen: cranium MM MaPhQ 211 (described and figured by Stehlin, 1912, as "Montauban 2").

Leptadapis leenhardti Stehlin, 1912

Type specimen: cranium "Montauban 3" described by Stehlin (1912), now lost. Neotype: cranium YPM PU 11481, Yale Peabody Museum (Figures 12.3, 12.5, 12.6).

Horizon and locality: unknown; old collections, Quercy region, south France.

Diagnosis: *Leptadapis* smaller than *L. magnus*, having a very low sagittal crest, weaker anterior zygomatic root than in *L. magnus*; frontal plane lying just above the braincase; skull anteroposteriorly short; P4/ simple and especially narrow lingually.

Leptadapis filholi, new species

Type specimen: cranium UM ACQ 209, Montpellier University (Figures 12.2, 12.5–12.7, 12.9).

Derivatio nomini: in honor of Henri Filhol, who named the species *L. magnus* and made substantial contributions to our knowledge of fossil mammals.

Horizon and locality: unknown; old collections, Quercy region, south France.

Diagnosis: *Leptadapis* with a higher muzzle and a broader palate than in *L. magnus* and *L. leenhardti*; sagittal and nuchal crests moderate in size (well developed but smaller than in *L. magnus*); distance between braincase and frontal plane intermediate between *L. magnus* and *L. leenhardti* (closer to the latter); the three molars are broader lingually than in the two other species; M1/ has an especially broad hypocone.

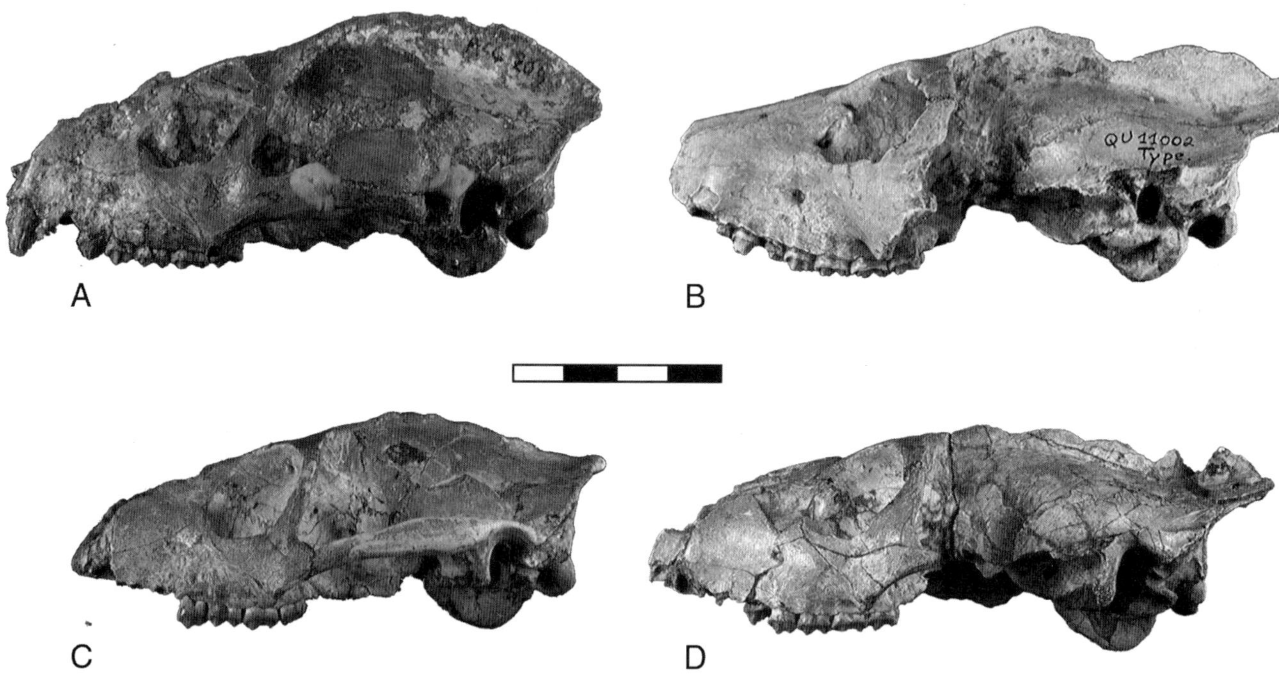

FIGURE 12.6. The four crania of *Leptadapis* in lateral views at the same scale (bar is 4 cm). Type specimen of *L. filholi*, MU ACQ 209 (A); type specimen of *L. magnus*, QU 11002 (cast, B) showing the highest sagittal and nuchal crests, and a thickened postorbital bar; the neotype of *L. leenhardti*, YPM PU 11481 (C) has a very low sagittal crest, and its posterior extremity and zygomatic arch are reconstructed with plaster; another *L. magnus*, MaPhQ 211 (D).

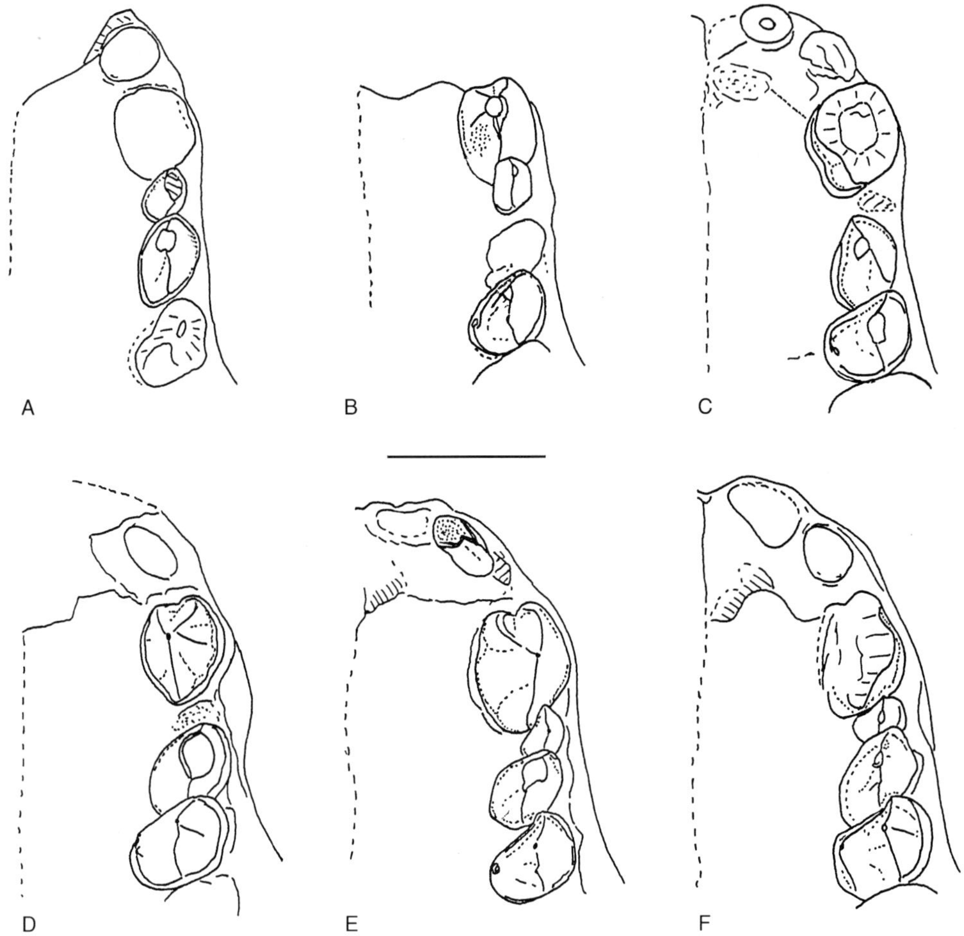

FIGURE 12.7. Drawings of the left anterior part of the muzzles of large adapines, showing the canine or its alveolus, P1/ to P3/ or alveoli, and incisor remnants or alveoli when preserved. Placing complete canines as vertical as possible inclines the nearby premolars, their labial sides being favored and their lingual side being diminished. All drawn at the same scale (bar is 1 cm). A is QU 11002, type of *Leptadapis magnus*; B is ACQ 209, type of *L. filholi*, n. sp.; C is QU 10872, type of *Magnadapis laurenceae*, n. gen. n. sp.; D is QU 10875, type of *M. quercyi* n. gen. n. sp. (inverted left-right; right side); E is ACQ 214, *L. intermedius* n. gen. n. sp.; F is MaPhQ 210, the type of *M. fredi*, n. gen. n. sp. Note that canines vary in size and outline (that in B is partially worn). Rare preservation of the incisor region shows that the alveolus of I2/ is close to that of the canine in *Leptadapis* (A), whereas incisors or alveoli are separated by a short diastema and slightly shifted medially in *Magnadapis* species (C–F); incisor alveoli are best preserved in F (canine is better preserved on the other side, but incisor alveoli are less clear); a partially worn I2/ is present in E and the root of I1/ is present in C.

12.3.5 Phylogeny of *Leptadapis* Species

Our systematic interpretation of the four *Leptadapis* skulls is to classify them in three different species. Can a phylogenetic interpretation of those be proposed? In the most evident character, cranial superstructures and robusticity, there is a clear morphocline. *L. leenhardti* has a very low sagittal crest, the braincase lying just beneath the frontal depression, the anterior zygomatic arch is lower than on the other specimens. Next is *L. filholi*, which has an intermediate sized sagittal crest and a braincase still anterior but several millimeters beneath the frontal rims; the anterior zygomatic arch is as high as in *L. magnus*. Lastly, *L. magnus* has a strongly developed sagittal crest, the sagittal and nuchal crests project much farther beyond the occipital plane, and the braincase is pushed backward (distinctly more isolated from the anterior part of the skull). The postorbital bar is clearly thicker on the *L. magnus* type specimen than on the other specimens.

In terms of superstructures and robusticity, there is a clear morphocline *leenhardti-filholi-magnus*. The increased development of the masticatory musculature is accompanied by a lengthening of the space between the frontal region and the braincase (the postorbital narrowing being stronger) and an accentuation of the posterior projection of the nuchal crests. Both aspects produce a marked lengthening of the whole skull in *L. magnus* in comparison with the two other species. However, some of the other observable skull characters do not fit into a similar morphocline. This is especially true of the breadth of the palate and muzzle, which is greater in *L. filholi* than in the two other species. This also seems to be true of muzzle height. This is visible despite the incompleteness of the nasal region on *L. leenhardti*. Little is missing between the two maxillae, and the height of its muzzle on the right side must be very close to the original state. Clearly the muzzle is much lower on *L. leenhardti* than on *L. filholi*: again *L. filholi* does not appear to be intermediate between the two other species in terms of muzzle morphology.

Concerning dental characters, there are no big differences between *L. leenhardti* and *L. magnus*, however those present can be interpreted. The lingual narrowness of M3/ and the double-rooted P2/ in *L. leenhardti* can be primitive (or variability?). On the P4/ of *L. leenhardti*, the absence of the crista obliqua and the less equally-sized labial cusps make this tooth appear simpler, and labially less molarized. *L. leenhardti* may well be primitive relative to *L. magnus* for dental and cranial characters, leading to the most parsimonious hypothesis that it is a possible ancestor of the latter (see Figure 12.11). We mentioned that characters of the M1/ in *L. filholi* (length, very large hypocone) are derived in comparison with the two other species. The P4/ however is not more derived than that of *L. magnus*. It could be at the same time derived in its transverse shortness and lingual breadth, and primitive in retaining a very small metacone. As is true of cranial characters, dental characters of *L. filholi* suggest a lineage independent of *L. magnus*. This leads to the view that two lineages of *Leptadapis* are present in the Quercy region. In this context, we cannot say anything concerning possible sexual dimorphism in *Leptadapis* species. We cannot say for sure if the *L. magnus* type specimen is a male and the *L. filholi* is a female, differences in cranial superstructures between the two species might be less accentuated. However we cannot show this through canine size. The two specimens attributed to *L. magnus* have small differences in cranial superstructures, which could reflect either a small amount of dimorphism or a small evolutionary distance. Available specimens do not give positive evidence of sexual dimorphism in *Leptadapis* species. Do some of the skull characters of *L. leenhardti* indicate that it might be a juvenile? The specimen, however, has a definitive P4/ and M3/, and the *L. magnus* type specimen has P4/ and M1/ which have only a small degree of wear. The difference in age at death between these two specimens must have been small, indicating that growth cannot account for most of the differences found in cranial superstructures. On the whole, *Leptadapis* species as redefined here are incompletely known: their anterior dentition is poorly documented; only one of them, *L. filholi*, shows a complete zygomatic arch, a canine and a P1/. No specimen has an associated mandible. It will be important to search in the biochronologically situated assemblages for mandibles which can be attributed to *Leptadapis* species.

12.4 Second Group of Generic Value: *Magnadapis* n. gen.

The other group of specimens includes the skull QU 11035-11036, the crania QU 10870, QU 10872, QU 10875, MaPhQ 210, the two muzzles PLV 6, ACQ 214, and specimens from Euzet-les-Bains. As explained above, they represent several species sharing a number of significant characters. These species need to be distinguished from the preceding group at the generic level, and we coin a new genus for them. Because this group has never been properly recognized before, we name several new species, for the skull morphologies which appear to us to warrant such systematic recognition. We first give names and diagnoses for the clearly recognizable species, and explain our choices and hesitations concerning other specimens in the following discussion.

12.4.1 Systematics

Magnadapis, new genus

Type-species: *Magnadapis quercyi*, new species.

Derivatio nomini: from *magnus*, large, and *Adapis*, because this group includes the largest known adapine species.

Diagnosis: species of *Magnadapis* differ primarily from species of *Leptadapis* by a broader interorbital breadth, associated with a broader muzzle; the palate is also broad, and the dental rows are less convergent anteriorly than in *Leptadapis* species (except *L. filholi*, which also has a broad palate); there is no posterior palatal spine. I2/ is small and isolated from the canine by a small diastema; the canines are very large and marked by deep vertical grooves; the P1/ seems comparatively smaller than in *Leptadapis*; P4/ often lingually broader than on *Leptadapis* species, and never presenting a crista obliqua interrupting the trigon basin; upper molars with hypocones generally smaller than in *Leptadapis* species, sometimes absent on M2/; M3/ transversely broad, with the trigon basin open posteriorly (no crista obliqua, usually no posterior cingulum). Several *Magnadapis* specimens are somewhat larger than *Leptadapis* specimens.

Other included species: *Magnadapis fredi* n. sp., *M. laurenceae* n. sp, and *M. intermedius* n. sp.

Magnadapis quercyi, new species

Type specimen: MNHN QU 10875, cranium from the Muséum National d'Histoire Naturelle, Paris (Figures 12.4, 12.5, 12.7–12.9). Cranium figured by Genet-Varcin (1963), Saban (1963) and Simons (1972).

Derivatio nomini: in reference to the South-France province where all the large adapine skulls were found in a well known paleokarst.

Horizon and locality: unknown; old collections, Quercy region, south France.

Diagnosis: cranial superstructures weakly developed, root of zygomatic arch higher than in *M. laurenceae*; large orbits and narrow postorbital bar; height of muzzle tapering anteriorly

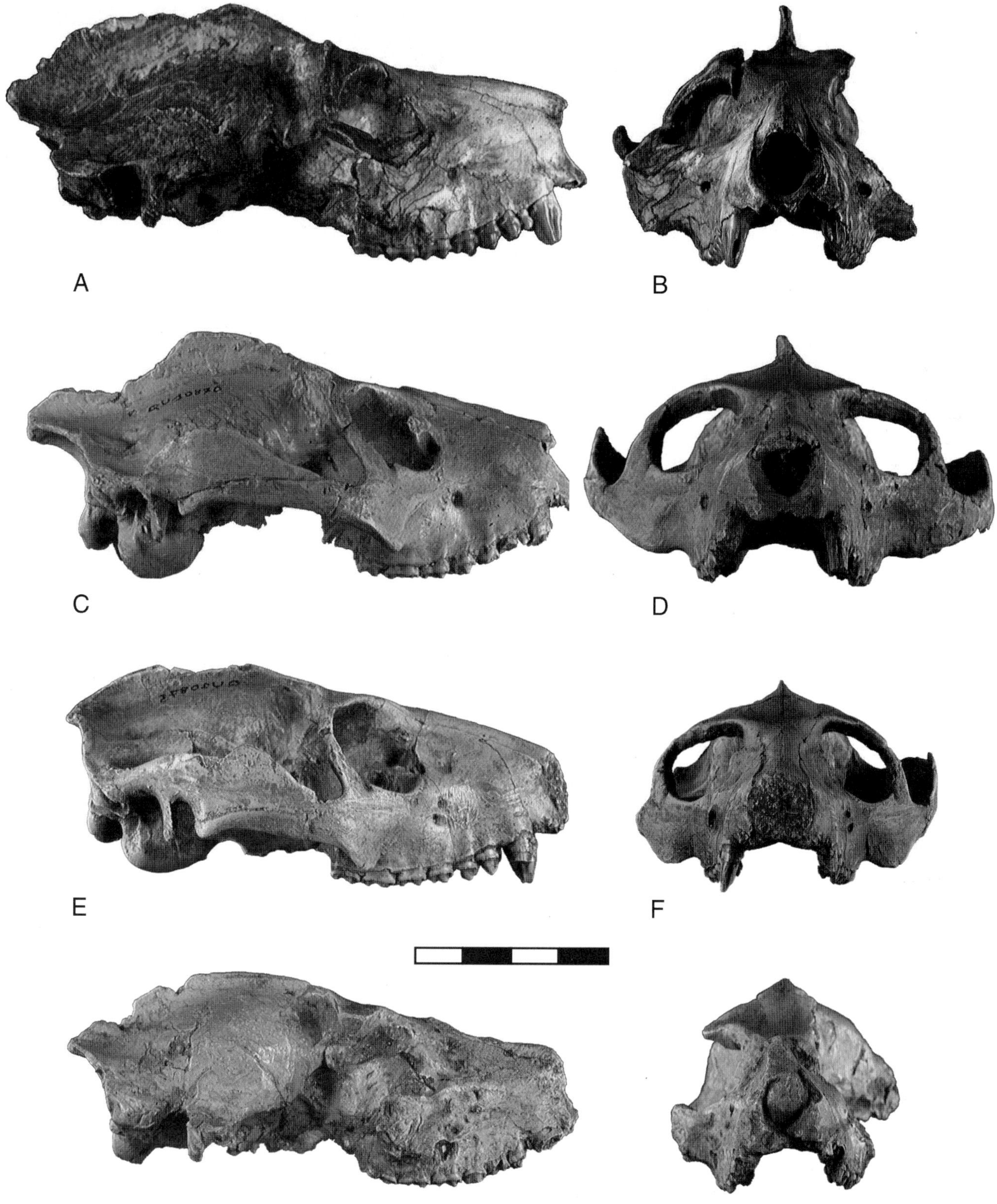

FIGURE 12.8. Four crania illustrating the four different *Magnadapis* species, in lateral (A, C, E, G) and in anterior (B, D, F, H) views, all at the same scale (bar is 4 cm). Type specimen of *M. fredi* n. gen. n. sp., MaPhQ 210 (A, B); type specimen of *M. intermedius* n. gen. n. sp., MNHN QU 10870 (C, D); type specimen of *M. quercyi*, MNHN QU 10875 (E, F); type specimen of *M. laurenceae*, MNHN QU 10872 (G, H). They are arranged with cranial superstructures increasingly developed from bottom to top, but the nuchal projection is more accentuated in G than in E. Note an accompanying increase in facial and anterior muzzle height.

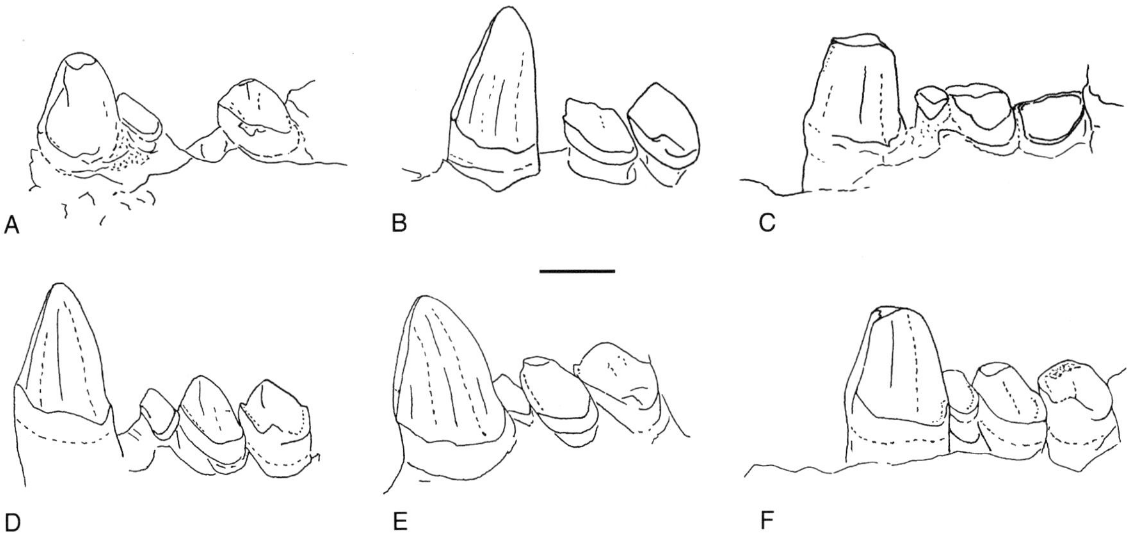

FIGURE 12.9. Schematic drawings of left upper canine to P3/ in lingual view in several large adapines, all at the same scale (bar is 5 mm). A is ACQ 209, type of *Leptadapis filholi*, n. sp.; B is QU 10875, type of *Magnadapis quercyi* n.gen. n. sp.; C is QU 10870, type of *M. intermedius* n. gen. n. sp.; D is PLV 6, referred to as *M.* aff. *quercyi*; E is ACQ 214, referred to *M. intermedius*; F is the type of *M. fredi*, n. gen. n. sp. Although the canine in A is partially worn, its anterior vertical groove and its lingual cingulum are partially preserved, showing a clear difference in size and shape between the only upper canine preserved in a *Leptadapis* species (A) and the more numerous upper canines of *Magnadapis* species. Note also increasing molarization of P3/ from A to B, E, to D, and to F, where the protocone lobe is broad and high.

much more than in *M. fredi*; canine moderate in size; P2-3-4/ less developed lingually, less molarized than in *M. fredi*; M2/ with small recognizable, cuspidate hypocone.

Referred specimen: muzzle PLV 6 from Leuven University, Belgium.

Magnadapis laurenceae, new species

Type specimen, MNHN QU 10872, incomplete cranium from the Muséum National d'Histoire Naturelle, Paris (Figures 12.7, 12.8, 12.10).

Derivatio nomini: in honor of Laurence Lanèque, in recognition of her important dissertation work on adapine skulls.

Horizon and locality: unknown; old collections, Quercy region, south France.

Diagnosis: weakly developed cranial superstructures; very low sagittal crest; frontal lines joining more posteriorly than on all other large adapine skulls; correlatively braincase closer to the anterior part of the skull than in other *Magnadapis* species; nuchal crests projecting further posteriorly than in *M. quercyi*; muzzle narrower, orbits smaller and anterior zygomatic root lower than in *M. quercyi*; premolars similar to those of *M. quercyi* in terms of molarization; canine relatively large and unusual in its great labio-lingual breadth (crown subcircular instead of anteroposteriorly elongated as in other species); M2/ with complete lingual cingulum and no hypocone.

Magnadapis fredi, new species

Type specimen: MM MaPhQ 210, cranium of the Montauban Natural History Museum (Figures 12.5, 12.7–12.10); reconstruction figured as "*Adapis magnus*, Montauban 1" by Stehlin (1912).

Derivatio nomini: in honor of Frederick S. Szalay, Fred to his close colleagues and friends, in recognition for his extensive contributions to primate paleontology and his commitment to theoretical questions.

Horizon and locality: unknown; old collections, Quercy region, south France.

Diagnosis: *Magnadapis* with enormous cranial superstructures, braincase slightly more than 1 cm below frontal rims, and pushed posteriorly relative to the anterior part of the skull; postorbital narrowing correlatively strong; very high sagittal crest in the posterior part of the skull, more than 1.5 cm high; cranium extended posteriorly further beyond the external auditory meatus than in other species; anterior zygomatic root higher than in all other species, and muzzle higher in its anterior part than in all other species of *Magnadapis*; nasofrontal suture shorter, much less posteriorly wedged between the frontals; P2-3-4/ lingually well developed, more molarized than in the three other *Magnadapis* species; M2/ with recognizable and cuspidate hypocone.

Magnadapis intermedius, new species

Type specimen: MNHN QU 10870/1, cranium and associated mandible (10871) from the Muséum National d'Histoire Naturelle, Paris (Figures 12.4, 12.5, 12.8, 12.9); figured by Grandidier (1905).

Derivatio nomini: to reflect the fact that its morphological characters are in several respects intermediate between those of *M. quercyi* and *M. fredi*.

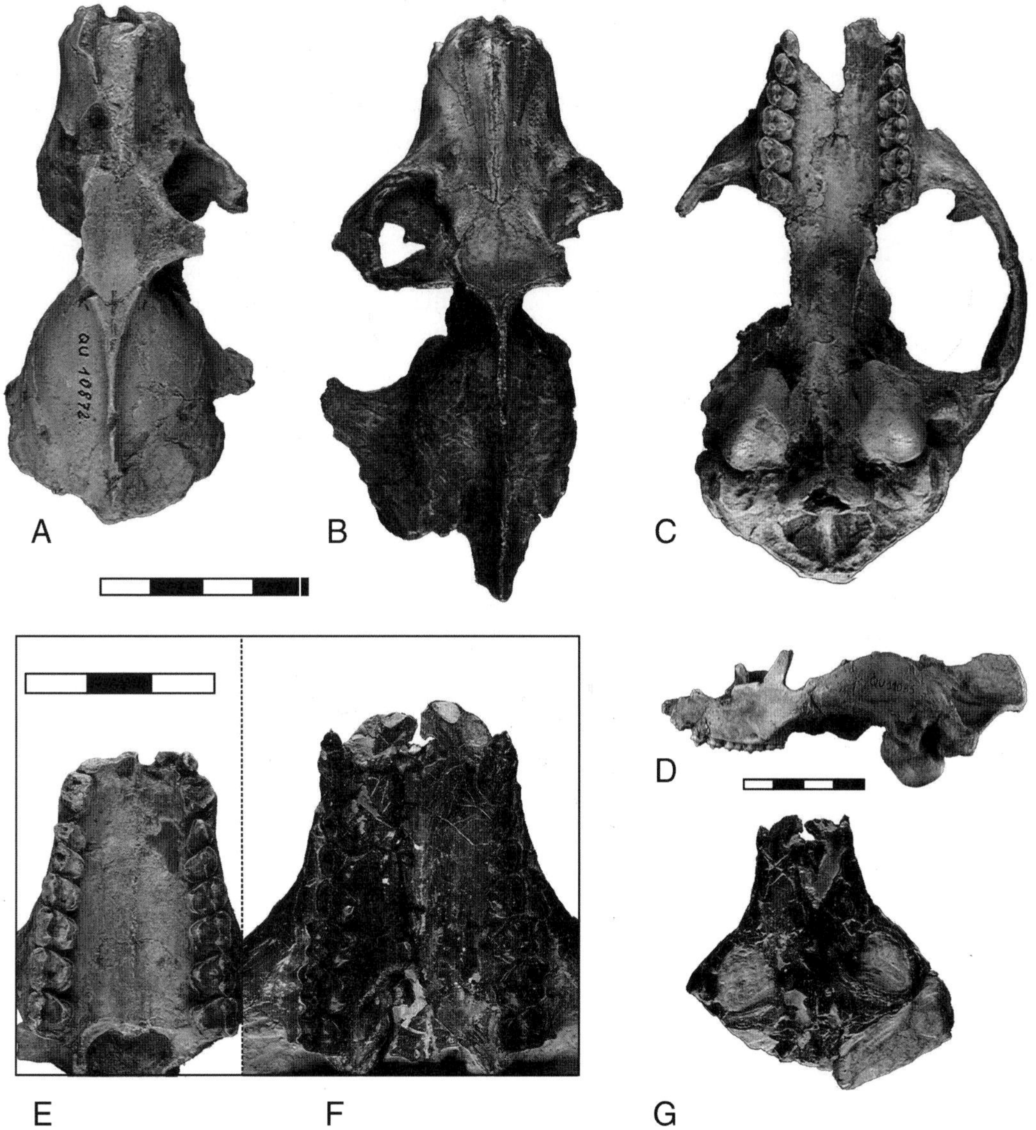

FIGURE 12.10. Two crania, one partial cranium and one muzzle of different *Magnadapis* species. Dorsal views of the type specimen of *M. laurenceae* n. gen. n. sp., MNHN QU 10872 (A), and the type specimen of *M. fredi* n. gen. n. sp., MaPhQ 210 (B); ventral view of MNHN QU 11035, *M.* aff. *intermedius* (C), and lateral view of the same (D); ventral views of the palates of *M. laurenceae* QU 10872 (E) and of the crushed muzzle BM St. H. 1634 from Euzet, referred to *M. fredi* (F); dorsal view of the same muzzle St.H. 1634 (G). At different scales: A–C and D, G, bars are 4 cm; E, F, bar is 3 cm. The two crania in A and B illustrate minimal (A) and maximal (B) cranial superstructure development in *Magnadapis* species. The posterior part of QU 11035 (D) probably approximates how the broken posterior part of *M. fredi* (B) may have looked like; notice the extremely high posterior part of the zygomatic arch.

Diagnosis: *Magnadapis* with well developed cranial superstructures, flaring zygomatic arches, sagittal and nuchal crests projecting far behind the foramen magnum, making a triangular posterior projection in dorsal view; postorbital constriction more accentuated than in *M. quercyi*; skull longer than in *M. quercyi*; cranial superstructures generally less extreme than in *M. fredi*; canines larger than in *M. quercyi*.

Referred specimens: partial cranium MNHN QU 11035 (Figure 12.10) and associated mandible (11036), muzzle ACQ 214 (Figures 12.7, 12.9).

The four crania used to name these four species can be arranged in a morphocline according to the general development of cranial superstructures (Figure 12.8). However, several peculiarities of *M. laurenceae* set it apart, and later we will discuss its possible significance. The partial cranium QU 11035 is somewhat intermediate between the type specimens of *M. intermedius* and *M. fredi*, and the muzzles PLV 6 and ACQ 214 also pertain to the same group. A parsimonious phylogenetic interpretation of this evidence is that these species constitute a lineage marked by increasing robusticity and cranial superstructure development, starting with *M. quercyi* and ending with *M. fredi* (an hypothesis which implies a chronological succession). However, the number of species to be distinguished between these two extremes is not straightforward, due to the possible role of sexual dimorphism in the morphological differences between the specimens.

12.4.2 Comparison Between QU 10870 and QU 10875

The most interesting case is QU 10870/1, the type specimen of *M. intermedius*. It is slightly larger than the *M. quercyi* type specimen QU 10875, and presents markedly more developed sagittal and nuchal crests, an increased nuchal projection, and its zygomatic arches are more flared laterally (Figure 12.4). These specimens show quite similar degrees of tooth wear, indicating that age cannot be responsible for their morphological differences. The difference between them in canine size is difficult to evaluate because the two canines of QU 10870 are incomplete and partially worn; the left one is more complete and preserves remnants of its basal cingulum: it was slightly larger than that of QU 10875 (Figure 12.9). If there were sexual dimorphism in large adapines, these two specimens could be considered a female (QU 10875) and a male (QU 10870) of the same species. Their orbital morphologies are remarkably similar. Size differences in sagittal and nuchal crests could exist between males and females as in some living anthropoids, however to what degree could they exist in Eocene strepsirhines? In fact, close inspection shows that they differ by some important skull characters; looking at these skulls in a posterolateral view, one sees that the minimal distance between the anterior part of the braincase and the frontal plane (the two converging frontal rims) is between 2 and 3 mm on QU 10875, and is 7–8 mm on the left side of QU 10870 (around 4 mm on the right side, where a bony plate artificially lengthens the braincase anteriorly). In dorsal or lateral view, the braincase of QU 10870 is posteriorly shifted behind the frontal region, accentuating its postorbital narrowing, and also increasing its skull length (Figures 12.4, 12.8). Related to its posterior shift, the posterior part of QU 10870 shows a more accentuated posteroventral inclination of the braincase relative to the anterior part of the skull (i.e., a higher degree of basal flexure or klinorhynchy). There is also a difference in the anterior part of the skull, the muzzle of QU 10870 being higher in its posterior part than on QU 10875 (Figure 12.8). This difference resides in the infraorbital height, greater on QU 10870. This difference in height is also present in the anterior part of the muzzle, thinner on *M. quercyi*. Strangely, in this context one would suspect that the orbit might be more vertical on QU 10870, and in fact the contrary is true: the orbit has a more posterior inclination on the higher of the two skulls, QU 10870 (Figure 12.8). On the whole, the huge difference in the size of their sagittal and nuchal crests, the marked nuchal projection on QU 10870, only incipient on QU 10875 (Figure 12.8), the increased flaring of the zygomatic arches on QU 10870, and its higher anterior zygomatic roots, are accompanied by a marked anteroposterior lengthening of the skull, an increased distance between braincase and frontal plane, and a clear difference in muzzle height. Such differences markedly exceed those found in species of *Alouatta*, the most dimorphic platyrrhine (which is roughly similar in size). If *Magnadapis* species had a higher degree of sexual dimorphism than highly dimorphic anthropoids, one would expect this to be reflected by very big canines in males. However, the difference in size of the canines observed between the two specimens is small (Figure 12.9). In this context, an evolution of cranial superstructures linked to dietary adaptation seems much more likely than an extreme degree of unusual sexual dimorphism. Even if a small part of the marked morphological differences between these two crania might be due to sexual dimorphism, we feel that they need to be placed in two different species.

12.4.3 Comparisons of QU 11035 and the Muzzles PLV 6 and ACQ 214

At this moment, it is difficult for us to find dental characters which would clearly delineate species of *Magnadapis*. The teeth of the *M. quercyi* type specimen seem primitive, due to the square outline of M1/, the simple, lingually narrow P4/, and the very small protocone of P3/. Its canine is also relatively slender. However, the muzzle PLV 6, which would fit very well with *M. quercyi* because it has a similar orbit and a similarly slender anterior zygomatic root, has less square M1/, clearly broader in its anterior than in its posterior half. This muzzle also retains a well preserved canine, more robust than that of the *M. quercyi* type. Is this evidence of possible canine dimorphism, usual intraspecific variability, or slight evolution toward more robust upper canines? We cannot clearly answer this question without more information on intraspecific dental variability in large adapines. We leave PLV 6 in open nomenclature as *M.* aff *quercyi*.

The incomplete skull QU 11035/6 raises another interesting question. In its preserved parts, the crests are extremely developed and in this way it resembles MaPhQ 210, the *M. fredi* type specimen. However, the anterior root of its zygomatic arch is much lower than on MaPhQ 210. Its P3/ is also less molarized than on MaPhQ 210. Its P2/ is remarkably little extended lingually and is also less molarized than on MaPhQ 210. Hence this skull is clearly closer to QU 10870 than to the *M. fredi* type specimen. We cannot find clear dental characters separating QU 11035 and QU 10870, whereas such characters exist for QU 11035 and MaPhQ 210. QU 11035 is slightly larger than QU 10870, the *M. intermedius* type specimen. Its zygomatic arches flare more widely and its nuchal crests and the posterior extremity of its sagittal crest are more extended than on QU 10870. In dorsal view, QU 10870 shows a salient posterior triangle due to these crests (Figure 12.4), whereas QU 11035 has a more regular outline (Figure 12.10), due to the increased lateral development of its nuchal crests. In QU 11035, the zygomatic arch is higher, remarkably robust in its posterior part (Figure 12.10), and it seems to have even greater basicranial flexure than QU 10870. The profile view also shows the very steep departure of the sagittal crest above the nuchal projection, not preserved on QU 10870. These differences concern crests, muscular attachments, and their possible correlates. The anterior zygomatic roots have only a small difference (slightly higher on QU 11035), easily accommodated in intraspecific variability. Hence these two specimens, QU 10870 and QU 11035 could possibly illustrate sexual dimorphism in *M. intermedius*, the type specimen being a female and QU 11035 being a male. However, other differences between the two specimens might also point toward some evolutionary distance between them instead. The base of the skull is slightly more extended posteriorly on QU 11035 than on QU 10870. In ventral view, the area posterolateral to the bullae and surrounded by a ribbon-like area continuous with the nuchal plane (presumably for a more extensive area of nuchal musculature attachment), is larger on QU 11035 than on QU 10870. Another difference, more clearly seen, is the much larger bullae on QU 11035 than on QU 10870 (visible even in lateral views, Figures 12.8, 12.10). Whereas we cannot definitely rule out that sexual dimorphism would explain some of these differences, we consider likely that QU 11035 represents a more advanced stage in a lineage affected by developing cranial superstructures, an increasing basicranial flexure and inflating bullae. Because the interpretation is difficult and we do not want to excessively multiply species names, we leave QU 11035 in open nomenclature as *M.* aff *intermedius*.

Canine size was not directly used in the preceding comparison because QU 11035 only retains a partial alveolus for its left canine. This seems to have accommodated a large tooth, and the muzzle ACQ 214 may add some information here. Its anterior zygomatic root is more similar to that of QU 10870 than to that of QU 11035 (thus it is very distinct from the *M. fredi* type specimen), and this muzzle bears a very large canine (Figure 12.9). This canine is clearly larger than those of QU 10870: it could fit in the alveolus of QU 11035. This agrees with both interpretations of the last specimen: in the case of sexual dimorphism, it would illustrate a marked canine size dimorphism in *M. intermedius* (both QU 11035 and ACQ 214 being males); in the case of different evolutionary stages, it would add to the preceding evolutionary changes an increase in canine size (in agreement with the very robust canine of *M. fredi*). In the other parts of its dentition, one can see that the P3/ protocone is larger than that of QU 11035 and QU 10870 (and more marked on the right than on the left side). This favors an evolutionary stage going toward *M. fredi*, the species which can be recognized by its clearly more molarized premolars. Because it is far from having the enormous zygomatic root of the *M. fredi* type specimen, we also refer ACQ 214 to *M. intermedius*, noting that, if the zygomatic root height were a very reliable character, ACQ 214 would be closer to the *M. intermedius* type specimen than to QU 11035. It might then illustrate a marked canine size dimorphism. If P3/ molarization were a better systematic indicator than zygomatic root height, ACQ 214 would be a more derived evolutionary stage, confirming at the same time the value of canine size as an indicator of evolution in this *Magnadapis* lineage.

12.4.4 Comparison of MaPhQ 210

The Montauban cranium MaPhQ 210, chosen as the type of *M. fredi*, has cranial superstructures further increased over the state seen in *M. intermedius*. The braincase is slightly more distant from the frontal plane (around 10 mm), and the two frontal lines converge sharply, which results in a short and very narrow frontal triangle (Figure 12.10). The postorbital narrowing is again slightly increased over that of QU 10870. The posterior part of the sagittal crest is enormous (around 18 mm visible) and seems more extended than on QU 10870 (though the latter is broken). The missing posterior part is not very extensive: the nuchal projection may be the authentic one (to be verified on the original), and the right lateral extremity of the nuchal plane is preserved. The most posterior part of the skull is more extended than on QU 10870. In ventral view, the osseous area posterior to the bulla is clearly more extensive on MaPhQ 210 than on QU 10870, and slightly more than on QU 11035. Likewise in lateral view, the part of the skull posterior to the external auditory meatus is very short on QU 10870 (as on *M. quercyi*), and is clearly more extended on MaPhQ 210. QU 11035 is similar to MaPhQ 210 in this view, and its lateral aspect probably gives a relatively good approximation of how the crests of *M. fredi* would have looked (Figure 12.10). Among the peculiarities of MaPhQ 210 are that the nasofrontal sutures protrude posteriorly into the frontals much less, the increased height of the anterior part of its muzzle, and especially the incredibly exaggerated height of its anterior zygomatic root (Figure 12.8). In lateral view, it seems also that MaPhQ 210 has a smaller orbit than the others. This is difficult to show, due to deformation: the postorbital bar has two pieces that are displaced and badly adjusted. However, in this view, the very high zygomatic root of MaPhQ 210 leaves a strong impression that the orbit was smaller and less posteriorly inclined than on the other skulls.

Many of the above-mentioned characters clearly distinguish the types of *M. fredi* and *M. intermedius*, but they cannot be compared with QU 11035 because the dorsal part of QU 11035 is crushed. However, dental characters also clearly separate MaPhQ 210 from all the preceding specimens: its P3/ is clearly broader and more extended lingually than on other specimens, having a more voluminous protocone (Figures 12.7, 12.9); the whole premolar series is more molarized and confirms a probably more derived evolutionary stage. In this context, there is no reason to wonder if QU 11035 might have been a female of *M. fredi*, and the latter species appears well justified.

The crushed muzzle NMB St.H. 1634 from Euzet-les-Bains has a broad interorbital region, a broad palate, large canines and its premolar series is highly molarized, making it very similar to the type of *M. fredi* (Figure 12.10F). Its hypocone is quite small on M2/, whereas this cusp is well developed and more voluminous on MaPh Q 210. However, this specimen from Euzet has markedly worn molars, rendering cusp size estimation difficult. It will be important to make a detailed study of dental variations in the Euzet assemblage, and compare it with the morphology of the M. *fredi* type specimen. Pending such a study, we provisionally refer the Euzet assemblage to *M. fredi*. This is important because Euzet is placed in the European biochronological scale (MP 17a), which gives a good idea of the age of the *Magnadapis* lineage, close to MP 16 – MP 17.

12.4.5 Comparisons of QU 10872, *M. Laurenceae*

The very peculiar *M. laurenceae* has the thinnest anterior zygomatic root and the lowest sagittal crest of all *Magnadapis* crania. Paradoxically, its nuchal crests project posteriorly beyond the foramen magnum further than in QU 10875. The only way it could fit in the preceding lineage of species would be as a species more primitive than *M. quercyi*, with even weaker cranial superstructures. However, other characters suggest a more complex relationship. Concerning cranial superstructures, QU 10872 has an anteriorly very low sagittal crest, as in *M. quercyi*, and its two frontal lines join even further posteriorly than in that species (Figure 12.10). This gives the impression that the braincase is slightly closer to the anterior part of the skull in *M. laurenceae* than in *M. quercyi* (correcting for the slight deformation of QU 10872 would possibly increase that impression). For that very posterior frontal junction, this specimen is an extreme within the large adapines, and it resembles *Notharctus* and other fossil primates. Could it be primitive for adapines? Posteriorly, the sagittal and nuchal crests project relatively far beyond the occipital plane (around 8 mm), clearly further back than in *M. quercyi*. Some variability or dimorphism can be expected there. However, it is strange to have the projection more accentuated on the slightly smaller, and possibly more primitive specimen. The anterior root of the zygomatic arch is slightly less robust in *M. laurenceae*, in proportion with its slightly smaller size. For these cranial superstructures, *M. laurenceae* appears possibly more primitive and in line with the preceding lineage, however the other cranial characters appear more problematic. In dorsal view, its muzzle is narrower than in *M. quercyi*. In anterior and lateral views, QU 10872 clearly appears to have a smaller orbit than *M. quercyi* and QU 10870 (Figure 12.8). In lateral view, the profile of the muzzle seems slightly concave, which would be unusual for adapines. However, this part of the skull is badly preserved and it is sediment instead of real bone which suggests this profile. In lateral view, the alveolar rim appears ventrally convex and markedly curving upward anteriorly, as in other *Magnadapis*. The ventral view confirms that the skull is proportionately shorter than in *M. quercyi*, the braincase and the anterior part being closer to each other. The palate is narrower and also seems somewhat shorter. On the whole, all these differences in skull characters show *M. laurenceae* to be quite distinct from other *Magnadapis* species. Because for several of these differences, it resembles *Leptadapis* species, further interpretation of its characters is relevant to the relationships between the two genera. These are discussed below.

For dental characters, *M. laurenceae* also shows differences from other *Magnadapis* species. The hypocone of its molars is especially small, being completely absent on M2/, which has a continuous lingual cingulum well separated from the postprotocrista, and small on M1/. It seems that M1/ and M2/ have a straighter centrocrista than on QU 10875 (M2/ is partly worn labially on the left side, where the molars are best preserved). The P4/ appears to be slightly more transverse, or slightly less anteroposteriorly elongated (with paracone and metacone closer, less equal) than on *M. quercyi*. P3/ appears quite similar on both specimens. The same is true for P2/. There seems to be a marked difference on the canines, seen in the unusual outline of the left canine in *M. laurenceae*. This tooth is broken, however the outline of its base is intact, and one can see the posterior part of the lingual cingulum, with a much more accentuated concavity. On the right side, the base of the canine is lingually worn, and the outline of the tooth is unclear. It is possible to place side by side the left canine of QU 10872 and the right canine of QU 10875. Their outlines are different (Figure 12.7). In *M. laurenceae*, the canine is broader in its anterior part (below the anterior groove). It is also linguo-labially broader in its median part, and it is less posteriorly extended (less oval). The canine of *M. quercyi*, more oval in occlusal outline, is similar to the canines of other large adapines. In contrast, the canine of *M. laurenceae* has a more circular outline (similar grooves can be deduced from the outline and the base at least anteriorly and posterolingually). The root of the left I1/ can be seen; it seems relatively small in comparison with the alveoli as preserved in *M. fredi*. We are cautious with regard to the significance of this because no I1/ is preserved in the large adapines studied here, and well preserved alveoli are also rare in this group. In sum, differences in dental characters reflect the uniqueness of *M. laurenceae*, and, like cranial characters, raise questions about character polarities. If a small hypocone on M1/ and no hypocone on M2/ were constant in *M. laurenceae*, it would be an autapomorphic character. The P4/ and the relatively straight centrocrista of M1-2/ would suggest primitive dental character states, possibly the relatively quadrangular M1/ also. We feel that these dental characters will need a more detailed and quantified study including assemblages from known localities, to get an idea of dental character variability. Perhaps we are overemphasizing small dental characters which may vary a lot within large adapine species.

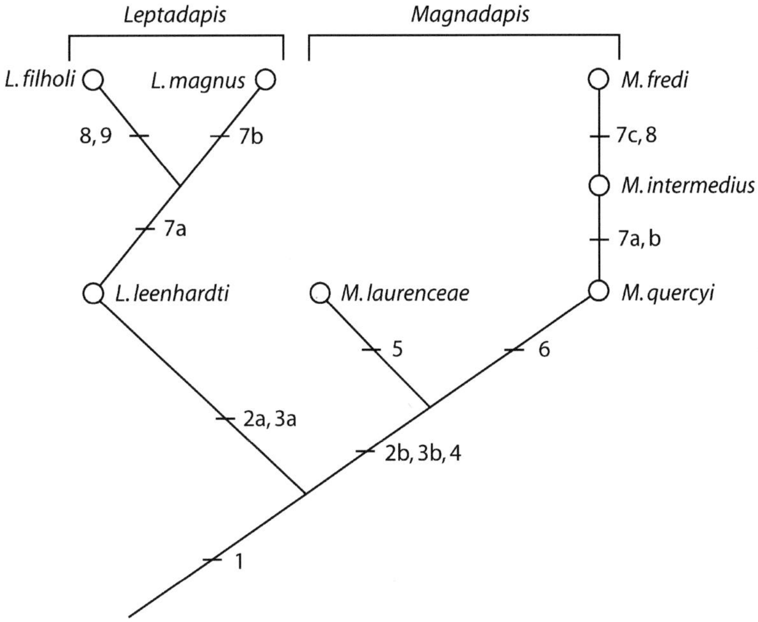

FIGURE 12.11. Schema showing our working hypothesis for large adapine phylogenetic relationships primarily based on cranial characters, with the addition of some dental traits. The characters corresponding to numbers are: orbits larger than in *Adapis* (1); weak interorbital breadth (2a) and greater interorbital breadth (2b); moderate hypocone size on molars (3a), and a decrease in hypocone size (3b); enlargement of M3/ (4); decrease in orbit size (5); broadening of the muzzle (6); moderate (7a) and more marked (7b) development of cranial superstructures, and very strong cranial superstructures including an exaggerated height of the anterior zygomatic root (7c); molarization of the upper premolars (8); lengthening and increase in size of the hypocone on M2-3/ (9).

12.5 Character Polarity and Relationships Between *Leptadapis* and *Magnadapis*

Comparing species of *Leptadapis* and *Magnadapis* might suggest something about character polarity and possible ancestral states. We found in both genera a morphocline from smaller animals having less developed cranial superstructures and an anteroposteriorly shorter skull (primitive) to species of larger size, more developed superstructures and long skulls (derived). These morphoclines would be *M. laurenceae-quercyi-fredi* and *L. leenhardti-filholi-magnus*. For reasons explained above, these morphoclines cannot be simple lineages of species. However, by comparing the most primitive species in each genus, or species having characters intermediate between the two genera, we might try to suggest hypotheses for other characters. *L. filholi* has a broad palate and a broader and higher muzzle than other *Leptadapis* species. Could this be a link with *Magnadapis*? We do not think so. As seen quite clearly with dental characters, especially the M1/ (Figure 12.5), we consider *L. filholi*, which has moderate cranial superstructures, as autapomorphic for its dental and muzzle characters (Figure 12.11).

Comparison of the primitive *L. leenhardti* to *M. laurenceae* does not suggest a place for the latter. Both have a very low sagittal crest, generally weak cranial superstructures, and a relatively narrow muzzle and palate. Both are anteroposteriorly short. For all these characters, they are likely primitive. However, the two genera were already well separated, as shown by the broad interorbital breadth, the higher muzzle, and the alveolar rim arcuate and curving upward anteriorly in *M. laurenceae. M. laurenceae* has a transversely elongated M3/ and a lingually broad P4/, as do other species of the genus. The bizarre character of *M. laurenceae* is its smaller orbit, which is also smaller than in other *Magnadapis* species (except possibly *M. fredi*). By comparison with *Adapis*, and because it is associated with the very posterior frontal line junction, it could be primitive for large adapines. However, in that case, convergence would have occurred for orbit size increase in both genera. This is not parsimonious. Branching of *M. laurenceae* before the split between *Magnadapis* and *Leptadapis* would appear even less parsimonious, implying convergence in all the derived characters defining *Magnadapis*. More probably, a secondary decrease in orbit size, a reversal, appeared in the lineage leading to *M. laurenceae* (Figure 12.11).

We did not fully elaborate on the polarity of dental characters. This appears as a complex task. The inclusion of *Paradapis ruetimeyeri*, which is the oldest large adapine species, might suggest polarities opposed to those chosen by us for some dental characters. This species has molars with very large hypocones, and the P4/ is highly molarized (Stehlin, 1916). However, these teeth differ from those of the crania studied here. They are not close enough to suggest a reversal of our dental polarities, which would imply a reversal in cranial trends. The resulting cranial gracilization required in large adapines seems very unlikely to us. Figure 12.11 shows our working hypothesis for the phylogeny of these species. It is primarily based on cranial characters, with the addition of some probably meaningful dental evidence. It is not yet placed in a stratigraphic frame, and future work might make it more complex (e.g., if the Euzet assemblage turned out to differ from *M. fredi*).

12.6 Preliminary Morphometric Study

The preceding systematic analysis is based on one quantitative analysis, the allometric approach of interorbital breadth by Lanèque (1993), and otherwise entirely on qualitative observations. In order to more quantitatively assess our material, we tried two different approaches using geometric morphometrics: (1) one using the seven fossil crania possessing a postorbital bar, to see if a quantitative approach would confirm our groups or not; and (2) a second one including our fossil sample and living species, to see if, by comparison with intraspecific variability in skull shape, we might have overestimated the number of species.

All data acquisition was done by taking three-dimensional coordinates of landmarks, using an Immersion Microscribe, three-dimensional point digitizer. For the first analysis, 38 points were digitized on the fossil skulls (Figure 12.12). Some points were missing on the most incomplete specimens. Some methods now are available for morphometrics with missing data and fossil reconstruction (Gunz et al., 2004). However, these methods require a reference-specimen (while none of our specimens can be considered as a reference), and are not appropriate for small samples. As the method used for this study does not work with missing points, we chose, when breakage of specimens did not allow real measurements, to visually estimate where the missing landmarks would have been. To attenuate the subjectivity introduced by these estimates, all measurements and estimates were done twice and each specimen used as two different individuals. This on one hand artificially increases the sample, and on the other hand reveals uncertainties due to estimation by showing, for the most incomplete skulls, an increased distance between the two points of one specimen (Figures 12.13, 12.14). Data were treated by Generalized Procrustes Adjustment, and Procrustes residuals were used in a principal component analysis (PCA, Gower, 1975). The first three axes of the PCA explain 33.86% (first axis), 25.39% (second axis) and 11.23% (third axis) of the total variance. Scatter diagrams of the specimens along the first two axes are given in Figure 12.13. We tested the possible effect of autocorrelation introduced by using each specimen two times. The same analysis was performed with one point for each specimen. It gave similar results and scatter diagrams. The difference in the variance explained by the axes in the two analyses is between 2.4% and 4.9% of the total variance. With an effect lower than 5%, we consider that autocorrelation is not a serious problem of our analysis. A hierarchical classification analysis was performed in order to evaluate the influence of estimated landmarks. Both measurements of each specimen were always grouped together, testifying that error due to landmark estimation is very low in comparison with interspecimen morphological variation. Lastly, we realized that the distribution along the two principal axes appeared driven by two specimens, MaPhQ 210 and MaPhQ 211, which are clearly less well-preserved than the others. The two points for each specimen are more separated than for other specimens, suggesting the influence of missing point reconstruction. Hence we performed an analysis without these two specimens (Figure 12.13B). It is discussed below.

In order to understand the meaning of the first axes, we extracted the variables driving them (weight over 0.7) and we studied their correlation with size. In the first analysis including the two damaged Montauban specimens, the first axis shows a significant correlation with size (correlation coefficient of 0.70; $p < 5\%$), whereas the second axis is not significantly correlated with size (0.32). For the first axis, the negative pole is influenced among others by three landmarks linked to the breadth of the posterior part of the palate (13, 16, 17), and two landmarks linked to interorbital breadth (19, 24), suggesting that *M. fredi* differs from other *Magnadapis* in being extreme for these characters. The positive pole raises the question of a possible difference in the height of the posterior part of the skull (33, 34, 35), which needs to be confirmed. The second axis separates *Magnadapis* specimens, below, from the specimens ascribed to *Leptadapis* above (with MaPhQ 211, distorted specimen, probably pushed more distantly by missing points). The negative pole of the second axis is influenced by the height between palate and orbits (12, 16, 17, 21, 18, 28), and by the anteroposterior length of the bullar region (9, 34, 35, 36). The positive pole seems influenced by the overall breadth of the skull (3, 4, 5) and also raises a question concerning a possible difference in height of the frontal line (3, 4, 23).

The scatter of specimens along the third axis (not shown here) spreads individuals between the *L. magnus* type specimen (positive pole) and QU 10870 (negative pole), specimens which are not distorted (however the former has no intact zygomatic arch). This axis better separates *L. leenhardti* and *L. filholi*, which were close to each other on the preceding diagram; it also widely separates QU 10875 and QU 10870, which were very close on the other diagram. There is probably some interesting signal here. Among the characters influencing this axis are, for the negative pole, M1/-M2/ length (16, 17), height of the posterior plane (6, 7), and for the positive pole canine projection (15), breadth of the postbullar region (34, 36).

The analysis performed without the two most damaged specimens (Figure 12.13B) shows an overall similar scatter, but with some interesting differences. Along axis one, QU 10875 is now well separated from QU 10870 (differences between *M. quercyi* and *M. intermedius*). Two other speci-

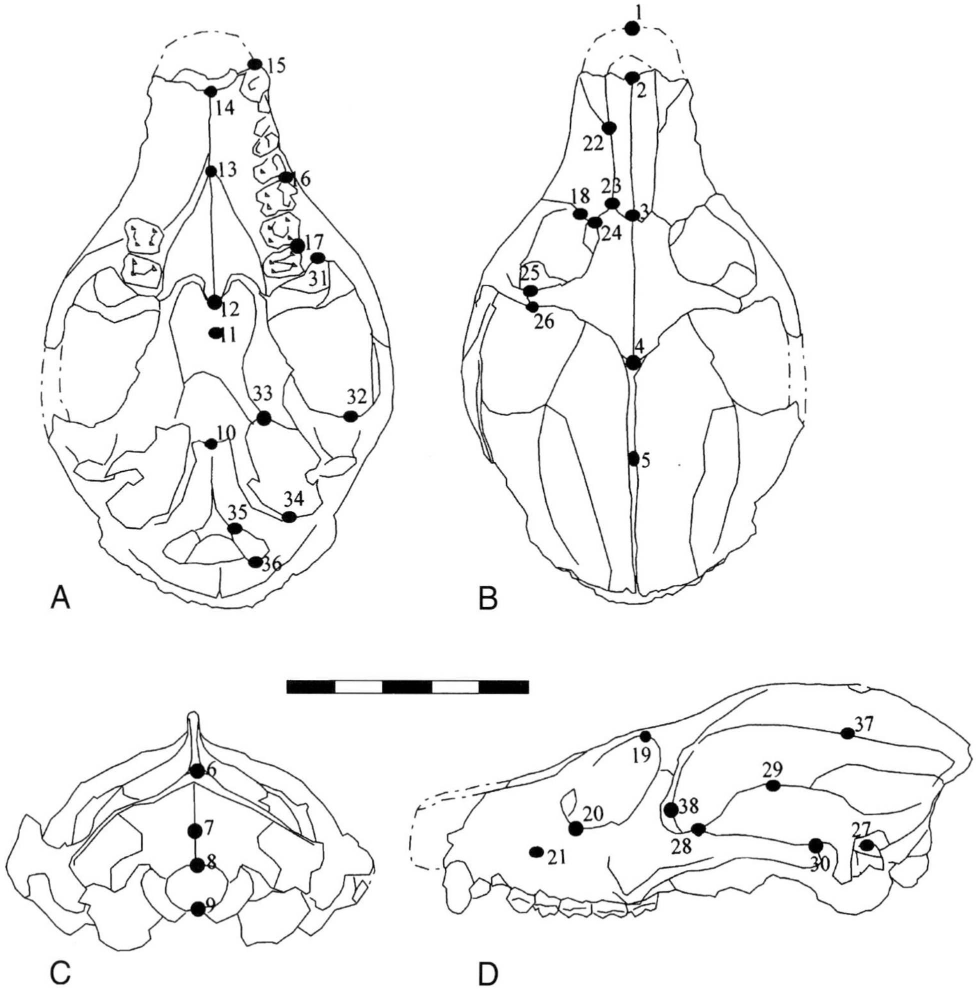

FIGURE 12.12. Schematic drawings of a *Leptadapis* cranium in ventral (A), dorsal (B), posterior (C), and lateral (D) views, showing all the landmarks used in the morphometric study (listed in Table 12.2). Scale bar is 5 cm.

TABLE 12.2. Listing of the 38 landmarks digitized for each cranium (all the measurements were performed on the left side of the cranium). Asterisks indicate the 21 points which are used in the analysis of adapines and *Alouatta* species.

Number	Definition	Number	Definition
1*	Interincisors	21*	Infraorbital foramen
2*	Summit of the nasal opening	22	Suture nasal/maxilla/premaxilla
3*	Nasion	23	Suture nasal/maxilla/frontal
4*	Bregma	24	Suture maxilla/frontal on the rim of the orbit
5	Summit of the sagittal crest	25	Suture frontal/zygomatic on the lateral rim of the orbit
6*	Lambda	26	Suture frontal/zygomatic on the rim of the temporal fossa
7*	Inion	27*	Porion
8*	Opisthion	28*	Suture zygomatic/squamosal on the upper rim of the zygomatic arch
9*	Basion		
10	Suture basioccipital/basisphenoid on the midline	29	Summit of the zygomatic arch
11	Suture basisphenoid/ presphenoid on the midline	30*	Suture zygomatic/squamosal on the lower rim of the zygomatic arch
12*	Suture of the palatines on the midline		
13*	Suture maxillas/palatines	31	Anterior point of the temporal fossa
14	Suture maxilla/premaxilla on the palate	32	Posterior point of the temporal fossa
15	Suture maxilla/premaxilla on the face	33*	Anterior point of the tympanic bulla
16	Proximo-vestibular point of the P4/ alveolus	34*	Posterior point of the tympanic bulla
17	Proximo-vestibular point of the M3/ alveolus	35*	Anterior point of the occipital condyle
18*	Suture maxilla/zygomatic on the rim of the orbit	36*	Posterior point of the occipital condyle
19*	Summit of the orbit	37	Median point on the braincase (same height as point 5)
20*	Base of the orbit	38	Maximum of constriction between the braincase and the face

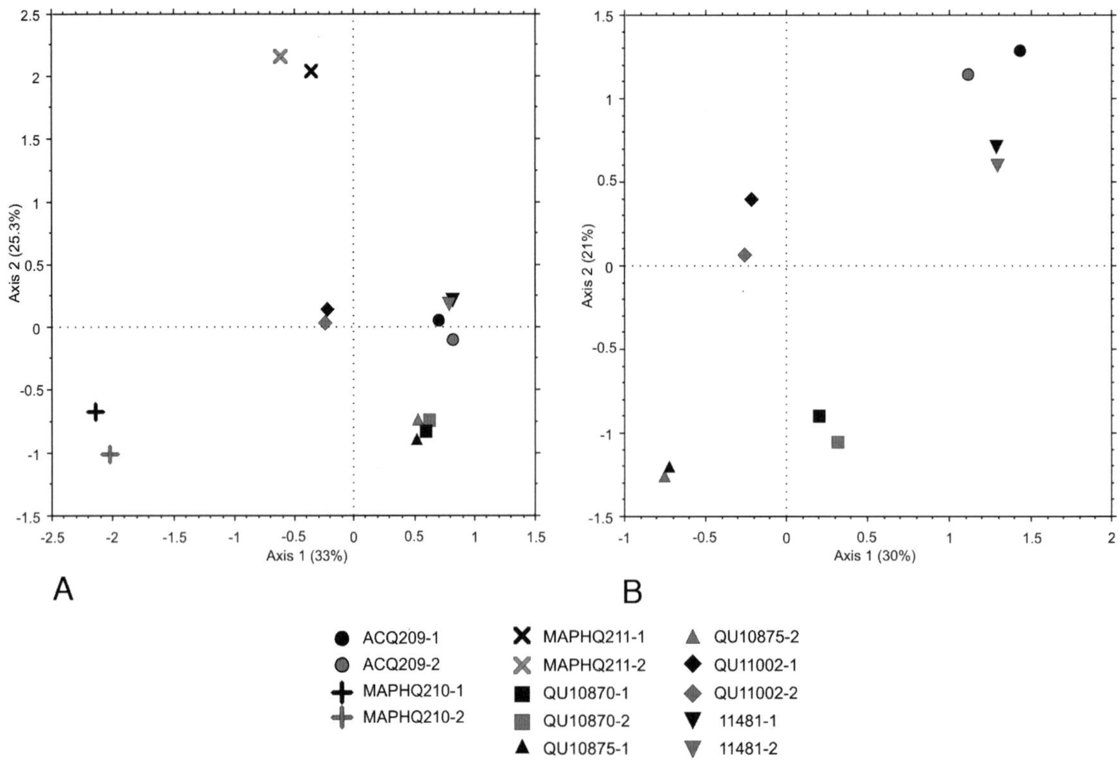

FIGURE 12.13. Scatter diagrams of large adapine crania along the two first axes of the principal component analysis of Procrustes residuals (38 landmarks); each cranium is represented by two points corresponding to two different sets of measurements. Analysis with the seven best preserved crania (A), in which the two points for MaPhQ 210 and MaPhQ 211 are relatively distant (influence of estimated missing points). Analysis without the two most damaged specimens (B). In both analyses, axis 2 separates *Leptadapis* specimens (top) from *Magnadapis* specimens (bottom).

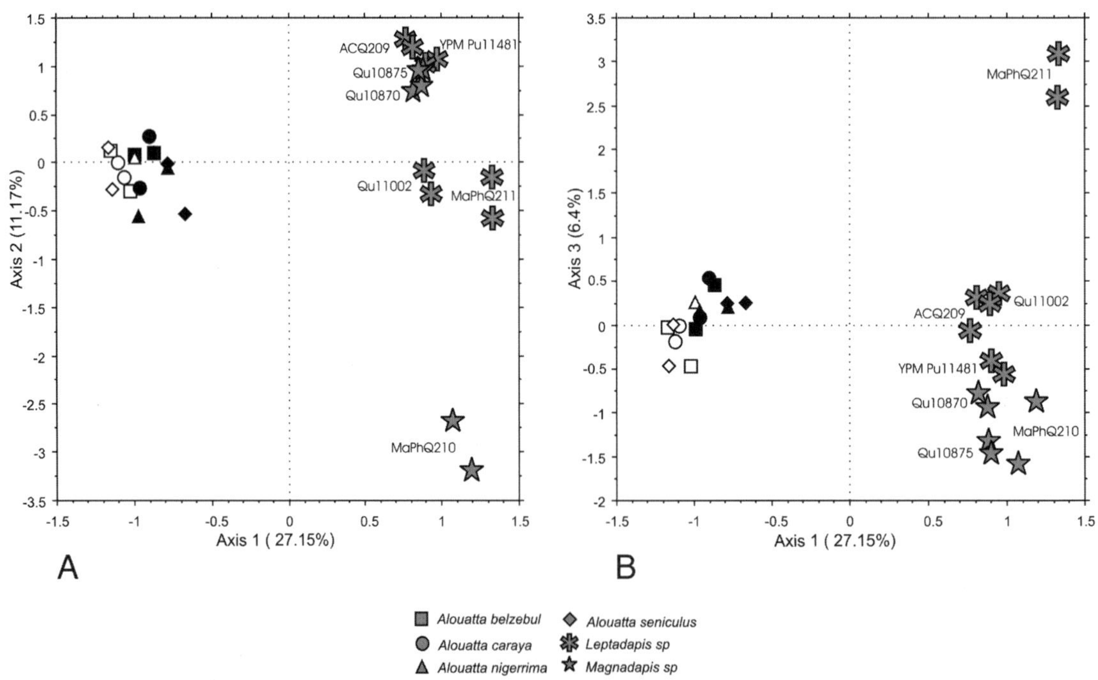

FIGURE 12.14. Scatter diagrams of a set of crania of four species of *Alouatta* and of the large adapine crania along the three first axes of the principal component analysis of Procrustes residuals (21 landmarks); first and second axes (A) and first and third axes (B). Female *Alouatta* are figured in white, and male *Alouatta* are figured in black; fossils are of unknown sex and are shown in grey.

mens now show a distance between their two measurements, ACQ 209 along axis one, and QU 1102 along axis two, revealing that estimated points still play a role. The two first axes explain a somewhat inferior percentage of the total variance, respectively 30% and 21%, and this time the first axis does not appear correlated with size. The variables driving the distribution were extracted (weight over 0.7). The negative pole of the first axis (*M. quercyi*) is again driven by a majority of seven variables in breadth (9, 10, 11, 13, 16, 17, 19), and also five variables in length. The positive pole better isolates two *Leptadapis* specimens. It is influenced by a great majority of eight variables in length (2, 5, 6, 13, 15, 23, 27, 35), against two in breadth and two in height. Axis two separates more clearly than in the preceding analysis the two *Magnadapis* specimens (below) from the three *Leptadapis* specimens (above). This is very interesting. The negative pole, toward *Magnadapis*, is influenced by seven variables in height, again suggesting differences between the palate and the orbit and infraorbital foramen (landmarks 12, 14, 15, 16, 17, 18, 21), five in length (including three posterior landmarks 9, 34, 36) and three in breadth. The positive pole, toward *Leptadapis*, gives an even clearer signal, with six variables in height, two in length and two in breadth. Almost all the variables in height are between the orbit and the frontal (landmarks 3, 4, 19, 20, 23). This analysis completes the distinction of the two genera. The latter were distinguished based on interorbital and muzzle breadth. This analysis adds shape differences in height between the dorsal frontal line, the base of the orbit and the palate. Visual inspection confirms marked differences in the anterior part of the cranium. Facial height is greater in *Magnadapis*, lower and tapering more anteriorly in *Leptadapis*. This is clearly seen in comparing specimens of similar length and cranial superstructures, as *L. magnus* (Figure 12.6B) and *M. intermedius* (Figure 12.8C).

Our second study attempts to compare these fossils with some living primates. Using the data base of one of us (S.C.) on living platyrrhines, we selected the genus *Alouatta* as being the most dimorphic platyrrhine genus, and the closest to large adapines in terms of overall size. Landmarks in common between the study of adapines and that of platyrrhines include 21 points (Table 12.2). In order to equilibrate the two samples, we used a sample of 15 adult *Alouatta*. A Generalized Procrustes Adjustment was done and a PCA analysis of Procrustes residuals. Scatter diagrams of specimens along the three first axes are given in Figure 12.14. In fact, we did a first analysis, not shown here, with a sample of 14 adult *Alouatta seniculus* (7 males and 7 females). The results were very similar. The only difference was that the *Alouatta* sample was more concentrated than in Figure 12.14, less spread along axis two. In order to increase the morphological variation in the living genus, we performed a second analysis with 15 individuals (7 females and 8 males) pertaining to four different species (Figure 12.14). Axis one clearly separates howling monkeys from large adapines. One third of the variables have a significant weight, almost all points are affected. This is not surprising: the morphologies of the two groups differ profoundly. Along this axis, female *Alouatta* (on the left) are almost separated from the males (on the right), whereas species of *Alouatta* are not isolated. The distances between the extremes within the two groups are similar. Axis two separates large adapines into three groups which are clearly spread more distantly than the sample of four different howler monkey species (not separated on this axis either). Even taking into account that MaPhQ 210 is damaged, and the distance between its two points indicate some influence of reconstructed landmarks, there seems to be a clear signal: the distinction of *M. fredi* from all the others confirms our analysis of this species as being extreme in the morphocline of cranial superstructures and/or muzzle height, and suggests that it may warrant generic distinction. However, this axis does not separate our two proposed generic groups. Interestingly, despite the shift toward the right of MaPhQ 211 (deformed specimen), this specimen groups with the *L. magnus* type specimen, which agrees with our systematic choice. Axis three again shows the species of *Alouatta* grouped together, and the large adapines more regularly spread (with the exception of MaPhQ 211, probably linked to missing points). Along this axis (Figure 12.14B), the three *Magnadapis* specimens appear on one side, toward the bottom of the diagram, and the four *Leptadapis* specimens on the other side, above, suggesting a possible systematically meaningful signal. These results are appealing, however we need to be cautious because there is clearly a strong influence of missing points (MaPhQ 211 very isolated, and the two points for MaPhQ 210 quite apart from each other). We defer a more precise morphological interpretation to future analyses avoiding the influence of missing points.

On the whole, these first geometrical morphometric attempts show two probably significant results. The comparison of the large adapine skulls with those of several *Alouatta* species suggests that the large adapines show a higher morphological disparity than several living species of one genus, giving further quantitative confirmation that two genera can be distinguished among them. The two studies also suggest that the two proposed genera can be quantitatively separated by shape variables, along axis 3 in the second study, and along axis 2, independent of size, in the study including only adapines. The results helped the recognition of major differences in facial height between the two genera. Other results call for further examination of other quantitative characters. However, Procrustes methods are poorly suited to locate shape differences. Further elaboration on these results will require more precise analyses avoiding missing points, a search for new landmarks which would better reflect some of the peculiar morphological differences analyzed here (e.g., distance between braincase and frontal rims), and the use of other methods.

12.7 Summary and Perspective

The large adapine skulls are for the first time subjected to a global study. We propose to distinguish among them two genera, *Leptadapis* and *Magnadapis*, and seven species. We also propose a first phylogenetic hypothesis for these species (Figure 12.11). The proposed lineage *M. quercyi – M. fredi* is parsimonious in minimizing the number of cladogenetic events. A different chronological succession would imply a higher number of lineages.

Our proposed lineage *M. quercyi – M. intermedius – M. fredi* stands in marked opposition with earlier hypotheses explaining cranial differences between the large adapines by sexual dimorphism (Gingerich, 1981; Gingerich and Martin, 1981). We emphasize that differences in this lineage affect not only cranial superstructures, but also facial height, which must result from phyletic evolution. Also, the differences in the canines of these species mainly reflect an evolution in canine robustness (length and breadth). There is no clear evidence between putative male and female cranial pairs of the strong canine size difference, including canine height, that should be present if large differences in cranial superstructures had been due to sexual dimorphism. This view agrees with the absence of canine dimorphism found in the largest adapine assemblage known from one locality, Euzet-les-Bains (Gingerich, 1977). We interpret the increase in canine size and robustness in this lineage as a phyletic trend paralleling the other cranial trends. However, some intraspecific canine size variability is suggested by the two muzzles. Canine dimorphism is still a possibility in large adapines, but in our opinion it would not be sufficient to explain the marked cranial characters that we used to define different species.

Our comparisons lead us to suspect that *"L." assolicus* is closely related to *Cryptadapis* and to return *"L." stintoni* to genus *Adapis*. This gives an interesting indication concerning the possible age of the largest *Adapis* species created by Stehlin (1912), and emphasizes the diversity of *Adapis* species in the latest Eocene (MP 19-20). It also restricts the known distribution of large adapines to older levels (MP 14-18).

The lineage *M. quercyi – M. fredi* may have been of relatively short duration. Its adaptive evolution and that of *Leptadapis* species possibly was a reaction to environmental change. The Perrière fauna (MP 17a) includes a large adapine and reveals the first signs of aridity, probable dry seasons, and less forested environments in the Quercy region (Legendre, 1987, 1989).

Our phylogenetic hypothesis will have to be confronted with other possible character interpretations. It will be important to study the possible effect of growth and aging on characters of the cranial superstructures, and also to further scrutinize the possible effect of sexual dimorphism on these characters. Until now, lack of provenance of the skulls of the old Quercy collections prevented a sound estimation of these factors, because we do not know how many samples they represent. Further scrutiny of the historical provenance of some might lead us to delineate possible assemblages (the Moscow skull and Paris specimens for example). Also, more complete phylogenetic analyses including *Adapis*-sized skulls should be done in the future. They might alter some of the character polarities which have been endorsed here.

The evolutionary history of large adapines appears as complex as that of the smaller adapines which diversified at the end of the Eocene (Lanèque, 1992a, b, 1993; Bacon and Godinot, 1998). Study of the dental material of the new Quercy collections and from some stratified localities will help to estimate the intraspecific variability and thus the systematic value of dental characters. It will also help us to understand the polarity of dental characters.

The dental record in its biochronological framework cannot provide a real Popperian test of hypotheses based on skull characters, because we are interpreting historical, and not experimental, data. Nevertheless, it should provide crucial arguments for the elaboration of understandable and parsimonious historical narratives of large adapine evolution during the late Eocene. This evolution already appears as a history of diversification, size increase, cranial superstructure development and dental specialization linked to diet. An increase in adapine lower molar cresting through time is documented, suggesting a folivorous adaptation, however the enormous cranial superstructures of some of them have yet to be fully explained. Certainly, more will be extracted from the study of this beautiful collection of Eocene primate skulls when analyses are extended to a richer dental record.

Acknowledgments. We thank the curators of several institutions for access to specimens in their care, B. Engesser in Basel, W. Joyce in Yale, E. Ladier in Montauban, L. Ginsburg and P. Tassy in Paris, colleagues in Leuwen and Montpellier. Some of these specimens were forwarded to M.G. by P. Gingerich. One of us (M.G.) is indebted to F. Szalay as a precursor and inspiration in many aspects of paleontological research; when he was working on his Ph.D. in Montpellier, some Montauban skulls were on Fred's desk in New York! We thank M. Dagosto and E. Sargis for their invitation to contribute to a volume honoring this remarkable colleague. D. Geffard-Kuriyama helped with photography, drawing Figure 12.11, and mounting of all the figures. N. Martin-Duverneuil did the CT scan for us at the Hôpital de la Pitié-Salpêtrière in Paris. B. Battail helped translating Russian. Jonathan Perry made a thorough job of improving the English of a complete first draft of this paper. Mary Silcox also suggested corrections. She and an anonymous reviewer raised good questions. Ellen Miller and Marian Dagosto gave help on the English of the final version of our paper.

References

Bacon, A.-M., Godinot, M., 1998. Analyse morphofonctionnelle des fémurs et des tibias des "*Adapis*" du Quercy: mise en évidence de cinq types morphologiques. Folia Primatologica, 69, 1–21.

BiochroM'97, 1997. Synthèses et tableaux de corrélations. In: Aguilar, J.-P., Legendre, S., Michaux, J. (Eds.), Actes du Congrès BiochroM'97. Mém. Trav. E.P.H.E., Inst. Montpellier, 21, 769–805.

Crochet, J.-Y., Hartenberger, J.-L., Rage, J.-C., Rémy, J. A., Sigé, B., Sudre, J., Vianey-Liaud, M., 1981. Les nouvelles faunes de vertébrés antérieures à la "Grande Coupure" découvertes dans les phosphorites du Quercy. Bulletin du Muséum National d'Histoire Naturelle, 4e Ser. C, 3, 245–265.

Crusafont-Pairo, M., 1967. Sur quelques prosimiens de l'Eocène de la zone préaxiale pyrénaïque et un essai provisoire de reclassification. In: Evolution des Vertébrés. Colloque International du CNRS N° 163. Editions du CNRS, Paris, pp. 611–632.

Depéret, C., 1917. Monographie de la faune de mammifères fossiles du Ludien Inférieur d'Euzet-les-Bains (Gard). Annales de l'Université de Lyon, Nouvelle Série (I. Sciences, Médecine), 40, 1–290.

Filhol, H., 1874. Nouvelles observations sur les mammifères des gisements de phosphates de chaux (Lémuriens et Pachylémuriens). Annales des Sciences Géologiques, 5, 1–36.

Fleagle, J. G., 1999. Primate Adaptation and Evolution, 2d Ed. Academic Press, San Diego, CA.

Franzen, J. L., 1994. The Messel primates and anthropoid origins. In: Fleagle, J.G., Kay, R.F. (Eds.), Anthropoid Origins. Plenum Press, New York, pp. 99–122.

Gebo, D. L., 2002. Adapiformes: Phylogeny and adaptation. In: Hartwig, W. C. (Ed.), The Primate Fossil Record. Cambridge University Press, Cambridge, pp. 21–43.

Genet-Varcin, E., 1963. Les singes actuels et fossiles. Boubée, Paris.

Gervais, P., 1876. Zoologie et paléontologie générale, 2e Ed. Bertrand, Paris.

Gingerich, P. D., 1977. Radiation of Eocene Adapidae in Europe. Géobios, Mémoire Spécial 1, 165–182.

Gingerich, P. D., 1981. Cranial morphology and adaptations in Eocene Adapidae. I. Sexual dimorphism in *Adapis magnus* and *Adapis parisiensis*. American Journal of Physical Anthropology 56, 217–234.

Gingerich, P. D., Martin, R. D., 1981. Cranial morphology and adaptations in Eocene Adapidae. II. The Cambridge skull of *Adapis parisiensis*. American Journal of Physical Anthropology 56, 235–257.

Godinot, M., 1984. Un nouveau genre témoignant de la diversité des Adapinés (Primates, Adapidae) à l'Eocène terminal. Comptes Rendus de l'Académie des Sciences de Paris, Série II 299, 1291–1296.

Godinot, M., 1998. A summary of adapiform systematics and phylogeny. Folia Primatologica 69 (Supplement 1), 218–249.

Gower, J. C., 1975. Generalized procrustes analysis. Psychometrica 40, 33–51.

Grandidier, G., 1905. Recherches sur les lémuriens disparus, et en particulier ceux qui vivaient à Madagascar. Nouvelles Archives du Muséum, 4e Sér., 7, 1–143.

Gregory, W. K., 1920. On the structure and relations of *Notharctus*, an American Eocene primate. Memoirs of the American Museum of Natural History, New Series 3, 49–243.

Gunz, P., Mitteroecker, P., Bookstein, F. L., Weber, G. W. 2004. Computer aided reconstruction of incomplete human crania using statistical and geometrical estimation methods. In: BAR Int., Enter the Past: Computer Applications and Quantitative Methods in Archaeology, Series 1227, Oxford, pp. 92–94.

Lanèque, L., 1992a. Variation in the shape of the palate in *Adapis* (Eocene, Adapiformes) compared with living primates. Human Evolution 7, 1–16.

Lanèque, L., 1992b. Analyse de distance de matrice euclidienne de la région du museau chez *Adapis* (Adapiformes, Eocène). Comptes Rendus de l'Académie des Sciences de Paris, Série II 314, 1387–1393.

Lanèque, L., 1993. Variation of orbital features in adapine skulls. Journal of Human Evolution 25, 287–317.

Legendre, S., 1987. Les communautés de mammifères d'Europe occidentale de l'Eocène supérieur et Oligocène: structures et milieux. Münchner geowissenschaftliche Abhandlungen. Reihe A, Geologie und Paläontologie 10, 301–312.

Legendre, S., 1989. Les communautés de mammifères du Paléogène (Eocène supérieur et Oligocène) d'Europe occidentale: structures, milieux et évolution. Münchner geowissenschaftliche Abhandlungen. Reihe A, Geologie und Paläontologie 16, 1–110.

Pavlova, M. B., 1910. Catalogue of the Collections of the Geological Cabinet of the Moskow Imperial University. V.1, Section II 2, Mammals. Moskow Imperial University Press, Moskow [in Russian].

Piveteau, J., 1957. Traité de Paléontologie. Tome VII. Primates, Paléontologie Humaine. Masson, Paris.

Remy, J. A., Crochet, J.-Y., Sigé, B., Sudre, J., Bonis, L. de, Vianey-Liaud, M., Godinot, M., Hartenberger, J.-L., Lange-Badré, B., Comte, B., 1987. Biochronologie des Phosphorites du Quercy: Mise à jour des listes fauniques et nouveaux gisements de mammifères fossiles. In: Schmidt-Kittler, N. (Ed.), International Symposium on Mammalian Biostratigraphy and Paleoecology of the European Paleogene. Münchner geowissenschaftliche Abhandlungen. Reihe A, Geologie und Paläontologie, 10, 169–188.

Richard, M., 1940. Nouveaux mammifères fossiles dans le Ludien supérieur de Pont d'Assou. Bulletin de la Société d'Histoire Naturelle de Toulouse, 75, 252–259.

Saban, R., 1963. Contribution à l'étude de l'os temporal des Primates. Mémoires du Muséum National d'Histoire Naturelle, Nouvelle Série, A 29, 1–377.

Schmidt-Kittler, N., Godinot, M., Franzen, J. L., Hooker, J. J., Legendre, S., Brunet, M., Vianey-Liaud, M., 1987. European reference levels and correlation tables. In: Schmidt-Kittler, N. (Ed.), International Symposium on Mammalian Biostratigraphy and Paleoecology of the European Paleogene. Münchner geowissenschaftliche Abhandlungen. Reihe A, Geologie und Paläontologie, 10, 13–31.

Simons, E. L., 1972. Primate Evolution. An Introduction to Man's Place in Nature. Macmillan, New York.

Stehlin, H. G., 1912. Die Säugetiere des Schweizerischen Eocaens. Siebenter Teil, erste Hälfte, *Adapis*. Abhandlungen der Schweizerischen paläontologischen Gesellschaft 38, 1163–1298.

Stehlin, H. G., 1916. Die Säugetiere des Schweizerischen Eocaens. Siebenter Teil, zweite Hälfte. Abhandlungen der Schweizerischen paläontologischen Gesellschaft 41, 1299–1552.

Szalay, F. S., Delson, E., 1979. Evolutionary History of the Primates. Academic Press, New York.

13. Primate Tibiae from the Middle Eocene Shanghuang Fissure-Fillings of Eastern China

Marian Dagosto*
Department of Cell and Molecular Biology
Feinberg School of Medicine, Northwestern University
303 E. Chicago Ave.
Chicago, IL 60611-3008, USA
m-dagosto@northwestern.edu

Daniel L. Gebo
Department of Anthropology
Northern Illinois University
DeKalb, IL 60115, USA
dgebo@niu.edu

Xijun Ni
Department of Vertebrate Paleontology
American Museum of Natural History
79th St. at Central Park West
New York, NY 10024, USA
nixj@amnh.org

Tao Qi
Institute of Vertebrate Paleontology and Paleoanthropology
Chinese Academy of Sciences
Beijing, 100044
China

K. Christopher Beard
Section of Vertebrate Paleontology
Carnegie Museum of Natural History
4400 Forbes Avenue
Pittsburgh, PA 15213, USA
beardc@CarnegieMNH.org

13.1 Introduction

Since this is a volume in celebration of the work of Frederick S. Szalay, we think it is entirely appropriate to open with an appreciation. We gratefully acknowledge Dr. Szalay's innovative efforts to bring the study of mammalian postcranial remains to the forefront of evolutionary morphology, a development that has inspired all of our research. MD thanks Dr. Szalay for being a supportive mentor, for instilling a broad and deep understanding of evolutionary biology, for generously allowing a naive graduate student access to important fossil specimens, and for providing the most stimulating environment for research. DLG thanks Dr. Szalay for his many kindnesses

* Address for correspondence: m-dagosto@northwestern.edu

E.J. Sargis and M. Dagosto (eds.), *Mammalian Evolutionary Morphology: A Tribute to Frederick S. Szalay*, 315–324.

and thoughtful discussions over the years, and celebrates his intuitive ability to demonstrate how postcranial morphology can be used to decipher important evolutionary events in mammalian evolution. KCB acknowledges the intellectual debt he owes to Dr. Szalay, whose comprehensive studies of the systematics, phylogenetic relationships, and functional anatomy of Paleogene primates and other mammals has inspired subsequent generations to continue that legacy. XN and TQ congratulate Dr. Szalay on a long and productive career. Although this essay does not exhibit the breadth and depth typical of Fred's work, it does in its own small way build upon themes evident in his own: the important contribution postcranial remains make to the interpretation of primate and mammalian evolution, systematics, and functional morphology (Szalay et al., 1975, 1987; Szalay, 1977, 1981a, b, 1984, 2007; Szalay and Drawhorn, 1980; Szalay and Lucas, 1993; Szalay and Sargis, 2001).

13.2 The Shanghuang Primates

The joint field expeditions of Carnegie Museum of Natural History and the Institute of Vertebrate Paleontology and Paleoanthropology of the Chinese Academy of Sciences have recovered numerous mammalian fossil remains from fillings in Triassic limestone fissures near the village of Shanghuang, southern Jiangsu province China. Five fissures, labeled A-E have been sampled, and are biostratigraphically correlated with the Irdinmanhan and early Sharamurunian Land Mammal Ages approximately 45 Ma (Qi et al., 1996). A broad array of mammals has been sampled and described from these localities (Beard et al., 1994; Wang and Dawson, 1994; Qi and Beard, 1996; Qi et al., 1996; Dawson and Wang, 2001). The primates include typically Eocene forms such as adapids, and a single omomyid, *Macrotarsius*, but also the earliest record of a tarsier and a previously unknown group of primates, the Eosimiidae, which are basal anthropoids (Beard et al., 1994, 1996; Ross et al., 1998; Beard, 2002; Kay et al., 2004). The affinities of the latter group have not been without controversy (e.g., Szalay, 2000; Gunnell and Miller, 2001). The discovery of more nearly complete dental remains and postcranial remains have answered some of the early criticism that *Eosimias* is not a primate, and the diversity of postcranial remains demonstrate that not all the primates from Shanghuang fit comfortably under the umbrella of tarsiid or omomyid (Gebo et al., 2001).

The tali and calcanei of several primate groups were described by Gebo and colleagues (Gebo et al., 2000a, b, 2001; Gebo and Dagosto, 2004). The dental remains and tarsal bones suggest the presence of at least five groups of primates, including adapids, an "unnamed haplorhine family" morphologically most similar to omomyids, tarsiids, and two kinds of anthropoids (eosimiids and "new protoanthropoids"). There are several size classes within each group. Here, we describe some less numerous, but still informative limb bone elements.

To clarify the following discussion, readers should note that in this paper we follow the classification given by Gunnell and Rose (2002). The family name Omomyidae refers to lower level taxa included in the Anaptomorphinae, Omomyinae, and Microchoerinae. Tarsiiformes includes the families Omomyidae and Tarsiidae. Following Szalay and Delson (1979), we include both Adapiformes and Lemuriformes in the taxon Strepsirhini, and Tarsiiformes and Anthropoidea (including eosimiids and protoanthropoids) in the taxon Haplorhini. The informal term "prosimian" is used for the group of non-anthropoid primates, e.g., Strepsirhini, Omomyidae, and Tarsiidae.

13.3 Tibiae

Five distal tibiae have been recovered, three from Quarry D and two from Quarry E. The bones are recognized as primate on the basis of the conformation of the articular surfaces for the talus which is unique to Primates among mammals (Dagosto, 1985). Figure 13.1 illustrates once again the point that most of the primate postcranial remains

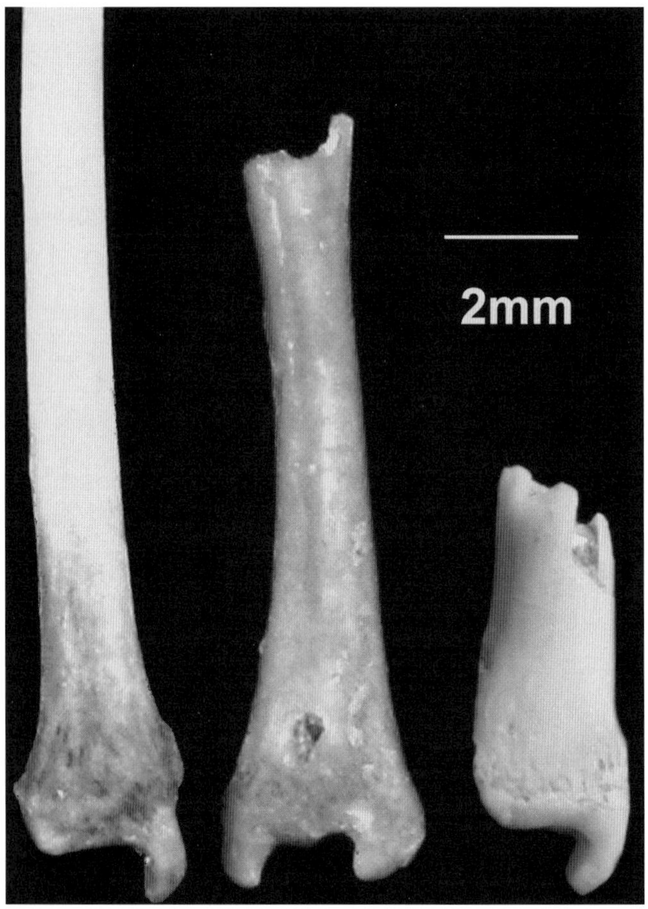

FIGURE 13.1. Size comparison. From left to right, anterior view of distal tibiae of *Microcebus berthae* (FMNH unnumbered) (~30 g); V13020 (fused morph), and V13033 (unfused morph).

TABLE 13.1. Measurements of the distal tibia (mm) in Shanghuang fossils and comparisons to living primates. Measurements were taken with a Reflex microscope. The measurements are illustrated in Figure 13.2, with the exception of measurement 7, which is the mediolateral width across the fused tibiofibula.

	Shanghuang fused morph		Shanghuang unfused morph			*M. berthae* (30 g)	*M. rufus* (50 g)	*Galagoides sp.* (70–100g)	*T. syrichta* (125–150 g)
Measurement	V13019	V13020	V13032	V13033	V13034	N = 1	N = 1	N = 4	N = 4
1	0.714	0.84	0.899	1.001	0.864	0.953	0.821	1.799	1.900
2	0.623	0.674	0.629	0.735	0.685	0.653	0.530	1.143	1.334
3	1.223	1.21	1.049	1.011	1.081	1.076	1.197	2.088	2.426
4	1.406	1.399	1.359		1.309	1.810	1.551	2.184	3.256
5	1.468	1.599	2.007	1.911	2.128	2.109	2.231	3.177	3.431
6			1.972		2.122	2.075	2.250	3.256	
7	2.952	3.039							6.499
AP	1.444	1.463	1.721	1.530	1.801	1.688	1.711	2.778	3.320
ML	1.502	1.278	1.417		1.484	1.566	1.537	2.199	3.210
Malleolar rotation	14	11	22	24	24			21	14
AP/ML	96.14	114.97	121.45		121.36	107.8	111.32	126.76	104.0

TABLE 13.2. Shanghuang primate taxa known from dental or postcranial remains. (Data from Beard et al., 1994; Gebo et al., 2001).

	Mass estimated from teeth	Mass estimated from tarsal remains	Number of specimens	Quarries
Adapoides troglodytes	~200–300 g	NA		B, D
Macrotarsius macrorhysis	~1,000 g	NA		D
Tarsius eocaenus	<70 g			A, C, D
Eosimias sinensis	67–137 g			B
Unnamed haplorhines	NA		Calcanei–2	D, E
(size class 2)		30–60 g	Tali–3	A, D
Tarsiidae	NA		Calcanei–4	D
(size class 1)	NA	20–30 g	Tali–0	D
(size class 2)		70 g	Calcanei–1	
			Tali–1	
Eosimiidae	NA		Calcanei–5	C, D
(size classes 1–3)		17–75 g	Tali–2	D, E
Protoanthropoids	NA	28–80 g	Calcanei–6	A, D, E
(size classes 1–3)			Tali–2	C

NA = not available

found at Shanghuang come from very small animals (Gebo et al., 2000). We estimate that all of the Shanghuang tibiae belong to primates weighing 50 g or less; in absolute measurements they are as small as or smaller than most individuals of *Microcebus* (30–60 g; Table 13.1). In terms of potential allocations, this immediately rules out the two adapids, which are estimated at 200–400 g on the basis of tooth size, the only currently recognized omomyid, *Macrotarsius*, estimated at 1,000 g, and *Eosimias sinensis*, estimated at 67–137 g (Beard et al., 1994). Considering size alone, the tibiae could belong to the smaller size classes within the "new haplorhine", tarsiid, or either of the protoanthropoid groups, all of which have representatives within the 20–60 g size range at quarries D and E (Table 13.2).

All of the Shanghuang tibiae described here belong to haplorhine primates as evidenced by the moderate rotation of the anterior part of the medial malleolus (10–24 degrees, Table 13.1); the flat, laterally facing posterior part of the medial malleolus; and the parallel anterior and posterior edges of the inferior tibial surface which make a relatively square shaped articular surface. These are features typical of tarsiers, omomyids, and anthropoids (Dagosto, 1985; Table 13.3, Figure 13.3). Strepsirhine primates (lemurs, lorises, and adapids) exhibit a very different conformation of the distal tibia with a more strongly rotated medial malleolus (20–40 degrees); no flat laterally facing part of the malleolus; and anterior and posterior edges that diverge laterally making a triangular shaped articular surface for the talus (Dagosto, 1985). That all the tibiae found so far are haplorhine is not surprising given that the vast majority of the tarsal bones are also haplorhine (Gebo et al., 2001). The tibiae, however, clearly represent two different kinds of haplorhine primates, one type in which the tibia is fused to the fibula, and another in which these bones remain separate.

Table 13.3. Distribution of tibial character states in primates utilized in Figure 13.3. The filled and diagonal-lined boxes are presumed to be derived conditions; the open boxes, primitive conditions, but the polarity of some of these features (especially 5–7) is not yet certain.? = character state is unknown. For feature 7, anthropoids exhibit all three character states.

Open box	Filled box	Shaded box
1. Lesser degree of rotation of medial malleolus	1. Strong degree of rotation of medial malleolus	
2. Posterolateral surface of medial malleolus flat, laterally facing	2. Posterolateral surface of medial malleolus curved, anteriorly facing	
3. Anterior and posterior edges of distal tibia parallel, rectangular outline	3. Anterior and posterior edges of distal tibia divergent, triangular outline	
4. AP/ML index low	4. AP/ML index high	
5. Tibial and fibular malleoli are parallel	5. Fibular malleolus slopes laterally = talofibular facet on talus slopes laterally	
6. Tibial and fibular malleoli are of equal length	6. Fibular malleolus shorter due to lateral slope	6. Fibular malleolus is shorter than tibial
7. Distal tibiofibular joint synovial	7. Tibia and fibula fused	7. Syndesmosis
8. Medial malleolus long, U-shaped, no strongly marked pit for deltoid ligament	8. Medial malleolus short, rectangular, pit for deltoid ligament	

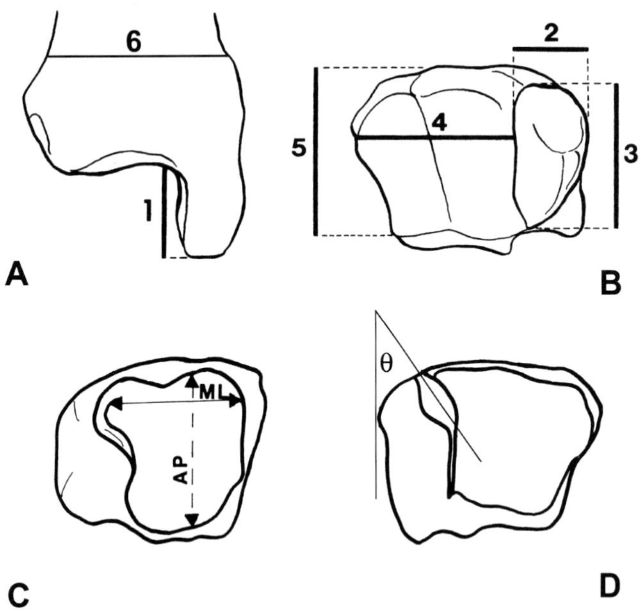

Figure 13.2. Measurements of the tibia used in this paper. 1. proximodistal height of malleolus; 2. mediolateral width of malleolus; 3. Anteroposterior depth of malleolus; 4. width of inferior tibial surface; 5. Anteroposterior depth of tibia; 6. width across the tibia just above the distal epiphysis; AP, anteroposterior depth of tibial facet; ML, mediolateral width of tibial facet; θ, angle of malleolar rotation. A and B, anterior and inferior view of distal tibia after Figure 27.8 of Meldrum and Kay (1997). C and D, inferior views of distal tibia in the strepsirhine *Eulemur* (C), and an omomyid (haplorhine) (D), modified from Dagosto (1985) figure 5.

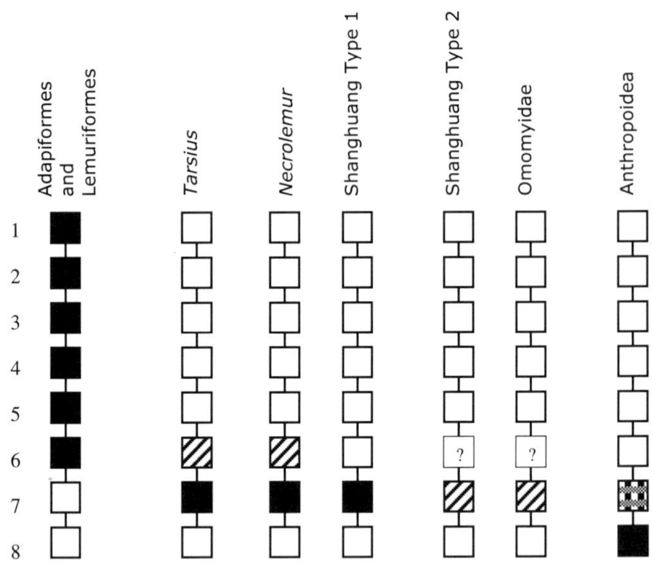

Figure 13.3. Distribution of tibial character states in primates (see also Table 13.3). The filled and diagonally lined boxes are presumed to be derived conditions; the open boxes, primitive conditions, but the polarity of some of these features (especially 5–7) is not yet certain.? = character state is unknown. For feature 7, anthropoids exhibit all three character states.

13.3.1 Type 1: Shanghuang Primates With Fused Tibiae-fibulae

Type 1 is represented by two specimens, V13020 (Quarry E) and V13019 (Quarry D) (Figure 13.4). In terms of absolute size these bones are smaller than any measured individual of *Microcebus*, including the 30 g *M. berthae* (Table 13.1). The two specimens are similar enough, both in size and morphology, to belong to the same or very closely related species. In both of these specimens the distal part of the fibula is completely fused to the tibia. V13020 has almost no hint of a suture line. Although the bones are solidly fused in V13019, the suture line is clearly visible. This individual is possibly not fully adult, as the fibular malleolus also exhibits a clear epiphyseal suture line. In the high degree of fusion, these specimens are more similar to *Tarsius* and differ from *Necrolemur*, in which a clearly visible suture line remains, even in adults (Schlosser, 1907; Dagosto, 1985).

As in most primates, there is a small pointed process on the anterior edge of the distal tibia; however, it does not appear to have a smooth surface for articulation with the talus. This

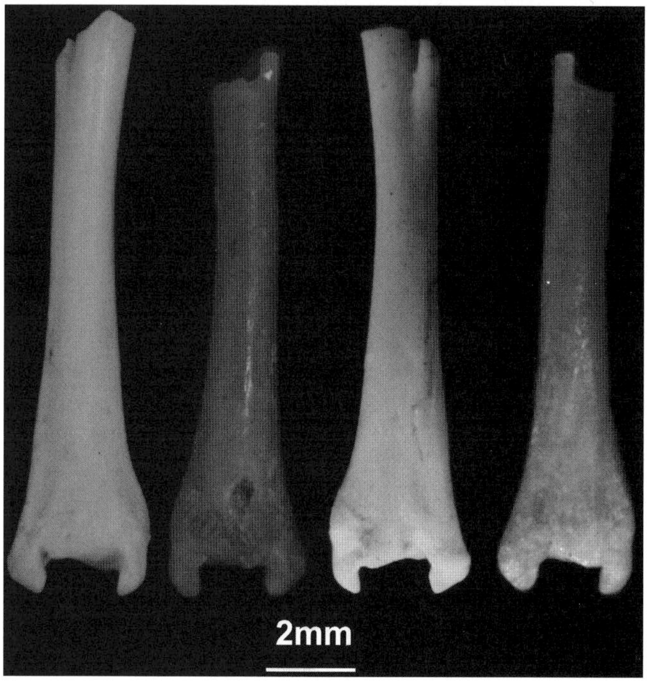

FIGURE 13.4. Anterior (left) and posterior (right) views of fused morph. From left to tight, anterior view of V13019, anterior view of V13020, posterior view of V13019 and posterior view of V13020.

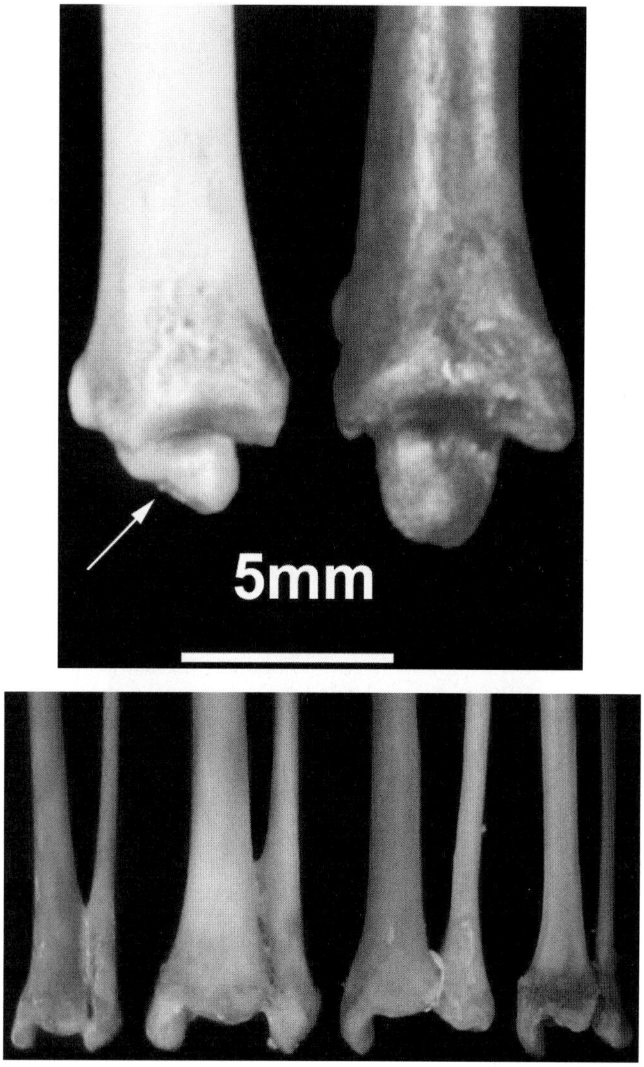

FIGURE 13.5. Differences between anthropoid and strepsirhine distal tibiae. A, lateral view of the tibial malleolus in *Callithrix* (left) and *Galago* (right). Note the shorter, smaller malleolus in the anthropoid. The arrow points to the pit for the deltoid ligament. B, anterior view of tibiofibular mortise in (from left to right) *Cebuella, Callithrix, Eulemur*, and *Galagoides* (not to scale). Note the symmetrical form of the mortise in the anthropoids due to the proximodistally straight fibular malleolus which extends as far distally as the medial malleolus, contrasted with the asymmetrical form of the mortise in strepsirhines due to the short, laterally flared fibular malleolus.

differs from most primates, but is a similarity to *Tarsius, Necrolemur*, and *Neosaimiri* (Dagosto, 1985; Meldrum and Kay, 1997). The shaft just above the articular surface is compressed anteroposteriorly, as in *Tarsius* and *Necrolemur*. The groove for the tendon of tibilias posterior is shallow and curves around the medial edge of the malleolus, as in *Tarsius, Necrolemur*, and the majority of anthropoids. In Bridger Basin omomyids, *Shoshonius* and strepsirhines the groove usually runs more inferiorly (Dagosto, 1985; Ford, 1986; Dagosto et al., 1999).

The ratio of the anteroposterior and mediolateral dimensions of the tibial facet (AP/ML; Table 13.1) is low, as is typical of most haplorhines. The degree of medial malleolar rotation is low (11–14 degrees) like that of *Tarsius*. Like "prosimian" primates, the medial malleolus is fairly long (proximodistally) and U-shaped, and it does not have a particularly well marked pit on its inferior surface (presumably for part of the deltoid ligament (Meldrum and Kay, 1997)). Most anthropoids generally have shorter, wider, more rectangular shaped malleoli, with a marked indentation on the inferior surface for the deltoid ligament, making a stepped shape (Figure 13.5A). *Cebuella*, however, appears to be an exception to this generality, having a malleolus shaped more like that of a prosimian.

Fusion is one obvious similarity of these specimens to *Tarsius* or *Necrolemur*, as is the anteroposterior compression at the distal end of the shaft, and the lack of an articular tibial "stop". These specimens are also of the appropriate size to belong to *Tarsius eocaenus* and the calcanei and talus attributed to the smaller size class of Tarsiidae in Gebo et al. (2001), and thus we provisionally attributed V13020 to this species (Dagosto et al., 1996). There are however, some noteworthy anatomical differences between extant tarsiers and the Shanghuang specimens (Figures 13.6 and 13.7) making other attributions equally possible.

Despite the marked degree of fusion of the bones in the Shanghuang primate, the point of separation of the two bones does not appear to extend as far proximally as in extant tarsiers (Figure 13.6). Although the bones are not complete, we estimate that in the Shanghuang specimens, the bones

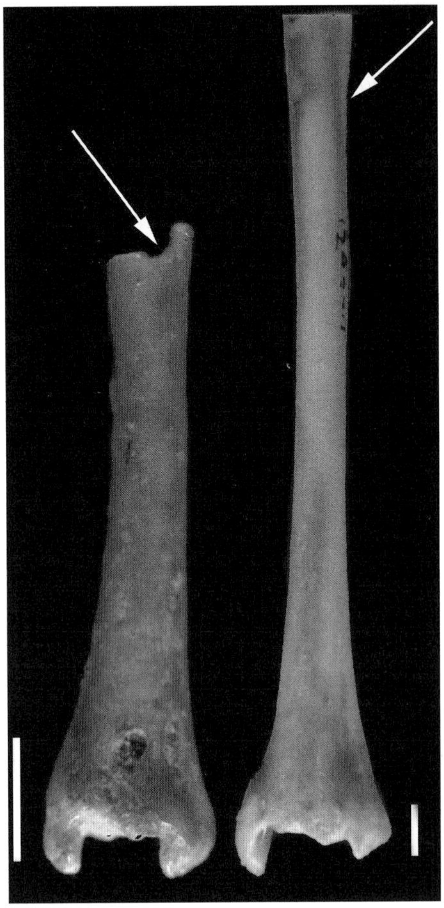

FIGURE 13.6. Anterior view of tibiofibula in V13020 (left) with *Tarsius syrichta* (right). The arrows indicate the site of tibiofibular fusion. Scales are 2 mm.

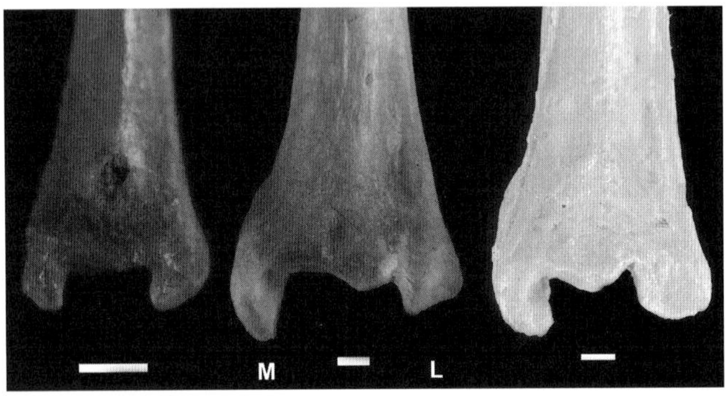

FIGURE 13.7. Anterior view of left tibiofibular mortise of Shanghuang fused morph (V13020; left), *Tarsius syrichta* (middle) and *Necrolemur* (right). M = the medial (tibial) malleolus. L = lateral (fibular) malleolus. Scales are 1 mm. Compare with Figure 13.5B.

were fused for only 40–50% of their length, while in *Tarsius* the comparable figure is 60%. In this feature, the fossils are more similar to *Necrolemur* than *Tarsius* (Schlosser, 1907; Dagosto, 1985). In *Tarsius*, the site where the fibula separates from the tibia occurs about 1 cm above the most distal point of the cnemial crest, while in the fossils the site is closer to the distal point of the cnemial crest. If one wanted to entertain the hypothesis that the extent of fusion is tarsier-like (e.g., 60% of tibia length) in the Shanghuang specimens, it would follow that the fossil tibia are relatively short. This would be a different, but still significant difference from *Tarsius*.

In the fused Shanghuang tibiae-fibulae, the tibial malleolus is relatively small, while the fibular malleolus is a much more substantial feature. It is as wide and as long as the medial malleolus, making a symmetrical frame for the talus. This contrasts with *Tarsius*, in which the fibular malleolus, although wide, does not extend as far distally as its medial counterpart (Jouffroy et al., 1984) making an incomplete, asymmetrical frame for the talus (Figure 13.7). *Necrolemur* is similar to *Tarsius*, but has a slightly longer fibular malleolus (Figure 13.7). In these features, V13019 and V13020 differ from *Tarsius* and *Necrolemur* but are similar to small monkeys, especially *Callithrix, Cebuella, Saimiri, Pithecia*, and the fossil anthropoid *Apidium*, all of which likewise have malleoli of equal length (Gregory, 1920; Fleagle and Simons, 1983; Fleagle and Meldrum, 1988; Meldrum and Kay, 1997).

Strepsirhines, even those with closely appressed tibiae and fibulae (e.g., *Microcebus, Galago, Galagoides*), have a very different profile for the tibiofibular mortise which differs greatly from that of any haplorhine (Figure 13.5B). In contrast to all haplorhines, in which the medial and lateral malleolar surfaces for the talus are parallel to each other, regardless of relative length, in strepsirhines the fibular malleolus slopes strongly laterally, matching the slope of the articulating facet on the talus (Gregory, 1920; Beard et al., 1988) so that the mortise is asymmetrical to an even more exaggerated degree and in a different way than in *Tarsius* and *Necrolemur*.

13.3.2 Type 2: Shanghuang Primates With Unfused Tibiafibulae

The other type of tibia is represented by three specimens, two from fissure D (V13032 and V13033), and one from fissure E (V13034). Based on absolute dimensions, these tibiae also belong to primates in the 30–60 g size range (Figure 13.1, Table 13.1), and therefore could belong to the smaller size classes of any of the Shanghuang haplorhine groups. The three specimens are similar enough in size and morphology to represent the same or closely related species (Figure 13.8).

Although the fibula was clearly not fused to the tibia, these tibiae all belonged to primates with closely appressed bones having a strong syndesmosis between them. Crests for the anterior and posterior tibiofibular ligaments end just above the joint, and there is no evidence of fibular apposition proximal to this point, as is observed, for example, in *Shoshonius* and *Absarokius* (Covert and Hamrick, 1993; Dagosto et al., 1999). In this, these tibiae are more similar to Bridger Basin omomyids or small anthropoids (Dagosto, 1985; Meldrum and Kay, 1997). The crests for the anterior

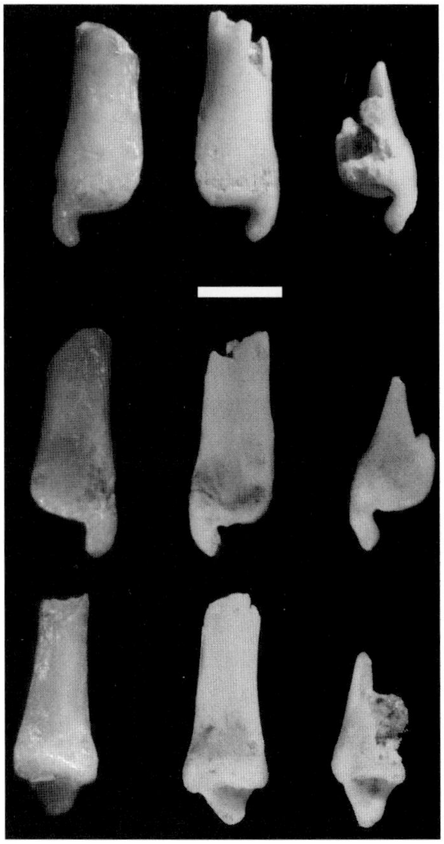

FIGURE 13.8. Comparative views of unfused morph. From left to right V13032, V13033, V13034. Top row, anterior view. Middle row, posterior view. Bottom row, lateral view. Scale is 2 mm.

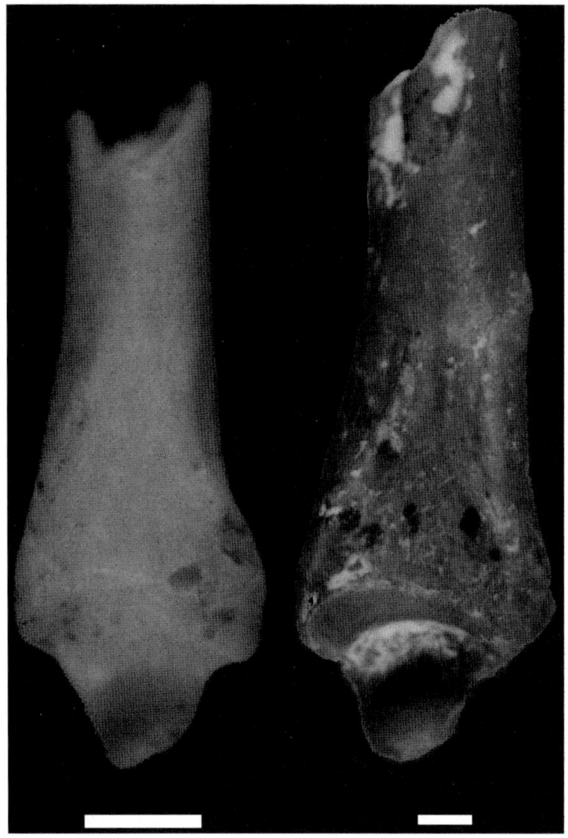

FIGURE 13.9. Comparison of lateral view of tibia of Shanghuang unfused morph (V13032) and Bridgerian omomyid (USNM 336189). Scales are 1 mm.

and posterior tibiofibular ligaments are not as well developed in the Shanghuang tibiae as they are in Bridger omomyids or *Neosaimiri* (Dagosto, 1985; Meldrum and Kay, 1997).

Like known omomyids, these tibiae appear to lack any distal articular facet for the fibula (Figure 13.9). This contrasts with most anthropoids, which have a small articular facet extending most of the anteroposterior length of the lateral surface of the distalmost aspect of the tibial shaft (Dagosto, 1985; Ford, 1986, 1990), although *Neosaimiri (Laventiana) annectens* also lacks a clear facet (Meldrum and Kay, 1997), and the facet is small in *Apidium* (Fleagle and Simons, 1983, 1995). It is often, however, extremely difficult to identify the tibial articular facet in *Microcebus*, galagos, and small platyrrhines (e.g., some callitrichids), even though the corresponding facet on the fibula is usually discernible. Therefore, like omomyids with unfused tibiofibulae, the best that can be said about these Shanghuang specimens is that the tibiofibular joint was largely syndesmotic and any synovial articulation was small, perhaps absent.

In addition to the lack of tibiofibular fusion, this type differs from the previous one in having a longer and narrower tibial articular surface (higher AP/ML ratio; Table 13.1) and greater rotation of the medial malleolus (22–24 degrees). In these respects, this morph more closely resembles omomyines, anaptomorphines, and anthropoids than tarsiers or *Necrolemur*. The anterior process is not strongly developed in V13032 and V13033, but is more salient in V13034. It does not appear to be faceted. The groove for the tibialis posterior is shallow but runs more inferiorly than medially. The tibial malleolus is relatively long, narrow, and triangular in shape and is without a marked indentation for the deltoid ligament (Figure 13.9).

13.4 Discussion and Summary

Five distal tibiae of haplorhine primates have been recovered from the middle Eocene Shanghuang fissure-fillings. There are two types of tibial morphology represented; in type 1 the tibia is fused to the fibula and in type 2 the bones are separate. Both types exhibit features (lesser degree of malleolar rotation; tibiofibular joint fused or syndesmotic; restricted mortise shape) which imply that flexion-extension movement between the crus and the tarsus was accompanied by less conjunct rotation than in the majority of strepsirhine primates or larger anthropoids. These characteristics are commonly found among small leaping primates (Hafferl, 1932; Fleagle and Simons, 1983,

TABLE 13.4. An index of medial height of the talus divided by lateral height of the talus. Only a selection of extant primates is shown, but almost all living taxa of strepsirhines and platyrrhines were measured. In no case did any extant primate have a mean index as high as *Tarsius*.

	Mean	sd	n	Min	Max
Tarsiidae					
Tarsius syrichta	136.0	7.4	7	125.1	146.0
Tarsius bancanus	133.9	12.2	8	111.6	148.6
Omomyidae					
Bridger B omomyid	113.0	5.4	7	103.5	118.4
Bridger C&D omomyid	119.3	6.2	6	112.7	129.1
Tetonius sp.	111.8	3.0	2		
Shoshonius cooperi	89.7		2		
Necrolemur sp.	122.2		1		
Adapidae					
Adapis sp.	110.9		2	107.5	114.3
Leptadapis sp.	95.4		5	90.5	111.3
Notharctus tenebrosus	107.4		2	104.8	110.0
Smilodectes gracilis	111.5	2.5	6	107.4	114.8
Extant primates					
Cebuella pygmaea	105.8	11.6	5	88.2	120.8
Saguinus	114.0	1.1	3	112.8	115.0
Saimiri sciureus	120.6	6.6	15	107.9	130.4
Microcebus murinus	111.0	5.5	15	100.0	122.7
Galago moholi	111.9	3.0	5	108.8	115.6
Eulemur fulvus	118.9	7.1	16	106.6	130.3
Shanghuang primates					
Eosimiid tali	111.2	8.2	6	105.0	126.8
New protoanthropoid	111.0		1		
New haplorhine	86.7		1		

1995; Dagosto, 1985; Fleagle and Meldrum, 1988). In terms of assessing possible phylogenetic affiliations, we first note that it is extremely difficult to distinguish between omomyids (aside from *Necrolemur*) and anthropoids solely on the basis of distal tibial morphology. One salient feature may be the shape of the medial malleolus in lateral view which in anthropoids (with the exception of *Cebuella*) tends to be shorter (at least posteriorly) with a more pronounced indentation for the deltoid ligament. With respect to these features, the shape of the tibial malleolus is more prosimian-like than anthropoid-like in both kinds of Shanghuang tibiae.

Given the fusion of the tibia and fibula exhibited by Type 1, it is reasonable to propose that these tibiae belong to the appropriately sized *Tarsius eocaenus* or an allied species. Regarding the conformation of the tibiofibular mortise (the relative size, distal extent, and degree of sloping of the malleoli), however, the most striking resemblance of the Shanghuang fused morph is to anthropoids, and not to *Tarsius* or *Necrolemur*. However, we cannot say whether the mortise shape exhibited by anthropoids is derived for crown-group anthropoids or primitive for haplorhines (with the *Tarsius-Necrolemur* condition being more derived in the latter case). We do not know the mortise shape of the unfused morph or of omomyids other than *Necrolemur*. The short fibular malleolus of *Tarsius* is likely correlated with the strong asymmetry in height between the medial and lateral sides of its talus. In the majority of primates, the medial side of the talus is taller than the lateral, but this asymmetry is most exaggerated in *Tarsius* (Table 13.4). If we are correct about the relationship between talar body and mortise asymmetry, the values for omomyid tali (Table 13.4) suggest that they, like anthropoids, did not exhibit the mortise asymmetry seen in *Tarsius* or even the less derived condition seen in *Necrolemur*. This suggests that the symmetrical mortise of omomyids and anthropoids is primitive for haplorhines.

The "equal length-nonasymmetrical" type of mortise shape may even be primitive for euprimates, since the alternate condition of an asymmetrical mortise shared by Lemuriformes and Adapiformes, made by the flare of the lateral malleolus, is almost certainly there to accommodate the flared talofibular facet, a feature considered a derived strepsirhine apomorphy (Beard et al., 1988; Dagosto and Gebo, 1994). Neither tree shrews (including *Ptilocercus*), nor dermopterans exhibit such an asymmetry of the mortise, nor does it seem to be characteristic of the plesiadapiforms *Ignacius* or micromomyids (Bloch and Boyer, 2007).

If the type 1 Shanghuang tibia does belong to a tarsiid, the differences in mortise morphology suggest that the postcranium of the Shanghuang tarsiids is more different from extant *Tarsius* than are the molars or the skull (Beard et al., 1994; Rossie et al., 2006) and that tibiofibular fusion was attained independently from extant *Tarsius* or *Necrolemur*. Only one talus from Shanghuang has been provisionally allocated to a tarsiid (on the basis of its relatively low and wide talar body and short talar neck) although there are other significant differences between it and extant tarsiers (e.g., talar head shape). This particular talus is too large to be from the same species as the tibiae discussed here (for example, the space between the malleoli in these specimens measures less than 1 mm, and the trochlear width of this talus is 2.13 mm; too large to fit in the mortise). Unfortunately the preservation of this specimen does not permit an assessment of talar asymmetry, but it seems unlikely that it was as exaggerated as in *Tarsius*. Although this actually means that in terms of morphology (but not size), this talus could functionally articulate with the Shanghuang tibiae, it points out again the significant differences between the Shanghuang form and extant tarsiers.

In addition, the dental and postcranial remains of Shanghuang tarsiids are relatively rare compared to the anthropoids or "unnamed haplorhines" (Table 13.2). Therefore, we think it equally likely that this morph belongs to one of the smaller size classes of the better represented Eosimiidae or protoanthropoid group and may provide further evidence of their anthropoid affinities. An eosimiid is the more likely option given that the tarsal morphology of the protoanthropoid group suggests less developed leaping abilities (Gebo et al., 2001). Given the strong degree of tibiofibular apposition in *Apidium, Cebuella*, and *Callithrix*, and the fusion (possibly convergent) in *Necrolemur* and *Tarsius*, it is not unreasonable to assume that tibiofibular fusion might occur convergently in other small anthropoids or haplorhines.

The fused tibiofibula from the Fayum region of Egypt that has been referred to *Afrotarsius* (Rasmussen et al., 1998) actually does not belong to a primate (White and Gebo, 2004). These Shanghuang tibiae, however, might provide evidence of tibiofibular fusion in Eosimiidae, a group that is sometimes linked to *Afrotarsius* (e.g., Ross et al., 1998; Gunnell and Miller, 2001). We stress, however, that (1) these tibiae may very well belong to a haplorhine group other than Eosimiidae; (2) these tibiae are too small to belong to *Eosimias sinensis*; (3) tibiofibular fusion is likely a compliant feature that is closely related to hindlimb function; by itself it is not enough to provide strong support for any potential phylogenetic relationship with *Tarsius*; and (4) the presence of such a derived character in one taxon, even if it proves to be an eosimiid, means only that this particular taxon is unlikely to be *directly ancestral* to anthropoids; it cannot refute an hypothesis that the whole clade is the sister group of crown anthropoids.

The affinities of the unfused morph (Type 2) are also uncertain. The greatest overall phenetic resemblance is to omomyids other than *Necrolemur* (e.g., *Omomys, Hemiacodon, Shoshonius*, and *Absarokius*). One significant difference from anthropoids may be the absence of an articular facet for the fibula in the unfused Shanghuang morph and omomyids, and the presence of such a facet in almost all anthropoids. Malleolar shape is also more like prosimians than anthropoids. Therefore, attribution to some as yet unrecognized Shanghuang omomyid or the "unnamed haplorhine family" at Shanghuang are both reasonable hypotheses. On the other hand, this morphology is not strikingly different from that of anthropoids and is probably primitive for haplorhines as a whole. Therefore attribution to either of the anthropoid groups is also possible. None of the morphological differences among the tali of the Shanghuang groups makes one attribution more likely than another.

Although we are unable to confidently allocate these bones to any specific group of Shanghuang primates, these tibiae, like the tarsal bones from Shanghuang, demonstrate the diversity of primates that were present at this locality, clearly show that these primates are phylogenetically haplorhine, support the recognition evidenced by the dentitions and tarsal bones of at least two major clades among the haplorhine primates at Shanghuang, and reveal the existence of anthropoid-like morphology among some of these primates.

Acknowledgments. Financial support for this research was provided by grants from the Leakey Foundation, the National Science Foundation (including BCS 9615557 and 0309800), and the Chinese NSF. We thank S. Goodman of the Field Museum of Natural History Chicago for access to uncatalogued *Microcebus* material, the staff of the Mammalogy department at FMNH for access to specimens, and Wang Yuanqing, Li Chuankui, Wang Banyue, Guo Jianwei, and numerous other colleagues at the IVPP for their generous assistance. Two reviewers made very helpful suggestions for improving the manuscript.

References

Beard, K. C., 2002. Basal anthropoids. In: Hartwig, W. C. (Eds.), The Primate Fossil Record. Cambridge University Press, New York, pp. 133–149.

Beard, K. C., Dagosto, M., Gebo, D. L., Godinot, M., 1988. Interrelationships among primate higher taxa. Nature 331, 712–714.

Beard, K. C., Qi, T., Dawson, M. R., Wang, B., Li, C., 1994. A diverse new primate fauna from middle Eocene fissure-fillings in southeastern China. Nature 368, 604–609.

Beard, K. C., Tong, Y. S., Dawson, M. R., Wang, J. W., Huang, X. S., 1996. Earliest complete dentition of an anthropoid primate from the Late Middle Eocene of Shanxi Province, China. Science 272, 82–85.

Bloch, J. I., Boyer, D. M., 2007. New skeletons of Paleocene–Eocene plesiadapiformes: A diversity of arboreal positional behaviors in early primates. In: Ravosa, M., Dagosto, M. (Eds.), Primate Origins: Adaptation and Evolution. Kluwer, New York, pp. 535–582.

Covert, H. H., Hamrick, M. W., 1993. Description of new skeletal remains of the early Eocene anaptomorphine primate *Absarokius* (Omomyidae) and a discussion about its adaptive profile. Journal of Human Evolution 25, 351–362.

Dagosto, M., 1985. The distal tibia of primates with special reference to the Omomyidae. International Journal of Primatology 6, 45–75.

Dagosto, M., Gebo, D. L., 1994. Postcranial anatomy and the origin of the Anthropoidea. In: Fleagle, J. G., Kay, R. F. (Eds.), Anthropoid Origins. Plenum, New York, pp. 567–594.

Dagosto, M., Gebo, D. L., Beard, K. C., Qi, T., 1996. New primate postcranial remains from the middle Eocene Shanghuang fissures, southeastern China. American Journal of Physical Anthropology Supplement 22, 92–93.

Dagosto, M., Gebo, D. L., Beard, K. C., 1999. Revision of the Wind River Faunas, early Eocene of Central Wyoming. Part 14. Postcranium of *Shoshonius cooperi* (Mammalia, Primates). Annals of Carnegie Museum 68, 175–211.

Dawson, M. R., Wang, B. Y., 2001. Middle Eocene Ischyromyidae (Mammalia: Rodentia) from the Shanghuang fissures, Southeastern China. Annals of Carnegie Museum 70(3), 221–230.

Fleagle, J. G., Meldrum, D. J., 1988. Locomotor behavior and skeletal morphology of two sympatric pitheciine monkeys, *Pithecia pithecia* and *Chiropotes satanas*. American Journal of Primatology 16, 227–249.

Fleagle, J. G., Simons, E. L., 1983. The tibio-fibular articulation in *Apidium phiomense*, an Oligocene anthropoid. Nature 301, 238–239.

Fleagle, J. G., Simons, E. L., 1995. Limb skeleton and locomotor adaptations of *Apidium phiomense*, an Oligocene anthropoid from Egypt. American Journal of Physical Anthropology 97, 235–289.

Ford, S. M., 1986. Systematics of the New World Monkeys. In: Swindler, D., Erwin, J. (Eds.), Comparative Primate Biology. Alan R. Liss, New York, pp. 73–135.

Ford, S. M., 1990. Locomotor adaptations of fossil platyrrhines. Journal of Human Evolution 19, 141–173.

Gebo, D. L., Dagosto, M., 2004. Anthropoid origins: Postcranial evidence from the Eocene of Asia. In: Ross, C. F., Kay, R. F. (Eds.), Anthropoid Origins: New Visions. Kluwer/Plenum, New York, pp. 369–380.

Gebo, D. L., Dagosto, M., Beard, K. C., Qi, T., 2000a. The smallest primates. Journal of Human Evolution 38, 585–594.

Gebo, D. L., Dagosto, M., Beard, K. C., Qi, T., Wang, J. W., 2000b. The oldest known anthropoid postcranial fossils and the early evolution of higher primates. Nature 404, 276–278.

Gebo, D. L., Dagosto, M., Beard, K. C., Qi, T., 2001. Middle Eocene primate tarsals from China: Implications for haplorhine evolution. American Journal of Physical Anthropology 116(2), 83–107.

Gregory, W. K., 1920. On the structure and relations of *Notharctus*: an American Eocene primate. Memoirs of the American Museum of Natural History 3, 51–243.

Gunnell, G. F., Miller, E. R., 2001. Origin of Anthropoidea: Dental evidence and recognition of early anthropoids in the fossil record, with comments on the Asian anthropoid radiation. American Journal of Physical Anthropology 114(3), 177–191.

Gunnell, G. F., Rose, K. D., 2002. Tarsiiformes: Evolutionary history and adaptation. In: Hartwig, W. C. (Eds.), The Primate Fossil Record. Cambridge University Press, New York, pp. 45–82.

Hafferl, A., 1932. Bau und funktion des Affenfusses. Ein beitrag zur Gelenk und Muskelmechanic. II. Die prosimier. Zeitschrift fur Anatomie und Entwicklung Gesellschaft 99, 63–112.

Jouffroy, F. K., Berge, C., Niemitz, C., 1984. Comparative study of the lower extremity in the genus *Tarsius*. In: Niemitz, C. (Eds.), Biology of Tarsiers. G. Fischer, New York, pp. 167–190.

Kay, R. F., Williams, B. A., Ross, C. F., Takai, M., Shigehara, N., 2004. Anthropoid origins: A phylogenetic analysis. In: Ross, C. F., Kay, R. F. (Eds.), Anthropoid Origins: New Visions. Kluwer/Plenum, New York, pp. 91–135.

Meldrum, D. J., Kay, R. F., 1997. Postcranial skeleton of Laventan platyrrhines. In: Kay, R. F., Madden, R. H., Cifelli, R. L., Flynn, J. J. (Eds.), Vertebrate paleontology in the Neotropics. Smithsonian Institution Press, Washington, DC, pp. 459–472.

Qi, T., Beard, K. C., 1996. *Nanotitian shanguhangensis*, gen. et sp. nov.: The smallest known bronothere (Mammalia:Perissodactyla). Journal of Vertebrate Paleontology 16, 578–581.

Qi, T., Beard, K. C., Wang, B. Y., Dawson, M. R., Guo, J. W., Li, C. K., 1996. The Shanghuang mammalian fauna, Middle Eocene of Jiangsu: history of discovery and significance. Vertebrata Palasiatica 34, 202–214.

Rasmussen, D. T., Conroy, G. C., Simons, E. L., 1998. Tarsier-like locomotor specializations in the Oligocene primate *Afrotarsius*. PNAS 95, 14848–14850.

Ross, C., Williams, B., Kay, R. F., 1998. Phylogenetic analysis of anthropoid relationships. Journal of Human Evolution 35, 221–306.

Rossie, J. B., Ni, X., Beard, K. C., 2006. Cranial remains of an Eocene tarsier. Proceedings of the National Academy of Sciences USA 103(12), 4381–4385.

Schlosser, M., 1907. Bietrag zur Osteologie und systematischen Stellung der Gattung *Necrolemur*, sowie zur Stammesgeschicte der Primaten ueberhaupt. Neues Jahrbuch für Mineralogie, Geologie und Paläontologie Monatshefte 1907, 199–226.

Szalay, F. S., 1977. Phylogenetic relationships and a classification of the eutherian Mammalia. In: Hecht, M., Goody, P., Hecht, B. (Eds.), Major Patterns in Vertebrate Evolution. Plenum, New York, pp. 315–374.

Szalay, F. S., 1981a. Functional analysis and the practice of the phylogenetic method as reflected in some mammalian studies. American Zoologist 21, 37–45.

Szalay, F. S., 1981b. Phylogeny and the problem of adaptive significance: the case of the earliest primates. Folia Primatologica 36, 157–182.

Szalay, F. S., 1984. Arboreality: Is it homologous in metatherian and eutherian mammals. In: Hecht, M., Wallace, B., Prance, G. (Eds.), Evolutionary Biology. Plenum, New York, pp. 215–258.

Szalay, F. S., 2000. Eosimiidae. In: Delson, E., Tattersall, I., Van Couvering, J., Brooks, A. S. (Eds.), Encyclopedia of Human Evolution and Prehistory. Garland, New York, pp. 235.

Szalay, F. S., 2007. Ancestral locomotor modes, placental mammals, and the origin of Euprimates: Lessons from history. In: Ravosa, M. J., Dagosto, M. (Eds.), Primate Origins: Adaptations and Evolution. Springer, New York, pp. 457–487.

Szalay, F. S., Decker, R. L., Tattersall, I., 1975. Phylogenetic relationships of *Plesiadapis*: postcranial evidence. In: Szalay, F. (Eds.), Approaches to Primate Paleobiology. Karger, New York, pp. 136–166.

Szalay, F. S., Delson, E., 1979. Evolutionary History of the Primates. New York, Academic.

Szalay, F. S., Drawhorn, G., 1980. Evolution and diversification of the Archonta in an arboreal milieu. In: Luckett, W. (Eds.), Comparative Biology and Evolutionary Relationships of Tree Shrews. Plenum, New York, pp. 133–169.

Szalay, F. S., Lucas, S. G., 1993. Cranioskeletal morphology of archontans, and diagnoses of Chiroptera, Volitania, and Archonta. In: MacPhee, R. (Eds.), Primates and their Relatives in Phylogenetic Perspective. Plenum, New York, pp. 187–226.

Szalay, F. S., Sargis, E. J., 2001. Model based analysis of postcranial osteology of marsupials from the Paleocene of Itaborai (Brazil) and the phylogenetics and biogeography of Metatheria. Geodiversitas 23, 139–302.

Szalay, F. S., Rosenberger, A. L., Dagosto, M., 1987. Diagnosis and differentiation of the Order Primates. Yearbook of Physical Anthropology 30, 75–105.

Wang, B., Dawson, M. R., 1994. A primitive cricetid (Mammalia, Rodentia) from the Middle Eocene of Jiangsu Province, China. Annals of the Carnegie Museum 63(3), 239–256.

White, J. L., Gebo, D. L., 2004. Unique proximal tibial morphology in strepsirrhine primates. American Journal of Primatology 64(3), 293–308.

14. *Rooneyia*, Postorbital Closure, and the Beginnings of the Age of Anthropoidea

Alfred L. Rosenberger*
Department of Anthropology and Archaeology
Brooklyn College, CUNY
2900 Bedford Ave, Brooklyn, NY 11210, USA
Department of Anthropology, The Graduate Center, CUNY
New York Consortium in Evolutionary Primatology (NYCEP)
American Museum of Natural History/Mammalogy
alfredr@brooklyn.cuny.edu

Russell Hogg
Department of Anthropology, The Graduate Center, CUNY
365 Fifth Avenue, New York, NY 10016, USA
New York Consortium in Evolutionary Primatology (NYCEP)
Hard Tissue Research Unit, New York University
hogg@nycep.org

Sai Man Wong
Department of Anthropology and Archaeology
Brooklyn College, CUNY
2900 Bedford Ave, Brooklyn, NY 11210, USA
vmax137@yahoo.com

"…it is particularly striking that no fossil prosimians show postorbital closure, yet all early anthropoids show a walled-off orbit. Where did the anthropoid condition come from? Or the tarsier condition, for that matter?"

Fleagle and Kay (1994:693)

"If and when we are compelled to conclude that the two [septa] are not homologous, it will only be because a convincing analysis of haplorhine phylogeny has given us convincing reasons for thinking that the last common ancestor of tarsiers and anthropoids lacked a postorbital septum."

Cartmill (1994:563)

"A large flange of the frontal descends behind the orbits [of *Rooneyia*]. Judged from the postorbital constriction of the skull, part of the major mass of the temporalis muscle extended slightly anteriorly above the orbits. In the case of *Rooneyia* the orbital partition, perhaps the homologue of that part of the postorbital funnel in *Tarsius*, platyrrhines and catarrhines, appears to be the bony wall which kept the muscles from intruding into the orbit. Possibly this partition is the initial adaptation responsible for the role of protecting the eyeballs and associated structures from the contraction of the temporalis."

Szalay (1976:349).

* Address for correspondence: alfredr@brooklyn.cuny.edu

E.J. Sargis and M. Dagosto (eds.), *Mammalian Evolutionary Morphology: A Tribute to Frederick S. Szalay*, 325–346.

14.1 Prologue

In the Age of Anthropoidea, the higher primates came to dominate primate evolution – at least since the Oligocene and probably even before that. In his research on the origins of anthropoids during the 1970s, F.S. Szalay set the stage for the present paper in three ways: he established its overarching phylogenetic framework; he promoted a methodology that emphasized the integration of phylogenetics and adaptational analysis in the reconstruction of evolutionary history; and, thankfully for us, he made a key morphological observation that produced the line of inquiry that this paper has followed up.

Regarding phylogeny, Szalay championed three big ideas that are crucial to an understanding of anthropoid origins. First, he helped convince primatologists to embrace the fossil record in applying Pocock's (1918) concept of Haplorhini (Szalay, 1975a), crafted originally in response to the phylogenetic puzzle of a single living genus, *Tarsius*. Second, he promoted the idea that anthropoids are monophyletic (Szalay, 1975b) at a time when the anatomical similarities between modern platyrrhines and catarrhines were seen by more senior authorities (e.g., Simpson, 1961; Gazin, 1958) as effects of parallel evolution and evidence of a dual origin. Third, Szalay developed the notion that omomyids are *the* model of pre-anthropoid anatomy (Szalay, 1976) while another equally authoritative school of thought (e.g., Gingerich, 1975, 1980) preferred adapids, a group with more obvious superficial morphological similarities to many anthropoids. Szalay thus established the modern version of the omomyid-anthropoid hypothesis (OAH), which remains the most widely accepted working hypothesis regarding the affinities and potential ancestry of higher primates (see Gregory, 1922; Le Gros Clark, 1934; Rosenberger and Szalay, 1980; Ross and Kay, 2004). The most viable single alternative to the OAH is the tarsier-anthropoid hypothesis (TAH; see Ross and Kay, 2004, for a brief history), which has strong promoters, too.

Szalay has advocated an approach to systematics – powerful and perhaps even more vital than the foregoing concepts because it is a tool – that builds on hypothetical *transformation series* of characters as phylogenetic evidence, which in turn generates readily testable hypotheses of the twin elements of phylogeny: sister-group (cladistic) and ancestral-descendant relationships (Szalay, 1977). His method for inferring a transformation series has been both dynamic and multifaceted, most often following a line of reasoning wherein character state A is posited to have evolved into state B because that is the most likely sequence suggested by the fossil record, and/or because that is the most logical direction selection would have taken to alter the evolutionary adaptation of a particular feature and its biological roles.

Also to influence this paper was the clue he left, buried in his seminal monograph on the systematics of the Omomyidae. Szalay (1976) elaborately confirmed Wilson's (1966) prior observation on the skull of the late Eocene fossil from Texas, *Rooneyia viejaensis* – it reveals an incipient form of postorbital closure. This passage, cited above, is a morphological keystone of our analysis.

We proceed by reopening questions of homology, phylogeny and classification that have critical bearing on the matter of anthropoid origins, followed by an examination of the morphology of the haplorhine skull as a context for interpreting the affinities of tarsiers and the enigmatic *Rooneyia*. Pursuing Szalay's lead (1976), and his interest in bringing classificatory rigor to higher phylogeny, we have identified several other features of the orbits which strongly indicate that *Rooneyia* belongs to a lineage that is the sister-group of Anthropoidea, a theory we have expressed by reclassifying *Rooneyia* and revising the higher classification of haplorhines (Rosenberger, 2006). In removing *Rooneyia* from the conventional grouping of Omomyidae, acknowledged by many to be paraphyletic (e.g., papers in Ross and Kay, 2004), the fossil tarsiiforms become somewhat more homogeneous in their morphology and adaptations, and more tarsier-like. This enables us to extend prior cranio-skeletal studies which show that some genera, European and North American, are cladistically allied with *Tarsius*, which again requires a rethinking of tarsiiform classification.

14.2 Setting the Agenda

14.2.1 Homology, Character Analysis, Adaptation and Origins of the Alisphenoid Septum

Simply put, anthropoids reinvented the primate skull. Because of the complexity of anatomical modifications this group experienced around its inception, apparently, researchers are bound to explain this reinvention in more ways than one.

A cardinal feature of the anthropoid cranium, postorbital closure made possible by a highly modified postorbital septum, has attracted enormous attention in recent years as a phylogenetic character (e.g., Cartmill and Kay, 1978; Cartmill, 1980; Rosenberger, 1986; Ross, 1994; Ross et al., 1998, Ross, 2000; Kay and Kirk, 2000; Simons, 2003; Hogg et al., 2005; see also chapters in Ross and Kay, 2004). An equally important feature is orbital orientation, and this too has been intensively studied (Ross, 1993, 1994, 1995). In some quarters, particularly among advocates of the TAH, the results of these parallel inquiries are interestingly asymmetrical. While the occurrence of a septum in taxa outside Anthropoidea is held to be phylogenetically and functionally informative, similarities in orbital convergence and frontation tend to be seen as functionally significant but phylogenetically moot. This duality, labeling characters as to their "functional" or "phylogenetic" value, reflects another key facet of Szalay's philosophy (1981) – he finds the distinction overblown and artificial. It also speaks to the core inferential issues of phylogenetics that he advocates, the search for homologies and the importance of weighting characters. These factors are laced throughout this paper, and crucial to evaluating competing hypotheses about anthropoid interrelationships.

As our opening quotes from Cartmill, Fleagle and Kay suggest, understanding the evolution of the postorbital septum is not straightforward. There are two schools of thought regarding its origins. One regards it as a decisive homology linking anthropoids and tarsiers, whereas the other sees it as convergently evolved in anthropoids and tarsiers. Until the advent of computerized parsimony analyses based on superabundant samples of taxa and characters, the argument came to pivot increasingly on the homology of a slip of the alisphenoid bone which has been put forth as *the* defining attribute of the postorbital septum, unique to tarsiers and anthropoids. Only recently, Cartmill (1994), for example, continued to discuss the alisphenoid problem at great length, concluding that there was no logical way to employ conventional character analysis of the septum in order to resolve the matter "in advance of our phylogenetic analyses" (p. 563). Whether or not one agrees with Cartmill, it is evident that undue attention to the question of the alisphenoid poses a larger danger, for the faces of Anthropoidea and of tarsiers probably have longer, more complex, and potentially more informative histories than the story of the alisphenoid in and of itself. To be fair, advocates of the TAH, and the alisphenoid's role in supporting it, have also invoked the morphology of the auditory bulla as evidence of tarsier–anthropoid monophyly (see Cartmill and Kay, 1978; MacPhee and Cartmill, 1986). While this region is beyond the scope of our paper, we refer the reader to Beard and MacPhee (1994), wherein one of the architects of the bulla analysis retreats from his earlier position.

The surest way to test the homologies of the tarsier and anthropoid alisphenoid postorbital septa would be to find at least one transformation series through time which revealed directly how it evolved in one group or the other. This is the phylogenetic gauntlet that Cartmill (1994) laid out to resolve the alisphenoid debate. However, there are no euprimate fossils that present anything like an alisphenoid precursor to the septum, which greatly limits the ways in which the anatomy can be studied and assessed. On the other hand, there are more than a half-dozen tarsiiform genera that offer *other cranial features* amenable to character analysis and phylogeny reconstruction, of tarsiers explicitly and of anthropoids by implication. Based on these fossils and characters, as we discuss below, one can see that the assumption of the alisphenoid plate as the final arbiter in a tarsier-anthropoid comparison poses an uncalled-for risk; this tiny plate of bone does not pass the threshold of a high-weight character in this context.

Some may argue that the question of tarsier-anthropoid alisphenoid homology has already been well-tested cladistically by the extensive series of parsimony (PAUP*) studies of anthropoid interrelationships that have been conducted (see chapters in Ross and Kay, 2004). While we agree that such analyses are useful in some ways, their results have been notoriously inconsistent for particular questions (Rosenberger, 2005; see further below) – usually the hard ones – and they are replete with unresolved polytomies. Almost all of the various alternative cladograms generated in these studies (e.g., Ross et al., 1998) were unable to root and/or sort the relationships of fossil tarsiiforms. This raises severe questions about pivotal conclusions regarding *Tarsius*. For if tarsiers are not most closely related to anthropoids they must surely be related to some set of fossil tarsiiforms, yet the interrelationships of this group would appear to be the only haplorhines whose affinities cannot be adequately addressed by these data and methods. In other words, if the cladistic relationships of the animals most similar phenetically to tarsiers (all sharing a "tarsiiform morphology," for lack of any other useful generalization) prove to be utterly confounding as a research outcome, why believe the particular results spun out for one small sample of them – genus *Tarsius*? If these studies return suspect or irresolvable phylogenetic relationships, it follows that the homologies and polarities upon which those results are based must be equally dubious. But which ones?

While the phylogeny test can shed light on homologies *post hoc*, there are other pointed reasons why the homologization of tarsier and anthropoid alisphenoid septa is not to be trusted in advance of a cladistic result. After all, this is a two-point comparison conducted exclusively using morphologically derived terminal taxa. (Fayum anthropoids, notwithstanding their geological age, are utterly modern in this regard, making them essentially equivalent to a living *Saimiri* or *Cercopithecus* in this context.) There are no plesiomorphic fossils (ignoring *Rooneyia* for the moment; see Szalay, 1976; Rosenberger, 2006) with the requisite anatomy and there is no meaningful, detailed morphocline among the living forms, meaning the *a priori* risk of a homology error is quite high. As baseline conditions, this does not bode well for homology inferences involving a question of deep-time origins. This situation is exacerbated by the fact that tarsiers, no matter what opinion one has about their origins, remain a vestigial phylogenetic twig as well as a morphological outlier. The risk of homology error is compounded when the morphological congruence between presumptive homologues occurs in taxa that are so vastly different, objectively, that scholars universally agree to distinguish them taxonomically at near-ordinal levels for the morphology of the character complex in question – the orbits – in addition to a myriad of other phenetic issues. And the risk level rises higher still when the septum is assumed to serve the same functional adaptation – preventing mechanical interference from chewing muscles – as we know intuitively that hardly anything in the tiny tarsier head could avoid coming under the selective and morphogenetic regime dictated by eyesight and eyeballs. This evolutionary/anatomical milieu is certainly unlike that which propels the small-eyed anthropoids.

Thus it is not surprising that the proposed homologization has met strong criticism. While the focal point of today's debate centers around the case as it has been most fully fleshed out by Cartmill and colleagues (e.g., Cartmill, 1980, 1994; Cartmill and Kay, 1978; Kay et al., 1997, 2004; Ross, 1993, 1994, 1996), the essence of their point follows the reasoning of earlier workers articulated at a time when the morphology of fossil tarsiiforms was poorly sampled, when morphologists were quite limited in terms of justifiable comparisons, explanations and alternative hypotheses. For example, Duckworth (1915:104) noted, "...the postorbital wall (to which the alisphenoid makes a distinct contribution) constitutes a resemblance to the Anthropoidea, and severs *Tarsius* (*sic*) from the Lemurs." Pocock (1918:51) agreed but in more general terms, saying that for "...the presence of the postorbital partition, and other well-known features, it seems that Hubrecht was quite right in removing *Tarsius* from the Lemurs and placing it in the higher grade of Primates." Le Gros Clark (1934:64) essentially concurred: "The orbits of the pithecoid skull are...almost completely cut off from the temporal region by a bony wall formed by the malar and alisphenoid (an advanced character which, it has been seen, occurs to a slight degree in *Tarsius*)."

For the early advocates of this school of thought, tarsiers represent an intermediate state of a series leading to anthropoid closure wherein a postorbital septum, deriving from the still more primitive euprimate postorbital bar, is enlarged but does not fully seal off the orbital fossa behind the eyeball (see Hershkovitz, 1977). Adding modernity to the argument that the alisphenoid component of the septum "proves" that the partition is homologous with anthropoids, Cartmill (1980 *et seq.*) and his colleagues offered a covering adaptive explanation to enhance the logic of the case. They proposed that a single adaptive reason for compartmentalizing the orbit in the tarsier-anthropoid group, to protect its contents from mechanical interference originating in the adjacent temporal fossa, where contraction of the temporalis muscle would otherwise disrupt the vision of these animals that place a high premium on pinpoint visual acuity (but see Ross, 2004, on the moot homologies of haplorhine foveae). A sizable literature has sought to establish this hypothesis, a variant of the visual predation hypothesis (e.g., Cartmill, 1972), by examining allometric and masticatory contingencies relating to eye size, orbit size and biomechanics (see reviews in Ross and Kay, 2004; Ravosa and Hogue, 2004; Ross, 1994; Ross, 2000). While important in their own right, these studies seek to corroborate by correlation and association. They do not doubt the supposition that the postorbital septa of tarsiers and anthropoids are homologous, and rarely challenge the interference explanation.

An empirical behavioral test of the interference/visual predation hypotheses has not yet been conducted, to our knowledge. If the septum does successfully insulate the eyeball, do tarsier eyes not wobble when the temporalis is stimulated? Do their eyes wobble less than a galago's, where there is no postorbital septum? Can it be shown that tarsiers have, need, or benefit from foveal, pinpoint vision as a motion detection device? Or, does a foveal retina primarily benefit hand-eye coordination, i.e., prey capture and manipulation, which would be another form of the visual predation hypothesis? Do tarsiers actually scan for prey and calculate takeoff coordinates while masticating? They ought to if the interference hypothesis is correct. Or, do they finish a meal before hunting again? As hold-and-feed animals, doesn't the logic of the interference hypothesis suggest that selection for the septum in tarsier ancestors favored populations with the fickle habit of chewing a live victim while clutching it and also being able to *simultaneously* take off again in order to...drop the first and grab a new one? Testing hypotheses of functional evolutionary adaptation is always complex and none of these questions alone would prove much if they were answered individually. But solutions would probably advance our knowledge of the issues to a new state and perhaps challenge the functional rationale of the homology hypothesis, which is tied to the proposition of visual predation as a causal explanation.

Such difficulties notwithstanding, the primary morphological substance of the hypothesis has also been challenged by Simons and Russell (1960; see also Simons and Rasmussen, 1989; Simons, 2003) and Rosenberger and Szalay (1980), who independently argued it is more likely that the slips of alisphenoid contributing to the postorbital wall of tarsiers and anthropoids are not homologous. This means that the evolution of the anthropoid eye socket and the tarsier postorbital septum were coincidental, convergent events. The general hypothesis advanced by these authors is this: in tarsiers the small alisphenoid rampart belongs to a series of lip-like orbital superstructures that are correlated autapomorphies, none of which occur in anthropoids. In adult tarsiers, the constituents of this pattern are evident superiorly, in the form of an everted superior orbital margin; inferiorly, by a shelf-like posterior extension of the maxillary orbital floor; posteriorly, by a broadened wing of the frontal bone that is continuous with a narrow horizontal process of the alisphenoid; and, laterally, by an enlargement of the surface of the maxillary-zygomatic complex (see below). In this view there is no simple "tarsier postorbital septum." Rather, tarsiers have a periorbital structural system whose principal biological role is related to eyeball hypertrophy and position, again distinguishing it fundamentally from the smaller-eyed anthropoids where the major biological role of the alisphenoid is not related to enlarged eyeballs. Additional support for this notion can be found in their different ontogenies. In anthropoid neonates, the alisphenoid plate forms a readily visible, proportionately large "wing," while in tarsier newborns there is little more than a nubbin of bone evident where the alisphenoid process arises. It appears to develop postnatally, in concert with the other periorbital flanges. As discussed below, one part of this derived pattern is already evident among fossil tarsiiforms in a mosaic distribution that suggests the alisphenoid of the tarsier condition is a "final" element of the design uniquely evolved in the genus.

While descriptively dissecting anatomical parts in this way involves some arbitrariness, it is instructive to consider briefly another major facial element of the orbital surround, discussed further below. This is the laterally flaring and essentially horizontal paralveolar extension of the tarsier face, which encompasses the anterior root of the zygomatic arch and forms the lowest and most lateral portion of the bony ring around the eyeball. Enlargement of the surface of the maxillary-zygomatic complex in *Tarsius*, which essentially everts the lateral face of the maxilla, has not figured as a character in discussions of tarsier and anthropoid orbits, yet it seems to make the case emphatically that the periorbital components of tarsiers are all functionally tied to the large-eye syndrome. Its purpose must be to enlarge the orbital floor laterally, extending it beyond the margin of the toothrow in order to accommodate hypertrophic eyeballs in a skull where there is no place to grow bone but outward. Thus tarsier faces have enormous bony facial extensions anteriorly and laterally, displacing the lateral orbital margin away from the midline and braincase. With the obvious highly derived exception of *Aotus*, hardly at all a mirror for the pattern, anthropoid skulls are nothing like this.

The upshot of this extensive integration of unique tarsier features is that it becomes difficult to isolate the septum from the others and ascribe to it a unique functional explanation apart from the rest. Rather than being fundamentally related to closing off the eye from the temporal fossa as the interference hypothesis claims, for both large- and small-eyed haplorhines (i.e., anthropoids and by extension *Rooneyia*; see Rosenberger, 2006, and below), the tarsier septum appears to represent an entirely different adaptational history and transformation series. It is difficult to say if it is essential to mechanically supporting the eye and its attachments as opposed to being simply an epigenetic reflection of orbital hypertrophy, which may be a distinction without a difference. In any event, this does not negate the interpretation that the alisphenoid septum provides bony insulation from interference as preferred by Cartmill and colleagues. But if this is a secondarily acquired biological role of a larger morphological pattern related to eyeball enlargement, it means that the tarsier morphology is less likely to be a homology shared with anthropoids.

Simons and Rasmussen (1989) offered a second challenge to the premise that the evolutionary essentials of anthropoid postorbitum pivots on the alisphenoid element. They pointed out, instead, that in anthropoids the ascending ramus of the zygomatic is what provides the principle separation of orbital and temporal fossae, not the alisphenoid. This contrasts with the *Tarsius* condition, where the ascending frontal process of the zygomatic bone is not so enlarged. To the contrary, it may seem surprisingly narrow given the size of its zygomatic and frontal roots, and the other superstructures described above. In other words, tarsiers are seen as retaining a primitive albeit modified postorbital bar. Anthropoids, in contrast, show a dramatically transformed postorbital bar predicated on a unique size and shape of the ascending process of the zygomatic bone, which was modified into a spoon-like shape, to use Simons' terminology, from a bar-like process. In all anthropoid skulls this laterally positioned lamina of the zygomatic is what makes for postorbital closure, with only a small fraction of the partition being formed by alisphenoid medially. In this view, the tarsier-anthropoid alisphenoid comparison turns out to be a red herring.

Arguing from another perspective, Rosenberger (1985) opposed the phylogenetic aspect of the TAH and the homologization of the alisphenoid flange in tarsiers and anthropoids. Building on Simons and Russell (1960), he suggested there is a series of uniquely derived features of the basicranium that align *Tarsius* more closely with European microchoerine tarsiiforms, which we now regard as tarsiids (Table 14.1; see Simons, 1972). Beard et al. (1991) and Beard and MacPhee (1994) then showed that newly discovered skulls of the North American tarsiiform *Shoshonius* also present this same suite of features (see also Dagosto et al., 1999). These data and arguments, along with the presence of definite Eocene tarsiids (Beard, 1998; Rossie et al. 2006), indicates that tarsiers were part of a larger, tricontinental radiation already well

TABLE 14.1. A provisional classification of non-anthropoid haplorhines that forms the basis of this study. Tarsioids and tarsiids are distinguished from other tarsiiforms as likely monophyletic groups sharing a suite of cranial characters relating to relatively large and hypertrophic eyes, in conjunction with postcranial features related to leaping, such as extensive apposition of the tibiofibula (see review in Dagosto et al., 1999). Tarsiines and microchoerines are known to show highly advanced postcranial adaptations, such a tibiofibular fusion (*Tarsius, Necrolemur, Pseudoloris*) and enhanced anterior calcaneal elongation (*Tarsius, Necrolemur, Microchoerus*) as well as a derived tubular auditory meatus (*Tarsius, Necrolemur, Microchoerus*). The *incertae sedis* tarsiids are known to share mosaics of the primitive and derived cranio-skeletal states of these features, so they may be referable to either Tarsiinae or Microchoerinae on cladistic grounds upon further study. Some microchoerines, such as *Pseudoloris*, may prove to be justifiably included in the tarsiines. *Xanthorhysis* is allocated to Tarsiidae based on Beard's (1998) analysis of the dentition. It is likely that other genera now regarded as omomyids will be classified as tarsioids when they are reconsidered. *Teilhardina* is kept outside the tarsiid group, as an anaptomorphid, because of its primitive craniodental morphology. (With the nominate genus *Omomys* removed to the Tarsiidae, the family-level term Omomyidae cannot be applied to non-tarsid tarsiiforms, and the first available name becomes Anaptomorphidae Cope, 1883 based on chronological priority.) The classification of Rooneyia is discussed further elsewhere (Rosenberger, 2006), where the new higher taxa are formally proposed based in part on the analysis presented herein.

Suborder Haplorhini
Semisuborder Tarsiiformes
  Superfamily Tarsioidea
    Family Tarsiidae
      Subfamily Tarsiinae
        Tarsius
      Subfamily Microchoerinae
        Hemiacodon, Microchoerus, Nannopithex, Necrolemur, Pseudoloris
    Family Tarsiidae *incertae sedis*
        Absarokius, Omomys, Shoshonius, Tetonius, Xanthorhysis
  Superfamily *incertae sedis*
    Family Anaptomorphidae
        Teilhardina
Semisuborder Simiiformes
Hyporder Protoanthropoidea
    Family Rooneyiidae
      Rooneyia
Hyporder Anthropoidea
  Infraorder Platyrrhini
  Infraorder Catarrhini

established in the Eocene (see also Rosenberger and Pagano, in press), which eliminates the genus from having a sister-group relationship with anthropoids.

While this particular phylogenetic point, which is further developed below, weakens the phylogenetics of the TAH and the underlying character analyses pertaining to the alisphenoid, it does not refute it entirely. The fallback position might be that anthropoids are still more closely related to a *greater tarsier clade* than to any other tarsiiforms (see Ross et al., 1998). While we consider this unlikely, it is worth noting that several variations of the cladistic interrelationships of tarsiers, "omomyids" and anthropoids may be said to be currently in play if one subscribes to the array of parsimony (PAUP*) analyses performed in the past decade by Kay and colleagues (e.g., Kay et al., 2004).

14.2.2 Toward A New Classification of Tarsiiforms

Thus in our view the alisphenoid postorbital septum has already been over-interpreted by those who regard it as a homology shared with anthropoids. But this does not explain why these points, several of which have been made before in other ways, have not sealed away the argument. We surmise that in a subtle way, this is because the problem has been cast too deeply in neontological terms, bound up in a heuristically outmoded taxonomy that fails to integrate paleontology. Cartmill (1994), for example, in his extended explication of the alisphenoid problem, makes almost no mention of fossil evidence. How is this possible in tracing the evolution of such a structure, or a lineage like Anthropoidea? Only part of the answer rests with the fact that an alisphenoid postorbital septum has not been observed in non-anthropoid fossils. But another part of the answer surely is that the *status quo* has long considered *Tarsius* a genus apart from fossil tarsiiforms, adaptationally and phylogenetically, and this, in turn, helped promote a limiting approach as to how tarsiers tend to be classified, compared and understood.

We would argue that the concept of Tarsiidae, as implemented in the literature in recent decades, has been too narrow. This is evident in formal classifications and the less formal ways that taxonomic terms are used and/or extended conceptually in various works. For instance, it has been rare for primate classifications published during the twentieth century to include any other genus besides *Tarsius* in the Tarsiidae. Osman Hill (1955) and Simons (1972) present the significant counterexamples. The only other case where this rule seems to have been broken recently involves the allocation of a new Chinese Eocene genus, *Xanthorhysis*, to Tarsiidae by Beard (1998); a bold move given today's aversion to recognizing modern primate families during epochs before the Miocene. It is noteworthy also that Simons, (1972; 2003), influenced by Teilhard de Chardin (1921), had previously discussed the genus *Pseudoloris* as the fossil most closely related to modern tarsiers and called it a tarsiid, but his argument has not been carefully assessed and so his reasoning has not been extended to other tarsiiform genera. A case in point: in placing *Xanthorhysis*, Beard (1998) did not consider Simon's points about *Pseudoloris*, which is also Eocene, nor did he integrate other highly pertinent phylogenetic analyses (e.g., Rosenberger, 1985; Beard et al, 1994; Dagosto et al., 1999) which suggest strongly that other tarsiiform genera are close cladistic relatives of modern *Tarsius* as well. Following from this, to present an illustration of a different sort, Jablonski (2003) discussed the origins of the tarsier ecological niche, specifying only *Xanthorhysis* and the Egyptian *Afrotarsius* (see Simons and Bown, 1985; Rasmussen et al., 1998) as fossil tarsiid genera and concluding that the animals must have originated in eastern Asia. There would be a far more complex case to be evaluated if one were to acknowledge European microchoerines and North American forms such as *Shoshonius* (see Beard et al., 1991) as being part of a monophyletic family of tarsiids. While Beard has motivated some welcome movement to expand the concept of Tarsiidae, as was the case with *Homo*/Hominidae for decades (see Simpson, 1961), the gradistic consensus of *Tarsius*/Tarsiidae as a category of its own has supported a reluctance to group tarsiers with potential or demonstrable cladistic relatives in an integrative way.

There is another set of forces at work which calls for a shift in how tarsiers, and tarsiiforms, ought to be classified. It begins with the gradual breakdown of Szalay's concept of Omomyidae (1976), which is steeped in a deeper history, most notably the synthetic works of Gregory (1922) and Le Gros Clark (1934), and his view that no fossil tarsiiforms are close enough to tarsiers phylogenetically to warrant expansion of the one-genus concept of Tarsiidae. In addition to the phyletic arguments already alluded to, fossil tarsiiforms are becoming better known adaptively. There is a host of genera for which we have information on cranial and postcranial morphology, as well as dentitions. Several show that advanced leaping adaptations *and* cranial features associated with relatively enormous eyes were present in combination, as we emphasize here. Thus the supposed ecomorphological differences between modern tarsiers and Eocene tarsiiforms is diminishing, and the facile argument that parallelism explains away suites of anatomical similarities between them is no longer compelling. As implied above, Beard et al. (1994) has even allocated an Eocene species, dentally similar and with good indications of having large eyes, to genus *Tarsius*.

For these reasons we provide a provisional classification that takes into account recent findings (Table 14.1), emphasizing the taxa that are relevant to our discussion of the postorbital septum. We recognize the incompleteness of this exercise and expect this iteration to be useful only as an interim step. However, to us it seems to be an effective way to promote necessary changes in the systematics and classification of Eocene tarsiiforms in particular, which holds the key to tarsier – and possibly anthropoid – origins. From a taxonomic standpoint, our intention is to maintain a monophyletic family Tarsiidae. Following Simons (1972), we keep *Tarsius* in a distinct subfamily and allocate other tarsiiforms that can be shown to be probably monophyletically related to it by cranial and/or postcranial characters to Subfamily Microchoerinae. This move was anticipated by Rosenberger (1985), who used the informal term "necrolemurs" to refer to this group, which then included only the classic microchoerines, *Necrolemur, Microchoerus, Nannopithex*, and *Pseudoloris*.

14.2.3 Questions and Goals

The forgoing should make it clear that, in our view, the structural antecedents of the transformed anthropoid skull is an unsettled matter in spite of a prodigious effort to understand the history of the postorbital septum and forward-facing orbits. The neontological work that has dominated debate must be extended more effectively to accommodate early relevant fossils if we are to get beyond the current stalemate of ideas. How anatomically, why adaptively, when temporally, and whom taxonomically was involved as the anthropoid orbital complex was reconfigured by natural selection? Even murkier is the question of phylogenetic transformation: what anatomical prelude was preadaptive to postorbital closure?

Our goal is to address the origins of the anthropoid skull by expanding the focus of inquiry, starting with a rethinking of the anatomical and spatial relationships of important components of the orbit relative to the face and neurocranium in early haplorhines, especially tarsiiforms. The skulls of pertinent Eocene tarsiiforms are still relatively scarce and understudied, but they are reasonably known in varying states of preservation from about seven genera: *Necrolemur, Microchoerus, Nannopithex, Pseudoloris, Shoshonius, Teilhardina* and *Tetonius*. Only a few of the important observations can be made on *Teilhardina*, which has been reconstructed via high resolution CT imaging (Ni et al., 2004).

In addition to these forms, we emphasize the late Eocene fossil from Texas, *Rooneyia viejaensis*, a controversial taxon (e.g., Szalay, 1976; Ross et al., 1998; Gunnell and Rose, 2002; Kay et al., 2004; Rosenberger, 2006) still known from only one relatively complete, undistorted and little damaged specimen (Wilson, 1966). The centrality of *Rooneyia* to the question of anthropoid origins is contextualized by the OAH: *Rooneyia* has most often been considered an omomyid for about 30 years now (see Gunnell and Rose, 2002, for a recent review). A different view promoted by some advocates of the TAH is that the systematics of *Rooneyia* is fundamentally un-interpretable in that there are several viable phylogenetic solutions. To wit, paraphrasing Ross et al. (1998:255) *Rooneyia* is: (1) not an omomyid; (2) related to extant strepsirhines; (3) related to an adapid/strepsirhine clade; (4) related to anthropoids; (5) the sister-taxon of all primates; (6) related to a parapithecine-*Aegyptopithecus* group; (7) the sister-taxon of an omomyid/tarsier/anthropoid clade. Here we consider *Rooneyia* a member of the Protoanthropoidea (Rosenberger, 2006), a group formally defined as a non-tarsiiform sister-group of Anthropoidea. The species has seldom been considered in detail in connection with anthropoid origins (e.g., Simons, 1972; Hogg et al., 2005; Rosenberger 2006) even though its skull stands well apart from fossil tarsiiforms in overall morphology, as shown by Fleagle (1999:376) in a rare comparative illustration. This is somewhat surprising given the clarity with which Szalay (1976), as quoted above, discussed the morphology of its postorbitum and the Cartmillian rationale he then offered to explain the adaptive benefits of postorbital closure.

14.3 Comparative Morphology

14.3.1 Haplorhines and *Rooneyia*

Using *Rooneyia* as a starting point, we draw on 3-D digitizations based on laser surface scanning to clarify how the orbits of haplorhines are packaged in the skull. *Notharctus* sp. was chosen as a comparative model for early strepsirhine cranial morphology (see Szalay and Delson, 1979; Gebo, 2002). As noted above, we attempt to refocus the discussion of the origins of the anthropoid orbit away from a narrow emphasis on the postorbital plate toward a balance of several factors. Our most important conclusions are these: (1) Haplorhine orbits are derived among primates in having a posterior-medially shifted orbital fossa and a mediolaterally extensive, relatively horizontal orbital floor. (2) The functional concern about spatial adjacency of the orbital and temporal fossae in foveate tarsiers and anthropoids is probably exaggerated. (3) *Rooneyia viejaensis* is unique among known Paleogene non-anthropoids in having a pattern of attributes that may foreshadow the evolution of an anthropoid eye socket, including: a funnel-shaped orbital fossa deeply recessed below the forebrain; a dorsoventrally and laterally extensive frontal process that forms a partial postorbital septum and implies, albeit tenuously, the existence of a relatively large ascending processes of the zygomatic bone (postorbital bar); a relatively large frontal bone with a fused metopic suture (see Figure 14.3), that extends roof-like above the orbit; highly convergent and frontated orbits. Simultaneously, *Rooneyia* is more primitive than fossil tarsiiforms for which skulls are known in having relatively small, anthropoid-sized eyeballs and in lacking numerous features that are correlated with eyeball hypertrophy, immediately around the orbital fossae, in the organization of the face, and in the morphology of the posterior palate and nasopharyngeal region that relates to enlarged eyes.

14.3.1.1 Orbital Fossa

There are profound differences in the size and placement of the orbital fossae in *Notharctus* and *Rooneyia* (Figure 14.1). We hypothesize that the *Notharctus* morphology represents the ancestral euprimate and strepsirhine pattern and that the *Rooneyia* arrangement models the ancestral condition of haplorhines. In the former, the orbital floor is situated far forward of the braincase and it is located quite laterally on the snout, nestled in the space formed by the junction of the anterior root of the zygomatic arch and the rostrum (Figure 14.1a). This is the common condition among mammals and must have been ancestral in euprimates. The orientation of the floor among strepsirhines can vary in the transverse plane. It may be horizontal or pitched upward, antero-dorsally, for example. However, the restricted size of the orbital floor is maintained among strepsirhines even in cases where the eyeballs are relatively large, as in lorises.

In contrast, in *Rooneyia* and other haplorhines, the orbital fossa is situated posteriorly in the face, essentially at the craniofacial junction (Figure 14.1b). The floor is greatly expanded, especially in its transverse dimension, and tends to be built largely from a horizontal lamina formed within the maxilla. The large size of the floor can be explained as the lamina's medial incursion into the space of the rostrum. In some cases, the floor is also enlarged laterally as a paralveolar expansion that is confluent with the root of the zygomatic (see below). In horizontal section (Figure 14.1c, d), the large orbital floor is clearly seen in connection with the typically haplorhine reduction of the nasal fossa in the transverse dimension, and approximation of the medial walls of the orbits. In superior view, the relatively large size of the orbital floor of haplorhines is also evident (Figure 14.2), whether the eyeballs are relatively large (e.g., *Necrolemur*) or small (*Rooneyia*).

14.3.1.2 Frontal Bone, Craniofacial Junction and Temporal Fossa

The complex morphology of the frontal bone and craniofacial junction is markedly different in *Rooneyia* and *Notharctus*. To begin with, the metopic suture is fused in *Rooneyia* (Figure 14.3; *contra* Ross et al., 1998). It tends to be fused in *Notharctus* and in the majority of living strepsirhines, contrary to conventional wisdom. (Rosenberger and Pagano, in press). The type specimen of *Rooneyia* is a young adult, judging by its little-worn molar teeth, suggesting that frontal fusion did not occur as bone was remodeled during aging. Unlike *Tarsius*, on the external surface of the frontal there is no indication of a longitudinal ridge or a sagittal canal (see Rosenberger and Pagano, in press). The frontal bone is also large in overall size and extends shelf-like above the orbital fossae (Rosenberger, 2006; Hogg et al., 2005). This is well illustrated by comparing the positions of the anterior margins relative to a line defining the transverse axis of postorbital constriction in *Rooneyia* and *Necrolemur* (Figure 14.2).

In *Notharctus*, the frontal bone is smaller and, because the degree of convergence and frontation is less and the orbital fossa is positioned further forward on the snout, the superior margin of the orbit does not overhang the orbital fossa (Figure 14.2). It is most likely that this typically strepsirhine condition is the primitive euprimate pattern. Tarsiiforms such as *Tetonius, Microchoerus*, and *Necrolemur* also tend to have laterally facing orbital margins rather than a forward-projecting superior rim. Thus their orbits are not roofed by the frontal, as in *Rooneyia*.

Ross (1995) has shown that the orientation of the orbital plane in *Rooneyia* is essentially anthropoid (Figure 14.4), i.e., its forward facing orbits just fall at the boundary (of a minimum convex polygon) of a bivariate plot of the angles of convergence and frontation, a geometry that is rare among non-anthropoids. Rosenberger (2006) argued that this is unlikely to be a homoplastic similarity shared with anthropoids; rather, it may be homologously derived. *Notharctus*, in presenting what must be the primitive condition for primates (e.g., Le Gros Clark, 1934), has laterally facing, relatively divergent orbits typical of most strepsirhines, fossil tarsiiforms, and modern tarsiers, quite unlike *Rooneyia* and anthropoids. It is the spread along the convergence axis of the bivariate plot describing the orbital plane (Figure 14.4) that most clearly distinguishes these forms from more primitive euprimates.

Regarding the vertical tilt of the plane, frontation, the anthropoids and *Rooneyia* accomplish this similarly by combining several factors: prolongation of the frontal to form a roof-like extension over the orbital fossae, combined with the deep recession of the orbits toward the braincase and a somewhat reduced interorbitum. This flattens the angle of tilt fixed by the upper and lower orbital margins in lateral view. Tarsiers may resemble *Rooneyia* and anthropoids in their metrics but not anatomically. The superior margin of the tarsier orbit is everted dorsally like a pitched awning rather than prolonged horizontally as a roof, and the inferior margin is extended anteriorly as part of the paralveolar expansion (see below). But their angles are similar because the tarsier facial skull is uniquely bent downward relative to the basicranial axis (Spatz, 1969; Starck, 1975), displacing the ventral margin of the orbit inferiorly and tilting the plane of the orbit into an anthropoid-like orientation.

An important consequence of the orbit's location within the cranium is the funnel-like shape of the orbital fossae in *Rooneyia*, as seen in the cutaway of Figure 14.5. This is a product of the subcerebral position of the orbit (which in turn contributes to the bony orbital roof), the medial incursion of the orbital floor and the convergence of the orbital apices toward the midline. That is, the anterior wall of the braincase effectively becomes part of the back wall of the orbital fossa, while angulation of the medial walls is conditioned by the width differential between the interorbitum and the span between the optic foramina. This pattern approximates the cone-shaped "eye socket" that defines Anthropoidea, differing

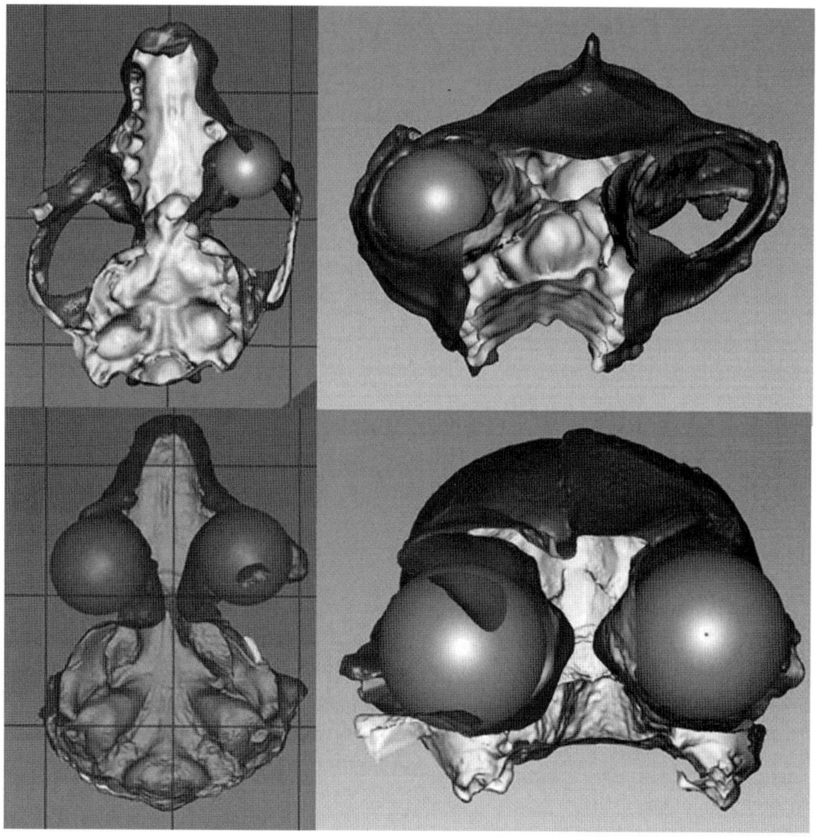

FIGURE 14.1. Images captured from three-dimensional laser scans of *Notharctus* (top) and *Rooneyia* (bottom) in horizontal and coronal sections taken near the lacrimal canal. The size and position of virtual eyeballs are based on contours of the orbital fossae, and are meant for illustrative purposes. Two alternatives are shown for *Rooneyia*, where the right eyeball is colliding (see irregular splotches) slightly with the back of the orbit and with matrix on the orbital floor. These images show the primitive antero-lateral placement of the orbital fossa in strepsirhines as compared with the derived postero-medial position in haplorhines, which is related to reduction of the posterior nasal fossa. The large size of the orbital floor in *Rooneyia*, like all haplorhines, is evident.

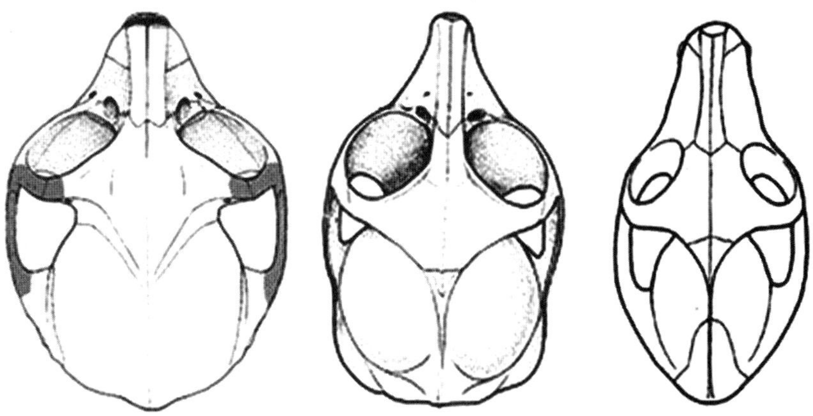

FIGURE 14.2. Dorsal views of *Rooneyia, Necrolemur* and *Notharctus*, left to right, brought to the same length (adapted from Szalay, 1976). The relatively large size of the haplorhine orbital floor is evident, as are recession of the orbits toward the braincase and the relatively wide shape of the braincase. Notice the relatively smaller size of the frontal bone in *Notharctus*, its intermediate size in *Necrolemur* and its relatively large size in *Rooneyia*, where the superior margin of the orbit is prolonged to partially roof over the fossa. In *Necrolemur*, the strongly tapering, concave profile of the snout and the relative narrowness of the interorbital region are aspects of eyeball enlargement and sagittally shifted medial orbital walls, part of the derived transformation series leading to the extensively modified arrangement of *Tarsius* where paralveolar expansion and fused medial orbital walls are part of the hypertrophic eyeball pattern.

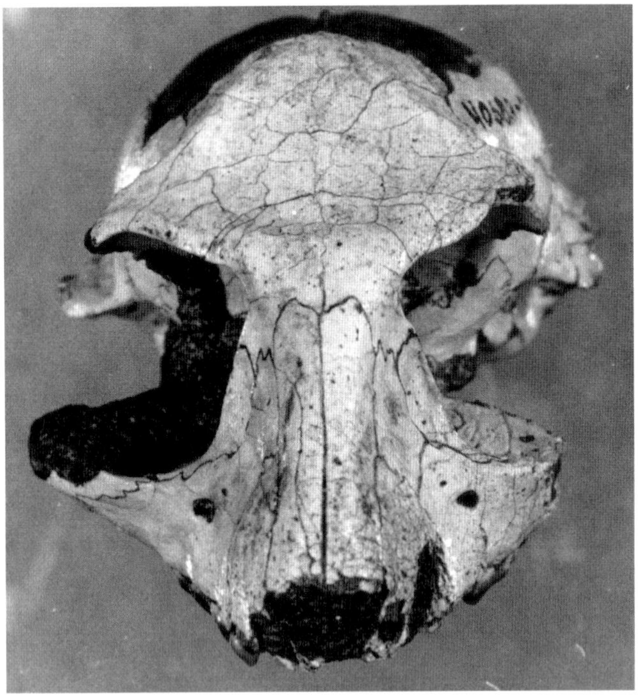

FIGURE 14.3. Anterodorsal view of *Rooneyia* (courtesy of Dr. Timothy Rowe, Vertebrate Paleontology Laboratory, Texas Memorial Museum, University of Texas at Austin) before the right side of the frontal bone was removed to expose the forebrain endocast. The metopic suture is fully fused and obliterated, except perhaps for a line of a few millimeters continuous with internasal suture, which is most likely a postmortem crack. The dorsal margins of the orbits are not everted and there are no anterior paralveolar extensions as in the large-eyed tarsiids.

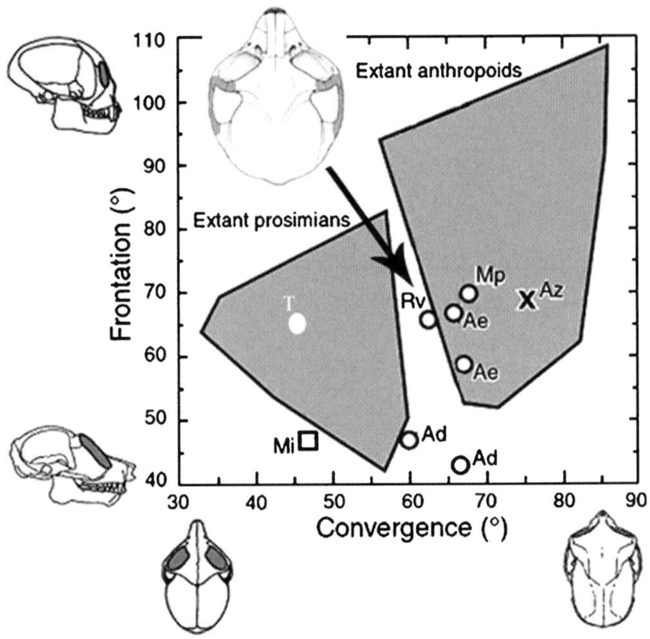

FIGURE 14.4. Orientation of the orbital plane in primates (modified from Ross, 2000, and Szalay, 1976). Minimum convex polygons outline the widest spatial distributions of points for extant species of anthropoids and non-anthropoids. Symbols: T, *Tarsius* spp.; Mi, *Microchoerus* sp.; Ad, *Adapis* sp.; Ae, *Archaeolemur edwardsi*; Az, *Aegyptopithecus zeuxis*; Mp. *Mesopropithecus pithecoides*; Rv, *Rooneyia viejaensis*. Approximate midpoint position of three *Tarsius* species based on plot in Ross (1995). In spite of differences in relative eye size, the orbital plane of *Tarsius* is laterally directed as in *Microchoerus*, both resembling strepsirhines. The higher degree of orbital frontation in *Tarsius* is a product of the unique downward tilting of the face on the neurocranium coupled with the extensively everted supraorbital flange, thus producing a superficial resemblance to anthropoids in this measure. Absent these specializations, and with a much more primitive overall cranial design, *Rooneyia* resembles higher primates more than any other non-anthropoids because of its prolonged frontal bone plane and recessed orbital fossae.

only in the absence of a structure that closes off the fossa laterally, i.e., the spoon-shaped zygomatic. Since the orbital fossae of *Notharctus* and other strepsirhines are placed so far forward on the snout, well away from the braincase, there is nothing comparable to this in their morphology (Figure 14.1a, b).

The posterior envelopment of the eye by the frontal-alisphenoid complex at the craniofacial junction is related to the width of the postorbital constriction, which tends to be larger relative to braincase width in modern haplorhines than in strepsirhines (Figure 14.6). With the exception of *Victoriapithecus*, all anthropoids in our plot fall above the slope of the line fit through our combined sample of strepsirhine and haplorhines. Another distinction is that the relationship between postorbital breadth and braincase width is somewhat more complex in strepsirhines than in haplorhines. In the strepsirhines, the correlation coefficient between these variables is 0.55, for an r^2 of only.31. In the haplorhines, the coefficient is 0.95, resulting in an r^2 of.90. Thus, among haplorhines the constriction is more tightly constrained by braincase width, and vice versa.

It is noteworthy that among the Eocene euprimates, *Notharctus* and the other adapids consistently fall well below the regression line. This corresponds with the notion that early strepsirhines are more primitive than early haplorhines in having a relatively narrow craniofacial junction, although this condition is probably exaggerated in the large-jawed, heavily muscled and small-brained (e.g., Martin, 1990) *Adapis* and *Leptadapis*. While the much wider postorbitum of *Tarsius* is also unusual for a haplorhine of its body size, this is undoubtedly a function of several associated features: hypertrophic eyeballs, an unusually wide forebrain (Starck, 1975), and the bent craniofacial axis (Spatz, 1969; Starck, 1975). When this outlier is eliminated, it is evident that the relatively wide postorbitum of typical haplorhines, which is also related to brain shape – their relatively broad frontal and temporal lobes (e.g., Radinsky, 1970) – is derived for euprimates.

When these features are considered together, a new picture of the spatial relationships of orbital and temporal fossae

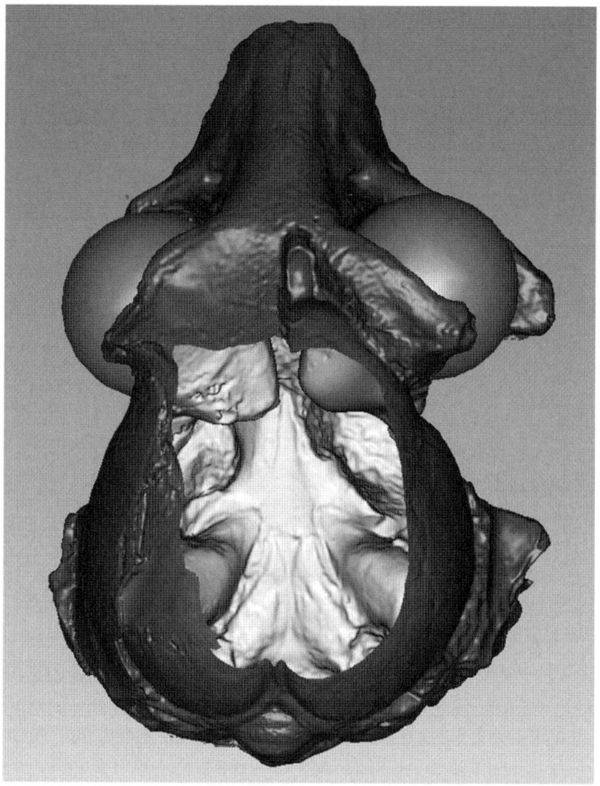

FIGURE 14.5. Cutaway of a three-dimensional model (same as in Figure 14.1) of the braincase of *Rooneyia*, with the basicranium shown in obverse for orientation. Note the V-shaped conformation of the orbital fossa, with it posterolateral wall formed by the braincase, the recessed virtual eyeballs, and the anterior overhang of the frontal bone.

emerges, different from that generally depicted in the literature. The chief determinants of their separation are the orbit's anteroposterior position relative to the braincase and the width differential between face and braincase as manifest by the degree of postorbital constriction at the craniofacial junction. Thus, as shown in Figure 14.1, in *Notharctus*, with eyes in the forward and lateral position and the constriction narrow, orbital and temporal fossae are adjacent. In *Rooneyia*, with the eyes pulled back toward the braincase and situated medially, and the constriction broad, a large part of the globe is shielded from the temporal fossae by the braincase. The literature (e.g., Cartmill, 1980; Ross 1993) seems to assert that these spaces would be broadly continuous in strepsirhines and haplorhines were it not for a *de novo* architectural partition formed of a bony screen, the postorbital plate. This view is at least partly inaccurate because it does not recognize that the temporal fossa has been lateralized in haplorhines by a broadening of the braincase, while eyeball position also differs from strepsirhines by having been shifted medially and posteriorly.

These fundamental differences hold even for larger-eyed strepsirhines and haplorhines. As shown in Figure 14.7, even though the orbits of the large-eyed loris have also shifted medially by virtue of the expanded transverse diameter of the eyeball, the narrow postorbital constriction is retained; therefore, much of the temporal fossa is located just behind a large segment of the eye. In *Tarsius*, on the other hand, the anteriorly wide braincase backs approximately half the diameter of the eyeball and displaces the temporal fossa far laterally as well. The small size of the tarsier temporal fossa is again evident Figure 14.7c, which also raises doubts about its physical impact on orbital contents.

14.3.1.3 Zygomatic

As the lateral segment of the postorbital bar is built from the ascending frontal process of the zygomatic (FPZ), this element is important to the interpretation of the early evolution of postorbital closure. Unfortunately, there is no way of making an accurate reconstruction of this feature in *Rooneyia*, for it is completely gone. Too much bone is also missing on both sides of the skull where the maxilla meets the root of the zygomatic, so the morphology cannot be established. However, the remains of the lateral process of the frontal (LPF) and comparisons with other primates enable us to clarify some details and propose several points for consideration.

Notharctus has a typical euprimate postorbital bar like that of most strepsirhines and fossil tarsiiforms: a uniformly narrow, flattened shaft of bone connecting the FPZ with the LPF. All the tarsiiforms are similar. However, as noted by Szalay (1976), the configuration of the LPF differs in *Rooneyia*, and this suggests that the "postorbital bar" of *Rooneyia* also differs. It is a large, flange-like process that we surmise is part and parcel of the overall enlargement of the frontal bone. However, in our view, Szalay's (1976) reconstruction of the postorbital bar and anterior zygomatic arch in *Rooneyia* is unnecessarily conservative. Figure 14.8 shows his diagrammatic reconstruction of this area and its appearance in two living strepsirhines, a galago and a loris. Szalay's *Rooneyia* differs little from the galago. But the clearly enlarged LPF would better match an equally well developed FPZ, perhaps as exemplified by the loris. While we do not suggest that the shape of the FPZ of *Rooneyia* was quite that similar to a loris, where hypertrophic eyeballs have played a large role in shaping this region, there are no obvious reasons requiring *Rooneyia* to have a slender postorbital bar as depicted. Rather, given the size of the LPF flange, it may have been considerably wider.

14.3.1.4 Frontal Process and Postorbital Flange

Figure 14.9 examines the dorsoventral extent of the LPF and its configuration as a postorbital flange. A partial, laterally broken flange with a distinct vertical lamina is present on both the right and left sides of the specimen. The right side probably preserves more of its bone overall but the left preserves undamaged the flange's inferior junction with the braincase. On the left side, the postorbital flange extends vertically downward until the line of the frontal-sphenoid suture, i.e.,

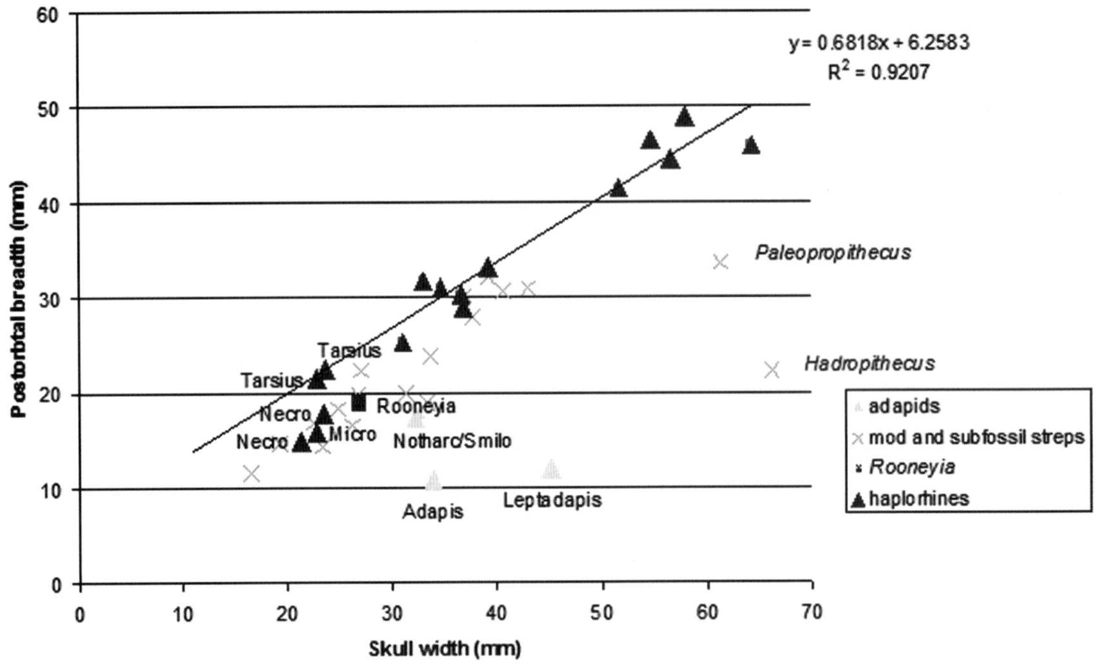

FIGURE 14.6. Bivariate plot of braincase width and postorbital breadth. Haplorhines have a relatively wider postorbitum than strepsirhines, and this holds true even for Eocene forms. Consequently, adjacency of temporal and orbital fossae is reduced since the orbits are situated at the craniofacial junction, and shielded medially, while the temporal fossa is located more laterally. Tarsiers have an unusually wide postorbitum, as their forebrains are distinctly broad, thus effecting the largest transverse spatial separation of orbital and temporal fossae. The regression is based on the anthropoid sample.

as far ventrally as possible without crossing a bone boundary. This is the area where in anthropoids the zygomatic sutures to the sidewall of the braincase. It is where in some platyrrhines there is a lateral orbital foramen (in *Saimiri* and *Cebus* possibly transmitting a branch of the superficial temporal artery, pers. obs.). In other words, the frontal process in *Rooneyia* is broadly similar in its spatial extent to the configuration of platyrrhines. Below this point, however, *Rooneyia* differs markedly for there is no alisphenoid component joining the frontal or zygomatic.

However, the ventral depth of the LPF in *Rooneyia* is extensive. In Figure 14.9b we have reoriented the skull of *Rooneyia* from the way it is usually depicted (e.g., Szalay, 1976) and into the Frankfurt plane, aligning it with *Necrolemur*, which tends to resemble *Notharctus* and other euprimates. Line “a” marks the lower horizon of the LPF in *Necrolemur*; line “b” marks it in *Rooneyia*. It is evident that the LPF in *Rooneyia*, as Szalay (1976) emphasized, partitions a proportionately larger amount of the orbital fossa from behind. Figure 14.9c makes this point by illustrating the right side, where the LPF is broken ventrally as well as laterally but still covers a proportionately large segment of a virtual eyeball fit into the orbit. We know of no other Eocene primate, strepsirhine or haplorhine, which matches this pattern.

14.3.2 Tarsiers and Tarsiids

There is an increasing body of evidence supporting the notion that tarsiers are most closely related to a collection of Eocene tarsiiform genera, which we have moved to classify as tarsiids (Table 14.1). Most active workers who disagree with this hypothesis believe that tarsiers are more closely related to anthropoids, the TAH (e.g., Cartmill and Kay, 1977; MacPhee and Cartmill, 1986; Ross, 1994; Kay et al. 1997; Ross et al., 1998; Kay et al., 2004; Ross and Kay, 2004). Therefore, placement of tarsiers is crucial to an understanding of the origins of anthropoids and the anthropoid orbit, as implied by the Cartmill (1994) quote that opens this paper. Of course, the proposed link between living tarsiers and fossils designated as tarsiiforms is not a novel hypothesis. It was widely (though not always dogmatically) assumed generations ago, albeit stated in less modern terms and argued without today’s cladistic formalisms (e.g., Gregory, 1922; Le Gros Clark, 1934, 1959; Simons, 1972). In the past, genera often singled out as having a close relationship with *Tarsius* included *Tetonius, Necrolemur* and *Pseudoloris*. For example, influenced by Teilhard de Chardin (1921), Le Gros Clark said (1934:269): “…it seems not unlikely that *Pseudoloris* represents the direct Eocene precursors of the modern *Tarsius*.” This roster of relatives has been enlarged recently following new character analyses of the skull and postcranium (Rosenberger, 1985; Beard et al., 1991; Beard and MacPhee 1994; MacPhee et al. 1995; Dagosto and Gebo, 1994; Dagosto et al., 1999), including some parsimony-based (PAUP*) studies. Among the postcranial synapomorphies identified in these studies as derived homologies shared by the fossils and *Tarsius* are features of the knee, calcaneus, tibio-fibula and, in the skull, several involving the basicranium and bulla, the glenoid fossa, pterygoid plates and the choanae (see summary in Dagosto et al., 1999).

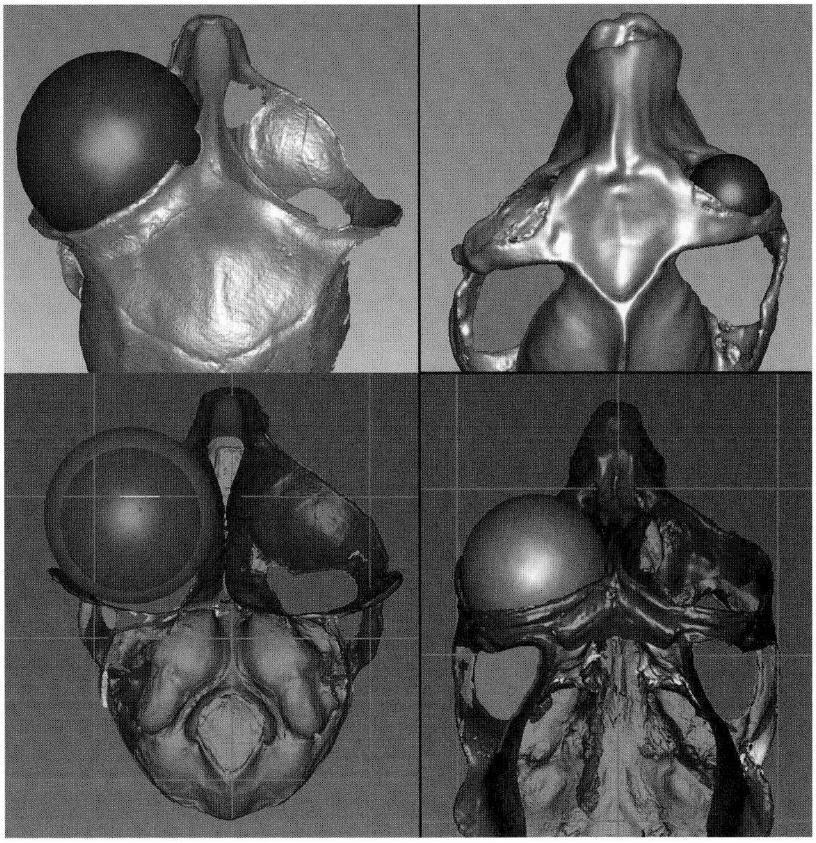

FIGURE 14.7. Images captured from three-dimensional laser scans of *Tarsius* (left top and bottom), *Notharctus* (top right) and *Loris* sp. (bottom right) to illustrate spatial relationships of orbital and temporal fossae and virtual eyeballs, and influence of postorbital breadth, with regard to the interference hypothesis of the postorbital septum. Rostrad placement of the orbital fossae in the strepsirhines is apparent irrespective of relative eyeball size. With a relatively wide anterior braincase, tarsiers have very small, laterally displaced temporal fossae. In dorsal view, the margin immediately surrounding the eyeball can be seen as an everted rim, continuous with the lateral process of the frontal. Fossil tarsiids with eyes that are probably roughly similar to *Tarsius* in their proportions, e.g., *Shoshonius*, and those with a less exaggerated size, e.g., *Necrolemur*, also show a superior everted margin, indicating this is a transformation series exclusive to the large-eyed tarsioids.

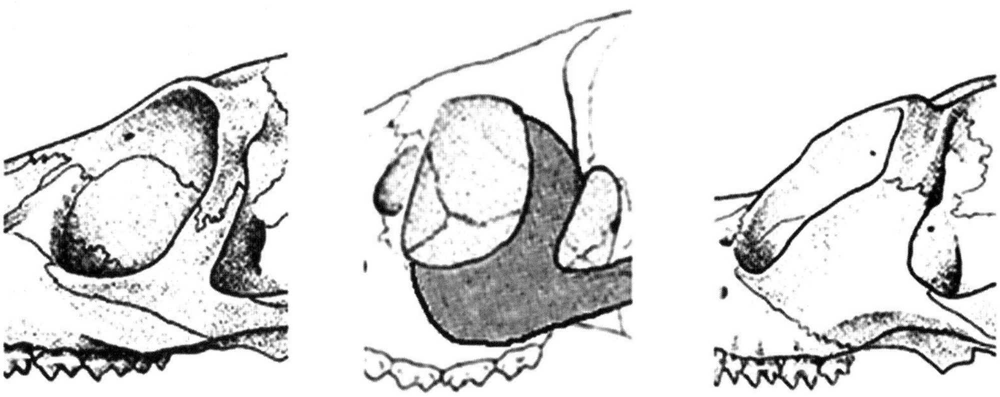

FIGURE 14.8. Lateral views of the postorbital bar in *Lemur* sp., *Rooneyia* and *Loris* sp. (left to right). The narrow ascending process of the zygomatic is the norm among strepsirhines while the loris condition illustrates how sizeable the zygomatic can become in response to eyeball enlargement. The small-eyed anthropoids are a different, parallel example of zygomatic hypertrophy. This portion of the *Rooneyia* postorbital bar may have been reconstructed too conservatively by Szalay (1976; see Figure 14.8), as its dorsal area of attachment in large, as depicted, raising the possibility that *Rooneyia* may have had a more loris-like pattern, predisposing it to a more extensive lateral closure of the orbit by the zygomatic that could approximate anthropoids. Note that the image of *Rooneyia* has been modified by stippling to better show the full extent of missing bone from the zygomatic/orbital region of the original, and to better reveal the conservatism of the original reconstruction. (Adapted from Szalay, 1976, Mahe, 1976).

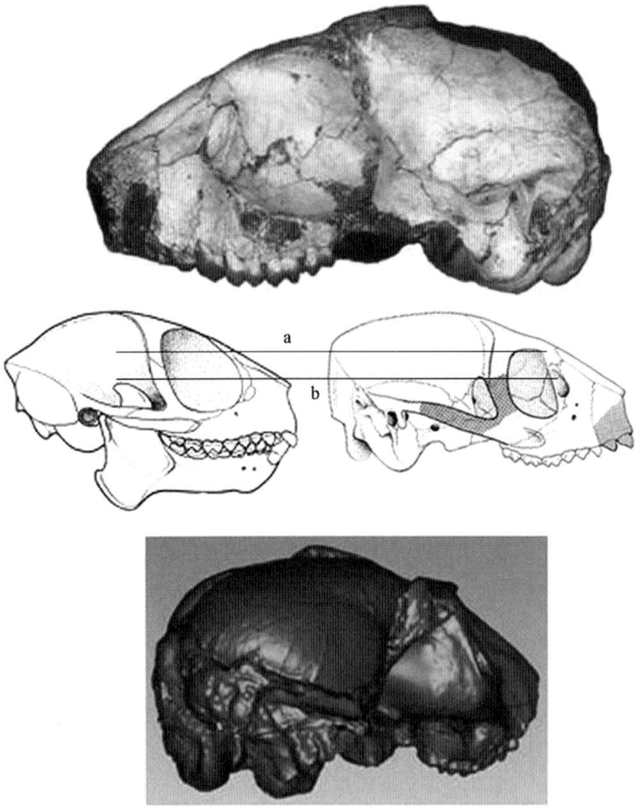

FIGURE 14.9. Top. Left lateral view of *Rooneyia* showing the great vertical depth of the lateral process of the frontal bone that forms the flange-like upper root of the postorbital bar, effecting partial postorbital closure superiorly (courtesy of Eric Delson). Middle. Lateral views of *Necrolemur* (left) and *Rooneyia* (right), drawn to same approximate cranial length and oriented on the Frankfurt plane, comparing the ventral horizon of the lateral frontal process in *Necrolemur* (line a) and *Rooneyia* (line b). Bottom. Posterior view of the broken frontal flange on the right side of *Rooneyia*, with a virtual eyeball set in place to illustrate how much of the eye is closed off from behind by the process and how relatively large the zygomatic process may have been in its area of attachment to the flange (middle and bottom adapted from Rosenberger, 2006).

It is rare for these studies to list cranial characters connected with the most striking morphological adaptation of the tarsier skull, the enormous orbits, as evidence of a phyletic link between fossil tarsiiforms and *Tarsius*. This is consistent with metric assessments that have attempted to infer relative orbit/eyes size among tarsiiforms and other fossil primates (e.g., Kay and Cartmill, 1977; Martin, 1990; Kay and Kirk, 2000). Although various Eocene tarsiiforms have relatively large eyes in the nocturnal euprimate range (e.g., Kay and Kirk, 2000; Heesy and Ross, 2001; Ravosa and Savakova, 2004), with the possible exception of one genus none of these studies have demonstrated that any fossils have eyes as anywhere nearly as large relatively as a tarsier's – "Most omomyiforms do not exhibit the enormously enlarged orbits (and thus eyes) characteristic of both extant tarsiers and owl monkeys..." (Kirk and Kay, 2004:582). This conclusion was wrought in consideration of Kay and Kirk's (2002) metric demonstration that orbits are relatively large in *Necrolemur, Microchoerus* and *Tetonius*, and unusually large relative to body size in *Shoshonius*.

Advocates of the TAH might explain the presence of large-eyed fossil tarsiiforms as evidence of parallelism, or a haplorhine last common ancestor which had large eyes that became reduced subsequently in anthropoids. Either way, it is reasonable for PAUP* users to code eye size in Eocene tarsiiforms and other euprimates (e.g., Ross et al., 1998 *et seq.*) in a three-state scheme as follows: "0" equals small; "1" equals large; "2" equals extremely large, the latter found only in *Tarsius*. (*Shoshonius* has only recently been added to the dataset used in the studies cited.) However, the morphological organization of tarsiid dentitions and skulls, as presented here, suggests that these necessarily reductionistic methods of metric assessment and parsimony analysis underestimate how relatively large the eyes of some fossil tarsiiforms actually are. Here we consider several cranial features belonging to a pattern which points to eyeball hypertrophy in many of them.

Before proceeding, however, we elect to elaborate on what should be obvious. Evolutionary biologists would probably all agree that there is no reason to expect that even a sister-genus of the tarsier *must* have an eye as large as a *Tarsius*. Meaning, even if its eyes were smaller, that would not negate a close phylogenetic relationship. To the contrary, it is expected. And that provides a rationale for homologizing and weighting heavily evidence of relatively large eyes in fossil tarsiiforms, although this is rarely done.

Sprankel's figures (1965) for juvenile *T. bancanus* indicate an eyeball:brain size weight ratio of 90%. While it is prudent to assume as a working hypothesis (but not an axiom) that this ratio is likely to be utterly unique among all primates living and extinct, as phylogenetic evidence the state coded (relatively) "large" in another taxon is fully acceptable as an "ordered" synapomorphy when the anatomical patterns associated with it suggests that the large-eyed similarity to *Tarsius* is homologous and that the ancestral condition of the larger group in question is thought to have unenlarged eyes and orbits. Thus, with reference to Kay and Kirk's (2002) careful metrical study, it becomes difficult to interpret their data on relative orbit size to reflect anything other than a phylogenetically meaningful transformation series, with *Necrolemur, Microchoerus* and *Tetonius* reflecting one shared, homologously derived state, large, relative to primitive euprimates; *Shoshonius* reflecting a more derived/more enlarged state; and, *Tarsius* reflecting the most modified state *if indeed its eyes were larger relative to body size than Shoshonius*. In the absence of a way to resolve this last caveat, the parsimonious interpretation would assume that *Tarsius* and *Shoshonius* share a derived version of the "large" condition. And, if it turns out that the living tarsier is a variant of that state, the tarsier condition would be seen as an autapomorphy derived from the *Shoshonius-Tarsius* condition. Lest the point not be clear, this mode of argument also implies that the other three genera mentioned share "large" as a derived state with *Shoshonius-Tarsius* as well, but one node removed from their common ancestor, which would have had the "small" condition.

What is vitally important here is that there are other features indicating that an advanced state of eyeball enlargement or hypertrophy obtained broadly among fossil tarsiids, especially for taxa lacking orbits sufficiently complete to be measured by conventional means. This reinforces the supposition that the derived metrics of the eyes in all these tarsiiforms is homologous. As a start, our anatomical perspective can be reduced to these points of reference: (1) arcade shape; (2) paralveolar morphology; (3) osseous interorbital septum; (4) choanal shape; and, (5) everted dorsal orbital margins.

14.3.2.1 Arcade Shape

Modern tarsiers tend to have what might be called a modified bell-shaped dental arcade (Figure 14.10). We add the qualifying term "modified," because it is best to look at this feature transformationally. This shape reflects an extreme narrowness of the anterior snout, i.e., closely set antemolar teeth, coupled with an exceptionally broad posterior palate, i.e., width across the molars. The postcanines diverge so dramatically toward the rear that bimolar breadth almost equals the maximum width of the braincase. The narrow anterior snout, sometimes described as tubular (e.g., Rosenberger, 1985), is not indicative of diminished function in the anterior teeth. Rather, it relates to a robust premaxilla with tall, stout, well rooted medial incisors and strong canines arranged in a particular way, which probably concentrates muscular force to enhance the efficacy of these teeth in puncturing prey while working against the lower anterior teeth (e.g., Thalmann, 1994). The posterior breadth of the arcade relates to megadontia and hypertrophic eyeballs. Tarsiers, and most likely some fossil tarsiiforms, also have large molars for their body size (Gingerich et al., 1982; Dagosto and Terranova, 1992). Additionally, with each eye approaching the volume of the whole brain, it stands to reason that the breadth of the orbital floor and the palate, to which the latter is fused, has been grossly modified to reflect transverse eyeball diameter.

In all fossil tarsiids for which the anatomy is known, in contrast to all anthropoids and strepsirhines, the arcade is pinched in the middle and even more precisely bell-shaped, more than in *Tarsius*. The anterior snout is narrow, the molars spaced far apart and the transition from premolar to molar is contoured to bridge the width differential. It is easy to visualize the differences between the fossil tarsiids and *Tarsius* as a transformation series where differences are related to simple contrasts in premolar-molar tooth widths, tooth proportions and, eventually, the massively enlarged molars in tarsiers. The bell-shaped silhouette is muted in some forms because the premolar-molar shape transition also conforms to another novelty of the tarsiiform face, the anteriorly extended paralveolar region of tarsiers as discussed below. Therefore, we interpret the formative bell-shaped arcade that is widespread

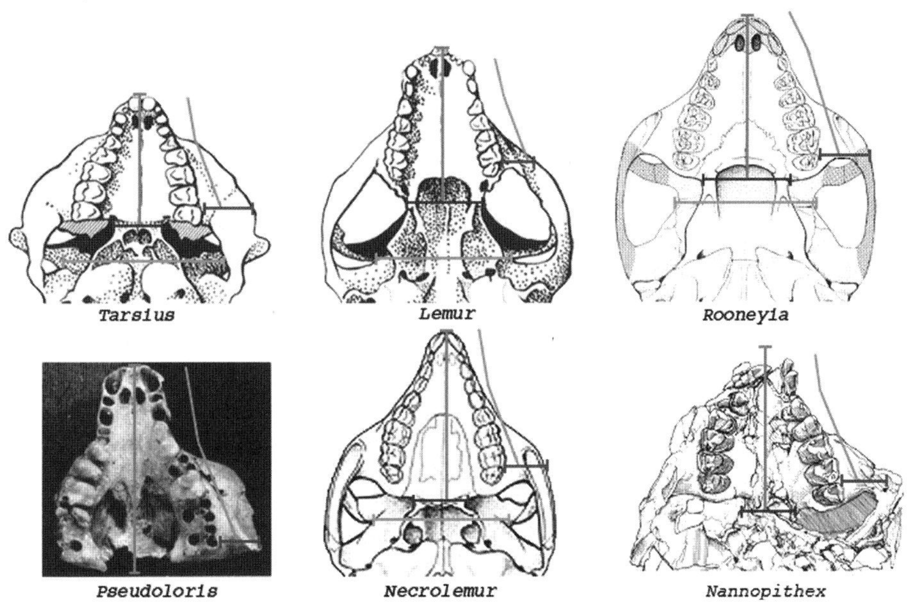

FIGURE 14.10. Basal views of (clockwise from top left) *Tarsius*, sp., *Lemur*, sp., *Rooneyia, Pseudoloris, Necrolemur*, and *Nannopithex* showing variations in the tarsiiform "bell-shaped" dental arcade, paralveolar extension of maxilla beyond the toothrow, relatively narrow choanal breadth and wide postorbital breadth. *Lemur* is typical of strepsirhines and taken to represent the ancestral euprimate pattern. *Rooneyia* retains the primitively wide choanae. Paralveolar extension and choanal narrowing are postero-lateral and postero-medial factors connected with eyeball hypertrophy, which is most exaggerated in *Tarsius, Pseudoloris* and perhaps *Nannopithex*. (Adapted from Cartmill, 1980, Szalay, 1976, Thalmann, 1994).

among tarsiids as a derived feature among euprimates relating to an increase in the transverse diameter of the orbits. As noted below, when the orbits become further exaggerated in genus *Tarsius*, the phenomenon is also played out in the anteroposterior axis in a localized way that makes the bell-shape less pronounced.

In *Rooneyia*, there is no sharp transition from premolar to molar so there is no indication of a bell-shaped arcade behind the canines. Since the tip very front of the snout is not preserved, we are limited in terms of what can be inferred about it. However, there is no reason to suppose it was narrow or tubular, as in the tarsiids. To the contrary, judging by the postcanine contour, and guided by the frugivorous cheek teeth, it would seem more likely that the snout was not set up to have tarsier-like piercing teeth and that it accommodated relatively wide incisors as Szalay (1976) believed. A wide anterior snout would resemble the condition of adapids, and is therefore likely to be ancestral in euprimates.

14.3.2.2 Paralveolar Morphology

As noted, a shape feature of *Tarsius* that is part of the bony underpinning of orbital hypertrophy involves the paralveolar surface of the face (Figure 14.10). To some degree, one might consider our delineation of this feature as a redundancy, a correspondent to the bell-shaped arcade. This may be a legitimate point, but we believe there is descriptive value in separating the two, which are surely functionally correlated – and that is our point: all five of the features discussed here are functionally correlated with one another and with hypertrophic eyes. But the anatomical expressions differ among the genera, and this permits and provokes evolutionary character analyses.

With the orbital floor grossly expanded transversely and fused to the hard palate medially, there is essentially no "suborbital" facial depth to speak of in *Tarsius* (see Seiffert et al., 2005). Rather, lateral to the dental arcade, the maxillary-zygomatic complex is extended anteriorly and transversely to form an enlarged overhanging plane which conforms to a massively expanded orbital floor (Figure 14.10). Since the equator of the eyeball lies high above the orbital floor, the line of maximum bi-orbital width actually lies well above the junction between zygomatic arch and maxilla, so that the surface of the latter has essentially become everted.

Fossil tarsiids do not uniformly present such an exaggerated condition (Figure 14.10) and none (as well as can be documented by measuring or estimating external orbital diameters) appear to have such hypertrophic orbits, except perhaps *Shoshonius* (see Beard et al., 1991; Kay and Kirk, 2000). However, the pattern is well developed in *Pseudoloris*, which genus may exhibit the closest resemblances to *Tarsius* in having an anteriorly and laterally expanded paralveolar shelf. The crushed skull of *Nannopithex* may have a more developed expansion than *Necrolemur*, where it is modest but better developed than the extent seen in *Rooneyia*. Others, such as *Tetonius, Microchoerus* and *Necrolemur*, of varying body sizes, show varying degrees of laterally expanded orbital shelves but none appear to be anteriorly expanded, and none appear to have grossly everted zygomatic roots. The lack of anterior expansion may mean that these taxa did not evolve the elongate, tube-shape eyeballs of *Tarsius* (Castenholtz, 1984), but retain a more primitive, spherical eye, albeit an impressively large one. It may also simply signal a smaller relative eye size.

With a raised orbital floor and small orbits, the maxillary morphology of anthropoids is fundamentally different. *Rooneyia*, however, has a modest paralveolar expansion. We attribute this to the combination of a broad and shallow face, and a relatively wide anterior braincase and (inferentially) a wide zygomatic arch, which makes for a laterally positioned anterior zygomatic root. Since this is associated with unenlarged orbits and is closest to the non-haplorhine condition, we believe that the relatively small paralveolar shelf of *Rooneyia* represents the ancestral state for haplorhines.

14.3.2.3 Osseous Interorbital Septum

The enormity of tarsier eyes requires a medial shift in the position of the medial orbital walls, thus producing an extensive fusion of these surfaces (Le Gros Clark, 1934) into an osseous interorbital septum (see Starck, 1975). The septum occupies a large segment of the anteroposterior dimension of the orbit from its posterior apex to the anterior-inferior margin. As a consequence, the nasal fossa is restricted to the front of the snout and, as Starck (1975) and others have shown, the olfactory nerves are thus routed to the nasal fossa via a long olfactory tube above the septum. High resolution CT scans may determine how extensive interorbital septa are in the fossil tarsiids, and if they have an olfactory tube like this, but the morphology of several, e.g., *Tetonius, Necrolemur, Microchoerus* and *Pseudoloris* (Cartmill and Kay, 1978; Ross, 1994), for example, is similar enough to be highly suggestive of an osseous septum, or at least orbits closely approximated over an long antero-posterior span. Breaks in several specimens of the last three genera show that fusion definitely occurred according to Ross (1994, Figure 14.14), but it is important to determine not only the presence of fusion, which may be common at the orbital apex, but also the anterior extent of it. One *Pseudoloris* specimen (see Teilhard de Chardin, 1921) shows quite dramatically a combination of features that appears to be indisputably tarsier-like in this regard: a remnant of extensively fused orbital walls can be seen demarcating an anteriorly isolated nasal fossa. Thus it appears to be a rather solid conclusion that the eyeballs of *Pseudoloris* were exceptionally large, as Simons (e.g., 2003) has emphasized. In *Rooneyia* (Rosenberger, 2006; Figure 14.5), as with anthropoids and most strepsirhines, the medial orbital walls remain separated (only sometimes contacting posteriorly at the apex in smaller species), suggesting the more widespread and primitive condition with the olfactory nerves entering the nasal fossa through a typically positioned cribriform plate, posteriorly, at the interface with the anterior cranial fossa.

14.3.2.4 Choanae

Ross (1994) and Szalay (2000) have related the small, peaked choanae in *Tarsius* to hypertrophic eyes. Connected with this is the correspondingly narrow distance between the pyramidal processes. These features are fixed by the mediolateral position of the orbital walls. Because the walls have been drawn to fuse in the midline of the skull, the posterior nasal aperture is narrow and small, and the pterygoid plates which extend behind them are also drawn to the middle. This is a derived condition among euprimates and it does not occur in strepsirhines, *Rooneyia* or anthropoids (Rosenberger, 1985; Beard and MacPhee, 1994; Ross, 1994). All of the tarsiids that preserve either the pyramidal processes or the choanae resemble *Tarsius* (Figure 14.10), i.e., *Necrolemur, Microchoerus, Nannopithex, Tetonius* and *Shoshonius*. This is also indirect evidence that they all have an osseous interorbital septum, at least apically, within the orbit. *Teilhardina* appears not to have this pattern.

14.3.2.5 Everted Dorsal Margin

By comparison with the other features discussed here, the often subtle lipping (not so in *Tarsius*, where it is dramatic) of the superior margin of the orbit seems less trenchant. However, the comparative evidence suggests that it is correlated consistently with eyeball hypertrophy. For example, other extant primates that show at least a slight eversion of the superior margin are *Loris* and *Aotus*. With the eyeballs of *Tarsius* jutting beyond the perimeter of the orbital fossa (Castenholtz, 1984), it seems reasonable to infer that the strongly everted superior margins are a direct correlate, as discussed above. Similar margins appear in *Necrolemur, Microchoerus, Hemiacodon*, and *Shoshonius*. In the smaller-eyed *Rooneyia* and most strepsirhines, the lip is not everted, suggesting that the tarsiid condition is derived.

To summarize, these features all appear to be associated with large eyes in tarsiids, suggesting a functional pattern that is part of the orbital hypertrophy syndrome, a derived euprimate adaptation that is most highly modified in *Tarsius*. Several of these features are graphically compared in chart form (Figure 14.11), derived from the images shown in Figure 14.10. In contrast, *Rooneyia* differs in nearly every detail, and is more primitive than any of the tarsiiforms (see Rosenberger, 1985). The morphology of *Teilhardina* is still not fully described but it appears not to manifest many of these features. Thus, *Teilhardina* may serve as a good model for the ancestral, small-eyed, haplorhine condition.

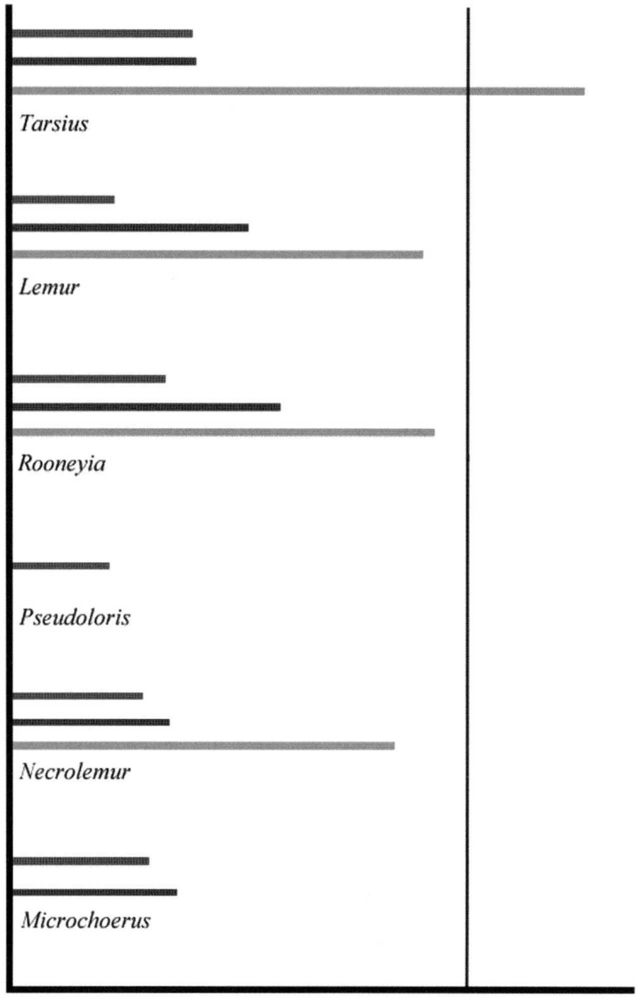

FIGURE 14.11. Graphic representation of features marked in Figure 14.10 to show their proportions. The position of the vertical line along the horizontal axis of the chart represents a standardized facial length, generated by scaling all the skulls to the same palatal length (see Figure 14.10). The horizontal bars represent the ratios, measured against this standard palatal length, of paralveolar extension of the maxilla (red), choanal breadth (blue), and postorbital breadth (teal). The relatively wide choanae of *Lemur* and *Rooneyia* are apparent, in contrast to the large-eyed tarsiids. *Tarsius, Necrolemur*, and *Nannopithex* have strikingly similar proportions of choanal breadth and paralveolar breadth.

14.4 Discussion

14.4.1 Systematics Of *Rooneyia viejaensis*

To our knowledge, *Rooneyia* is the only non-anthropoid primate, with the exception of *Tarsius*, that has been mentioned for its potential as being structurally "on the way" to evolving postorbital closure. We have made it central to our morphological discussion, but its systematics must also be "right" if our hypothesis is to have extended credibility. As reviewed above, Szalay's (1976) interpretation of *Rooneyia* as an omomyid has been the most influential (see Szalay and Delson, 1979; Fleagle, 1999; Gunnell and Rose, 2002), and his overall view of haplorhine evolution was also important in establishing what he called omomyids as a likely ancestral stock for Anthropoidea (Rosenberger and Szalay, 1980). Taken literally, that notion is itself a complex proposition, for it implies that Szalay's omomyids are paraphyletic. In the decades since his landmark monograph, workers have come to be more convinced that this is indeed the case (e.g., Rosenberger, 1985; Beard et al., 1991; Beard and MacPhee, 1994; Dagosto et al. 1999; Ross et al., 1998). So, if Omomyidae *sensu* Szalay is not monophyletic, is *Rooneyia* one of the taxa whose inclusion forced the family's paraphyly – and obscured the close phylogenetic links between many "omomyids" and tarsiers?

The most recent treatments bearing on the systematics of *Rooneyia* are pertinent here. Gunnell and Rose, for example, classify *Rooneyia* within a subfamily (Omomyinae) of the Omomyidae but state (2002:70), "The taxonomic position of *Rooneyia* is in question…and further evidence is required to elucidate its phylogenetic relationships." Still, they do not go so far as to remove the genus from the family in their classification, as was done with several other questionable forms (*Ekgmowechashala, Altanius, Kohatius, Afrotarsius, Altiatlasius*; see Delson et al., 2000) – these they label "Problematic Taxa."

A series of parsimony (PAUP*) studies (e.g., Ross, 1994; Kay et al., 1997; Ross et al., 1998; Kay and Kirk, 2000; Kay et al., 2004) more directly challenged the *status quo*, stunningly removing *Rooneyia* from both of the accepted major ordinal-level taxonomic groups of primates on account of its tendency to link cladistically with disparate taxa (see above) depending on how the input character/taxon matrices were constructed and coded. Thus, as a new taxonomic alternative, *Rooneyia* has been separated from both Semiorder Strepsirhini and Semiorder Haplorhini and placed in Semiorder *incertae sedis* (e.g., Kay et al., 2004). In a different judgment, *Rooneyia* was installed even more remotely, into a taxonomic gehenna – "Plesion indet." (Kay and Kirk, 2000).

While a detailed examination of the characters, character coding and missing-data cells used in the large character/taxon data matrices at the heart of these projects (e.g., Ross, 1994; Kay et al., 1997; Ross et al., 1998; Kay and Kirk, 2000; Kay et al., 2004) may shed light on why so many divergent solutions to the *Rooneyia* problem were found, several larger points are worth mentioning here. In one sense, the mere fact that the cladistic position of *Rooneyia* was highly unstable means only that specific cladistic hypotheses could not be corroborated using these particular methods and conditions. It does not mean that the relationships of *Rooneyia* are imponderable, unknowable or to be found among some heretofore unrecognized group of primates. From another perspective, all of these proposed solutions are conjectural. Other than in Ross's (1994) robust character analysis, which dealt with a relatively small list of the full data set used in this series of studies, little effort was given to refute the evidence and analyses provided in other studies except in a *post hoc* fashion, after the cladograms were automatically constructed. Thus, even if Omomyidae *sensu* Szalay is paraphyletic, the phylogenetic ambiguity promoted in this string of papers is not a compelling rationale for abandoning the still unfalsified interpretation of *Rooneyia* as a haplorhine: there is a difference between saying "Wrong." and saying "We can't tell." But what kind of haplorhine is it? Our concurring analysis of the morphology of the orbital region also suggests that the affinities of *Rooneyia* are outside of Szalay's omomyids. We have proposed that *Rooneyia* is more closely related to Anthropoidea than any other euprimate genus for which we have cranial evidence (Rosenberger, 2006). But, importantly, the refinement we offer of the hypothesis Szalay began to conceptualize decades ago (Szalay, 1976; Rosenberger and Szalay 1980) is that the ancestral stock or nearest relatives of Anthropoidea that he anticipated would be found among the "omomyids" in fact appears to be one that he included in the family – *Rooneyia*.

The orbital morphology discussed above presents several correlated derived characters that *Rooneyia* appears to share exclusively with anthropoids. These may be reduced to anatomical shapes, proportions and spatial relationships of the frontal bone, and they relate to how the orbit is constructed on nearly all its sides – the aperture (high convergence plus frontation), the roof (large size and anteriorly extended frontal), its posterior (large, dorsoventral and laterally extensive lateral frontal process or flange), and, conjecturally, its lateral segment (potentially somewhat enlarged ascending process of zygomatic). Postero-inferiorly, the orbital floor is still open; there is a large inferior orbital fissure. The orbital fossa is still not fully drawn in beneath the forebrain, which is still almond-shaped but noticeably less so than in strepsirhines, and the space is probably not sealed by anything more than an incipiently large zygomatic. But, as Wilson (1966) and Szalay (1976) noted, the orbital morphology of *Rooneyia* fulfills what was then seen as the primary structural requirement of an antecedent to anthropoid postorbital closure. Dorsally, perhaps half of the height of the opening was closed by the transversely aligned frontal flange. How to "get there from here?" Rosenberger (2006) suggested that further widening of the forebrain in anthropoids may have induced a rotation of the "postorbital flange" into a parasagittal plane, thus making it possible to (1) widen the frontal-zygomatic suture complex and (2) reshape the zygomatic into the spoon-like enclosure that is universal in anthropoids.

We have rendered the characters this way in part to emphasize that they are likely to be interconnected. However, it is legitimate to question our character analysis, specifically the homologies of the unit characters making up this suite. For example, why interpret the high levels of orbital convergence plus frontation as homologous in *Rooneyia* and anthropoids? Or, why isn't the lateral frontal process making the postorbital flange of *Rooneyia* regarded as the homologue of the postorbital alisphenoid-septum of *Tarsius*? Our reasoning regarding the orbital aperture is partly philosophical and partly based on related hypotheses. Apart from a *few* of the ambiguous results of the PAUP* studies, there are simply no alternative cladistic hypotheses offered in the literature to compete with the notion of *Rooneyia* as haplorhine. *Rooneyia* shares many primitive euprimate features with strepsirhines and haplorhines (Rosenberger, 1985; Beard and MacPhee, 1994), along with others that show it to be a haplorhine (Szalay, 1976) and a protoanthropoid (Rosenberger, 2006). Therefore, since there is no reason to suspect that any particular resemblance to anthropoids is a likely product of analogy, and the orientation of its orbits, for example, fits the universal expectation of a pre-anthropoid pattern, why not assume it is homologous as a working hypothesis until that hypothesis fails?

The debate about the homologies of the anthropoid and tarsier septa has been carried out with no mention of *Rooneyia*, but for Szalay's (1976) comment. Concerning the possibility that the frontal flange of *Rooneyia* is homologous with the tarsier septum, it appears that the evidence is strongly against it. *Rooneyia* is relatively small-eyed; *Tarsius* eyes are hypertrophic. As discussed above, fossil tarsiids with eyes that may approach a tarsier's in size (*Shoshonius*), or are relatively smaller (*Necrolemur*), do not have a flange of any kind, suggesting that the *Tarsius* pattern is neomorphic. Thus, this particular similarity of *Tarsius* and *Rooneyia* appears to be a classic case of convergence.

14.4.2 Origins Of Postorbital Closure

Based on the forgoing, to limit semantic confusion we suggest a redefinition of the character states relating to the separation of orbital and temporal fossae in euprimates. A frontal-alisphenoid septum evolved in *Tarsius, sui generis*. A postorbital flange of the frontal evolved in the common ancestor of *Rooneyia* and the anthropoids. Postorbital closure, distinguished and defined by a large contribution of the zygomatic, evolved in anthropoids, in connection with enlargement and rotation of the flange. While we regard both the interference and structural support hypotheses as having merit as adaptive explanations, and sometimes difficult to distinguish, the specific selective conditions that may have influenced the morphology of tarsiers and anthropoids were probably quite different. Common to both, and what is likely to have set the stage for this example of functional parallelism, is that haplorhines since their inception had an orbital fossa that was recessed posterior-medially into the face and positioned at the craniofacial junction, unlike the more primitive strepsirhines. This may have added a new range of developmental influences which came into play as tarsiid eyeballs evolved their large size and as anthropoid eyes and faces shifted further beneath an enlarging anterior cranial fossa.

14.5 Epilogue

Fleagle and Kay asked, "Where did the anthropoid condition come from?" (1994:693). At a time when primatologists were looking at "prosimian grade" primates for the answer, Szalay (1976) saw that the haplorhine concept was more powerful and turned our attention to tarsiiforms for ideas about the anthropoid's ancestral stock. He recognized that *Rooneyia viejaensis* offered a structural example of pre-anthropoid morphology, with an incipient postorbital septum. We have extended his analysis and found that the orbital morphology of *Rooneyia*, now conceived more broadly, indeed suggests that *Rooneyia* is more closely related to Anthropoidea than any other early euprimate, meaning that postorbital closure could have evolved from the group we have called Protoanthropoidea.

And "…the tarsier condition, for that matter?" – Fleagle and Kay (1994:693). To paraphrase Cartmill (1994:563): Are we now compelled to conclude that the septa of tarsiers and anthropoids are not homologous, because a convincing analysis of haplorhine phylogeny has given us convincing reasons for thinking that the last common ancestor of tarsiers and anthropoids lacked a postorbital septum? Our answer is: Yes. Unlike Szalay (1976), who believed that the enlarged frontal-alisphenoid flange of tarsiers was equivalent to *Rooneyia's* frontal postorbital flange, we suggest that an impressive suite of cranial features among fossil tarsiiforms argue otherwise. For several fossil tarsiiforms these demonstrate an exclusive monophyletic relationship shared with tarsiers, while also showing that the morphology of their last common ancestor was too derived – in a tarsier direction – to have been ancestral to anthropoids as well. The postcranial evidence, basically from the same collection of species, shows the same thing. This means that there is a powerful case for modern tarsiers evolving from an array of faunivorous, small, nocturnal, leaping quadrupeds, Eocene tarsiiforms which already had members that presaged the ultra-large eyed, hyper-leaping mode of life that is the *Tarsius bona fides*.

Did the two postorbital partitions of anthropoids and tarsiers evolve for the same reason, as Szalay (1976) and Cartmill (1980) suggested, for "…protecting the eyeballs and associated structures from the contraction of the temporalis (Szalay, 1976:349)"? From the perspective of a character analysis, which has been our vantage point, this matter can only be addressed indirectly. But it tends to indicate there is little common ground in the local and peripheral morphologies or transformational histories of the two patterns, making it unlikely that they share the same underlying causality. While postorbital closure in anthropoids is largely effected by framing the eyeball laterally with an enlarged zygomatic and backing it medially by the braincase, tarsiers are different. They have radically expanded the frontal into a plate-like shield, with a small splint of alisphenoid tethered to it. Large-eyed fossil tarsiids anticipate this pattern in developing a brim-like frontal differing only in degree from the morphology of *Tarsius*. What the fossil tarsiids lack are the maxillary process and the alisphenoid splint, both of which would seem to provide additional support for the eyeball. In other words, the ancestors of tarsiers were already developing a scaffold to cup giant eyeballs before the alisphenoid-septum was selected, either for its support or insulation value. There is no evidence that the small-eyed anthropoids passed through such a stage. Rosenberger and Pagano (in press) suggested that fusion of the frontal bone was influential as a preadaptation in basal anthropoids by producing a new biomechanical regime that shunts masticatory strains laterally towards the frontal-zygomatic sutures as opposed to being dissipated by a mobile, open interfrontal (metopic) suture.

Acknowledgments. We are grateful to the museums and individuals who have allowed us access to specimens and facilities over the years, especially to our colleagues at the American Museum of Natural History. Financial assistance was provided by the Brooklyn College Fund for New Faculty, a Tow Faculty Travel Fellowship from Brooklyn College, a PSC CUNY research award and an NSF grant, SEI (SBE) 0513660. For help with this study in particular, we also thank Dr. R.D.E. MacPhee, American Museum of Natural History; Drs. Pascal Tassy and Marc Godinot, Muséum national d'Histoire Naturelle; Paula Jenkins, Natural History Museum, London; Dr. Walter Joyce, Yale Peabody Museum; Drs. Tom Demere and Stephen Walsh, and Kesler Randall, San Diego Natural History Museum; Dr. Pat Holroyd, Museum of Paleontology, University of California. Dr. K. Christopher Beard, Carnegie Museum of Natural History, and Dr. Timothy Rowe, Vertebrate Paleontology Laboratory, Texas Memorial Museum, University of Texas at Austin, provided important illustrations, for which we are grateful. We are thankful to Xijun Ni for helpful discussions, sharing observations and rescuing us from error, and we are beholden to Marian Dagosto and Eric Sargis for their kind forbearance, generosity and clarity of purpose in developing this tribute to Fred Szalay. Eric Delson helped us immeasurably, again; thanks barely conveys how much. But our deepest gratitude goes to Fred, a great teacher, omnipresent source of inspiration and much more than a good friend.

References

Beard, K. C., 1998. A new genus of Tarsiidae (Mammalia: Primates) from the middle Eocene of Shanxi Province, China, with notes on the historical biogeography of tarsiers. Bulletin of the Carnegie Museum of Natural History 34, 260–277.

Beard, K. C., Krishtalka, L., Stucky, R. K., 1991. First skulls of the early Eocene primate *Shoshonius cooperi* and the anthropoid-tarsier dichotomy. Nature 349, 64–67.

Beard, K. C., MacPhee, R. D. E., 1994. Cranial anatomy of *Shoshonius* and the antiquity of Anthropoidea. In: Fleagle, J. G., Kay, R. F. (Eds.), Anthropoid Origins. Plenum, New York, pp. 55–97.

Beard, K.C., Qi, T., Dawson, M.R., Wang, B., Li, C., 1994. A diverse new primate fauna from middle Eocene fissure-fillings in southeastern China. Nature 368: 604–609.

Cartmill, M., 1972. Arboreal adaptations and the origin of the order Primates. In: Tuttle, R. (Ed.), The Functional and Evolutionary Biology of Primates. Aldine, Chicago, pp. 97–122.

Cartmill, M., 1980. Morphology, function and evolution of the anthropoid postorbital septum. In: Ciochon, R., Chiarelli, B. (Eds.), Evolutionary Biology of the New World Monkeys and Continental Drift. Plenum, New York, pp. 243–274.

Cartmill, M., 1994. Anatomy, antinomies, and the problem of anthropoid origins. In: Fleagle, J. G., Kay, R. F. (Eds.), Anthropoid Origins. Plenum, New York, pp. 549–566.

Cartmill, M., Kay, R. F., 1978. Cranio-dental morphology, tarsier affinities, and primate suborders. In: Chivers, D., Joysey, J. (Eds.), Recent Advances in Primatology, Volume 3, Evolution, D. Academic, London, pp. 205–214.

Castenholtz, A., 1984. The eye of *Tarsius*. In Niemitz, C. (Ed.), Biology of Tarsiers. Gustav-Fisher Verlag, Stuttgart, pp. 303–318.

Dagosto, M., Gebo, D. L., 1994. Postcranial anatomy and the origin of the Anthropoidea. In: Fleagle, J. G., Kay, R. F. (Eds.), Anthropoid Origins. Plenum, New York, pp. 567–594.

Dagosto, M., Gebo, D. L., Beard, K. C., 1999. Revision of the Wind River faunas, early Eocene of central Wyoming. Part 14. Postcranium of *Shoshonius cooperi* (Mammalia: Primates). Annals of the Carnegie Museum 68, 175–211.

Dagosto, M., Terranova, C.J., 1992. Estimating the body size of Eocene primates: A comparison of results from dental and postcranial variables. International Journal of Primatology 13: 307–344.

Delson, E., Tattersall, I., Van Couvering, J. A., Brooks, A. S., 2000. Encyclopedia of Human Evolution and Prehistory, 2nd Ed. Garland Publishing, New York.

Duckworth, W. L. H., 1915. Morphology and Anthropology Publisher. Cambridge University Press, Cambridge.

Fleagle, J. G., 1999. Primate Adaptation and Evolution. Academic, San Diego.

Fleagle, J. G., Kay, R. F. (Eds.), 1994. Anthropoid Origins. Plenum, New York

Gazin, C. L., 1958. A review of the middle and upper Eocene primates of North America. Smithsonian Miscellaneous Collection 136, 1–112.

Gebo, D. L., 2002. Adapiformes: phylogeny and adaptation. In: Hartwig, W. C. (Ed.), The Primate Fossil Record. Cambridge University Press, Cambridge, pp. 21–44.

Gingerich, P. D., 1975. Dentition of *Adapis parisiensis* and the evolution of lemuriform primates. In: Tattersall, I., Sussman, R. W. (Eds.), Lemur Biology. Plenum, New York, pp. 65–80.

Gingerich, P., 1980. Eocene Adapidae: paleobiology and the origin of South American Platyrrhini. In: Ciochon, R., Chiarelli, B. (Eds.), Evolutionary Biology of the New World Monkeys and Continental Drift. Plenum, New York, pp. 123–138.

Gingerich, P.D., Smith, B.H., Rosenberg, K., 1982. Allometric scaling in the dentition of primates and prediction of body weight. American Journal of Physical Anthropology 58: 81–100.

Gregory, W. K., 1922. The Origin and Evolution of the Human Dentition. Williams & Wilkins, Baltimore, MD.

Gunnell, G.F., Rose, K.D., 2002. Tarsiiformes: evolutionary history and adaptation. In: Hartwig, W. (Ed.), The Primate Fossil Record. Cambridge University Press, Cambridge, pp. 45–82.

Heesy, C. P., Ross, C. F., 2001. Evolution of activity patterns and chromatic vision in primates: morphometrics, genetics, & cladistics. Journal of Human Evolution 40, 111–149.

Hershkovitz, P., 1977. Living New World Monkeys (Platyrrhini), volume 1. University of Chicago Press, Chicago.

Hill, W. C. O., 1955. Primates, Comparative Anatomy and Taxonomy, volume 2: Haplorhini: Tarsioidea. University of Edinburgh Press, Edinburgh.

Hogg, R. T., Rosenberger, A. L., Wong, S. M., 2005. Eye on haplorhine orbits. Abstract. American Journal of Physical Anthropology 126, S40, 117.

Jablonski, N., 2003. The evolution of the tarsiid niche. In: Wright, P.C., Simons, E.L., Gursky, S. (Eds.), Tarsiers: Past, Present, and Future. Rutgers University Press, New Brunswick, pp. 35–49.

Kay, R. F., Cartmill, M., 1977. Cranial morphology and adaptation of *Palaechthon nacimienti* and other Paromomyidae (Plesiadapoidea/Primates), with a description of a new genus and species. Journal of Human Evolution 6, 19–35.

Kay, R. F., Ross, C. F., Williams, B. A., 1997. Anthropoid Origins. Science 275, 797–804.

Kay, R. F., Kirk, E. C., 2000. Osteological evidence for the evolution of activity pattern and visual acuity in primates. American Journal of Physical Anthropology 113, 235–262.

Kay, R. F., Williams, B. A., Ross, C. F., Takai, M., Shigehara, N., 2004. Anthropoid origins: a phylogenetic analysis. In: Ross, C., Kay, R. F. (Eds.), Anthropoid Origins: New Visions. Plenum, New York, pp. 91–135.

Kirk, E. C., Kay, R. F., 2004. The evolution of high visual acuity in the Anthropoidea. In: Ross, C., Kay, R. F. (Eds.), Anthropoid Origins: New Visions. Kluwer/Plenum, New York, pp. 523–586.

Le Gros Clark, W. E., 1934. Early Forerunners of Man. Bailliere, London.

Le Gros Clark, W. E., 1959. The Antecedents of Man. Harper, New York.

MacPhee, R. D. E., Cartmill, M., 1986. Basicranial structures and primate systematics. In: Swindler, D. R. Erwin, J. (Eds.), Comparative Primate Biology, Volume I. Alan R. Liss, New York, pp. 219–276.

MacPhee, R. D. E., Beard, K. C., Qi, T., 1995. Significance of primate petrosal from Middle Eocene fissure-fillings at Shanghuang, Jiangsu Province, People's Republic of China. Journal of Human Evolution 29, 501–513.

Mahe, J., 1976. Craniométrie des lémuriens: analyses multivariables-phylogénie. Mémoires du Musée National d'Histoire Naturelle de Paris, Ser. C. 32. pp. 1–342.

Martin, R. D., 1990. Primate Origins and Evolution. Princeton University Press, Princeton, NJ.

Ni, X., Wang, Y., Hu, Y., Li, C., 2004. A euprimate skull from the early Eocene of China. Nature 427, 65–68.

Pocock, R. I., 1918. On the external characters of the lemurs and *Tarsius*. Proceedings of the Zoological Society of London 1918, 19–53.

Radinsky, L. B., 1970. The fossil evidence of prosimian brain evolution. In: Noback, C. R., Montagna, W. (Eds.), The Primate Brain. Appleton-Century-Crofts, New York, pp. 209–224.

Ravosa, M. J., Hogue, A. S., 2004. Function and fusion of the mandibular symphysis in mammals: a comparative and experimental perspective. In: Ross, C. F., Kay, R. F. (Eds.), Anthropoid Origins: New Visions. Kluwer/Plenum, New York, pp. 299–448.

Ravosa, M. J., Savakova, D. G., 2004. Euprimate origins: the eyes have it. Journal of Human Evolution 46, 357–364.

Rosenberger, A. L., 1985. In favor of the *Necrolemur-Tarsius* hypothesis. Folia Primatologica 45, 179–194

Rosenberger, A.L., 1986. Platyrrhines, catarrhines and the anthropoid transition. In: Wood, B, Martin, L. Andrews, P. (Eds.), Major Topics in Human and Primate Evolution. Cambridge University Press, New York, pp. 66–88.

Rosenberger, A.L., 2005. The hunt for the dawn monkey: Unearthing the origins of monkeys, apes, and humans. Journal of Mammalian Evolution 12, 513–516.

Rosenberger, A. L., 2006. Protoanthropoidea (Primates, Simiiformes): A new primate higher taxon and a solution to the *Rooneyia* problem. Journal of Mammalian Evolution 13, 139–146.

Rosenberger, A. L., Szalay, F. S., 1980. On the tarsiiform origins of Anthropoidea. In: Ciochon, R. L., Chiarelli, A. B. (Eds.), Evolutionary Biology of the New World Monkeys and Continental Drift. Plenum, New York, pp. 139–157.

Rosenberger, A. L., Pagano, A., Anatomical Record. Frontal fusion: Collapse of another anthropoid synapomorphy in press?

Ross, C. F., 1993. Allometric and functional influences on primate orbit orientation and the origins of the Anthropoidea. Journal of Human Evolution 29, 201–227.

Ross, C. F., 1994. The craniofacial evidence for anthropoid and tarsier relationships. In: Fleagle, J. G., Kay, R. F. (Eds.), Anthropoid Origins. Plenum, New York, pp. 469–547.

Ross, C. F., 1995. Allometric and functional influences on primate orbit orientation and the origins of the Anthropoidea. Journal of Human Evolution 29, 201–227.

Ross, C. F., 1996. Adaptive explanation for the origins of the Anthropoidea (Primates). American Journal of Primatology 40, 205–230.

Ross, C. F., 2000. Into the light: the origins of Anthropoidea. Annual Review of Anthropology 29, 147–194.

Ross, C.F., 2004. The tarsier fovea: functionless vestige or nocturnal adaptation? In: Ross, C.F., Kay, R.F. (Eds.), Anthropoid Origins: New Visions. Kluwer/Plenum Press, New York, pp. 477–537.

Ross, C. F., Williams, B., Kay, R. F., 1998. Phylogenetic analysis of anthropoid relationships. Journal of Human Evolution 35, 221–306.

Ross, C. F., Kay, R. F. (Eds.), 2004. Anthropoid Origins: New Visions. Kluwer/Plenum, New York.

Rossie, J. B., Ni, X., Beard, K. C., 2006. Cranial remains of an Eocene tarsier. Proceedings of the National Academy of Sciences USA 103, 4381–4385.

Seiffert, E. R., Simons, E. L., Clyde, W. C., Rossie, J. B., Attia, Y., Bown, T. M., Chatrath, P., Mathison, M. E., 2005. Basal anthropoids from Egypt and the antiquity of Africa's higher primate radiation. Science 310, 300–304.

Simons, E. L., 1972. Primate Evolution. MacMillan, New York.

Simons, E. L., 2003. The fossil record of tarsier evolution. In: Wright, P. C., Simons, E. L., Gursky, S. (Eds.), Tarsiers: Past, Present, and Future. Rutgers University Press, Piscataway, NJ, pp. 9–34.

Simons, E. L., Russell, D. E., 1960. Notes on the cranial anatomy of *Necrolemur*. Breviora 127, 1–14.

Simons, E. L., Bown, T. M., 1985. *Afrotarsius chatrathi*, first tarsiiform primate (? Tarsiidae) from Africa. Nature 313, 475–477.

Simons, E. L., Rasmussen, D. T., 1989. Cranial anatomy of *Aegyptopithecus* and *Tarsius* and the question of the tarsier-anthropoidean clade, American Journal of Physical Anthropology 79, 1–23.

Simpson, G. G., 1961. Principles of Animal Taxonomy. Columbia University Press, New York.

Spatz, W. B., 1969. An interpretation of the sagittal shape of the skull of higher primates, based on observations on the skull of *Tarsius*. Proceedings of the 2nd International Congress in Primatology, Volume 2. Karger, Basel, pp. 187–191.

Sprankel, H., 1965. Studies on Tarsius. I: Morphology of the tail along with ethological observations. Folia Primatologica (Basel) 3, 153–188.

Starck, D., 1975. Phylogeny of the primate higher taxa: the basicranial evidence. In: Luckett, W. P., Szalay, F. S. (Eds.), Phylogeny of the Primates: A Multidisciplinary Approach. Plenum, New York, pp. 91–125.

Szalay, F. S., 1975a. Phylogeny of primate higher taxa: the basicranial evidence. In: Luckett, W. P., Szalay, F. S. (Eds.), Phylogeny of the Primates: a Multidisciplinary Approach. Plenum, New York, pp. 357–404.

Szalay, F. S., 1975b. Haplorhine relationships and the status of the Anthropoidea. In: Tuttle, R. H. (Ed.), Primate Functional Morphology and Evolution. Mouton Publishers, The Hague, pp. 3–22.

Szalay, F. S., 1976. Systematics of the Omomyidae (Tarsiiformes, Primates): Taxonomy, phylogeny, and adaptations. Bulletin of the American Museum of Natural History 156, 159–449.

Szalay, F. S., 1977. Ancestors, descendants, sister groups, and testing of phylogenetic hypotheses. Systematic Zoology 26, 12–18.

Szalay, F. S., 2000. Tarsiiformes. In: Delson, E., Tattersall, I., Van Couvering, J. A., Brooks, A. S. (Eds.), Encyclopedia of Human Evolution and Prehistory, 2nd Ed. Garland Publishing, New York, pp. 691–693.

Szalay, F. S., Delson, E., 1979. Evolutionary History of the Primates. Academic, New York.

Teilhard de Chardin, P., 1921. Les mammifères de l'Eocene inférieur Francais et leurs gisements. Annales. De Paléontologie 10, 171–176, Annales De Paléontologie 11, 9–116.

Thalmann, U., 1994. Die Primaten aus dem eozänen Geiseltal bei Halle/Saale (Deutschland). Courier Forschungsinstitut Senckenberg 175, 1–161.

Wilson, J. A.,1966. A new primate from the earliest Oligocene, west Texas, preliminary report. Folia Primatologica 4, 227–240.

15. Epitensoric Position of the Chorda Tympani in Anthropoidea: a New Synapomorphic Character, with Remarks on the Fissura Glaseri in Primates

Wolfgang Maier*
Lehrstuhl Spezielle Zoologie
Universität Tübingen
Auf der Morgenstelle 28
D-72076 Tübingen, Germany
wolfgang.maier@uni-tuebingen.de

15.1 Introduction

As is well known from human anatomy, the chorda tympani detaches from the facial nerve before its exit from the stylomastoid foramen; recurving around the hyoid it enters the tympanic cavity via a small foramen in the facial canal (canaliculus chordae tympani). Then, embedded in an epithelial fold, it passes between the crus longum of the incus and the malleus across the tendon of the tensor tympani muscle. Anteriorly, it is situated at the medial side of the anterior process of the malleus (Folianus) before it exits from the tympanic cavity through the petrotympanic fissure or fissura Glaseri (cf. Warwick and Williams, 1973, figure 286; Henson, 1974, figure 21A; and many other textbooks of human anatomy). The anterior process consists in fact of a small dermal bone called the gonial in the German anatomical literature (Gaupp, 1908, 1911, 1913), which is synonymous with the prearticular of vertebrate paleontology and English comparative anatomy (Williston, 1903; De Beer, 1937).

The chorda tympani nerve transmits fibers from the taste buds and to the salivary glands of the lower jaw. The complicated course of the nerve is tied to the evolutionary history of the mammalian middle ear which was carefully analysed by Gaupp (1913) and Goodrich (1930), to name only the most important early researchers. In fact, the chorda is a posttrematic branch of the facial nerve, which attains its dorsal and recurrent position due to the secondary evolutionary ascent of the mammalian middle ear from the angle of the lower jaw to the basicranium (Allin, 1975, Maier, 1990; Allin and Hopson, 1992).

* Address for correspondence: wolfgang.maier@uni-tuebingen.de

E.J. Sargis and M. Dagosto (eds.), *Mammalian Evolutionary Morphology: A Tribute to Frederick S. Szalay,* 347–360.

The mammalian chorda tympani was first analyzed from a comparative point of view by Bondy (1907). His study was based on histological sections of adult specimens. Although he was mainly interested in the pars flaccida of the tympanic membrane, he also presented many details on the complete course of the chorda tympani in more than 50 species of mammals representing most orders. Among these taxa were several primates, i.e., *Macaca, Ateles, Hylobates* and *Homo*. He realized that these 'higher' primates are all similar in the relationship of the chorda tympani and the tensor tympani muscle, i.e., the nerve passes above this muscle. This is in clear contrast to the majority of mammals from monotremes through marsupials, to 'insectivores', bats, rodents, carnivores and cetartiodactyls, where the nerve runs ventral to the muscle. Next to the studied primates only a few other species studied by Bondy showed a dorsal course of the nerve or a course that pierced the muscle tendon: *Myoxus* and *Sciurus* (rodents), *Herpestes* (carnivores), and *Equus* (perissodactyls).

These observations of Bondy (1907) were recognized by a number of authors, namely by Gaupp (1911, 1913) and Goodrich (1930). However, when the latter author wrote: "the relation of its (-m. tensor tympani, W.M.-) ligament to the chorda tympani varies. Usually it reaches the malleus below the nerve; but in *Sciurus* and *Equus* it is pierced by the nerve, and in some (including Man and the Apes) it passes above" (p. 467), he obviously misinterpreted Bondy completely.

Saban (1963) provided valuable information on the relationship between the chorda tympani and the tensor tympani muscle in Primates and he gave detailed drawings of various strepsirhine taxa. In his figure 41 he depicted the middle ear of *Daubentonia* with the chorda tympani passing underneath the insertion of the tensor tympani muscle. He also documented this ventral position of the nerve for other lemuriforms and lorisiforms. With respect to *Tarsius*, he remarked on the relationship between the chorda tympani and other structures of the middle ear: "La corde du tympan pénètre dans la cavité tympanique entre l'enclume et le marteau, fait une boucle très serrée autour du muscle du marteau, puis s'engage dans la gouttière malléolaire" (p. 299) ("the chorda tympani enters the cavum tympani between anvil and hammer, forms a sharp loop around the muscle of the hammer and then enters the groove for the anterior process of the malleus"; translation W.M.); although the relationship of the chorda with the tensor tympani is not completely clear from this description, it probably means that the nerve runs underneath the muscle. However, his descriptions are based on macroscopic preparations of adult specimens only, and hence several of the anatomical data he presented seem not completely reliable. In his figures 21 and 83A he depicted a complicated m. tensor tympani in *Tupaia sp*., while tupaiids are definitively missing that muscle according to all evidence of microscopic anatomy (see below); in his figure 83H he showed a chorda tympani passing below the tensor muscle in *Homo*, while all available evidence testifies that the nerve runs above it. Of platyrrhines he only considered *Ateles* in passing, but his figure 83E seems to show a chorda running above the tensor tympani. Methodologically, Saban (1963) was interested in defining different evolutionary morphotypes of the tensor tympani muscle rather than an interpretation of his data in terms of phylogenetic systematics in the sense of Hennig (1966).

It has long been realized that ontogenetic information is essential for understanding the details of cranial morphology (Gaupp, 1906; Goodrich, 1930; De Beer, 1937; Starck, 1967; MacPhee, 1981; Kuhn and Zeller, 1987; Maier, 1993; Novacek, 1993). Relationships of soft-tissue and skeletal structures are more easily recognized at earlier ontogenetic stages and very often become obscured in adult crania. However, time-consuming preparation and interpretation of histological serial sections is a handicap for broad systematic comparisons. I have been able to build up a fairly comprehensive collection of serial sections of the heads of many primate taxa over the last 30 years, and I am therefore in a fortunate position to investigate the relationship between the chorda tympani nerve and the tensor tympani muscle based on relatively extensive material.

Voit (1909) provided one of the first detailed descriptions of a mammalian fetal skull (chondrocranium) based on histological serial sections. In this study of *Oryctolagus* (rabbit) he also documented the position of the chorda tympani; he noticed the difference from human conditions and provided a drawing that is the basis of Figure 15.1 in this work. Goodrich (1930, figures 488, 489) presented very clear semi-diagrammatic pictures of the ear region of a young specimen of the marsupial *Trichosurus* that also illustrates the positioning of the relevant soft tissue structures including the chorda tympani. However, this is not the place to review the great number of monographic studies on craniogenesis in mammals where remarks on the chorda tympani are found; only the few works of specific relevance to Primates are reviewed:

Starck (1960, p.595, figure 18) stated for the chorda tympani in a fetus of *Pan troglodytes* "Sie verläuft wie bei *Homo* über die Sehne des Musc. tensor tympani…" ("It runs across the tendon of the tensor tympani muscle as in *Homo*"; translation W.M.). In his study of the fetal cranium of *Propithecus sp*., Starck (1962) did not comment on the relationship of the chorda tympani with the tensor tympani muscle; however, we are able to present observations on the serial sections of the same specimen (see below).

15.2 Materials and Methods

The histological serial sections that have been studied for this paper are listed in Table 15.1. The ontogenetic stages of the sectioned specimens as well as their state of preservation are quite different; brief information on the age or size is given in column 2.

In order to put my data in an appropriate theoretical framework, I also consider briefly the probable outgroups. This is necessary to establish the ancestral morphotypes at different levels in order to define the polarity of character distribution. The Archonta concept was posed by Gregory (1910) on very doubtful morphological arguments; consequently it has been discredited for many decades. Some modern systematic studies, however, have revived the concept (MacPhee and Cartmill, 1986; Simmons, 1993; Novacek, 1990; Wible and Martin, 1993; Murphy et al., 2001; Springer et al., 2005), and some authors even consider Dermoptera as the living sister-group of Primates (Beard, 1993) or even as member of an expanded order Primates (McKenna and Bell, 1997). Bats have been removed from the Archonta by most molecular systematists, leaving only Euarchonta (Scandentia, Dermoptera, and Primates) as a natural group (Waddell et al., 1999; Murphy et al., 2001; Silcox et al., 2005). Surprisingly, Primates and other Euarchonta have recently been united with Glires (rabbits and rodents) in a supraordinal clade Euarchontoglires on molecular data by some authors (Adkins and Honeycutt, 1991; Waddell et al., 1999; Murphy et al., 2001; Sargis, 2002; Bloch and Boyer, 2002; Springer et al., 2005), and therefore a few remarks on Glires may be useful.

As to primate systematics I refer to the studies of Szalay (1975b), Cartmill and Kay (1978), Beard and MacPhee (1994), Allard et al. (1996), Soshani et al. (1996), Kay et al. (1997), Ross et al. (1998), and Delpero et al. (2001) as working hypotheses. For taxonomy I refer to Wilson and Reeder (1993).

15.3 Results

15.3.1 GLIRES

In *Oryctolagus* (Lagomorpha) and in the majority of rodents the chorda tympani passes below the insertion of the tensor tympani muscle. This morphotype is documented by a (modified) illustration of Voit (1909) (Figure 15.1). I define this topographic arrangement, which is most probably the plesiomorphic mammalian condition (see discussion), as ***hypotensoric***.

TABLE 15.1. List of histological serial sections that have been examined for the relationship between the chorda tympani and the tensor tympani muscle. All sectional series are presently housed at the Department of Zoology, University of Tuebingen. CRL = crown-rump-length. HL = head length.

Species	CRL/stage	Ventral	Intermediate	Dorsal
Oryctolagus cuniculus	41 mm	+	–	–
Ochotona pusilla	34 mm	?	?	?
Rattus norvegicus	Several	+	–	–
Tupaia belangeri	Several	?	?	?
Cynocephalus volans	145 mm	+	–	–
Daubentonia madagascariensis	98 mm	+	–	–
Microcebus murinus	Fetal	+	–	–
Lemur catta	Several	+	–	–
Lepilemur mustelinus	54 mm	+	–	–
Propithecus sp.	26 mm HL	+	–	–
Indri indri	66.5 mm	+	–	–
Galagoides demidoff	Neonate	+	–	–
Galago senegalensis	Fetal	+	–	–
Otolemur crassicaudatus	Neonate	+	–	–
Loris tardigradus	Neonate	+	–	–
Tarsius bancanus	Neonate	+	–	–
Saimiri sciureus	80 mm	–	–	+
Cebus albifrons	Fetal	–	–	+
Callimico goeldii	62 mm	–	–	+
Callithrix pygmaea	60 mm	–	–	+
Callithrix jacchus	20,31,80 mm	–	–	+
Oedipomidas oedipus	Neonate	–	–	+
Callicebus cupreus	Fetal	–	–	+
Aotus trivirgatus	Fetal	–	–	+
Pithecia monachus	90 mm	–	–	+
Alouatta caraya	36 mm	–	–	+
Ateles geoffroyi	51 mm	–	–	+
Macaca fuscata	58 mm	–	–	+
Papio hamadryas	115 mm	–	–	+
Theropithecus gelada	88 mm	–	–	+
Chlorocebus aethiops	60 mm	–	–	+
Nasalis larvatus	100 mm	–	–	+
Trachypithecus vetulus	Fetal	–	–	+
Hylobates sp.	100 mm	–	–	+
Symphalangus syndactylus	Neonate	–	–	+
Gorilla gorilla	Fetal	–	–	+
Pan troglodytes	80 mm	–	–	+
Homo sapiens	Several	–	–	+

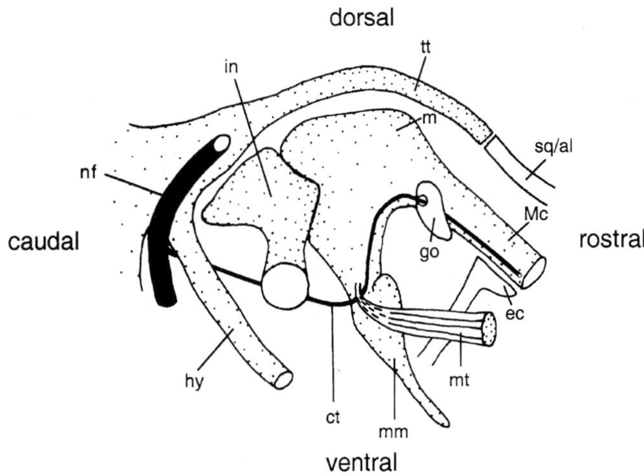

FIGURE 15.1. Diagrammatic drawing of the course of the chorda tympani through a left middle ear viewed from the medial side; the drawing is modified from the study of Voit (1909) on *Oryctolagus cuniculus*. In this species the anterior exit of the tympanic cavity is a fissura squamosotympanica. The following histograms are dorsoventral cross-sections through the insertion of the tensor tympani muscle or through the fissura squamosotympanica. Abbreviations for all figures: al – alisphenoid, as – arteria stapedia, c – capsula otica, ct – chorda tympani, ec – ectotympanicum, GF – Glaserian fissure, go – gonial (prearticular), hy – hyoid, in – incus, m – malleus, Mc – Meckel's cartilage, mm – manubrium mallei, mt – musculus tensor tympani, nf – nervus facialis, sq – squamosum, tt – tegmen tympani.

In older fetuses of *Oryctolagus*, the anterior portion of the gonial shows a medial process that articulates with the ventral side of the tegmen tympani; for a short distance the chorda is squeezed between gonial and tegmen (goniopetrosal fissure). More rostrally, the chorda reaches the retroarticular space between the medial flange of the ectotympanic and the alisphenoid; hence there exists no petrotympanic fissure (Glaseri) in the sense of human anatomy. The remnant of Meckel's cartilage shows no close contact with any of the bones in its vicinity.

It may be mentioned in passing that *Ochotona* has lost the tensor tympani muscle secondarily, as documented by a serial section in our collection. At least in *Oryctolagus*, the chorda tympani pierces the gonial which constitutes the anterior process of the malleus, and we have good reasons to assume that this also represents a plesiomorphic mammalian condition (cf. Gaupp, 1911). The distribution of this character complex in Glires is presently under study by Ruf, Frahnert and Maier (in press).

15.3.2 Scandentia

Because the tensor tympani muscle is lacking in both *Tupaia* and *Ptilocercus*, its relation with the chorda tympani cannot be judged (Spatz, 1964; Zeller, 1983, 1986). The chorda tympani runs in close contact with the medial side of the gonial, but this ossicle is not pierced by the nerve (Spatz, 1964; Zeller, 1983). I speculate that the reduction of the tensor tympani muscle is due to the large anterior portion of the stapedial artery (Bugge, 1974), which is partly protected by a ventral lamella of the tegmen tympani (Spatz, 1964); this ventral process of the tegmen would prevent a functioning access of the tensor tympani muscle at the base of the manubrium mallei (Figure 15.2A).

Doran (1878) had wrongly remarked that in tree shrews "the tensor-tympani tendon is inserted into the root of the manubrium" (p. 441). On the basis of macroscopic dissection, Saban (1963, figures 21 and 83) described and depicted in his *Tupaia* four bundles of a tensor tympani muscle that alledgedly insert on a quite distinctive muscular process of the malleus. Because histological serial sections do not show this

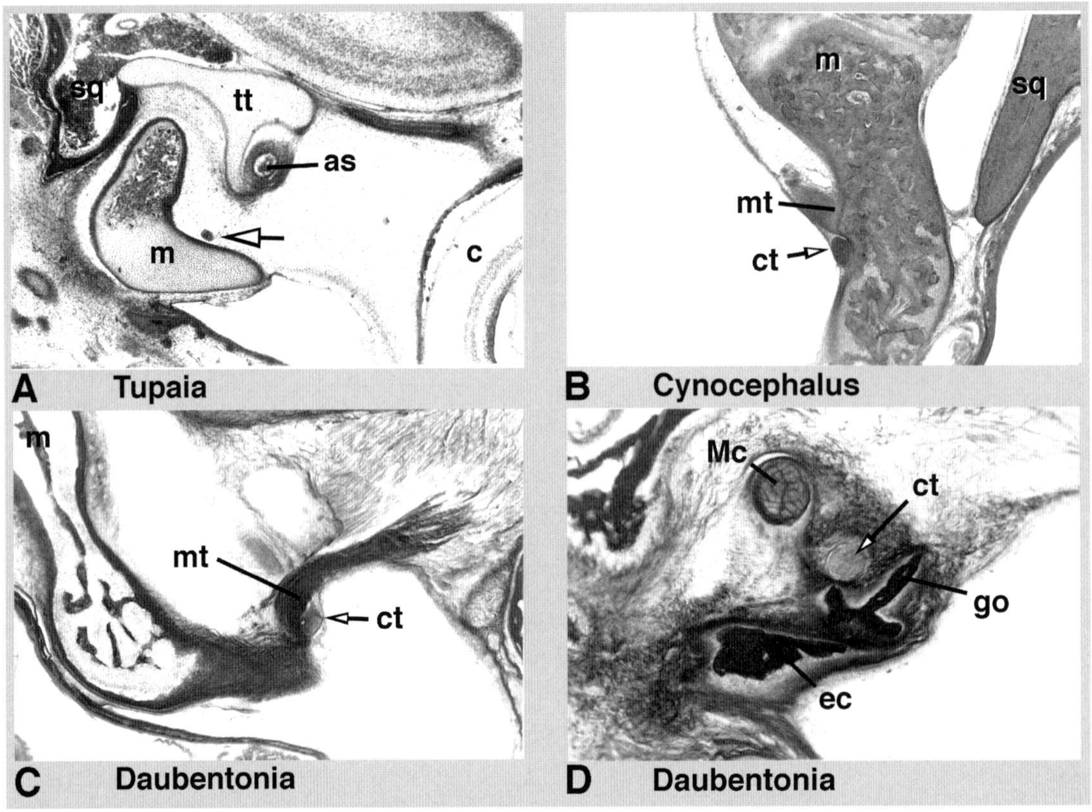

FIGURE 15.2. **A**, *Tupaia glis* (neonate; section 336-2); malleus without the tensor tympani muscle, which is absent; ventral process of the tegmen tympani protects the stapedial artery. **B**, *Cynocephalus volans* (juvenile; section 1290, left side); chorda tympani passes below the insertion of the tensor tympani muscle. **C**, *Daubentonia madagascariensis* (fetal stage; section 171-2); chorda tympani passes below the insertion of the tensor tympani muscle. **D**, *Daubentonia madagascariensis* (same specimen; section 162-2); chorda tympani runs between the gonial and Meckel's cartilage; the gonial is not pierced.

muscle, I suspect that this author misinterpreted the anterior mallear ligaments as slender muscle slips. MacPhee (1981), knowing this controversy, judged that the presence of a tensor tympani muscle may be variable among the tree shrews, but Zeller (1986) argued that the absence of this muscle, which is also missing in *Ptilocercus*, is a synapomorphy of ptilocercines and tupaiines. The ear ossicles of *Tupaia* are also shown by Fleischer (1973; figure 17); he only noted that a muscular process and a collum mallei are missing. We consider the absence of a tensor tympani as an autapomorphy of the Scandentia.

Tupaia is characterized by a laterally situated but expanded tegmen tympani as well as by a prominent posterior extension of the epitympanic wing of the alisphenoid (Spatz, 1964; MacPhee, 1981). In the adult skull, neither the tegmen tympani nor the epitympanic wing of the petrosal do reach the anterior end of the bulla; instead the alisphenoid forms a considerable part of the anterior roof of the tympanic cavity (Saban, 1963; MacPhee, 1981, figure 29). In my neonate, the chorda tympani appears in the retroarticular space between the protruding gonial and the epitympanic wing of the alisphenoid, close to the suture with the overlying squamosal. With the formation of the entotympanic bulla, the conditions become more complex at later stages (Saban, 1963; MacPhee, 1981). In sum, *Tupaia* does not possess a fissura petrotympanica or fissura Glaseri sensu stricto.

15.3.3 Dermoptera

In a juvenile *Cynocephalus volans*, the tensor tympani muscle is relatively small, and its thin tendon, which runs in a straight posterior direction, inserts just above the chorda tympani (Figure 15.2B). More rostrally, the chorda tympani is embedded in a longitudinal groove at the medial side of the ossified malleus. This blade-like root of the anterior process of the malleus most likely represents the gonial; however, no perforation of the gonial is observed. Anteriorly, the chorda tympani leaves the cavum tympani squeezed between the ectotympanic and the greatly expanded squamosal – far away from any part of the petrosal, i.e., there is no fissura Glaseri sensu stricto. The ear ossicles of *Cynocephalus (Galeopithecus)* are described and compared at length by Doran (1878, p. 442 ff.).

15.3.4 Primates

15.3.4.1 Strepsirhini

Daubentonia madagascariensis: The morphology of the chorda tympani in strepsirhine primates is introduced by the example of *Daubentonia*. The arrangement of the chorda tympani looks plesiomorphic in most aspects (cf. Figure 15.2C). It passes on the lateral side of the hyoid and squeezes beetween the posterior limb of the ectotympanic and the hyoid to reach the dorsolateral wall of the tympanic cavity. It then runs rostrally on the lateral side of the crus longum of the incus, just near its ventral end. The lenticular process of the incus of *Daubentonia* forms a small rostral apophysis, which can also be recognized in Doran (1878; Plate 58, figure 29); it is through the notch formed by this small process (still cartilaginous in our fetal specimen) that the chorda gets guided medially. The chorda is not adjacent to the medial side of the collum of the malleus, but lies more medially in the roof of the primary tympanic cavity. The medially projecting muscular process brings the chorda in close proximity to the malleus below the insertion point of the tensor tympani tendon.

Immediately in front of the tendon, the chorda curves dorsally at right angles to reach the gonial which is already completely fused with the anterior process of the malleus proper. The chorda comes to lie in a groove on the medial side of the gonial, but no perforation occurs. More anteriorly, the dorsal portion of the anterior process is revealed as the cartilaginous element of Meckel, and the nerve is situated between the gonial and the cartilaginous rod (Figure 15.2D). Further forward, the chorda runs along the dorsal side of the gonial, which becomes detached from Meckel's cartilage but is tightly fixed to the anterior limb of the ectotympanic. In front of the gonial, the chorda runs for a short distance between the ectotympanic and the medial process of the squamosal near its suture with the tegmen tympani; therefore in *Daubentonia* the chorda tympani does not exit from the tympanic cavity by a petrotympanic fissure or fissura Glaseri sensu stricto but between the squamosal on the one side and the gonial or ectotympanic respectively on the other (Figure 15.8B). Still more rostrally, the chorda passes laterally between the ectotympanic and the cartilage of Meckel to eventually join the lingual nerve. The cartilage of Meckel is embedded in a groove at the medioventral edge of the squamosal.

Microcebus murinus: The anatomy of the middle ear of the fetal and adult ear region of the mouse lemur have been carefully described by Saban (1963) and MacPhee (1981). In our fetal specimen, the chorda tympani enters the tympanic cavity between the ectotympanic and the posttympanic process of the squamosal. It then runs almost transversely to the medial side of the malleus, where it passes under the tendon of the tensor tympani muscle (Figure 15.3A) before ascending to the gonial; however, the nerve does not come near this bony element and does not perforate it.

Lepilemur mustelinus: The chorda tympani is relatively thin in the weasel lemur. The nerve enters the tympanic cavity across the upper end of the posterior crus of the ectotympanic; it passes underneath the tensor tympani muscle, but at some distance from the insertion (Figure 15.3B). It bends dorsally in front of the muscle but does not come close to the gonial, and only gradually approaches the medial side of Meckel's cartilage. The chorda enters the retroarticular space of the lower jaw through a gap between the ectotympanic and squamosal, i.e., definitely not through a petrotympanic fissure.

Lemur catta: The ring-tailed lemur shows a relationship between chorda tympani and tensor tympani muscle that closely resembles that in *Lepilemur*; however, the chorda seems to contain more fibres (Figure 15.3C). It only gradually approaches the anteriormost portion of the gonial but always remains at some distance from it. Finally it reaches the retroarticular space together with the cartilage of Meckel between the ectotympanic and the alisphenoid. Later on in ontogeny, the morphology of this region in Lemuriformes is complicated by the formation of a petrosal bulla and by the inclusion of the ectotympanic ring (cf. MacPhee, 1977, 1981); therefore, this region is not strictly comparable with that of anthropoids.

Propithecus sp. and *Indri indri*: The chorda tympani in the sifaka corresponds largely to that in other Lemuriformes; however, it passes underneath the tensor tympani muscle close to its insertion at the malleus (Figure 15.3D). Our fetus of *Indri indri* is not very well preserved, but the sectional series clearly shows that the chorda tympani runs below the tendon of the tensor tympani muscle. In *Propithecus sp.* the tegmen tympani reaches far forward (Starck, 1962; Figures 15.2 and 15.3), but at the anterior end of the ectotympanic ring the tegmen has ended and the chorda tympani together with the cartilage of Meckel are dorsally covered by a narrow medial wing of the squamosal.

Galago senegalensis, Galagoides demidoff, Otolemur crassicaudatus: The relatively young fetus of *Galago senegalensis* shows a tensor tympani muscle that runs almost transversely from the lower side of the epitympanic wing of the petrosal to insert at the muscular process of the malleus; the chorda tympani crosses the muscle on its ventral side, but at some distance from the insertion point. In a somewhat older specimen of *Galagoides demidoff*, the nerve also passes underneath the tensor tympani muscle somewhat medial to its insertion (Figure 15.4A). In both galagos the chorda tympani pierces the gonial, a character state not otherwise encountered in primates!

In the neonatal *Otolemur crassicaudatus*, the chorda tympani runs almost transversely and curves around the lower side of the tendon of the tensor tympani (Figure 15.4B). Although the penetration hole is not directly shown in our sectional series, the nerve runs inside the gonial more anteriorly. Saban (1963; figure 83C) postulated a special lorisiform-type of the tensor tympani muscle; his figure shows the chorda tympani passing below the muscle insertion.

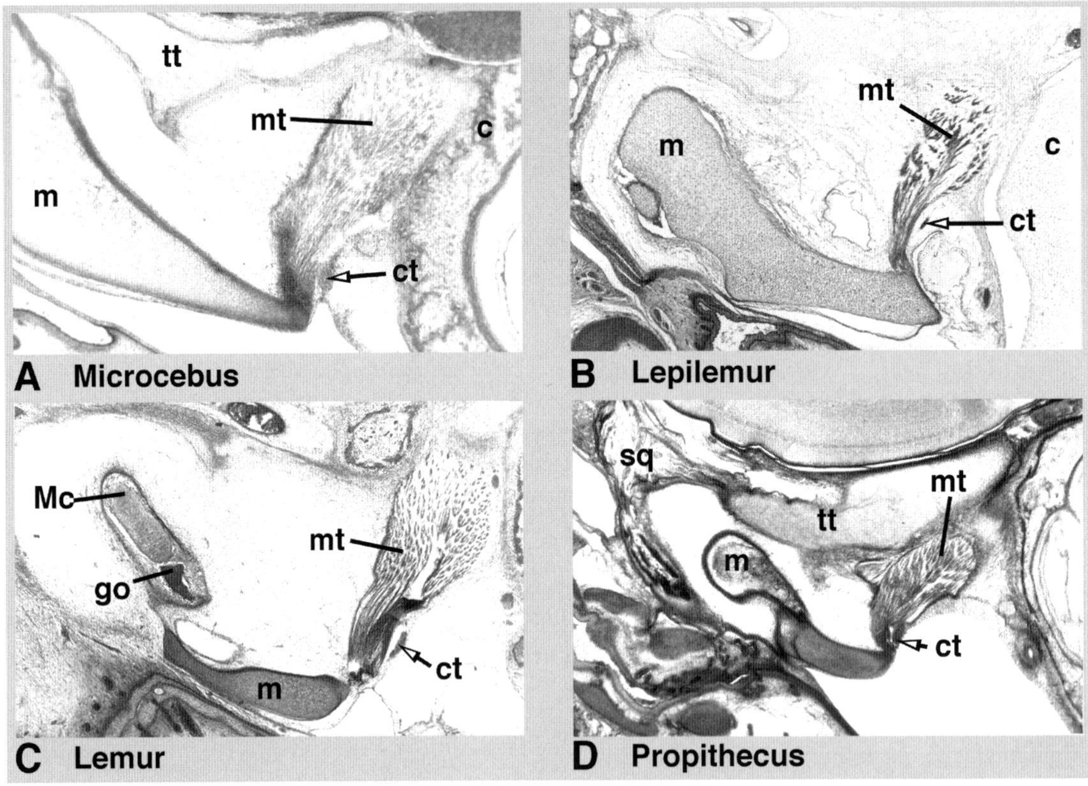

FIGURE 15.3. **A**, *Microcebus murinus* (fetal stage; section 107-1-3); chorda tympani passes below the tensor tympani tendon close to its insertion. **B**, *Lepilemur mustelinus* (fetal stage; section 99-2-4); thin chorda tympani passes below the tensor tympani muscle at some distance from its insertion. **C**, *Lemur catta* (fetal stage; section 278-2-1); chorda tympani passes below the tensor tympani muscle somewhat medial to its insertion. **D**, *Propithecus sp.* (fetal stage; section 161-2); chorda tympani passes below the tensor tympani muscle close to its insertion at the malleus.

Loris tardigradus: In a neonatal slender loris the chorda tympani enters the tympanic cavity between the ectotympanic and the squamosal, then runs almost transversely to pass below the tensor tympani – muscle (Figure 15.4C). The nerve sharply bends around the strong tendon and steeply ascends in front of it. The tensor tympani muscle runs in a posterior-anterior direction, and the nerve passage is not close to its insertion at the malleus. Quite different from galagids, the anterior part does not come close to the anterior process or gonial respectively, and in this feature it resembles some lemuriforms.

15.3.4.2 Haplorhini

Tarsiiformes

Tarsius bancanus: In a serial section of a newborn *T. bancanus* it can be clearly seen that the chorda tympani passes below the tendon of the tensor tympani muscle, i.e., it is of the primitive hypotensoric morphotype (Figure 15.4D). Because this head was sectioned sagittally, it proved to be impossible to follow the chorda posteriorly and anteriorly.

The ear region of adult *Tarsius sp.* was carefully described by Saban (1963). As to the tensor tympani muscle he stated:"De type lémuriforme, il se compose d'un étroit ruban musculaire logé dans le toit de la caisse du tympan. Devant la fenêtre ovale, les fibres, ceintureé par la corde du tympan se redressent pour se fixer par un tendon sur le manche du marteau" (p. 296) ("Being of the lemuriform type, it is composed of a straight muscle slip fit into the roof of the cavum tympani. In front of the oval window its fibres, embraced by the chorda tympani, take a new direction to insert by a tendon at the neck of the malleus"; translation W.M.). His figure 83 D shows that the chorda tympani passes underneath the insertion of the tensor tympani muscle; consequently he classified it with the lemuriform type (p. 321).

Anthropoidea

Platyrrhini, Saimiri sciureus: The chorda tympani detaches from the facial nerve, then turns around the proximal end of the hyoid cartilage and runs along the ventral margin of the squamosal into the middle ear. The area of detachment is covered by the posterior limb of the ectotympanic ring. The nerve turns medially and slightly dorsally in front of the crus longum of the incus. The chorda passes well above the tensor tympani muscle and runs in the middle of the epitympanic recess (Figure 15.5A). I call this condition, which is almost certainly derived, ***epitensoric*** (see discussion).

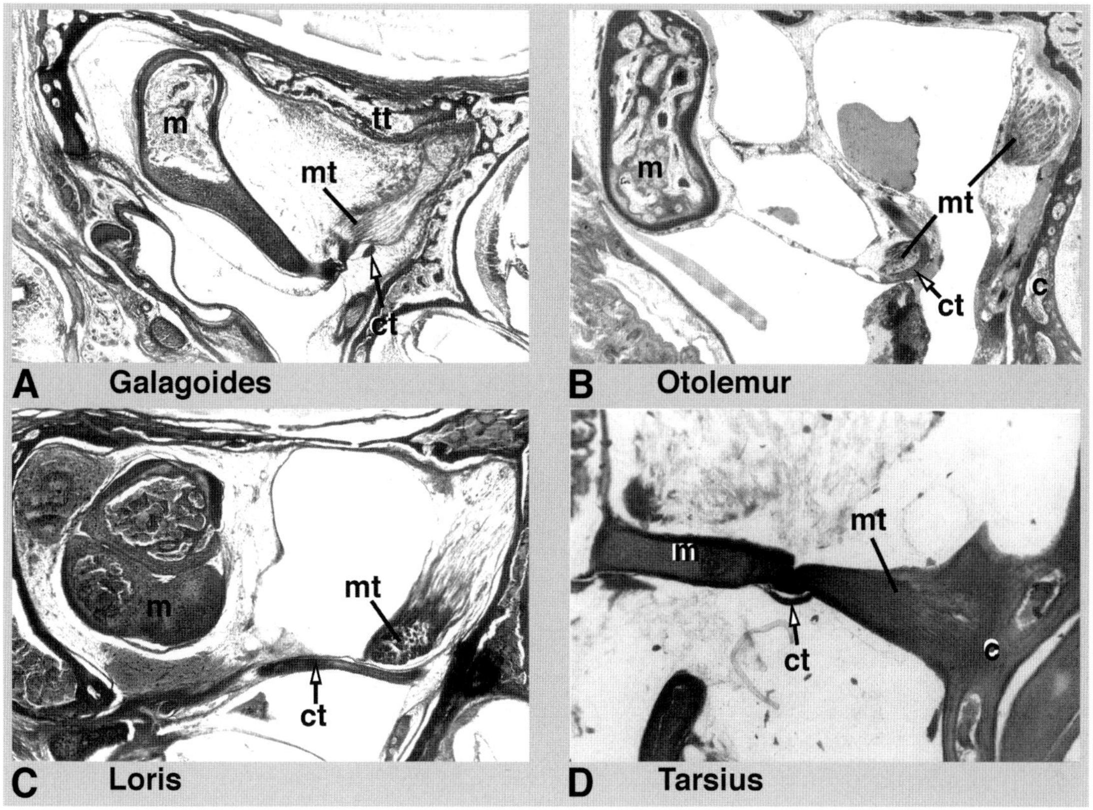

FIGURE 15.4. **A**, *Galagoides demidoff* (neonatal; section 40-2-1); chorda tympani passes below the tensor tympani muscle somewhat medial to its insertion. **B**, *Otolemur crassicaudatus* (neonatal; section 1290); tendon of the tensor tympani runs more longitudinally and the chorda tympani is shown curving underneath the tendon. **C**, *Loris tardigradus* (neonatal; section 133-2-1); chorda tympani runs transversely and turns around the tendon of the tensor tympani muscle. **D**, *Tarsius bancanus* (neonate; section 347-2); chorda tympani passes just below the insertion of the tensor tympani muscle, i.e., it shows a hypotensoric condition.

The tensor tympani muscle in *Saimiri* originates more rostrally and is deflected at right angles by a tendinous loop fixed to a processus cochleariformis before inserting at the processus muscularis of the malleus; this muscle originates in the gap between the cochlear capsule and the tegmen tympani, but it does not reach as far forward as the auditory tube. Rostrally, the chorda tympani is not near the anterior process of the malleus or Meckel's cartilage, but rather runs medially to adjoin the belly of the tensor tympani muscle. Due to the lateral and anterior expansion of the tegmen tympani (as part of the petrosal), the chorda enters the retroarticular space by a gap between the ectotympanic and this part of the petrosal, i.e., by a true petrotympanic fissure (Glaseri).

Cebus albifrons, Callimico goeldii, Ateles geoffroyi. From a fairly long list of platyrrhine species (see Table 15.1) only a few have been chosen for illustration. Although the topography and size of the tensor tympani muscle seems to vary considerably, the chorda tympani invariably crosses well above its tendon, i.e., it is always epitensoric (Figures 15.5 B–D). The approximation of the nerve to the malleus varies in the different taxa, but it never pierces the anterior process of this ear ossicle.

Catarrhini: Macaca fuscata, Papio hamadryas doguera, Trachypithecus vetulus, Nasalis larvatus. All investigated cercopithecoids, both colobids and cercopithecids, (see Table 15.1) invariably show a chorda tympani passing well above the tensor tympani muscle and at a distance from the malleus. Only in *Trachypithecus* does the nerve lie relatively close to the insertion of the tendon of the tensor muscle (Figures 15.6 A–D). Anteriorly, the chorda tympani does not come close to the gonial, which is quite diminutive. In the fetal macaque the chorda enters the retroarticular space between the anterior end of the tegmen tympani and Meckel's cartilage; only after the resorption of this relatively massive cartilage rod does the ectoympanic secondarily form the ventrolateral margin of the Glaserian fissure. In the baboon, the chorda lies more medially and indeed exits from the tympanic cavity by the gap formed between the tegmen tympani (still being cartilaginous) and the medial lamella of the ectotympanic, i.e., by a true petrotympanic fissure (Figure 15.8B). The same conditions are seen in the *Trachypithecus* specimen.

Symphalangus syndactylus, Gorilla gorilla, Pan troglodytes, Homo sapiens. All studied hominoids have a chorda tympani that runs above the tendon of the tensor tympani muscle, i.e.,

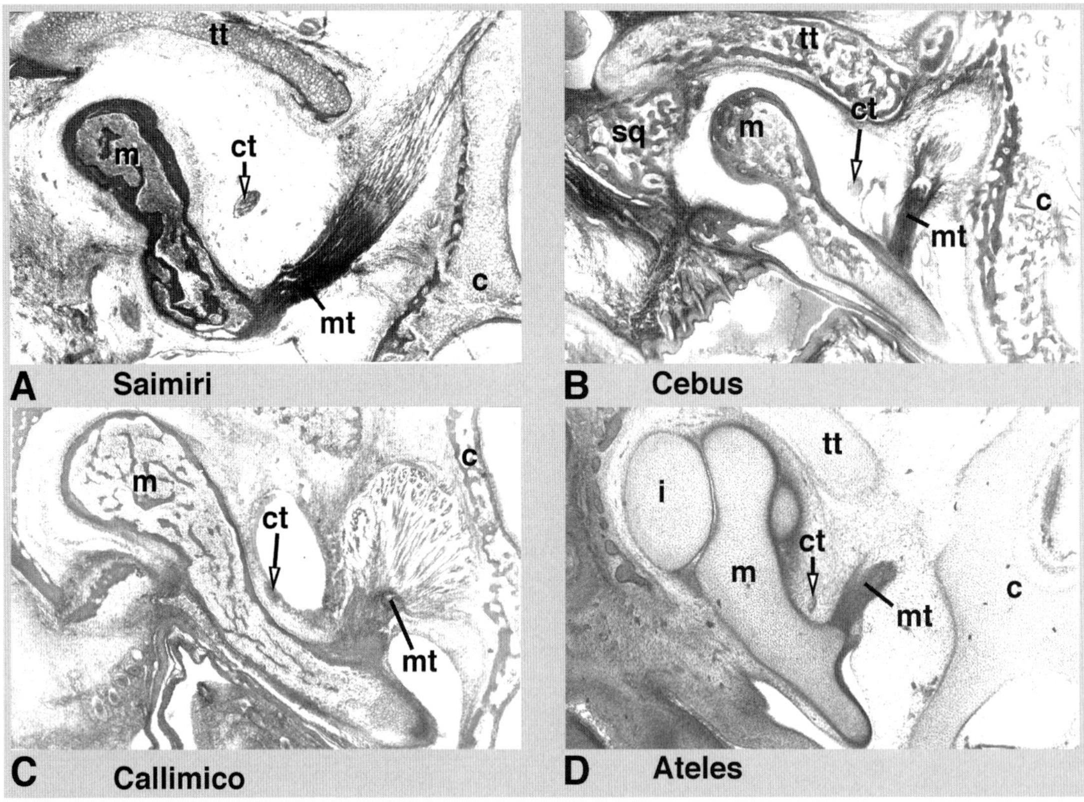

FIGURE 15.5. **A**, *Saimiri sciureus* (fetal stage; section 278-1); chorda tympani crosses above the tensor tympani muscle; it shows no close approximation to the malleus. **B**, *Cebus albifrons* (late fetal stage; section 263); chorda tympani passes above the tensor tympani muscle. **C**, *Callimico goeldii* (late fetal stage; section 131-21); chorda tympani passes across the tensor tympani muscle; due to pneumatization of the epitympanic recessus, the nerve is brought closer to the malleus. **D**, *Ateles geoffroyi* (young fetal stage; section 79-1-2); chorda tympani runs above the insertion of the tensor tympani muscle.

they are all epitensoric. As compared with the majority of cercopithecoids, the chorda lies closer to the collum mallei and closer to the insertion point of the tensor tympani muscle (Figures 15.7 A–D). It is clear in the *Hylobates* fetus that the chorda leaves the tympanic cavity squeezed between the ectotympanic and the still cartilaginous tegmen tympani, i.e., by a petrotympanic fissure. The chorda has no close relationship with the gonial or cartilage of Meckel in this taxon. In all the hominoids the anterior exit of the chorda tympani is by a petrotympanic fissure, i.e., by a true Glaserian fissure.

15.4 Discussion

The monophyly of the Haplorhini and of the Anthropoidea is well founded (Szalay, 1975b; Ross, 1994; Ross et al., 1998), but the number of supporting morphological characters is not overwhelmingly great (Mickoleit, 2004). Middle ear anatomy has been most relevant for establishing this monophyly (Szalay, 1975a), but it has mostly concentrated on the specific course of the promontory branch of the internal carotid artery. The chorda tympani has not yet been addressed in this context although some information on this nerve is dispersed in the literature as shown above. To my knowledge, this nerve has never been studied under a strict phylogenetic-systematic aspect. One reason for this neglect may be that this problem cannot be properly addressed on a pure osteological basis but requires histological sections of fetal stages. With a collection of more than 40 serial sections of representatives of most primate families at hand, a comparative study of the microscopic anatomy of this nerve and its relations with other middle ear structures was feasible on a broader scale.

It can be clearly demonstrated that in all anthropoid species the chorda tympani passes across the mallear insertion of the tensor tympani muscle, i.e., it is epitensoric. In contrast, in all studied strepsirhines it passes below that insertion, i.e., it is hypotensoric. *Tarsius* is also hypotensoric.

Of course this character distribution raises the question of the ancestral morphotype in order to define the plesiomorphic and apomorphic condition. There exists ample evidence, however, that the hypotensoric state is plesiomorphic for mammals: Bender (1906) and Bondy (1907) have shown that the chorda has a hypotensoric position in monotremes and in a number of marsupials. Zeller (1989) does not explicitly

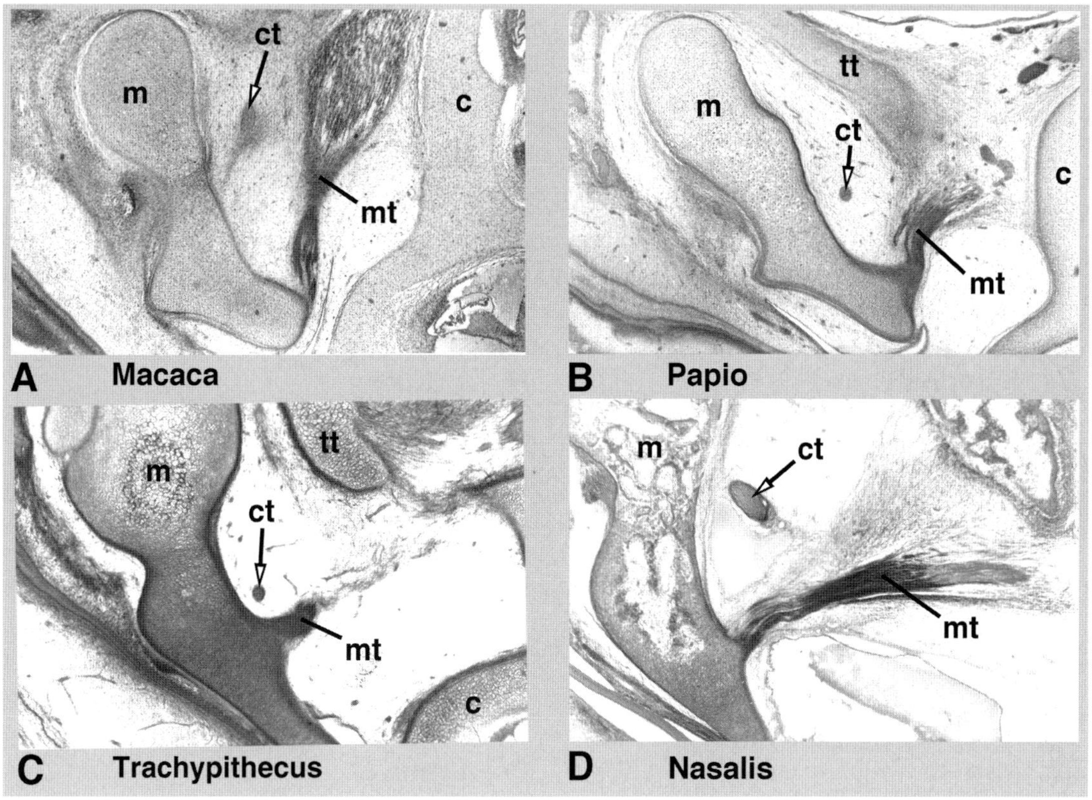

FIGURE 15.6. **A**, *Macaca fuscata* (young fetal stage; section 92-2); chorda tympani crosses way above tensor tympani muscle. **B**, *Papio hamadryas* doguera (young fetal stage; section 107-2); chorda runs well above the tensor tympani muscle. **C**, *Trachypithecus vetulus* (fetal stage; section 179-2-1); chorda tympani passes across the tensor tympani muscle, but in proximity of its insertion. **D**, *Nasalis larvatus* (late fetal stage; section 295-1-1); chorda runs above the tendon of the tensor tympani muscle.

comment on the relationship of the chorda with the tensor tympani in *Ornithorhynchus*, but his figure 47 clearly shows that it passes below that muscle. Cords (1915) for *Perameles* and Toeplitz (1920) for *Didelphis* have stated that the chorda tympani in these marsupials passes below the tendon, and no different information has come forth so far from any other marsupial taxon. Figures 488 and 489 in Goodrich (1930), which also show the ventral position of the chorda tympani, are based on *Trichosurus*. We have checked a number of marsupial taxa, and all have a hypotensoric chorda tympani.

If we accept the cladogram of Murphy et al. (2001) as working hypothesis, basal afrotherians and xenarthrans are of special importance for reconstructing the ancestral morphotype of eutherians. MacPhee (1981, figure 63 a, c) showed that at least in the tenrecs *Hemicentetes* and *Setifer* the chorda tympani passes below the tensor tympani muscle; I can confirm the hypotensoric position of the chorda in *Microgale pusilla. Elephantulus myurus* and *Macroscelides proboscideus* of my collection are also hypotensoric. Reinbach (1952, p. 395) described in his most careful study of the fetal skull of *Dasyurus novemcinctus* (xenarthran) the chorda tympani running underneath the tensor tympani muscle. These scattered data together with the evidence from many other eutherian orders render it most likely that the hypotensoric condition represents the ancestral morphotype of Eutheria and Boreoeutheria.

Glires are more difficult to interpret: *Oryctolagus* (Leporidae) and the majority of rodents are hypotensoric, but a few taxa (*Aplodontia*, all glirids, some sciurids, *Castor, Anomalurus, Pedetes*) are epitensoric (Ruf, Frahnert and Maier, in press). Of the euarchontans, *Cynocephalus* is hypotensoric whereas in scandentians the tensor tympani muscle is secondarily lost and the chorda cannot be typified. Because all strepsirhine primates are also hypotensoric, it is most parsimonious to conclude that the ancestral morphotypes of Euarchontoglires, Euarchonta, and Primates are hypotensoric as well.

Obviously only a small number of taxa acquired the apomorphic epitensoric condition independently. This character distribution seems to prove that this feature is strongly constrained developmentally and that it is not easily modified. The invariable epitensoric condition in Anthropoidea can therefore be considered with confidence to be a valid new synapomorphy for this group. It is interesting that within the haplorhines *Tarsius* has still retained the plesiomorphic state.

The question remains why the course of the chorda tympani is so conservative in most mammals and why it is altered in

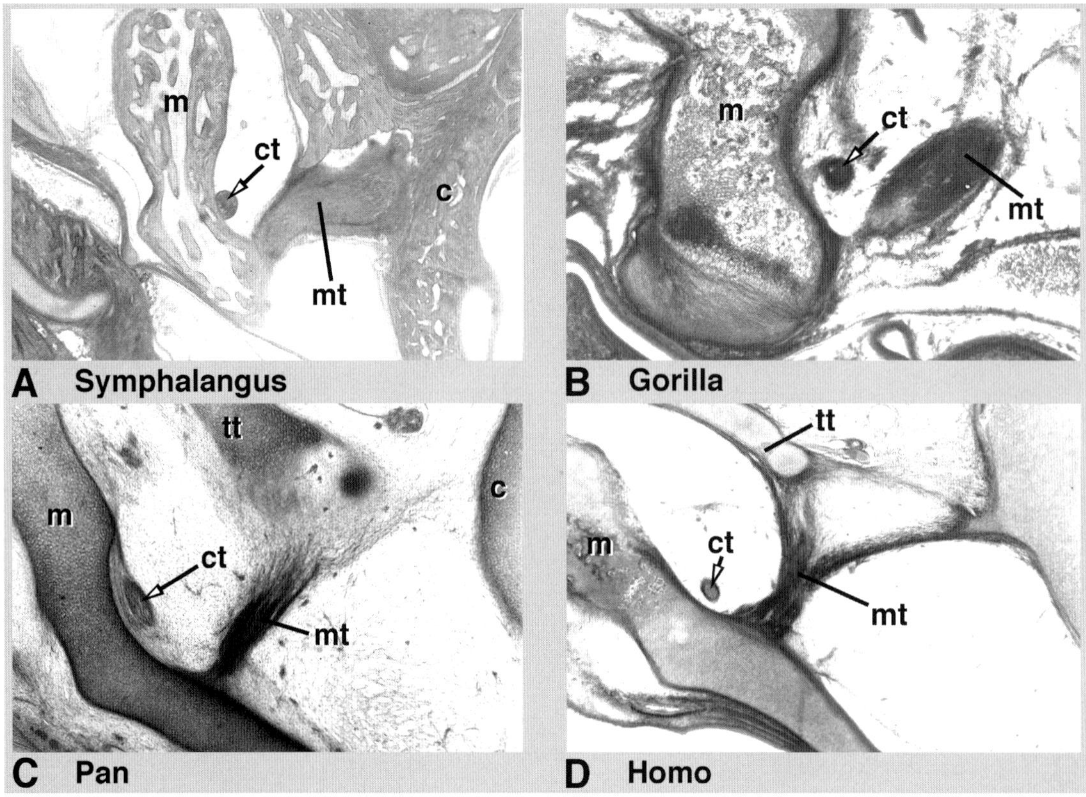

FIGURE 15.7. **A**, *Symphalangus syndactylus* (late fetal stage; section 634); chorda tympani crosses above the insertion of the tensor tympani muscle. **B**, *Gorilla gorilla* (newborn; section 366); chorda tympani passes above the insertion of the tensor tympani muscle. **C**, *Pan troglodytes* (younger fetal stage; section 193-2); the chorda runs above the insertion of the tensor tympani muscle. **D**, *Homo sapiens* (late fetal stage; section 641); chorda tympani passes across the insertion of the tensor tympani muscle.

a few groups. The conservative arrangement of the chorda is the more surprising as it normally makes a remarkable detour ventrally to pass at the underside of the tensor tympani and then ascends again to reach the gonial. This suggests that the relationship of these two heterogeneous structures is established early in ontogeny when the chorda usually shows a more straight course; when the muscle is displaced ventrally later in development, the nerve is trapped by the muscle and has to follow its shift. Anatomists have other well- known examples of such recapitulatory relationships: cervical roots of the accessory nerve, recurrent branch of the vagus nerve, sympathetic and parasympathtic nerves for the heart etc. Very probably, the ventral displacement of the insertion point of the tensor tympani muscle – and with it the chorda – is connected with the evolutionary development of a neck region (collum) of the malleus; Fleischer (1973) has shown that a neck only occurs in functionally derived morphotypes of the malleus.

Why in a few taxa – among them the anthropoids – the chorda tympani has escaped the developmental trap is not yet understood. Perhaps, a heterochronous development of nerve and muscle allows a dorsal transgression of the chorda. Sciurid embryos could probably provide a good model for studying this correlation, because the nerve course not only varies interspecifically (as far as known) in this rodent family but also intraspecifically in *Sciurus vulgaris*. In some taxa (*Herpestes, Equus*) the nerve pierces the muscle tendon thus representing an intermediate stage. At least it could be argued that the epitensoric condition with the straight course of the chorda tympani should be slightly more economic, i.e., selectively advantageous once the ontogenetic constraints are overcome.

There exists another curious structural correlation between the chorda tympani and a bony element of the lower jaw, the gonial or prearticular. Gaupp (1908, 1911, 1913) carefully studied the peculiar relationship of the two structures. He clearly demonstrated that a homologous perforation of an exocranial bony element lying at the medial side of the articular and mandibular cartilage is found in many amphibians, sauropsids and mammals. He used this relationship, together with other arguments, to homologize the bony element and named it the gonial (Gaupp, 1908). He did not know at that time that Williston (1903) had already called this same bone the prearticular in some extant and fossil sauropsids. Formerly the bone had been known under a variety of names which were based on incorrect homologizations with dermal bones of the lower jaw of osteognathostome fish. Meanwhile, the prearticular (gonial) is well known in various extant and fossil fishes (Jarvik, 1980).

Although the term prearticular of Williston (1903) has temporal priority, the concept of the gonial is based on a much broader and much more subtle morphological analysis in Gaupp (1911). Whereas Goodrich (1930, figure 483) still used both terms alternatively, meanwhile the term prearticular dominates, especially in paleontology; the term gonial is only used in German embryology. Starck (1962) discussed the problems of homology and terminology of the gonial at some length.

The perforation of the gonial/prearticular by the chorda tympani varies among mammals. Already in monotremes there exists a difference: *Ornithorhynchus* shows a perforation, *Tachyglossus* does not. Most marsupials also show a perforated gonial as do many eutherians (Gaupp, 1911). It is certainly most unlikely that this peculiar relationship between the chorda tympani and the gonial is established several times independently in osteognathostomes; therefore, this character state is generally supposed to be plesiomorphic. It is of some interest that the perforation of the gonial (prearticular) is found only in galagos (not in lorisids) among euarchontans. In strepsirhines the anterior part of the chorda shows varying relationships with the gonial: in some it runs close to the medial side of this bone while in others it remains at some distance; the same holds true for the epitensoric anthropoids.

J.H. Glaser (1680) described in man a 'hiatus membranae tympani', "der von späteren Anatomen irriger Weise auf die zwischen der Fossa glenoidalis und dem Meatus auditorius externus osseus befindliche Spalte bezogen wurde (Fissura Glaseri)" (Hyrtl, 1845, p. 53); ("which erroneously by later anatomists was referred to the fissure between the glenoid fossa and the bony meatus externus"; translation W.M.) Despite this misunderstanding the fissure was later on named after him as fissura Glaseri. According to Hyrtl (1873, p. 253) it was J. Henle who recognized that the fissura Glaseri in fact is not situated between the tympanic and squamosal but between the tympanic and the 'tegmentum tympani', which in Eutheria is an expansion of the petrosal. The statement of van der Klaauw (1931) that "the fissura Glaseri is at first the aperture for the cartilage of Meckel, which disappears later in development" (p. 164), is misleading; the existence of a cartilage of Meckel was only discovered in the first half of the nineenth century by Meckel (1815), when embryology had become an important field of research (Russell, 1916; Nyhart, 1995). The first modern textbook of comparative embryology explicitly mentions that in the human embryo the chorda tympani together with the cartilage of Meckel pass through the "Fissura petrotympanica oder (or) Glaseri" (Hertwig, 1888, p. 461). Therefore, the fissura Glaseri sensu stricto is by definition only the gap between the (ecto-) tympanic and the tegmen tympani as it is found in man (see also Gegenbaur, 1892; p. 214). It should be mentioned that the exact position of the anterior exit of the cavum tympani is often not well defined and difficult to study in serial sections of fetuses; morever, its morphology may change considerably during ontogeny.

It was soon recognized by some authors that the anterior exit of the chorda tympani is often somewhat different from the human condition and may not be fully equivalent in different mammals (van Kampen, 1905; van der Klaauw, 1931; MacPhee, 1981). Other skeletal elements were found framing the gap in various mammals, namely the squamosal and the alisphenoid. For example, monotremes and marsupials have no petrotympanic fissure, because they do not possess a tegmen tympani (contra van Kampen, 1905); in the latter clade, tiny tubercles at the suprafacial commissure or bony crests of the petrosal, which are the result of secondary excavations by pneumatization, have been interpreted as tegmen tympani, but this homology is doubtful and these structures are far away from the 'Glaserian fissure sensu lato'. In a juvenile of *Monodelphis domestica* (head-rump-length 63 mm) the chorda tympani exits just medial to the postglenoidal process through the gap between ectotympanic and alisphenoid bulla; this confirms the notion of van der Klaauw (1931, p. 165) on *Notoryctes*. This exit is certainly no fissura Glaseri as defined in man and in anthropoids.

If again we take the cladogram of Murphy et al. (2001) as reference, the anatomical conditions of afrotherians and xenarthrans are of special relevance for the reconstruction of the ancestral morphotype of Eutheria. As far as I am aware, no study of the anterior exit of the chorda tympani exists for primitive taxa close to the basal branching of the eutherians. However, it can be concluded from figures 56 and 57 of MacPhee (1981) that in tenrecs only squamosal and sphenoid wings but no part of the petrosal are near the rostral end of the tympanic cavity. The situation is different in macroscelidids, where the expanded tegmen tympani reaches the rostral end of the ectotympanic and contacts a short squamosal (MacPhee, 1981, figure 44). In my specimen of *Elephantulus*, the chorda enters the retroarticular space by a gap whose dorsal margin is formed by the squamosal and alisphenoid and the lower by the gonial/prearticular.

As to the Glires, it has been mentioned above that in *Oryctolagus* the chorda exits between the ectotympanic and the alisphenoid wing. *Ochotona* is difficult to interpret, because the sidewall of the cranium shows a wide fontanella in the relevant area. In rodent fetuses the region of the anterior exit of the chorda tympani looks special, because parts of the tegmen tympani are shifted outside the primary sidewall of the chondrocranium; moreover, the glenoid fossa is modified, and it seems difficult to compare this region with that of other mammals (Ruf Frahnert and Maier, in press).

The anatomy of *Tupaia* and *Cynocephalus* has been described above; it is evident that in none of them does a fissura Glaseri sensu stricto exist. The same holds true for the studied strepsirhines. *Tarsius* has not yet been studied for this aspect of its anatomy. We can therefore conclude that in the groundplan of Euarchonta and Primates there existed no petrotympanic or Glaserian fissure.

The medial position of the anterior portion of the chorda tympani, the long and broad tegmen tympani as well as the

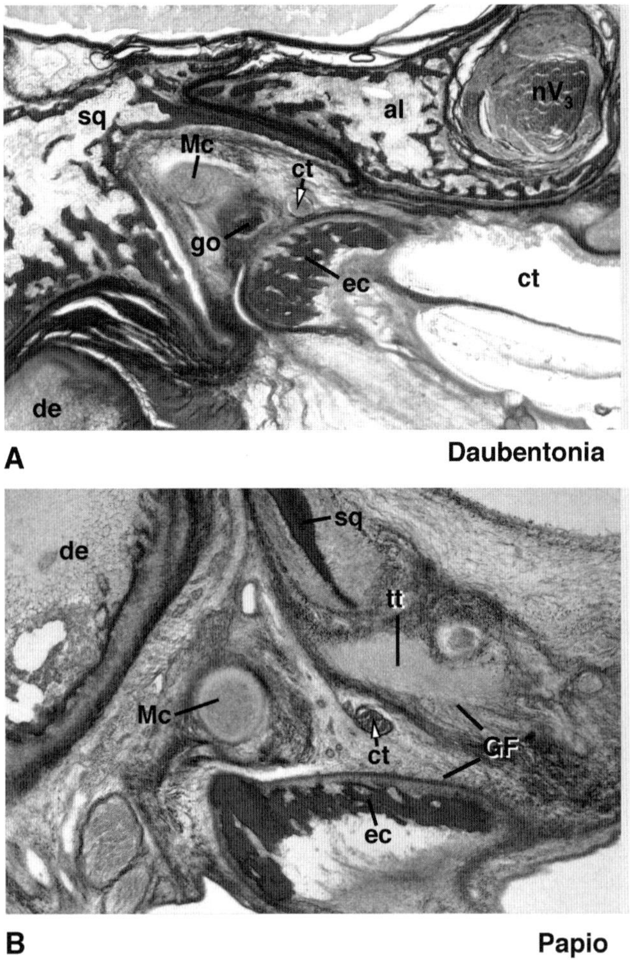

FIGURE 15.8. Dorsoventral cross-section at the fissura Glaseri. **A**, In the strepsirhine *Daubentonia* the chorda exits between the ectotympanic, gonial (prearticular), and squamosum or alisphenoid respectively, i.e., through a fissura Glaseri sensu lato. **B**, In the anthropoid *Papio hamadryas doguera*, the chorda tympani exits anteriorly from the tympanic cavity between the ectotympanic and the tegmen tympani, which is a part of the otic capsule, i.e., through the petrotympanic or fissura Glaseri sensu stricto (GF).

narrow epitympanic wing of the squamosal seem to be conditions for the formation of a true petrotympanic fissure of Glaser. Therefore I presume that this character may well be a further apomorphy of the Anthropoidea (Figure 15.8). In cleaned skulls the contribution of the petrosal by means of the tegmen tympani is not always easy to establish, but CT-sections should be helpful. The diagnosis of the petrotympanic fissure is potentially useful for the study of fossil skulls. Already van Kampen (1905), van der Klaauw (1931) and a few others had recognized the problem of an exact homologization of the Glaserian fissure. MacPhee (1981) suggested to avoid the term petrotympanic fissure and to use 'Glaserian fissure' in a wider sense instead, but this suggestion would mean to ignore a craniological character of possible systematic value.

15.5 Summary

The course of the chorda tympani nerve within the cavum tympani has been studied on the basis of more than 40 histological serial sections representing all major taxa of Primates and some presumed outgroups. It is shown that in all strepsirhines the chorda passes underneath the tensor tympani muscle; this hypotensoric character state, which is plesiomorphic for mammals and for primates, is also found in *Tarsius*. In all Anthropoidea, however, the chorda tympani passes above the tensor tympani muscle; this derived epitensoric state is considered as new synapomorphy of this clade. There is good evidence that the anterior exit of the chorda tympani from the cavum tympani through a petrotympanic fissure (Glaserian fissure sensu stricto) is also an apomorphic state of anthropoids, which is not yet found in strepsirhines and euarchontan outgroups.

References

Adkins, R. M. Honeycutt, R. L., 1991. Molecular phylogeny of the superorder Archonta. Proceedings of the National Academy of Sciences USA 88, 10317–10321.

Allard, M. W., McNiff, B. E., Miyamoto, M. M., 1996. Support for interordinal eutherian relationships with an emphasis on primates and their archontan relatives. Molecular Phylogenetics and Evolution 5, 78–88.

Allin, E. F., 1975. Evolution of the mammalian middle ear. Journal of Morphology 147, 403–438.

Allin, E. F., Hopson, J. A., 1992. Evolution of the auditory system in Synapsida (Mammal-like reptiles and primitive mammals) as seen in the fossil record. In: Webster, D. B., Fay, R. R., Popper, A. N. (Eds.), The Evolutionary Biology of Hearing. Springer, NewYork, pp. 587–614.

Beard, K. C., 1993. Origin and evolution of gliding in early Cenozoic Dermoptera (Mammalia, Primatomorpha). In: MacPhee, R. D. E. (Ed.), Primates and their Relatives in Phylogenetic Perspective. Plenum, New York, pp. 63–90.

Beard, K. C., MacPhee, R. D. E., 1994. Cranial anatomy of *Shoshonius* and the antiquity of the Anthropoidea. In: Fleagle, J. G., Kay, R. F. (Eds.), Anthropoid Origins. Plenum, New York, pp. 55–97.

Bender, O., 1906. Die Schleimhautnerven des Facialis, Glossopharyngeus und Vagus. Studien zur Morphologie des Mittelohres und der benachbarten Kopfregion der Wirbeltiere. Denkschriften der Medizinisch-Naturwissenschaftlichen Gesellschaft zu Jena, Volume VII, pp. 341–454.

Bloch, J. I., Boyer, D. M., 2002. Grasping primate origins. Science 298, 1606–1610.

Bondy, G., 1907. Beiträge zur vergleichenden anatomie des Gehörorgans der Säuger. (Tympanicum, Membrana Shrapnelli und Chordaverlauf). Anatomische Hefte 35, 293–408.

Bugge, J., 1974. The cephalic arterial system in insectivores, primates, rodents and lagomorphs, with special reference to the systematic classification. Acta Anatomica 87(Supplement 62), 1–160.

Cartmill, M., Kay, R. F., 1978. Cranio-dental morphology, tarsier affinities and primate suborders. In: Chivers, D. J., Joysey, K. A. (Eds.), Recent Advances in Primatology. Academic, New York, pp. 205–213.

Cords, E., 1915. Ueber das Primordialcranium von *Perameles spec.?* unter Beruecksichtigung der Deckknochen. Anatomische Hefte 52, 1–82.

De Beer, G. R., 1937. The Development of the Vertebrate Skull. Clarendon, Oxford.

Delpero, M., Masters, J. C., Cervella, P., Crovella, S., Ardito, G., Rumpler, Y., 2001. Phylogenetic relationships among the Malagasy lemuriforms (Primates: Strepsirrhini) as indicated by mitochondrial sequence data from the 12S rRNA gene. Zoological Journal of the Linnean Society 133, 83–103.

Doran, A. H. G., 1878. Morphology of the mammalian Ossicula auditus. Transactions of the Linean Society London, series 2, 1, 371–497.

Fleischer, G., 1973. Studien am Skelett des Gehörorgans der Säugetiere, einschließlich des Menschen. Säugetierkundliche Mitteilungen 21, 131–239.

Gaupp, E., 1906. Die Entwickelung des Kopfskelettes. In: Hertwig, O. (Ed.), Handbuch der vergleichenden und experimentellen Entwickelungslehre der Wirbeltiere III (2), 573–873. Gustav Fischer, Jena.

Gaupp, E., 1908. Zur Entwickelungsgeschichte und vergleichenden Morphologie des Schädels von *Echidna aculeata* var. typica. Semonsche Zoologische Forschungsreisen III, Part II, 2, 539–788. Jena.

Gaupp, E., 1911. Beiträge zur Kenntnis des Unterkiefers der Wirbeltiere. I. Der Processus anterior (Folii) des Hammers der Säuger und das Goniale der Nichtsäuger. Anatomischer Anzeiger 39, 97–135.

Gaupp, E., 1913. Die Reichertsche Theorie (Hammer-, Aboss- und Kieferfrage). Archiv für Anatomie und Entwicklungsgeschichte 1912 (Supplement), 1–416.

Gegenbaur, C., 1892. Lehrbuch der Anatomie des Menschen. 5th edition. W. Engelmann, Leipzig.

Glaser, H., 1680. Tractatus Posthumus de Cerebro.- Basil. (quoted after Hyrtl, 1845).

Goodrich, E. S., 1930. Studies on the Structure and Development of Vertebrates. Macmillan, London.

Gregory, W. K., 1910. The orders of mammals. Bulletin of the American Museum of Natural History 27, 1–524.

Hennig, W., 1966. Phylogenetic Systematics. University of Illinois Press, Urbana, IL.

Henson, O. W., 1974. Comparative anatomy of the middle ear. In: Keidel, W. D., Neff, W. D. (Eds.), Handbook of Sensory Physiology Springer, Berlin, pp. 39–109.

Hertwig, O., 1888. Lehrbuch der Entwicklungsgeschichte des Menschen und der Wirbelthiere. Fischer, Jena.

Hyrtl, J., 1845. Vergleichend-anatomische Untersuchungen ueber das innere Gehoerorgan des Menschen und der Saeugethiere. F. Ehrlich, Prague.

Hyrtl, J., 1873. Lehrbuch der Anatomie des Menschen. 12th edition. W. Braumueller, Vienna.

Jarvik, E., 1980. Basic Structure and Evolution of Vertebrates. Academic, London.

Kay, R. F., Ross, C., Williams, B., 1997. Anthropoid origins. Science 275, 797–804.

Kuhn, H.-J., Zeller, U., (Eds.) 1987. Morphogenesis of the Mammalian Skull - Mammalia Depicta 13. Parey Verlag, Hamburg, pp. 1–144.

MacPhee, R. D. E., 1977. Ontogeny of the ectotympanic -petrosal plate relationship in strepsirhine prosimians. Folia primatologica 27, 245–283.

MacPhee, R. D. E., 1981. Auditory regions of primates and eutherian insectivores. Contributions to Primatology 18, 1–282.

MacPhee, R. D. E., Cartmill, M., 1986. Basicranial structures and primate systematics. In: Swindler, D., Erwin, J. (Eds.), Comparative Primate Biology, Volume 1. Alan R. Liss, New York, pp. 219–275.

Maier, W., 1990. Phylogeny and ontogeny of mammalian middle ear structures. Netherlands Journal of Zoology 40, 55–74.

Maier, W., 1993. Cranial morphology of the therian common ancestor, as suggested by the adaptations of neonate marsupials. In: Szalay, F. S., Novacek, M. J., McKenna, M. C., Mammal Phylogeny. Mesozoic differentiation, Multituberculates, Monotremes, Early Therians, and Marsupials. Springer, New York, pp. 165–181.

McKenna, M. C., Bell, S. K., 1997. Classification of Mammals above the Species Level. Columbia University Press, New York.

Meckel, J. F., 1815. Handbuch der menschlichen Anatomie. Buchhandlung des Halleschen Weisenhauses, Halle/Berlin.

Mickoleit, G., 2004. Phylogenetischen Systematik der Wirbeltiere. Verlag Dr. Friedrich Pfeil, Muenchen.

Murphy, W. J., Eizirik, E., O'Brien, J., Zhang, Y. P., Madsen, O., Scally, M., Douady, C. J., Teeling, E., Ryder, O. A., Stanhope, M. J., DeJong, W. W., Springer, M. S., 2001. Resolution of the early placental mammal radiation using Bayesian phylogenetics. Science 294, 2348–2351.

Novacek, M. J., 1990. Morphology, paleontology, and the higher clades of mammals. In: Genoways, H. (Ed.), Current Mammalogy Volume 2, Plenum, New York, pp. 507–543.

Novacek, M. J., 1993. Patterns of diversity in the mammalian skull.- In: Hanken, J., Hall, B. K. (Eds.), The skull, Volume 2, The University of Chicago Press, Chicago, pp. 438–545.

Nyhart, L. K., 1995. Biology Takes Form. Animal Morphology and the German Universities 1800–1900. Chicago University Press, Chicago.

Reinbach, W., 1952. Zur Entwicklung des Primordialcraniums von *Dasypus novemcinctus* LINNÉ (*Tatusia novemcincta* LESSON). Part I.- Zeitschrift für Morphologie und Anthropologie 44, 375–444.

Ross, C., 1994. The craniofacial evidence for anthropoid and tarsier relationships. In: Fleagle, J. G. and Kay, R. F. (Eds.), Anthropoid Origins. Plenum, New York, pp. 469–547.

Ross, C., Williams, B., Kay, R. F., 1998. Phylogenetic analysis of anthropoid relationships. Journal of Human Evolution 35, 221–306.

Ruf, I., Frahnert, S., Maier, W., in press. The chorda tympani and its contribution to rodent phylogeny. Mammalian Biology.

Russell, E. S., 1916. Form and Function. A Contribution to the History of Animal Morphology. J. Murray, London (Reprint Chicago University Press 1982).

Saban, R., 1963. Contribution à l'étude de l'os temporal des Primates. Mémoires du Muséum national d'histoire naturelle, ns, A 39, 1–378.

Sargis, E. J., 2002. Primate origins nailed. Science 298, 1564–1565.

Soshani, J., Groves, C. P., Simons, E. L., Gunnell, G. F., 1996. Primate phylogeny: Morphological vs molecular results. Molecular Phylogenetics and Evolution 5, 102–154.

Silcox, M. T., Bloch, J. I., Sargis, E. J., Boyer, D. M., 2005. Euarchonta (Dermoptera, Scandentia, Primates). In: Rose, K. D., Archibald, J. D. (Eds.), The Rise of Placental Mammals. Johns Hopkins University Press, Baltimore, MD, pp. 127–144.

Simmons, N. B., 1993. The importance of methods: Archontan phylogeny and cladisitic analysis. In: MacPhee, R. (Eds.), Primates and their Relatives in Phylogenetic Perspective. Plenum, New York, pp. 1–62.

Spatz, W., 1964. Beitrag zur Kenntnis der Ontogenese des Cranium von *Tupaia glis* (Diard 1820). Morphologisches Jahrbuch 106, 321–416.

Springer, M. S., Murphy, W. J., Eizirik, E., O'Brien, J., 2005. Molecular evodence for major placental clades. In:. Rose, K. D., Archibald, J. D. (Eds.), The Rise of Placental Mammals. Johns Hopkins University Press, Baltimore, MD, pp. 37–49.

Starck, D., 1960. Das cranium eines Schimpansenfetus (*Pan troglodytes* Blumenbach, 1799) von 71 mm SchStlge. Morphologisches Jahrbuch 100, 559–647.

Starck, D., 1962. Das cranium von *Propithecus spec.* (Prosimiae, Lemuriformes, Indriidae). Bibliotheca primatologica 1, 163–196.

Starck, D., 1967. Le crâne des Mammifères. In: Grassé, P.-P.(Ed.), Traité des Zoologie, Volume XVI, Masson, Paris. pp. 405–549.

Szalay, F. S., 1975a. Phylogeny of primate higher taxa: the basicranial evidence. In: Luckett, W., Szalay, F. (Eds.), Phylogeny of the Primates. Plenum, New York, pp. 91–125.

Szalay, F. S., 1975b. Haplorhine phylogeny and the status of the Anthropoidea. In: Tuttle, R. H. (Ed.), Primate Functional Morphology and Evolution. Mouton Publishers, The Hague, The Netherlands, pp. 3–22.

Toeplitz, C., 1920. Bau und Entwicklung des Knorpelschaedels von *Didelphis marsupialis*. Zoologica 27, 1–84.

van Kampen, P. N., 1905. Die Tympanalgegend des Säugetierschädels. Morphologisches Jahrbuch 34, 321–722.

van der Klaauw, C. J., 1931. The auditory bulla in some fossil mammals, with a general introduction to this region of the skull. Bulletin of the American Museum of Natural History 62, 1–352.

Voit, M., 1909. Das Primordialcranium des Kaninchen unter Berücksichtigung der Deckknochen. Anatomische Hefte, 1.Abt. 38, 425–616.

Waddell, P. J., Okada, N., Hasegawa, M., 1999. Towards resolving the interordinal relationships of placental mammals. Systematic Biology 48, 1–5.

Warwick, R., Williams, P. L., (Eds.), 1973. Gray's Anatomy. 35th edition. Longman, London.

Wible, J. R., Martin, J. R., 1993. Ontogeny of the tympanic floor and roof in archontans. In: MacPhee, R. D. E. (Ed.), Primates and their Relatives in Phylogenetic Perspective. Plenum, New York, pp. 111–146.

Williston, S. W., 1903. North American Plesiosaurs. Part I.- Publ. Field Columbian Museum 73, Geological series II/1, 1–77.

Wilson, D. E., Reeder, D. M. 1993. Mammal Species of the World. 2nd edition. Smithsonian Institution Press, Washington, DC.

Zeller, U. A., 1983. Zur Ontogenese und Morphologie des Craniums von *Tupaia belangeri* (Tupaiidae, Scandentia, Mammalia). Medical Dissertation, University Goettingen, 308 pp.

Zeller, U., 1986. The systematic relations of tree shrews: evidence from skull morphogenesis. In: Else, J. G., Lee, P. C. (Eds.), Primate Evolution. Cambridge University Press, Cambridge, pp. 273–280.

Zeller, U., 1989. Die Entwicklung und Morphologie des Schädels von *Ornithorhynchus anatinus* (Mammalia: Prototheria: Monotremata). Abhandlungen (Senckenbergische Naturforschende Gesellschaft) 545, 1–188.

16. Evolutionary Morphology of the Guenon Postcranium and Its Taxonomic Implications

Eric J. Sargis*
Department of Anthropology
Yale University
P.O. Box 208277
New Haven, CT 06520, USA
*Eric.Sargis@yale.edu**

Carl J. Terranova
Department of Anatomy
Touro College of Osteopathic Medicine
230 West 125th St.
New York, NY 10027, USA
carl.terranova@touro.edu

Daniel L. Gebo
Department of Anthropology
Northern Illinois University
Dekalb, IL 60115, USA
dgebo@niu.edu

16.1 Introduction

Guenons (Primates, Cercopithecini) are relatively small-bodied (Table 16.1) Old World monkeys endemic to Africa. They exhibit a variety of substrate preferences, spanning from arboreal to semiterrestrial to terrestrial (Table 16.2). The ancestral guenon was likely arboreal; indeed, the postcranial morphology of semiterrestrial guenons, including the basal *Allenopithecus* (Tosi et al., 2004, 2005), resembles that of their arboreal relatives (Gebo and Sargis, 1994). Morphological modifications attributable to terrestriality are only found in three guenon taxa (Gebo and Sargis, 1994): patas monkeys (*Erythrocebus patas*), the lhoesti group (*Cercopithecus lhoesti, C. preussi*, and *C. solatus*), and vervet monkeys (*Cercopithecus aethiops*; Manaster, 1979; Anapol and Gray, 2003; Anapol et al., 2005).

16.1.1 Functional Morphology

Gebo and Sargis (1994) conducted both qualitative and univariate analyses on the postcranium of guenons. They found that *Erythrocebus patas, Cercopithecus lhoesti*, and *C. preussi* exhibit numerous, but variable, terrestrial adaptations, whereas *C. aethiops* is characterized by far fewer modifications for terrestriality (Table 16.3). They combined Lernould's (1988) classification of guenons with the consensus tree from Ruvolo's (1988) electrophoretic analysis of 14 proteins, and mapped terrestriality onto this cladogram. From this, Gebo and Sargis (1994) hypothesized that terrestriality and the postcranial adaptations for this substrate preference evolved independently three times among guenons. In this paper, we re-evaluate this proposal in light of new molecular

* Address for correspondence: *Eric.Sargis@yale.edu*

E.J. Sargis and M. Dagosto (eds.), *Mammalian Evolutionary Morphology: A Tribute to Frederick S. Szalay*, 361–372.

TABLE 16.1. Number of specimens per taxon, taxon abbreviations used for all plots, and mean body weight (in g) of each taxon.

Taxon	Abbreviation	*n*	Body weight*
Allenopithecus nigroviridis	AN	2	4678
Cercopithecus aethiops	CAe	11	4656
Cercopithecus ascanius	CAs	10	3426
Cercopithecus cephus	CC	4	3355
Cercopithecus diana	CD	6	4643
Cercopithecus hamlyni	CH	1	4638
Cercopithecus lhoesti	CL	1	4663
Cercopithecus mitis	CM	14	5656
Cercopithecus neglectus	CN	6	5701
Cercopithecus pogonias	CPo	1	3187
Cercopithecus preussi	CPr	2	?
Erythrocebus patas	EP	6	7661
Miopithecus talapoin	MT	5	1795

*Mean body weights from Delson et al. (2000) appendix 2; see their appendix 1 for sources of raw body weights.

TABLE 16.2. Guenon substrate preferences.

Taxon	Substrate preferences	Sources
Allenopithecus nigroviridis	Semiterrestrial	Gautier-Hion, 1988
Cercopithecus aethiops	**Semiterrestrial***	Rose, 1979
Cercopithecus ascanius	Arboreal	Gebo and Chapman, 1995
Cercopithecus campbelli	Semiterrestrial	McGraw, 1998, 2000
Cercopithecus cephus	Arboreal	Gautier-Hion, 1988
Cercopithecus diana	Arboreal	McGraw, 1998, 2000
Cercopithecus hamlyni	Semiterrestrial	Gebo and Sargis, 1994
Cercopithecus lhoesti	**Terrestrial**	Kaplin and Moermond, 2000
Cercopithecus mitis	Arboreal	Gebo and Chapman, 1995
Cercopithecus neglectus	Semiterrestrial	Gautier-Hion, 1988
Cercopithecus nictitans	Arboreal	Gautier-Hion, 1988
Cercopithecus petaurista	Arboreal	McGraw, 2000
Cercopithecus pogonias	Arboreal	Gautier-Hion, 1988
Cercopithecus preussi	**Terrestrial**	Gautier-Hion, 1988
Cercopithecus solatus	**Terrestrial**	Gautier-Hion, 1988
Erythrocebus patas	**Terrestrial**	Isbell et al., 1998
Miopithecus talapoin	Arboreal	Gautier-Hion, 1988

* *C. aethiops* may be terrestrial; this species exhibits some postcranial modifications like those of the terrestrial species. (From Gebo and Sargis, 1994.)

evidence on the phylogeny of guenons and we perform a re-analysis of the morphological data.

Although vervet monkeys do not exhibit as many terrestrial adaptations as l'Hoest's or patas monkeys, they can still be distinguished from arboreal guenons in studies of postcranial morphology (Gebo and Sargis, 1994). For example, in his qualitative analysis of the humerus and femur, Nakatsukasa (1994) was able to differentiate *C. aethiops* from the arboreal *C. mitis* and *C. mona*. In addition to osteological differences, the fiber architecture of forelimb and hind limb muscles differs between the more terrestrial *C. aethiops* and the arboreal *C. ascanius* (Anapol and Barry, 1996; Anapol and Gray, 2003). Specifically, the muscle fiber architecture in *C. aethiops* allows higher velocity/excursion for terrestrial running, whereas that of *C. ascanius* is better suited for storage of passive elastic strain energy for use in the compliant canopy (Anapol and Barry, 1996; Anapol and Gray, 2003). Furthermore, Anapol and Gray (2003) showed that the distribution of force potential in the shoulder and arm muscles of *C. aethiops* is better for transitioning between the ground and canopy and for braking during terrestrial running, while that of *C. ascanius* is better suited for stabilization during quadrupedal descent and propulsion during arboreal running. These myological differences led them to propose that "semiterrestrial" is a discrete locomotor category, intermediate between arboreal and terrestrial (Anapol and Gray, 2003; Anapol et al., 2005).

16.1.2 Scaling

Martin and MacLarnon (1988) conducted several logarithmic bivariate regression (major axis) analyses to examine the scaling of various craniodental variables to body size. They explored scaling of 18 cranial variables relative to prosthion-inion (skull) length. Of these, five cranial variables scale isometrically, nine are negatively allometric, and four are positively allometric. Martin and MacLarnon (1988) also examined the scaling trends of seven dental variables relative to body weight, all of which scaled isometrically. Although it is possible that there is a real difference in scaling between cranial and dental measures in guenons, it is perhaps more likely that the variable patterns of scaling seen in the two data sets are a result of Martin and MacLarnon's (1988) use of different body size surrogates for the cranial and dental analyses (Shea, 1992). In this paper, we will explore scaling patterns in the postcrania of guenons.

Shea (1992) also studied scaling in guenons, specifically ontogenetic scaling of *M. talapoin* and *C. cephus*. He analyzed 11 cranial and 6 postcranial measurements; the latter are of particular interest for our study. In his bivariate regression (least-squares and reduced major axis) analyses, he found that the scaling of humerus and tibia length relative to femur length is negatively allometric (Shea, 1992); the same is true when regressing individual limb elements to pelvic height. Here, we examine adult interspecific data, which, unlike the ontogenetic data examined by Shea (1992), do not allow us to examine process. However, scaling patterns are important to explore in our effort to elucidate both the functional and evolutionary patterns of association among guenons.

16.1.3 Multivariate Analyses

Further evidence for the distinction of *C. aethiops* from arboreal guenons comes from Manaster's (1979) multivariate analysis of 67 postcranial variables in 7 guenon species. She conducted a canonical analysis, which separated three groups: (1) *C. aethiops*; (2) *C. mitis* and *C. neglectus*; and (3) *C. mona, C. cephus, C. nictitans*, and *C. diana*. In this analysis, canonical axis 1 separated the terrestrial *C. aethiops* from the six arboreal and semiterrestrial taxa (Manaster, 1979).

TABLE 16.3. Postcranial adaptations present in terrestrial taxa (from Gebo and Sargis, 1994).

Character	Taxa
1. Elongated fore- and hind limbs (Hurov, 1987)	*patas*
2. High intermembral index	*patas*
3. High brachial index	*patas*
4. Long, narrow scapula (short vertebral border)	*patas*
5. Short infraspinous fossa	*patas*
6. Square glenoid fossa	*patas*
7. Strongly retroflexed humerus	*patas*
8. Narrow posterior humeral trochlea with high medial edge	*patas*
9. Straight ulna and radius	*patas*
10. Strongly retroflexed femur	*patas*
11. Deep (a-p) knee with high lateral patellar rim	*patas*
12. Digitigrade feet	*patas*
13. Short supraspinous fossa	*patas, lhoesti*
14. Oblique angle of humeral head	*patas, lhoesti, preussi*
15. Medially curved humeral shaft	*patas, lhoesti, preussi*
16. Deep radial and ulnar fossae on humerus	*patas, lhoesti, preussi*
17. Small radial facet on ulna	*patas, lhoesti, preussi*
18. Distal radial facets flat (not concave)	*patas, lhoesti, preussi*
19. Small femoral head articular surface	*patas, lhoesti, preussi*
20. Small/reduced anterior calcaneal facet (distal part)	*patas, lhoesti, preussi*
21. Shallow calcaneocuboid pivot (on calcaneus)	*patas, lhoesti, preussi*
22. Narrow and smooth (no lateral ridge) humeral trochlea	*lhoesti, preussi*[a]
23. Medially twisted distal ulnar shaft	*lhoesti, preussi*
24. Femoral head/neck perpendicular to shaft (not oblique)	*lhoesti, preussi*
25. Short tibial crest	*lhoesti, preussi*
26. Small peroneal tubercle (on calcaneus)	*lhoesti, preussi*
27. Narrow and tall talar head	*lhoesti, preussi, aethiops*
28. Acromion process/spine not angled cranially	*lhoesti*
29. Small infraspinatus flange	*lhoesti*
30. Narrow humeral head	*lhoesti*
31. Greater/lesser tuberosities extend above humeral head	*lhoesti*
32. Short deltopectoral crest	*lhoesti*
33. Vertical medial trochlear rim (posterior humerus)	*lhoesti*
34. Small radial head and articular surface	*lhoesti*
35. Greater trochanter extends above femoral head	*lhoesti*
36. Posterior position of lesser trochanter	*lhoesti*
37. Narrow patellar groove (with sharp medial rim)	*lhoesti*
38. Deep (a-p) distal tibial facet with small medial part	*lhoesti*
39. Deep, pronounced tibial cup (medial talar facet for tibia)	*lhoesti*
40. Less medially angled (straighter) calcaneal tuber	*lhoesti*
41. Reduced attachment areas on calcaneus for ligaments	*lhoesti*
42. Low humeral mid-shaft cortical areas	*lhoesti*[b]
43. Long cuboid	*lhoesti, aethiops*
44. Narrow humeral facet on ulna	*aethiops*
45. Long navicular	*aethiops*
46. Short third metatarsal	*aethiops*
47. Long calcaneus (Kingdon, 1988)	*aethiops, patas*[c]
48. Short manual and pedal digits (Kingdon, 1988)	*aethiops, patas*[c]

[a] *C. preussi* has a small lateral trochlear ridge.
[b] *C. lhoesti* was the only terrestrial taxon to be scanned.
[c] More extreme in *E. patas* than *C. aethiops* (Kingdon, 1988).

Martin and MacLarnon (1988) also performed multivariate analyses on 12 guenon species, although their study focused on craniodental morphology. They conducted cluster and multidimensional scaling analyses on 18 cranial and 7 dental variables, as well as analyses on a combined data set that included the 7 dental variables and 11 of the 18 cranial variables. In their cranial analysis, *C. lhoesti* was united to *Miopithecus talapoin* and *E. patas* clustered with these two taxa, while *Allenopithecus nigroviridis* and *C. neglectus* were distant. The dental analysis united *E. patas* with *M. talapoin* and *C. lhoesti* clustered with these two genera. Again, *Allenopithecus* was distant, but it formed a cluster with *C. aethiops* in this analysis. The combined craniodental analysis produced results similar to the cranial analysis in some ways and to the dental analysis in other ways. Specifically, *C. lhoesti* joined *M. talapoin* and *E. patas* joined these two species, as in the cranial analysis. Furthermore, *Allenopithecus* was, once again, distant, and, as in the dental analysis, it clustered with *C. aethiops*. In summary, *C. lhoesti, M. talapoin*, and *E. patas* are phenetically similar craniodentally, whereas *Allenopithecus* is dissimilar. Although *C. aethiops* was united to *Allenopithecus* in the dental analysis, it clustered with arboreal *Cercopithecus* species in the cranial analysis. Based on these results, Martin and MacLarnon (1988) stated that the topologies from their cluster analyses were similar to those from the karyotype analyses of Dutrillaux et al. (1988), which supported a terrestrial group including *C. lhoesti, E. patas*, and *C. aethiops*. However, this group was not precisely supported in Martin and MacLarnon's (1988) multivariate analyses, as *M. talapoin* replaced *C. aethiops* in a cluster with *C. lhoesti* and *E. patas*. An additional similarity between the craniodental and chromosomal studies is the distant position of *Allenopithecus*. Based on these similarities, Martin and MacLarnon (1988) proposed that their phenetic analysis was more indicative of phylogeny than function even though this was not expected at the outset of their study. They also proposed that taxonomically restricted analyses, such as their examination of guenons, might better elucidate phylogenetic patterns, while more inclusive analyses may emphasize functional convergences (Martin and MacLarnon, 1988). Here, we compare our results from a cluster analysis of guenon postcrania to those from Martin and MacLarnon's (1988) craniodental analyses.

16.1.4 Molecular Phylogenetics

Both the karyotype analysis of Dutrillaux et al. (1988) and the Y-chromosome sequence data of Tosi et al. (2002, 2003, 2004) support a terrestrial clade that includes *E. patas, C. aethiops*, and *C. lhoesti*. Additional sequence data from an X-chromosome intergenic region also support this terrestrial clade, as well as the basal divergence of *Allenopithecus* (Tosi et al., 2004, 2005). Within the terrestrial clade, the X-chromosome evidence supports a *C. aethiops-E. patas* sister taxon relationship (Tosi et al., 2004, 2005), which is also supported by three cranial synapomorphies (Groves, 2000). Based on these phylogenetic results, Tosi et al. (2002, 2003, 2004) proposed that terrestriality is derived and evolved only once among guenons. This contradicts Gebo and Sargis' (1994) proposal that terrestriality evolved three times among guenons, which was based on a phylogeny from Ruvolo (1988) where *E. patas, C. aethiops*, and *C. lhoesti* were not closely related. We re-evaluate this proposal in light of recent molecular evidence (see Tosi et al., 2002, 2003, 2004, 2005).

16.1.5 Taxonomy

The inclusion of *C. aethiops* and *C. lhoesti* in a clade with *E. patas* to the exclusion of other *Cercopithecus* species makes the genus *Cercopithecus* paraphyletic (Tosi et al., 2002, 2003, 2004). Four different solutions to this taxonomic problem have been proposed (Tosi et al., 2002, 2003, 2004): (1) patas monkeys could be sunk into *Cercopithecus*; (2) patas monkeys could be left in *Erythrocebus* (Trouessart, 1897), vervet monkeys could be restored to *Chlorocebus* (Gray, 1870), and l'Hoest's monkeys could be reinstated to *Allochrocebus* (Elliot, 1913); (3) vervet, patas, and l'Hoest's monkeys could be placed in *Chlorocebus*, which has priority over *Erythrocebus*, to formally recognize the close relationship between these taxa in the terrestrial clade; (4) l'Hoest's monkeys could be restored to *Allochrocebus* and vervet and patas monkeys could be placed in *Chlorocebus*, which would formally recognize the sister taxon relationship between these two species (Tosi et al., 2004). Proposal #3 is the favored classification of Tosi et al. (2002, 2003, 2004). Here, we use the postcranial evidence to assess each of these four proposals.

16.2 Materials and Methods

16.2.1 Sample

The number of guenon specimens of each taxon included in this study is summarized in Table 16.1. Gebo and Sargis (1994) reported 37 measurements (their table 3) and 26 indices (their table 4), which we used in our multivariate and scaling analyses.

16.2.2 Scaling Analyses

Scaling patterns were assessed using reduced major axis (RMA) regression of natural logarithm transformed species mean data. RMA (Bohonak, 2002) was used to estimate regression parameters. All 13 taxa were included in analyses focusing on the scaling of long bone lengths with other long bone lengths. We also develop scaling equations for all long bones on species mean body weight (see Table 16.1). No weight data are available for *Cercopithecus preussi* (Delson et al., 2000), so only 12 taxa are included in this portion of the scaling analysis.

16.2.3 Multivariate Analyses

A cluster analysis (unweighted pair-group average [UPGA]) was performed on the raw species means of the 26 indices reported by Gebo and Sargis (1994, table 4). The tree is presented with Euclidean distances (see Table 16.1 for abbreviations).

A principal components analysis (PCA) was performed on a correlation matrix computed from the natural logarithm transformed species means of the 37 measurements reported by Gebo and Sargis (1994, table 3). No additional rotations (e.g., Varimax) were performed (see Neff and Marcus, 1980, p. 104). The first two factors of the PCA, which were the only factors with eigenvalues greater than one, were compared in a bivariate plot (see Table 16.1 for abbreviations), and this is included with the eigenvalues, percent of total variance, factor scores, and factor loadings below. Statistica (StatSoft Inc., Tulsa, OK) was used to perform these multivariate analyses.

16.3 Results

16.3.1 Cluster Analysis

The cluster analysis of 26 postcranial indices shows that three terrestrial taxa, *Cercopithecus preussi, C. lhoesti*, and *Erythrocebus patas*, are linked to the semiterrestrial *Allenopithecus* (Figure 16.1). The other terrestrial taxon, *C. aethiops*, is linked to four arboreal taxa in a separate cluster. A third cluster is formed by an arboreal taxon and two semiterrestrial species. The arboreal *C. pogonias*, a relatively small-bodied species (see Table 16.1), is quite distant from all other guenons. It is worth noting that *E. patas* is linked to much smaller taxa and *Miopithecus talapoin* is linked to much larger species, thereby indicating that the clusters are not simply related to body size.

16.3.2 Principal Components Analysis

Figure 16.2 shows a bivariate plot of the first two factors from the principal components analysis (see Tables 16.4–16.6 for eigenvalues, factor scores, and factor loadings, respectively). Factor 1, which accounts for 88.3% of the total variance (Table 16.4), is likely a size vector because it separates large- (*Erythrocebus patas*), medium- (*Cercopithecus, Allenopithecus*), and small- (*Miopithecus*) bodied guenons (see Figure 16.2; Table 16.1). Factor 2, however, separates *C. lhoesti* from all other guenons, including *C. preussi* (Figure 16.2).

Some of the measurements with the highest loadings for factor 2 (Table 16.6) reflect the qualitative (and univariate) differences between the terrestrial *C. lhoesti* and the arboreal *C. mitis* outlined by Gebo and Sargis (1994). For instance, *C. lhoesti* has a relatively narrow humeral head (Figure 16.3a) and short supraspinous fossa (Figure 16.3b) compared to those of *C. mitis* (Gebo and Sargis, 1994), and humeral head width (measurement #1) and supraspinous fossa height (measurement #4) have relatively high loadings for factor 2 (Table 16.6). Similarly, *C. lhoesti* has a very small ulnar radial facet (Figure 16.3c) relative to that of *C. mitis* (Gebo and Sargis, 1994), and ulnar radial facet width (measurement #8) has the highest loading for factor 2 (Table 16.6). Finally, *C. lhoesti* has a relatively short olecranon process and a very narrow patellar groove (Figure 16.3d) relative to those of *C. mitis* (Gebo and Sargis, 1994), and ulnar olecranon process length (measurement #9) and femoral patellar groove width (measurement #19) have relatively high loadings for factor 2 (Table 16.6).

16.3.3 Scaling

Table 16.7 presents the reduced major axis equations for the entire sample. Correlations are uniformly high, indicating that the relationship uncovered is unlikely to be much affected by model selection. Each talapoin-to-patas monkey equation is characterized by an isometric slope. This indicates no change in gross shape of the fore- and hind limbs with size change among guenons.

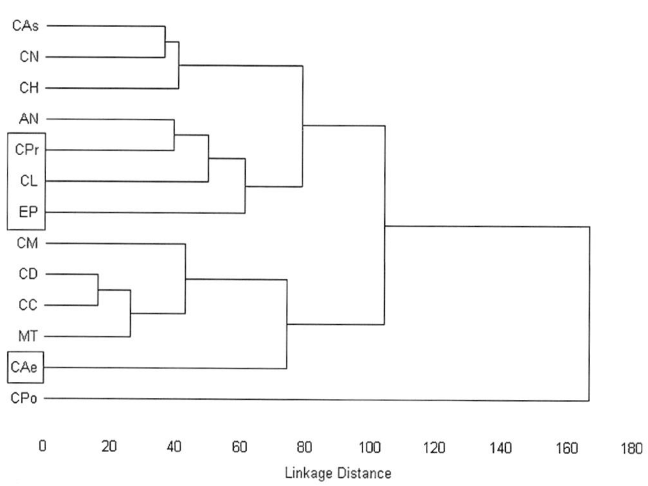

FIGURE 16.1. Cluster analysis of 26 postcranial indices. See Table 16.1 for abbreviations. Note that three terrestrial taxa (CPr, CL, EP) are linked to the semiterrestrial *Allenopithecus* (AN) rather than the terrestrial *Cercopithecus aethiops* (CAe).

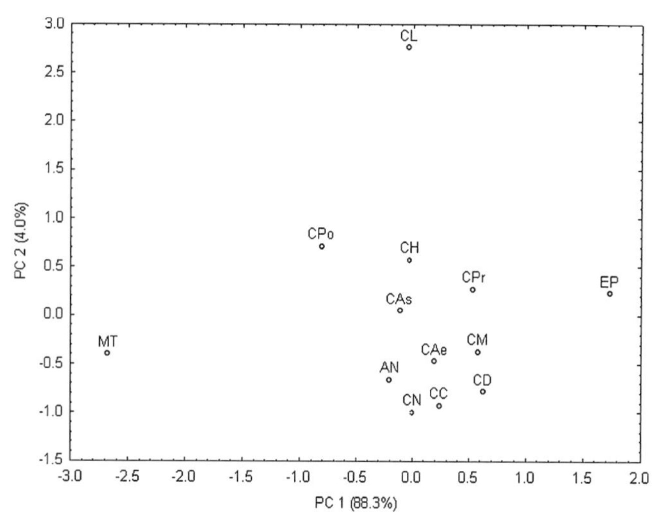

FIGURE 16.2. Bivariate plot of Factors 1 and 2 from the principal components analysis. Note that Factor 1 is likely a size vector and Factor 2 separates *Cercopithecus lhoesti* (CL) from all other guenons. See Table 16.1 for abbreviations.

Table 16.4. Eigenvalues from principal components analysis.

Factors	Eigenvalues	% of Total variance	Cumulative eigenvalues	Cumulative % of total variance
1	32.7	88.3	32.7	88.3
2	1.5	4.0	34.2	92.3

Table 16.5. Factor scores from principal components analysis.

Taxon	Factor 1	Factor 2
Allenopithecus nigroviridis	−0.2032	−0.6587
Cercopithecus aethiops	0.1893	−0.4726
Cercopithecus ascanius	−0.1166	0.0512
Cercopithecus cephus	0.2395	−0.9238
Cercopithecus diana	0.6260	−0.7833
Cercopithecus hamlyni	−0.0303	0.5737
Cercopithecus lhoesti	−0.0518	2.7605
Cercopithecus mitis	0.5778	−0.3736
Cercopithecus neglectus	−0.0009	−0.9922
Cercopithecus pogonias	−0.8159	0.7125
Cercopithecus preussi	0.5332	0.2742
Erythrocebus patas	1.7315	0.2346
Miopithecus talapoin	−2.6786	−0.4023

Table 16.8 presents the reduced major axis scaling equations of long bone length on mean body weight. The femur and tibia scale isometrically with body weight. The forelimb (humerus, radius, and ulna) is characterized by slight positive allometry. However, the correlation between any two variables does not exceed 0.80. Although all are significant, it is difficult to have much confidence in these slope (or y-intercept) point estimates. As such, we regard the scaling patterns identified here as tentative.

Table 16.6. Factor loadings from principal components analysis*.

Measurement	Factor 1	Factor 2
1. Humeral head width	−0.931	**0.263**
2. Humeral head height	−0.979	0.041
3. Humerus length	−0.986	−0.086
4. Supraspinous fossa height	−0.963	**0.193**
5. Supraspinous fossa length	−0.950	−0.016
6. Ulnar sigmoid notch length	−0.944	0.153
7. Ulnar humeral facet width	−0.885	0.170
8. Ulnar radial facet width	−0.416	**0.865**
9. Ulnar olecranon process length	−0.934	**0.298**
10. Ulna length	−0.962	−0.061
11. Radius length	−0.962	−0.087
12. Pisiform length	−0.974	0.133
13. Innominate length	−0.990	−0.093
14. Ischium length	−0.974	0.094
15. Pubis length	−0.874	−0.146
16. Ilium length	−0.977	−0.149
17. Femur length	−0.982	−0.052
18. Femoral greater trochanter height	−0.948	0.055
19. Femoral patellar groove width	−0.910	**0.292**
20. Femoral epicondylar width	−0.990	0.011
21. Femoral condylar height	−0.982	−0.084
22. Femoral medial condyle width	−0.985	0.081
23. Femoral lateral condyle width	−0.937	−0.108
24. Tibia length	−0.973	−0.085
25. Tibial patellar crest length	−0.787	0.093
26. Calcaneus length	−0.982	−0.078
27. Distal calcaneal length	−0.947	−0.250
28. Posterior calcaneal facet length	−0.956	−0.225
29. Calcaneal heel width	−0.982	0.072
30. Talar trochlea width	−0.992	−0.092
31. Talar head width	−0.947	−0.235
32. Talar head height	−0.960	−0.109
33. Talus length	−0.974	−0.092
34. Talar neck length	−0.945	−0.133
35. Navicular length	−0.790	0.061
36. Cuboid length	−0.953	−0.057
37. Third metatarsal length	−0.961	−0.084

*Values in bold are discussed in the text.

16.4 Discussion

16.4.1 Scaling

We focused on the scaling of long bone lengths because they have been examined previously by Shea (1992). Our data differ from those examined by Shea in several important ways. He examined two taxa and his data were ontogenetic. Our sample is restricted to adults, although we included the entire range of body size within Cercopithecini. From these analyses, we can suggest that limb bone lengths are geometrically similar across all guenons.

However, some interesting patterns are evident when species mean body weights are used (Table 16.8). Hind limb long bones (femur and tibia) are isometric with body weight among guenons. On the other hand, all of the forelimb long bones scale with slight positive allometry. This pattern reflects the relatively elongated forelimb (Figure 16.4), high intermembral index, and high brachial index (Figure 16.5) of *E. patas* (Table 16.3).

16.4.2 Locomotor Evolution

By mapping substrate preference onto the consensus tree from Ruvolo's (1988) electrophoretic analysis of proteins, Gebo and Sargis (1994) proposed that terrestriality evolved three times among guenons. Alternatively, when this character is mapped onto the maximum likelihood (ML) tree from the analysis of X-chromosome sequence data in Tosi et al. (2005), terrestriality appears to have evolved only once in this group (Figure 16.6; Tosi et al., 2002, 2003, 2004). Their ML topology is congruent with their maximum parsimony tree (Tosi et al., 2005), and the terrestrial clade is also supported by their Y-chromosome data (Tosi et al., 2002, 2003, 2004), as well as the karyotype data of Dutrillaux et al. (1988). Based on the congruence of these data sets, their topology appears to represent a robust phylogenetic hypothesis (i.e., species tree) of guenon relationships (Tosi et al., 2004). We (DLG and EJS) therefore retract our previous proposal that

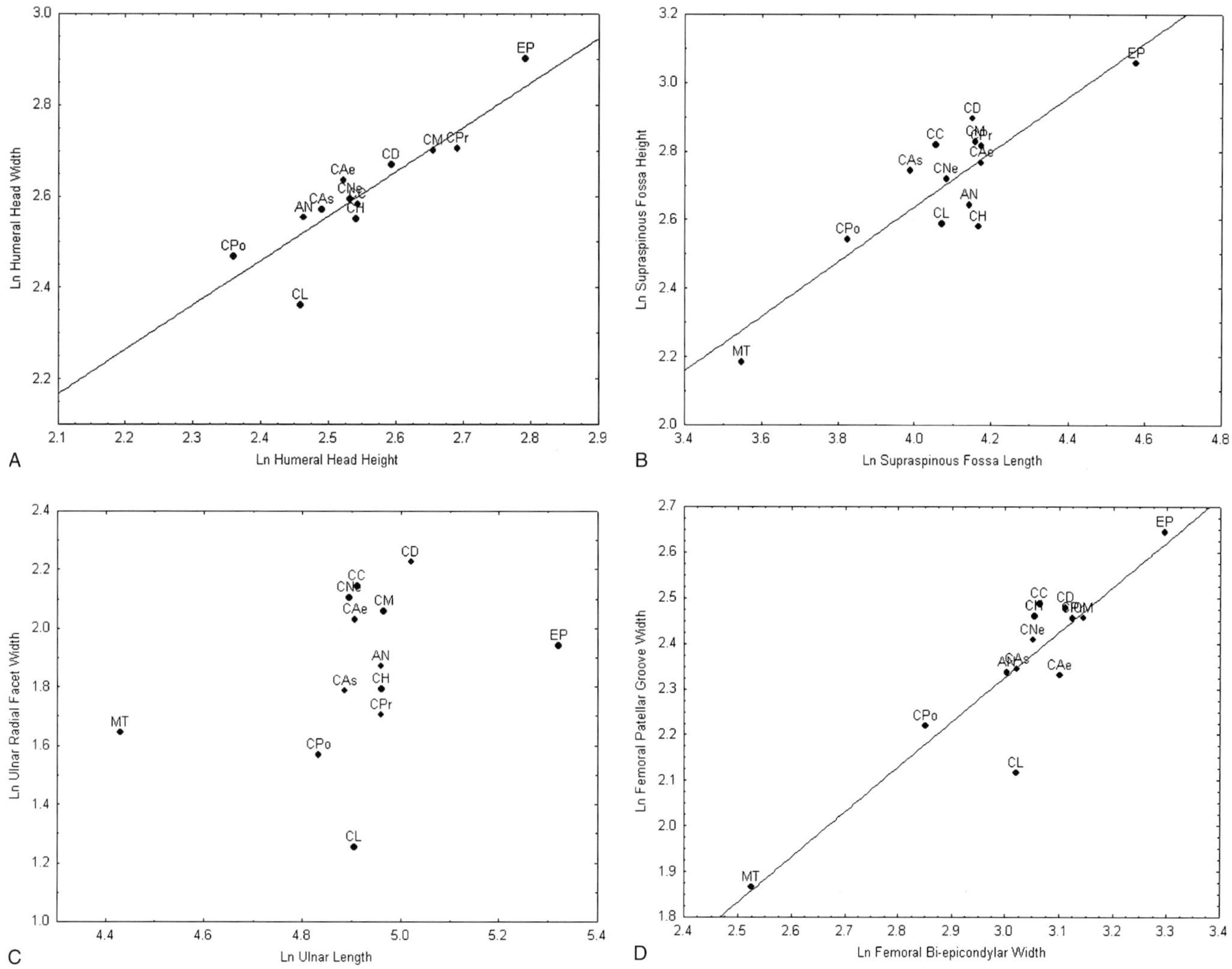

FIGURE 16.3. Bivariate plots of variables with high loadings for Factor 2 from the principal components analysis (see Table 16.6). A, Note the relatively narrow humeral head of *Cercopithecus lhoesti* (CL). B, Note the relatively short supraspinous fossa of *C. lhoesti* (CL). C, Note the relatively small ulnar radial facet of *C. lhoesti* (CL). A best fit line is not depicted because linear regression is not significant. D, Note the relatively narrow patellar groove of *C. lhoesti* (CL).

TABLE 16.7. Reduced Major Axis regression parameters and scaling patterns for long bone length regressed on long bone or innominate length.

	Slope	95% CI, Slope	y-Intercept	R^2	*n*[a]	Scaling
Humerus * Radius	0.96	0.82–1.1	0.23	0.95	13	Isometric
Humerus * Ulna	0.95	0.78–1.1	0.13	0.92	13	Isometric
Humerus * Femur	0.99	0.84–1.2	−0.20	0.94	13	Isometric
Femur * Tibia	1.05	0.93–1.2	−0.25	0.97	13	Isometric
Tibia * Radius	0.91	0.72–1.1	0.64	0.91	13	Isometric
Tibia * Ulna	0.91	0.69–1.1	0.54	0.88	13	Isometric
Humerus * Innominate	1.08	0.95–1.2	−0.33	0.97	13	Isometric
Radius * Innominate	1.13	0.90–1.4	−0.59	0.91	13	Isometric
Ulna * Innominate	1.14	0.92–1.4	−0.48	0.92	13	Isometric
Femur * Innominate	1.08	0.93–1.2	−0.14	0.96	13	Isometric
Tibia * Innominate	1.03	0.86–1.2	0.11	0.95	13	Isometric
Forelimb * Hind limb	1.05	0.86–1.2	−0.21	0.93	13	Isometric

[a] *n* = Number of taxa.

TABLE 16.8. Reduced Major Axis regression parameters and scaling patterns for long bone length regressed on body weight*.

	Slope	95% CI, Slope	y-Intercept	R^2	n^{a}	Scaling
Humerus * Weight	0.50	0.34–0.66	0.61	0.74	12	Positive
Radius * Weight	0.54	0.35–0.72	0.33	0.76	12	Positive
Ulna * Weight	0.57	0.37–0.70	0.44	0.80	12	Positive
Femur * Weight	0.50	0.32–0.69	0.79	0.74	12	Isometric
Tibia * Weight	0.48	0.30–0.67	0.95	0.71	12	Isometric

*Mean body weights from Table 16.1.
[a] n = Number of taxa.

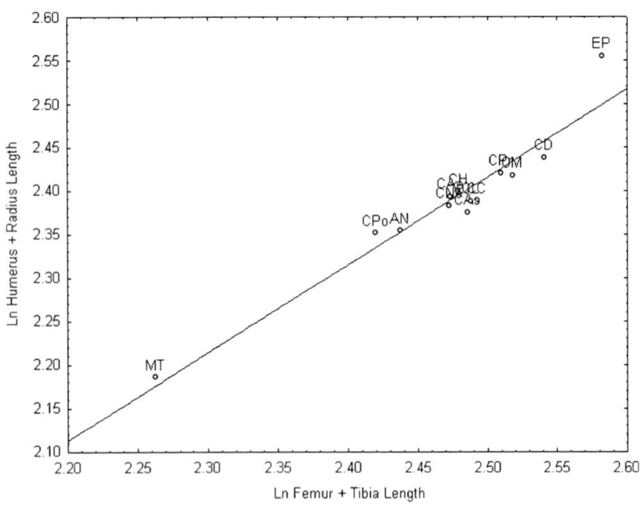

FIGURE 16.4. Bivariate plot of forelimb and hind limb length. Note the relatively long forelimb of *Erythrocebus patas* (EP). This illustrates the high intermembral index of *E. patas* (Table 16.3).

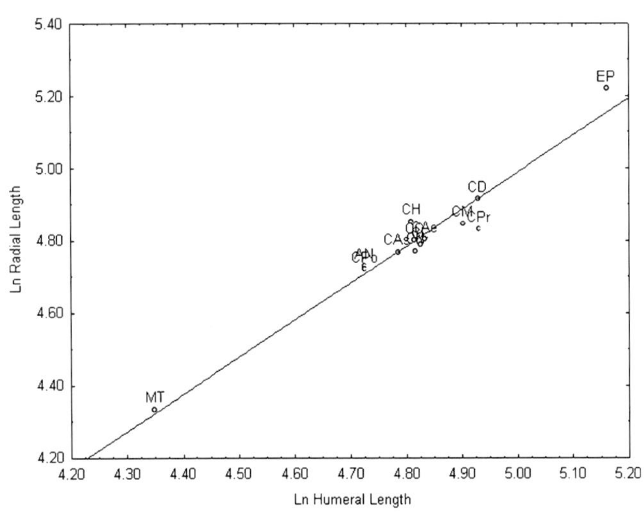

FIGURE 16.5. Bivariate plot of radial and humeral length. Note the relatively long radius of *E. patas* (EP). This illustrates the high brachial index of *E. patas* (Table 16.3).

terrestriality evolved three times in Cercopithecini (Gebo and Sargis, 1994) since Ruvolo's (1988) phylogenetic hypothesis has subsequently been rejected in analyses of a larger and more diverse data set (see Tosi et al., 2004). However, if terrestrial locomotion evolved only once among extant guenons, does this mean that a single suite of terrestrial adaptations also evolved once in the ancestor of this terrestrial clade?

16.4.3 Character Evolution

Surprisingly, the terrestrial species do not form a group in our cluster analysis (Figure 16.1). Although *Cercopithecus preussi*, *C. lhoesti*, and *Erythrocebus patas* cluster together, they do so with *Allenopithecus* rather than with *C. aethiops*; the latter appears to be quite different from the other terrestrial taxa given this phenetic assessment of postcranial morphology. This is particularly surprising because one would have predicted that the terrestrial taxa would cluster together because of postcranial adaptations for terrestriality, whereas one would not necessarily have predicted that they would form a clade in a molecular phylogenetic analysis. Alternatively, it is not surprising that *C. aethiops* is distant from the terrestrial taxa in the cluster analysis given Gebo and Sargis' (1994) earlier assessment that it lacks many of the qualitative terrestrial features present in the other terrestrial taxa (see Table 16.3). Furthermore, Martin and MacLarnon's (1988) cluster analyses also recovered a group that included *C. lhoesti* and *E. patas*, although with *Miopithecus* rather than *Allenopithecus*, to the exclusion of *C. aethiops*.

In contrast to the cluster analysis, *C. lhoesti* and *E. patas* are distinct in the principal components analysis (Figure 16.2). *E. patas* is separated on Factor 1, whereas *C. lhoesti* is separated on Factor 2. The separation of *E. patas* on Factor 1 is likely due to its large body size, just as the separation of *M. talapoin* is likely due to its small body size (see Table 16.1). Alternatively, the separation of *C. lhoesti* on Factor 2 is not related to body size. In fact, many of the highest loadings for Factor 2 (see Table 16.6) correspond to qualitative traits that differentiate *C. lhoesti* from other guenons (see above; Gebo and Sargis,

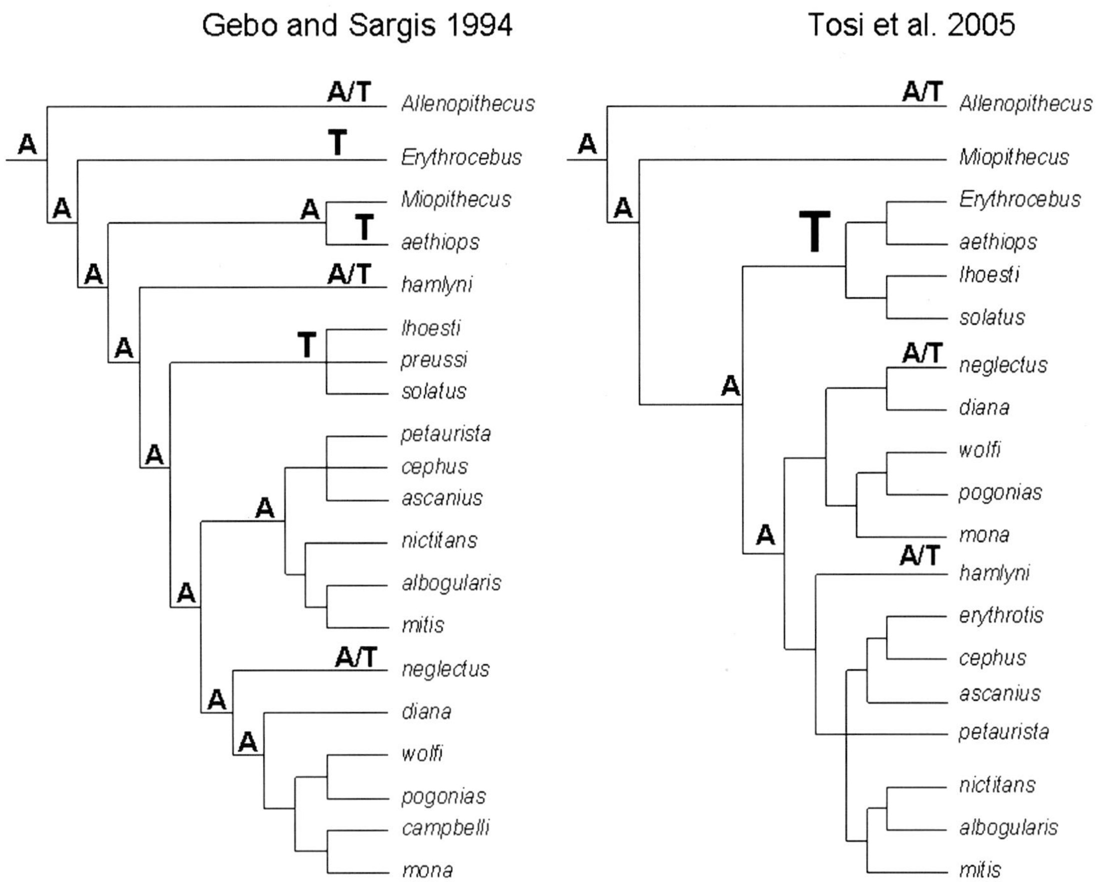

FIGURE 16.6. Cladogram used by Gebo and Sargis (1994) with substrate preference mapped on (left). Cladogram from Tosi et al. (2005) with substrate preference mapped on (right). A: arboreal; T: terrestrial; A/T: semiterrestrial. Note that terrestriality is derived in both cases, but either evolved three times (left) or only once (right); see text for discussion. Also note that semiterrestrial guenons are postcranially similar to arboreal guenons (Gebo and Sargis, 1994).

1994). In summary, the cluster analysis differentiates *C. aethiops* from the other terrestrial taxa, whereas the principal components analysis differentiates *C. lhoesti* and *E. patas*.

The results of these multivariate analyses show that terrestrial guenons are quite different from one another in their postcranial morphology (Figures 16.1–16.2), a conclusion also reported by Gebo and Sargis (1994). In fact, there do not appear to be any terrestrial adaptations that are shared by all three terrestrial groups (*E. patas, C. aethiops*, and *C. lhoesti/preussi*; see Table 16.3; Gebo and Sargis, 1994). Although there are terrestrial features that are shared by two of the three taxa, many appear to be unique to *E. patas* or *C. lhoesti* (see Table 16.3, nos. 1–12 and 28–42, respectively). This leads us to conclude that although terrestrial locomotion may have evolved only once among guenons (Tosi et al., 2002, 2003, 2004), the postcranial morphology of the three terrestrial taxa must have diverged significantly from the condition found in the common ancestor of the terrestrial clade. This is particularly true of *E. patas* and *C. lhoesti*, whose numerous postcranial differences indicate independent acquisition of many terrestrial adaptations. This view was stated previously by Gebo and Sargis (1994) when they proposed that terrestriality evolved three times among guenons, and it remains probable even if terrestrial locomotion evolved only once. The contrast is simply continued postcranial divergence of *E. patas* and *C. lhoesti* from a terrestrial ancestor (i.e., within the terrestrial clade) rather than independent evolution of terrestriality in these taxa.

Despite their numerous postcranial differences, *E. patas* and *C. lhoesti/preussi* share many more terrestrial features (Table 16.3, nos. 13–21) than *C. lhoesti/preussi* shares with *C. aethiops* (Table 16.3, nos. 27 and 43) or *E. patas* shares with *C. aethiops* (Table 16.3, nos. 47–48). This is surprising considering that the X-chromosome data support a *C. aethiops-E. patas* sister taxon relationship within the terrestrial clade (Tosi et al., 2004, 2005). This relationship was predicted by cranial morphology (three synapomorphies; Groves, 2000), but not by postcranial morphology. Table 16.3 lists eight characters shared by *E. patas* and *C. lhoesti/preussi* (nos. 14–21), but only two are shared by *E. patas* and *C. aethiops* (nos. 47–48) and both are more extreme in *E. patas* (Kingdon, 1988). In addition to differences in limb morphology, Hurov (1987) outlined a number of differences in axial skeleton morphology and sagittal back motion between *E. patas* and *C. aethiops*. For example, *C. aethiops* exhibits an increased range of sagittal back motion when galloping speed increases, but *E. patas* does not. Sagittal back mobility is increased in *C. aethiops* by the presence of thicker intervertebral disks and reduced dorsoventral diameter of the thoracic cage relative to the conditions found in *E. patas*. Hurov (1987) summarized these differences between vervet and patas monkeys by concluding that *C. aethiops* increases stride length by increasing its range of sagittal back motion, whereas *E. patas* increases stride length via its elongated fore- and hind limbs.

Reconstructing the ancestral condition for the terrestrial clade is difficult because no terrestrial features are ubiquitous among the terrestrial taxa (Table 16.3). The terrestrial clade is subdivided into the *Lhoesti*-group (*C. lhoesti, C. preussi*, and *C. solatus*) and the *C. aethiops-E. patas* sister group (Figure 16.6; Tosi et al., 2004, 2005). It is therefore tempting to use postcranial features shared by members of these two groups to characterize the last common ancestor of the terrestrial clade (e.g., Table 16.3, nos. 13–21, 27, and 43). For example, it is possible that the ancestor of the terrestrial clade exhibited more terrestrial adaptations than *C. aethiops*, possibly some of those shared by *E. patas* and *C. lhoesti*, and that *C. aethiops* subsequently lost some of these features. We do not consider this to be a likely scenario. It is also possible that this ancestor was most similar to *C. aethiops*, which is considered the least specialized of the terrestrial taxa because it exhibits the fewest terrestrial adaptations (Table 16.3; Gebo and Sargis, 1994). We consider the latter possibility more likely. Consequently, we propose that the ancestral condition of the terrestrial clade was postcranially similar to *C. aethiops* and that *E. patas* and *C. lhoesti* became increasingly specialized through the independent acquisition of their numerous terrestrial adaptations.

16.4.4 Taxonomy

As noted above, the genus *Cercopithecus* is paraphyletic because *C. aethiops* and *C. lhoesti* are more closely related to *E. patas* than they are to other species of *Cercopithecus* (Tosi et al., 2002, 2003, 2004). Tosi et al. (2004) outlined four solutions to this taxonomic problem, all of which were summarized above. Here, we evaluate each of these four proposals in light of the postcranial evidence discussed above.

We reject the proposal (#1) to sink patas monkeys into *Cercopithecus* because it fails to formally recognize the terrestrial clade, thereby ignoring the various terrestrial adaptations found among *E. patas, C. aethiops*, and *C. lhoesti* (Tosi et al., 2002, 2004). Alternatively, each of the terrestrial taxa exhibit a different suite of terrestrial adaptations, and none of these characters are present in every terrestrial taxon (Table 16.3; see above). We also reject the proposal (#4) to reinstate l'Hoest's monkeys to *Allochrocebus* and place vervet and patas monkeys in *Chlorocebus*, thereby formally recognizing their sister taxon relationship. This view overemphasizes the derived nature of l'Hoest's monkeys' postcrania, while simultaneously undervaluing the autapomorphous postcranial morphology of patas monkeys (Tosi et al., 2004). Although our principal components analysis separates l'Hoest's monkeys (Figure 16.2), patas monkeys are certainly as derived based on qualitative data (Table 16.3; Gebo and Sargis, 1994). Concerning proposal #3, we agree with Tosi et al. (2002, 2003, 2004) that vervet, patas, and l'Hoest's monkeys should be placed in *Chlorocebus* to formally recognize the terrestrial clade. *Chlorocebus* (Gray, 1870) has priority over *Erythrocebus* (Trouessart, 1897), and *Cercopithecus* cannot be used for the terrestrial clade because *Cercopithecus diana*, an arboreal guenon that is not included in this clade, is the type species of the genus (International Commission on Zoological Nomenclature, 1954). We also see some value in proposal #2, which places patas monkeys in *Erythrocebus*, vervet monkeys in *Chlorocebus*, and l'Hoest's monkeys in *Allochrocebus* (Tosi et al., 2002, 2003, 2004). Ultimately, we reject this option because it does not formally recognize the terrestrial clade (Tosi et al., 2002, 2003, 2004), but we appreciate that it acknowledges the numerous differences found in the postcrania of vervet, patas, and l'Hoest's monkeys (Table 16.3; Figures 16.1–16.2; Gebo and Sargis, 1994). In the end, we propose a compromise classification of the terrestrial taxa (Table 16.9) that both formally recognizes the terrestrial clade by placing all of these taxa in *Chlorocebus*, but also acknowledges the postcranial diversity within this clade by placing them in separate subgenera. This arrangement further acknowledges the relatively long Y-chromosome lineages of the three terrestrial cercopithecin taxa, which are as long or longer than those of papionin genera (Tosi et al., 2003, 2004).

TABLE 16.9. Classification of terrestrially adapted guenons advocated here (see text)

Taxonomic names	Species group	Common names
Chlorocebus (Chlorocebus) aethiops	*Aethiops*-Group	Vervet, Grivet, Green monkeys
Chlorocebus (Erythrocebus) patas		Patas monkeys
Chlorocebus (Allochrocebus) lhoesti	*Lhoesti*-Group	l'Hoest's monkeys
Chlorocebus (Allochrocebus) preussi	*Lhoesti*-Group	Preuss's monkeys
Chlorocebus (Allochrocebus) solatus	*Lhoesti*-Group	Sun-tailed monkeys

Our proposal to use subgenera to distinguish vervet, patas, and l'Hoest's monkeys is similar to one of Groves' (2000) three taxonomic proposals regarding vervet and patas monkeys. He suggested using the subgenera *Chlorocebus* and *Erythrocebus* for vervet and patas monkeys, respectively, although he proposed these as subgenera of *Cercopithecus* rather than *Chlorocebus* and he favored his alternative proposal of recognizing them as distinct genera. Groves' (2000) third proposal would place both vervet and patas monkeys in *Chlorocebus*. He considered this option to be confusing because patas and vervet monkeys are so different morphologically, citing six distinctive features in patas monkeys and five in vervets. We consider this additional evidence for distinguishing these taxa at the level of subgenus, in an attempt to acknowledge the many differences among the terrestrial taxa. Furthermore, we disagree with Groves' (2000) favored proposal of separating patas and vervet monkeys as distinct genera because this arrangement fails to formally recognize the terrestrial clade.

In summary, we agree with Tosi et al. (2002, 2003, 2004) that terrestriality evolved only once among guenons. However, vervet, patas, and l'Hoest's monkeys exhibit many postcranial differences indicating that they have diverged significantly from the common ancestor of the terrestrial clade. We have recognized this terrestrial clade taxonomically by placing all three taxa in the genus *Chlorocebus* (see Tosi et al., 2002, 2003, 2004). We have also acknowledged their numerous postcranial differences by placing them in separate subgenera (Table 16.9).

Although terrestrial locomotion evolved only once among guenons, the "terrestrial" taxa, particularly patas and l'Hoest's monkeys, all exhibit different terrestrial adaptations (Table 16.3). This conclusion is significant for the proposal made by Tosi et al. (2003) to use substrate preference as a character. They consider this to be additional evidence in support of the terrestrial clade because a single transition to terrestriality would be more parsimonious than three independent transitions. However, we would caution against the use of such a character because vervet, patas, and l'Hoest's monkeys may engage in terrestrial locomotion in different ways (see Hurov, 1987). They have certainly acquired distinct postcranial adaptations for this substrate preference.

In conclusion, our study has demonstrated significant postcranial variation within the terrestrial clade of guenons. Although our sample permitted an examination of variation among patas, vervet, and l'Hoest's monkeys, we were unable to assess variation within the *Lhoesti*-group. Any future analysis that focuses on this group should increase the sample size of *Chlorocebus (Allochrocebus) lhoesti* and *Chlorocebus (Allochrocebus) preussi*, as well as incorporate samples of *Chlorocebus (Allochrocebus) solatus*. Such a study has the potential to reveal yet more evolutionarily significant variation within the terrestrial clade. This would further demonstrate that the evolution of terrestriality among guenons was not simply a single transition, but included continued divergence in multiple lineages.

Acknowledgments. Thanks to Eric Delson, John Oates, and Steve Frost for their insights on guenon taxonomy, and to Marian Dagosto and two anonymous reviewers for helpful comments. We thank Fred Szalay for being an inspirational and consummate mentor (graduate advisor to EJS, postdoctoral advisor to CJT, and senior colleague to DLG) and for his friendship over the years.

References

Anapol, F., Barry, K., 1996. Fiber architecture of the extensors of the hindlimb in semiterrestrial and arboreal guenons. American Journal of Physical Anthropology 99, 429–447.

Anapol, F., Gray, J. P., 2003. Fiber architecture of the instrinsic muscles of the shoulder and arm in semiterrestrial and arboreal guenons. American Journal of Physical Anthropology 122, 51–65.

Anapol, F., Turner, T. R., Mott, C. S., Jolly, C. J., 2005. Comparative postcranial body shape and locomotion in *Chlorocebus aethiops* and *Cercopithecus mitis*. American Journal of Physical Anthropology 127, 231–239.

Bohonak, A. J., 2002. RMA: software for reduced major axis regression. Version 1.17. http://www.bio.sdsu.edu/pub/andy/RMA.html

Delson, E., Terranova, C. J., Jungers, W. L., Sargis, E. J., Jablonski, N. G., Dechow, P. C., 2000. Body mass in Cercopithecidae (Primates, Mammalia): estimation and scaling in extinct and extant taxa. Anthropological Papers of the American Museum of Natural History 83, 1–159.

Dutrillaux, B., Muleris, M., Couturier, J., 1988. Chromosomal evolution of Cercopithecinae. In: Gautier-Hion, A., Bourliere, F., Gautier, J.-P., Kingdon, J. (Eds.), A Primate Radiation: Evolutionary Biology of the African Guenons. Cambridge University Press, Cambridge, pp. 150–159.

Elliot, D. G., 1913. A Review of the Primates, 3 volumes. American Museum of Natural History, New York.

Gautier-Hion, A., 1988. Polyspecific associations among forest guenons: ecological, behavioural, and evolutionary aspects. In: Gautier-Hion, A., Bourliere, F., Gautier, J.-P., Kingdon, J. (Eds.), A Primate Radiation: Evolutionary Biology of the African Guenons. Cambridge University Press, Cambridge, pp. 452–476.

Gebo, D. L., Sargis, E. J., 1994. Terrestrial adaptations in the postcranial skeletons of guenons. American Journal of Physical Anthropology 93, 341–371.

Gebo, D. L., Chapman, C. A., 1995. Positional behavior in five sympatric Old World monkeys. American Journal of Physical Anthropology 97, 49–76.

Gray, J. E., 1870. Catalogue of Monkeys, Lemurs, and Fruit-eating Bats in the Collection of the British Museum. British Museum Trustees, London.

Groves, C. P., 2000. The phylogeny of the Cercopithecoidea. In: Whitehead, P. F., Jolly, C. J. (Eds.), Old World Monkeys. Cambridge University Press, Cambridge, pp. 77–98.

Hurov, J. R., 1987. Terrestrial locomotion and back anatomy in vervets (*Cercopithecus aethiops*) and patas monkeys (*Erythrocebus patas*). American Journal of Primatology 13, 297–311.

International Commission on Zoological Nomenclature, 1954. Opinion 238. Validation, under the plenary powers, of the generic name *Cercopithecus* as from Linnaeus, 1758 (Class Mammalia). Opinions and Declarations of the International Commission on Zoological Nomenclature 4, 25–26.

Isbell, L. A., Pruetz, J. D., Lewis, M., Young, T. P., 1998. Locomotor activity differences between sympatric patas monkeys (*Erythrocebus patas*) and vervet monkeys (*Cercopithecus aethiops*): implications for the evolution of long hindlimb length in *Homo*. American Journal of Physical Anthropology 105, 199–207.

Kaplin, B. A., Moermond, T. C., 2000. Foraging ecology of the mountain monkey (*Cercopithecus lhoesti*): implications for its evolutionary history and use of disturbed forest. American Journal of Primatology 50, 227–246.

Kingdon, J., 1988. Comparative morphology of hands and feet in the genus *Cercopithecus*. In: Gautier-Hion, A., Bourliere, F., Gautier, J.-P., Kingdon, J. (Eds.), A Primate Radiation: Evolutionary Biology of the African Guenons. Cambridge University Press, Cambridge, pp. 184–193.

Lernould, J.-M., 1988. Classification and geographical distribution of guenons: a review. In: Gautier-Hion, A., Bourliere, F., Gautier, J.-P., Kingdon, J. (Eds.), A Primate Radiation: Evolutionary Biology of the African Guenons. Cambridge University Press, Cambridge, pp. 54–78.

Manaster, B. J., 1979. Locomotor adaptations within the *Cercopithecus* genus: a multivariate approach. American Journal of Physical Anthropology 50, 169–182.

Martin, R. D., MacLarnon, A. M., 1988. Quantitative comparisons of the skull and teeth in guenons. In: Gautier-Hion, A., Bourliere, F., Gautier, J.-P., Kingdon, J. (Eds.), A Primate Radiation: Evolutionary Biology of the African Guenons. Cambridge University Press, Cambridge, pp. 160–183.

McGraw, W. S., 1998. Comparative locomotion and habitat use of six monkeys in the Tai Forest, Ivory Coast. American Journal of Physical Anthropology 105, 493–510.

McGraw, W. S., 2000. Positional behavior of *Cercopithecus petaurista*. International Journal of Primatology 21, 157–182.

Nakatsukasa, M., 1994. Morphology of the humerus and femur in African mangabeys and guenons: functional adaptation and implications for the evolution of positional behavior. African Study Monographs Supplement 21, 1–62.

Neff, N. A., Marcus, L. F., 1980. A Survey of Multivariate Methods for Systematics. Privately Published and AMNH, New York.

Rose, M. D., 1979. Positional behavior of natural populations: some quantitative results of a field study of *Colobus guereza* and *Cercopithecus aethiops*. In: Morbeck, M. E., Preuschoft, H., Gomberg, N. (Eds.), Environment, Behavior, and Morphology: Dynamic Interactions in Primates. Gustav Fischer, New York, pp. 75–93.

Ruvolo, M., 1988. Genetic evolution in the African guenons. In: Gautier-Hion, A., Bourliere, F., Gautier, J.-P., Kingdon, J. (Eds.), A Primate Radiation: Evolutionary Biology of the African Guenons. Cambridge University Press, Cambridge, pp. 127–139.

Shea, B. T., 1992. Ontogenetic scaling of skeletal proportions in the talapoin monkey. Journal of Human Evolution 23, 283–307.

Tosi, A. J., Buzzard, P. J., Morales, J. C., Melnick, D. J., 2002. Y-chromosomal window onto the history of terrestrial adaptation in the Cercopithecini. In: Glenn, M. E., Cords, M. (Eds.), The Guenons: Diversity and Adaptation in African Monkeys. Kluwer/Plenum, New York, pp. 15–26.

Tosi, A. J., Disotell, T. R., Morales, J. C., Melnick, D. J., 2003. Cercopithecine Y-chromosome data provide a test of competing morphological evolutionary hypotheses. Molecular Phylogenetics and Evolution 27, 510–521.

Tosi, A. J., Melnick, D. J., Disotell, T. R., 2004. Sex chromosome phylogenetics indicate a single transition to terrestriality in the guenons (tribe Cercopithecini). Journal of Human Evolution 46, 223–237.

Tosi, A. J., Detwiler, K. M., Disotell, T. R., 2005. X-chromosomal window into the evolutionary history of the guenons (Primates: Cercopithecini.) Molecular Phylogenetics and Evolution 36, 58–66.

Trouessart, E. L. 1897. Catalogus Mammalium tam Viventium quam Fossilium. R. Friedländer & Sohn, Berlin.

17. Analysis of Selected Hominoid Joint Surfaces Using Laser Scanning and Geometric Morphometrics: A Preliminary Report

William E. H. Harcourt-Smith*
Department of Vertebrate Paleontology
American Museum of Natural History
Central Park West & 79th Street
New York, NY 10024, USA
willhs@amnh.org

Melissa Tallman
Anthropology PhD Program
CUNY Graduate School
New York, NY 10016, USA
& NYCEP
tallman@nycep.org

Stephen R. Frost
Department of Anthropology
University of Oregon
Eugene, OR 97403, USA
sfrost@uoregon.edu

David F. Wiley
Institute for Data Analysis and Visualization
University of California
Davis, CA 95616, USA
wiley@cs.ucdavis.edu

F. James Rohlf
Department of Ecology and Evolution
Stony Brook University
Stony Brook, NY 11794, USA
NYCEP
rohlf@life.bio.sunysb.edu

Eric Delson
Department of Anthropology
Lehman College/CUNY
& Department of Vertebrate Paleontology
American Museum of Natural History
Central Park West & 79th Street
New York, NY 10024, USA
& NYCEP
eric.delson@lehman.cuny.edu

* Address for correspondence: willhs@amnh.org

E.J. Sargis and M. Dagosto (eds.), *Mammalian Evolutionary Morphology: A Tribute to Frederick S. Szalay*, 373–383.

17.1 Foreword

Fred Szalay is a polymath of evolutionary morphology. From the mid-1960s (Szalay, 1968) to at least the mid-1980s (Szalay et al., 1987), he was acknowledged as the leading researcher on non-anthropoid fossil primates, complementing the anthropoid expertise of Elwyn Simons who had just laid the foundations for paleoprimatology as a distinct field through his revisions and field success in the Fayum of Egypt. Fred edited two volumes in this area in 1975, in both of which Delson (as a very junior colleague) was most pleased to be included (Szalay, 1975; Luckett and Szalay, 1975), which in turn led to their collaborative review of the whole order (Szalay and Delson, 1979). Fred was also an early critic of the cladistic approach to phylogenetic reconstruction, passionately opposed to the narrowness of evolutionary thinking that he felt stemmed from cladistic thinking, so he developed (often with Walter Bock) a refinement of the "evolutionary" taxonomic approach (e.g., Szalay, 1977b, 1993; Szalay and Bock, 1991). Although his earliest work focused on crania and dentitions, Fred was equally interested in postcranial morphology from both functional and phylogenetic viewpoints (Decker and Szalay, 1974; Szalay and Decker, 1974; Szalay et al., 1975); he almost single-handedly made morphology of the postcranium relevant to mammalian phylogeny reconstruction, and he argued that the distinction between functionalism and non-functional thinking was entirely artificial. His long 1977(a) paper on mammalian phylogeny was almost entirely based on the evidence from foot bones, and this focus continued in his work on primates (Szalay and Dagosto, 1980, 1988; Szalay and Langdon, 1986) as well as eutherian and especially metatherian mammals (Szalay and Drawhorn, 1980; Szalay, 1984, 1994; Szalay and Lucas, 1993, 1996; Szalay and Sargis, 2001). Fred argued that taxa could just as readily be distinguished by their postcrania as by their teeth, ear regions or facial structure. Many of his colleagues rejected this idea or thought that (at best), postcranial morphology might delineate families or subfamilies, but not genera or species. Fred's most recent contribution (Szalay, 2007; in a book whose co-editor is the same as this volume's) reviewed the locomotor adaptations of the earliest primates and their predecessors, comparing several entrenched hypotheses unfavorably to his own prior interpretations.

We hope Fred will appreciate this paper, which attempts to distinguish genera and even individuals from the morphology of their ankles, one of his favorite anatomical regions. On the other hand, we realize that Fred never cared much for complex statistical analyses (though he adopted a variety of relatively high-tech approaches after their usefulness was demonstrated to him), and we hope he will not be too put off by this work on that basis.

17.2 Introduction

Allocating fossil specimens to a particular taxon, and even in some cases to a particular individual, is a primary problem in paleontology. For many mammals, including primates, fossil taxa are defined from craniodental morphology, and postcranial elements can often be allocated only when directly associated with cranial parts. Moreover, when multiple individuals are recovered, it is usually important to associate elements of a single individual in order to help determine functional adaptations, overall size and proportions, as well as other factors, such as the number of individuals preserved.

In addition to careful taphonomic analysis, most studies attempting to allocate isolated postcranial elements of unknown association, particularly of fossil hominins, have used a combination of visual estimation of morphological similarity and linear measurements (whether analyzed in a uni-, bi-, or multivariate manner). Central to this study is the concept of joint congruence, that is, the closeness of fit between the articulating elements of a joint complex. Two previous publications (Aiello et al., 1998; Wood et al., 1998) reported the results of preliminary analyses using a laser surface scanner (LSS) to compare congruence in the talo-crural joint in hominoids and one human fossil. Their work represents an important new research direction, but they were unable to continue along this line because the statistical techniques, the computer software and hardware, and the models they developed, were unable to address the complex 3D relationships between the reciprocal shapes involved.

It is clear that the effect of soft tissue (cartilage, ligament, tendon and musculature) is an important part of congruence in most joints (see, e.g., Hamrick, 1999). However, fossil and archaeological material does not usually preserve soft tissue structures, and researchers nearly always have to work with hard tissue joint surfaces preserved as subcondral bone. As we are ultimately interested in matching previously unassociated fossil and archaeological elements, we considered it desirable to use the hard-tissue joint surfaces of extant specimens housed in museum collections. In that context, we present preliminary results from a new approach to joint congruence. This new approach to matching is an indirect one that only requires a strong covariation between the shapes of the matching pairs of structures. We combine laser surface scans of opposing joint surfaces with geometric morphometrics and multivariate statistical analyses to examine ways to differentiate taxa and match elements from the same individual, utilizing an initial sample of extant hominoids. We use the tibial and talar components of the ankle joint because it is a relatively "tight" and predominantly uniaxial joint. As a result, there is a reasonable expectation that the reciprocal surface should be relatively congruent.

17.3 Background

There have been a number of studies on individual joint articular surfaces, as well as reciprocal joint surface geometry and congruence of such complexes as the gleno-humeral, humero-ulnar, radio-ulnar, carpo-metacarpal, metacarpo-phalangeal, femoro-tibial, patello-femoral, and tarso-metatarsal joints. These studies have mainly focused on documenting the normal anatomy of the skeletal elements involved in each joint complex (Leardini et al., 1999; Matsuda et al., 1998; Medley et al., 1983; Rostlund et al., 1989; Shiba et al., 1988; Siu et al., 1996; Soslowsky et al., 1992; Staron et al., 1994; Tamai et al., 1988; Yoshioka et al., 1988).

Many of the biomedical studies have dealt with joint geometry and mechanics and how this information is utilized in prosthesis design (Hertel and Lehmann, 2001; Hertel et al., 2002; Leardini, 2001; Roberts et al., 1991; Swieszkowski et al., 2001). Knowledge of normal and pathological joint morphology and kinematics allows for the design and implantation of prostheses that accurately approximate normal range and quality of motion (Bullough, 1981; Frost, 1999; Haut et al., 1998; Hlavacek and Vokoun, 1998; Kauer and de Lange, 1987; Kelkar et al., 2001; Pretterklieber, 1999; Soslowsky et al., 1992; Waide et al., 2000). Anthropological studies include those that have examined patterns of joint size dimorphism in the elbow and knee of catarrhine primates (Lague, 2003), patterns of sexual dimorphism in hominoid humeri (Lague and Jungers, 1999), patellar articular proportions of recent and Pleistocene humans (Trinkaus, 2000), and the relationship between hip joint congruence and function (MacLatchy and Bossert, 1996; MacLatchy, 1996).

Various methodologies have been employed in these studies to understand (and in some cases to visualize) joint geometry and morphology. Methods utilized include: digitization of joint facets (Dykyj et al., 2001; Yoshioka et al., 1988); Magnetic Resonance imaging (Matsuda et al., 1998; Staron et al., 1994; Staubli et al., 1999); Merchants's Skyline views & Axial Computed Arthro-tomography (CTA) (Walker et al., 1993); Stereophotogrammetry (SPG) (Ateshian et al., 1992; Huiskes et al., 1985; Kelkar et al., 2001; Soslowsky et al., 1992); three-dimensional Coordinate Digitizing System (3-DCDS) (Haut et al., 1998); computed tomography (CT) (Siu et al., 1996); X-ray absorptiometry scans (Mikhail et al., 1996); and classical morphometrics (e.g., Rostlund et al., 1989). Most recently, Tocheri et al. (2003) have studied modern and fossil trapezia with the aid of LSS data, but their methodology was not designed to take full advantage of those data in terms of estimating joint congruence.

17.4 Materials

The sample consists of the left tibia and talus of 22 extant *Homo sapiens*, 20 *Pan troglodytes* and 12 *Gorilla gorilla*. All individuals are adults and numbers of males and females are as even as was possible to arrange. The modern human sample includes individuals representing Alaskan Inuit from Point Hope (n = 7), ancient Egyptians (n = 3), archaeological Native Americans from Canyon Del Muerto, Arizona (n = 5) and modern African- (n = 5) and Euro-Americans (n = 2) from New York medical school collections, studied by courtesy of the Department of Anthropology, American Museum of Natural History, New York; the *Pan* and *Gorilla* samples are composed of wild-shot adults housed in collections within the museum's departments of Anthropology and Mammalogy.

17.5 Methods

Whole tibiae and tali were scanned with either a Cyberware 3030 laser surface scanner (Cyberware Inc., 2110 Del Monte Av., Monterey, CA 93940) or a portable Minolta Vivid 910 laser surface scanner (Konica Minolta Photo Imaging USA Inc., P.O. Box 92253, Chicago, IL 60675). The Cyberware 3030 is capable of scanning to a resolution of ~300 microns in the z plane, and the Minolta Vivid 910 scanner to ~30 microns. Resulting data files were edited and processed using *CyDir* (Cyberware Inc.) or *Geomagic Studio* 8.0 (Geomagic Inc., 3200 Chapel Hill-Nelson Rd., Research Triangle Park, NC 27709) respectively. The output files were saved in the .ply polygon model file format capable of storing additional information such as color and surface normals, thus providing highly accurate surface renditions of the actual object scanned, in this case whole bones. The .ply files were then entered into *Landmark Editor* (Wiley, 2006), a software package written for our team's LSS research, where the user can visualise the images as NURBS (Nonuniform rational B spline) surfaces.

A grid of points was then placed on the "virtual" joint surface of each laser scan. This grid was anchored by eight discrete homologous landmarks placed at identifiable features along the perimeter of the articular surface, and one in the center, of the distal tibia and talar trochlea. These landmarks were adapted from talar landmarks devised by Harcourt-Smith (2002) and tibial landmarks devised by Harcourt-Smith et al. (2004) and Garcia and Harcourt-Smith (2006). Table 17.1 summarizes the definitions for positioning these landmarks. Levels of landmark homology were determined after Bookstein (1991) and O'Higgins (2000).

The grid-defined articular areas were then re-sampled into a three-dimensional mesh of 361 (19 x 19) evenly-spaced points. Using an odd number of semi-landmark points makes it possible to correctly place a middle line of points along a linear anatomical structure, in this case the trochlear groove of the talus. Moreover, it was determined by visual inspection that the 19 x 19 grid created a very dense distribution of points on structures as small as the distal tibial articular surface and talar trochlea. In that respect, this number of points was deemed sufficient to capture the shape of the articular surfaces.

Landmark Editor automatically spaces these points by calculating the distance between two landmarks along the surface of the polygon model and then evenly spacing points along that line according to a number specified by the user. This methodology can be applied to either a line or a grid of points. These points are called semilandmarks because their relative location on the surface is arbitrary, and only their variation in directions orthogonal to the surface reflect differences in the shape of the surface (see Bookstein, 1997; Delson et al., 2001; Gunz et al., 2005;

TABLE 17.1 Homologous landmarks taken on the talar trochlear surface and distal tibial articular surface. Talar landmarks are after Harcourt-Smith (2002) and tibial landmarks adapted from Harcourt-Smith et al. (2004) and Garcia and Harcourt-Smith (2006). The types of landmark (i.e., level of homology) follow Bookstein (1991) and O'Higgins (2000).

Talar Landmarks		
Number	Type	Description
1	III	Most distal point of the trochlear groove.
2	II	Most distal point of contact between the medial malleolar facet and the trochlear surface.
3	III	Most dorsal point on the medial facet margin.
4	II	Most proximal point of contact between the medial malleolar facet and the trochlear surface.
5	III	Most proximal point of the trochlear groove.
6	II	Most proximal point of contact between the lateral malleolar facet and the trochlear surface.
7	III	Most dorsal point on the lateral facet margin.
8	II	Most distal point of contact between the lateral malleolar facet and the trochlear surface.
9	III	Most dorsal point on the trochlear groove.
Tibial Landmarks		
Number	Type	Description
1	II	Point where anterior and lateral facet margins meet.
2	III	Midpoint between landmarks 1 and 3 along the lateral side of the articular surface.
3	II	Point where posterior and lateral facet margins meet.
4	III	Midpoint between landmarks 3 and 5 on the posterior facet margin.
5	II	Point where posterior and medial facet margins meet. Point should be just before surface rises to become the medial malleolus.
6	III	Midpoint on the medial side of the articular surface between landmarks 5 and 7. Landmark should be on the edge just before it rises for the medial malleolus.
7	II	Point where medial and aneterior facet margins meet. Point should be just before it rises to become the medial malleolus.
8	III	Midpoint between landmarks 9 and 1 on the anterior facet margin.
9	III	Middle of articular surface.

and Figure 17.1). The same number of semilandmarks was used on all specimens.

The surfaces, as characterized by the sets of landmarks and semilandmark points, were then registered using Generalized Procrustes Analysis (GPA), which removes the effects of variation in orientation, location, and size (Rohlf and Slice, 1990). In addition, the effects of variation due to the somewhat arbitrary spacing in different specimens of the semilandmarks over the sampled surface were minimized using a "sliding" process. During registration, the semilandmarks are slid along planes tangent to the sampled surface around each semilandmark so as to minimize the Procrustes distance among configurations (Figure 17.2), so that the effects of the arbitrary positioning of semilandmarks on the surface are mitigated. This technique, originally applied to 2D curves (Bookstein, 1997), has recently been extended to surfaces (Gunz et al., 2005) using bending energy as the sliding criterion. In this study, however, the criterion being minimized was Procrustes distance rather than bending energy (Rohlf, 2005). Results obtained with bending energy or Procrustes distance as the minimized criterion were recently shown to be comparable (Perez et al., 2006). The sliding algorithm did not work well for the points on the grids of semilandmarks representing the malleolar facets of the tibia and talus, as it caused considerable erroneous deviation of a number of semilandmarks. This is being further investigated, but as a result only analyses of the trochlear surfaces of the talus and tibia are presented. Both GPA and sliding were completed using code written by F.J.R

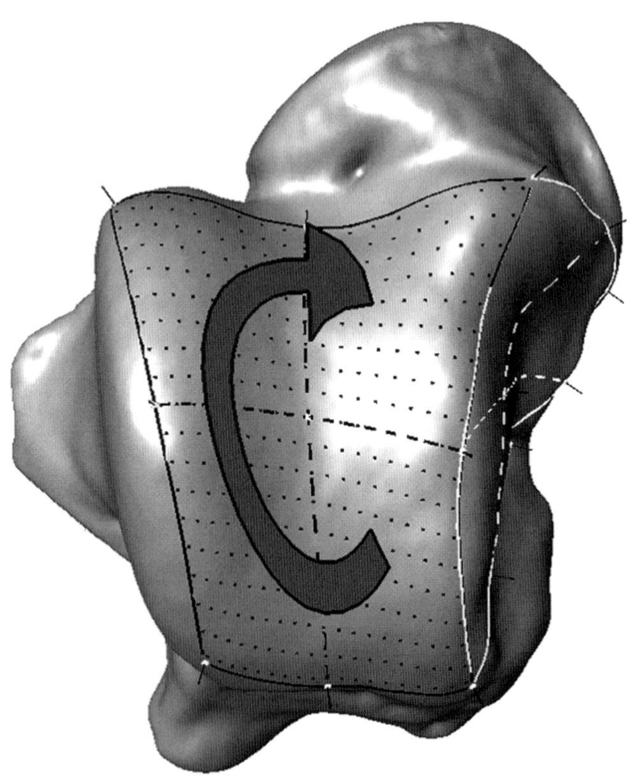

FIGURE 17.1. Screen capture from *Landmark Editor* software, showing the dorsal view of a laser scan of a modern human talus. On the trochlear surface is the dense 19 × 19 grid of semilandmark points. The arrow represents the order in which the points are collected.

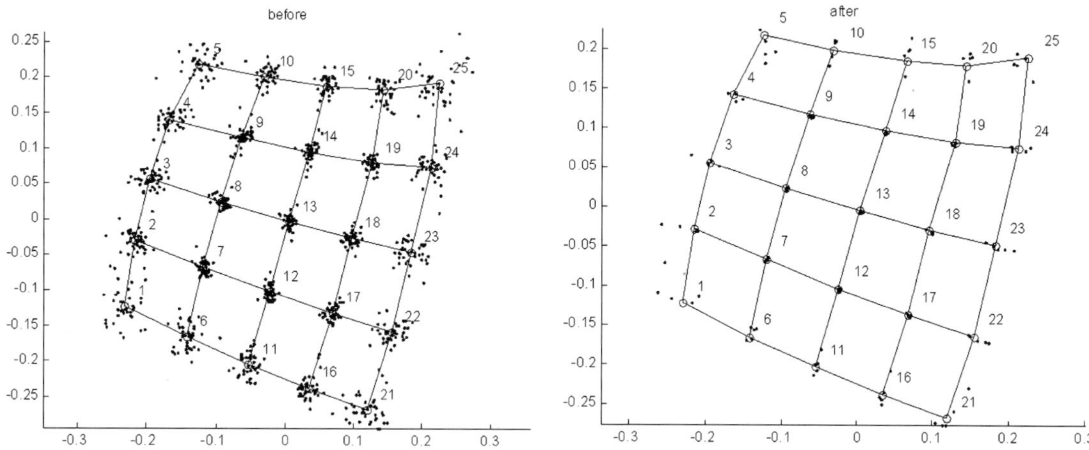

FIGURE 17.2. Unslid (left) versus slid (right) grids of semilandmarks. In this example, only 5 × 5 landmark grids are shown for ease of visualization. Black dots represent semilandmarks, the grid is for visualization only. Sliding was achieved using Procrustes distance as the criterion. Note positioning of semilandmarks on the surface is equalized, but their elevation normal to the surface is preserved.

for MATLAB software (The Mathworks, Inc., Natick, MA). After GPA and sliding, size was restored to the data (i.e., backscaled) by multiplying each coordinate by centroid size using SAS (SAS Institute, Cary, NC).

17.6 Data Analysis

All data analysis was conducted using SAS. Two Block Partial Least Squares (2B-PLS) or Singular Warps Analysis was used to analyze the relationships between the morphologies of the opposing joint surfaces. This technique analyzes the covariation between two sets of variables (Rohlf and Corti, 2000; Bookstein et al., 2003; Frost et al., 2003). In this study, these are the shape variables (i.e., the 3D coordinates of landmarks and semilandmarks) of the opposing joint surfaces, with the tibia comprising one matrix and the talus the other. This is a novel application of the technique, using the information on covariation between the shapes of opposing surfaces to detect whether a pair of structures has in fact been correctly associated.

This technique has a potential advantage over alternative approaches that rely on directly aligning the reciprocal joint surfaces to evaluate their congruence (e.g., assessing the volume separating "optimally" positioned surfaces). Such approaches require assumptions as to the distribution of the thickness of joint cartilage as well as the habitual posture and range of motion of those joints in the organisms involved (i.e., assumptions are made about how the opposing joint surfaces should articulate). None of these are required here. Instead, this approach simply assumes that a relationship exists between the shapes of opposing joint surfaces (i.e., they covary), and that the shapes of mismatched joint surfaces deviate from this baseline. More specifically, a matrix of covariances between shape variables based on correctly paired structures is decomposed using a singular-value decomposition to yield linear combinations of variables from each of the structures that best account for the observed covariation between the paired structures (Rohlf and Corti, 2000; Bookstein et al., 2003). Pairs of scores are then computed for the correctly paired structures. The scores for the first singular vector from each structure are expected to show a tight pattern of correlation. Pairs of scores are also computed for the incorrectly matched structures using the matrices of singular vectors based on the matched data. Scores for mismatched specimens are expected to fall outside of the pattern of covariation for correctly matched surfaces. That is, the fit of a particular pair of joint surfaces is compared to the pattern of covariation of matched pairs. As noted, no assumptions are made about how the opposing joint surfaces should articulate (as one might do by manually fitting them together).

Finally, the PLS scores of the first ten singular warps were subjected to a canonical discriminant function (DF) analysis, also called a canonical variates analysis (see Hubberty, 1994) to explore whether pairs of reciprocal joint surfaces could be correctly classified as matched or mismatched with any degree of certainty. To compute the reclassification rates, a cross validation procedure was used, where each pair was removed in turn from the data, the discriminant functions recomputed from the remaining pairs, and then that pair reclassified. All analyses were conducted on landmark data that were subjected to sliding and backscaled using centroid size as a multiplier, because the results from the DF analysis were considerably better using sliding and backscaled data than without.

17.7 Results

Sliding the semilandmarks gives considerably better results than not sliding them. Specifically, where the data were slid, but there was no backscaling, 84% of pairs were correctly classified. For unslid data, where backscaling (using centroid size as a multiplier) was included, the rate of correct classification rose to 92%. Finally, the backscaled and slid data produced the highest rate of correct classification, at 95%. Therefore, only the results of analyses performed after sliding are shown here. For the full dataset, PLS of the slid and unslid semilandmark grids does not produce any significant differences between matched and unmatched individuals when size is adjusted for using GPA. However, results from the backscaled data were more positive. Figure 17.3 shows a bivariate plot of PLS axis 1 for both the tibia (x-axis) and the talus (y-axis) using the backscaled data. The matched specimens form a relatively narrow diagonal "band" running from the intersect of the x and y axes at roughly a 45° angle. The intergenerically mismatched specimens predominantly fall away from this line, either above or below it; in each case, the name of the genus providing the tibia precedes that yielding the talus. The *Pan-Homo* and *Pan-Gorilla* pairs fall exclusively above the line of matched specimens. For the *Homo-Pan* and *Homo-Gorilla* pairs, most specimens fall away from the band of matched individuals, but a number of these mismatches do fall within the range of matches. The *Gorilla-Pan* and *Gorilla-Homo* mismatches also fall partially within the range of matched individuals, and partially outside it. Overall, 76% of intergeneric mismatches fell outside the range of variation of the matched individuals.

In the above analysis, matched and mismatched humans did not significantly differ. However, when the analysis was restricted to *Homo* alone (Figure 17.4), there was a strong separation between pairs from the same or different individuals: 73% of modern human mismatched individuals fell outside the range of variation of matched specimens, either above or below the latter's diagonal band.

Finally, canonical discriminant function (DF) analysis was conducted on the first 10 PLS scores for the tibia and talus where size had been restored as above, using centroid size as a multiplier. Table 17.2 summarizes the rates of correct allocations for each set of matched and unmatched pairings. For the intrageneric pairs, 100% of *Gorilla*, 95% of *Pan* and 93.75% of *Homo* were correctly re-allocated to genus. For the intergeneric mismatched pairings, 90.9% of *Gorilla-Pan* (i.e., *Gorilla* tibia and *Pan* talus) mismatches were correctly assigned to that group, while 100% of the *Pan-Gorilla* pairs were correctly assigned. All of the 3 *Homo -Gorilla* mismatches were correctly assigned, 8 of the 9 *Homo -Pan* mismatches, and all of the 12 *Pan-Homo* pairings. Overall the percentage of correct assignments was high for all combinations of matched and unmatched individuals.

Figure 17.5 shows the canonical scores for DF 1 (x-axis) versus DF 2. Most of the pre-assigned groups form tight, distinct clusters separated from each other. Both the *Homo* and *Pan* intrageneric groups have very little or no overlap with any other group. DF 1 separates the intra-*Homo* group and the *Pan-Homo* group (on the negative end of the axis) from all others. There is also some separation between the *Homo-Pan, Gorilla-Pan* and *Pan* groups and the *Pan-Gorilla, Homo-Gorilla* and *Gorilla* groups. The relative position of any group on DF1 seems to be predominantly influenced by the second taxon of the pair, i.e., that represented by talar landmarks, with *Homo* at

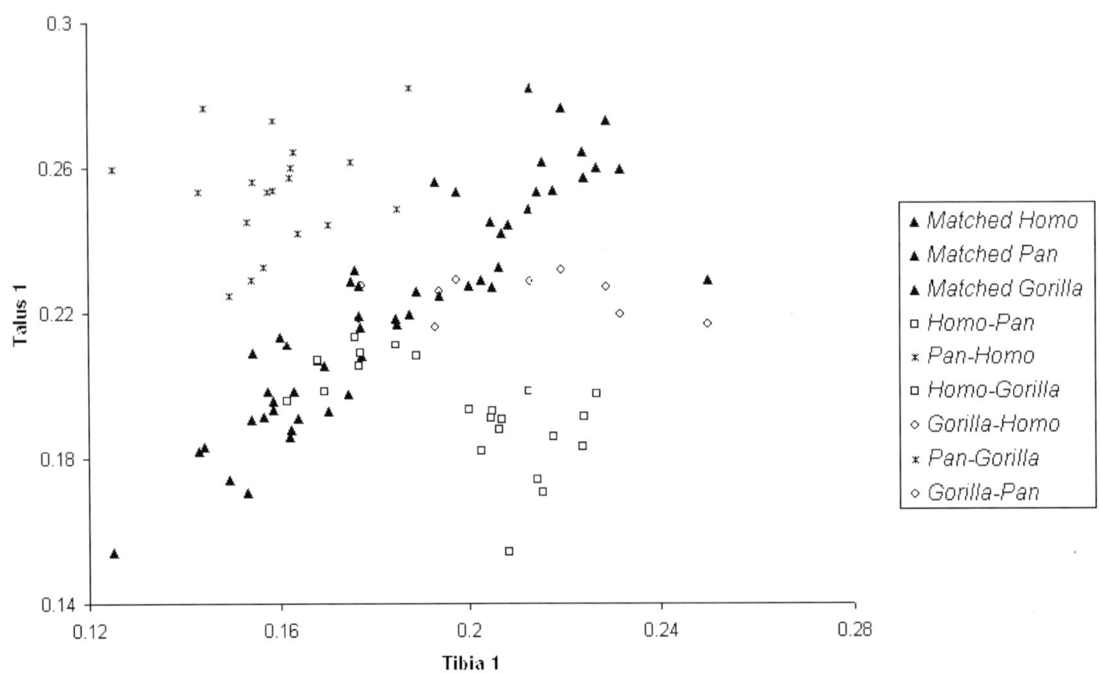

FIGURE 17.3. PLS (singular) vector 1 scores (after sliding and backscaling) for tibio-talar pairs; tibia on the x-axis and talus on the y-axis. All intraspecific pairs, whether from the same or different individuals, are given the same labels (solid triangles) for visual clarity. For each interspecific pairing, the first taxon in the legend refers to the tibial surface, the second to the talar.

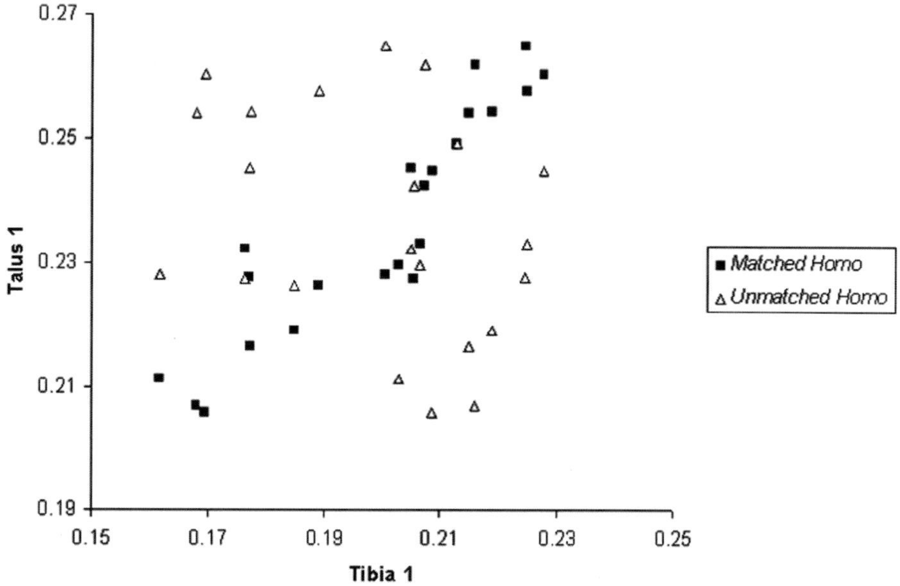

FIGURE 17.4. PLS (singular) vector 1 scores (after sliding and backscaling) for tibio-talar pairs; tibia on the x-axis and talus on the y-axis. Results for modern humans only, pairs from same individuals (matched) vs. random combinations.

TABLE 17.2. Frequency of correct and incorrect allocations from discriminant function analysis of PLS scores. For this DF analysis, a cross validation procedure was used. The actual pairing is shown on the left, the categories into which pairs were classified are listed across the top. For each pairwise comparison, the upper value is the number of allocations for that category, and the lower value (**bold**) is the corresponding percentage. Values along the major axis report correct reclassifications, while off-axis values are misclassifications where, for example, a *Gorilla* tibia and *Pan* talus was reported as *Pan* + *Pan*. H = *Homo*, G = *Gorilla*, P = *Pan*. First letter of a pair represents the taxon of the tibial surface, and the second the talar.

	Classified as								
	G + G	G + P	H + G	H + H	H + P	P + G	P + H	P + P	Total
Actual G + G	11 100%								11
G + P		10 90.9%						1 9.1%	11
H + G			3 100%						3
H + H				30 93.75%			2 6.25%		32
H + P					8 88.9%			1 11.1%	9
P + G						8 100%			8
P + H							12 100%		12
P + P					1 5%			19 95%	20
Total correct									101 95.3%
Total incorrect									5 4.7%
Grand total									106 100%

the negative end of the axis, *Pan* intermediate, and *Gorilla* at the positive extreme. It is interesting to note that these are not arranged according to size, as the *Pan* talar and tibial surfaces are smaller than those of *Homo*. On DF2, the principal separation is between the *Homo-Pan, Gorilla-Pan* and *Pan* groups and all others. The intrageneric groups all fall closer to the middle of DF2, suggesting that it is intergeneric mismatches that are mainly driving variation on this axis. In other words, the intergeneric mismatches essentially fall outside the range of variation seen within the intrageneric matches.

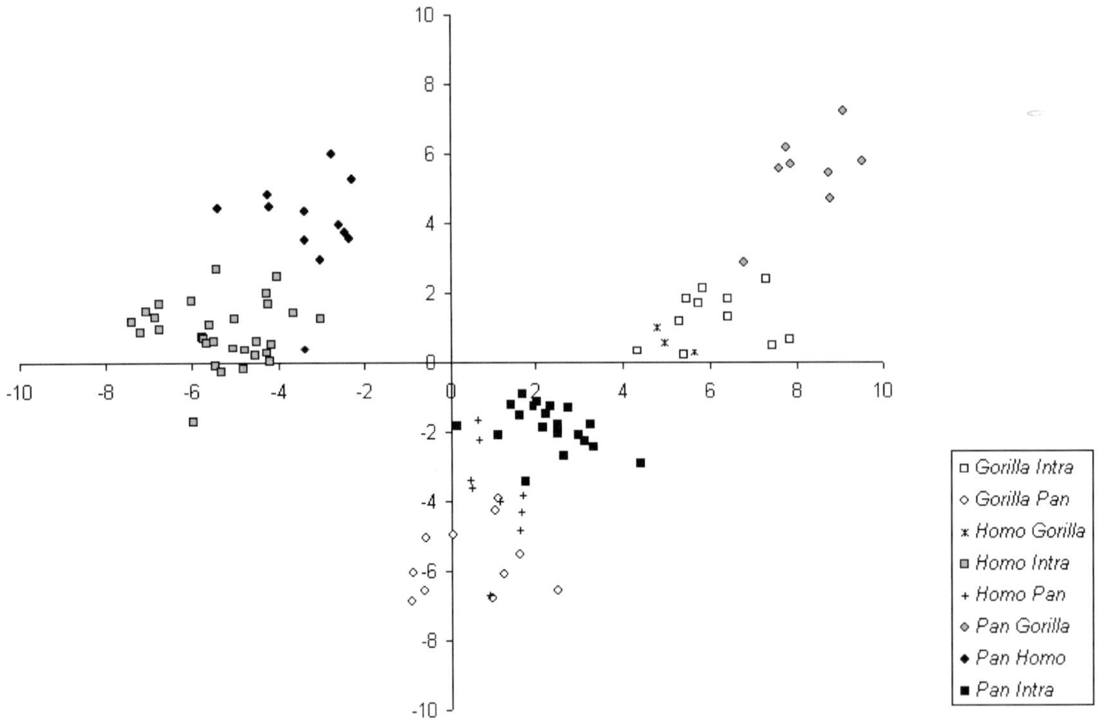

FIGURE 17.5. Scores from discriminant function analysis of the first ten PLS (singular) vectors (after sliding and backscaling) for tibio-talar pairs. DF1 scores on x-axis, DF2 on y-axis. For each interspecific pairing, the first taxon in the legend refers to the tibial dataset, the second to the talar dataset. In the key, "Intra" refers to both matched and mismatched intra-generic pairings.

17.8 Summary and Conclusions

2B-PLS analysis of GPA aligned and slid (semi)landmark data from laser scans of *Homo, Pan* and *Gorilla* tibiae and tali was undertaken in order to test a new approach for distinguishing between taxa and individuals. When centroid size is restored to the data, the preliminary results of our joint congruence study are encouraging. Thus it appears clear that both size and shape, rather than shape alone, are critical in determining the statistical degree of congruence between the tibial and talar joint surfaces. This may seem intuitive, since when trying to match joints using visual estimation alone, one naturally would discard elements incompatible due to size differences. Shape alone, at least for the surfaces used here, is not a sufficient factor in estimating joint congruence.

In a comparison of all three genera using tibial and talar PLS axis 1 scores, there was reasonably good visual separation (~75%, see Figure 17.3) of intrageneric matches from intergeneric mismatches. Within *Homo*, individual matches were not distinguished from individual mismatches when the apes were included. However, in a second analysis where apes were not included, 73% of individual mismatches within *Homo* were separated from individual matches (Figure 17.4). It is likely that intergeneric differences swamp the subtler distinctions expected within a genus (or species). Therefore, when considering potential hominin fossil material in an analysis such as this one, the important factor is whether the values for that pair of fossils fall far away from the band of matched individuals, or are closer to (or even within) that band.

The DF analysis of the first 10 pairs of singular warp scores yielded good reclassification results, with 101 of 106 pairs correctly identified (over 95%, as per Table 17.2). Given that DFA considers variation across all variables and compares differences between groups relative to variation within groups, it is not surprising that discrimination is better in Figure 17.5 than in Figure 17.3. However, the basic pattern of group distribution is similar.

This statistical treatment of the PLS scores could therefore be useful for discriminating between matched and unmatched isolated fossil elements (where reciprocal joint surfaces are present). This is particularly true given that the tibial malleolar surface was not included in this analysis, and that the results were nonetheless positive; its inclusion should improve the results. This technique, therefore, has considerable potential for sorting isolated elements recovered from paleontological or archeological sites, in particular, where it is hypothesized that two or more closely related taxa may co-occur.

The results from this study are preliminary, and a number of different directions are currently being explored to further refine our techniques. In particular, we are evaluating different ways of sliding the semilandmarks, so that more complex surfaces such as the malleolar facets can be used in future analyses. Further analyses will also benefit from larger sample sizes, which are currently being collected. Most importantly, our next phase of research will incorporate fossil elements into the analysis for the first time.

Acknowledgments. We thank Eric Sargis and Marian Dagosto for inviting us to participate in this volume (and for waiting relatively patiently while we completed our analyses), and Ian Tattersall, Nancy Simmons and Eileen Westwig for access to specimens in their care at the American Museum of Natural History. Thanks also to Ratha Ry, Anthony Pagano, Emily Henderson and Miriam Ruiz for help in laser-scanning and data collection. Comments by two anonymous reviewers and the editors on a previous version helped to improve the manuscript. This research was supported by the National Science Foundation (BCS-0452539). This is NYCEP morphometrics contribution number 18.

References

Aiello, L. C., Wood B. A., Key, C., Wood, C., 1998. Laser scanning and palaeoanthropology: an example from Olduvai Gorge, Tanzania. In: Strasser, E., Fleagle, J. G., Rosenberger, A. L., McHenry, H. M. (Eds.), Primate Locomotion: Recent Advances. Plenum, New York, pp. 223–236.

Ateshian, G.A., Rosenwasser, M.P., Mow, V.C., 1992. Curvature characteristics and congruence of the thumb carpometacarpal joint: Differences between female and male joints. Journal of Biomechanics 25, 591–607

Bookstein, F. L., 1991. Morphometric Tools for Landmark Data: Geometry and Biology. Cambridge University Press, Cambridge.

Bookstein, F. L., 1997. Landmark methods for forms without landmarks: localizing group differences in outline shape. Medical Image Analysis 1, 225–243.

Bookstein, F. L, Gunz, P., Mitteroecker, P., Prossinger, H., Schaefer, K., Seidler, H., 2003. Cranial integration in *Homo*: singular warps analysis of the midsagittal plane in ontogeny and evolution. Journal of Human Evolution 44, 167–187.

Bullough, P. G. 1981. The geometry of diarthrodial joints, its physiologic maintenance, and the possible significance of age-related changes in geometry-to-load distribution and the development of osteoarthritis. Clinical Orthopaedics and Related Research 156, 61–66.

Decker, R. L., Szalay, F. S., 1974. Origins and function of the pes in the Eocene Adapidae (Lemuriformes, Primates). In: Jenkins, F. A. (Ed.), Primate Locomotion. Academic, New York, pp. 261–291.

Delson, E., Harvati, K., Reddy, D., Marcus, L. F., Mowbray, K., Sawyer, G. J., Jacob, T., Márquez, S., 2001. The Sambungmacan 3 *Homo erectus* calvaria: a comparative morphometric and morphological analysis. Anatomical Record 262, 360–377.

Dykyj, D., Ateshian, G. A., Trepal, M. J., MacDonald, L. R., 2001. Articular geometry of the medial tarsometatarsal joint in the foot: comparison of metatarsus primus adductus and metatarsus primus rectus. Journal of Foot and Ankle Surgery 40, 357–365.

Frost, H. M., 1999. Joint anatomy, design, and arthroses: insights of the Utah paradigm. Anatomical Record 255, 162–174.

Frost, S. R., Marcus, L. F., Reddy, D. P., Bookstein, F., Delson, E., 2003. Cranial allometry, phylogeography and systematics of large bodied papionins (Primates: Cercopithecinae) inferred from geometric morphometric analysis of landmark data. Anatomical Record 275A, 1048–1072.

Garcia, G., Harcourt-Smith, W.E.H., 2006. A geometric morphometric analysis of the distal tibia of *Homo habilis*. American Journal of Physical Anthropology Supplement 42, 92.

Gunz, P., Mitteroecker, P., Bookstein, F. L., 2005. Semilandmarks in three dimensions. In: Slice, D. E. (Ed.), Modern Morphometrics in Physical Anthropology. Kluwer, Dordrecht, The Netherlands, pp. 73–98.

Hamrick, M. W., 1999. A chondral modeling theory revisited. Journal of Theoretical Biology 201, 201–208.

Harcourt-Smith, W.E.H., 2002. Form and Function in the hominoid tarsal skeleton. PhD Thesis. University College London: London.

Harcourt-Smith, W. E. H., Tallman, M., Frost, S. R., Kim, J., Delson, E., 2004. Estimating hominoid reciprocal joint congruence: A comparison of two morphometric techniques. American Journal of Physical Anthropology Supplement 38, 108–109.

Haut, T. L., Hull, M. L., Howell, S. M., 1998. A high-accuracy three-dimensional coordinate digitizing system for reconstructing the geometry of diarthrodial joints. Journal of Biomechanics 31, 571–577.

Hertel, R., Lehmann, O., 2001. Glenohumeral joint. Anatomical aspects and implications for prosthesis design. Orthopade 30, 363–369.

Hertel, R., Knothe, U., Ballmer, F. T. 2002. Geometry of the proximal humerus and implications for prosthetic design. Journal of Shoulder and Elbow Surgery 11, 331–338.

Hlavacek, M., Vokoun, D., 1998. The influence of articular surface incongruity on lubrication and contact pressure distribution of loaded synovial joints. Proceedings of the Institute of Mechanical Engineering [H] 212, 11–22.

Hubberty, C. J., 1994. Applied Discriminant Analysis. Wiley, New York.

Huiskes, R., Kremers, J., de Lange, A., Woltring, H. J., Selvik, G., van Rens, T. J., 1985. Analytical stereophotogrammetric determination of three-dimensional knee-joint geometry. Journal of Biomechanics 18, 559–570.

Kauer, J. M., de Lange, A., 1987. The carpal joint. Anatomy and function. Hand Clinics 3, 23–29.

Kelkar, R., Wang, V. M., Flatow, E. L., Newton, P. M., Ateshian, G. A., Bigliani, L. U., Pawluk, R. J., Mow, V. C., 2001. Glenohumeral mechanics: A study of articular geometry, contact, and kinematics. Journal of Shoulder and Elbow Surgery 10, 73–84.

Lague, M. R., 2003. Patterns of joint size dimorphism in the elbow and knee of catarrhine primates. American Journal of Physical Anthropology 120, 278–297.

Lague, M. R., Jungers, W. L., 1999. Patterns of sexual dimorphism in the hominoid distal humerus. Journal of Human Evolution 36, 379–399.

Leardini, A., 2001. Geometry and mechanics of the human ankle complex and ankle prosthesis design. Clinical Biomechanics 16, 706–709.

Leardini, A., O'Connor, J. J., Catani, F., Giannini, S., 1999. A geometric model of the human ankle joint. Journal of Biomechanics 32, 585–591.

Luckett, W. P., Szalay, F. S. (Eds.), 1975. Phylogeny of the Primates: a Multi-disciplinary Approach. Plenum, New York.

MacLatchy, L. M., 1996. Another look at the australopithecine hip. Journal of Human Evolution 31, 455–476.

MacLatchy, L. M., Bossert, W. H., 1996. An analysis of the articular surface distribution of the femoral head and acetabulum in anthropoids, with implications for hip function in Miocene hominoids. Journal of Human Evolution 31, 425–453.

Matsuda, S., Matsuda, H., Miyagi, T., Sasaki, K., Iwamoto, Y., Miura, H., 1998. Femoral condyle geometry in the normal and varus knee. Clinical Orthopaedics and Related Research 349, 183–188.

Medley, J. B., Dowson, D., Wright, V., 1983. Surface geometry of the human ankle joint. Engineering Medicine 12, 35–41.

Mikhail, M.B., Vaswani, A.N., Aloia, J.F., 1996. Racial differences in femoral dimensions and their relation to hip fracture, Osteoporosis International 6(1), 22–24.

O'Higgins, P., 2000. Advances in approaches to the study of morphological variation in the hominid fossil record: Biology, landmarks and geometry. Journal of Anatomy 197, 103–120.

Perez, S. I., Bernal, V., Gonzalez, P. N., 2006. Differences between sliding semilandmark methods in geometric morphometrics: with an application to human craniofacial and dental variation. Journal of Anatomy 208, 769–784.

Pretterklieber, M. L., 1999. Anatomy and kinematics of the human ankle joint. Radiologe 39, 1–7.

Roberts, S. N., Foley, A. P., Swallow, H. M., Wallace, W. A., Coughlan, D. P., 1991. The geometry of the humeral head and the design of prostheses. Journal of Bone and Joint Surgery [Br] 73, 647–650.

Rohlf, F. J., 2005. TpsRelw, v. 1.42. Ecology and Evolution. Stony Brook University, New York.

Rohlf, F. J., Corti, M., 2000. Use of two-block partial least-squares to study covariation in shape. Systematic Biology 49, 740–753.

Rohlf, F. J., Slice, D., 1990. Extensions of the Procrustes method for the optimal superimposition of landmarks. Systematic Zoology 39, 40–59.

Rostlund, T., Carlsson, L., Albrektsson, B., Albrektsson, T., 1989. Morphometrical studies of human femoral condyles. Journal of Biomedical Engineering 11, 442–448.

Shiba, R., Sorbie, C., Siu, D. W., Bryant, J. T., Cooke, T. D., Wevers, H. W., 1988. Geometry of the humeroulnar joint. Journal of Orthopaedic Research 6, 897–906.

Siu, D., Rudan, J., Wevers, H. W., Griffiths, P., 1996. Femoral articular shape and geometry. A three-dimensional computerized analysis of the knee. Journal of Arthroplasty 11, 166–173.

Soslowsky, L. J., Flatow, E. L., Bigliani, L. U., Mow, V. C., 1992. Articular geometry of the glenohumeral joint. Clinical Orthopaedics and Related Research 285, 181–190.

Staron, R. B., Feldman, F., Haramati, N., Singson, R. D., Rosenwasser, M., Esser, P. D., 1994. Abnormal geometry of the distal radioulnar joint: MR findings. Skeletal Radiology 23, 369–372.

Staubli, H. U., Durrenmatt, U., Porcellini, B., Rauschning, W., 1999. Anatomy and surface geometry of the patellofemoral joint in the axial plane. Journal of Bone and Joint Surgery [Br] 81, 452–458.

Swieszkowski, W., Skalski, K., Pomianowski, S., Kedzior, K., 2001. The anatomic features of the radial head and their implication for prosthesis design. Clinical Biomechanics 16, 880–887.

Szalay, F. S., 1968. The beginnings of primates. Evolution 5, 19–36.

Szalay, F. S. (Ed.) 1975 Approaches to Primate Paleobiology. Karger, Basel.

Szalay, F. S., 1977a. Phylogenetic relationships and a classification of the eutherian Mammalia. In: Hecht, M. K., Goody, P. C., Hecht, B. M. (Eds.), Patterns of Vertebrate Evolution. Plenum, New York, pp. 315–374.

Szalay, F. S., 1977b. Ancestors, descendants, sister groups and testing of phylogenetic hypotheses. Systematic Zoology 26, 12–18.

Szalay, F. S., 1984. Arboreality: Is it homologous in metatherian and eutherian mammals? Evolutionary Biology 18, 215–258.

Szalay, F. S., 1993. Species concepts: the tested, the untestable and the redundant. In: Kimbel, W. H., Martin, L. B. (Eds.), Species, Species Concepts, and Primate Evolution. Plenum, New York, pp. 21–42.

Szalay, F. S., 1994. Evolutionary history of the marsupials and an analysis of osteological characters. Cambridge University Press, Cambridge.

Szalay, F. S., 2007. Ancestral locomotor modes, placental mammals, and the origin of Euprimates: Lessons from history. In: Ravosa, M. J., Dagosto, M. (Eds.), Primate Origins: Adaptations and Evolution. Springer, New York, pp. 457–487.

Szalay, F. S., Bock, W. J., 1991. Evolutionary theory and systematics: Relationships between process and patterns. Zeitschrift für zoologische Systematik und Evolutions-Forschung 29, 1–39.

Szalay, F. S., Dagosto, M., 1980. Locomotor adaptations as reflected on the humerus of Paleogene primates. Folia Primatologica 34, 1–45.

Szalay, F. S., Dagosto, M., 1988. Evolution of hallucial grasping in the primates. Journal of Human Evolution 17, 1–33.

Szalay, F. S., Decker, R. L., 1974. Origins, evolution and function of the tarsus in late Cretaceous eutherians and Paleocene Primates. In: Jenkins, F. A. (Ed.), Primate Locomotion. Academic, New York, pp. 223–259.

Szalay, F. S., Delson, E., 1979. Evolutionary History of the Primates. Academic, New York.

Szalay, F. S., Drawhorn, G., 1980. Evolution and diversification of the Archonta in an arboreal milieu. In: Luckett, W. P. (Ed.), Comparative Biology and Evolutionary Relationships of Tree Shrews. Plenum, New York, pp. 133–169.

Szalay, F. S., Langdon, J., 1986. The foot of *Oreopithecus bambolii*: An evolutionary assessment. Journal of Human Evolution 15, 585–621.

Szalay, F. S., Lucas, S. G., 1993, Cranioskeletal morphology of archontans, and diagnoses of Chiroptera, Volitantia, and Archonta. In: MacPhee, R. D. E. (Ed.), Primates and their Relatives in Phylogenetic Perspective. Plenum, New York, pp. 187–226.

Szalay, F. S., Lucas, S. G., 1996 The postcranial morphology of Paleocene *Chriacus* and *Mixodectes* and the phylogenetic relationships of archontan mammals. Bulletin of New Mexico Museum of Natural History and Science 7, 1–47.

Szalay, F. S., Rosenberger, A. L., Dagosto, M., 1987. Diagnosis and differentiation of the Order Primates. Yearbook of Physical Anthropology 30, 75–105.

Szalay, F. S., Sargis, E. J., 2001. Model-based analysis of postcranial osteology of marsupials from the Paleocene of Itaborai (Brazil) and the phylogenetics and biogeography of Metatheria. Geodiversitas 23, 139–302.

Szalay, F. S., Tattersall, I., Decker, R., 1975. Phylogenetic relationships of *Plesiadapis* – postcranial evidence. In: Szalay, F. S. (Ed.), Approaches to Primate Paleobiology. Karger, Basel, pp. 136–166.

Tamai, K., Ryu, J., An, K. N., Linscheid, R. L., Cooney, W. P., Chao, E. Y., 1988. Three-dimensional geometric analysis of the metacarpophalangeal joint. Journal of Hand Surgery [Am] 13, 521–529.

Tocheri, M. W., Marzke, M. W., Liu, D., Bae, M., Jones, G. P., Williams, R. C., Razdan, A., 2003. Functional capabilities of modern and fossil hominid hands: Three-dimensional analysis of trapezia. American Journal of Physical Anthropology 122, 101–122.

Trinkaus, E., 2000. Human patellar articular proportions: recent and Pleistocene patterns. Journal of Anatomy 196, 473–483.

Waide, D. V., Lawlor, G. J., McCormack, B. A., Carr, A. J., 2000. The relationship between surface topography and contact in the elbow joint: development of a two-dimensional geometrical model in the coronal plane. Proceedings of the Institute of Mechanical Engineering [H] 214, 413–423.

Walker, C., Cassar-Pullicino, V. N., Vaisha, R., McCall, I. W., 1993. The patello-femoral joint – a critical appraisal of its geometric assessment utilizing conventional axial radiography and computed arthro-tomography. British Journal of Radiology 66, 755–761.

Wiley, D. F., 2006. *Landmark Editor* 3.0. Institute for Data Analysis and Visualization, University of California, Davis (http://graphics.idav.ucdavis.edu/research/EvoMorph)

Wood, B. A., Aiello, L. C., Wood, C., Key, C. A., 1998. A technique for establishing the identity of 'isolated' fossil hominid limb bones. Journal of Anatomy 193, 61–72.

Yoshioka, Y., Siu, D. W., Cooke, T. D., Bryant, J. T., Wyss, U., 1988. Geometry of the first metatarsophalangeal joint. Journal of Orthopaedic Research 6, 878–885.

18. Comparative Primate Bone Microstructure: Records of Life History, Function, and Phylogeny

Johanna Warshaw*
Hard Tissue Research Unit
Departments of Biomaterials & Basic Sciences, Rm 817-S
New York University College of Dentistry
NYU Mail Code: 9448
345 East 24th Street
New York, NY 10010-4086, USA
jw143@nyu.edu

18.1 Introduction

The study of comparative osteology has yielded an abundant literature regarding the form, function and biological roles of bones and whole skeletons. These data have been variably applied to the fossil record, contributing to the reconstruction of locomotion, life history features and the evolutionary histories of the better-known extinct animals, yet many questions regarding the paleobiology of fossil taxa remain. The vast majority of work to date has examined whole bones and their macroscopically defined parts, rather than their microstructure. However, such inquiry often suffers from superficial analogies with living taxa, as well as from the frequent incompleteness of fossil material (e.g., Szalay, 2000; Szalay and Sargis, 2001). These shortcomings in turn limit phylogenetic interpretations.

At the microstructural level, bone is hypothesized to record aspects of the mechanics and life history of vertebrates. As bones grow, the growth remodeling process (cf. Enlow, as differentiated from secondary, intracortical or Haversian remodeling) leaves a record of changes in length, diameter and shape, and drift through anatomical space, all of which are potentially visible microscopically in the pattern of tissues at the level of the cross-section (Enlow, 1962a, b, 1963, 1976, 1982; Enlow and Hans, 1996). Thus variations in skeletal developmental patterns, integrally related to organismal life history, are encoded within the cortical tissue. In addition, because bones form within the context of their functional adaptation, their micro- as well as macrostructures must accommodate mechanical demands. It follows that significant information with bearing upon the biology and behavior of fossil taxa is potentially signaled *within*, and not just at the surface of, bones. The possibility for recovery of this information is particularly exciting in light of recent methodological advances (e.g., automated capturing of high resolution images of entire bone cross-sections). Yet the additional insights to be offered from an exploration of bone microstructure have been relatively neglected. As de Ricqlès noted in 1992, "thin section techniques…have been used more rarely and somehow reluctantly by vertebrate palaeontologists…even up to the present day" (p. 40). The situation has somewhat ameliorated in the years since de Ricqlès made this statement, in particular with respect to the study of dinosaur and therapsid paleohistology. For example, the histological studies of researchers such as Chinsamy, Erikson, Horner, de Ricqlès, and their colleagues have provided dramatic support for the reevaluation of the physiology and life history in a number of fossil reptilian and mammal-like reptile species (Curry, 1998; Botha and Chinsamy, 2000; Erickson and Tumanova, 2000; Horner et al., 2000; de Ricqlès et al., 2000; Erickson et al., 2001; Padian et al., 2001; Horner and Padian, 2004). However, with a few notable exceptions, there remains a dearth of data regarding both extant and extinct mammalian taxa, including primates. This is particularly the case for studies of primary tissue type, while the situation is somewhat better with regard to the investigation of variation in intracortical (Haversian) remodeling.

I report here on variation in primary bone tissues in midshaft cortices from the postcrania of a range of living primates. Before interpretations of microstructure in extinct species are possible, the extent of variability that exists in living groups, and the relationship of this variability to biological and behavioral features, must be understood. Employing conventional and polarized light microscopy techniques, my research goals are threefold. First, I qualitatively and quantitatively describe the variation in primary tissues types that exists at the midshaft of long bones within a carefully assembled sample of primates. This study represents a departure from most earlier comparative mammalian bone microstructural studies in that the sample is focused, yet broad enough to provide a phylogenetic context and to explore a wide range of variability.

* Address for correspondence: johanna364@yahoo.com

E.J. Sargis and M. Dagosto (eds.), *Mammalian Evolutionary Morphology: A Tribute to Frederick S. Szalay,* 385–426.

Second, I assess ideas regarding several principal factors affecting the intra- and intertaxonomic variability in primary bone tissue types: ontogenetic processes, organismal growth rates, body size, mechanical adaptations, and phylogeny. Third, I explore the possibilities for use of primary tissue type patterns to test specific ideas about primate biology: given a cross section of bone, what can we learn of the life ways of an individual?

Ultimately, the goal is to obtain results from the extant models that will be applicable to the interpretation of adaptation and evolution among fossil specimens. In anthropology, many controversial questions endure, for example: who are the specific ancestors to living strepsirhines and haplorhines (e.g., Krishtalka and Schwartz, 1978; Rosenberger and Szalay, 1980; Szalay et al., 1987; Szalay and Dagosto, 1988; Dagosto, 1988, 1994; Rasmussen, 1994; Simons et al., 1995; Shoshani et al., 1996; Simons, 1997)? Ideally, microstructural tools could be employed in addressing such questions with respect to fossil material.

18.2 Background

18.2.1 Previous Comparative Bone Microstructure Studies

18.2.1.1 General

Since the invention of the microscope there have been many investigations into the structure and composition of the bony tissue of vertebrates. The seventeenth to nineteenth centuries witnessed the description of its microscopic constituent parts and varied organization by such workers as Leeuwenhoek, the renowned Havers (employing a magnifying glass), and others (see Enlow, 1963; Martin and Burr, 1989; Bromage, 1986 for reviews). By the mid nineteenth century, scholars had produced broadly comparative works in bone histology (e.g., Quekett 1849, 1855; Foote, 1913, 1916; Crawford, 1940). In these publications, the objective was primarily taxonomic: it was believed that bone histology could be used by paleontologists to identify otherwise intractable specimens. However, in later works such as de Ricqlès (1968, 1969, 1975, 1976, 1977a, b, 1978a, b), Amprino and Godina (1947), and Enlow and Brown (1956, 1957, 1958), a more modern understanding of bone was applied to broad comparative samples. These seminal publications emphasized the diversity of tissue types that can be found among vertebrates and, while recognizing that group patterns do exist, did not place primary importance on taxonomy. There was an appreciation that convergences, either mechanical or with respect to life history, can lead to near identity in bone histology among disparate groups of vertebrates (de Ricqlès, 1992). Therefore, today the emphasis is on function, and the possibility that bone microstructure may offer a record of such adaptive aspects of organismal biology as patterns of growth, size, age, physiological adaptations, mineral homeostasis, and biomechanics (de Ricqlès, 1992, p. 41) (see Warshaw, 2007 for a more comprehensive review of previous studies).

It follows that comparative bone histology should take into account a variety of interdependent factors, and that comprehensive accounts of comparative bone microstructure – reaching beyond description to attempt functional analysis – are thus limited in number and scope. Nevertheless, despite the inherent difficulties, a number of important works have been produced within the context of this contemporary understanding of bone (e.g., Amprino and Godina, 1947; Burr, 1979, 1992; Enlow and Brown, 1958; Jowsey 1966, 1968; de Ricqlès (1968–1978); Singh et al., 1974).

18.2.1.2 Primates

Studies which include descriptions of primate bone histology span works of the nineteenth century through those undertaken in recent years. The earliest of these can be used principally for their descriptions and illustrations of bone cross-sections (e.g., Quekett, 1855; Foote, 1916; Demeter and Mátyás, 1928). More pertinent to the current study, Amprino and Godina (1947) have described the microstructure observed in postcranial elements from a range of vertebrates, including Old World monkeys, New World monkeys, and apes. They discuss variation in primary tissue type distribution, among other features (patterns of intracortical remodeling and, to a limited degree, variation in collagen fiber patterns). The data are wholly qualitative, and Amprino and Godina consider variation to be largely determined by age, body size and ontogenetic processes. Similarly, the descriptions in Enlow and Brown (1958) are qualitative in their reference to differences in tissue type distributions. For these authors, it is the process of growth and development that are considered to be largely responsible for the variable patterns observed in all vertebrates (see also Enlow, 1963, 1982). Enlow and Brown do not differentiate among species of living primates, but rather describe the order as a whole, as well as the somewhat different tissue type pattern observed in the dentary of a single Paleocene plesiadapiform primate, *Plesiolestes*.

Other comparative mammalian studies include those of Jowsey (1966, 1968), and Singh et al. (1974). The first author has undertaken an investigation of intracortical remodeling and degree of mineralization in a limited number of mammalian species from diverse orders (including rhesus macaques), and identifies some taxonomic differences attributed to variability in longevity and body size. Singh et al. (1974) published the first in what was to be a series of papers documenting variability in mammalian bone histology, including in this study strepsirhine and haplorhine primates. Variables that Singh et al. examine include tissue type variation, degree of vascularity of the cortical bone, extent of intracortical remodeling, and density of osteocyte lacunae, both within and between taxa. The effects of mechanics, diet, age, and specimen preparation are mentioned. While fairly comprehensive in its coverage of mammals, taxa were selected opportunistically (from zoo and laboratory collections), rather than with the intent to elucidate systematic patterns. Therefore, the work ultimately failed to plumb the potential of the comparative approach despite the quantification of some variables, remaining a noteworthy piece of research inspiring further investigation.

Collectively, the work outlined above provides bone microstructural data for a wide range of vertebrates, including primates. However, it is only with some measure of phylogenetic control that the causal factors accounting for variability can begin to be identified. Therefore, research that focuses on a smaller number of closely related taxa, while still maintaining a broad taxonomic and adaptive context, is a necessary follow-up to the more extensive review studies of the past (de Ricqlès, 1992). With three notable exceptions in the primate literature (Burr, 1979; Paine and Godfrey, 1997; Schaffler and Burr, 1984), such studies are veritably non-existent.

Burr (1979; 1992) and Schaffler and Burr (1984) have explored variation in mineral density and the degree of intracortical remodeling among primates. Results point primarily to a relationship between microanatomical variability and mechanics. The 1984 paper on intracortical remodeling is particularly significant in that it is the first study to examine postcranial bone microstructure in a purposefully selected primate sample with the intent to explain the diversity found therein. For this reason, and as the authors suggest, this paper serves as "a baseline for other studies" (p. 196). An equally important paper by Paine and Godfrey (1997), addresses the same variable in seven primate species, but obtains contrasting results. In this case the principal factor thought to affect remodeling is body size, with a subsidiary locomotor effect. These papers, while pioneering in their approach and significant in their findings, address only a limited range of microstructural features – excluding primary tissue types – and only in the femur and humerus. Therefore, Ruff and Runestad's 1992 statement that "microstructural analysis is a potentially informative but currently underexplored method of studying primate limb bone adaptations" (p. 427) holds true today, for both primates and non-primate mammals.

18.2.2 Tissue Typology of Compact Bone

> The laws of Nature are undeviating in the construction of the skeleton of vertebrate animals: the same regularity in structure, the same method of arrangement of the bone-cells, has existed from the time when the surface of our planet was first inhabited by a vertebrate animal up to the present period…The bones of the Mastodon and the huge Megatherium, the giants of the land, are no more remarkable for the coarseness of their structure than are those of the smallest mammiferous quadrupeds, the mouse, and such has been the prevailing law from the commencement of the earth's existence, and such, no doubt, will continue to the end of time (Quekett, 1849, p. 57–78).

It appears that bone tissues of living vertebrates have been composed of the same basic units of construction and, with presumably similar function, since the appearance of earliest vertebrate skeletal fossils (Enlow and Brown, 1958; Moss, 1964; de Ricqlès, 1978). Bone is a composite of organic and non-organic components, comprising a collagen and mineral matrix, cells that are associated with this matrix, and a system of surrounding and penetrating blood vessels and nerves. These components come together in a range of relatively stereotyped patterns. The differentiation between, and naming of, tissue types is inherently problematic given that there exist, at least within some categories, seemingly countless variations. In addition, one tissue type may grade imperceptibly into another, so that at times it becomes impossible to decide where one ends and the other starts, or whether to apply yet another term to the intervening area. Nevertheless, because the features of bone tissue that are used to differentiate between types likely have biological (in terms of life history, physiology and mechanics) as well as descriptive value, it is essential to arrive at a satisfactory means by which to discuss them. In addition, such categorizations are necessary for the quantification of tissue type areas.

Ultimately, the most successful descriptive categorizations will consider many, if not all, of the meaningful features that differ between bone tissues, and there have been many attempts to construct workable systems of tissue typology. The most comprehensive, most commonly applied classification that exists in the literature is one proposed by de Ricqlès in the 1970s, expanding upon earlier typologies (e.g., Enlow and Brown, 1956; Enlow, 1966), and modified by him, together with colleagues, in the intervening years (de Ricqlès, 1975; Francillon-Vieillot et al., 1990; de Ricqlès et al., 1991; see Warshaw 2007 for an outline of this classification).

The typology of compact bone requires two levels of description. First, the fibrillar structure of the bone is identified. This is classically organized into one of three major patterns: (1) woven-fibered (= woven), (2) parallel-fibered, or (3) lamellar, depending on the degree of regularity in the organization of the collagen fibers and the shape, distribution and orientation of the osteocyte lacunae. These fibrillar matrix types are found in both compact and cancellous bone, and may be primary or secondary in origin. While secondary bone is often (including here) used to refer solely to the tissue deposited intracortically around Haversian canals subsequent to resorption of existing bone, some consider secondary bone to be all those tissues that have replaced previously deposited bone, identified by the presence of a reversal line. However, as de Ricqlès and Colleagues have pointed out, the concept of primary versus secondary bone has a limited meaning, and can only be used as a record of relative osteogenesis time at the very local level (de Ricqlès et al., 1991, Castanet and de Ricqlès, 1986–87).

Subsequent to the characterization of the fibrillar structure, the identification of vascular patterns and the spatial distribution of fibrillar matrix types, as well as the ontogenetic origin of the tissue, assist in the second level of differentiation between primary bone types. Bone tissues of all fibrillar structures can be avascular. When vascular canals are present, these may be simple – no lamellae circumferentially surround the canal

– or they may be in the form of primary osteons. The latter are surrounded by lamellae that have been circumferentially deposited around the blood vessels in the canals. Such canals may course longitudinally through a bone, or may follow perpendicular, oblique or radial directions. In addition, a single canal may change directions within the cortex, and may anastomose with other canals (Robling and Stout, 1999).

18.2.3 Existing Hypotheses Regarding the Factors Determining Primary Tissue type Variability

Differences in primary tissue type patterns can be found between taxa, individuals, bones and parts of bones. The extent to which, and manner in which, these differences are mediated by genetically determined and environmental factors remains largely unknown. As de Ricqlès aptly comments, "histodiversity has…a complex, multifactorial causality which is mainly of an ontogenetic, functional origin, but which cannot generally be subsumed to any unique, single efficient cause" (1992, p. 47). Inroads have been made in this area, so that it has become possible to gain some understanding of the relationship between tissue types and a number of other biological variables. Four potentially interrelated factors are frequently suggested to impact upon the distribution of primary tissue types: developmental processes, organismal growth rate, body size and mechanical adaptations. Reviewed briefly below, each will be treated in more detail in the discussion section of this chapter. The phylogenetic status of a species is another factor that must be considered in conjunction with all of these. Tissue types and distributions may be group specific inasmuch as growth patterns, life histories, and/or positional behaviors are constrained by heritage. While the current literature does not offer strong support for the prospect that some microstructural features may be reflective of subordinal phylogenetic relationships – independent of proximate, adaptive factors – the possibility that this may occur should not be rejected *a priori*.

> It is true that *some* bone histological characteristics may sometimes appear as taxa-specific and, if they are not caused by homoplasy, used, at least tentatively, as synapomorphies…Many more may be discovered in the future using finer quantitative histological surveys, as permitted by image analysis (de Ricqlès, 1992, p. 53).

18.2.3.1 Tissue Types and Development

It is the work of Enlow (1963, 1976, 1982; Enlow and Hans, 1996) that is most responsible for bringing the process of osteogenesis to the forefront and taking this into account when assessing variation in tissue type distribution within and between bones, and among taxa. As animals grow, bones must change size while maintaining mechanically functional shapes and relationships with adjacent bones and soft tissues. However, because the osteocytes within calcified matrix are not capable of movement or mitotic division, this relocation of bone requires deposition (or apposition, cf. Enlow) of new matrix at some surfaces and resorption of existing matrix at others; in this way the bone organ accomplishes "drift" through anatomical space. This process of growth remodeling leaves behind a record of its progression in the patterning of tissue types. For example, the presence of compacted coarse cancellous tissue at the midshaft may indicate that the bone in this region was once metaphyseal and trabecular. Even when at the periosteal surface of a cortex, this tissue type indicates original endosteal deposition. Particularly with the assistance of lines of discontinuity marking pauses in growth or reversals between resorption and deposition, differences in tissue types offer a means to reconstruct the growth remodeling process of the whole bone.

18.2.3.2 Tissue Types, Growth Rate and Body Size

Amprino (1946) was the first to suggest a relationship between the rate of growth of bone and the tissue type that results; this was later coined "Amprino's rule" (de Buffrénil and Pascal, 1984). His observations have been supported by subsequent research (e.g., Newell-Morris and Sirianni, 1982; Burr et al., 1989; de Ricqlès et al., 1991; Castanet et al., 1996; Castanet et al., 2000; de Margerie et al., 2002; de Margerie et al., 2004) and have been applied to the palaeobiological reconstruction of fossil reptile growth patterns (e.g., Horner et al., 2000; de Ricqlès et al., 2000; Sander, 2000; but see Starck and Chinsamy, 2002 for a discussion of potential pitfalls).

In this two-level scheme, the degree of organization of the fibrillar structure is rate-dependent. Woven bone is deposited most quickly (e.g., on average 9.8 µm/day in early macaque fetuses), parallel-fibered tissues have an intermediate deposition rate (e.g., 4 µm/day in early postnatal macaques), and lamellar bone is deposited most slowly (e.g., 1.5 µm/day in a juvenile macaque) (Newell-Morris and Sirianni, 1982). The second determinant of rate is the intensity (Castanet et al., 1996, 2001; de Margerie et al., 2002) and perhaps patterning (de Margerie et al., 2004), of vascularization. More vascularization is associated with greater rate of deposition (e.g., Castanet et al., 1996). Such a scheme suggests that variation in deposition rates will account for many of the patterns observed between species, importantly reflecting the interaction between body size and growth rates. It follows that tissue types and distributions are group specific inasmuch as growth patterns, or life histories, are group specific.

18.2.3.3 Tissue Types and Mechanical Properties of Bone

Because the supporting function of the skeleton is a principal selected characteristic, tissue deposition must proceed within the context of mechanical imperatives. This can be seen in the maintenance of strain patterns throughout ontogeny (Biewener et al., 1986; Biewener, 1991). Not all tissue types possess the same mechanical properties, due to variation in such compositional and organizational features as degree of calcification, collagen fiber organization, porosity, or presence of cement lines (Evans and Bang, 1967; Simkin and Robin, 1974; Burr,

1980; Martin, 1991; Currey, 1999). For example, it has been shown that bone cortices with more secondary osteons have reduced tensile strength – and reduced fatigue life – relative to cortex with more interstitial primary lamellar bone (Amprino, 1948; Evans and Bang, 1967; Simkin and Robin, 1974; Vincentelli and Grigorov, 1985; Rho et al., 1999) or fibro-lamellar bone (Martin and Ishida, 1989; Schaffler et al., 1990); these results are not corroborated in all studies however (Martin and Ishida, 1989; Riggs et al., 1993b).

The difference in the mechanical properties among most primary bone tissue types for the most part remains to be explored, although a small literature hints that differences do exist. It has been suggested, for example, that woven bone is weaker (i.e., will fail under smaller loads) than lamellar bone (Martin, 1991). Furthermore, the increase in porosity found in fibrolamellar bone relative to less vascularized lamellar-zonal tissues is expected to reduce strength and stiffness (Martin, 1991; Schaffler and Burr, 1988; Lieberman and Pearson, 2001). Finally, de Margerie and colleagues have recently proposed that a laminar vascular organization of fibro-lamellar tissue is an adaptation to torsional loads (de Margerie, 2002; de Margerie et al., 2005). There exists no body of work, however, examining the relative mechanical properties of lamellar and parallel-fibered matrix types, both ubiquitous among primates and other vertebrates. Given the histocompositional differences between these two types, the potential for mechanical significance should not be overlooked.

18.2.4 Objectives of the Current Study

Taking into account the research outlined above, the objectives of the current study are to address the following. First and foremost, the presence or absence of variation in primary tissue types among primates, and patterning of this variation, must be asertained. While it can be assumed that diversity will be found (given the few previous primate studies and the more comprehensive database which exists for other vertebrate groups), and that it is non-random, this has never before been systematically determined. Second, any observed variation is to be considered in light of the hypotheses outlined above regarding factors that are determintive of it: ontogenetic processes, growth rate, body size, positional behavior and phylogeny. It is predicted that variation will be patterned within and among skeletal elements, and among taxa, reflective of these life history and mechanical factors, as well as, potentially, phylogenetic status.

18.3 Materials and Methods

18.3.1 Sample Selection

In selecting a sample range for a study of variation in bone microstructure, there are two useful, complementary levels of investigation, both of which focus on a subset of closely related mammals. One approach examines one or a few species in great depth, with relatively large sample sizes for each taxon (e.g., Burr, 1979; Paine and Godfrey, 1997; McFarlin, 2006). This offers the opportunity to assess intraspecific variability and determine species patterns, including during ontogeny when juveniles are included. The other approach – one which serves as a bridge between intra-taxon studies and extensive vertebrate reviews – is broad-based, but nevertheless restricts the bulk of the effort to taxa which are phylogenetically constrained, such as the members of a single order (e.g., Schaffler and Burr, 1984). This is the method adopted here, with the intent to identify broad, biologically meaningful, patterns.

Using a carefully constructed comparative sample allows for correlation and assessment of similarities and differences, and thus the elucidation of possible causal relationships between morphological attributes and adaptations. The order Primates is well suited to this research for three reasons. First, primates are well studied, and there exists a vast literature on their phylogeny, behavior, ecology, physiology, and functional skeletal morphology. Second, it is a group including species that vary greatly in size, other aspects of life history, and positional behavior. Finally, while many primates likely share similar growth and behavioral life histories due to common ancestry, there are also many instances of convergence in adaptations.

The sample is composed of a selection of primates including members of five strepsirhine families (Galagidae, Lorisidae, Cheirogaleidae, Indriidae, Lemuridae), three platyrrhine families (Callitrichidae, Cebidae, Atelidae), and the family Tarsiidae (Table 18.1). The sample encompasses representatives of several broad primate positional behavior categories, as well as a range of body sizes and life history strategies.

In some cases the taxonomic level of analysis is the genus (e.g., *Otolemur, Callithrix*), although more than one species may be represented. In other cases, several genera from a single family are included as a single taxon unit. Such generalization is only applied in cases where, in terms of positional behavior and broad life-history features, the genera are very similar to one another. For example, *Nycticebus, Loris* and *Perodicticus* are referred to as a single unit for analysis: the Lorisidae. These three genera display very comparable (albeit not identical) ecological, mechanical and life history adaptations (Gebo, 1987; Demes et al., 1990; Müller, 1982; Rasmussen and Izard, 1988; Kappeler, 1996; Weisenseel et al., 1998). With the exception of the galagids, whenever the genus is the level of analysis the nominal taxon is the sole representative of its family. So, for example, *Saimiri* is the only genus from the family Cebidae included in the study. Effectively therefore, in all but one case, families are the units of analysis, with some containing one representative genus, and some more than one. The galagids have been divided into two subgroups (greater and lesser galagos) because of the distinctive variation in positional behavior between the two. This offers an opportunity to examine a smaller range of variation within a single family.

For the purposes of the current study, I suggest that the allocation of sample species to families or subfamilies (e.g., Lorisidae vs. Lorisinae, or Cebinae vs. Cebidae) is less important than the monophyletic grouping of closely related genera. I am aware of no current research that would invalidate the 10 groups I use (with the possible exception of paraphyly among lorisids; see Yoder et al., 2001; Poux and Douzery, 2004), in terms of both the taxa I place within the groups, and the separation between them, despite the likelihood that some would quibble with the use of the family level (e.g., Rosenberger, 2002). Therefore, there is no further consideration here of competing taxonomies.

Sample composition was designed so that the following kinds of comparisons could be made: (a) between closely related taxa with differing positional behavior in terms of frequencies of locomotor modes (e.g., *Lemuridae/Indriidae*), (b) between phylogenetically distant taxa with similar positional behavior (e.g., *Tarsiidae/lesser galagos*), (c) between species of differing size and/or growth rates but similar positional behavior (e.g., *lesser galagos/Indriidae*), and (d) between species with similar size and positional behavior but differing growth rate (e.g., *Cheirogaleidae/Callithrix*).

All individuals are adults, as determined by the fusion of all epiphyses and eruption of all permanent teeth (when skulls were available). While adulthood – as defined by the attainment of stable body mass – does not necessarily coincide with dental and skeletal maturity (e.g., Smith and Jungers, 1997), these are the only features that can be used to identify adulthood in the absence of body mass records. In addition, epiphyseal fusion is an appropriate measure given that the features of interest are related to skeletal growth and development. Although changes in proportions of the cortex composed of intracortically remodeled bone may occur with aging, the primary tissue types deposited during growth are not typically expected to vary substantially after attainment of adult size (barring pathology). In addition, non-human primate studies suggest that, although percent secondary osteonal bone may increase during growth, it does not change in a predictable fashion in the adult (Schaffler and Burr, 1984; Burr, 1992; Paine and Godfrey, 1997; Jerome and Peterson, 2001). Geriatric individuals, identified from tooth wear and overall condition of the skeletal material, were avoided.

18.3.2 Element Selection and Section Sampling Location

Midshaft sections were taken from the femur, humerus, tibia, radius and ulna on one side of the body. While the majority of previous comparative bone microstructure studies have concentrated on the femur, all the elements of the skeleton will differentially reflect posture and locomotion, as well as growth (Erickson and Tumanova, 2000). The midshaft was selected, rather than more distal or proximal locations, as this has long been a standard in the literature and provides for comparability among studies (Schaffler and Burr, 1984; Paine and Godfrey, 1997; Goldman, 2001). In addition, beam theory has provided a superior understanding of the mechanical loads at the midshaft relative to the potentially more complicated mechanical environments closer to the joints (Lanyon and Rubin, 1985).

Diaphyseal rather than maximum length was used in order to maintain consistency with colleagues examining juvenile specimens, which may not have epiphyses. For each bone, landmark standards were established to determine the proximal and distal extent of the diaphysis (see McFarlin, 2006; Warshaw, 2007). In total, 124 sections were included in the study, two of which were ultimately excluded from the analyses because of apparent pathology. Due to the labor intensive nature of specimen preparation (as well as the difficulty in obtaining specimens for destructive work) the large number of taxa necessitates small sample sizes for each taxon.

18.3.3 Specimen Collection, Processing, Preparation and Imaging

Midshaft blocks (0.3–1 cm long) were extracted from each long bone shaft using a Buehler Isomet low-speed diamond saw or a custom portable circular saw, cleaned of organics, dehydrated and defatted, and embedded in clear blocks of poly-methylmethacrylate, according to the methods detailed in Warshaw (2007). To record orientation relative to the whole bone, a small groove was sawn at the anterior midline on the end of the block opposite the midshaft surface.

The embedded midshaft surface of each block was exposed, ground with graded carbide papers (on a Buehler Handimet II hand grinder, to a final 1200 grit finish) and mounted to a glass slide using dental adhesives according to the methods developed in the HTRU (Goldman et al., 1998; Goldman, 2001; Warshaw, 2007). Mounted blocks were sectioned on the Isomet to produce 120 ±4 μm thick sections and ground to 100 ± 5 μm, finishing with 1200 grit paper (see Warshaw, 2007, for a discussion of section thickness determination).

Image acquisition was digital, with data transferred directly from cameras to Pentium-based computers. Sections were imaged at low magnification (5X objective) in conventional transmitted light (LM) and circularly polarized transmitted light (CPL) using a Leica-Leitz DMRX/E Universal Microscope, configured with a Marzhauser motorized stage. The benefits of CPL over linearly polarized light (LPL), particularly for quantification of collagen fiber orientation, are discussed in Bromage et al., 2003. While collagen fiber orientation is not considered in this chapter (see Warshaw, 2007) CPL images were employed qualitatively to assist in the identification of tissue types. LM and CPL images of entire sections were acquired with a JVC KY-F55B color video camera using Syncroscopy Montage Explorer software (Synoptics, Ltd.). The Syncroscopy software allows for real-time montaging during image acquisition, resulting in high-resolution images of entire sections. All work was performed in the Hard Tissue Research Unit and the Analytical Microscopy and Imaging Center in Anthropology, Hunter College of the City University NewYork now relocked to the New York University College of Dentistry

18.3.4 Image Processing and Quantification of Tissue Types

For each bone, CPL and LM images were overlain and aligned in Abode Photoshop (8.0), and non-bone areas were masked out. To allow for quantification of the relative proportions of tissue types within a cortex, areas for each of the major bone tissue types were traced and assigned identifying colors (Figures 18.1 and 18.2). Tissue type maps were converted to grayscale and imported into Optimas (Version 6.51; Media Cybernetics, Inc.) and the areas (in pixels) were calculated for each tissue type in each section, using a macro developed for Optimas (Goldman, 2001). For use in the majority of analyses, absolute values were converted to proportions, relative to the total cortical area (TCA) or relative to the endosteal or periosteal portion of the cortex (for the endosteal and periosteal tissues respectively).

18.3.5 Identification of Tissue Types

Following is a description of the individual tissue types identified in the current study and the methods for identification of these tissues. The typology generally follows the scheme described by de Ricqlès and colleagues, adapted for maximal utility with this primate sample and the specific questions I address. Standards developed in the HTRU were employed for the identification of tissues and the demarcation of boundaries between tissues, following the typology outlined below. Tissue types are described as they are observed in LM as well as CPL. The former allows for the identification of vascular canal and osteocyte lacunar morphologies and distributions, while the latter provides information regarding collagen fibrillar matrix patterns. In addition, the sections were examined real-time in LM and CPL to clarify type identification when the captured images did not provide sufficient resolution.

Tissues were first categorized as either endosteal, periosteal or intracortically remodeled (i.e., composed of secondary osteons, or Haverisan systems). Within the endosteal and periosteal areas, tissues were further assigned to one of five primary types: (1) woven (WOV), (2) fibrolamellar (FBL), (3) parallel-fibered (PF), (4) lamellar (LAM) and (5) coarse

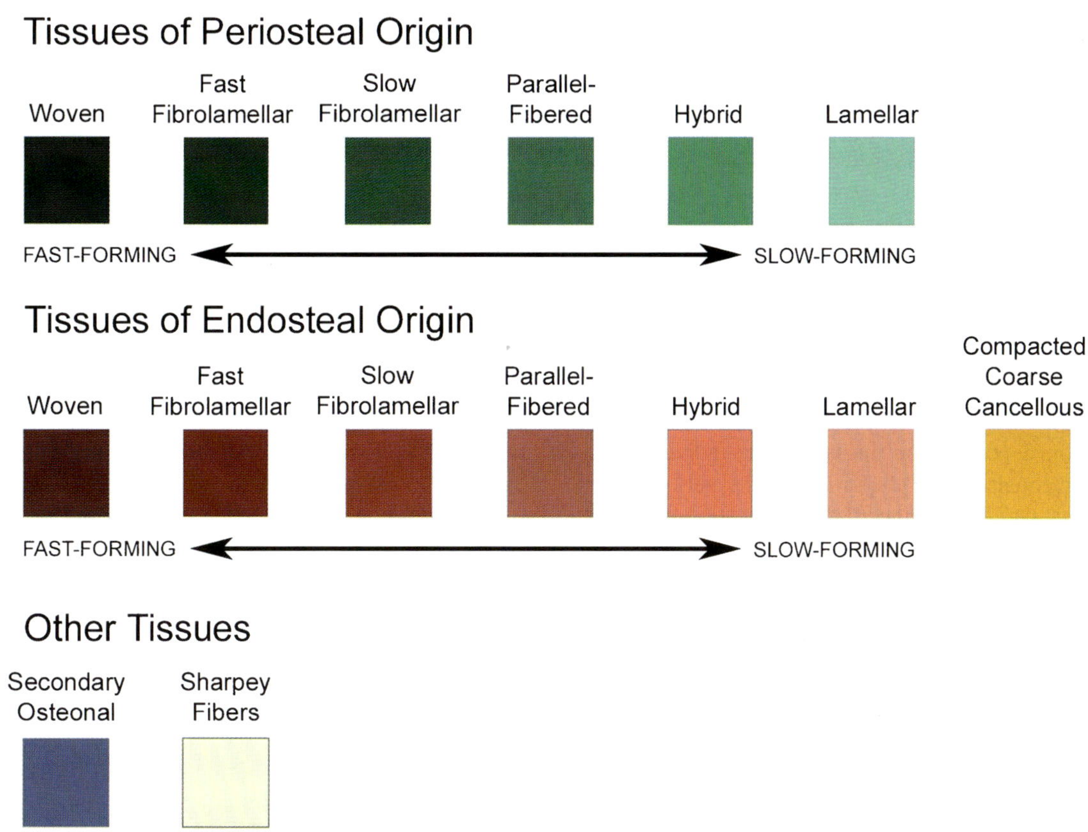

FIGURE 18.1. Tissue mapping color code. See the text and Warshaw (2007) for a discussion of the "hybrid", "fast fibrolamellar" and "slow fibrolamellar" types.

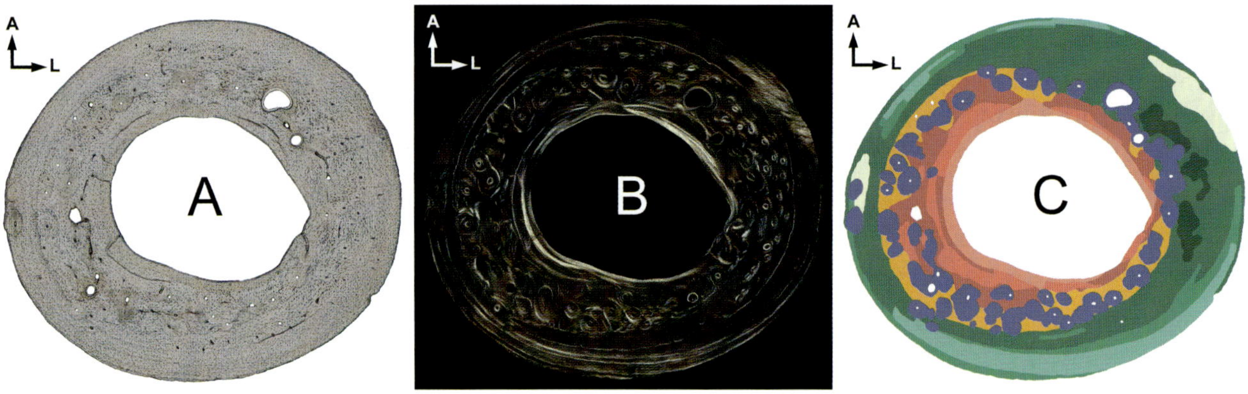

FIGURE 18.2. Sample cortex in three modes. Midshaft *Nycticebus coucang* (HC27) radius: A, imaged in conventional transmitted light; B, imaged in circularly polarized light; and C, translated into a tissue map.

compacted cancellous (CCC). The last of these appears only in the endosteal cortex. With the exclusion of CCC bone, the tissues have previously been described as falling on a formation-rate continuum, as per their numbering above (e.g., Amprino, 1946; Castanet et al., 1996; Newell-Morris and Sirianni, 1982; de Ricqlès et al., 1991), with woven-fibered and fibrolamellar bone at the fast end of the range, and lamellar bone at the slow end. As a general rule, faster-forming tissues display less organized fibrillar structure and rounder osteocyte lacunae (e.g., in woven bone) and/or more vascularization (e.g., in fibrolamellar bone), while the slower tissues are characterized by highly organized fibrillar patterns, flattened lacunae, and less vascularization (e.g., in lamellar bone). However, in practice, a range of variation in these features may occur within each category. When dense Sharpey fibers prevented the identification of one of the other tissue types, the bone was labeled as Sharpey fiber bone (SF). In addition, intracortically remodeled bone was labeled as such (HAV).

Because intracortical remodeling effectively replaces primary tissue, the type that was initially deposited and present previous to remodeling is unknown. Therefore, no primary tissue type was assigned to areas occupied by HAV bone. It follows that, depending on the taxon, the individual, and the skeletal element, some proportion of the periosteal and endosteal area was not assigned a primary tissue type. In other words, in the analyses that follow, total periosteal area is equal to real total periosteal area minus periosteal HAV area, and total endosteal area is equal to real total endosteal area minus endosteal HAV area. For quantitative analysis, SF bone is included with periosteal tissues as it is only found in periosteally deposited bone. CCC bone is of endosteal origin, but because of its distinct developmental trajectory it is examined in isolation, as it may provide information regarding taxon- and element-specific records of growth.

It should be noted that de Ricqlès and colleagues often do not differentiate between PF and LAM bone within a lamellar-zonal continuum (i.e., circumferentially formed regions of lamellar and parallel-fibered bone displaying a depositional periodicity). However, because much of the primate cortices in this sample are composed of bone in these categories, and because it has been suggested that there are differences in growth rates between them (e.g., Amprino, 1946; Newell-Morris and Sirianni, 1982), the two types are considered independently, rather than lumped in a single lamellar-zonal category. The degree of vascularization within lamellar-zonal tissues is not quantified, as vascularization is generally sparse (but see the discussion section).

Two modifications to previously published typologies were introduced in the HTRU. First, the identification of a hybrid type (see figure 18.1) intermediate in morphology between PF bone and LAM bone is an innovation to describe variation observable in primate bones that would not otherwise be accounted for, and which may be of biologically significance (i.e., may indicate rates of growth intermediate between LAM and PF tissues). However, during the analysis stage of this work, it became evident that proportions of hybrid bone did not vary in a patterned fashion, while PF and LAM proportions did. Hybrid tissue was therefore collapsed into the LAM category (although still visible as a separate color in the tissue maps). This was chosen, rather than a combining with PF bone, because of the presence of lamellae in hybrid bone, albeit as a reduced proportion of the tissue relative to LAM bone.

The second innovation, a differentiation between more and less heavily vascularized fibrolamellar tissue ("fast" versus "slow" fibrolamellar, see Figure 18.1), was also ultimately deemed not useful to the examination of the particular taxa included in this study. Therefore this will not be further considered here, but is discussed in Warshaw (2007). Because woven bone occurs very infrequently, it was combined in the analysis with fibrolamellar bone (together labeled FAST bone), the other rapidly depositing tissue.

Due to the continuous nature of tissue type variation, it is at times difficult to confidently assign boundary lines between types. There is therefore a measure of subjectivity, and resultant error, in the creation of these colorized tissue maps. To assess the degree of error that may be produced, a repeated measures test was performed. An area of cortex from a single section containing a range of tissues was mapped three times. Variation in the proportion of tissue types, relative to TCA, did not exceed 3% of the TCA.

18.3.5.1 Woven-Fibered Bone

WOV bone is identified on the basis of rounded or stellate-shaped, randomly distributed osteocyte lacunae (visible in LM images) and a multidirectional fibrillar texture (appearing "woven" in CPL images). This tissue, most often associated with infant and juvenile cortices, was absent in the majority of the sections in this study. When present, it was found in very small isolated areas, and in no cases did it comprise a significant proportion of the cortex.

18.3.5.2. Parallel-Fibered Bone (Figure 18.3B)

In PF bone, the osteocyte lacunae are typically more flattened and less randomly distributed than is the case for WOV bone. This is associated with regions within which collagen fibers are all similarly aligned, appearing smooth and featureless in CPL. In practice, tissue that appears to be parallel-fibered in terms of the collagen patterning may contain a range of osteocyte shapes (from rounded to more flattened) and distributions (from random to aligned in rows). In the current study, bone is considered to be PF based on the characteristic collagen patterning visible in CPL, that is, when 75% or more of the tissue in a given area (assessed by eye) is free of lamellar structure (see below). This tissue type is frequently located within areas of "lamellar-zonal" bone as described by de Ricqlès and others.

18.3.5.3 Lamellar Bone (Figure 18.3C)

In this study, bone is recognized as LAM when more than 25% of the tissue comprises visible lamellae. In some cases, lamellar structure is not visible in a CPL image, but can be seen in the LM image or when looking at the section real-time via the microscope oculars. Lamellar bone is typically characterized by flattened osteocyte lacunae, aligned with their long axes coursing with the lamellae. Again, however, in practice the shape of the lacunae may vary, from very flattened to lozenge, or peach-pit shaped. In addition, the lamellae may be thick or thin, distinct along their edges, or less clearly delineated. The lamellae may appear in discrete circumferential rings or series of rings, interspersed with rings of PF bone, or may be fragmented and more randomly distributed within an area otherwise composed of PF bone. As with PF bone, LAM bone, when running in circumferential swaths, would be classified as part of the lamellar-zonal type by de Ricqlès and others.

18.3.5.4 Fibrolamellar Bone (Figure 18.3A)

FBL bone comprises woven areas punctuated by vascular canals that have become surrounded by LAM tissue. Typically, the vascularization here will be denser than what is seen in lamellar-zonal bone. The FBL arrangement is formed when spaces in rapidly forming woven bone are filled in around the vascular canals with lamellar matrix, establishing a more compact tissue (Currey, 1960; Francillon-Vieillot et al., 1990). As is the case with lamellar-zonal bone, the vascular canals in this type may vary in density, may course in all directions, may be randomly distributed, or may display a regular patterning. Some patterns are distinct enough to have been assigned names, such as reticular, radiating, plexiform, and laminar. Recently it has been recognized that FBL bone may also be zonal, forming with the kind of periodicity previously associated solely with lamellar-zonal tissue (Castanet et al., 2001).

FBL is distinguished from vascular LAM or PF bone (i.e., lamellar-zonal types) by the following criteria. First, 75% or more of the vascularization must consist of primary osteons rather than simple vascular canals, with at least two lamellae around the canal. Second, the distance between the primary osteons must be no more than twice the diameter of the osteons. Third, at least 25% of the extra-osteonal matrix must be composed of WOV tissue, as identified by rounded or stellate osteocyte lacunae that are distributed in a random, clumping pattern. Fourth, no more than 75% of the osteocyte lacunae may be flattened and coursing in arched lines above and below the primary osteons (indicating LAM or PF tissue).

18.3.5.5 Coarse Compacted Cancellous Bone (Figure 18.3D)

CCC bone (type name *sensu* Enlow, 1963, 1982), is a uniquely endosteal tissue that forms when LAM or PF tissue fills in the spaces between trabeculae of cancellous bone, leaving only the vascular canals open. CCC bone is characterized by odd-shaped convolutions of LAM or PF bone demarcated in some cases by resting or reversal lines. However, unlike the cement lines of secondary osteons, they do not form closed circles or ellipses.

18.3.5.6 Sharpey Fibers (Figure 18.3E)

Sharpey fibers, or extrinsic fibers (*sensu* Boyde and Jones, 1998), mark the insertion of muscle, tendon, and ligament to the bone surface (Francillon-Vieilliot et al., 1990). They can be found deep within the cortex, marking soft tissue attachments at earlier stages of growth remodeling, as well as at the surface, marking current attachment sites. SF bone, whether abutting the periosteum or buried deeper in the cortex, is identified by the presence of collagen fiber bundles that run at an angle to the fiber patterns in the surrounding or underlying bone tissue, and at and angle, or even perpendicular to the periosteal edge of the bone. These are easily visible as parallel dark lines in LM imaging, and parallel bright lines in CPL. Tissue is

TABLE 18.1. Sample composition and provenance.

Taxonomic level	Genus	Species	Provenance	F	H	T	R	U
Lesser galagos	*Galago*	*senegalensis*	MNHU 60601	■	■			
	Galago	*moholi*	DPC 22b		■	■	■	■
	Galago	*moholi*	DPC 33		■	■	■	■
	Galagoides	*demidovii*	HC 25	■	■			
Greater galagos	*Otolemur*	*crassicaudatus*	DPC 75	■	■	■	■	■
	Otolemur	*crassicaudatus*	CT OA	■	■	■		
	Otolemur	*crassicaudatus*	HC 25	■	■	■	■	■
	Otolemur	*crassicaudatus*	USNM 1	■	■	■		
Lorisidae, *sensu stricto*	*Nycticebus*	*coucang*	MCZ 36116	■	■			
	Nycticebus	*coucang*	MCZ 422	■	■	■	■	■
	Nycticebus	*coucang*	HC 27	■	■	■	■	■
	Nycticebus	*coucang*	USNM 7	■	■			
	Nycticebus	*coucang*	DPC 12	■	■	■	■	■
	Loris	*tardigradus*	DPC 14	■	■	■	■	■
	Perodicticus	*potto*	DPC 90	■	■		■	■
Cheirogaleidae	*Cheirogaleus*	*major*	MCZ 44850	■	■			
	Cheirogaleus	*major*	DPC 22c	■	■			
	Mirza	*coquereli*	DPC 39	■	■	■	■	
	Microcebus	*murinus*	DPC	■	■	■	■	■
Lemuridae	*Varecia*	*variegata*	MCZ 18740	■	■	■	■	■
	Lemur	sp.	HC 30	■	■			
Indriidae	*Propithecus*	sp.	MNHU 44684	■	■	■	■	■
	Propithecus	sp.	MNHU 44685	■	■			
	Indri	*indri*	HC 19	■	■			
Tarsiidae	*Tarsius*	*spectrum*	MNHU An11318	■	■			
	Tarsius	*spectrum*	HC 16	■	■	■	■	■
	Tarsius	*syrichta*	DPC 8	■	■	■	■	■
	Tarsius	sp.	HC 23	■	■			
Callitrichidae	*Callithrix*	*geoffroyi*	MNHU 34381	■	■		■	■
	Callithrix	sp.	HC 28	■	■			
Cebidae	*Saimiri*	*sciureus*	HC 12	■				
	Saimiri	sp.	HC 13	■	■	■	■	■
	Saimiri	sp.	HC 14	■	■	■	■	
	Saimiri	sp.	MNHU 85197	■	■	■	■	■
Atelidae	*Ateles*	*fusciceps*	USNM 68	■	■			
	Ateles	sp.	MNHU 44079	■	■			

CT = Carl Terranova, DPC = Carl Terranova, HC = Hunter College of the City University of New York, FMNH = Field Museum of Natural History, MCZ = Museum of Comparative Zoology at Harvard University, MNHU = Museum für Naturkunde of Humboldt University, USNM = National Museum of Natural History; F = midshaft femur section, H = midshaft humerus section, T = midshaft tibia section, R = midshaft radius section, U = midshaft ulna section.

categorized as SF only when 50% or more of the area is composed of Sharpey fibers, masking the underlying histological structure.

18.3.5.7 *Intracortically Remodeled Bone (Figure 18.3F)*

Secondary osteons are identified by the presence of a reversal line (cement line) at their perimeter, generally appearing dark in LM, and often (although not always) bright in CPL. The reversal line is often scalloped in appearance. Within this perimeter are found concentric lamellae. Secondary osteons are classically round or oval, but in practice can drift into elongated and convoluted shapes (Robling and Stout, 1999), although they will always, by definition, be completely encircled by a reversal line, unlike the convolutions of CCC bone. Unlike primary osteons (which will have no reversal line) they will show a discontinuity with the primary interstitial bone around them.

18.3.6 Key to Abbreviations Used in the Following Sections

LAM = Lamellar + hybrid bone
PF = Parallel-fibered bone
FAST = Woven + fibrolamellar bone
CCC = Compacted coarse cancellous bone
SF = Sharpey fiber bone
P = Bone of periosteal origin
E = Bone of endosteal origin
TCA = Total cortical area
PHYLO = Phylogenetic status
PB = Positional behavior
SUS = Suspension/brachiation
AQ = Arboreal quadrupedalism/grasp-leaping
SC = Slow climbing/grasp-clinging and creeping
VCL = Vertical-clinging and leaping

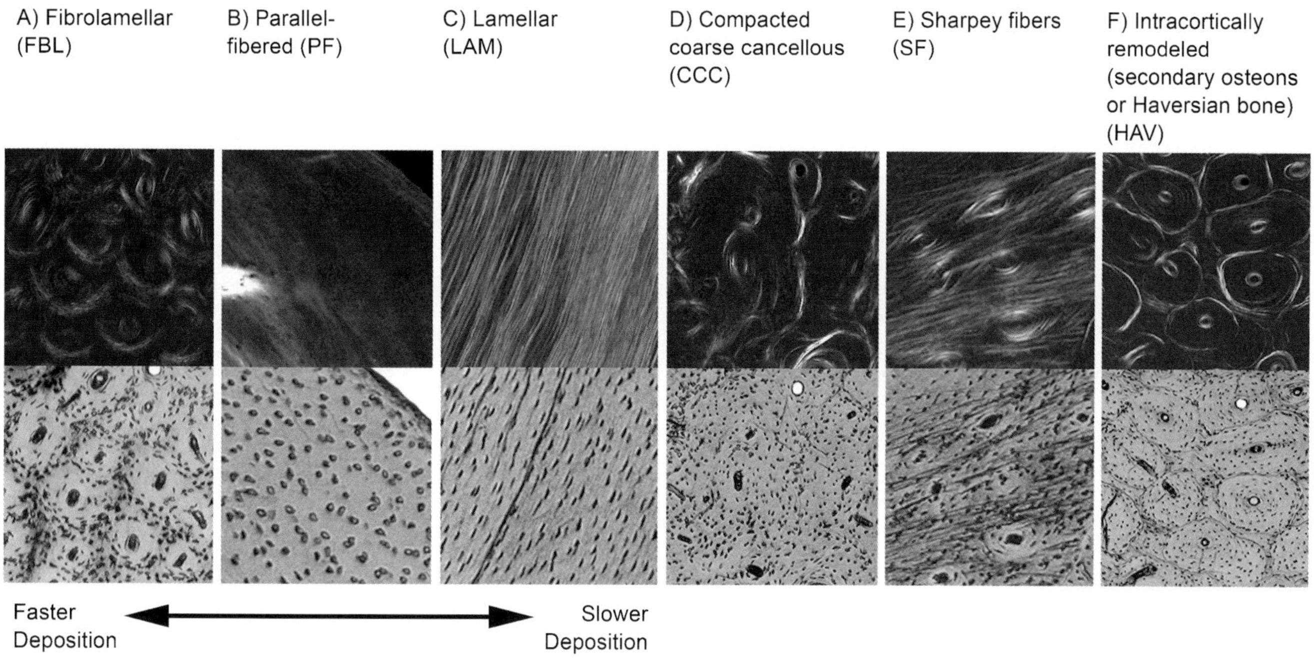

FIGURE 18.3. Tissue type examples. For each pair, the upper is a circularly polarized light image and the lower is a conventional transmitted light image.

18.3.7 Statistical and Descriptive Analyses

Sample sizes are small, with a maximum of 34 (for all sample femora, and all sample humeri) and in some cases as few as a single specimen (for distal skeletal elements in some taxa). In many instances the data are not normally distributed and cannot be successfully normalized via transformation. This may be due to sample size and composition, or perhaps the nature of the variable itself. Because some sample distributions remained non-normal after transformation, non-parametric statistics are applied where possible. When p values are highly significant, as is often the case (e.g., $p<0.01$), results are considered to be more reliable, despite the non-normal distributions. However, because of small sample sizes in many of the analyses, significances may be low or results may be insignificant despite visible differences in graphs or the sections themselves. In such cases, an observed pattern is considered to be more robust when it is repeated (for example among different skeletal elements, or among taxa). Even when results are statistically significant, identified patterns are verified using graphical methods. In the absence of visible and consistent patterns, the results and conclusions must remain tentative. Finally, in many cases, sample sizes are so small as to preclude significance testing altogether. In those cases, only descriptive methods are used.

Non-parametric Sign tests and Wilcoxon Matched Pairs tests are used to identify differences in the proportion of tissue types in the sample as a whole, examining each skeletal element independently. Subsequently, multiple regression analyses are employed to determine the importance of three independent variables in accounting for the observed variation in primary tissue type proportions. The variables – total cortical area (TCA), phylogenetic status (PHYLO), and positional behavior (PB) – were chosen as best able address the questions outlined in the objectives section, and are described below. A summary of taxon features is provided in Table 18.2. Given the frequent non-normality of the samples, the multiple regression analysis serves only as a rough guide to the link between dependent and independent variables, and suggested relationships are subsequently examined via descriptive graphical methods and basic statistics. Differences among PB or PHYLO categories are tested for significance using Kruskall-Wallis tests for multiple independent variables. If significance is identified with these tests ($p \leq 0.05$), pairwise comparisons (Mann-Whitney U tests) are utilized to pinpoint the source of significant variation. Descriptive analysis is provided for intra-taxonomic variation in periosteal tissue-type proportions within and between skeletal elements. Finally, femoral mean primary tissue type proportions in select pairs of taxa are qualitatively compared in light of the multiple regression results.

18.3.8 Independent Variables Included in the Multiple Regression Analyses

18.3.8.1 Total Cortical Area

TCA is used as a proxy measure of body size rather than body mass, because only estimated body mass values, which were culled from the literature, are available for this sample. Long bone length could also be used as a body size indicator, but including this in addition to cortical area would introduce

TABLE 18.2. Taxon life history and positional behavior information.

Taxonomic level	Species	Adult mass (g)	Scaled growth rate score	Positional behavior
	Galago moholi			
Lesser galagos	*Galago senegalensis*	173–315	43[w]	VCL
	Galagoides demidovii	60–63	–	VCL
Greater galagos	*Otolemur crassicaudatus*	1110–1190	36[w]	VCL/AQ/SC[*]
Cheirogaleidae	*Cheirogaleus major*	362–438	53[k]	AQ
	Mirza coquereli	304–326	–	AQ
	Microcebus murinus	59–63	48[k]	AQ
Lorisidae, *sensu stricto*	*Nycticebus coucang*	626–1100	55[w]	SC
	Perodicticus potto	830–1250	–	SC
	Loris tardigradus	192–269	77[w]	SC
Lemuridae	*Varecia variegata*	3520–3630	31[w]–68[k]	AQ
	Lemur sp.	1080–2510	112[w]	AQ
Indriidae	*Propithecus* sp.	2950–3250	–	VCL
	Indri indri	5830–6840	–	VCL
Tarsiidae	*Tarsius spectrum*			
	Tarsius syrichta	104–134	–	VCL
	Tarsius sp.			
Callitrichidae	*Callithrix geoffroyi*	330–360	165[w]–234[k]	AQ
	Callithrix sp.			
Cebidae	*Saimiri sciureus*	662–1020	210[k]–262[w]	AQ
	Saimiri sp.			
Atelidae	*Ateles fusciceps*	7290–9410	225[w]	SUS
	Ateles sp.			

Adult body mass ranges are from Smith and Jungers (1997). Where there are more than one species per genus, the range for all species combined is provided. For *Propithecus*, although the species is unknown for these specimens, the weights provided are for *P. verreauxi*. Scaled growth rate scores coded with a *k* are from Kirkwood (1985). The values are derived from the number of days (D) between age at 25% of adult mass and attainment of 75% of adult mass (Wa), following the equation D x $Wa^{1/4}$. A larger number indicates slower growth rate (i.e., greater number of days to increase mass by the same amount). Scaled growth rate scores coded with a *w* are those compiled by the author using the same equation. However, in these instances, D represents the number of days between birth and attainment of adult mass, as data for age at 25% and 75% of adult mass were not available. [*] *O. crassicaudatus* is classified as VCL for the multiple regression ananysis.

Positional behavior data were compiled from the following sources. **Lesser galagos:** Napier and Walker, 1967; Charles-Dominique, 1971b; Stevens et al., 1971, 1972; Hall-Craggs, 1974; Jolly and Gorton, 1974; Jouffroy et al., 1974; Walker, 1974, 1979; Jouffroy and Lessertisseur, 1979; Oxnard et al., 1981a, b; Burr et al., 1982; Gebo, 1987; Dagosto, 1988; Gebo and Dagosto, 1988; Demes and Günther, 1989; Ruff and Runestad, 1992; Terranova, 1995a; Aerts, 1998; Preuschoft et al., 1998. **Greater galagos:** Napier and Walker, 1967; Walker, 1974; Walker, 1979; Jungers, 1985; Gebo, 1987; Ruff and Runestad, 1992; Terranova, 1995a, b; Preuschoft et al., 1998; Runestad Connour et al., 2000. **Cheirogaleidae:** Walker, 1974, 1979; Jouffroy, 1975; Jouffroy and Lessertisseur, 1979; Oxnard et al., 1981b, Jungers 1985; Gebo, 1987; Ford, 1988; Preuschoft et al., 1995, 1998. **Lorisidae:** Suckling et al., 1969; Walker, 1969, 1979; Petter and Hladik, 1970; Charles-Dominique, 1971a, b; Hall-Craggs, 1974; Jouffroy and Lessertisseur, 1979; Dykyj, 1980; Szalay and Dagosto, 1980; Jouffroy et al., 1983; Glassman and Wells, 1984; Jungers, 1985; Gebo, 1987; Demes and Jungers, 1989; Demes et al., 1990; Ishida et al., 1990, 1992; Jouffroy and Petter, 1990; Jouffroy and Stern, 1990; Gomez, 1992; Preuschoft et al., 1995; Schulze and Meier, 1995; Runestad, 1997; Preuschoft et al., 1998. **Lemuridae:** Walker, 1974, 1979; Jouffroy, 1975; Jouffroy and Lessertisseur, 1979; Gebo, 1987; Dunbar, 1988; Gebo and Dagosto, 1988; Godfrey, 1988; Dagosto, 1994; Terranova, 1996; Preuschoft et al., 1998; Runestad Connour et al., 2000; Godfrey et al., 2004. **Indriidae:** Walker, 1974, 1979; Jouffroy 1975; Jouffroy and Lessertisseur, 1979; Szalay and Dagosto, 1980; Oxnard et al., 1981b; Jungers, 1985; Gebo, 1987; Gebo and Dagosto, 1988; Godfrey, 1988; Demes and Gunther, 1989; Demes et al., 1991; Godfrey et al., 2004. **Tarsiidae:** Walker, 1974, 1979; Jouffroy and Lessertisseur, 1979; Szalay and Dagosto, 1980; Oxnard et al., 1981b; Niemitz, 1983; Jouffroy et al., 1984; Dagosto, 1988; Ford, 1988; Gebo and Dagosto, 1988; Demes and Günther, 1989; Rasmussen et al., 1998; Runestad Connour et al., 2000. **Callitrichidae:** Szalay, 1981; Sussman and Kinzey, 1984; Ford, 1988; Falsetti and Cole, 1992; Preuschoft et al., 1995, Garber and Leigh, 1997. **Cebidae:** Fleagle and Mittermeier, 1980; Szalay and Dagosto, 1980; Boinski, 1989; Fontaine, 1990; Rose, 1993; Garber and Leigh, 1997; Nakatsukasa et al., 1997; Johnson and Shapiro, 1998. **Atelidae:** Jenkins et al., 1978; Mittermeier, 1978; Fleagle and Mittermeier, 1980; Ford, 1988; Swartz et al., 1989; Fontaine, 1990; Johnson and Shapiro, 1998; Turnquist et al., 1999; Runestad Connour et al., 2000; Cant et al., 2001.

multicollinearity. The correlation between cortical area and bone length is significant and positive (depending on the element, r ranges between .9182 and .9536), therefore either can be used. Of these two indirect measures of body size, cortical area is particularly appropriate to this study, given that the microstructural variables are examined as proportions relative to the midshaft cross-sectional area. In addition, it has been shown that the cross sectional properties of bones can function as reasonable predictors of body size (e.g., Delson et al., 2000; Ruff, 2003).

18.3.8.2 Phylogenetic Status

For the multivariate analyses, the sample was split into three taxonomic groups: strepsirhines, *Tarsius*, and platyrrhines. Separation at this level, rather than at a lower one, was chosen because the multivariate analysis is designed to capture broad patterns of variation. Subsequent pairwise analyses are used to identify differences among more closely related taxa.

When PHYLO and the scaled growth rate data available for this analysis were compared, they were found to be significantly correlated with one another (r = 0.9250). This is not surprising given that strepsirhines typically display accelerated growth relative to platyrrhines (Cubo et al., 2002; Kirkwood, 1985), i.e., phylogeny and growth rate are not independent of each other. Therefore, despite the relevance of growth rate to primary tissue type variation, this variable was not included in the multivariate analysis, as both growth rate and PHYLO could not included together in the multiple regressions, again to avoid the problem of multicollinearity and also to maximize the power of the regression analysis. Because data points exist for all individuals with regard to phylogenetic status, and given that these two variables are so highly correlated, PHYLO was selected over scaled growth rate to include in this portion of the analysis.

18.3.8.3 Positional Behavior

Positional behaviors have been allocated to four broadly-defined categories: brachiation/suspension (SUS), grasp leaping/generalized arboreal quadrupedalism (AQ), grasp-clinging and creeping/slow climbing (SC), and vertical-clinging and leaping (VCL). These follow the categories traditionally used to describe primate locomotion (e.g., Napier and Napier, 1967; Napier and Walker 1967; Szalay and Dagosto, 1980, Runestad Connour et al., 2000; Kimura, 2002). While the naming of these categories may vary, the association of individual taxa with each of them, and the descriptions of the *characteristic* behaviors with which they are associated, generally do not. As many researchers have pointed out, there is variation within these positional behavior groups, for example in the frequency and styles of locomotor modes associated with the category, in their morphological correlates, between sexes, with age, and with body size (e.g., Oxnard, 1973; Walker, 1979; Jouffroy and Lessertisseur, 1979; Oxnard et al., 1981a; 1981b; Terranova, 1995a, 1996; Fontaine, 1990). Variation in behavior may occur within species, and even individuals, dependent on factors such as habitat, time of day, or activity (Walker, 1979; Fleagle and Mittermeier, 1980). In addition, it is likely that certain positional behavior types evolved more than once among primates (e.g., vertical-clinging and leaping, see Szalay and Dagosto, 1980) with convergent or different morphologies and mechanics. Nevertheless, because the sample is constructed to identify broad patterns of variation and, because ultimately it is designed for functional comparison with fossil specimens, such categorization is useful (see Gebo, 1987; Godinot, 1990). And while primates of a given taxon may engage in a range of postures and locomotions, the frequencies with which they perform certain behaviors makes it possible to generalize about what is habitual for that taxon (Jouffroy and Lessertisseur, 1979). There is a precedent for this kind of broad categorization in past investigations of comparative bone microstructure, as well as other bone morphological features (e.g., Schaffler and Burr, 1984; Paine and Godfrey, 1997; Runestad, 1997).

18.4 Results

18.4.1 Periosteal/Endosteal Comparisons

The taxon means for relative proportions of P, E and CCC bone were examined for each skeletal element. Without exception, in all taxa and elements, there is a greater average proportion of P than CCC (Figures 18.4 and 18.5). Likewise, the proportion of P surpasses E, with the exception of the ulna and radius in *Cebidae* and the humerus in Cheirogaleidae. Indeed, this periosteal/endosteal disparity is so distinct that it is readily discernable with visual inspection of the cortices (Figure 18.5). There are patterned differences in the relationship among P, E and CCC bone between elements (Figures 18.4 and 18.5). First, in the sample as a whole, the proportion of E tissues is greater in the elements of the forelimb than in the hindlimb. In addition, the tibia is characterized by an absence or near absence in all individuals (with one exception) of CCC bone.

The distribution of primary tissue types is less variable endosteally than periosteally. This has also been found to be the case by de Margerie et al. (2002), in a sample of mallard long bones. Because of this, and because endosteal bone comprises less of the cortex than periosteal bone (and may be absent altogether in some sections), the remainder of the study focuses on the periosteal portion of the cortex.

18.4.2 Sample-wide Patterns of Tissue Types in Periosteal Bone

For each skeletal element, mean proportions of primary tissue types (relative to the periosteal cortical area) for all individuals combined were examined in order to identify sample-wide patterns. In the femur, values are lowest for

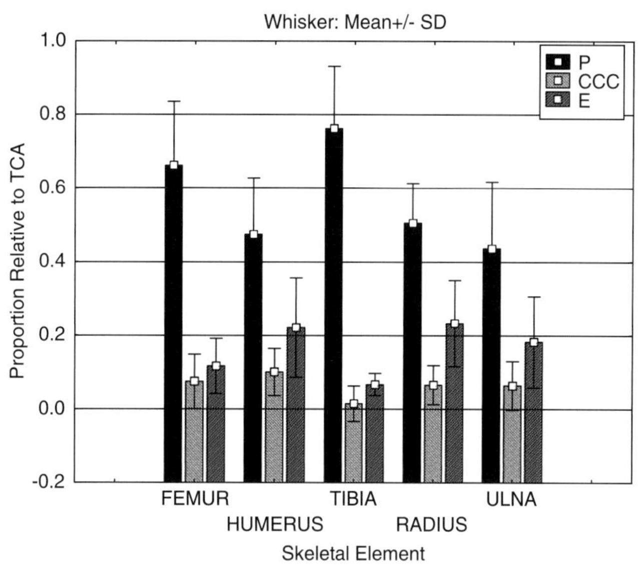

FIGURE 18.4. Proportions (relative to TCA) of periosteal, endosteal and coarse compacted cancellous bone in each element for all individuals combined.

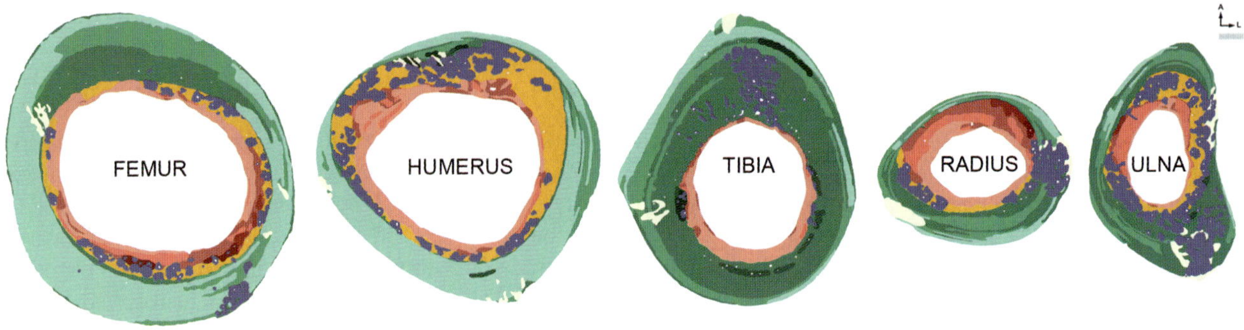

FIGURE 18.5. Midshaft sections from one *Otolemur crassicaudatus* individual (HC31). Proportions of slow-forming tissues (lighter greens) in the periosteal portion of the cortex are higher in the femur and humerus than in the other three elements. Also evident is the higher proportion of endosteal tissues (pinks and yellow) in the elements of the forelimb. Patterns of element-specific growth are evident. For example, the humerus has a strong posterior-ward pattern of drift, whereas growth in the femur proceeds in a more evenly circumferential fashion. The tibia is characterized by the absence of compacted coarse cancellous bone (yellow). Scale indicates 0.5 mm.

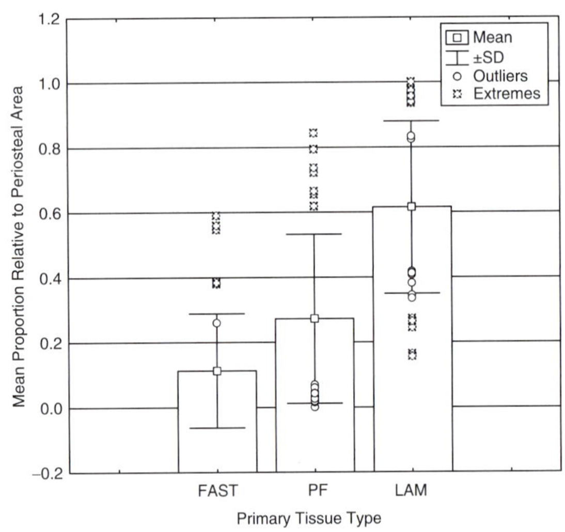

FIGURE 18.6. Proportions (relative to periosteal area) of FAST, PF and LAM bone in the femur for all individuals combined.

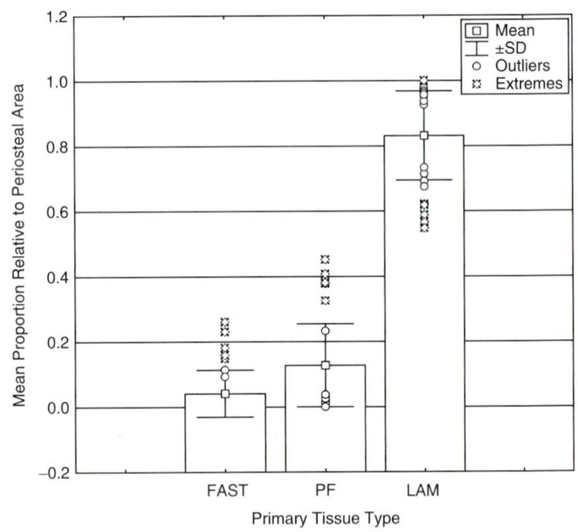

FIGURE 18.7. Proportions (relative to periosteal area) of FAST, PF and LAM bone in the humerus for all individuals combined.

FAST bone, higher for PF bone, and higher still for LAM bone (Figure 18.6). The differences among types are significant ($p < 0.01$ for all pairwise comparisons). As a whole then, the primate femur is characterized by an emphasis on tissues at the slow-deposition end of the spectrum. However, examination of the outliers and extremes in Figure 18.6 highlights the variability within tissue categories. In the humerus, the pattern is similar, although with an even greater emphasis on the slow-deposition LAM bone relative to PF and FAST bone (Figure 18.7). The difference between LAM bone and the other two types is highly significant ($p < 0.0001$).

The pattern in the distal bones of both the forelimb and hindlimb contrasts with what is seen in the proximal elements. As a whole, the radius, ulna and tibia show relatively more PF bone, and less LAM bone than the femur and humerus. In other words, there is a diminished emphasis on the slow-forming tissues. In all three bones, differences between LAM and PF tissues are insignificant, while values for FAST bone are significantly lower than the other two types ($p < 0.001$) (Figure 18.8). Figure 18.9 provides a visual summary of the contrast among elements. These patterns are consistent throughout the sample; in each individual, the proximal bones comprise larger proportions of LAM bone than distal bones (Figures 18.10 and 18.11).

18.4.3 Factors Accounting For Variation in Primary Tissue Types Across the Entire Sample

Multiple regression analyses confirm that the independent variables (TCA, PHYLO, PB), chosen based on their purported relationship with primary tissue type diversity, do account for a significant proportion of the variation observed in this sample,

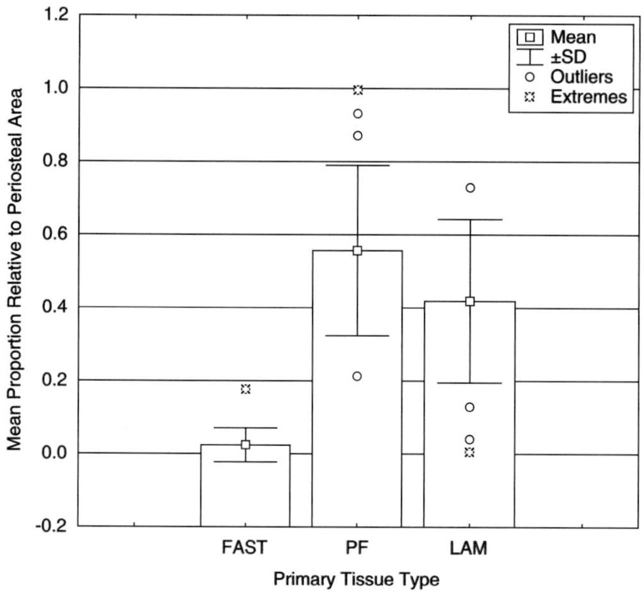

FIGURE 18.8. Proportions (relative to periosteal area) of FAST, PF and LAM bone in the ulna for all individuals combined.

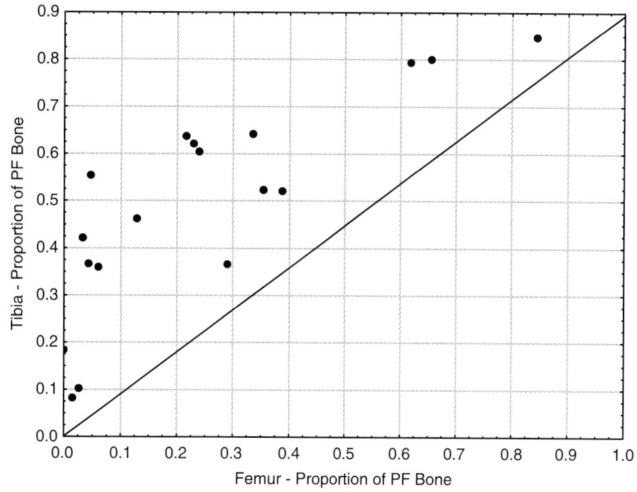

FIGURE 18.10. Comparison of individual values for PF bone in the femur and tibia. The values for each individual for the proportion of PF bone in the femur and tibia are plotted against each other. The diagonal line indicates where data points would fall should the proportion of PF bone be equivalent in both elements within an individual. For all individuals, the tibia contains a greater proportion of PF bone than the femur.

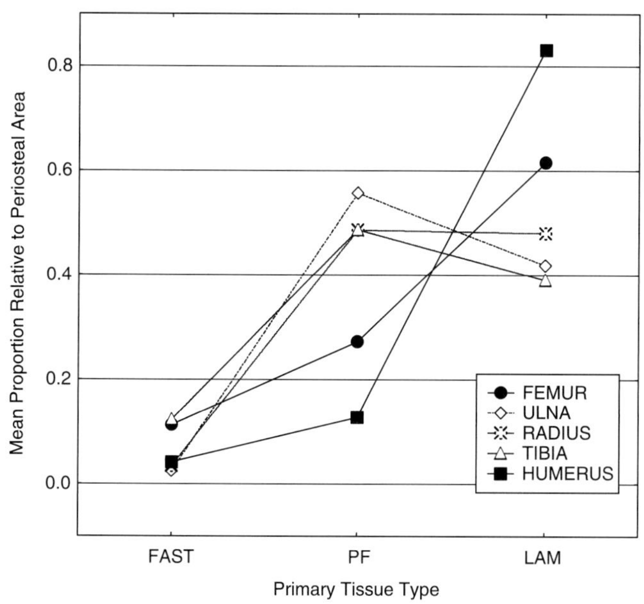

FIGURE 18.9. Comparison of mean periosteal tissue type values for all five skeletal elements for all individuals. Proportions (relative to periosteal area) of FAST, PF and LAM bone. Femoral and humeral patterns of tissue type proportions in the periosteal cortex are similar, with highest values for the LAM category. In contrast, the distal bones for both forelimb and hindlimb contain as much, or more, PF tissue as LAM tissue.

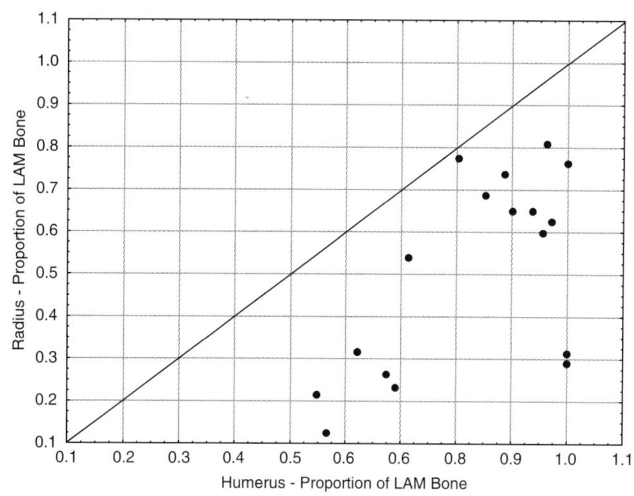

FIGURE 18.11. Comparison of individual values for LAM bone in the humerus and radius.The values for each individual for the proportion of LAM bone in the humerus and radius were plotted against each other. The diagonal line indicates where data points would fall should the proportion of LAM bone be equivalent in both elements within an individual. For all individuals, the humerus contains a greater proportion of LAM bone than the radius.

although some variation clearly remains unexplained. The models yielding the highest adjusted R^2 may include all three predictor variables, only two of the three, or in some cases just a single variable (Table 18.3). Overall, significant R^2 values range between 0.22 and 0.77 with p values of <0.05 to <0.0001. Depending on the tissue type, and to a lesser degree the element, the relationship with the independent variables differs, both in the magnitude of the total variance explained, as well as in the independent variables accounting for the variation. Therefore, in the following results, FAST, PF and LAM tissues are discussed individually.

TABLE 18.3. Multiple regression results for the entire sample.

Tissue type	Element	Adjusted R^2	Model
FAST	Femur	0.71^{***}	PHYLO, PB, TCA
	Humerus	0.22^{*}	PHYLO, TCA
	Tibia	0.55^{**}	PB, TCA
	Radius	0.73^{**}	PHYLO, PB, TCA
	Ulna	0.74^{**}	PHYLO, PB
PF	Femur	0.77^{***}	PHYLO, PB, TCA
	Humerus	0.41^{**}	PHYLO, PB, TCA
	Tibia	0.40^{*}	PHYLO, PB
	Radius	0.63^{**}	PHYLO, PB, TCA
	Ulna	0.64^{***}	PB
LAM	Femur	0.67^{***}	PHYLO, PB, TCA
	Humerus	0.55^{***}	PHYLO, PB, TCA
	Tibia	0.60^{**}	PHYLO, PB, TCA
	Radius	0.75^{***}	PHYLO, PB
	Ulna	0.69^{***}	PB

For each tissue type, the model with the highest adjusted R^2 was selected. The number of independent variables included in the model differs depending on the tissue type and element. Note that betas for some of the independent variables in these models may not be significant, as per the descriptions of results below. * indicates $p < 0.05$; ** indicates $p < 0.01$; ***indicates $p < 0.001$.

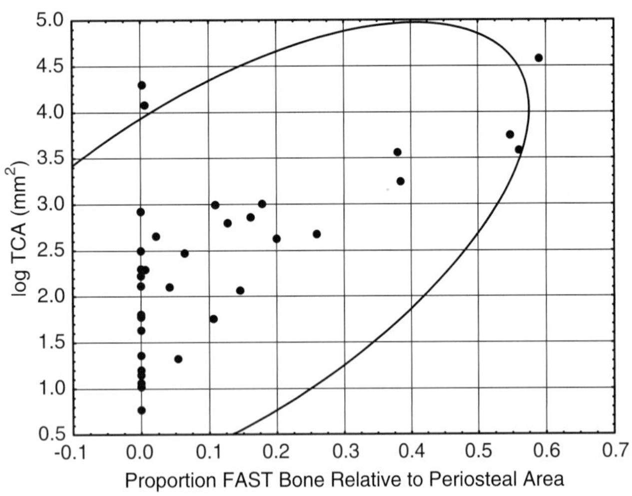

FIGURE 18.12. Correlation between FAST bone and TCA in the femur. There is a positively allometric relationship between the proportion of FAST bone (relative to periosteal area) and TCA ($r = 0.64$, $p<0.001$).

18.4.3.1 *FAST Bone*

In all elements except the ulna, TCA is the strongest predictor for the proportion of the cortex, relative to periosteal area that is composed of FAST bone. In the femur, humerus, tibia and radius, the relationship is positively allometric: the greater the cortical area, the greater the proportion of it that is composed of these fast-forming tissues (Figure 18.12). Given the consistency among these four elements, it is most likely that the absence of this pattern in the ulna is due to sample composition, rather than a difference in growth or functional adaptation in this bone relative to the others.

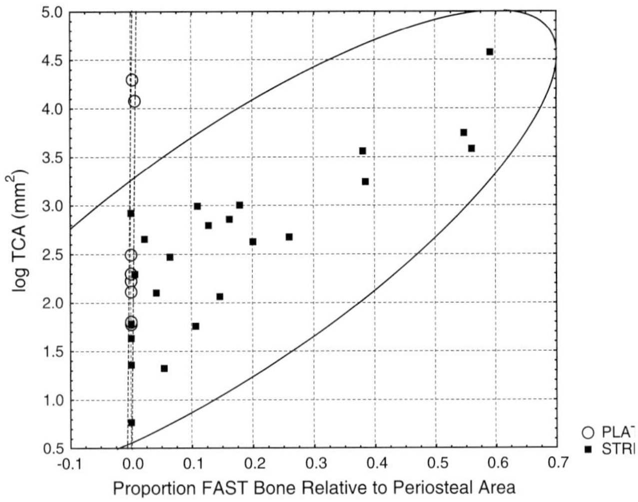

FIGURE 18.13. Correlation between FAST bone and TCA in the femur, categorized by PHYLO.

This relationship in Figure 18.12 is strengthened when the data are categorized by PHYLO (Strepsirhini $r = 0.82$, $p < 0.001$; Platyrrhini $r = 0.83$, $p < 0.05$). Tarsiids are not included as all tarsiid values for both axes are approximately equivalent.

For the femur and ulna, the multiple regression analyses indicate that taxonomic status is also a factor accounting for variation in the proportion of FAST bone. This is best appreciated in the femur, where a bivariate plot of FAST bone against TCA yields higher correlation coefficients when it is categorized by PHYLO (Figure 18.13). Small individuals (<500 g) have very low values for FAST bone in the femur (<10% of the total periosteal area), and often have none of this tissue at all. While the amount of FAST bone increases with TCA in strepsirhines, this is not the case for platyrrhines; values for larger individuals remain close to zero.[1] The significance of PHYLO as a factor is corroborated by non-parametric tests. For all elements, means for percentage of FAST bone in the periosteal cortex are higher in strepsirhines than in platyrrhines, significantly so in the femur, humerus and tibia ($p<0.05$ to <0.01). Strepsirhine means are also higher than tarsiid means for all elements, although only significantly so for the femur ($p<0.01$). The consistency in this pattern suggests that PHYLO, along with TCA, is indeed an important predictor for the amount of FAST bone in the periosteal cortex.

[1] The large platyrrhines (*Ateles* individuals) are highly secondarily remodeled, and it is possible that the secondary osteons (Haversian systems) may have obscured areas previously occupied by FAST bone. This potential problem is mitigated by the fact that in the larger strepsirhines individuals (e.g., *Indri indri*), the cortex is also highly remodeled, yet the proportion of FAST bone that is still identifiable is high relative to what is seen in *Ateles*.

In the femur, radius and ulna, multiple regression analysis indicates that PB is also a factor accounting for a significant proportion of variation in FAST bone. However, the lack of concordance between elements[2] – combined with the generally marginal significance or lack of significance for pairwise comparisons between PB categories for all elements – suggests that the relationship between PB and FAST bone as revealed in the multiple regression should be viewed with skepticism.

18.4.3.2 PF Bone

In contrast to the multiple regression results for FAST bone, TCA does not appear to be an important predictor of the proportion of the periosteal cortex composed of PF tissue. A bivariate plot of the femur (Figure 18.14) indicates a weak negatively allometric relationship between the proportion of PF bone and TCA. For all other elements, bivariate plots show no relationship between TCA and the proportion of PF bone relative to the periosteal cortex.

More important are the variables PHYLO and PB. For all bones except the ulna, multiple regressions indicate that phylogenetic status accounts for a significant proportion of the variation in the sample, where tarsiids display more PF bone in the femur, humerus, tibia and radius than do either strepsirhines and platyrrhines (Figure 18.15). Differences are non-significant in most cases. Despite this, the consistency of these patterns in conjunction with the multiple regression results does suggest a taxonomic patterning in the proportion of the periosteal cortex composed of PF bone.

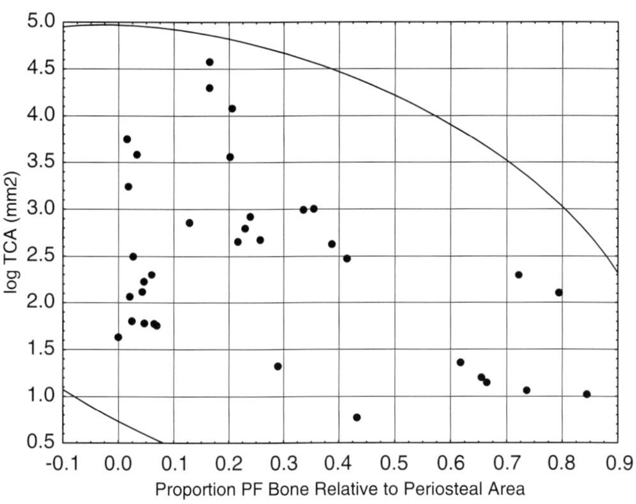

FIGURE 18.14. Correlation between PF bone and TCA in the femur. There is a weak negatively allometric relationship between PF bone (as a proportion of periosteal area) and TCA ($r = -0.43$, $p < 0.01$).

[2] For the radius, slow climbers have the greatest mean proportion of FAST bone, while in the ulna arboreal quadrupeds have the highest mean values.

For all elements except the tibia, multiple regression results indicate that PB is also a significant factor. In all cases (including the tibia), arboreal quadrupeds display the smallest mean proportions of PF bone, while slow-climbers include a greater mean proportion of PF bone than vertical-clinging leapers and arboreal quadrupeds (except in the femur, where vertical-clinging leapers and slow-climbers display equivalent proportions of PF bone) (Figure 18.16). Differences between arboreal quadruped and slow-climber

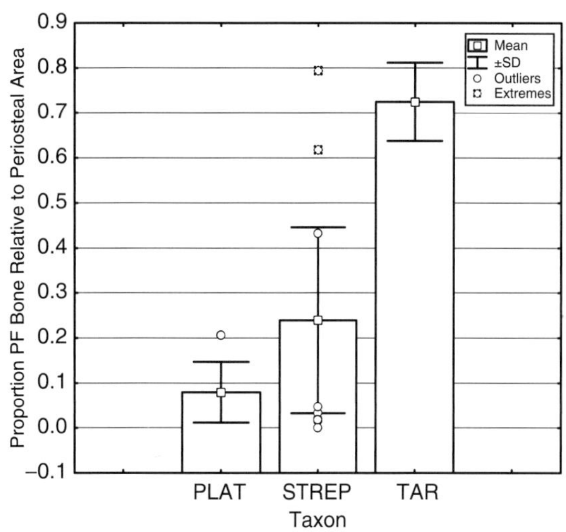

FIGURE 18.15. Femoral proportions of PF bone, categorized by PHYLO.
The proportion of PF bone (relative to periosteal area) is significantly higher in tarsiids than in strepsirhines or platyrrhines. The proportion of PF bone is also higher in strepsirhines than in platyrrhines.

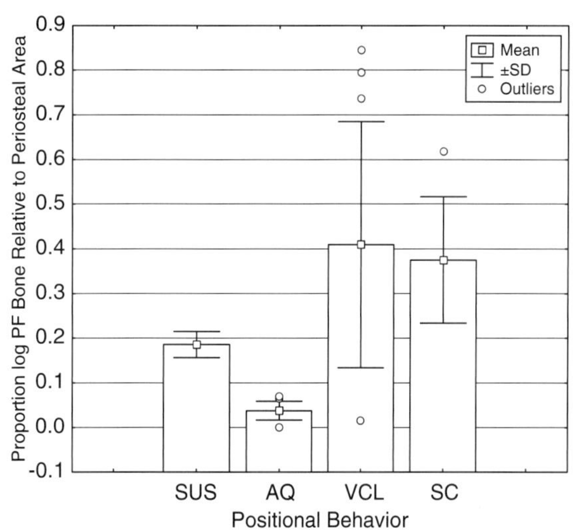

FIGURE 18.16. Femoral proportions of PF bone, categorized by PB.
The proportion of PF bone (relative to periosteal area) is significantly lower in arboreal quadrupeds than in slow climbers or vertical-clinger leapers.

values are significant for all elements ($p < 0.05$ to < 0.001). In all cases except the femur, vertical clingers and leapers display an intermediate amount of PF bone, with statistically insignificant differences between vertical-clinging leapers and other PB categories in most cases. The values for the suspensory spider monkeys, which are only available for the femur and humerus, are conflicting (the humerus value is almost as high as slow climbers, but the value for the femur is lower than both SC and VCL), perhaps due to the small sample size ($n = 2$).

18.4.3.3 LAM Bone

Similar to PF bone, and unlike FAST bone, multiple regression analyses indicate that TCA is subsidiary to PHYLO and PB in its importance as a predictor of the proportion of LAM bone relative to periosteal area. Corroborating this, correlations between LAM bone and TCA are insignificant (Figure 18.17) for all elements. A categorization by PHYLO results in significant negatively allometric relationships between TCA and the proportion of LAM bone for platyrrhines and strepsirhines, but only in a couple of elements (Figure 18.18). In these cases, the larger the cortex, the less slow-forming tissue present. The pattern is the inverse of that seen for FAST bone. As the humeral and femoral cortical areas increase, strepsirhines reduce the proportion of LAM bone more than the platyrrhines do, again the inverse of what is seen for FAST bone.

For all elements except the ulna, multiple regression results point to PHYLO as a significant factor. In all elements (including the ulna) the mean value for LAM bone is higher in platyrrhines than in strepsirhines or *Tarsius* (Figure 18.19), although significantly so only in a few instances. In addition, in all cases except the ulna, mean values for *Tarsius* are lower than for strepsirhines (significant only for the femur).

Multiple regressions show that PB also accounts for a significant proportion of variation in LAM bone in the periosteal cortex. In all elements, arboreal quadrupeds have proportionately more of this tissue than vertical-clinger leapers, as well as slow climbers (Figure 18.20) – significantly so

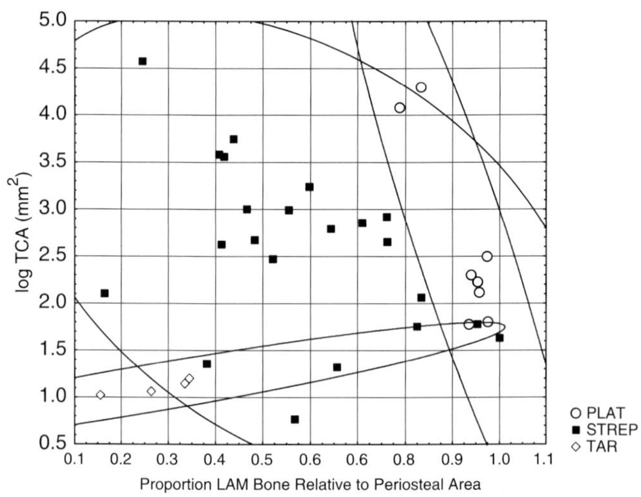

FIGURE 18.18. Correlation between LAM bone and TCA in the femur, categorized by PHYLO. When the data in Figure 18.17 are categorized by PHYLO there is a significant relationship for platyrrhines ($r = -0.91$, $p < 0.01$), as well as strepsirhines ($r = -0.42$, $p < 0.05$).

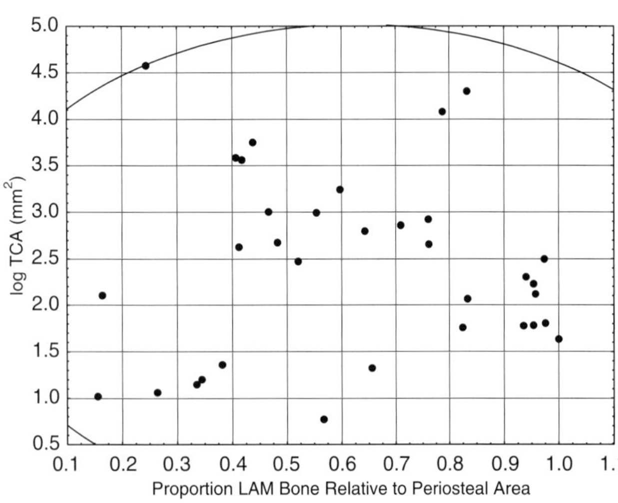

FIGURE 18.17. Correlation between LAM bone and TCA in the femur. There is no relationship between LAM bone (as a proportion of periosteal area) and TCA ($r = 0.00$, $p > 0.05$).

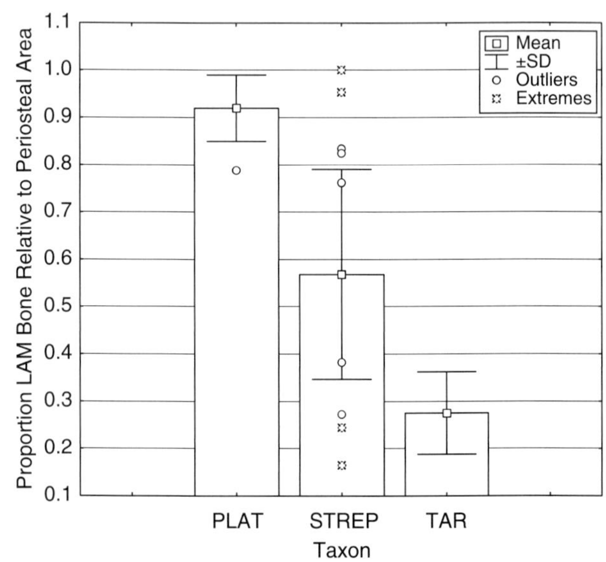

FIGURE 18.19. Femoral proportions of LAM bone, categorized by PHYLO. See text for description.

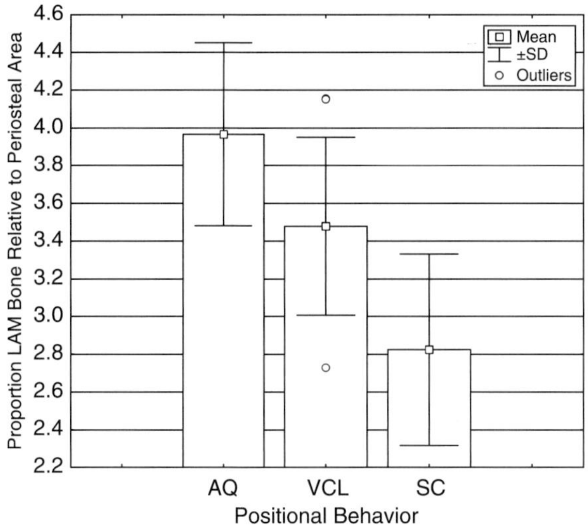

FIGURE 18.20. Tibial proportions of LAM bone, categorized by PB. See text for description.

in only a few instances. As with the independent variable PHYLO, this is the inverse of what is seen in PF bone; not a surprising result given that most of the cortex in the majority of the specimens is composed of a combination of LAM and PF tissues. In all cases except the femur, slow climbers have the least LAM bone, with vertical-clinger leapers intermediate between slow climbers and arboreal quadrupeds. In the femur, vertical-clinger leapers have the least LAM bone although the mean values for VCL and SC are close. The difference between vertical-clinger leapers and SC is significant for the tibia, radius and ulna ($p < 0.05$ to < 0.01). In both the femur and humerus, mean values for the suspensory category are lower than those for arboreal quadrupedalism. Overall, there are several consistent PB patterns among skeletal elements for the proportion of the periosteal cortex composed of LAM bone, corroborated by both multiple regressions and comparisons between means.

18.4.4 Characterization of Individual Taxa and Targeted Comparisons Amongst Taxa

18.4.4.1 Intra- and Intertaxonomic Diversity

Investigations of variation within the sample as a whole highlight patterned relationships between primary tissue types and body size, phylogeny and/or growth, and positional behavior. However, within the broad patterns described above, there is evidence of both intra- and intertaxonomic diversity (see for example the standard deviation for some of the means in Figure 18.16). Following is an exploration of this variability. Because within-taxon sample sizes are small, particularly for the tibia, radius and ulna, analyses remain descriptive.

18.4.4.2 Consistency[3] Within Taxa: Intrataxonomic Variation for Individual Elements

There is an overall pattern of intrataxonomic consistency in proportions of primary tissue types for individual elements. However, some skeletal elements appear to be more constrained in their distribution of tissues than others. With a couple of taxon exceptions, femoral and humeral patterns are remarkably uniform within most taxa. For the tibia and radius, reduced sample sizes make assessments of variability less reliable, but overall these bones also display a uniformity of tissue type patterning. In contrast, patterns of tissue type in the ulna are more variable within taxa. Figures 18.21–18.23 are examples of consistent and inconsistent intrataxon patterning for the femur.

18.4.4.3 Consistency Within Taxa: Comparing Means for Skeletal Elements Within Taxa

For the sample as a whole there is considerable variation among elements. When examining means within an individual taxon the disparities are again apparent. Only in the three platyrrhine taxa do all skeletal elements show a similar pattern of tissue type distribution, with LAM bone comprising the majority of the cortex, followed by PF bone and finally FAST bone (although all five bones are available only for the cebids). In all other taxa there is variation among elements in the tissue that predominates across the

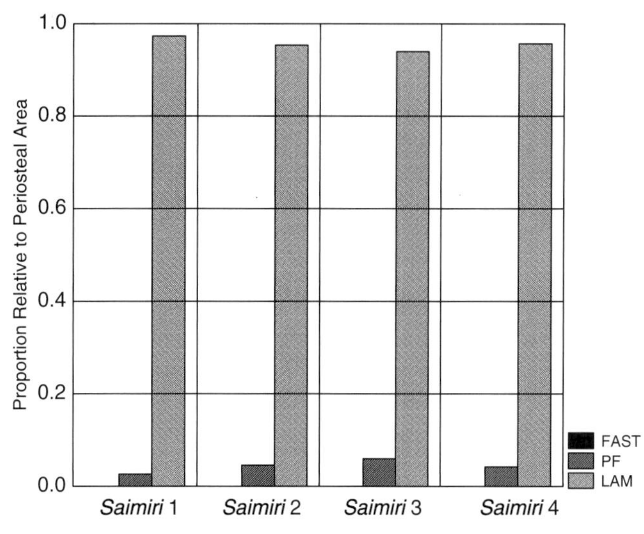

FIGURE 18.21. Individual values for the proportions of FAST, PF and LAM bone in the femora of *Saimiri* (Cebidae). *Saimiri* is an example of a taxon where femoral proportions of primary tissue types are highly uniform among individuals.

[3] Patterns are considered to be consistent when all individuals within a taxon have approximately the same proportion of the three primary tissue types: FAST, PF and LAM. This is determined through examination of graphical depictions of tissue type values.

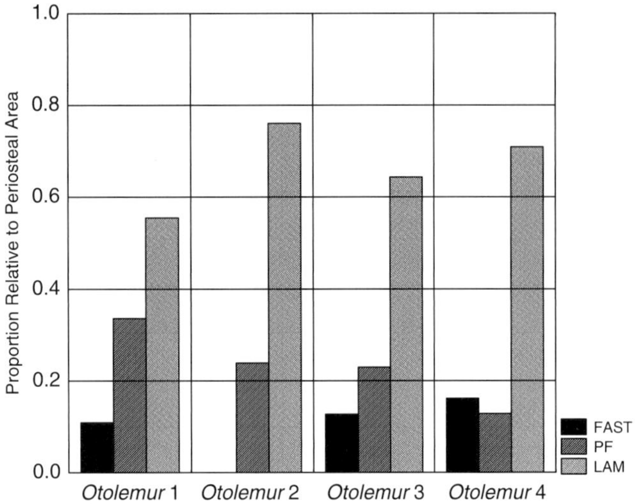

FIGURE 18.22. Individual values for the proportions of FAST, PF and LAM bone in the femora of *Otolemur crassicaudatus* (greater galagos). In *O. crassicaudatus*, proportions are still consistent, although somewhat less so than in *Saimiri*.

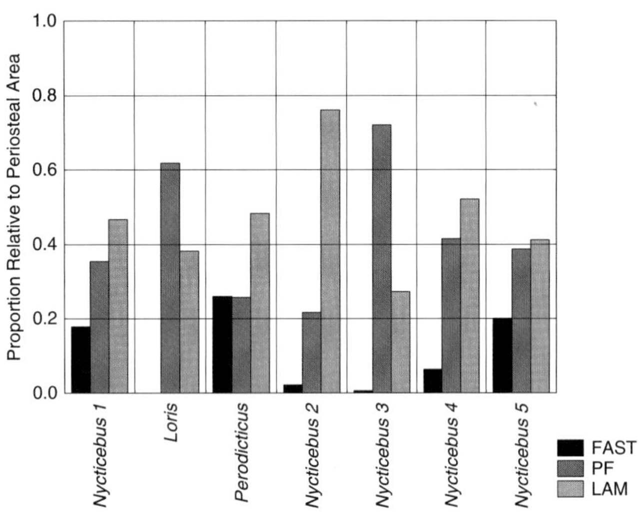

FIGURE 18.23. Individual values for the proportions of FAST, PF and LAM bone in the femora of lorisids. Patterns of tissue type variation among the lorisid femora are inconsistent from one individual to the next, although several individuals do show similar values. *Nycticebus* 3 was excluded from the analyses due to apparent pathology.

cortex, with the larger-bodied strepsirhine indriids and lemurids displaying the most diversity.

18.4.4.4 Consistency Between Taxa: Means For Skeletal Elements Across Taxa

If element means are compared across taxa it is clear that humeri are the least variable. In all taxa, LAM bone is the predominant tissue in the humerus, followed by PF and then FAST bone, or by similar values for PF and FAST. In all other elements, the order of tissues in terms of their predominance in the cortex varies among taxa (but, as described above, not in a consistent fashion). The femur and tibia appear to be the most variable, with the radius and ulna intermediate between humeri and the hindlimb bones.

18.4.4.5 Characterization of Femoral Patterns of Periosteal Primary Tissue Type Distributions for Individual Taxa

Because tissue types in the humerus appear to be more constrained, and because sample sizes for the tibia, humerus, and ulna are smaller than those for the femur (yielding less reliable means), description of taxon patterns and comparisons among individual taxa are here restricted to the femur.

Lesser Galagos (Figure 18.24). There is some heterogeneity in this group, with two of the three individuals (*Galago moholi* and *Galagoides demidovii*) displaying a predominance of LAM bone (>55%), and the third individual (*Galago senegalensis*) with more PF bone (80%) in the periosteal portions of the cortex. Small amounts (5%) of FAST tissue are present in the two *Galago* individuals in the anterior quadrant (the thickest cortical region), close to the boundary between endosteal and periosteal bone. The LAM and PF tissues are sparsely vascularized,[4] with canals that are simple and course longitudinally. The bone may be avascular in the medial, lateral and posterior quadrants of the cortices.

Greater Galagos (Figure 18.25). In all *Otolemur crassicaudatus* individuals, LAM tissue makes up the largest proportion of the periosteal cortex (55–75%), with PF and FAST tissues generally (although not exclusively) concentrated in the anterior quadrant of the cortex. This quadrant is the thickest in three of four individuals. In addition, in three of four femora, faster-forming tissues are located close to the boundary with endosteal bone. There is sparse to moderate vascularization of the PF and LAM tissues, with denser vascularization in the anterior cortex, and areas of avascular bone most frequently in the posterior quadrant and close to the most peripheral regions of the cortex. Vascular canals are generally of the simple type, coursing longitudinally, with one individual displaying short, radially oriented vascular canals.

[4] Vascularization in the sample has not been quantified. Therefore, the relative densities of vascular canals are assessed visually and described qualitatively. The following three terms are used: (1) sparse vascularization = vascular canals are irregularly distributed, with avascular stretches between them that generally exceed three times the diameter of the canals, (2) moderate vascularization = vascular canals are more regularly distributed, with the distance between canals frequently less than three times the diameter of the canals, (3) intense vascularization = vascular canals are separated from each other by less than the diameter of two canals.

FIGURE 18.24. *Galago senegalensis* (MNHU 60601) midshaft femur. Conventional transmitted light image on left. Color coded tissue map on right. Scale bar indicates 0.5 mm. See text for description.

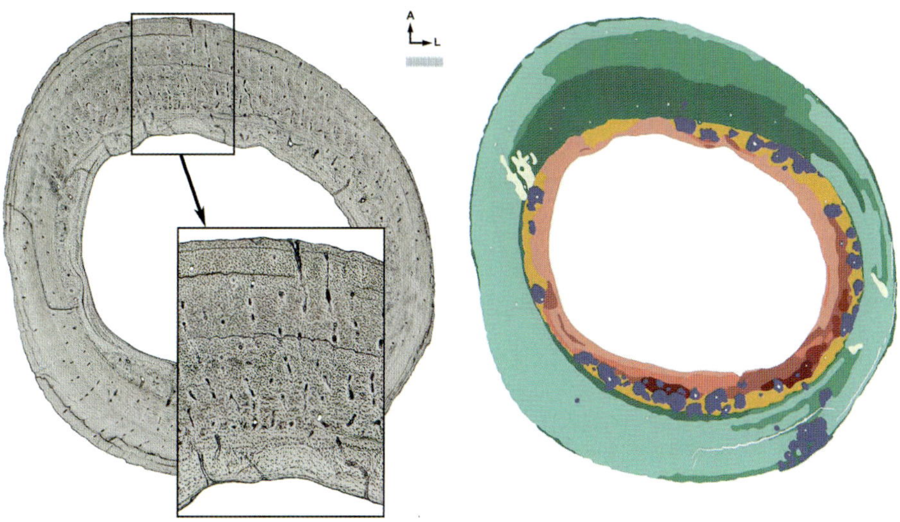

FIGURE 18.25. *Otolemur crassicaudatus* (HC 31) midshaft femur. Legend as per Figure 18.24, with magnified conventional transmitted light inset.

Lorisidae, sensu stricto (Figure 18.26). Similar to the lesser galagos, there is some heterogeneity in this group. In four of the six individuals, LAM bone is the predominant tissue (47–77%), while in one individual, proportions of LAM and PF are equivalent (>40%). In the sole specimen of *Loris tardigradus*, PF is found in the majority of the periosteal cortex (>62%). In most instances, LAM tissue is found closer to the periosteal margin, while FAST tissue is near the boundary with endosteally-deposited bone. FAST bone, present in all individuals except the noted specimen of *L. tardigradus*, is concentrated posteriorly in the four individuals of *Nycticebus coucang*, and laterally in the sole specimen of *Perodicticus potto*. The majority of the LAM and PF bone is sparsely vascularized, or avascular, with a single individual of *N. coucang* displaying denser, but still moderate vascularization. Where present, vascularization is not clustered in a single quadrant of the bone, but evenly distributed around the midshaft cortex.

Cheirogaleidae (Figure 18.27). The cheirogaleids are characterized by a predominance of LAM bone in the periosteal cortex (>80%). Very little PF bone is present (<10%). The two *Cheirogaleus major* individuals display some FAST bone (>10–15%) posteriorly in the thickest part of the cortex, close to the endosteal region of the bone, as well as laterally in one individual. Neither the *Mirza coquereli* nor *Microcebus*

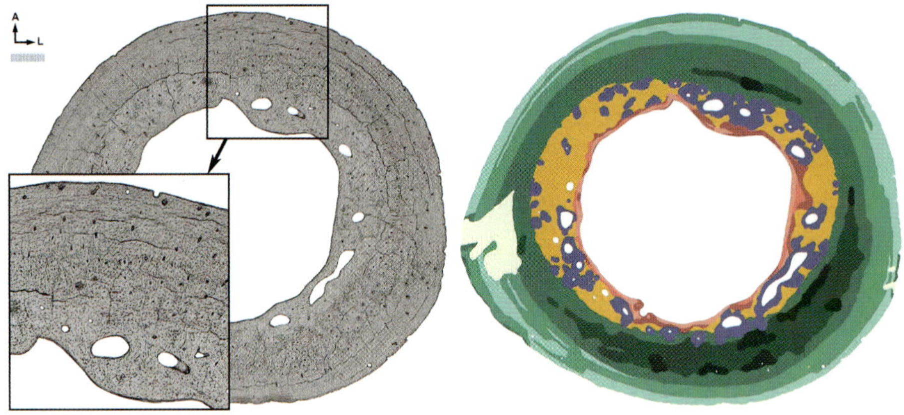

FIGURE 18.26. *Nycticebus coucang* (DPC 12) midshaft femur. Legend as per Figure 18.24, with magnified conventional transmitted light inset.

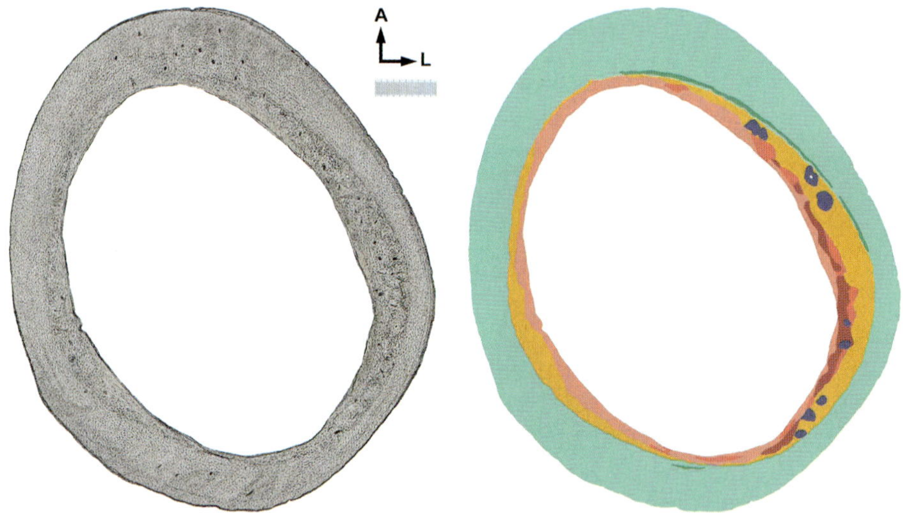

FIGURE 18.27. *Mirza coquereli* (DPC 39) midshaft femur. Legend as per Figure 18.24.

murinus individuals exhibit FAST bone in the periosteal cortex. The LAM bone is largely avascular, with few areas of sparse vascularization by simple longitudinal canals.

Indriidae (Figure 18.28). FAST bone makes up the greatest proportion of the periosteal cortex in one specimen of *Propithecus* sp. and in *Indri indri* (>50%), followed by LAM bone (<45%). In the second specimen of *P.* sp., LAM bone comprises approximately 5% more of the cortex than FAST bone. In all individuals, PF is the least common tissue (≤20%). FAST bone is found in all quadrants of the section. The largest uninterrupted areas are close to the endosteal boundary, but smaller strips, punctuated by circumferential swathes of LAM or PF bone are present closer to the periphery. The cortex as a whole shows a periodic alternation of tissue types in the progression from endosteum to the periosteal margin. Vascularization in LAM and PF areas (by longitudinal canals) ranges from sparse to intense.

Lemuridae (Figure 18.29). The two lemurids display near equivalent proportions of FAST and LAM tissues, with *Varecia variegata* exhibiting a greater proportion of FAST bone (55% of the periosteal cortex), and *Lemur* sp. more LAM bone (60% of the periosteal cortex). The proportion of PF bone is low (<5% of the periosteal cortex) in both individuals. FAST bone is usually, although not always, concentrated close to the endosteal boundary, particularly the larger uninterrupted areas of this tissue type, and is found in all quadrants of the cortex. Although FAST bone may not necessarily be the first tissue found abutting endosteally-deposited bone, it is never present at the periphery of the cortex. As with the indriids, narrow circumferential strips of FAST bone, interspersed with strips of LAM, are also present, generally closer to the periosteal margin than the larger swathes of FAST bone. The LAM bone is sparsely to moderately vascularized. Vascular canals are of the simple

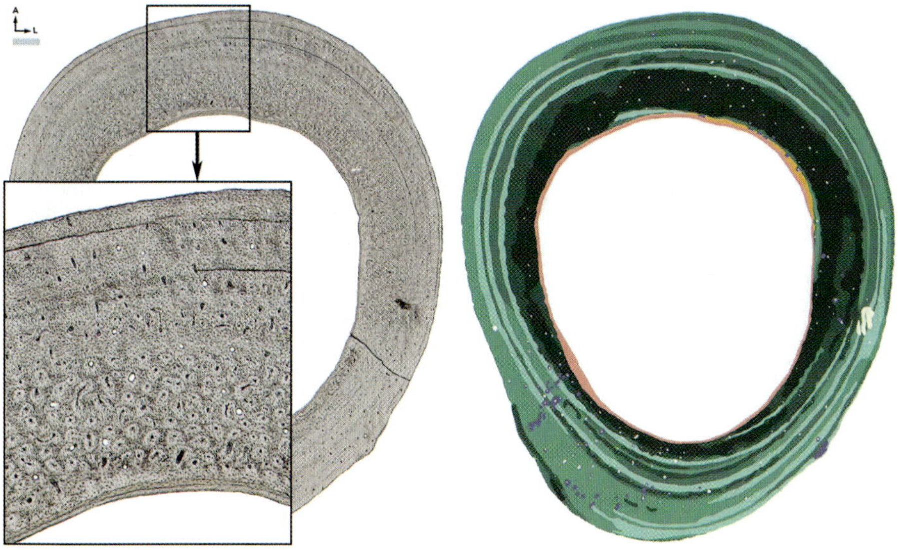

FIGURE 18.28. *Propithecus* sp. (MNHU 44685) midshaft femur. Legend as per Figure 18.24, with magnified conventional transmitted light inset.

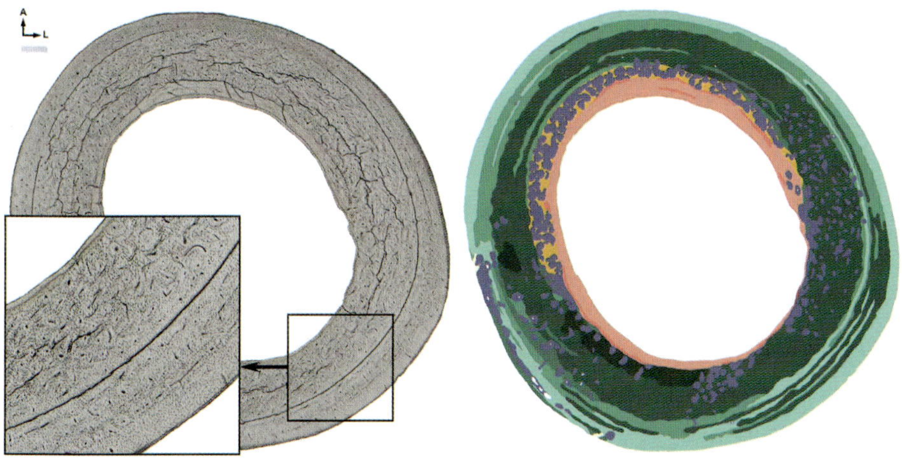

FIGURE 18.29. *Varecia variegata* (MCZ 18740) midshaft femur. Legend as per Figure 18.24, with magnified conventional transmitted light inset.

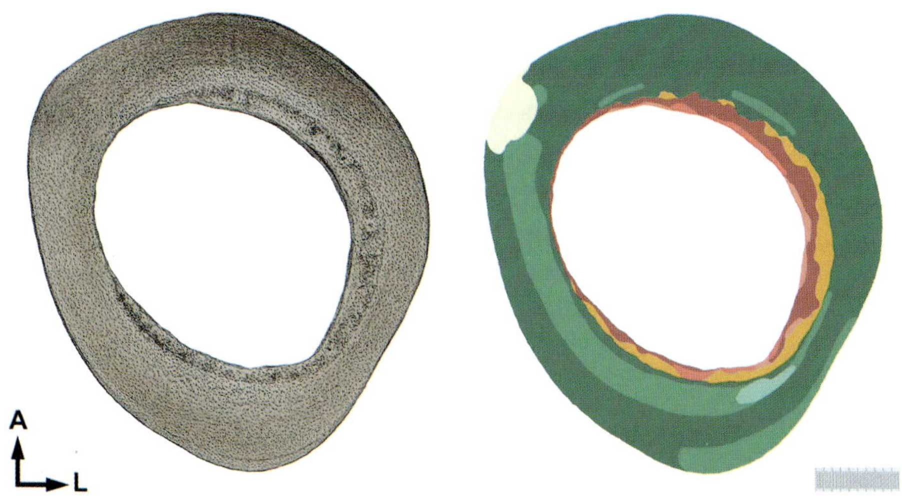

FIGURE 18.30. *Tarsius spectrum* (MNHU An11318) midshaft femur. Legend as per Figure 18.24.

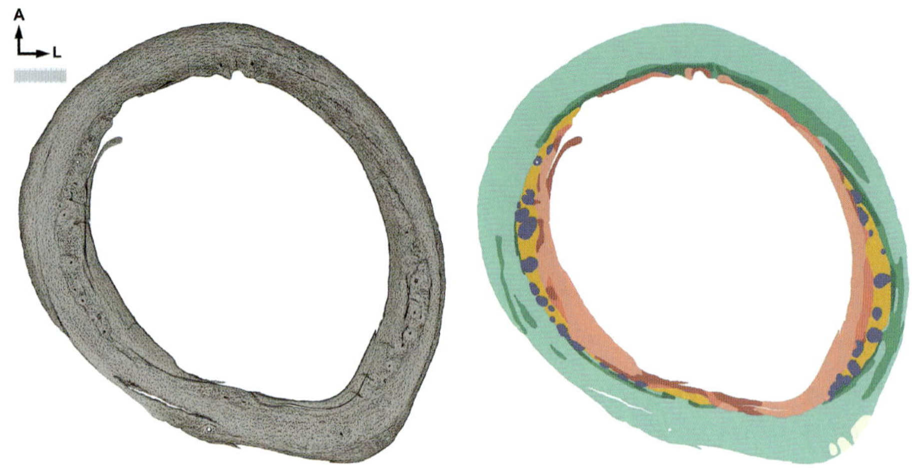

FIGURE 18.31. *Callithrix geoffroyi* (MNHU 34381) midshaft femur. Legend as per Figure 18.24.

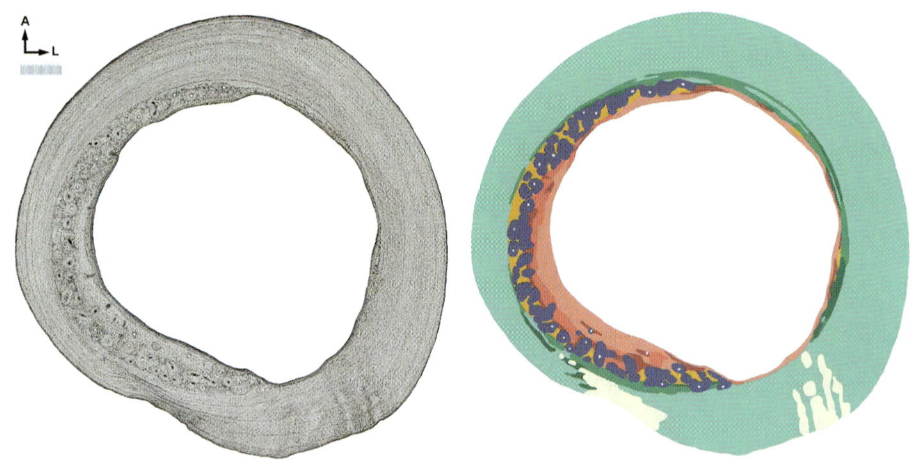

FIGURE 18.32. *Saimiri* sp. (MNHU 85197) midshaft femur. Legend as per Figure 18.24.

longitudinal type, with some short radially coursing canals in the *V. variegata* individual.

Tarsiidae (Figure 18.30). Tarsiids are characterized by a preponderance of PF bone in the periosteal cortex (~65–85%). The remainder of the tissue is of the LAM type. No FAST bone is present in the four individuals examined here. When present, the LAM bone is often concentrated in the medial and lateral quadrants, where the cortex is thinnest. The periosteal cortex in tarsiids is almost entirely avascular; in only one of the four individuals are a few vascular canals found, (in the posterior cortex). These are of the simple, longitudinal or short radial type.

Callitrichidae (Figure 18.31). In *Callithrix*, the vast majority of the cortex (>95% in both individuals) is composed of LAM bone, with the remainder of the PF type. PF bone is concentrated close to the endostreal boundary. There is no FAST bone present in the two individuals examined here. The periosteal cortex is largely avascular, with each individual displaying only a few simple longitudinal canals. In one of the two specimens of *Callithrix*, these are concentrated in a region of soft-tissue attachment (most likely the femoral adductor and extensor muscles).

Cebidae (Figure 18.32). As is the case with callitrichids, the periosteal cortex in cebid specimens is primarily composed of LAM bone (>95% in all four individuals). The remainder of the bone is PF, with no FAST tissue in any of the four specimens. Almost all PF bone occurs close to the endosteal boundary, and may be in any of the cortical quadrants. In three of the four individuals, the periosteal cortex is entirely avascular, except for the occasional secondary osteon. The cortex of the fourth individual is also largely avascular, with a few short, simple, radial canals anteromedially.

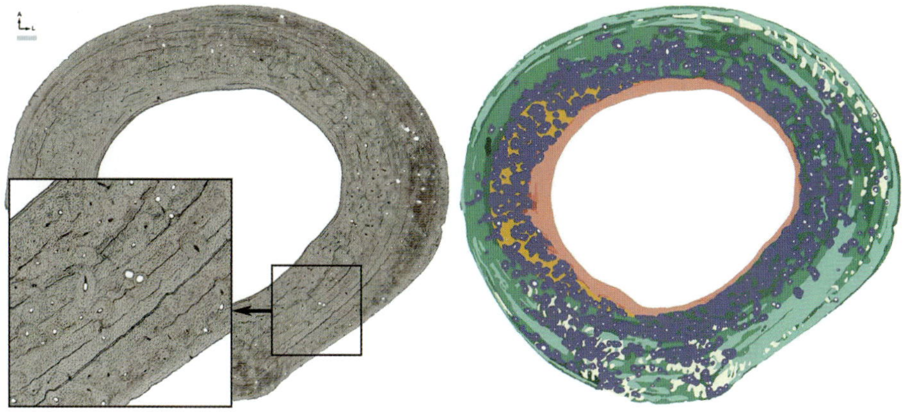

FIGURE 18.33. *Ateles* sp. (MNHU 44079) midshaft femur. Legend as per Figure 18.24, with magnified conventional transmitted light inset.

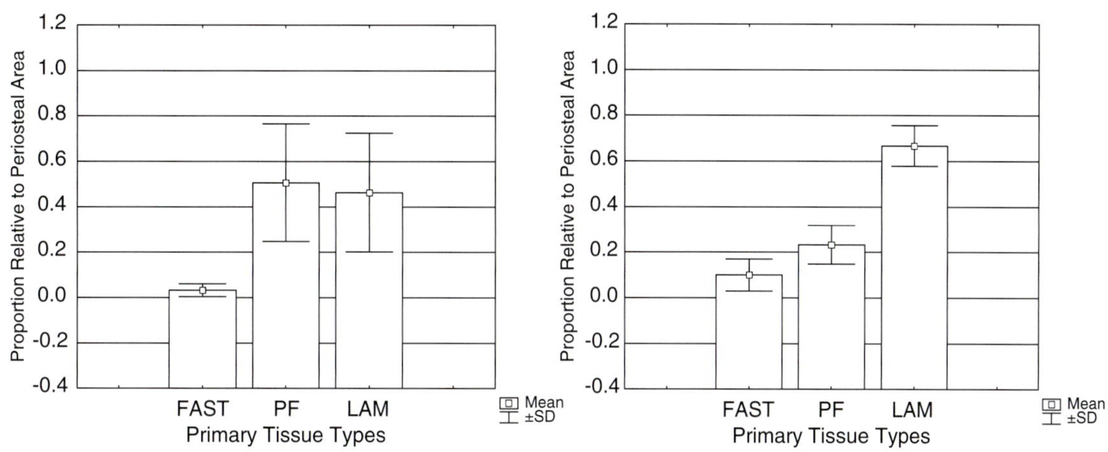

FIGURE 18.34. Mean proportions of primary tissue types in lesser galagos (left) and tarsiids (right).

Atelidae (Figure 18.33). The large atelid individuals also show a predominance of LAM primary tissue in the periosteal cortex (>75% in both specimens). The remainder of the tissue is mostly PF (distributed in patches throughout the periosteal cortex, with less frequency in the posterior quadrant). FAST bone comprises less than 2% of the periosteal primary bone cortical area. Excluding secondary osteons, the LAM and PF bone areas are avascular to moderately vascularized, with simple longitudinal canals and the occasional longitudinal primary osteon.

18.4.4.6 Comparisons of Femoral Means Between Select Pairs of Taxa

The multivariate analyses of patterns of variation were limited to broad taxonomic and positional behavior categories. In addition, growth rate was not included as an independent variable. The following targeted comparisons are therefore designed to determine whether: (a) the patterns revealed in the multiple regression analyses with respect to both function and life history are repeated at a more detailed phylogenetic and functional level, and (b) growth rate can be separated from phylogeny as a factor that can predict variation in tissue type patterns. Although comparisons can be made between more pairs of taxa than are offered here (see Warshaw, 2007), the following examples suffice to highlight the patterns of variation that would be apparent in a more thorough review. Within-strepsirhine or within-haplorhine patterns of phylogenetic difference (e.g., lemuroids vs. lorisoids) – independent of growth, size or positional behavior – are difficult to assess in this sample given the confluence of evolutionary heritage with positional behavior and/or body size and/or growth adaptation.

Lesser Galagos vs. Tarsiidae (Figure 18.34). These two taxa show a similar[5] proportion of FAST bone in the periosteal cortex (≤5%), as would be expected given their comparable

[5] Mean values are considered to be similar when the difference between them is <10% of the total periosteal area.

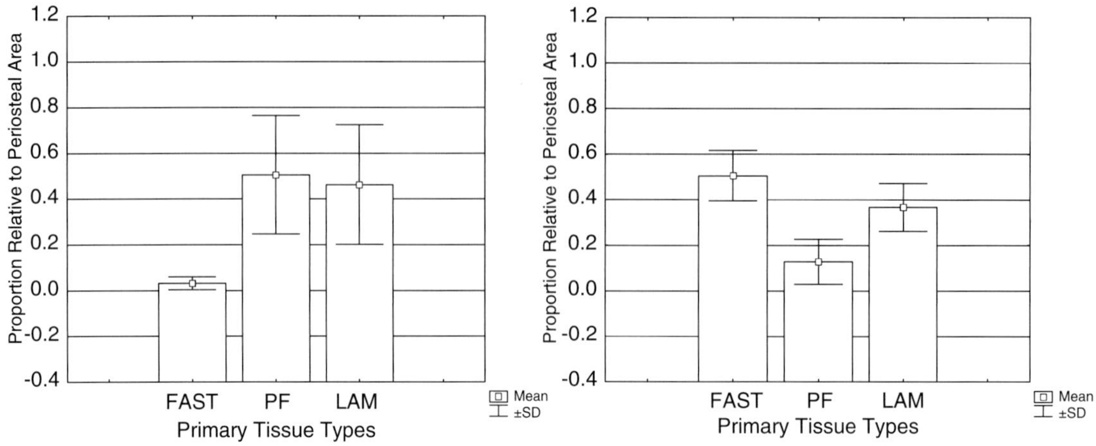

FIGURE 18.35. Mean proportions of primary tissue types in lesser galagos (left) and indriids (right).

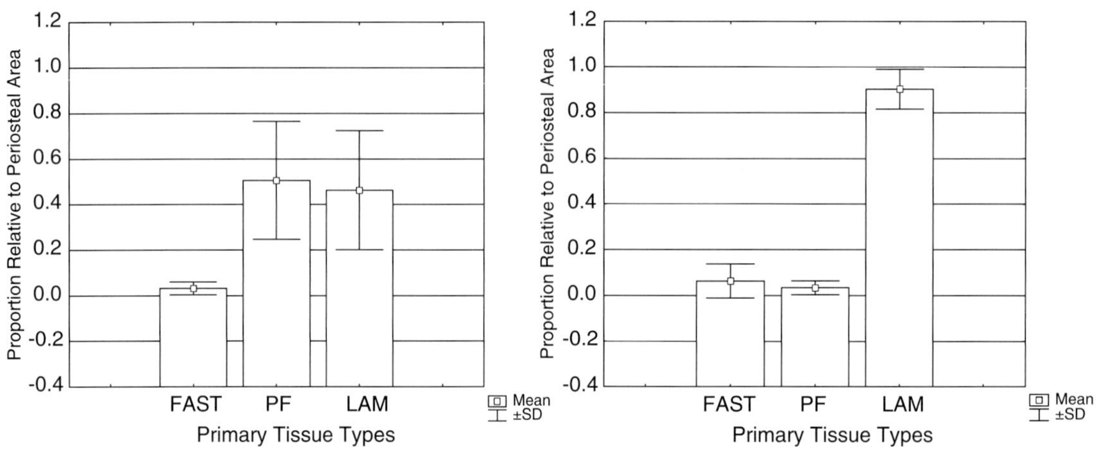

FIGURE 18.36. Mean proportions of primary tissue types in lesser galagos (left) and cheirogaleids (right).

small body size. The smallest individuals (all tarsiids, and *Galagoides demidovii*) have no FAST bone at all. No differences are expected based on positional behavior, as the two taxa share very similar, if independently derived, locomotory patterns. Phylogenetic distance predicts that divergence in the proportions of LAM and PF tissues should be found. This is indeed the case, with *Tarsius* femora on average exhibiting more PF bone.[6] The pattern is not consistent with predictions based on growth rates. The slower-growing tarsiids (as suggested by infant growth data; Kappeler, 1998) would be expected to show more LAM bone and, concomitantly, a lower proportion of PF bone than the galagos.

Lesser Galagos vs. Indriidae (Figure 18.35). Indriids have proportionately much more FAST bone than the lesser galagos, consistent with their vastly larger body sizes. Scaled infant growth rates are lower for indriids than galagos; the prediction is therefore that a larger proportion of LAM bone will be found in the indriids. This is not reflected in the observed proportions; mean values for LAM bone are close in the two taxa. Given their similarity in positional behavior, this parity conforms to expectation. However, in the galagos, the proportion of PF bone is higher, perhaps reflecting the reduced amount of FAST bone relative to the indriids.

Lesser Galagos vs. Cheirogaleidae (Figure 18.36). In this case, both groups are heterogeneous in terms of body size, with *Microcebus murinus* and *Galagoides demidovii* having roughly the same body mass, but with the larger cheirogaleids weighing more than the average *Galago*. In terms of sample averages, the cheirogaleids are larger. Therefore the expectation is that the cheirogaleids would display more FAST bone than the lesser galagos. This is not borne out; values for this tissue type are very close for the two groups. However, the highest values for the cheirogaleids (~15%) are greater than

[6] The lesser bushbaby sample is variable, with one *Galago* showing more PF bone than three out of the four *Tarsius* specimens.

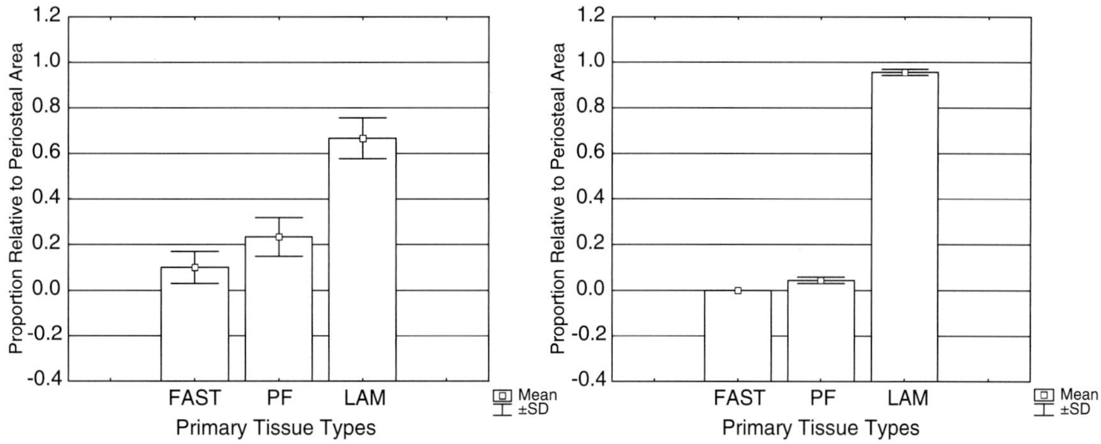

FIGURE 18.37. Mean proportions of primary tissue types in greater galagos (left) and cebids (right).

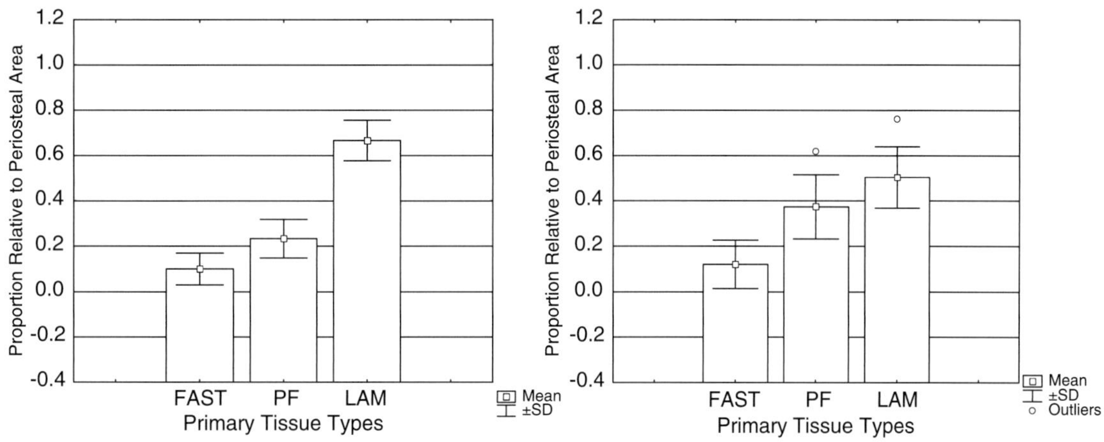

FIGURE 18.38. Mean proportions of primary tissue types in greater galagos (left) and lorisids (right).

the highest values for the lesser galago group (~5%). Adult growth scores are slightly higher for *Cheirogaleus major* and *M. murinus* than *Galago* (rates unavailable for *G. demidovii*). These data predict more LAM and less PF bone in the cheirogaleids. This is the observed pattern, although it is more marked than might be expected given that growth rates are not highly divergent. Habitual positional behaviors are different between the two groups, with the cheirogaleids more quadrupedal and less saltatory than the galagos. Proportions of LAM and PF are consistent with the contrast in locomotor adaptation.

Cebidae vs. Greater Galagos (Figure 18.37). These two taxa share roughly the same body size. However, as with other platyrrhine/strepsirhine comparisons, *Otolemur crassicaudatus* displays more FAST bone than *Saimiri* (on average approximately 10% of the cortex in the former, versus 0% in the latter). This is in accordance with both phylogenetic relationship, and associated faster growth rates in *O. crassicaudatus* (both in infancy and later in ontogeny). Likewise, *O. crassicaudatus* exhibits less LAM bone, and more PF bone, also consistent with growth differences. These same patterns are also in accord with the positional behavior differences between the two taxa, with arboreal quadrupeds displaying more LAM bone and less PF bone than vertical-clinger leapers. This is somewhat complicated by the variable positional behavior of *O. crassicaudatus*, as the greater galagos locomote frequently in non-leaping quadrupedal modes. Nevertheless, vertical clinging and leaping still comprises a greater proportion of the *O. crassicaudatus* locomotor repertoire than is the case for *Saimiri* (Napier and Walker, 1967; Walker, 1974; Fleagle and Mittermeier, 1980; Gebo, 1987; Boinski, 1989; Fontaine, 1990; Ruff and Runestad, 1992; Terranova, 1995a, 1995b; Preuschoft et al., 1998).

Greater Galagos vs. Lorisidae (Figure 18.38). In accord with expectation, these two similarly sized and closely related taxa show roughly the same proportion of FAST bone in the cortex. Because of the close phylogenetic relationship, any differences in LAM and PF types should in this case reflect differences in growth rates. Again, however, the observed

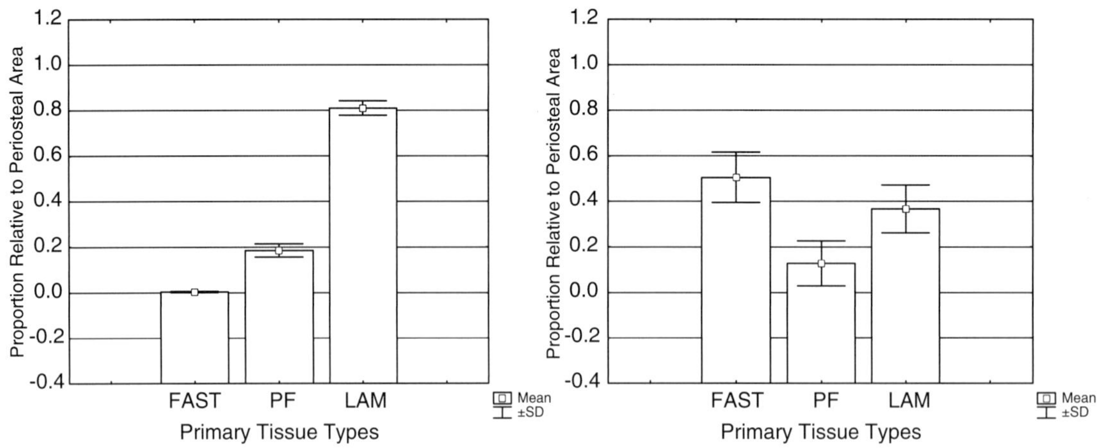

FIGURE 18.39. Mean proportions of primary tissue types in atelids (left) and indriids (right).

patterns do not correspond with prediction based on observations of growth and development. The scaled growth rate is slower for the lorisids than *O. crassicaudatus*, as are scaled daily growth rates for lorisid infants relative to *O. crassicaudatus* infants, yet *O. crassicaudatus* shows more of the slow-forming LAM bone and less of the faster-forming PF bone than the lorisids. Here, as above, a positional behavior comparison does correspond with what is seen in the distribution of tissue types if *Otolemur* is categorized as an arboreal quadruped, intermediate in its locomotion between vertical clinging and leaping and arboreal quadrupedalism, or even as a part-time slow climber.

Indriidae vs. Atelidae (Figure 18.39). The larger size of *Ateles* predicts for more FAST bone. However, in this case (comparable to the comparison between *O. crassicaudatus* and *Saimiri*), phylogeny and/or associated differences in growth rate appear to relate to primary tissue proportions. The slower-growing platyrrhine (as described by scaled infant growth rates; there are no post-weaning growth rates available for the indriids) displays more LAM bone, while more FAST bone is found in the indriids. The proportion of PF bone is similar in the two taxa. Multiple regression results suggest that the vertical clinging and leaping indriids should display more PF and less LAM bone than *Ateles*, and this is borne out here for LAM bone, but not PF. The indriids have less PF bone than expected relative to *Ateles*, but this may be due to the large proportion of the cortex composed of FAST tissue.

18.4.5 Summary of Paired Comparisons of Means

Overall, when looking at femora in selected pairs of taxa, the proportion of FAST bone in the femoral cortex appears to coincide with predictions based on the results of the multiple regression analyses, where larger cortices (and larger animals) contain more FAST bone. This is evident, for example, in a comparison between the lesser galagos and indriids, two taxon groups with vastly divergent body size but sharing a commitment to vertical clinging and leaping. In contrast, when two similar-sized individuals from families such as the indriids and lemurids are compared, the values for FAST bone are similar. The relationship between TCA and FAST bone even extends to genus level comparisons within taxon groups (e.g., *Galago* vs. *Galagoides*). The subsidiary relationship between FAST bone and phylogeny is also borne out, with strepsirhines showing more FAST bone than platyrrhines at a given body size, in all pairwise comparisons (e.g., greater galagos vs. cebids, or the indriids vs. atelids).

The distribution of proportions of LAM bone and PF bone is much more difficult to characterize at this more focused level of taxon comparison. First, with the exception of contrasts between the distantly related strepsirhines and platyrrhines, differences in these two tissue types do not match the expectations based on relative growth rates. When similar-sized but phylogenetically disparate taxa such as *O. crassicaudatus* and *Saimiri* are compared, it is the slower-growing cebid that displays more LAM, and less PF bone, as expected. Indeed, with a single exception (discussed below) all platyrrhines in the sample display more LAM bone than strepsirhines of similar body size. In contrast, within strepsirhines, the faster growing taxa do not necessarily display less LAM bone than the slower-growing taxa. This can be seen in the comparison of the closely related *O. crassicaudatus* and lorisids.

When pairs of taxa are compared, there appears to be a positional behavior pattern in the relative proportions of PF, and particularly LAM, bone. This observation is most salient among the strepsirhines, where sufficient numbers and taxon diversity representative of different positional behaviors are present. Arboreal quadrupeds display more LAM, and less PF bone than similar-sized vertical-clinger leapers or slow climbers. For example, the vertical clinging and leaping indriids display less LAM and more PF bone than the more generalized quadrupedal lemurids. Similarly, the lesser galagos, which are habitual and committed vertical-clinger leapers, exhibit less LAM and more PF than *O. crassicaudatus*, intermediate in its locomotion between a vertical-clinger and leaper and an arboreal quadruped or slow climber.

Two peculiarities relative to the patterns identified above are present in the sample. First, the cheirogaleids display more LAM and less PF bone than all other strepsirhines, equivalent to what is found in similar-sized platyrrhines (e.g., *Callithrix*). Second, while differences between the distantly related platyrrhine and strepsirhine groups correspond with the disparity in growth adaptations between the two, the pattern among the phylogenetically distinct tarsiids does not match what is known of their growth and development. The predominance of PF bone in tarsiids is unusual relative to all other primates in this sample.

18.5 Discussion

18.5.1 Primary Tissue Types and Element Specific Ontogenies

As Enlow (e.g., 1962a, 1962b, 1963, 1976, 1982; Enlow and Hans, 1996) has so clearly demonstrated, one of the most important factors determining the tissue types that will be found across a cortical cross-section is the developmental history of a bone. The shape that it must ultimately become in order to accommodate functional demands – given phylogenetic and constructional constraints – will leave a record throughout the hard tissue, varying from region to region dependent upon local growth remodeling. Every cross section, then, encodes part of the story of element-specific development in length and diameter and it's interaction with surrounding soft tissue and other bones, in addition to the story that it may tell about the whole organism.

Comparison among skeletal elements in this study shows that not all long bone midshafts are built the same. In all sections (with two exceptions) there is a greater proportion of periosteally deposited bone than either CCC or endosteally deposited bone. This is in part a consequence of sampling at the midshaft, rather than in metaphyseal regions of the shaft. However, there is a distinctive separation between hindlimb bones (the femur and tibia) and forelimb bones (the humerus, radius and ulna). Mean values for the proportion of the cortex composed of endosteal tissues (not including CCC bone) are higher in the humerus, radius and ulna than in the femur and tibia. In other words, relative to the hindlimb bones, forelimb midshafts achieve adult cortical proportions either via more endosteal deposition, less periosteal deposition, or both. Studies of cortical expansion during development might differentiate among these options.

The humerus is almost always distinguished by a posteriorly directed pattern of drift, accounting in part for the increase in endosteal tissue relative to the femur and tibia. That is, the humeral midshaft moves posteriorly through anatomical space while growing. Lorisids are the exception to this rule, where humeri exhibit a more even circumferential deposition of tissue in all quadrants of the cortex, much like all femora in the sample.[7] The humeral drift observed in this study was also identified in a vital labeling study of macaque skeletal development by Newel-Morris and Sirianni (1982), as well as in descriptions of the macaque humerus provided by Enlow (1963). Greater deposition rates in the posterior cortex relative to the anterior cortex for the prenatal period were found by Newell-Morris and Sirianni. Later in growth, they observed very slow anterior deposition relative to the posterior cortex, or none at all. Interestingly, in that study, tissue in both parts of the bone could be lamellar despite the disparity in deposition rates (see section on growth rate below).

Humeri are also distinguished by a relative paucity of variation in tissue types: diversity at the midshaft in this element is reduced relative to all others, with a predominance of LAM bone in all taxa. What might explain this unique pattern? On the one hand it is well established that primate humeral growth usually continues longer than other long bones (i.e., the proximal humerus is the last or close to the last epiphysis to fuse) (Galliari, 1998; Kohn et al., 1997; King 2003). However, it cannot be assumed that this necessitates a longer period of circumferential bone deposition at the midshaft. Might this highly organized tissue be indicative of some mechanical imperatives, universal among the humeri examined here? In the absence of studies to test these hypotheses, ideas regarding the constraints that might be acting to produce this pattern must remain speculative.

In almost all cases the tibia is characterized by a lack of, or reduced amount of, CCC bone. This suggests that the midshaft compacta of the tibia is not incorporating previously cancellous portions of bone, which would likely have been in the metaphyseal region in a previous growth phase. Rather, the tibia would have been lengthening at both ends to the same degree, as has been described by Enlow (1962b), and midshaft of the juvenile would remain midshaft in the adult.[8] In contrast, the midshafts of all other elements would have previously been parts of the bone that were either distal or proximal to the current midshaft, indicating that long bone extension likely proceeded more quickly on one end than the other. Serial sections would be required to determine

---

[7] This finding is interesting in light of the unique positional behavior among the lorisids, where humeri and femora function in a more equivalent fashion than is seen in other primates (Ishida et al., 1990; Jouffroy and Stern, 1990), and where other features of humeral morphology also point to the distinctiveness of this group (Szalay and Dagosto, 1980).

[8] It is possible that cancellous, or CCC bone is present at the midshaft tibia at an earlier stage of growth, and that it is remodeled away as the bone reaches adult proportions. However, given the patterns observed in this sample, it is still likely that the tibial midshaft drifted less during growth, either proximally or distally, than the midshafts of other elements.

growth directionality (but see the discussion of Bateman, 1954, below). In the femur, radius and ulna, unlike the humerus, this uneven extension would not have been accompanied by substantial drift; there is no consistent pattern of such evident in these bones.

Studies of primate epiphyseal closure sequences support these observation, although not unequivocally. For example, in *Saimiri boliviensis*, there is a significant lag time between distal and proximal humeral fusions (up to 37 months) (Galliari, 1988). This suggests considerable distal-ward relocation of the juvenile midshaft. In the tibia this lag is much reduced, but still present (up to 9 months). In the radius, ulna, and femur the lag is also reduced relative to the humerus, but reversed, with the proximal end fusing first (up to 22 months earlier). Kohn et al. (1997) find similar results for *Callithrix* and *Saguinas*, and these authors suggest that there is a "general pattern in Primate skeletal development" (p. 129). Indeed King's (2003) comparative study of epiphyseal fusion sequences shows that, despite some variation among taxa, there are consistent patterns. For example, among the long bones examined there, the distal humerus is always the first growth plate to fuse, while the proximal humerus fuses last or close to last. In contrast, the fusion times for the proximal and distal ends of the tibia are generally close to each other. In all cases the proximal radii and ulnae fuse before the distal ends. Fusion at the femoral head occurs before fusion at the distal femur in almost all cases. While these data are mostly consistent with the distribution of primary bone tissues in the primate midshaft cortex, the observation of lag time between distal and proximal epiphyseal unions in the tibia is not.

Relative epiphyseal union times, however, do not tell the whole story, as they cannot provide information regarding the rate at which the epiphyseal region serves to lengthen the bone. A given fusion may occur subsequent to extensive lengthening, moderate lengthening, or none at all. These would have different ramifications for the pattern found at the midshaft of the adult cortex. Although there exist no comparable data among primates, a study of the pattern of long bone lengthening in grey-lethal and microphthalmic mutant mice (Bateman, 1954) provides results that are entirely consistent with the pattern observed in the current study. Bateman found that during growth, 80% of femoral lengthening occurs at the distal end, with some proximal lengthening. In the radius and ulna, the pattern is similar, although with less growth at the proximal end. In the humerus, lengthening occurs proximally, with almost no distal growth. (The forelimb pattern is one where lengthening occurs minimally at the elbow joint, and extensively where the humerus joins the shoulder and the radius and ulna join the carpus). Finally tibial and fibular growth is almost equivalent in both directions. These observations of mouse ontogeny match patterns of compacted coarse cancellous distribution at the midshaft in the sample primates. Either serial sectioning along the length of the shaft or observations of growth would serve to confirm the similarities.

In addition to differences in patterns of drift among skeletal elements, it is also evident that the tibia, ulna and radius generally display a smaller proportion of the slower-forming LAM tissues, relative to the humerus and femur. Given the smaller cortices of the distal elements (except for the tibia in some cases), the initial expectation is that they would display larger proportions of LAM bone, having more time during growth to develop a smaller cortex. Several explanations can be put forth to account for this lack of concordance between TCA and tissue type within individuals.

First, growth of skeletal elements – including their circumferential expansion at midshaft – does not necessarily occur at the same rate, or for the same duration, in all cases. This can be seen at its extreme in the disparity between hindlimb and forelimb growth rates in marsupials (e.g., Hamrick, 1999). For primates, limb growth allometry studies also highlight differences in growth rates between skeletal elements (e.g., Falsetti and Cole, 1992).

Second, tissue deposition may not proceed at the same rate in all elements, even for the same type. For example, Castanet et al. (2000) note that the tissue formation in ratite bird hindlimbs is faster than in the forelimbs, and that bony "elements (i.e., femur, humerus) may provide different estimations of overall skeletal growth dynamics, even when similar bone tissue types are observed" (p. 549). Within the fibro-lamellar tissues category they identify a tremendous range of deposition rates, from 10 to 80 μ/day. de Margerie et al. (2004) confirm these observations in their fluorochrome study of king penguin chick long bones. While all skeletal elements in their study contain a range of fibrolamellar vascular organizations, rates in the radius were 38% lower than in the femur, humerus and tibiotarsus. Likewise, Starck and Chinsamy (2002) identify differences among skeletal elements: fibrolamellar bone deposition rates range between 10–50 μ/day, and cross-sections in the femur and humerus increase their thickness more rapidly than the cortices in more distal long bones. They suggest that "individual skeletal elements respond differently to environmental conditions, thus producing different patterns in bone microstructure" (p. 244). It does not appear that rates are as variable in the PF and LAM categories, but ranges still may overlap. (See the section on growth rates below for more discussion of this issue.)

Lastly, there is the possibility that, within individuals, differences in growth rates between elements are not substantial enough to influence tissue type, and that variation in the organization of the tissue laid down is dependent upon other factors, such as difference between elements in mechanics at the midshaft (see section below on positional behavior).

18.5.2 Primary Tissue Types and Body Size

Results of this study indicate that TCA, as a proxy for body size, is a significant factor in determining the proportion of highly vascularized fast-forming bone (woven and fibro-lamellar types = FAST bone) within a primate long bone cortex. This is not surprising; the larger the individual and the larger the cortical cross-section of its long bones, the more bone tissue it must deposit to achieve functional competence during growth and development. FAST bone is often found in the thickest quadrants of the cortex, further evidence of the relationship with cortical area. The data indicate that there may be a threshold to this size related pattern. Individuals with body masses below 500 g display very little (i.e., less than 10% of the periosteal cortex) or no FAST bone tissue. This is consistent with findings of other researchers for disparate groups of vertebrates. For example, Bonis et al. (1972) found that small tetrapods have cortices composed of lamellar bone, while larger ones contain fibrolamellar tissues or remodeled bone. Castanet et al. (2001), in a study of bird long bones, find that small birds (< 100 g) do not have densely vascularized cortices. In their review of Enlow and Brown (1956) and Klevezal (1996), they find this to hold true for all tetrapods, regardless of phylogenetic status and metabolic adaptation, where species up to 1 kg are have poorly vascularized bones. They conclude that, in general, the larger the species, the more densely vascularized its bones will be, and the more likely they will be composed of fibrolamellar, rather than lamellar-zonal tissues. This is in accord with the results of the present study. Although the density of vascularization was not quantified, the presence of FBL bone indicates denser vascularization than LAM or PF bone.

This relationship between FAST bone and body size is not repeated to the same extent with the other, slower-forming, primary tissue types examined here: PF and LAM. A larger midshaft cortex does not mean there will be more PF bone, hypothesized to be faster depositing than LAM tissue. Similarly, smaller cortices do not necessarily comprise more slow-forming LAM bone than PF bone. In addition, when values for FAST bone are high, this does not preferentially occur at the expense of either PF or LAM values. In some cortices, if the proportion of FAST bone is large, there is proportionately less PF tissue, while in others it is the proportion of LAM that is diminished. Finally, in some cases, both LAM and PF tissues are reduced relative to their proportions in bones that contain less FAST tissue.

A potential explanation for the lack of a strong relationship between body size and PF or LAM bone is the subtle nature of the separation in growth rates between these two, at least as described by Newell-Morris and Sirianni (1982), relative to the much faster rates of woven and heavily vascularized tissues (see discussion of growth rates below). Alternatively, or perhaps in addition to this, LAM and PF bone tissues may be responding to other, non size related features, as I suggested may be the case for explaining the difference in tissue types among skeletal elements within an individual.

18.5.3 Primary Tissue Types and Growth Rate

As would be predicted from previous studies of primary bone formation rates, and in accordance with "Amprino's rule" (Amprino, 1946; de Buffrénil and Pascal, 1984), the results of this study suggest a broad relationship between growth rate and primary tissue types. However, this pattern breaks down when individual primate taxa are compared.

As a whole, the primates in this sample have a large proportion of the adult midshaft periosteal cortex (as well as endosteal) composed of slow-forming, sparsely vascularized tissues such as LAM bone. Comparative reviews (e.g., Enlow and Brown, 1958) show that many non-primate mammals (e.g., carnivores, rodents) are likely to display much higher proportions of faster tissues such as the highly vascularized fibro-lamellar bone (see *Mustela, Bassariscus, Canis, Dasypus* and *Lepus* in Enlow and Brown, 1958). The preponderance of slower-forming bone tissue types in primates is consistent with the generally lengthened life histories (slow growth, delayed maturity) of members of the primate order (e.g., Cubo et al., 2002; Watts, 1990; Ross, 1998), but also with the small to medium body sizes of the species in this study (all species < 10 kg).

One indication of the relationship between growth rates and tissue types is that FAST tissue is often found close to the boundary between periosteal and endosteal bone, in older parts of the periosteal cortex of femora. These areas, having been deposited earlier in a developmental trajectory than more peripheral areas, are more likely to have formed when growth was more rapid. Another sign of the relationship between growth rate and tissue type is the difference between platyrrhines and strepsirhines in proportions of all tissue types. The slower growing platyrrhines (reflected in scaled growth scores, and reported previously by Kirkwood, 1985; see also Cubo et al., 2002) exhibit more slow-forming LAM bone, less PF bone and less FAST bone than the strepsirhines. This is the case both for strepsirhine and platyrrhine means, and also when femoral values for pairs of comparably sized lower-level taxa are compared (with the exception of the comparison between the cebids and cheirogaleids). Within suborders, however, the relationship between tissue type and growth rate is less evident (at least for the femur). So, for example, the lorisids, with slower growth rates relative to the comparably sized greater galagos, display less, rather than more LAM bone, and concomitantly more PF bone.

Even within infraorders there are two exceptions to the general concordance between growth rate and primary tissue type distributions. First, the cheirogaleids display a surprisingly high, plartyrrhine-like proportion of LAM bone at the midshaft femur. One could speculate that this may relate to the periods of lethargy, torpor or hibernation – a unique adaptation among primates – found in this family (Ortmann et al., 1997; Schmid, 1998, 2000; 2001). Growth curves are not know for cheirogaleids, but if periods of reduced activity, metabolic rate and body temperature occur during the later phases of growth when the adult cortex is forming, their rates of tissue deposition and primary tissue type proportions would be affected. *Microcebus murinus*, for example, has been observed to enter a period of dry season hibernation about four months after birth, remaining in hibernation for up to several months (Schmid and Kappeler, 1998). In a study of lines of arrested growth in the tibia of captive *M. murinus*, Castanet et al. (2004) found a diaphyseal cortical pattern suggestive of depressed rates of growth during the first seasonal hibernation (approximately 0.1 μm/day) after a more rapid phase of summer growth (with deposition rates of up to 8 μm/day). Thereafter, for a period of up to 6 years, bone was slowly deposited circumferentially. The transition from faster vascular tissue to slower avascular tissue concurrent with the first hibernation suggests that the unique life history adaptation of the cheirogaleids my indeed be reflected in its cortical bone microstructure. While *Mirza coquereli* is not known to have any daily or seasonal periods of reduced activity, it also displays a preponderance of slow-growth LAM bone at the midshaft femur. This would be explained if a pattern of growth similar to that found in *Cheirogaleus* and *Microcebus* were a family-specific trait.

A second surprising finding is that the tiny tarsiers, reported to have a slow life history relative to strepsirhines of their size (corroborated by infant growth rates), display a large proportion of the cortex composed of PF bone (a mean of about 70% of the periosteal portion of the cortex). Not only are mean values of PF bone for the femur higher than those found in the lesser galagos, the taxon with the best match in terms of size and positional behavior, but they are substantially higher (by at least 20% of the total periosteal area, and usually much more) than in any other taxon examined in this study.

There are several potential explanations for the cheirogaleid and tarsiid patterns, as well as the general lack of agreement with scaled growth rate scores when pairwise comparisons are made of closely related taxa. First, it may be that changes in primary tissue type are not necessary when the difference in deposition rates needed to construct bones in an adequate span of time is small. So, for example, while the scaled growth rate data indicate that it will take a *Saimiri* individual approximately four times as long to achieve the same increase in mass as a lorisid, the scaled growth rate values for lorisids are approximately two times, or less, the value for *O. crassicaudatus*. A related possibility is that the disparity in bone deposition rates between PF and LAM is too subtle to be reflected among species where somatic growth rate differences are not extreme, or that the deposition rates for these tissues vary and/or overlap with each other, as has been observed to be the case for fibrolamellar tissues. For example, Castanet et al. (2000), de Margerie et al. (2004), and Starck and Chinsamy (2002) have all identified a range of tissue deposition rates within the fibrolamellar pattern in vital labeling studies of ratite birds, king penguins, and Japanese quails respectively. Starck and Chinsamy in particular warn of the plastic and highly responsive nature of bone in its reaction to variation in environmental factors such as diet and mechanical stress. While data for rates among lamellar-zonal tissues are more limited, there are indications that in these types as well, variability and overlap are significant. Results of the vital labeling study of Newell-Morris and Sirianni (1982) suggest that there is only an average 2 μm/day difference in deposition rates between PF and LAM tissue (4 μm/day and 2 μm/day respectively), with overlaps in the ranges of rates between these two tissues.

Second, there are, in some cases, subtle differences in the density of vascularization of the LAM and PF tissues that are not quantified in this study, and which may reflect differences in growth rates. de Margerie et al. (2002), and Castanet et al. (1996) have shown that vascularized tissues deposit more rapidly than non-vascular tissues, that the inclusion of primary osteons indicates faster deposition than simple vascular canals, and that as the size of primary osteons increase, so does the growth rate.

Third, there is the possibility that the growth rate data available for the taxa included in this study are inappropriate for comparison to patterns of tissue types found in the adult midshaft cortex. As described in many studies (e.g., Leigh and Terranova, 1998) somatic growth rates are not constant throughout ontogeny. The scaled rates used here represent averages derived from the entire growth period from birth to attainment of adult mass, or averages of the rates of increase in mass between 25% of adult weight and 75% of adult weight. In both cases, these may not accurately reflect the latter periods of bone deposition found in an adult cortex, where much of the bone formed early in development has been resorbed away. In addition, age at attainment of adult mass was not available for all taxa, and in some cases was estimated from published growth curves (e.g., data for *Otolemur crassicaudatus* and *Loris tardigradus* estimated from Rasmussen and Izard, 1988), or based on rough estimates from the literature when the species was not known (e.g., data for *Lemur* sp. estimated from Leigh and Terranova, 1998).

18.5.4 Primary Tissue Types and Positional Behavior

There exists a voluminous literature on the mechanical properties of secondary osteons, both as individual units, and relative to primary bone tissue (Chamay and Tschantz, 1972; Bouvier and Hylander, 1981; Layon et al., 1982; Portigliatti-Barbos et al., 1983; Layon, 1984, 1993; Burr et al., 1985; Layon and Rubin, 1985; Schaffler et al., 1989, 1990; Burr, 1993; Martin, 1993, 2000; Lieberman and Crompton, 1998). A substantial and growing literature also considers the relationship between collagen fiber orientation and functional adaptation, although this is most frequently in reference to intracortically remodeled bone (Ascenzi and Bonucci, 1967, 1968, 1972; Vincentelli and Evans, 1971; Ascenzi et al., 1973; 1987; Portigliatti-Barbos et al., 1983; Ascenzi and Benvenuti, 1986; Carando et al., 1989, 1991; Martin and Ishida, 1989; Boyde and Riggs, 1990; Bromage, 1992; Riggs et al., 1993a, 1993b, McMahon et al., 1995; Goldman, 2001; Warshaw, 2007). The relative properties of various primary bone tissue types remain relatively unexplored. Interestingly, recent investigations of primary tissue types and whole bone mechanics have led to the suggestion that vascular patterning in primary bone may vary as a function of habitual loads on bone. de Margerie (2005) found that, in the mallard, the bones most likely to be subjected to torsional loads (e.g., most of the bones of the wing skeleton, as well as the femur) exhibited fibrolamellar bone complexes with a more laminar organization of primary osteons (primary vascular canals running in circumferential rings around the cortex) than found in other long bones. He hypothesized that this vascular scheme is adapted to resist the shearing that torsional loads impose, and that the "putative adaptation must have an ultrastructural basis, perhaps through particular collagen fibre orientation in the primary osteons of laminar bone" (p. 523).

In a subsequent study of turkey ulnae, Skedros and Hunt (2004), found "significant differences in [collagen fiber orientation] and/or other microstructural characteristics…between 'tension', 'compression' and 'shear' (neutral axis) regions of subadult bones" (p. 122) although these differences were less evident in adult bones. Portions of the cortex containing greater laminarity had more oblique to transverse collagen fibers, and these were most often found where compression was the predominant load. In addition, they determined that bones displaying more laminarity throughout the cortex were correlated with a more uniform distribution of preferred collagen around the cortex. In this case, the suggestion is that both the vascular and collagen fiber patterns may be adapting to torsion, although the authors caution that the vascular patterns may be more directly a product of variation in growth rate than loading regime.

The data from those studies suggest that collagen fiber orientation is not the only microstructural feature of primary bone that may be related to mechanical adaptation. It follows that there may be functional significance to variable distributions of LAM and PF bone, as observed in this study. While in some cases (e.g., the lorisids) differences linked to positional behavior are impossible to separate from those due to other taxon-specific features (e.g., growth, phylogeny), members of other locomotor categories are more heterogeneous with respect to phylogeny and life history. In particular, the arboreal quadrupedalism category displays consistency in terms of the proportion of the femoral midshaft cortex composed of LAM bone, relative to other positional behavior types. This broadly defined group includes both strepsirhines and platyrrhines, and a range of taxa in terms of body size and life history.

The identification of the pattern, however, does not easily lead to an explanation of its significance. Given the histocompositional divergences between PF and LAM bone, it is likely that there are differences in mechanical properties between the two types. However, there are no data to support this proposition or to suggest what specific mechanical variables might differ (e.g., strength, stiffness, fracture toughness). Second, no comparative strain gauge data exist for the non-human primate midshaft femur. And while in the past it has been assumed that the cross-sectional shape of a bone might reflect habitual patterns of strain, more recent studies have shown that this is often not the case (e.g., Demes et al., 2001). So, for example, while it has been reported that forces generated and absorbed in the limbs of generalized arboreal quadrupeds are greater than those in size matched specialized vertical clingers and leapers performing equal locomotor tasks (Demes et al., 1999), the actual loading patterns at the level of the femoral midshaft section are unknown.

Finally, it is important to keep in mind another potential explanation for the apparent mechanical patterns observed in this sample. Although the relative proportions of LAM and PF bone are not highly correlated with total cortical area (or periosteal cortical area), it is possible that cortical thickness (not measured here) may impact upon the distribution of tissue types. Two sections with a similar TCA may have different diameters, with the smaller of the two displaying thicker cortices than the larger. This bears future investigation.

18.5.5 Primary Tissue Types and Phylogeny

Related to all factors above, and difficult to separate from them, is the question of phylogenetic signals in bone microstructure, independent from more proximate, adaptive life history and mechanical features. Excluding the earliest comparative bone histology works (e.g., Foote, 1913; Quekett, 1849, 1855), previous studies have not often identified phylogenetically determined, taxon-specific bone microstructural features. For example, Castanet et al. (2001) find that only 1.3% of variation in the intensity of vascularization in the long bone cortices of birds is due to a phylogenetic effect, independent of adaptive features. Bonis et al. (1972) determine that most variation is explicable by differences in body size and growth rates. There are exceptions to this apparent rule cited in the contemporary literature; for example the presence of acellular bone and Williamson canaliculi in some groups of fishes (Enlow and Brown, 1956; Castanet et al., 2001). Starck and Chinsmay (2002) suggest that the absence of lines of arrested growth in extant birds, even when experimentally subjected to periods of food restriction, represents the loss of this adaptation in the stem lineage leading to ornithurine birds. Certainly group-specific patterns may be identifiable. For the most part, however, it is likely is that they can be explained via an understanding of growth, physiological, and mechanical adaptations.

Here, taxon-specific patterns are clearly present. For example, platyrrhines exhibit less FAST and PF bone and more LAM bone than strepsirhines. Tarsiids have more PF bone at the midshaft femur than any other taxa in the sample. Cheirogaleids display high proportions of LAM bone relative to all other strepsirhines. The question remains: can these patterns be explained as reflections of other features that serve to segregate taxa along phylogenetic lines? In the case of the separation between platyrrhines and strepsirhines, it seems likely that this is related to the divergent growth rate adaptations in the two groups. The pattern seen in the tarsiids may reflect a phylogenetically determined peculiarity with no clearly adaptive explanation (at least with respect to extant members of the tarsiid lineage). Alternatively, the almost total lack of vascularization in the midshaft cortex – not the case even for the smallest galagid – may be compensated for by the slightly higher deposition rates of PF bone. With respect to the cheirogaleids, both life history (as outlined above) and phylogenetic explanations are speculative. Then again, there may be some as-yet unrecognized pattern related to cortical thickness that accounts for the seemingly uncharacteristic primary tissue type distributions in this taxon. The data presented here are insufficient to answer this question satisfactorily, and the prudent conclusion for the time being – given the overwhelming evidence elsewhere in the sample – would be that a life history and/or mechanical explanation is likely.

18.6 Summary and Conclusions

Returning to objectives outlined earlier, I offer the following conclusions. First, the primates in this study are characterized overall by a high proportion of slow-forming tissues (lamellar-zonal types). This emphasis on LAM and PF tissues is consistent with the observed "slow" life-history adaptation in the order as a whole. Second, within this generalized primate pattern there exists significant and non-random variation, both among skeletal elements, and among taxa. Such variation is partly explicable via consideration of ontogenetic processes at the level of the cross-section, reflecting whole bone development. Individual elements display typical patterns of periosteal and endosteal tissue distribution, with features that differentiate between the primate forelimb and hindlimb as well as between proximal and distal limb element ontogenies, and that highlight the distinctive growth trajectories of humeri and tibiae.

The range of primary tissue type patterns is also partly accounted for by diversity in body size and growth rate – as has been previously suggested – as well as habitual positional behavior. Dependent upon the tissue type, different variables appear to play more or less of a significant role in accounting for the observed patterns (despite the obviously correlated nature of proportions among tissue types within a cortex[9]). Proportions of FAST bone relate primarily to body size (and therefore the need for more rapid bone formation), with a subsidiary phylogenetic effect likely related to subordinal differences in somatic growth rates. In contrast, LAM and PF tissues exhibit little or no relationship with body size, but do appear to track broad phylogenetic and/or growth differences, as well as positional behaviors. Needless to say, the separation between factors such as body size, growth and habitual locomotion is somewhat artificial, as all are interrelated adaptive (and phylogenetically informed) features of a species. Nevertheless, it does appear that different primary tissue types may preferentially track different demands on the growing bone.

Both the multiple regression analyses and simpler qualitative and quantitative exploration of the data make clear that some of the observed variation remains to be explained. Given the apparent ability of bone to respond so efficiently to both organismal and environmental demands, it is possible that a range of factors not explored here may be relevant to a more full explanation of variation in primary tissue types at the midshaft. Sex, parity, diet, cyclical or catastrophic climate changes, physiological perturbations both normal and pathological; all have the potential to impact upon a growing bone, leaving records in the tissues of its cross-section. What I have shown, however, is that several factors commonly referred to in the literature – body size, growth rate, mechanical environment – do indeed leave signals in primate bone.

What can we learn of the life history, positional behavior, and phylogenetic affiliation of an individual from an analysis of bone tissue organization? At present I propose that we can glean important, albeit tentative, clues. We can see, for example, the difference between a platyrrhine and a strepsirhine. We know that the former generally has an extended period of growth relative to the latter, and this is evident in the relative paucity of fast forming, and preponderance of slow forming, bone tissue. Familial differences in growth rate are not, however, reflected in the data. Among strepsirhines, we might identify a pattern indicating that an individual is an habitual arboreal quadruped rather than a vertical clinger and leaper. The distinctive tarsiid tissue type scheme, presently an adaptive puzzle, is unique and may be an autapomorphic phylogenetic marker. These clues offer a tantalizing glimpse of the potential that primary tissue type diversity has to inform the analysis of fossil material. They also make clear that more work needs to be done, both comparative and experimental, before microstructural features in primates can be interpreted with confidence.

Acknowledgments. I would like to acknowledge the following people and institutions: Tim Bromage, my doctoral dissertation advisor, for his support, enthusiasm, and inspiration, and for access to the one-of-a-kind laboratory facilities that he developed (the Hard Tissue Research Unit and the Analytical Microscopy and Imaging Center in Anthropology, both previously housed at Hunter College, CUNY, and currently at the New York University College of Dentistry), Carl Terranova, Donald Enlow, Jim McMahon, David Thomas (for development of the Optimas macro), John Wahlert, Mitchell Schaffler, Shannon McFarlin, Haviva Goldman, Stuart Marquis, Stephen McJonathan, Jan Hinsch, and Joseph Luders for the perspective, assistance, and expertise that they have provided, and two anonymous reviewers for their helpful comments. I am lucky to have known Justine Salton, and miss her energy, talent and, most profoundly, her friendship. I am grateful to Carl Terranova, the Field Museum of Natural

[9] When values for one tissue type increase, something else must decrease.

History and Lawrence R. Heaney and William Stanley, the Museum of Comparative Zoology at Harvard University and Maria Rutzmoser, the Museum für Naturkunde of Humboldt University and Manfred Ade, the National Museum of Natural History and Richard Thorington and Helen Kafka, for access to specimens. Finally, I thank Marian Dagosto and Eric Sargis for inviting me to contribute to this volume. It provides me with the opportunity to honor Fred Szalay, who inspired me with his voracious intellect and interest in the natural world. His sparkling delight in, and profound knowledge of, all things biological have always served to remind me how rewarding the fields of vertebrate morphology and evolutionary biology can be. As an advisor and mentor, he has shown unfailing patience and care, and I am very fortunate to have had classroom and field time with him, and also to consider him a friend. This research was supported by a graduate fellowship from the New York Consortium in Evolutionary Biology (NYCEP), a Research Grant from The Leakey Foundation, and a National Science Foundation Dissertation Improvement Award (DDI-0205185).

References

Aerts, P., 1998. Vertical jumping in *Galago senegalensis*: the quest for an obligate mechanical power amplifier. Philosophical Transaction of the Royal Society B 353, 1607–1620.

Amprino, R., 1946. La structure du tissue osseux envisagee comme expression de differences dans la vitesse de l'accroissement. Archives de Biologie LVII, 315–330.

Amprino, R., 1948. A contribution to the functional meaning of the substitution of primary by secondary bone tissue. Acta Anatomica 5, 291–300.

Amprino, R., Godina, G., 1947. La struttura delle ossa nei vertebrati: ricerche comparative negli anfibi e negli amnioti. Commentationes. Pontificia Accademia Delle Scienze 11(9), 329–464.

Ascenzi, A., Bonucci, E., 1967. The tensile properties of single osteons. Anatomical Record 158, 375–386.

Ascenzi, A., Bonucci, E., 1968. The compressive properties of single osteons. Anatomical Record 161, 377–392.

Ascenzi, A., Bonucci, E., 1972. The shearing properties of single osteons. Anatomical Record 172, 499–510.

Ascenzi, A., Benvenuti, A., 1986. Orientation of collagen fibers at the boundary between two successive osteonic lamellae and its mechanical interpretation. Journal of Biomechanics 19, 455–463.

Ascenzi, A., Bonucci, E., Simkin, A., 1973. An approach to the mechanical properties of single osteonic lamellae. Journal of Biomechanics 6, 227–235.

Ascenzi, A., Boyde, A., Portigliatti Barbos, M., Carando, S., 1987. Micro-biomechanics vs. macro-biomechanics in cortical bone. A micromechanical investigation of femurs deformed by bending. Journal of Biomechanics 20, 1045–1053.

Bateman, N., 1954. Bone growth: a study of the grey-lethal and microphthalmic mutants of the mouse. Journal of Anatomy 88(2), 212–262.

Biewener, A., 1991. Musculoskeletal design in relation to body size. Journal of Biomechanics 24, 19–29.

Biewener, A. A., Swartz, S. M., Bertram, J. E. A., 1986. Bone modeling during growth: dynamic strain equilibrium in the chick tibiotarsus. Calcified Tissue International 39, 390–395.

Boinski, S., 1989. The positional behavior and substrate use of squirrel monkeys: ecological implications. Journal of Human Evolution 18, 659–677.

Bonis, L. de, Lebeau, M., de Ricqlès, A., 1972. Etude de la répartition des types de tissues osseux chez les vértébrés tétrapodes au moyen de l'analyse factorielle des correspondences. Comptes rendus de l'Académie des sciences D 274, 3084–3087.

Botha, J., Chinsamy, A., 2000. Growth patterns deduced from the bone histology of the cynodonts *Diademodon* and *Cynognathus*. Journal of Vertebrate Paleontology 20(4), 705–711.

Bouvier, M., Hylander, W. L., 1981. Effect of bone strain on cortical bone structure in macaques (*Macaca mulatta*). Journal of Morphology 167, 1–12.

Boyde, A., Jones, S., 1998. Aspects of anatomy and development of bone. In: Zaidi, M., Adebanjo, O. A., Huang, C. L. H. (Eds.), Advances in Organ Biology: Molecular and Cellular Biology of Bone, Part 5A. JAI Press, Stamford, pp. 3–44.

Boyde, A., Riggs, C. M., 1990. The quantitative study of the orientation of collagen in compact bone slices. Bone 11, 35–39.

Bromage, T. G., 1986. A comparative scanning electron microscope study of facial growth and remodeling in early hominids. Ph.D. dissertation, University of Toronto, Ontario, Canada.

Bromage, T. G., 1992. Microstructural organization and biomechanics of the macaque circumorbital region. In: Smith, P., Tchernov, E. (Eds.), Structure, Function and Evolution of Teeth. Freund Publishing House, London, pp. 257–272.

Bromage, T. G., Goldman, H. M., McFarlin, S. C., Warshaw, J., Boyde, A., Riggs, C. M., 2003. Circularly polarized light standards for investigations of collagen fiber orientation in bone. Anatomical Record. Part B New Anatomist 274(1), 157–68.

Burr, D. B., 1979. Percentage ash content of nonhuman primate long limb bones. American Journal of Physical Anthropology 51, 361–364.

Burr, D. B., 1980. The relationships among physical, geometrical and mechanical properties of bone, with a note on the properties of nonhuman primate bone. Yearbook of Physical Anthropology 23, 109–146.

Burr, D. B., 1992. Estimated intracortical bone turnover in the femur of growing macaques: implications for their use as models in skeletal pathology. Anatomical Record 232, 180–189.

Burr, D. B., 1993. Remodeling and the repair of fatigue damage. Calcified Tissue International 53, S75–S81.

Burr, D. B., Martin, R. B., Schaffler, M. B., Radin, E. L., 1985. Bone remodeling in response to *in vivo* fatigue microdamage. Journal of Biomechanics 18, 189–200.

Burr, D. B., Piotrowski, G., Martin, R. B., Cook, P. N., 1982. Femoral mechanics in the lesser bushbaby (*Galago senegalensis*): structural adaptations to leaping in primates. Anatomical Record 202, 419–429.

Burr, D. B., Schaffler, M. B., Yang, K. H., Lukoschek, M., Sivaneri, N., Blaha, J. D., Radin, E. L., 1989. Skeletal change in response to altered strain environments: is woven bone a response to elevated strain? Bone 10, 223–233.

Cant, J. G. H., Youlatos, D., Rose, M. D., 2001. Locomotor behavior of *Lagothrix lagothricha* and *Ateles belzebuth* in Yasuni National Park, Ecuador: general patterns and nonsuspensory modes. Journal of Human Evolution 41, 141–166.

Carando, S., Portigliatti-Barbos, M., Ascenzi, A., Boyde, A., 1989. Orientation of collagen in human tibial and fibular shaft and possible correlation with mechanical properties. Bone 10, 139–142.

Carando, S., Portigliatti-Barbos, M., Ascenzi, A., Riggs, C. M., Boyde, A., 1991. Macroscopic shape of, and lamellar distribution within, the upper limb shafts, allowing inferences about mechanical properties. Bone 12, 265–269.

Castanet, J., de Ricqlès, A., 1986–87. Sur la relativité de la notion d'osteones primaires et secondaires et de tissus osseux primaire et secondaire en général. Annales des Sciences Naturelles, Zoologie 13(8), 103–109.

Castanet, J., Croci, S., Aujard, F., Perret, M., Cubo, J., de Margerie, E., 2004. Lines of arrested growth in bone and age estimation in a small primate: *Microcebus murinus*. Journal of Zoology 263(1), 31–39.

Castanet, J., Cubo, J., de Margerie, E., 2001. Signification de l'histodiversité osseuse: le message de l'os. Biosystema – Systématique et Paléontologie 19, 133–147.

Castanet, J., Grandin, A., Abourachid, A., de Ricqlés A., 1996. Expression of growth dynamic in the structure of the periosteal bone in the mallard *Anas platyrhynchos*. Comptes Rendus de l'Académie des Sciences III-Vie 319, 301–308.

Castanet, J., Rogers, K. C., Cubo, J., Boisard, J. J., 2000. Periosteal bone growth rates in extant ratites (ostriche and emu). Implications for assessing growth in dinosaurs. Comptes rendus de l'Académie des sciences III-Vie 323(6), 543–550.

Chamay, A., Tschantz, P., 1972. Mechanical influences in bone remodeling. Experimental research on Wolff's law. Journal of Biomechanics 5, 173–180.

Charles-Dominique, P., 1971a. Eco-ethologie des prosimiens du Gabon. Revue Biologia Gabonica 7, 121–228.

Charles-Dominique, P., 1971b. Eco-ethologie et vie sociale des prosimiens du Gabon. Bulletin de la Société d'Ecologie II, 318–326.

Crawford, G. N. C., 1940. The evolution of the Haversian pattern. Journal of Anatomy 74, 284–299.

Cubo, J., Berge, C., Quilhac, A., Margerie, E. de, Castnet, J., 2002. Heterochronic patterns in primate evolution: evidence from endochondral ossification. European Journal of Morphology 40(2), 81–88.

Currey, J. D., 1960. Differences in the blood supply of bone of different histological types. Quarterly Journal of Microscopical Science 101, 351–370.

Currey, J. D., 1999. The design of mineralized hard tissues for their mechanical functions. Journal of Experimental Biology 202, 3285–3294.

Curry, K., 1998. Histological quantification of growth rates in *Apatosaurus*. Journal of Vertebrate Paleontology 18(3), 36A–37A.

Dagosto, M., 1988. Implications of postcranial evidence for the origin of euprimates. Journal of Human Evolution 17, 35–56.

Dagosto, M., 1994. Testing positional behavior of Malagasy lemurs: a randomized approach. American Journal of Physical Anthropology 94, 189–202.

Dagosto, M., Gebo, D. L., 1994. Postcranial anatomy and the origin of the Anthropoidea. In: Fleagle, J. G., Kay, R. F. (Eds.), Anthropoid Origins. Plenum, New York, pp. 567–594.

de Buffrénil, V., Pascal, M., 1984. Croissance et morphogenèse postnatales de la mandibule du vison (*Mustela vison* Schreibner): donnes sur la dynamique et l'interprétation fonctionelle des depots osseux mandibulaires. Canadian Journal of Zoology 62, 2026–2037.

Delson, E., Terranova, C. J., Jungers, W. L., Sargis, E. J., Jablonski, N. G., Dechow, P. C., 2000. Body mass in Cercopithecidae (Primates, Mammalia). Anthropological Papers of the American Museum of Natural History 83, 1–159.

de Margerie, E. 2002. Laminar bone as an adaptation to torsional loads in flapping flight. Journal of Anatomy 201(6), 521–526.

de Margerie, E., Cubo, J., Castanet, J., 2002. Bone typology and growth rate: testing and quantifying 'Amprino's rule' in the mallard (*Anas platyrhynchos*). Comptes Rendus Biologies 325, 221–230.

de Margerie, E., Robin, J. P., Verrier, D., Cubo, J., Groscolas, R., Castanet, J., 2004. Assessing a relationship between bone microstructure and growth rate: a fluorescent labelling study in the king penguin chick (*Aptenodytes patagonicus*). Journal of Experimental Biology 207(Pt 5), 869–879.

de Margerie, E., Sanchez, S., Cubo, J., Castanet, J., 2005. Torsional resistance as a principal component of the structural design of long bones: comparative multivariate evidence in birds. Anatomical Record Part A 282A, 49–66.

Demes, B., Fleagle, J. G., Jungers, W. L., 1999. Takeoff and landing forces of leaping strepsirhine primates. Journal of Human Evolution 37, 279–292.

Demes, B., Gunther, M. M., 1989. Biomechanics and allometric scaling in primate locomotion and morphology. Folia Primatologica 53, 125–141.

Demes, B., Jungers, W. L., 1989. Functional differentiation of long bones in lorises. Folia Primatologica 52, 58–69.

Demes, B., Jungers, W. L., Nieschalk, U., 1990. Size-and speed-related aspects of quadrupedal walking in slender and slow lorises. In: Jouffroy, F., Stack, M., Niemitz, C. (Eds.), Gravity, Posture and Locomotion in Primates. Firenze, Il Sedicesimo, pp. 175–197.

Demes, B., Jungers, W. L., Selpien, K., 1991. Body size, locomotion, and long bone cross-sectional geometry in indriid primates. American Journal of Physical Anthropology 86, 537–547.

Demes, B., Qin, Y.-X., Stern, J. T. J., Larson, S. G., Rubin, C. T., 2001. Patterns of strain in the macaque tibia during functional activity. American Journal of Physical Anthropology 116, 257–265.

Demeter, G., Mátyás, J., 1928. Mikroskopisch verleichend-anatomische Studien an Röhrenknochen mit besonderer Rucksicht auf die Unterscheidung menschlicher und tierischer Knochen. Zeitschrift fur Anatomie und Entwicklungsgeschichte 87, 45–99.

de Ricqlès, A., 1968. Recherches paleohistologiques sur les os longs des tetrapodes. I – Origine du tissue osseux plexiforme des dinosauriens sauropodes. Annales de Paleontologie (Vertebres) 54, 131–145.

de Ricqlès, A., 1969. Recherches paleohistologiques sur les os longs des tetrapodes. II – Quelques observations sur la structure des os longs des theriodontes. Annales de Paleontologie (Vertebres) 55, 1–52.

de Ricqlès, A., 1975. Recherches paleohistologiqus sur es os longs des tetrapodes. VII – Sur la classification, la signification fonctionelle et l'histoire des tissues osseux des tetrapodes (premiere partie). Annales de Paleontologie (Vertebres) 61, 51–129 (plus plates).

de Ricqlès, A., 1976. Recherches paleohistologiqus sur es os longs des tetrapodes. VII – Sur la classification, la signification fonctionelle et l'histoire des tissues osseux des tetrapodes (deuxieme partie). Annales de Paleontologie (Vertebres) 62, 71–126.

de Ricqlès, A., 1977a. Recherches paleohistologiqus sur es os longs des tetrapodes. VII – Sur la classification, la signification fonctionelle et l'histoire des tissues osseux des tetrapodes (deuxieme partie, fin). Annales de Paleontologie (Vertebres) 63, 133–160.

de Ricqlès, A., 1977b. Recherches paleohistologiqus sur es os longs des tetrapodes. VII – Sur la classification, la signification fonctionelle et l'histoire des tissues osseux des tetrapodes (deuxieme partie, suite). Annales de Paleontologie (Vertebres) 63, 33–56.

de Ricqlès, A., 1978a. Recherches paleohistologiqus sur es os longs des tetrapodes. VII – Sur la classification, la signification fonctionelle et l'histoire des tissues osseux des tetrapodes (troisieme partie). Annales de Paleontologie (Vertebres) 64, 85–111.

de Ricqlès, A., 1978b. Recherches paleohistologiqus sur es os longs des tetrapodes. VII – Sur la classification, la signification fonctionelle et l'histoire des tissues osseux des tetrapodes (troisieme partie, fin). Annales de Paleontologie (Vertebres) 64, 153–184.

de Ricqlès, A., 1992. Some remarks on palaeohistology from a comparative evolutionary point of view. In: Grupe, G., Garland, A. N. (Eds.), Histology of Ancient Human Bone: Methods and Diagnosis. Springer, Berlin, pp. 37–77.

de Ricqlès, A., Meunier, F., Castanet, J., Francillon-Vieillot, H., 1991. Comparative microstructure of bone. In: Hall, B. (Ed.), Bone–Volume 3: Bone Matrix and Bone Specific Products. CRC, Boca Raton, FL, pp. 1–78.

de Ricqlès, A., Padian, K., Horner, J. R., Francillion-Vieillot, H., 2000. Palaeohistology of the bones of pterosaurs (Reptilia: Archosauria): anatomy, ontogeny, and biomechanical implications. Zoological Journal of the Linnean Society, London 129, 349–385.

Dunbar, D. C., 1988. Aerial maneuvers of leaping lemurs: the physics of whole-body rotations while airborne. American Journal of Physical Anthropology 16, 291–303.

Dykyj, D., 1980. Locomotion of the slow loris in a designed substrate context. American Journal of Physical Anthropology 52, 577–586.

Enlow, D. H., 1962a. Functions of the Haversian system. American Journal of Anatomy 110, 269–305.

Enlow, D. H., 1962b. A study of the postnatal growth and remodeling of bone. American Journal of Anatomy 110, 79–101.

Enlow, D. H., 1963. Principles of Bone Remodeling. Charles C. Thomas, Springfield.

Enlow, D. H., 1966. An evaluation of the use of bone histology in forensic medicine and anthropology. In: Evans, F. G. (Ed.), Studies on the Anatomy and Function of Bone and Joints. Springer, New York, pp. 93–112.

Enlow, D. H., 1976. The remodeling of bone. Yearbook of Physical Anthropology 20, 19–34.

Enlow, D. H., 1982. Handbook of Facial Growth. W. B. Saunders, Philadelphia.

Enlow, D. H., Brown, S. O., 1956. A comparative histological study of fossil and recent bone tissues. Part I. Texas Journal of Science VIII, 405–443.

Enlow, D. H., Brown, S. O., 1957. A comparative histological study of fossil and recent bone tissues. Part II. Texas Journal of Science IX, 186–214.

Enlow, D. H., Brown, S. O., 1958. A comparative histological study of fossil and recent bone tissues. Part III. Texas Journal of Science X, 187–230.

Enlow, D. H., Hans, M. G., 1996. Essentials of Facial Growth. W.B. Saunders, Philadelphia.

Erickson, G. M., Rogers, K. C., Yerby, S. A., 2001. Dinosaurian growth patterns and rapid avian growth rates. Nature 412(6845), 429–433.

Erickson, G. M., Tumanova, T. A., 2000. Growth curve of *Psittacosaurus mongoliensis* Osborn (Ceratopsia: Psittacosauridae) inferred from long bone histology. Zoological Journal of the Linnean Society of London 130, 551–566.

Evans, G., Bang, S., 1967. Differences and relationships between the physical properties and the microscopic structure of human femoral, tibial and fibular cortical bone. American Journal of Anatomy 120, 79–88.

Falsetti, A. B., Cole, T. M. I., 1992. Relative growth of the postcranial skeleton in callitrichines. Journal of Human Evolution 23, 79–92.

Fleagle, J. G., Mittermeier, R. A., 1980. Locomotor behavior, body size, and comparative ecology of seven Surinam monkeys. American Journal of Physical Anthropology 52, 301–314.

Fontaine, R., 1990. Positional behavior in *Saimiri boliviensis* and *Ateles geoffroyi*. American Journal of Physical Anthropology 82, 485–508.

Foote, J., 1913. The comparative histology of the femur. Smithsonian Miscellaneous Collections 61, 1–9.

Foote, J., 1916. A contribution to the comparative histology of the femur. Smithsonian Contributions to Knowledge 35(3), iii–ix, 1–242.

Ford, S. M., 1988. Postcranial adaptations of the earliest platyrrhine. Journal of Human Evolution 17, 155–192.

Francillon-Vieillot, H., de Buffrénil, V., Castanet, J., Geraudie, J., Meunier, F. J., Sire, J. Y., Zylberberg, L., de Ricqlès, A., 1990. Microstructure and mineralization of vertebrate skeletal tissues. In: Carter, J. G. (Ed.), Skeletal Biomineralization: Patterns, Processes and Evolutionary Trends. Van Nostrand Reinhold, New York, pp. 471–530.

Garber, P. A., Leigh, S. R., 1997. Ontogenetic variation in small-bodied new world primates: Implications for patterns of reproduction and infant care. Folia Primatologica 68, 1–22.

Galliari, C. A., 1988. A study of postnatal appendicular maturation in captive-born squirrel monkeys (*Saimiri boliviensis*). American Journal of Primatology 16, 51–61.

Gebo, D. L., 1987. Locomotor diversity in prosimian primates. American Journal of Primatology 13, 271–281.

Gebo, D. L., Dagosto, M., 1988. Foot anatomy, climbing, and the origin of the Indriidae. Journal of Human Evolution 17, 135–154.

Glassman, D., Wells, J., 1984. Positional and activity behavior in a captive slow loris: a quantitative assessment. American Journal of Primatology 7, 121–132.

Godfrey, L. R., 1988. Adaptive diversification of Malagasy strepsirrhines. Journal of Human Evolution 17, 93–134.

Godfrey, L. R., Samonds, K. E., Jungers, W. L., Sutherland, M. R., Irwin, M. T., 2004. Ontogenetic correlates of diet in Malagasy lemurs. American Journal of Physical Anthropology 123, 250–276.

Godinot, M., 1990. An introduction to the history of primate locomotion. In: Jouffroy, F. K., Stack, M. H., Niemitz, C. (Eds.), Gravity, Posture and Locomotion in Primates. Firenze, Il Sedicesimo, pp. 45–60.

Goldman, H. M., 2001. Histocomposition and geometry at the human midshaft femur. Ph.D. dissertation, The City University of New York, New York.

Goldman, H. M., Kindsvater, J., Bromage, T. G., 1998. Correlative light and backscattered electron microscopy of bone–Part I: specimen preparation methods. Scanning 21, 40–43.

Gomez, A., 1992. Primitive and derived patterns of relative growth among species of Lorisidae. Journal of Human Evolution 23, 219–233.

Hall-Craggs, E. C. B., 1974. Physiological and histochemical parameters in comparative locomotor studies. In: Martin, R. D., Doyle, G. A., Walker, A. C. (Eds.), Prosimian Biology. Duckworth, London, pp. 829–845.

Hamrick, M., 1999. Development of epiphyseal structure and function in *Didelphis virginiana* (Marsupiala, Didelphidae). Journal of Morphology 239, 283–296.

Horner, J. R., Padian, K., 2004. Age and growth dynamics of *Tyrannosaurus rex*. Proceedings of the Royal Society London, Series B Biology 271(1551), 1875–1880.

Horner, J. R., de Ricqlès, A., Padian, K., 2000. Long bone histology of the hadrosaurid dinosaur *Maiasaura peeblesorum*: growth dynamics and physiology based on an ontogenetic series of skeletal elements. Journal of Vertebrate Paleontology 20, 115–129.

Ishida, H., Jouffroy, F., Nakano, Y., 1990. Comparative dynamics of pronograde upside down horizontal quadrupedalism in the slow loris (*Nycticebus coucang*). In: Jouffroy, F., Stack, M., Niemitz, C. (Eds.), Gravity, Posture and Locomotion in Primates. Firenze, Il Sedicesimo, pp. 209–220.

Ishida, H., Hirasaki, E., Matano, S., 1992. Locomotion of the slow loris between discontinuous substrates. In: Matano, S., Tuttle, R., Ishida, H., Goodman, M. (Eds.), Topics in Primatology, Volume 3: Evolutionary Biology, Reproductive Endocrinology, and Virology. University of Tokyo Press, Tokyo, pp. 139–152.

Jenkins, F. A. Jr., Domlnowshi, P. J., Gordon, E. P., 1978. Analysis of the shoulder in brachiating spider monkeys. American Journal of Physical Anthropology 48, 65–76.

Jerome, C. P., Peterson, P. E., 2001. Nonhuman primate models in skeletal research. Bone 29(1), 1–6.

Johnson, S. E., Shapiro, L. J., 1998. Positional behavior and vertebral morphology in atelines and cebines. American Journal of Physical Anthropology 105, 333–354.

Jolly, C. J., Gorton, A. T., 1974. Proportions of the extrinsic foot muscles in some lorisid prosimians. In: Martin, R. D., Doyle, G. A., Walker, A. C. (Eds.), Prosimian Biology. Duckworth, London, pp. 801–815.

Jouffroy, F. K., 1975. Osteology and myology of the lemuriform postcranial skeleton. In: Tattersal, I., Sussman, R. E. (Eds.), Lemur Biology. Plenum, New York, pp. 149–192.

Jouffroy, F. K., Berge, C., Niemitz, C., 1984. Comparative study of the lower extremity in the genus *Tarsius*. In: Niemitz, C. (Ed.), Biology of the Tarsiers. Gustav Fischer Verlag, New York, pp. 167–190.

Jouffroy, F. K., Gasc, J. P., Decombas, M., Oblin, S., 1974. Biomechanics of vertical leaping from the ground in *Galago alleni:* a cineradiographic analysis. In: Martin, R. D., Doyle, G. A., Walker, A. C. (Eds.), Prosimian Biology. Duckworth, London, pp. 817–827.

Jouffroy, F. K., Lessertisseur, J., 1979. Relationships between limb morphology and locomotor adaptations among prosimians: an osteometric study. In: Morbeck, M. E., Preuschoft, H., Gomberg, N. (Eds.), Environment, Behavior, and Morphology: Dynamic Interactions in Primates. Gustav Fischer, New York, pp. 143–181.

Jouffroy, F. K., Petter, A., 1990. Gravity-related kinematic changes in lorisine horizontal locomotion in relation to position of the body. In: Jouffroy, F., Stack, M., Niemitz, C. (Eds.), Gravity, Posture and Locomotion in Primates. Firenze, Il Sedicesimo, pp. 199–208.

Jouffroy, F. K., Renous, S., Gasc, J., 1983. Etude cinéradiographique des deplacements du membre antérieur du potto de Bosman (*Perodicticus potto*, P.L.S. Muller, 1766) au cours de la marche quadrupède sur une branche horizontale. Annales des Sciences Naturelles, Zoologie, Paris 5, 75–87.

Jouffroy, F. K., Stern, J. T. J., 1990. Telemetered EMG study of the antigravity versus propulsive actions of knee and elbow muscles in the slow loris. In: Jouffroy, F., Stack, M., Niemitz, C. (Eds.), Gravity, Posture and Locomotion in Primates. Firenze, Il Sedicesimo, pp. 221–236.

Jowsey, J., 1966. Studies of Haversian systems in man and some animals. Journal of Anatomy 100, 857–864.

Jowsey, J., 1968. Age and species differences in bone. Cornell Veterinarian 58, 74–94.

Jungers, W. L., 1985. Body size and scaling of limb proportions in primates. In: Jungers, W. L. (Ed.), Size and Scaling in Primate Biology. Plenum Press, New York, pp. 345–381.

Kappeler, P. M., 1996. Causes and consequences of life-history variation among strepsirhine primates. American Naturalist 148, 868–891.

Kappeler, P. M., 1998. Nests, tree holes, and the evolution of primate life histories. American journal of Primateology 46, 7–33.

Kimura, T., 2002. Primate limb bones and locomotor types in arboreal or terrestrial environments. Zeitschrift fur Morphologie und Anthropologie 83, 201–219.

King, S. J., 2003. An evolutionary perspective on differential craniodental and postcranial growth and development in primates. Ph.D. dissertation, University of Massachusetts, Massachusetts.

Kirkwood, J. K., 1985. Patterns of growth in primates. Journal of Zoology, London (A) 205, 123–136.

Klevezal, G. A., 1996. Recording Structures of Mammals. Determination of Age and Reconstruction of Life History. A. A. Balkema, Rotterdam, The Netherlands.

Kohn, L. A. P., Olson, P., Cheverud, J. M., 1997. Age of epiphyseal closure in tamarins and marmosets. American Journal of Primatology 41, 129–139.

Krishtalka, L., Schwartz, J. H., 1978. Phylogenetic relationships of plesiadapiform-tarsiiform primates. Annals of the Carnegie Museum 47, 515–540.

Lanyon, L. E., 1984. Functional strain as a determinant for bone remodeling. Calcified Tissue International 36, S56–S61.

Lanyon, L. E., 1993. Osteocytes, strain detection, bone modeling and remodeling. Calcified Tissue International 53, S102–S107.

Lanyon, L. E., Goodship, A. E., Pye, C. J., MacFie, J. H., 1982. Mechanically adaptive bone remodelling. Journal of Biomechanics 15, 141–154.

Lanyon, L. E., Rubin, C. T., 1985. Functional adaptation in skeletal structures. In: Hildebrand, M., Bramble, D. M., Liem, K. F., Wake, D. B. (Eds.), Functional Vertebrate Morphology. Harvard University Press, Cambridge, pp. 1–25.

Lieberman, D. E., Crompton, A. W., 1998. Responses of bone to stress: constraints on symmorphosis. In: Weibel, E. R., Taylor, C. R., Bolis, L. (Eds.), Principles of Animal Design: The Optimization and Symmorphosis Debate. Cambridge University Press, Cambridge, pp. 78–86.

Lieberman, D. E., Pearson, O. M., 2001. Trade-off between modeling and remodeling responses to loading in the mammalian limb. Bulletin of the Museum of Comparative Zoology 156, 269–282.

Leigh, S. R., Terranova, C. J., 1998. Comparative perspectives on bimaturism, ontogeny, and dimorphism in lemurid primates. International Journal of Primatology 19, 723–749.

Martin, R. B., 1991. Determinants of the mechanical properties of bones [published erratum appears in 1992, Journal of Biomechanics 25(10), 1251]. Journal of Biomechanics 24(Supplement 1), 79–88.

Martin, B., 1993. Aging and strength of bone as a structural material. Calcified Tissue International 53, S34–S40.

Martin, R. B., 2000. Toward a unifying theory of bone remodeling. Bone 26, 1–6.

Martin, R. B., Burr, D. B., 1989. Structure, Function, and Adaptation of Compact Bone. Raven, New York.

Martin, R. B., Ishida, J., 1989. The relative effects of collagen fiber orientation, porosity, density, and mineralization on bone strength. Journal of Biomechanics 22, 419–426.

McFarlin, S. C., 2006. Ontogenetic variation in long bone microstructure in catarrhines and its significance for life history research. Ph.D. dissertation, The City University of New York, New York.

McMahon, J. M., Boyde, A., Bromage, T. G., 1995. Pattern of collagen fiber orientation in the ovine calcaneal shaft and its relation to locomotor-induced strain. Anatomical Record 242, 147–158.

Mittermeier, R. A., 1978. Locomotion and posture in *Ateles geoffroyi* and *Ateles paniscus*. Folia Primatologica 30, 161–193.

Moss, M. L., 1964. The phylogeny of mineralized tissues. International Review of General and Experimental Zoology 1, 297–331.

Müller, E. F., 1982. Heart rate in the slow loris (*Nycticebus coucang*). Folia Primatologica 38, 250–3258.

Nakatsukasa, M., Takai, M., Setoguchi, T., 1997. Functional morphology of the postcranium and locomotor behavior of *Neosaimiri fieldsi*, a Saimiri-like middle Miocene platyrrhine. American Journal of Physical Anthropology 102(4), 515–544.

Napier, J. R., Napier, P. H., 1967. A Handbook of Living Primates. Academic, New York.

Napier, J. P., Walker, A. C., 1967. Vertical clinging and leaping – a newly recognized category of locomotor behavior of primates. Folia Primatologica 6, 2047–2219.

Newell-Morris, L., Sirianni, J. E., 1982. Parameters of bone growth in the fetal and infant macaque (*Macaca nemestrina*) humerus as documented by trichromatic bone labels. In: Dixon, A. D., Sarnat, B. G. (Eds.), Factors and Mechanisms Influencing Bone Growth. Alan R. Liss, New York, pp. 243–258.

Niemitz, C., 1983. New results on the locomotion of *Tarsius bancanus*, Horsfield, 1821. Annales des Sciences Naturelles, Zoologie, Paris Series 13, 89–100.

Ortmann, S., Heldmaier, G., Schmid, J., Ganzhorn, J. U., 1997. Spontaneous daily torpor in Malagasy mouse lemurs. Naturwissenschaften 84, 28–32.

Oxnard, C. E., 1973. Some locomotor adapations among lower primates: implications for primate evolution. Symposia of the Zoological Society of London 33, 255–299.

Oxnard, C. E., German, R., Jouffroy, F. K., Lessertisseur, J., 1981a. A morphometric study of limb proportions in leaping prosimians. American Journal of Physical Anthropology 54, 421–430.

Oxnard, C. E., German, R., McArdle, J., 1981b. The functional morphometrics of the hip and thigh in leaping prosimians. American Journal of Physical Anthropology 54, 481–498.

Padian, K., de Ricqlès, A. Horner, J. R., 2001. Dinosaurian growth rates and bird origins. Nature 412(6845), 405–408.

Paine, R. R., Godfrey, L. R., 1997. The scaling of skeletal microanatomy in non-human primates. Journal of Zoology, London 241, 803–821.

Petter, J. J., Hladik, C. M., 1970. Observations sure le domaine vital et la densité de population de *Loris tardigradus* dans les forêts de Ceylan. Mammalia 34, 394–409.

Portigliatti-Barbos, M., Bianco, P., Ascenzi, A., 1983. Distribution of osteonic and interstitial components in the human femoral shaft with reference to structure, calcification and mechanical properties. Acta Anatomica 115, 178–186.

Poux, C., Douzery, E. J. P., 2004. Primate phylogeny, evolutionary rate variations, and divergence times: a contribution from the nuclear gene IRBP. American Journal of Physical Anthropology 124(1), 1–16.

Preuschoft, H., Günther, M. M., Christian, A., 1998. Size dependence in prosimian locomotion and its implications for the distribution of body mass. Folia Primatologica 69, 60–81.

Preuschoft, H., Witte, H., Fischer, M., 1995. Locomotion in nocturnal primates. In: Alterman, L., Doyle, G., Izard, M. (Eds.), Creatures of the Dark: The Nocturnal Prosimians. Plenum, New York, pp. 453–472.

Quekett, J., 1849. On the intimate structure of bone, as composing the skeleton in the four great classes of animals, viz., mammals, birds, reptiles, and fishes, with some remarks on the great value of the knowledge of such structure in determining the affinities of minute fragments of organic remains. Transactions of the Microscopical Society of London 2, 46–458.

Quekett, J., 1855. Descriptive and Illustrated Catalogue of the Histological Series Contained in the Museum of the Royal College of Surgeons of England. Prepared for the Microscope. Royal College of Surgeons, London.

Rasmussen, D. T., 1994. The different meanings of a *Tarsius*-anthropoid clade and a new model of anthropoid origins. In: Fleagle, J. F., Kay, R. F. (Eds.), Anthropoid Origins. Plenum, New York, pp. 335–360.

Rasmussen, D. T., Izard, M. K., 1988. Scaling of growth and life history traits relative to body size, brain size, and metabolic rate in lorises and galagos (Lorisidae, Primates). American Journal of Physical Anthropology 75, 357–367.

Rho, J.-Y., Zioupos, P., Currey, J. D., Pharr, G. M., 1999. Variations in the individual thick lamellar properties within osteons by nanoindentation. Bone 25, 295–300.

Riggs, C. M., Lanyon, L. E., Boyde, A., 1993a. Functional associations between collagen fiber orientation and locomotor strain direction in cortical bone of the equine radius. Anatomy and Embryology 187, 231–238.

Riggs, C. M., Vaughan, L. C., Evans, G. P., Lanyon, L. E., Boyde, A., 1993b. Mechanical implications of collagen fibre orientation in cortical bone of the equine radius. Anatomy and Embryology 187, 239–248.

Robling, A. G., Stout, S. D., 1999. Morphology of the drifting osteon. Cells Tissues Organs 164, 192–204.

Rose, M. D., 1993. Functional anatomy of the elbow and the forearm in primates. In: Gebo, D. L. (Ed.), Postcranial Adaptation in Nonhuman Primates. Northern Illinois University Press, Dekalb, IL, pp. 70–95.

Rosenberger, A. L., 2002. Platyrrhine paleontology and systematics: the paradigm shifts. In: Hartwig, W. C. (Ed.), The Primate

Fossil Record. Cambridge University Press, Cambridge, pp. 151–159.

Rosenberger, A. L., Szalay, F. S., 1980. On the tarsiiform origins of Anthropoidea. In: Ciochon, R. L., Chiarelli, A. B. (Eds.), Evolutionary Biology of the New World Monkeys and Continental Drift. Plenum, New York, pp. 139–157.

Ross, C., 1998. Primate life histories. Evolutionary Anthropology 6(2), 54–63.

Ruff, C. B., 2003. Long bone articular and diaphyseal structure in Old World monkeys and apes. II: Estimation of body mass. American Journal of Physical Anthropology 120(1), 16–37.

Ruff, C., Runestad, J., 1992. Primate limb bone structural adaptations. Annual Review of Anthropology 21, 407–433.

Runestad, J. A., 1997. Postcranial adaptations for climbing in Loridae (Primates). Journal of Zoology, London 242, 261–290.

Runestad Connour, J., Glander, K., Vincent, F., 2000. Postcranial adaptations for leaping in primates. Journal of Zoology, London 251, 79–103.

Sander, P. M., 2000. Long bone histology of the Tendaguru sauropods: implications for growth and biology. Paleobiology 26, 466–488.

Schaffler, M. B., Burr, D. B., 1984. Primate cortical bone microstructure: relationship to locomotion. American Journal of Physical Anthropology 65, 191–197.

Schaffler, M. B., Burr, D. B., 1988. Stiffness of compact bone: effects of porosity and density. Journal of Biomechanics 21(1), 13–16.

Schaffler, M. B., Radin, E. L., Burr, D. B., 1989 Mechanical and morphological effects of strain rate on fatigue of compact bone. Bone 10, 207–214.

Schaffler, M. B., Radin, E. L., Burr, D. B., 1990. Long-term fatigue behavior of compact bone at low strain magnitude and rate. Bone 11, 321–326.

Schmid, J., 1998. Daily torpor in mouse lemurs, *Microcebus* spp.: metabolic rate and body temperature (abstract). Folia Primatologica 69, 403.

Schmid, J., 2000. Daily torpor in the gray mouse lemur (*Microcebus murinus*) in Madagascar: energetic consequences and biological significance. Oecologia 123, 175–183.

Schmid, J., 2001. Daily torpor in free-ranging gray mouse lemurs (*Microcebus murinus*) in Madagascar. International Journal of Primatology 22, 1021–1031.

Schmid, J., Kappeler, P. M., 1998. Fluctuating sexual dimorphism and differential hibernation by sex in a primate, the gray mouse lemur (*Microcebus murinus*). Behavioral Ecology and Sociobiology 43, 125–132.

Schulze, H., Meier, B., 1995. Behavior of captive *Loris tardigradus nordicus*: a qualitative description, including some information about morphological bases of behavior. In: Alterman, L., Doyle, G., Izard, M. (Eds.), Creatures of the Dark: The Nocturnal Primates. Plenum, New York, pp. 221–249.

Shoshani, J., Groves, C. P., Simons, E. L., Gunnell, G. F., 1996. Primate phylogeny: morphological vs molecular results. Molecular Phylogenetics and Evolution 5(1), 102–154.

Simkin, A., Robin, G., 1974. Fracture formation in differing collagen fiber pattern of compact bone. Journal of Biomechanics 7, 183–188.

Simons, E. L., 1997. Discovery of the smallest Fayum Egyptian primates (Anchomomyini, Adapidae). Proceedings of the National Academy of Sciences USA 94, 180–184.

Simons, E. L., Rasmussen, D. T., Gingerich, P. D., 1995. New cercomoniine adapid from Fayum, Egypt. Journal of Human Evolution 29, 577–589.

Singh, I. J., Tonna, E. A., Gandel, C. P., 1974. A comparative histological study of mammalian bone. Journal of Morphology 144, 4213–4438.

Skedros, J. G., Hunt, K. J., 2004. Does the degree of laminarity correlate with site-specific differences in collagen fibre orientation in primary bone? An evaluation in the turkey ulna diaphysis. Journal of Anatomy 205(2), 121–134.

Smith, R. J., Jungers, W. L., 1997. Body mass in comparative primatology. Journal of Human Evolution 32, 523–559.

Starck, J. M., Chinsamy, A., 2002. Bone microstructure and developmental plasticity in birds and other dinosaurs. Journal of Morphology 254(3), 232–246.

Stevens, J., Edgerton, V., Mitton, S., 1971. Gross anatomy of hindlimb skeletal system of the *Galago senegalensis*. Primates 12, 313–321.

Stevens, J., Mitton, S., Edgerton, V., 1972. Gross anatomy of hindlimb skeletal muscles of the *Galago senegalensis*. Primates 13, 83–101.

Suckling, J. A., Suckling, E. E., Walker, A. C., 1969. Suggested function of the vascular bundles in the limbs of *Perodicticus potto*. Nature 221, 379–380.

Sussman, R. W., Kinzey, W. G., 1984. The ecological role of the Callitrichidae: a review. American Journal of Physical Anthropology 64, 419–449.

Swartz, S. M., Bertram, J. E. A., Biewener, A. A., 1989. Telemetered *in vivo* strain analysis of locomotor mechanics of brachiating gibbons. Nature 342, 270–272.

Szalay, F. S., 1981. Phylogeny and the problem of adaptive significance: the case of the earliest primates. Folia Primatologica 36, 157–182.

Szalay, F. S., 2000. Function and adaptation in paleontology and phylogenetics: why do we omit Darwin? Palaeontologia Electronica 3(2), 1–25.

Szalay, F. S., Dagosto, M., 1980. Locomotor adaptations as reflected on the humerus of paleogene primates. Folia Primatologica 34, 1–45.

Szalay, F. S., Dagosto, M., 1988. Evolution of hallucial grasping in the primates. Journal of Human Evolution 17, 1–33.

Szalay, F. S., Rosenberger, A. L., Dagosto, M., 1987. Diagnosis and differentiation of the Order Primates. Yearbook of Physical Anthropology 30, 75–105.

Szalay, F. S., Sargis, E. J., 2001. Model-based analysis of postcranial osteology of marsupials from the Palaeocene of Itaboraí (Brazil) and the phylogenies and biogeography of Metatheria. Geodiversitas 23(2), 139–302.

Terranova, C. J., 1995a. Functional morphology of leaping behaviors in galagids: associations between landing limb use and diaphyseal geometry. In: Alterman, L., Doyle, G. A., Izard, M. K. (Eds.), Creatures of the Dark: The Nocturnal Prosimians. Plenum, New York, pp. 473–493.

Terranova, C. J., 1995b. Leaping behaviors and the functional morphology of strepsirhine primate long bones. Folia Primatologica 61, 181–201.

Terranova, C. J., 1996. Variation in the leaping of lemurs. American Journal of Primatology 40, 145–165.

Turnquist, J. E., Schmitt, D., Rose, M. D., Cant, J. G. H., 1999. Pendular motion in the brachiation of captive *Lagothrix* and *Ateles*. American Journal of Physical Anthropology 48(4), 263–281.

Vincentelli, R., Evans, F. G., 1971. Relations among mechanical properties, collagen fibers, and calcification in adult human cortical bone. Journal of Biomechanics 4, 193–201.

Vincentelli, R., Grigorov, M., 1985. The effects of Haversian remodeling on the tensile properties of human cortical bone. Journal of Biomechanics 18, 201–207.

Walker, A., 1969. The locomotion of lorises, with special reference to the potto. East African Wildlife Journal 7, 1–5.

Walker, A., 1974. Locomotor adaptations in past and present prosimian primates. In: Jenkins, F. A. Jr. (Ed.), Primate Locomotion. Academic, New York, pp. 349–381.

Walker, A., 1979. Prosimian locomotor behavior. In: Martin, R. D., Doyle, G. A. (Eds.), The Study of Prosimian Behavior. Academic, New York, pp. 543–565.

Warshaw, J., 2007. Primate bone microstructural variability: relationships to life history, mechanical adaptation and phylogeny. Ph.D. dissertation, The City University of New York, New York.

Watts, E. S., 1990. Evolutionary trends in primate growth and development. In: Rousseau, C. J. de (Ed.), Primate Life History and Evolution. Wiley-Liss, New York, pp. 89–104.

Weisenseel, K. A., Izard, M. K., Nash, L. T., Ange, R. L., Poorman-Allen, P., 1998. A comparison of reproduction in two species of *Nycticebus*. Folia Primatologica 69, 321–324.

Yoder, A. D., Irwin, J. A., Payseur, B. A., 2001. Failure of the ILD to determine data combinability for slow loris phylogeny. Systematic Biology 50(3), 408–424.

Taxonomic Index

Subject Index

Vertebrate Paleobiology and Paleoanthropology

Published Titles:

Neanderthals Revisited: New approaches and perspectives
Edited by K. Harvati and T. Harrison
ISBN: 978-1-4020-5120-3, 2006
A Volume in the Max-Planck Institute Sub-series in Human Evolution

The Evolution and Diversity of Humans in South Asia: Interdisciplinary studies in archaeology, biological anthropology, linguistics and genetics
Edited by M. Petraglia and B. Allchin
ISBN: 978-1-4020-5561-4, 2007

Dental Perspectives on Human Evolution: State of the art research in dental paleoanthropology
Edited by S.E. Bailey and J-J. Hublin
ISBN: 978-1-4020-5844-8, 2007
A Volume in the Max-Planck Institute Sub-series in Human Evolution

Hominin Environment in the East African Pliocene: An Assessment of the Faunal Evidence
Edited by R. Bobe, Z. Alemseged and A.K. Behrensmeyer
ISBN: 978-1-4020-3097-0, 2007

Deconstructing Olduvai: A Taphonomic Study of the Bed I Sites
By M. Domínguez-Rodrigo, C.P. Egeland and R. Barba Egido
ISBN: 978-1-4020-6150-9, 2007

Mammalian Evolutionary Morphology: A Tribute to Frederick S. Szalay
Edited by Eric J. Sargis and Marian Dagosto
ISBN: 978-1-4020-6996-3, 2008